T0383273

Fish Physiology

The 50th Anniversary Issue of Fish Physiology: Physiological Systems and Development

Volume 40A

This is Volume 40A in the

FISH PHYSIOLOGY series

Edited by Anthony P. Farrell, Colin J. Brauner and Erika J. Eliason

Honorary Editors: William S. Hoar and David J. Randall

A complete list of books in this series appears at the end of the volume

Fish Physiology

The 50th Anniversary Issue of Fish Physiology: Physiological Systems and Development

Volume 40A

Edited by

David J. Randall
Department of Zoology, The University of British Columbia, Vancouver, British Columbia, Canada

Anthony P. Farrell
Department of Zoology, and Faculty of Land and Food Systems, The University of British Columbia, Vancouver, British Columbia, Canada

Colin J. Brauner
Department of Zoology, The University of British Columbia, Vancouver, British Columbia, Canada

Erika J. Eliason
Department of Ecology, Evolution, and Marine Biology, University of California - Santa Barbara, Santa Barbara, CA, United States

Academic Press is an imprint of Elsevier
50 Hampshire Street, 5th Floor, Cambridge, MA 02139, United States
525 B Street, Suite 1650, San Diego, CA 92101, United States
The Boulevard, Langford Lane, Kidlington, Oxford OX5 1GB, United Kingdom
125 London Wall, London, EC2Y 5AS, United Kingdom

First edition 2023

ISBN: 978-0-443-13733-4
ISSN: 1546-5098

For information on all Academic Press publications
visit our website at https://www.elsevier.com/books-and-journals

Publisher: Zoe Kruze
Acquisitions Editor: Mariana Kuhl
Editorial Project Manager: Palash Sharma
Production Project Manager: Abdulla Sait
Cover Designer: Christian Bilbow

Typeset by STRAIVE, India

Contents

Contributors

C. Albers (367)

M.V.L. Bennett (677)

J.H.S. Blaxter (483)

Colin J. Brauner (361), University of British Columbia, Vancouver, BC, Canada

Lauren Chapman (557), Department of Biology, McGill University, Montréal, QC, Canada

Robert H. Devlin (401), Fisheries and Oceans Canada, West Vancouver, BC, Canada

Edward M. Donaldson (253)

Anthony P. Farrell (91), Department of Biological Sciences, Simon Fraser University, Burnaby, British Columbia, Canada

W.N. Holmes (253)

G.M. Hughes (9), Research Unit for Comparative Animal Respiration, University of Bristol, Bristol, United Kingdom

Takashi Iwamatsu (401), Aichi University of Education, Kariya, Aichi, Japan

Kjell Johansen (187)

David R. Jones (91), Department of Zoology, University of British Columbia, Vancouver, British Columbia, Canada

Kiyoshi Kikuchi (401), Fisheries Laboratory, University of Tokyo, Shizuoka, Japan

Daiani Kochhann (557), Laboratory of Behavioural Ecophysiology, Center of Agrarian and Biological Sciences, Acaraú Valley State University, Sobral, CE, Brazil

Duncan Leitch (669), Department of Integrative Biology and Physiology, University of California, Los Angeles, CA, United States

J. Adam Luckenbach (401), Environmental Physiology Program, Northwest Fisheries Science Center, National Marine Fisheries Service, National Oceanic and Atmospheric Administration, Seattle; Center for Reproductive Biology, Washington State University, Pullman, WA, United States

Casey A. Mueller (475), Department of Biological Sciences, California State University San Marcos, San Marcos, CA, United States

Yoshitaka Nagahama (401), National Institute for Basic Biology, Okazaki; College of Science and Engineering, Kanazawa University, Kanazawa, Japan

David J. Randall (361), University of British Columbia, Vancouver, BC, Canada

Peter J. Rombough (567), Zoology Department, Brandon University, Brandon, Manitoba, Canada R7A 6A9

Holly A. Shiels (79), Division of Cardiovascular Sciences, Faculty of Biology, Medicine, and Health, University of Manchester, Manchester, United Kingdom

Tobias Wang (173), Section for Zoophysiology, Department of Biology, Aarhus University, Aarhus C, Denmark

Nicholas C. Wegner (1), Fisheries Resources Division, Southwest Fisheries Science Center, National Marine Fisheries Service, National Oceanic and Atmospheric Administration, La Jolla, CA, United States

Chris M. Wood (235), Department of Zoology, University of British Columbia, Vancouver, BC; Department of Biology, McMaster University, Hamilton, ON, Canada; Rosenstiel School of Marine and Atmospheric Science, University of Miami, Miami, FL, United States

Toki-O Yamamoto (421)

Alex M. Zimmer (235), Department of Biological Sciences, University of New Brunswick, Saint John, NB, Canada

Preface

In the mid-1960s, Bill Hoar and Dave Randall prepared a six-volume treatise entitled "Fish Physiology" to be published with the hope that it would serve biologists of the 1970s as Margaret Brown's "The Physiology of Fishes" had served readers earlier. They had no intention of continuing the series beyond six volumes, but the response was such that additional volumes were developed, covering in-depth analysis of specific areas of fish biology. Advice from the scientific community when developing volumes was an important factor. The interval between volumes depended on developments in the field and how ready they were to spend time on the editing process. Bill Hoar eventually retired, and Dave Randall had had enough! At an international meeting, he discussed this with several colleagues and was persuaded to continue with guest editors to spread the load. Tony Farrell agreed to join Dave Randall as an editor in this enterprise. Dave Randall retired, and Colin Brauner assumed the challenge. Tony Farrell retired recently and was replaced by Erika Eliason. The series "Fish Physiology" recently celebrated its 50th anniversary.

The middle of the 20th century saw the rapid growth of comparative physiology. The huge success of studies of the squid giant axon in unraveling the nature and transmission of the action potential and ideas of model systems in biological research engendered interest in comparative physiology. The Fisheries Research Board (now the Department of Fisheries and Oceans) was established in Canada. Fred Fry at the University of Toronto had established a framework for the environmental study of fish. He was also reputed to have an electric suite that he could plug in and survive the cold Canadian winters. The increased research activity in fish physiology created a market for the series. Contributing authors were from everywhere, and the books were used worldwide. In total, the series has produced a total of 47 books (several volumes have two books) that contain almost 500 chapters since the inaugural volume was published in 1969.

The treatise "Fish Physiology" is the result of the many contributions of the members of the international consortium of fish physiologists and other interested persons. There are many subgroups, but no overarching structure or organization to this broad collection of like-minded people who have found it useful to have this summary of their efforts and hopes. Academic Press, the original publishers, imposed length and layout and decided the books would have a green cover. Content was entirely left to the editors. Each volume represents the views of individual specialists on the subject under discussion.

However, while discussing functional processes, the authors have referred to a wealth of comparative material so that the treatise has become more than an account of the physiology of fishes; it contains many fundamental concepts and principles important in the broad field of comparative animal physiology. The treatise has survived where others (e.g., Physiology of Reptiles) have long disappeared. We think the longevity is because the volumes serve and belong to the community of "Fish Physiologists." During the life of "Fish Physiology," publishers have always been supportive, despite changes in the business with the advent of the Internet. In addition, the world has become a much more bureaucratic and conservative place, with sound bites, fake news, and less regard for the rigor of scientific study. Careful analysis of data and ideas has been central to the presentation of "Fish Physiology," avoiding celebrity status, promotion, and politics. The volumes have maintained their initial focus to serve a broad range of fish physiologists.

The content of the "Fish Physiology" volumes has evolved over time. The initial volumes were devoted to understanding the basic mechanisms and principles of fish physiology, with a focus on a few model species and some application to natural environmental conditions. Then, as the field better understood mechanisms, the approach was broadened to not only delve deeper into system physiology (e.g., chapters in early volumes were expanded to become books), but interspecific differences in physiology were explored, permitting a more evolutionary framework. Finally, as interspecific physiological mechanisms were further resolved, it became possible to discuss physiology in light of a changing world. Thus, physiology can now inform conservation, sustainability, and management, as exemplified by the most recent volumes.

Narrowing down the 50th anniversary issue to less than 20 chapters was a challenge. In general, the majority of chapters written throughout the series' history are very strong, expertly written, and could be candidates. Therefore, we limited ourselves to the first 14 volumes in the series to highlight the very important early work in the field that was published in the series. In particular, we wished to (re)introduce new researchers (such as graduate students) to this research that has stood the test of time and that has shaped the field. We want students and early-career researchers to gain an appreciation for the history of the field and also see how earlier research remains applicable. Too often, research conducted in an earlier era is forgotten or not thought relevant simply due to the date of publication. However, this is far from the case, and this anniversary issue aims to address this for the field in general and the series in particular. Second, within this time window, we carefully reviewed selected chapters that were particularly novel or impactful on the field of fish physiology. The standout chapters that we selected in no way should undermine the quality and impact of the many chapters that were not selected. In the first 25 years of the series, there were 155 chapters published. To appropriately highlight this tremendous body of work, we required two volumes. Each volume includes 9–10 republished chapters. We recruited acknowledged experts in the field of each republished chapter to write a short review, which precedes each of the republished chapters in their original form. Volume A focuses on physiological systems and development, while Volume B focuses on physiological applications to altered environments and growth. Thus, Volumes A and B reflect the evolution of the series as described above.

In the Preface to Volume 1, Bill Hoar and Dave Randall expressed the hope that "Fish Physiology" would prove as valuable in fisheries research laboratories as in university reference libraries and that it would be a rich source of detailed information for the comparative physiologist and the zoologist as well as the specialist in fish physiology. It has fulfilled this hope because of the efforts of many, many people. We thank them all for their hard work and continued support, and we feel fortunate to be part of such a welcoming and supportive community.

David J. Randall, Anthony P. Farrell, Colin J. Brauner and Erika J. Eliason
August 2023
Vancouver, Canada

Chapter 1

The fish gill: Where fish physiology begins

Nicholas C. Wegner*
Fisheries Resources Division, Southwest Fisheries Science Center, National Marine Fisheries Service, National Oceanic and Atmospheric Administration, La Jolla, CA, United States
**Corresponding author: e-mail: nick.wegner@noaa.gov*

Nicholas C. Wegner discusses the G. Hughes chapter "General Anatomy of the Gills" in *Fish Physiology*, Volume 10A, published in 1984. The next chapter in this volume is the re-published version of that chapter.

George Hughes' (1984) seminal book chapter titled "General Anatomy of the Gills" in *Fish Physiology Volume 10A* is a comprehensive review of fish gill functional morphology that continues to serve as a common reference in the field of fish respiratory science. Hughes' chapter summarizes the work of his productive career and the status of fish respiratory morphology at the time, with particular emphasis on the diversity of gross and fine gill structure across fishes that vary in evolutionary lineage, life history, habitat, and metabolic requirements. Specifically, Hughes discusses the multiple factors sculpting gill dimensions on evolutionary time scales and shows how gill structure and function can be adapted to meet the diverse physiological demands in fishes ranging from fast-swimming ram ventilators to fishes that breathe air through auxiliary air-breathing organs. His discussions on a vast range of important topics, including gill diffusion capacity, fish ventilatory mechanics, and allometric scaling, continue to inspire the field today.

It can be said that the study of fish physiology begins at the gills. The gills generally comprise the primary exchange surface through which fish acquire environmental oxygen to support aerobic metabolism, release carbon dioxide, exchange ions with the surrounding water to maintain and balance internal osmolarity and pH, and through which nitrogenous waste excretion occurs. Due to their multiple functions, the gills thus provide an easily accessible window for looking into the biology and physiological requirements of a fish. This was clearly recognized by George M. Hughes, who most consider to be the founder of the modern era of fish respiratory science (Randall, 2014; Wegner and Graham, 2010); Hughes stated concerning the gills, "Their adaptation to particular environments provides a fascinating study in comparative functional anatomy and physiology" (Hughes, 1984). Hughes' (1984) book

Fish Physiology, Vol. 40A. https://doi.org/10.1016/bs.fp.2023.09.003

chapter in the *Fish Physiology* series titled "*General Anatomy of the Gills*" serves as a seminal work summarizing the knowledge of fish gill function and morphology at the time. In fact, much of the material contained within the chapter's original 72 pages are derived directly from Hughes' productive career, which ultimately included over 250 scientific papers published over the span of more than 50 years (1952–2003). Not surprisingly, Hughes' chapter has been cited nearly 500 times, and while now 40 years old, continues to serve as a common reference in modern studies of gill morphology and function, and still gets pulled off my book case with regularity.

I remember first sitting down at a lab bench when I was a young undergraduate student volunteering in a laboratory at Scripps Institution of Oceanography with a copy of the *Fish Physiology Volume 10A* opened to George Hughes's chapter alongside a copy of *Sharks, Skates, and Rays: The Biology of Elasmobranch Fishes* opened to Patrick Butler's chapter on the "Respiratory System" (Butler, 1999), while I carefully dissected a leopard shark, *Triakis semifasciata*, to expose the gills. Thinking that I had discovered some inconsistencies between my dissection and the anatomical illustrations and images of the gills in these book chapters, I hastily alerted my future graduate advisor, Jeffrey B. Graham, who carefully looked at my dissection and the figures and simply replied, "I don't see anything inconsistent here. Keep reading." And so, I did. Hughes' chapter (although probably not designed for this purpose) became my self-taught guide to gill morphology and physiology, which was further augmented with other works by Hughes (e.g., Hughes, 1966; Hughes and Morgan, 1973; Muir and Hughes, 1969) that are largely summarized in his 1984 chapter.

The power of George Hughes' chapter is in its exhaustive and scrupulous summary of the state of knowledge of gill morphology. Beginning with gill development, Hughes describes the origin of fish gills, their relationship to other respiratory organs, and their basic structure composed of primary and secondary folds, historically referred to as primarily and secondary lamellae, but now typically referred to as gill filaments and lamellae, respectively, to avoid confusion. These structures are generally conserved across fish species, although, as Hughes elaborates in this and other important works, the size and thickness of these gill components can vary widely among taxa from different habitats or species exhibiting differences in activity level and hence oxygen demand (Hughes, 1966, 1972, 1984; Hughes and Morgan, 1973).

While the recognition that gill dimensions correlate with both fish habitat and activity level largely predates the work of Hughes (e.g., Gray, 1954), his prolific publication record examining the gills from diverse fish species (and other aquatic animals) greatly refined our understanding of the diversity of gill structure and function. He was particularly interested in the total respiratory surface area of the gills (gill surface area) as well as diffusion distances, which together determine the diffusion capacity of the gills for both gas and ion exchange. In Hughes' chapter, he categorizes fishes into simple ecomorphotypes based largely on these dimensions, describing: 1. Fast-swimming oceanic species, 2. Fishes of intermediate activity, 3. Sluggish fishes, and 4. Air breathers (i.e., fishes that use an auxiliary air-breathing organ which could include, but is not limited to, modified gill tissue). He uses these groups to help illustrate how gill morphology can be sculpted on evolutionary time scales to meet physiological demands. For example, a fast-swimming oceanic fish, such as a tuna (family Scombridae), can have a gill surface area that is an order of magnitude larger and a thickness of the gill epithelium (the water-blood barrier) that can be an order of magnitude thinner than those of a "sluggish" marine fish. The simplicity of comparing such fish ecomorphotypes to demonstrate their diversity and the various factors shaping gill functional morphology has been repeated in subsequent reviews of gill dimensions (e.g., Palzenberger and Pohla, 1992; Wegner, 2011). In my review chapter in the *Encyclopedia of Fish Physiology* (Wegner, 2011), I purposely included these same groups from Hughes (1984) with additional categories including 5. Freshwater fishes and 6. Hypoxia dwellers.

Inherent to the differences among the respiratory morphology of different fish groups are the multiple factors that shape gill dimensions. Ultimately, the gills are a balance of having a large enough gill surface area to meet oxygen demands for aerobic metabolism, while not having gills excessively large that they incur unnecessary osmoregulatory costs, all while optimizing branchial resistance to water flow to maximize oxygen extraction efficiency while minimizing ventilatory costs. Hughes had an innate ability to simplify fluid dynamics in a way that allows the non-engineer (such as me) to quickly grasp how changing certain gills dimensions (e.g., the spacing of the gill lamellae) not only affects gill surface area, but also the resistance to branchial waterflow, and hence the energy required to ventilate the gills (e.g., see Hughes' Fig. 20). For a fast-swimming ram ventilator, such as a tuna, high branchial resistance associated with closely spaced lamellae appears to be an acceptable or perhaps even advantageous morphological configuration for water hydrodynamics and gas exchange, while also allowing for a high gill surface area (Hughes, 1984; Muir and Hughes, 1969; Wegner et al., 2010a). Conversely, for a less-active (i.e., "sluggish") fish, a smaller gill surface area with widely-spaced lamellae drastically decreases the costs of active ventilation while also limiting osmoregulatory costs (Hughes, 1984; Hughes and Gray, 1972).

Hughes' comparative approach in documenting the diversity of gill functional morphology greatly impacted my development as a graduate student in the laboratory of Jeffrey Graham, where Hughes' works were a source of inspiration for a research group primarily focused on fishes that push the limits of cardio-respiratory function (e.g., "high-performance fishes," such as tunas and lamnid sharks, as well as air-breathing fishes). Indeed, my PhD dissertation titled "Morphology, function, and evolution of the gills of high-performance fishes" documented gill adaptations used by scombrids and billfishes to enhance gill surface area, minimize diffusion distances, streamline branchial flow, and strengthen the gills for ram ventilation (Wegner et al., 2010a, 2013), all of which drew heavily from ideas first explored by Hughes and colleagues and summarized in Hughes' (1984) chapter. My interests were further piqued by Hughes' discussion of differences in the gill interbranchial septum among fish lineages (see Figs. 5, 6, and 23), particularly how the elasmobranch gill has a potential added source of branchial resistance as the water that passes the gill lamellae must subsequently flow through septal channels that are absent or greatly reduced in most teleosts. Building upon Hughes' hypotheses we confirmed that the interbranchial septum of the elasmobranch gill does, in fact, provide an added source of branchial resistance and appears to preclude fast-swimming sharks, such as the shortfin mako, *Isurus oxyrinchus* (family Lamnidae), from having the same closely spaced lamellae as observed in tunas (which would further increase resistance to water flow). This ultimately appears to limit the relative gill surface area of lamnid sharks in comparison to tunas and potentially the evolutionary convergence between these two groups for high levels of aerobic performance (Wegner et al., 2010b, 2012).

On the other end of the fish spectrum, Hughes was also at the forefront of documenting the diversity of respiratory morphology in air-breathing fishes. Pioneering work by Hughes and colleagues such J.S. Datta Munshi in the 1960s and 1970s (e.g., Hughes and Munshi, 1973; Hughes et al., 1973) helped to document the independent evolution of air breathing among different fish groups and their associated myriad of air-breathing organs, ranging from lungs and respiratory gas bladders, to suprabranchial chambers and modified gills, to respiratory digestive tracts. While Hughes' chapter contains only a relatively short summary of this work, the diversity of such fish air-breathing organs adds to the rich tapestry of morphological and physiological specialization that excites researchers in the field of comparative physiology. Indeed, Hughes's works not only inspired my career trajectory, but also that of my previously mentioned graduate advisor, Jeffrey Graham, who became one of the world's authorities on the respiratory physiology of air-breathing fishes (Graham, 1997, 2006; Graham and Wegner, 2010).

Along with documenting the diversity of fish respiratory morphology, Hughes was also instrumental in establishing our current understanding of how fish ventilate their gills with a nearly continuous and unidirectional stream of water for oxygen uptake (see Figs. 18 and 22). While the study of fish ventilation has a long history (Gudger, 1946; Wegner and Graham, 2010), the work of Hughes and his student, Graham Shelton, largely formalized our current understanding of how gill ventilation is driven by the contraction and expansion of a dual pumping mechanism comprised of the buccal and opercular chambers in teleosts and the orobranchial and parabranchial chambers in elasmobranchs (Hughes, 1960a, 1960b; Hughes and Shelton, 1958, 1962). Hughes' depiction of the expansion and contraction of these chambers as pistons (Fig. 18), further adds to the effective visualization he employed to help describe relatively complex hydrodynamic processes in simple fashion. In such work, Hughes again drew from the diversity of fish cranial and branchial morphology and swimming modes to understand common features of fish ventilation. Such work laid the foundation for future research on fish ventilatory modes, biomechanics, and more comprehensive modeling of respiratory gas exchange (Randall, 2014; Wegner and Graham, 2010).

Finally, Hughes was also one of the first to emphasize the importance of the allometric scaling of gill surface area and related dimensions (i.e., the rate of how gill surface area changes ontogenetically with body mass) in order to more accurately compare gill size and the respiratory capacity among fishes of different body size (see Figs. 28 and 31, Table 6). Because gill surface area does not usually grow isometrically with body mass, Hughes' emphasis on scaling largely helped to correct less accurate comparison methods such as simply dividing by body mass (i.e., determining mass-specific gill surface area) of fishes of different size (e.g., as done originally by Gray, 1954). Hughes thus advocated for the determination of gill surface area and other dimensions over a wide range of body sizes in individual species and cautioned about the extrapolation of such relationships beyond the data available. The importance of this topic has resurfaced in recent years with the resurgence of the Gill Oxygen Limitation (GOL) hypothesis, which argues that a potential mismatch in the scaling of oxygen supply (through gill surface area) and demand (i.e., aerobic metabolism) potentially limits organismal metabolism and growth, and that this may explain decreases in fish size and fishery yields associated with climate warming (Cheung et al., 2013; Pauly, 1981, 2021). It seems from Hughes' writing within his chapter and elsewhere that he was intrigued by the idea of the gills potentially constraining fish metabolism and growth. However, more recent work suggests that there is little physiological basis for the GOL hypothesis (Lefevre et al., 2017, 2018), with several lines of evidence arguing against it (Bigman et al., 2023; Prinzing et al., 2023; Somo et al., 2023). If there is a potential mismatch between oxygen supply and demand under warmer temperatures, this does not appear associated with the gills (Wegner and Farrell, 2023). Addressing such complex issues associated with oxygen uptake at the gills begins with the basic framework and understanding that chapters such as Hughes (1984) provide.

It is thus clear that the works of George Hughes, as summarized in his 1984 chapter "General Anatomy of the Gills," are still at the forefront of fish respiratory science. In fact, this is the second review I have been fortunate to write on Hughes' impacts on the field (Wegner and Graham, 2010). From his comparative approach in describing gill morphology among representative fish species, it is apparent that George Hughes appreciated the August Krogh principle that "for a large number of problems there will be some animal of choice, or a few such animals, on which it can be most conveniently studied." Indeed, in his chapter, Hughes captures the diversity of gill morphology from different fish ecomorphotypes in order to draw common and generalizable conclusions as to the role of the gills in most fishes and how specialized adaptations work within the structural framework of the gills to meet diverse physiological requirements. As I reread Hughes' chapter in preparation for writing this review, I actually started a new list of questions that I would like to investigate—a sign of an interesting and timeless chapter on a topic that still has much to be explored! Perhaps this was said best by Hughes himself while closing his chapter, "To summarize, the whole study of gill morphometry has only just begun…".

References

Bigman, J.S., Wegner, N.C., Dulvy, N.K., 2023. Gills, growth and activity across fishes. Fish Fish. 24, 730–743.

Butler, P.J., 1999. Respiratory system. In: Hamlett, W.C. (Ed.), Sharks, Skates, and Rays: The Biology of Elasmobranch Fishes. John Hopkins University Press, Baltimore, pp. 174–197.

Cheung, W.W., Sarmiento, J.L., Dunne, J., Frölicher, T.L., Lam, V.W., Palomares, M.D., Watson, R., Pauly, D., 2013. Shrinking of fishes exacerbates impacts of global ocean changes on marine ecosystems. Nat. Clim. Chang. 3, 254–258.

Graham, J.B., 1997. Air-Breathing Fishes: Evolution, Diversity, and Adaptation. Academic Press, San Diego, p. 299.

Graham, J.B., 2006. Aquatic and aerial respiration. In: Evans, D.H., Claiborne, J.B. (Eds.), The Physiology of Fishes, third ed. CRC Press, Boca Raton, pp. 85–117.

Graham, J.B., Wegner, N.C., 2010. Breathing air in water and in air: the air breathing fishes. In: Nilsson, G.E. (Ed.), Respiratory Physiology of Vertebrates: Life with and without Oxygen. Cambridge University Press, Cambridge, pp. 174–221.

Gray, I.E., 1954. Comparative study of the gill area of marine fishes. Biol. Bull. 107, 219–225.

Gudger, E.W., 1946. Oral breathing valves in fishes. J. Morphol. 79, 263–285.

Hughes, G.M., 1960a. A comparative study of gill ventilation in marine teleosts. J. Exp. Biol. 37, 28–45.

Hughes, G.M., 1960b. The mechanism of gill ventilation in the dogfish and skate. J. Exp. Biol. 37, 11–27.

Hughes, G.M., 1966. The dimensions of fish gills in relation to their function. J. Exp. Biol. 45, 177–195.

Hughes, G.M., 1972. Morphometrics of fish gills. Respir. Physiol. 14, 1–25.

Hughes, G.M., 1984. General anatomy of the gills. In: Hoar, W.S., Randall, D.J. (Eds.), Fish Physiology. vol. 10A. Academic Press, Orlando, pp. 1–72.

Hughes, G.M., Gray, I.E., 1972. Dimensions and ultrastructure of toadfish gills. Biol. Bull. 143, 150–161.

Hughes, G.M., Morgan, M., 1973. The structure of fish gills in relation to their respiratory function. Biol. Rev. 48, 419–475.

Hughes, G.M., Munshi, J.S.D., 1973. Nature of the air-breathing organs of the Indian fishes *Channa*, *Amphipnous*, *Clarias* and *Saccobranchus* as shown by electron microscopy. J. Zool. 170, 245–270.

Hughes, G.M., Shelton, G., 1958. The mechanism of gill ventilation in three freshwater teleosts. J. Exp. Biol. 35, 807–823.

Hughes, G.M., Shelton, G., 1962. Respiratory mechanisms and their nervous control in fish. Adv. Comp. Physiol. Biochem. 1, 275–364.

Hughes, G.M., Dube, S.C., Munshi, J.S.D., 1973. Surface area of the respiratory organs of the climbing perch, *Anabas testudineus* (Pisces: Anabantidae). J. Zool. 170, 227–243.

Lefevre, S., Mckenzie, D.J., Nilsson, G.E., 2017. Models projecting the fate of fish populations under climate change need to be based on valid physiological mechanisms. Glob. Chang. Biol. 23, 3449–3459.

Lefevre, S., Mckenzie, D.J., Nilsson, G.E., 2018. In modelling effects of global warming, invalid assumptions lead to unrealistic projections. Glob. Chang. Biol. 24, 553–556.

Muir, B.S., Hughes, G.M., 1969. Gill dimensions for three species of tunny. J. Exp. Biol. 51, 271–285.

Palzenberger, M., Pohla, H., 1992. Gill surface area of water-breathing freshwater fish. Rev. Fish Biol. Fish. 2, 187–216.

Pauly, D., 1981. The relationships between gill surface area and growth performance in fish: a generalization of von Bertalanffy's theory of growth. Berichte der Deutschen Wissenchaftlichen Kommission für Meeresforschung 28, 251–282.

Pauly, D., 2021. The gill-oxygen limitation theory (GOLT) and its critics. Sci. Adv. 7, eabc6050.

Prinzing, T.S., Bigman, J.S., Skelton, Z.R., Dulvy, N.K., Wegner, N.C., 2023. The allometric scaling of oxygen supply and demand in the California Horn Shark, *Heterodontus francisci*. J. Exp. Biol. 226, jeb246054.

Randall, D., 2014. Hughes and Shelton: the fathers of fish respiration. J. Exp. Biol. 217, 3191–3192.

Somo, D.A., Chu, K., Richards, J.G., 2023. Gill surface area allometry does not constrain the body mass scaling of maximum oxygen uptake rate in the tidepool sculpin, *Oligocottus maculosus*. J. Comp. Physiol. B. 193, 425–438.

Wegner, N.C., 2011. Gill respiratory morphometrics. In: Farrell, A.P. (Ed.), Encyclopedia of Fish Physiology: From Genome to Environment. Academic Press, San Diego, pp. 803–811.

Wegner, N.C., Farrell, A.P., 2023. Plasticity in gill morphology and function. In: Alderman, S.L., Gillis, T. (Eds.), Encyclopedia of Fish Physiology, second ed. Elsevier Press.

Wegner, N.C., Graham, J.B., 2010. George Hughes and the history of fish ventilation: from Du Verney to the present. Comp. Biochem. Physiol. 157A, 1–6.

Wegner, N.C., Sepulveda, C.A., Bull, K.B., Graham, J.B., 2010a. Gill morphometrics in relation to gas transfer and ram ventilation in high-energy demand teleosts: scombrids and billfishes. J. Morphol. 271, 36–49.

Wegner, N.C., Sepulveda, C.A., Olson, K.R., Hyndman, K.A., Graham, J.B., 2010b. Functional morphology of the gills of the shortfin mako, *Isurus oxyrinchus*, a lamnid shark. J. Morphol. 271, 937–948.

Wegner, N.C., Lai, N.C., Bull, K.B., Graham, J.B., 2012. Oxygen utilization and the branchial pressure gradient during ram ventilation of the shortfin mako, *Isurus oxyrinchus*: is lamnid shark–tuna convergence constrained by elasmobranch gill morphology? J. Exp. Biol. 215, 22–28.

Wegner, N.C., Sepulveda, C.A., Aalbers, S.A., Graham, J.B., 2013. Structural adaptations for ram ventilation: gill fusions in scombrids and billfishes. J. Morphol. 274, 108–120.

Chapter 2

GENERAL ANATOMY OF THE GILLS☆

G.M. HUGHES
Research Unit for Comparative Animal Respiration, University of Bristol, Bristol, United Kingdom

Chapter Outline

I. INTRODUCTION

The gills form a highly characteristic feature of fishes, and their presence has a marked effect on the anatomy and functioning of the rest of the animal. Although their origin among early chordates may have been with particular reference to feeding, nevertheless, it is to exchanges with the environment, particularly O_2 and CO_2, that they have become most adapted. Their evolution may be considered in relation to this function, and their adaptation to

☆This is a reproduction of a previously published chapter in the Fish Physiology series, "1984 (Vol. 10A)/General Anatomy of the Gills: ISBN: 978-0-12-350430-2; ISSN: 1546-5098".

Fish Physiology, Vol. 40A. https://doi.org/10.1016/bs.fp.2023.09.001

particular environments provides a fascinating study in comparative functional anatomy and physiology. During this evolution it is apparent that the existence of a large surface exposed to the external medium with a thin barrier separating the internal and external media inevitably leads to exchanges in addition to O_2 and CO_2, and consequently, these have had a modifying influence on the evolution of these organs and in certain circumstances other functions have become more important than that of respiration.

It is only in the early stages of most living fish that the gills are clearly visible: external gills form important respiratory and indeed nutritive organs in some elasmobranchs (Needham, 1941). In adults the red gills are normally covered up and enclosed within elaborations of the gill slits. These slits originate in the pharyngeal region as perforations between the alimentary canal and the lateral body surface, the enclosed mesoderm developing into the gill arches. The large number of gill slits among agnathan fishes becomes reduced to a more or less constant number among true fishes, in which typically there are five slits on each side of the animal. Anterior to these branchial organs there are the hyoid, mandibular, and premandibular gill arches of the primitive head, which can be recognized during embryonic development (Goodrich, 1930; Horstadius, 1950). The slit between the premandibular and mandibular arches becomes incorporated in the mouth, and only in certain orders is a slit present between the mandibular and hyoid arches. In most elasmobranchs and some primitive actinopterygians this is confined to the dorsal region where it forms the spiracle. The remaining gill slits are of the same general type and consist of increased surface foldings of the epithelium, which is perforated between the alimentary canal and outside. Spacing between these slits in primitive forms is greater than in more advanced groups, and from an anatomic point of view involves a reduction in the interbranchial septum and is one of the most important changes that has occurred in the external anatomy of gills. As a consequence there has been a condensation along the longitudinal axis in the extent of the gilled portion of the pharynx. These changes are associated with further modifications of the head skeleton leading to the evolution of gill covers of various kinds, which reach their maximum development in the evolution of the opercular bones that cover the gills of the teleost fishes.

Another important selection pressure has been in relation to the swimming habits of these animals and the necessity for the maintenance of streamlining at the head end of the fish. Economy of space and accommodation within the head are clearly important aspects of external gill morphology. The most advanced developments of this kind are found in oceanic forms such as tunas, in which streamlining of the head has been maintained in spite of the evolution of a gill system with a very large surface exposed to the water. Ventilation of these gills is associated with the swimming movements, the water entering the mouth by ram ventilation, and leaving by the opercular slits as the fish swims through the water. The proportion of water ventilating the gill system is regulated by control of the openings of the mouth and the opercular slits. It is of interest that such ramjet ventilation has also evolved among the cartilaginous fishes (Hughes, 1960a), as some sharks show this form of ventilation when they swim beyond a certain velocity. The condition in this group is completely different from that in tunas, as the operculum has not developed and externally the gill slits are clearly visible. In these forms the extent of the external openings is restricted in comparison with some primitive elasmobranchs, in which the whole gill slit is open to the outside, whereas in more advanced forms the external openings are restricted to approximately one-third at the ventral end of each gill.

A. Relationship of Gills to Lungs

Typically gills are the gas exchange organs of water-breathing fishes, but in some species they are also involved in gas exchange with the air (Schlotte, 1932; Singh, 1976; Graham, 1976). Furthermore, many modifications of the gills have occurred during evolution, notably of the teleosts, in which they have become important aerial gas exchange organs. In these fishes, as in those in which a diverticulum of the pharynx forms a lung homologous with that of tetrapods, the accessory organs generally have the same basic structure of a hollow intucking of the gut lining that is ventilated tidally (Fig. 1). Such a method of ventilation inevitably leads to certain structural and functional complications, and contrasts with the more continuous flow of water through the gills (Table I). Problems of support for the much enlarged surface are different in water and air. In the "true" lungs there has been a development

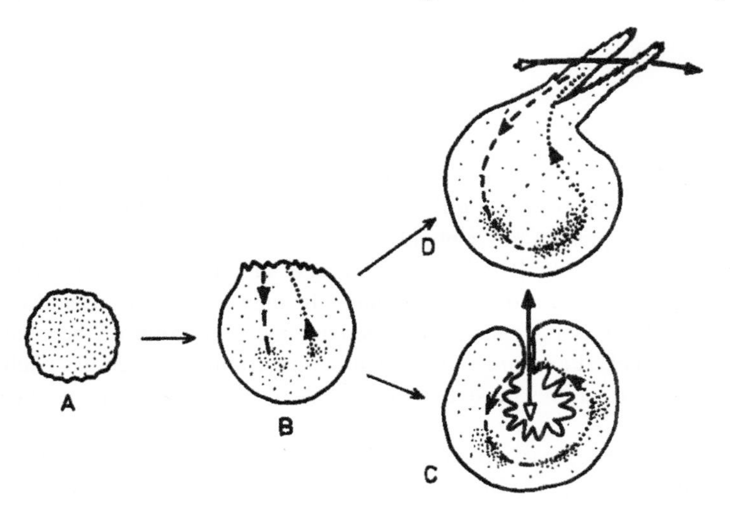

Fig. 1. Diagrams to illustrate general course of evolution from a small spherical organism (A) depending solely on diffusion for its O_2 supply to larger organisms in which localized areas of the external surface become specialized for gas exchange (B) and later differentiate into either external expansions (gills) with continuous flowthrough of water (D) or (C) air-breathing organs (lungs), which are "intuckings" of the external surface relying on tidal ventilation. In (B), (C), and (D), convection systems have evolved between the sites of gaseous exchange and the regions where the oxygen is utilized and CO_2 released. (After Hughes, 1984a.)

Table I Summary of Main Types of Ventilation Found among Vertebrates[a]

Medium	Respiratory organ	Ventilatory flow	Example	References
Water	Gill	Continuous	Tuna	Brown and Muir (1970)
		(Ram jet)	Shark	Hughes (1960a)
	Gill	+ or –	Trout	Hughes and Shelton (1958)
		Continuous		
Water	Gill	Tidal	Dogfish	Hughes (1960a)
			Lamprey	Hughes and Shelton (1962)
Air	Lung	Tidal	Lungfish	McMahon (1969)
			Tetrapods	Gans (1976)
Air/water/air	Suprabranchial chamber	Tidal	Gourami	Hughes and Peters (1984)
Air	Lung	Continuous	Birds	Brackenbury (1972); Scheid (1979)

[a]From Hughes (1978b).

of a special lining layer or surfactant (Pattle, 1976, 1978; Hughes and Weibel, 1978) that is absent from the gills. Nevertheless, such a surfactant layer does not appear to be present in air-breathing organs that have developed independently of the true lung (Hughes, 1978a), and support for the gas exchange surface is achieved in a different manner. In many cases the basic lamellar structure of the gills is apparent, and perhaps the proportion of the total surface of the intucked exchange organ that is directly involved in gas exchange is less than that of the most evolved lungs, but few quantitative data are available.

In spite of the great structural variety of the parallel evolution of the air-breathing organs, a number of physiological generalizations suggested many years ago (Hughes, 1966a; Rahn, 1966) are now sufficiently well documented to justify acceptance. Thus, in most so-called bimodal breathers, the gas exchange with the water is more important with respect to CO_2 release, whereas the air-breathing organ is more responsible for oxygen uptake. The consequent difference in the gas exchange ratio at the two organs must lead to some differences in basic function at a molecular or fine-structural level, but these have scarcely been investigated. The relative balance in the importance of the two organs is variable and related to the particular mode of life of the fish concerned. The ventilation of some air-breathing organs has been shown to involve quite a different mechanism, in which the gas exchange surface, although in contact with air for most of the cycle, becomes flushed with water during the uptake of air at the water surface (Peters, 1978; Hughes, 1977). Such a mechanism is clearly intermediate between the typical ventilation mechanism of the gill and that of organs that are only in contact with air. Perhaps it is this feature of gas exchange organs modified from gills that has precluded the development of surfactant lining and the associated cytological structures that have been noted previously. It also supports the view that fish gas exchange organs show the greatest range in ventilatory mechanisms (Table I).

II. DEVELOPMENT OF GILLS

Fish eggs vary in the amount of yolk they contain and consequently the timing of different developmental stages. In all cases, however, the development of the gills forms an important part of the whole process. From a respiratory point of view, it is essential that they should reach a sufficient level of morphological and functional development that they can take over gas exchange functions when the surface area: volume ratio becomes too small and gas exchange across the body surface is insufficient to meet the growing demands of the developing fish (Figs. 1 and 2). In a number of cases, development contains a "critical" period (May, 1974) during which much mortality occurs and can be a serious problem in aquaculture. One important aspect is clearly the change in nutrition of the embryo as the yolk sac is used up and other sources of energy become more important. One other feature of this problem, however, is that there are also changes in the surface area: volume ratio, particularly in the developing gill region. A morphometric study (Iwai and Hughes, 1977) has shown, for example, that the critical period in the Black Sea bream coincides with the stage at which the surface area:volume ratio of the developing gills reaches a minimum (Fig. 2). After this stage, the increasing number of lamellae ensures that the ratio increases, and hence sufficient surface is available for oxygen uptake. Coincident with development of the gills, there is also development of other parts of the respiratory and cardiovascular system and coordination of the pumping systems for water and blood flow through the gills (Holeton, 1971; Morgan, 1974).

Although such functional aspects have not been investigated a great deal until recently, there has always been much interest in the detailed morphological development of the gill system, as it constitutes one of the diagnostic characters of chordates. In some primitive forms (e.g., *Polypterus*) there are patches of cilia on these epithelia (Hughes, 1980a, 1982) that may represent the more continuous ciliation of ancestral chordates, which depend on ciliary currents for feeding and respiration. Early studies concerned themselves with

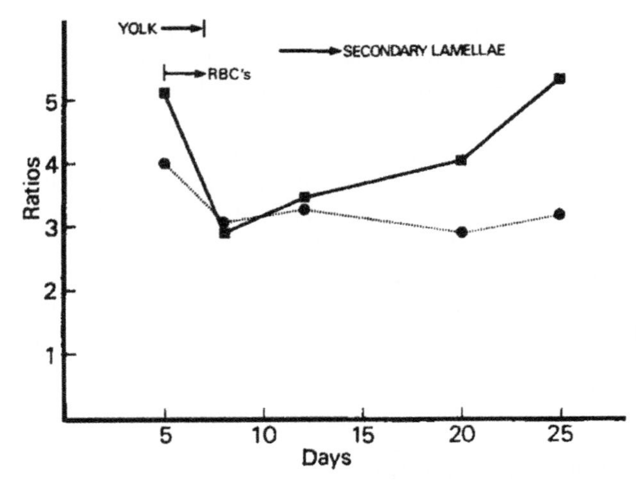

Fig. 2. Changes in the ratio of the intersection count to the point count for the gill arches (dotted line) and gill filaments (solid line) during developmental stages of Black Sea bream. These ratios are proportional to the surface area: volume ratio and show that for the gill filaments it is at a minimum after about 7 days, and this coincides with the sensitive period. The surface area: volume ratio for the arches declines steadily during development, because there is no further increase in surface folding. (From Iwai and Hughes, 1977.)

the morphological nature of the gill epithelium and especially the development of the branchial blood vessels (Sewertzoff, 1924). During development the whole sequence is continuous; the epithelium that forms the surface of the gill clefts later becomes the surface of the filament, and still later the surface of the lamellae. Later differentiation occurs in different regions of the adult in relation to the particular microenvironmental and functional conditions. It is, however, convenient to discuss sequentially the development of the main constituents.

A. Branchial Arches

Differentiation of the gills proceeds from the anterior end of the embryo soon after gastrulation as gill pouches begin to evaginate from the endoderm and meet ectodermal invaginations. These two processes are followed by piercing and formation of the gill clefts, which isolate interbranchial bars between them. Much discussion has entered around the nature of the epithelium that finally develops into the gills, and some of the earlier workers (Moroff, 1904; Goette, 1878) thought that the gills developed mainly from ectoderm, whereas subsequent scholars such as Bertin (1958) and others believe that they are endodermal in origin. Some morphologists have distinguished "endobranchiate" Cyclostomata from "ectobranchiate" Gnathostomata. The precise nature of the germ layer is not now considered to be of importance, and there appears to be no consistent morphological or functional difference between ectodermal and endodermal regions of the gills.

The interbranchial bars also isolate mesodermal regions from which the coelom becomes obliterated and the mesoderm develops into musculature. At this stage, six or more primary branchial vessels arise from the ventral aorta and pass around the pharynx to become connected with the dorsal aorta. Each primary vessel becomes differentiated into afferent and efferent branchial arteries (Fig. 3A). The precise pattern of this differentiation varies

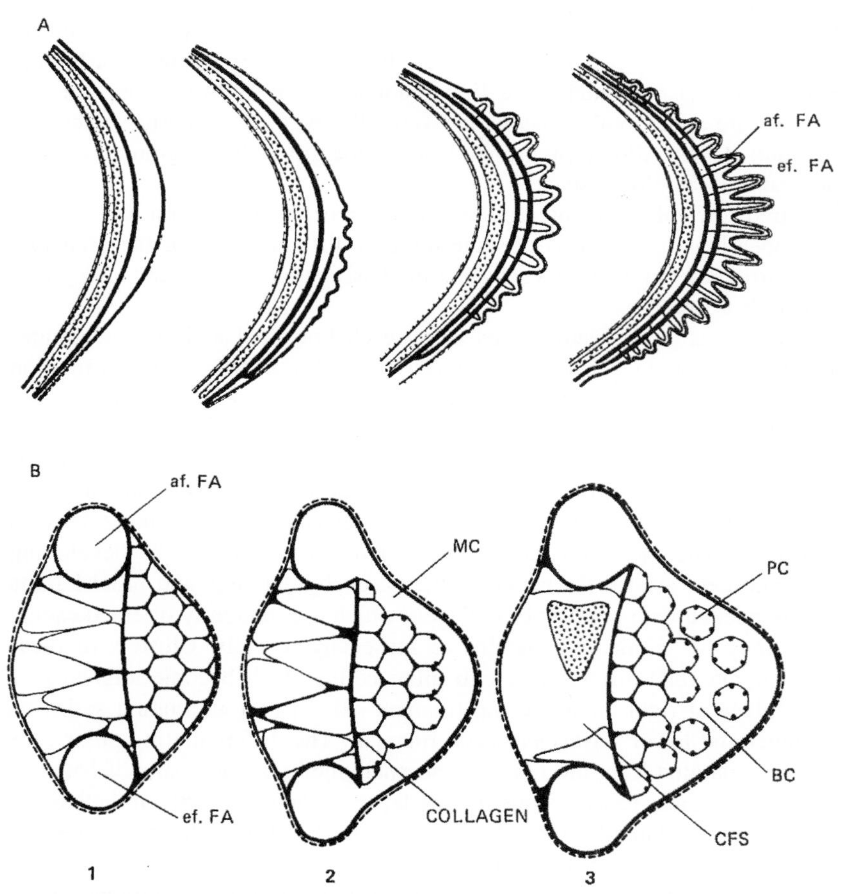

Fig. 3. Rainbow trout (*Salmo gairdneri*). (A) Diagrams showing the formation of the afferent and efferent branchial arteries from the primary branchial artery of a gill arch. Note the beginning of the development of the filament arteries between afferent and efferent branchial arteries. (B) Diagrams showing stages in the development of a trout lamella. Formation of the marginal channel (MC) and development of pillar cells (PC) in the main body of the lamella are illustrated. Blood channels (BC) between the pillar cells and along the margin connect the afferent (af.FA) and efferent filament arteries (ef.FA). In **1** there is an accumulation of mesenchyme cells, and in **2** the formation of the marginal channel and columns in pillar cells. In **3**, formation of the blood spaces and pillar cells is shown. Note the presence of the central venous sinus (CVS). Note marginal channel endothelial lining. (From Hughes and Morgan, 1973.)

between elasmobranchs and teleosts, and leads ultimately to the well-known difference between selachians—with their paired efferent branchials that join to form single epibranchial arteries that enter the dorsal aorta—and teleosts, in which there is usually a single efferent and a single afferent branchial. Modifications in this general pattern occur among different teleost groups (Muir, 1970; Adeney and Hughes, 1977), and some functional significance may be ascribed to it.

The filaments start to develop as the gill bar epithelium begins to bulge outwards and becomes invaded by a series of vascular loops, which connect between the existing afferent and efferent branchial arteries (Fig. 3A). In both teleosts and elasmobranchs, the first filaments form midway along the gill arch and are added both dorsally and ventrally over a long period extending into adult stages. Allometric growth leads to the gill arches beginning to bend backwards to an elbow shape.

Paired rows of filaments grow out in a laterocaudal direction, the anterior row appearing first. At a later stage there is differentiation of the branchial arch cartilages, so that in each septum basi-, hypo-, epi-, and pharyngobranchials arise.

The corpus cavernosum forms along each branchial arch and penetrates into the developing gill filaments (Acrivo, 1938). Details of the formation of the blood vessels vary among different groups and species, but eventually the filament loops lengthen as the afferent and efferent branchial arteries grow apart. During this early stage of filament development, the blood vessels are surrounded by mesenchyme and there is a well-defined basal lamina below the surface epithelium (Fig. 3B). The underlying collagen layer is well defined but later becomes more diffuse. Toward the tips of the developing filaments there are spaces between the mesenchymal cells and basal lamina that are packed with bundles of collagen, which are randomly oriented except where they lie within indentations of mesenchyme cells. Clusters of these mesenchyme cells differentiate into pillar cells, which become aligned in a single plane at right angles to the filament axis in two alternating series on the upper and lower surfaces of each filament. The differentiating pillar cells become separated from the filament mesenchyme by a thick band of collagen that joins the inner basement membrane layers of opposite sides at the base of each developing lamella (Fig. 25). Distal to this band of collagen, the mesenchyme cells continue to differentiate into pillar cells, particularly by the formation of collagen columns, which join the two basement membrane layers of the developing lamellae (Fig. 3B). Collagen columns lie within but are external to pillar cells, because the intuckings that contain the collagen are lined with pillar cell membrane.

New lamellae are added toward the tips of the filaments during the entire period of growth. Spaces within the mesenchyme around the edge of the lamellae merge to form a continuous channel joining the afferent and efferent filament arteries, and this gives rise to the marginal channels of the adult.

Once blood flow through this channel begins, the cells lining its outer border develop the typical membrane-bound Weibel–Palade bodies (Morgan, 1974), which is typical of the adult marginal channel (Tovell *et al.*, 1971). During growth the shape of the secondary lamella changes from ovoid to triangular, with the apex toward the efferent filament artery. New pillar cells are formed by division of the proximal row of mesenchyme cells, this row being completely isolated from the main body of the filament by a thick collagen sheet outside of which is the so-called pillar cell system (Hughes and Perry, 1976). The proximal cell layer is usually about one cell thick, and these mesenchyme cells sometimes have collagen columns around their outer borders. As more columns develop, the cells divide and move away to leave the smaller undifferentiated cells (Fig. 3B). In the early stages the external surface of the epithelium is folded and has short microvilli, the height and frequency of which increases to that found in the adult condition at a later stage. To begin with, the water–blood barrier is relatively thick—for example, 11 μm at 31 days in rainbow trout (Morgan, 1974), whereas after 102 days it is 7 μm. The higher proportion of chloride cells in young trout may be related to the greater osmotic problems of smaller animals having the greater surface area:volume ratio (Wagner *et al.*, 1969).

B. Hyoid Arch

The hyoid arch is situated immediately anterior to the first branchial arch, in front of which is the mandibular arch. In most fishes it becomes reduced and is represented by the hyomandibular and basihyal cartilages. It often plays an important role in the suspension of the jaw, which is derived from the mandibular arch. Between the hyoid arch and the first branchial arch, there is a complete gill slit; however, a prehyoidean slit, between the mandibular and hyoid arches, is only complete in a fossil group, the Aphetohyoidea (Watson, 1951). Among modern fish, it only persists as a spiracle in the dorsalmost portion and is especially well developed in bottom-living rays, where it forms the main entry for water into the orobranchial cavity and is actively closed by a valve (Hughes, 1960a). The posterior hemi-branch attached to the hyoid arch persists in most elasmobranchs and some primitive bony fish as a complete hemibranch, for example in sturgeons and *Latimeria* (Table II). Among the cartilaginous fishes, it forms a distinct hemibranch that lines the anterior part of the first gill slit. Morphologically the hyoidean posterior hemibranch seems to be identical to other hemibranchs at both gross and fine-structural levels. In most fishes, development of an operculum is associated with a reduction in the spiracle, and a hyoidian hemibranch is usually absent. In such fishes as sturgeons, where the hyoid hemibranch is found attached to the inner side of the operculum, the presence of a spiracle is perhaps surprising as the pseudobranch represents the gill of the slit anterior to the hyoid arch and is often associated with the spiracle, which is well developed in elasmobranchs.

Table II Presence of Gills on Different Visceral Arches of Fishes[a]

Genus	Mandibular arch	Spiracle	Hyoid arch	Branchial arches						
				I	*II*	*III*	*IV*	*V*	*VI*	*VII*
Scyliorhinus	ps	+	ph	H	H	H	H			
Hexanchus	ps	+	ph	H	H	H	H	H		
Heptanchus	ps	+	ph	H	H	H	H	H	H	
Raia	ps	+	ph	H	H	H	H	.		
Chimaera	.	–	ph	H	H	H	ah	.		
Latimeria	.	–	ph	H	H	H	H	.		
Neoceratodus	.	–	ph	H	H	H	H	.		
Protopterus	.	–	ph	.	.	H	H	ah		
Acipenser	ps	+	ph	H	H	H	H	.		
Huso	ps	–	ph	H	H	H	H	.		
Lepisosteus	ps	–	ph	H	H	H	H	.		
Amia	ps	–	.	H	H	H	H	.		
Polypterus	.	+	.	H	H	H	ah	.		
Ophiocephalus (*Channa*)	.	–	.	H	H	H	ah	.		
Anabas	.	–	.	H	H	H	.	.		
Amphipnous	.	–	.	.	H	.	.	.		
Opsanus	.	–	.	H	H	H	.	.		
Dibranchus	.	–	.	.	H	H	.	.		

[a]Abbreviations: ah, anterior hemibranch; H, holobranch; ph, posterior hemibranch; ps, pseudobranch (external).

C. Pseudobranch

The pseudobranch represents the gill of the slit that lies between the mandibular and hyoid arches. In fact, it is the posterior hemibranch of the mandibular arch. It has been recognized for a long time that this gill is often supplied with blood that has already been oxygenated in the hemibranch of the hyoid or first branchial arch; therefore, it was considered to be non-respiratory and hence was named "pseudo." Many different functions have been ascribed to the pseudobranch (e.g., chemoreceptor, mechanoreceptor, secretory), but as yet there is no certain indication of any function common to all fishes. Muller (1839) distinguished two morphological types, the first being free pseudobranchs in which lamellae are distinguishable and in direct contact with the water. The second type is the covered pseudobranch where the surfaces do not come into close contact with the water. Of this latter kind, the glandular type of pseudobranch is perhaps the most developed, but even in the structure of this organ it is possible to recognize the typical pillar cell organization of the lamellae (Munshi and Hughes, 1981). Some authors have described at least four types of pseudobranchs that vary in their degree of isolation from the external environment (Leiner, 1938; Bertin, 1958). It should be emphasized that the carotid labyrinths of bony fishes (Siluriformes) and the carotid labyrinth and body of tetrapods have quite a different origin and structurally show no trace of pillar cells.

One of the interesting features of the pseudobranch is that the efferent blood from it passes to the brain and in many cases to the eyes. It has been suggested that in some cases it may result in these organs being supplied with blood of even higher than normal oxygen levels. Certainly, damage to the pseudobranch soon produces blindness in a number of fishes, but it is difficult to extirpate the pseudobranch surgically and not interfere with the ophthalmic artery.

III. GILL ORGANIZATION

The general organization of the gills is based on a system of progressive subdivision, first of all giving rise to the individual pouches separated by the gill septa, which vary in extent from group to group, being most extensive in the more primitive forms. Along their full length and on anterior and posterior surfaces of each arch are found the gill filaments, which may be regarded as an increase in surface of the interbranchial septum (Fig. 4). The filaments themselves have their surface further increased by being folded into a series of lamellae. The many variations in detail of this plan are related both to the systematic position of the fish and its mode of life. In some cases the primary epithelium of both the filaments and lamellae becomes modified.

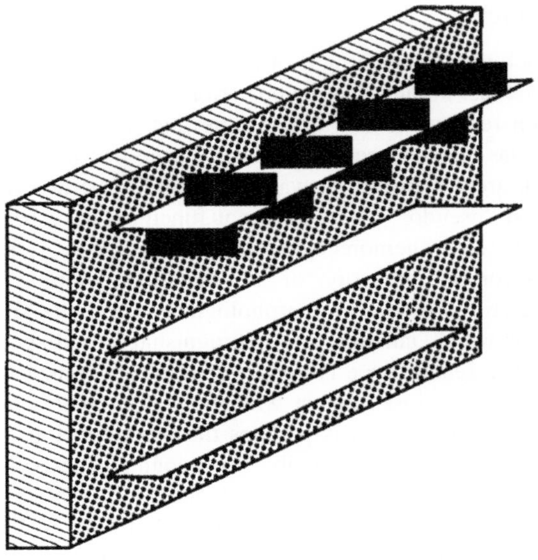

Fig. 4. Diagrammatic three-dimensional representation of a single gill bar of a developing fish showing the three main epithelia: (A) the gill cleft (coarsely dotted area), which becomes folded into (B) the gill filaments (white areas), which in their turn become folded into (C) the lamellae (finely dotted areas).

A. Gill Septum

A gill septum separates two adjacent gill pouches, and to its surface is attached a series of filaments. In the most primitive groups (e.g., Elasmobranchii) the septum forms a complete partition between the pharynx and the outer body wall. Its extension forms a flap-valve for the next posterior slit. In more advanced groups there is a progressive reduction in the septum and the consequent freeing of the filaments at their tips (Fig. 6), so that water can pass more freely between the filaments of a given hemibranch and so enter the opercular cavity. All the filaments attached to one side of a gill septum form a hemibranch, and the two hemibranchs attached to a given branchial arch constitute a holobranch. It is clear that, where the septum is more complete, there will be some interference with the passage of water across the filament. This is less than was initially supposed, as more detailed anatomic and physiological studies have demonstrated that on the inner side of the attachment of the filaments there is a so-called septal channel (Kempton, 1969; Grigg, 1970a,b) along which the water can flow outward (Figs. 5 and 23B). Blood and water flows are countercurrent (Hills and Hughes, 1970; Piiper and Scheid, 1982), although a more crosscurrent

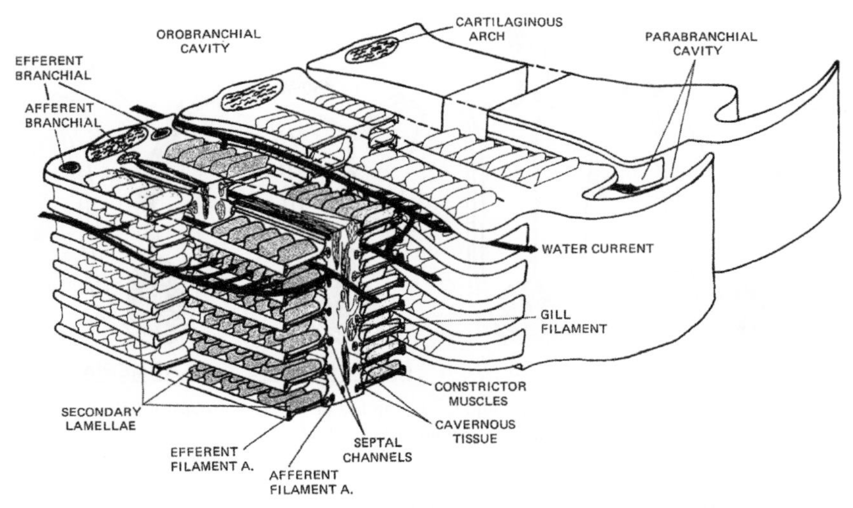

Fig. 5. Stereogram to show the structure of three gill arches in a dogfish. Each arch is shown in different degrees of detail. In the anteriormost arch is shown the structure of the secondary lamella and septum; in the last arch the main organization and shape of the filaments are visible, as well as the way in which extensions of the septa overlap the next posterior arch and form the valve at the exit of the parabranchial cavity. The direction of water flow between the secondary lamellae along the septal channels is illustrated in thick arrows.

arrangement has also been considered (Piiper and Schumann, 1967). Before leaving through the external gill slit, water passes into a series of parabranchial cavities, which are analogous to the opercular cavity of the teleost system. In the elasmobranchs, Woskoboinikoff (1932) distinguished the cavity before the gills as the orobranchial cavity, comprising the main cavity of the mouth together with parts of the gill slits from the internal slit to the point where the water flows between the filaments. In teleost fish the orobranchial cavity is more commonly referred to as the buccal cavity; furthermore, as already mentioned, the opercular cavities are partly homologous with the combined parabranchial cavities, as in teleosts the septa have become much reduced and the development of the operculum has produced a separate cavity into which the water from all of the gill slits flows. Functionally the separation between these cavities is the gill resistance, mainly constituted by the narrow gap between lamellae.

The degree of development of gill septa is best understood from a series of diagrammatic sections across individual gill arches (Fig. 6). The two extremes indicated earlier are exemplified in the elasmobranchs, with their complete septum, and in the teleosts, where the septum is very significantly reduced. In some of the intermediate groups there are varying degrees of septal development. These include, for example, the condition in *Neoceratodus* (Dipnoi)

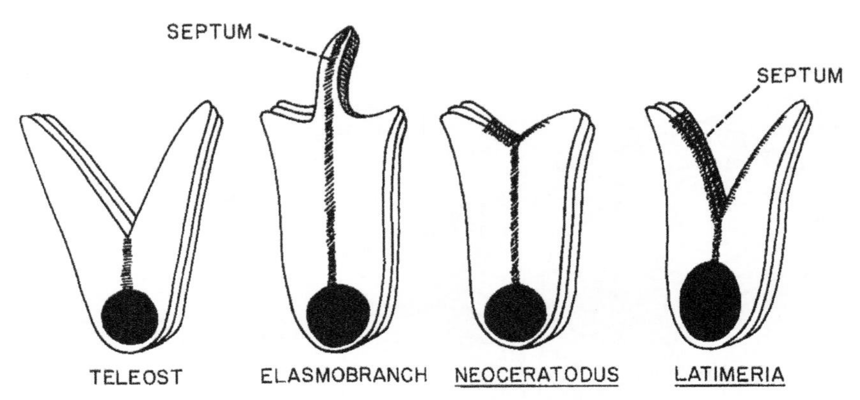

Fig. 6. Diagrammatic cross section through the gill arches of different groups of fishes showing varying degrees of development of the gill septum. (After Hughes, 1980c.)

where the gill filaments of a given holobranch are held together except near their tips. A similar situation has been described for the gill of *Latimeria* (Hughes, 1980c), where the main interbranchial septum splits proximally but continues and maintains contact between the filaments of individual hemibranchs except at their tips. *Latimeria* has an operculum that is not so well developed, and a similar situation is also found in the chimaeroid fish. Among primitive actinopterygians such as *Amia*, the interbranchial septum is well developed proximally, but the gill filaments are free for about one-third of their length. Among different groups of teleosts, there is variation in the degree of development of the septum, that has been used as a basis for gill classification (Miscalencu and Dornescu, 1970). In some forms the septum is almost complete, whereas in the more advanced perciform type the filaments are free for most of their length.

These differences are associated with other features of the anatomy shown diagrammatically in Fig. 7. Some of the associated features concern the development of the filament adductor muscles, which serve to draw together the tips of the filaments of the adjacent hemibranchs, a movement that occurs rhythmically and with particular force during coughing (Bijtel, 1949). Normally the tips of the filaments are in close contact with one another because of the elastic properties of the gill filament skeleton. However, it would seem that the different degrees of development of the septa are very much related to the systematic position of the fish.

Dornescu and Miscalencu (1968a, b) distinguished three types (Fig. 7) of gills on the basis of their studies of over 50 species from six different orders of Teleostei. The clupeiform type has a well-developed interbranchial septum, so that the filaments are only free for about one-third of their length. These workers recognized the so-called "blebs" at about one-third of the length along the efferent filament artery, which would appear to be similar to those

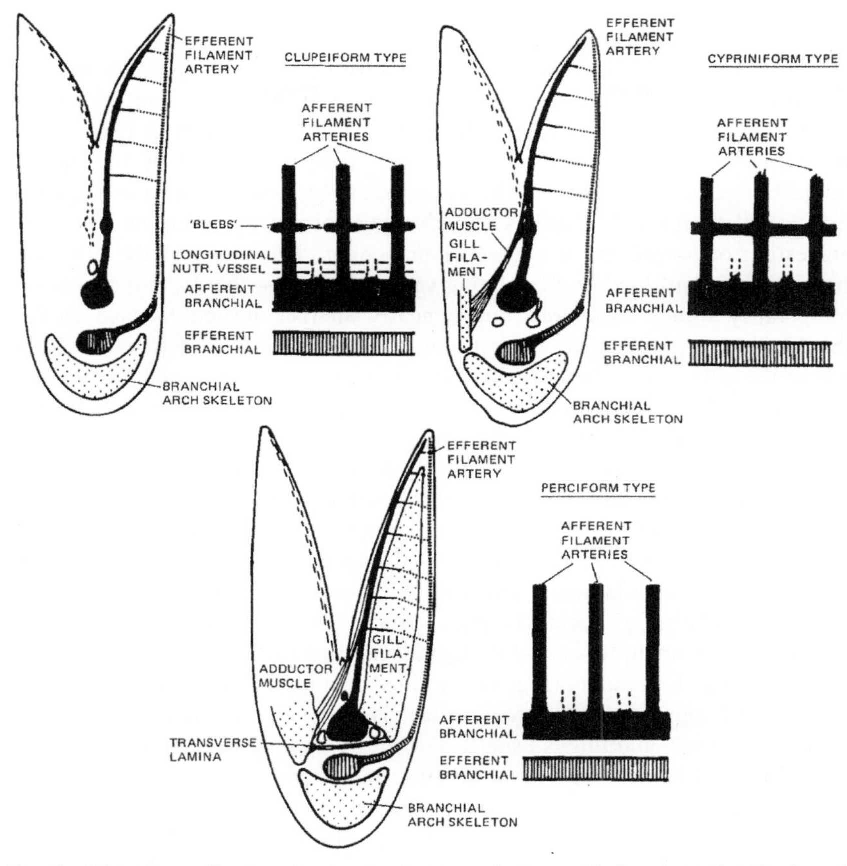

Fig. 7. Diagrams to illustrate the three main types of teleost gills based on classification of Dornescu and Miscalencu (1968a). A transverse section across a gill arch is shown for each type together with a medial longitudinal section to indicate the main blood vessels of the arch and gill filaments. (From Hughes, 1980c.)

described by Fromm (1976). The cypriniform type has a more expanded bleb, and the filament adductor muscles are mainly intrafilamental. The most advanced is the perciform type, which has a much shorter gill septum, with much of the adductor muscles running along the gill filament cartilage. It would seem that these different types with differing degrees of septal development among teleosts will influence the path of the water current. Little attention has been paid to the possible pathway of water once it has traversed the interlamellar spaces of those lamellae proximal to the end of the gill septum, it must be supposed that water flow continues along a channel analogous to the septal channel of the elasmobranch gill.

B. Filaments

Filaments form the most distinctive respiratory structure of fish gills and are sometimes referred to as primary lamellae. However, some authors have referred to the original surface lining the interbranchial septum (Fig. 4) as the primary lamella, in which case the gill filaments formed by its folding would be secondary lamellae. The most generally used and least ambiguous term is gill filament. It also has the advantage of not involving any specific suggestion of homology or degree of branching. The shape of the gill filaments varies considerably, from being very filamental—that is, elongated—to being fairly stubby structures, but in nearly all fish the length exceeds the breadth. Although they give the appearance of being paired structures on either side of the gill septum, they usually alternate with one another, and the number of filaments in the two hemibranchs of a given holobranch are not always the same. In fact, it is more common for the number of filaments attached to the two hemibranchs that line a given gill slit to be closer in number.

In adult fish the number of filaments does not increase so markedly as during the juvenile growth period, but there is a very significant increase in the length of each of them as the fish grows. This leads to an increase in the total length (*L*) of all the filaments, which is an important morphometric dimension used in calculating gill area (Hughes, 1966b, 1984b). When measuring the length of the filament it is usual to begin from the position where the filament joins the gill arch to the tip (Fig. 8), but this is not always the full length of the filament that supports secondary lamellae. The length of the filaments along a given arch varies, and this is usually studied by making measurements of the first and last filament together with filaments at regularly spaced intervals such as every fifth, tenth, or twentieth filament, the intervals at which these measurements are made being related to the total number of filaments along a given hemibranch. The number may range from less than 50 to several hundred in large and active fish. In many cases there is a gradual increase in length of filaments from the dorsal end of the arch to about one-third of the filament number and then a gradual decrease to the ventral end of the arch. A common variation on this plan is that there is often a peak in filament length just before the main angle of the gill arch, which then increases to another maximum shortly afterwards, whereupon the filament length decreases to the last filament (Fig. 9). In general the length of the filaments is measured on each of the hemibranchs on one side of a fish, but in some species (e.g., flatfishes) there are differences on the two sides and it is necessary to measure the length of the filaments of all the hemibranchs on both sides of the fish. There are also differences in length of corresponding filaments on the same arch. For the anterior arches, the filaments of the posterior hemibranch are usually longer, whereas the anterior hemibranch filaments are longer in arches I and II. These differences may vary along a given arch, and this modification is very striking in the gills of *Latimeria* (Fig. 10).

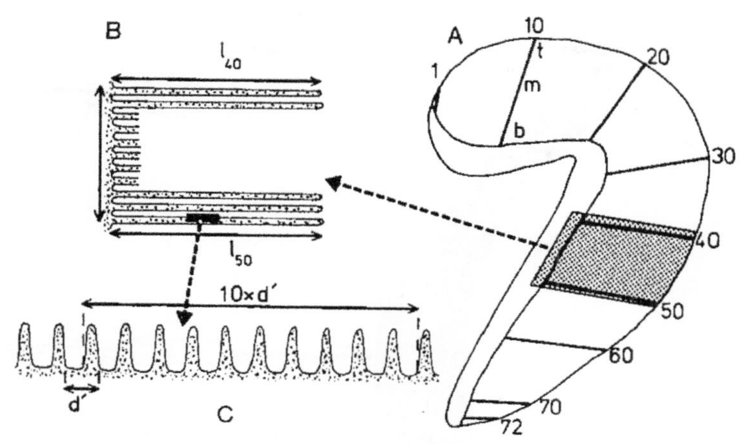

Fig. 8. (A) Diagram of a single hemibranch (72 filaments) of a teleost fish showing the position of filaments selected for measurement at regular intervals around the arch. Lengths of the first and last filaments are also determined. The positions for secondary lamellae selected for determination from the tip, middle, and base of each selected filament are indicated. (B) Diagram showing method for measurement of length of filaments 40 (l_{40}) and 50 (l_{50}) and the distance between filaments 40 and 50. (C) Diagram illustrating the method for measurement of secondary lamella frequency $1/d, = n/2$ by measuring the distance for 10 secondary lamellae. (From Hughes, 1984b.)

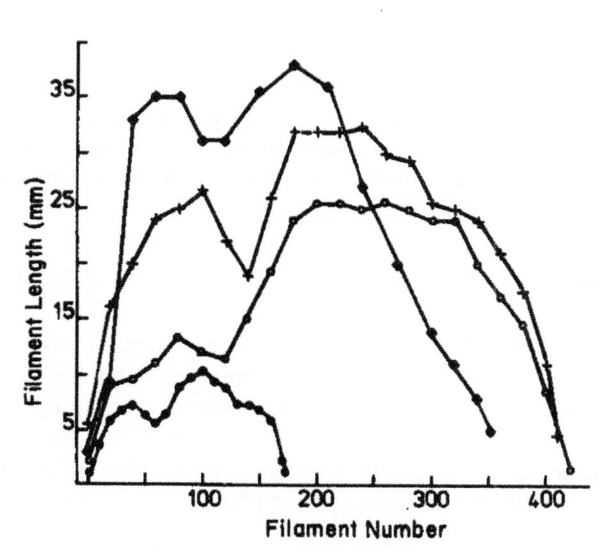

Fig. 9. Diagram showing variation in filament length when plotted for hemibranchs of different species of fish: mackerel, *Scomber scomber* (●), (After Hughes and Iwai, 1978.) Skipjack (◆), grouper (+), barracuda (○). (After Hughes, 1980c.)

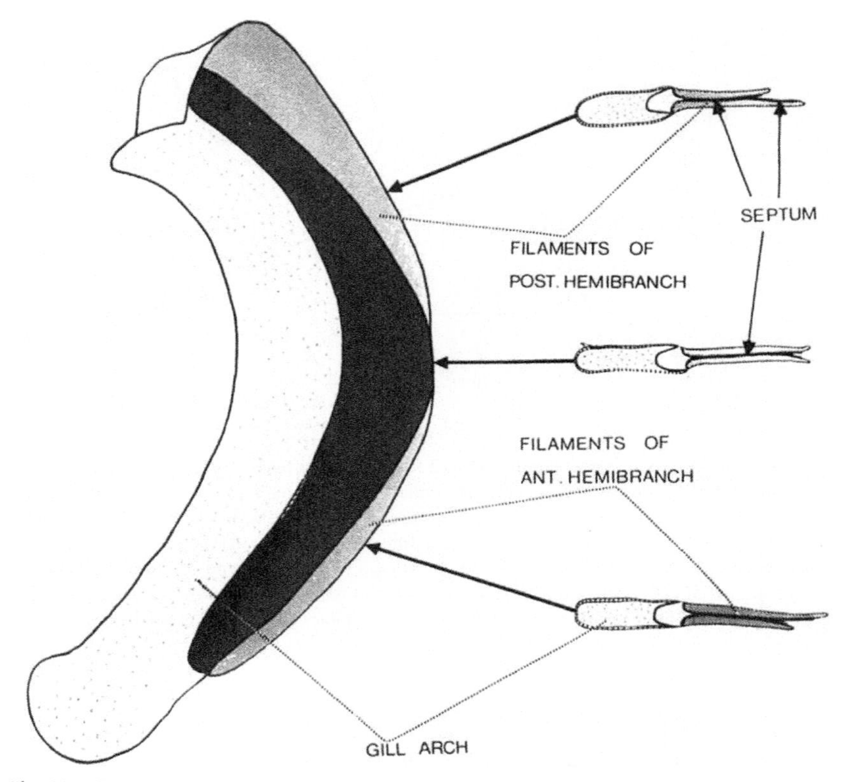

Fig. 10. Diagram of the gill of a single arch of *Latimeria*. Dorsally the filaments of the posterior hemibranch are longer than those of the anterior hemibranch, but ventrally the filaments of the anterior hemibranch are the longer. Sections across the arch at three levels show these differences in filament length. (From Hughes, 1976.)

Total filament length (*L*) is probably the most readily determined overall dimension of the gills and should be achieved with most accuracy. As indicated earlier, the tips of the filaments of adjacent hemibranchs are usually in close contact with one another, and this forms an important feature of the gill sieve, ensuring that little water will be shunted past the tips of the filaments. Such a shunting occurs, however, during hyperventilation, with a consequent fall in percentage utilization (Hughes, 1966b; Hughes and Umezawa, 1968). During coughing, contraction of the filament adductor muscles reduces the resistance of this pathway and enables the water current to be reversed in certain patterns of coughing movements (Hughes and Adeney, 1977). The filaments are supported by gill rays and connective tissue and vascular spaces, the detailed nature of which varies from species to species. In the elasmobranchs the vascular spaces form a special cavernous tissue on the afferent filament side only (Acrivo, 1938), which probably has a supporting hydrostatic function (Hughes, 1980b), and gill rays support the interbranchial septum.

Among teleosts there is a reduction of the afferent arterial cavernous tissue, which may be represented by the "blebs," and each filament is supported by a single gill ray with its expanded edge in the plane of the filament.

The arrangement and form of the gill rays varies from species to species (e.g., between rainbow and brown trout) and has been used as a systematic character; this is also true (Iwai, 1963, 1964; Kazanski, 1964) of the gill rakers that line the gill arches mainly on the pharyngeal side. The form of the gill rays provides greatest support to the trailing edge of the filament rather than to the leading edge that faces the ventilatory water current. This position of the gill rays is related to their function as skeletal supports for the insertion of the filament adductor muscles. Expansion of the gill rays in the plane of the filament is especially notable in large-gilled fish such as tunas, where their support during ramjet ventilation is especially important. Oceanic fishes also show the development of thickened supporting tissue on the leading and trailing edges of the filaments, which may coalesce to form interfilamental junctions (Table III). Thickenings without such coalescence

Table III Presence (+) or Absence (−) of Fusions between Filaments and/or Lamellae

Species	Filamentar fusion	Lamellar fusion
Coryphaena hippurus	−	−
Scomber scombrus	−	−
Mola mola	−	−
Katsuwonus pelamis	−	+
Euthynnus affinis	−	+
Thunnus albacares	+	+
Thunnus atlanticus	+	+
	(outflow surface only)	
Thunnus obesus	+	+
Thunnus thynnus	+	+
Acanthocybium solandri	+	−
Terrapturus audax	+	+
Xiphias gladius	+	−
Istriophorus sp.	+	−
Amia calva	−	+

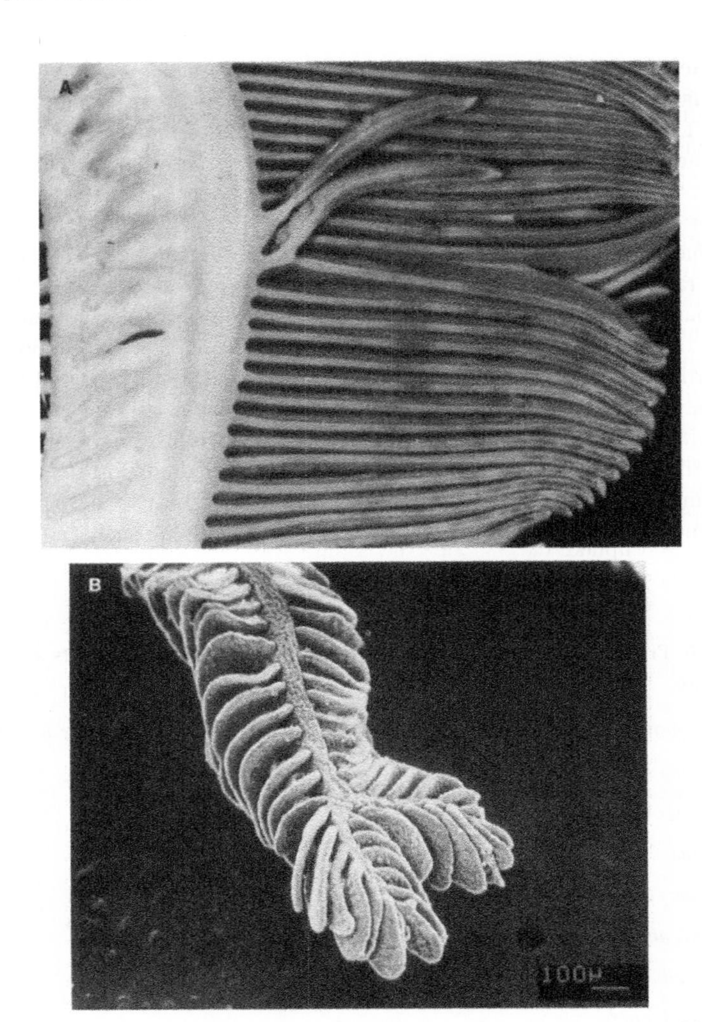

Fig. 11. Photographs of gill filaments that are divided at the tip. (A) Whole gill of *Huso huso* with a single divided filament; (B) scanning electron micrograph of tip of a single filament of a mudskipper.

(Fig. 16C) are found in the sunfish, *Mola mola* (Adeney and Hughes, 1977), which is not a rapid swimmer, confirming that such thickenings are related to the size of the gill. Similarly, the interfilamental junctions are only found in specimens above a certain size in some species.

The filaments are sometimes divided at their tips (Fig. 11), a condition that is more common in some species than others. It probably arises as an anomaly during development, perhaps as a result of physical damage.

C. Lamellae

The lamellae are the most important units of the gill system from the point of view of gas exchange. The rest of the basic anatomy is directed to providing a suitable support for these structures and to enable the water and blood to come into close proximity. Essentially each lamella comprises two epithelia that are kept separated by a series of pillar cells between which the blood can flow. The direction of blood flow is opposite to that of the water and thus facilitates gas exchange (van Dam, 1938; Hughes and Shelton, 1962). Although lamellae have apparently the same basic structure among all groups of fishes, there are nevertheless important differences in detailed anatomy. The extent to which these involve details at an electron-microscopic level is discussed in Chapter 2, especially in so far as they involve differences in relation to blood flow.

From a design point of view, lamellae are required to have a large surface area, where gas exchange can be facilitated without any excessive exchange of ions and water. A close contact between water and blood must be achieved so that oxygen uptake can occur in the limited period (contact time = about 1 sec, Hughes *et al.*, 1981) during which the water and blood are passing the lamellae. At the same time, the finer the pores in the gill sieve, the greater will be the resistance to water flow and hence the greater the expenditure of energy by the fish in moving the viscous and dense respiratory medium. In classical descriptions of the gill system, it has been thought of as being homogeneous; in fact, this superficial impression is shown to be misleading from both a structural and functional points of view. As with most respiratory systems, heterogeneity is more typical (Hughes, 1973b).

1. *Number*

The number of lamellae in a fish is very large and increases steadily with body size; it may reach more than 5 million in a very active fish of 1 kg body weight. In contrast, inactive species such as the toadfish may have a very much smaller number of lamellae (Fig. 12).

2. *Shape*

The heterogeneity of the gill system is well illustrated by the variation in shape of lamellae from a single gill arch (Fig. 13). Heterogeneity produces problems for quantitative analyses both in relation to surface area and also of the distribution of water flow through the gill sieve. There are many variations in shape of lamellae from similar regions of a wide variety of teleost species (Fig. 14). Some elasmobranchs (e.g., *Raia*) may have horn-shaped projections at the anterior end of the lamellae, and this arrangement seems to be found in some other elasmobranchs (Cooke, 1980). In general, however, it is clear that the greatest proportion of the surface of a lamella is found toward the leading edge, that is, the edge at which the water enters the gill from the buccal cavity.

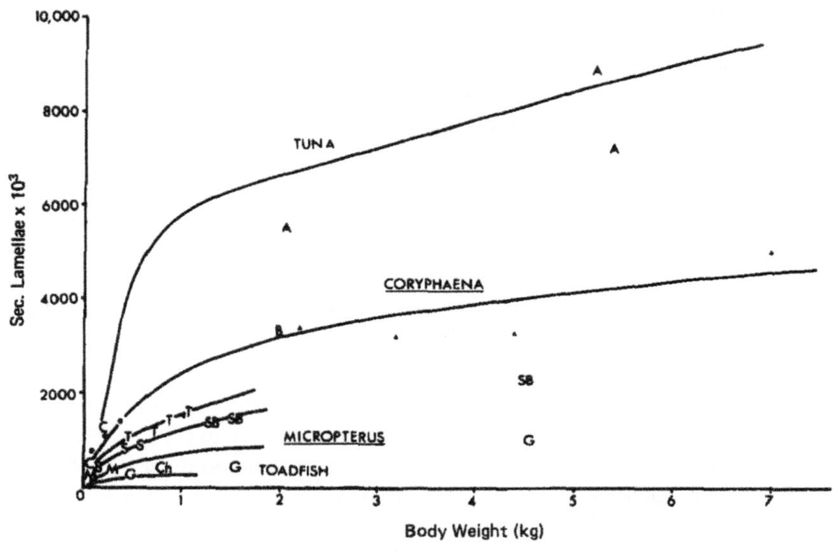

Fig. 12. Relationship between total number of lamellae and body mass for a variety of fishes. Lines are drawn for six of them (tuna, *Coryphaena* & mackerel, sea trout & mullet, striped bass and scup, *Micropterus*, and toadfish). In addition, points for specimens of individual fish are given for the following: A = false albacore, B = bonito, C = *Caranx*, Ch = *Chaenocephalus*, G = goosefish, M = mullet, S = scup, SB = striped bass, T = sea trout.

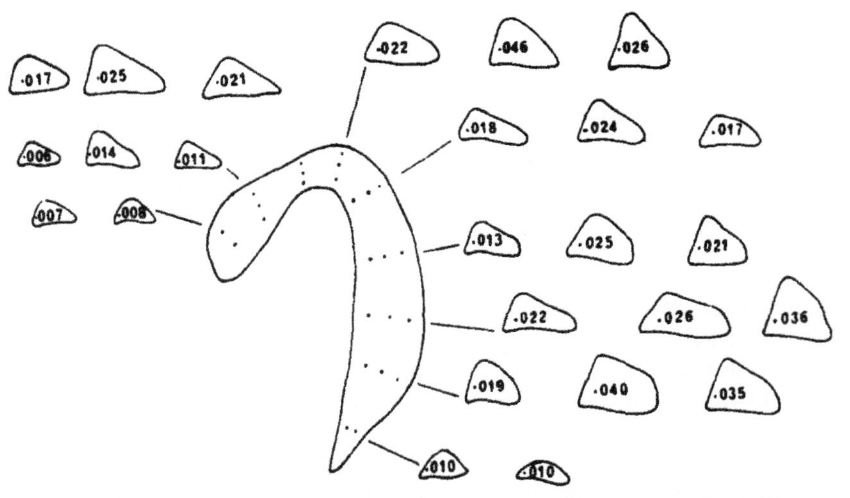

Fig. 13. Diagram of the anterior hemibranch of the first gill from the left side of an 8-cm trout showing regional variation in the surface area of the secondary lamellae. (After Morgan, 1971.)

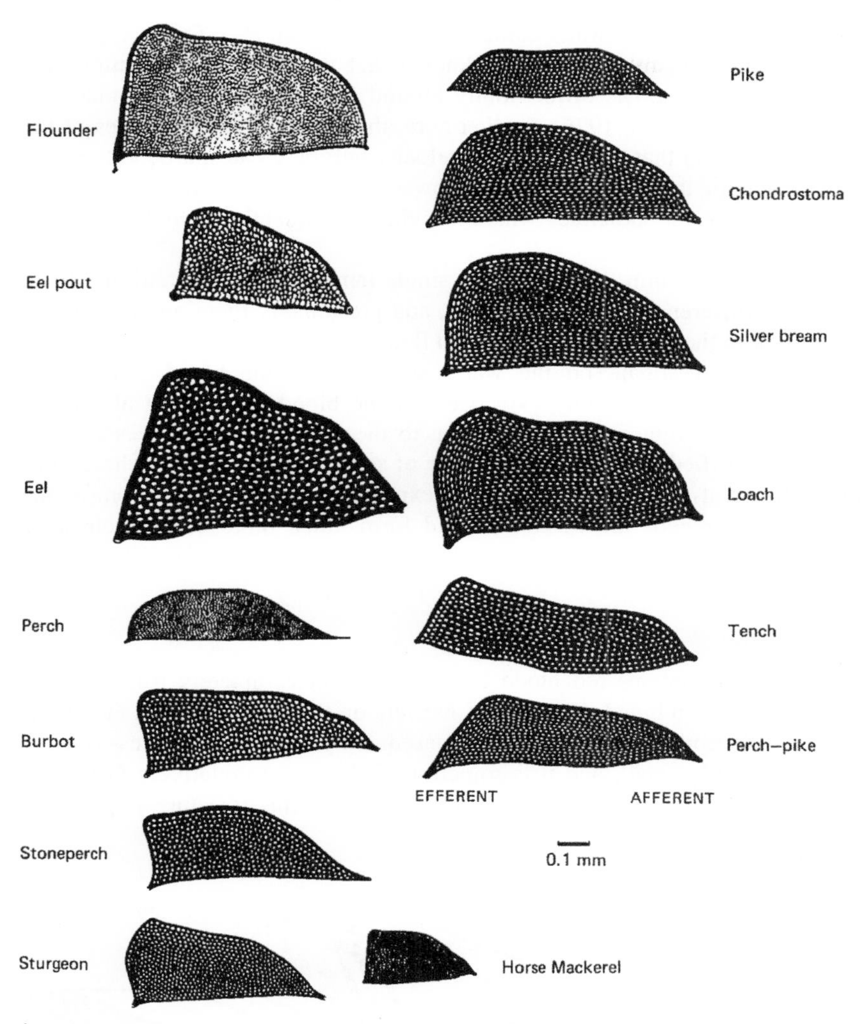

Fig. 14. Tracings of individual secondary lamellae of a variety of fish based on drawings by Byczkowska-Smyk (1957–1962), Nowogrodska (1943), and Oliva (1960). (Modified from Hughes, 1970a.)

Some of the consequences of the different shapes of lamellae have been discussed by reference to gas exchange in the gill of icefish, where complications due to the oxygen–hemoglobin dissociation curve are not present (Hughes, 1972b). A general conclusion has been reached from such analyses that the distribution that maximizes lamellar area at the inlet side produces the best gas exchange situation both in counter- and cocurrent flow. For some of the shapes it ensures that blood in the marginal channels at the inlet side will come into contact with water containing the maximum P_{O_2}. This is especially relevant, as in many species it is known that under resting conditions the blood sometimes flows differentially around only the marginal channels of the lamellae (Hughes, 1976). Furthermore, the general hydrodynamics of the situation suggest that the water flow velocity will be greatest around the marginal edges of the lamellae and be slowest in the crypt region between the bases of adjacent lamellae, where it will consequently have the greatest contact time.

Thus, even within the limits of a single lamella, it is apparent that there would be differences in gas exchange, and presumably there are adaptations in relation to the microflows of the two fluids. Such differences are extremely difficult to investigate on the water side, but on the blood side of the exchanger there is certainly evidence that the blood flow is not always by the most direct route from the afferent to the efferent filament vessel. This has been clarified for the large lamellae of some tunas (Fig. 15), where most blood flows at a definite angle to the expected water flow direction, as the adjacent pillar cells are in contact and form much more definite channels along which the red blood cells are directed (Muir and Kendall, 1968).

3. *Support*

Each lamella contains red blood cells, which can completely fill the blood channels; the combined thickness of barriers on the two sides of a lamella is 3–20 μm depending on the species. Consequently, the total thickness of each lamella is fairly small, and it is sometimes subjected to relatively fast flows of water. The main skeletal supporting system of this structure is probably

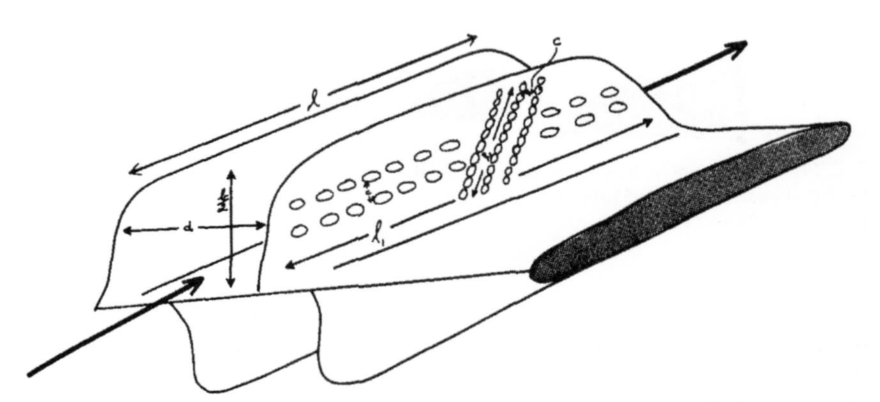

Fig. 15. Diagram of a single lamella of a tuna showing the blood pathways at an angle to the direction of the water flow. The relationship between water flow and dimensions for the water is $\dot{V}_G = (N \cdot \Delta P \cdot 5d^3b)/24l$ (Hughes, 1966) and for the blood is $\dot{Q} = (Ag \cdot \Delta p \cdot \pi c^3)/256\mu l_1{}^2$ (Muir and Brown, 1971).

attributable to the blood contained in the channels and to the combined effects of hydrostatic pressure, viscosity, and stiffness of the red blood cells. There is also an intrinsic skeletal system of the pillar cell system, formed by the basement membrane together with the extracellular columns of collagen (Hughes and Weibel, 1972), which run in membrane-lined channels within the pillar cells. This system will also serve to resist outward distortion due to the arterial blood pressure (Bettex-Galland and Hughes, 1973) and seems to be antagonized by a contractile mechanism within the pillar cells themselves (Hughes and Grimstone, 1965; Smith and Chamley-Campbell, 1981). It would appear that the lamellae are well supported and effectively retain their orientation within the water current. In general it is believed that the orientation of the lamellae on the filaments is such that the direction of water flow is parallel to that of the lamellar surfaces, and the flow is laminar. To increase this possibility there are differences in the orientation of lamellae at different positions along the axis of the gill filaments (Wright, 1973). This would appear to ensure that there is less resistance to the water flow provided by the lamellae in their normal orientation.

IV. MODIFICATIONS IN RELATION TO HABIT

As with other systems of the fish body, the gills show many modifications that are adaptive to the particular physicochemical conditions under which they must operate. Furthermore, as fishes from a variety of taxonomic groups frequently become adapted to similar conditions, they illustrate extremely well some of the principles of convergence and parallel evolution; perhaps the air-breathing fish provide the best examples of the latter.

A. Fast-Swimming Oceanic Species

These fish are characterized by the presence of a large gill surface, which is the result of both an increase in total filament length and a high frequency of lamellae, and consequently very large numbers of lamellae. The lamellae themselves may, however, be relatively small and the barrier between the water and blood less than 1 μm in certain tunas (see Table IV). It is well known that these fish show ramjet ventilation, in which the main energy forcing water through the gill sieve is provided by the body musculature (Hughes, 1970b; Roberts, 1975). It is probable that this represents an economy in the overall energetics of the fish. These fish may develop into relatively large specimens and swim at high speeds, and it is inevitable that sometimes foreign matter may enter the mouth and cause physical damage to the gill system (e.g., Fig. 16A). It is perhaps in response to such conditions that the various additional structures on the filaments have evolved, as they serve to consolidate the hemibranchs. In some cases (Fig. 16B) water only

Table IV The Water–Blood Barrier: Measurements Showing Range Found for Length of the Three Main Parts of the Barrier[a,b]

Fish species	Epithelium (μm)	Basal lamina (μm)	Pillar cell flange (μm)	Total water–blood distance (μm)	Mean[c] (μm)
Elasmobranchs and benthic teleosts					
Scyliorhinus canicula	2.38–18.48	0.3–0.95	0.37–0.71	5.24–19.14	11.27
Scyliorhinus stellaris	3.0–9.3	0.25–1.32	0.37–0.62	4.29–11.88	9.62
Squalus acanthias	3.0–22.5	0.3–0.6	0.12–0.6	3.43–29.55	10.14
Galeus vulgaris	1.5–22.5	0.11–0.66	0.05–0.2	2.31–24.0	9.87
Raia montagui	0.8–14.8	0.2–1.65	0.08–1.2	1.2–15.6	4.85
Raia clavata	0.5–11.5	0.13–0.63	0.03–1.13	3.0–11.6	5.99
Pleuronectes platessa	0.35–10.4	0.09–1.7	0.48–0.95	0.93–17.28	3.85
Solea solea	1.3–3.4	0.03–0.41	0.03–0.18	1.97–3.98	2.80
Solea variegata	1.0–12.3	0.13–0.33	0.03–0.63	2.06–23.1	5.55
Limanda limanda	0.13–7.0	0.1–0.41	0.063–1.0	0.88–8.4	2.53
Microstomus kitt	0.25–16.7	0.1–0.69	0.013–0.1	0.71–44.25	3.23
Pelagic teleosts					
Thunnus albacares	0.017–0.625	0.048–0.166	0.025–0.875	0.166–1.125	0.533
Euthynnus affinis	0.05–0.475	0.063–0.125	0.025–0.25	0.313–1.063	0.596
Katsuwonis pelamis	0.013–0.625	0.075–1.875	0.017–0.375	0.24–1.906	0.598
Scomber scombrus	0.165–1.875	0.066–1.0	0.033–1.75	0.600–3.625	1.215
Trachurus trachurus	0.38–3.4	0.063–0.125	0.30–2.33	0.25–3.13	2.221
Salmo gairdneri	2.075–9.25	0.125–1.25	0.20–2.4	3.32–9.6	6.37

[a]Measurements for elasmobranchs and benthic teleosts from Hughes and Wright (1970); those for pelagic teleosts from Hughes (1970a).
[b]The electron micrograph sections used for the measurements were not always perpendicular to the lamellar surface.
[c]The mean of about 50 measurements for the total distance.

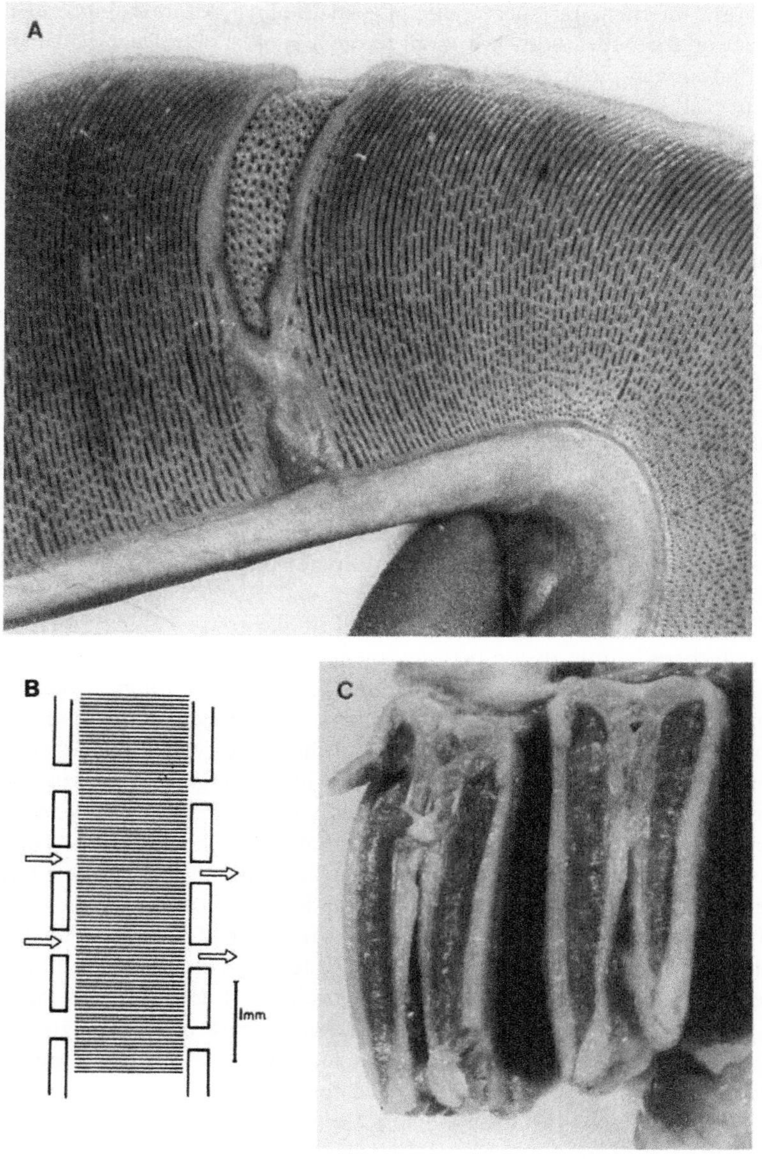

Fig. 16. (A) Photograph of a single gill arch from a marlin. Several filaments have been damaged and show regeneration. The formation of interfilamental junctions leaving pores for the flow of water are clearly visible and correspond to those shown diagrammatically in (B). (B) Diagram illustrating the flow pathway of water across a filament of a tuna gill in which interfilamental junctions are developed. On the inflow side, water enters through small pores, then passes through the interfilamental space containing secondary lamellae and leaves by a corresponding series of slightly smaller outlet holes into the opercular cavity. (After Hughes, 1970b.) (C) Section of two gill arches of *Mola mola*, showing two rows of hemibranchs in each arch with a dense layer of hardened tissue covering the leading and trailing edges. (After Adeney and Hughes, 1976.)

enters the interlamellar spaces through small pores before being collected together on the outlet side once more through another set of pores. Although a supporting and protective function has naturally been emphasized for these structures, there is also the important possibility that they effectively slow down the water velocity past the lamellae and thus extend contact time. An analysis based on the dimensions of both the water and blood parts of the gas exchanger (Fig. 15) has indicated the adaptive features in this system (Brown and Muir, 1970).

In view of the development of the interfilamental junctions in tunas, the presence of similar consolidation of the hemibranchs in the holostean *Amia* is surprising, as the habits of these fish are entirely different (Bevelander, 1934). There can be no doubt that they are independently evolved structures and apparently have a different function. A possible suggestion is that these structures serve to reduce the collapse of the gill system during air breathing of these freshwater fishes (Hughes, 1966a,b).

Another oceanic fish with special developments of the gill is the sunfish (*Mola mola*), which may reach sizes exceeding 1000 kg. The leading and trailing edges of the gill filaments are covered by a dense layer of hard tissue (Fig. 16C) in which are embedded many spines originating from a layer just beneath the outermost epithelium (Adeney and Hughes, 1977). Growth of the filaments can lead to some restriction of the space between filaments, although there is little fusion between them. One of the most interesting features of *Mola* gills is the extent of the internal gill slits through which water enters the spaces between hemibranchs. This is restricted to about one-third of the total length of the gill, and consequently, the typical teleost flow of water through the filament can only occur in this region; otherwise water entering through the slit passes either dorsally or ventrally into pouches from which the water can pass between the filaments of the hemibranchs. This situation closely resembles that of elasmobranchs, rather than the typical teleost system. The opening of the internal gill slits is in fact the only region where the branchial arches are present; the fusion together of the hemibranches is due to connective tissue that thus prevents water from entering the gill chamber except in the middle region. The gill ray is well developed on the inlet side and forms a flattened structure that extends across the full length of the lamellae. In the regions where the branchial arch skeleton is absent (i.e., ventrally and dorsally), there is a special fusion of the gill rays of adjacent filaments, but not with filaments of the opposite hemibranchs. However, the bases of gill rays are connected to filaments of the neighboring hemibranch by a thin layer of connective tissue. These structures apparently support the hemibranch and replace the gill arch in these regions, as there are no cartilaginous connections between the gill arch and the base of the gill rays in the middle region.

B. Fishes of Intermediate Activity

For a long time it has been recognized that it is difficult but desirable to classify fishes according to their activity, the most controllable method being to measure swimming speed in relation to oxygen consumption (Brett, 1972). Such a quantitative basis has not yet been completed. Nevertheless, a broad band of species regarded as "intermediate" in activity by Gray (1954) were as follows: *Sarda sarda*, *Mugil cephalus*, *Caranx crysos*, *Roccus lineatus*, *Archosargus probatocephalus*, *Chilomyceterus schoepfi*, *Stenotomus chrysops*, *Tatutogo onitus*, *Prionotus strigatus*, *Poronotus triacanthus*, *Cynoscion regalis*; *Palinurichthyes perciformis*, *Echeneis naucrates*, *Spheroides maculatus*, *Centropristis striatus*, *Peprilus alepidatus*, *Prionotus carolinus*, *Trichiurus lepturus*, *Paralichthys dentatus*. Such fishes represent the major groups of teleosts, and their respiratory systems have the following typical structure.

The gill arches are well developed and support filaments of average length. The operculum covers over the whole of the gill system and communicates with the exterior via a more or less continuous slit. The lamellae are average in size, and their frequency on each side of a filament is usually about 18 to 25 per millimeter. Ventilation of gills usually involves rhythmic ventilatory movements, although under certain conditions (e.g., high swimming speeds) some of them might show ram ventilation. The buccal and opercular suction pumps are equally important in maintaining the ventilatory current and would be classified in group I of Baglioni's scheme (Hughes, 1960b).

C. Sluggish Fishes

There are many fish, usually of benthic habits, that are relatively inactive, but this "sluggish" group would also include some fish that maintain a constant position in a condition of neutral buoyancy and make only occasional darting movements connected with feeding and mating. The latter category includes many deep-water fishes such as groupers and *Latimeria*; the icefish, lacking hemoglobin, is another example. All of these species are expected to have low oxygen consumption in relation to their body size, and their gills are relatively poorly developed. In some instances gills are restricted to three arches, as in the toadfish (Table II). The filaments are short especially in some deep-sea fishes (Marshall, 1960; Hughes and Iwai, 1978). Individual lamellae are relatively large in area, and their spacing (10–15 per millimeter) is wider than that of other species. The resistance to water flow through the gill sieve is therefore low. The opercular pumps are well developed usually because of the good development of the branchiostegal apparatus. In bottom-living forms, exit from the opercular cavity is often by a limited part of the opercular slit, which is directed dorsally. Similar restrictions of the opercular openings are found in midwater hoverers such as trigger and puffer fish. Ventilatory frequency is low, the opercular expansion phase being slow, but ejection of water from the respiratory system occupies a short portion of the whole cycle.

Flatfishes show some special adaptations related to their mode of life, especially the way in which they come to rest on one of the two sides of the body. In most cases the fish rest on the morphological right side, which is pale-colored, whereas the upper surface is darker and contains many chromatophores. Ventilation is achieved with a well-developed suction pump mainly because of the extensive branchiostegal apparatus (Schmidt, 1915; Henschel, 1941). In fact, recordings of pressure changes in the opercular cavities (Hughes, 1960b) showed slight asymmetry and the absence of an apparent reversal phase that is characteristic of many teleost fishes. It was suggested that this resulted from the active mechanism closing the opercular openings, as this would reduce the possible influx of sand or other particles from the benthic habitat. The gills themselves are well developed on both sides, although there are some differences in their detailed morphology. The upper opercular cavity is more convex and longer and accommodates somewhat better developed gills than the flatter and smaller lower opercular cavity with its moderately developed gills. Measurements of the gills on the two sides have been made by a number of authors and are indicated in Table IV.

There appear to be changes in the path of the ventilatory current especially during development (Al-Kadhomiy, 1984). The premetamorphic stages are free living, and their body configuration is similar to that of typical teleosts. In these forms the opercular cavities, gills, and bony elements concerned with gill ventilation are equally developed on both sides. Following metamorphosis there is gradual development of the asymmetry of the adult form. The life of some flatfishes seems to alternate between swimming freely in midwater and periods when they are on the sea bottom, especially during migrations (Harden-Jones, 1980). It seems possible that during the midwater phases, the ventilation is equal on both sides of the respiratory system, whereas when on the bottom they make greater use of the special channel (Yazdani and Alexander, 1967; Al-Kadhomiy, 1984), which enables water from the ventral opercular cavity to be passed into the dorsal opercular cavity and thence to pass out via the dorsal opercular opening. In such a situation, however, both sets of gills are well ventilated; thus, the observation does not necessarily lead to the expectation that they are unequally involved in gas exchange.

D. Air Breathers

Air breathing has evolved many times among different groups of fishes and in most cases seems to have been related to a worsening of conditions for aquatic respiration. In their classical studies on the Paraguayan chaco, Carter and Beadle (1931) drew attention to groups of fishes that live near the surface and ventilate their gills with these waters of relatively high oxygen tension. Fish that habitually live in deeper waters must swim to the surface to gulp air, with a consequent increase in oxygen consumption (Singh, 1976). In more extreme conditions the aquatic environment dries up and the fish must use some form of air breathing to survive, either during migration to some other aquatic environment or as a preliminary to some resting stage. In all cases, as has been pointed out (Carter, 1957), the modifications that have evolved concern parts of the alimentary canal, where gas exchange with the air takes place. In each of these sites there is an increased vascularization associated usually with a reduction in the tissue barrier separating the air and blood.

1. Air-Breathing Organs

The precise anatomy ranges from suprabranchial cavities and their enclosed labyrinthine organs to forms in which special regions of the rectum have become the place where bubbles of air taken in at the mouth and passed through the alimentary canal come to rest and where gas exchange occurs (e.g., loaches, *Misgurnus*). The increase in surface for gas exchange often involves intucking of the endodermal epithelium, as in the lungs of *Polypterus* and dipnoans, which are homologous with the tetrapod lung. Striking convergences are found, as in some catfishes (e.g., *Saccobranchus*), where a backward projection from the suprabranchial chamber forms an air sac on either side of the vertebral column (Munshi, 1962). When the "lungs" of *Saccobranchus* were investigated morphologically, it was apparent that there were similarities to the basic structure of gills, and this was substantiated by more detailed studies using electron microscopy (Hughes and Munshi, 1973a). The surface structure as revealed by scanning electron microscopy showed similarities between the microvillous surfaces of the lamellae and of the respiratory islets, whereas the "lanes" between the lamellar structures have microridged surfaces comparable to those of the gill filament epithelium (Fig. 17). Electron microscopy also showed that the labyrinthine organs of anabantoid fish, which were supposed to be homologous with gill lamellae, are not derived so directly from the gills themselves (Hughes and Munshi, 1968, 1973b). The suggested homology with gills (Munshi, 1968) rested on the recognition of the typical pillar cell structure in the labyrinthine plates, but electron microscopy revealed that these so-called pillars are intucked epithelial cells that have a similar position separating adjacent blood channels. Electron micrography also showed that the air–blood barrier was extremely thin (<1 μm), and consequently, such structures could have a relatively high diffusing capacity, although their surface might be fairly restricted (Table V). In other cases, however, the air-breathing organs—though well vascularized—have fairly thick barriers between the air and blood and hence could not form such important gas exchange organs.

From a physiological point of view, it has been recognized that the air-breathing organs of fish are especially important in relation to the uptake

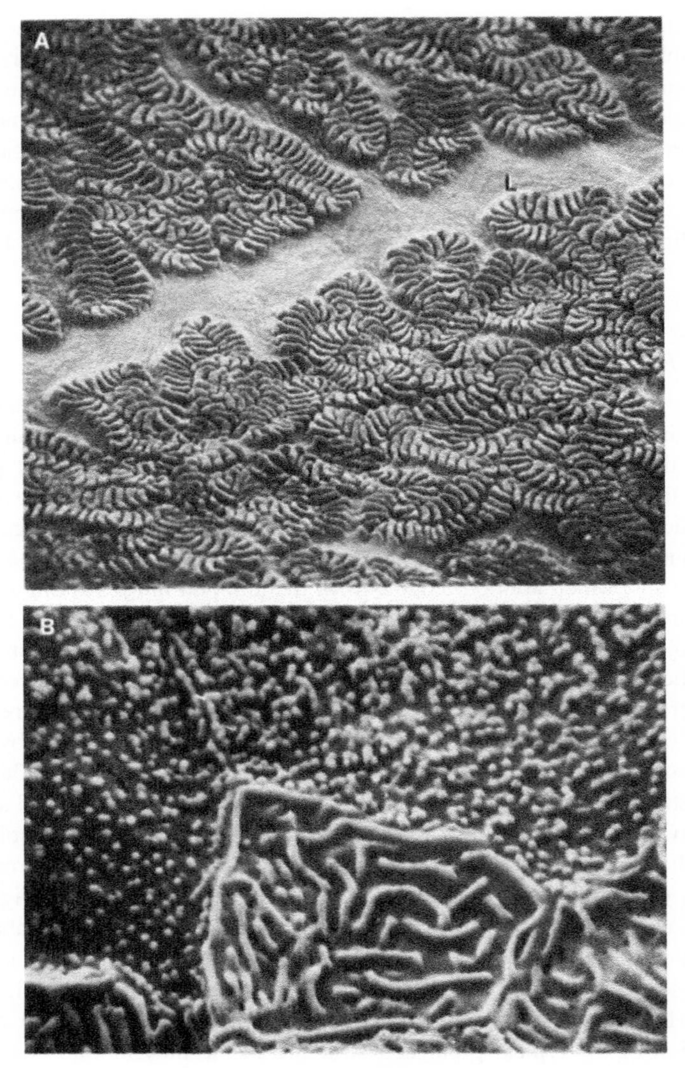

Fig. 17. (A) Scanning electron micrographs (SEM) of respiratory islets from the air tube of *Saccobranchus*. Notice the whorl-like appearance of the islets and the presence of a "lane" (L) separating groups of islets in which a biserial arrangement can be identified. (×120) (B) High-power SEM showing the microridged surface of the flattened cells of the lanes, which change abruptly into the microvillus surfaces of the respiratory islets. (×12,000) (After Hughes and Munshi, 1978.)

of oxygen, as this gas is in relatively short supply in the water (Singh, 1976). The release of carbon dioxide is not such an important function of the air-breathing organ, and this is much more easily facilitated in the aquatic medium because of its high solubility in water. Consequently, the gas

Table V Diffusing Capacity of the Tissue Barrier (D_t) and Component Measurements for Respiratory Surfaces of *Channa* and *Anabas*[a]

Fish species	Respiratory surface for 1 g fish (mm^2)	Thickness of tissue barrier (μm)	Area g wt 100 g $fish^{-1}$ (mm^2)	Diffusing capacity (ml min^{-1} mm Hg^{-1} kg^{-1})	References
Channa punctata					
Total gills	470.39	2.0333	71.8229	0.0530	Hakim *et al.* (1978)
Suprabranchial chamber	159.08	0.7800	39.1705	0.0753	Hakim *et al.* (1978)
Anabas testudineus					
All gill arches	278.00	10.0000	47.2000	0.0071	Hughes *et al.* (1973)
Suprabranchial chamber	55.40	0.2100	7.6500	0.0539	Hughes *et al.* (1973)
Labyrinthine organ	80.70	0.2100	32.0000	0.2286	Hughes *et al.* (1973)

[a]From Hakim *et al.* (1978).

exchange ratio for air-breathing organs and gills may be quite different, although the overall ratio is 0.8–1.0, as in many other fishes.

2. *The Gills*

The gills of air-breathing fishes are therefore important in gas exchange, although not so much in oxygen uptake, and in general their surface area is less than typical aquatic breathers of the same body size. In some species the gills themselves are important organs not only in water breathing but also when the fish comes out into the air. This is especially true, for example, in some mudskippers (*Periophthalmus*), which live on mudflats and mangrove swamps in many different parts of the world. When they come out of the water such fishes expand the buccopharyngeal cavity and enclose a separate volume of air with which gas exchange appears to take place. In their gill structure different degrees of stiffening of the system have been recognized (Schlotte, 1932). In many of these species cutaneous respiration is also of great importance when the fish is out of water. During the life history of some air-breathing species it has often been found that in the earlier stages the fish are almost entirely dependent on aquatic respiration, but with an increase in size and greater development of the air-breathing organ relative to the gills, there is a transition toward greater dependence on air breathing (Hughes *et al.*, 1974a).

V. GILL VENTILATION AND ROLE OF BRANCHIAL MUSCLES

Because of the greater density and viscosity of water with respect to air and its low content of oxygen (Hughes, 1963; Dejours, 1976), it has generally been accepted that the problem of ventilation of respiratory surfaces that faces a water-breathing species is greater than that of those that breathe air. Because of this it is usually considered that a greater portion of the standard metabolism of the fish is required for maintaining sufficient water flow, but estimates of the particular percentage are quite variable. They range from less than 1 to more than 40%, and no definite value has been established. The consensus of many studies would seem to support a figure of 5 to 10% (Hughes, 1973a; Jones and Schwarzfeld, 1974; Holeton, 1980). Similar values have been attributed to the work of the cardiac pump (Hughes, 1973a; Jones and Randall, 1978). Regardless of the precise figure, there can be little doubt that mechanisms have evolved that are extremely economical from an energetic point of view, and this leads to the great fascination of attempts to elucidate their detailed functioning. Fish would appear to have a greater range of ventilation mechanisms than any other group of vertebrates (Hughes, 1978a), the basic one being the double pump whereby a more or less continuous flow of water is maintained across the gill surfaces.

A. Water Pumps

Early anatomic studies of the respiratory system were based on fixed material. From a functional point of view, such studies have the serious disadvantage that the gills have a completely abnormal orientation. Consequently, standard diagrams used to illustrate fish ventilation showed the gills with large spaces between filaments through which much water could flow without having any oxygen removed from it. The largely morphological studies by Bijtel (1949) and Woskoboinikoff (1932), who emphasized the position of the gill filaments during normal ventilation, were very important, and these conclusions were confirmed by physiological measurements (Hughes and Shelton, 1958; Hughes, 1960a,b) that demonstrated that the respiratory system could be divided into two functional cavities separated by a gill resistance (Fig. 18A and B). In fact, there are three cavities in teleosts—a single buccal cavity and an opercular cavity on each side—whereas in elasmobranchs the gills separate a single orobranchial cavity from five or more parabranchial cavities on each side of the fish. The double-pumping hypothesis was proposed because pressure changes recorded on the two sides of the gill resistance had different time courses and varied in relation to volume changes of the cavities. Apart from showing the differences in time course of pressures in the cavities, these studies using electromanometers also showed how small (about 1 cm H_2O) were the pressure changes during the normal ventilatory cycles. The small pressure changes contrast with the relatively large pressures recorded during feeding and some other activities involving the same basic apparatus (Alexander, 1970; Lauder, 1980, 1983). The low level of the pressure changes is indicative of the muscular activity involved and the consequent energetic economy during ventilation.

Analyses of cine films and pressure wave forms also indicated that the entrance to the typical system, the mouth, is guarded by valves (mandibular/maxillary), but often there are active changes in the size of the mouth opening. Similarly, in most cases movements of the opercular valves follow the pressure changes and indicate their largely passive function. At both the entrance and exit to the system, however, there are fishes in which active closing occurs, and this is important both in the normal rhythm and also during specil respiratory maneuvers such as coughing. Thus, bottom-living fishes such as flatfish have well-developed branchiostegal apparatus, which produces active closing of the opercular openings and prevents possible influx of sand particles before the expansion of the cavity occurs. Also, it has been shown that in rapidly swimming oceanic fishes such as tuna, regulation of the volume flow through the respiratory system is probably best achieved by the degree of opercular adduction. In these fish the dimensions of the elongated opercular slit also help to maintain good laminar flow of water across the body surface.

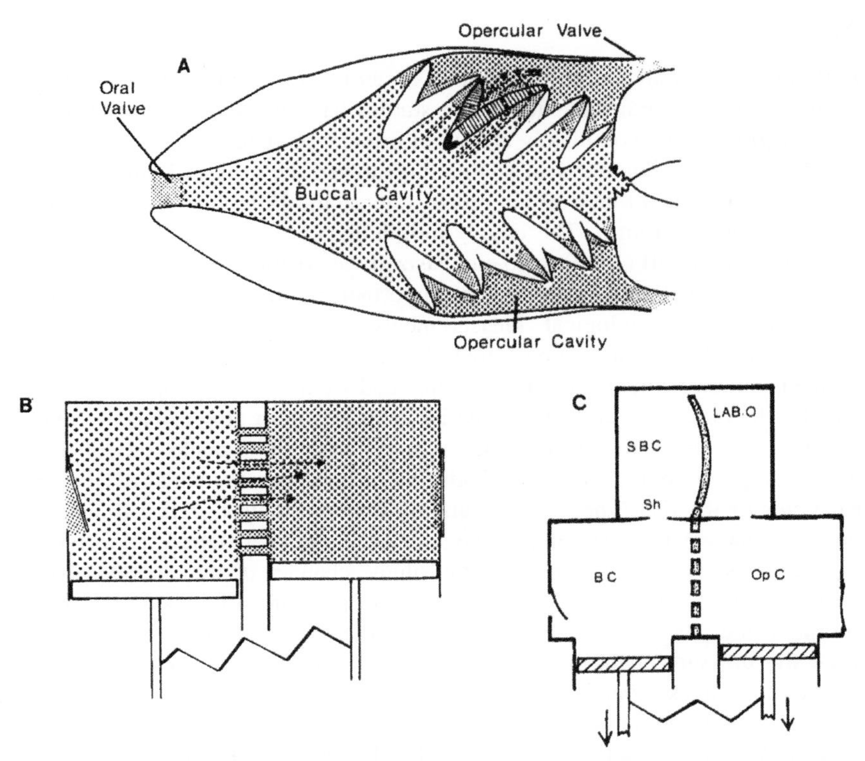

Fig. 18. (A) and (B) Diagrams to illustrate the double-pumping mechanism for the ventilation of fish gills. The equivalent of a simplified hydrodynamic model of the fish is shown by shading of corresponding parts. Communication between the buccal and opercular cavities is via a gill resistance, and the system is powered by changes in volume of the buccal and opercular cavities. It should be emphasized that changes in volume of these two cavities are not independent, and this mechanical coupling is indicated by a spring connecting two pumps of the hydrodynamic model. (From Hughes, 1976.) (C) Diagram of modified double-pumping model in *Osphronemus.* SBC, suprabranchial cavity; LAB O, labyrinthine organ; Sh, shutter.

From a respiratory point of view, the most important flow concerns the water current across the gill surface, and it has been extremely difficult to obtain information concerning its detailed nature. One biological method has been to utilize the attachment mechanism of small glochidia larvae of freshwater mussels experimentally introduced into the water inspired at the mouth of the fish. Paling (1968) carried out such experiments and found fixation of the glochidia on certain hemibranchs that were related to the distribution of gill area. Similar experiments using tench showed that the distribution was not in direct proportion to the gill area but tended to be greater than expected for the second and third slits, whereas it was less than directly proportional to the surface area on the first and fifth slits (Fig. 19)

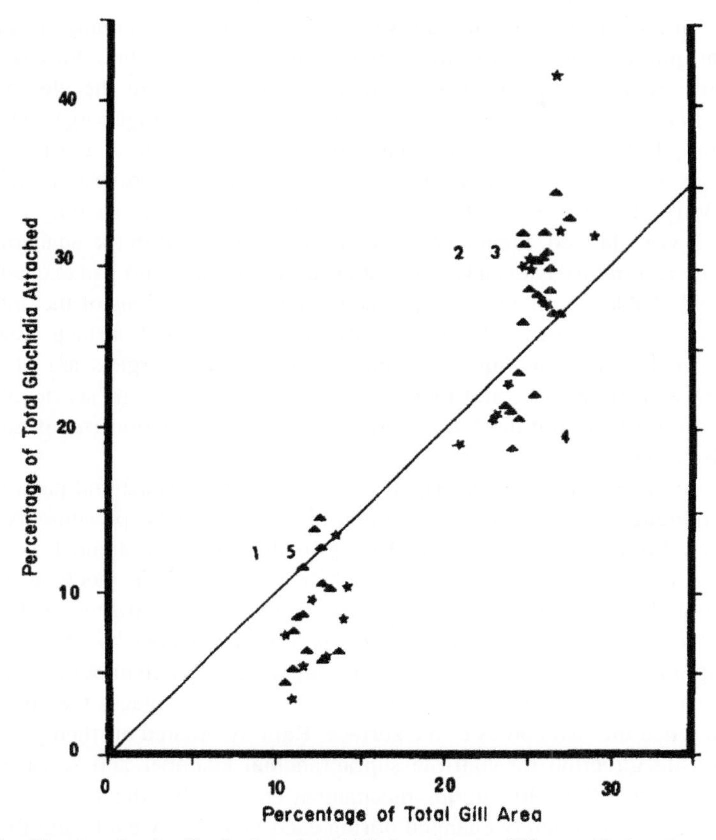

Fig. 19. Plot showing the distribution of glochidia larvae in the different gill slits of a tench. The percentage of the total numbers attached to a given fish are plotted in relation to the percentage of the total gill area that borders each of the gill slits (1–5). Closed triangle, experiments carried out in 1969; star; experiments carried out in 1970.

B. Ventilation of Air-Breathing "Gills"

The gills of a number of species of fish, notably intertidal gobies and mudskippers, and eels when traversing the land, are used, and the buccal and opercular cavities are largely filled with air. The ventilation mechanism here is mainly to inflate the cavities with air and to keep them closed by active contraction of the valves at the entrance and exit of the respiratory system. Measurements of the content of the cavities have been made in eels, but few other species have been investigated (Berg and Steen, 1965). There are also cases where typical aquatic fish come to the surface, possibly under hypoxic conditions, to gulp air, much of which probably remains in the buccal cavity and presumably can serve to help increase the P_{O_2} of the water as ventilation continues in the normal way. There are, however, many species of fish that gulp air and press it into contact with specific air-breathing organs, and some of these may be in the roof of the mouth, as in the electric eel (*Electrophorus*), where it forms an important gas exchange organ (Farber and Rahn, 1970). As the air will naturally stay in the roof of the mouth, it is perhaps not surprising that extensions of these cavities form the suprabranchial cavity of many air-breathing species. The surface of this cavity may be very well vascularized. Such a condition is well developed in the anabantoids, where the respiratory islets have very short tissue barriers and gas exchange is facilitated (Table V). In addition, anabantoids have expansions of the anterior branchial arches to form labyrinthine plates with increased surface area and are enclosed within the suprabranchial chamber. These organs are of great importance in these fish, and their mechanism of ventilation has developed modification of the normal buccal pressure–opercular suction pump mechanism (Fig. 18C).

In some genera (e.g., *Anabas*), air is gulped at the surface and passes into the suprabranchial chamber in the normal direction, its flow presumably controlled by differences in resistance of the gills themselves as against the resistance of the entrance to the suprabranchial chamber. This mechanism has been termed "monophasic," distinguished from the specialized mechanism called "diphasic" by Peters (1978), and is well illustrated by *Ostronemus*, the gourami. In this fish, the air within the suprabranchial chamber is replaced by water that enters from the opercular cavities and so displaces the air that is exhaled when the fish comes to the surface. Rapidly inhaled air then passes in the opposite direction through the suprabranchial chamber and so displaces the water (Fig. 18C). This unique mechanism ensures that the suprabranchial chamber gas is completely changed during each ventilatory cycle and that the respiratory surfaces in the suprabranchial cavity are bathed in water during this period. As a consequence of the high solubility of CO_2 in water and the thinness of barriers, it seems probable that carbon dioxide release into the water will also be aided by this novel ventilatory mechanism.

Many other mechanisms for ventilation of the air-breathing cavities of fish have been discussed by Randall *et al.* (1981), and in some cases there also seems to be the possibility of the air-breathing surface being in contact with water for at least part of the cycle. Other mechanisms, including ventilation of modified swim bladders, have suggested a whole variety of mechanisms including that of "jet streaming," which has been invoked in some interpretations of the ventilation of the anuran lung (Gans *et al.*, 1969).

VI. GILL MORPHOMETRY

Measurements of the dimensions of respiratory organs have been made for a long time, and analogies such as that between the surface of the human lung and a tennis court vividly portray the large increase in surface that is provided for gas exchange functions. Fish gills are very much more accessible than mammalian lungs, and during recent years the investigation of these structures has proceeded very rapidly.

The term morphometry has become much more commonly used and is now frequently applied to all studies that concern the measurement of dimensions of structures within living organisms. It is also sometimes applied to gross dimensions (e.g., body weight–length relationships). Originally, however, it was used more in connection with the application of stereological methods to the investigation of gill dimensions. Another term sometimes used for nonstereological studies (Hughes, 1970a) is morphological measurements, but the shorter form, morphometry, is preferable and has now come into common usage for this whole area of study. It is impossible to give an adequate summary of all the basic techniques of stereology, and anyone wishing to extend their knowledge of this field should refer to some of the more standard texts (e.g., Underwood, 1970; Weibel, 1979). Many of the techniques used for gill and air sac morphometry are relatively simple, and a detailed knowledge of stereological principles is not always required (Hughes and Weibel, 1976). Measurement of structural entities is time-consuming, and therefore, it is important to be sure which measurements are most appropriate for the particular analysis that is envisaged. Before embarking on such a study it is often advisable to discuss the problem with a physical scientist and then choose which measurements to make; engineers often show great interest in these problems.

Many measurements are possible, but most are directed toward two main aspects. The first concerns the resistance to the flow of water and blood through the gills, and the second analyzes conditions for gas exchange, particularly the area and thickness of the tissue barriers. Although many measurements are now available, there is a great need for standardization of techniques. However, from a comparative point of view the adoption of the same technique by groups of workers for a range of different species does provide useful comparative data. For the future development of this subject more standardization is desirable, especially to obtain absolute values for comparison with other animal groups.

One of the most general problems concerns the condition of the material from which the measurements are made. Obviously it is preferable to use fresh material wherever possible, but the actual measurements are usually easier on fixed material, despite the consequent post mortem and processing changes that this introduces.

A second general problem concerns the heterogeneity of the gill system and the consequent need for representative sampling. Much more is known about the range of variation, and weighting techniques are generally used when trying to obtain overall values for the whole gill system. Fixation certainly leads to some change in dimensions, and it is important that the degree of this change should be determined for any particular material.

Measurements made soon after fixation are probably the most accurate, and those that are carried out on fixed material following embedding and sectioning with an ultramicrotome do not produce the same degree of change. Fixation followed by sectioning in paraffin wax produces far greater shrinkage. Some of these problems have been discussed in more detail elsewhere (Hughes, 1984b; Hughes *et al.*, 1984).

A. Water and Blood Flow Dimensions

During their passage through the gills both the water and blood are channeled into relatively small spaces, the dimensions of which will have an important effect on the flow characteristics, notably the relationship between pressure difference and flow. From a design point of view, the flow must not be too rapid for the exchanges required, but the resistance to that flow must not be excessive and hence the dimensions must not be too fine in relation to the power available from the ventilatory and cardiac pumps. Icefish gills illustrate this principle very well (Holeton, 1972; Hughes, 1972b).

1. *Resistances to Flow*

The concept of gill resistance was first developed in relation to water flow through the gills (Hughes and Shelton, 1958), and as measurements in that study showed differential pressures across the resistance of about 1 cm of water, the question naturally arose whether such pressures were sufficient to ventilate the narrow spaces between the lamellae. Accordingly, measurements of these dimensions were undertaken, in particular the height ($b/2$) of the lamellae, their length (l) across the filament, and the distance between them (d), thus defining a rectangular water channel. As adjacent filaments are very close to one another, a simplification was made by doubling this space, as indicated in Fig. 20A. In fact the lamellae alternate with one another, and the possibility was envisaged that under certain conditions these spaces might be reduced in size (Fig. 20B), but no observations have been made that confirm such a possibility. The Poiseuille equation for laminar flow is usually defined in relation to a cylindrical tube, and a modification of this formula for a rectangular cross section (Hughes, 1966b) gave the following relationship:

$$q = \frac{P_1 - P_2}{\eta} \times \frac{5d^3b}{24l}$$

where q is the flow through a rectangular tube of length l, width d, and height b; $P_1 - P_2$ is the pressure difference, and η the water viscosity. Application of this formula to fish (e.g., tench) that had both their gill dimensions and differential pressures measured showed that the mean differential was certainly sufficient to ventilate the gills with a volume of water of the same

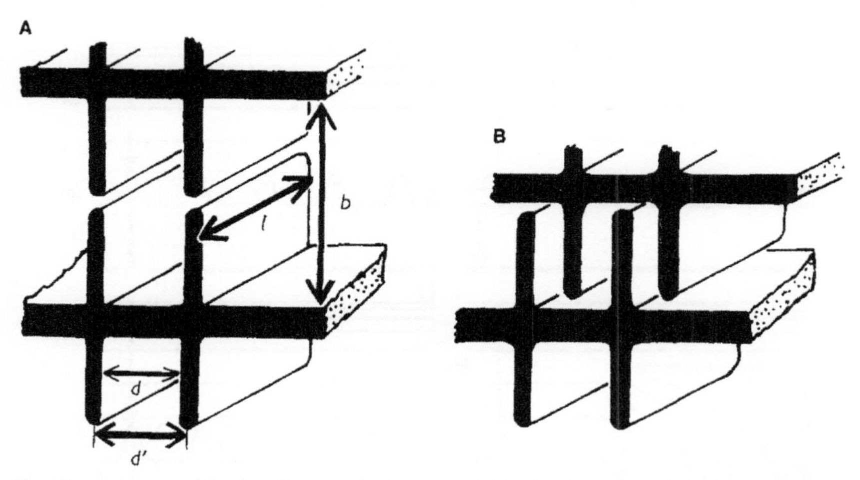

Fig. 20. Diagram of the lamellae attached to two neighboring filaments of the same gill arch to show dimensions used in calculations of gill resistance to water flow. In (A) the filaments are shown far apart with the lamellae opposite one another. In (B) the minimum possible size of pore is shown to result from the close interdigitation of the lamellae. (From Hughes, 1966b.)

order of magnitude as that measured (Hughes, 1966b). Measurements of these dimensions during the development of the gills in trout (Morgan, 1971) and small-mouthed bass (Price, 1931) provided valuable data not only for determining the resistance to water flow at different stages of the life cycle but also in relation to the scaling of these structures during development. Application of methods of dimensional analysis also proved fruitful (Hills and Hughes, 1970; Hughes, 1977), supporting the view that flow through the system is laminar at least up to certain sizes. The question of the nature of the water flow between the lamellae has been discussed on a number of occasions, but most authors agree that the low velocity through the channels in relation to their dimensions results in Reynolds numbers that are very small (<10) and make any nonlaminar flow extremely unlikely except in so far as it might be a result of intermittent movements of the filaments themselves. Although laminar flow is clearly beneficial from the point of view of water resistance, some authors have suggested that turbulent flow would produce greater mixing of the interlamellar water and hence be advantageous for gas exchange (Steen, 1971).

This question has become of further interest in relation to recent more detailed knowledge of the lamellar surfaces. Scanning electron micrographs have revealed the presence of many microridges and microvilli on the surfaces. The patterns of these surface sculpturings are variable in different species and in different parts of individual lamellae (Hughes, 1979b). It is an interesting but unanswered question whether these sculpturings have any

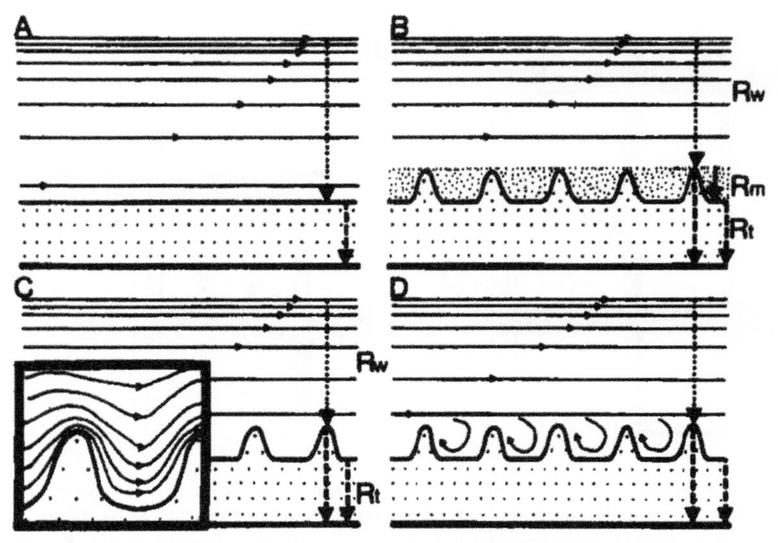

Fig. 21. Schematic diagrams to indicate possible relationships between water flow and the lamellar surface. (A) Laminar flow across a flat lamellar surface showing that the two main resistances to gas transfer are those in water and tissue barriers. (B) Laminar flow across a microridged surface in which thin layers of mucus fill up the spaces between the ridges. The resistance to gas transfer is here divided into either three portions between ridges (R_w, R_m, R_t) or two portions over the ridges (R_w, R_t). Subscripts w, m, and t refer to water, mucus, and tissue, respectively. (C) Laminar flow across a microridged surface with no mucus. The inset shows how the boundary layer would probably be maintained over all the ridged surface in view of the low Reynolds number that would operate at the dimensions of the lamella. (D) Laminar flow together with microturbulences between microridges of a surface without mucus. In this case, as in (C), the overall resistance to gas transfer consists of two components, that for the tissue (R_t) varying in length according to the position with respect to the ridged surface. (From Hughes, 1979b.)

hydrodynamic significance or whether in fact the hydrodynamic forces present during development play a part in determining the pattern of the ridging, and so on. The possibility that the sculpturing produces microturbulence at the boundary between water and lamella has been seriously considered (Lewis and Potter, 1976), but the presence of a layer of mucus that probably fills in the spaces between the ridges (Hughes and Wright, 1970) tends to make it more likely that the actual surface across which the water flows is smooth and covered with a very fine layer of mucus that will be thicker in the spaces between the ridges (Fig. 21)

2. *Subdivision of Water and Blood Flows*

On entering the mouth the water flow is largely laminar and soon subdivides as it enters different slits on the two sides of the buccal cavity. In teleosts one part passes anterior to the first gill arch and along the inner margin of the operculum. This pathway is usually very narrow, but during opercular expansion it will tend to increase in size. This increase does not lead to an enormous increase in water flow, however, because of the total movement of the whole branchial system, which expands during the expansion phase. This activity involves contractions of the intrinsic musculature of the gill arches (Hughes and Ballintijn, 1965; Ballintijn, 1968; Osse, 1969) and also to some extent those of the filaments themselves (Pasztor and Kleere-koper, 1962). Contraction of the abductor muscles tends to rotate the filaments relative to the longitudinal axis of each arch and help to maintain continuity of the gill sieve (Fig. 22). Most water passes between the gill slits of the first to fourth arches, but again there are differences in the last gill slit, and as at the front of the system there are various modifications in particular groups of fish (Table II). In teleosts typically there is only one hemibranch for the last slit, but in elasmobranchs there are two. Such differences between the individual slits produce variations in the amount of oxygen that is removed from the water as it passes along these major subdivisions of the flow pathway. The same variability extends to the flow of water between a typical pair of arches, since some of the water passes between the tips of the gill filaments, although

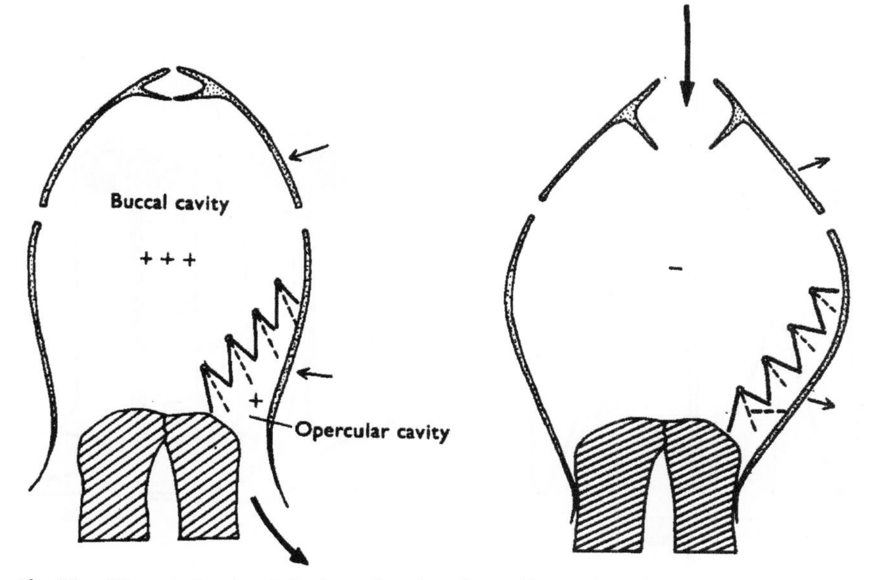

Fig. 22. Diagram showing, in horizontal section, the position of the gill arches and filaments on one side during the two main phases of the ventilatory cycle as described by Pasztor and Kleerekoper (1962). The anterior part of the buccal cavity with the oral valves is shown in vertical section. Directions of movement of the walls of the respiratory cavities are shown by thin arrows. Dashed lines indicate orientation of the gill arches. Pressures in the buccal and opercular cavities are indicated relative (+ −) to that of the outside water. The pressure difference across the gills results in flow from the buccal to the opercular cavities during both phases of the cycle. Thick arrows indicate water flow. (From Hughes and Morgan, 1973.)

they remain in close contact for much of the ventilatory cycle. This shunting of water has been likened to an anatomic dead space (Hughes, 1966b). During hyperventilation a larger proportion of the water flows through this pathway with a consequent lowering in the percentage utilization of oxygen from the water. The major portion of the water passes between the secondary lamellae of the gill filaments of the individual hemibranchs, and thence it is collected again into the opercular cavities.

As has been indicated previously, there are differences in the dimensions of the lamellae at different positions along the filaments, and in general we may distinguish those from the tip, middle, and base of each filament. The path of water flow for these different parts of the filament vary, and, most importantly, water passing between the basal lamellae must meet the septum especially in those teleosts where this remains relatively well developed. In the perciform gills, however, this is a relatively small portion of the water. Water traversing between lamellae at the tip has a less impeded flow pathway. From such morphological features of the water pathway it is clear that there will be different resistances to water flow depending on which pathway one considers. These different resistances are indicated in Fig. 23, where

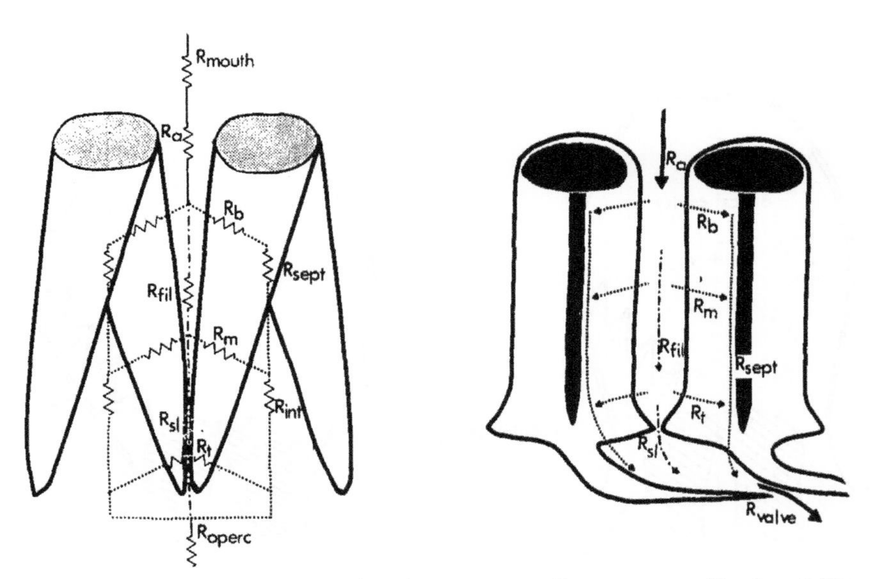

Fig. 23. Diagrams to illustrate the main resistances to water flow across two gill arches of (A) a teleost and (B) an elasmobranch gill. The resistances to water flow between lamellae at the base (R_b), middle (R_m), and tip (R_t) regions of the gill filaments are indicated. The septal resistance (R_{sept}) of this teleost gill is far less than in the elasmobranch gill. Resistance to water flow through the mouth of the fish (R_{mouth}) and between the adjacent gill arches (R_a) and filaments (R_{fil}) and outwards through the slit between tips of adjacent hemibranchs (R_{sl}) and finally out into the water via the opercular (R_{operc}) and external gill slits (R_{valve}) are also indicated. (—·—·), Axial flow; (····), interlamellar flow. (After Hughes, 1972c, 1973b.)

differences between the elasmobranch and teleost gill are indicated. The same principles apply to all gills, but the relative flows in different portions of the filament will vary.

The dimensions of spaces between lamellae vary according to the habit of the fish. In all cases there would be a part of the water flow that is more axially placed, where the distance for diffusion of any oxygen molecules to the lamellar surfaces is greater and the water flow velocity higher compared to that of water in the immediate boundary layers. Thus, we can consider the axial flow of water as representing a sort of physiological dead space (Fig. 24), the proportion of which will depend on the dimensions of the interlamellar spaces and the velocity of water flow. The extent of the physiological dead space, as with the morphological dead space, will increase with increasing ventilation. Flow velocities in the interlamellar spaces can be estimated from measurements of the total ventilation volume and morphometry, which indicates the total number of interlamellar spaces and their dimensions. Calculations give velocities of the order of 5 cm sec^{-1}. Such values have not been determined accurately, although measurements of water flow velocity at least in the neighborhood of the gills indicate velocities of 10 to 20 cm sec^{-1} (Hughes, 1976, 1978b; Holeton and Jones, 1975). Although the resistances of the secondary lamellar and filamental pathways are most important during ventilation, it has long been recognized that resistance of the gill arches and rakers are very important during feeding, and this has been shown in some elegant studies using rapidly responding pressure manometers (Lauder, 1983).

The situation on the blood side of the exchanger is more complex than was originally supposed, and for a detailed account of vascular pathways, reference should be made to Chapter 2. Originally it was supposed that, as with water flow, the system was homogeneous, but emphasis was given to the probability of greater flow around the marginal channels (Hughes and Grimstone, 1965) and the possibility of flow in this pathway and the proximal channels (Fig. 25) embedded in the filament being regulated relative to flow in most of the lamellar channels as a result of contractile activity within pillar cells. The efficacy of such a regulation was emphasized by Wood (1974), who calculated that a passive increase in diameter of the marginal channels of only 25% would result in the whole of the blood flow passing by this route. The absence of the contractile mechanism of the pillar cells would tend to produce a disproportionate increase in the marginal channel diameter with increase in blood pressure. Evidence for such a change in distribution of intralamellar blood flow has been obtained using morphometric methods (Soivio and Tuurala, 1981). Furthermore, it is also possible for blood flow to be restricted to certain lamellae depending on the conditions (Booth, 1978, 1979). Thus, on both the water and the blood sides there is recruitment of water and blood channels under different conditions. Each of these channels, however, can vary in the extent to which ventilation and/or perfusion occurs (Hughes, 1972c).

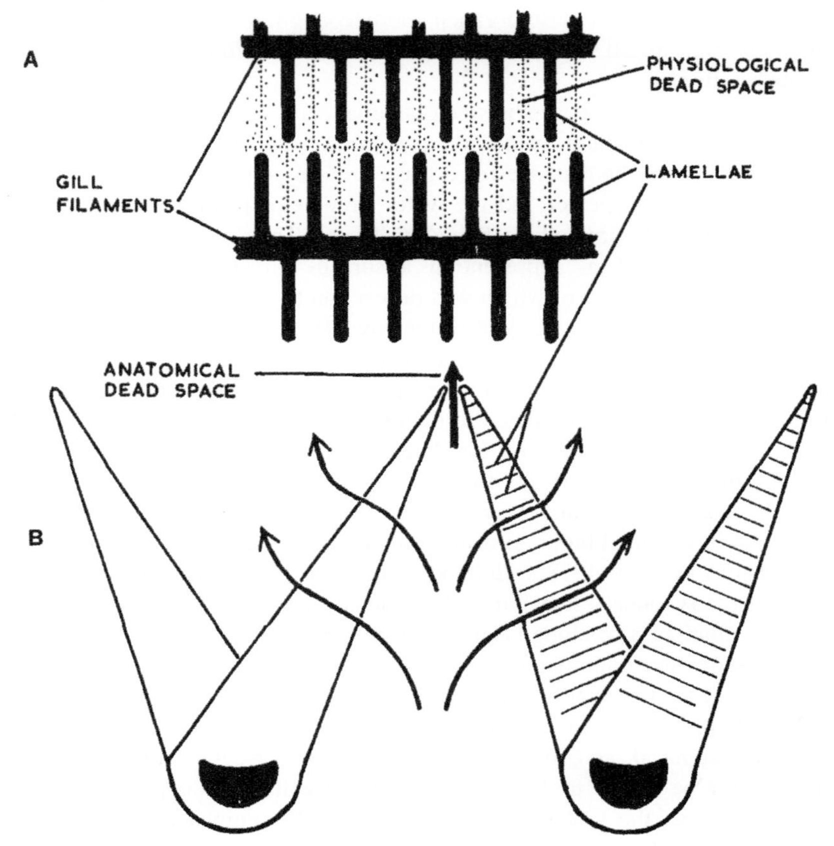

Fig. 24. Diagram of two gill arches of a teleost fish indicating the paths of different parts of the respiratory current. (A) Longitudinal section of two filaments. The water flows at right angles to the page. The concentration of oxygen in the interlamellar water (equivalent to alveolar air) is indicated by the density of the stippling. The water that passes through without losing any oxygen is equivalent to the physiological dead space. (B) Diagram of transverse section through two adjacent gill arches with their rows of attached filaments. The filaments normally touch one another at their tips, and most water passes between the lamellae (interlamellar water). With excessive pressure gradients across the gills, the volume of water shunted between the tips of the filaments increases. This portion of the respiratory water is equivalent to the anatomic dead space of the lung. (From Hughes, 1966b.)

There are many variations in the proportion of blood that passes through the respiratory or intralamellar pathway, as distinct from pathways not in close contact with the water and consequently less involved in gas and ionic exchanges. Earlier suggestions of shunting of blood between the afferent and efferent blood pathways are present in the literature (Muller, 1839), but this concept was first given prominence by Steen and Kruysse (1964).

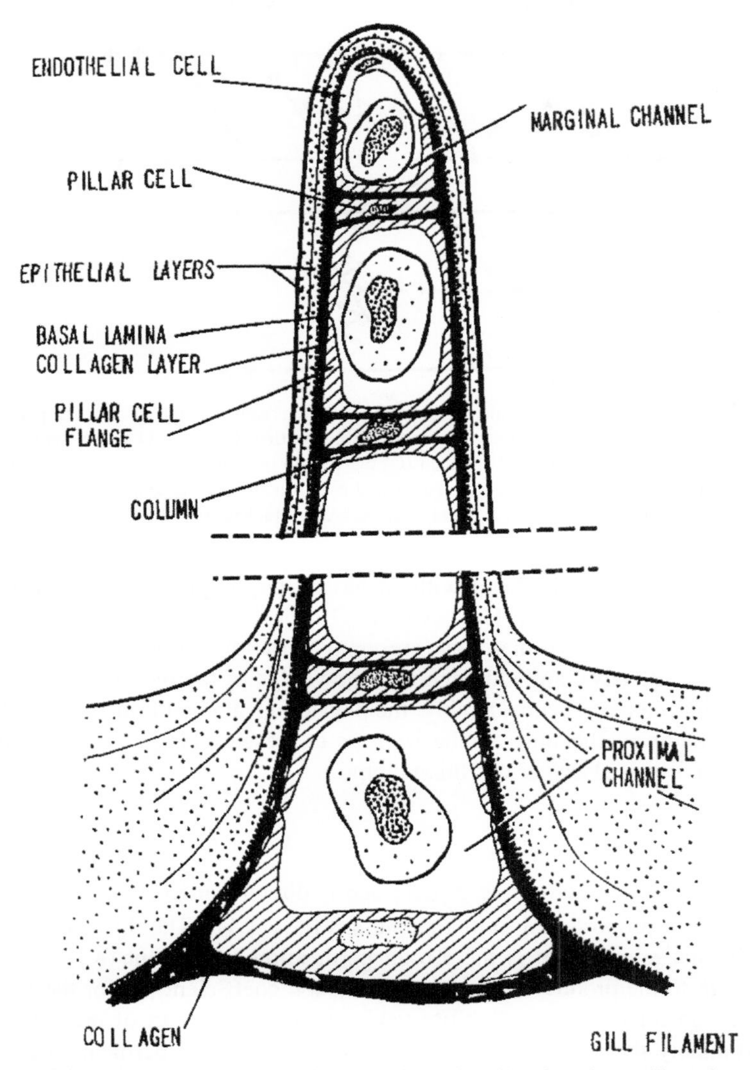

Fig. 25. Diagrammatic transverse section through the lamella of a teleost gill to show basic organization. The lamella is covered in two epithelial layers, within which is found the pillar cell system (PCS) consisting of the basal lamina and pillar cells, whose flanges enclose the main blood channel. The outer portion of the marginal channel is lined by endothelial cells. Note how the proximal or basal channel is embedded in the main tissue of the gill filaments. (From Hughes, 1980b.)

Their suggestion of a direct shunt is shown diagrammatically in Fig. 26B and compares with the classical situation. It is now accepted that there is evidence for some of the other pathways such as that shown in Fig. 26C. There is also evidence that, in at least some species, a relatively large portion of the cardiac

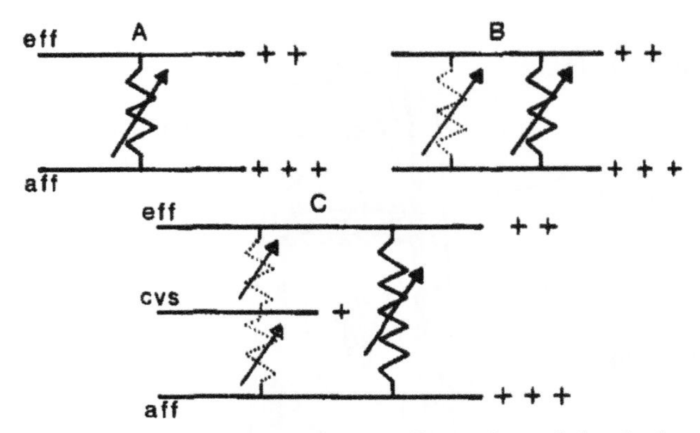

Fig. 26. Electrical analog diagrams to illustrate different views of the circulatory pathways between afferent and efferent filament arteries in fish. The first diagram (A) shows the classical circulation; a shunt is inserted in (B), but in (C) there is no true shunt between the afferent and efferent filament arteries as the pressure (indicated by +) in the central venous sinus (CVS) is lower than in either the efferent or afferent arteries. (After Hughes, 1979a.)

output may return to the heart without having any oxygen removed from it (e.g., up to 30% in the eel; Hughes *et al.*, 1982).

Thus, it is concluded that there are many variations in both the water and blood flow pathways, and although our knowledge of the morphological possibilities has increased considerably, the physiological conditions under which different pathways become emphasized are at present being actively studied. Details are given elsewhere in this volume.

B. Gas Exchange

Transfer of oxygen across the gill surface is directly proportional to its area and inversely proportional to its thickness: $\dot{V}_{O_2} = KA/t$. The diffusion coefficient in this relationship is the Krogh permeation coefficient (K), which is equivalent to the product of the "true" diffusion coefficient (d) for the particular gas molecule and its solubility (α) in the particular part of the pathway. Unfortunately there are few, if any, measurements of the permeation coefficient in fish tissues, and in most estimations, values obtained by Krogh (1919) for frog connective tissue have been used. Thus, much of the quantitative estimates may well need to be modified in the future if methods become available for determining the permeation coefficient of different parts of the water–blood barrier. For this reason it is important that details of different sections of the pathway should be measured wherever possible, but this involves a detailed study of transmission electron micrographs, and these are not always available. From such studies it is possible to subdivide the tissue barrier into several sections (Fig. 27). It is not known which of these is the most limiting.

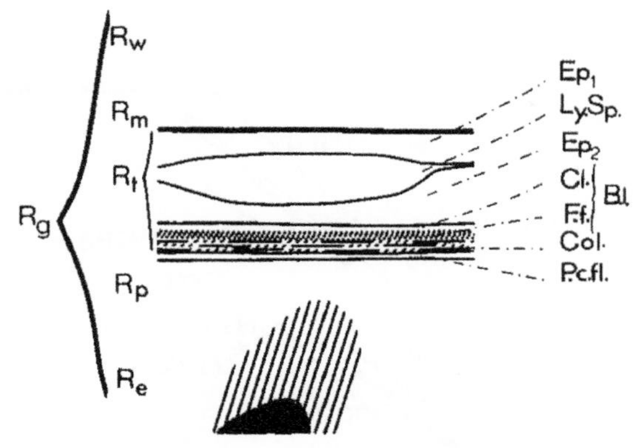

Fig. 27. Diagram to show different components of the resistance to gas transfer from water to blood in a fish gill lamella. The resistances are for water (R_w), the mucus layer (R_m), tissue (R_t), plasma (R_p), and within the erythrocyte (R_e). The tissue resistance (R_t) consists of seven different layers, namely outer and inner epithelial layers (Ep_1 and Ep_2), which in some regions are separated by lymphoid spaces (Ly.Sp). The basement membrane is composed of an outer clear layer (Cl), the middle, fine fibrous layer (F.f.), and an inner collagen layer (Col). The outer two form the basal lamina (B.I.), and the innermost layer of the tissue barrier is formed by a flange, of the pillar cell (P.c.fl.). (After Hughes, 1972c.)

1. *Surface Area*

As has been indicated earlier, the surface area of fish gills is greatly increased because of the subdivision of the primary epithelium into that of the gill filaments and of the lamellae. Measurements of surface area usually involve estimation of the total surface of all the lamellae of a given specimen. The techniques involved require sampling of the filaments and lamellae with special reference to the frequency of these structures along the filaments (Fig. 8) and their average surface area. The latter measurement is the most difficult and can give rise to significant errors, especially if no account is taken of the heterogeneity of the system and a weighting technique adopted. From these basic measurements the area is given by the following relationship: gill area $= Lnbl$, where L is the total filament length, n is the number of lamellae per millimeter on both sides of the filaments, and bl is the bilateral area of an average lamella.

It is usually this total area that is used in calculations and plots relating gill area to body mass. Some discussion has taken place regarding the extent to which the total lamellar area is equivalent to the respiratory area. It seems probable that some gas exchange can take place over this whole surface. It must be remembered, however, that the diffusion distance between the water and the nearest red blood cell will vary considerably according to its position directly over a blood channel or a pillar cell. Estimates of the proportion of the lamellar area directly above the pillar cells have given values up to

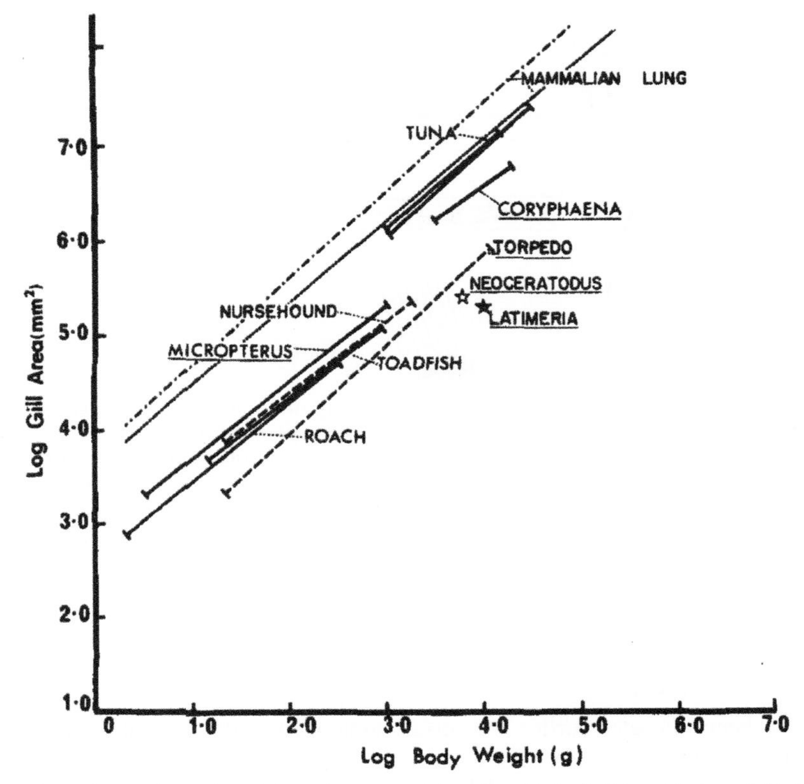

Fig. 28. Bilogarithmic plot of the surface area of the gills of fish of different body weight. The dotted lines show the relationship for mammalian lungs of "free-living" and captive species. (After Hughes, 1978b.)

30%, and on this basis only 70% of the total area is directly applicable to the thinner barriers of the regions that have been referred to as the diffusion channels.

The total area of the lamellae is much greater in more active fishes such as tuna and is far less in sluggish specimens such as toadfish (Fig. 28). Because the relationship between lamellar area and body mass is not a direct one, it is important to take account of the body size when making such comparisons. Accordingly, data are usually given for different body sizes from a regression line and can be calculated for any given body mass using the relationship, $X_1 = X_2\ (W_1/W_2)b$, where X_1 is the value required at weight W_1, and X_2 is the known value at weight W_2; b is the slope of the log/log regression line.

For a range of medium-size fish, the gill area of a 200-g fish has sometimes been used (Jager and Dekkers, 1975); however, where the size range is very great (e.g., 10 g–10 kg) it is perhaps better to compare fish over similar ranges, especially in cases where the regression coefficient (*b*) has not been determined and it is necessary to to utilize an average figure. The errors introduced by making such extrapolations may be quite large (Hughes, 1984b).

In fish that have accessory air-breathing organs the area of the gill surface is also small, and this is also true in some deep-water species that probably have a relatively sluggish existence. When assessing the importance of a respiratory surface for gas exchange it is important, however, to take into account not only the surface but also diffusion distances. This is because the diffusion of gas from the water to the blood, although directly proportional to surface area, is also inversely proportional to barrier thickness. Surfaces have sometimes been supposed to be important in gas exchange because of the degree of vascularization, although distances separating the blood and water may be very great, as in the cutaneous circulation of a number of fish, such as the icefish (Holeton, 1970). In comparison with the areas of lungs, no fish gills have such extensive surfaces as that of a mammal of comparable body mass. However, this is presumably compensated for by the greater ventilation of the gill surface and the lack of dead space and tidal ventilation.

Area measurements of lunglike air-breathing organs have not so far taken into account increases in surface due to the bulging capillaries. As has been suggested (Hughes and Munshi, 1978), the effective increase in area would be partly compensated by the nonrespiratory areas such as the "lanes" (Fig. 17). The need to take this heterogeneity into account when calculating diffusing capacity has been emphasized (Hughes, 1980d) and is especially important in fish (e.g., *Monopterus*) where the nonrespiratory regions are very extensive. In such cases, published values (Hughes *et al.*, 1974a) must be considerably reduced, but in fish with "lanes" (Table III) the effect may be compensated by projections of the respiratory islets (modified lamellae) above the air sac lining surface.

2. *Thickness*

As indicated earlier, the thickness of the barrier separating the water from the blood is an important factor in gas exchange. Measurements of these distances have been carried out in a number of fish gills, but unfortunately the methods adopted have varied quite considerably. In a number of cases the measurements have been confined to the "diffusion channel" that overlies the blood channels in the lamella. The minimum distances may be measured at regular intervals along the outside epithelium and an average value taken. In other cases only the minimum distance in the diffusion channel has been measured. One method that has been used for measuring the arithmetic mean thickness, particularly in lungs, is to obtain values for the surface area (intersection counting) of the air and capillary surfaces together with an estimate of the volume of the barrier using point counting. Thus, the arithmetic mean thickness of the barrier is the volume of the barrier divided by the mean area of the air and blood surfaces. Such a method has scarcely been applied to fish gills, but a method developed for measurement of the harmonic mean thickness in tetrapod lungs has given useful results. In this method, values for distances between the external surface and the nearest point on the blood capillary or the nearest blood corpuscle are measured using a special rule, and the numbers of occurrences between certain distances are calculated and, from their distribution, the harmonic mean thickness is calculated. For details of this method see Weibel (1971) and Perry (1983). In only a very few cases has the harmonic diffusion distance been measured for gills (Hughes, 1972c; Hughes and Perry, 1976). In this type of analysis the longer distances are of less importance, because in the estimation it is the reciprocal of the distance that is calculated. Where the harmonic diffusion distance has been used, then the appropriate area measurement is that of the total outer lamellar surface (Fig. 29). In cases where only the diffusion channel measurements have been used with a consequent smaller value, then the smaller area overlying the blood channels is the appropriate value for surface area (Hughes, 1982).

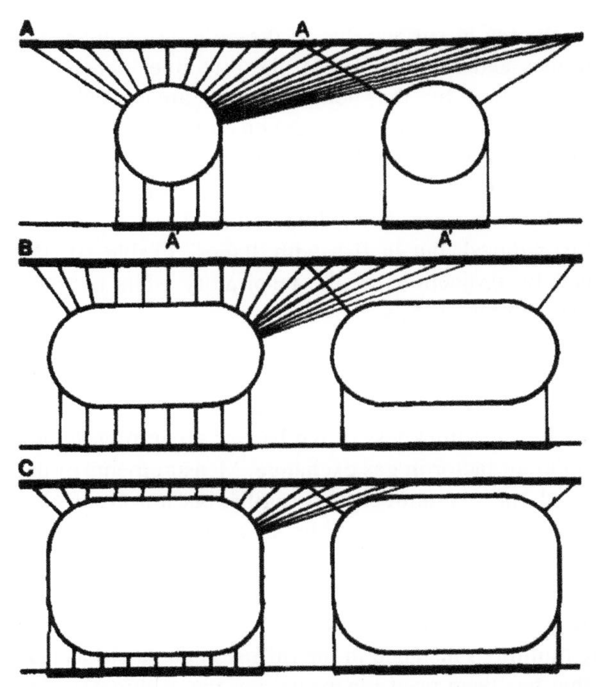

Fig. 29. Diagrams to illustrate two methods for measuring the thickness and surface area of a lamella for use in estimating diffusing capacity. The upper surface has an area (A), and the lines indicate distances for measurement of harmonic means. The lower surfaces show areas (A') over blood channels and lines for measurements used to obtain diffusion distances across the blood channels alone. The blood channels increase in relative size from a to c. (From Hughes, 1982.)

In all these discussions there has often been neglect of the water that is in immediate contact with lamellae. At least two different types of studies (Hills and Hughes, 1970; Scheid and Piiper, 1976) have indicated that this in fact forms an important part of the barrier to diffusion. Another problem relates to the thin layer of mucus that must overlie the lamellae during normal function, but the extent of this layer has so far proved impossible to determine in the living condition.

3. *Diffusing Capacity*

Respiratory physiologists introduced the term diffusing capacity to express the capability of a gas exchange organ to transmit gases, and it is the ratio of volume of gas transferred across the barrier in unit time to the mean difference in partial pressures of that gas on the two sides of the barrier. Diffusing capacities with respect to oxygen and also carbon monoxide have been determined in a number of cases. It is directly proportional to surface area and inversely proportional to the thickness. Gill diffusing capacity $\left(D_{\mathrm{g}}\right) = \dot{V}_{\mathrm{O_2}}/\Delta P_{\mathrm{O_2}}$. The diffusion constant is the Krogh permeation coefficient (K), which has not been determined for many fish tissues. Such calculations give values for diffusing capacity of the tissue barrier alone (D_{t}), but of course the transfer of oxygen from water to the hemoglobin molecules involves other stages, and these may be indicated by the following relationship:

$$1/D_{\mathrm{g}} = 1/D_{\mathrm{w}} + 1/D_{\mathrm{m}} + 1/D_{\mathrm{t}} + 1/D_{\mathrm{p}} + 1/D_{\mathrm{e}}.$$

where the subscripts w, m, t, p, and e indicate water, mucus layer, tissue barrier, plasma, and erythrocytes, respectively.

Although the term diffusing capacity is preferable because of its frequent usage in mammalian respiratory physiology and morphology, the term transfer factor introduced by Cotes (1965) for the mammalian lung has often been used for the fish gill. There are some drawbacks, however, as transfer factor was the term originally used in relation to blood platelet studies, and in mammalian respiratory physiology it usually refers to the transfer of carbon monoxide across the lung surface. A disadvantage of the term diffusing capacity, however, is that physiologically it concerns all the resistances between the oxygen contained in the water and its final combination with the hemoglobin molecules. It is thus greatly influenced by factors other than diffusion, notably convection within the water and plasma, and the kinetics of the reaction in the red blood corpuscles. Nevertheless, it is probably preferable to maintain the use of this term rather than confuse the literature still further. Accordingly, it would seem that for comparisons between different gill systems, data based on the diffusing capacity of the tissue barrier provide a useful guide to the systems' effectiveness (Table V). The greatest error probably arises because of the lack of information regarding the nature of the resistance in the water film. Once again this particular resistance will vary according to the degree of ventilation. Indeed, from a morphometric point of view, only diffusive resistances are considered with the exception of the reaction with hemoglobin.

Many of the different resistances to oxygen transfer between water and red cell are indicated in Fig. 27. Apart from the water, the main resistances are in the tissue, plasma, and erythrocytes, and these are the components included in morphometric estimates of diffusing capacity. Hence, $R_g = R_t + R_p + R_e$. Written as conductances, this becomes: $1/D_g = 1/D_t + 1/D_p + 1/D_e$, where D_g is the gill diffusing capacity.

Estimates of these relative contributions in the lung of *Lepidosiren*, for example, suggest that the tissue component forms 75% of the morphometrically determined diffusing capacity, and 13% is made up by the resistance in the erythrocytes. However, at least two studies (Hills and Hughes, 1970; Scheid and Piiper, 1976) have concluded that a large contribution to the overall resistance is contained within the water. Consequently, for the total resistance, assuming 50% is in the water, this suggests that only 37% is contained within the tissue barrier and about 7% in the erythrocyte. That the major resistance is diffusive and not chemical has also been confirmed (Hills *et al.*, 1982) by analysis of in vivo measurements using isolated lamellae. There are of course many difficulties in estimating these resistances, and at present it would be wrong to place too much certainty on any of the figures, although they are a good guide. Perhaps the most certain are those for the morphometric part of the tissue barrier ($D_t = KA/t$), but even here there are difficulties. The need for caution is particularly great when discussing absolute values for diffusing capacity and using them in comparisons with other animals. On a relative basis, however, the values available have proved to be of considerable usefulness. For example, in studies of the effects of pollutants on the gills (Hughes *et al.*, 1979) it has been shown that the concept of relative diffusing capacity (Hughes and Perry, 1976) enables comparisons to be made between experimental and control fish without making absolute measurements of either gill area or barrier thickness. The assumption was made, however, that the permeation coefficient would be the same under the two conditions. Soivio and Tuurala (1981) have extended some earlier studies (Hughes *et al.*, 1978) to show that there are changes not only in the barrier thickness but also in other dimensions of the lamellae following exposure to hypoxia. These studies also indicated the value of simple stereological methods using point and intersection counting to assess the volume and surfaces to particular parts of a structure relative to other parts using sectioned material. Figure 30 illustrates the typical method, which makes it possible to compare the relative volumes of plasma and pillar cells within a lamella. The superimposed grid is rectilinear, which is preferred when the sections are at random to the surface of the lamellae. Despite the regular orientation of lamellae, it is difficult to ensure that sectioning is always at right angles to their surface. Careful control can make this possible, in which case a Merz grid can be used (Hughes and Perry, 1976).

To summarize, the whole study of fish gill morphometry has only just begun, and methods are becoming more standardized. In future this will lead

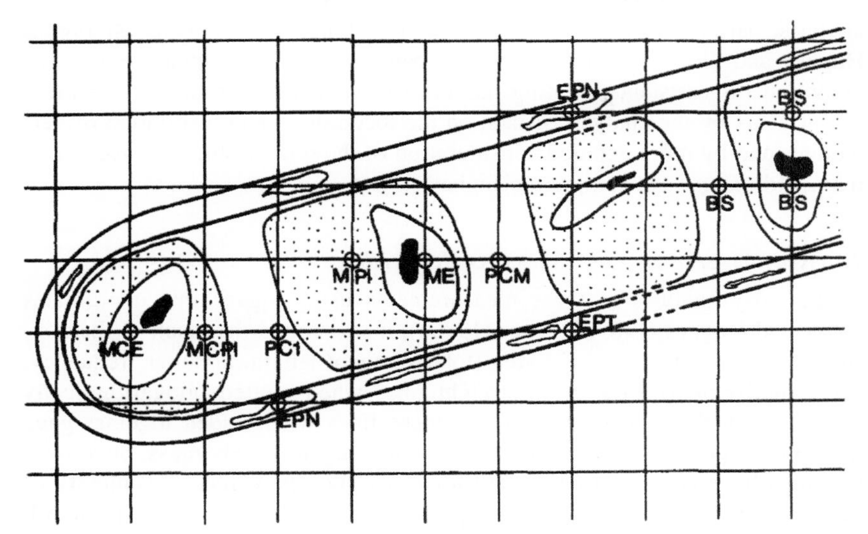

Fig. 30. Diagrammatic section of a lamella with superimposed rectilinear grid. Some of the points counted are indicated. E, erythrocyte cytoplasm; Nu, erythrocyte nucleus; Pl, plasma; PC, pillar cell; EPT, epithelial tissue; EPN, epithelial nontissue. (From Hughes 1979a.)

to a situation where greater certainty can be laid on the results of comparisons of different types of material. Although the basic methodology is simple, it is advisable that anyone wishing to carry out such studies should take some of the precautions that have been shown to be necessary by these early exploratory studies.

C. Scaling

This term has become generally used in relation to the way in which changes in dimensions of a given structure or function are related to differences in body mass. It is thus a special form of allometry, which was first introduced by Huxley and Teissier (1936) and has provided a valuable synthesis of much morphological and physiological data. Stimulating studies by D'Arcy Thompson (1917) and Huxley (1932) as experimental biology began to expand must be given much credit for the early influence of allometry. Since then there has always been an interest in allometry, but this has been renewed considerably in recent years perhaps because of the greater interest in functional morphology in relation to physiology and the ready availability of facilities that greatly speed up the computation and analysis of data. Methods for processing morphometric measurements similar to those outlined earlier have also been greatly improved by these developments. For example, Gray (1954) had made some pioneering measurements of the gill area of many species of fish covering a range of body sizes, but they were not subjected to regression analysis, partly because of the time required. Later studies (Hughes, 1966b; Hughes and Gray, 1972) have greatly benefited by these measurements of Gray and his assistants. Scaling has been brought into prominence in more recent years by the analysis of data on the locomotion and respiration of vertebrates and by more detailed consideration of the relationships between body mass and oxygen consumption (Hughes, 1984a).

1. RELATIONSHIP OF GILL AREAS TO BODY MASS

In the early analysis of these measurements using logarithmic coordinates (Muir and Hughes, 1969; Hughes, 1970a), it was found that each of the constituent measurements of gill area (total filament length, lamellar frequency, and lamellar area) all showed straight lines when plotted bilogarithmically (Fig. 31). Furthermore, the slopes of these lines when added together gave the same slope as that for the plot of total area against body mass. Such relationships have now been confirmed for many species of fish. A summary of the a and b values in the relationship, gill area $= aW^b$ is given in Table VI.

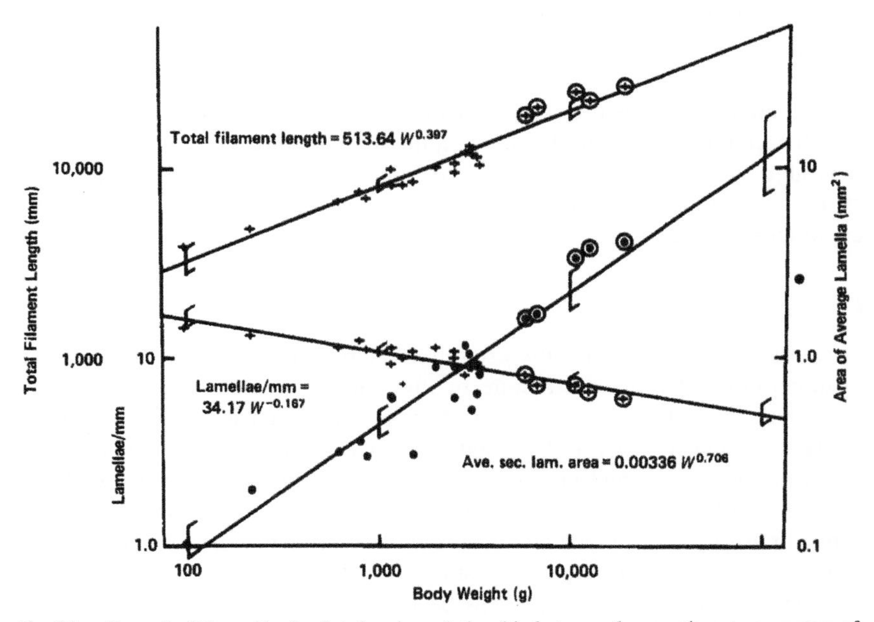

Fig. 31. *Torpedo.* Bilogarithmic plot showing relationship between the constituent parameters of the gill area and body mass for 22 specimens. The upper crosses represent total filament length (mm), the lower crosses show the number of lamellae per millimeter, and filled circles are for the area of an average lamella (mm^2). Most points plotted refer to *T. marmorata*; five of the measurements (enclosed with circles) were made on large specimens of *T. nobiliana*. Bars indicating 95% confidence limits for each of the regression lines are shown at body weights of 100, 1000, and 10,000 g. (From Hughes, 1978c.)

Table VI Summary of Results of Regression Analysis for Gill Dimensions of a Range of Fish Species[a]

Species	Gill measurement	Body weight			Equation for total gill area $Y = aW_b$	References
		10 g	*100 g*	*1000 g*		
Thunnus albacares, Thunnus thynnus	L	13,481	32,488	78,293	$3151W^{0.875}$	Muir and Hughes (1969)
	bl	0.0176	0.0674	0.2581		
	$1/d$	49.56	40.35	32.85		
Coryphaena hippurus	L	5069	13,675	36,369	$5208W^{0.713}$	Hughes (1970a)
	bl	0.0804	0.1707	0.3626		
	$1/d$	31.12	28.65	26.37		
Scomber scombrus	L	4054	10,434	26,849	$424.1W^{0.997}$	Hughes (1970b)
	bl	0.0187	0.0672	0.2417		
	$1/d$	28.56	30.14	31.81		
Salmo gairdneri	L	2434	6541	17,579	$314.8W^{0.932}$	Hughes (1980b)
	bl	0.0229	0.0853	0.3172		
	$1/d$	23.78	20.53	17.72		
Tinca tinca	L	3012	7674	19,540	$867.2W^{0.698}$	Hughes (1970b)
	bl	0.0263	0.0601	0.1373		
	$1/d$	23.76	22.17	20.69		
Opsanus tau	L	923.7	2825	8638	$560.7W^{0.79}$	Hughes (1970a)
	bl	0.1412	0.3347	0.7879		
	$1/d$	13.37	11.27	9.51		

Continued

Table VI Summary of Results of Regression Analysis for Gill Dimensions of a Range of Fish Species—Cont'd

Species	Gill measurement	Body weight			Equation for total gill area $Y = aW_b$	References
		10 g	*100 g*	*1000 g*		
Scyliorhinus canicula	L	1623	3644	8181	$262.3W^{0.961}$	Hughes (1970b)
	bl	0.04998	0.2412	1.1655		
	$1/d$	14.56	12.36	10.49		
Torpedo marmorata	L	1280.3	3191.1	79544	$117.5W^{0.937}$	Hughes (1978c)
	bl	0.0171	0.0869	0.4414		
	$1/d$	23.29	15.87	10.82		
Anabas testudineus	L	559	1209	2615	$556W^{0.615}$	Hughes and Munshi (1973a)
	bl	0.0398	0.106	0.2828		
	$1/d$	25.72	18.11	12.76		
Saccobranchus fossilis	L	832.2	2267	6172	$186.1W^{0.746}$	Hughes *et al.* (1974b)
	bl	1.02455	0.06276	0.1604		
	$1/d$	25.45	20.46	16.46		
Channa punctata	L	1485.6	3955	—	$470.4W^{0.592}$	Hakim *et al.* (1978)
	bl	0.0236	0.0475	—		
	$1/d$	26.24	19.12	—		

[a]Values for total filament length (L, mm), bilateral area of an average lamella (bl, mm^2), and frequency of lamellae on one side of a filament ($1/d$, mm) are given for body weights of 10, 100, and 1000 g. The relationship for total gill area ($Y = aW^b$) is also given for each species.

It is now apparent that no common slope b can be used for all species of fish. However, a value of about 0.8 is fairly common, although b may range from 0.5 to 1.0. Thus, it has been concluded that individual species show differences in the scaling of respiratory parameters, and some of the possible factors have been discussed but as yet are not understood.

The importance of emphasizing differences at an intraspecific level becomes apparent when comparisons are extended to relationships between oxygen consumption and body mass. Again the general relationship with a slope of 0.75 to 0.8 has been known for many years, and for a wide range of fish species Winberg (1956) suggested an average value of 0.82, which has often been adopted. Just as with the scaling of the respiratory surfaces, so there are variations between species with respect to the $\dot{V}_{O_2}/W$ relationship. Initially it was supposed (Muir and Hughes, 1969) that the two slopes would be approximately the same, suggesting that gill surface area was the main factor governing oxygen consumption. When measurements for more species were analyzed, however, differences between the slopes were recognized and it was further appreciated that the total gill area was not normally functional during resting conditions. Perhaps for this reason the slope of the gill area line should be considered more in relation to that for $\dot{V}_{O_2,\text{active}}$. As only two or three fish species have had their active metabolism measured over a range of body sizes, it was suggested that, in the absence of definite data, the slope of the gill area line might be taken to be indicative of the slope for active metabolism (Hughes, 1972c, 1977). Under these circumstances, it was appreciated that in some species the lines relating to resting metabolism and gill area (= active metabolism) would diverge, and consequently, the difference between these two (= scope for activity; Fry, 1957) would increase with body mass. In fact such a relationship has been demonstrated for salmon (Brett, 1965). In some fish, however, the two regression lines would be parallel, so that there would be no increase in scope with body size, and in other instances they might converge, indicating a decrease in scope with increase in body mass. The size where the two lines cross would suggest that no activity was possible above such a size, and the respiratory surface could be considered the major factor governing growth of the fish. Until sufficient data are available comparing the active and resting metabolism of fish with morphometric measurements of gill area and diffusing capacity, these must represent tantalizing hypotheses that may or may not prove to be helpful in further generalizations.

2. *Dimensional Analysis*

Engineers have long used a technique in which the dimensions of different features of a machine are analyzed in relation to the particular size of the whole structure and in this way to gain information regarding their functional relationships. A similar method has been applied to the gills of fish for which

Table VII Exponents for the Relationships between Body Mass and Different Parameters Relating to the Gill Dimensions of *Micropterus* and the Rainbow Trout, *Salmo gairdneri*

Parameter	*Micropterus*	*Salmo gairdneri*
Distance between secondary lamellae	0.022	0.09
Number of pores	0.54	0.23
Pore length	0.21	0.24
Ventilation volume	0.93	0.73
Gill area	0.95	0.78

measurements during their development are available. The data for the small-mouthed bass, *Micropterus dolomieu* (Price 1931), proved invaluable for making such an analysis (Hills and Hughes, 1970), and a similar technique was applied later to measurements made for the rainbow trout by Morgan (1971). Exponents for the relationships between body mass and different parameters relating to the gill dimensions were established as shown in Table VII. In spite of differences in the detailed exponents for these two species, the final conclusions were almost identical and have given useful general information regarding the changes in the structures during development. From this information the following conclusions were reached:

1. Water flow past the lamellae is probably laminar.
2. Relative to the blood, countercurrent flow of water is more probable than cocurrent.
3. The perfusion of gill surfaces with blood is constant at all body sizes.
4. Resistance to overall oxygen transfer provided by the water is between 5 and 10 times greater than that due to the tissue barrier, and this remains almost constant at all body sizes.
5. The water velocity between secondary lamellae increases slightly with body size; in the case of the trout it is proportional to $W^{0.17}$, whereas in *Micropterus* it is proportional to $W^{0.10}$.

Although the whole gill sieve may be considered as a large number of pores of the dimensions just discussed, which are in parallel to one another, and consequently, the volume flow per unit time through each pore is relatively small, nevertheless there are other paths for water flow between the buccal and opercular cavity that may be considered as part of an anatomic and physiological dead space, the total forming a water shunt that has been estimated to be as much as 60% of the total ventilation volume in trout (Randall, 1970). Randall further subdivided the physiological dead space into diffusion and distribution components. There have, however, been very few studies on the resistance properties of the whole gill network in relation to differences in hydrostatic pressure across the gill resistance. In some early studies (Hughes and Shelton, 1962) using anesthetized fish and by changing the pressure difference, it was shown that flow increased with increasing pressure difference until a point came when the flow for a given increment of pressure suddenly increased. It was suggested that perhaps this indicated the instant at which the gill resistance falls as the tips of the filaments separate. Such studies have also indicated (Hughes and Umezawa, 1968) that the resistance from the opercular to buccal cavities is greater than that in the normal direction of water flow.

The conditions regarding blood flow through lamellae have been established by Muir and Brown (1971), and dimensional analysis suggests that blood viscosity must increase with body mass (Hughes, 1977).

VII. CONCLUSIONS

It is apparent from investigations at a gross morphological level that there are many variations in the organization of fish gills. Nevertheless, the same basic structure is recognizable even when it has become modified to produce organs so diverse as the glandular pseudobranch of *Anabas* and the air tubes of *Saccobranchus*, which are so different from the highly evolved gills of tuna and other oceanic species. In all these situations the presence of pillar cells has proved to be an invaluable guide to the morphological nature of such highly modified organs. Despite much investigation, a great deal remains to be learned about the functioning of these interesting cells, which clearly can function not only in support but also may have important immunological functions (Chilmonczyk and Monge, 1980), and their similarity to some structures in the reticuloendothelial system of mammals is most striking (Hughes and Weibel, 1976). The filaments of fish gills also show a wide range in their structure and development, again related to the habits of the particular fishes.

The detailed nature of the vascular pathways at the filament level is still in course of investigation, and the control mechanisms involved remain to be elucidated. At the filament level it would appear that nervous mechanisms are involved (see Chapter 2, this volume), but for lamellae little information is available concerning either the sensory or the motor side of their functioning. It would appear that much of the motor activity of lamellae is regulated by blood-borne substances. These different control mechanisms must affect the overall balance between the respiratory and other functions of the gills and still presents a challenge to the fish physiologist. Much has been learned by the use of isolated gill preparations, but until more is known of the blood composition and a suitable perfusion medium is developed, it is important to compare constantly the condition of such preparations with those of the intact gill system (see Chapter 10, Volume XB, this series).

Comparison of the structure and function of fish gills with the vertebrate lung suggests that the gill is at least as well adapted to its respiratory function in water as is the lung to air. Their positions and roles in the circulation of the body have many similarities, although differences due to the single and double circulation have some influence. In particular, blood is supplied to the gill at a relatively high pressure compared with the lungs. The precise way in which this affects the filtration of plasma through the capillary beds is not yet fully understood, but the existence of low capillary permeability (Hargens *et al.*, 1974) may be related to these circulatory differences. Indeed there are indications that the whole functioning of the capillary system in fish may be quite different from what is accepted as normal for mammals. Vogel (1981) and Vogel and Claviez (1981) have drawn attention to these differences and have identified a secondary circulation that differs from the lymphatic system of higher vertebrates. Thus, perhaps some of the most fascinating studies of fish gills concern the air-breathing fishes, where the gills are modified to a certain extent while a lung or other air-breathing organ takes over some of the functions that they normally perform. Investigation of the wide variety of these fishes continues to reveal important general principles as, for example, the presence of something equivalent to a double circulation in some species of *Channa* (Ishimatsu and Itazawa, 1983).

REFERENCES

Acrivo, C. (1938). Sur l'organisation et la structure du corps caverneaux chez *Scyllium canicula* Cuv. *Bull. Histol. Appl. Tech. Microsc.* **12,** 362–372.

Adeney, R., and Hughes, G. M. (1977). Observations on the gills of the sunfish *Mola mola. J. Mar. biol. Assor. U.K.* **57,** 825–837.

Alexander, R. McN. (1970). Mechanics of the feeding action of various teleost fishes. *J. Zool.* **162,** 145–156.

Al-Kadhomiy, N. (1984). Gill development, growth, and respiration of the flounder, (*Platichthys flesus*). Thesis, Bristol University.

Ballintijn, C. M. (1968). The respiratory pumping mechanism of the carp, *Cyprinus carpio* (L). Thesis, University of Grøningen, Netherlands.

Berg, T., and Steen, J. B. (1965). Physiological mechanisms for aerial respiration in the eel. *Comp. Biochem. Physiol.* **15,** 469–484.

Bertin, L. (1958). Organes de la respiration aquatique. *In* "Traité de zoologie" (P. P. Grassé, ed.), vol. 13, pp. 1303–1341. Masson, Paris.

Bettex-Galland, M., and Hughes, G. M. (1973). Contractile filamentous material in the pillar cells of fish gills. *J. Cell Sci.* **13,** 359–366.

Bevelander, G. (1934). The gills of *Amia calva* specialised for respiration in an oxygen-deficient habitat. *Copeia* **3,** 123–127.

Bijtel, J. H. (1949). The structure and the mechanism of movements of the gill filaments in Teleostei. *Arch. Neerl. Zool.* **8,** 267–288.

Booth, J. H. (1978). The distribution of blood flow in the gills of fish: application of a new technique to rainbow trout (*Salmo gairdneri*). *J. Exp. Biol.* **73,** 119–129.

Booth, J. H. (1979). The effects of oxygen supply, epinephrine, and acetylcholine on the distribution of blood flow in trout gills. *J. Exp. Biol.* **83,** 31–39.

Brackenbury, J. H. (1972). Lung-air-sac anatomy and respiratory pressures in the bird. *J. Exp. Biol.* **57,** 543–550.

Brett, J. R. (1965). The relation of size to rate of oxygen consumption and sustained swimming speed of sockeye salmon (*Oncorhynchus nerka*). *J. Fish. Res. Board Can.* **22,** 1491–1501.

Brett, J. R. (1972). The metabolic demand for oxygen in fish, particularly salmonids, and a comparison with other vertebrates. *Respir. Physiol.* **14,** 151–170.

Brown, C. E., and Muir, B. S. (1970). Analysis of ram ventilation of fish gills with application to skipjack tuna (*Katsuwonus pelamis*). *J. Fish. Res. Board Can.* **27,** 1637–1652.

Byczkowska-Smyk, W. (1957). The respiratory surface of the gills in teleosts. 1. The respiratory surface of the gills in the flounder (*Pleuronectes platessa*) and perch (*Perca fluviatilis*). *Zool. Pol.* **8,** 91–111.

Byczkowska-Smyk, W. (1958). The respiratory surface of the gills in teleosts. 2. The respiratory surface of the gills in the eel (*Anguilla anguilla*), the loach (*Misgurnus fossilis* L.) and the perch-pike (*Lucioperca lucioperca* L.). *Acta Biol. Cracov. Ser. Zool.* **1,** 83–97.

Byczkowska-Smyk, W. (1959a). The respiratory surface of the gills in teleosts. 3. The respiratory surface of the gills in the tench (*Tinca tinca* L.), the silver bream (*Blicca bjoerkna*), and the chondrostomas (*Chondrostoma nasus* L.). *Acta Biol. Cracov.* **2,** 73–88.

Byczkowska-Smyk, W. (1959b). The respiratory surface of the gills in teleosts, 4. The respiratory surface of the gills in the pike (*Esox lucius*), stone-perch (*Acerina cernus*) and the burbot (*Lota lota* L.). *Acta Biol. Cracov.* **2,** 113–127.

Byczkowska-Smyk, W. (1962). Vascularisation and size of the respiratory surface in *Acipenser stellatus. Acta Biol. Cracov.* **5,** 304–315.

Carter, G. S. (1957). Air breathing. *In* "Physiology of Fishes" (M. E. Brown, ed.), Vol. 1, pp. 65–79. Academic Press, New York.

Carter, C. S., and Beadle, L. C. (1931). The fauna of the Paraguayan Chaco in relation to its environment. II. Respiratory adaptations in the fishes. *J. Linn. Soc. London, Zool.* **37,** 327–366.

Chilmonczyk, S., and Monge, D. (1980). Rainbow trout gill pillar cells: Demonstration of inert particle phagocytosis and involvement in viral infection. *J. Reticuloendothel. Soc.* **28,** 327–332.

Cooke, I. R. C. (1980). Functional aspects of the morphology and vascular anatomy of the gills of the Endeavour dogfish, *Centrophorus scalpratus* (McCulloch) (Elasmobranchii: Squalidae). *Zoomorphologie* **94,** 167–183.

Cotes, J. E. (1965). "Lung Function." Blackwell, Oxford.

D'Arcy, Thompson, W. (1917). "On Growth and Form." Cambridge Univ. Press, London and New York.

de Jager, S., and Dekkers, W. J. (1975). Relations between gill structure and activity in fish. *Neth. J. Zool.* **25,** 276–308.

Dejours, P. (1976). Water versus air as the respiratory media. *In* "Respiration of Amphibious Vertebrates" (G. M. Hughes, ed.), pp. 1–15. Academic Press, New York.

Dornescu, G. T., and Miscalencu, D. (1968a). Cele trei tipuri de branchii ale teleosteenilor. *Ann. Univ. Bucuresti, Ser. S. Nat. Biol.* **17,** 11–20.

Dornescu, G. T., and Miscalencu, D. (1968b). Etude comparative des branchies de quelques espèces de l'ordre clupéiformes. *Morphol. Jahrb.* **122**(H2), 261–276.

Farber, J., and Rahn, H. (1970). Gas exchange between air and water and the ventilation pattern in the electric eel. *Respir. Physiol.* **9,** 151–161.

Fromm, P. P. (1976). Circulation in trout gills: Presence of 'Blebs' in afferent filament vessels. *J. Fish. Res. Board Can.* **31,** 1793–1796.

Fry, F. E. J. (1957). The aquatic respiration of fish. *In* "Physiology of Fishes" (M. E. Brown, ed.), Vol. 1, pp. 1–63. Academic Press, New York.

Gans, C. (1976). Ventilatory mechanisms and problems in some amphibious aspiration breathers (*Chelydra, Caiman*—Reptilia). *In* "Respiration of Amphibious Vertebrates" (G. M. Hughes, ed.), pp. 357–374). Academic Press, New York.

Gans, C., DeJohngh, H. J., and Farber, J. (1969). Bullfrog (*Rana catesbeiana*) ventilation: how does the frog breathe? *Science* **163,** 1223–1225.

Goette, A. (1878). Zur Entwicklungsgeschichte der Teleosteirkieme. *Zool. Anz.* **1,** 52.

Goodrich, E. S. (1930). "Studies on the Structure and Development of Vertebrates." Macmillan, New York.

Graham, J. B. (1976). Respiratory adaptations of marine air-breathing fishes. *In* "Respiration of Amphibious Vertebrates" (G. M. Hughes, ed.), pp. 165–187. Academic Press, New York.

Gray, I. E. (1954). Comparative study of the gill area of marine fishes. *Biol. Bull. (Woods Hole, Mass.)* **107,** 219–225.

Grigg, G. C. (1970a). Water flow through the gills of Port Jackson sharks. *J. Exp. Biol.* **52,** 565–568.

Grigg, G. C. (1970b). The use of gill slits for water intake in a shark. *J. Exp. Biol.* **52,** 569–574.

Hakim, A., Munshi, J. S. D., and Hughes, G. M. (1978). Morphometrics of the respiratory organs of the Indian green snake-headed fish, *Channa punctata. J. Zool.* **184,** 519–543.

Harden-Jones, F. R., Arnold, G. P., Grier-Walker, M., and Scholes, P. (1979). Selective tidal stream transport and the migration of plaice (*Pleuronectes platessa* L.) in the southern North Sea. *J. Cons., Cons. Int. Explor. Mer.* **38,** 331–337.

Hargens, A. R., Millard, R. W., and Johansen, K. (1974). High capillary permeability in fishes. *Comp. Biochem. Physiol. A.* **48A,** 675–680.

Henschel, J. (1941). Neue Untersuchungen über den Atemmechanismus mariner Teleosteer. *Helgol. Wiss. Meeresunters.* **2,** 244–278.

Hills, B. A. (1974). "Gas Transfer in the Lung." Cambridge Univ. Press, London and New York.

Hills, B. A., and Hughes, G. M. (1970). A dimensional analysis of oxygen transfer in the fish gill. *Respir. Physiol.* **9,** 126–140.

Hills, B. A., Hughes, G. M., and Koyama, T. (1982). Oxygenation and deoxygenation kinetics of red cells in isolated lamellae of fish gills. *J. Exp. Biol.* **98,** 269–275.

Holeton, G. F. (1970). Fish respiration without haemoglobin. Ph.D. Thesis, Bristol University.

Holeton, G. F. (1971). Respiratory and circulatory responses of rainbow trout larvae to carbon monoxide and to hypoxia. *J. Exp. Biol.* **55,** 683–694.

Holeton, G. F. (1972). Gas exchange in fish with and without hemoglobin. *Respir. Physiol.* **14,** 142–150.

Holeton, G. F. (1980). Oxygen as an environmental factor of fishes. *In* "Environmental Physiology of Fishes" (M. A. Ali, ed.), pp. 7–32. Plenum, New York.

Holeton, G. F., and Jones, D. R. (1975). Water flow dynamics in the respiratory tract of the carp (*Cyprinus carpio* L.). *J. Exp. Biol.* **63,** 537–549.

Horstadius, S. (1950). "The Neural Crest. Its Properties and Derivatives in the Light of Experimental Research." Oxford Univ. Press, London and New York.

Hughes, G. M. (1960a). The mechanism of gill ventilation in the dogfish and skate. *J. Exp. Biol.* **37,** 11–27.

Hughes, G. M. (1960b). A comparative study of gill ventilation in marine teleosts. *J. Exp. Biol.* **37,** 28–45.

Hughes, G. M. (1963). "Comparative Physiology of Vertebrate Respiration," (Heinemann, London (2nd ed., 1974).

Hughes, G. M. (1966a). Evolution between air and water. *In* "Development of the Lung" (A.V.S. de Reuck and R. Porter, eds.), pp. 64–80. Churchill, London.

Hughes, G. M. (1966b). The dimensions of fish gills in relation to their function. *J. Exp. Biol.* **45,** 177–195.

Hughes, G. M. (1970a). Morphological measurements on the gills of fishes in relation to their respiratory function. *Folia Morphol. (Prague)* **18,** 78–95.

Hughes, G. M. (1970b). A comparative approach to fish respiration. *Experientia* **26,** 113–122.

Hughes, G. M. (1972a). Gills of a living coelacanth, *Latimeria chalumnae*. *Experientia* **28,** 1301–1302.

Hughes, G. M. (1972b). Distribution of oxygen tension in the blood and water along the secondary lamella of the icefish gill. *J. Exp. Biol.* **56,** 481–492.

Hughes, G. M. (1972c). Morphometrics of fish gills. *Respir. Physiol.* **14,** 1–25.

Hughes, G. M. (1973a). Respiratory responses to hypoxia in fish. *Am. Zool.* **13,** 475–489.

Hughes, G. M. (1973b). Comparative vertebrate ventilation and heterogeneity. *In* "Comparative Physiology" (L. Bolis, K. Schmidt-Nielsen, and S. H. P. Maddrell, eds.), pp. 187–220. North-Holland Publ., Amsterdam.

Hughes, G. M. (1976). Fish respiratory physiology. *In* "Perspectives in Experimental Biology" (P. Spencer Davies, ed.), Vol. 1, pp. 235–245. Pergamon Press, Oxford and New York.

Hughes, G. M. (1977). Dimensions and the respiration of lower vertebrates. *In* "Scale Effects in Animal Locomotion" (T. J. Pedley, ed.), pp. 57–81. Academic Press, New York.

Hughes, G. M. (1978a). A morphological and ultrastructural comparison of some vertebrate lungs. *Colloq. Sci. Fac. Med. Univ. Carol. [Pap.] , 21st, 1978* pp. 393–405.

Hughes, G. M. (1978b). Some features of gas transfer in fish. *Bull. Inst. Math. Appl.* **14,** 39–43.

Hughes, G. M. (1978c). On the respiration of *Torpedo marmorata*. *J. Exp. Biol.* **73,** 85–105.

Hughes, G. M. (1979a). The path of blood flow through the gills of fishes—some morphometric observations. *Acta Morphol. Sofia* **2,** 52–58.

Hughes, G. M. (1979b). Scanning electron microscopy of the respiratory surface of trout gills. *J. Zool.* **188,** 443–453.

Hughes, G. M. (1980a). Structure of fish respiratory surfaces. *Int. Congr. Anat., 11th, 1980* Abstract 36.

Hughes, G. M. (1980b). Functional morphology of fish gills. *In* "Epithelial Transport in the Lower Vertebrates" (B. Lahlou, ed.), pp. 15–36. Cambridge Univ. Press, London and New York.

Hughes, G. M. (1980c). Ultrastructure and morphometry of the gills of *Latimeria chalumnae*, and a comparison with the gills of associated fishes. *Proc. R. Soc. London, Ser. B* **208,** 309–328.

Hughes, G. M. (1980d). Morphometry of fish gas exchange organs in relation to their respiratory function. *In* "Environmental Physiology of Fishes" (M. A. Ali, ed.), pp. 33–56. Plenum, New York.

Hughes, G. M. (1981). Fish gills—past, present and future. *Biol. Bull. India* **3**(2), 69–87.

Hughes, G. M. (1982). An introduction to the study of gills. *In* "Gills" (D. F. Houlihan, J. C. Rankin, and T. J. Shuttleworth, eds.), pp. 1–24. Cambridge Univ. Press, London and New York.

Hughes, G. M. (1984a). Scaling of respiratory areas in relation to oxygen consumption of vertebrates. *Experientia* (in press).

Hughes, G. M. (1984b). Measurement of gill area in fishes: Practices and Problems. *J. Mar. Biol. Assoc. U. K.* (in press).
Hughes, G. M., and Adeney, R. J. (1977). Variations in the pattern of coughing in rainbow trout. *J. Exp. Biol.* **68,** 109–122.
Hughes, G. M., and Ballintijn (1965). The muscular basis of the respiratory pumps in the dogfish (*Scyliorhinus canicula*). *J. Exp. Biol.* **43,** 363–383.
Hughes, G. M., and Gray, I. E. (1972). Dimensions and ultrastructure of toadfish gills. *Biol. Bull. (Woods Hole, Mass.)* **143,** 150–161.
Hughes, G. M., and Grimstone, A. V. (1965). The fine structure of the secondary lamellae of the gills of *Gadus pollachius*. *Q. J. Microsc. Sci.* **106,** 343–353.
Hughes, G. M., and Iwai, T. (1978). A morphometric study of the gills in some Pacific deep-sea fishes. *J. Zool.* **184,** 155–170.
Hughes, G. M., and Morgan, M. (1973). The structure of fish gills in relation to their respiratory function. *Biol. Rev. Cambridge Philos. Soc.* **48,** 419–475.
Hughes, G. M., and Munshi, J. S. D. (1968). Fine structure of the respiratory surfaces of an air-breathing fish, the climbing perch, *Anabas testudineus* (Bloch). *Nature (London)* **219,** 1382–1384.
Hughes, G. M., and Munshi, J. S. D. (1973a). Fine structure of the respiratory organs of the climbing perch, *Anabas testudineus* (Pisces: Anabantidae). *J. Zool.* **170,** 201–225.
Hughes, G. M., and Munshi, J. S. D. (1973b). Nature of the air-breathing organs of the Indian fishes, *Channa, Amphipnous, Clarias* and *Saccobranchus* as shown by electron microscopy. *J. Zool.* **170,** 245–270.
Hughes, G. M., and Munshi, J. S. D. (1978). Scanning electron microscopy of the respiratory surfaces of *Saccobranchus* (= *Heteropneustes*) *fossilis* (Bloch). *Cell Tissue Res.* **195,** 99–109.
Hughes, G. M., and Munshi, J. S. D. (1979). Fine structure of the gills of some Indian air-breathing fishes. *J. Morphol.* **160,** 169–194.
Hughes, G. M., and Perry, S. F. (1976). Morphometric study of trout gills: A light microscopic method suitable for the evaluation of pollutant action. *J. Exp. Biol.* **64,** 447–460.
Hughes, G. M., and Peters, H. M. (1984). Ventilation of the gills and labyrinthine organs with air and water in the gouramy (*Osphronemus goramy*). (In preparation.)
Hughes, G. M., and Shelton, G. (1958). The mechanism of gill ventilation in three fresh water teleosts. *J. Exp. Biol.* **35,** 807–823.
Hughes, G. M., and Shelton, G. (1962). Respiratory mechanisms and their nervous control in fish. *Adv. Comp. Physiol. Biochem.* **1,** 275–364.
Hughes, G. M., and Umezawa, S. (1968). On respiration in the dragonet, *Callionymus lyra* L. *J. Exp. Biol.* **49,** 565–582.
Hughes, G. M., and Weibel, E. R. (1972). Similarity of supporting tissue in fish gills and the mammalian reticulo-endothelium. *J. Ultrastruct. Res.* **39,** 106–114.
Hughes, G. M., and Weibel, E. R. (1976). Morphometry of fish lungs. *In* "Respiration of Amphibious Vertebrates" (G. M. Hughes, ed.), pp. 213–232. Academic Press, New York.
Hughes, G. M., and Weibel, E. R. (1978). Visualization of layers lining the lung of the South American lungfish (*Lepidosiren paradoxa*) and a comparison with the frog and rat. *Tissue Cell* **10,** 343–353.
Hughes, G. M., and Wright, D. E. (1970). A comparative study of the ultrastructure of the water-blood pathway in the secondary lamellae of teleost and elasmobranch fishes—benthic forms. *Z. Zellforsch. Mikrosk. Anat.* **104,** 478–493.
Hughes, G. M., Dube, S. C., and Munshi, J. S. D. (1973). Surface area of the respiratory organs of the climbing perch, *Anabas testudineus* (Pisces: Anabantidae). *J. Zool.* **170,** 227–243.

Hughes, G. M., Singh, B. R., Thakur, R. N., and Munshi, J. S. D. (1974a). Areas of the air-breathing surfaces of *Amphipnous cuchia* (Ham.). *Proc. Indian Natl. Sci. Acad., Part B* **40,** No. 4, 379–392.

Hughes, G. M., Singh, B. R., Guha, G., Dube, S. C., and Munshi, J. S. D. (1974b). Respiratory surface areas of an air-breathing siluroid fish, *Saccobranchus (Heteropneustes) fossilis*, in relation to body size. *J. Zool.* **172,** 215–232.

Hughes, G. M., Tuurala, H., and Soivio, A. (1978). Regional distribution of blood in the gills of rainbow trout in normoxia and hypoxia: A morphometric study with two fixatives. *Ann. Zool. Fenn.* **15,** 226–234.

Hughes, G. M., Perry, S. F., and Brown, V. M. (1979). A morphometric study of effects of nickel, chromium and cadmium on the secondary lamellae of rainbow trout gills. *Water Res.* **13,** 665–679.

Hughes, G. M., Horimoto, M., Kikuchi, Y., Kakiuchi, Y., and Koyama, T. (1981). Blood flow velocity in microvessels of the gill filaments of the goldfish (*Carassius auratus* L). *J. Exp. Biol.* **90,** 327–331.

Hughes, G. M., Peyraud, C., Peyraud-Waitzenegger, M., and Soulier, P. (1982). Physiological evidence for the occurrence of pathways shunting blood away from the secondary lamellae of eel gills. *J. Exp. Biol.* **98,** 277–288.

Hughes, G. M., Perry, S. F., and Piiper, J. (1984). Quantitative anatomy of the gills of the larger spotted dogfish (*Scyliorhinus stellaris*). (In preparation.)

Huxley, J. S. (1932). "Problems of Relative Growth." Methuen, London.

Huxley, J. S., and Teissier, G. (1936). Terminology of relative growth. *Nature (London)* **137,** 780–781.

Ishimatsu, A., and Itazawa, Y. (1983). Difference in blood oxygen levels in the outflow vessels of the heart of an air-breathing fish, *Channa argus:* Do separate blood streams exist in a teleostean heart? *J. Comp. Physiol.* **149,** 435–440.

Iwai, T. (1963). Taste buds on the gill rakers and gill arches of the sea catfish *Plotosus anguillaris* (Lacapède). *Copeia* **2,** 271–274.

Iwai, T. (1964). A comparative study of the taste buds in gill rakers and gill arches of teleostean fishes. *Bull. Misaki Mar. Biol. Inst. Kyoto Univ.* **7,** 19–34.

Iwai, T., and Hughes, G. M. (1977). Preliminary morphometric study on gill development in Black Sea bream (*Acanthopagrus schlegeli*). *Bull. Jpn. Soc. Sci. Fish.* **43,** 929–934.

Jones, D. R., and Randall, D. J. (1978). The respiratory and circulatory systems. *In* "Fish Physiology" (W. S. Hoar and D. J. Randall, eds.), Vol. 7, p. 425. Academic Press. New York.

Jones, D. R., and Schwazfeld, T. (1974). The oxygen cost to the metabolism and efficiency of breathing in trout (*Salmo gairdneri*). *Respir. Physiol.* **21,** 241–254.

Kazanski, V. I. (1964). Specific differences in the structure of gill rakers in Cyprinidae. *Vopr. Ikhtiol.* **4,** 45–60.

Kempton, R. T. (1969). Morphological features of functional significance in the gills of the spiny dogfish *Squalus acanthias. Biol. Bull. (Woods Hole, Mass.)* **136,** 226–240.

Krogh, A. (1919). The rate of diffusion of gases through animal tissue. *J. Physiol. (London)* **52,** 391–408.

Lauder, G. V. (1980). The suction feeding mechanism in sunfishes (*Lepomis*): An experimental analysis. *J. Exp. Biol.* **88,** 49–72.

Lauder, G. V. (1983). Prey capture hydrodynamics in fishes: experimental tests of two models. *J. Exp. Biol.* **104,** 1–13.

Leiner, M. (1938). "Die Physiologie der Fischatmung." Akad. Verlagsges., Leipzig.

Lewis, S. V., and Potter, I. A. (1976). A scanning electron microscope study of the gills of the lamprey *Lampetra fluviatilis* (L). *Micron* **7,** 205–211.

McMahon, B. R. (1969). A functional analysis of the aquatic and aerial respiratory movements of an African lungfish, *Protopterus aethiopicus*, with reference to the evolution of the lung-ventilation mechanism in vertebrates. *J. Exp. Biol.* **51,** 407–430.

Marshall, N. B. (1960). Swimbladder structure of deepsea fishes in relation to their systematics and biology. *'Discovery' Rep.* **31,** 1–122.

May, R. C. (1974). Larval mortality in marine fishes and the critical period concept. *In* "The Early Life History of Fish" (J. H. S. Blaxter, ed.), pp. 2–19. Springer-Verlag, Berlin and New York.

Miscalencu, D., and Dornescu, G. T. (1970). Etude comparative des branchies de quelques perciformes marins. *Zool. Anz.* **184,** 67–74.

Morgan, M. (1971). Gill development, growth and respiration in the trout, *Salmo gairdneri.* Ph.D. Thesis, Bristol University.

Morgan, M. (1974). Development of secondary lamellae of the gills of the trout, *Salmo gairdneri* (Richardson). *Cell Tissue Res.* **151,** 509–523.

Moroff, T. (1904). Uber die Entwicklung der Kiemen bei Fischen. *Arch. Mikrosk. Anat. Entwicklungsmech.* **64,** 189–213.

Muir, B. S. (1970). Contribution to the study of blood pathways in teleost gills. *Copeia* **1,** 19–28.

Muir, B. S., and Brown, C. E. (1971). Effects of blood pathway on the blood pressure drop in fish gills with special reference to tunas. *J. Fish. Res. Board Can.* **28,** 947–955.

Muir, B. S., and Hughes, G. M. (1969). Gill dimensions for three species of tunny. *J. Exp. Biol.* **51,** 271–285.

Muir, B. S., and Kendall, J. I. (1968). Structural modifications in the gills of tunas and some other oceanic fishes. *Copeia* **2,** 388–398.

Muller, J. (1839). Vergleichende Anatomie der Myxinoiden. III. Uber das Gefassystem. *Abh. Akad. Wiss., Berlin* pp. 175–303.

Munshi, J. S. D. (1962). On the accessory respiratory organs of *Heteropneustes fossilis* (BL.). *Proc. R. Soc. Edinburgh, Sect. B: Biol.* **68,** 128–146.

Munshi, J. S. D. (1968). The accessory respiratory organs of *Anabas testudineus* (Bloch) (Anabantidae, Pisces). *Proc. Linn. Soc. London* **170,** 107–126.

Munshi, J. S. D. (1976). Gross and fine structure of the respiratory organs of air-breathing fishes. *In* "Respiration of Amphibious Vertebrates" (G. M. Hughes, ed.), pp. 73–104. Academic Press, New York.

Munshi, J. S. D., and Hughes, G. M. (1981). Gross and fine structure of the pseudobranch of the climbing perch, *Anabas testudineus* (Bloch). *J. Fish Biol.* **19,** 427–438.

Needham, J. (1941). "Biochemistry and Morphogenesis." Cambridge Univ. Press, London and New York.

Nowogrodska, M. (1943). The respiratory surface of gills of Teleosts. VII. *Trachurus trachurus. Acta Biol. Cracov.* **6,** 147–158.

Oliva, O. (1960). The respiratory surface of the gills in teleosts. 5. The respiratory surface of the gills in the viviparous blenny (*Zoarces viviparous*). *Acta Biol. Cracov.* **3,** 71–89.

Osse, J. W. M. (1969). Functional morphology of the head of the perch (*Perca fluviatilis*)—an electromyographic study. *Neth. J. Zool.* **19,** 290–392.

Paling, J. E. (1968). A method of estimating the relative volumes of water flowing over the different gills of freshwater fish. *J. Exp. Biol.* **48,** 533–544.

Pasztor, V. M., and Kleerekoper, H. (1962). The role of the gill filament musculature in teleosts. *Can. J. Zool.* **40,** 785–802.

Pattle, R. E. (1976). The lung surfactant in the evolutionary tree. *In* "Respiration of Amphibious Vertebrates" (G. M. Hughes, ed.), pp. 233–255. Academic Press, New York.

Pattle, R. E. (1978). Lung surfactant and lung lining in birds. *In* "Respiratory Function in Birds, Adult and Embryonic" (J. Piiper, ed.), pp. 23–32. Springer-Verlag, Berlin and New York.

Perry, S. F. (1983). "Reptilian Lungs. Functional Anatomy and Evolution." Springer-Verlag, Berlin and New York.

Peters, H. M. (1978). On the mechanism of air ventilation in anabantoids (Pisces: Teleostei). *Zoomorphology* **89,** 93–124.

Piiper, J., and Scheid, P. (1982). Physical principles of respiratory gas exchange in fish gills. *In* "Gills" (D. F. Houlihan, J. C. Rankin, and T. J. Shuttleworth, eds.), pp. 45–61. Cambridge Univ. Press, London and New York.

Piiper, J., and Schumann, D. (1967). Efficiency of O_2 exchange in the gills of the dogfish, *Scyliorhinus stellaris*. *Respir. Physiol.* **2,** 135–148.

Price, J. W. (1931). Growth and gill development in the small-mouthed black bass, *Micropterus dolomieu* Lacapede. *Contrib. Stone Lab. Ohio Univ.* **4,** 1–46.

Rahn, H. (1966). Gas transport from the external environment to the cell. *In* "Development of the Lung" (A. V. S. de Reuck and R. Porter, eds.), pp. 3–23. Churchill, London.

Randall, D. J. (1970). Gas exchange in fish. *In* "Fish Physiology" (W. S. Hoar and D. J. Randall, eds.), Vol. 4, pp. 253–292. Academic Press, New York.

Randall, D. J., Burggren, W. W., Farrell, A. P., and Haswell, M. S. (1981), "The Evolution of Air Breathing in Vertebrates." Cambridge Univ. Press, London and New York.

Roberts, J. L. (1975). Active branchial and ram gill ventilation in fishes. *Biol. Bull. (Woods Hole, Mass.)* **148,** 85–105.

Scheid, P. (1979). Mechanisms of gas exchange in bird lungs. *Rev. Physiol., Biochem. Pharmacol.* **86,** 138–185.

Scheid, P., and Piiper, J. (1976). Quantitative functional analysis of branchial gas transfer: Theory and application to *Scyliorhinus stellaris* (Elasmobranchii). *In* "Respiration of Amphibious Vertebrates" (G. M. Hughes, ed.), pp. 17–38. Academic Press, New York.

Schlotte, E. (1932). Morphologie und Physiologie der Atmung bei wassers-schlamm-und land-lebenden Gobiiformes. *Z. Wiss. Zool.* **140,** 1–113.

Schmidt, P. (1915). Respiratory adaptations of Pleuronectids. *Bull. Acad. Sci. St.-Petersbourg* [6] **9,** 421–444 (in Russian).

Sewertzoff, A. (1924). Die Entwicklung der Kiemen und Kiemenbogen-gefasse der Fische. *Z. Wiss. Zool.* **121,** 494–556.

Singh, B. N. (1976). Balance between aquatic and aerial respiration. *In* "Respiration of Amphibious Vertebrates" (G. M. Hughes, ed.), pp. 125–164. Academic Press, New York.

Smith, D. G., and Chamley-Campbell, J. (1981). Localisation of smooth-muscle myosin in branchial pillar cells of snapper (*Chrysophys auratus*) by immuno-fluorescence histochemistry. *J. Exp. Zool.* **215,** 121–124.

Soivio, A., and Tuurala, H. (1981). Structural and circulatory responses to hypoxia in the secondary lamellae of *Salmo gairdneri* gills at two temperatures. *J. Comp. Physiol.* **145,** 37–43.

Steen, J. B. (1971). "Comparative Physiology of Respiratory Mechanisms." Academic Press, New York.

Steen, J. B., and Kruysse, E. (1964). The respiratory function of teleostean gills. *Comp. Biochem. Physiol.* **12,** 127–142.

Tovell, P. W., Morgan, M., and Hughes, G. M. (1971). Ultrastructure of trout gills. *Electron Microsc., Proc. Int. Congr., 7th, 1970* Vol. III, p. 601.

Underwood, E. E. (1970). "Quantitative Stereology." Addison-Wesley, Reading, Massachusetts.

van Dam, L. (1938). On the utilisation of oxygen and regulation of breathing in some aquatic animals. Dissertation, University of Grøningen, Netherlands.

Vogel, W. O. P. (1981). Struktur und Organisationsprinzip im Gefassysten der Knochenfische. *Morphol. Jahrb.* **127,** 772–784.

Vogel, W. O. P., and Claviez, M. (1981). Vascular specialisation in fish, but no evidence for lymphatics. *Z. Naturforsch., C: Biosci.* **36C,** 490–492.

Wagner, H., Conte, F. P., and Fessler, J. (1969). Development of osmotic and ionic regulation in two races of juvenile chinook salmon, *O. tschawytscha. Comp. Biochem. Physiol.* **29,** 325–342.

Watson, D. M. S. (1951). "Palaeontology and Modern Biology," Chapter 1. Yale Univ. Press, New Haven, Connecticut.

Weibel, E. R. (1971). Morphometric estimation of pulmonary diffusion capacity. I. Model and method. *Respir. Physiol.* **11,** 54–75.

Weibel, E. R. (1979). "Stereological Methods," Vol. I. Academic Press, New York.

Winberg, G. G. (1956). (1) Rate of metabolism and food requirements of fishes. (2) New information on metabolic rate in fishes. *Fish. Res. Board Can.* (Transl. Ser. 194 and 362).

Wood, C. M. (1974). A critical examination of the physical and adrenergic factors affecting blood flow through the gills of the rainbow trout. *J. Exp. Biol.* **60,** 241–265.

Woskoboinikoff, M. M. (1932). Der Apparat der Kiemenatmung bei den Fischen. Ein Versuch der Synthese in der Morphologie. *Zool. Jahrb., Abt. Anat. Ontag. Tiere* **55,** 315–488.

Wright, D. E. (1973). The structure of the gills of the elasmobranch, *Scyliorhinus canicula* L. *Z. Zellforsch. Mikrosk. Anat.* **144,** 489–509.

Yazdani, G. M., and Alexander, R. McN. (1967). Respiratory currents of flatfish. *Nature (London)* **213,** 96–97.

Chapter 3

The oldies are the goodies: 30 years on "The Heart" still sets the pace

Holly A. Shiels*
Division of Cardiovascular Sciences, Faculty of Biology, Medicine, and Health, University of Manchester, Manchester, United Kingdom
*Corresponding author: e-mail: holly.shiels@manchester.ac.uk

Holly A. Shiels discusses the A.P. Farrell and D.R. Jones chapter "The Heart" in *Fish Physiology*, Volume 12A, published in 1992. The next chapter in this volume is the re-published version of that chapter.

"The Heart" remains the greatest collection of comparative information on the anatomical, morphological, and physiological properties of the fish heart. It is a work of depth and breath, and a lesson in how the comparative approach provides physiological insight beyond the boundaries of the knowledge base from which it is generated. My introduction focuses on key areas from this chapter that are dear to my own heart including structure-function relationships, cardiac remodeling and regulation of heart function in an era of climate change. However, countless introductions could flow from Farrell and Jones' chapter as it touches on nearly every aspect of fish cardiac physiology. I very much enjoyed re-reading this chapter and recognizing its impact on my work. I hope the re-release of this chapter engages the next generation the way it did mine.

1 Introduction to "The Heart"

"The Heart" synthesized decades of comparative research providing the foundational treatise on the form and function of the fish heart. The chapter takes a comparative and quantitative view of fish heart anatomy, morphology, and physiology, presenting the extremes of diversity whilst drawing together generalities to arrive at a set of principles governing fish heart function. The chapter is data rich. The details are made accessible through a fantastic collection of tables that include the norms and the boundaries of cardiovascular function across fishes. The physiological relevance of this collection of functional parameters is borne out through extensive considered discussions. The result is a compendium of fish cardiac physiology based on detailed

Fish Physiology, Vol. 40A. https://doi.org/10.1016/bs.fp.2023.09.005

morphology and in vivo and ex vivo experimentation, synthesized to provide functional insight that goes well beyond the state of knowledge of the component parts.

I remember reading this chapter as an undergraduate student and thinking Farrell must have been working in fish physiology for decades to have produced the ~45 publications which underpin many of the observations and interpretations laid out in the chapter. Soon after, I went to study with Tony at Simon Fraser University, British Columbia, Canada, first as an MSc student and then a PhD student. Here I realized the breadth and depth of work underpinning the chapter was not the results of a lengthy career (Tony was still an associate professor at the time) but rather the result of a burning passion for quantitative understanding of physiological processes. I think Tony would have made a similar deep impression on whatever field he decided to go into. The fish biology community is eternally grateful he turned his attention here. David Jones was already a full professor at the University of British Columbia and a member of the Royal Society of Canada at the time "The Heart" was published. His broad, long-term perspective on the form and function of the fish heart, particularly as it relates to oxygen transport systems during exercise and hypoxia, can be felt throughout the chapter. I didn't have the opportunity to work with Dave directly, but I knew him, enjoyed his lectures, and had the chance to walk around VanDusen's Gardens in Vancouver with him once, talking about art.

"The Heart" starts with a detailed anatomical overview of each chamber of the fish heart. The descriptions are quantitative, comparing what is know across species, stating norms and exceptions. The description of the ventricle builds on the categories and heart types developed by Santer (Santer, 1985; Santer and Greer Walker, 1980) and Tota (Agnisola and Tota, 1994; Basile et al., 1976; Tota, 1989). Farrell and Jones then discuss the proportions and relative roles of the spongy and compact ventricular layers, the diversity of organization across species, and their changing role during ontogeny within a species. They draw the conclusion that a larger proportional compact layer is associated with a more active lifestyle in fish, with regional endothermy in sharks, and with seasonal temperature change, but not with exercise training, in salmonids. These observations spurred a flurry of research into ventricular remodeling in fish with many labs, including Tony's and later my own, trying to understand what drives remodeling and what the consequence of remodelling is for performance. Work from Todd Gillis's lab and others, revealed some of the cellular drivers for cold-hypertrophy of the spongy layer and cold-atrophy of the compact layers (e.g., Ding et al., 2022; Keen et al., 2016b). That the number and size of individual ventricular myocytes increased in the cold had been known before "The Heart" was published (e.g., Farrell et al., 1988). Since, studies have tried to explain the mechanisms driving this remodeling in the ventricle (Clark and Rodnick, 1998; Keen et al., 2016a; Sun et al., 2009; Vornanen et al., 2005), the atria (Keen et al., 2018), and the bulbus arteriosus (Dimitriadi et al., 2021; Foddai et al., 2023; Keen et al., 2021). Recent work has also highlighted the importance of concurrent remodeling of the connective tissue that supports the contractile cardiomyocytes (Johnston and Gillis, 2022). Indeed, the realization that collagen is laid down in the cold to support the hypertrophied heart but regresses in the warm as the tissue atrophies (Shaftoe et al., 2023), presents a tantalizing paradigm for the regression of maladaptive fibrosis in the human heart, which is known to occur with age and with a range of cardiomyopathies (for review see Horn and Trafford, 2016). Evidently, Farrell and Jones recognized the potential of thermal remodeling of the fish myocardium as a model for cardiac remodeling in response to other stressors in other animals. The early insights from "The Heart" provided novel pathways for investigation of cardiac remodeling in fish (Dimitriadi et al., 2021; Filice et al., 2021), humans, and other animals in response to stressors such as exercise, pregnancy, disease, and the environment (for recent review see Gibb and Hill, 2018).

The chapter is exhaustive in breadth as well as depth and devotes considerable time to the description of the coronary vasculature of fishes as this was a key interest of both Tony and Dave. Drawing on their earlier work (Davie and Farrell, 1991; Farrell, 1987, 1991; Jones, 1971) and that of Tota (Tota, 1989; Tota et al., 1983) and others, they discuss the implications of all elasmobranchs having a coronary circulation that penetrates to both myocardial layers but only ~30% of teleosts having a coronary circulation (Axelsson, 1995), mostly associated with the compact myocardium (Farrell, 1996; Farrell et al., 1992). Differences in coronary vascular organization between heart types and fish species is reviewed, and the authors first conclude that oxygenated blood from either or both the cranial and pectoral circuits arrive at the fish heart through the coronary circulation at the highest possible post-brachial pressure. They then discuss the advent of coronary lesions in migrating salmon or in rapidly growing fish, highlighting both the resilience and vulnerability of these vessels during extreme performance stress (Saunders et al., 1992). Work from the Farrell lab added to this literature in the 30 years since the chapter was publish (e.g., Farrell, 2002) and today these observations resonate strongly with aquaculture and fish welfare biologists as the prevalence and severity of cardiac abnormalities including arteriosclerosis are a feature of stressed and rapidly growing farmed fish (Brijs et al., 2020; Frisk et al., 2020; Johansen et al., 2011).

Farrell and Jones were unable to determine a single factor which adequately explained the presence or absence of a coronary circulation in a given species. They recognized an association with an active lifestyle and hypoxia tolerance (see Bushnell and Jones, 1994; Davie and Farrell, 1991; Farrell, 1987; Jones et al., 1986), but admit this did not explain all observations. They point out adaptations that might overcome the "problem" of an avascular myocardium including elevated myocardial myoglobin, hypoxia and acidosis tolerance, and slow heart rates. Trying to determine the factors driving the presence or absence of a coronary circulation, or the performance advantage conferred by a coronary circulation, became a theme of work in the Farrell lab (Cox et al., 2016; Farrell, 2007; Farrell et al., 2012; Simonot and Farrell, 2009). This work culminated in another book chapter (Farrell et al., 2012) which involved Tony's graduate students as well as his daughter Nicole Farrell who sought insight from the vast fish collections of the Natural History Museum in London, England. In this chapter, they suggest the coronary circulation likely emerged with the evolution of the first jawed vertebrates as no cyclostome had a coronary circulation. They note the extensive coronary system in elasmobranchs supplying blood to both the spongy and compact layers of the ventricle and describe its loss across teleost lineages only to reappear multiple times in different groups through convergent evolution, usually limited to the compacta. Finally, they review the few teleost species with a coronary system that supplies both the spongy and compact myocardium. However, despite applying morphological, ecological, and phylogenetic lenses, the drivers for the evolution of the coronary circulation in teleosts remain unclear. It is likely that the evolution of the delivery of branchial blood to the teleost heart has been shaped by multiple drivers including myocardial architecture, the environment, and the cardiovascular requirements of the organism. This "need" sentiment may have been summed up best by Tota and Gattuso (1996) (who cite "The Heart" by Farrell and Jones) and who elegantly quote Gould's suggestion that "The evolution of most major groups is not the story of ecological variations..., but a history of mechanical improvement" (Gould, 1970).

Another important mechanical and structure-function relationship considered in the "The Heart" is the role of the ventricular trabeculations of the spongy myocardium. Farrell and Jones expand on an idea from Davie and Farrell (1991) which suggests that the diameter of the spongy trabeculae is a compromise between maximizing the cross-sectional area for force production and minimizing it for oxygen diffusion. Farrell and Jones recognized that the hearts of chondrichthyans (i.e. sharks and rays) were poised to shed light on this question by virtue of their robust coronary vascularization which should effectively uncouple trabecular size from oxygen diffusion limitations. The chapter encouraged further research in this area and in 2016, Cox and Farrell (Cox et al., 2016) used microCT scanning to demonstrate the extent of coronary vascularization in shark (*Squalus suckle*) versus teleost (*Oncorhynchus mykiss*) spongy trabeculae. They showed that trabeculae less than 20 μm in cross-section where not vascularized and that in larger trabecular, capillaries were found in the interior meaning that the outer part of the tissue still acquired oxygen via venous luminal blood. Recently Bjarke Jensen's group has expanded our understanding of the role of spongy trabeculations in the fish heart (Jensen et al., 2016) by revealing their contribution to conduction of the electrical impulse through the myocardium in the absence of a defined fast-conduction system like the His-Purkinje system of mammals (Boukens et al., 2019). The role of the trabeculated versus compacted myocardium was also considered by Farrell and colleagues in the perspective section of their chapter on the evolution of the coronaries in the vertebrate heart (Farrell et al., 2012). Thus, like all good reviews, Farrell and Jones' "The Heart" stimulated (and still stimulates) research in the field and beyond; their legacy is how many of their ideas were confirmed with later experimentation.

The intimate connection between structure, function, and requirement, is recognized and discussed throughout "The Heart" across all levels of biological organization. But is particularly forthcoming when "The Heart" moves from the organ and tissue level to the cardiomyocyte. Here, this insight, and their synthesis across species, revealed some important fundamental "truths." The first was the observation that the long thin fish cardiomyocytes with peripherally located myofibrils facilitated excitation-contraction coupling in the absence of a transverse tubular system and a reduced sarcoplasmic reticulum (in relation to mammals). Farrell and Jones develop these structure-function observations even further, outlining how subcellular organization and the complement of various organelles provides insight into strategies and limitations in cellular calcium cycling and cellular energetics. I was taken with these ideas as a student. Indeed, investigating the implications of the lack of transverse tubules and the low complement of sarcoplasmic reticulum on maximal heart rate and temperature tolerance in fishes became the focus of my post graduate studies under Tony's supervision. Tony and I published our first paper together on this topic in 1997 (Shiels and Farrell, 1997). I am still trying to understand these interrelationships today (Shiels, 2022; Shiels and Galli, 2014; Shiels and Sitsapesan, 2015)! Indeed, the influence of the structure-function paradigm championed by Farrell and Jones in "The Heart" has resonated throughout my career (e.g., Vornanen et al., 2002). Just last year I co-edited a special issue of the Philosophical Transactions of the Royal Society B entitled "The cardiomyocyte: new revelations on the interplay between architecture and function in growth, health, and disease" (for an introduction to the issue see Rajagopal et al., 2022). As this issue attests, the fundamental truths that Farrell and Jones gleaned from the comparative literature in 1992 have held fast and formed a foundation for current work across the life sciences. For example, their quantitative comparative approach revealed maintenance of myocyte size and shape with heart growth in fish. By comparing myocyte size (length and width and calculating volume) across different fish during ontogeny they conclude that myocyte volume increases ~6-fold for an ~100-fold increase in ventricular mass. Implicit in this estimation is that hyperplasia dominates over hypertrophy for cardiac growth in fish. We now know that this phenomenon is true of all non-mammalian vertebrates, but this is certainly one of the earliest published observations. Understanding what happens postnatally in mammalian hearts that leads to loss of proliferative potential restricting growth to hypertrophy, is a very active research area as it is concurrent with the loss of cardiac regenerative potential (Gan et al., 2020; Hirose et al., 2019; Sirsat et al., 2016).

Thus, once again, the comparative and the quantitative synthesis offered by Farrell and Jones in 1992 presented observations that underpin biological and biomedical research today.

The final section of "The Heart" is entitled "Cardiac Physiology" and describes the electrical, contractile, and power output of the fish heart. Here, Farrell and Jones integrate the structural and functional details provided earlier in the chapter with intrinsic (e.g. mechanical, endocrine, and neuronal) and extrinsic (e.g., temperature, hypoxia, acidosis) regulators of function. The intrinsic property of a cardiac muscle to either increase or decrease the force of contraction in response to an increment in pacing rate, first described by Bowditch (1871), is thoroughly reviewed in this section. Farrell and Jones surmised that by in large, fish hearts demonstrate a negative force-frequency relationship, meaning contractile force decreases with increased pacing frequency. The force-frequency relationships was also a hot topic in mammalian cardiac physiology in 1992 because although the positive force-frequency response (where contractile force increases with contraction frequency) dominated mammalian cardiac observations, rodents, and humans and animal models in end-stage heart failure, presented with a flat or even a negative force-frequency response (Rumberge and Reichel, 1972; Shattock and Bers, 1989). The directional implication of this relationship for function are obvious as stress and exercise increase heart rate, a feature considered necessary for increasing cardiac output to allow the body to meet or overcome the stressor. An intrinsic loss of contractility at higher heart rates in animals and patients with heart failure could compound circulation difficulties (Palomeque et al., 2004). We now know that the positive force-frequency phenomenon is attributed to an elevation in the amplitude of the intracellular calcium transient, primarily caused by augmented sarcoplasmic reticulum calcium cycling capacity during higher stimulation frequencies. This stems from a rise in extracellular calcium influx and a decrease in extracellular calcium efflux between beats (Balcazar et al., 2018). However, these cellular drivers were unknow when "The Heart" was published. Farrell and Jones employed the comparative approach in fishes to reveal mechanistic insight that extended to all animals. They noticed that generally (though exceptions exist) high performing fishes had a flat or less negative force-frequency relationship compared with more sedentary species, as did atrial tissue compared with ventricular tissue within a species (Keen et al., 1992; Vornanen, 1989). Finally, they noticed that while increased extracellular calcium or adrenergic stimulation (both which increase calcium influx into the cardiomyocyte) alleviated the negative-force frequency response to some extent, it predominantly shifted the relationships to a different range of contraction frequencies (Driedzic and Gesser, 1985; Hove-Madsen and Gesser, 1989). Unpicking the cellular mechanisms that determined the relationship between force and frequency in fish hearts, and the interplay between intrinsic (e.g., adrenergic stimulation) and extrinsic (e.g., temperature) factors on this relationship, was another focus of my graduate work in the Farrell lab (see Shiels et al., 2002).

The impact of temperature and its direct and interactive effects on heart rate and cardiac contractility has had a resurgence recently as fish physiologists strive to understand the implications of climate change and extreme weather events on fish health and survival. Thermal tolerance of the heart has emerged as a lynchpin for surviving or succumbing to environmental warming in both humans (Liu et al., 2022) and fishes (Farrell et al., 2009). Recent studies have implicated the force-frequency response and species-specific densities and activities of the pumps and ion channels that control the cellular cycling of calcium in fish cardiomyocytes in contributing to thermal tolerance (Gamperl and Syme, 2021; Gamperl et al., 2022; Vornanen, 2021). Of course, contraction follows electrical activation, and Matti Vornanen's lab has shown how contractile failure at high temperatures can be due to failure of the action potential to effectively propagate, leading to heart block and other serious arrhythmias (Haverinen and Vornanen, 2020; Kuzmin et al., 2022; Vornanen, 2020). There are many other physiological systems and processes that affect the thermal tolerance of fish (see Ern et al., 2023; Farrell et al., 2009). But the thorough consideration of heart rate regulation in fishes and the mechanisms that limit maximum heart rate laid out in "The Heart" provide the framework of physiological knowledge upon which current studies are built.

The field of fish cardiac physiology has moved forward in the >30 years since the publication of "The Heart"; spurred on by environmental realities like climate change and by the development of new technologies and methods for studying fish cardiovascular physiology. Indeed, direct measurement of myocardial, arterial, and venous oxygen content was not possible in 1992, nor was bio-telemetry advanced enough to provide physiological insight into cardiac function in free-swimming individuals. However, by using a quantitative comparative approach, emphasizing the connection between myocardial architecture and performance physiology, "The Heart" set us on a path of discovery that prepared us for addressing the pressing questions of today. Indeed, before climate change and conservation physiology were disciplines, Farrell and Jones showed how the fish cardiovascular system was fundamental to fish behavior, ecology, and evolution. The chapter broaden the range of questions that researchers were asking in the field and provided insights with implications for other fields of study. The oldies really are the goodies. No wonder I still have the paper reprint Tony gave me when I arrived on his doorstep in 1994.

References

Agnisola, C., Tota, B., 1994. Structure and function of the fish cardiac ventricle—flexibility and limitations. Cardioscience 5, 145–153.

Axelsson, M., 1995. The coronary circulation: a fish perspective. BJMBR = Revista brasileira de pesquisas medicas e biologicas 28, 1167–1177.

Balcazar, D., Regge, V., Santalla, M., Meyer, H., Paululat, A., Mattiazzi, A., Ferrero, P., 2018. Serca is critical to control the Bowditch effect in the heart. Sci. Rep. 8.

Basile, C., Goldspink, G., Modigh, M., Tota, B., 1976. Morphological and biochemical characterisation of the inner and outer ventricular myocardial layers of adult tuna fish (*Thunnus thynnus* L.). Comp. Biochem. Physiol. B 54, 279–283.

Boukens, B.J.D., Kristensen, D.L., Filogonio, R., Carreira, L.B.T., Sartori, M.R., Abe, A.S., Currie, S., Joyce, W., Conner, J., Opthof, T., Crossley 2nd, D.A., Wang, T., Jensen, B., 2019. The electrocardiogram of vertebrates: evolutionary changes from ectothermy to endothermy. Prog. Biophys. Mol. Biol. 144, 16–29.

Bowditch, H., 1871. Über die Eigentümlichkeiten der Reizbarkeit, welche die Muskelfasern des Herzens zeigen. Ber. K. Sächs. Ges. Wiss. Math.-phys. Kl 652.

Brijs, J., Hjelmstedt, P., Berg, C., Johansen, I.B., Sundh, H., Roques, J.A.C., Ekström, A., Sandblom, E., Sundell, K., Olsson, C., Axelsson, M., Gräns, A., 2020. Prevalence and severity of cardiac abnormalities and arteriosclerosis in farmed rainbow trout (*Oncorhynchus mykiss*). Aquaculture 526, 735417.

Bushnell, P.G., Jones, D.R., 1994. Cardiovascular and respiratory physiology of tuna: adaptations for support of exceptionally high metabolic rates. Environ. Biol. Fish 40, 303–318.

Clark, R.J., Rodnick, K.J., 1998. Morphometric and biochemical characteristics of ventricular hypertrophy in male rainbow trout (*Oncorhynchus mykiss*). J. Exp. Biol. 201, 1541–1552.

Cox, G.K., Kennedy, G.E., Farrell, A.P., 2016. Morphological arrangement of the coronary vasculature in a shark (*Squalus sucklei*) and a teleost (*Oncorhynchus mykiss*). J. Morphol. 277, 896–905.

Davie, P.S., Farrell, A.P., 1991. The coronary and luminal circulations of the myocardium of fishes. Can. J. Zool. 69, 1993–2001.

Dimitriadi, A., Geladakis, G., Koumoundouros, G., 2021. 3D heart morphological changes in response to developmental temperature in zebrafish: more than ventricle roundness. J. Morphol. 282, 80–87.

Ding, Y., Johnston, E.F., Gillis, T.E., 2022. Mitogen-activated protein kinases contribute to temperature-induced cardiac remodelling in rainbow trout (*Oncorhynchus mykiss*). J. Comp. Physiol. B. 192, 61–76.

Driedzic, W.R., Gesser, H., 1985. Ca2+ protection from the negative inotropic effect of contraction frequency on teleost hearts. J. Comp. Physiol. B. 156, 135–142.

Ern, R., Andreassen, A.H., Jutfelt, F., 2023. Physiological mechanisms of acute upper thermal tolerance in fish. Phys. Ther. 38, 141–158.

Farrell, A.P., 1987. Coronary flow in a perfused rainbow trout heart. J. Exp. Biol. 129, 107–123.

Farrell, A.P., 1991. From hagfish to tuna—a perspective on cardiac-function in fish. Physiol. Zool. 64, 1137–1164.

Farrell, A.P., 1996. Features heightening cardiovascular performance in fishes; with special reference to tunas. Comp. Biochem. Physiol. A Physiol. 113, 61–67.

Farrell, A.P., 2002. Coronary arteriosclerosis in salmon: growing old or growing fast? Comp Biochem Physiol A Mol Integr Physiol 132, 723–735.

Farrell, A.P., 2007. Tribute to P. L. Lutz: a message from the heart—why hypoxic bradycardia in fishes? J. Exp. Biol. 210, 1715–1725.

Farrell, A.P., Davie, P.S., Franklin, C.E., Johansen, J.A., Brill, R.W., 1992. Cardiac physiology in tunas. 1. In vitro perfused heart preparations from yellowfin and skipjack tunas. Can. J. Zool. Revue Canadienne De Zoologie 70, 1200–1210.

Farrell, A.P., Farrell, N.D., Jourdan, H., Cox, G.K., 2012. A Perspective on the Evolution of the Coronary Circulation in Fishes and the Transition to Terrestrial Life. Can. J. Zool. 75–102.

Farrell, A.P., Hammons, A.M., Graham, M.S., 1988. Cardiac growth in rainbow trout, *Salmo gairdneri*. Can. J. Zool. 66 (11), 2368–2373. https://doi.org/10.1139/z88-351.

Farrell, A.P., Sandblom, E., Clark, T.D., 2009. Fish cardiorespiratory physiology in an era of climate change. Can. J. Zool. 87, 835–851.

Filice, M., Barca, A., Amelio, D., Leo, S., Mazzei, A., Del Vecchio, G., Verri, T., Cerra, M.C., Imbrogno, S., 2021. Morpho-functional remodelling of the adult zebrafish (*Danio rerio*) heart in response to waterborne angiotensin Ii exposure. Gen. Comp. Endocrinol. 301, 113663.

Foddai, M., Carter, C.G., Anderson, K., Ruff, N., Wang, S., Wood, A.T., Semmens, J.M., 2023. Impact of temperature and dietary replacement of fishmeal on cardiovascular remodelling and growth performance of adult Atlantic salmon (*Salmo salar* L.). Aquaculture 573, 739590.

Frisk, M., Høyland, M., Zhang, L., Vindas, M.A., Øverli, Ø., Johansen, I.B., 2020. Intensive smolt production is associated with deviating cardiac morphology in Atlantic salmon (*Salmo salar* L.). Aquaculture 529, 735615.

Gamperl, A.K., Syme, D.A., 2021. Temperature effects on the contractile performance and efficiency of oxidative muscle from a eurythermal versus a stenothermal salmonid. J. Exp. Biol. 224.

Gamperl, A.K., Thomas, A.L., Syme, D.A., 2022. Can temperature-dependent changes in myocardial contractility explain why fish only increase heart rate when exposed to acute warming? J. Exp. Biol. 225.

Gan, P., Patterson, M., Sucov, H.M., 2020. Cardiomyocyte polyploidy and implications for heart regeneration. Annu. Rev. Physiol. 82, 45–61.

Gibb, A.A., Hill, B.G., 2018. Metabolic coordination of physiological and pathological cardiac remodeling. Circ. Res. 123, 107–128.

Gould, S.J., 1970. Evolutionary paleontology and the science of form. Earth Sci. Rev. 6, 77–119.

Haverinen, J., Vornanen, M., 2020. Atrioventricular block, due to reduced ventricular excitability, causes the depression of fish heart rate in fish at critically high temperatures. J. Exp. Biol. 223, jeb225227.

Hirose, K., Payumo, A.Y., Cutie, S., Hoang, A., Zhang, H., Guyot, R., Lunn, D., Bigley, R.B., Yu, H., Wang, J., Smith, M., Gillett, E., Muroy, S.E., Schmid, T., Wilson, E., Field, K.A., Reeder, D.M., Maden, M., Yartsev, M.M., Wolfgang, M.J., Grützner, F., Scanlan, T.S., Szweda, L.I., Buffenstein, R., Hu, G., Flamant, F., Olgin, J.E., Huang, G.N., 2019. Evidence for hormonal control of heart regenerative capacity during endothermy acquisition. Science 364, 184–188.

Horn, M.A., Trafford, A.W., 2016. Aging and the cardiac collagen matrix: novel mediators of fibrotic remodelling. J. Mol. Cell. Cardiol. 93, 175–185.

Hove-Madsen, L., Gesser, H., 1989. Force frequency relation in the myocardium of Rainbow-Trout: effects of K+ and adrenaline. J. Comp. Physiol. B Biochem. Syst. Environ. Physiol. 159, 61–69.

Jensen, B., Agger, P., De Boer, B.A., Oostra, R.J., Pedersen, M., Van Der Wal, A.C., Nils Planken, R., Moorman, A.F., 2016. The hypertrabeculated (noncompacted) left ventricle is different from the ventricle of embryos and ectothermic vertebrates. Biochim. Biophys. Acta 1863, 1696–1706.

Johansen, I.B., Lunde, I.G., Rosjo, H., Christensen, G., Nilsson, G.E., Bakken, M., Overli, O., 2011. Cortisol response to stress is associated with myocardial remodeling in salmonid fishes. J. Exp. Biol. 214, 1313–1321.

Johnston, E.F., Gillis, T.E., 2022. Regulation of collagen deposition in the trout heart during thermal acclimation. Curr. Res. Physiol. 5, 99–108.

Jones, D.R., 1971. Theoretical analysis of factors which may limit the maximum oxygen uptake of fish: the oxygen cost of the cardiac and branchial pumps. J. Theor. Biol. 32, 341–349.

Jones, D.R., Brill, R.W., Mense, D.C., 1986. The influence of blood-gas properties on gas tensions and ph of ventral and dorsal aortic blood in free-swimming tuna; *Euthynnus affinis*. J. Exp. Biol. 120, 201–213.

Keen, J.E., Farrell, A.P., Tibbits, G.F., Brill, R.W., 1992. Cardiac physiology in tunas. 2. Effect of ryanodine, calcium, and adrenaline on force frequency relationships in atrial strips from skipjack tuna, *Katsuwonus Pelamis*. Can. J. Zool. 70, 1211–1217.

Keen, A.N., Fenna, A.J., McConnell, J.C., Sherratt, M.J., Gardner, P., Shiels, H.A., 2016a. The dynamic nature of hypertrophic and fibrotic remodeling of the fish ventricle. Front Physiol 6, 427. https://doi.org/10.3389/fphys.2015.00427. PMID: 26834645; PMCID: PMC4720793.

Keen, A.N., Klaiman, J.M., Shiels, H.A., Gillis, T.E., 2016b. Temperature-induced cardiac remodeling in fish. J. Exp. Biol. 220, 147–160.

Keen, A.N., Fenna, A.J., McConnell, J.C., Sherratt, M.J., Gardner, P., Shiels, H.A., 2018. Macro- and micromechanical remodelling in the fish atrium is associated with regulation of collagen 1 alpha 3 chain expression. Pflugers Arch. - Eur. J. Physiol. 470, 1205–1219.

Keen, A.N., Mackrill, J.J., Gardner, P., Shiels, H.A., 2021. Compliance of the fish outflow tract is altered by thermal acclimation through connective tissue remodelling. J. R. Soc. Interface. 18 (184), 20210492.

Kuzmin, V., Ushenin, K.S., Dzhumaniiazova, I.V., Abramochkin, D., Vornanen, M., 2022. High temperature and hyperkalemia cause exit block of action potentials at the atrioventricular junction of rainbow trout (*Oncorhynchus mykiss*) heart. J. Therm. Biol. 110, 103378.

Liu, J., Varghese, B.M., Hansen, A., Zhang, Y., Driscoll, T., Morgan, G., Dear, K., Gourley, M., Capon, A., Bi, P., 2022. Heat exposure and cardiovascular health outcomes: a systematic review and meta-analysis. Lancet Planet. Health 6, e484–e495.

Palomeque, J., Vila Petroff, M.G., Mattiazzi, A., 2004. Pacing staircase phenomenon in the heart: from Bodwitch to the XXI century. Heart Lung Circ. 13, 410–420.

Rajagopal, V., Pinali, C., Shiels, H.A., 2022. New revelations on the interplay between cardiomyocyte architecture and cardiomyocyte function in growth, health, and disease: a brief introduction. Philos. Trans. R. Soc. B 377 (1864), 20210315.

Rumberge, E., Reichel, H., 1972. Force-frequency relationship—comparative study between warm-blooded and cold-blooded animals. Pflugers Archiv Eur. J. Physiol. 332.

Santer, R.M., 1985. Morphology and innervation of the fish heart. Adv. Anat. Embryol. Cell Biol. 89, 1–102.

Santer, R.M., Greer Walker, M., 1980. Morphological studeis onteh ventricle of teleost and elasmobranch hearts. J. Zool. 190, 259–272.

Saunders, R.L., Farrell, A.P., Knox, D.E., 1992. Progression of coronary arterial lesions in atlantic salmon (*Salmo salar*) as a function of growth rate. Can. J. Fish. Aquat. Sci. 49, 878–884.

Shaftoe, J.B., Manchester, E.A., Gillis, T.E., 2023. Cardiac remodeling caused by cold acclimation is reversible with rewarming in zebrafish (*Danio rerio*). Comp. Biochem. Physiol. A Mol. Integr. Physiol. 283, 111466.

Shattock, M.J., Bers, D.M., 1989. Rat vs. rabbit ventricle: Ca flux and intracellular Na assessed by ion-selective microelectrodes. Am. J. Phys. 256, C813–C822.

Shiels, H.A., 2022. Avian cardiomyocyte architecture and what it reveals about the evolution of the vertebrate heart. Philos. Trans. R. Soc. Lond. Ser. B Biol. Sci. 377, 20210332.

Shiels, H.A., Farrell, A.P., 1997. The effect of temperature and adrenaline on the relative importance of the sarcoplasmic reticulum in contributing Ca2+ to force development in isolated ventricular trabeculae from rainbow trout. J. Exp. Biol. 200, 1607–1621.

Shiels, H.A., Galli, G.L., 2014. The sarcoplasmic reticulum and the evolution of the vertebrate heart. Physiology (Bethesda) 29, 456–469.

Shiels, H.A., Sitsapesan, R., 2015. Is there something fishy about the regulation of the ryanodine receptor in the fish heart? Exp. Physiol. 100, 1412–1420.

Shiels, H.A., Vornanen, M., Farrell, A.P., 2002. The force-frequency relationship in fish hearts—a review. Comp. Biochem. Physiol. A Mol. Integr. Physiol 132, 811–826.

Simonot, D.L., Farrell, A.P., 2009. Coronary vascular volume remodelling in rainbow trout *Oncorhynchus mykiss*. J. Fish Biol. 75, 1762–1772.

Sirsat, S.K., Sirsat, T.S., Price, E.R., Dzialowski, E.M., 2016. Post-hatching development of mitochondrial function, organ mass and metabolic rate in two ectotherms, the American alligator (*Alligator mississippiensis*) and the common snapping turtle (*Chelydra serpentina*). Biol Open 5, 443–451.

Sun, X.J., Hoage, T., Bai, P., Ding, Y.H., Chen, Z.Y., Zhang, R.L., Huang, W., Jahangir, A., Paw, B., Li, Y.G., Xu, X.L., 2009. Cardiac hypertrophy involves both myocyte hypertrophy and hyperplasia in anemic zebrafish. PLoS One 4 (8), e6596.

Tota, B., 1989. Myoarchitecture and vascularization of the elasmobranch heart ventricle. J. Exp. Zool. 252, 122–135.

Tota, B., Gattuso, A., 1996. Heart ventricle pumps in teleosts and elasmobranchs: a morphodynamic approach. J. Exp. Zool. 275, 162–171.

Tota, B., Cimini, V., Salvatore, G., Zummo, G., 1983. Comparative study of the arterial and lacunary systems of the ventricular myocardium of elasmobranch and teleost fishes. Am. J. Anat. 167, 15–32.

Vornanen, M., 1989. Regulation of contractility of the fish (*Carassius carassius* L) heart ventricle. Comp. Biochem. Physiol. C Comp. Pharmacol. Toxicol. 94, 477–483.

Vornanen, M., 2020. Feeling the heat: source–sink mismatch as a mechanism underlying the failure of thermal tolerance. J. Exp. Biol. 223, jeb225680.

Vornanen, M., 2021. Effects of acute warming on cardiac and myotomal sarco(endo)plasmic reticulum ATPase (SERCA) of thermally acclimated brown trout (*Salmo trutta*). J. Comp. Physiol. B. 191, 43–53.

Vornanen, M., Shiels, H.A., Farrell, A.P., 2002. Plasticity of excitation–contraction coupling in fish cardiac myocytes. Comp. Biochem. Physiol. A Mol. Integr. Physiol. 132, 827–846.

Vornanen, M., Hassinen, M., Koskinen, H., Krasnov, A., 2005. Steady-state effects of temperature acclimation on the transcriptome of the rainbow trout heart. Am. J. Phys. Regul. Integr. Comp. Phys. 289, R1177–R1184.

Chapter 4

THE HEART☆

ANTHONY P. FARRELL
Department of Biological Sciences, Simon Fraser University, Burnaby, British Columbia, Canada

DAVID R. JONES
Department of Zoology, University of British Columbia, Vancouver, British Columbia, Canada

Chapter Outline

I. INTRODUCTION

Since the descriptions of the cardiovascular system and the cardiovascular changes associated with exercise that appeared in earlier volumes of this series (Randall, 1970; Jones and Randall, 1978), several informative reviews, monograms, and perspectives have appeared, dealing with either general or specific aspects of cardiovascular physiology and anatomy. These works include those of Johansen and Burggren (1980), Tota (1983), Nilsson (1983), Laurent *et al.* (1983), Farrell (1984), Santer (1985), Farrell (1985), Wood and Perry (1985), Butler (1986), Johansen and Gesser (1986), Butler and Metcalfe (1988), Tota (1989), Forster *et al.* (1991), Davie and Farrell (1991a), Brill and Bushnell (1991), Farrell (1991a), and Satchell (1991). This

☆This is a reproduction of a previously published chapter in the Fish Physiology series, "1992 (Vol. 12A)/The Heart: ISBN: 978-0-12-350435-7; ISSN: 1546-5098"

Fish Physiology, Vol. 40A. https://doi.org/10.1016/bs.fp.2023.09.002

chapter focuses on the heart, whereas the primary arterial circulation is dealt with in the following chapter. Cardiac anatomy and morphology are described in Section II. The cardiac cycle and the control of cardiac output ($\dot{Q}$) are described in Section III. Since the heart also generates sufficient pressure to overcome vascular resistance, pressure development and myocardial power output are also considered in Section III. Accessory hearts, which assist venous return, are described in Chapter 3.

II. CARDIAC ANATOMY AND MORPHOLOGY

The fish heart is a four-chambered organ contained within a pericardial sac. Together, the sinus venosus, atrium, ventricle, and either an elastic bulbus arteriosus (in teleosts), or a contractile conus arteriosus (in elasmobranchs) raise the potential and kinetic energy of the blood. Detailed descriptions of fish hearts in the recent literature include Santer and Greer Walker (1980), Yamauchi (1980), Tota (1983), Santer (1985), Greer Walker *et al.* (1985), Emery *et al.* (1985), Sanchez-Quintana and Hurle (1987), and Tota (1989). What follows is a brief functional description of the anatomy, fine structure, and morphometrics of the heart.

A. Sinus Venosus

The sinus venosus has been described as a thin-walled chamber usually 60–90 μm thick (Santer, 1985). Its volume is similar to that of the atrium in teleosts, but less than that of the atrium in elasmobranchs (Satchell, 1971). The sinus venosus receives venous blood via paired Cuverian ducts, hepatic veins, and anterior jugular veins. The sinoatrial ostium is guarded by a large valve, and the openings of the hepatic veins into the sinus venosus are guarded by muscular sphincters. However, the Cuverian ducts are not valved; they freely communicate with the major systemic veins, which are particularly capacious in elasmobranchs (see Chapter 3).

The principal tissue component of the sinus wall is connective tissue with bounding inner endothelial and outer epicardial linings. The amount of cardiac muscle present in the sinus venosus varies considerably among species. The European eel (*Anguilla anguilla*) has an almost complete layer of cardiac muscle (Yamauchi, 1980). In contrast, cardiac muscle is limited to a sparse arrangement of small bundles of 4 to 5 cells in plaice (*Pleuronectes platessa*) and bull rout (*Myoxocephalus scorpius*). The sinus venosus is virtually amuscular in loach (*Misgurnus anguillicaudatus*), brown trout (*Salmo trutta*), and zebra fish (*Zebra danio*). Goldfish (*Carassius auratus*) and carp (*Cyprinus carpio*) have only smooth muscle elements (Yamauchi, 1980).

The major functional role of the sinus venosus is related to the initiation and control of the heart beat. The sinus venosus is the site of specialized cardiac pacemaker tissue in many fish. Histological and neurophysiological evidence for the location of pacemaker tissue has been reviewed previously (Laurent, 1962; Yamauchi, 1980; Laurent *et al.*, 1983; Santer, 1985;

Satchell, 1991). What emerges from these reviews is that a ring of specialized myocardial cells (nodal tissue) is usually located at the base of the sinoatrial ostium and connects with atrial myocardium. Nodal tissue can have other locations. In the eel (*A. vulgaris*) and lungfish (*Protopterus ethiopicus*), the pacemaker is located at the junction of the sinus venosus and Cuverian ducts. Specialized pacemaker cells have also been reported in the atrium of plaice, brown bullhead (*Ictalurus nebulosus*), and Pacific hagfish (*Eptatretus stouti*), as well as in the atrioventricular region. Even in the absence of specialized pacemaker tissue, myogenic activity of the atrium, ventricle, or atrioventricular region can also initiate contraction, but the contraction rate is likely to be slower and more irregular than the pacemaker rate.

B. Atrium

Water-breathing fish have a single atrium. The atrium is an irregularly shaped chamber with thin trabecular walls (Santer, 1985). Atrial volume is similar to or larger than ventricular volume (see Section III,B,2). Atrial mass generally constitutes 8–25% of ventricular mass and 0.01–0.03% of body mass (Table I). However, exceptionally large atria are reported for the New Zealand hagfish (*Eptatretus cirrhatus*), tunas, and two species of red-blooded Antarctic fishes (Table I). In *E. cirrhatus, Pagothenia bernacchii*, and *P. borchgrevinki*, atrial mass is an unusually large proportion (33–50%) of ventricular mass. Skipjack tuna (*Katsuwonus pelamis*) have an exceptionally large ventricle, and the atrial to ventricular ratio is similar to that in other fish. As a result, their unusually large (0.06%) atrial mass relative to body mass is greater than the relative ventricle mass of some benthic fish (Table I).

The variable form of the atrium in part reflects the variable distance that exists between the atrioventricular and ventricular-bulbar orifices and the extent of the S-shaped folding of the heart. Inside the atrium, there are two arcuate fans of muscular trabeculae in many teleosts (Santer, 1985). The trabeculae (19–35 μm in diameter) arise at the atrioventricular ostium and form a meshlike network. When they contract, they help pull the roof and sides of the atrium toward the atrioventricular ostium (Satchell, 1971). The atrioventricular valve is supported by a ring of cardiac tissue. In benthopelagic teleosts, the more spacious arrangement of trabeculae gives the atrium a frail appearance (Greer Walker *et al*., 1985).

C. Ventricle

The ventricle shows considerable species variability with respect to its relative mass, gross morphology, histology, and vascularity. Santer (1985) emphasizes the fact that there is no "typical" shape to the ventricle in fishes. It is equally appropriate to state that there is no typical ventricular mass, histology, or vascular network. Instead, certain patterns are useful to categorize fish hearts. These categories reflect functional rather than phylogenetic

Table I Cardiac Morphometrics in Selected Fishes

	Relative ventricular mass (%)	Compacta (%)	Relative atrial mass (%)	Source[a]
Teleosts with compacta				
Katsuwonus pelamis (skipjack tuna)	0.38	65.6	0.061	(a)
Thunnus albacares (yellowfin tuna)	0.29	55.4	0.056	(a)
Thunnus thynnus (northern bluefin tuna)	0.31	39.1	—	(b,c)
Thunnus maccoyii (southern bluefin tuna)	0.29	48.5	0.053	(d)
Thunnus obesus (bigeye tuna)	—	73.6	—	(c)
Makaira nigricans (Pacific blue marlin)	0.087	48.0	0.013	(e)
Scomber scombrus (mackerel)	0.18	43.0	—	(c)
Engraulis encrasicolus (anchovy)	—	61.7	—	(c)
Clupea harengus (herring)	—	24.8	—	(c)
Seriola grandis (kingfish or yellowtail)	0.11	42.3	0.030	(f)
Salmo gairdneri (rainbow trout)	0.08–0.13	30–39	0.017	(g)
Cyrinus carpio (carp)	—	37.0	—	(h)
Anguilla anguilla (European eel)	—	34.0	—	(c)
Anguilla dieffenbachii (longfinned eel)	0.03–0.10	40.9	0.007–0.013	(i)
Conger conger (conger eel)	—	16.0	—	(c)
Elasmobranchs				
Carcharodon carcharias (great white shark)	—	36.0	—	(j)
Isurus oxyrinchus (shortfin mako shark)	0.14	41.5	0.028	(e)

Table I Cardiac Morphometrics in Selected Fishes—Cont'd

	Relative ventricular mass (%)	Compacta (%)	Relative atrial mass (%)	Source
Galeocerdo cuvieri (tiger shark)	—	19.8	—	(j)
Prionance glauca (blue shark)	—	16.7	—	(j)
Squalus acanthias (dogfish)	0.086	22.1	0.017	(k)
Squatina squatina (monkfish)	—	34.5	—	(c)
Raja clavata (thornback ray)	—	35.8	—	(c)
Raja hyperborea (Arctic skate)	—	17.3	—	(c)
Chimaera monstrosa	—	5.0	—	(c)
Teleosts with only spongiosa				
Eptatretus cirrhatus (New Zealand hagfish)	0.096	none	0.039	(d)
Chaenocephalus aceratus (hemoglobin-free icefish)	0.30	none	—	(l)
Chionodraco humatus (hemoglobin-free icefish)	0.39	none	—	(m)
Pagothenia borchgrevinki (red-blooded Antarctic fish)	0.16	none	0.040	(n)
Pagothenia bernacchi (red-blooded Antarctic fish)	0.11	none	0.050	(n)
Hemitripterus americanus (sea raven)	0.07	none	—	(o)
Chelidonichthys kumu (gurnard)	0.05	none	0.017	(d)
Pleuronectes platessa (flounder)	0.035	none	—	(p)
Coryphaenoides rupestris (deep-sea fish)	0.06[b]	none	—	(q)

[a](a) Farrell *et al.* (1992); (b) Poupa *et al.* (1981); (c) Santer and Greer Walker (1980); (d) P. S. Davie and co-workers (unpublished data); (e) Davie (1987); (f) Davie and Hutchinson (unpublished data); (g) Farrell *et al.* (1988a); (h) Bass *et al.* (1973); (i) Davie *et al.* (1992); (j) Emery *et al.* (1985); (k) Davie and Farrell (unpublished data); (l) Holeton (1970); (m) Tota *et al.* (1991); (n) Axelsson *et al.* (1992); (o) Farrell *et al.* (1985); (p) Santer *et al.* (1983); (q) Greer Walker *et al.* (1985)
[b]Total heart weight.

correlates, and some of the categories have hazy boundaries. Two features used to categorize fish hearts are (1) ventricular form, and (2) the relative development of an outer layer of compacta tissue and its associated coronary circulation.

1. Ventricle Form

Santer (1985) proposed three major categories of ventricular form in fishes:

1. A saclike ventricle, which is rounded in shape with an indistinct apex. This form is the most common shape in elasmobranchs and many marine teleosts, and the only shape found in 29 species of benthopelagic teleosts (Santer *et al.*, 1983; Greer Walker *et al.*, 1985).
2. A tubular ventricle with a cylindrical cross-section. This form is found only in fish with an elongate body shape, but all fish with an elongate body shape do not necessarily have a tubular heart.
3. A pyramidal ventricle with a triangular base forming the caudal aspect. The pyramidal ventricle is restricted to species with an active life style, e.g., the salmonid and scombrid families. A pyramidal ventricle always has an outer compacta layer, but not all fish with compacta have a pyramidal ventricle, e.g., elasmobranchs.

The functional significance of these shapes, beyond the tubular ventricle reflecting body form, is not entirely clear. While the saclike ventricle is clearly the most common form, the possession of a pyramidal ventricle may be advantageous in terms of pressure development, because the apices have small radii of curvature as well as an unusual arrangement of fiber bundles.

2. Spongiosa and Compacta

Tota (1989) proposed that fish hearts can be divided into four major categories based on the possession and arrangement of spongiosa and compacta. The ventricles have two basic forms: whereas one form has an arrangement of muscle trabeculae that span the entire ventricle to form a spongelike network, the spongiosa, the other form has an outer compacta layer enclosing an inner spongiosa. The ventricle is bounded externally by epicardium and internally by endocardium. The compacta is always associated with a coronary circulation. The principal features distinguishing the four categories of ventricles are the extent of spongiosa versus compacta and the pattern of vascularization (Fig. 1). The four categories reflect increasingly more complex arrangements.

The majority of fish species have a type I venticle. The type I ventricle has only spongiosa myocardium (also given terms such as the inner, spongy, trabecular, or trabeculated layer, the subendocardium, and the endocardium in the literature). Variations of the type I ventricle exist and reflect differences in vascular development, although there are no capillaries *per se* in the myocardium of any type I hearts. Most fish have a subtype Ia ventricle in which venous blood contained in the lumen and intertrabecular spaces of

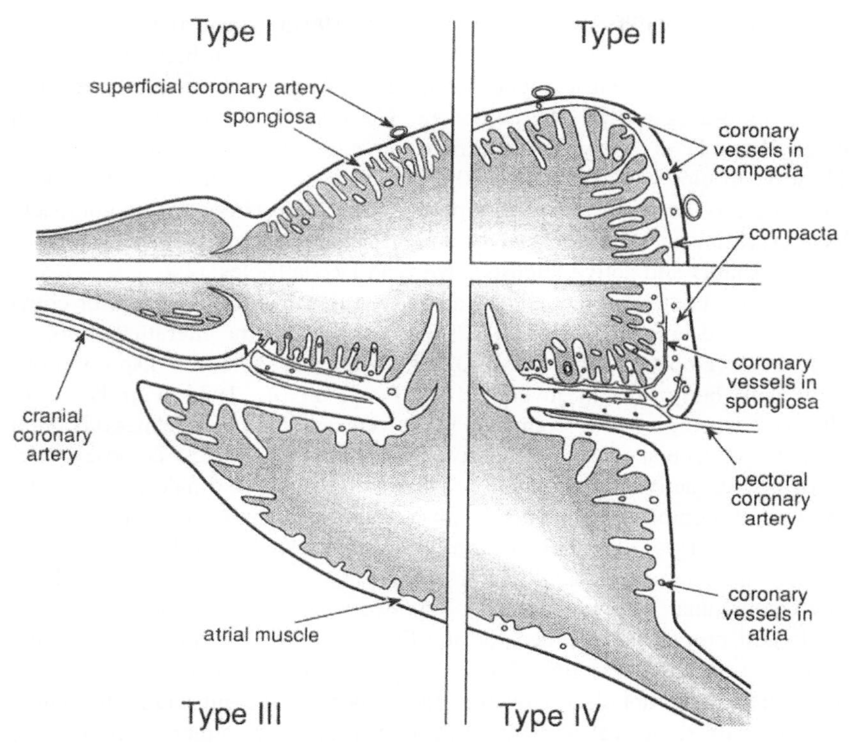

Fig. 1. A schematic illustration of the main features that distinguish the four ventricle types in fishes. The basic anatomical design of the fish heart has been divided into quadrants with each quadrant illustrating particular features of a given ventricle type. Type I is characterized by having a single myocardial type (spongiosa) and no capillaries in the ventricular muscle. Most type I hearts have no coronary vessels whatsoever. Type II is characterized by two muscle layers in the ventricle (inner spongiosa and outer compacta), a coronary circulation, and capillaries only in the outer compacta. Type III is similar to type II, but capillaries are found in both the spongiosa and compacta, and to a limited extent in the atrium. Type IV has a larger percentage of the ventricle as compacta (>30%) and a more extensive capillarization of the atrium. The coronary circulation, regardless of type, is derived from a cranial supply, as depicted in the lower left quadrant, and an additional pectoral supply in some fishes, as depicted in the lower right quadrant. Adapted with permission from Davie and Farrell (1991a).

the ventricle (luminal blood) provides the only blood supply (hence the terms venous, lacunary, or avascular hearts). Subtypes Ib and Ic possess a superficial coronary artery, but arterial vessels either are confined to the epicardium (subtype Ib), e.g., plaice, or connect directly to the intertrabecular spaces (subtype Ic), e.g., hemoglobin-free Antarctic fish (Tota, 1989).

The three other categories of ventricles (types II, III, and IV) have compacta and spongiosa myocardial tissue as well as capillaries in the myocardium (Tota, 1989; Davie and Farrell, 1991a). Type II ventricles are characterized by the capillaries of the coronary circulation being found only in the compacta. In ventricle types III and IV, the coronary circulation reaches the spongiosa as well as the compacta. The distinguishing feature between types III and IV is that type III ventricles have less than 30% compacta tissue. Those ventricles with more than 30% compacta and coronaries in the spongiosa are arbitrarily described as type IV. Most teleosts possessing compacta have type II ventricles. Most elasmobranchs have type III ventricles. Endothermic sharks and active teleosts have type IV ventricles.

Ontogenetically, the compacta (also given terms outer, compact, or cortical layer, the epicardium, and the subepicardium in the literature) appears early in development (Kitoh and Oguri, 1985) and increases disproportionately with body size in juvenile fish (Poupa *et al.*, 1974; Farrell *et al.*, 1988a). The proportion of compacta in the ventricle (and its thickness) is quite variable between species. The relative proportion of compacta is loosely correlated with activity. Endothermic sharks have approximately twice the amount of compacta compared with ectothermic sharks (36–40% versus 15–24%; Emery *et al.*, 1985). Among the active teleosts that have compacta, the most athletic ones generally have higher proportions of compacta (Table I). The bigeye tuna (*Thunnus obesus*) is reported to have 73.6% compacta tissue, the largest proportion recorded for any fish. Beyond a genetic constraint, the factors setting the upper limit for the proportion of compacta in the ventricle of mature fish are not clear. However, the proportion of compacta in the ventricle does vary somewhat with smoltification in Atlantic salmon (*Salmo salar*) and with changing seasons in rainbow trout (*Oncorhynchus mykiss*), but not with exercise training (Ostadl and Schiebler, 1972; Poupa *et al.*, 1974; Farrell *et al.*, 1988a; Farrell *et al.*, 1990a).

The greater the proportion of compacta, the more layers of different fiber bundles it has, each layer having a different architecture (Santer, 1985; Sanchez-Quintina and Hurle, 1987). Dogfish (*Galeorhinus galeus*) has 15% compacta, and blue shark (*Prionace glauca*) has 17% compacta; both species have a single layer of looped fiber bundles in the compacta. The more athletic Atlantic shortfin mako shark (*Isurus oxyrinchus*) has 28% compacta arranged in two layers. The outer layer is looped (Fig. 2A), whereas the inner layer forms a sac of circular fibers. Similarly, there are two layers in swordfish (*Xiphias gladius*) and albacore (*Thunnus alalunga*), which have 29 and 26% compacta, respectively. Again the outer layer is arranged in loops (Fig. 2B), but these teleosts have a pyramidal rather than a saclike ventricle, and the inner layer forms rings encircling the vertices of the ventricular pyramid. Three layers of fiber bundles are present in Atlantic bluefin tuna (*Thunnus thynnus*), which has 39% compacta. The inner (Fig. 2C) and outer layers have a similar fiber architecture as swordfish and albacore, whereas the additional layer between inner and outer layers has the shape of a stirrup that surrounds the cranial apex, the lateral borders, and the caudal face of the ventricle.

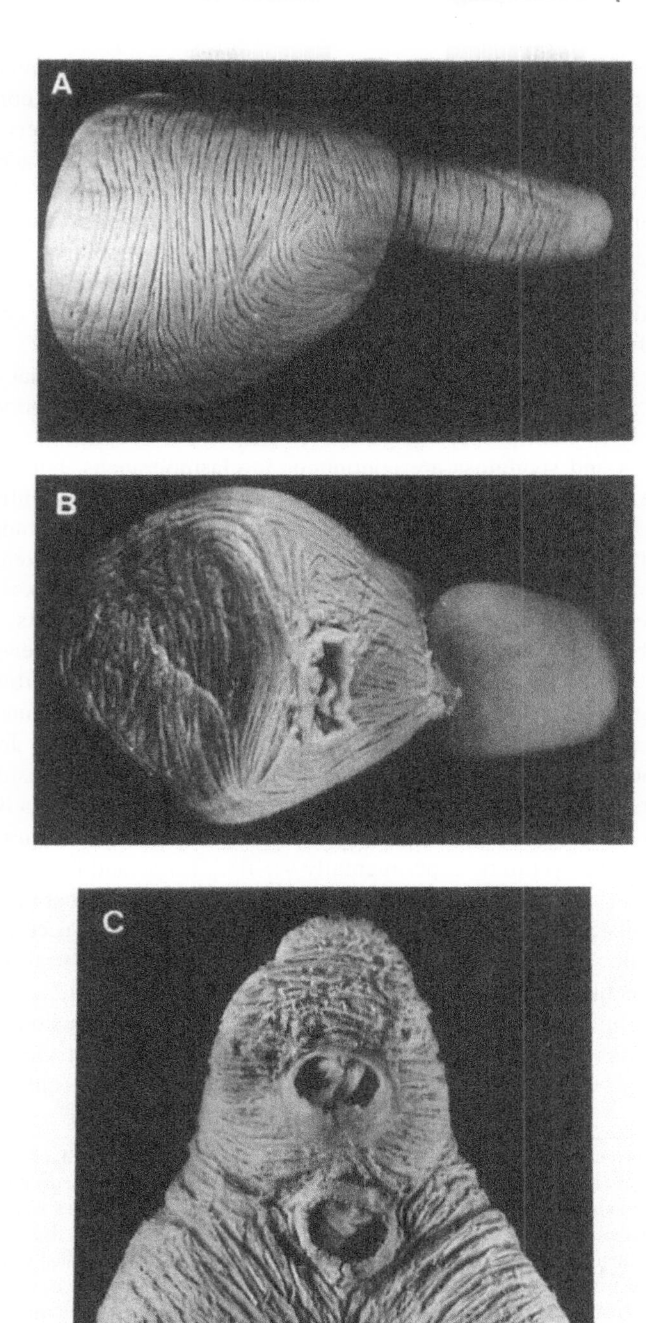

Fig. 2. See figure legend on next page.

These observations suggest that up to three layers can be found, with looped fibers being the most fundamental fiber arrangement in the compacta. In teleosts, the dominant longitudinal arrangement of looped fibers on the ventral faces and a dominant transverse arrangement on the caudal face would act to reduce the longitudinal and transverse diameters of the ventricle during systole (Sanchez-Quintina and Hurle, 1987). The addition of inner fibers either as a circular arrangement around the saclike elasmobranch ventricle, or as coils around the vertices of the pyramidal teleost ventricle is clearly a characteristic of fish that are more active, and perhaps provides a mechanical advantage for developing higher blood pressures.

Certain features distinguish the elasmobranch ventricle from that in teleosts. First, all elasmobranchs have compacta and a coronary circulation, whereas only a small proportion of active teleosts do. Second, the fibers of the compacta and spongiosa are continuous in elasmobranchs, but are separated by a layer of connective tissue in teleosts (Sanchez-Quintina and Hurle, 1987). Third, the elasmobranch spongiosa has a precise arrangement of arcuate trabeculae running from the atrioventricular to the conoventricular ostia. In contrast, teleosts have an anarchial array of trabeculae except near the bulboventricular region, where a longitudinal arrangement exists.

The trabecular arrangement of spongiosa accounts for the greater proportion of ventricular mass in almost all fishes (Table I). The idea that this trabecular arrangement allows the individual lacuna of the spongiosa to function as "many small hearts" within the larger ventricle was put forward by Johansen (1965). Since the radius of each lacuna is small, LaPlacian relationships dictate that pressure will be generated at a considerable mechanical advantage compared with that of hearts consisting solely of compact myocardium. However, this idea has yet to be experimentally verified. Davie and Farrell (1991a) suggested that the diameter of the trabeculae is a compromise between minimizing the distance for O_2 diffusion and maximizing the cross-sectional area for tension development. The basis for this suggestion is that atrial and ventricular trabeculae of most fish derive O_2 from luminal blood, and a trabecular muscle structure increases the surface area and reduces the diffusion distance for O_2 transfer compared with a compact muscle structure. In view of this suggestion, it would be interesting to learn whether the trabeculae of the

Fig. 2. The myocardial fiber architecture in the compacta of fish hearts. (A) A lateral view of the ventricle and conus arteriosus of the mako shark. Note the transverse arrangement of superficial fibers that loop around the sacular heart. An additional inner layer of fiber bundles in the compacta (not shown here) has a more circular arrangement. (B) A dorsal view of the swordfish ventricle and bulbus arteriosus showing the superficial fiber layers of the pyramidal ventricle. An inner layer consists of rings encircling the vertices, similar to but not so pronounced as those shown in C. (C) A cranial view of the Atlantic blue fin tuna ventricle showing the deep fiber layer of the compacta. The rings of fibers encircling the vertices of the ventricle are very pronounced. The atrioventricular and bulboventricular valves are evident. With permission from Sanchez-Quintina and Hurle (1987).

atrium and the spongiosa in elasmobranchs and certain active teleosts (types III and IV) have larger diameters, since these trabeculae have a coronary circulation.

3. *Relative Ventricle Mass*

Cardiac mass in fish scales almost in direct proportion to body mass (Poupa and Lindstrom, 1983), a situation which is similar to that in other vertebrates. However, this allometric relationship tends to obscure the fact that relative ventricular mass (ventricle mass divided by body mass) varies considerably among fishes and even within a species. Relative ventricular mass varies over a 10-fold range, being 0.035% of body mass in flounder to 0.38% of body mass in skipjack tuna (Table I).

The following generalizations can be made regarding relative ventricle mass.

1. Active species have larger ventricles. Active teleosts (e.g., Clupidae, 0.249%) have a larger relative ventricular mass than do active elasmobranchs (e.g., sharks, 0.196%) (Poupa and Lindstrom 1983). Even at the species level, larger hearts are observed in anadromous compared with lake varieties of *O. mykiss* (Graham and Farrell, 1992). However, benthic elasmobranchs (e.g., Selachii, 0.098%) have a larger relative ventricular mass than do benthic teleosts (e.g., Pleuronectids, 0.061%) (Poupa and Lindstrom, 1983).
2. Endothermic sharks have larger ventricles than ectothermic sharks (Emery *et al.*, 1985).
3. Exercise training has variable effects on cardiac growth. Some training regimens produce small increases in relative ventricle mass, whereas other regimens produce only isometric cardiac growth (Davison, 1989; Farrell *et al.*, 1990a).
4. Cold-acclimation increases relative ventricular mass in rainbow trout, carp, and goldfish (Tsukuda *et al.*, 1985; Farrell, 1987; Goolish, 1987; Graham and Farrell, 1989). In rainbow trout, ventricle mass increases by 70% when fish are acclimated to temperatures 10°C apart. Seasonal variations in relative ventricle mass of rainbow trout also correlate well with changes in water temperatures (Farrell *et al.*, 1988a). Antarctic fishes have relatively large ventricles (Holeton, 1970; Johnston *et al.*, 1983; Tota *et al.*, 1991).

These observations implicate a large relative ventricular mass as being important in (1) developing higher blood pressures in active fishes, (2) compensating for the negative inotropic effect of low temperature, and (3) accommodating large cardiac stroke volumes. These physiological considerations are covered in subsequent sections.

D. Innervation Patterns

The literature on the innervation patterns of fish hearts and pacemaker tissue has been comprehensively reviewed by Laurent *et al.* (1983), Nilsson (1983), and Santer (1985). The following summarizes the cardiac distribution of cholinergic and adrenergic fibers. Other aspects of cardiac innervation are described in Part B of this volume, Chapters 5 and 6.

Vagal cardiac innervation is found in all fishes except the hagfishes. In all cases the vagus carries small, myelinated, cholinergic efferent fibers. These efferents stimulate inhibitory muscarinic cholinoceptors in all species except lampetroids, which have excitatory nicotinic cholinoceptors. Only the sinus venosus is innervated in lampreys. In elasmobranchs and teleosts, the sinoatrial region is richly innervated, and it is possible that every muscle cell has at least one nerve profile. The muscle fibers of the atrium and the atrioventricular region are innervated to a lesser extent whereas those of the ventricle are only sparsely and partially innervated. Large myelinated fibers are found in the vagal trunks and throughout the atrium and ventricle. These fibers are thought to be sensory, cholinergic afferent fibers, but their role has not been extensively studied (see Laurent, 1962).

Adrenergic cardiac innervation is found in the majority of teleosts studied, but not in cyclostomes and elasmobranchs. Unmyelinated adrenergic efferents have various sizes, and their distribution pattern varies between species. Typically, the sinus venosus and atrium, especially the sinoatrial and atrioventricular regions, are well innervated. The compacta of the ventricle, especially near the atrioventricular region, the coronary vessels, and the bulbus arteriosus also receive adrenergic fibers. Benthic teleosts, such as plaice, lack adrenergic innervation.

From these distribution patterns, we can expect significant cholinergic and adrenergic neural influences on pacemaker rates, atrioventricular conduction, and atrial contractility. In addition, adrenergic influences may affect ventricular contractility to a greater degree than cholinergic influences, and adrenergic coronary vasoactivity is to be expected.

E. Myocytes

A number of features distinguish fish cardiac muscle cells (myocytes) from those in mammals. These include: (1) small size (1–12.5 μm) compared with 10–25 μm in mammals, (2) limited anatomical development of the sarcoplasmic reticulum (SR), (3) peripheral arrangement of myofibrils, and (4) variable amount of intracellular myoglobin. The fine structure of myocardial cells is described in considerable detail by Yamauchi (1980), Helle (1983), and Santer (1985).

The morphometric characteristics of myocytes from various fish atria and ventricles were compiled by Santer (1985). Myocyte diameters are most often between 2.5 and 6.0 μm and range from 1.0 to 12.5 μm. The differences that exist in the diameters of atrial and ventricular myocytes are small relative to the overall range in myocyte diameters. Tota (1989) notes that myocyte diameters are less in elasmobranchs than in teleosts. The largest myocyte diameters occur in *A. vulgaris*, *Cyprinus carpio*, *Misgurnus anguillicaudatus*, and two species of osteoglossids; all are species that tolerate aquatic hypoxia. Myocytes in the compacta have larger diameters than in the spongiosa in *Arapaima gigas* (Hochachka *et al.*, 1978) and albacore (Breisch *et al.*, 1983). Among the functional advantages of a narrow myocyte are a shorter diffusion distance from the outside to the center of the cell, and a higher ratio of sarcolemma area to intracellular volume. However, smaller diameters probably mean higher electrical resistance and slower conduction velocity. Poupa and Lindstrom (1983) proposed that myocyte diameter in homoeotherms is proportional to stroke volume, inversely proportional to heart rate, and inversely proportional to the maximum activity state of the animal. Thus, the "high speed" cardiac pump of small mammals is characterized by small myocytes. The small myocytes of the "low speed" fish heart clearly contrast with this dogma (Farrell, 1991a).

The effect of cardiac growth on myocyte size has been assessed for perch (*Perca perca;* Karttunen and Tirri, 1986) and rainbow trout (Farrell *et al.*, 1988a). In perch, mean myocyte diameter (4.2 μm) and length (133 μm) were independent of fish size (50–250 g), implying that cardiac growth is entirely hyperplastic over this range of body mass for perch. In rainbow trout, myocyte length increased from 44 μm to 103 μm, and myocyte diameter increased from 6.6 to 9.0 μm for a larger range of body mass (12–1750 g). From these dimensions it was estimated that myocyte volume increased 6-fold for a 100-fold increase in ventricle mass (Fig. 3A). Hence, hyperplastic cardiac growth occurs to a much greater extent than hypertrophic growth in rainbow trout. Hyperplastic growth of fish myocytes contrasts with the situation in mammals in which myocyte growth is hypertrophic, except in the embryonic period and sometimes in the first few weeks of neonatal life. Hyperplastic growth means that relatively small myocyte diameters are maintained during ontogeny.

The SR is sparse in fish hearts and the sarcolemma (SL) lacks the T-tubule system typical of all skeletal muscle and mammalian cardiac muscle. Instead, there are small (150 nm diameter), flask-shaped infoldings of the SL termed calveolae (Santer, 1985). Tota (1989) notes that elasmobranch myocytes have less SR compared with teleosts. Four types of SR are recognized (Fig. 3B). Subsarcolemmal cisternae are in contact with tubules that encircle the myofibrils. Longitudinal tubules connecting the circular tubules are particularly sparse in fish. Patches of reticular lattices are also limited, but are more prominent in active species of fishes (Santer, 1985). The impact of small myocyte size, absence of T-tubules, and limited anatomical development of the SR with respect to excitation–contraction coupling is discussed more fully in Chapter 6.

Ventricular myocytes contain more myofibrils than atrial myocytes and can occupy 55% of myocyte volume in active species, approaching proportions found in mammalian ventricles (Yamauchi and Burnstock, 1968; Midtun, 1980; Breisch *et al.*, 1983). Myofibrillar volume is particularly low in Antarctic

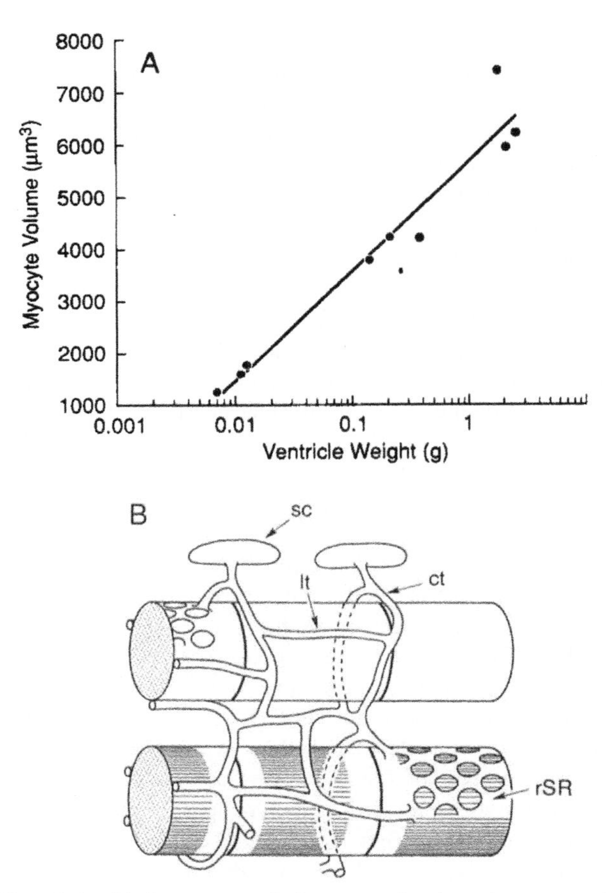

Fig. 3. (A) The relationship between ventricular mass and estimated myocyte volume for rainbow trout weighing between 10 and 1800 g. Since myocyte volume increases at only about one sixth the rate of ventricular mass, considerable hyperplastic myocyte growth is likely. From Farrell *et al.* (1988). (B) The sarcoplasmic reticulum (SR) is much reduced in fish myocardium. This schematic illustrates the four types of SR that have been identified in fishes and their anatomical relationship to two myofibrils: subsarcolemmal cisterna (sc); longitudinal tubules (lt); circular tubules (ct); reticular SR (rSR). Adapted from Santer (1985).

fishes (Johnston *et al.*, 1983). Myofibrils of some fish species are located in the peripheral part of the cell, close to the SL, with mitochondria being more centrally located. An advantage of peripherally located myofibrils is that, in the absence of a T-tubule system, calcium diffusion from the extracellular space to troponin-C is facilitated, which may be more critical for cardiac muscle performance than is the diffusion of oxygen to mitochondria (see Farrell, 1991b).

Myoglobin is particularly variable in fish hearts. Some myocytes appear white as a result of being essentially devoid of myoglobin, e.g., ocean pout

(*Macrozoarces americanus*), which has 5.2 nmol per g ventricular mass. However, most fish species possess myoglobin in the range of 50–150 nmol per g ventricular mass. The highest concentrations of myoglobin are found in active fishes, such as striped bass (400 nmol per g ventricular mass; Driedzic, 1983) and Atlantic bluefin tuna (up to 1300 nmol per g ventricular mass; Giovane *et al.*, 1980), and hypoxia-tolerant species, such as the North American eel (*A. rostrata*, 239 nmol per g ventricular mass; Bailey *et al.*, 1990). Ventricular mass increases with body size in Atlantic bluefin tuna, and myoglobin concentrations do likewise (Poupa *et al*, 1974). Clearly, high myoglobin concentrations in large ventricles (activity related, hypoxia related, and growth related) reflect a balance between fine architecture of the cell and adequate O_2 diffusion. Chapter 5 by Driedzic discusses the functional importance of myoglobin in facilitating O_2 delivery to the myocardium of hypoxic, but not normoxic, fish hearts.

Other cardiac cell inclusions, such as atrial naturetic factors, glycogen, and fat, are discussed in Chapter 5 and Part B, Chapter 3.

F. Conus and Bulbus Arteriosus

A cyclindrical-, pear-, or onion-shaped chamber lying within the pericardium is interposed between the ventricle and ventral aorta in chondrichthyes and osteichthyes. In selachians this chamber is rhythmically contractile and referred to as the conus arteriosus, whereas in teleosts it is elastic and called the bulbus arteriosus. Structurally, the bulbus resembles the ventral aorta, and this, combined with the fact that more ancient fishes have both conal and bulbar regions, has given rise to the notion that phylogenetically the primitive conus has been telescoped into the ventricle by backward growth of the ventral aorta. However, this view is not subscribed to by all, and Priede (1976), for instance, suggests that the teleost bulbus is of cardiac origin. Certainly, Senior's (1909) study of heart development in the shad (*Alosa sapidissima*) implies that the conus develops within the pericardium. Nevertheless, the nomenclature of the bulb in cyclostomes as a "conus", when its structure is aortic and it resides outside the pericardium, reflects the view of the conus as phylogenetically more primitive.

Embryologically, the conus has four endocardial ridges, which are capable of occluding its lumen during contraction, although the ventral one is usually vestigial (Goodrich, 1930). Pocket valves, with the concavity to the front, develop from the dorsal and lateroventral ridges in transverse rows or tiers. Up to six tiers can be present, although in "advanced" selachians, there are only two tiers of well-developed valves. In contrast, multiple tiers of valves occur in the conus of primitive teleosts (*Acipenser* and *Amia*). In *Polypterus* and *Lepidoseus*, the conus is elongated, and there are multiple rows and tiers of valves (Goodrich, 1930). In most teleosts, a narrow muscular ring with two valves at the ventriculobulbar junction represents the only vestige of the conus. However, in the sticklebacks, *Gasterosteus* and *Pungitius*, there may be an exceptional arrangement in that valves have been described at both the ventriculobulbar and bulboaortic junctions (Benjamin *et al.*, 1983). In most teleosts, the luminal wall of the bulbus is smooth or ridged although in trout and tuna (fish with high arterial blood pressures) there is a series of chordae-like structures (longitudinal elements), which run along the internal wall. Over most of their length, the obliquely running radial fibers connect the longitudinal elements to the bulbar wall (Priede, 1976).

Deep sea fishes (*Macrouridae*) also have a unique bulbar anatomy. At the ventriculobulbar junction, there is a layer of gelatinous and collagenous tissue that forms an "inner tube" within the bulbus running toward the ventral aorta (Fig. 4A; Greer Walker *et al.*, 1985). This "inner tube" even contains hyaline cartilage in *Chalinura profundicola*. The "inner tube" becomes continuous with spaces between anastomosing trabeculae in the main cavity of the bulbus.

In cyclostomes the "conus" has no cardiac muscle in its walls and is not rhythmically contractile. The hagfish conus is much thinner than that in the lamprey, which is probably a reflection of much lower blood pressures in hagfishes than in lampreys (Wright, 1984; Davie *et al.*, 1987).

The selachian conus contains cardiac muscle, and teleost bulbus contains smooth muscle in the walls. Hence, the nomenclature of wall structure for the conus is cardiac (i.e., epi-, myo-, and endocardium), whereas the corresponding

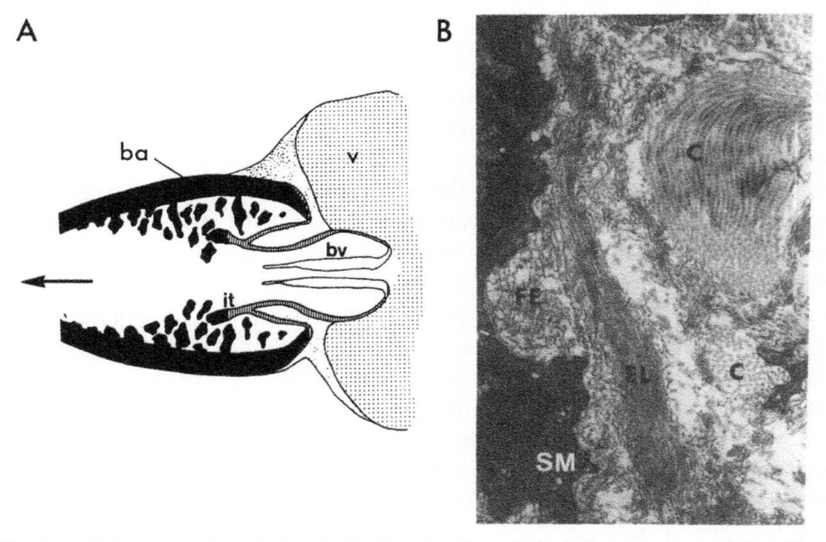

Fig. 4. (A) Reconstruction of a longitudinal section through the bulbus arteriosus of *Ventrifossa occidentalis*, based on a study of serial transverse sections. Ventricle (v), bulbus arteriosus (ba), inner tube (it), bulboventricular valves (bv). From Greer Walker *et al.* (1985), with permission. (B) Electronmicrograph (×20,000) of the bulbar wall in a yellowfin tuna. C, collagen; FE, fibrillar elastin; EL, elastic lamina; SM, smooth muscle. Jones, Wheaton, and Brill, 1991.

three layers of the bulbus are described using blood vessel terminology (i.e., adventitia, media, and intima). Priede (1976) has argued that blood vessel terminology is inappropriate since it implies arterial homology, which exists only in the sense that bulbus and heart are both embryologically derived from the subintestinal blood vessel.

The luminal surface (intimal or endocardial) consists of squamous, cuboid, or elongate (tall) cells, which do not appear to be aligned along the major axis of blood flow (Benjamin *et al.*, 1983; Langille *et al.*, 1983; Leknes, 1981, 1985). In the dogfish *Scyliorhinus stellaris*, there is a subendocardial layer of chromaffin cells associated with myelinated and unmyelinated nerve endings. It has been suggested that the chromaffin cells are part of a neurosecretory system that influences cardiac performance (Saetersdal *et al.*, 1975; Zummo and Farina, 1989). In teleosts, smooth muscle and collagen fibers may be associated with the subintimal layer (Santer, 1985).

The outermost layer of both bulbus and conus is usually described as visceral pericardium. Collagen is concentrated in this layer along with blood vessels and ganglion cells. Sympathetic fibers appear to innervate smooth muscle of the medial layer of the bulbus (Gannon and Burnstock, 1969; Watson and Cobb, 1979). As the nerves in the bulbar wall are generally associated with blood vessels, species lacking a well-developed coronary blood supply have few or no signs of neural innervation (i.e., plaice, *Pleuronectes platessa*).

The myocardial layer of the conus is made up of circularly arranged cardiac myocytes, some collagen fibers, and many capillaries. The medial layer looks very much like the compact myocardium of the heart (Santer, 1985; Zummo and Farina, 1989). In contrast, the bulbar medial wall consists of smooth muscle cells and a fibrillar elastic reticular network (elastica). Smooth muscle cells may be arranged either in a spiral of flat sheets (Watson and Cobb, 1979) or circumferentially around an inner layer of interdigitizing smooth muscle fasciles (Licht and Harris, 1973). The latter arrangement is reminiscent of the division of the myocardium into compact and trabeculate muscular regions. In rainbow trout and salmon, three muscle layers are recognized: outer and inner, which run longitudinally to the main axis of the bulbus, sandwiching a layer of circumferentially orientated cells (Serafini-Fracassini *et al.*, 1978).

Elastic fibrils (20–50 nm in diameter) are found in close association with smooth muscle cells (Fig. 4B). These fibrils form an isotropic, three-dimensional network and are not usually in sheets or lamellae (Fig. 4B). Occasionally fibrils anastomose to form a fibrillar core, although amorphous elastin of the type seen in higher vertebrates is conspicuously lacking in fish. Some form of microfibrillar (4–15 nm in diameter) network may be associated with all elastic elements (Isokawa *et al.*, 1988, 1990) as the skeleton on which they are laid. However, cyclostomes are a unique case because the "conus" contains microfibrils (Wright, 1984), that are not even made of elastin (Sage and Gray, 1979, 1980).

G. Coronary Circulation

The anatomical origin of the coronary circulation is well described in the older literature (e.g., Parker, 1886; Parker and Davis, 1899; Grant and Regnier, 1926; O'Donoghue and Abbot, 1928; Foxon, 1950). Developmental and evolutionary perspectives of the coronary circulation are presented by Halpern and May (1958). The distribution patterns of the coronary arteries are described for lamnid shark ventricles by De Andres *et al.* (1990) and for tuna by Tota (1978). Terminal distribution patterns within the ventricular wall and the functional significance of the luminal versus coronary circulations are covered in Tota *et al.* (1983), Tota (1989), and Davie and Farrell (1991a).

The coronary circulation has two sides of origin: a cranial (cephalad) circulation (e.g., rainbow trout), and in some fish an additional pectoral (caudal) circulation (e.g., eel and swordfish). One to three pairs of arteries branch from the efferent branchial arch arteries and meet medially to form the hypobranchial artery from which the cranial coronary circulation is derived as one or more arteries. Cranial coronary arteries reach the ventricle across the surface of the bulbus arteriosus or conus arteriosus. The pectoral coronary circulation arises from the first branch of dorsal aorta, the coracoid artery, and vessels penetrate the pericardium, reaching the apical and sometimes lateral aspects of the ventricle via pericardial ligaments. The cranial and pectoral circulations are functionally interconnected at the arterial level (Davie and Farrell, 1991b). The anatomical origins of both the cranial and pectoral coronary circuits are such that oxygenated blood is delivered directly from the gills to the heart at the highest possible postbranchial blood pressure.

In elasmobranchs, multiple coronary arteries are found on the conus arteriosus, often with a dominant ventral or dorsal vessel. On the ventricle, the coronary arteries form a visible, subepicardial (extramural) network and short perforating branches supply the compacta and spongiosa (Tota, 1989; De Andres *et al.*, 1990). The compacta is well vascularized with capillaries connecting to veins that drain into the atrial chamber close to the atrioventricular region. The trabeculae of the spongiosa in type III and IV ventricles have centrally located arteries, terminal arterioles, a few capillaries, and no venous system (Voboril and Schiebler, 1970; Ostadal and Schiebler, 1971). Terminal arterioles and capillaries open directly to the lumen, thereby forming arteriolacunary connections analogous to the Thebesian system of the mammalian heart (Tota, 1989). In the type IV ventricle of the lamnid shark, the arterial branches that go to the inner spongiosa are formed in the region of the conoventricular and atrioventricular grooves rather than from the subepicardial network (De Andres *et al.*, 1990), and the atrium receives small branches from arteries located in the atrioventricular region. Numerous very small, perpendicular branches circle and supply the conal cardiac muscle. Coronary veins also run parallel to the main coronary arteries on the conus arteriosus of elasmobranchs and presumably drain the conal cardiac muscle (A. P. Farrell, 1990).

In teleosts, the cranial coronary artery is often a single vessel running along the dorsal (e.g., skipjack tuna) or ventral (e.g., rainbow trout) surface of the ventral aorta. The extramural network, so prominent in elasmobranchs, is largely lacking in teleost ventricles. The arterial distribution pattern has not been studied with respect to the various layers of fiber bundles in the compacta. Capillaries are limited to the compacta in type II ventricles, and venous drainage is to the atrium in the atrioventricular region. A Thebesian system, similar to that in elasmobranchs, is found in type IV teleost ventricles (Tota *et al.*, 1983).

In view of this anatomical diversity, it is most likely that the coronary circulation evolved more than once (Davie and Farrell, 1991a) with the cranial origin being the most primitive (Tota, 1989). All of the elasmobranchs and about one quarter of the teleosts have coronaries, and no single factor adequately accounts for the presence of a coronary circulation. Typically, teleosts that either tolerate environmental hypoxia or are active swimmers have both a coronary circulation and high myoglobin levels. Santer and Greer Walker (1980) postulated that the cost of swimming was higher in elasmobranchs because of the absence of a swimbladder, which may explain why all elasmobranchs have a coronary circulation: a continuous mode of swimming may lower the luminal O_2 supply to a critical level for long periods (see Section II,E).

The histological structure of the fish coronary artery is characterized by an external parenchyma surrounding a medial layer of vascular smooth muscle; an internal elastic lamina separates the media from the intima, which normally has a single layer of endothelial cells (Fig. 5A). Robertson *et al.* (1961) observed arteriosclerotic lesions in and confined to the coronary vessels of mature Pacific salmon. These lesions have since been characterized as intimal proliferations of vascular smooth muscle with a disrupted elastic lamina (Fig. 5B). They resemble the early forms of mammalian arteriosclerotic lesions, and although the proliferation of intimal smooth muscle is just as extensive as in mature mammalian lesions, calcium and lipid inclusions typically found in mature mammalian lesions are absent (Moore *et al.*, 1976; McKenzie *et al.*, 1978; House and Benditt, 1981).

Lesions are particularly prevalent and severe in mature migratory salmonids (Pacific salmonids, genus *Oncorhynchus*), Atlantic salmon, and steelhead trout (*O. mykiss*), but are less severe or absent in other fish species (Vastesaeger *et al.*, 1965; Santer, 1985). Coronary lesions are typically found in more than 95% and often 100% of a sample population of migratory salmonids (Robertson *et al.*, 1961; Maneche *et al.*, 1973; Farrell *et al.*, 1986a; Farrell *et al.*, 1990b). In addition, severe lesions are located on 66–80% of the length of the main coronary artery, often obstructing the arterial lumen by 10 to 30%. Since the lesions are in a main arterial conduit, the potential exists for a substantial increase in vascular resistance as a result of lesion formation. Whether or not this leads to an ischemic state is unclear (Farrell *et al.*, 1990b).

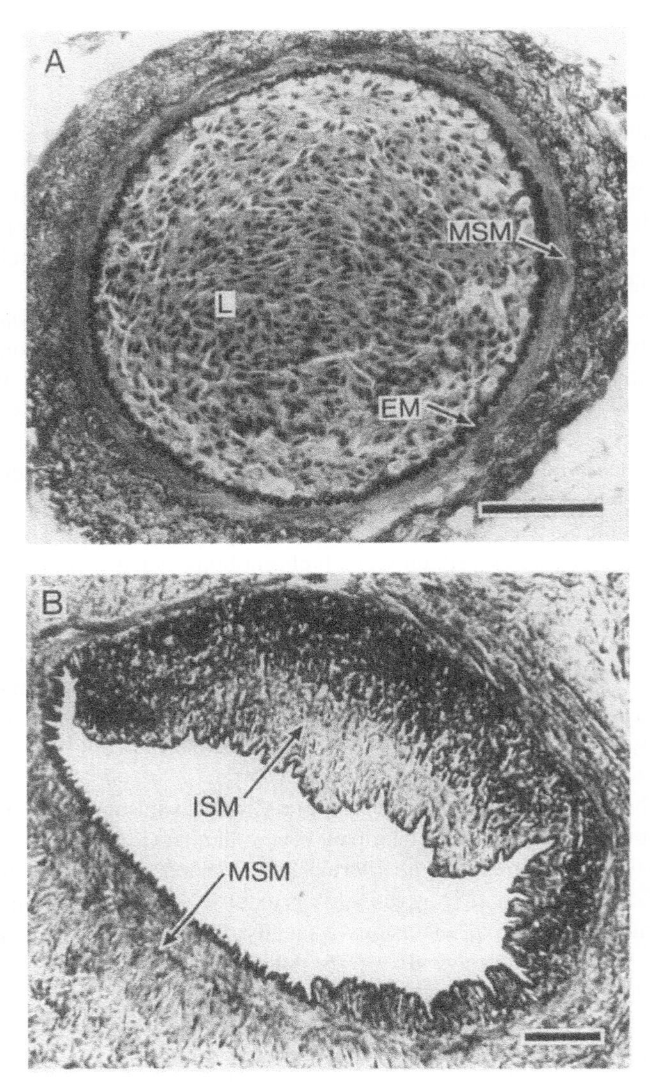

Fig. 5. Histological cross-sections of the main coronary artery of salmonids. (A) A normal coronary artery from an adult Atlantic salmon. The internal elastic membrane (EM) is differentially stained to demarcate the inner limit of the medial vascular smooth muscle layer (MSM). The vessel lumen (L) is filled with nucleated red blood cells. Bar = 50 μm. (B) A coronary artery from a spawning chinook salmon showing a severe arteriosclerotic lesion, which significantly occludes the vessel lumen (devoid of red blood cells). The main tissue component of the lesion is intimal vascular smooth muscle (ISM), which has a different fiber arrangement than the MSM. Bar = 50 μm.

The etiology of coronary lesions in salmonids is unclear. Sexual maturation has been implicated based on the observations that lesions are found in mature fish (Robertson *et al.*, 1961; Schmidt and House, 1979) and that injections of sex hormones promote lesion formation (House *et al.*, 1979). However, lesions are well developed at least 1 year before maturation in Atlantic salmon (Farrell *et al.*, 1986a). Factors related to growth are also implicated in lesion development. The prevalence and severity of coronary lesions accelerates in parallel with rapid bodily growth in the ocean (Saunders *et al.*, 1992; Kubasch and Rourke, 1990). The correlation between growth rate and lesion formation may explain the observation that lesions are fewer and less severe in the slower-growing, land-locked species of salmonids compared with the migratory varieties. Diet also influences lesion formation (Farrell *et al.*, 1986), and correlations are found between plasma low density lipoproteins and lesions (Eaton *et al.*, 1984; Farrell *et al.*, 1986). The possibility exists that lesions form because of mechanical stresses imposed on the coronary artery by unusually large expansions and contractions of the compliant bulbus arteriosus on which the coronary artery lies. Preliminary studies show no lesions in coronary arteries that lie on the conus arteriosus of elasmobranchs (A. P. Farrell, 1992).

Van Citters and Watson (1968), for steelhead trout, and Maneche *et al.* (1973), for Atlantic salmon, concluded that lesions regress when individuals return to the sea for repeated spawning. This idea was challenged by Saunders and Farrell (1988), who reported that coronary lesions in Atlantic salmon continue to develop and parallel the resumption of growth after spawning.

III. CARDIAC PHYSIOLOGY

A. Cardiac Cycle

The cardiac cycle refers to synchronized contractions and relaxations of the cardiac chambers (Randall, 1970). The cycle consists of isometric and isotonic contractions (systole), during which the ventricle is emptied, and a period of relaxation (diastole), during which the chamber is refilled. Diastasis is a period when there is no blood flow into or out of a cardiac chamber. The electrical events (waves of muscular depolarization and repolarization), recorded as the electrocardiogram (ECG) can be related to contractile events, as revealed by intracardiac pressure and ventral aortic flow measurements. When comparisons are made between fish species (Fig. 6), important differences emerge. This section considers general and species-specific aspects of the ECG, cardiac contractility, and cardiac work.

1. *Electrocardiogram*

The ECG terminology is the same in fishes as in mammals. The P wave represents atrial contraction, the QRS complex, ventricular contraction, and the

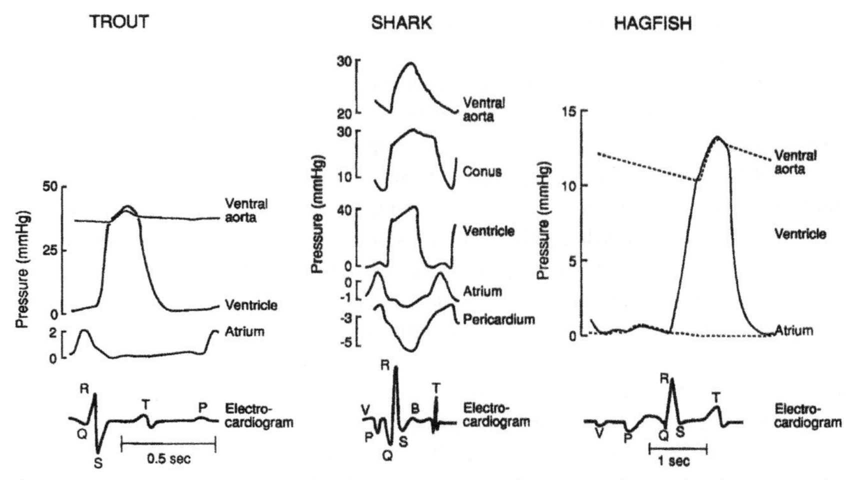

Fig. 6. Electrocardiograms and blood pressure traces during the cardiac cycle of representative fishes. Trout heart, adapted from Randall (1970) and Farrell (1991c). Shark heart, adapted from Satchell (1971). Hagfish heart, adapted from Davie *et al.* (1987).

T wave, ventricular relaxation. However, additional waves are present in fish compared with mammalian ECGs. Contraction of conal cardiac muscle in elasmobranchs produces a B wave during the S–T interval (Fig. 6). Contraction of cardiac muscle in the sinus venosus appears as a V wave during the P–T interval and is prominent in *A. anguilla* and hagfishes. Given the anatomical factors presented previously (a paucity of valves to prevent backflow of blood into the veins and the absence or limited amount of cardiac muscle in many species), it seems unlikely that the sinus venosus functions effectively as a contractile chamber to propel venous blood into the atrium, even though such a role has been ascribed to it in the past.

In the mammalian heart, coordination between the atrial and ventricular contractions is provided by specialized cardiac fibers, which first delay conduction to the ventricle (atrioventricular junctional fibers) and then speed up its propagation over the ventricle (fast-conducting fibers). An atrioventricular delay, as indicated by the P–Q interval, is evident in fish ECGs. This delay is important in allowing the complete contraction of the atrium before ventricular contraction starts. Atrioventricular conduction in some fish can be seriously compromised at low temperatures (Peyraud-Waitzenegger *et al.*, 1980). In perfused rainbow trout hearts, atrioventricular desynchrony can occur at 5 to 7°C (Bennion, 1968; Graham and Farrell, 1989), but is corrected with adrenergic stimulation (10 nmol adrenaline). This contrasts with the situation in *A. anguilla* at 8°C, in which stimulation of α-adrenoceptors with high concentrations of adrenaline (>5 μmol) actually causes atrioventricular desynchrony (Pennec and Peyraud, 1983).

The propagation of the action potential from the atrioventricular node to the various parts of the ventricle is coordinated and synchronized despite a lack of histological evidence for fast-conducting fibers in fish ventricles (Satchell, 1991). Whether the various layers of fiber bundles or the different myocyte diameters in the ventricle are important in the spread of excitation is unknown. At the level of individual myocytes, current flow between cells is affected by the degree of folding and number of gap junctions in the intercalated discs. Davie *et al.* (1987) noted that teleosts have intercalated discs that are less convoluted and have fewer gap junctions than those of higher vertebrates. Gap junctions may be absent or few in number in hagfish myocytes.

2. *Contractility*

Cardiac contractility is a term that expresses the vigor of contraction or, more specifically, the change in developed force at a given resting fiber length (Berne and Levy, 1988). An increase in fiber length above resting increases the force of contraction (Frank–Starling mechanism), but it does not increase contractility. Measures of contractility *in vitro* include peak isometric tension and maximal shortening velocity at a fixed initial length. Measurement of contractility *in vivo* is less precise. The dP/dt during the isovolumic phase of the cardiac cycle (isometric contraction) and initial velocity of blood flow in the aorta are used as indirect indices. Davie *et al.* (1987) noted that dP/dt values for ventricular contractions in teleosts ranged from 370 mm Hg/sec to 480 mm Hg/sec, about five times slower than those of mammalian hearts. On the other hand, tunas have higher heart rates and ventral aortic pressures, and dP/dt values are much higher compared with other teleosts, falling into the mammalian range (Jones *et al.*, 1992). In contrast, ventricular dP/dt for *Myxine glutinosa* and *Eptatretus cirrhatus* (approximately 22 mm Hg/sec) is more than 10 times slower than that for teleosts (Davie *et al.*, 1987). Ventricular dP/dt in the leopard shark (*Triakis semifasciata*) is also slow (25 mm Hg/sec), increasing to only 36 mm Hg/sec during swimming (Lai *et al.*, 1990a). The slow dP/dt in hagfish and elasmobranchs probably reflects low contractility and perhaps slow rates of electrical conduction between myocytes. Slow conduction rates would be expected in hagfish because of the lack of deeply penetrating T tubules in the SL and poor electrical couplings between myocytes (Davie *et al.*, 1987).

Neural (vagal and adrenergic nerve fibers), humoral, and local factors can increase (positive inotropy) and decrease (negative inotropy) cardiac contractility. Among the specific factors known to increase contractility of the atrium and/or ventricle in fishes are the following: increased temperature (Ask *et al.*, 1981; Bailey and Driedzic, 1990), β-adrenergic stimulation (see Ask *et al.*, 1981; Laurent *et al.*, 1983; Nilsson, 1983; Farrell, 1984; Vornanen, 1989), the peptides arginine vasotocin and oxytocin (Chui and Lee, 1990), adenosine (Lennard and Huddart, 1989), prostacyclin (Acierno *et al.*, 1990), and histamine

(Temma *et al.*, 1989). Negative inotropic effects on atrial or ventricular tissue occur with the following: hypoxia and acidosis (Gesser and Poupa, 1983; Farrell, 1984), acetylcholine (Randall, 1970; Holmgren, 1977; Cameron and Brown, 1981), α-adrenergic stimulation in some species (Tirri and Lehto, 1984), purinergic agents (Meghi and Burnstock, 1984), and adrenaline in combination with adenosine (Lennard and Huddart, 1989). Isolated atrial and ventricular tissues can have different sensitivities to pharmacological agents. For example, acetyl-choline has negative inotropic effects on the Atlantic cod atrium but not on the ventricle, whereas the ventricle is more sensitive to adrenergic stimulation (Holmgren, 1977). This implies that vagal innervation is confined to the atrial region. Interestingly, the affinity for β-adrenergic atrial stimulation is greater in *Platichthys flesus* and *Squalus acanthias* (fish without adrenergic innervation) than in rainbow trout (Ask, 1983).

Contractility, as measured by the force of contraction of cardiac muscle strips *in vitro*, is dependent on the duration of the active state (the period of contraction and relaxation) and its intensity (rate of contraction). The relationship between maximal isometric force and contraction frequency is documented for several species. In ventricular and atrial strips from hagfish (*Myxine glutinosa*) and a variety of teleost species, an increase in contraction frequency reduces the duration of the active state, decreases maximal isometric tension, and, at higher frequencies, the rate of contraction (Ask, 1983; Driedzic and Gesser, 1985; Vornanen, 1989). This inverse relationship between maximal isometric tension and contraction frequency for these teleosts and hagfish is referred to as a negative staircase effect. In other vertebrates, including ventricular tissue from mammals and several elasmobranch species (Driedzic and Gesser, 1988) and atrial tissue from skipjack tuna (Keen *et al.*, 1992), there is a positive relationship between contraction frequency and maximal isometric tension at low frequencies, whereas at higher frequencies the relationship is again negative. Thus, the force–frequency relationship has an apex, i.e., a contraction frequency at which maximal isometric tension reaches a peak. In elasmobranchs, the apices occur at contraction frequencies (0.3–0.4 Hz) that correspond to *in vivo* heart rates. Similarly, the apex for skipjack tuna (1.4–1.6 Hz) also corresponds to *in vivo* heart rates. Moreover, skipjack atrial tissue can contract up to a maximal frequency of 3.4 Hz (Keen *et al.*, 1992), a frequency that is well beyond those reported for other fish species. The relevance of these *in vitro* observations to the relationships between maximal isometric tension, contraction frequency, and the duration of the active state to the *in vivo* situation is further brought home by Vornanen's observation that the duration of the action potential is very similar to the duration of contraction.

In electrically paced cardiac strips *in vitro*, increasing temperature increases the rate of contraction, thereby increasing contractility (Ask, 1983). However, the duration of the active state is shorter, with increasing temperature through shorter contraction and relaxation times (Vornanen, 1989; Bailey and Driedzic, 1990), and this reduces maximal isometric tension.

Nonetheless, because the active state is shorter, the muscle strips can contract at higher frequencies at high temperatures. An additional effect of temperature is revealed in spontaneously beating cardiac strips *in vitro*; the negative staircase effect is less pronounced at high than at low temperatures because the higher heart rates result in a shorter active state. Some of the changes in cardiac contractility associated with temperature acclimation were identified by Bailey and Driedzic (1990). Among these were a pronounced reduction in the duration of relaxation (e.g., yellow perch, *Perca flavescens*) and a shorter-duration contraction (e.g., smallmouth bass, *Micropterus dolomieui*, and yellow perch) in cold-acclimated compared with warm-acclimated fish.

The positive inotropic effect of β-adrenergic stimulation is the result of increases in both the rate and duration of contraction in crucian carp and rainbow trout (Vornanen, 1989). Again positive chronotropy also caused by β-adrenergic stimulation offsets, but does not overcome, the positive inotropic effects (Ask, 1983). In contrast, adrenaline increased maximal tension independent of frequency in atrial strips from skipjack tuna; the duration of contraction increased with no change in the rates of contraction and relaxation (Keen *et al.*, 1992).

Cardiac contractility in fishes is dependent on extracellular calcium concentrations, reflecting the overall importance of transsarcolemmal calcium movements in the availability of activator calcium for excitation–contraction coupling (see Chapter 6). In several species of elasmobranchs and teleosts, including skipjack tuna, maximal isometric force increases severalfold with increasing extracellular calcium (1–9 m*M*) (Driedzic and Gesser, 1985; Driedzic and Gesser, 1988; Keen *et al.*, 1992). One consequence of the positive inotropic effect of extracellular calcium is to shift the relationship between maximal isometric tension and contraction frequency to the right (Driedzic and Gesser, 1988). Despite marked *in vitro* effects on contractility by extracellular calcium, *in vivo* effects may be quite limited because a reasonably good homeostatic mechanism maintains extracellular calcium levels above the threshold for major cardiac effects. An increase from 1 m*M* to 2 m*M* extracellular calcium substantially increased $\dot{Q}$ and pressure development in isolated perfused Atlantic cod hearts (Driedzic and Gesser, 1985). However, increases in extracellular calcium above the threshold level of around 1.5 m*M* resulted in very modest increases in maximal cardiac power output of *in situ* perfused heart preparations (Farrell *et al.*, 1986a).

3. Cardiac Stroke Work and Power Output

Pressure–volume loops are presented in Fig. 7 for rainbow trout and leopard shark at rest and during exercise. The area of the pressure–volume loop is proportional to the external work performed by the ventricle. This is termed ventricular stroke work (units of joules). During exercise the ventricle clearly performs more work as a result of higher pressures and stroke volumes.

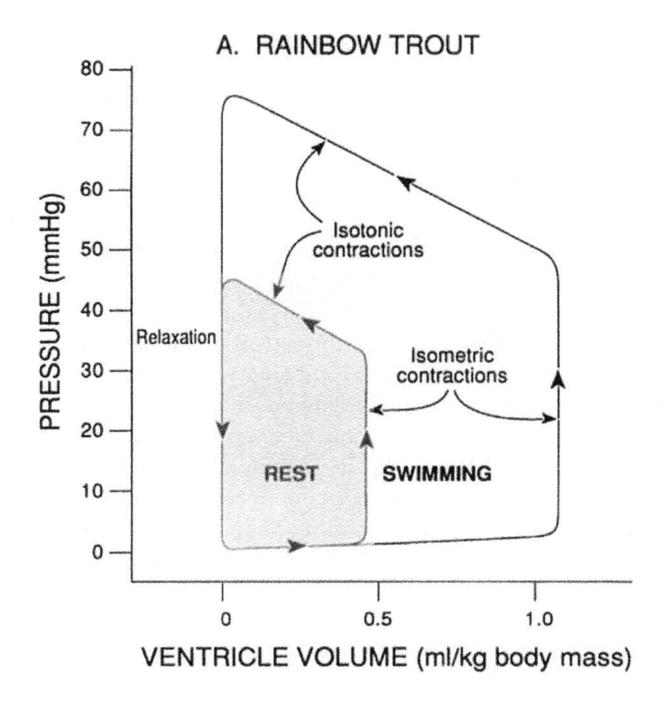

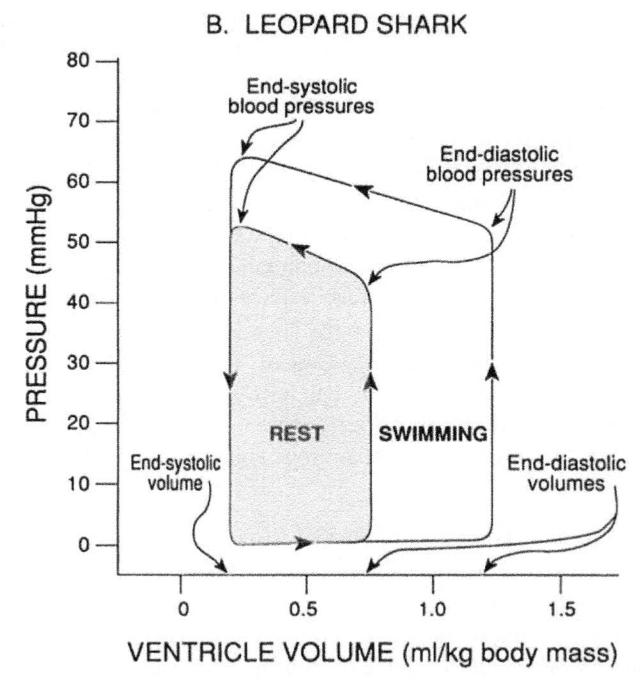

Fig. 7. Pressure–volume loops based on data from resting and exercising rainbow trout and leopard sharks. The area of the pressure–volume loops reflects cardiac work. With swimming, substantial increases in both stroke volume and blood pressure raise the work performed by the heart.

However, usually stroke work is not obtained from pressure–volume loops but estimated as the product of stroke volume times mean driving pressure (ventral aortic pressure minus central venous blood pressure). If information on central venous blood pressure is absent, using a venous blood pressure of zero probably introduces an error of no more than 5% in the calculation.

Myocardial power output is the product of stroke work times heart rate, or alternatively, the product of $\dot{Q}$ and ventral aortic pressure (units of watts). Stroke work and power output are often expressed per gram ventricular mass in fishes because ventricular mass *per se* is quite variable.

Comparisons of power output, such as those presented in Table II, are useful because power output measures the integrated performance of the heart (flow work and pressure work). A number of general points can be drawn from these data. The highest and lowest myocardial power output values are for tuna and hagfish, respectively, with the difference being over sixfold. With the exception of hagfish and a high value for *Scyliorhinus canicula*, most temperate water species have resting power output values in the range 0.9–1.8 mW/g ventricular mass a range which is similar to that for Antarctic species. Power output can increase two- to fourfold with exercise or adrenergic stimulation. The maximal power output observed for elasmobranchs and sluggish species of teleosts is about half that reported for the more active salmonids (3.2 mW/g versus 6–7 mW/g). Maximal power output in rainbow trout is only 30–50% greater than the resting value for tuna. A threefold increase in the power output values reported for tuna in Table II would not be unexpected (i.e., a twofold increase in $\dot{Q}$ and a 50% increase in ventral aortic pressure). This would mean that maximal power output values for tuna are of the order of 15–18 mW/g as compared with 7.0 mW/g in rainbow trout, 3.3 mW/g in leopard shark, and 0.27 mW/g in Atlantic hagfish. The physiological and biochemical basis for this large range in power output per unit mass of ventricle warrants attention.

The slow heart rates of Antarctic fish result in the highest values for resting stroke work (Table II). Conversely, the fast heart rates in tuna result in a stroke work that is not so different from that of other fishes, even though power outputs in tuna are the highest reported.

The impact of a large relative ventricular mass on overall cardiac performance can be assessed by expressing cardiac power output as a function of body mass rather than ventricular mass and comparing data for skipjack tuna and rainbow trout. The resting power output for skipjack tuna heart (~26 mW/kg body mass) is more than 10 times the resting power output for rainbow trout heart (~2 mW/kg body mass) (Farrell *et al.*, 1992). Since a large ventricular mass increases total myocardial power output, a large ventricular mass will compensate for reduced contractility associated with cold temperatures. This sort of compensation may explain why cardiac hypertrophy is associated with cold-acclimation in fish and why Antarctic fishes have relatively large ventricles (Table I).

Table II Cardiac Work and Power Output in Selected Fishes at Rest and while Swimming

	Stroke work (mJ/g)		Power output (mW/g)		Temperature (°C)	Body mass (kg)	Source[a]
	Rest	*Exercise*	*Rest*	*Exercise*			
Temperate water fishes							
Myxine glutinosa[b]	0.21	0.62	0.08	0.27	11	0.08	(a)
Triakis semifasciata	2.00	3.59	1.71	3.30	14–24	1.93	(b)
Scyliorhinus stellaris	2.00	3.21	1.43	2.46	19	2.6	(c)
Scyliorhinus canicula	6.50	—	2.07	—	14	0.5–1.0	(d)
Gadus morhua	2.47	3.87	1.77	3.29	10.5	0.4–0.8	(e)
Ophiodon elongatus	2.36	4.28	1.18	3.21	10	4.2	(f)
Hemitripterus americanus	1.54	3.54	1.16	3.13	10	1.2	(g)
Anguilla australis[b]	1.09	1.09	0.96	0.98	16–17	0.62	(h)
Anguilla australis	1.29	1.88	1.08	2.19	16–20	0.9–1.1	(i)
Oncorhynchus mykiss	2.41	8.27	1.53	7.03	11	1.0	(j)
Oncorhynchus kisutch	2.44	8.96	1.22	5.97	5	1.4	(k)
Tropical fishes							
Katsuwonus pelamis	2.23	—	4.70	—	26	1–2	(l)
Thunnus albacares	3.46	—	5.60	—	26	1–2	(l)

Antarctic fishes							
Chaenocephalus aceratus	4.02	—	0.98	—	0.5–2	1.0	(m)
Chaenocephalus aceratus	6.60	—	1.80	—	1.2	2.0	(n)
Pseudochaenichthys georgianus	5.32	—	0.72	—	0.5–2	1.0	(m)
Chionodraco hamatus[c]	—	4–6	—	1.6–3.4	0–2	0.29–0.47	(o)
Pagothenia borkgrevinki	5.58	6.0	1.05	2.00	0	0.06	(p)
Pagothenia bernacchii	4.69	—	0.82	—	0	0.05	(p)

[a](a) Axelsson *et al.* (1990) and Forster *et al.* (1991); (b) Lai *et al.* (1989a,b, 1990a); (c) Piiper *et al.* (1977); (d) Short *et al.* (1977); (e) Axelsson and Nilsson (1986); (f) Farrell (1982); (g) Axelsson *et al.* (1989) and Farrell *et al.* (1985); (h) Davie and Forster (1980); (i) Hipkins (1985); (j) Kiceniuk and Jones (1977); (k) Davis (1966); (l) Bushnell (1988); (m) Hemmingsen and Douglas (1977); (n) Hemmingsen *et al.* (1972); (o) Tota *et al.* (1991); (p) Axelsson *et al.* (1992).

[b]Maximum values for postadrenaline infusion.

[c]Maximum values in perfused heat preparations. Where ventricle mass is not known 0.2 g/kg body and 0.08 g/kg body were assumed for elasmobranch and teleost species.

4. *Efficiency of Cardiac Contraction*

Efficiency is the quotient of external work divided by the total energy transformed and, like that of most mechanical and biological systems, the efficiency of the fish heart is low (Farrell *et al.*, 1985; Farrell and Milligan, 1986). Hence, only a small proportion of cardiac O_2 consumption appears as external work and, *up to a point*, functional or design strategies that improve external work performance will have relatively little impact on cardiac energetics. The rider, "up to a point," is included in the above statement because, theoretically, both mechanical and biological systems increase efficiency with increasing power output from idle or rest. Hence, it is possible for more external work to be performed with little or no reflection of this in increased cardiac O_2 uptake. In this respect, Houlihan *et al.* (1988) report no increase in mechanical efficiency from the resting value of 21% for perfused rainbow trout hearts. In contrast, others have reported resting mechanical efficiency to be 11–15%, increasing to a maximum of 25% at higher power outputs (Farrell *et al.*, 1985; Graham and Farrell, 1990). Really, it depends on where you start, because at subnormal power outputs, mechanical efficiency is greatly reduced (4% in sea raven; Driedzic *et al.*, 1983). Hence, given that efficiency changes in fish hearts are small, then a general prediction is that more active fishes with larger hearts, higher $\dot{Q}$'s and higher blood pressures also have a greater myocardial oxygen consumption per unit time $(\dot{V}O_2)$. Furthermore, reasonably accurate predictions of myocardial $\dot{V}O_2$ will be possible from values of $\dot{Q}$ and ventral aortic pressure.

The majority of myocardial $\dot{V}O_2$ is utilized for developing tension to generate blood pressure, so anything that decreases cardiac tension (e.g., a fall in ventricular volume or blood pressure) or the time over which the heart is contracting (e.g., decrease in heart rate) will bring about an improvement in cardiac efficiency. As a consequence, if ventricular work is increased in mammals by raising mean aortic pressure (pressure work), at constant stroke volume and rate, then the increase in external work is paralleled by an increase in myocardial $\dot{V}O_2$. Obviously, efficiency is unaltered. In contrast, if work is increased a like amount by increasing stroke volume alone (flow work), then myocardial $\dot{V}O_2$ increases only slightly owing to a large increase in efficiency (Sarnoff, 1958a,b). However, in fish, both pressure and flow work increase myocardial $\dot{V}O_2$ by similar amounts for similar increases in external work (Farrell, 1984), so the increase in wall tension in enlarged fish ventricles must offset gains in efficiency due to increased external work, *per se*.

As noted previously, external work is obtained as the area enclosed by the ventricular pressure–volume loop. More usually, this value is approximated as the product of mean aortic pressure and flow, which gives values within 20% of those obtained from pressure–volume loops. Using a colorific equivalent, power outputs can be converted to O_2 cost, and comparing this value to measured values of myocardial $\dot{V}O_2$ yields the efficiency of cardiac contraction. Unfortunately, making measurements of myocardial $\dot{V}O_2$ in intact fish is next to impossible owing to the variable development of the coronary system, lack of a coronary sinus, and difficulty in assessing O_2 uptake from venous blood flowing through the heart.

Consequently, studies with working perfused fish hearts have provided much of the basis for our understanding of myocardial $\dot{V}O_2$ in fish. Since myocardial $\dot{V}O_2$ is directly proportional to cardiac power output under aerobic conditions (Fig. 8), then myocardial $\dot{V}O_2$ appears to be similar in all hearts, being around 0.3 μl O_2/sec per mW of cardiac power output (Davie and Farrell, 1991a). Sea raven and *Eptatretus cirrhatus* hearts (type I) use 0.29 and 0.42 μl O_2/sec per mW, respectively (Farrell *et al.*, 1985; Forster, 1991). Rainbow trout hearts (type II) use 0.26–0.28 μl O_2/sec per mW (Houlihan *et al.*, 1988; Graham and Farrell, 1990). *Squalus acanthias* hearts (type III) use 0.30 μl O_2/sec per mW (Davie and Franklin 1992).

Assuming a myocardial $\dot{V}O_2$ of 0.3 μl/sec per mW for other species, it is possible to calculate the O_2 cost of cardiac pumping in tuna and Antarctic hemoglobin-free fishes, two species with high $\dot{Q}$ values. Using a total resting $\dot{V}O_2$ value of 242 μl O_2/sec kg body mass for skipjack tuna (Bushnell and Brill, 1992) and a cardiac power output of 4.7 mW/g ventricular mass

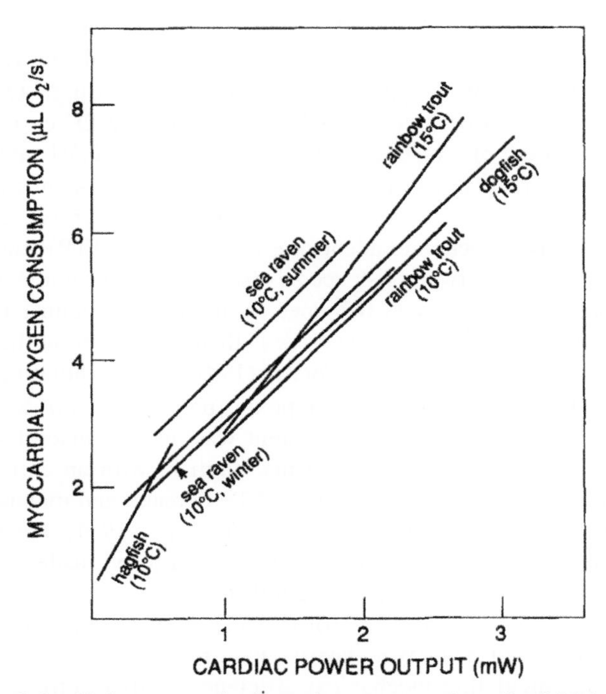

Fig. 8. The relationship between cardiac $\dot{V}_{O_2}$ and power output for perfused fish hearts. The data were obtained from the following sources: dogfish, Davie and Franklin (1992); hagfish, Forster (1991); sea raven, Farrell *et al.* (1985); rainbow trout, Graham and Farrell (1989).

(Table II), the 4-g ventricle in a 1-kg skipjack tuna would develop 18.8 mW and consume 5.64 μl O_2/sec, representing 2.3% of total $\dot{V}O_2$. For a 1-kg *Chaenocephalus aceratus* using a total $\dot{V}O_2$ of 5.8 μl O_2/sec (MacDonald *et al.*, 1987) and a total cardiac power output of 3.0 mW (1.0 mW/g for a 3-g ventricle; Table II), myocardial $\dot{V}O_2$ is 1.2 μl O_2/sec, which in contrast is 23% of resting metabolism. This high O_2 cost of cardiac pumping may limit activity in *C. aceratus*, particularly since its oxygen supply must be obtained from the venous blood.

Estimates of the O_2 cost of cardiac pumping have been made for other species. Cameron (1975) estimated O_2 cost of cardiac pumping as between 0.08 and 4.0% of total $\dot{V}O_2$. Jones and Randall (1978) estimated myocardial $\dot{V}O_2$ in resting rainbow trout as 3.5% of total $\dot{V}O_2$, increasing to 4.5% at maximal prolonged swimming speed (U_{crit}). These early predictions are close to more recent values based on myocardial $\dot{V}O_2$ measurements. In sea raven, myocardial $\dot{V}O_2$ is approximately 0.6% of total resting $\dot{V}O_2$ (Farrell *et al.*, 1985). The O_2 cost of cardiac pumping represents 4.6% of total $\dot{V}O_2$ in rainbow trout, but *decreases* to 1.9% at U_{crit} (Farrell and Steffensen, 1987a) and is unlikely to limit swimming performance in rainbow trout, as was suggested earlier (Jones, 1971). The reasons for this decrease in O_2 cost are that the increase in $\dot{Q}$ provides around half of the increase in internal O_2 convection and that myocardial efficiency increases. In *E. cirrhatus*, myocardial $\dot{V}O_2$ for a resting power output of 0.2 mW is approximately 3% of total resting $\dot{V}O_2$ (2.77 uL O_2/s.kg body mass; Forster *et al.*, 1992). Therefore, the O_2 cost of cardiac pumping in a variety of temperate fishes ranges between 0.5% and 5.0%.

It is also important to recognize that anaerobic metabolism can contribute significantly to cardiac work in certain fishes under hypoxic conditions. Since the capacity to generate adenosine triphosphate (ATP) anaerobically does not vary greatly between fish species (see Chapter 5), but power output can vary considerably between fish species, the relative contribution made by anaerobic metabolism to cardiac work is then dependent on the total cardiac power output (Farrell, 1991a). Forster (1991), after examining the myocardial $\dot{V}O_2$ and lactate release by hypoxic, perfused *E. cirrhatus* hearts, reported that most of the ATP requirement was met by anaerobic metabolism (see also Forster *et al.*, 1991). Thus, hagfish, with an extremely low cardiac power output, have a low cardiac ATP demand and maintain resting cardiac function through glycolysis (Forster, 1989, 1991; Forster *et al.*, 1991). In most other fish, glycolysis would appear to be inadequate to support the much higher power outputs (Farrell, 1991a). For example, anaerobic metabolism contributes at best 25% of the ATP requirement during hypoxic exposure in perfused sea raven hearts. Failure to account for anaerobic metabolism in calculating mechanical efficiency leads to erroneously high estimates, e.g., Farrell *et al.* (1985).

B. Control of Stroke Volume

Cardiac output is the product of stroke volume and heart rate. A number of important differences exist between fish and mammalian hearts with respect to the control of stroke volume. Foremost, many fish can increase stroke volume to a much greater degree than heart rate (Farrell, 1991a; see also Section III,D). Second, atrial contraction is the primary, if not sole determinant of ventricular filling. Third, atrial filling is achieved through *vis-à-fronte* and *vis-à-tergo* mechanisms. (*Vis-à-fronte* filling directly utilizes some of the energy of ventricular contraction, "force from in front," whereas *vis-à-tergo* filling utilizes energy remaining in the venous circulation, "force from behind.") Fourth, end-systolic volume of the ventricle is quite small. Because of these differences, it is important to fully appreciate the control of stroke volume in fish. The following describes in some detail the Frank–Starling mechanism, atrial filling, and other factors that affect stroke volume.

In mammalian hearts, atrial contraction provides around 25% of ventricular filling, the majority occurring directly from the veins, with a small proportion due to elastic recoil of the ventricle. In contrast, it is generally held that atrial contraction is the sole determinant of ventricular filling in fish (Randall, 1970; Johansen and Burggren, 1980). This being the case, the fish heart needs to be viewed as two pumps (atrium and ventricle) in series. The volume pumped by each chamber is set by the end-diastolic volume minus the end-systolic volume (Fig. 7).

End-diastolic volume of the atrium is determined by the time available for filling and the volume and distensibility of the atrium. With *vis-à-tergo* filling, central venous pressure and perhaps sinus contraction provide the energy for atrial filling. With *vis-à-fronte* filling, the force of ventricular contraction will be an additional factor because of the effect it has on the negative (suction) pressure within the pericardium. Atrial end-systolic volume is determined by (1) the Frank–Starling mechanism, which will be described; (2) atrial contractility; (3) the atrioventricular delay, which sets ventricular filling time; and (4) ventricular volume and distensibility.

End-systolic volume of the ventricle is determined by the force of contraction, the diastolic output pressure in the aorta, and the timing of the closure of the ventricular outflow valves. Force of contraction is altered either through the Frank–Starling mechanism, or by factors that modulate contractility (see preceding sections). End-systolic volume of the ventricle is much lower in fish than in mammals. Normal contraction appears to almost completely empty the ventricle. Mean ejection fraction is estimated at 80% in the leopard shark (Lai *et al.*, 1990a). In rainbow trout, the ejection fraction is normally near 100% (Franklin and Davie, 1992). Thus, ventricular end-diastolic volume approximates atrial stroke volume. Clearly, modulation of atrial contractility and the Frank–Starling mechanism are particularly important in determining ventricle filling and hence cardiac stroke volume. In terms of the pressures developed, the atrium can be viewed as a preamplifier and the ventricle as a power amplifier.

An intrinsic property of cardiac muscle is that end-systolic volume of the ventricle is normally independent of mean aortic output pressure. This is termed homeometric regulation. Homeometric regulation has been quantified in perfused fish hearts, by measuring the heart's intrinsic ability to maintain resting stroke volume over a range for aortic output pressure (Farrell, 1984). As expected, the range is species specific and correlates well with *in vivo* ventral aortic pressures, i.e., hearts from more active fishes are capable of sustaining stroke volume at higher aortic pressures (Fig. 9A). However, at extremely high pressures homeometric regulation breaks down, the heart does not empty completely, end-systolic volume increases (Franklin and Davie, 1992), and stroke volume decreases (Fig. 9A). Factors that alter cardiac contractility also affect the maximal output pressure against which stroke volume can be maintained. The Antarctic fish *Chionodraco* is unusual in that even small increases in output pressure decrease stroke volume substantially in perfused heart preparations (Tota *et al.*, 1991).

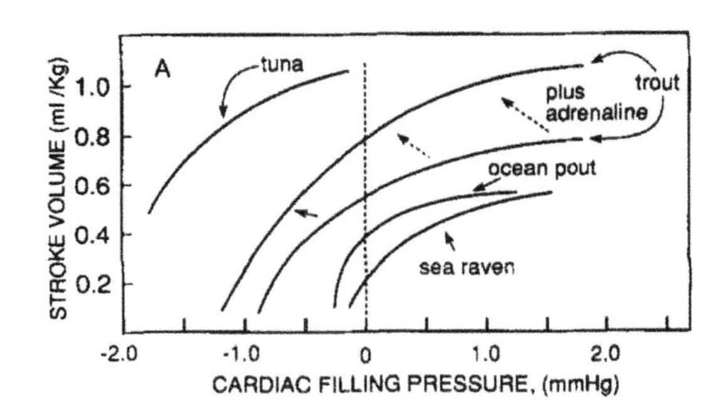

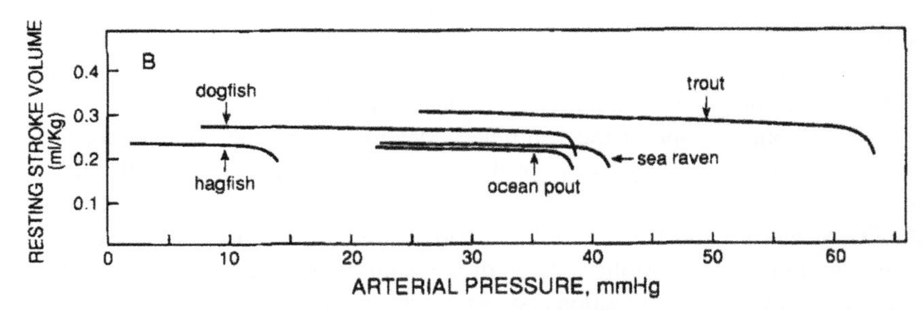

Fig. 9. (A) The effect of increased output pressure on the ability of fish hearts to maintain constant stroke volume (homeometric regulation). At resting $\dot{Q}$ levels, increases in output pressure have little effect on resting stroke volume over a physiological pressure range. The maximum pressure is species specific, being higher in active fish. (B) A general comparison of Starling curves obtained for perfused hearts from several teleost species. Note that, through *vis-à-fronte* filling, a large portion of the curve for rainbow trout is at subambient pressure. The positive inotropic effect of adrenaline is to shift the Starling curve for rainbow trout upward and to the left. From Farrell (1991a).

An anatomical constraint on stroke volume is the size of the ventricular chamber. This anatomical constraint is presumably reflected in the maximal stroke volume. Maximal stroke volume varies between fish species in the general range of 0.7 to 1.4 ml/kg body mass (Fig. 9B). Such a small interspecific range is perhaps surprising because fish hearts develop different blood pressures, and the size of the ventricle relative to body mass varies by more than 10–fold (Table I). Antarctic fishes are important exceptions in that hemoglobin-free and red-blooded species have unusually high stroke volumes, ranging from 2 to 10 ml/kg (Holeton, 1970; Hemmingsen *et al.*, 1972; Tota *et al.*, 1991; Axelsson *et al.*, 1992).

Stroke volume is not entirely independent of heart rate (Farrell, 1984, 1985). In isolated hearts of rainbow trout, increasing cardiac pacing frequency within the physiological range reduced maximal stroke volume to such a degree that $\dot{Q}$ did not increase at higher heart rates (Farrell *et al.*, 1989). This observation can be interpreted as either cardiac filling time being compromised, or the force of contraction being reduced at high heart rates (an expression of the negative staircase effect). Furthermore, there seems to be an antagonistic rather than synergistic interaction between increasing inotropic and chronotropic forces in yellowfin tuna. An increase in stroke volume of 30%, caused by injecting phenylephrine into the circulation of restrained fish, was accompanied by a 50% increase in the length of the systolic ejection phase. In fact, in one tuna, stroke volume reached 1.2 ml/kg, but heart rate fell from 112 to 105 beats per minute (bpm) (Jones *et al.*, 1992).

1. *Frank–Starling Mechanism*

With greater cardiac filling, the force of contraction must be greater to achieve complete cardiac emptying. According to the Frank–Starling mechanism of the heart, "the energy of contraction is a function of the length of the muscle fibre." Thus, increased end-diastolic volume results in a greater force of contraction and a greater stroke volume. This intrinsic property of muscle applies to the atrium and ventricle. Since an increase in cardiac filling pressure increases end-diastolic volume, the relationship between cardiac filling pressure and stroke volume is referred to as a Starling curve.

The Starling curve has functional significance in fish because stroke volume varies appreciably. Figure 9B shows a series of Starling curves for perfused hearts from teleosts. Similar curves have been obtained with *E. cirrhatus* (Forster *et al.*, 1991) and *S. acanthias* (Davie and Farrell, 1991b). These curves illustrate four important points. First, the *in vitro* range for stroke volume corresponds well with the *in vivo* range (see Table V). Second, the cardiac filling pressures *in vitro* are low, even subambient on occasions, and are comparable with central venous pressures *in vivo*. Third, only small changes in cardiac filling pressures (about 1–2 mm Hg) are needed to move along the Starling curve from resting to near maximal stroke volume. In this regard, perfused fish hearts are much more sensitive to filling pressure than perfused mammalian hearts, which often require filling pressures 10 times higher than those in fish to evoke maximal stroke volume. The greater sensitivity of the fish heart to filling pressure may reflect a combination of a highly distensible, thin-walled atrium and atrial contraction determining ventricular filling. Fourth, inotropic actions of neurohumoral factors create a family of Starling curves, making the heart more or less sensitive to filling pressure. For example, the positive inotropic effect of adrenaline increases the sensitivity of the rainbow trout heart to filling pressure, shifting the Starling curve to the left and upward (Fig. 9B). The negative inotropic action of acetylcholine probably shifts the Starling curve to the right and makes it flatter.

What is not clear in terms of the regulation of stroke volume *in vivo* is whether venous filling pressure, cardiac contractility at a constant filling pressure, or some combination of both are utilized to adjust stroke volume. Increased performance due to the Frank–Starling mechanism causes a shift along a given curve, whereas altered contractility causes a shift to another curve. Blood pressure in the cardinal vein increased significantly with swimming in the leopard shark (Lai *et al.*, 1990b). In rainbow trout, the increase in venous pressure at U_{crit} (maximal prolonged swimming speed), though not statistically significant (1.9 mm Hg; Kiceniuk and Jones, 1977), corresponds to the upper arm of the Starling curve. Hence these data suggest that increased venous pressure, alone, may make an important contribution.

2. Cardiac Filling

a. Atrial Filling. Atrial filling is achieved by one of two mechanisms: *vis-à-tergo* and *vis-à-fronte*. *Vis-à-tergo* atrial filling utilizes potential and kinetic energy that either remains in the venous circulation, or is generated by sinus contraction. Intuitively, the large unvalved communications between the Cuverian ducts and the sinus venosus should render contraction of the sinus cardiac muscle, when present, relatively ineffective at filling the atrium. Therefore, *vis-à-tergo* atrial filling is determined primarily by the pressure gradient set by the central venous blood pressure.

In contrast, *vis-à-fronte* filling uses some of the energy of ventricular contraction directly to distend the atrium and bring about atrial filling. This is achieved by subambient intrapericardial pressures, which create a negative atrial transmural pressure gradient. A rigid pericardial cavity is necessary to develop subambient intrapericardial pressures. The anatomical arrangements for a rigid pericardial cavity include a thick pericardial membrane (e.g., elasmobranchs) or a rigid musculoskeletal framework over which a relatively thin pericardial membrane is attached (e.g., active teleosts). An important consequence of *vis-à-fronte* filling is that central venous blood pressures will be subambient.

The relative importance of the *vis-à-fronte* and *vis-à-tergo* atrial filling mechanisms in fish has been examined with *in situ* perfused hearts. What has emerged from these studies is that benthic teleosts such as the sea raven and ocean pout use only *vis-à-tergo* filling. Above-ambient filling pressures are required for normal and elevated stroke volumes (Farrell, 1984, 1985). In contrast, elasmobranchs and active teleosts employ both *vis-à-fronte* and *vis-à-tergo* filling, with the *vis-à-tergo* mechanism increasing in importance at high stroke volumes. For perfused hearts from rainbow trout (Fig. 8A), *S. acanthias*, yellowfin tuna (Farrell *et al.*, 1992), *A. dieffenbachii* (Franklin and Davie, 1991), and *Seriola grandis* (P. S. Davie and C. E. Franklin, personal communication), the lower half of the range for stroke volume can be generated with subambient filling pressures (*vis-à-fronte* filling); but maximal stroke volume requires positive filling pressures (*vis-à-tergo* filling). Subambient intrapericardial pressures are reported for rainbow trout (Farrell *et al.*, 1988b), various elasmobranchs (Satchell, 1971; Shabetai *et al.*, 1985), and tuna (Lai *et al.*, 1987). Typically, intrapericardial pressure oscillates during the cardiac cycle with a sharp decrease in pressure during ventricular systole coincident with atrial filling (Fig. 10A). The observation that intrapericardial pressures increased when stroke volume increased in the leopard shark (Lai *et al.*, 1989a) is also consistent with the idea of transition from *vis-à-fronte* to *vis-à-tergo* filling at higher stroke volumes. When the pericardium is either punctured or is absent in perfused hearts from rainbow trout, *S. acanthias*, and *A. dieffenbachii*, intrapericardial pressure remains at ambient pressure, and the Starling curve is right shifted (Fig. 10B); above-ambient filling pressures are then needed for *vis-à-tergo* filling (Farrell *et al.*, 1988b; Davie and Farrell, 1991b; Franklin and Davie, 1991). Satchell and Jones (1967) report that puncturing the pericardium had no effect on $\dot{Q}$ in *Heterodontis portusjacksoni*.

Hence, the potential for *vis-à-fronte* filling exists, but whether this potential is realized, especially during activity, is unclear. For example, Lai *et al.* (1990a) suggest that atrial filling is *vis-à-tergo* in the leopard shark. They reported a triphasic filling pattern for the atrium, involving atrial relaxation, ventricular ejection, and sinus venosus contraction. Also cardinal sinus blood pressures were higher than pericardial pressures, which contrast with the typically subambient venous pressures recorded in other elasmobranchs (see Chapter 3, this volume). Furthermore, recent recordings from tuna show no evidence of large subambient intrapericardial pressures (Jones *et al.*, 1992).

Potential advantages and disadvantages of *vis-à-fronte* atrial filling are worth considering. First, subambient venous pressures, which *vis-à-fronte* filling permits, will tend to keep the stressed venous blood volume (see Part B, Chapter 3) at a minimum (Farrell *et al.*, 1988b). This may be particularly advantageous in fish with either fast circulation times (e.g., active teleosts), or capacious venous blood vessels (e.g., elasmobranchs). Second, higher aortic pressures may be possible with *vis-à-fronte* filling. Puncturing the pericardium to prevent *vis-à-fronte* filling reduced the maximal pressure that could be developed by perfused rainbow trout and *A. dieffenbachii* hearts

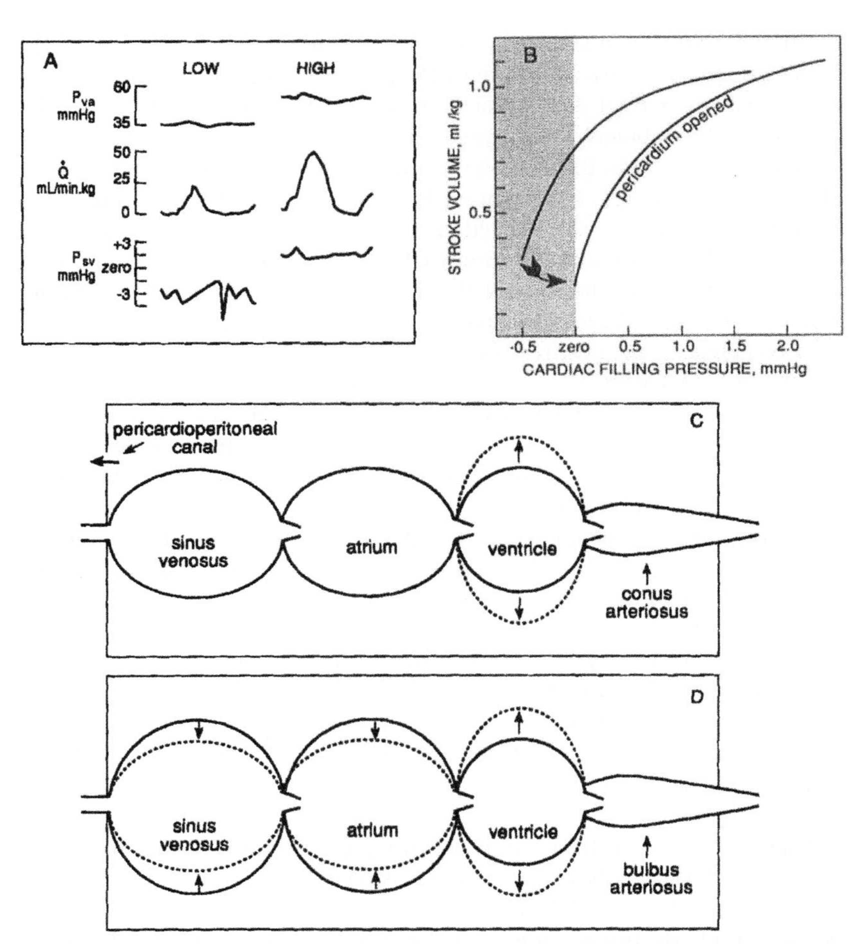

Fig. 10. (A) Cardiovascular measurements from a perfused rainbow trout heart during low (equivalent to a resting *in vivo* status) and high (elevated $\dot{Q}$ and diastolic output pressure) work conditions. The pressure in the sinus venosus (P_{SV}) is below zero with the low-work condition, but above zero with the high-work condition. From A. P. Farrell, unpublished observations. (B) The effect of puncturing the pericardium on the Starling curve of a perfused rainbow trout heart. Normally, filling pressures that are below zero account for 50% of the Starling curve, indicating a *vis-à-fronte* cardiac filling mechanism. When the pericardium is open, there is a right shift in the Starling curve, and filling pressures are above zero. Adapted with permission from Farrell *et al.* (1988b). (C) and (D) illustrate how certain fish modulate stroke volume even though they have a rigid pericardial cavity that facilitates *vis-à-fronte* filling. In each case the solid lines represent the end-diastolic volume when $\dot{Q}$ and stroke volume increase. (C) The situation in elasmobranchs. Atrial volume can be matched with ventricular volume, and, as ventricle end-diastolic volume and intrapericardial pressure increase, pericardial fluid is displaced from the pericardium into the peritoneum through the pericardio-peritoneal canal. (D) The probable situation in trout. The volumes of the sinus venosus and atrium are larger than the volume of the ventricle and act as variable-volume reservoirs. When ventricular end-diastolic volume increases, the end-diastolic volumes of the atrium and sinus venosus decrease. From Farrell (1991a).

(Farrell *et al.*, 1988b; Franklin and Davie, 1991). However, the ventricle has to generate pressure against a small negative intrapericardial pressure, which must reduce efficiency. Third, *vis-à-fronte* filling may be more rapid than *vis-à-tergo* filling since filling time is set by the duration of isometric ventricular contraction, rather than by the longer diastolic period. This may permit higher heart rates (Farrell, 1991b). In terms of disadvantages, the rigid pericardial cavity needed for *vis-à-fronte* filling has a finite volume, which theoretically should limit any increase in stroke volume. The fact that teleosts and elasmobranchs switch to *vis-à-tergo* filling at high stroke volumes is undoubtedly a reflection of this problem. Nonetheless, two- to threefold increases in stroke volume are common in fishes, and up to 50% of this change is possible with *vis-à-fronte* filling (Fig. 10B). There appear to be two solutions to the problem of a finite pericardial volume. Elasmobranchs have a pericardio–peritoneal canal, which allows a controlled and rapid removal of fluid from the pericardial cavity to peritoneal cavity (Shabetai *et al.*, 1985; Lai *et al.*, 1989a). Thus, when stroke volume increases, the ventricle can occupy a greater proportion of the pericardial volume by displacing pericardial fluid into the peritoneum (Fig. 10C). Pericardial fluid is replaced at a slower rate by secretion. Teleosts do not have a pericardio-peritoneal canal, and the near zero end-systolic volume of the ventricle exacerbates the problem (i.e., increases in stroke volume cannot draw on an end-systolic reserve of the ventricle). To accommodate an increase in stroke volume within the rigid pericardial cavity, the mechanism suggested for rainbow trout is that the atrium and sinus act as variable-volume reservoirs; their end-diastolic volumes are larger under low than under high stroke volume conditions (Fig. 10D). For this to occur, maximal end-diastolic volume of the sinus and atrium must be larger than that of the ventricle (i.e., maximal stroke volume), and this is the case (Farrell, 1991a). Tuna could circumvent the problem of a rigid pericardial cavity by increasing heart rate rather than increasing stroke volume to any significant degree (see the following), although recent recordings *in vivo* suggest that stroke volume can change by a factor of two (Jones *et al.*, 1992). Another problem with *vis-à-fronte* filling relates to the atrial and sinus venosus distensibility being quite similar, with the result that ventricular contraction would produce *vis-à-fronte* filling of the sinus venosus and would tend to short-circuit *vis-à-fronte* filling of the atrium. Tonic contraction of the cardiac muscle in the sinus venosus, or a sinus venosus with a low volume, as in elasmobranchs and tuna, could minimize this problem (Farrell, 1991a).

b. Ventricular Filling. Evidence for atrial filling being the sole determinant of ventricular filling comes from blood pressure measurements, which show an unfavorable pressure gradient for filling directly from either the sinus venosus or the veins during atrial diastole, and from angiographic images, which show no filling of the ventricle before atrial contraction (See Randall, 1970; Satchell, 1971; Johansen and Burggren, 1980). However, Lai *et al.* (1990a), working with the leopard shark and using echo-Doppler and angiographic imaging techniques, challenged the concept of atrial contraction as the sole determinant of ventricular filling. They provided evidence for biphasic ventricular filling. Blood first entered the ventricle during early diastole, probably as a result of elastic recoil, and was followed by a period of diastasis. A second surge of flow accompanied atrial contraction. The velocity–time integrals were similar for these two filling phases, suggesting that the relative flow contributions were similar.

C. Control of Heart Rate

Heart rate is set by the intrinsic rhythm of the sinoatrial pacemaker in the absence of extrinsic modulation. At least four mechanisms modulate the pacemaker rhythm to some degree and set the *in vivo* heart rate. These are stretch of the pacemaker cells, cholinergic nerve fibers, adrenergic nerve fibers, and hormones (Randall, 1970). The interplay of these effector mechanisms is species specific and temperature dependent. As will be evident from the following, temperature directly affects the pacemaker rate, alters the responses to modulators, and produces acclimatory changes. Positive chronotropy and negative chronotropy refer to increases (tachycardia) and decreases (bradycardia) in heart rate, respectively.

1. Intrinsic Heart Rate

It is very difficult to remove all extrinsic modulation of heart rate *in vivo* and measure intrinsic heart rate. However, based on an analysis of the various modulators (see the following), it is clear that the adrenergic and cholinergic controls are by far the most important. Thus, heart rate *in vivo* after β-adrenoceptor blockade (e.g., propranolol or sotalol) and muscarinic blockade (e.g., benzetimide or atropine) provides a good approximation of intrinsic heart rate. In fact there is good agreement between estimates of intrinsic heart rate obtained in this way and measurements of intrinsic heart rate in spontaneously beating isolated heart preparations. For example, a heart rate of 38 bpm in sea raven at 10°C after pharmacological blockade (Table III) compares well with an intrinsic heart rate of 39 bpm for a working perfused heart preparation (Farrell *et al.*, 1983). In view of this, the data base on heart rate after muscarinic and adrenergic blockade (Table III) is used to provide general insights into the factors that might affect intrinsic heart rate.

On the basis of the data presented in Table III, intrinsic heart rate is lowest in cyclostomes, higher in elasmobranchs, and highest in certain teleosts. In fact, a sevenfold difference in heart rate exists between hagfish and tuna. Furthermore, active fish tend to have higher heart rates within a phylogenetic grouping. There is not enough information to determine whether intrinsic heart rate in fish scales with a negative exponent of body mass. Dissimilar rates have been measured for closely related and similar sized (500 g) chinook salmon (*Oncorhynchus tshawytscha*) and rainbow trout, whereas heart rate in larger (3–6 kg) coho salmon (*O. kitsutch*) is lower than both of the former, indicating a potential for species as well as size-related differences.

Table III Estimates of the Intrinsic Heart Rate and the Relative Levels of Cholinergic and Adrenergic Tone which Set the Resting Heart Rate in Various Fish Species

Species	Cholinergic tone (%)	Adrenergic tone (%)	Ratio	Intrinsic[a] heart rate (bpm)	Temp. (°C)	Source[b]
Cyclostomes						
Myxine glutinosa	—[c]	50.8	—	17	10–11	(a)
Eptatretus cirrhatus	—[c]	86.1	—	14	17–18	(b)
Elasmobranchs						
Squalus acanthias	30.8	27.3	1.2	19	7–8	(c)
Squalus acanthias	—	—	—	40	15	(d)
Teleosts						
Pagothenia borchgrevinki	55.4	3.2	17.2	23	0–0.5	(e)
Pagothenia bernacchii	80.4	27.5	3.9	22	0–0.5	(e)
Chionodraco hamatus	—	—	—	26	0.6	(f)
Oncorhynchus kisutch	55.2	60.0	0.9	30	11–13	(g)
Oncorhynchus tshawytscha	15.4	38.8	0.4	54	8–9	(h)
Oncorhynchus mykiss	—	—	—	45	10	(i)
Gadus morhua	37.7	21.0	1.9	39	10–11	(j)
Hemitripterus americanus	34.8	39.6	0.9	38	10–12	(k)
Hemilepidotus hemilepidotus	17.4	19.9	0.9	47	8–10	(l)

Continued

Table III Estimates of the Intrinsic Heart Rate and the Relative Levels of Cholinergic and Adrenergic Tone which Set the Resting Heart Rate in Various Fish Species—Cont'd

Species	Cholinergic tone (%)	Adrenergic tone (%)	Ratio	Intrinsic heart rate (bpm)	Temp. (°C)	Source
Pollachius pollachius	19.7	33.2	0.6	38	11–12	(k)
Labrus mixtus	14.2	14.7	1.0	52	11–12	(l)
Labrus berggylta	33.9	15.4	2.2	48	11–12	(m)
Ciliata mustela	14.5	29.6	0.5	55	11–12	(m)
Raniceps raninus	12.5	28.8	0.4	28	11–12	(m)
Zoarces viviparus	18.9	67.1	0.3	39	11–12	(m)
Myoxocephalus scorpius	11.5	25.6	0.4	42	11–12	(m)
Carassius auratus	66.0	22.0	0.3	57	20–25	(n)
Macrozoarces americanus	—	—	—	56	10	(o)
Anguilla dieffenbachii	—	—	—	41	15	(p)
Seriola grandis	—	—	—	80	19	(q)
Katsuwanus pelamis	—	—	—	138	25	(r)
Thunnus albacares	—	—	—	123	25	(r)

[a]Intrinsic heart rate was estimated *in vivo* after pharmacologic blockade with atropine and a B-adrenergic antagonist. These values are indicated by the accompanying values for cholinergic and adrenergic tone. Other values of intrinsic heart rate are from perfused heart preparations.
[b](a) Axelsson *et al.* (1990); (b) Forster *et al.* (1991); (c) Holmgren *et al.* (1992); (d) Davie and Farrell (1991b); (e) Axelsson *et al.* (1992); (f) Tota *et al.* (1991); (g) Axelsson and Farrell (unpub.); (h) Thorarensen and Farrell (unpub.); (i) Farrell *et al.* (1991); (j) Axelsson (1988); (k) Axelsson *et al.* (1989); (l) Axelsson *et al.* (1992); (m) Axelsson *et al.* (1987); (n) Cameron *et al.* (1979a); (o) Farrell *et al.* (1983); (p) Davie *et al.* (1992); (q) Davie *et al.* (1992); (r) Farrell *et al.* (1992).
[c]No vagal innervation of the heart.

Intrinsic pacemaker rate varies directly with temperature (Table III) as a result of a direct effect on the membrane permeability of pacemaker fibers (Randall, 1970). Moreover, the factor by which rates change for an increase in temperature of 10°C (Q_{10}) for intrinsic heart rate is 2.0 or greater, indicating conformer-type responses to acute temperature changes. For *in vitro* working perfused sea raven hearts, the Q_{10} is 2.0 (Graham and Farrell, 1985) and is 2.0–2.3 in nonworking, isolated goldfish hearts (Tsukuda *et al.*, 1985). A similar Q_{10} of 2.0 is reported for *P. bernacchi* over a temperature range of 0°–5°C after β-adrenergic and muscarinic blockade (Axelsson *et al.*, 1992). In fact, similar Q_{10} values of 2.1 for *Scyliorhinus canicula* (Butler and Taylor, 1975), 2.1 for 15°C- and 25°C-acclimated *A. anguilla* (Seibert, 1979), and 2.5–2.9 for 10°C- and 24°C-acclimated sole (*Solea vulgaris*) (Sureau *et al.*, 1989) are also reported after only muscarinic blockade. Likewise, the Q_{10} for heart rate is 2.1 for 6.5°C- and 15°C-acclimated rainbow trout after bilateral vagotomy (Priede, 1974).

Temperature acclimation results in a compensatory change in pacemaker rate, so that cold-acclimated fish have a higher pacemaker rate than warm-acclimated fish at a given test temperature. For example, at the same test temperature pulsation rates *in vitro* (Tsukuda *et al.*, 1985) and heart rates *in vivo* (after muscarinic blockade) are lower in warm-acclimated compared with cold-acclimated *Carassius auratus, Solea vulgaris* and *A. anguilla* (Seibert, 1979; Sureau *et al.*, 1989). As a result, a Q_{10} value derived from the heart rates at the two acclimation temperatures is typically less than 2.0 ($Q_{10} = 1.46$ for bilaterally vagotomized rainbow trout, Priede, 1974; $Q_{10} = 1.73$ for muscarinically blocked *A. anguilla*, Seibert, 1979; $Q_{10} = 1.64$ for isolated goldfish hearts, Tsukuda *et al.*, 1985; $Q_{10} = 1.4$–1.9 for perfused sea raven hearts, Graham and Farrell, 1985; $Q_{10} = 1.4$ for perfused rainbow trout hearts, Graham and Farrell, 1989). The mechanism underlying the adjustment of pacemaker rate in response to temperature acclimation is not understood.

2. Resting Heart Rate

Resting heart rate is rarely the intrinsic pacemaker rate and, at a given temperature, is determined primarily by a "push-pull" type modulation as set by the relative levels of adrenergic and cholinergic tone on the heart (Farrell, 1991a). Numerous studies show that atropine injections in resting fish increase heart rate, from as little as 10 to 50% in rainbow trout, Atlantic cod, and dogfish (Priede, 1974; Short *et al.*, 1977; Taylor *et al.*, 1977; Wood *et al.*, 1979; Axelsson *et al.*, 1987), to as much as 175% in *A. anguilla* (Seibert, 1979) and 300% in red-blooded Antarctic fish (Axelsson *et al.*, 1992). All species examined so far, except for cyclostomes, have a resting cholinergic tone to the heart (Table III). In addition, all species have a resting adrenergic tone as revealed by β-adrenergic antagonists. In some species, e.g., goldfish, cholinergic tone is greater than the adrenergic tone, resulting in slow heart rates, whereas in others, the adrenergic tone is greater than the cholinergic tone, resulting in higher heart rates, e.g., *Zoarces viviparus*. Cholinergic tone and adrenergic tone are similar in some species, e.g., sea raven (Table III). Thus, resting heart rate can be slower than, faster than, or the same as the intrinsic pacemaker frequency, depending on the relative contributions of adrenergic and cholinergic tone.

3. *Cholinergic Control of Heart Rate*

Cholinergic fibers carried in the vagus are responsible for lowering heart rate. Increased cholinergic tone is the primary mechanism for abrupt bradycardia, as is often observed during exposure to hypoxia, at the initiation of burst swimming, and with visual and olfactory stimuli. Vagotomy or atropine injections abolish bradycardia. It is likely that acetylcholine, released from the large number of cholinergic nerve terminals in the sinoatrial region, hyperpolarizes pacemaker cells and slows the pacemaker rate (Saito, 1973). However, the specific action of neurotransmitter release on pacemaker cells in fish probably needs reexamining, since Hirst *et al.* (1991) find that in amphibian and mammalian cardiac cells, acetylcholine applied as a pharmacological agent causes pacemaker hyperpolarization whereas acetylcholine released as a result of vagal stimulation does not.

A cholinergic inhibitory tone exists in resting fish, so a reduction in the cholinergic tone (vagal release) causes tachycardia. Obviously, the greater the resting cholinergic tone, the greater the potential for tachycardia from vagal release. During swimming a twofold increase in heart rate is observed in *P. borchgrevinki* as a result of vagal release of resting cholinergic tone (Axelsson *et al.*, 1992), whereas in rainbow trout and *Scyliorhinus stellaris*, vagal release produces a much smaller tachycardia during exercise (Priede, 1974; Piiper *et al.*, 1977).

The cholinergic mechanisms that control heart rate are affected by temperature acclimation. For example, the level of cholinergic inhibition of heart rate is greater in cold-acclimated compared with warm-acclimated *Scyliorhinus canicula* and rainbow trout (Taylor *et al.*, 1977; Wood *et al.*, 1979). In contrast, the level of cholinergic inhibition of heart rate is lower in cold-acclimated *A. anguilla* (Siebert, 1979). Hydrostatic pressure also affects cholinergic control of heart rate. Heart rate in *A. anguilla* is elevated at a hydrostatic pressure of 101 bar. In addition, unlike the bradycardia observed at 1 bar hydrostatic pressure, tachycardia is normally observed with motor activity at a hydrostatic pressure of 101 bar, unless water temperature is >24.5°C, and then bradycardia occurs (Sebert and Barthelomy, 1985). These observations suggest a direct cardiac effect of hydrostatic pressure which is temperature dependent. Indeed, Gennser *et al.* (1990) report that twitch tension and frequency of atrial strips *in vitro* are both decreased by hyperbaric conditions at 8°–10°C, but twitch tension increases and frequency is unchanged with hyperbaric conditions at 16°–24°C.

4. ADRENERGIC CONTROL

The cardiac response to adrenergic stimulation depends to a large degree on the relative density of SL adrenoceptor subtypes, the sensitivity of the adrenoceptor, and the concentration and type of ligand (adrenaline and noradrenaline). The relative density of adrenoceptor subtypes affect the chronotropic response in a qualitative way, because α- and β-adrenoceptor subtypes, both of which are found in fish hearts, have opposite chronotropic effects; cardiac α-adrenoceptors mediate bradycardia, and β-adrenoceptors mediate tachycardia. The absolute density and sensitivity of the adrenoceptors, as well as the concentration and type of ligand, affect the chronotropic response in a quantitative way. For example, adrenaline and noradrenaline have different potencies for adrenoceptor subtypes and are coreleased in different ratios. Circulating levels of adrenaline and noradrenaline (nanomolar concentrations) are often similar (e.g., Butler *et al.*, 1978; Nilsson, 1983; Ristori and Laurent, 1985; Butler, 1986; Primmett *et al.*, 1986; Milligan *et al.*, 1989), but with stress and disturbance, these levels increase 10- to 100-fold (Nilsson, 1983; Butler, 1986). In cyclostomes, elasmobranchs, and dipnoans, either noradrenaline becomes the dominant hormone, or adrenaline and noradrenaline increase to similar levels. In contrast, in teleost fish, with the exception of *Cyprinus carpio*, adrenaline becomes the predominant circulating hormone. Adrenaline and noradrenaline are also coreleased from adrenergic nerve endings in fish hearts (in contrast to mammalian hearts), with adrenaline being the primary neurotransmitter (Cameron, 1979b; Pennec and Le Bras, 1984).

The majority of studies indicate that adrenergic stimulation of fish hearts causes positive chronotropy (Nilsson, 1983; Laurent *et al.*, 1983; Farrell, 1984; Butler and Metcalfe, 1988). Studies with *in vitro* cardiac muscle strips and perfused hearts suggest that positive chronotropy is mediated by β-adrenoceptors, probably the β_2 subtype (Ask *et al.*, 1980, 1981; Cameron and Brown, 1981; Farrell *et al.*, 1986b; Temma *et al.*, 1986; Davie *et al.*, 1992). There are, however, a few species in which α-adrenoceptors mediate negative chronotropy, and these include *S. acanthias* (Capra and Satchell, 1977), carp (Laffont and Labat, 1966; Temma *et al.*, 1989), perch (Tirri and Ripatti, 1982), and eel (Forster, 1976; Chan and Chow, 1976; Peyraud-Waitzenegger *et al.*, 1980). In perch, α-adrenoceptor-mediated negative chronotropy dominates over a very weak β-adrenoceptor-mediated positive chronotropy (Tirri and Ripatti, 1982). In *C. carpio*, however, α-adrenoceptor-mediated negative chronotropy is revealed only at high concentrations of noradrenaline, with positive chronotropy occurring at lower concentrations (Temma *et al.*, 1989). Cardiac α-adrenoceptors are apparently not involved in adrenergically mediated chronotropy in either rainbow trout or *Pleuronectes platessa* (Farrell *et al.*, 1986b; Temma *et al.*, 1986). Thus, α- and β-adrenoceptor subtypes may both be found in fish hearts, but both receptor types are not necessarily found in a single species. An additional confounding factor, which is not yet fully understood, is the impact of seasonal changes in temperature on adrenoceptor function. In rainbow trout, for example, the number of cardiac α- and

β-adrenoceptors and adrenoceptor sensitivity, but apparently not the concentrations of circulating catecholamines, vary with acclimation temperature (Ask, 1983; Graham and Farrell, 1989; Milligan *et al.*, 1989).

Adrenergic control of heart rate is possible through three means, which can be grouped phylogenetically (Laurent *et al.*, 1983; Nilsson, 1983): (1) endogenously, as in cyclostomes and dipnoans, which have chromaffin tissue within the heart (endogenous catecholamine stores); (2) exogenously, for all fish that have chromaffin tissue outside the heart (exogenous catecholamine stores), release of catecholamines being under sympathetic control in elasmobranchs and teleosts; and (3) neurally, as in the teleosts and holosteans that have adrenergic cardiac innervation. Bearing in mind these groupings, the relative importance of these three adrenergic control mechanisms will be examined.

a. *Endogenous Catecholamine Stores*. Of the three adrenergic mechanisms, the role of cardiac catecholamine stores is the most poorly understood (see Laurent *et al.*, 1983; Dashow and Epple, 1985). Nonetheless, based on the observations that *in vivo* injection of β-adrenergic antagonists reduces heart rate significantly in hagfish (Axelsson *et al.*, 1990; Forster *et al.*, 1992) and that depletion of catecholamine stores in hagfish with reserpine also depresses heart rate (Bloom *et al.*, 1961), a suggestion worth further study is whether endogenous catecholamines play a role in tonic cardiac stimulation in cyclostomes (Forster *et al.*, 1991).

b. *Circulating Catecholamines*. The low levels of circulating catecholamines in resting fish are probably responsible for the excitatory adrenergic tone observed in fish, with the possible exception of cyclostomes (see above). Nilsson *et al.* (1976) first suggested that the adrenergic tone on the heart was a result of circulating catecholamines, after noting that the threshold concentration for adrenergic stimulation of isolated cardiac muscle strips was similar to the blood concentrations of catecholamines. Since then several studies have found that nanomolar concentrations of adrenaline and noradrenaline stimulate perfused teleost and elasmobranch hearts and make them less prone to deterioration with time (Graham and Farrell, 1989; Davie and Farrell, 1991). Tonic adrenergic stimulation in rainbow trout may be particularly important for maintaining atrioventricular conduction at low temperature (Graham and Farrell, 1989).

Because the sensitivity of cardiac tissue to catecholamines matches the concentration range for circulating catecholamines, the 10- to 100-fold increase in circulating catecholamine concentrations under stressful conditions should produce tachycardia. Thus, elevated levels of circulating catecholamines may be responsible for the tachycardia following burst swimming (Farrell, 1982). They may also oppose the vagally mediated bradycardia observed with exposure to hypoxia, and may prevent the negative chronotropy produced by extracellular acidosis.

During prolonged swimming, circulating catecholamine levels do not increase appreciably except at the maximal prolonged swimming speed (U_{crit}), when there is a small component of burst swimming (Nilsson, 1983; Butler, 1986). The implication of this observation is that tachycardia observed during aerobic exercise is probably determined largely by vagal release rather than by additional adrenergic stimulation beyond the tonic level observed in resting fish. Even though circulating catecholamine levels increase at U_{crit}, heart rate has already reached a plateau (Priede, 1974; Kiceniuk and Jones, 1977). An anomalous observation is that in bilaterally vagotomized rainbow trout heart rate increased by 8% and 83% at 6.5°C and 15°C, respectively, with exercise (Priede, 1974). The tachycardia after vagotomy could reflect an involvement of adrenergic innervation other than that contained in the vagal trunk, an abnormal increase in circulating catecholamines during prolonged swimming, or some other positive chronotropic mechanism.

c. *Cardiac Adrenergic Innervation.* One of the major difficulties in separating the role of adrenergic nerves from that of circulating catecholamines is that catecholamine injections *in vivo* produce vasoconstriction and cross the blood–brain barrier, either of which can elicit cardiac reflexes that can mask any direct cardiac effect (Peyraud-Waitzenegger *et al.*, 1979; Wood and Shelton, 1980a,b; Jones and Milsom, 1982; Farrell, 1986). However, these roles have been separated in one study on the Atlantic cod (Axelsson and Nilsson, 1986). Blockade of adrenergic nerve activity with bretylium reduced heart rate in both resting and swimming fish by only 3 bpm (7%). Thus, chronotropic effects mediated by adrenergic nerves are minimal in Atlantic cod. Nonetheless, it would be worth studying fishes that have a more active lifestyle before making firm conclusions regarding the limited involvement of adrenergic nerves in terms of chronotropic effects. In this regard, it is interesting to note that in fishes lacking adrenergic cardiac innervation (e.g., elasmobranchs), tachycardia associated with swimming is usually quite small; a change of often less than 10 bpm in elasmobranchs (Lai *et al.*, 1989a,b).

5. *Mechanical Stretch of the Pacemaker Cells*

Pacemaker cells are stretched with increased cardiac filling and mechanical stretch of the pacemaker cells increases the intrinsic heart rate. In the functionally aneural heart of hagfish, stretch of the pacemaker cells may be an important mechanism to increase heart rate (Jensen, 1960, 1961). Even so, normal heart rate in hagfish probably varies by no more than 10–20% (Forster *et al.*, 1991, 1992). Mechanical stretch of the pacemaker has even less physiological significance in terms of modulating heart rate in elasmobranchs and teleosts. Physiological increases in cardiac filling pressure of perfused hearts have little, if any, effect on the intrinsic pacemaker rate and, furthermore, the effect is to decrease rather than increase heart rate by 1 to 2 bpm

(Farrell, 1984; Farrell *et al.*, 1986; Davie and Farrell, 1991a). Thus, the mechanism for mechanical modulation of heart rate has been superseded by neurohumoral control of pacemaker rate in phylogenetically advanced fishes. The mechanical effects on heart rate, reported previously for lampreys, elasmobranchs, and teleosts (Jensen, 1969, 1970) were observed with isolated perfused hearts performing a subphysiological work load and beating either at rates slower than the sinoatrial pacemaker rate (8 bpm and 24 bpm at 21°C in rainbow trout, and 12 bpm at 13°C in elasmobranchs), or not beating at all (lampreys). In view of these observations, it is possible that either depolarization rates of cardiac tissues other than the sinoatrial pacemaker are more sensitive to stretch than is the sinoatrial pacemaker, or the sinoatrial pacemaker can be *kick-started* by stretch.

6. *Other Factors Affecting Heart Rate*

In *A. japonica*, arginine vasotocin, oxytocin, metosotocin, and isotocin, but not arginine vasopressin, increase heart rate in isolated atria (Chui and Lee, 1990). Both metosotocin and isotocin are more effective at 28°C. Moderate levels of hypoxia do not have a direct affect on pacemaker frequency, but severe hypoxia and anoxia can lead to cardiac failure. Heart rate is unchanged when the hearts from sea raven, rainbow trout, *A. dieffenbachii*, and *S. acanthias* are perfused with hypoxic saline (Farrell *et al.*, 1985; Farrell *et al.*, 1989; Davie and Farrell, 1991b; Davie *et al.*, 1992). High extracellular calcium reduces intrinsic heart rate in sea raven and rainbow trout (Farrell *et al.*, 1983; Farrell *et al.*, 1986b).

7. *Maximal Heart Rate*

In mammals, an allometric relationship exists between heart rate and body mass so that resting and maximal heart rates are higher in smaller mammals. Presently there is no evidence for or against a similar allometric relationship in fishes. Instead, there is evidence to suggest that lower vertebrates, with the exception of tunas, have a range for heart rate which does not exceed 120 bpm (Farrell, 1991a). Examples of the highest heart rates reported for fish at high temperatures include 112 bpm for rainbow trout at 20°C (Wood *et al.*, 1979), 120 bpm for spangled perch, *Leipotherapon unicolor* at 30°C (Gehrke and Fiedler, 1988), 30 bpm for *C. carpio* at 25°C (Moffit and Crawshaw, 1983), 56 bpm for *A. japonica* at 22°C (Chan, 1986), and 80 bpm for *Seriola grandis* at 20°C (P. S. Davie and C. E. Franklin, 1991, personal communication). Thus, if an allometric relationship for heart rate does exist for fish, it would have to be compressed into a range that has a maximal value of 120 bpm or lower, depending on the species and water temperature.

Tuna have heart rates that exceed 120 bpm, the highest rates among lower vertebrates. Heart rates are 117–126 bpm in spinally blocked skipjack tuna (Bushnell, 1988; Bushnell *et al.*, 1989; Bushnell and Brill, 1991) and range from 91 to 172 bpm in perfused hearts from skipjack tuna and yellowfin tuna (Farrell *et al.*, 1992). Rates of 180 to 240 bpm have been measured in swimming tuna (Kanwisher *et al.*, 1974). What might be the modifications that allow tuna hearts to beat faster? Possibilities include the following:

1. Membrane permeabilities in ion-pumping activities. The SL membrane events associated with the action potential (the sodium leak, channel opening and closing, and pumps that restore ionic status) obviously match a given heart rate, but whether any of these SL membrane events are unusual in tuna is unknown. Bailey and Driedzic (1990) noted that cold-acclimation in yellow perch was associated with a more rapid rate of relaxation, suggesting some modification of SL calcium ATPase to improve maximal contraction frequency.
2. Calcium diffusion during excitation–contraction (E–C) coupling. Since the majority of calcium that binds to troponin-C (activator calcium) is derived from the extracellular space and passes through L-type calcium channels during excitation (see Chapter 6) the rate of diffusion of extracellular calcium to and from the innermost troponin-C could set a limit for contraction frequency in fishes. Skeletal muscle (mammalian and fish) and mammalian cardiac muscle, all of which contract at high frequencies, have intracellular calcium stores in the SR, bringing most of the activator calcium nearer to the troponin C. In tuna, SR calcium is involved to a significant degree in E–C coupling (Keen *et al.*, 1992). In contrast, hagfish, with exceptionally slow heart rates, appear to "protect" the extracellular calcium with an unusually thick glycocalyx, which surrounds the myocyte (Poupa *et al.*, 1985).
3. Conduction of electrical activity. Every myocardial fiber is activated with each heart beat, and so the anatomical factors affecting the transmission of the action potential may be different in tuna.
4. Oxygen supply. A high dP/dt in tunas coupled with a high heart rate increases not only the absolute myocardial O_2 demand but also the rate at which O_2 must diffuse through the myocardium. The presence of very high cardiac myoglobin concentrations, and the presence of and wider distribution pattern for the coronary circulation in tuna may mean that diffusion of O_2 from blood to mitochondria is not limiting heart rate. In fact, Poupa and Brix (1984) note that cardiac capillary density, blood P_{50}, and mitochondrial density are all high in fast-beating hearts.

D. Cardiac Output

Cardiac output ($\dot{Q}$) is the product of stroke volume and heart rate. Since the mechanisms controlling heart rate and stroke volume have been described, this section focuses on interspecific differences in $\dot{Q}$ and changes in $\dot{Q}$ in response to environmental perturbations.

The importance of $\dot{Q}$ in terms of internal oxygen convection is evident from a rearrangement of the Fick equation:

$$\dot{Q} = \dot{V}_{O_2}/A - \dot{V}_{O_2}$$

where $\dot{V}_{O_2}$ is the oxygen consumption and $A - \dot{V}_{O_2}$ is the difference in arterial–venous blood O_2 content. Differences in $\dot{Q}$ account for a major proportion of interspecific differences in $\dot{V}_{O_2}$. For example, there is a 75-fold difference between the routine $\dot{V}_{O_2}$ of *Eptatretus cirrhatus* and *Katsuwonus pelamis*, which is reflected in a 15-fold difference for $\dot{Q}$ and a five fold difference for $A - \dot{V}_{O_2}$ content (Farrell, 1991a; Table IV). Despite the obvious importance of $\dot{Q}$ in terms of O_2 transport in fishes, *in vivo* measurements of $\dot{Q}$ are far from comprehensive. The available $\dot{Q}$ values, arranged according to measurement technique, species, and temperature, are presented in Table IV. In order to evaluate this information, it is essential to first appreciate the technical limitations of these $\dot{Q}$ measurements.

1. Measurement

$\dot{Q}$ can be measured by direct or indirect techniques. Of the indirect methods used for measuring $\dot{Q}$, the Fick equation is the most common. In fact, two thirds of the $\dot{Q}$ values presented in Table IV are based on the Fick principle. Unfortunately, Fick estimates in fish are unreliable because errors introduced by the method can substantially (20–40%) overestimate or underestimate $\dot{Q}$ (Metcalfe and Butler, 1982; Dax-boeck *et al.*, 1982; Hughes *et al.*, 1982; Butler, 1988; Peyraud-Waitzenegger and Soulier, 1989). However, in some cases, the sum of the errors can be such that the Fick principle may accurately estimate $\dot{Q}$ (Randall, 1985).

Direct techniques most commonly use cuff-type or cannulating electromagnetic or Doppler-flow probes on the ventral aorta. In teleosts, the ventral aorta is often accessible, but limited space in active species favors the use of miniature Doppler probes over the more bulky electromagnetic probes. In elasmobranchs, posterior afferent branchial arteries form immediately after the conus arteriosus reaches the anterior pericardial wall. Therefore $\dot{Q}$ can be estimated only by using assumptions regarding flow distribution to the various gill arches because a flow probe must be placed on an unbranched segment of the ventral aorta between branchial arteries (Satchell *et al.*, 1970; Short *et al.*, 1977; Lai *et al.*, 1989b). Doppler and electromagnetic techniques have different technical merits. Electromagnetic flow probes can be calibrated accurately, but a zero level must be established regularly *in vivo* (because of the sizeable diastolic blood flow in the ventral aorta) by means of either a cuff-type occluder, if anatomy permits, or bradycardia (evoked by cholinergic antagonists or by frightening the fish). Therefore, the accuracy of the zero

Table IV Cardiac Output ($\dot{Q}$) for Fishes at Rest

Species	$\dot{Q}$ (ml/min · kg)	Temp (°C)	Method[a] (source)[b]	Species	$\dot{Q}$ (ml/min · kg)	Temp °C	Method[a] (source)[b]
Cyclostomes				*Pseudopleuronectes americanus*	15.5	5	i (30)
Eptatretus cirrhatus	15.8	17	d (1)		23.1	10	i
Myxine glutinosa	8.7	10–11	d (2)		41.8	15	i
Elasmobranchs				*Anguilla anguilla*	12.2	8.5–10.5	d (31)
Dasyastis sabina	47	21.5–24.2	i (3)		8.7	8.5–10.5	i
Raja rhina[c]	21.2–23.3	10	d (4)		11.5	15	d (32)
Scyliorhinus canicula	19.2	7	i (5)		9.3	15	i
	23.2	12			16.6–25.7	18–20	d,i (33)
	39.8	17			18.8	21–24	i (34)
	43.7	15	i (6)	*Anguilla australis*	9.1	15.5–18.5	d (35)
	32.1	14–16	i (7)		6.2–10.2	16–20	d (36)
Scyliorhinus stellaris	22.3	16	i (8)		10.4	16–20	d (37)
	52.5	18.5–19.3	i (9)	*Anguilla australis schmidtii*	10.9–11.3	16.6–17.0	i (38)
Squalus acanthias	24.8	—	i (10)	*Anguilla japonica*	12.8	21–23	i (39)
	25.5	—	i (11)	*Cyprinus carpio*	18.3	9–11	i (40)
	26.7	—	i (12)		20.7	15–16	i (34)

Continued

Table IV Cardiac Output ($\dot{Q}$) for Fishes at Rest—Cont'd

Species	$\dot{Q}$ (ml/min · kg)	Temp (°C)	Method (source)	Species	$\dot{Q}$ (ml/min · kg)	Temp °C	Method (source)
Squalus suckleyii	21	6–7	i (13)		36.5	24–25	i (41)
	32	9–10	i		24.6–34.2	25.1–25.7	i (42)
Triakis semifasciata[c]	33.1	14–24	d (14)	*Oplegnathus fasciatus*	36.6	—	i (34)
Antarctic teleosts				*Salmo gairdneri*	15–30	4–8	i (43)
Chaenocephalus aceratus[c]	61	0–1	i (15)		19.8	6	i (44)
Chaenocephalus aceratus	20–30	1–2	d (16)		38.7	12	i
	119	1–2	i		62.5	18	i
	66	0.5–2	d (17)		18.3	8.5–8.9	i (45)
	94–104	1–2	i		18.6	8.5–8.9	i (46)
Pseudochaenichthyes	50–87	0.5–2	d		17.6	9.0–10.5	i (47)
Pagothenia bernacchii	29.6	0	d (18)		34	13–16	d (48)
Pagothenia borchgrevinki	17.6	0	d		31.2–36.7	13–16	d (49)
Temperate water teleosts					65–100	14–15	i (50)

Gadus morhua	17–26	9–10	d (19)		45.9	14.8–15.2	i (51)
	29.1	10	d (20)		16	15–16	i (34)
	17.3	10–11	d (21)	*Tinca tinca*	14–18	12–14	i (52)
	19.2	10–12	d (22)	**Tropical teleosts**			
Hemitripterus americanus	10.8	7	d (23)	*Katsuwonus pelamis*	50–80	23–25	i (53)
	14.6	10.5	d		132.3	24–26	d (54)
	18.8	10–12	d (24)	*Thunnus albacares*	115	24–26	d
Myoxocephalus scorpius	27.7	15–18	i (25)	**Air-breathing fish**			
Ophiodon elongatus	10.9	10	d (26)	*Protopterus aethiopicus*	20	18	d (55)
	11.2	10	d (27)	*Electrophorus electricus*	40–70	28–30	d
	5.9	13	d (28)	*Amphipnous cuchia*	80	20–30	i (56)
Platichthys stellatus	45.1	9–11	i (29)	*Hoplerythrinus unitaeniatus*	27.6–32.2	26–30	d (57)
	93.5	19.5–20.0	i				

[a]Method of measurement: i, indirectly measured, d, directly measured.
[b](1) Forster *et al.* (1992); (2) Axelsson *et al.* (1990); (3) Cameron *et al.* (1971); (4) Satchell *et al.* (1970); (5) Butler and Taylor (1975); (6) Taylor *et al.* (1977); (7) Short *et al.* (1979); (8) Baumgarten-Schumann and Piiper (1968); (9) Piiper *et al.* (1977); (10) Robin *et al.* (1964); (11) Robin *et al.* (1966); (12) Murdaugh *et al.* (1965); (13) Hanson and Johansen (1970); (14) Lai *et al.* (1989a); (15) Holeton (1970); (16) Hemmingsen *et al.* (1972); (17) Hemmingsen and Douglas (1977); (18) Axelsson *et al.* (1992); (19) Jones *et al.* (1974); (20) Peterssen and Nilsson (1980); (21) Axelsson and Nilsson (1986); (22) Axelsson (1988); (23) Farrell (1986); (24) Axelsson *et al.* (1989); (25) Goldstein *et al.* (1964); (26) Farrell (1981); (27) Farrell (1982); (28) Stevens *et al.* (1972); (29) Watters and Smith (1973); (30) Cech *et al.* (1976); (31) Hughes *et al.* (1982); (32) Peyraud-Waitzenegger and Soulier (1989); (33) Motais *et al.* (1969); (34) Itazawa (1970); (35) Hipkins *et al.* (1986); (36) Hipkins and Smith (1983); (37) Hipkins (1985); (38) Davie and Forster (1980); (39) Chan (1986); (40) Garey (1970); (41) Itazawa and Takeda (1978); (42) Takeda (1990); (43) Stevens and Randall (1967); (44) Barron *et al.* (1987); (45) Cameron and Davis (1970); (46) Davis and Cameron (1971); (47) Kiceniuk and Jones (1977); (48) Wood (1974); (49) Wood and Shelton (1980a,b); (50) Holeton and Randall (1967); (51) Neumann *et al.* (1983); (52) Eddy (1974); (53) Stevens (1972); (54) Bushnell (1988); (55) Johansen *et al.* (1968); (56) Lomholt and Johansen (1976); (57) Farrell (1978).
[c]Assumptions made regarding distribution of blood flow.

level is a major concern. In contrast, Doppler-flow probes have a reliable "built-in" zero, but calibration is less accurate even when extreme care is taken. Thus, Doppler-flow probes are particularly useful when measuring relative changes in $\dot{Q}$. The recent development of ultrasonic flow probes may resolve some of these technical difficulties.

An additional concern is the state of the animal, which is not always clearly defined in the literature. Stress, spontaneous activity, and blood loss all will tend to elevate resting $\dot{Q}$. Also, opening the pericardium will be of concern given the importance of *vis-à-fronte* cardiac filling in certain teleosts. Hence, the following description of some of the factors that influence $\dot{Q}$ is at best tentative.

2. Body Mass

$\dot{V}_{O_2}$, $\dot{Q}$ and end-diastolic volume scale to body mass in mammals with an exponent of 0.75–0.77, whereas heart rate scales with an exponent of −0.25 (see Schmidt-Nielsen, 1989). This dogma for endothermic mammals may not be true for ectothermic fish. For example, the exponent for the allometric relationship between resting metabolic rate and body mass in the catfish (*Silurus meridionalis*) decreases with increasing water temperature, being 0.94 at 10°C and 0.75 at 30°C (Xiaojun and Ruyung, 1990). The significance of this in terms of $\dot{Q}$ needs to be established.

3. Activity

Resting $\dot{Q}$ varies by 15-fold among fish species ranging from hagfish to tuna (Table IV). Fish that display high levels of activity often have a higher resting $\dot{Q}$ than sluggish forms. For example, benthic fishes such as sea raven and lingcod (*Ophiodon elongatus*) have a resting $\dot{Q}$ near 10 ml/min/kg at 10°C, whereas 17 ml/min/kg is reported for rainbow trout at similar temperatures. Tuna have the highest $\dot{Q}$ values among fishes. In fact, the 132 ml/min/kg at 26°C reported for spinally blocked skipjack tuna is about half that of resting $\dot{Q}$ values reported for similar sized mammals at 37°C. Most of the 15-fold range in $\dot{Q}$ between hagfish and tuna is accounted for by the seven fold difference in heart rate (Table II).

Fish exercise anaerobically (burst exercise) for short periods. Burst exercise is often accompanied by decreases in heart rate, $\dot{Q}$, and blood pressure (Stevens *et al.*, 1972; Farrell, 1982). The bradycardia may begin even before an exercise bout begins (Farrell, 1982). The depression of cardiac activity during burst exercise may be related to preventing hypertension as the skeletal muscle contracts violently and closes off peripheral blood flow. Recovery from burst exercise is characterized by increases in $\dot{Q}$, heart rate, and ventral aortic blood pressure (P_{va}).

With aerobic swimming, increases in $\dot{Q}$ and $A-V_{O_2}$ increase internal O_2 convection by similar amounts (Kiceniuk and Jones, 1977; Jones and Randall, 1978; Farrell, 1991a). In response to prolonged swimming, $\dot{Q}$ increases by 47% in Atlantic cod, 64% in sea raven, 70% in *Scyliorhinus stellaris*, 70% in leopard shark, and threefold in rainbow trout (Table V). Brill and Bushnell (1991) estimate that tuna increase $\dot{Q}$ twofold. An anomalous observation is that $\dot{Q}$ does not change from resting values in swimming eels (Table V).

Information on maximal values for $\dot{Q}(\dot{Q}_{max})$ is available from work with *in situ* perfused heart preparations. The $\dot{Q}_{max}$ is determined by elevating heart rate with adrenaline in the perfusate and elevating stroke volume by mechanically increasing the filling pressure of the heart (e.g., Farrell *et al.*, 1988b; Milligan and Farrell, 1991). These data suggest that the $\dot{Q}_{max}$ for *in vitro* perfused heart studies is similar to the highest $\dot{Q}$ value observed with exercise *in vivo;* $\dot{Q}_{max}$ has been measured at 50 to 70 ml/min/kg for rainbow trout and 25 ml/min/kg for sea raven (Farrell *et al.*, 1983; Farrell *et al.*, 1986b; Farrell *et al.*, 1991), and 39 ml/min/kg in *S. acanthias* (Davie and Franklin, 1992). However, in *A. dieffenbachii*, $\dot{Q}_{max}$ (22 ml/min/kg) is about twice that observed *in vivo* (Davie *et al.*, 1992).

Whereas interspecific differences in resting $\dot{Q}$ can be largely accounted for by differences in heart rate, changes in stroke volume are important in bringing about increased $\dot{Q}$ during aerobic exercise. When both stroke volume and heart rate have been measured during aerobic exercise, the relative contribution of stroke volume is usually similar to or greater than that of heart rate (Table V). For example, during prolonged swimming the percentage change in stroke volume versus heart rate is, respectively, 200% versus 50% in rainbow trout (Kiceniuk and Jones, 1977), 55–63% versus 7–15% in *Scyliorhinus stellaris* (Piiper *et al.*, 1977), 33% versus 7% in leopard shark (Lai *et al.*, 1989b), 26% versus 18% in the Atlantic cod (Axelsson and Nilsson, 1986), and 25% versus 31% in the sea raven (Axelsson *et al.*, 1989). Farrell (1991a) notes that cardiac pumping in mammals, birds, reptiles, and amphibians is primarily frequency modulated and not volume modulated. Thus, the predominantly stroke volume-modulated increase in $\dot{Q}$ in response to exercise is different from other vertebrates. However, two notable exceptions to this trend exist among fishes; they are tunas and certain Antarctic fishes. The reported range for heart rate (90–240 bpm) in tuna would adequately account for the predicted twofold increase in $\dot{Q}$ (Brill and Bushnell, 1991; Farrell, 1991). In the red-blooded Antarctic notothenid (*P. borchgrevinki*), swimming in a swim-tunnel results in a twofold increase in heart rate with little change in stroke volume (Axelsson *et al.*, 1992). Interestingly, these two fish live at opposite ends of the temperature spectrum for fishes.

Table V Effect of Swimming Activity on Cardiac Output ($\dot{Q}$), Heart Rate, Stroke Volume, Ventral Aortic Blood Pressure (P_{va}), and Oxygen Uptake ($\dot{V}_{O_2}$)[a]

Species (mass)	Activity state	$\dot{Q}$ (ml/min · kg)	Heart rate (beats/min)	Stroke volume (ml/kg)	P_{va} (mm Hg)	$\dot{V}_{O_2}$ (ml O_2/min · kg)	Method (temp.)	Source[b]
Anguilla australis schmidtii (0.62 kg)	Rest	11.3	53.5	0.21	40	—	Indirect	(a)
	15 cm/sec	11.3	53.9	0.21	40	—	(17°C)	
Gadus morhua (0.35–0.73 kg)	Rest	19.2	30.5	0.49	—	—	Direct	(b)
	2/3 bl · sec^{-1}	30.2	42.7	0.61	—	—	(11°C)	
Gadus morhua (0.4–0.8 kg)	Rest	17.3	43.2	0.39	38	—	Direct	(c)
	2/3 bl · sec^{-1}	25.4	51.2	0.49	48	—	(10.5°C)	
Hemitripterus americanus (0.67–1.40 kg)	Rest	18.8	37.3	0.51	29	—	Direct	(d)
	30 cm/sec	30.9	49.1	0.64	36	—	(10–12°C)	
Salmo gairdneri (0.9–1.5 kg)	Rest	17.6	37.8	0.46	38	0.56	Fick	(e)
	41–63% U_{crit}	28.4	42.7	0.62	40	1.52	(9–10°C)	
	70–78% U_{crit}	34.8	49.0	0.70	50	1.90		
	81–91% U_{crit}	42.9	51.3	0.86	60	3.12		
	Maximum	52.6	51.4	1.03	—	4.34		

Salmo gairdneri (1.4–1.7 kg)	Rest	45.9	—	—	—	2.2	Fick	(f)
	30 min postexercise	71.9	—	—	—	3.1	(15°C)	
Scyliorhinus stellaris (2.8 kg)	Rest	52.5	43	1.21	25	1.30	Fick	(g)
	0.27 bl · sec^{-1}	89.2	46	1.94	26	1.90	(19°C)	
Triakis semifasciata (1–93 kg)	Rest	33.1	51.3	0.77	48	—	Direct	(h)
	0.3–0.7 bl · sec^{-1}	56.2	55.1	1.02	58	—	(14–24°C)	
	postexercise	60.4	49.5	1.22	55	—		
Pagothenia borchgrevinki (0.064 kg)	Rest	29.6	11.3	2.16	28	—	Direct	(i)
	1 bl · sec^{-1}	51.8	21.0	2.16	28	—	(0–1°C)	

[a]Swimming speeds are presented as cm/sec, bodylengths per second (bl · sec^{-1}), or a percentage of the maximum prolonged swimming speed (U_{crit}).
[b](a) Davie and Forster (1980); (b) Axelsson (1988); (c) Axelsson and Nilsson (1986); (d) Axelsson *et al.* (1989); (e) Kiceniuk and Jones (1977); (f) Neumann *et al.* (1983); (g) Piiper *et al.* (1977); (h) Lai *et al.* (1989a, 1990a); (i) Axelsson *et al.* (1992).

4. Temperature

Acute and chronic (or seasonal) changes in temperature are expected to affect $\dot{Q}$. Temperature effects can be assessed using Q_{10} values with a Q_{10} value of around 2.0 indicating temperature conformity and a Q_{10} value of 1.0 indicating complete temperature compensation. When subjected to an acute 5°C increase in temperature, Q_{10} values for resting $\dot{Q}$ were reported as 1.56, 2.12 and 2.35 for 5°C-, 10°C-, and 15°C-acclimated winter flounder, respectively (Cech *et al.*, 1976). In *S. acanthias* the Q_{10} values are 2.10 for both $\dot{Q}$ and heart rate for an acute temperature change from 7° to 17°C (Butler and Taylor, 1975). A Q_{10} value of 2.00 was reported with perfused hearts from the sea raven for the temperature range 3°–13°C (Graham and Farrell, 1985). In contrast, an acute increase in temperature from 0°–5°C produces only a 20% increase in $\dot{Q}$ in *P. bernacchi* (Axelsson *et al.*, 1992). Following temperature acclimation of winter flounder, a Q_{10} value of 2.70 was reported for resting $\dot{Q}$ values over the temperature range 5° to 15°C (Cech *et al.*, 1976). Following temperature acclimation of rainbow trout Barron *et al.* (1987) report a similar Q_{10} of 2.6 for resting $\dot{Q}$ over the temperature range 6° to 18°C. In contrast, the $\dot{Q}_{max}$ for perfused hearts from temperature-acclimated rainbow trout had a Q_{10} of 1.36 over the temperature range 5° to 15°C (Graham and Farrell, 1989). This observation raises the possibility that resting $\dot{Q}$ and $\dot{Q}_{max}$ are affected differently with temperature acclimation.

5. Acidosis

Extracellular acidosis occurs in a number of situations including exposure to hypercapnic and acid environments, and after burst exercise. Under these conditions, extracellular pH can decrease by as much as 0.5 pH units (see Wood *et al.*, 1977; Ruben and Bennett, 1981; Graham *et al.*, 1982; Milligan and Wood, 1986). The effects of extracellular acidosis on isolated cardiac muscle strips and perfused hearts have been reviewed previously (Gesser and Poupa, 1983; Farrell, 1984; Gesser, 1985). In general, extracellular acidosis has negative chronotropic and inotropic effects, with some species being more acidosis sensitive than others. An extreme extracellular pH of 6.8 to 7.1 reduces the force developed by ventricular muscle strips by as much as 55% in rainbow trout, perch, Atlantic cod, plaice, eel, and flounder (Gesser *et al.*, 1982; Gesser and Poupa, 1983; Gesser, 1985). Perfused hearts exposed to pH 7.4, as is more usual in extracellular acidosis, show a small decrease (10–20%) in resting $\dot{Q}$, primarily through negative chronotropy (Farrell, 1984). The negative inotropic effect of acidosis is better revealed at higher cardiac work loads. Figure 11A illustrates that at pH 7.4, $\dot{Q}_{max}$ for the perfused sea raven heart is reduced by more than 50% compared with that of controls at pH 7.9. Perfused rainbow trout hearts are more tolerant to a similar acidotic challenge (Fig. 11B); $\dot{Q}_{max}$ is higher and is reduced by only 15%. In the absence

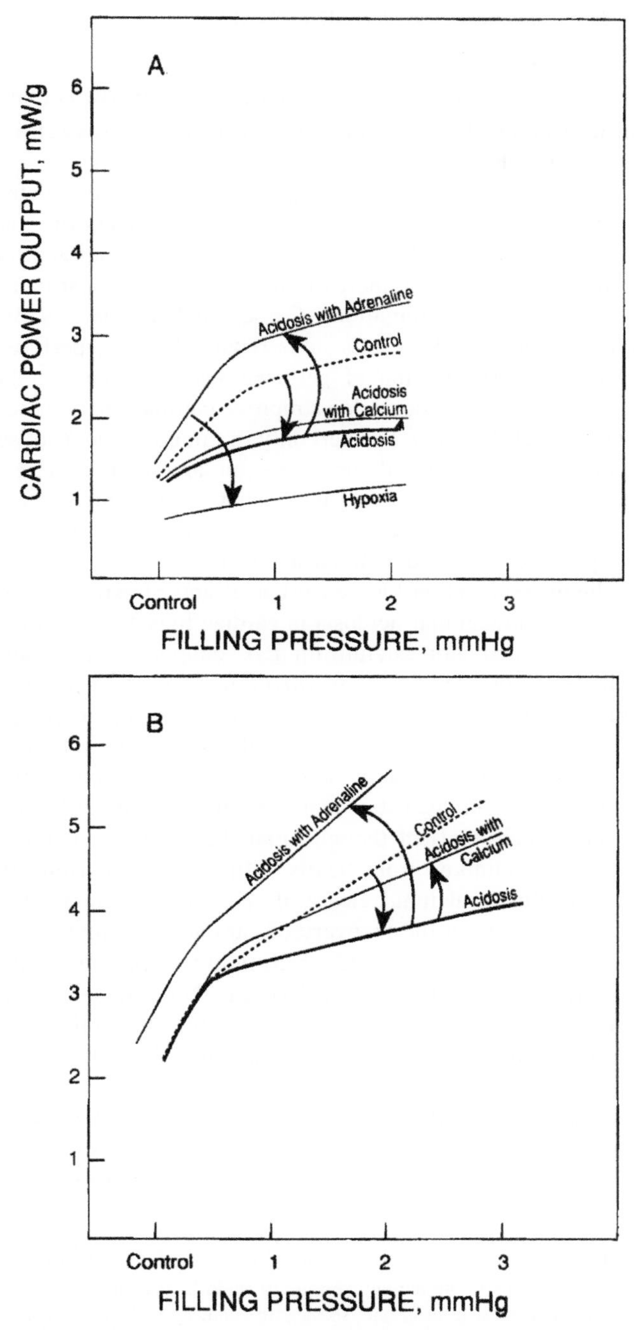

Fig. 11. The effect of hypercapnic acidosis on cardiac performance in perfused sea raven (A) and rainbow trout (B) hearts. Curves for cardiac power output versus cardiac filling pressure are presented. Since mean output pressure and heart rate change only slightly with the increases in filling pressure, the changes for $\dot{Q}$ and stroke volume were quantitatively similar. The sea raven heart is clearly more sensitive than the rainbow trout heart to extracellular acidosis, as indicated by the greater percentage

(Continued)

of *in vivo* data, a direct effect of extracellular acidosis *in vivo* is likely to be a reduction in maximum $\dot{Q}$, being more significant in acidosis-sensitive compared with acidosis-tolerant fishes.

An important observation made with perfused hearts and isolated muscle strips is that the negative inotropic and chronotropic effects of extracellular acidosis can be ameliorated. In isolated muscle strips, contractility can be restored during acidosis with an increase in either extracellular calcium concentration or adrenergic stimulation (Gesser and Poupa, 1983; Gesser, 1985). Somewhat different results are obtained with perfused hearts (Fig. 11). Adrenergic stimulation of perfused sea raven and rainbow trout hearts during extracellular acidosis does improve cardiac performance beyond even the control level (Fig. 11). Further, only in rainbow trout (not sea raven) does increasing extracellular calcium have an ameliorative effect during extracellular acidosis. Thus, it seems likely that *in vivo*, increased levels of circulating catecholamines protect the heart against the negative inotropic and chronotropic effects of extracellular acidosis.

The most likely explanation for the cardiac effects of extracellular acidosis is that it causes an intracellular acidosis in cardiac muscle, which results in a functional deficit of activator calcium for E–C coupling (Gesser and Poupa, 1983). This idea is supported by the *in vitro* effects on isolated muscle strips of increasing extracellular calcium. Furthermore, following adrenergic stimulation, there can be a greater influx of extracellular calcium into the myocardium since by activating cardiac β-adrenoceptors, the probability that SL calcium channels are in an open state is increased. [The possibility that adrenaline exerts its protective effect though intracellular pH regulation was discounted in perfused rainbow trout hearts (Milligan and Farrell, 1986).] In addition to extracellular calcium, release of calcium from intracellular stores is thought to be involved in the recovery of cardiac performance that occurs in acidosis-tolerant species during a 15- to 20-minute exposure to extracellular acidosis (Gesser, 1985). Sensitivity to acidosis is therefore set by factors that include the sensitivity of the myofilaments to calcium availability, the level of adrenergic stimulation, and the potential for calcium release from intracellular stores.

A problem does exist with the activator calcium deficit hypothesis, and it relates to the fact that myocardial intracellular acidosis is not necessarily observed *in vivo* even though intracellular acidosis is observed *in vitro* during

Fig. 11—Cont'd reduction in maximal power output. When physiological levels of adrenaline (0.1 μM) are added to the perfusate during exposure to acidosis, maximal cardiac performance is improved beyond the control level in both species. In contrast, doubling the concentration of extracellular calcium in the perfusate restores cardiac performance only for rainbow trout hearts. Exposure to hypoxic perfusate has a similar negative inotropic action on sea raven hearts. Data adapted from Farrell (1985) and Farrell *et al.* (1986b).

extracellular acidosis (Gesser and Jorgensen, 1982; Farrell and Milligan, 1986). For example, a marked extracellular acidosis is produced by strenuous activity in sea raven, rainbow trout, and starry flounder (*Platichthys stellaris*). However, during the first 30 min of this extracellular acidosis, ventricular intracellular pH increased significantly in sea raven, was not significantly changed in rainbow trout, and decreased in starry flounder (Milligan and Wood, 1986; Wood and Milligan, 1987). Based on these limited data, the level to which myocardial intracellular acidosis occurs *in vivo* appears to be greater in the acidosis-tolerant species of fish as measured by *in vitro* approaches.

6. Hypoxia

During environmental hypoxia, many fish can maintain $\dot{Q}$ if levels are moderate (water $P_{O_2} > 70$ torr). Moderate to severe hypoxia (water $P_{O_2} = 40 - 70$ torr) causes reflex bradycardia, but in some fishes, $\dot{Q}$ is maintained because of a compensatory increase in stroke volume (e.g., rainbow trout, Holeton and Randall, 1967; *Scyliorhinus canicula*, Butler and Taylor, 1975) whereas in others, $\dot{Q}$ decreases even though stroke volume increases somewhat (e.g., lingcod, Farrell, 1982; *A. japonica*, Chan, 1986; *A. anguilla*, Peyraud-Waitzenegger and Soulier, 1989). Tuna are extremely hypoxia sensitive and, at a water P_{O_2} between 85 and 104 torr, hypoxic bradycardia develops, but there is no compensatory increase in stroke volume to maintain $\dot{Q}$ (Bushnell *et al.*, 1990; Bushnell and Brill, 1991). Hagfishes are exceptional in terms of hypoxic tolerance. *Myxine glutinosa* maintain $\dot{Q}$ and heart rate under severely hypoxic (water $P_{O_2} = 13 - 20$ torr) conditions (Axelsson *et al.*, 1990), whereas *Eptatretus cirrhatus* increase $\dot{Q}$ during hypoxia (water $P_{O_2} = 40$ torr) (Forster *et al.*, 1992).

The direct effect of hypoxia on cardiac tissue is negative inotropy (Gesser *et al.*, 1982). Factors that probably influence the overall cardiac response to hypoxia, but have not been fully examined, include the following:

1. The anaerobic ATP generating potential of the heart relative to total cardiac power output (see preceding sections).
2. The intrinsic ability of the contractile machinery to tolerate hypoxia. The eel myocardium, for example, has a remarkable intrinsic tolerance to hypoxia (Gesser *et al.*, 1982). In fact, the maximal capacity of perfused eel hearts is reduced by only 20% with a perfusate P_{O_2} of 11.5 torr (Davie *et al.*, 1992); rainbow trout hearts fail at a perfusate P_{O_2} of 40 torr (Farrell *et al.*, 1988).
3. The presence of a coronary circulation. Increased coronary blood flow during exposure to moderate to severe hypoxia may be important in maintaining oxygen delivery to the myocardium as partial pressure of oxygen in venous blood P_{vO_2} decreases (Farrell, 1984). In this way, $\dot{Q}$ could be maintained during hypoxic exposure.
4. Temperature and the normoxic level of cholinergic tone. At a high water temperature, hypoxia-induced bradycardia is greater in *S. canicula*

(Butler and Taylor, 1975). This may reflect increased myocardial O_2 demand and a lower normoxic level of cholinergic tone at higher temperatures. In *P. bernacchi* at 0°C, the normoxic level of cholinergic tone is near maximal, and bradycardia does not develop during hypoxia (water P_{O_2}=40 torr). Since the heart is already under an unusually high resting cholinergic tone, a further increase associated with hypoxia may be difficult (Axelsson *et al.*, 1992).

5. Involvement of higher cerebral centers. Despite the high hypoxic tolerance of the eel myocardium *in vitro*, $\dot{Q}$ is not maintained *in vivo* at water P_{O_2} below 40 torr (Chan, 1986). Therefore, it would appear that eels down-regulate resting $\dot{Q}$ during hypoxic exposure.
6. Humoral factors. Hypoxia is often accompanied by extracellular acidosis. The effects of hypoxia and acidosis are additive in terms of negative inotropy (Gesser *et al.*, 1982). However, circulating catecholamines increase during hypoxia to levels that should prevent or at least moderate this negative inotropy (Gesser *et al.*, 1982).

E. Myocardial O_2 Supply, and the Threshold Venous P_{O_2}

Since the majority of fish have a type I ventricle and spongiosa accounts for the major portion of type II ventricles, luminal O_2 is the predominant myocardial O_2 supply in fish. Coronary arteries supply O_2 to the compacta and to some extent to the spongiosa in ventricle types III and IV. The extent to which luminal blood obviates the need for a coronary circulation and the extent to which the coronary circulation supplements the luminal O_2 supply is not entirely clear (Davie and Farrell, 1991a).

Luminal blood provides such a high volume of O_2 relative to the myocardial O_2 demand (because the entire $\dot{Q}$ passes through the heart) that a measurable decrease in the O_2 content of blood leaving the heart, compared with that entering, is unlikely (Jones and Randall, 1978; Farrell, 1984). This means that venous O_2 content is unlikely to limit myocardial O_2 delivery. Instead, luminal P_{vO_2} which, to a large degree, sets the gradient for O_2 diffusion across the spongiosa, might limit the luminal O_2 supply under certain conditions. Whether or not a threshold P_{vO_2} exists has not been rigorously tested, but a number of observations that bear on the issue were examined by Davie and Farrell (1991a). They suggested a threshold P_{vO_2} of around 10 torr; the exact value of which would vary somewhat according to the exact distribution pattern of the coronary circulation, the level of O_2 demand as set by myocardial power output, the blood residence time in the heart as set by heart rate, the intracellular myoglobin concentration, and the hemoglobin O_2 dissociation curve. For example, at U_{crit}, when myocardial $\dot{V}_{O_2}$ is near maximal, the range of P_{vO_2} values is 9–16 torr (e.g., 16 torr in rainbow trout, Kiceniuk and Jones, 1977; 9 torr in leopard shark, Lai *et al.*, 1990b). During hypoxia, when cardiac performance is lower, P_{vO_2} is correspondingly lower at the point when $\dot{Q}$

decreases below a resting value (e.g., 7 torr in *S. canicula*, Short *et al*., 1977; 6–10 torr in rainbow trout, Holeton and Randall, 1967; Wood and Shelton, 1980b; 6 torr in *A. dieffenbachii*, Forster, 1985). As a means of experimentally lowering P_{vO_2} in a progressive fashion and to establish a threshold P_{vO_2} for rainbow trout, Steffensen and Farrell (1992) exposed coronary-ligated and sham-operated fish to stepwise decreases in the water P_{O_2} while they were swimming at 50% of U_{crit}. Ventral aortic blood pressure (P_{va}) was used as a measure of cardiac performance. The sham-operated fish swam until P_{vO_2} was 8.6 torr and the coronary-ligated fish that had only a luminal myocardial O_2 supply swam until P_{vO_2} was 6.8 torr (Steffensen and Farrell, 1992). Furthermore, at a water P_{O_2} of 60 torr when P_{vO_2} was 14.3 torr and 13.6 torr in sham-operated and coronary-ligated, respectively, P_{va} increased with the increasing levels of hypoxia in sham-operated fish, but decreased in coronary-ligated fish, suggesting earlier cardiac failure when there was no coronary circulation to supplement the luminal O_2 supply.

The problem of a P_{vO_2} threshold is particularly acute in the hemoglobin-free Antarctic fishes. The myocardial $\dot{V}_{O_2}$ of 1.2 μl O_2/sec.kg body mass in *Chaenocephalus aceratus* requires a high O_2 extraction from luminal plasma. Assuming a plasma O_2 solubility of 0.056 μl O_2/ml torr at 0°C, a P_{vO_2} of 30 torr and $\dot{Q}$ of 100 ml/min.kg body mass, venous plasma will supply 2.8 μl O_2/sec. kg body mass. Myocardial O_2 demand therefore uses 43% of the luminal O_2 supply. Clearly, myocardial O_2 supply in these fish will be jeopardized in situations when the luminal O_2 supply is reduced, e.g., a decrease in either P_{vO_2}, or $\dot{Q}$ or some combination.

A coronary circulation either reduces or eliminates the reliance of the heart on the luminal O_2 supply. In view of the possibility of a P_{vO_2} threshold, it is not surprising, therefore, that fish that encounter environmental hypoxia or are active swimmers (two situations in which P_{vO_2} decreases) have a coronary circulation. Nevertheless, there is no clear estimate of the relative contributions made by the luminal and coronary O_2 supplies to the O_2 requirement of the heart. Indirect information has been obtained from *in vivo* studies in which the coronary circulation is interrupted, from coronary perfusion studies *in vitro*, and from theoretical estimates of O_2 delivery via the coronaries. Three important points have emerged from studies with salmonids in which coronary blood flow has been stopped by surgical ablation or ligation of the main coronary artery (Daxboeck, 1982; Farrell and Steffensen, 1987b; Farrell *et al*., 1990a). First, the coronary circulation in rainbow trout and chinook salmon is not essential for short-term survival in captivity. Coronary-ligated fish can survive for several months. Second, the coronary artery can regrow around the ligation or ablation site. Angiogenesis is quite rapid, a matter of weeks, which presumably indicates a long-term survival value. Third, swimming is still possible after stopping the coronary O_2 supply, but U_{crit} is reduced by 10–33% (Farrell and Steffensen, 1987b; Farrell *et al*., 1990a) provided there is no coronary regrowth (see Daxboeck, 1982).

Four adaptations alleviate the problem of P_{vO_2} threshold.

1. A coronary circulation. This well-oxygenated blood supply to the heart supplements the luminal O_2 supply in all elasmobranchs, as well as active and hypoxia-tolerant teleosts.
2. High myocardial myoglobin concentration. Myoglobin facilitates O_2 diffusion under hypoxic conditions in fish hearts and is found in high concentrations in very active and hypoxia-tolerant species.
3. Hypoxia (and acidosis) tolerance. The myocardia of some teleosts and elasmobranchs are particularly tolerant to hypoxia and the attendant acidosis produced by anaerobic metabolism. Hagfish hearts *in vitro* can function under anoxic conditions, and *in vivo* P_{vO_2} is 5 torr after exercise (Wells *et al.*, 1986), or 2 torr during exposure to progressive hypoxia (Forster *et al.*, 1992).
4. Slow heart rates. Slow heart rates in Antarctic fishes, for example, increase the residence time of blood in the lumen of the heart.

Of the above adaptations, the development of a coronary circulation is particularly interesting. Fish are, in effect, at the evolutionary interface in the development of the vertebrate coronary circulation. The coronary O_2 supply in salmonids supplements rather than replaces the luminal O_2 supply, and appears to be important for supporting normal pressure development by the ventricle. Experimental support for this suggestion is provided by the observation that when coronary-ligated trout swim at 50% of their critical swimming speed and are exposed to progressive hypoxia to systematically lower venous P_{O_2}, they cannot develop normal ventral aortic pressures below a venous P_{O_2} of approximately 15 torr and, as a result, cardiac power output was reduced by an estimated 37% (Steffensen and Farrell, 1992). Similarly perfused hearts from rainbow trout, *S. acanthias* and *A. dieffenbachii*, cannot generate high output pressures when pumping hypoxic saline without coronary perfusion, but can maintain normal or elevated flow rates (Farrell *et al.*, 1989; Davie and Farrell, 1991b; Davie *et al.*, 1992). A higher cardiac performance is possible when the coronary circulation is perfused with either red blood cell suspensions under a condition of luminal hypoxia (Davie and Farrell, 1991; Davie *et al.*, 1992) or with saline under aerobic conditions (Houlihan *et al.*, 1988). These *in vitro* observations are consistent with the coronary O_2 supply having a greater impact on homeometric regulation than on the Frank–Starling mechanism.

Whereas it is unlikely that luminal O_2 can supply sufficient O_2 to meet the entire myocardial O_2 demand in salmonids because of a diffusion limitation, it is equally unlikely that the coronary circulation normally satisfies the entire myocardial O_2 demand. Based on a coronary flow of 0.43 ml/min.kg body mass and a simultaneously measured myocardial power output of 4.7 mW in coho salmon, the coronary O_2 supply would be 1.42 μl O_2/sec to the heart of a 1-kg fish, assuming an O_2 content of 11 vols% (Axelsson and Farrell, 1992). This would represent approximately half the myocardial O_2 requirement for this particular cardiac work load.

Tuna may be different from salmonids in terms of requirement for coronary O_2 supply. This conclusion is based on limited success perfusing tuna hearts using techniques previously successful with other fish. Resting cardiac power outputs could be achieved in perfused yellowfin tuna but not in perfused skipjack tuna hearts (Farrell *et al.*, 1992). Furthermore, the hearts released large quantities of lactate into the perfusate even though luminal P_{O_2} exceeded 600 torr. The most likely conclusion to be drawn from these observations is that the myocardial $\dot{V}_{O_2}$ of skipjack tuna is predominantly supplied by the arterial blood in the coronary circulation.

F. Control of Coronary Blood Flow

Cameron (1975) estimated coronary flow as 0.56 to 0.65% of $\dot{Q}$ for burbot (*Lota lota*) and sucker (*Catastomus catastomus*) using microspheres. These values should be treated with some caution because microspheres injected into the dorsal aorta should not reach the coronary circulation in appreciable numbers. Axelsson and Farrell (1992) simultaneously measured coronary artery blood flow, dorsal aortic blood pressure, and $\dot{Q}$ in coho salmon (Fig. 12). Coronary flow was 0.20 ml/min.g ventricular mass (0.43 ml/min.kg body mass), which represented 1.1% of $\dot{Q}$. Coronary flow in two anaesthetized school sharks (*Galeorhinus australis*) was measured as 0.64 ml/min.g ventricular mass (0.37 ml/min.g body mass) (P. S. Davie and C. E. Franklin, 1992, personal communication). Pressure–flow relationships for the coronary circulation have been

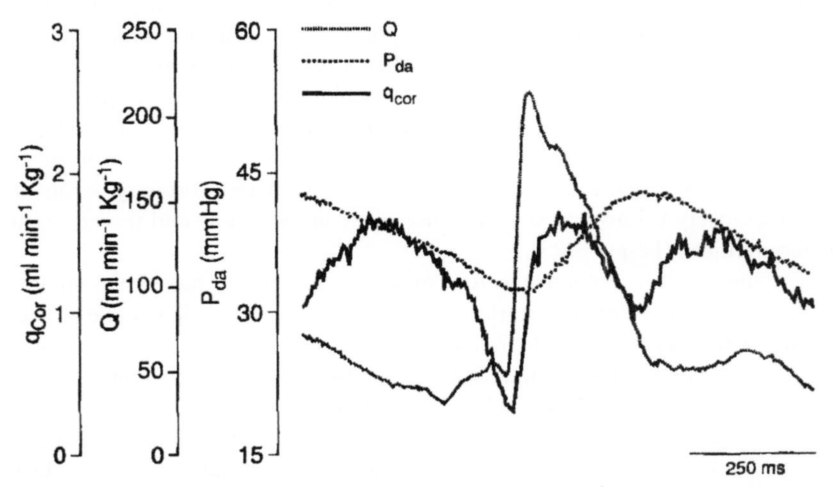

Fig. 12. Simultaneous measurements of ventral aortic blood flow ($\dot{Q}$), dorsal aortic blood pressure (P_{da}) and coronary blood flow (q_{cor}) in a coho salmon (4 kg, 10°–11°C). The continuous but phasic nature of blood flow in the main coronary artery during one cardiac cycle is clearly illustrated. With permission, Axelsson and Farrell (1992).

used to estimate coronary flow as 0.38 and 0.67 ml/min.g ventricular mass in rainbow trout and skipjack tuna (Farrell, 1987; Farrell *et al.*, 1992), which are 1.5% and 1.9% of resting $\dot{Q}$, respectively. These flows are higher than *in vivo* measurements, perhaps because a large coronary vasodilatory reserve exists *in vivo* and is lost with *in vitro* saline perfusion.

At U_{crit} in rainbow trout, $\dot{Q}$ increases threefold, ventral aortic blood pressure increases by 50% (Kiceniuk and Jones, 1977), and mechanical efficiency of the heart doubles (Graham and Farrell, 1989), so myocardial $\dot{V}_{O_2}$ must quadruple. To meet this increase in myocardial O_2 demand, total coronary blood flow must also increase as much as fourfold if there is no change in O_2 extraction. How a change in coronary flow of this magnitude is achieved is not entirely clear.

Coronary vasoactivity and changes in dorsal aortic blood pressure affect coronary blood flow. Recent *in vivo* measurements of coronary blood flow support the idea of coronary vasoactivity altering coronary blood flow (Axelsson and Farrell, 1992). In coho salmon, coronary vascular resistance decreased when coronary flow increased by 114% with hypoxia, 60% with adrenaline injections, and 40% with isoproterenol injections. Axelsson and Farrell (1992) concluded that α-adrenergic constriction and β-adrenergic dilatation, as well as increases in dorsal aortic pressure are involved in regulating coronary flow in coho salmon. Analysis of the pressure–flow relationships for rainbow trout coronaries suggests that the small increase in dorsal aortic blood pressure associated with sustained swimming would, by itself, increase total coronary blood flow by 50% (Farrell, 1987), which means that there must be a large coronary vasodilatory reserve. That is, under resting conditions there is a tonic vasoconstriction of the coronary circulation, which can be released to increase coronary flow. Satchell (1971) noted that the coronary arteries in fish, being derived from postbranchial arteries, have a blood pressure significantly less than the ventricular pressure. This may be a good reason why coronary blood flow in fish hearts is not so tightly linked to coronary blood pressure as in the mammalian heart, where coronary perfusion pressure closely follows pressure developed by the ventricle and therefore, its O_2 requirement (Feigl, 1983).

The mechanisms responsible for coronary vasoactivity have been studied *in vitro* at the arteriolar and arterial levels. The picture emerging is one of species diversity. Based on perfusion studies of the entire coronary circulation, α-adrenergic contraction dominates β-adrenergic relaxation in Atlantic salmon (Farrell and Graham, 1986), rainbow trout (Farrell, 1987), and marlin (Davie and Daxboeck, 1984). In the conger eel, β-adrenergic relaxation predominates (Belaud and Peyraud, 1971). α-Adrenergic vasoconstruction is more pronounced in 15°C- than 5°C-acclimated rainbow trout (Farrell, 1987). Acetylcholine causes contractions in the coronary circulation of the conger eel (Belaud and Peyraud, 1971). Even though most circulatory control is normally exerted at the arteriolar level, there is significant vasoactivity at the arterial

Table VI A Summary of the Vasoactive Responses of Arterial Rings from the Main Coronary Artery in Fishes[a]

Agonist	Rainbow trout[b]	Skate[c]	Mako shark[d]
Acetylcholine	+	(+)/−	(−)
Noradrenaline	−	−	−
Adrenaline	− & +	− & +	−
Isoproterenol	−	−	−
Adenosine	+	+	−
ATP	+	+/−	+/−
ADP	+	+/−	+/−
Histamine	0	(−)	×
Bradykinin	0	(−)	×
Serotonin	(−)	−	0
Prostaglandin $F_{2\alpha}$	+ +	+ +	+ +
Nitroglycerine	−	−	×
Nitroprusside	−	×	×
Endothelin	+ +	×	×
Thrombin	+	×	×

[a]+ +, very strong contractions; +, contractions; −, relaxations; 0, no response; ×, not tested; (), (weak response); /, two responses observed at low/high concentrations; &, both responses observed but on different rings.
[b]Small *et al.* (1990).
[c]Farrell and Davie (1992).
[d]Farrell and Davie (1991b)

level in mammalian coronaries (Kalsner, 1982). The vasoactive responses of isolated coronary vascular rings from three species of fishes are summarized in Table VI, and some of the responses are different from those of the whole coronary circulation. β-Adrenergic relaxations predominate over α-adrenergic contractions, and postaglandin $F_{2\alpha}$ produces strong contractions in all three species. In contrast, the responses to purinergic agents and acetylcholine are species specific. Clearly, receptor subtypes, their relative density, and location in the coronary network are quite variable between fish species. For example, in salmonids the α-adrenoceptors dominate in arterioles, whereas β-adrenoceptors dominate in the main artery. The mechanism underlying the proposed vasodilatory reserve in fish has not been established, but the high sensitivity to and potent constrictor effects of endothelin and prostaglandin $F_{2\alpha}$ warrant further study as tonic vasoconstrictor agents.

Physical factors also affect coronary flow. Contraction of the mammalian left ventricle can reverse coronary flow for up to 20% of the cardiac cycle. The effects of vascular compression are less dramatic in fish hearts. There are beat-by-beat oscillations in coronary perfusion pressure in perfused rainbow trout and *S. acanthias* hearts, and alterations to distolic output pressure also affect coronary perfusion pressure (Farrell, 1987; Davie and Farrell, 1991b). *In vivo* blood flow in the main coronary artery of coho salmon is phasic but continuous (Fig. 12). The systolic rise in coronary flow lags behind ventral aortic flow and is almost coincident with the rise in dorsal aortic blood pressure. The lowest flow always just precedes the onset of ventral aortic flow (coincident with ventricular contraction), and a second rise in coronary flow occurs during the early phase of diastole. However, in the anaesthetized school shark, coronary blood flow in the main artery is briefly reversed, coincident with ventricular systole (P. S. Davie and C. E. Franklin, 1992, personal communication). This difference may be related to conal contraction and the fact that the coronary circulation goes to the spongiosa in sharks but not in rainbow trout.

ACKNOWLEDGMENTS

Work by the authors was supported by grants from the Natural Sciences and Engineering Research Council of Canada, the British Columbia and Yukon Heart Foundation, and the British Columbia Health Care Research Foundation. We are indebted to Joanne Harrington for her excellent typing skills in preparing the many drafts of this manuscript and to Alan Kolok for his comments on early drafts.

REFERENCES

Acierno, R., Agnisola, C., Venzi, R., and Tota, B. (1990). Performance of the isolated and perfused working heart of the teleost *Conger conger:* Study of the inotropic effect of prostacyclin. *J. Comp. Physiol. B.* **160,** 365–371.

Ask, J. A. (1983). Comparative aspects of adrenergic receptors in the hearts of lower vertebrates. *Comp. Biochem. Physiol.* **76A,** 543–552.

Ask, J. A., Stene-Larsen, G., and Helle, K. B. (1980). Atrial β_2-adrenoceptors in the trout. *J. Comp. Physiol.* **139,** 109–116.

Ask, J. A., Stene-Larsen, G., and Helle, K. B. (1981). Temperature effects on the β_2-adrenoceptors of the trout atrium. *J. Comp. Physiol.* **143,** 161–168.

Axelsson, M. (1988). The importance of nervous and humoral mechanisms in the control of cardiac performance in the Atlantic cod *Gadus morhua* at rest and during non-exhausting exercise. *J. Exp. Biol.* **137,** 287–303.

Axelsson, M., and Farrell, A. P. (1992). Coronary blood flow *in vivo* in the coho salmon (*Oncorhynchus kisutch*). Society for Exptal. Biol. Meeting. Abstract A12.1.

Axelsson, M., and Nilsson, S. (1986). Blood pressure control during exercise in the Atlantic cod, *Gadus morhua. J. Exp. Biol.* **126,** 225–236.

Axelsson, M., Ehredstrom, F., and Nilsson, S. (1987). Cholinergic and adrenergic influence on the teleost heart *in vivo. Exp. Biol.* **46,** 179–186.

Axelsson, M., Driedzic, W. R., Farrell, A. P., and Nilsson, S. (1989). Regulation of cardiac output and blood flow in the sea raven, *Hemitripterus americanus. Fish Physiol. Biochem.* **196,** 2–12.

Axelsson, M., Farrell, A. P., and Nilsson, S. (1990). Effect of hypoxia and drugs on the cardiovascular dynamics of the Atlantic hagfish, *Myxine glutinosa*. *J. Exp. Biol.* **151,** 297–316.

Axelsson, M., Davison, W., Forster, M. E., and Farrell, A. P. (1992). Cardiovascular responses of the red-blooded Antarctic fishes, *Pagothenia bernacchii* and *P. borchgrevinki*. *J. Exp. Biol.* In press.

Bailey, J. R., and Driedzic, W. R. (1990). Enhanced maximum frequency and force development of fish hearts following temperature acclimation. *J. Exp. Biol.* **149,** 239–254.

Bailey, J. R., Sephton, D. H., and Driedzic, W. R. (1990). Oxygen uptake by isolated fish hearts with differing myoglobin concentrations under hypoxic conditions. *J. Mol. Cell. Cardiol.* **22,** 1125–1134.

Barron, M. G., Tarr, B. D., and Hayton, W. L. (1987). Temperature-dependence of cardiac output and regional blood flow in rainbow trout, *Salmo gairdneri* Richardson. *J. Fish. Biol.* **31,** 735–744.

Basile, C., Goldspink, G., Modigh, M., and Tota, B. (1976). Morphological and biochemical characterization of the inner and outer myocardial layers of adult tuna fish. *Comp. Biochem. Physiol.* **54B,** 279–283.

Bass, A., Ostadal, B., Pelouch, V., and Vitek, V. (1973). Differences in weight parameters, myosin ATP-ase activity and the enzyme pattern of energy supplying metabolism between the compact and spongious cardiac musculature of carp and turtle. *Pflugers Arch.* **343,** 65–77.

Baumgarten-Schumann, D., and Piiper, J. (1968). Gas exchange in the gills of resting unanaesthetized dogfish (*Scyliorhinus stellaris*). *Resp. Physiol.* **5,** 317–325.

Belaud, A., and Peyraud, C. (1971). Étude preliminaire du debit coronaire sur coeur perfusé de poisson. *J. Physiologie (Paris)* **63,** 165A.

Benjamin, M., Norman, D., Santer, R. M. and Scarborough, D. (1983). Histological, histochemical and ultrastructural studies on the bulbus arteriosus of the stickle-backs, *Gasterosteus aculeatus* and *Pungitius pungitius* (Pisces: Teleostei). *J. Zool. Lond.* **200,** 325–346.

Bennion, G. R. (1968). The control of the function of the heart in teleost fish. M.S. thesis, University of British Columbia, Vancouver, Canada.

Berne, R. M., and Levy, M. N. (1988). "Physiology." C. V. Mosby Co., Toronto, Canada.

Bloom, G., Ostlund, E., Von Euler, U. S., Lishajko, F., Ritzen, M., and Adams-Ray, J. (1961). Studies on catecholamine-containing granules of specific cells in cyclostome hearts. *Acta Physiol. Scand.* **63,** 1–34.

Breisch, E. A., White, F., Jones, H. M., and Laurs, R. M. (1983). Ultrastructural morphometry of the myocardium of *Thunnus alalunga*. *Cell Tiss. Res.* **233,** 427–438.

Brill, R. W., and Bushnell, P. G. (1991). Metabolic scope of high energy demand teleosts—the tunas. *Can. J. Zool.* **69,** 2002–2009.

Bushnell, P. G. (1988). Cardiovascular and respiratory responses to hypoxia in three species of obligate ram ventilating fishes, skipjack tuna, *Katsuwonus pelamis*, yellowfin tuna, *Thunnus albacares*, and bigeye tuna, *Thunnus obesus*. Ph.D. thesis, University of Hawaii, Honolulu. pp. 276.

Bushnell, P. G., and Brill, R. W. (1991). Responses of swimming skipjack (*Katsuwonus pelamis*) and yellowfin (*Thunnus albacares*) tunas to acute hypoxia, and a model of their cardiorespiratory system. *Physiol. Zool.* **64,** 787–811.

Bushnell, P. G., and Brill, R. W. (1992). Oxygen transport and cardiovascular responses in skipjack tuna (*Katsuwonus pelamis*) and yellowfin (*Thunnus albacares*) tuna exposed to acute hypoxia. *J. Comp. Physiol. B* in press.

Bushnell, P. G., Brill, R. W., and Bourke, R. W. (1990). Cardiorespiratory responses of skipjack tuna (*Katsuwonus pelamis*), yellowfin tuna (*Thunnus albacares*), and big-eye tuna (*T. obesus*) to acute reductions in ambient oxygen. *Can. J. Zool.* **68,** 1857–1863.

Butler, P. J. (1986). Exercise. *In* "Fish physiology: Recent advances" (S. Nilsson, S. Holmgren, eds.), pp. 102–118. Croom Helm, London.

Butler, P. J. and Metcalfe, J. D. (1988). Cardiovascular and respiratory systems. *In* "Physiology of Elasmobranch Fishes" (T. V. Shuttleworth, ed.), pp. 1–47. Springer-Verlag, New York.

Butler, P. J., and Taylor, E. W. (1975). The effect of progressive hypoxia on respiration in the dogfish (*Scyliorhinus canicula*) at different seasonal temperatures. *J. Exp. Biol.* **63,** 117–130.

Butler, P. J., Taylor, E. W., Capra, M. F., and Davison, W. (1978). The effect of hypoxia on levels of circulating catecholamines in the dogfish *Scyliorhinus canicula*. *J. Comp. Physiol.* **127,** 325–330.

Cameron, J. N. (1975). Morphometric and flow indicator studies of the teleost heart. *Can. J. Zool.* **53,** 691–698.

Cameron, J. N., and Davis, J. C. (1970). Gas exchange in rainbow trout (*Salmo gairdneri*) with varying blood oxygen capacity. *Fish. Res. Board Can.* **27,** 1069–1085.

Cameron, J. N., Randall, D. J., and Davis, J. C. (1971). Regulation of the ventilation–perfusion ratio in the gills of *Dasyatis sabina* and *Squalus suckleyi*. *Comp. Biochem. Physiol.* **39A,** 505–519.

Cameron, J. S. (1979a). Autonomic nervous tone and regulation of heart rate in the goldfish, *Carassius auratus*. *Comp. Biochem. Physiol.* **63C,** 341–349.

Cameron, J. S. (1979b). Effect of temperature on the availability of alpha- and beta-adrenergic receptors in goldfish hearts. *Am. Zool.* **19,** 1974.

Cameron, J. S., and Brown, S. E. (1981). Adrenergic and cholinergic responses of the isolated heart of the goldfish *Carassius auratus*. *Comp. Biochem. Physiol.* **70C,** 109–116.

Cameron, J. S., and O'Connor, E. F. (1979). Liquid chromatographic demonstration of catecholamine release in fish heart. *J. Exp. Zool.* **209,** 473–479.

Capra, M. F., and Satchell, G. H. (1977). The differential hemodynamic responses of the elasmobranch, *Squalus acanthias*, to the naturally occurring catecholamines, adrenaline and noradrenaline. *Comp. Biochem. Physiol.* **58C,** 41–47.

Cech, J. J., Bridges, R. W., Rowell, D. M., and Balzer, P. J. (1976). Cardiovascular responses of winter flounder, *Pseudopleuronectes americanus*, to acute temperature increase. *Can. J. Zool.* **54,** 1383–1388.

Chan, D. K. O. (1986). Cardiovascular, respiratory, and blood adjustments to hypoxia in the Japanese eel, *Anguilla japonica*. *Fish Physiol. Biochem.* **2,** 179–193.

Chan, D. K. O., and Chow, P. H. (1976). The effects of acetylcholine, biogenic amines and other vasoactive agents on the cardiovascular functions of the eel, *Anguilla anguilla*. *J. Exp. Zool.* **196,** 13–26.

Chui, K. W., and Lee, Y. C. (1990). The cardiac effects of neurohypophysial hormones in the eel, *Anguilla japonica* (Temminck and Schlegel). *J. Comp. Physiol.* **160,** B213–B216.

Cobb, J. L. S., and Santer, R. M. (1973). Electrophysiology and cardiac function in teleosts: Cholinergically mediated inhibition and rebound excitation. *J. Physiol.* (*London*) **230,** 561–573.

Dashow, L., and Epple, A. (1985). Plasma catecholamines in the lamprey: Intrinsic cardiovascular messengers? *Comp. Biochem. Physiol.* **82C,** 119–122.

Davie, P. S. (1987). Coronary supply to the myocardium of the hearts of very active fishes. *Proc. Physiol. Soc. N. Z.* **6,** 36.

Davie, P. S., and Daxboeck, C. (1984). Anatomy and adrenergic pharmacology of the coronary vascular bed of the Pacific blue marlin (*Makaira nigricans*). *Can. J. Zool.* **62,** 1886–1888.

Davie, P. S., and Farrell, A. P. (1991a). Cardiac performance of an isolated heart preparation from the dogfish (*Squalus acanthias*): The effects of hypoxia and coronary artery perfusion. *Can. J. Zool.* **69,** 1822–1828.

Davie, P. S., and Farrell, A. P. (1991b). The coronary and luminal circulations of the myocardium of fishes. *Can. J. Zool.* **69,** 1993–2001.

Davie, P. S., and Forster, M. E. (1980). Cardiovascular responses to swimming in eels. *Comp. Biochem. Physiol.* **67A,** 367–373.

Davie, P. S., and Franklin, C. E. (1992). Myocardial oxygen consumption and mechanical efficiency of a perfused dogfish heart preparation. In preparation.

Davie, P. S., Forster, M. E., Davison, B., and Satchell, G. H. (1987). Cardiac function in the New Zealand hagfish, *Eptatretus cirrhatus*. *Physiol. Zool.* **60,** 233–240.

Davie, P. S., Franklin, C. F., and Farrell, A. P. (1992). The effect of hypoxia and coronary perfusion on the maximum performance of the hypoxic eel heart. *J. Exp. Zool.* In press.

Davis, J. C. (1966). Influence of water and blood flow on gas exchange at the gills of rainbow trout, *Salmo gairdneri*. M. S. thesis, University of British Columbia, Vancouver, Canada.

Davis, J. C., and Cameron, J. N. (1971). Water flow and gas exchange at the gills of rainbow trout, *Salmo gairdneri*. *J. Exp. Biol.* **54,** 1–18.

Davison, W. (1989). Training and its effects on teleost fish. *Comp. Biochem. Physiol.* **94A,** 1–10.

Daxboeck, C. R. (1982). Effect of coronary artery ablation on exercise performance in *Salmo gairdneri*. *Can. J. Zool.* **60,** 375–381.

Daxboeck, C. R., Davie, P. S., Perry, S. F., and Randall D. J. (1982). Oxygen uptake in a spontaneously ventilating, blood-perfused trout preparation. *J. Exp. Biol.* **101,** 35–46.

DeAndres, A. V., Munoz-Chapuli, R., Sans-Coma, V., and Garcia-Garrido, L. (1990). Anatomical studies of the coronary system in elasmobranchs: I. Coronary arteries in lamnoid sharks. *Am. J. Anatomy* **187,** 303–310.

Dizon, A. E., Brill, R. W. and Yuen, H. S. H. (1978). Correlations between environment, physiology, and activity and the effects on thermoregulation in skipjack tuna. *In* "The Physiological Ecology of Tunas." (G. D. Sharp, and A. E. Dizon, eds.), pp. 223–259. Academic Press, New York.

Driedzic, W. R. (1983). The fish heart as a model system for the study of myoglobin. *Comp. Biochem. Physiol.* **76A,** 487–493.

Driedzic, W. R., and Gesser, H. (1985). Ca^{2+} protection from the negative inotropic effect of contraction frequency on teleost hearts. *J. Comp. Physiol.* **156B,** 135–142.

Driedzic, W. R., and Gesser, H. (1988). Differences in force–frequency relationships and calcium dependency between elasmobranch and teleost hearts. *J. Exp. Biol.* **140,** 227–242.

Driedzic, W. R., Scott, D. L., and Farrell, A. P. (1983). Aerobic and anaerobic contributions to energy metabolism in perfused isolated sea raven (*Hemitripterus americanus*) hearts. *Can. J. Zool.* **61,** 1880–1883.

Eaton, R. P., McConnell, T., Hnath, J. G., Black, W., and Swartz, R. E. (1984). Coronary myointimal hyperplasia in fresh water Lake Michigan salmon (Genus *Oncorhynchus*): Evidence for lipoprotein-related atherosclerosis. *Am. J. Pathol.* **116,** 311–318.

Eddy, F. B. (1974). Blood gases of the tench (*Tinca tinca*) in well-aerated and oxygen-deficient waters. *J. Exp. Biol.* **60,** 71–83.

Emery, S. H., Mangano, C., and Randazzo, V. (1985). Ventricle morphology in pelagic elasmobranch fishes. *Comp. Biochem. Physiol.* **82A,** 635–643.

Farrell, A. P. (1978). Cardiovascular events associated with air breathing in two teleosts, *Hoplerythrinus unitaeniatus* and *Arapaima gigas*. *Can. J. Zool.* **56,** 953–958.

Farrell, A. P. (1981). Cardiovascular changes in the lingcod (*Ophiodon elongatus*) following adrenergic and cholinergic drug infusions. *J. Exp. Biol.* **91,** 293–305.

Farrell, A. P. (1982). Cardiovasclar changes in the unanaesthetized lingcod (*Ophiodon elongatus*) during short-term progressive hypoxia and spontaneous activity. *Can. J. Zool.* **60,** 933–941.

Farrell, A. P. (1984). A review of cardiac performance in the teleost heart: Intrinsic and humoral regulation. *Can. J. Zool.* **62,** 523–536.

Farrell, A. P. (1985). Cardiovascular and hemodynamic energetics of fishes. *In* "Circulation, Respiration, and Metabolism." (R. Gilles, ed.), pp. 377–385. Springer-Verlag, Berlin, Germany.

Farrell, A. P. (1986). Cardiovascular responses in the sea raven, *Hemitripterus americanus*, elicited by vascular compression. *J. Exp. Biol.* **122,** 65–80.

Farrell, A. P. (1987). Coronary flow in a perfused rainbow trout heart. *J. Exp. Biol.* **129,** 107–123.

Farrell, A. P. (1991a). From hagfish to tuna—a perspective on cardiac function. *Physiol. Zool.* **64,** 1137–1164.

Farrell, A. P. (1991b). Introduction to cardiac scope in lower vertebrates. *Can. J. Zool.* **69,** 1981–1984.

Farrell, A. P. (1991c). Circulation of body fluids. *In* "Comparative Animal Physiology" (C. L. Prosser, ed.), pp. 509–558. Wiley-Liss, Inc., New York.

Farrell, A. P., and Davie, P. S. (1991a). Coronary artery reactivity in the mako shark, *Isurus oxyrinchus. Can. J. Zool.* **69,** 375–379.

Farrell, A. P., and Davie, P. S. (1991b). Coronary artery reactivity in the rough skate, *Raja nasuta. Comp. Biochem. Physiol.* **99C,** 555–560.

Farrell, A. P., and Daxboeck, C. (1981). Oxygen uptake in the lingcod, *Ophiodon elongatus*, during progressive hypoxia. *Can. J. Zool.* **59,** 1272–1275.

Farrell, A. P., and Graham, M. S. (1986). Effects of adrenergic drugs on the coronary circulation of Atlantic salmon (*Salmo salar*). *Can. J. Zool.* **64,** 481–484.

Farrell, A. P., and Milligan, C. L. (1986). Myocardial intracellular pH in a perfused rainbow trout heart during extracellular acidosis in the presence and absence of adrenaline. *J. Exp. Biol.* **125,** 347–359.

Farrell, A. P., and Steffensen, J. F. (1987a). An analysis of the energetic cost of the branchial and cardiac pumps during sustained swimming in trout. *Fish Physiol. Biochem.* **4,** 73–79.

Farrell, A. P., and Steffensen, J. F. (1987b). Coronary ligation reduces maximum sustained swimming speed in Chinook salmon, *Oncorhynchus tshawytscha. Comp. Biochem. Physiol.* **87A,** 35–37.

Farrell, A. P., MacLeod, K. R., Driedzic, W. R., and Wood, S. (1983). Cardiac performance during hypercapnic acidosis in the *in situ* perfused fish heart. *J. Exp. Biol.* **107,** 415–429.

Farrell, A. P., Hart, T., Wood, S., and Driedzic, W. R. (1985). Myocardial oxygen consumption in the sea raven, *Hemitripterus americanus:* The effects of volume loading, pressure loading and progressive hypoxia. *J. Exp. Biol.* **117,** 237–250.

Farrell, A. P., Saunders, R. L., Freeman, H. C., and Mommsen, T. P. (1986a). Arterio-sclerosis in Atlantic salmon: Effects of dietary cholesterol and maturation. *Arterio-sclerosis* **6,** 453–461.

Farrell, A. P., MacLeod, K. R., and Chancey, B. (1986b). Intrinsic mechanical properties of the perfused rainbow trout heart and the effects of catecholamines and extracellular calcium under control and acidotic conditions. *J. Exp. Biol.* **125,** 319–345.

Farrell, A. P., Hammons, A. M., Graham, M. S., and Tibbits, G. F. (1988a). Cardiac growth in rainbow trout, *Salmo gairdneri. Can. J. Zool.* **66,** 2368–2373.

Farrell, A. P., Johansen, J. A., and Graham, M. S. (1988b). The role of the pericardium in cardiac performance of the trout (*Salmo gairdneri*). *Physiol. Zool.* **61,** 213–221.

Farrell, A. P., Small, S., and Graham, M. S. (1989). Effect of heart rate and hypoxia on the performance of a perfused trout heart. *Can. J. Zool.* **67,** 274–280.

Farrell, A. P., Johansen, J. A., Steffensen, J. F., Moyes, C. D., West, T. G., and Suarez, R. K. (1990a). Effects of exercise training and coronary ablation on swimming performance, heart size and cardiac enzymes in rainbow trout, *Oncorhynchus mykiss. Can. J. Zool.* **68,** 1174–1179.

Farrell, A. P., Johansen, J. A., and Saunders, R. L. (1990b). Coronary lesions in Pacific salmonids. *J. Fish Dis.* **13,** 97–100.

Farrell, A. P., Johansen, J. A., and Suarez, R. K. (1991). Effects of exercise-training on cardiac performance and muscle enzymes in rainbow trout, *Oncorhynchus mykiss. Fish. Physiol. Biochem.* **9,** 303–312.

Farrell, A. P., Davie, P. S., Franklin, C. E., Johansen, J. A., and Brill, R. W. (1992). Cardiac physiology in tunas: I. *In vitro* perfused heart preparations from yellowfin and skipjack tunas. *Can. J. Zool.* In press.

Feigl, E. O. (1983). Coronary physiology. *Physiol. Rev.* **63,** 1–205.

Forster, M. E. (1976). Effects of adrenergic blocking drugs on the cardiovascular system of the eel, *Anguilla anguilla* (L). *Comp. Biochem. Physiol.* **55C,** 33–36.

Forster, M. E. (1985). Blood oxygenation in shortfinned eels during swimming and hypoxia: Influence of Root effect. *N. Z. J. Mar. Freshwater Res.* **19,** 247–251.

Forster, M. E. (1989). Performance of the heart of the hagfish, *Eptatretus cirrhatus. Fish Physiol. Biochem.* **6,** 321–327.

Forster, M. E. (1991). Myocardial oxygen consumption and lactate release by the hypoxic hagfish heart. *J. Exp. Biol.* **156,** 583–590.

Forster, M. E., Axelsson, M., Farrell, A. P., and Nilsson, S. (1991). Cardiac function and circulation in hagfishes. *Can. J. Zool.* **69,** 1985–1992.

Forster, M. E., Axelsson, M., Davison, W., and Farrell, A. P. (1992). Hypoxia in the New Zealand hagfish, *Eptatretus cirrhatus. Respir. Physiol.* In press.

Foxon, G. E. H. (1950). A description of the coronary arteries in Dipnoan fishes and some remarks on their importance from the evolutionary standpoint. *J. Anat.* **84,** 121–131.

Franklin, C. E., and Davie, P. S. (1991). The pericardium facilitates pressure work in the eel heart. *J. Fish Biol.* **39,** 559–564.

Franklin, C. E., and Davie, P. S. (1992). Cardiac volumes in rainbow trout, *Oncorhynchus mykiss*. In preparation.

Gannon, B. J. and Burnstock, G. (1969). Excitatory adrenergic innervation of the fish heart. *Comp. Biochem. Physiol.* **29,** 765–773.

Garey, W. (1970). Cardiac output of the carp (*Cyprinus carpio*). *Comp. Biochem. Physiol.* **33,** 181–189.

Gehrke, P. C., and Fielder, D. R. (1988). Effects of temperature and dissolved oxygen on heart rate, ventilation rate and oxygen consumption of spangled perch, *Leipotherapon unicolor* (Gunther, 1985) (Percoidei, Teraponidae). *J. Comp. Physiol.* **157B,** 771–782.

Gennser, M., Karpe, F., and Ornhagen, H. C. (1990). Effects of hyperbaric pressure and temperature on atria from ectotherm animals (*Rana pipiens* and *Anguilla anguilla*). *Comp. Biochem. Physiol.* **95A,** 219–228.

Gesser, H. (1985). Effects of hypoxia and acidosis on fish heart performance. *In* "Respiration and Metabolism" (R. Gilles, ed.), pp. 402–410. Springer-Verlag, Berlin, Germany.

Gesser, H., and Jorgensen, E. (1982). pH, contractility and Ca^{2+} balance under hypercapnic acidosis in the myocardium of different vertebrate species. *J. Exp. Biol.* **96,** 405–412.

Gesser, H., and Poupa, O. (1983). Acidosis and cardiac muscle contractility: Comparative aspects. *Comp. Biochem. Physiol.* **76A,** 559–566.

Gesser, H., Andresen, P., Brams, P., and Sund-Laursen, J. (1982). Inotropic effects of adrenaline on the anoxic or hypercapnic myocardium of rainbow trout and eel. *J. Comp. Physiol.* **147,** 123–128.

Giovane, A., Greco, G., Maresca, A., and Tota, B. (1980). Myoglobin in the heart ventricle of tuna and other fishes. *Experientia* **36,** 219–220.

Goldstein, L., Forster, R. P., and G. M., Fanelli Sr. (1964). Gill blood flow and ammonia excretion in the marine teleost *Myoxcephalus scorpius*. *Comp. Biochem. Physiol.* **12,** 489–499.

Goodrich, E. S. (1930). "Studies on the Structure and Development of Vertebrates." Macmillan, London.

Goolish, E. M. (1987). Cold-acclimation increases the ventricle size of carp, *Cyprinus carpio*. *J. Thermal Biol.* **12,** 203–206.

Graham, J. B. and Diener, D. R. (1978). Comparative morphology of the central heat exchangers in the skipjacks *Katsuwonus* and *Euthynnus*. *In* "The Physiological Ecology of Tunas" (G. D. Sharp and A. E. Dizon, eds.), pp. 113–134. Academic Press, New York.

Graham, J. B., Dewar, H., Lai, N. C., Lowell, W. R., and Arce, S. M. (1990). Aspects of shark swimming performance using a large water tunnel. *J. Exp. Biol.* **151,** 175–192.

Graham, M. S., and Farrell, A. P. (1985). Seasonal intrinsic cardiac performance of a marine teleost. *J. Exp. Biol.* **118,** 173–183.

Graham, M. S., and Farrell, A. P. (1989). The effect of temperature acclimation and adrenaline on the performance of a perfused trout heart. *Physiol. Zool.* **62,** 38–61.

Graham, M. S., and Farrell, A. P. (1990). Myocardial oxygen consumption in trout acclimated to 5° and 15°C. *Physiol. Zool.* **63,** 536–554.

Graham, M. S., and Farrell, A. P. (1992). Environmental influences on cardiovascular variables in rainbow trout, *Oncorhynchus mykiss* (Walbaum). *J. Fish Biol.* In press.

Graham, M. S., Wood, C. M., and Turner, J. D. (1982). The physiological responses of the rainbow trout to strenuous exercise: Interactions of water hardness and environmental acidity. *Can. J. Zool.* **60,** 3153–3164.

Grant, R. T., and Regnier, M. (1926). The comparative anatomy of the cardiac coronary vessels. *Heart* **14,** 285–317.

Greer Walker, M., Santer, R. M., Benjamin, M., and Norman, D. (1985). Heart structure of some deep-sea fish (Teleostei-Macrouridae). *J. Zool.* (*London*) **205,** 75–89.

Halpern, M. H., and May, M. M. (1958). Phylogenetic study of the extracardiac arteries to the heart. *Am. J. Anat.* **102,** 469–480.

Hanson, D., and Johansen, K. (1970). Relationship of gill ventilation and perfusion in Pacific dogfish, *Squalus suckleyii*. *J. Fish. Res. Board Can.* **27,** 551–564.

Helle, K. B. (1983). Structures of functional interest in the myocardium of lower vertebrates. *Comp. Biochem. Physiol.* **76A,** 447–452.

Hemmingsen, E. A., Douglas, E. L., Johansen, K., and Millard, R. W. (1972). Aortic blood flow and cardiac output in the hemoglobin-free fish *Chaenocephalus aceratus*. *Comp. Biochem. Physiol.* **43A,** 1045–1051.

Hemmingsen, E. A. and Douglas, E. L. (1977). Respiratory and circulatory adaptations to the absence of hemoglobin in Chaenichthyid fishes. *In* "Adaptations within Antarctic Ecosystems" (G. A. Llano, ed.), pp. 479–487. Smithsonian Institute, Washington, D.C.

Hipkins, S. F. (1985). Adrenergic responses of the cardiovascular system of the eel, *Anguilla australis, in vivo*. *J. Exp. Zool.* **235,** 7–20.

Hipkins, S. F., and Smith, D. G. (1983). Cardiovascular events associated with spontaneous apnea in the Australian short-finned eel, *Anguilla australis*. *J. Exp. Zool.* **227,** 339–348.

Hipkins, S. F., Smith, D. G., and Evans, B. K. (1986). Lack of adrenergic control of dorsal aortic blood pressure in the resting eel, *Anguilla australis*. *J. Exp. Zool.* **238,** 155–166.

Hirst, G. D. S., Edwards, F. R., Bramich, N. J., and Klemm, M. F. (1991). Neural control of cardiac pacemaker potentials. *News Physiol. Sci.* **6,** 185–190.

Hochachka, P. W., Guppy, M., Guderley, H., Storey, K. B., and Hulbert, W. C. (1978). Metabolic biochemistry of water vs. air-breathing osteoglossids: Heart enzymes and ultrastructure. *Can. J. Zool.* **56,** 759–768.

Holeton, G. F. (1970). Oxygen uptake and circulation by a hemoglobinless Antarctic fish, (*Chaenocephalus aceratus* Lonnberg) compared with three red-blooded Antarctic fish. *Comp. Biochem. Physiol.* **34,** 457–471.

Holeton, G. F., and Randall, D. J. (1967). The effect of hypoxia upon the partial pressure of gases in the blood and water afferent and efferent to the gills of rainbow trout. *J. Exp. Biol.* **46,** 317–328.

Holmgren, S. (1977). Regulation of the heart of a teleost, *Gadus morhua*, by autonomic nerves and circulating catecholamines. *Acta Physiol. Scand.* **99,** 62–74.

Holmgren, S., Axelsson, M., and Farrell, A. P. (1992). The effect of catecholamines, substance P and vasoactive intestinal polypeptide (VIP) on blood flow to the gut in the dogfish, *Squalus acanthias. J. Exp. Biol.* In press.

Houlihan, D. F., Agnisola, C., Lyndon, A. R., Gray, C., and Hamilton, N. M. (1988). Protein synthesis in a fish heart: Responses to increased power output. *J. Exp. Biol.* **137,** 565–587.

House, E. W., and Benditt, E. P. (1981). The ultrastructure of spontaneous coronary arterial lesions in steelhead trout (*Salmo gairdneri*). *Am. J. Pathol.* **104,** 250–257.

House, E. W., Dornauer, R. J., and Van Lenten, B. J. (1979). Production of coronary arteriosclerosis with sex hormones and human chorionic gonadotrophin (HCG) in juvenile steelhead and rainbow trout, *Salmo gairdneri. Atherosclerosis* **34,** 197–206.

Hughes, G. M., Peyraud, C., Peyraud-Waitzenneger, M., and Soulier, P. (1982). Physiological evidence for the occurrence of pathways shunting blood away from the secondary lamellae of eel gills. *J. Exp. Biol.* **98,** 277–288.

Isokawa, K., Takagi, M. and Toda, Y. (1988). Ultrastructural cytochemistry of trout arterial fibrils as elastic components. *Anat. Rec.* **220,** 369–376.

Isokawa, K., Takagi, M. and Toda, Y. (1989). Ultrastructural cytochemistry of aortic microfibrils in the arctic lamprey, *Lampetra japonica. Anat. Rec.* **223,** 158–164.

Isokawa, K., Takagi, M. and Toda, Y. (1990). Ultrastructural and cytochemical study of elastic fibers in the ventral aorta of a teleost, *Anguilla japonica. Anat. Rec.* **226,** 90–95.

Itazawa, Y. (1970). Heart rate, cardiac output and circulation time of fish. *Bull. Jap. Soc. Sci. Fish* **36,** 926–931.

Itazawa, Y., and Takeda, T. (1978). Gas exchange in the carp gills in normoxic and hypoxic conditions. *Respir. Physiol.* **35,** 263–269.

Jensen, D. (1961). Cardioregulation in an aneural heart. *Comp. Biochem. Physiol.* **2,** 181–201.

Jensen, D. (1965). The aneural heart of the hagfish. *Ann. N. Y. Acad. Sci.* **127,** 443–458.

Jensen, D. (1969). Intrinsic cardiac rate regulation in the sea lamprey, *Petromyzon marinus*, and rainbow trout, *Salmo gairdneri. Comp. Biochem. Physiol.* **30,** 685–690.

Jensen, D. (1970). Intrinsic cardiac rate regulation in elasmobranchs: The horned shark, *Heterodontus prancisci*, and the thornback ray, *Platyrhinoidis triseriata. Comp. Biochem. Physiol.* **34,** 289–296.

Johansen, K. (1965). Cardiovascular dynamics in fishes, amphibians and reptiles. *Ann. N. Y. Acad. Sci.* **127,** 414–442.

Johansen, K. and Burggren, W. W. (1980). Cardiovascular function in the lower vertebrates. *In* "Heart and Heartlike Organs" (G. H. Bourne, ed.), pp. 61–117. Academic Press, New York.

Johansen, K. and Gesser, H. (1986). Fish cardiology: Structural, haemodynamic, electromechanical and metabolic aspects. *In* "Fish Physiology: Recent Advances." (S. Nilsson, S. Holmgren, eds.), pp. 71–85. Croom Helm, London.

Johansen, K., Lenfant, C., and Hanson, D. (1968). Cardiovascular dynamics in lung-fishes. *Z. vergl. Physiol.* **59,** 157–186.

Johnston, I. A., Fitch, N., Zummo, G., Wood, R. E., Harrison, P., and Tota, B. (1983). Morphometric and ultrastructural features of the ventricular myocardium of the haemoglobinless ice fish *Chaenocephalus aceratus*. *Comp. Biochem. Physiol.* **76,** 475–480.

Jones, D. R. (1971). Theoretical analysis of factors which may limit the maximum oxygen uptake of fish: The oxygen cost of the cardiac and branchial pumps. *J. Theor. Biol.* **32,** 341–349.

Jones, D. R. (1992). Cardiac energetics and the design of vertebrate arterial systems. *In* "Efficiency and Economy in Animal Physiology" (R. W. Blake, ed.), pp. 159–168. Cambridge University Press, New York.

Jones, D. R., and Milsom, W. K. (1982). Peripheral receptors affecting breathing and cardiovascular function in non-mammalian vertebrates. *J. Exp. Biol.* **100,** 59–91.

Jones, D. R. and Randall, D. J. (1978). The respiratory and circulatory systems during exercise. *In* "Fish Physiology" (W. S. Hoar and D. J. Randall, eds.), pp. 425–501. Academic Press, New York.

Jones, D. R., Brill, R. W., and Bushnell, P. G. (1992). Ventricular and arterial dynamics of anesthetized and swimming tuna. In review *J. Exp. Biol.*

Jones, D. R., Langille, B. W., Randall, D. J., and Shelton, G. (1974). Blood flow in dorsal and ventral aortae of the cod, *Gadus morhua. Am. J. Physiol.* **226,** 90–95.

Kalsner, S. (1982). Vasoconstrictors, spasm and acute myocardial events. *In* "The Coronary Artery" (S. Kalsner, ed.), pp. 551–595. Cambridge University Press, New York.

Kanwisher, J., Lawson, K., and Sundness, G. (1974). Acoustic telemetry from fish. *Fish. Bull., U.S.* **72,** 251–255.

Karttunen, P., and Tirri, R. (1986). Isolation and characterization of single myocardial cells from the perch, *Perca fluviatilis. Comp. Biochem. Physiol.* **84A,** 181–188.

Keen, J. E., Farrell, A. P., Tibbits, G. F., and Brill, R. W. (1992). Cardiac dynamics in tunas. II. Effect of ryanodine, calcium and adrenaline on force–frequency relationships in arterial strips from skipjack tuna, *Katsuwonus pelamis. Can. J. Zool.* In press.

Kiceniuk, J. W., and Jones, D. R. (1977). The oxygen transport system in trout (*Salmo gairdneri*) during sustained exercise. *J. Exp. Biol.* **69,** 247–260.

Kitoh, K., and Oguri, M. (1985). Differentiation of the compact layer in the heart ventricle of rainbow trout. *Bull. Jpn. Soc. Sci. Fish* **51,** 539–542.

Kubasch, A., and Rourke, A. W. (1990). Arteriosclerosis in steelhead trout, *Oncorhynchus mykiss* (Walbaum): A developmental analysis. *J. Fish Biol.* **37,** 65–69.

Laffont, J., and Labat, R. (1966). Action de l'adrénaline sur la fréquence cardiaque de la corpe commune. Effect de la température du millieu sur l'intensite de la réaction. *J. Physiol.* (*Paris*) **58,** 351–355.

Lai, N. C., Graham, J. B., Lowell, W. R., and Laurs, R. M. (1987). Pericardial and vascular pressures and blood flow in the albacore tuna, *Thunnus alalunga. J. Exp. Biol.* **146,** 187–192.

Lai, N. C., Graham, J. B., and Lowell, W. R. (1989a). Elevated pericardial pressure and cardiac output in the leopard shark *Triakis semifasciata* during exercise: The role of the pericardioperitoneal canal. *J. Exp. Biol.* **147,** 263–277.

Lai, N. C., Graham, J. B., Bhargava, V., Lowell, W. R., and Shabetai, R. (1989b). Branchial blood flow distribution in the blue shark (*Prionace glauca*) and the leopard shark (*Triakis semifasciata*). *Exp. Biol.* **48,** 273–278.

Lai, N. C., Shabetai, R., Graham, J. B., Holt, B. D., Sunnerhagen, K. S., and Bhargava, V. (1990a). Cardiac function in the leopard shark, *Triakis semifasciata. J. Comp. Physiol.* **160,** 259–268.

Lai, N. C., Graham, J. B., and Burnett, L. (1990b). Blood respiratory properties and the effect of swimming on blood gas transport in the leopard shark *Triakis semifasciata. J. Exp. Biol.* **151,** 161–173.

Langille, B. L., Stevens, E. D., and Anantaraman, A. (1983). Cardiovascular and respiratory flow dynamics. *In* "Fish Biomechanics" (P. W. Webb and D. Weihs, eds.), pp. 92–139 Praeger, New York.

Laurent, P. (1962). Contribution à létude morphologique et physiologique de l'innervation du coeur des téléostéens. *Arch. Anat. Microsc. Morphol. Expt.* **51,** 337–458.

Laurent, P., Holmgren, S., and Nilsson, S. (1983). Nervous and humoral control of the fish heart: Structure and function. *Comp. Biochem. Physiol.* **76A,** 525–542.

Leknes, I. L. (1981). On the ultrastructure of the endothelium in the bulbus arteriosus of the three teleostean species. *J. Submicrosc. Cytol.* **13(1),** 41–46.

Leknes, I. L. (1985). A scanning electron microscopic study on the heart and ventral aorta in a teleost. *Zool. Anz.* **214,** 142–150.

Lennard, R., and Huddart, H. (1990). Electrophysiology of the flounder heart (*Platichthys flesus*)—The effect of agents which modify transmembrane ion transport. *Comp. Biochem. Physiol. C.* **94C,** 499–509.

Licht, J. H. and Harris, W. S. (1973). The structure, composition and elastic properties of the teleost bulbus arteriosus in the carp, *Cyprinus carpio. Comp. Biochem. Physiol.* **46A,** 669–708.

Lomholt, J. P., and Johansen, K. (1976). Gas exchange in the amphibious fish, *Amphipnuos cuchia. J. Comp. Physiol.* **107B,** 141–158.

MacDonald, J. A., Montgomery, J. C., and Wells, R. M. G. (1987). Comparative physiology of Antarctic fishes. *Adv. Mar. Biol.* **24,** 321–388.

Maneche, H. C., Woodhouse, S. P., Elson, P. F., and Klassen, G. A. (1973). Coronary artery lesions in Atlantic Salmon (*Salmo salar*). *Exp. Mol. Pathol.* **17,** 274–280.

McKenzie, J. E., House, E. W., McWilliam, J. G., and Johnson, D. W. (1978). Coronary degeneration in sexually mature rainbow and steelhead trout, *Salmo gairdneri. Atherosclerosis* **29,** 431–437.

Meghji, P., and Burnstock, G. (1984). Actions of some autonomic agents on the heart of the trout (*Salmo gairdneri*) with emphasis on the effects of adenyl compounds. *Comp. Biochem. Physiol.* **78C,** 69–76.

Meltcalfe, J. D., and Butler, P. J. (1982). Differences between directly measured and calculated values for cardiac output in the dogfish: A criticism of the Fick Method. *J. Exp. Biol.* **99,** 255–268.

Midtun, B. (1980). Ultrastructure of atrial and ventricular myocardium in the pike *Esox lucius* L. and mackerel *Scomber scombrus* L. (Pisces). *Cell. Tissue Res.* **211,** 41–50.

Milligan, C. L., and Farrell, A. P. (1986). Extra- and intracellular acid–base status following enforced activity in the sea raven (*Hemitripterus americanus*). *J. Comp. Physiol.* **156,** 583–590.

Milligan, C. L., and Farrell, A. P. (1991). Lactate utilization by an *in situ* perfused trout heart: Effects of workload and lactate transport blockers. *J. Exp. Biol.* **155,** 357–373.

Milligan, C. L., and Wood, C. M. (1986). Tissue intracellular acid–base status and the fate of lactate after exhaustive exercise in the rainbow trout. *J. Exp. Biol.* **123,** 123–144.

Milligan, C. L., Graham, M. S., and Farrell, A. P. (1989). The response of trout red cells to adrenaline during seasonal acclimation and changes in temperature. *J. Fish. Biol.* **35,** 229–236.

Moffitt, B. P., and Crawshaw, L. I. (1983). Effects of acute temperature changes on metabolism, heart rate, and ventilation frequency in carp *Cyprinus carpio* L.. *Physiol. Zool.* **56,** 397–403.

Moore, J. F., Mayr, W., and Hougie, C. (1976). Ultrastructure of coronary arterial changes in spawning Pacific salmon genus *Onchorhynchus* and steelhead trout *Salmo gairdneri. J. Comp. Pathol.* **86,** 259–267.

Motais, R., Isaia, J., Rankin, J. C., and Maetz (1969). Adaptive changes of the water permeability of the teleostean gill epithelium in relation to external salinity. *J. Exp. Biol.* **51,** 529–546.

Muir, B. S., and Brown, C. E. (1971). Effects of blood pathway on the blood-pressure drop in fish gills, with special reference to tunas. *J. Fish. Res. Board Can.* **28,** 947–955.

Murdaugh, H. V., Robin, E. D., Millen, J. E., and Drewry, W. F. (1965). Cardiac output determinations by the dye-dilution method in *Squalus acanthius. Am. J. Physiol.* **209,** 723–726.

Neumann, P., Holeton, G. F., and Heisler, N. (1983). Cardiac output and regional blood flow in gills and muscles after strenuous exercise in rainbow trout (*Salmo gairdneri*). *J. Exp. Biol.* **105,** 1–14.

Nilsson, S. (1983). "Autonomic Nerve Function in the Vertebrates." Springer-Verlag, Berlin, Germany.

Nilsson, S., Abrahamson, T., and Grove, D. J. (1976). Sympathetic nervous control of adrenaline release from the head kidney of the cod *Gadus morhua. Comp. Biochem. Physiol.* **55,** 123–127.

O'Donogue, C. H., and Abbot, E. (1928). The blood vascular system of the spiny dogfish *Squalus acanthias* and *Squalus suckleyii. Trans. R. Soc. Edinb.* **55,** 823–890.

Ostadal, B., and Schiebler, T. H. (1971). Über die terminale Strombahn in Fishherzen. *Z. Anat. Entwick* **134,** 101–110.

Ostadal, B., and Schiebler, T. H. (1972). Blood supply of the heart in different species of fish. *Physiol. Bohemoslov.* **21,** 103.

Parker, G. H., and Davis, F. K. (1899). The blood vessels of the heart in *Carcharias, Raja* and *Amia. Proc. Boston Soc. Nat. Hist.* **29,** 163–178.

Parker, T. J. (1886). On the blood vessels of *Mustelus antarticus. Phil. Trans. R. Soc. Lond., B.* **177,** 685–732.

Pennec, J. P., and LeBras, Y. M. (1984). Storage and release of catecholamines by nervous endings in the isolated heart of the eel (*Anguilla anguilla* L.). *Comp. Biochem. Physiol.* **77C,** 167–172.

Pennec, J. P., and Peyraud, C. (1983). Effects of adrenaline on isolated heart of the eel (*Anguilla anguilla*) during winter. *Comp. Biochem. Physiol.* **74C,** 477–480.

Perry, S. F. and Farrell, A. P. (1989). Perfusion preparations in comparative respiratory physiology. *In* "Techniques in Comparative Respiratory Physiology: An Experimental Approach. Society for Experimental Biology Seminar Series" (C. R. Bridges and P. J. Butler, eds.), pp. 223–257. Cambridge University Press, London.

Petersson, K., and Nilsson, S. (1980). Drug induced changes in cardiovascular parameters in the Atlantic cod, *Gadus morhua. J. Comp. Physiol.* **137B,** 131–138.

Peyraud-Waitzenegger, M., and Soulier, P. (1989). Ventilatory and circulatory adjustments in the European eel (*Anguilla anguilla* L.) exposed to short-term hypoxia. *Exp. Biol.* **48,** 107–122.

Peyraud-Waitzenegger, M., Savina, A., Laparra, J., and Morfin, R. (1979). Blood–brain barrier for epinephrine in the eel (*Anguilla anguilla* L.). *Comp. Biochem. Physiol.* **63C,** 35–38.

Peyraud-Waitzenegger, M., Barthelemy, L., and Peyraud, C. (1980). Cardiovascular and ventilatory effects of catecholamines in unrestrained eels (*Anguilla anguilla*). A study of seasonal changes in reactivity. *J. Comp. Physiol.* **138B,** 367–375.

Piiper, J., Meyer, M., Worth, H., and Willmer, H. (1977). Respiration and circulation during swimming activity in the dogfish *Scyliorhinus stellaris. Respir. Physiol.* **30,** 221–239.

Poupa, O., and Brix, O. (1984). Cardiac beat frequency and oxygen supply: A comparative study. *Comp. Biochem. Physiol.* **78A,** 1–3.

Poupa, O., and Lindstrom, L. (1983). Comparative and scaling aspects of heart and body weights with reference to blood supply of cardiac fibres. *Comp. Biochem. Physiol.* **76A,** 413–421.

Poupa, O., Gesser, H., Jonsson, S., Sullivan, L. (1974). Coronary-supplied compact shell of ventricular myocardium in salmonids: growth and enzyme pattern. *Comp. Biochem. Physiol.* **48A,** 85–95.

Poupa, O., Lindstrom, L., Maresca, A., and Tota, B. (1981). Cardiac growth, myoglobin, proteins and DNA in developing tuna (*Thunnus thynnus thynnus*). *Comp. Biochem. Physiol.* **70A,** 217–222.

Poupa, D., Helle, K. B., and Lomsky, M. (1985). Calcium paradox from cyclostome to man—A comparative study. *Comp. Biochem. Physiol.* **81A,** 801–806.

Priede, I. G. (1974). The effects of swimming activity and section of the vagus nerves on heart rate in rainbow trout. *J. Exp. Biol.* **60,** 305–319.

Priede, I. G. (1976). Functional morphology of the bulbus arteriosus of rainbow trout (*Salmo gairdneri* Richardson). *J. Fish. Biol.* **9,** 209–216.

Primmett, D. R. N., Randall, D. J., Mazeaud, M., and Boutilier, R. G. (1986). The role of catecholamines in erythrocyte pH regulation and oxygen transport in rainbow trout (*Salmo gairdneri*) during exercise. *J. Exp. Biol.* **122,** 139–148.

Randall, D. J. (1968). Functional morphology of the heart in fishes. *Am. Zool.* **8,** 179–189.

Randall, D. J. (1970). The circulatory system in fish physiology. *In* "Fish Physiology" (W. S. Hoar and D. J. Randall, eds.), pp. 132–172. Academic Press, New York.

Randall, D. (1985). Shunts in fish gills. *In* "Cardiovascular shunts—Phylogenetic, Ontogenetic and Clinical Aspects. Alfred Benson Symposium XXI" (K. Johansen, W. Burggren, ed.), pp. 71–87. Munksgaard, Copenhagen, Denmark.

Ristori, M. T., and Laurent, P. (1985). Plasma catecholamines and glucose during moderate exercise in the trout: Comparison with bursts of violent activity. *Exp. Biol.* **44,** 247–253.

Robertson, O. H., Wexler, B. C., and Miller, B. F. (1961). Degenerative changes in the cardiovascular system of the spawning Pacific salmon (*Oncorhynchus tshawytscha*). *Cir. Res.* **9,** 826–834.

Robin, E. D., Murdaugh, H. E., and Millen, J. E. (1964). Gill gas exchange in dogfish sharks. *Fed. Proc.* **23 I,** 469.

Robin, E. D., Murdaugh, H. V., and Millen, J. E. (1966). Acid–base, fluid and electrolyte metabolism in the elasmobranch. III. Oxygen, CO_2, bicarbonate and lactate exchange across the gill. *J. Cell Physiol.* **67,** 93–100.

Ruben, J. A., and Bennett, A. F. (1981). Intense exercise, bone structure, and blood calcium levels in vertebrates. *Nature* **291,** 411–413.

Saetersdal, T. S., Sorensen, E., Myklebust, R. and Helle, K. B. (1975). Granule-containing cells and fibres in the sinus venosus of elasmobranchs. *Cell Tissue Res.* **163,** 471–490.

Sage, E. H. and Gray, W. R. (1979). Studies on the evolution of elastin. I. Phylogenic distribution. *Comp. Biochem. Physiol.* **64B,** 313–327.

Sage, E. H. and Gray, W. R. (1980). Studies on the evolution of elastin. II. Histology. *Comp. Biochem. Physiol.* **66B,** 13–22.

Saito, T. (1973). Effects of vagal stimulation on the pacemaker action potentials of carp heart. *Comp. Biochem. Physiol.* **44A,** 191–199.

Sanchez-Quintana, D., and Hurle, J. M. (1987). Ventricular myocardial architecture in marine fishes. *Anatomical Rec.* **217,** 263–273.

Santer, R. M. (1985). Morphology and innervation of the fish heart. *Adv. Anat. Embryology Cell Biology* **89,** 1–102.

Santer, R. M., and Greer Walker, M. (1980). Morphological studies on the ventricle of teleost and elasmobranch hearts. *J. Zool. Lond.* **190,** 259–272.

Santer, R. M., Greer Walker, M. G., Emerson, L., and Witthames, P. R. (1983). On the morphology of the heart ventricle in marine teleost fish (Teleostei). *Comp. Biochem. Physiol.* **76A,** 453–457.

Sarnoff, S. J., Braunwald, E., G. H., Welch, Jr., Case, R. B., Stansby, W. N. and Macruz, R. (1958a). Haemodynamic determinants of oxygen consumption of the heart with special reference to the tension–time index. *Am. J. Physiol.* **192,** 148–56.

Sarnoff, S. J., Case, R. B., G. H., Welch Jr., Braunwald, E. and Stainsby, W. N. (1958b). Performance characteristics and oxygen debt in a nonfailing, metabolically supported, isolated heart preparation. *Am. J. Physiol* **192,** 141–47.

Satchell, G. H. (1971). "Circulation in Fishes." Cambridge Monographs in Experimental Biology." Cambridge University Press, Cambridge, England.

Satchell, G. H. (1991). "Physiology and Form of Fish Circulation." Cambridge University Press, Cambridge, England.

Satchell, G. H., and Jones, M. P. (1967). The function of the conus arteriosus in the Port Jackson shark, *Heterodontus portusjacksoni. J. Exp. Biol.* **46,** 373–382.

Satchell, G. H., Hanson, D., and Johansen, K. (1970). Differential blood flow through the afferent branchial arteries of the skate, *Raja rhina. J. Exp. Biol.* **52,** 721–726.

Saunders, R. L., and Farrell, A. P. (1988). Coronary arteriosclerosis in Atlantic salmon: No regression of lesions after spawning. *Arteriosclerosis* **8,** 378–384.

Saunders, R. L., Farrell, A. P., and Knox, D. E. (1992). Progression of coronary arterial lesions in Atlantic salmon as a function of growth rate. *J. Fisheries Aquatic Sci.* In press.

Schmidt, S. P., and House, E. W. (1979). Time study of coronary myointimal hyperplasia in precocious male steelhead trout, *Salmo gairdneri. Atherosclerosis* **34,** 375–381.

Schmidt-Nielsen, K. (1984). "Scaling: Why Is Animal Size So Important." Cambridge University Press, Cambridge.

Schmidt-Nielsen K. (1989). "Animal Physiology: Adaptation and Environment." 3rd Ed. Cambridge University Press.

Sebert, P. H., and Barthelemy, L. (1985). Hydrostatic pressure and adrenergic drugs (agonists and antagonists): Effects and interactions in fish. *Comp. Biochem. Physiol.* **82C,** 207–212.

Senior, H. D. (1909). The development of the heart in shad (*Alosa sapidissima* Wilson). *Am. J. Anat.* **9,** 211–262.

Serafini-Fracassini, A., Field, J. M., Spina, M., Garbisa, S. and Stuart, R. J. (1978). The morphological organization and ultrastructure of elastin in the arterial wall of trout (*Salmo gairdneri*) and salmon (*Salmo salar*). *J. Ultrastruc. Res.* **65,** 1–12.

Shabetai, R., Abel, D. C., Graham, J. B., Bhargava, V., Keyes, R. S., and Witztum, K. (1985). Function of the pericardium and pericardioperitoneal canal in elasmobranch fishes. *Am. J. Physiol.* **248,** H198–H207.

Seibert, H. (1979). Thermal adaptation of heart rate and its parasympathetic control in the European eel *Anguilla anguilla* (L.). *Comp. Biochem. Physiol.* **64C,** 275–278.

Short, S., Butler, P. J., and Taylor, E. W. (1977). The relative importance of nervous, humoral, and intrinsic mechanisms in the regulation of heart rate and stroke volume in the dogfish (*Scyliorhinus canicula* L.). *J. Exp. Biol.* **70,** 77–92.

Short, S., Taylor, E. W., and Butler, P. J. (1979). The effectiveness of oxygen transfer during normoxia and hypoxia in the dogfish (*Scyliorhinus canicula* L.) before and after cardiac vagotomy. *J. Comp. Physiol.* **132,** 289–295.

Small, S. A., Macdonald, C., and Farrell, A. P. (1990). Vascular reactivity of the coronary artery in rainbow trout (*Oncorhynchus mykiss*). *Am. J. Physiol.* **258,** R1402–R1410.

Steffensen, J. F., and Farrell, A. P. (1992). Cardiac and swimming performance of coronary-ligated rainbow trout, *Oncorhynchus mykiss*, exposed to hypoxia. Submitted.

Stevens, E. D. (1972). Some aspects of gas exchange in tuna. *J. Exp. Biol.* **56,** 809–823.

Stevens, E. D., and Randall, D. J. (1967). Changes of gas concentrations in blood and water during moderate swimming activity in rainbow trout. *J. Exp. Biol.* **46,** 329–338.

Stevens, E. D., Bennion, G. R., Randall, D. J., and Shelton, G. (1972). Factors affecting arterial blood pressures and blood flow from the heart in intact, unrestrained ling cod, *Ophiodon elongatus. Comp. Biochem. Physiol.* **43,** 681–695.

Sureau, D., Lagardere, J. P., and Pennec, J. P. (1989). Heart rate and its cholinergic control in the sole (*Solea vulgaris*) acclimatized to different temperatures. *Comp. Biochem. Physiol.* **92A,** 49–51.

Takeda, T. (1990). Ventilation, cardiac output and blood respiratory parameters in the carp, *Cyprinus carpio*, during hyperoxia. *Resp. Physiol.* **81,** 227–240.

Taylor, E. W., Short, S., and Butler, P. J. (1977). The role of the cardiac vagus in the response of the dogfish *Scyliorhinus canicula* to hypoxia. *J. Exp. Biol.* **70,** 57–75.

Temma, K., Kishi, H., Kitazawa, T., Kondo, H., Ohta, T., and Katano, Y. (1986). Are beta-adrenergic receptors in ventricular muscles of carp heart (*Cyprinus carpio*) mostly beta-2 type? *Comp. Biochem. Physiol.* **83,** C261–C264.

Temma, K., Akamine, M., Shimizu, T., Kitazawa, T., Kondo, H., and Ohita, T. (1989). Histamine directly acts on β-adrenoceptors as well as H_1-histaminergic receptors, and causes positive inotropic effects in isolated ventricular muscles of carp heart (*Cyprinus carpio*). *Comp. Biochem. Physiol.* **92C,** 143–148.

Thomas, S., Poupin, J., Lykkeboe, G., and Johansen, K. (1987). Effects of graded exercise on blood gas tensions and acid–base characteristics of rainbow trout. *Respir. Physiol.* **68,** 85–97.

Tibbits, G. F., Hove-Madsen, L., and Bers, D. M. (1991). Calcium transport and the regulation of cardiac contractility in teleosts: A comparison with higher vertebrates. *Can. J. Zool.* **69,** 2014–2019.

Tirri, R., and Lehto, H. (1984). Alpha- and beta-adrenergic control of contraction force of perch heart. *Comp. Biochem. Physiol.* **77C,** 301–304.

Tirri R., and Ripatti, R. (1982). Inhibitory adrenergic control of heart rate of perch (*Perca fluviatilis*) *in vitro*. *Comp. Biochem. Physiol.* **73,** C399–C402.

Tota, B. (1978). Functional cardiac morphology and biochemistry in Atlantic bluefin tuna. *In* "The Physiological Ecology of Tunas." (G. D. Sharp and A. E. Dizon, eds.), pp. 89–112. Academic Press, New York.

Tota, B. (1983). Vascular and metabolic zonation in the ventricular myocardium of mammals and fishes. *Comp. Biochem. Physiol.* **76A,** 423–437.

Tota, B. (1989). Myoarchitecture and vascularization of the elasmobranch heart ventricle. *J. Exp. Zool.* (Suppl.) **2,** 122–135.

Tota, B., Cimini, V., Salvatore, G., and Zummo, G. (1983). Comparative study of the arterial and lacunary systems of the ventricular myocardium of elasmobranch and teleost fishes. *Am. J. Anat.* **167,** 15–32.

Tota, B., Acierno, R., and Agnisola, C. (1991). Mechanical performance of the isolated and perfused heart of the haemoglobinless antarctic icefish *Chionodraco hamatus* (Lönnberg): Effects of loading conditions and temperature. *Phil. Trans. R. Soc. Lond. (B.)* **332,** 191–198.

Tsukuda, H., Liu, B., and Fujii, K. I. (1985). Pulsation rate and oxygen consumption of isolated hearts of the goldfish, *Carassuis auratus*, acclimated to different temperatures. *Comp. Biochem. Physiol.* **82A,** 281–283.

Van Citters, R. L., and Watson, N. W. (1968). Coronary disease in spawning steelhead trout, *Salmo gairdnerii*. *Science* **159,** 105–107.

Vastesaeger, M. M., Delcourt, R. and Gillot, P. H. (1965). Spontaneous atherosclerosis in fishes and reptiles. *In* "Comparative Atherosclerosis" (J. C. Roberts and R. Straus, eds.), pp. 129–149. Harper & Row, New York.

Voboril, Z., and Schiebler, T. H. (1970). Zur Gefaβversorgung von fischherzen. *Z. Anat. Entwicklungsgesch* **130,** 1–8.

Vornanen, M. (1989). Regulation of contractility of the fish (*Carassius carassius* L.) heart ventricle. *Comp. Biochem. Physiol.* **94C,** 477–483.

Watson, A. D., Cobb, J. L. S. (1979). A comparative study on the innervation and the vascularization of the bulbus arteriosus in teleost fish. *Cell Tissue Res.* **196,** 337–346.

Watters, K. W., and Smith, L. S. (1973). Respiratory dynamics of the starry flounder, *Platichthys stellatus*, in response to low oxygen and high temperature. *Mar. Biol.* **19,** 133–148.

Wells, R. M. G., Forster, M. E., Davison, W., Taylor, H. H., Davie, P. S., and Satchell, G. H. (1986). Blood oxygen transport in the free-swimming hagfish, *Eptatretus cirrhatus*. *J. Exp. Biol.* **123,** 43–53.

Wood, C. M. (1974). A critical examination of the physical and adrenergic factors affecting blood flow through the gills of the rainbow trout. *J. Exp. Biol.* **60,** 241–265.

Wood, C. M., and Milligan, C. L. (1987). Adrenergic analysis of extracellular and intracellular lactate and H^+ dynamics after strenuous exercise in the starry founder, *Platichthys stellatus*. *Physiol. Zool.* **60,** 69–81.

Wood, C. M., and Perry, S. F. (1985). Respiratory, circulatory, and metabolic adjustments to exercise in fish. *In* "Circulation, Respiration, and Metabolism" (R. Gilles, ed.), pp. 2–22. Springer-Verlag, Berlin, Germany.

Wood, C. M., and Shelton, G. (1980a). Cardiovascular dynamics and adrenergic responses of the rainbow trout *in vivo*. *J. Exp. Biol.* **87,** 247–270.

Wood, C. M., and Shelton, G. (1980b). The reflex control of heart rate and cardiac output in the rainbow trout: Interactive influences of hypoxia, haemorrhage and systemic vasomotor tone. *J. Exp. Biol.* **87,** 271–284.

Wood, C. M., McMahon, B. R., and McDonald, D. G. (1977). An analysis of changes in blood pH following exhausting activity in the starry flounder (*Platichthys stellatus*). *J. Exp. Biol.* **69,** 173–185.

Wood, C. M., Pieprzak, P., and Trott, J. N. (1979). The influence of temperature and anaemia on the adrenergic and cholinergic mechanisms controlling heart rate in the rainbow trout. *Can. J. Zool.* **57,** 2440–2447.

Wright, G. M. (1984). Structure of the conus arteriosus and ventral aorta in the sea lamprey, *Petromyzon marinus*, and the atlantic hagfish, *Myxine glutinosa:* Microfibrils, a major component. *Can. J. Zool.* **62,** 2445–2456.

Xiaojun, X., and Ruyung, S. (1990). The bioenergetics of the southern catfish (*Silurus meridionalis chen*). I. Resting metabolic rate as a function of body weight and temperature. *Physiol. Zool.* **63,** 1181–1195.

Yamauchi, A. (1980). Fine structure of the fish heart. *In* "Hearts and Heartlike Organs" (G. H. Bourne, ed.), pp. 119–148. Academic Press, New York.

Yamauchi, A., and Burnstock, G. (1968). Electron microscopic study on the innervation of the trout heart. *J. Comp. Neurol.* **132,** 567–588.

Zummo, G. and Farina, F. (1989). Ultrastructure of the conus arteriosus of *Scyliorhinus stellaris*. *J. Exp. Zool.* (Suppl.) **2,** 158–164.

Chapter 5

How the evolution of air-breathing shaped the form and function of the cardiorespiratory systems

Tobias Wang*
Section for Zoophysiology, Department of Biology, Aarhus University, Aarhus C, Denmark
**Corresponding author: e-mail: tobias.wang@bio.au.dk*

Tobias Wang discusses the K. Johansen chapter "Air Breathing in Fishes" in *Fish Physiology*, Volume 4, published in 1970. The next chapter in this volume is the re-published version of that chapter.

The evolution of air-breathing in water-breathing fishes was a major transition in the history of vertebrates that enabled the first tetrapods to conquer terrestrial environments. Breathing of air involved modifications of existing structures (buccal cavity, esophagus and even stomach or intestines) to facultatively or obligately replace gills for efficient uptake of oxygen, while CO_2 excretion and ion-regulation persisted across the gills. Breathing of air allowed survival in oxygen-depleted water and first forays of vertebrates into terrestrial environments. However, the evolution of these novel gas exchange surface also required major modifications of the cardiovascular system to provide adequate perfusion with blood of these new respiratory surfaces. It was the study of these cardiovascular adaptations in air-breathing fish that Kjell Johansen pioneered during a sojourn to Brazil in 1964. His studies over the next 5–6 years formed the basis of his influential review "*Air Breathing in Fishes,*" which appeared in 1970 in volume 4 of the *Fish Physiology* series. My primary goal of this short perspective is to provide some biographical information on Kjell Johansen's early career with focus on his studies on air-breathing fish. Johansen's review had a profound impact, and marked the start of a new era where the entire oxygen transport cascade was framed in air-breathing fish for the first time. Johansen also provided the first synthesis on the regulation of the marked interactions between ventilation of the air-breathing organ and its perfusion, mostly due to his own contributions over the four preceding 4 years. The study of air-breathing fish remain a vibrant and exciting topic, and Johansen's review continue to inspire researchers in this research area.

Fish Physiology, Vol. 40A. https://doi.org/10.1016/bs.fp.2023.09.009

"Few events in the life history of vertebrates stand out in importance like the transition from life in water to life on land. Such transition is inseparable from the development of accessory organs of respiration making possible the direct utilization of O_2 in the atmospheric air. The fishes occupy a crucial phylogenetic position in this respect, and they demonstrate numerous and extremely varied examples of how the fundamental process of breathing with gills in water is modified or assisted by accessory organs for aerial respiration." Thus starts Kjell Johansen's first publication (1966) on air-breathing fish where he describes the cardiorespiratory changes associated with air-breathing in the South American swamp eel (*Synbranchus marmoratus*). Characteristic for Johansen's many research contributions, this study was performed on an expedition—in this case to Brazil—and he later became known for expanding the iconic Krogh principle—"*For a large number of problems there will be some animal of choice, or a few such animals, on which it can be most conveniently studied* (Krogh, 1929)," which he qualified by adding: "*if this species occur in Cleveland, choose another problem.*" Quite fitting to this hedonic revision of the Krogh principle, the title of Johansen's August Krogh Lecture at The International Union of Physiological Sciences (IUPS) in Vancouver was "*The World as a Laboratory: Physiological Insights from Nature's Experiments.*"(Johansen, 1987).

As evident from the lyric start of Johansen's, 1966 publication, it is not surprising that many of the early founders of comparative anatomy in the 19th century took interest in the describing the air-breathing organs (ABOs) in various air-breathing fish that were collected mostly in tropical regions. Also, as soon as physiological methods had been sufficiently developed to study how the ABOs function, comparative physiologists started to measure the air-breathing behavior including breathing frequency and depth as well as the respiratory properties of the blood in the various species of air-breathing fish. However, until Kjell Johansen's seminal chapter in 1970 on "*Air Breathing in Fishes*" in volume 4 of the newly founded book series *Fish Physiology*, only the British zoologist Carter (1957) and the Indian zoologist Das (1927) had synthesized the available information on air-breathing into actual reviews that sought to delineate general patterns. Both reviews, however, were mostly confined to descriptions of the structures of the ABOs as well as some preliminary information on overall gas exchange and blood oxygen binding properties (e.g., Carter, 1935; Carter and Beadle, 1931). Johansen's review, in contrast, included sophisticated cardiovascular measurements that allowed for hemodynamic interpretations on the relationships between blood flows, vascular resistance and blood pressure and how these variables changed during the intermittent bouts of air-breathing that characterize most air-breathing fish. The review also contained modern analyses of gas exchange and detailed interpretations of arterial and venous blood gases.

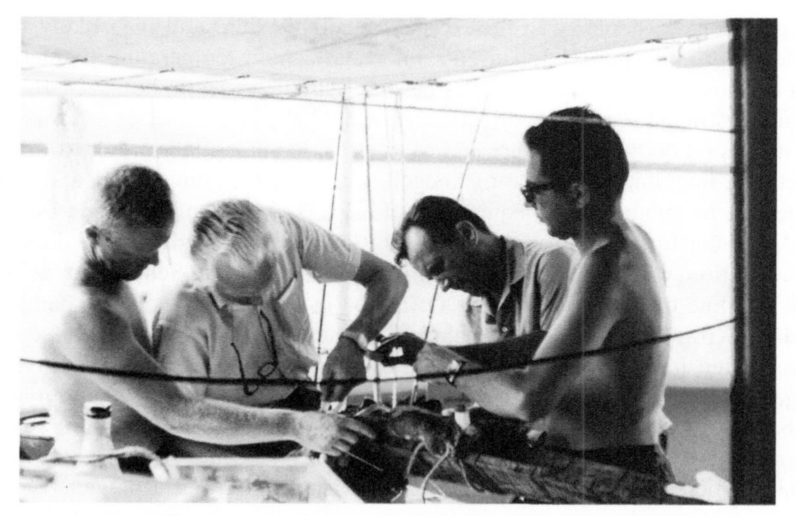

FIG. 1 Kjell Johansen, Knut Schmidt-Nielsen, Claude Lenfant and possibly Jorge Petersen on the Alpha-Helix expedition in the Amazon in 1967. It is not easy to discern whether they are working on a South American lungfish or an electrical eel, but both species were studied on this expedition. Schmidt-Nielsen, chief-scientist on the expedition, described that "*the collaboration with Kjell made me feel like a student again. I admired his surgical competence and his excellent skills as a team leader. I felt privileged to be included in his research*" (Schmidt-Nielsen, 1998).

Johansen's review, therefore, marked the start of a new era with an understanding of the entire oxygen transport cascade in air-breathing fish. He also provided the first synthesis on the regulation of the marked interactions between ventilation of the air-breathing organ, its perfusion (Fig. 1).

Kjell Johansen was 38 years old when he published *Air Breathing in Fishes* in *Fish Physiology* (1970), and established himself as a leading expert on air-breathing fish within a mere 4 years.

1 Kjell Johansen's background and entry to the study of air-breathing fish

Kjell Johansen was born in Oslo (Norway) on September 9, 1932. Upon completion of his university degree in Oslo in 1956, he was appointed at the Institute for Experimental Medical Research at the University Hospital in Oslo (Ullevål) to develop and implement the new heart-lung machines that were taken into use to secure adequate oxygen delivery during invasive surgeries in the thorax. Many of these surgeries were performed on hypothermic patients to reduce metabolism and lessen the demands for adequate oxygenation, particularly in the oxygen-sensitive brain, and it is likely that these conditions spurred Johansen's interests in both temperature regulation and cardiovascular physiology. His academic interests were, however, by no means, restricted to human patients, and he published a number of experimental studies on body temperature and heart rates, including interpretation of the electrocardiogram, in hibernating hedgehog and northern birch mice, and soon thereafter he submitted the equivalent of a Ph.D. thesis on the diving bradycardia in the European grass snake. Many of the early studies were done in collaboration with John Krogh, a former student of Per Scholander at the new Institute of Zoophysiology in Oslo, and in addition to their comparative studies, Krog and Johansen developed methods to determine intravascular oxygen tensions using Teflon-covered polarographic oxygen electrodes. Together with Dr. Ragnar Hol at the Radiological Department, Johansen also embarked on cinematographic investigations of the circulation and gill ventilation in *Myxine* and other fish using injections of contrast media, such as small particles of barium, that could be visualized using Røntgen. Later, Johansen would be amongst the first to use such techniques to study the circulatory physiology of air-breathing fish. Consequently, his ability to incorporate modern techniques of investigation—often through extensive collaboration with medical schools—became one of the reasons for his enormous impact in the field of comparative physiology.

In 1959, Johansen accepted a trainee position at with Dr. Loren Carlson at Department of Physiology and Biophysics in the Medical School at University of Seattle, but he also managed to spend time at the University in Alaska in Fairbanks. During the first years in United States, he upheld close contact with Oslo University, and defended his Doctor of Science (Dr. scient., a Scandinavian degree recognizing a substantial and independent academic contribution beyond what is required for a PhD) on the cardiovascular dynamics in *Amphiuma tridactylum*; a large aquatic salamander that is native to the southwest of North America (Johansen, 1962).

Johansen was employed as trainee on a research program supported by the American Air Force under Dr. Loren Carlson, who was well-aware of Johansen's interest in comparative physiology, and prompted a meeting with Dr. Art W. Martin in the Department of Zoology. Martin was a well-known invertebrate physiologist with a wide interest in physiological problems including osmoregulation, and together with Johansen they now started to investigate the circulatory physiology of various invertebrates including octopus and giant earthworms. It was their shared interest in giant earthworms that would bring Johansen to Brazil where he embarked on his influential studies on air-breathing fish that would become an integral part of the rest of his academic career (Fig. 2).

FIG. 2 Kjell Johansen brought aestivating lungfish (*Protopterus annectans*) to Aarhus and one specimen—endeared with the Swahili name Kamongo—stood on his desk for seven years. One of these aestivating lungfish now resides in the office of the author.

2 Studies of air-breathing fish at University of Sao Paulo

Martin and Johansen spend 5 months at Sao Paulo University (USP) in 1964 with support from NSF and a Fulbright stipend to the young Johansen. 2 years earlier, Dr. Paolo Sawaya, who had been appointed as the first professor of general animal physiology at USP already in 1937, had visited Martin to give lectures at a course at Friday Harbor and together they planned to continue hemodynamic studies of the giant earthworms, that now were renowned for their high pressures in the hemolymph. The findings from the 1964 expedition were published in *Journal of Experimental Biology* (Johansen and Martin, 1965a,b, 1966), but Johansen also found time to study cardiac shunting in toads, blood pressures in sloths and the cardiorespiratory physiology of the air-breathing marbled swamp eel (*Synbranchus marmoratus*). The iconic marbled swamp eel is native to swamps and ponds across central and South America and received its Latin name because its operculae are fused on the ventral side, such that expired respiratory water must leave the branchial cavity through a common exit. This anatomical adaptation probably enables symbranchid eels (there are sister species in Asia, such as *Monopterus*

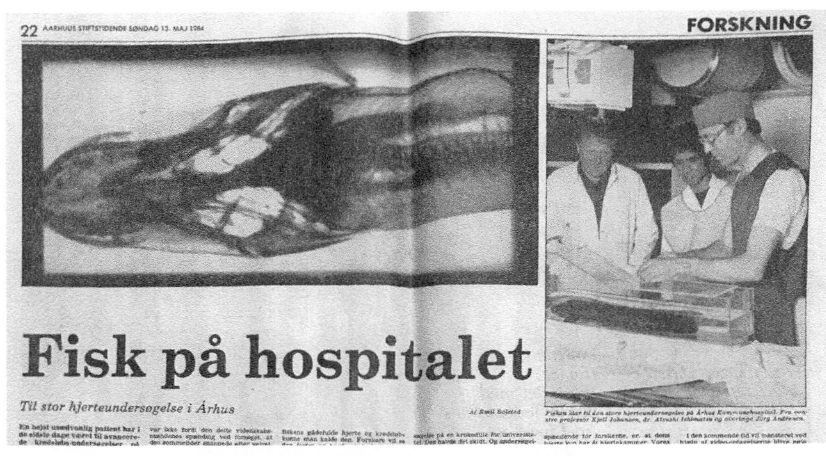

22 AARHUUS STIFTSTIDENDE SØNDAG 13. MAJ 1984

FORSKNING

Fisk på hospitalet

Til stor hjerteundersøgelse i Århus

FIG. 3 The radiographic studies of heart function in snake head (*Channa argus*) attracted interest from the local newspapers in Denmark, and the research was featured in the title "Fish in the hospital." The picture to the right shows radiologist Jorg Andreasen, Dr. Atsushi Ishimatsu and Professor Kjell Johansen. I the article Johansen explain how the hearts of air-breathing fish and other ectotherms have resemblance to cardiac malformations in humans and that the study of exotic animal may inspire novel treatment of such disease-states.

albus) to hold large, engulfed air-bubbles inside the pharynx for the purpose of respiratory gas exchange occur over a richly vascularized buccal epithelium—air-breathing. In *Synbranchus*, significant aerial gas exchange presumably also occurs over the rather sturdy gills that appear not to collapse in air exposure, unlike the gills of purely water-breathing fishes. Johansen (1966) reported on colossal elevations in heart rate at unaltered blood pressure when the buccal cavity was inflated with air at the water surface, and he took arterial blood samples as well as samples from the ABO to demonstrate how oxygen and CO_2 were exchanged. He correctly surmised that stretch receptors in the buccal cavity elicit a release of vagal tone on the heart to explain the rise in heart rate and he speculated that a barostatic response is responsible for the maintenance of arterial blood pressure. By and large, these interpretations have been substantiated in symbranchids and other air-breathing fish over the years (Graham et al., 1995; McKenzie et al., 2007; Iversen et al., 2011), and it is also clear that increased venous tone, possibly through adrenergic innervation, must play a role in maintaining cardiac filling when filling time is reduced as heart rate increases (Skals et al., 2006). It was not until recently, however, that the barostatic regulation of heart rate in response to altered blood pressures were characterized in air-breathing fish (Armelin et al., 2021) (Fig. 3).

3 Lungfish studies with Claude Lenfant and Gordon Grigg on three continents

Lungfish (*Dipnoi*) feature prominently in Johansen's 1970 review. This is a natural choice given that lungfish represent a sister taxa to the air-breathing vertebrates, and have derived their name because they, unlike the air-breathing teleosts, have real paired lungs that originate ventrally from the esophagus. However, not much had been known about this important taxa, and Johansen and Martin (1965a,b) lamented this dearth of knowledge in their chapter in Handbook of Physiology titled "comparative aspects of cardiovascular function in vertebrates": "*The interesting group of lungfishes, the Dipnoi, has unfortunately not yet received any noticeable attention from physiologists.*" Johansen would certainly change that upon his return from the inaugural visit to Brazil, where he joined forces with Dr. Claude Lenfant, a French surgeon and respiratory physiologists that was recruited to The University of Seattle in the mid-1960s. Lenfant later became the director for The National Institute of Health (NIH) in Bethesda. Johansen and Lenfant had befriended each other a few years prior to becoming colleagues in Seattle, and they now combined strengths in cardiovascular physiology and expertise in gas exchange. Fortunately, none of the extant dipnoid species occur in Cleveland, so the newly formed and dynamic duo was obliged to travel to Brazil to study *Leipdosiren*, to Uganda to study *Protopterus*, and to Brisbane, Australia, where they studied *Neoceratodus* in collaboration with Dr. Gordon Grigg. As is well-known today, the three extant species of lungfish differ profoundly in their dependence on aerial gas exchange, with *Protopterus* being an obligate air-breather that will drown if denied aerial access, while *Neoceratodus* has retained very well-developed gills and is a facultative airbreather. As reviewed in 1970, Johansen and Lenfant provided a detailed description of the cardiorespiratory physiology of these three species through very impressive and simultaneous measurements of blood flows and pressures as well gas composition in various blood vessels and lungs. The data were interpreted in terms of classic diagrams for lung physiology that were being developed around that time, making these papers an absolute must on the reading list for any serious comparative cardiorespiratory physiologist. Lenfant and Johansen also collaborated on similarly integrative studies on electric eels on the Alpha-Helix expedition in the Amazon in 1966 where Knut Schmidt-Nielsen and the young Brazilian zoologist Jorge Petersen figure as co-authors (Johansen et al., 1968; Fig. 4).

Johansen would return to the Amazon on the to study air-breathing fishes when Drs. David Randall and Peter Hochachka arranged for the Alpha Helix to return to the Amazon near Manaus in 1977. This expedition resulted in a book on the evolution of air-breathing (Randall et al., 1981) and an entire

FIG. 4 Kjell Johansen in his office at Aarhus University. The specific occasion is unfortunately not known to the author, but could be a favorable grant evaluation or an invitation to an expedition to an exotic place.

issue of research articles in the Canadian Journal of Zoology (volume 56, 1978).

Johansen continued to study lungfish for the rest of his career and he even brought an aestivating Protopterus through customs from Kenya and placed it on his desk in Aarhus. Still within in its native mud, Kamongo—the Swhahili name for the African lungfish and the title of one of Homer Smiths books (Smith, 1932)—was equipped with a Fleisch tube connected to a pneumotachygraph (Glass et al., 1978), so the low ventilatory activity could be monitored during the suspended animation of aestivation. Seven (!) years later, a foul smell in Johansen's office revealed that Kamongo had given up the long wait for the rain that would have enabled Kamongo to resume normal lungfish business under natural conditions. The reduction in metabolism during aestivation had been described 50 years earlier by Smith (1930), kidney physiologist and founder of comparative renal physiology, and the attending reduction in lung ventilation and the ventilatory responses to hypercapnia were characterized decades later by Perry et al. (2008).

4 Air-breathing fish at Aarhus University

Shortly after the publication of his 1970-review in *Fish Physiology*, Johansen was called to Aarhus University (i.e., he was offered the position of full professor and founding head without any applications for the position!). Here, Johansen was given the possibility to establish an institute for Zoophysiology as well as freedom to design three floors in the building where I and other comparative physiologists still reside, although relegated to a section through various organizational changes over the past 50 years. The name zoophysiology was undoubtedly chosen with inspiration from August Krogh's department of Zoophysiology at Copenhagen University and the term is now synonymous with comparative animal physiology. Johansen was also offered ample funds to purchase scientific equipment as well as the possibility to hire technicians and post docs. The basement quickly became a state-of-the-art animal care facility with climatic chambers and the possibility of holding many different exotic species, including air-breathing fish. In the following 15 years, Johansen continued to study air-breathing fish and many international visitors were trained in the physiology of air-breathing fishes under his tutelage. These included Tadeu Rantin, Eduardo Bicudo, and Augusto Abe from Brazil, all visiting on fellowships from DANIDA, as well as Atsushi Ishimatsu from Japan. With Drs. Roy Weber, Gunnar Lykkeboe and Geoff Maloy from Nairobi, the blood oxygen binding of whole blood and hemoglobins were described in awake and aestivating lungfish (Johansen et al., 1976; Weber et al., 1977). Also, the local Mads Lomholt and Mogens Glass studied air-breathing fish in Aarhus, and Mogens Glass would later focus on the respiratory physiology of the South American lungfish when appointed to a full professorship at the medical school in Riberao Preto, Brazil (Nunan et al., 1989; Wang and Wood, 2020). With Ishimatsu, Glass and Jorgen Andresen, a radiologist at Aarhus University Hospital, Johansen continued the cinematographic studies to yield dynamic pictures of blood flows and cardiac chamber function in *Channa argus* (Andresen et al., 1987). The central hemodynamics of this south East Asian species of snakehead fish is particularly interesting because the inflow of oxygen-rich blood from the ABO is separated from the oxygen-poor blood that returns to the heart from the systemic veins (Ishimatsu and Itazawa, 1983). This is a rather spectacular finding because both blood streams enter a single atrium with no anatomical specialization to separate blood flows and an equally normal piscine-like ventricle; the mechanistic basis of this remarkable flow separation in these actively contracting chambers remains to be fully understood (Ishimatsu, 2012; Ishimatsu et al., 1979). Thus, relationship between functional morphology and the separation of blood flows in the heart and its inflow and outflow tracts remain to be clarified in air-breathing fish (Ishimatsu, 2012) and it would seems that a merger of modern imaging techniques and model description of fluid dynamics would be a way forward to understand this complex problem.

5 What are the next research question on air-breathing fish?

While much fundamental insight is provided by Johansen in 1970, there are still many fundamental aspects of air-breathing fish that beg to be resolved, and I believe these questions span from on overall evolutionary questions on the main drivers of the evolution of air-breathing to more mechanistic physiological questions and priorities of physiological regulation under different conditions. Because air-breathing evolved independently on many occasions, the evolutionary questions lend themselves to be resolved by proper phylogenetic approaches. Thus, in many ways, the air-breathing fish provide a rather exciting means to examine whether similar evolutionary trajectories are chosen and such patterns could be framed in the context of "Tinbergens four questions" pertaining to function, evolution, causation and development (e.g., Bateson and LaLand, 2013). It is interesting, in this context, that Johansen, and contemporary comparative physiologists, often used an evolutionary perspective in the writings, but it was only recently that the different modes of ABOs were mapped on a phylogenetic tree (Damsgaard et al., 2020).

Over the past couple of decades, a number of excellent reviews provide excellent overviews of the existing knowledge of ABO structure, regulation of blood vessel responses, gill structure in air-breathing fish, as well as physiological responses to environmentally relevant situations such as hypercapnia, hypoxia, elevated temperature and physical exercise (e.g., Bayley et al., 2019; Damsgaard et al., 2020; Florindo et al., 2018; Graham, 1997; Ishimatsu, 2012; Milsom, 2018; Milsom et al., 2022; Pelster and Wood, 2018; Shartau and Brauner, 2014; Turko et al., 2021; Zaccone et al., 2018). Given the enormous amount of information accrued since Johansen's 1970-review, we are therefore in the very fortunate position of being to understand how the various homeostatic responses are prioritized. For example, the gills of most air-breathing fish are small with reduced surface area and thick filaments, which is normally attributed to being a morphological adaptation to avoid branchial oxygen loss in aquatic hypoxia (e.g., Aaskov et al., 2022; Brauner et al., 2004). This is likely correct, but to what extent does the reduced gills impair all the other branchial functions, such as acid-base balance and ion homeostasis, in air-breathing fish? These types of questions have started to become addressed in the past decade (e.g., Brauner et al., 2004; Pelster and Wood, 2018) and their solutions require an integrated approach and the ability to understand the responses in the context of the natural history of the species. I think this approach is in line with Kjell Johansen's research philosophy.

6 Kjell Johansens impact

It is hard to overestimate Kjell Johansen's enormous contribution to our present understanding of the cardiorespiratory physiology of air-breathing fish. His many primary publications and his 1970-review have stood the test of time and continue to inspire. In informal discussions, some contemporary comparative physiologists nonetheless argue that Johansen and Lenfant "skimmed the cream" by limiting their investigations to relatively few individual animals and often providing representative traces of physiological measurements rather than statistically solid data with small error bars and equally small stars highlighting statistically significant differences. The low n-values probably represent a relic from the past, but I venture to argue that their conclusions and interpretations remain solid. It should also be recognized that many of the surgical preparations were both demanding and innovative. Indeed, Dipnoi, in particular, are notoriously difficult to surgically operate upon and very few have succeeded to reach the level achieved by Johansen and Lenfant. Rather, the Viking and Physiologist Kjell Johansen should be remembered as the pioneer who first truly tilled the field of cardiorespiratory physiology of air-breathing fishes and provided its first synthesis. As a direct result, the study of air-breathing fishes became a fertile research field for many others to explore, even if many questions remain unexplored or many responses need to be explained.

Kjell Johansen died in March 1987, and while I started my biology degree in September 1985 at Aarhus University, I never had the privilege of meeting him in person. Johansen, however, has opened many doors for me as graduate student, post-doc and young faculty; Whenever I presented myself as being from Aarhus at a conference or in other academic contexts, I was approached by an older established physiologist who admired their friend Kjell Johansen and immediately engaged in a friendly conversation. In that way, Johansen has almost been a guardian angel in my own career.

Acknowledgments

This short article is based on many conversations and extended correspondence with many of Johansen's colleagues and friends over three decades. I am particularly indebted to Steve Wood (supervised by Kjell Johansen in the 1960s and temporarily employed at Aarhus University immediately after Johansen had established the institute for zoophysiology) for providing access to an unpublished description of the "Seattle days" written by Dr. Arthur W Martin in connection with the honorary conference at the August Krogh Institute at Copenhagen University to celebrate Kjell Johansen's career in June 1989 (see Wood et al., 1992).

References

Aaskov, M.L., Jensen, R.J., Skov, P.V., Wood, C.M., Wang, T., Malte, H., Bayley, M., 2022. *Arapaima gigas* maintains gas exchange separation in severe aquatic hypoxia but does not suffer branchial oxygen loss. J. Exp. Biol. 225, jeb243672.

Andresen, J.H., Ishimatsu, A., Johansen, K., Glass, M.L., 1987. An angiocardiographic analysis of the central circulation in the air breathing teleost *Channa argus*. Acta Zool. (Stockh.) 68, 165–171.

Armelin, V., Braga, V.H., Teixeira, M., Mariana, I., Guagnoni, I.N., Wang, T., Florindo, L.H., Henrique, L., 2021. The baroreflex in aquatic and amphibious teleosts: does terrestriality represent a significant driving force for the evolution of a more effective baroreflex in vertebrates? Comp. Biochem. Physiol. 255A, 110916.

Bateson, P., LaLand, K.N., 2013. Tinbergen's four questions: an appreciation and an update. Trends Ecol. Evol. 28, 712–718.

Bayley, M., Damsgaard, C., Thomsen, M., Malte, H., Wang, T., 2019. Learning to air-breathe: the first steps. Phys. Ther. 34, 14–29.

Brauner, C.J., Matey, V., Wilson, J.M., Bernier, N.J., Val, A.L., 2004. Transition in organ function during the evolution of air-breathing; insights from *Arapaima gigas*, an obligate air-breathing teleost from the Amazon. J. Exp. Biol. 207, 1433.

Carter, G.S., 1935. Respiratory adaptations of the fishes of the forest waters, with descriptions of the accessory respiratory organs of *Electrophorus electricus* (Linn.) (=*Gymnotus electricus auctt.*) and *Plecostomus plecostomus* (Linn.). J. Linn. Soc. 39, 219–233.

Carter, G.S., 1957. Air-breathing. In: Brown, M. (Ed.), The Physiology of Fishes. Academic Press, pp. 65–79.

Carter, G.S., Beadle, L.C., 1931. The fauna of the Paraguayan Chaco in relation to its environment. II. Respiratory adaptations in the fishes. J. Linn. Soc. 37, 327–368.

Damsgaard, C., Baliga, V.B., Bates, E., Burggren, W., McKenzie, D.J., Taylor, E., Wright, P.A., 2020. Evolutionary and cardio-respiratory physiology of air-breathing and amphibious fishes. Acta Physiol. 228, e13406.

Das, B.K., 1927. The bionomics of certain air-breathing fishes of India, together with an account of the development of their air-breathing organs. Philos. Trans. R. Soc. B 216, 183–219.

Florindo, L.H., Armelin, V.A., McKenzie, D.J., Rantin, F.T., 2018. Control of air-breathing in fishes: central and peripheral receptors. Acta Histochem. 120, 642–653.

Glass, M.L., Wood, S.C., Johansen, K., 1978. The application of pneumotachography on small unrestrained animals. Comp. Biochem. Physiol. 59A, 425–427.

Graham, J.B., 1997. Air-Breathing Fishes: Evolution, Diversity and Adaptation. Academic Press, San Diego. 1997, 299 pp.

Graham, J.B., Lai, N.C., Chiller, D., Roberts, J.L., 1995. The transition to air breathing in fishes V. Comparative aspects of cardiorespiratory regulation in *Synbranchus marmoratus* and *Monopterus albus* (*Synbranchidae*). J. Exp. Biol. 198, 1455–1467.

Ishimatsu, A., 2012. Evolution of the cardiorespiratory system in air-breathing fishes. Aqua BioSci. Monogr. 5, 1–28.

Ishimatsu, A., Itazawa, Y., 1983. Difference in blood oxygen levels in the outflow vessels of the heart of an air-breathing fish, *Channa argus*: do separate blood streams exist in a teleostean heart? J. Comp. Physiol. 149, 435–440.

Ishimatsu, A., Itazawa, Y., Takeda, T., 1979. On the circulatory systems of the snakeheads *Channa maculata* and *C. argus* with reference to bimodal breathing. Jpn. J. Ichthyol. 26, 167–180.

Iversen, N.K., Huong, D.T.T., Bayley, M., Wang, T., 2011. Autonomic control of the heart in the Asian swamp eel (*Monopterus albus*). Comp. Biochem. Physiol. 2011 (158A), 485–489.

Johansen, K., 1962. Double circulation in the amphibian, *Amphiuma tridactylum*. Nature 194, 991–992.

Johansen, K., 1966. Air breathing in the teleost, *Symbranchus marmoratus*. Comp. Biochem. Physiol. 18, 910–918.

Johansen, K., 1987. The August Krogh lecture: the world as a laboratory—physiological insights from nature's experiments. In: McLennan, H., Ledsome, J.R., McIntosh, C.H.S., Jones, D.R. (Eds.), Advances in Physiological Research. Plenum Press, New York, pp. 277–396.

Johansen, K., Lenfant, C., Schmidt-Nielsen, K., Petersen, J.A., 1968. Gas exchange and control of breathing in the electric eel, *Electrophorus electricus*. Z. Vgl. Physiol. 61, 137–163.

Johansen, K., Lykkeboe, G., Weber, R.E., Maloiy, G.M., 1976. Respiratory properties of blood in awake and estivating lungfish, *Protopterus amphibious*. Respir. Physiol. 27, 335–345.

Johansen, K., Martin, A.W., 1965a. Circulation in a giant earthworm, *Glossoscolex giganteus*. I. Contractile processes and pressure gradients in the large blood vessels. J. Exp. Biol. 43, 333–347.

Johansen, K., Martin, A.W., 1965b. Comparative aspects of cardiovascular function in vertebrates. Handbook of physiology II. Circulation 3, 2583–2614.

Johansen, K., Martin, A.W., 1966. Circulation in a giant earthworm, *Glossoscolex giganteus*. II. Respiratory properties of the blood and some patterns of gas exchange. J. Exp. Biol. 45, 165–172.

Krogh, A., 1929. The progress in physiology. Am. J. Physiol. 90, 243–251.

McKenzie, D.J., Campbell, H.A., Taylor, E.W., Micheli, M., Rantin, F.T., Abe, A.S., 2007. The autonomic control and functional significance of the changes in heart rate associated with air breathing in the jeju, *Hoplerythrinus unitaeniatus*. J. Exp. Biol. 210, 4224–4232.

Milsom, W.K., 2018. Central control of air-breathing in fish. Acta Histochem. 120, 691–700.

Milsom, W.K., Kinkead, R., Hedrick, M., Gilmour, K., Perry, S.F., Gargaglioni, L., Wang, T., 2022. Control of breathing in ectothermic vertebrates. Compr. Physiol. 12, 3869–3988.

Nunan, B.L.C.Z., Silva, A.S., Wang, T., da Silva, G.S.F., 1989. Respiratory control of acid-base status in lungfish. Comp. Biochem. Physiol. 237A, 110533.

Pelster, B., Wood, C.M., 2018. Ionoregulatory and oxidative stress issues associated with the evolution of air-breathing. Acta Histochem. 120, 667–679.

Perry, S.F., Euverman, R., Wang, T., Loong, A.M., Chew, S.F., Ip, Y.K., Gilmour, K.M., 2008. Control of breathing in African lungfish (*Protopterus dolloi*): a comparison of aquatic and cocooned (terrestrialized) animals. Respir. Physiol. Neurobiol. 160, 8–17.

Randall, D.J., Burggren, W.W., Farrell, A.P., Haswell, M.S., 1981. The Evolution of Air Breathing in Vertebrates. Cambridge University Press, Cambridge. 133 pp.

Schmidt-Nielsen, K., 1998. The Camels Nose—Memoirs of a Curious Scientist. Island Press, Washington DC. 339 pp.

Shartau, R.B., Brauner, C.J., 2014. Acid-base and ion balance in fishes with bimodal respiration. J. Fish Biol. 84, 682–704.

Skals, M., Skovgaard, N., Taylor, E.W., Leite, C.A.C., Abe, A.S., Wang, T., 2006. Cardiovascular changes under normoxic and hypoxic conditions in the air-breathing teleost *Synbranchus marmoratus*: importance of the venous system. J. Exp. Biol. 209, 4167–4173.

Smith, H.W., 1930. Metabolism of the lungfish, *Protopterus æthiopicus*. J. Biol. Chem. 88, 97–130.

Smith, H.W., 1932. Kamongo or the Lungfish and the Padre. Viking Press NY.

Turko, A.J., Rossi, G.S., Wrigth, P.A., 2021. More than breathing air: evolutionary drivers and physiological implications of an amphibious lifestyle in fishes. Physiology (Bethesda) 36, 307–314.

Wang, T., Wood, S.C., 2020. Mogens Lesner Glass (1946-2018). Braz. J. Med. Biol. Res. 53, e10838.

Weber, R.E., Johansen, K., Lykkeboe, G., Maloiy, G.M.O., 1977. Oxygen-binding properties of hemoglobins from estivating and active African lungfish. J. Exp. Zool. 199, 85–96.

Wood, S.C., Weber, R.E., Hargens, A.R., Millard, R., 1992. Physiological Adaptations in Vertebrates: Respiration, Circulation, and Metabolism. Marcel Dekker Inc, New York.

Zaccone, G., Lauriano, E.R., Capillo, G., Kuciel, M., 2018. Air- breathing in fish: air- breathing organs and control of respiration: nerves and neurotransmitters in the air-breathing organs and the skin. Acta Histochem. 120, 630–641.

Chapter 6

AIR BREATHING IN FISHES*,☆

KJELL JOHANSEN

Chapter Outline

I. OCCURRENCE AND BIONOMICS OF AIR-BREATHING FISHES

Natural selection of air-breathing habits in fishes has occurred many times and in diverse ways during vertebrate evolution.

The majority of extant fishes showing structural adaptations for direct use of atmospheric oxygen are tropical freshwater or estuarine forms. Most air-breathing fishes are teleosts but typically the dipnoan, chondrostean,

*This chapter was written while the author was supported by grants GB 7166 from the National Science Foundation and HE 12071 from the National Institutes of Health.
☆This is a reproduction of a previously published chapter in the Fish Physiology series, "1970 (Vol. 4)/Air Breathing in Fishes: ISBN: 978-0-12-350404-3; ISSN: 1546-5098".

Fish Physiology, Vol. 40A. https://doi.org/10.1016/bs.fp.2023.09.012

and holostean fishes are dominated by air-breathing forms. Reports of air-breathing habits among elasmobranch fishes have not been adequately confirmed (George, 1953).

A smaller number of air-breathing fishes live in temperate regions, and in exceptional cases such as the bowfin, *Amia calva,* they occupy waters seasonally frozen over.

A shortage of dissolved oxygen constitutes the primary environmental condition which has stimulated the development of air-breathing in fishes (Barrell, 1916; Carter and Beadle, 1931a), although such adaptations may also have been correlated with behavior or activity patterns. A few forms, inhabiting oxygen-deficient tropical swamps, have acquired the faculty of sustained air breathing when entirely removed from water during droughts. Such periods are passsed in an estivating condition. Air breathing has also evolved among fishes living in torrential mountain streams in Asia. Here, the air-breathing habit is an adaptation to drought only since these streams carry well-oxygenated water during the wet season. These fishes either estivate or migrate over land to neighboring streams (Das, 1927, 1940; Saxena, 1963). Many tropical swamps and stagnant tributaries to larger rivers have very acid and CO_2-rich water. Adverse conditions for aquatic breathing with gills also result when water gets excessively turbid with suspended material which may cover the gill surfaces.

The principal causes of oxygen depletion in tropical freshwaters are related to the abundance of organic matter which provides a substrate for rapid oxygen consumption by microorganisms. The prevailing high temperatures of the tropics accelerate oxygen depletion by increasing the O_2 requirements of the microfauna as well as the other inhabitants of the swamps. Diurnal temperature variations are small, and thus little vertical water movement occurs to promote reentry of atmospheric O_2 and the swamps become stagnant. Oxygen release from photosynthetic activity is very slight because of the low intensity of light below the dense foliage of the prolific vegetation in regions of tropical swamps and rain forests. The same foliage cover prevents wind disturbance from effectively stirring and thus aerating the water.

Naturally tropical waters show a vertical gradient in O_2 concentration owing to a more well-oxygenated surface layer. However, as close as 1 cm to the surface, O_2 concentrations can be as low as 2% of saturation value. Several inhabitants of tropical swamps show behavioral adaptations for utilization of the more O_2-rich surface layer by breathing at or near the surface. Often O_2 availability is so limiting that all species not adapted for direct air breathing must utilize the surface water as a source for oxygen (Carter and Beadle, 1931a,b).

Among air-breathing fishes, those occupying brackish and marine habitats are rare because of the well-stirred turbulent conditions of these waters. However, several members of marine gobiid and blennid families show structural adaptations for air breathing (Schöttle, 1931; Oglialoro, 1947). Estuarine waters may show large spatial, diurnal, and seasonal fluctuations in oxygen availability (Todd and Ebeling, 1966). More exceptionally, fishes like *Periopthalmus* have developed air-breathing habits in support of behavioral acts such as feeding and escape reactions.

II. NATURE OF THE STRUCTURAL ADAPTATIONS FOR AIR BREATHING

Structural adaptations for air breathing in fishes are extremely diverse. The extent of modifications from structures typical of an aquatic breather often reflects the relative role of air breathing in overall gas exchange. Some species rely predominantly on water breathing with gills and only supplement gas exchange with air breathing when adverse respiratory qualities in the water make aquatic breathing insufficient or too costly for extraction of the required oxygen. Others are obligate air breathers and succumb if denied access to air for a short time (Rauther, 1910; Johansen, 1968).

All air-breathing fishes have a hollow compartment as a receptacle or holding space for air. The air chamber may have evolved as a modification of an already existing structure or it may be a neomorphic structure. It is richly supplied with blood vessels, and fine capillaries or lacunar spaces invest the membranes bordering the air space. The interface between blood and air is specialized to reduce the diffusion distance and other diffusion barriers between air and blood. Frequently, various forms of structural changes such as foldings, papillations, or arborizations of the gas exchange membranes have resulted in a considerable surface expansion that enhances the diffusion exchange. Such structural specialization is more prominent in species utilizing air breathing as a major means of gas exchange. A concurrent trend is for the primary aquatic gas exchange organs, the gills, to degenerate progressively to few, coarse, and sparsely perfused gill filaments as air breathing assumes a more important role (Carter and Beadle, 1931b; Dubale, 1951). In some species, such as the electric eel, *Electrophorus electricus,* the gills have become vestigial (Evans, 1929; Johansen *et al.,* 1968). Air-breathing fishes also show structural specializations for rhythmic ventilation of the air-breathing organs, while the muscle and skeletal parts involved with purely aquatic breathing have become reduced or modified for the new purpose of breathing air.

Table I lists representative species of air-breathing fishes arranged according to the type of their structural adaptation. The list is not exhaustive, and the interested reader is referred to more specialized morphological texts (Rauther, 1910; Carter and Beadle, 1931b; Das, 1940; Saxena, 1963).

A. Structural Derivatives of the Mouth and Pharynx as Air-Breathing Organs

By far the most common and yet the most diversified structural adjustments for air breathing are those associated with the branchial, opercular, and pharyngeal cavities.

Table I A List of Representative Air-Breathing Fishes Arranged According to the Type of Their Structural Adaptation[a]

Type of air breathing	Species	Habitat	Relative importance of air breathing	Remarks	Reference
Gills Pharyngeal and opercular walls	*Symbranchus marmoratus*	Swamps and rivers, South America	Essential in O_2-deficient water and during estivation	Gills capable of both aerial and aquatic breathing	Taylor (1913) Johansen (1966)
Skin	*Hypopomus brevirostris*	Swamps and rivers, South America		Gills capable of both aerial and aquatic breathing	Carter and Beadle (1931b)
	Anguilla anguilla	Rivers in Europe, Asia, Africa, North America	Accessory during migrations over land	Rhythmic inflation of branchial chamber with air, when air exposed	Krogh (1904) Berg and Steen (1965)
	Periopthalmus sp.	Marine tropical shores, estuarine and brackish water	Essential during voluntary air exposure	Air breathing appears more related to behavior than to environmental conditions	Schöttle (1931)
	Gillichtys mirabilis	Marine shallow tidal flats	Important in oxygen-deficient water	Remarkable ability to change from aquatic to aerial breathing	Todd and Ebeling (1966)
Pharyngeal epithelium, papillated and vascular	*Electrophorus electricus*	Rivers and swamps of South America	Air breathing obligatory	Air breathing dominates O_2 absorption; CO_2 elimination via vestigial gills and skin	Johansen *et al.* (1968)

Air chambers as diverticula from the opercular and pharyngeal cavities (opercular and pharyngeal lungs)	*Clarias* sp.	Tropical pools, Asia	Important in oxygen-deficient water	Capable of estivation, migrates overland	Rauther (1910) Moussa (1956)
	Anabas scandens	Tropical pools, Asia	Air breathing accessory but necessary in O_2-deficient water	Estivates, migrates out of water	Rauther (1910) Das (1927)
	Amphipnous cuchia	Rivers in northern India and Burma	Facultative air breather	Estivates, gills more reduced than in other air-breathing fishes	Das (1940)
	Ophiocephalus punctatus	Tropical pools, Asia and Africa	Air breathing obligatory	Estivates	Rauther (1910) Das (1927) Qasim *et al.* (1966)
Stomach or intestine modified for air breathing	*Hoplosternum* sp.	Swamps, South America	Obligate air breather	Intestinal breather	Carter and Beadle (1931b)
	Misgurnus fossilis	Temperate and tropical rivers, Europe	Air breathing accessory	Intestinal breather	Calugareanu (1907)
	Ancistrus anisitsi	Swamps, South America	Air breathing accessory	Stomach breather	Carter and Beadle (1931b)
Air bladder functions as lung	*Polypterus* sp.	Lakes and rivers in Africa	Survives on aquatic breathing in well-aerated water. Air breathing obligatory under most natural conditions	Activity markedly increases air breathing	Budgett (1900) Magid (1966)

Continued

Table I A List of Representative Air-Breathing Fishes Arranged According to the Type of Their Structural Adaptation—Cont'd

Type of air breathing	Species	Habitat	Relative importance of air breathing	Remarks	Reference
Air bladder functions as lung	*Amia calva*	Rivers and lakes, North America	Air breathing accessory, increased importance at increased temperatures and/or during activity	Capable of estivation	Johansen *et al.* (1970)
	Arapaima gigas	Rivers and swamps, South America	Air breathing obligatory, dies quickly if denied access to air	Very large fish, exceeds 3 meters	Sawaya (1946)
	Neoceratodus forsteri	Rivers, East Australia	Air breathing accessory	Increased importance in O_2-deficient water and during activity	Grigg (1965) Lenfant *et al.* (1966) Johansen *et al.* (1967)
	Lepidosiren paradoxa	Rivers and swamps, South America	Obligatory air breather, succumbs if denied access to air	Structurally most advanced of all air-breathing fishes, estivates	Johansen and Lenfant (1967) Lenfant *et al.* (1970)
	Protopterus sp.	Lakes and swamps, Africa	Air breathing obligatory	Estivates	Lenfant and Johansen (1968) Johansen and Lenfant (1968)

[a]For complete lists, see Rauther (1910), Carter and Beadle (1931b), Das (1940), and Saxena (1963).

Piscine gills are designed to function in water where lack of net gravitational forces prevents collapse of the fine filaments and lamellae. However, the collapse of these in air makes the gills of most fishes useless as organs of aerial gas exchange. A very few fishes, such as *Symbranchus* and *Hypopomus,* are able to use their gills for air breathing. They fill the buccal cavity with air regularly at the surface, and the used air is voided from the mouth or the opercular openings (Taylor, 1913; Carter and Beadle, 1931b; Johansen, 1966). The gills of these fishes are structurally modified to prevent their collapse in air. Most notably they show a high ratio between the breadth and height of the secondary lamellae. The ratio is more than twice that in purely aquatic fishes or fishes possessing other types of structural adaptations for air breathing (Carter and Beadle, 1931b). The epithelium of the buccal and pharyngeal cavities of these two fishes is richly vascularized and plays a supporting role in aerial gas exchange.

The common eel, *Anguilla anguilla* (Berg and Steen, 1965), and species of *Periopthalmus* (Schöttle, 1931) also employ gill breathing successfully during air exposure.

A further development of aerial mouth breathing is seen in fishes with foldings and papillations of the buccal epithelia projecting into the buccal cavity and forming extensive vascular surfaces for aerial gas exchange. The electric eel, *Electrophorus electricus,* offers a striking example (Evans, 1929; Böker, 1933; Richter, 1935; Carter, 1935; Johansen *et al.*, 1968). In this fish the papillated projections are distributed over both the floor and the roof of the mouth, while smaller prominences are present on the branchial arches and the lateral branchial walls. The system of papillae forms a labyrinth of air passages. Studies on formalin-fixed material revealed a surface area of the mouth respiratory organ about 15% of the total body surface (Johansen *et al.,* 1968). This is considerably less than for either gills in purely aquatic fishes or lungs of higher vertebrates. Yet the mouth organ in *Electrophorus* functions well enough to have permitted the gills of adult fishes to become almost vestigial (Fig. 1).

The next step in a progressive structural adjustment to air breathing includes the formation of an air chamber extending from the buccal or pharyngeal cavities. In the Ophiocephalidae with several air-breathing species in tropical marshes of Asia and Africa, the air chamber lies dorsal to the branchial area and becomes partly enclosed within the cavity of the skull (Munshi, 1962).

In *Amphipnous cuchia,* another air-breathing fish living in India, the paired air chambers referred to as pharyngeal lungs (Das, 1940) are more free and extend several centimeters behind the head. The internal walls of the air chamber are very vascular and may show surface expanding structures in the form of rosettelike arborizations protruding into the lumen. In others, internal trabeculation may give the air chamber a resemblance to an amphibian lung.

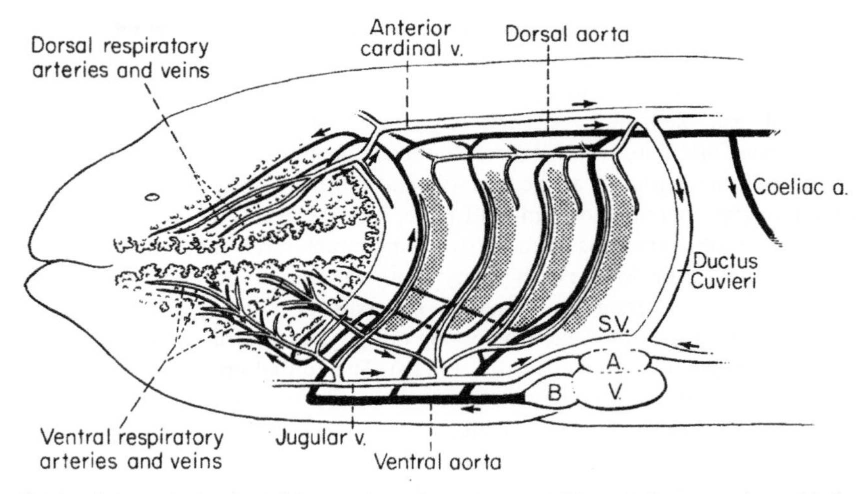

Fig. 1. Schematic drawing of the mouth respiratory organ and its vascular connections with the heart and central circulation in the electric eel. The arrows indicate the direction of the blood flow. From Johansen *et al.* (1968). A, Atrium; B, bulbusarteriosus; V, ventricle; a, artery; v, vein.

Another structural variation is seen in fishes where the opercular cavities have become richly vascularized and serve as receptacles for air. When air is lodged in such chambers they stand out in balloonlike fashion. The opercular walls have become thin and elastic and their bone structures have regressed. The opercular openings are either fused to one small channel ventrally or have otherwise become reduced in size. These modifications and others involving the intrinsic muscles of the opercular walls permit ventilatory movements and renewal of air inside the opercular chambers. Examples of such specialization are seen in *Hypopomus brevirostris, Monopterus* sp., and several species of gobiid fishes (e.g., *Pseudapocryptes lanceolatus*) (Wu and Kung, 1940).

More elaborate use of the opercular cavities in air breathing results when specialized diverticula develop from their dorsal side commonly between the hyoid and the first branchial arches. Two distinctly different types of such opercular air chambers or lungs are found. In some cases they remain free and extend posteriorly. Internally they possess ridges and trabeculations or other protrusions expanding the highly vascular internal walls. Often these air sacs extend half the length of the animal and lie embedded within the myotomes of the body wall, a fact which undoubtedly bears significance in the filling and emptying of the sacs. Indian fishes of the genus *Saccobranchus* are well-known examples of this type of structural adaptation (Rauther, 1910).

The other variety of opercular air chamber results when the opercular diverticulum extends inside the skull to take the form of a labyrinthlike organ

(*Anabantidae*) or corallike dendritic formations as seen in species of *Clarias* (Moussa, 1956; Munshi, 1961).

Members of the Anabantidae (*Anabas, Macropodus, Trichogaster,* and others) typically show the air labyrinths starting as outgrowths from the epibranchial section of the first branchial arch. Laminated projections contribute to the surface expansion inside the chambers. The detailed structure of the labyrinthine organ is extremely complicated (Munshi, 1968). The diffusion distance from air to blood in *Trichogaster* has been measured as the smallest of all recorded in respiratory organs, showing a minimal value of 600 Å, a medium range of 1300–1700 Å, and a maximum value of 1.2 μ (Schulz, 1960). Based on electron microscopy, Schulz reported that capillaries of some areas of the labyrinthine organ lack both basal membrane and endothelium, thus leaving only a thin epithelium intervening between blood and air.

In *Clarias* the two remarkable vascular rosettelike trees which fill the air chambers on each side receive structural support from gill arches 2 and 4 and their vascular supply from the first and fourth afferent branchial arteries (Moussa, 1956). Figure 2 shows the gross appearance of the labyrinth in *Clarias lazera*.

B. Structural Adaptations of the Gastrointestinal Tract for Air Breathing

In some fishes portions of the gastrointestinal tract, notably the stomach or part of the small intestine, have been modified to serve in gas exchange between the blood and rhythmically swallowed air. The used air is voided at the anus or the mouth. Members of the families Loricaridae, e.g., *Plecostomus plecostomus* (Carter, 1935), and Callichthyidae, e.g., *Callichthys* and *Hoplosternum* (Carter and Beadle, 1931b) offer typical examples of gastrointestinal air breathers. Most fishes employing the gastrointestinal tract for gas exchange use it as an accessory breathing organ when aquatic conditions become severely oxygen deficient. Others depend on it during short migrations over land (*Plecostomus*), and a few are obligate air breathers that need to supplement the aquatic gas exchange with air breathing at all times.

C. The Air Bladder as a Respiratory Organ

It is now generally conceded that the air bladder in fishes originally functioned as an accessory respiratory organ. In most extant actinopterygian fishes the air or swim bladder serves primarily in buoyancy control or sound production or sound detection. In dipnoan fishes the air bladder has retained and further evolved the respiratory function it allegedly possessed in the crossopterygian fishes that also gave rise to amphibians and land vertebrates. Members of the primitive orders Chondrostei (*Polypterus*) and Holostei (*Amia* and *Lepisosteus*) have also retained a respiratory function of the air bladder

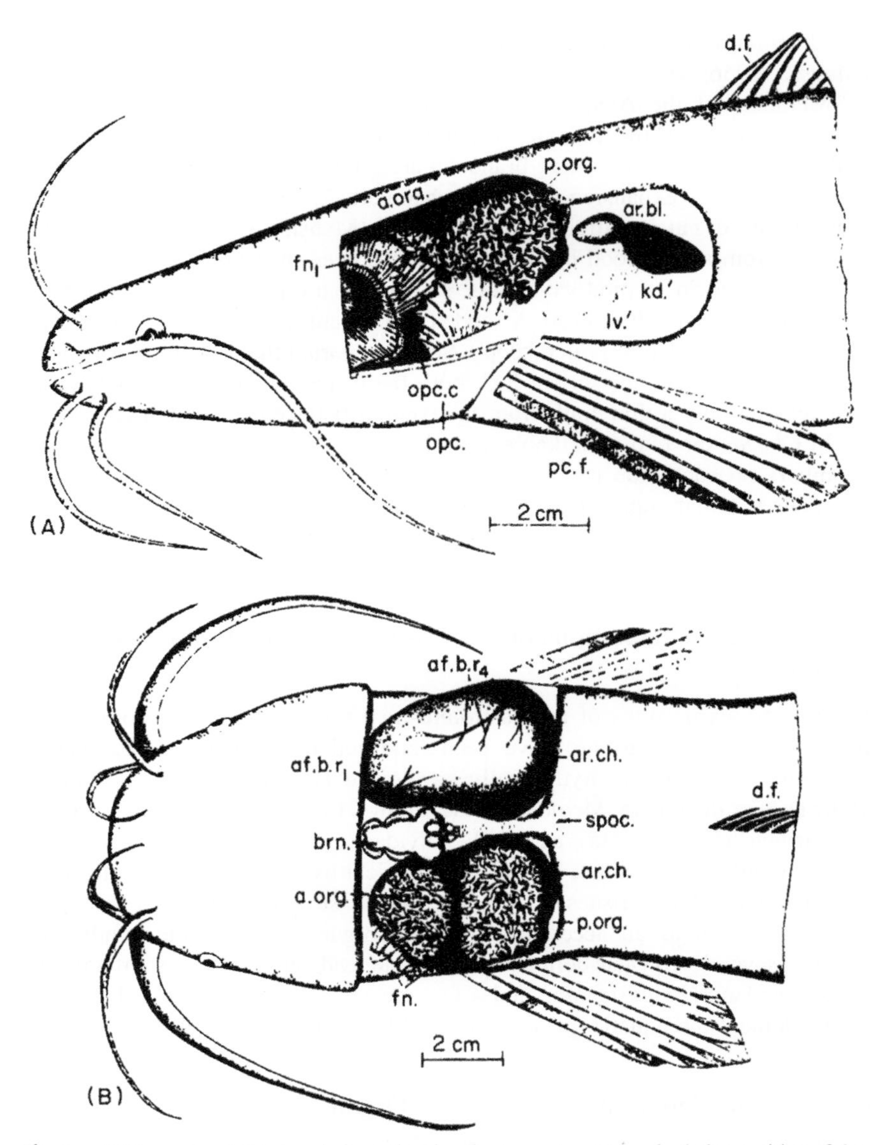

Fig. 2. (A) Lateral and (B) dorsal views showing the gross structure and relative position of the air chamber in *Clarias lazera*. In A and to the left in B the wall of the air chamber has been removed to expose the surface expanding arborescent organs. From Moussa (1956). af.b.r., 1 and 4, ramifications of the first and fourth afferent branchial vessels in the wall of the air-chamber; a.org., anterior arborescent organ; ar.bl., air or swim bladder; ar.ch., air-chamber; brn., brain; d.f., dorsal fin; $fn._{1\text{-}4}$, serial order of fans; kd'., part of kidney; lv'., part of liver; opc., operculum; opc.c., opercular cavity; pc.f., pectoral fin; p.org., posterior aborescent organ; spoc., supraoccipital.

but fall short of the refinement in structural adjustment for air breathing seen in the dipnoans (Klika and Lelek, 1967). Among modem teleosts some have secondarily developed the air bladder into a respiratory organ such as seen in the South American fishes *Erythrinus uniteniatus* and *Arapaima gigas* (Carter and Beadle, 1931b; Sawaya, 1946).

Only in the lungfishes, and of these *Protopterus* and *Lepidosiren* in particular, has the air bladder evolved structurally to resemble a lung as it exists in lower vertebrates. Internal septa, ridges, and pillars divide the air space into smaller compartments opening into a median cavity. Further subdivisions of these compartments by progressively smaller reticulated septa terminates in alveoli-like pockets richly covered with blood vessels. *Neoceratodus* has one lung, while *Protopterus* and *Lepidosiren* have two lungs fused in their anterior region where the pneumatic duct and the pneumatic sphincter connect the lungs with the esophagus. The short airway duct, which in no way resembles the air distributing system of the trachea and bronchii in terrestrial vertebrates, is richly supplied with smooth muscle as is the lung parenchyma itself. The smooth musculature is important for the mechanics of breathing and may exert additional influence in distributing the air inside the lung (Grigg, 1965; Johansen and Reite, 1967). Electron microscopy has revealed that the fine structure of the lung in *Protopterus* does not differ basically from that in higher vertebrates (DeGroodt *et al.*, 1960; Schulz, 1960). Klika and Lelek (1967), reporting on lung structure in *Protopterus,* state that the respiratory epithelium adjoins the basement membrane which in turn is separated from the endothelial basement membrane by a thin layer of collagen fibrils. The approximate blood to air diffusion distance in *Protopterus* is about 0.50 μ. The *Protopterus* lung is much more similar to the amphibian lung than that of *Polypterus* which has a smooth inner surface much less vascular than in the lungfishes. Figure 3A shows the gross structure of the *Protopterus* lung.

Structural specializations in the perfusion pattern through the gas exchange organs and the vascular linkage between them and the general circulation is of utmost importance for efficient gas exchange. In air-breathing fishes presumably the highest degree of efficiency will prevail when the air-breathing organ is perfused with systemic venous blood and when the respiratory efferent blood is dispatched directly for systemic arterial distribution. Most air-breathing fishes also employ aquatic breathing and efficient perfusion of the gills and/or skin is important since these areas are the primary sites for CO_2 elimination. The vascular system should hence ensure that both the air- and water-breathing respiratory organs are perfused before the blood is turned over to the nutritive circulation.

Figure 4 summarizes the various types of circulatory arrangements seen in air-breathing fishes. All types fall far short of matching the ideal perfusion pattern existing in purely aquatic fishes (type A) or in the highest vertebrates where the respiratory and systemic vascular beds are coupled in direct series. All air-breathing fishes show a parallel linkage between the two types of

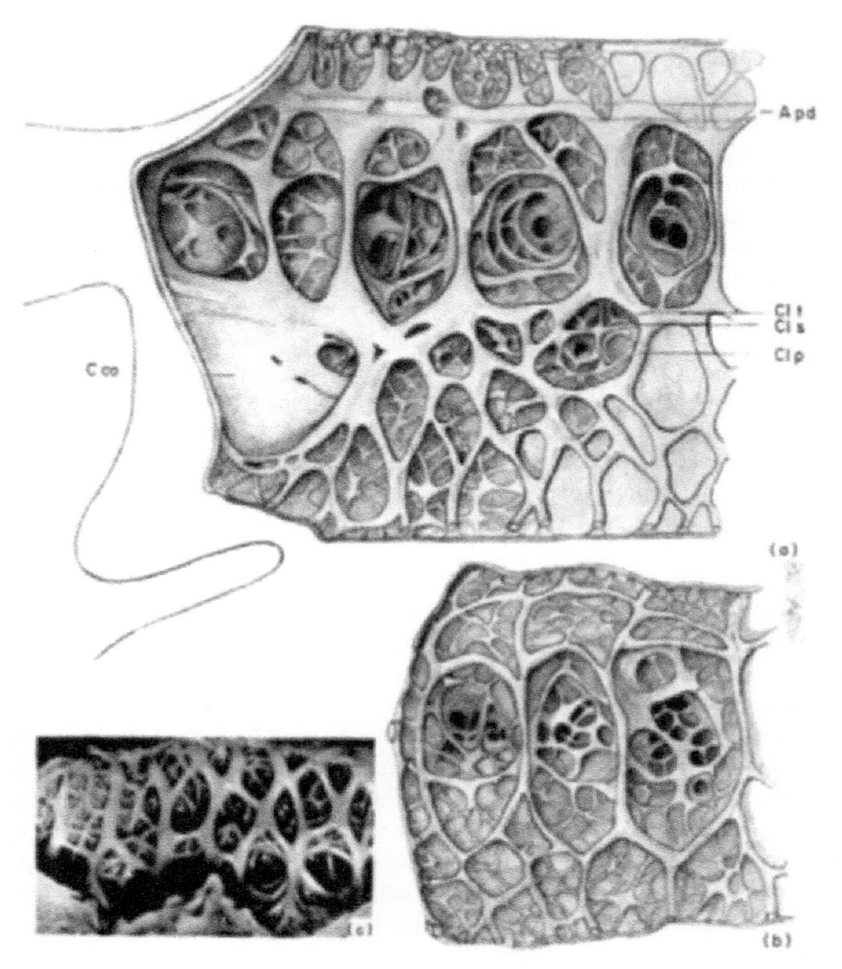

Fig. 3. Gross structure of the anterior region of the lung in *Protopterus aethiopicus* (Heckel). (b and c) Posterior region of the *Protopterus* lung showing extensive trabeculation. From Poll (1962). Apd, dorsal pulmonary artery; C co, communicating cavity between the two lungs; Cl p, primary septum; Cl s, secondary septum; Cl t, tertiary septum.

vascular beds, and only the lungfishes are structurally adapted to minimize shunting between these vascular beds. In the electric eel (type B), the entire gain in oxygenation resulting from perfusing the mouth is shunted back to the systemic veins before blood is redistributed to the systemic and respiratory arteries. This deficiency prevails to various degrees in all types except the very few fishes able to use their gills alternately for aerial and aquatic breathing [e.g., *Symbranchus* or the few fishes employing air breathing with specialized air chambers extending from the opercular cavities such as *Clarias* (type C)].

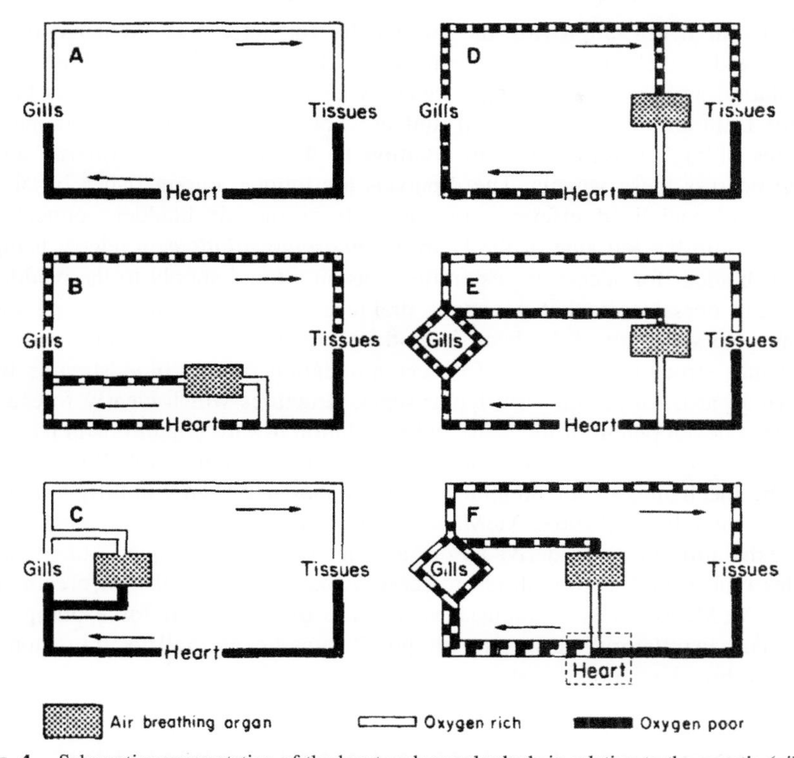

Fig. 4. Schematic representation of the heart and vascular beds in relation to the aquatic (gills) and aerial respiratory organs in air-breathing fishes. The amount of white and black represents only the approximate level of oxygenated and deoxygenated blood carried in the vessels. (A) General piscine arrangement with the branchial (gill) and systemic (tissues) vascular beds in direct series. (B) Air-breathing organ derived from pharyngeal and/or opercular mucosa. Afferent vessels to air-breathing organ arranged in parallel with branchial circulation and derived from afferent branchial vessels. Efferent vessels from air-breathing organ connected to systemic veins. Arrangement typical of *Monopterus, Ophiocephalus, Electrophorus, Amphipnous, Periopthalmus,* and *Anabas.* (C) Gills, buccal mucosa, or chambers extending from the opercular cavity serving as air-breathing organs. Afferent vessels to air-breathing organ arranged in parallel with the branchial vascular bed. Efferent circulation largely connected with the efferent branchial circulation. Systemic arterial blood highly oxygenated. *Clarias, Saccobranchus,* and in part, *Symbranchus* and *Hypopomus* typical of type C. (D) Air-breathing organ associated with the gastrointestinal tract. Afferent circulation derived from the dorsal aorta (systemic arteries). Efferent circulation connected to systemic veins. Arterial blood mixed. Arrangement typical in *Hoplosternum, Plecostomus,* and *Ancistrus.* (E) Air bladder serving as air-breathing organ. Specialized afferent vessel derived from the most posterior epibranchial arteries. Efferent circulation connected to systemic veins. Arterial blood mixed. Arrangement typical of *Polyterus, Amia,* and *Lepisosteus.* (F) Air bladder structurally advanced to resemble amphibian lung. Specialized afferent and efferent circulation in parallel with systemic arterial circulation. Partial septation of the heart allows a high degree of selective perfusion in the respiratory and systemic circuits. Arrangement seen in lungfishes and in *Protopterus* and *Lepidosiren* in particular.

Only in fishes employing the air bladder for aerial gas exchange have specialized vessels developed as connections between the primary branchial circulation and the gas exchange organ (types E and F). These vessels take origin from the sixth aortic arch and are homologue with the pulmonary arteries of higher vertebrates. In addition to the dipnoans, *Polypterus* and *Amia* possess these vessels. The dipnoans are unique among fishes by also having a specialized efferent circulation from the air bladder connected directly with the left side of the heart. *Gymnarchus niloticus,* a teleost using the air bladder for accessory respiration, has an arterial supply to the bladder from the dorsal aorta but a venous drainage connecting directly with the atrium. No septation of the heart is evident. However, the heart in dipnoans is in part structurally adapted for accommodation of two bloodstreams by showing partial atrial and ventricular septa, structures which clearly forecast a further development of the heart and circulation toward conditions in terrestrial vertebrates (Johansen and Hanson, 1968; Johansen and Hol, 1968).

The mechanics of air breathing in lungfishes has been carefully studied (Bishop and Foxon, 1968; McMahon, 1969). The mechanics of inhalation and exhalation of air is derived from a series of basically aquatic breathing cycles typical of fishes modified to serve a specific step in the air-breathing cycle. McMahon (1969) emphasizes the role of the buccal force pump as the main ventilatory mechanism in purely aquatic as well as transitional bimodal breathers among fishes and amphibians.

III. PHYSIOLOGICAL ADAPTATIONS IN AIR-BREATHING FISHES

A. Respiratory Properties of Blood

Studies of respiratory properties of blood from air-breathing fishes have disclosed adaptive features which correlate with environmental factors and the relative importance of air and water breathing (Willmer, 1934; Fish, 1956; Swan and Hall, 1966; Lenfant and Johansen, 1968).

Table II summarizes available information in representative species. A large variability is apparent in hemoglobin concentration and O_2 capacity, with no adaptive trend evident in air-breathing fishes as a group. However, a comparison of tropical fishes studied by Willmer (1934) disclosed a definite tendency for O_2 capacities to be higher in the air-breathing forms. This difference probably reflects the nature of the fishes' habitats, particularly the degree of O_2 deficiency rather than the habit of air breathing per se. The fast flowing rivers and creeks where the purely aquatic breathers *Myelus setiger* and *Hoplias malabaricus* live are well aerated, whereas the habitat of the air-breathing fishes *Hoplosternum* and *Electrophorus* can be severely O_2 deficient.

Table II Some Respiratory Properties of Blood in Representative Air-Breathing Fishes[a]

Species	Hemoglobin (%)	Hematocrit (%)	O_2 capacity (vol. %)	P_{10} at pH_a	Bohr effect	Root effect	Haldane effect	Standard bicarbonate	Buffering capacity	Reference
Electrophorus	11.2	41.0	13.90	10.7 (7.57; 28°C)	−0.680	None	4.8 (30)[b]	33.5	4.7	Johansen *et al.* (1968)
Electrophorus	15.9[c]		19.75	15						Willmer (1934)
Lepidosiren[d] (juvenile)	5.4[c]	19.0	6.80	10.8 (7.6; 23°C)	−0.234	None	3.8 (27)	15.0	1.49	Johansen and Lenfant (1967)
Lepidosiren[d] (adult)	6.6	28.0	8.25		−0.295	None	1.5 (25)	27.2	2.43	Lenfant *et al.* (1970)
Neoccratodus[d]	7.2[c]	35.0	9.00	12.5 (7.6; 18°C)	−0.620	None	4.0 (10)	11.0	1.33	Lenfant *et al.* (1966)
Hoplosternum			18.1	15	Slight	None				Willmer (1934)
Myelus setiger			10.7	20	Large	Large				Willmer (1934)
Hoplias malabaricus			6.5	20	Large	Large				Willmer (1934)
Amia calva	5.4	23.0								
Protopterus[d]	7.6	30.5	9.50		−0.470	None	1 (25)	20	1.52	Lenfant and Johansen (1968)
Protoplerus				27.3 (7.6; 23°C)	−0.335					Swan and Hall (1966)

Continued

Table II Some Respiratory Properties of Blood in Representative Air-Breathing Fishes—Cont'd

Species	Hemoglobin (%)	Hematocrit (%)	O_2 capacity (vol. %)	P_{10} at pH_a	Bohr effect	Root effect	Haldane effect	Standard bicarbonate	Buffering capacity	Reference
Protoplerus										Fish (1956)
Symbranchus[d]	13.0	47	17.30	7.2 (7.60; 25°C)	–0.400	None	2.1 (25)	24.1	2.70	Lenfant *et al.* (1970)
Protopterus[e]	7.8[c]	30.1	9.73	25.0 (7.64; 25°C)	–0.280	None	2.30 (27)	34.2	2.60	Lahari *et al.* (1968)

[a]Myelus setiger and Hoplias malabaricus are purely aquatic breathers included for comparison.
[b]Calculated from true plasma.
[c]Derived from O_2 capacity using O_2 combining power of 1.25 ml O_2/g Hb.
[d]Maximum values of Hb, hematocrit, and O_2 capacity.
[e]Average given by authors.

In evaluating this type of adaptive adjustment, one must recognize the relative importance of air and aquatic breathing in the total gas exchange of these fishes. If water breathing is the dominate mode and air breathing is only accessory or supplemental, adaptive adjustment of the blood to meet the conditions in the water may be expected. An early documentation by Krogh and Leitch (1919) showed that aquatic species tended to show higher O_2 capacities when living in O_2 deficient water. The obligate air breathers, among fishes, have an ample reservoir of atmospheric oxygen available. However, these fishes may be compared with the diving animals among higher vertebrates which, by means of high oxygen capacities, increase their O_2 stores and consequently extend the time between surfacings. Sluggish or active habits also may influence hemoglobin levels in fishes (Root, 1931). Still another type of selection pressure for development of high O_2 capacities in air-breathing fishes has been suggested and exemplified in the electric eel, *Electrophorus* (Johansen *et al.*, 1968), and applies to most other air-breathing fishes. This relates to the shunting of respiratory efferent blood into systemic veins which results in systemic arterial perfusion with mixed blood (Fig. 4). Thus arterial blood never gets fully saturated, a condition which obviously reduces the efficiency of the gas transport system. The situation in lungfishes differs considerably in that recirculation of the well-oxygenated blood from the lung is largely prevented by a separate pulmonary vascular circuit and a partial division of the heart and its outflow channels.

The general shape of O_2–Hb dissociation curves conform to those of other fishes. Oxygen affinity generally decreases in vertebrates with increasing dependence on aerial gas exchange (McCutcheon and Hall, 1937; Lenfant and Johansen, 1967). Another generalization relates an increased O_2 affinity to the ability of a species to survive in an O_2 poor medium (Krogh and Leitch, 1919). The applicability of these generalizations to air-breathing fishes depends on the relative importance of air breathing in their bimodal gas exchange. Among species emphasizing water breathing, there is a tendency for higher O_2 affinity to correlate with the most O_2-deficient habitats. No clear-cut adaptive trends are apparent from a comparison of the obligate air breathers and those practicing occasional supplemental air breathing. In fishes like *Electrophorus* in which the O_2-rich blood is admixed with O_2-deficient blood before perfusion of the arteries, a high O_2 affinity would improve the efficiency loss caused by the mixed condition of arterial blood.

Several authors have claimed evidence of a relationship between the CO_2 sensitivity of the O_2-Hb dissociation and the habitat and air-breathing habits of fishes. Willmer (1934) offered convincing data to show that tropical fishes from larger rivers and fast flowing creeks with well-aerated water of low CO_2 content have blood which is adversely affected by CO_2 and pH changes. Conversely, blood from fishes living in stagnant swamps and creeks with O_2-deficient, CO_2-rich water were distinctly less sensitive and often even insensitive to CO_2 changes (Fig. 5). For instance, if *Myelus setiger,* normally

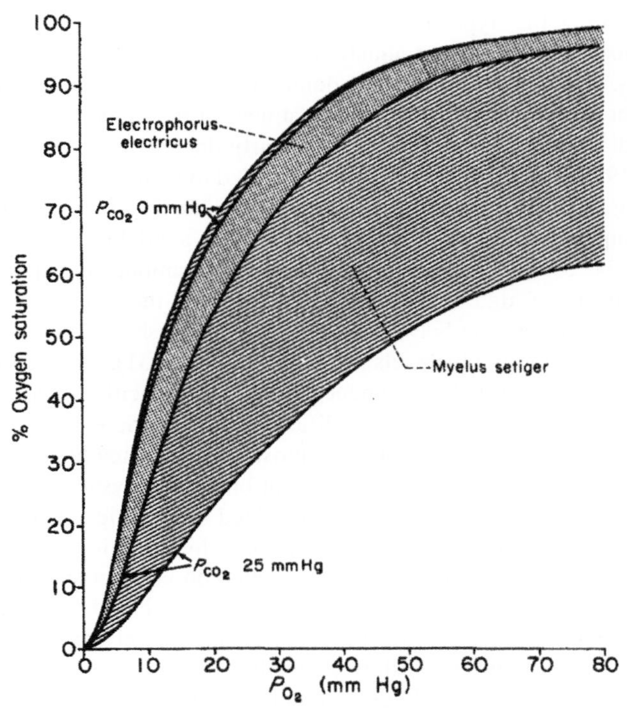

Fig. 5. Oxygen dissociation curves showing a conspicuous Bohr effect in *Myelus* living in well-aerated CO_2-free water. A slight Bohr effect is present in *Electrophorus* living in stagnant oxygen-deficient, CO_2-rich water. Redrawn from Willmer (1934).

living in well-aerated CO_2-free water, was transferred to a typical swamp habitat, its blood could never be more than 40% saturated with oxygen because of the large Root and Bohr shifts. It has been argued that the diminishing sensitivity to CO_2 in fishes normally living in a CO_2-rich habitat is an important adaptation needed for adoption of air breathing since this change inevitably will entail an elevated concentration of CO_2 at the gas exchange surfaces because of the bidirectional airway of air breathers (Carter, 1951).

The CO_2 combining power of blood shows a clear increase with increased importance of air breathing in overall gas exchange. Figure 6 shows a comparison of CO_2 dissociation curves with normal values of arterial P_{co_2} indicated for representative air-breathing fishes.

The increase in CO_2 combining power correlates with an increase in the average level of arterial P_{co_2}, thus meeting the obvious need for a higher capacity in CO_2 transport associated with the air-breathing habit. Figure 7, showing a composite plot of overall buffering capacity in air-breathing fishes, illustrates the same adaptive increase in buffering capacity with the increased importance of air breathing.

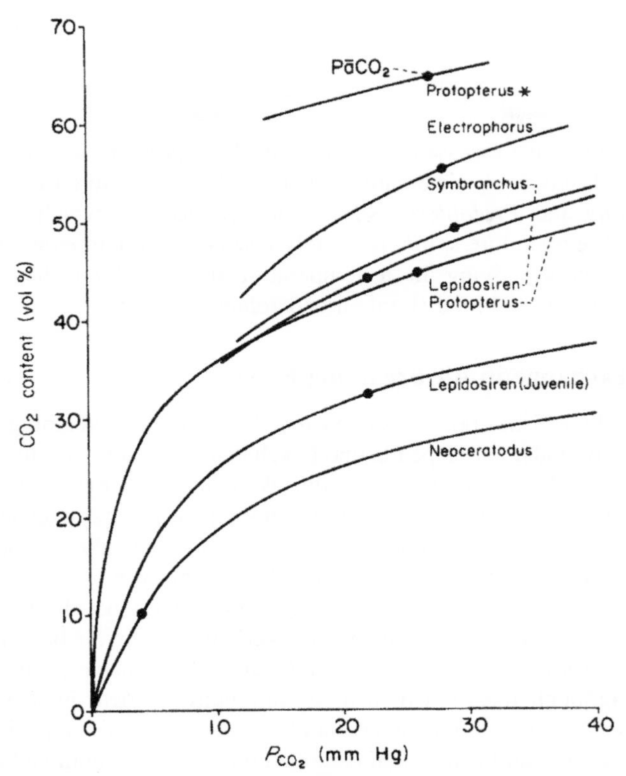

Fig. 6. Comparison CO_2 dissociation curves from representative species of air-breathing fishes. The black dots indicate the normal levels of arterial P_{CO_2}. Information selected from Lahiri *et al.* (1968), *Protopterus**; Johansen *et al.* (1968), *Electrophorus;* Lenfant *et al.* (1970), *Symbranchus;* Lenfant and Johansen (1968), *Lepidosiren and Protopterus;* Johansen and Lenfant (1967), *Lepidosiren* (juvenile); and Lenfant *et al.* (1966), *Neoceratodus.*

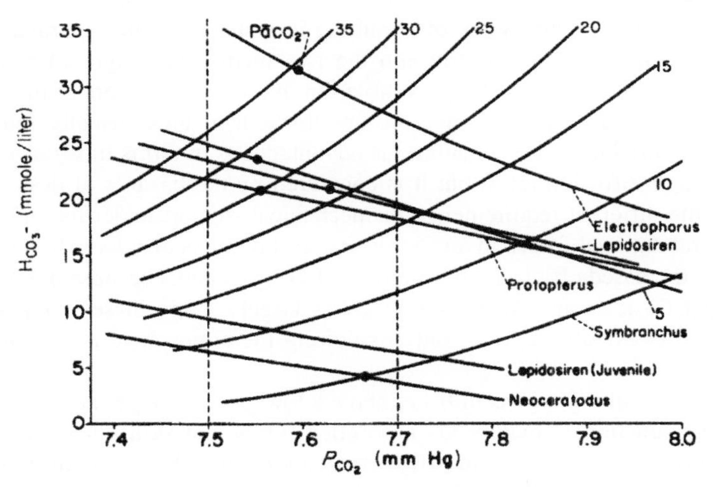

Fig. 7. A comparison of standard bicarbonate and buffering capacities in representative air-breathing fishes. Information selected from Johansen *et al.* (1968), Electrophorus; Lenfant *et al.* (1970), *Lepidosiren and Symbranchus;* Lenfant and Johansen (1968), Protopterus; Johansen and Lenfant (1967), *Lepidosiren* (juvenile); and Lenfant *et al.* (1966), *Neoceratodus.*

Few studies have assessed the influence of temperature on the respiratory properties of blood in air-breathing fishes. A comparison of blood from *Neoceratodus* and *Protopterus* shows that the former, which experiences large annual temperature shifts, is notably insensitive to temperature changes, while *Protopterus,* living in a temperature-stable habitat, has far more temperature-sensitive blood (Lenfant and Johansen, 1968).

B. Gas Exchange in Air-Breathing Fishes

All air-breathing fishes possess some capacity for aquatic breathing. Thus gas exchange is bimodal and depends upon exchange diffusion with the water via gills or skin and with air rhythmically taken into, and expelled, from, the air-breathing organ. The fish may employ both means of gas exchange simultaneously or may resort to one method at a time depending upon external circumstances. A very few fishes undertake voluntary migrations over land in moist or wet grass. Aquatic respiration must then be altogether halted except for some possible exchange via the skin. Estivating air-breathing fishes represent a special case. They must rely almost exclusively on aerial gas exchange although retention of a moist microenvironment in their cocoons or burrows in the substratum is essential for survival. This requisite may be as important for fluid balance and osmotic control as for continued cutaneous gas exchange. The literature is deficient of any attempt to study gas exchange in estivating fishes.

Air-breathing fishes experience two vastly different environments. A low solubility of O_2 in water provides an oxygen availability, in aerated water, about one-thirtieth of that in air in a tropical environment. Hence, aquatic breathers must ventilate much larger volumes than air breathers at similar O_2 uptakes. The greater work of breathing in water is further aggravated by the huge density difference between the two media. The high solubility of CO_2 in water along with the high ventilation of aquatic breathers brings about internal CO_2 tensions which may be one-thirtieth of those usually found in terrestrial vertebrates. Air breathing is advantageous in terms of O_2 availability and high diffusion rates, but it is countered by the hazards of desiccation and by the different requirements for mechanical support under the influence of net gravitational forces in air. Exclusive air breathers also have to meet the need for increased efficiency of internal buffering systems because of elevated internal CO_2 tensions. Air-breathing fishes largely avoid these problems by remaining essentially aquatic, only exploiting the atmosphere as a source of oxygen.

All air-breathing organs in fishes show a low gas exchange ratio indicating that their function is mainly O_2 absorption. Gills and skin conversely play their most important role in CO_2 elimination and show accordingly gas exchange ratios exceeding unity.

Table III offers a compilation of the relative role of aquatic and aerial gas exchange in the few fishes that have been adequately studied.

Table III Oxygen Uptake, CO_2 Production, and Gas Exchange Ratio for Aquatic and Air Breathing in Representative Air-Breathing Fishes

	Oxygen uptake (ml/min/kg)				CO_2 production (ml/min/kg)				Gas exchange ratio				
Species	*Aquatic*	*Air*	*Total*	*Aquatic/Air*	*Aquatic*	*Air*	*Total*	*Aquatic/Air*	*Aquatic*	*Air*	*Total*	**Remarks**	**Reference**
Protopterus	0.021	0.169	0.20	0.12	0.101	0.013	0.15	2.34	4.7	0.25	0.75	Fasting, free swimming with access to air, 20°C	Lenfant and Johansen (1968)
Lepidosiren (150 g)	0.90	0.51	1.40	0.18	1.21	0.37	1.58	3.27	1.34	0.72	1.12	Juvenile, 18°C	Johansen and Lenfant (1967)
Amia	0.316	1.20	1.52	0.31	1.54	0.48	2.02	3.2	4.9	0.37	1.33	19°C	Johansen *et al.* (1970)
	0.34	1.66	2.00	0.205	1.15	0.54	1.69	2.09	3.1	0.33	0.85	27°C	
Gillichtys			1.73							0.28		20°C	Todd and Ebeling (1966)
Neoceratodus	0.25		0.25		0.31					0.2	1.24	18°C	Lenfant *et al.* (1966)
Cobitis fossilis	0.24	0.99	1.23	0.24	0.39	0.80	1.19	0.5	1.6	0.81	0.98	25°C, adult	Calugareanu (1907)
Lepidosiren	0.03	0.67	0.70	0.04								20°C	Sawaya (1946)
Protopterus			0.87									20 °C feeding	Smith (1930)

Among obligate air breathers such as *Lepidosiren* and *Protopterus,* aquatic breathing is of little significance in O_2 absorption and contributes less than 10% of the total oxygen uptake in adult fish (Johansen and Lenfant, 1967; Lenfant and Johansen, 1968). Aquatic breathing remains very important in CO_2 elimination and is, in *Protopterus* (Lenfant and Johansen, 1968), about 2.5 times more effective than the lungs in removing CO_2. In many fishes the skin plays an important role in gas exchange, particularly in CO_2 elimination. Cunningham (1934) reported a gas exchange ratio in excess of 10 for the skin of *Lepidosiren* in water. During air exposure the skin is also important in gas exchange and marked cutaneous vasodilation has been reported for the common eel, *Anguilla* (Berg and Steen, 1965) and *Lepidosiren* (Johansen and Lenfant, 1967).

In *Cobitis fossilis,* an intestinal air breather (Calugareanu, 1907), and *Gillichtys mirabilis,* a marine gobiid fish practicing accessory air-breathing with the buccopharyngeal cavity (Todd and Ebeling, 1966), O_2 absorption is the dominant feature of air breathing (Table III).

The large variation of total O_2 uptake, apparent in Table III, may be related to feeding. Oxygen uptake in *Protopterus* has been reported to decline rapidly between 24 and 48 hr after the last food intake, and a constant metabolic rate was maintained only as long as the intake of food exceeded the prevailing oxidation rate (Smith, 1930).

A labile oxygen uptake was also apparent during air exposure of juvenile *Lepidosiren* (Johansen and Lenfant, 1967) and adult *Amia.* In *Lepidosiren* the O_2 uptake fell to 20% of the value in water in less than 6 hr. Similarly, Berg and Steen (1965), working on the common eel, reported on O_2 uptake in air half of the value measured in water. A different pattern is evident in *Protopterus* during air exposure. A normal level of O_2. uptake was maintained when rebreathing air until the ambient O_2 tension had fallen to 80 mm Hg and CO_2 tension had risen to 35–40 mm Hg.

Figure 8 shows the rate of change of gas composition inside air-breathing organs of a number of fishes based on samples obtained when the animals were resting in well-aerated water. The rate of O_2 absorption in the obligate air breathers is conspicuously higher than in the fishes employing accessory or supplemental air breathing. Note, also, the marked effect of temperature on the rate of O_2 depletion from the air bladder in *Amia.*

A comparison of the slopes for O_2 depletion and CO_2 accumulation reveals the former to be much steeper, suggesting a low gas exchange ratio. The slopes for CO_2 accumulation also show a more abrupt change to a gentler slope late in the interval between surfacings for air, a fact suggesting that the low gas exchange ratio becomes even lower, later in the interval between air breaths. In the case of *Protopterus* and *Electrophorus,* this change is particularly apparent (Figs. 9 and 12). Thus aquatic CO_2 elimination varies and may increase with time, possibly owing to a vasodilatory effect of increasing CO_2, in gills and skin vessels.

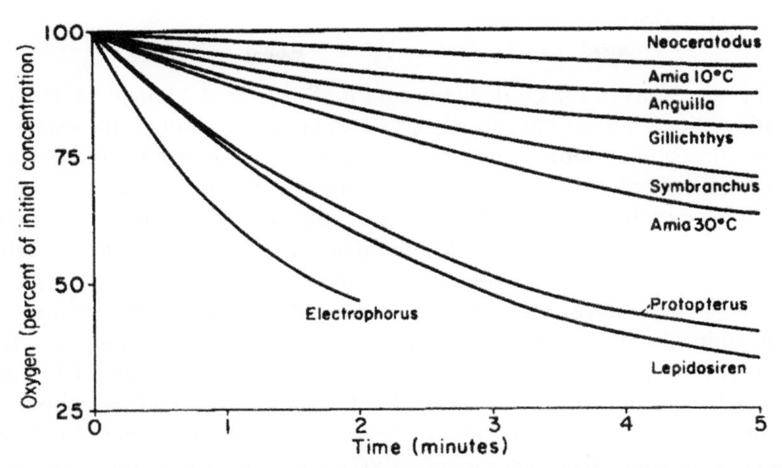

Fig. 8. Rate of O_2 depletion from air-breathing organs in various fishes. The slopes represent average values. Information selected from Lenfant *et al.* (1966), Neoceratodus; Johansen *et al.* (1970), *Amia;* Berg and Steen (1965), *Anguilla;* Johansen (1966), *Symbranchus;* Todd and Ebeling (1966), *Gillichtys;* Lenfant and *Johansen* (1968), *Protopterus;* Johansen and Lenfant (1967), Lepidosiren; and Johansen *et al.* (1968), *Electrophorus.*

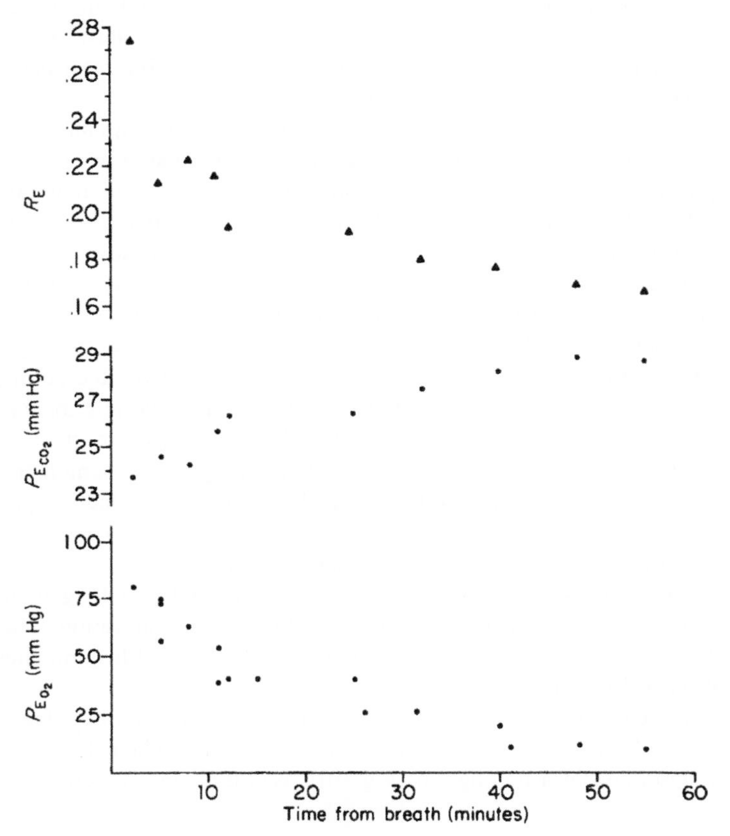

Fig. 9. Time course of changes in O_2 tension, CO_2 tension, and the gas exchange ratio (R_E) inside the lung of *Protopterus* between successive air breaths. From Lenfant and Johansen (1968).

The slopes given for O_2 depletion in Fig. 8 represent average values, and variations likely reflect changes in the blood perfusion through the exchange organ. The average slopes seem related with the breathing habits of the fishes. Thus, in *Electrophorus,* intervals between air breaths rarely exceeded 2 min, while in *Lepidosiren* and *Protopterus* the intervals varied but were usually between 3 and 6 min, whereas, *Symbranchus* and *Amia* both showed infrequent surfacings often exceeding 25–30 min between air breaths.

Few studies have assessed the total volume and tidal volume of air-breathing organs in fishes. Juvenile specimens of *Protopterus* (weight 57–78 g) are reported to show average lung volumes of 1.7 ml (or 25.3 ml/kg) measured by radiological techniques. Tidal volumes measured by spirometry averaged 1.45 ml (or 19.8 ml/kg) and, hence, represent a fraction of total lung volume as large as 80% (Jesse *et al.,* 1967).

C. Internal Gas Transport in Air-Breathing Fishes

Gas transport between the respiratory organs and the metabolizing tissues depends on the metabolic rate, the pattern and rate of blood circulation, and on the respiratory properties of blood. Very few air-breathing fishes have been adequately studied by frequent sampling of blood and gas to allow an evaluation of internal gas transport. Table IV summarizes available information on gas composition in blood and aerial exchange organs.

The Australian lungfish, *Neoceratodus,* is special among the lungfishes by being predominantly a water breather. During rest, in aerated water, the gills are responsible for the entire O_2 absorption which suffices to maintain the arterial blood 95% saturated with O_2. Comparisons of gas tensions in samples from the pulmonary artery, pulmonary vein, and the lung disclose a near equilibrium indicating no role of the lung in gas exchange. These samples were obtained between the sporadic breaths for air which often fell more than one hour apart (Lenfant *et al.,* 1966). The two other species of lungfishes emphasize air breathing. In *Lepidosiren paradoxa,* blood from the pulmonary artery was only 40–50% saturated with O_2 while that in the pulmonary vein was almost completely saturated (Table IV). The difference in gas tensions between blood in the dorsal aorta and the pulmonary artery demonstrates a partial selective passage of blood through the heart. This selective passage tends to minimize recirculation of blood from the pulmonary vein back to the lungs (Johansen and Lenfant, 1967; Lenfant *et al.,* 1970).

In *Protopterus aethiopicus* passage of blood through the lung increases the O_2 saturation more than 30% to exceed 90% of full saturation. There is a concurrent drop in CO_2 tensions but aquatic respiration still dominates the elimination of CO_2 (Table IV).

The marked differences in the importance and efficiency of air breathing in *Neoceratodus* and *Protopterus* can best be discussed when the blood gas data are described in a nomogram (Fig. 10). While resting in aerated water,

Table IV Normal Values of Blood Gas Tensions Sampled from Free Swimming Unrestrained Species of Air-Breathing Fishes

Species	Systemic venous blood		Systemic arterial blood			Afferent blood aerial organ (pulmonary artery)		Efferent blood aerial organ (pulmonary vein)		Temperature (°C)	Reference
	P_{O_2}	P_{CO_2}	P_{O_2}	% sat.	P_{CO_2}	P_{O_2}	P_{CO_2}	P_{O_2}	P_{CO_2}		
Lepidosiren	—	—	31.5	80	—	10.0	—	48–50	90	18–20	Johansen and Lenfant (1967)
Neoceratodus	14.3	6.5	38.9	95	3.60	38.9	3.6	36.3	3.8	20	Lenfant *el al.* (1966)
Protopterus	2	—	27.0	78	25.7	19.9	25.5	39.8	21.7	20	Lenfant and Johansen (1968)
Electrophorus	—	—	20.8	70	27.7	20.8	27.7	—	—	25–28	Johansen *et al.* (1968)
Lepidosiren	—	—	38.0	86	22	28	—	69	—	25–28	Lenfant *et al.* (1970)
Symbranchus	—	—	22	90	29	—	—	—	—	25–28	Lenfant *et al.* (1970)
Protopterus	—	—	36	—	—	—	—	—	—	25	Lahiri *et al.* (1968)
Electrophorus	—	—	13	50	29.0	—	—	—	—	—	Garey and Rahn (1968)

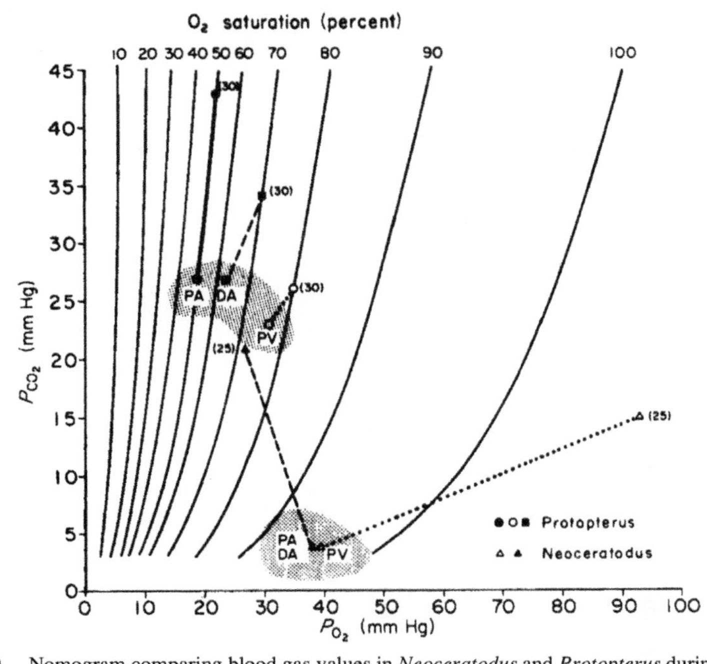

Fig. 10. Nomogram comparing blood gas values in *Neoceratodus* and *Protopterus* during undisturbed breathing in well-aerated water (values on stippled background) and following air exposure. (Duration of air exposure in minutes indicated by numbers in parentheses.) PA, pulmonary artery; DA, dorsal aorta; and PV, pulmonary vein. From Lenfant and Johansen (1968).

Neoceratodus practices water breathing only, whereas *Protopterus* practices frequent air breathing which results in a sizable gain in O_2 saturation of blood passing the lungs. A significant drop in blood CO_2 tension also accompanies perfusion of the lungs. The much higher CO_2 tensions in *Protopterus* than in *Neoceratodus* is a correlate of the increased importance of air breathing. The behavioral response to air exposure of these two lungfishes differs markedly. While both intensify their air breathing by increased frequency and depth, *Neoceratodus* becomes restless and splashes about in frantic efforts to get back into water. *Protopterus,* on the other hand, normally takes to air exposure with ease. Air exposure of *Neoceratodus* promptly engages the lung in O_2 absorption, and the O_2 tension in blood from the pulmonary vein increases from about 40 mm Hg to more than 90 mm Hg. Yet, it is apparent that this change fails to maintain the systemic arterial O_2 saturation which drops precipitously because of inadequate blood flow through the lung. Note that the changes in oxygenation are accompanied by a conspicuous rise in the blood CO_2 tension. *Protopterus* shows an increased oxygenation in both the pulmonary vein and in systemic arteries. The concurrent increase in CO_2 tension is related to a loss of the important escape route via gills and skin and the

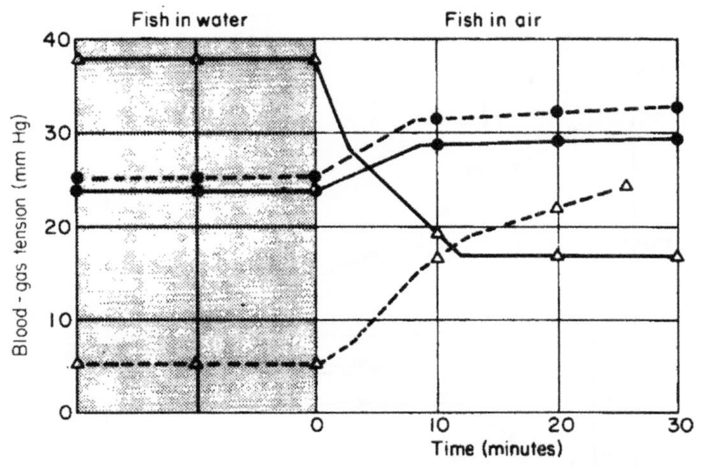

Fig. 11. A comparison of the time course of changes in O_2 and CO_2 tension of arterial blood during air exposure of *Neoceratodus* and *Protopterus*. Broken lines indicate CO_2 tensions. Solid lines indicate O_2 tensions. Adapted from Lenfant and Johansen (1968).

inevitable retention of CO_2 resulting from the anatomical dead space of the bidirectional airway. The CO_2 retention is minimized by the high tidal volumes relative to total lung volume. The time course of the blood gas changes during air exposure of the two lungfishes (Fig. 11) also shows the inadequacy of the *Neoceratodus* lung to cope with both O_2 absorption and CO_2 elimination, while the *Protopterus* lung can handle both tasks even though the internal CO_2 levels stabilize at a much higher level in air than in water (Lenfant and Johansen, 1968).

Gas exchange and gas transport in *Electrophorus,* which practices air breathing with the uniquely modified oral mucosa, is illustrated in Fig. 12. The O_2 tension inside the mouth drops from about 140 mm Hg right after a breath to 60 mm Hg at the end of an average 2-min breath interval. Concurrently, the CO_2 tension increases from about 7 to 30 mm Hg inside the mouth, with most of the increase taking place early in the breath interval. The gas exchange ratio drops from about 0.85 just after a breath to 0.20 just before the next breath. Hence, CO_2 is eliminated to a large extent by aquatic gas exchange through the skin or vestigial gills. The low O_2 tension corresponding to about 65% O_2 saturation in arterial blood results from the shunting of blood from the mouth organ back to the systemic veins before arterial distribution. Blood from the jugular vein draining the mouth organ and the anterior systemic veins showed higher O_2 tensions corresponding to about 90% O_2 saturation. This difference in saturation is caused by the additional admixture of systemic venous blood from the posterior end before the blood is distributed from the heart to the arteries (Fig. 1).

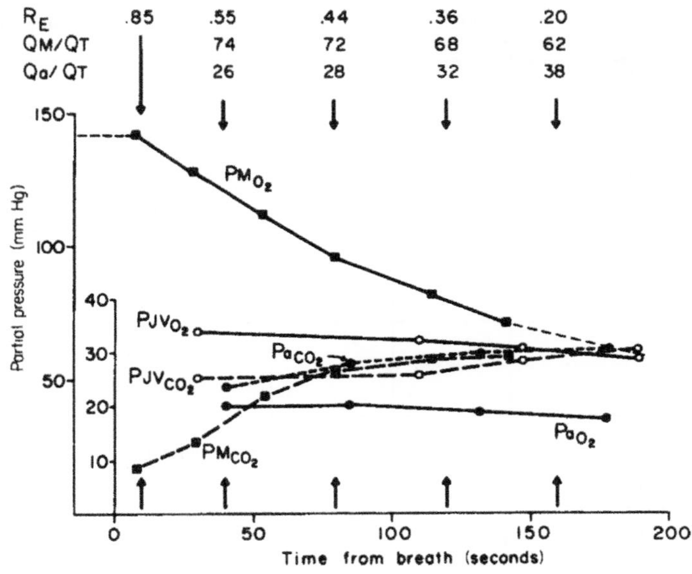

Fig. 12. Changes in O_2 and CO_2 tensions in the mouth gas (M), systemic arterial blood (a), and blood from the jugular vein (JV) in *Electrophorus* between consecutive air breaths. Listed on top are computed values of the gas exchange ratio (R_E), the fraction of cardiac output going to the mouth QM/QT and to the systemic arteries Qa/QT. The vertical arrows indicate the times at which these calculations were made from the corresponding values of gas partial pressures. From Johansen *et al.* (1968).

The higher level of CO_2 tension in the jugular vein than in the arterial blood suggests that systemic venous blood returning via the posterior cardinal veins, which receive the major cutaneous drainage, must have the lowest CO_2 tension; otherwise CO_2 must have been largely eliminated during passage through the vestigial gills.

The nomogram (Fig. 13) summarizes results obtained in *Electrophorus* and substantiates the efficiency loss in gas transport caused by the vascular shunt from the mouth organ (Johansen *et al.*, 1968).

D. Control of Breathing in Air-Breathing Fishes

Breathing in vertebrates is, in general, geared to satisfy changing metabolic requirements by maintaining stable internal conditions for gas exchange and gas transport. The factors affecting the control of breathing are markedly different for water breathers and air breathers. The aquatic animal is surrounded by an environment of changing gas composition in which O_2 and CO_2 tensions may change in the same or in opposite directions. Also, in aquatic animals the external and internal environments are in intimate diffusion contact at the gills, and external changes will quickly change the internal

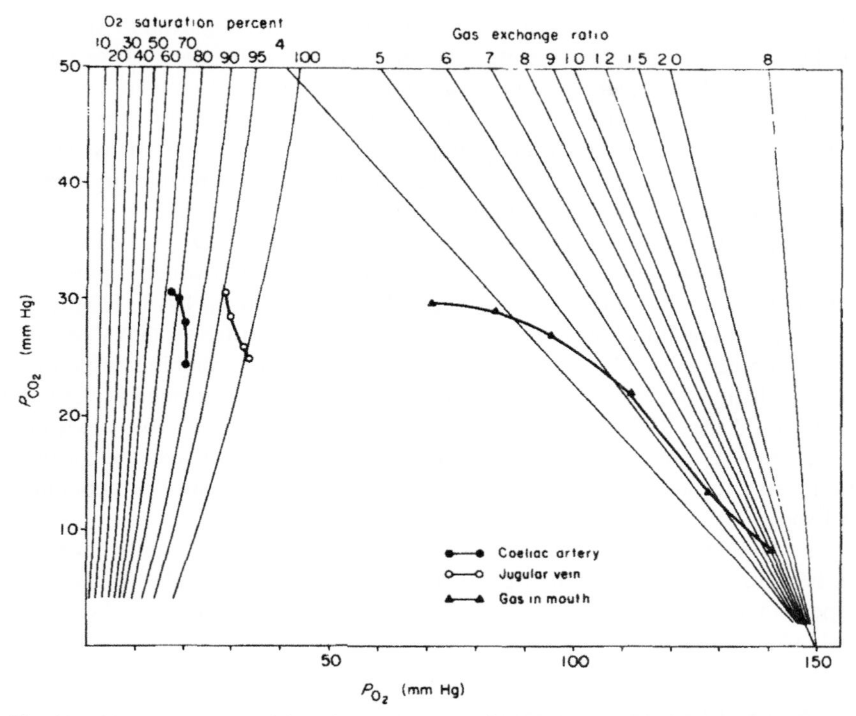

Fig. 13. Nomogram summarizing changes in gas tensions in blood and inside the mouth during an interval between air breaths in *Electrophorus*. From Johansen and Lenfant (1970).

environment. Air-breathing terrestrial animals, on the other hand, normally experience an external environment stable in its respiratory qualities; in addition, the presence of an internal gas compartment in the lungs will protect the internal environment against external changes. Control of breathing in the air breather can hence be based on negative feedback from receptors sensing internal parameters important to gas transport such as O_2 and CO_2 tension and pH of arterial blood. The aquatic animal has a more difficult problem in regulation because an internal factor like the arterial O_2 tension may change as a result of external, as well as, internal changes. If, for instance, internal hypoxia occurs when a fish enters severely O_2-deficient water, an increased breathing effort may further aggravate the internal hypoxia. It would be advantageous for the fish to move to water of higher O_2 content. Control of breathing should hence allow the fish to distinguish between changes occurring internally or externally. This complex scheme calls for chemosensitive mechanisms that can screen external as well as internal changes.

It has long been contended that fishes show preference reactions to external gradients in gas composition and that this is linked with an ability to detect the gas tensions in the water by external sensors (Höglund, 1961).

Many authors view the orientation of fish to the higher O_2 availability in a gradient as a nondirective escape reaction caused by incipient suffocation. On the other hand, there is little doubt that many fishes living in severely oxygen-deficient swamp water show a clear chemotaxis for the more oxygen-rich surface layers of the swamp. In fact, the existence of many fishes depends on frequent return to the surface layers to satisfy respiratory needs (Carter and Beadle, 1931b; Johansen, 1968).

The apparent variance among fishes in their response to O_2 gradients does not extend to avoidance reactions to P_{CO_2} and pH gradients which reportedly are common and much more acute (Shelford and Allee, 1913; Höglund, 1961). Such reactions are presumably triggered by external gustatory receptors (Scharrer *et al.,* 1947). Some authors consider the breathing responses to CO_2 change in fishes to be connected to chemoreceptors located in the gill region (Dijkstra, 1933; Powers and Clark, 1942; Jesse *et al.,* 1967). Meanwhile, no experimental work has been reported that conclusively identifies the location of and mechanism of chemosensitive reflex control of breathing in fishes.

A consideration of these problems takes on special significance in air-breathing fishes. From their practice of a dual mode of breathing one will expect that air and aquatic breathing are both individually controlled and mutually coordinated.

E. Normal Breathing Behavior

The amplitude and frequency of both air and aquatic breathing in air-breathing fishes are very labile. In fishes employing aquatic as well as air breathing the frequency of branchial pumping generally exceeds the frequency of air breathing. Commonly, the rate of branchial pumping is low just after an air breath, but it increases progressively before the next air breath (Johansen and Lenfant, 1967, 1968). The contribution to gas exchange from air breathing evidently reduces the chemoreceptor drive to branchial breathing. Conversely, this drive will progressively increase with time after an air breath with a resultant higher paced branchial pumping and finally evocation of another air breath. The two types of breathing are apparently under a common or mutual control. In many fishes, the rates of both branchial and air breathing show a notable increase during swimming activity. Young specimens of *Neoceratodus* change from sporadic air breaths during rest in aerated water to a steady higher rate of air breathing during swimming (Grigg, 1965). Similarly, *Polypterus senegalus* and *Amia calva,* both of which use the swim bladder for accessory air breathing, show an augmented rate of branchial and air breathing with the onset of exercise (Magid, 1966). Thus the resting rate of water breathing provides little reserve for increased metabolic activity. The energy expenditure of water ventilation is high, and a definite advantage attends the use of less costly alternative ways of gas exchange during periods of increased demand. Water breathing during low activity, however, permits the fish to exploit its aquatic environment without the need for frequent ascents to the surface for air.

It is generally held that structural adaptations for air breathing in fishes evolved in association with environmental oxygen lack. Yet fishes such as the Australian lungfish, the holostean fishes *Amia* and *Lepisosteus,* and many teleosts breathe air even though their habitats are not oxygen deficient. Thus, internal oxygen deficiencies resulting from activity may have provided a selection pressure for development of air breathing independent of the oxygen availability in the water. Certainly, the exceptional air-breathing habits of marine pelagic fishes such as the tarpon, *Megalops* (Schlaifer and Breder, 1940), support this contention. Recent work comparing the relative importance of water breathing and air breathing in *Amia* at different temperatures has emphasized the importance of air breathing for support of higher metabolic activity (Johansen *et al.,* 1970). The situation resembles the prompt stimulation of breathing in exercising mammals without an apparent initiating role of arterial chemoreceptors.

A few fishes, such as the electric eel, have come so far in the dominance of air breathing that the muscles used for water ventilation and the gills themselves have atrophied to a point where mechanical manifestation of branchial breathing has ceased altogether. However, aquatic gas exchange continues mainly as CO_2 elimination through the skin or the atrophied gills. This important part of overall gas exchange is mainly regulated by changes of blood perfusion at the interfaces between water and blood in the skin or gills.

A marked change in the respiratory behavior of air-breathing fishes likely attend their ontogenetic development. Some species such as the African lungfish are decidedly aquatic breathers during larval and juvenile stages, but air breathing takes on a dominating importance in the adult fish.

F. Breathing Responses to Changes in External Gas Composition

1. *Changes of Environmental Oxygenation*

Exposure to hypoxic water evokes different breathing responses among air-breathing fishes. The response patterns express an apparent relationship to the relative role of aquatic and air breathing in overall gas exchange.

The Australian lungfish is an almost exclusive water breather when at rest in aerated water. It reacts to hypoxic water by promptly increasing its branchial breathing efforts. Water ventilation increased as much as five times when the O_2 tension in the water was reduced to about 80 mm Hg. Figure 14 shows the result of a typical experiment. Ventilation doubled, while the extraction of O_2 decreased as did the overall O_2 uptake. The increased ventilation was caused by an increase of the volume propelled by each branchial pumping movement and not by an increased frequency of the branchial pumping. The pattern of air breathing changed from no air breaths in the hour

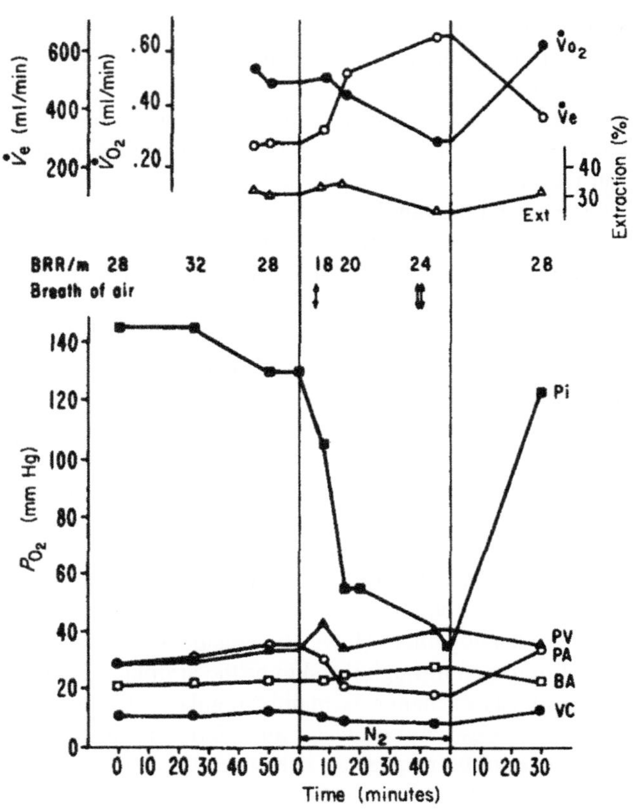

Fig. 14. Responses to breathing deoxygenated water in *Neoceratodus*. $\dot{V}_e$, ventilation of water; $\dot{V}_{O_2}$, oxygen uptake from the water; Ext, percent extraction of oxygen from the water; and BRR, branchial respiratory rate. Arrows mark breaths of air. Oxygen tensions followed in: Pi, inspired water; PV, pulmonary vein; Pa, pulmonary artery; BA, an afferent branchial artery; and VC, vena cava. From Johansen *et al.* (1967).

preceding the hypoxia to more than three air breaths in the first half-hour of hypoxia. The stimulation of air breathing was much slower in onset than the increase of branchial breathing. The commencement of air breathing, however, initiated important changes in the pattern of blood perfusion. Most notably, the O_2 tensions increased in blood drawn from those branchial arteries, which supply the systemic circulation, while blood in the pulmonary arteries decreased in O_2 tension. The O_2 tension of blood in the pulmonary vein also increased. Thus the lung had assumed a role in O_2 uptake, and important circulatory changes attended this change (Lenfant *et al.,* 1966; Johansen *et al.,* 1967).

The yarrow, *Erythrinus* sp., is another fish utilizing the swim bladder for accessory air breathing (Carter and Beadle, 1931b; Willmer, 1934).

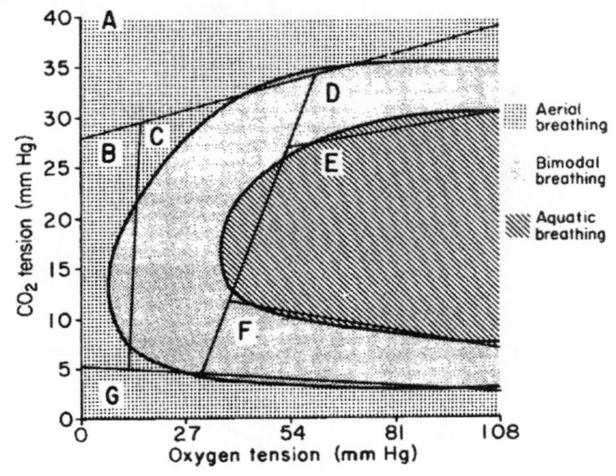

Fig. 15. Relationship between the mode of respiration in the yarrow *Erythrinus unitaeniatus*, and the O_2 and CO_2 tensions in the external water. Suggested reasons for the respiratory behavior: In area A, CO_2 tensions too high. Gill opercula actively closed. Area B, oxygen lost to water if gills remain active. Area C, gills inadequate to saturate the blood with oxygen. Area D, CO_2 too high, but kept within bounds by aerial respiration. Area E, aquatic breathing sufficient. Area F, gill movements insufficient to saturate blood. Area G, CO_2 content too low to stimulate gill movements. Figure and comments from Willmer (1934).

Figure 15 shows how the breathing behavior is related to the external gas composition for *Erythrinus*. Depending on the CO_2 content of the water (see later discussion) aquatic respiration suffices to satisfy the respiratory needs at decreasing O_2 tension down to about 40 mm Hg.

The bowfin, *Amia calva,* also utilizes the air bladder as an accessory gas exchanger. The relative importance of air breathing in this fish depends in a large measure on the temperature of the water (Fig. 16A). The fish occupies temperate regions of North America and experiences large annual temperature variations. At water temperatures of 20°–25°C the fish regularly ascends for air even in well-aerated water. Upon deoxygenation of the water the fish initially increases the branchial breathing rate moderately. Further deoxygenation elicits an increased rate of air breathing. When the latter occurs, both the rate and amplitude of branchial breathing drop below the level prevailing before deoxygenation. This depression of branchial breathing may be an adaptive response which reduces the loss of oxygen from the air bladder to the oxygen-deficient water via the gills (Fig. 16B).

Among obligate air breathers such as the African and South American lungfishes or the electric eel, the breathing responses to external changes are different. In these fishes the gills are variously degenerated, and the gill surface area is reduced. Thus there is a gradual reduction of the diffusion link between the external water and the blood as the fish shifts emphasis from water to air breathing. Deoxygenation of the water elicits no change in the breathing pattern of these fishes. In a sense they have become relieved of

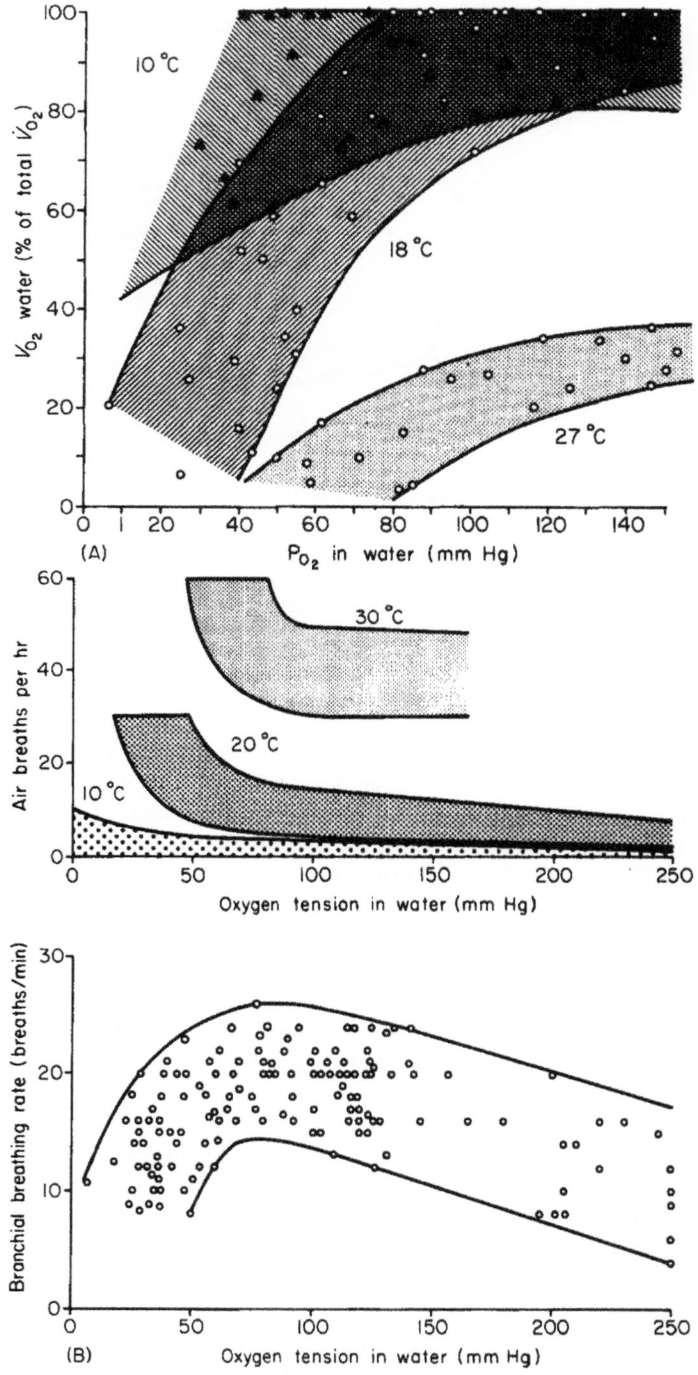

Fig. 16. (A) Oxygen uptake (V_{O_2}) of *Amia calva* from water in percent of the total oxygen uptake, related to the oxygen tension and temperature of the water. (B) Rate of air breathing in *Amia calva* (top) at different temperatures and rates of branchial breathing (temp. range, 14°–26°C) both as functions of oxygen tensions in the ambient water. From Johansen *et al.* (1970).

the environmental deoxygenation stress which originally provided the selection pressure favoring the air-breathing habit. Blood gas measurements have substantiated that no change in internal O_2 tension occurs in response to deoxygenation of the water, nor do the fishes exhibit any restlessness suggesting the presence of external chemosensation of O_2 that could guide the fish in avoidance or attraction to the degree of external oxygenation (Johansen and Lenfant, 1968).

Experiments on juvenile specimens of the African lungfish have produced data in conflict with those cited above (Jesse *et al.*, 1967). Deoxygenated water elicited an increase of both branchial and air breathing in the juveniles. However, these specimens survived on water breathing alone for several days, which contrasts markedly with other observations on adult fish. In addition, the experimental procedure evoked hypoxia simultaneously in water and in the air phase and consequently could stimulate the fish via the lung, as well as the gills. The interesting possibility emerges from these studies that control of breathing changes markedly during the ontogenetic development of the African lungfish.

A logical and simple experiment used to test the importance of air breathing in fish is to deprive them of access to air. Depending on the temperature and gas composition of the external water, such experiments have produced variable results, but they show that the degree of internal oxygenation clearly influences the breathing efforts. The attempts to breathe air increase in all air-breathing fishes, while branchial breathing appears to increase only in species normally dependent on branchial breathing for an important part of their oxygen uptake.

Experiments in which the fish have surfaced into atmospheres of controlled gas composition, while the conditions in the water have remained stable, have been more conclusive. Only in a few experiments have internal blood gas tensions been correlated with such procedures. In the African lungfish, introduction of hypoxic gas into the lung causes a prompt and sharp increase in the rate of air breathing. Simultaneous measurements of arterial O_2 tension show that air breathing is stimulated at much higher values than those recorded when the fish practices undisturbed breathing in normal air (Johansen and Lenfant, 1968). This implies that the chemosensitive areas which trigger the response to inhalation of hypoxic gas are more sensitive to the rate of change than to the actual level of oxygen tension in the hypoxic condition. Experiments on *Electrophorus* support such a contention by showing remarkably prompt responses to changes in composition of inhaled air. Figure 17a illustrates how the rate of air breathing was changed after a single breath of air following hypoxic breathing. This implies a location of the chemoreceptors in the buccal mucosa or in the blood path in intimate contact with the gas in the mouth. A location in the systemic arteries is ruled out because of the complete shunt of blood from the gas exchange organ to the systemic veins. The resulting changes in systemic arterial blood are much too slow to represent an important feedback stimulus. Figures 17b and 18

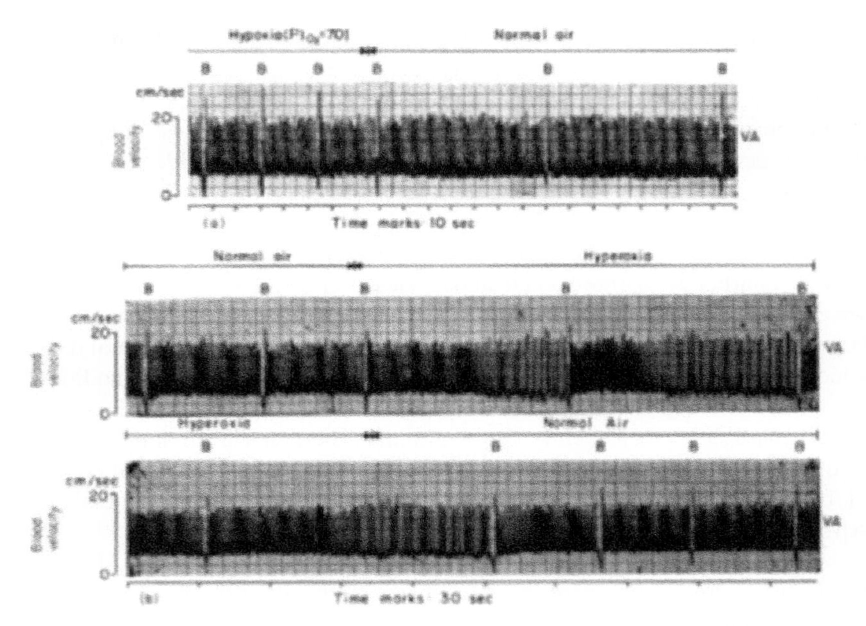

Fig. 17. Continuous tracings of blood velocity in the ventral aorta (VA) of *Electrophorus*. Each air breath shows as a distinct larger excursion (B). (a) shows the very prompt change in the rate of air breathing following the first breath of air after hypoxic breathing. (b) Inhalation of hyperoxic air immediately prolongs the interval between air breaths. From Johansen *et al.* (1968).

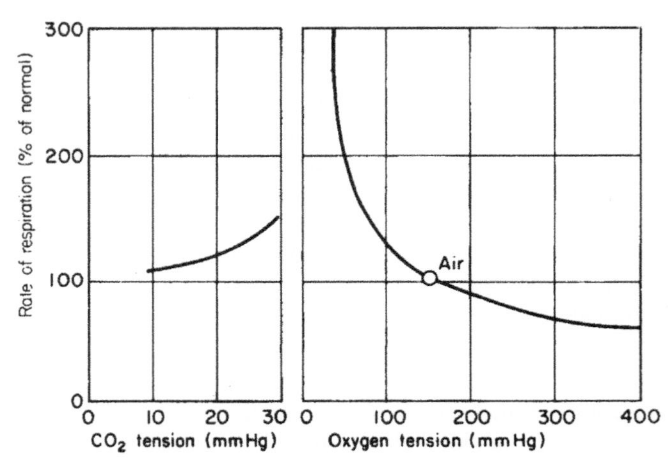

Fig. 18. Percent change in the rate of air breathing of *Electrophorus* in relation to the gas composition of the inhaled air. From Johansen *et al.* (1968).

show that inhalation of oxygen-enriched atmospheres causes a depression of air breathing and is important by suggesting the removal of a tonic P_{O_2}-dependent stimulus. This indicates that normal spontaneous breathing is governed by changes of oxygen tension in the air-breathing organ. The situation corresponds to conditions in higher vertebrates where oxygen inhalation is effective in removing the tonic activity of chemoreceptor cells stimulated by the normally prevailing levels of P_{O_2}. Hyperoxic breathing also reduces the rate of air breathing in *Protopterus* and other air-breathing fishes. Species depending largely on aquatic breathing show a depression of branchial breathing in hyperoxic water or following inhalation of hyperoxic air.

2. *Changes of Environmental CO_2 Tensions*

Internal CO_2 tensions are very low among purely water-breathing fishes living in well-aerated water (Lenfant and Johansen, 1966; Piiper and Schumann, 1967). A negative feedback control of breathing based on the level of metabolically produced CO_2 appears to be of little consequence in controlling breathing in such fishes. However, in stagnant tropical swamps, the CO_2 levels can build up high enough to adversely affect the dissociation of oxy-hemoglobin and to disrupt internal acid–base balance. Adoption of air breathing also inevitably leads to a retention and increase of internal CO_2 tensions. For tropical air-breathing fishes the external and internal levels of CO_2 may hence be important factors in the control of breathing.

Exposure to increasing CO_2 tensions in the external water evokes extremely varied breathing responses in different species of air-breathing fishes. This variety and the paucity of controlled experiments complicate the evaluation of the role of CO_2 in respiratory control. It appears that CO_2 at moderately high tensions stimulates both branchial and air breathing. At increasing tensions, however, a depression of branchial breathing occurs while air breathing continues at increased rates. The response pattern also seems to depend on the relative importance of air breathing and water breathing in the normal respiratory behavior of the fish.

In a few fishes studied adequately, such as the teleost *Symbranchus* and the Australian lungfish *Neoceratodus,* exposure to CO_2 concentration as low as 1–2% caused a prompt depression of branchial breathing while the rate of air breathing was accelerated; CO_2 concentrations below 1% were not tested (Johansen, 1966; Johansen *et al.*, 1967). In *Amia calva*, which employs accessory air breathing using the air bladder, CO_2 concentrations up to 3% stimulated both branchial and air breathing while higher concentrations depressed branchial breathing. Figure 19 demonstrates the rapid depression of branchial breathing in *Neoceratodus* following exposure to 2% CO_2 in the water. Upon removal of the hypercarbic stimulus, the changes were promptly reversed. Atropinization prevented the CO_2 response. This suggests a cholinergic link in the reflex chain eliciting the response. It is of

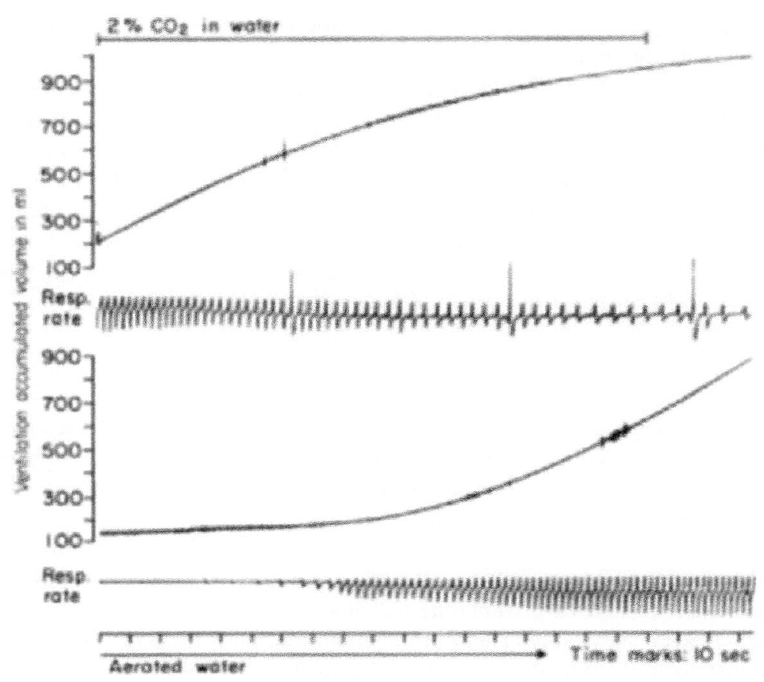

Fig. 19. Continuous recording of water ventilation and branchial respiratory rate in *Neoceratodus* during (top) and after (bottom) an increase in P_{co_2} of the ambient water. From Johansen *et al.* (1967).

considerable interest that filling the lung in *Neoceratodus* with a CO_2-rich gas mixture, which greatly elevates the CO_2 level in the pulmonary vein and in blood afferent to the gills, evokes no breathing response whatever. However, blood sampled distal to the gills was very low in CO_2 and attested to a remarkable efficiency of the gills in removing that gas. These data prompted the suggestion that the CO_2-sensitive receptors eliciting the depression of branchial breathing must be external sensors like gustatory or olfactory receptors or they must be located in the blood path efferent to the branchial circulation. Strong corroborative evidence was gained from studies of *Symbranchus*, which is capable of both air and aquatic breathing with its gills (Johansen, 1966). The fish shows inhibition of branchial breathing in CO_2-rich water which persists as CO_2 leaks into the bloodstream. However, when the fish is transferred to CO_2-free water, a high frequency of forceful branchial movements is established. Hence, the fish appears to possess a sensory mechanism that can differentiate between the external and internal CO_2 level. Similarly, in experiments with *Protopterus*, bubbling of 5% CO_2 in the water stopped branchial breathing before there were appreciable changes in CO_2 tension of the arterial blood.

The stimulation of air breathing by CO_2 may be a direct effect, or it may be secondary to the reduced oxygen uptake caused by the depression of branchial breathing. The response of *Protopterus* and *Electrophorus,* in which branchial breathing contributes little or nothing to the oxygen uptake, suggests a direct effect. Also, inhalation of CO_2-enriched gas stimulates air breathing in these fish while internal oxygen levels remain the same or increase. However, in predominantly water-breathing fishes such as *Neoceratodus* or *Amia,* the influence of CO_2 on air breathing may be both direct and indirect.

The variability found in the few studies on CO_2 response in air-breathing fishes was also apparent in the important work of Willmer (1934). Figure 15 summarizes his work on the factors controlling breathing in the yarrow, *Erythrinus,* a teleost which uses its air bladder for accessory breathing. Willmer contended that a certain CO_2 concentration in the blood is necessary to drive the branchial breathing. At very low O_2 contents of the water, branchial movements ceased, possibly to prevent loss to the water of O_2 gained by air breathing. The opercula also remain actively closed at higher external CO_2 concentrations, allegedly to prevent internal acidosis which would have resulted if gill breathing persisted. The figure also explains the conditions under which respiratory needs are satisfied by aquatic respiration alone and similarly the factors in the environment which cause the fish to employ bimodal breathing.

G. Breathing Responses to Mechanical Stimuli

A possible role of external mechanosensitive receptors in control of breathing rhythms is suggested by recent experiments showing a marked acceleration of branchial breathing rate in *Lepidosiren* and *Protopterus* when the water was stirred or agitated (Johansen and Lenfant, 1967, 1968). Such a role of mechanoreceptors in fish takes on special significance in the normally O_2-deficient stagnant tropical swamps, where stirring of surface water by wind movement or other mechanical agitation should improve the respiratory quality of the water and make branchial breathing more efficient.

Inflation and deflation of the air bladder in *Amia* definitely influence the breathing pattern. A deflation causes the fish to promptly ascend for air, while fishes approaching the surface for air breathing interrupt their ascent and settle to the bottom again if the bladder is artificially inflated. This response occurs whether air, oxygen, or nitrogen is injected. This inflation response shows striking similarity with the inflation reflex described long ago and known as the Hering-Breuer reflex in mammals. In higher vertebrates the inhibition of breathing caused by stretch in the lungs is no longer held to exert any important influence in control of breathing since such changes do not supply information to the respiratory center about the respiratory quality of the blood. In air-breathing fishes a stretch reflex may prove to play an important role since the volume and thus the distention of the air bladder will change during the long intervals between air breaths because of the low gas exchange ratio for the air-breathing organ. Since the air bladder is primarily an O_2 absorber, the change in volume with time provides a measure of the rate of O_2 depletion. If the reflex described in fishes such as *Amia* proves to be homologous with the vagus-mediated inflation reflex in higher vertebrates, the case may exemplify a basic reflex type that exerted a more important regulatory role early in its phylogenetic history.

An important role of the air which a fish takes below the surface is its effect on buoyancy and, indeed, in most teleosts the air bladder is engaged mainly in buoyancy control. In fishes which use the air bladder primarily as a respiratory organ, changes of buoyancy clearly influence breathing behavior. Ascent for air in such fishes may be provoked simply by tying a weight around them (Spurway and Haldane, 1963). The air-breathing fishes consequently face the additional task of coordinating the mechanisms for buoyancy control with those engaged in satisfying the respiratory needs.

H. Breathing Response to Air Exposure

Only a few of the fishes adapted for air breathing make voluntary excursions onto dry land. Estivating fishes are, of course, compelled to subsist on air breathing when entrapped in the substratum during drought.

It seems typical that fishes engaged in voluntary air exposure under natural conditions promptly commence frequent ventilatory movements and show no signs of distress and erratic behavior such as seen in purely aquatic breathers when air-exposed in the laboratory (Berg and Steen, 1965; Johansen and Lenfant, 1968). Only in the South American teleost *Symbranchus* and in the lungfishes have blood gas values been correlated with breathing behavior during air exposure (Johansen, 1966; Lenfant and Johansen, 1968). In *Symbranchus,* arterial P_{O_2} and P_{CO_2} both increased during air exposure. In *Protopterus*, the onset of the accelerated air breathing was too rapid to have been elicited by internal changes in blood gas composition. The sensory input evoking the response must have been the physical act of air exposure itself. It is of interest that *Protopterus,* while recovering from anesthesia, will attempt to breathe air when air-exposed, even while still unresponsive to tactile or painful stimuli. The blood gas values show an increase in both P_{O_2} and P_{CO_2} (Fig. 20); the latter obviously results from the interruption of aquatic CO_2 exchange. The sustained rate of air breathing during air exposure could be driven by the increased level of internal P_{CO_2}, although this stimulus could not play a role in the initial elicitation of the response.

I. Coupling of Respiratory and Circulatory Events

Cardiovascular and respiratory events are strongly interrelated in fishes. Cardiac output determination by the Fick principle in *Neoceratodus* indicated

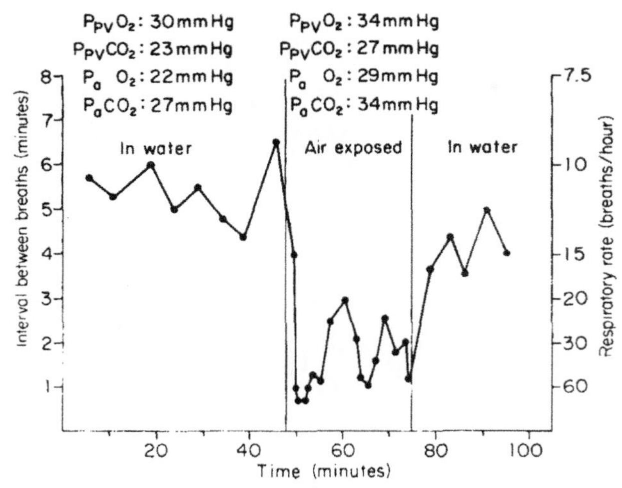

Fig. 20. Changes in rate of air breathing and blood gas tensions following air exposure in *Protopterus*. PV, pulmonary vein; PA, pulmonary artery; and a, systemic artery. From Johansen and Lenfant (1968).

a marked increase during exposure to hypoxic water. The increase in blood perfusion exceeded the relative increase in the ventilation of water. Consequently, the ventilation perfusion ratio of the gills changed from about 12 before hypoxia to about 4, after 45 min exposure to hypoxic water, then climbed to 17 during recovery. The ratio of pulmonary blood flow to total blood flow rose with increased rate of air breathing (Johansen *et al.,* 1967). In *Protopterus,* direct blood flow measurements have shown that spontaneous air breathing causes distinct changes in heart rate, total cardiac output, and regional blood flow. Cardiac output increased sharply with each air breath and declined slowly during the subsequent breath interval (Fig. 21A) (Johansen *et al.,* 1968). Branchial vascular resistance tended to decrease in conjunction with an air breath. The most consistent change was a marked increase of pulmonary blood flow which exceeded that in the vena cava and thus indicated a regional flow shift from the systemic to the pulmonary circuit. Thus, pulmonary flow could vary from less than 20% to more than 70% of the cardiac output.

In *Electrophorus* there is an equally strong interrelationship between circulatory and respiratory events. When the intervals between air breaths were relatively long (exceeding 1 min), cardiac output and heart rate decreased progressively in the late phase of the interval. At the next breath, both heart rate and blood velocity promptly increased and were reestablished at the levels prevailing just after the preceding breath (Fig. 21B). This cyclic phenomenon occurred as a normal event when the intervals between breaths were long, whereas no time-dependent changes were recorded at short

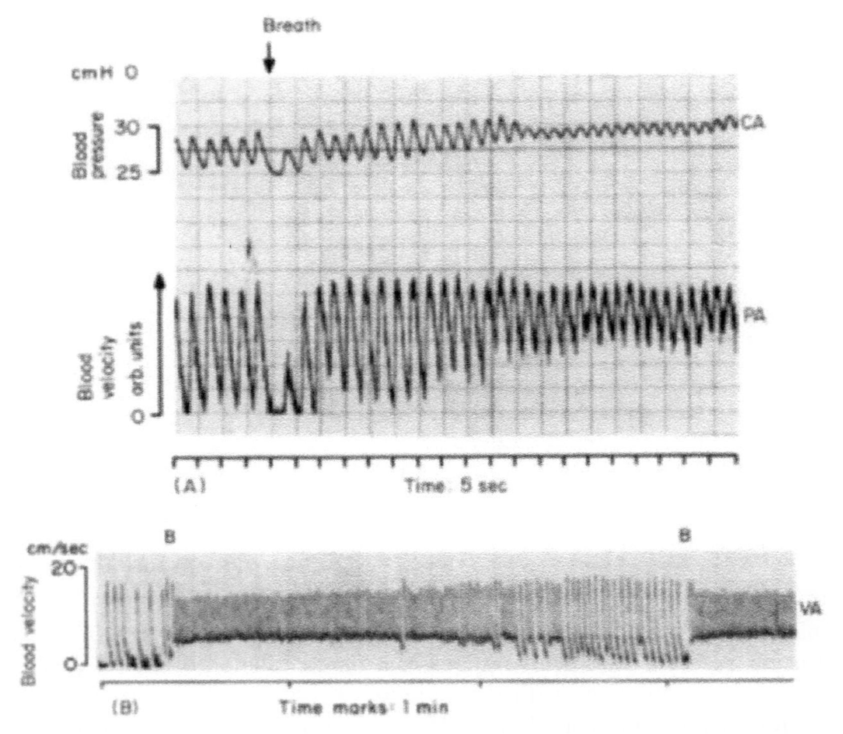

Fig. 21. (A) Change in blood velocity in the pulmonary artery associated with an air breath in *Protopterus*. CA, coeliac artery; and PA, pulmonary artery. From Johansen *et al.* (1968). (B) Changes in ventral aortic blood velocity related to the phase of the interval between air breaths in *Electrophorus*. VA, ventral aorta; and B, air breath. From Johansen *et al.* (1968).

intervals. The sudden changes in blood perfusion elicited by air breaths are at least, in part, controlled by the degree of mechanical distension of the air-breathing organ inside the mouth. Artificial inflation of the mouth via a catheter elicited an increase in heart rate and cardiac output whether nitrogen, air, or oxygen were injected; thus, the chemical composition of the gas in the mouth does not constitute the actual stimulus. A similar interrelation between cardiovascular events and distention of the air-breathing organ has been reported for *Symbranchus* (Johansen, 1966). Based on blood gas values the fractional distribution of cardiac output has been calculated in *Electrophorus*. Just after a breath the blood flow fraction going to the air-breathing organ is at its highest, and it declines steadily later in the breathing interval. Similar experiments during hypoxic and hyperoxic breathing showed that blood will shift toward the respiratory organ when the O_2 tension in the mouth is high (Fig. 22). This adjustment has the obvious rationale of promoting the matching process between blood and gas in the air-breathing organ.

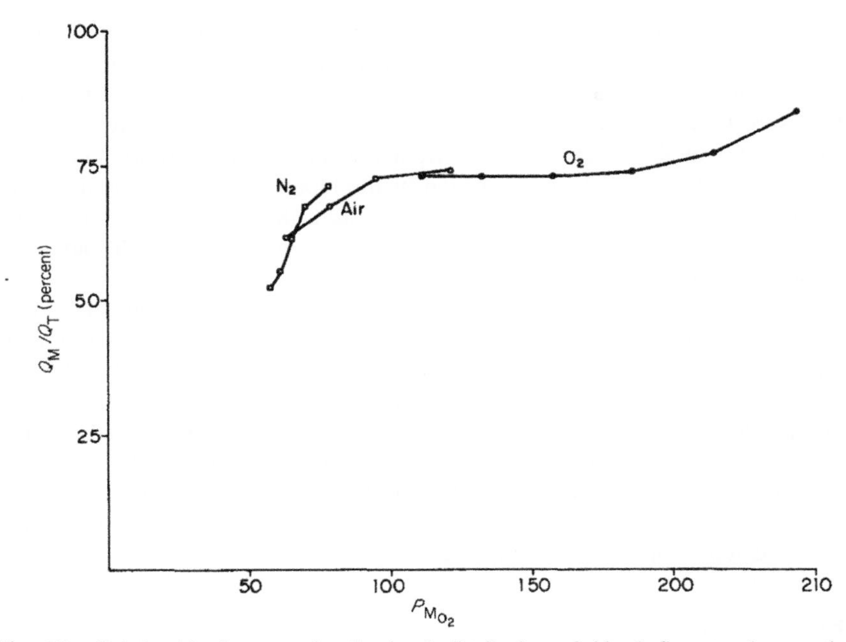

Fig. 22. Relationship between the fractional distribution of blood flow to the mouth air-breathing organ and the oxygen tension of the air inside the mouth of *Electrophorus*. The three curves show the relationship when breathing a hypoxic and a hyperoxic atmosphere and normal air, respectively. From Johansen *et al.* (1968). Q_M, Blood flow to mouth; Q_A, total cardiac output.

Increased perfusion through skin and mucosal linings of the buccal cavity has been frequently reported in air-exposed air-breathing fishes (Johansen and Lenfant, 1967; Berg and Steen, 1965). The marine gobiid fish, *Gillichtys*, displays some remarkable vascular changes when placed in hypoxic water. Under such conditions this fish habitually visits the surface and gulps air which is placed as a bubble between the tongue and the buccal roof. When stimulated to breath air by low O_2 tension in the water, the buccal epithelial capillaries promptly dilate and become engorged with blood, giving the epithelium a corrugated lunglike appearance, bright red from a proliferous circulation. This change in perfusion occurs within a few minutes and is totally absent in related fishes which are purely aquatic breathers (Todd and Ebeling, 1966). The close coupling between mechanisms controlling breathing and vasomotor responses seems basic to very diverse types of respiratory organs ranging from the gills in fishes to the lungs of both lower and higher vertebrates.

Only in the lungfishes among air-breathing fishes has the vascular circuit to the air-breathing organ gained enough anatomical separation from the systemic circuit to allow separate selective perfusion of the two circuits.

During rest in aerated water, the lung in *Neoceratodus* is of no consequence to gas exchange, and no apparent tendency for a preferential distribution of blood from the pulmonary vein to the systemic arteries is apparent. Exposure to hypoxic water evokes an active use of the lung and results in a clear preferential passage of blood from the lung to the anterior branchial arteries which carry the major portion of the blood to the dorsal aorta. One analysis showed that more than 83% blood from the pulmonary vein passes this way, whereas recirculation of systemic venous blood to the anterior arches was reduced to about 16%.

Lepidosiren and *Protopterus* are much more dependent on air breathing and show a consistent gradient of O_2 tension between blood in systemic and pulmonary arteries. This gradient reflects the degree of preferential perfusion which increases with an increased rate of air breathing (Table III). In *Protopterus* (Johansen *et al.*, 1968) the proportion of blood conveyed to the systemic arteries from the pulmonary circuit could be as high as 95% shortly after a breath and decline to about 65% just prior to the next air breath. The factors important in the control of the preferential passage of blood are hence also coupled to the breathing rhythm.

REFERENCES

Barrell, J. (1916). Influence of silurian-devonian climates on the rise of air-breathing vertebrates. *Bull. Geol. Soc. Am.* **27**, 387–436.

Berg, T., and Steen, J. B. (1965). Physiological mechanisms for aerial respiration in the eel. Comp. *Biochem. Physiol.* **15**, 469–484.

Bishop, I. R., and Foxon, G. E. H. (1968). The mechanism of breathing in the South American lungfish, *Lepidosiren paradoxa;* a radiological study. *J. Zool., Lond.* **154**, 263–271.

Boker, H. (1933). Uber einige neue Organe bei luftatmenden Fischen und im Uterus der Anakonda. *Anat. Anz.* **76**, 118–155.

Budgett, J. S. (1900). Observations on *Polypterus* and *Protopterus*. *Proc. Cambridge Phil. Soc.* **10**, 236.

Calugareanu, D. (1907). Die Darmatumung von *Cobitis fossilis.* II. Mitteilung: Über den Gaswechsel. *Arch. Ges. Physiol.* **120**, 425–450.

Carter, G. S. (1935). Reports of the Cambridge expedition to British Guiana 1933. Respiratory adaptations of the fishes of the forest waters, with descriptions of the accessory respiratory organs of Electrophorus electricus L. and Plecostomus plecostomus L. *J. Linnean Soc. London* **39**, 219–233.

Carter, G. S. (1951). "Animal Evolution." Sidgwick & Jackson, London.

Carter, G. S., and Beadle, L. C. (1931a). The fauna of the swamps of the Paraguayan Chaco in relation to its environment. I. Physico-chemical nature of the environment. *J. Linnean Soc. London* **37**, 205–258.

Carter, G. S., and Beadle, L. C. (1931b). The fauna of the swamps of the Paraguayan Chaco in relation to its environment. II. Respiratory adaptations in the fishes. *J. Linnean Soc. London* **37**, 327–366.

Cunningham, J. T. (1934). Experiments on the interchange of oxygen and carbon dioxide between the skin of *Lepidosiren* and the surrounding water, and the probable emission of oxygen by the male *Symbranchus*. *Proc. Zool. Soc. London* 875–887.

Das, B. K. (1927). The bionomics of certain airbreathing fishes of India, together with an account of the development of their airbreathing organs. *Phil. Trans. Roy. Soc. London* **216**, 183–219.

Das, B. K. (1940). Nature and causes of evolution and adaptation of the air-breathing fishes. *Proc. 27th Indian Sci. Congr.* Presidential address, No. 2, pp. 215–260.

DeGroodt, M., Lagasse A., and Sebruyns, M. (1960). Elektronenmikroskopische morphologie der lungenalveolen des *Protopterus* und *Amblystoma. Proc. 4th Intern. Conf. Electron Microscopy, Berlin, 1958* Vol. 1, pp. 418–421. Springer, Berlin.

Dijkstra, S. J. (1933). Über Wesen und Ursache der Notatmung. Z. *Vergleich Physiol.* **19**, 666–672.

Dubale, M. S. (1951). A comparative study of the extent of gill-surface in some representative Indian fishes, and its bearing on the origin of the airbreathing habit. *J. Univ. Bombay* [N. S.] **19**, 90–101.

Evans, M. (1929). Some notes on the anatomy of the electric eel, *Gymnotus electrophorus*, with special reference to a mouthbreathing organ and the swimbladder. *Proc. Zool. Soc. London* 17–23.

Fish, G. R. (1956). Some aspects of the respiration of six species of fish from Uganda. *J. Exptl. Biol.* **33**, 186–195.

Garey, W. F., and Rahn, H. (1968). Unpublished data.

George, J. C. (1953). Observations on the air-breathing habit in *Chiloscyllium. J. Univ. Bombay* [N.S.] **21**, 80.

Grigg, G. C. (1965). Studies of the Queensland lungfish, *Neoceratodus forsteri* (Krefft). *Australian J. Zool.* **13**, 243–253.

Höglund, L. B. (1961). The reactions of fish in concentration gradients. A comparative study based on fluviarium experiments with special reference to oxygen, acidity, carbon dioxide, and sulphite waste liquor (SWL). Rept. No. 43, pp. 1–147. Inst. Freshwater Res. Fish Board, Sweden.

Jesse, M. T., Shub, C., and Fishman, A. P. (1967). Lung and gill ventilation of the African lungfish. *Respiration Physiol.* **3**, 267–287.

Johansen, K. (1966). Airbreathing in the teleost *Symbranchus marmoratus. Comp. Biochem. Physiol.* **18**, 383–395.

Johansen, K. (1968). Airbreathing fishes. *Sci. Am.* **219**, 102–111.

Johansen, K., and Hanson, D. (1968). Functional anatomy of the hearts of lungfishes and amphibians. *Am. Zoologist* **8**, 191–210.

Johansen, K., and Hol, R. (1968). A radiological study of the central circulation in the African lungfish, *Protopterus aethiopicus. J. Morphol.* **126**, 333–438.

Johansen, K., and Lenfant, C. (1967). Respiratory function in the South American lungfish, *Lepidosiren paradoxa* (Fitz). *J. Exptl. Biol.* **46**, 205–218.

Johansen, K., and Lenfant, C. (1968). Respiration in the African lungfish. II. Control breathing. *J. Exptl. Biol.* **49**, 453–468.

Johansen, K., and Lenfant, C. (1970). Unpublished data.

Johansen, K., and Reite, O. B. (1967). Influence of acetylcholine and biogenic amines on pulmonary smooth muscle in the African lungfish, *Protopterus aethiopicus. Acta Physiol. Scand.* **71**, 248–252.

Johansen, K., Lenfant, C., and Grigg, G. C. (1967). Respiratory control in the lungfish, *Neoceratodus forsteri* (Krefft). *Comp. Biochem. Physiol.* **20**, 835–854.

Johansen, K. Lenfant, C., Schmidt-Nielsen, K., and Petersen, J. A. (1968). Gas exchange and control of breathing in the electric eel, *Electrophorus electricus. Z. Vergleich. Physiol.* **61**, 137–163.

Johansen, K., Hanson, D., and Lenfant, C. (1970). Respiration in a primitive air breather, *Amia calva*. *Respiration Physiol.* (in press).

Klika, E., and Lelek, A. (1967). A contribution to the study of the lungs of the *Protopterus annectens* and *Polypterus senegalensis*. *Folia Morphol.* **15**, 168–175.

Krogh, A. (1904). Some experiments on the cutaneous respiration of vertebrate animals. *Skand. Arch. Physiol.* **16**, 348–357.

Krogh, A., and Leitch, I. (1919). The respiratory function of the blood of fishes. *J. Physiol. London* **52**, 288–300.

Lahiri, S., Shub, C., and Fishman, A. P. (1968). Respiratory properties of blood of the African lungfish. (*Protopterus*) *Proc. 24th Intern. Union Physiol. Sci. Washington, D.C.* Vol. **7**, 251.

Lenfant, C., and Johansen, K. (1966). Respiratory function in the elasmobranch *Squalus suckleyi*. *Respiration Physiol.* **1**, 13–29.

Lenfant, C., and Johansen, K. (1967). Respiratory adaptations in selected amphibians. *Respiration Physiol.* **2**, 247–260.

Lenfant, C., and Johansen, K. (1968). Respiration in the African lungfish, *Protopterus aethiopicus*. I. Respiratory properties of blood and normal patterns of breathing and gas exchange. *J. Exptl. Biol.* ***49***, 437–452.

Lenfant, C., Johansen, K., and Grigg, G. C. (1966). Respiratory properties of blood and pattern of gas exchange in the lungfish *Neoceratodus forsteri* (Krefft). *Respiration Physiol.* **2**, 1–21.

Lenfant, C., Johansen, K., Petersen, J. A., and Schmidt-Nielsen, K. (1970). Gas exchange in the airbreathing fishes; *Lepidosiren paradoxa* and *Symbranchus marmoratus*. In preparation.

McCutcheon, F. H., and Hall, C. G. (1937). Hemoglobin in the Amphibia. *J. Cellular Comp. Physiol.* **9**, 191–197.

McMahon, B. R. (1969). A functional analysis of the aquatic and aerial respiratory movements of an African lungfish, *Protopterus aethiopicus*, with reference to the evolution of the lung-ventilation mechanism in vertebrates. *J. Exptl. Biol.* **51**, 407–430.

Magid, A. M. A. (1966). Breathing and function of the spiracles in *Polypterus senegalus*. *Animal Behaviour* **14**, 530–533.

Moussa, T. A. (1956). Morphology of the accessory airbreathing organs of the teleost, *Clarias lazera* (C & V). *J. Morph.* **98**, 125–160.

Munshi, J. S. D. (1961). The accessory respiratory organs of *Clarias batrachus* (Linn). *J. Morph.* **109**, 115–140.

Munshi, J. S. D. (1962). On the accessory respiratory organs of *Ophiocephalus punctatus* (bloch) and *Ophiocephalus striatus* (bloch). *J. Linnean Soc. London, Zool.* **44**, 616–624.

Munshi, J. S. D. (1968). The accessory respiratory organs of *Anabas testudineus* (Bloch) (Anabantidae, Pisces). *Proc. Linnean Soc. London* **179**, 107–126.

Oglialoro, C. M. (1947). Vascolarizzazione della mucosa bucco-faringea e respirazione accessoria attraverso tale regione in alcuni *Teleosti marini*. *Boll. Soc. Ital. Biol. Sper.* **23**, 990–992.

Piiper, J., and Schumann, D. (1967). Efficiency of O_2 exchange in the gills of the dogfish, *Scyliorhinus stellaris*. *Respiration Physiol.* **2**, 135–148.

Poll, M. (1962). Etude sur la structure adulte et la formation des sacs pulmonaires des Protopteres. *Ann. Reeks Zool. Wetenschap.* **108**, 131–172.

Powers, E. B., and Clark, R. T. (1942). Control of normal breathing in fishes by receptors located in the regions of the gills and innervated by the Xth and Xth cranial nerves. *Am. J. Physiol.* **138**, 104–107.

Qasim, S. Z., Qayyum, A., and Garg, R. K. (1966). The measurement of carbon dioxide produced by air-breathing fishes and evidence of the respiratory function of the accessory respiratory organs. *Proc. Indian Acad. Sci.* **52B**, 19–26.

Rauther, M. (1910). Die akzessorischen Atmungsorgane der Knochenfische. *Ergeb. Zool.* **2**, 517–585.

Richter, G. (1935). Die Luftatmung und die akzessorischen Atmungsorgane von *Gymnotus electricus* L. *Morphol. Jahrb.* **75**, 469–475.

Root, R. W. (1931). The respiratory function of the blood of marine fishes. *Biol. Bull.* **61**, 427–456.

Sawaya, P. (1946). Sobre abiologia de alguns peixes de respiracao aerea. *Univ. Sao Paulo, Fac. Filosof., Cienc. Letras, Zool.* **11**, 255–286.

Saxena, D. B. (1963). A review on ecological studies and their importance in the physiology of air-breathing fishes. *Ichthyologica* **2**, 116–128.

Scharrer, E., Smith, S. W., and Palay, S. L. (1947). Chemical sense and taste in the fishes, *Prionotus* and *Trichogaster*. *J. Comp. Physiol.* **86**, 183–198.

Schlaifer, A., and Breder, C. M. (1940). Social and respiratory behaviour of small tarpon. *Zoologica* **25**, 493–512.

Schöttle, E. (1931). Morphologie und Physiologie der Atmung bei wasser, schlamm und landlebenden Gobiiformes. Z. *Wiss. Zool.* **140**, 1–114.

Schulz, H. (1960). Die submikroskopische Morphologie des Kiemenepithels. *Proc. 4th Intern. Conf. Electron Microscopy, Berlin, 1958* Vol. 2, pp. 421–426. Springer, Berlin.

Shelford, V. E., and Allee, W. C. (1913). The reactions of fishes to gradients of dissolved atmospheric gases. *J. Exptl. Zool.* **14**, 207–266.

Smith, H. W. (1930). Metabolism of the lungfish, *Protopterus aethiopicus*. *J. Biol. Chem.* **88**, 97–130.

Spurway, H., and Haldane, J. B. S. (1963). The regulation of breathing in a fish, *Anabas testudineaus*. *In* "The Regulation of Human Respiration" (D. J. C. Cunningham and B. B. Lloyd, eds.), pp. 431–434. Blackwell, Oxford.

Swan, H., and Hall, F. G. (1966). Oxy-hemoglobin dissociation in *Protopterus aethiopicus*. *Am. J. Physiol.* **210**, 487–489.

Taylor, M. (1913). The development of *Symbranchus marmoratus*. *Quart. J. Microscop. Sci.* **59**, 1–51.

Todd, E. S., and Ebeling, A. W. (1966). Aerial respiration in the longjaw mudsucker *Gillichtys mirabilis* (Teleostei: Gobiidae). *Biol. Bull.* **130**, 265–288.

Willmer, E. N. (1934). Some observations on the respiration of certain tropical fresh water fish. J. *Exptl. Biol.* **11**, 283–306.

Wu, H. W., and Kung, C. C. (1940). On the accessory respiratory organ of *Monopterus*. *Sinensia* **11**, 59–67.

Chapter 7

Volume and composition of body fluids: The lasting impact of the first chapter of the Fish Physiology series and the value of summary tables

Alex M. Zimmer[a,*] and Chris M. Wood[b,c,d]
[a]*Department of Biological Sciences, University of New Brunswick, Saint John, NB, Canada*
[b]*Department of Zoology, University of British Columbia, Vancouver, BC, Canada*
[c]*Department of Biology, McMaster University, Hamilton, ON, Canada*
[d]*Rosenstiel School of Marine and Atmospheric Science, University of Miami, Miami, FL, United States*
[*]*Corresponding author: e-mail: alex.zimmer@unb.ca*

Alex M. Zimmer and Chris M. Wood discuss the W.N. Holmes and E.M. Donaldson chapter "The Body Compartments and the Distribution of Electrolytes" in *Fish Physiology*, Volume 1, published in 1969. The next chapter in this volume is the re-published version of that chapter.

This chapter written by Holmes and Donaldson (1969) is the first of the Fish Physiology series. In their chapter, the authors painstakingly summarized the literature on body fluid volumes and ionic compositions of diverse fish species and their work has stood the test of time as an invaluable reference for fish physiologists. In this overview, we discuss the impact that this chapter has had as a reference for typical body fluid volumes and ion compositions in fishes in several fields, some extending beyond fish physiology. In addition, we present 10 insights made by Holmes and Donaldson (1969), some of which seem to have anticipated important research areas, while others have gone largely overlooked. This chapter is clearly a classic in the Fish Physiology series which, nearly 6 decades later, still has the potential to catalyze new avenues of research in the field.

Fish Physiology, Vol. 40A. https://doi.org/10.1016/bs.fp.2023.09.006

1 Introduction

This classic review, the very first chapter in the *Fish Physiology* series, is a broad overview of body fluid volumes and compositions from diverse groups of fishes, including many species that have been poorly studied to date. The authors of the chapter point out that fishes have evolved to occupy many different ecological niches in freshwater and marine environments, with some species adopting life cycles involving "sojourns" into freshwater and seawater, to even terrestrial habitats for some part of their lives. These niches impose distinct ionoregulatory and osmoregulatory challenges with which species must cope in order to maintain, in the words of Claude Bernard, a constant "milieu intérieur" (Holmes and Donaldson, 1969). In addition, the volume and composition of body fluid compartments can differ substantially from one another, and these differences must be maintained within a relatively constant range because, as discussed by the authors in the introduction to their chapter, differences in the concentration of charged molecules between intracellular and extracellular fluids are critically important for nearly all physiological functions. Their chapter has served as an invaluable reference for researchers in the field to evaluate and compare body fluid volume and composition data. Importantly, the authors provide insightful context and highlight the significance of changes in body fluid volume and composition in response to different environmental variables (e.g., salinity, estivation), the differences between body compartments, and trends across species and groups.

Perhaps surprisingly for the first entry in the *Fish Physiology* series, the chapter was actually written by two experts in bird physiology. Neil Holmes had been an Assistant Professor at UBC, and Ed Donaldson was his Ph.D. student, investigating adrenocortical function in ducks. Holmes' group worked on both birds and fishes at the time, and produced some now classic papers on kidney function in salmonids (e.g. Holmes and McBean, 1963; Holmes and Stainer, 1966). By the time the chapter was published, Holmes had moved on to a position at the University of California at Santa Barbara where he had a distinguished career focused on osmoregulatory physiology and petroleum toxicity in ducks, among other research pursuits, while Donaldson had taken a research position at the West Vancouver Laboratory of the Dept. of Fisheries and Oceans Canada where he moved into fish endocrinology. Donaldson stayed there all his life, broadening his endocrinological interests to many aspects of salmonid culture and became a world leader in the reproductive physiology, sex determination, and growth enhancement of salmonids as applied to aquaculture. Indeed, he later guest-edited Volumes 9A and B ("*Reproductive Physiology*") and contributed several other chapters to the *Fish Physiology* series.

Prior to 1969, the "go to" review for those working in fish osmoregulation was the chapter of Virginia Safford Black (1957) "*Excretion and Osmoregulation*." in Margaret Brown's "*Physiology of the Fishes*" (Brown, 1957). But there was a lack of quantitative information all summarized in one document. It is this aspect of the Holmes and Donaldson chapter, more than anything else, that explains its enduring legacy. In conversation with the present authors, Dave Randall well remembers encouraging the authors in the production of the exhaustive tables of concentrations and volumes, which are the hallmarks of the chapter, and realizing their long-term value. But he does not remember who did most of the work in compiling them!

2 Importance of chapter

As of September 2023, the chapter by Holmes and Donaldson had been cited 336 times, ranking it in the top 25 most-cited of the 482 articles published in the *Fish Physiology* series to date, according to the Elsevier Scopus database (Fig. 1). This extensive body of citing literature highlights the utility of the chapter, which is most often cited to refer to typical values for blood or plasma volume, plasma osmolarity, and plasma ion concentrations in a given species for comparative purposes. The data summarized in the many extensive tables throughout the chapter have clearly been of great value to the fish physiology community. The work that has cited the chapter spans over 6 decades of research, highlighting its lasting impact and the continued interest of fish physiologists in the area of environmental and comparative osmoregulation and ionoregulation. Some of the most well-cited articles of the 336 citing publications include many chapters of the *Fish Physiology* series (Brett, 1979; Cowey and Sargent, 1979; McDonald and Milligan, 1992), reviews in specific research areas such as the comparative regulation of nervous system pH (Chesler, 1990), fish osmo- and ionoregulation (Evans et al., 1999), aquatic toxicology (Larsson et al., 1985), and some of the seminal papers that helped define our collective understanding of the mechanisms of ion regulation in fishes (e.g., Karnaky et al., 1976).

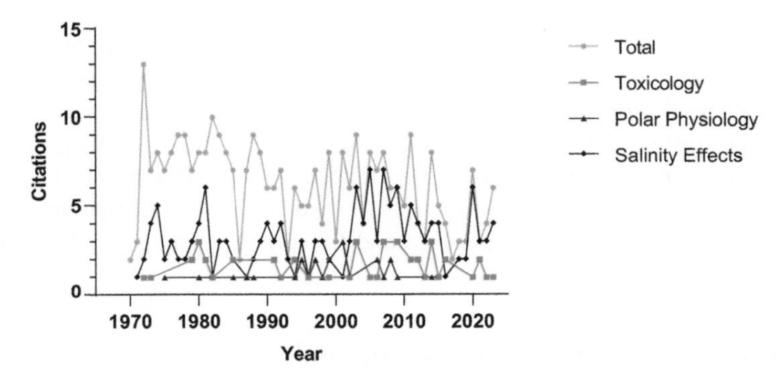

FIG. 1 Citations of the Holmes and Donaldson (1969) chapter over time. Citations reported by Scopus as of September 2023. Displayed are total citations (gray circles) and citations from research in toxicology (search terms "toxicity"; red squares), polar physiology (search terms "antifreeze OR cryoprotection OR cryoprotectant"; blue triangles), and salinity effects (search terms "salinity AND blood OR plasma"; black diamonds). See text for more details.

Many of the early citations of the Chapter in the 1970s and 1980s added to the characterization of blood volume and composition in diverse fish species. The Chapter was evidently also an important tool in other areas such as toxicology (48 Scopus search hits, including 17 original research articles, using search term "toxicity" from 1972 to 2022), being used to estimate fluid compartment volumes in order to determine the distribution and metabolism of methylmercury (Giblin and Massaro, 1973) and DDT (Dvorchik and Maren, 1974), and to construct models for predicting the toxicity of metal contaminants (Paquin et al., 2002). Research in the physiology of polar fishes has also frequently cited this Chapter (28 Scopus hits, including 17 original research articles, using search term "antifreeze OR cryoprotection OR cryoprotectant" from 1975 to 2015), usually as a reference to normal ranges of plasma ions and freezing points of teleost plasma (e.g., Denstad et al., 1987; Treberg et al., 2002). The most profound impact of this chapter, however, is its use as a reference to typical concentrations of inorganic ions in freshwater and seawater fishes (161 Scopus hits using search term "salinity AND blood OR plasma"), with the citing articles in this area generating over 5000 citations since 1970.

In our opinion, this chapter still stands as the most comprehensive summary of the distribution of fluids between body compartments in fishes. The subject matter was not reviewed again until the chapter of Olson (1992) in *Fish Physiology Volume 12*, which cites Holmes and Donaldson (1969) extensively. Although citations of the chapter have been declining over time (Fig. 1), we encourage researchers to continue to exploit this resource to compare and contextualize blood volume and composition data. The impressive number of species (>100) included in the 19 data tables makes this chapter an important reference for researchers working on non-model or poorly-studied species. Moreover, in addition to the compilation of blood chemistry parameters (Chapter Tables 12, 13, 14, 16, 17), the authors also provided summaries of the chemical composition of many other bodily fluids (Chapter Tables 15 and 19) including the cranial fluid, pericardial fluid, the aqueous humor of the eye, and the endolymph and perilymph of the inner ear. Notably, the latter has important implications in the formation of otoliths, which are widely used to assess age, growth, and the history of events in the life of a fish, and the regulation of endolymph chemistry is a field unto itself (Payan et al., 2004). These summarized data may therefore shed light on comparative and evolutionary trends in the composition and function of these specialized body compartments.

3 History and consideration of methods

It is worthwhile here to briefly review the methods used to collect these data; indeed, critical evaluation of methods was an important theme in the chapter itself. Prior to 1969, major cations were measured by flame photometry or atomic absorption spectrophotometry, and Cl^- by coulometric titration with Ag^+ ions ("chloridometers"); while modern instruments are often automated, the basic methodology has not changed, so "old" (pre-1969) and "new" (post-1969) measurements are directly comparable. For other anions (phosphate, sulphate, bicarbonate), inexpensive but time-consuming chemical methods have been largely replaced by more expensive instrument-based methods such as ion chromatography and acid-displacement followed by infrared CO_2 detection, so some caution should be used in comparing old and new data.

Holmes and Donaldson provided an overview of the methods used to measure the volume of different body compartments in fishes, highlighting that the only direct measurement that can be made is total body water, determined by the difference between wet and dry body weight. All other measurements of fluid volumes are estimates that can be greatly influenced by the methodology employed. The general approach is to inject or infuse a given marker and measure its concentration in the blood to determine a dilution factor which can be used to calculate volume. The distribution of specific markers into different fluid compartments allows researchers to estimate the volume of these compartments. However, estimates are complicated by the fact that different markers are endogenously produced (e.g., urea), metabolized (e.g., sulfanilamide), or excreted (e.g., antipyrine). Plasma volume can be estimated by injection of radiolabeled serum albumin or by T 1824 (Evans blue) which binds to plasma proteins, but these estimates are complicated by the fact that capillaries of fishes can be permeable to proteins, leading to an overestimation because labeled proteins would be diluted into the rest of the extracellular space. Red blood cell (RBC) space can also be determined using radiotracers that accumulate in red cells and total blood volume can be calculated using RBC volume (or hematocrit) and plasma volume. Complications can arise from the fact that RBCs may not be homogenously distributed throughout the vasculature (see Insight # 6 below), and blood volume is ultimately dependent upon the technique used to estimate plasma volume. Holmes and Donaldson note that, due to substantial method-dependent variability, fluid volumes are often cited as the volume of the distribution of a given compound, rather than a specific fluid space. In order to make meaningful comparisons between species and studies, researchers should compare measurements made with the same techniques, which appears to be the approach used by the authors who most often cite blood volume as "T 1824 space and hematocrit" determined by the same researcher (Thorson) across several studies. Notably, Olson (1992) provided an exhaustive and critical review of the methods to measure extracellular fluid volume (ECFV) described in this chapter, and others developed after its publication, and concluded "*It becomes readily apparent that the variability due to different methods is often as great, if not greater, than actual differences between fish.*"

3.1 Ten insights on body volume and ionic composition of fishes from Holmes and Donaldson, circa 1969

As discussed above, the biggest impact of this chapter has been its use as a reliable reference source for typical blood volume and chemistry values in fishes. However, upon reviewing this chapter, we have come to believe that the data tables therein may have overshadowed important insights by the authors that have either set the stage for key research advances in the field or have gone largely overlooked to date. As such, we now present "Ten Insights on Body Volume and Ionic Composition of Fishes from Holmes and Donaldson, circa 1969".

3.2 Insight #1: A Nernstian approach to ion distribution

In the Introduction to their chapter, the authors used the Nernst equation to illustrate the electrochemical forces associated with strong cation distribution across the intracellular (muscle)/extracellular membranes of the eel in different environments. With rare exceptions, this approach has seen little uptake in studies of whole animal physiology, probably reflecting the reluctance of most physiologists to deal with equations involving logarithms. Nevertheless, these exceptions prove the prescience of Holmes and Donaldson. A Nernstian approach has been used to successfully explain the distribution of ammonia across cell membranes and how it changes with exercise, reflecting the relatively high permeability ratio of cell membranes to NH_4^+ relative to NH_3 in ammoniotelic fishes, compared to other animals (Wright and Wood, 1988; Wright et al., 1988). It also explains how a reduction of this ratio correlates with the evolution of ureotelism and a reduction in muscle ammonia stores in amphibians (Wood et al., 1989) and could be an interesting avenue of research in the study of estivating lungfish that accumulate urea as a mechanism to conserve water, which was yet another topic discussed by the authors of this chapter. And more than 30 years later, in aquatic toxicology, Nernstian analysis explained the depolarization of transepithelial potential across the gills which is predictive of major ion toxicity in freshwater fishes (Wood et al., 2020).

3.3 Insight #2: Mitochondria as regulators of intracellular ion concentrations

As a caveat to some of the data presented in their tables, Holmes and Donaldson acknowledged that intracellular fluids are not homogenous solutions, but rather contain "islands" rich in specific ions and solutes. In particular, they point out the surprising lack of attention given to the possibility of the regulation of intracellular ion concentrations by mitochondria, highlighting that according to the endosymbiotic theory, the bacterial ancestor of mitochondria would have possessed active ion transport mechanisms that could contribute to the regulation of cytosolic ion concentrations. Today, it is clear that mitochondria possess complex mechanisms of ion regulation that are critical to the regulation of oxidative phosphorylation and can also contribute to apoptotic signaling pathways (O'Rourke, 2007). In addition to the active uptake of K^+ described in the chapter, mitochondria have long been recognized as sinks for Ca^{2+} and have the capacity to regulate a constant steady-state extramitochondrial Ca^{2+} concentration under a wide range of conditions (Nicholls, 1978). Mitochondria also maintain a transmembrane Na^+ gradient in cardiac (Jung et al., 1992) and smooth muscle (Poburko et al., 2007) cells that is regulated by Na^+/H^+ and Na^+/Ca^{2+} exchanges, indicating a potential role for mitochondria in the regulation of intracellular Na^+ concentration. Interestingly, an important unresolved issue in ionoregulatory physiology is the thermodynamics of freshwater Na^+ absorption across the apical membrane of mitochondrion-rich ionocytes against an "apparent" Na^+ concentration gradient that is directed outwards. Mitochondria could potentially contribute to maintaining a low cytosolic Na^+ concentration that would enable Na^+ absorption from fresh water, although it is important to note that mitochondria isolated from pig hearts maintained a Na^+ gradient whereby the intramitochondrial Na^+ concentration was lower than that of the extramitochondrial solution (Jung et al., 1992). Nevertheless, the role of mitochondria in overall ion homeostasis may be a fruitful avenue for future research which could be facilitated by application of Na^+ sensitive dyes that differentially accumulate in mitochondria and cytosol (Poburko et al., 2007).

3.4 Insight #3: Phylogenetic trends in blood volume

Throughout the tables in their chapter, Holmes and Donaldson summarize the blood volume of 23 species determined arithmetically using hematocrit and plasma volume measured by the T 1824 method. The authors highlight a phylogenetic trend whereby more derived species appear to have lower blood volumes (Fig. 2A) and posit that differences in blood volume may be related to differences in hemoglobin function in less derived species. The hemoglobin of lampreys and hagfishes are monomeric and display oxygen dissociation curves more similar to myoglobin than to the hemoglobin of teleost fishes, with a "flattened zone" over physiological ranges of PO_2 in these species. The authors speculate that this would perhaps limit oxygen delivery to the tissues in these species, suggesting that more research comparing hemoglobin oxygen binding properties across the fish phylogeny is needed. The summary tables presented by the authors also demonstrate an inverse relationship between blood volume and hematocrit among these species (Fig. 2B), which would suggest that less derived fishes (agnathans and chondrichthyans) may

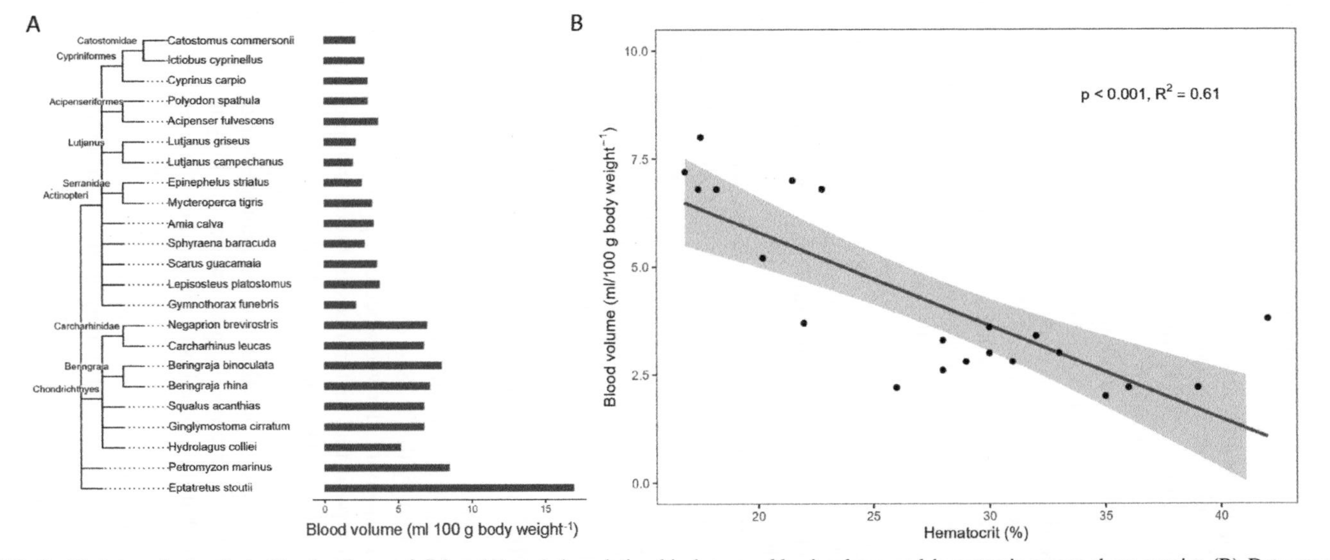

FIG. 2 Phylogenetic trends in blood volume of fishes (A) and the relationship between blood volume and hematocrit among these species (B) Data were extracted from tables in Holmes and Donaldson (1969) and represent blood volumes calculated using hematocrit and T 1824 (Evans Blue) space as an estimate of plasma volume. Phylogeny was created using NCBI Common Tree with species names (updated to reflect current species names on FishBase) as inputs. Note that for (B), hematocrit data for *Eptatretus stoutii* and *Petromyzon marinus* were not available in the original chapter and were therefore obtained from Perry et al. (2009) and Tufts (1991), respectively. Gray shading around the regression line in panel B represents 95% confidence intervals.

require a larger blood volume to deliver the same amount of O_2 compared to more derived teleost species. Given the positive relationship between hematocrit and blood viscosity, it is interesting to speculate whether the hearts of these less derived species, which typically have lower cardiac power outputs (Driedzic et al., 1987), may have placed a limitation on blood hematocrit (and hence viscosity), necessitating a larger blood volume to deliver sufficient O_2 to tissues. Notably, however, increased blood volume may place even greater demands on cardiac function, as exemplified by the hemoglobin-less icefishes that devote considerable amount of resting metabolic rate to cardiac function in order to circulate large volumes of blood lacking RBCs (Sidell and O'Brien, 2006). To our knowledge, this particular insight has gone largely overlooked to date, and the evolutionary relationships between hemoglobin oxygen affinity, blood volume, and cardiac function in fishes have yet to be explored in earnest. In mammals, some evidence suggests that hemoglobin affinity sets functional limits on the size of RBCs, such that species with low affinity hemoglobin have smaller RBCs, which would theoretically facilitate oxygen unloading in the tissues of species with high mass-specific metabolic rates (Udroiu, 2023). Such analyses in fishes may reveal the basis and functional significance of the apparent differences in blood volume and hematocrit across phylogenetically distant species.

3.5 Insight #4: Influences of growth, age, and smoltification on extracellular and intracellular fluid volumes and ion concentrations

The authors devoted considerable text on this topic, reviewing extensive earlier literature on this area, and raising questions about whether observed changes were a function of growth or smoltification. Our impression is that interest subsequently waned in the physiological community, despite an incisive review (Takei et al., 2014), but it has continued strongly in the aquaculture community. There, the focus has been mainly on plasma electrolytes and muscle water content (e.g., the classic "seawater challenge" test; Clarke, 1982) and whole-body ion composition at different stages in the life cycle (e.g., Shearer, 1984), often together with measurements of mRNA or enzyme activity of key ionoregulatory transporters in the gills (e.g., Handeland et al., 2014). The fluid volume regulation aspects have not been pursued, probably because of the methodological challenges discussed earlier. This is a topic that should be revisited by physiologists. An area that has, however, generated much interest and research in the physiological community is the hormonal control of ionoregulation and smoltification. Holmes and Donaldson discuss the role of various hormones in the regulation of osmotic and ionic balance in fishes, including the role of prolactin in smoltification, but noted that understanding was incomplete at the time. These topics have since been the subject of chapters in the *Fish Physiology* series (Hoar, 1988; McCormick, 1995), and our collective understanding of the neuroendocrine regulation of ion balance in fishes continues to improve (Yan and Hwang, 2019).

3.6 Insight #5: Hypertonic urine in killifish in seawater

The chapter highlighted experiments by Stanley and Fleming (1964) and Fleming and Stanley (1965) showing that the plains killifish (*Fundulus kansae*) produces a hypertonic urine for about a week after transfer to seawater. To our knowledge, with the exception of the report of Kowarsky (1973) on the osmolality of bladder urine in the marine catfish *Cnidoglanis macrocephalus*, these remain the only reports of hypertonic urine production in a marine teleost. Fish are generally believed to lack this ability because of the absence of a countercurrent exchange system in their nephrons. However, there have in fact been very few studies on Na^+ and Cl^- flux rates during the dynamic period of seawater challenge (e.g. Wood, 2011), and very few functional studies on the kidney of seawater fishes in the intervening 50 years. A wider range of species should be investigated. The advent of new molecular (Takvam et al., 2021), micro-dissection (Fehsenfeld and Wood, 2018) and micro-probe approaches (Fehsenfeld et al., 2019) for examining the function of teleost nephrons has opened up new possibilities.

3.7 Insight #6: Lymphatic system of fishes

In their description of different body fluid compartments, the authors describe the "lymphatic system" of fishes that has since been a subject of disagreement among fish biologists. In mammals, the lymphatic system regulates interstitial fluid volume, which is a function of the forces favoring filtration and absorption across capillaries. Based on mammalian literature, Holmes and Donaldson summarized these forces in fishes (Chapter Fig. 1), however at this time they were unaware that unlike mammals, oncotic pressure plays virtually no role in fluid filtration because fish capillaries are permeable to proteins (Olson et al., 2003). Historically, the vessels observed in fishes were originally identified as "lymphatics" due to their general lack of RBCs (Steffensen et al., 1986), morphological similarity to mammalian lymphatics, and to a "*modicum of mammalian ethnocentrism*". They were later reclassified as a "secondary circulatory system" and subsequent research questioned the very existence of a lymphatic system in fishes (Olson, 1996). In more recent work, molecular evidence in zebrafish (*Danio rerio*) suggests that these vessels may in fact be true lymphatics (Küchler et al., 2006), supporting the original mammalian ethnocentric view of Holmes and Donaldson. Interestingly, however, some research suggests that the perfusion of lymphatic vessels with RBCs can be modulated by neuroendocrine regulation, such that these vessels can switch their functional identity from lymphatics to blood vessels which might serve to increase tissue O_2 delivery under hypoxic conditions (Jensen et al., 2009).

3.8 Insight #7: Importance of studying the composition of specialized fluid compartments

This chapter is also an incredible resource for researchers studying the specialized fluid compartments of fishes. Determining the composition of these specialized fluids is often critical to understanding the function of the organ system in which they are contained. For instance, the ionic composition of the endolymph, high in K^+, is critical to mechanosensory signal transduction in the inner ear. Holmes and Donaldson demonstrate that high endolymph [K^+], a unique feature among extracellular fluids, is observed across virtually all fish species studied up to that point, and this feature appears to be an extremely well-conserved trait amongst vertebrates. As discussed above, the ionic composition of the endolymph is also undoubtedly critical to the precipitation of $CaCO_3$ in the formation of otoliths (Payan et al., 2004), and elucidating the mechanisms governing the regulation of endolymph composition may help understand and predict the effects of ocean acidification on otolith growth (Checkley et al., 2009; Kwan and Tresguerres, 2022). The ionic composition of specialized fluid compartments can also indicate the mechanisms by which they are formed. The authors discuss high [HCO_3^-] and/or [Na^+] in the aqueous humor of the eye, suggesting that secretion of either or both of these ions is responsible for its formation. In fact, administration of carbonic anhydrase inhibitors is a common treatment for glaucoma, a disease associated with increased intraocular pressure. Given the increasing use of fish models, particularly zebrafish, in human biomedical research including ocular diseases (Gestri et al., 2011), this insight by Holmes and Donaldson stretches beyond the field of fish physiology.

3.9 Insight #8: The regulation of K^+ concentrations

The authors devote as much attention to the regulation of [K^+] as they do to the regulation of [Na^+] and [Cl^-]. This occurs in diverse areas—Nernstian analysis of electrochemical forces (Insight #1), K^+ uptake by mitochondria (Insight #2), muscle [K^+] regulation during seawater challenge (Chapter Fig. 3), [K^+] in extravascular compartments (Insight #7), and plasma [K^+] regulation in hagfishes, lampreys, sharks, and teleosts. Regrettably, studies on K^+ regulation seem to have fallen out of favor in the last few decades, with much more focus on Na^+, Cl^-, and Ca^{2+} regulation; yet K^+ is the major intracellular cation. In the extracellular fluid, [K^+] must be kept low, but not too low. In both mammals and fish, if plasma [K^+] is allowed to deviate out of a relatively narrow range (e.g., 2.5–5 $mmol\,L^{-1}$), potentially fatal disturbances of cardiac function, and incapacitation of skeletal and gastrointestinal muscles may occur (Barratt and Huddart, 1979; Hanson et al., 2006; Nomura et al., 2019; Kuzmin et al., 2022). There is a need for a renewed focus on K^+ regulation in fishes, complementing some recent progress in understanding the role of ROMK ("renal outer medullary K^{+}") channels and NKCC ("Na^+, K^+, 2 Cl^- co-transporter") in K^+ excretion (Furukawa et al., 2014; Horng et al., 2017) and a newly discovered Na^+/K^+ exchange process responsible for sustaining branchial Na^+ uptake in zebrafish at low environmental pH (Clifford et al., 2022).

3.10 Insight #9: Sex differences in plasma Ca^{2+} concentrations

The chapter devoted an entire subsection to an overview of the differences in plasma [Ca^{2+}] between male and female teleost fishes. In general, females have higher plasma [Ca^{2+}] that increases significantly during spawning, a phenomenon which is generally attributed to increases in circulating vitellogenin, a Ca^{2+}-binding yolk precursor (for a review of vitellogenesis, see Mommsen and Walsh, 1988 in the *Fish Physiology Volume 11A*). The term "vitellogenin" was not coined until 1969 (Pan et al., 1969), but it undoubtedly accounts for the concomitant increases in plasma Ca^{2+} and protein concentrations observed during spawning, or in response to exogenous estrogen treatment, in the many studies cited in this section of the chapter. The incorporation of Ca^{2+}-bound vitellogenin into oocytes resulted in a greater than 10-fold increase in gonadal Ca^{2+} content in female Atlantic salmon returning to spawning grounds (Persson et al., 1998). The accumulation of Ca^{2+} in the gonads occurred only following the return to freshwater; therefore, salmon must rely on mobilization of Ca^{2+} stores from scales (Persson et al., 1998) and active Ca^{2+} absorption from their dilute environment. This important insight from Holmes and Donaldson has since been applied to the estimation of reproductive status in rainbow trout (Nagler et al., 1987), further demonstrating the utility of characterizing species- and sex-specific differences in blood ionic composition in fishes and warranting further research on such sex-specific differences in other species.

3.11 Insight #10: Cold effects on plasma osmolality and composition

The chapter devoted considerable attention to freezing point depression and apparent "supercooling" of the plasma in polar fishes. The authors postulated that "organic components in addition to inorganic components may be responsible for the lowering of plasma freezing point at low temperatures", thereby anticipating the landmark discovery by Art DeVries and colleagues of antifreeze proteins in notothenioid fishes that are endemic to the ice-cold waters of the Southern Ocean off the coast of Antarctica (DeVries and Wohlschlag, 1969; DeVries et al., 1970). The field subsequently exploded and remains an active area of research (e.g., Zhuang et al., 2019).

The authors also discussed apparent pathologies of plasma ion concentrations and osmolality at cold temperatures. This was probably instrumental in catalyzing the flood of research on the ionoregulatory effects of "cold-shock" over the following few decades. Out of this came a general realization that active transport processes (uptake of most ions in freshwater, and efflux of most ions in seawater fish) were more sensitive than passive diffusive processes, that extreme cold caused plasma composition to drift towards that of the environment, and that this osmoregulatory failure could contribute towards lethality (reviewed by Donaldson et al., 2008). In polar fishes, however, this drift in plasma osmolarity and ionic composition towards that of seawater is a hallmark of cold-water adaptation, attributed to a reduction in metabolic costs of ionic regulation in these species that display depressions in metabolic rate (Prosser et al., 1970) and/or to a depression of the freezing point of the plasma (Duman and Devries, 1975). More recently, because of global warming concerns, there has been a great deal of interest in high temperature effects on physiology, but potential ionoregulatory issues have been largely overlooked. A recent review (Ern et al., 2023) highlighted various physiological disturbances that could contribute to high temperature death, but osmoregulation was not specifically mentioned, except for the action of high temperature in increasing membrane fluidity, and therefore permeability. Ionoregulatory function and dysfunction at high temperature is an area ripe for investigation.

4 Holmes and Donaldson (1969) as a fish physiology "classic"

This chapter is undoubtedly a classic, standing out as one of the most cited chapters in the history of the series. The extensive summary of volume and composition data of many body compartments across diverse fish species has served as an important comparative tool in the field. While the authors provide valuable and meaningful context to these data, as exemplified by the 10 insights highlighted above, there are some concepts that the authors did not touch upon which are worth noting. For instance, the 1960s were the heyday of radio-isotopic flux measurements pioneered by the Villefranche (France) group—Jean Maetz, Rene Motais and many colleagues—as well as Bill Potts and David Evans in the U.K.. These seminal studies laid the groundwork for our modern understanding of ionoregulatory physiology in fishes, but were not discussed in this chapter. A critically important advancement in our understanding of ionoregulation by freshwater fishes was the confirmation of the idea of August Krogh (1939), by Garcia Romeu and Maetz (1964) and Maetz and Garcia Romeu (1964), that Na^+ and Cl^- uptake in freshwater fish are independent processes, with Na^+ being exchanged for acidic equivalents (H^+ or NH_4^+), and Cl^- being exchanged for basic equivalents (HCO_3^- or OH^-). These papers proved that ion and acid-base regulation were inextricably linked by the laws of electroneutrality, a concept that is widely accepted today but was unheard of at the time; the authors of this chapter devoted little text to the regulation of acid-base balance. In their discussion of body compartment volumes, the authors did not elaborate upon the orders of magnitude higher ion exchange rates in seawater fish, or the presence of exchange diffusion (Motais et al., 1966; Evans, 1967) or the incredibly high rates of diffusive water flux in freshwater fish that are reduced in marine fish (Potts et al., 1967). Finally, the critical importance of drinking for water acquisition in seawater teleosts (Potts et al., 1967; Evans, 1968) is not mentioned, apart from speculation on whether the coelacanth drinks. Importantly, some of these topics were considered in Chapter 3 "*Salt Secretion*" in this same volume (Conte, 1969), which could explain their absence in this chapter. Nevertheless, this chapter stands out as an invaluable compendium of reference values for fish physiologists, and we believe that the many insights made by its authors will act as catalysts for continued, renewed, and novel avenues of research in our field. We encourage readers to visit (or revisit) the impressive data tables of this classic which we hope will spark curiosity and lead to further research in comparative studies of ionoregulation and osmoregulation in fishes.

References

Barratt, L., Huddart, H., 1979. Spontaneous activity and related calcium movements of fish intestinal smooth muscle. The effect of depolarization, caffeine and lanthanum. Gen. Pharmacol. Vasc. S. 10, 21–30.

Brett, J.R., 1979. Environmental factors and growth. In: Hoar, W.S., Randall, D.J., Brett, J.R. (Eds.), Fish Physiology. vol. 8. Academic Press, pp. 599–675.

Brown, M.E. (Ed.), 1957. The Physiology of Fishes. vols. I and II. Academic Press, New York.

Checkley, D.M., Dickson, A.G., Takahashi, M., Radich, J.A., Eisenkolb, N., Asch, R., 2009. Elevated CO_2 enhances otolith growth in young fish. Science 324, 1683.

Chesler, M., 1990. The regulation and modulation of pH in the nervous system. Prog. Neurobiol. 34, 401–427.

Clarke, W.C., 1982. Evaluation of the seawater challenge test as an index of marine survival. Aquaculture 28, 177–183.

Clifford, A.M., Tresguerres, M., Goss, G.G., Wood, C.M., 2022. A novel K^+-dependent Na^+ uptake mechanism during low pH exposure in adult zebrafish (*Danio rerio*): new tricks for old dogma. Acta Physiol. 234, e13777.

Conte, F.P., 1969. Salt secretion. In: Hoar, W.S., Randall, D.J. (Eds.), Fish Physiology. vol. 1. Academic Press, pp. 241–292.

Cowey, C.B., Sargent, J.R., 1979. Nutrition. In: Hoar, W.S., Randall, D.J., Brett, J.R. (Eds.), Fish Physiology. vol. 8. Academic Press, pp. 1–69.

Denstad, J.-P., Aunaas, T., Börseth, J.F., Aarset, A.V., Zachariassen, K.E., 1987. Thermal hysteresis antifreeze agents in fishes from Spitsbergen waters. Polar Res. 5, 171–174.

DeVries, A.L., Wohlschlag, D.E., 1969. Freezing resistance in some Antarctic fishes. Science 163, 1073–1075.

DeVries, A.L., Komatsu, S.K., Feeney, R.E., 1970. Chemical and physical properties of freezing point-depressing glycoproteins from Antarctic fishes. J. Biol. Chem. 245, 2901–2908.

Donaldson, M.R., Cooke, S.J., Patterson, D.A., MacDonald, J.S., 2008. Cold shock and fish. J. Fish Biol. 73, 1491–1530.

Driedzic, W.R., Sidell, B.D., Stowe, D., Branscombe, R., 1987. Matching of vertebrate cardiac energy demand to energy metabolism. Am. J. Phys. Regul. Integr. Comp. Phys. 21, R930–R937.

Duman, J.G., DeVries, A.L., 1975. The role of macromolecular antifreezes in cold water fishes. Comp. Biochem. Physiol. 52A, 193–199.

Dvorchik, B.H., Maren, T.H., 1974. Distribution, metabolism, and elimination and toxicity of 14C-DDT in the dogfish, *Squalus acanthias*. Comp. Gen. Pharmacol. 5, 37–49.

Ern, R., Andreassen, A.H., Jutfelt, F., 2023. Physiological mechanisms of acute upper thermal tolerance in fish. Phys. Ther. 38, 141–158.

Evans, D.H., 1967. Sodium, chloride and water balance of the intertidal teleost, *Xiphister atropurpureus*: III. The roles of simple diffusion, exchange diffusion, osmosis and active transport. J. Exp. Biol. 47, 525–534.

Evans, D.H., 1968. Measurement of drinking rates in fish. Comp. Biochem. Physiol. 25, 751–753.

Evans, D.H., Piermarini, P.M., Potts, W.T.W., 1999. Ionic transport in the fish gill epithelium. J. Exp. Zool. 283, 641–652.

Fehsenfeld, S., Kolosov, D., Wood, C.M., O'Donnell, M.J., 2019. Section-specific H^+ fluxes in renal tubules of fasted and fed goldfish. J. Exp. Biol. 222, jeb200964.

Fehsenfeld, S., Wood, C.M., 2018. Section-specific expression of acid-base and ammonia transporters in the kidney tubules of the goldfish *Carassius auratus* and their responses to feeding. Am. J. Physiol. Ren. Physiol. 315, 1565–1582.

Fleming, W.R., Stanley, J.G., 1965. Effects of rapid changes in salinity on the renal function of a euryhaline teleost. Am. J. Phys. 209, 1025–1030.

Furukawa, F., Watanabe, S., Kakumura, K., Hiroi, J., Kaneko, T., 2014. Gene expression and cellular localization of ROMKs in the gills and kidney of *Mozambique tilapia* acclimated to fresh water with high potassium concentration. Am. J. Phys. Regul. Integr. Comp. Phys. 307, R1303–R1312.

Garcia Romeu, F., Maetz, J., 1964. The mechanism of sodium and chloride uptake by the gills of a fresh-water fish. *Carassius auratus*: I. Evidence for an independent uptake of sodium and chloride ions. J. Gen. Physiol. 47, 1195–1207.

Gestri, G., Link, B.A., Neuhauss, S.C.F., 2011. The visual system of zebrafish and its use to model human ocular diseases. Dev. Neurobiol. 72, 302–327.

Giblin, F.J., Massaro, E.J., 1973. Pharmacodynamics of methyl mercury in the rainbow trout (*Salmo gairdneri*): tissue uptake, distribution and excretion. Toxicol. Appl. Pharmacol. 24, 81–91.

Handeland, S.O., Imsland, A.K., Nilsen, T.O., Ebbesson, L.O.E., Hosfeld, C.D., Pedrosa, C., Toften, H., Stefansson, S.O., 2014. Osmoregulation in Atlantic salmon *Salmo salar* smolts transferred to seawater at different temperatures. J. Fish Biol. 85, 1163–1176.

Hanson, L.M., Obradovich, S., Mouniargi, J., Farrell, A.P., 2006. The role of adrenergic stimulation in maintaining maximum cardiac performance in rainbow trout (*Oncorhynchus mykiss*) during hypoxia, hyperkalemia and acidosis at 10 C. J. Exp. Biol. 209, 2442–2451.

Hoar, W.S., 1988. The physiology of smolting salmonids. In: Fish Physiology. vol. 11B. Academic Press, pp. 275–343.

Holmes, W.N., Donaldson, E.M., 1969. The body compartments and the distribution of electrolytes. In: *Fish Physiology*. vol. 1. Academic Press, pp. 1–89.

Holmes, W.N., McBean, R.L., 1963. Studies on the glomerular filtration rate of rainbow trout (*Salmo gairdneri*). J. Exp. Biol. 40, 335–341.

Holmes, W.N., Stainer, I.M., 1966. Studies on the renal excretion of electrolytes by the trout (*Salmo gairdneri*). J. Exp. Biol. 44, 33–46.

Horng, J.-L., Yu, L.-L., Liu, S.-T., Chen, P.-Y., Lin, L.-Y., 2017. Potassium regulation in medaka (*Oryzias latipes*) larvae acclimated to fresh water: passive uptake and active secretion by the skin cells. Sci. Rep. 7, 16215.

Jensen, L.D.E., Cao, R., Hedlund, E.-M., Söll, I., Lundberg, J.O., Hauptmann, G., Steffensen, J.F., Cao, Y., 2009. Nitric oxide permits hypoxia-induced lymphatic perfusion by controlling arterial-lymphatic conduits in zebrafish and glass catfish. Proc. Natl. Acad. Sci. U. S. A. 106, 18408–18413.

Jung, D.W., Apel, L.M., Brierley, G.P., 1992. Transmembrane gradients of free Na^+ in isolated heart mitochondria estimated using a fluorescent probe. Am. J. Phys. Cell Phys. 31, C1047–C1055.

Karnaky, K.J., Kinter, L.B., Kinter, W.B., Stirling, C.E., 1976. Teleost chloride cell. II. Autoradiographic localization of gill Na, K-ATPase in killifish *Fundulus heteroclitus* adapted to low and high salinity environments. J. Cell Biol. 70, 157–177.

Kowarsky, J., 1973. Extra-branchial pathways of salt exchange in a teleost fish. Comp. Biochem. Physiol. A 46, 477–486.

Krogh, A., 1939. Osmotic Regulation in Aquatic Animals. Cambridge University Press, Cambridge.

Küchler, A.M., Gjini, E., Peterson-Maduro, J., Cancilla, B., Wolburg, H., Schulte-Merker, S., 2006. Development of the zebrafish lymphatic system requires vegfc signaling. Curr. Biol. 16, 1244–1248.

Kuzmin, V., Ushenin, K.S., Dzhumaniiazova, I.V., Abramochkin, D., Vornanen, M., 2022. High temperature and hyperkalemia cause exit block of action potentials at the atrioventricular junction of rainbow trout (*Oncorhynchus mykiss*) heart. J. Therm. Biol. 110, 103378.

Kwan, G.T., Tresguerres, M., 2022. Elucidating the acid-base mechanisms underlying otolith overgrowth in fish exposed to ocean acidification. Sci. Total Environ. 823, 153690.

Larsson, Å., Haux, C., Sjöbeck, M.-L., 1985. Fish physiology and metal pollution: results and experiences from laboratory and field studies. Ecotoxicol. Environ. Saf. 9, 250–281.

Maetz, J., Garcia Romeu, F., 1964. The mechanism of sodium and chloride uptake by the gills of a fresh-water fish, *Carassius auratus*: II. Evidence for NH_4^+/Na^+ and HCO_3^-/Cl^-exchanges. J. Gen. Physiol. 47, 1209–1227.

McCormick, S.D., 1995. Hormonal control of gill Na^+, K^+-ATPase and chloride cell function. In: Wood, C.M., Shuttleworth, T.J. (Eds.), Fish Physiology. vol. 14. Academic Press, pp. 285–315.

McDonald, D.G., Milligan, C.L., 1992. Chemical properties of the blood. In: Fish Physiology. vol. 12B. Academic Press, pp. 55–133.

Mommsen, T.P., Walsh, P.J., 1988. Vitellogenesis and oocyte assembly. In: Hoar, W.S., Randall, D.J. (Eds.), Fish Physiology. vol. 11A. Academic Press, pp. 347–406.

Motais, R., Garcia Romeu, F., Maetz, J., 1966. Exchange diffusion effect and euryhalinity in teleosts. J. Gen. Physiol. 50, 391–422.

Nagler, J.J., Ruby, S.M., Idler, D.R., So, Y.P., 1987. Serum phosphoprotein phosphorous and calcium levels as reproductive indicators of vitellogenin in highly vitellogenic mature female and estradiol-injected immature rainbow trout (*Salmo gairdneri*). Can. J. Zool. 65, 2421–2425.

Nicholls, D.G., 1978. The regulation of extramitochondrial free calcium ion concentration by rat liver mitochondria. Biochem. J. 176, 463–474.

Nomura, N., Shoda, W., Uchida, S., 2019. Clinical importance of potassium intake and molecular mechanism of potassium regulation. Clin. Exp. Nephrol. 23, 1175–1180.

O'Rourke, B., 2007. Mitochondrial ion channels. Annu. Rev. Physiol. 69, 19–49.

Olson, K.R., 1992. Blood and extracellular fluid volume regulation: role of the renin-angiotensin system, kallikrein-kinin system, and atrial natriuretic peptides. In: Fish Physiology. vol. 12. Academic Press, pp. 135–254.

Olson, K.R., 1996. Secondary circulation in fish: anatomical organization and physiological significance. J. Exp. Zool. 275, 172–185.

Olson, K.R., Kinney, D.W., Dombkowski, R.A., Duff, D.W., 2003. Transvascular and intravascular fluid transport in the rainbow trout: revisiting Starling's forces, the secondary circulation and interstitial compliance. J. Exp. Biol. 206, 457–467.

Pan, M.L., Bell, W.J., Telfer, W.H., 1969. Vitellogenic blood protein synthesis by insect body fat. Science 165, 393–394.

Paquin, P.R., Zoltay, V., Winfield, R.P., Wu, K.B., Mathew, R., Santore, R.C., Di Toro, D.M., 2002. Extension of the biotic ligand model of acute toxicity to a physiologically-based model of the survival time of rainbow trout (*Oncorhynchus mykiss*) exposed to silver. Comp. Biochem. Physiol. C 133, 305–343.

Payan, P., De Pontual, H., Boeuf, G., Mayer-Gostan, N., 2004. Endolymph chemistry and otolith growth in fish. C. R. Palevol 3, 535–547.

Perry, S.F., Vulesevic, B., Braun, M., Gilmour, K.M., 2009. Ventilation in Pacific hagfish (*Eptatretus stoutii*) during exposure to acute hypoxia or hypercapnia. Respir. Physiol. Neurobiol. 167, 227–234.

Persson, P., Sundell, K., Björnsson, B.T.H., Lundqvist, H., 1998. Calcium metabolism and osmoregulation during sexual maturation of river running Atlantic salmon. J. Fish Biol. 52, 334–349.

Poburko, D., Liao, C.-H., Lemos, V.S., Lin, E., Maruyama, Y., Cole, W.C., van Breemen, C., 2007. Transient receptor potential channel 6-mediated, localized cytosolic [Na^+] transients drive Na^+/Ca^{2+} exchanger-mediated Ca^{2+} entry in purinergically stimulated aorta smooth muscle cells. Circ. Res. 101, 1030–1038.

Potts, W.T.W., Foster, M.A., Rudy, P.P., Howells, G.P., 1967. Sodium and water balance in the cichlid teleost, *Tilapia mossambica*. J. Exp. Biol. 47, 461–470.

Prosser, C.L., Mackay, W., Kato, K., 1970. Osmotic and ionic concentrations in some Alaskan fish and goldfish from different temperatures. Physiol. Zool. 43, 81–89.

Black, V.S., 1957. Excretion and osmoregulation. In: The Physiology of Fishes. Metabolism. vol. 1. Academic Press, New York, pp. 163–205.

Shearer, K.D., 1984. Changes in elemental composition of hatchery-reared rainbow trout, *Salmo gairdneri*, associated with growth and reproduction. Can. J. Fish. Aquat. Sci. 41, 1592–1600.

Sidell, B.D., O'Brien, K.M., 2006. When bad things happen to good fish: the loss of hemoglobin and myglobin expression in Antarctic icefishes. J. Exp. Biol. 209, 1791–1802.

Stanley, J.G., Fleming, W.R., 1964. Excretion of hypertonic urine by a teleost. Science 144, 63–64.

Steffensen, J.F., Lomholt, J.P., Vogel, W.O.P., 1986. In *vivo* observations on a specialized microvasculature, the primary and secondary vessels in fishes. Acta Zool. 67, 193–200.

Takei, Y., Hiroi, J., Takahashi, H., Sakamoto, T., 2014. Diverse mechanisms for body fluid regulation in teleost fishes. Am. J. Phys. Regul. Integr. Comp. Phys. 307, R778–R792.

Takvam, M., Wood, C.M., Kryvi, H., Nilsen, T.O., 2021. Ion transporters and osmoregulation in the kidney of teleost fishes as a function of salinity. Front. Physiol. 12, 664588.

Treberg, J.R., Wilson, C.E., Richards, R.C., Ewart, K.V., Driedzic, W.R., 2002. The freeze-avoidance response of smelt *Osmerus mordax*: initiation and subsequent suppression of glycerol, trimethylamine oxide and urea accumulation. J. Exp. Biol. 205, 1419–1427.

Tufts, B.L., 1991. Acid-base regulation and blood gas transport following exhaustive exercise in a Agnathan, the sea lamprey *Petromyzon marinus*. J. Exp. Biol. 159, 371–385.

Udroiu, I., 2023. Phylogeny and evolution of erythrocytes in mammals. J. Exp. Biol. 226, jeb245384.

Wood, C.M., 2011. Rapid regulation of Na^+ and Cl^- flux rates in killifish after acute salinity challenge. J. Exp. Mar. Biol. Ecol. 409, 62–69.

Wood, C.M., Munger, R.S., Toews, D.P., 1989. Ammonia, urea and H^+ distribution and the evolution of ureotelism in amphibians. J. Exp. Biol. 144, 215–233.

Wood, C.M., McDonald, M.D., Grosell, M., Mount, D.R., Adams, W.J., Po, B.H., Brix, K.V., 2020. The potential for salt toxicity: can the trans-epithelial potential (TEP) across the gills serve as a metric for major ion toxicity in fish? Aquat. Toxicol. 226, 105568.

Wright, P.A., Randall, D.J., Wood, C.M., 1988. The distribution of ammonia and H^+ between tissue compartments in lemon sole (*Parophrys vetulus*) at rest, during hypercapnia and following exercise. J. Exp. Biol. 136, 149–175.

Wright, P.A., Wood, C.M., 1988. Muscle ammonia stores are not determined by pH gradients. Fish Physiol. Biochem. 5, 159–162.

Yan, J.-J., Hwang, P.-P., 2019. Novel discoveries in acid-base regulation and osmoregulation: a review of selected hormonal actions in zebrafish and medaka. Gen. Comp. Endocrinol. 277, 20–29.

Zhuang, X., Yang, C., Murphy, K.R., Cheng, C.H.C., 2019. Molecular mechanism and history of non-sense to sense evolution of antifreeze glycoprotein gene in northern gadids. Proc. Natl. Acad. Sci. 116, 4400–4405.

Chapter 8

THE BODY COMPARTMENTS AND THE DISTRIBUTION OF ELECTROLYTES☆

W.N. HOLMES
EDWARD M. DONALDSON

Chapter Outline

I. INTRODUCTION

The cells contained within the integument of all but the simplest multicellular organisms are bathed in a relatively stable fluid medium. Basing his judgment on observations on the composition of blood, Claude Bernard perceived that the constancy of this medium, which he termed the "*milieu interieur*," was an essential condition of free and independent life. Water is the solvent of this fluid medium and in it are dissolved numerous inorganic and organic solutes. The organic solutes are mainly nutrients and the products of metabolism while the inorganic solutes consist of oxygen, carbon dioxide, and electrolytes occurring in ratios similar to those found in dilute seawater. From this

☆This is a reproduction of a previously published chapter in the Fish Physiology series, "1969 (Vol. 1)/The Body Compartments and the Distribution of Electrolytes: ISBN: 978-0-12-350401-2; ISSN: 1546-5098".

Fish Physiology, Vol. 40A. https://doi.org/10.1016/bs.fp.2023.08.001

medium the cells take up oxygen and nutrients and into it they discharge carbon dioxide and the products of metabolism.

Separating the extracellular environment from the internal environment of the cells is a highly organized functional boundary, the cell membrane. This membrane is approximately 75 thick and consists of a bimolecular lipoprotein layer which severely restricts the passive movement of solutes from the extracellular fluid into the cell. Again, water is the major solvent of the intracellular fluid but, although the number of solvent particles per unit volume of intracellular and extracellular fluid is roughly equal, the composition in terms of the relative abundance of particle species in each fluid is strikingly different. In the extracellular fluid the concentrations of Na^+, Cl^-, HCO_3-, and Ca^{2+} are relatively high and, with the exception of blood plasma, the protein concentration is relatively low when compared to the intracellular fluid. In contrast, the concentrations of K^+, PO_4^{3-}, Mg^{2+} and protein in the intracellular fluid are relatively high when compared to the extracellular fluids. Since some of the constituents of both the extracellular and the intracellular fluids possess electrical charges and some of these charged particles are either selectively accumulated in one of the fluids or are nondiffusible, the intracellular fluid becomes negatively charged with respect to the extracellular fluid. Such a charge tends to aid or mitigate against the passive movement of charged particles into or out of the cell body.

There exist, therefore, regulatory mechanisms, presumed to be situated in the cell membrane, which are repsonsible for the maintenance of the differential distribution of solutes between the extracellular and intracellular fluids. These mechanisms involve the expenditure of energy in order to move particle species against the concentration and/or the electrochemical gradient. Thus any tendency to change the concentrations of solutes in these fluids either by passive diffusion, albeit often slow, or by changes in the metabolic state of the individual are counteracted.

By definition such mechanisms are termed "active transport" mechanisms. Probably the mechanism associated with the transport of Na^+ out of the cell and K^+ into the cell is the one best understood at this time. Energy derived from metabolism is used to sustain these ion fluxes and, on the average, these active movements of Na^+ and K^+ just balance the diffusion of Na^+ into and K^+ out of the cell. The potential which exists across the membrane is believed to be related to the fact that K^+ tends to permeate the membrane more rapidly than does Na^+.

Among the Vertebrata we find animals which have adapted to a wide range of ecological niches and even within the class Agnatha and the various classes of gnathostomatous fishes there exist forms which have adapted to freshwater, brackish water, and marine environments. Indeed, others such as certain members of the Salmonoidei and Anguilloidei have adopted life cycles involving sojourns in both freshwater and seawater. Others, such as the lungfishes, have even become adapted to the terrestrial habitat for at least part of their lives. In general, however, despite the range of ecological niches which have been occupied by the vertebrates, a remarkable constancy in the compositions of the extracellular and intracellular fluids exists. Furthermore, among the Agnatha and the gnathostomatous fishes, only the members of the order Myxiniformes and the class Chondrichthyes show exceptions to the generalized pattern of electrolyte distribution between the two major body compartments. An approximation of the steady state ion concentrations and the attendant potential differences expected to occur in fish may be visualized from an examination of Table I.

So far the concept of body compartments has only been dealt with in general terms, and we must now proceed to consider more rigorous definitions of these spaces.

II. THE TOTAL BODY VOLUME

The total body volume is of course self-evident, but as a physiological parameter in most vertebrates it is somewhat meaningless when the distribution and movement of solutes are under consideration. Large portions of the total body volume are occupied by structures having very low turnover rates of metabolites and solutes. The integument and skeleton of many fishes for instance cannot be considered to be solute pools having significant short-term exchanges of water and solutes with the surrounding tissues and fluids. For this reason, therefore, the total body water content is more usefully related to the body compartments since it is the common solute of, and is apportioned between, the major compartments and their subdivisions. Although water is passively distributed according to the disposition of the solutes, it delineates the compartments of organisms in which the metabolic reactions may take place.

A. The Intracellular Compartment

The intracellular compartment of any tissue, organ, or organism may be defined as the sum of the cellular volumes contained within the limits of the cell membrane. This is of course an oversimplification of the actual state of affairs obtaining within each individual cell. The cells are structurally extremely complex and contain structures such as the nucleus, the nucleoli, the mitochondria, the endoplasmic reticulum, and so on.

Some of the organelles are themselves bounded by membranes, e.g., the nucleus and the mitochondria, and consequently they represent compartments within the cell body itself. Many of these membranes almost certainly possess specific active transport properties, and substances appear to be selectively accumulated within these organelles.

Probably, the most intensely studied organelle from this standpoint is the mitochondrion. Isolated mitochondria have been observed to actively take up K^+ from the surrounding medium. Respiratory substrates are necessary for this process, but in contrast to the Na^+ and K^+ active transport mechanisms in the plasma membrane, it is not inhibited by the presence of the cardiac glycoside, ouabain. Thus, the system can be clearly differentiated from that occurring in the cell membrance. Also isolated kidney mitochondria have

Table I The Approximate Steady State Ion Concentrations and Potentials Existing between the Muscle Cells and Their Interstitial Fluid in the Various Physiological and Environmental States of the Eel, *Anguilla anguilla*[a]

	Ion	Ion concentrations (mmole/liter)			
		Interstitial fluid (ion$_o$)	*Intra-cellular fluid (ion$_i$)*	$\frac{[ion_o]}{[ion_i]}$	ε_{ion}(mV)
Parietal muscle					
Freshwater yellow eel	Na^+	143.2	20.3	7.0542	+47.1
	K^+	2.26	129	0.0175	−99.3
Freshwater silver eel	Na^+	150.1	26.7	5.6217	+42.3
	K^+	1.75	112.0	0.0156	−102.1
Seawater yellow eel	Na^+	164.2	18.9	8.6878	+53.0
	K^+	3.35	146.0	0.0229	−92.7
Seawater silver eel	Na^+	183.3	22.4	8.1830	+51.6
	K^+	3.22	143.0	0.0225	−93.1
Tongue muscle					
Freshwater yellow eel	Na^+	143.2	15.4	9.2987	+54.7
	K^+	2.26	122.0	0.1852	−97.9
Freshwater silver eel	Na^+	150.1	21.3	7.0469	+47.9
	K^+	1.75	120.0	0.0146	−103.7
Seawater yellow eel	Na^+	164.2	14.1	11.6454	+60.2
	K^+	3.35	159	0.0211	−94.7
Seawater silver eel	Na^+	183.3	22.6	8.1106	+51.4
	K^+	3.22	154	0.0209	−94.9

[a]The membrane potentials were calculated using the Nernst equation for univalentions:

$$\mathcal{E}_{ion} = \frac{R \cdot T}{F \cdot Z} \log_e \frac{[ion_o]}{[ion_i]} \text{ V}$$

where R is the gas constant (8.31 joules/mole-deg absolute), T is the temperature (deg absolute), F is the Faraday constant (96,500 C/mole), and Z is the valency (1+). By converting to $\log_{10}$, expressing $\mathcal{E}$ in millivolts, and assuming the temperature of the tissues to be the same as that of the environment (12°C), the equation becomes $\mathcal{E}_{ion} = 56.5 \log_{10}[ion_o]/[ion_i]$ mV. Calculated from Chan *et al.* (1967).

been observed to show a 50-fold increase in their Ca^{2+} content during respiration *in vitro* (Vasington and Murphy, 1962). This active uptake of Ca^{2+} into the mitochondrion *in vitro* is dependent upon the presence of Mg^{2+}, inorganic phosphate, and adenosine triphosphate (ATP) in the medium. Furthermore, the influx of inorganic phosphate approximates the simultaneous influx of Ca^{2+}. The quantities of inorganic phosphate and Ca^{2+} entering the mitochondrion *in vitro*, however, far exceeds the solubility of the possible calcium salts (Lehninger *et al.*, 1963). These authors concluded that it would be necessary for at least one salt, possibly hydroxyapatite, to precipitate within the mitochondrion; such areas of precipitation may be associated with the dense osmophilic granules within the mitochondria (see Lehninger, 1965). The *in vivo* accumulation of Ca^{2+} within the mitochondrion is by no means so striking and indeed appears to be considerably less than that observed *in vitro*.

Surprisingly little consideration, however, has been given to the possible physiological implications of intramitochondrial accumulation of ions. Several years ago it was postulated that the mitochondria evolved from bacteria which had originally parasitized and ultimately become symbiotic within the aerobic cell. Such speculation considerably complicates our concept of the intracellular compartment. Nevertheless, the role of the mitochondrial membrane appears to be, at least in part, associated with maintaining a constant *intramitochondrial milieu* in which the enzyme systems may function. Further the active transport mechanisms in the membrane of the mitochondrion may function to regulate the solute composition of the hyaloplasm. Such a mechanism would of course only achieve a temporary sequestration of ions within the mitochondrion, but it may serve to temporarily regulate any localized concentrations of solutes. The existing membranes of the bacteria and their associated active transport mechanisms could well have been adapted to these purposes.

The foregoing serves to illustrate the complexity of the intracellular compartment in even the simplest cell type. Much of the experimental data concerning the role of the mitochondrion within the cell has been derived from mammalian tissues, but there is no reason to believe that the data derived from these tissues do not equally apply to the tissues of fishes. Indeed, the mammalian studies do, in fact, indicate the directions of future studies in fishes. For a precise and elegant survey of the role of the mitochondrion in cellular regulation the reader is referred to Lehninger (1965).

It is clear, therefore, that one must recognize the presence of islands, rich in specific ions and cell solutes, which occur within the intracellular compartment. Further, the compartment is not in reality a homogeneous solution within the aqueous phase of the cell. Even so, it is convenient and meaningful to consider the intracellular compartment as a single aqueous phase when discussing the movement of solutes, particularly electrolytes, between the cell and its surrounding medium.

B. The Extracellular Compartment

The extracellular compartment is that space which exists outside the plasma membranes of the cells, and it contains the fluid and the inclusions surrounding these cells. Anatomically the extracellular compartment can be divided into several subcompartments.

One group of extracellular spaces is anatomically characterized by the presence of a continuous layer of epithelial cells separating it from the remainder of the extracellular compartment. These spaces are collectively termed the "transcellular space." Fluids passing through the epithelial cell boundary into the transcellular compartment are invariably modified. In vertebrates generally they include the gastrointestinal and biliary secretions, the cerebrospinal, intraoccular, pericardial, peritoneal, synovial, and pleural fluids; the luminal fluid of the thyroid gland; the cochlear endolymph, the secretions of the sweat and other glands and the contents of the renal tubules and urinary tract.

The remainder of the extracellular compartment is composed of the *intravascular fluid* or blood plasma, the *interstitial fluid* and the lymph.

The intravascular fluid is circulated through a closed system consisting of arteries and veins which are connected by the capillary network. The capillary network is a system of narrow vessels, 10–20μ in diameter, with walls composed of a single layer of flattened endothelial cells. During the passage of blood through the capillary network a certain "leakage" of water and plasma solutes take place. This net fluid movement out of the capillaries occurs, according to the Starling hypothesis, as the resultant of the forces of filtration and the forces of absorption along the inside and outside of the capillaries. The forces tending to filter fluid out of the capillaries are greater at the arterial end of the capillary and consist of the capillary hydrostatic pressure (CHP) and the oncotic pressure of the interstitial fluid (IOP). Conversely, the forces tending to absorb fluid from the interstitial space are greater at the distal or venous end of the capillary and they include the hydrostatic pressure of the interstitial fluid (IHP) and the oncotic pressure of the plasma (POP). Thus, the net force of filtration (+*ve*) or absorption (−*ve*) at any point along the capillary = (CHP+IOP) − (IHP+POP). Values for the forces of filtration and absorption in fish have not been determined, but typical values taken from the mammalian literature are shown in Fig. 1.

Clearly, a net filtration of fluid out of the capillary system into the interstitial space occurs during the passage of blood through the capillary network. It is this continuous transcapillary efflux of water and solutes out of the blood which constitutes the intersitial fluid.

By virtue of the limited permeability of the capillary wall, much of the protein remains in the capillary and an ultrafiltrate of plasma emerges into the interstitial space. This phenomenon is illustrated by the progressively increasing oncotic pressure of the capillary blood as it passes from the arterial end to the venous end of the capillary loop. Further, the progressive reduction

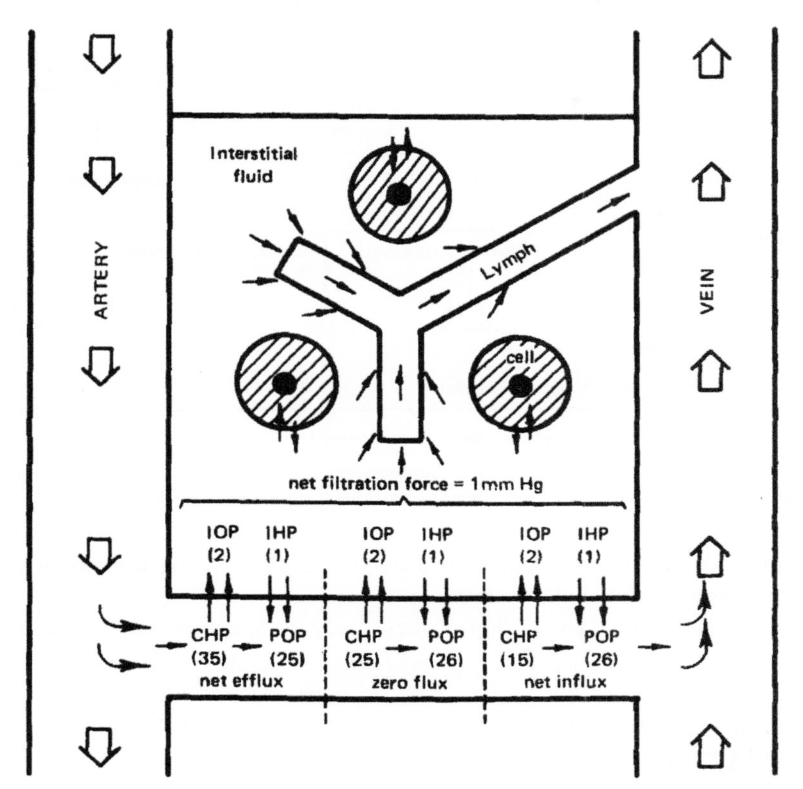

Fig. 1. A diagrammatic representation of the transcapillary movement of ultrafiltrate into the interstitial space and the return of this filtrate via the lymphatic system to the intravascular space. CHP = capillary hydrostatic pressure, IHP = interstitial hydrostatic pressure, POP = plasma oncotic pressure, and IOP = interstitial oncotic pressure. All values in parentheses present hypothetical pressures in mmHg.

in plasma volume as the filtration continues along the length of the capillary results in a progressive decrease in the capillary hydrostatic pressure.

The existence of drainage channels, which start as blind tubules surrounded by interstitial fluid, ensure that the interstitial fluid does not remain stagnant. These endothelial tubules collect into larger vessels, known as "lymphatic ducts," which finally empty into a large vein after passing through the *lymph nodes*. In the lymph nodes the fluid is filtered through a trabeculum of cells, and lymphocytes are added before it rejoins the venous circulation. This system of tubules, beginning with the blind ducts and ending in the large trunk vessels, is known as the "lymphatic system" and the fluid contained in the system is known as "lymph." Since it is derived from interstitial fluid, lymph may also be considered to be an ultrafiltrate of plasma. However, although the inorganic constituents occur at approximately the same concentrations as those found in plasma, the protein composition is quite variable

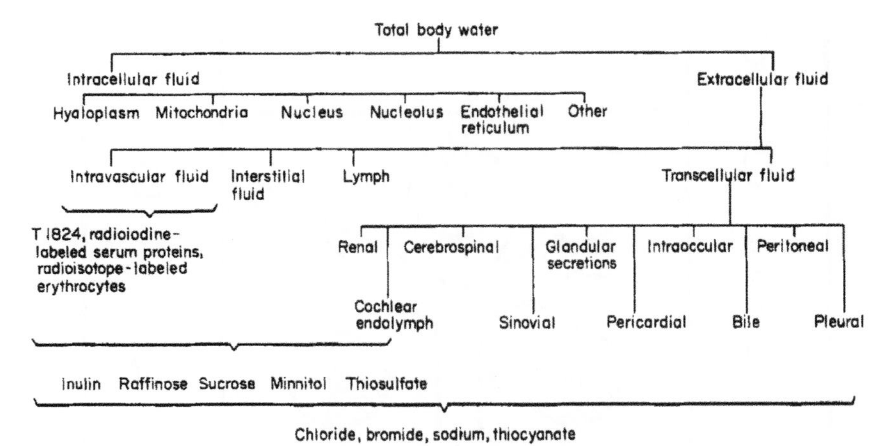

Fig. 2. A summary of the major body compartments in the vertebrates. The approximate volumes of distribution of some of the more common indicator substances are bracketed.

and even the largest molecules such as fibrinogen may be present in the lymph at some point in the system.

A summary and classification of these body compartments are illustrated in Fig. 2.

III. METHODS FOR THE DETERMINATION OF BODY COMPARTMENTS

With the exception of total body water, the compartmental volumes of an individual cannot be measured directly, and therefore indirect methods have been devised. These methods often involve the application of the dilution principle to some nontoxic indicator which is rapidly and homogeneously distributed throughout the compartment to be measured. The volume in which a known amount of this substance is homogeneously distributed may then be calculated after the concentration of the substance in the compartmental fluid has been determined. The general equation for this relationship is as follows:

$$V = Q/C \tag{1}$$

where V is the volume of distribution, Q is the quantity administered, and C is the concentration of the indicator in the compartmental fluid. Should this solute leave the compartment as a result of excretion or metabolism, then a correction must be applied for this loss. The amount lost from the compartment is subtracted from the amount administered and the general formula becomes:

$$V = (Q - E)/C \tag{2}$$

where E is the quantity excreted. The volume of distribution is therefore defined as the volume which would be necessary to accommodate all the indicator substances in the body at a specific time if it were distributed in the compartment at the observed plasma concentration of the substance at that time.

Two principal dilution methods exist for the determination of compartmental volumes. These are the *infusion-equilibrium method* and the *kinetic method*. The infusion-equilibrium method is most suited to use with indicator substances which are rapidly excreted from the organism, e.g., inulin. A priming dose, sufficient to saturate the tissues, is administered intravenously, and the substance is then slowly infused until a constant plasma concentration is achieved. At this point the infusion is stopped and excretory products are continuously collected until the plasma concentration of the indicator has declined to zero. The amount of indicator collected in the excretory products represents the amount which was present in the body (Q) at the time the infusion was stopped. By dividing this quantity (Q) by the plasma concentration at the end of infusion (C) the volume of distribution of the indicator is obtained [Eq. 1].

This method has serious drawbacks to its application in studies on fish. First, since the excretion of the indicator may extend over a considerable period, it is necessary to retain an intravenous cannula throughout the collection period. Such techniques are difficult in fish, particularly if the individuals are maintained in a free-swimming state. Second, many species of fish possess both renal and extrarenal pathways of excretion, and therefore both pathways must be monitored to determine the amount of indicator excreted. Third, in instances where the indicator is an inorganic substance such as chloride, bromide, deuterium oxide, or tritium oxide, the possibility exists that these substances may, after having been excreted by the kidney, reenter the circulation from the environment via the extrarenal uptake mechanisms in the gill epithelium. Therefore, cognizance must be taken of these possible sources of error when volumes of distribution in fish are determined according to the infusion-equilibrium method.

In the kinetic method a series of plasma samples are analyzed following the intravenous administration of a single dose of indicator substance. The log concentration of the indicator substance is then plotted against time. Initially the semilogarithmic decline in the plasma concentration of indicator substance is nonlinear, but thereafter it is linear. By extrapolation of the linear portion of the curve to zero time a theoretical value for the plasma concentration at this time is obtained. This value represents what the equilibrium concentration of indicator substance in the plasma would have been if instantaneous distribution had occurred following the injection. The volume of distribution is then obtained by dividing the amount of indicator administered (Q) by the zero-time equilibrium concentration (C) [Eq. (28–33)]. The volume of distribution may also be determined at any other time after injection by measuring the plasma concentration at this time and the amount of indicator substance excreted up to this time. These values, together with the amount of indicator initially injected, are then substituted in Eq. (158–159). Again, when the latter technique is applied to fish, the excretion via both the renal and extrarenal pathways must be monitored and consideration, in some instances, must be given to the possible reentry of excreted indicator into the fish.

We will now consider the methods available to the physiologist for the determination of specific compartmental volumes.

A. Total Body Water

The total body water content of any animal is easily and most accurately determined by a comparison of the wet and dry body weight of the animal following desiccation to constant weight. The terminal nature of this method does, of course, limit its usefulness.

Other less drastic and indirect techniques have therefore been devised; but in each case their accuracy must be determined by simultaneous comparison with the desiccation method.

Each of these methods involves the application of the dilution principle to indicator substances which become rapidly and homogeneously distributed throughout the total body water, including the transcellular spaces. The chemicals commonly used for this purpose include urea, thiourea, sulfanilamide, antipyrine, 4-acetyl-4-aminoantipyrine (NAAP), deuterium oxide, and tritium oxide. It is assumed that these chemicals become evenly distributed throughout the body, and the accuracy of each method is largely determined by the extent to which this assumption is true.

The use of urea is limited owing to the endogenous production of this substance, and thiourea has been shown to distribute unevenly in the body water (Winkler *et al.*, 1943). Sulfanilamide, at least in some mammals, may become conjugated in the liver although values derived from the use of this substance in the dog are in good agreement with desiccation values. Antipyrine is rapidly distributed throughout the body water, but it is also rapidly metabolized and excreted in the urine (Brodie, 1951). Although the metabolism and excretion of antipyrine appears to occur at a uniform rate (Soberman, 1950) and the appropriate corrections may be applied [Eq. 2], the use of the compound is probably more appropriate in acute experiments. However, the use of NAAP would seem to be more suitable under experimental and environmental circumstances where the rates of urine flow in teleosts are likely to vary (Holmes and McBean, 1963; Holmes and Stainer, 1966). This compound is only slowly metabolized by the tissues and is excreted extremely slowly. When deuterium oxide and tritium oxide are used to determine the total body water content of fish maintained in a closed environment, the possible reentry of the excreted compounds into the body must be examined. Studies on mammals also indicate that approximately 5% of the labile hydrogen atoms from both compounds will exchange with unlabeled hydrogen atoms of substances other than those in the body water (Elkinton and Danowski, 1955).

B. The Extracellular Volume

Since whole blood is itself a tissue consisting of the blood cells suspended in their extracellular environment the total *intravascular volume* may be determined by establishing (1) the total blood cell volume or (2) the plasma volume. In each case an estimate of the total intravascular volume may be obtained as follows if a simultaneous measurement of the hematocrit is made:

$$\text{Intravascular volume} = \text{plasma volume} \times \frac{100}{100 - \text{hematocrit}} \tag{3}$$

$$\text{Intravascular volume} = \times \frac{\text{cell volume}}{\text{hematocrit} \times 100} \tag{4}$$

Plasma volume may be determined according to the kinetic dilution method after the intravenous administration of known amounts of radioiodinated serum albumin (RISA), or T 1824 (Evans blue) which binds to the plasma proteins. Since serum albumin tends to leak out of the vascular space during passage through the capillaries, the estimates of plasma volume using RISA or T 1824 tend to be high. By using a larger protein molecule exclusively as the indicator, such as radioiodine-labeled fibrinogen, the transcapillary leakage is minimized and the plasma volume values obtained are some 2—12% lower than those obtained with albumin and other small protein molecules.

The alternative method of determining total intravascular volume depends upon the selective accumulation of certain compounds within the blood cells, principally the erythrocytes. When isolated erythrocytes are incubated in the presence of radioisotopes such as ^{52}Fe, ^{55}Fe, ^{32}P, ^{42}K and thorium B, the radioisotopes penetrate the cells and become bound to the hemoglobin or some other protein within the cell. The more firmly bound the isotope becomes, the lower the rate of loss of the radioactive indicator from the cells after it has been injected into the animal. In this regard ^{51}Cr, which becomes firmly bound to the globin portion of the hemoglobin molecule, shows the lowest rate of loss from the system. No appreciable loss occurs during the first 24 hr after injection; thus, only one blood sample is necessary to establish the dilution (Sterling and Gray, 1950; Gray and Sterling, 1950). On the other hand, ^{32}P is less firmly bound and up to 6% per hour may be lost from the circulating cells. Many blood samples must therefore be taken in order to establish the dilution curve for ^{32}P labeled cells (Gregersen and Rawson, 1959). A unique technique for the estimation of blood cell volume has recently been developed in the Pacific hagfish, *Eptatretus stoutii*, by McCarthy (1967). This method involves labeling the red blood cells of the hagfish with L-methioninemethyl-^{14}C. Blood from donor animals is incubated *in vitro* in the presence of the L-methionine-methyl - ^{14}C in a rotary incubator for 2 hr at 10°C. At the end of the incubation period the blood–isotope mixture is gently and rapidly centrifuged, the plasma supernatant is removed, and the cells are resuspended three times in cold physiological saline. The cells are finally suspended in a small amount of physiological saline and intravenously injected into the experimental animals. Approximately 3% per hour of the zero-hour blood concentration of isotope is lost from the hagfish circulation.

When plasma volumes or total intravascular volumes are calculated from dilution studies using labeled red blood cells, the values tend to be lower than those obtained by labeled plasma dilution techniques. This discrepancy arises from two sources. First, the red blood cells are not homogeneously distributed throughout the intravascular compartment, and therefore the hematocrit is not constant for all blood vessels. This is particularly true of the hagfish (Johansen *et al.*, 1962). Second, a certain amount of plasma is always trapped between the blood cells when they are separated from the plasma (Gregersen and Rawson, 1959). The most accurate estimation of intravascular volume therefore is obtained by simultaneous measurement of plasma and cell volumes (Armin *et al.*, 1952).

The distribution of several substances is purported to measure the combined intravascular, interstitial, and lymphatic spaces. These substances include inulin, raffinose, sucrose, mannitol, and thiosulfate, and the value which each substance gives increases in that order. The molecular weights of these substances, which are in a reverse order, largely determine the rates at which they diffuse throughout the various extracellular subcompartments, particularly the connective tissue and the transcellular spaces. For this reason the volume of distribution of the compound is often cited rather than the specific volume it is purported to measure. Inulin and sucrose appear to show two phases of distribution: A rapidly equilibrating phase is followed by a second phase where it is believed that these substances become selectively accumulated in the macrophages. These-substances are not, therefore, distributed homogeneously at this point and consequently erroneously high values may be obtained. Mannitol does not have this characteristic and is a more reliable index in chronic experiments.

An estimation of the volumes of distribution of body Na^+ and Cl^- may also be used to determine extracellular volume. Neither Na^+ nor Cl^- is homogeneously distributed throughout the extracellular compartment of any vertebrate. They are preferentially distributed between certain transcellular spaces, e.g., gastric secretion and renal tubular fluid; and some, albeit small, amounts of both ions occur within the intracellular compartment. Nevertheless, surprisingly accurate estimations of extracellular volume may be made if corrections are applied for the heterogeneous distribution of the ions. Indeed the methods are particularly useful when applied to tissue samples such as skeletal muscle. A refined method for calculating the Cl^- space of a tissue or organism was described by Manery (1954) as follows:

$$Cl^- \text{ space } (H_2O)_{E^{cl-}} = \frac{Cl_t - \times r_{cl} - \times H_2O_p}{Cl_p -} \text{ g water/kg wet weight} \quad (5)$$

where Cl_t- = tissue Cl^- concentration in millimoles per kilogram wet weight tissue,

H_2O_p = plasma water content in milliliter per kilogram wet weight plasma,

Cl_p- = plasma Cl^- concentration in millimoles/liter plasma, and

r_{Cl^-} = Gibbs-Donnan ratio for Cl^-.

The following analogous equation may also be derived for calculating the Na^+ space of an organism or tissue:

$$\text{Na}^+ \text{ space } (H_2O)_{E^{Na+}} = \frac{Na_t + \times H_2O_p}{Na_p + \times r_{Na} +} \text{ g water/kg tissue} \quad (6)$$

where $Na_t{}^+$ = tissue Na^+ concentration in millimoles per kilogram wet weight tissue,
H_2O_p = plasma water content in milliliter per kilogram wet weight plasma,
Na_{p^+} = plasma Na^+ concentration in millimoles per liter plasma, and
r_{Na^+} = Gibbs-Donnan ratio for Na^+.

The Gibbs-Donnan ratios for Na^+ and Cl^- used in Eqs. (5) and (6) have not been estimated for lower vertebrate tissues. Therefore, the values obtained from mammalian studies must be applied as approximations and these values are $r_{Na^+} = 0.942$ and $r_{cl^-} = 0.977$. For a critical evaluation of the methods used in the determination of extracellular volume from the distributions of Na^+ and Cl^-, the reader is referred to reviews by Manery (1954) and Cotlove and Hogben (1962).

C. The Intracellular Volume

The intracellular volume cannot be measured directly. As pointed out above, it is an extremely complex and certainly not a homogeneous compartment of the body. An estimate of intracellular volume may be estimated by subtracting simultaneous values for the total body water and the extracellular volume, but this estimate is clearly dependent upon the accuracy of the methods used for the determination of body water and extracellular space.

A summary of the classification of the body compartments is included in Fig. 2, and the range of distribution of some of the indicator substances is outlined.

IV. COMPARTMENTAL SPACES IN FISH

A. Class Agnatha

Order Myxiniformes and Order Petromyzontiformes

Until 1959 only one value for a compartmental volume in the Agnatha appeared in the literature. This value was the blood volume determined in a single specimen of the sea lamprey, *Petromyzon marinus*, by Welcker in 1858. The animal taken from the sea had a body weight of 1094 g, and the blood volume was reported to be 4.16% of the body weight. This value is extremely low when compared to the subsequent data obtained for the fresh-water form of this species by Thorson (1959). Although Thorson's data (Table II) are undoubtedly more reliable, the discrepancy is nevertheless large. According to Thorson (1959) the difference probably reflects an inverse

Table II Summary of the Available Data on Compartmental Volumes in Two Species of Cyclostomes[a]

Parameter	Pacific hagfish, *Eptatretus stoutii,* (McCarthy, 1967) (ml/100 g body wt)	Method of determination	Sea lamprey, *Petromyzon marinus,* (Thorson, 1959) (g/100 g body wt)	Method of determination
Total body water	74.6 ± 3.4 (5)	Desiccation	75.6 ± 0.15 (12)	Desiccation
Extracellular water	25.9 ± 5.1 (5)	Inulin-carboxyl-^{14}C	23.9 ± 0.23 (12)	Sucrose space
Interstitial water	10.3 ± 1.6 (5)	Inulin space minus T 1824 space	18.4 ± 0.19 (12)	Sucrose space minus T 1824 space
Intravascular volume	16.9 ± 2.1 (5)	T 1824 space plus red blood cell volume	8.5 + 0.12 (12)	T 1824 space and hematocrit
Plasma volume	13.8 ± 3.1 (10)	T 1824 space	5.5 ± 0.09	T 1824 space
Red blood cell volume	4.9 ± 2.0 (5)	L-Methionine-methyl-^{14}C space	3.0 (12)	Intravascular space minus plasma volume
Intracellular volume	48.7 ± 1.7 (5)	Intravascular volume minus inulin space	51.7 ± 0.26 (12)	Total body water minus sucrose space

[a]The values taken from Thorson (1959) were reported by the author in g/100 g body weight. These values may be converted to ml/100 g body weight by using the reported specific gravity values of 1.018 and 1.040 for plasma and blood, respectively. The numbers in parentheses indicate the number of individual fish used for the determination. All means ± S.E.

relationship between relative blood volume and body size similar to that which has been demonstrated in the elasmobranchs (Martin, 1950). The possibility also exists that there may be a difference in the distribution of body compartments between the freshwater and marine forms of this species. The blood volume of the Pacific hagfish, however, is even higher than that of the freshwater form of *Petromyzon marinus* (Table II).

The relative intravascular or blood volume of the Pacific hagfish, *Eptatretus stoutii*, is the highest reported for any vertebrate species. This high value is because of both a high plasma volume and a high red blood cell volume in this species when compared to *Petromyzon marinus* or other groups of fishes. Furthermore, the total extracellular volume of *Eptretatus stoutii* is similar to that of *Petromyzon marinus* (Table II), and the increase in intravascular volume appears to have occurred at the expense of the interstitial space (McCarthy and Conte, 1966; McCarthy, 1967).

B. Class Chondrichthyes

Subclass Elasmobranchii

An earlier set of data reporting the blood volumes of elasmobranchs (Table III; Martin, 1950) showed somewhat lower mean values than those which were later reported by Thorson (1958) (Table IV). The lower values found by Martin may, however, be attributable to the considerably larger fish used in his study. A single specimen of *Raja binoculata*, which has not been included in the synopsis of his data (Table III), had a body weight of 4750 g, which was within the body weight range of this species studied by Thorson. The blood volume for this individual was 7.3%; a value close to that reported by Thorson for this species (8.0%). Furthermore, Martin (1958) found that the best fit for his data from the heavier skate was obtained when the blood weights were plotted against the 0.75 exponent of the body weight. This correction would bring the mean blood weight values of the larger skate closer to those found in the smaller individuals by Thorson (cf. Tables III and IV).

A comparison of the blood volumes of several elasmobranch species with the blood volume of a single osteichthyan species, the lingcod (*Ophiodon elongatus*), also showed that the blood volumes of the elasmobranchs were significantly greater than those of the lingcod (Martin, 1960) (Table III). This trend has been further documented by Thorson (1958, 1961, 1962; Tables VIII and IX).

As in the case of many vertebrate species, the values obtained for the extracellular volume of the elasmobranchs depend upon the chemical indicator used for the determinations. The degree of penetration, and the resulting volume of distribution, is inversely related to the molecular weight of inulin, raffinose, and sucrose (Table V). On the other hand, sodium thiocyanate appears to be readily excreted via both the renal and extrarenal pathways

Table III Summary of the Data on Blood Volumes in Chondrichthian Fishes[a,b]

Species	Body wt (g)	Hematocrit (% cells)	Blood vol (% body wt)	*p* value
Chondrichthyes				
Ratfish *Chimaera colliei*	1225 ± 53	18.0 ± 0	2.6 ± .49	NS[c]
Dogfish *Squalus sucklii*	2168 ± 816	18.0 ± 2.5	9.0 ± 2.3	<0.02
Skate *Raja rhina*	3225 ± 751	18.5 ± 1.8	5.0 ± .66	<0.02
Skate *Raja binoculata*	♂ 1468 ± 129	18.3 ± 1.9	4.1 ± .35	<0.01
	♀ 3334 ± 337	17.1 ± 1.1	3.4 ± .23	NS[c]
			3.7 ± .21	<0.01
Osteichthyes				
Lingcod *Ophiodon elongatus*	4435 ± 896	35 ± 4.7	2.8 ± .22	—

[a]From Martin (1950).
[b]These values are compared to the blood volumes of an osteichthyan, *Ophiodon elongatus.* (All values are means ± S.E.; *p* values represent the significance with respect to the value for *Ophiodon elongatus.*)
[c]NS indicates not significant.

and consequently, when no correction is made, gives inordinately low values for the volume of distribution (Thorson, 1958).

Owing to the different degrees of penetration observed for inulin and sucrose, the calculated interstitial fluid spaces (inulin or sucrose space minus plasma space) of the elasmobranchs tend to be lower or higher depending upon whether the volume of distribution of inulin or sucrose is used in the calculation (Table IV). Conversely, the estimates of intracellular volume (total body water minus inulin or sucrose space) tend to be higher or lower for the same reason (Table IV).

The compartmental spaces of only one freshwater elasmobranch have been measured. This species was the Lake Nicaragua shark, *Carcharhinus nicaraguensis*, and none of the compartmental spaces, including total body water, differed significantly from the corresponding values in a variety of marine species (Table IV).

A limited amount of data is available on the transcellular spaces in elasmobranchs (Table VI). The relatively large volumes of peritoneal and cerebrospinal fluid in the long-nosed skate, *Raja rhina*, probably is reflected in the high values for total body water of this species (cf. Tables IV and VI). Since neither sucrose nor inulin was observed to penetrate the transcellular spaces measured, these high values for some of the transcellular spaces are therefore not reflected in the estimates of extracellular volumes (Thorson, 1958).

Table IV A Summary of the Body Compartmental Volumes in Chondrichthian Fishes[a-c]

	Marine species						Freshwater species
	Long-nosed skate, Raja rhina	*Big skate,* Raja binoculata	*Dogfish,* Squalus acanthias	*Lemon shark,* Negaprion brevirostris	*Nurse shark,* Ginglymostoma cirratum	*Ratfish,* Hydrolagus colliei	*Lake shark,* Carchashinus nicaraguensis
Body weight (g)	4933 (12) (1400–16550)	8184 (4) (2646–18100)	2631 (33) (1120–6350)	6400 (9) (3180–12270)	1603 (5) (1136–2270)	1067 (16) (520–1573)	48070 (10) (27700–57200)
Hematocrit (% cells)	16.8 (11) (12–21)	17.5 (4) (15–20)	18.2 (25) ± 0.55	21.5 (9) ± 0.37	17.4 (5) ± 1.3	20.2 (12) (15–25)	22.8 (10) ± 1.4
Pulse (beats/min)	11.1 (10) (6–18)	16.0 (4) (12–22)	31.0 (14) (18–40)	26 (9) (20–32)	22 (5) (16–24)	31.8 (11) (18–42)	12.2 (10) (8–18)
Plasma volume (T 1824 space)	5.9 (8) ± 0.53	6.5 (4) ± 0.60	5.5 (24) ± 0.29	5.4 (9) ± 0.11	5.7 (5) ± 0.27	4.2 (8) ± 0.32	5.1 (10) ± 0.22
Blood volume (T 1824 space and hematocrit)	7.2 (8) (4–9.5)	8.0 (4) (6.5–9.9)	6.8 (24) ± 0.37	7.0 (9) ± 0.14	6.8 (5) ± 0.33	5.2 (8) (4.1–7.4)	6.8 (10) ± 0.35
Extracellular fluid (inulin)	11.8 (8) ± 0.67	13.2 (2) ± 1.6	12.7 (13) ± 0.35	—	—	10.6 (8) ± 0.25	—
Extracellular fluid (sucrose)	—	—	21.2 (3) ± 1.4	21.2 (8) ± 0.47	21.9 (4) ± 1.2	—	19.7 (8) ± 0.54
Interstitial fluid (inulin or sucrose space minus T 1824 space)	5.5 (6) (4–8.6)	7.9 (2) (5.6–10.2)	15.7 (3)	15.8 (8)	16.2 (4)	6.7 (6) (5.9–7.7)	14.6 (8)

Continued

Table IV A Summary of the Body Compartmental Volumes in Chondrichthian Fishes[a-c]—Cont'd

	Marine species						Freshwater species
	Long-nosed skate, Raja rhina	*Big skate,* Raja binoculata	*Dogfish,* Squalus acanthias	*Lemon shark,* Negaprion brevirostris	*Nurse shark,* Ginglymostoma cirratum	*Ratfish,* Hydrolagus colliei	*Lake shark,* Carchashinus nicaraguensis
Intracellular fluid (total water minus extracellular space)	68.3	67.5	50.5	49.9	49.8	59.7	52.4
Total body water (desiccation)	82.0 (3) ± 0.41	82.6 (13) ± 0.40	71.7 (16) ± 0.48	71.1 (6) ± 0.14	71.7 (3) ± 0.22	71.4 (13) ± 0.47	72.1 (4) ± 0.37

[a]From Thorson (1958).
[b]All volumes are expressed as percentages of mean body weight.
[c]The range of measurements or the S.E. of the mean is given under each mean value. The number of individual fish used for each determination is given in parentheses after the mean value.

Table V Comparison of Inulin, Raffinose, and Sucrose Spaces of *Squalus acanthias*[a]

Indicator	No. of specimens	Av body weight (g)	Volume (% body weight)	Range
Inulin	13	2818	12.7	11.4–14.4
Raffinose	2	2605	15.2	15.1–15.4
Sucrose	3	2183	21.2	18.5–24.3

[a]Data from Thorson (1958).

Table VI The Fluid Volumes of Some Transcellular Spaces in Three Species of Chondrichthian Fishes[a,b]

Indicator	*Hydrolagus colliei*	*Raja rhina*	*Squalus acanthias*
Peritoneal fluid	Trace (7)	1.0 (6) (0.52 – 1.48)	0.33 (6) (0.11 – 0.51)
Cerebrospinal fluid	0.25 (7) (0.17–0.40)	0.77 (6) (0.58–1.0)	0.40 (5) (0.32–0.52)
Ocular fluid	0.85 (5) (0.78–0.99)	0.17 (3) (0.14–0.20)	0.38 (5) (0.30–0.57)

[a]From Thorson (1958).
[b]The values in parentheses below the mean values indicate the ranges, and the numerals in parentheses after the mean values indicate the number of fishes used for the determination.

C. Class Osteichthyes

1. Classes Sarcopterygii, Brachiopterygii, and Actinopterygii

The plasma and blood volumes of osteichthyan fishes have long been known to be low when compared to those of mammalian species. Until 1961, however, the available data were rather scant; indeed no data were recorded in the literature between 1858 and 1934. Most of the values reported by various investigators up to 1964 are recorded in Table VII. A more detailed taxonomic analysis of the blood volumes in osteichthyan fishes, however, was published by Thorson in 1961, and these data are summarized in Table VIII. The more primitive freshwater chondrostean and holostean fishes tend to have somewhat higher plasma and blood volumes than the more advanced teleostean fishes from freshwater. Comparison of the mean values

Table VII A Summary of the Data on Blood Volumes Presented by Authors Other than Thorson (1961)[a]

Species[b]	No. of fish	Body weight (g)	Blood vol (ml)	Blood wt (g)	Blood wt (% body wt)	Reference and method
Cyprinus tinea	1	269.5	4.81	5.04	1.87	Welcker (1858) (bleeding)
Perch *Perca fluviatilis*	1	122.7	1.26	1.32	1.07	Welcker (1858) (bleeding)
Perch *Perca fluviatilis*	1	98.2	1.26	1.32	1.34	Welcker (1858) (bleeding)
Tautog *Tautoga onitis*	3	—	—	—	1.5	Derrickson and Amberson (1934) (bleeding)
Bullhead *Ameiurus natalis*	6	171.4 ± 34.7	2.16 ± 0.33	—	1.76 ± 0.26	Prosser and Weinstein (1950) (T 1824 and hematocrit)
Lingcod *Ophiodon elongatus*	8	4435 ± 896	202 ± 50	211 ± 52	2.8 ± 0.22	Martin (1950) (T 1824 and hematocrit)
Rock fish *Sebastodes* sp.	1	2150	57	60	2.8	Martin (1950) (Vital red and hematocrit)
Sculpin *Cottidae* sp.	3	4020 ± 189	89.3 ± 10.2	93.7 ± 10.2	2.3 ± 0.19	Martin (1950) (Vital red and hematocrit)
Goldfish *Carassius carassius*	—	—	—	—	2.5–3.0	Korzhuyev and Nikolskaya (1951) (Hemoglobin washout)
Common sole *Solea vulgaris*	—	—	—	—	4.0–5.9	(Hemoglobin washout)
Common sucker *Catostomus commersoni*	23	200–999	—	—	1.5	Lennon (1954) (bleeding)
Rainbow trout—FW *Salmo gairdneri*	10	8–29	—	—	2.25	Schiffman and Fromm (1959) (T 1824 and hematocrit)

Rainbow trout—FW	10	—	—	—	3.5 ± 0.9	Conte *et al.* (1963) (T 1824 and hematocrit)
Rainbow trout—FW	5	—	—	—	3.3 ± 0.9	Albumin-^{131}I and hematocrit
Rainbow trout—FW	9	—	—.	—	2.9 ± 0.8	Simultaneous T 1824 and albumin-^{131}I and hematocrit
Rainbow trout—FW	6	—	—	—	2.8 + 1.0	T 1824 or albumin-^{131}I and ^{51}Cr-labeled red blood cell
Steelhead trout—SW *Salmo gairdneri*	13	553 ± 127	—	—	6.9 ± 1.8	Smith (1966) (T 1824 and hematocrit)
Rainbow trout—FW *Salmo gairdneri*	4	548 ± 49	—	—	2.4 ± 0.4	Smith (1966) (bleeding)
Sockeye salmon—FW *Oncorhynchus nerka*	8	1814	—	—	4.0 ± 0.6	Smith (1966) (T 1824 and hematocrit)
Coho salmon—SW *Oncorhynchus kisutch*	8	930 ± 104	—	—	4.5 ± 1.5	Smith (1966) (T 1824 and hematocrit)
	6	1012 ± 120	—	—	2.3 ± 0.4	Bleeding
Pink salmon—SW *Oncorhynchus gorbuscha*	—	—	—	—	7.8	Smith and Bell (1964) (T 1824 and hematocrit)
Atlantic cod *Gadus morhua*	—	—	—	—	2.4	Ronald *et al.* (1964) (Fluorescein and hematocrit)

[a]All mean values are recorded ± S.E.
[b]Here FW indicates freshwater and SW seawater.

Table VIII A Summary of the Data on Compartmental Volumes in Several Species of Osteichthyan Fishes[a,b,c]

	Freshwater Chondrostei (order Acipenseriformes)		Freshwater Holostei (order Lepisosteiformes)		Freshwater Teleostei		
Parameter	*Lake sturgeon, Acipenser fulvescens*	*Paddlefish, Polyodon spatula*	*Bowfin, Amia calva*	*Short-nosed gar, Lepisosteus platostomum*	*Common sucker, Catostomus commersoni*	*Carp, Cyprinus carpio*	*Bigmouth buffalo fish, Ictiobus cyprinellus*
Weight (g)	3058 (8) (2275–4530)	4679 (5) (3740–5910)	1963 (6) (1020–3265)	1185 (7) (855–1730)	617 (2) (580–655)	2412 (7) (1585–3190)	3395 (8) (1980–5440)
Pulse (beats/min)	49 (8) (44–52)	22 (5) (16–28)	20(6) (14–28)	19 (6) (14–24)	55 (2) (47–64)	28 (6) (18–39)	38 (8) (19–70)
Respiration (per min)	53 (8) (40–72)	17(5) (10–26)	14 (6) (9–20)	28 (5) (24–36)	47 (2) (45–50)	27 (3) (19–38)	22(4) (20–24)
Hematocrit (% cells)	22 (8) (19–29)	30 (5) (24–37)	32 (6) (32–34)	42(7) (33–50)	39 (2) (39–40)	33 (7) (23–24)	29 (8) (18–40)
Specific gravity, plasma	1.016 (3) (all 1.016)	1.017 (3) (1.016–1.018)	1.018 (3) (1.0175–1.0185)	1.016 (3) (1.015–1.017)	1.016 (3) (1.015–1.017)	1.019 (3) (1.018–1.0495)	1.016 (3) (all 1.016)
Specific gravity, blood	1.036 (3) (1.033–1.040)	1.040 (3) (1.039–1.041)	1.045 (3) (1.044–1.047)	1.051 (3) (1.050–1.052)	1.041 (3) (1.040–1.042)	1.040 (3) (1.039–1.0495)	1.042 (3) (1.041–1.043)
Plasma volume (T 1824 space)	2.8 ± 0.18 (8)	2.2 ± 0.09 (5)	2.2 ± 0.19 (6)	2.1 ± 0.14 (7)	1.2 ± 0.14 (2)	1.8 ± 0.10 (7)	1.9(8) ± 0.22
Blood volume	3.7 (8) (2.8–4.9)	3.0 (5) (2.4–3.6)	3.4 (6) (2.9–5.0)	3.8 (7) (3.0–5.2)	2.2 (2) (1.8–2.7)	3.0 (7) (2.4–3.5)	2.8 (8) (1.8–4.1)
Extracellular fluid (sucrose space)	20.1 ± 1.4 (8)	15.6 ± 0.42 (5)	18.9 ± 1.4 (6)	13.6 ± 0.33 (7)	12.2 ± 0.32 (2)	15.5 ± 1.3 (7)	13.2 (8) ± 0.45
Interstitial fluid (sucrose space minus plasma)	17.3	13.4	16.7	11.5	11	13.7	11.3
Total body water	72.7 ± 0.25 (8)	74.0 ± 0.48 (6)	74.5 ± 0.48 (6)	66.7 ± 0.70 (7)	74.4 ± 0.45 (2)	71.4 ± 0.45 (7)	70.6 (8) ± 1.2
Intracellular fluid (total water minus sucrose space)	52.6	58.4	55.6	53.1	62.2	55.9	57.4

[a]From Thorson (1961).
[b]Values are indicated as means ± S.E. or with the ranges reported in parentheses below.
[c]All volumes expressed as percentage of body weight.

Marine Teleostei						
Tiger rockfish, Mycteroperca tigris	*Nassau grouper, Epinephelus striatus*	*Red snapper, Lutianus campechanus*	*Gray snapper, Lutianus griseus*	*Green moray, Gymnothorax funebris*	*Great barracuda, Sphyraena barracuda*	*Rainbow parrot fish, Pseudoscarus guacamaia*
5885 (1)	1270 (2) (930–1610)	3765 (2) (3130–4400)	3711 (1900–4680)	4062 (6) (3050–4815)	2204 (11) (1432–4575)	4607 (19) (1650–6830)
51 (1)	52(1)	58 (2) (48–68)	54(3) (44–66)	70 (5) (60–88)	68 (11) (32–98)	48 (16) (30–98)
—	—	—	41 (4) (30–48)	22 (3) (18–28)	41 (7) (30–54)	40 (14) (22–64)
28 (1)	28 (2) (28–29)	36 (2)	35 (6) (28–40)	26 (6) (24–28)	31 (11) (25–36)	30 (27) (20–40)
Used av (1.017)	Used av (1.017)	Used av (1.017)	Used av (1.017)	Used av (1.017)	Used av (1.017)	Used av (1.017)
Used av (1.042)	Used av (1.042)	Used av (1.042)	Used av (1.042)	Used av (1.042)	Used av (1.042)	Used av (1.042)
2.3 (1)	1.8 ± 0.18 (2)	1.3 ± 0.04 (2)	1.3 ± 0.04 (2)	1.6 ± 0.26 (6)	1.9 ± 0.09 (10)	2.4 ± 0.11 (16)
3.3(1)	2.6 (2)	2.2 (2)	2.0 (6)	2.2 (6)	2.8 (10)	3.6 (16)
12.5 (1)	14.5 ± 1.0 (2)	14.0(2)	14.0 ± 0.40 (6)	15.8 ± 1.1 (6)	15.9 ± 0.73 (8)	16.6 ± 0.57 (8)
10.2	12.7	12.7	12,7	14.2	14.0	14.2
71.1 (1)	71.7 ± 0.95 (2)	71.3 ± 0.11 (2)	72.3 ± 0.42 (6)	63.7 ± 2.4 (6)	70.6 ± 0.65 (9)	73.1 ± 0.32 (14)
58.6	57.2	57.3	58.3	47.9	54.7	56.5

from most of the chondrostean and holostean species examined are in fact statistically higher than the corresponding values from the individual freshwater teleostean species. Furthermore, when the values from the plasma volumes from all the species of chondrostean fishes are pooled, the mean value is significantly greater than the pooled value for the freshwater teleost species (Table IX, $P<0.001$). This relationship does not hold, however, when the pooled mean value for the holostean species is compared to the mean pooled value for the freshwater teleost species. Also, a comparison of the plasma volumes of the freshwater and marine teleost species does not reveal any significant difference (Table IX).

Values obtained for the measurement of extracellular fluid volumes in the Osteichthyes vary according to the degree of penetration of the indicator substance, the usual inverse relationship being found between the molecular weight of the indicator and the recorded volume of distribution (Table X). Again the extracellular compartment, as indicated by the sucrose space, tends to be smaller in the more advanced teleosts.

The mean pooled extracellular volumes of both the chondrostean and the holostean species are significantly greater than that of the freshwater teleost species (Table IX, $p< 0.001$ and < 0.05, respectively). Also, the marine teleosts tend to have slightly greater extracellular volumes than do the freshwater teleosts (Table IX, $p< 0.05$).

The total body water composition of the chondrostean fishes is significantly higher than that of the freshwater teleost species (Table IX, $p < 0.02$), but there appears to be no significant difference between the body water content of the holosteans, freshwater teleosts, or the marine teleosts (Table IX). As a consequence of the extracellular space and the total water composition, the intracellular volumes of the more primitive members of the class Osteichthyes tend to be lower than intracellular volumes estimated for the teleost species (Table IX).

2. Blood Volume Changes Associated with the Evolution of the Fishes

Among the classes Agnatha, Chondrichthyes, and Osteichthyes there appears to be a correlation between the blood volume and the degree of primitiveness of the individual fish (Thorson, 1961), the trend being toward lower red blood cell (RBC) and plasma volumes in the Chondrichthyes and Osteichthyes. Although large RBC and plasma volumes may be, in general, considered to be primitive characteristics among the aquatic vertebrates, the reasons why this should be so are not immediately apparent. One is tempted to suggest that the circulatory system in the primitive fish may be less efficient. Unfortunately, there are no comparative data on the cardiac outputs of fish to substantiate this hypothesis. Indeed the available data on pulse and respiration rates of the agnathans, chondrichthyans, and osteichthyans do not seem to show any phlogenetic trend (Thorson, 1961; Table VIII). An examination of the cyclostome hemoglobins, however, suggests that they may represent the early

Table IX A Summary of the Mean Compartmental Spaces in the Various Taxonomic Groups of the Osteichthyes[a,b]

Parameter	Osteichthyes			
	Freshwater Chondrostei	*Freshwater Holostei*	*Freshwater Teleostei*	*Marine Teleostei*
Weight (g)	3681 (13)	1544 (13)	2664 (17)	3710 (47)
Pulse (beats/min)	39 (13)	20 (12)	37 (16)	57 (39)
Respiration (per min)	47 (13)	20 (11)	29 (9)	38 (28)
Hematocrit (% cells)	25 (13)	38 (13)	32 (17)	30 (55)
Specific gravity, plasma	1.0185 (6)	1.0185 (6)	1.017 (9)	—
Specific gravity, blood	1.048 (6)	1.048 (6)	1.041 (9)	—
Plasma volume (T 1824 space)	2.5 ± 0.11^{c} (13)	2.1 ± 0.12^{d} (13)	1.8 ± 0.11 (17)	$1.9 + 0.06^{d}$ (43)
Blood volume	3.5 (13)	3.6 (13)	2.8 (17)	2.9 (43)
Extracellular fluid (sucrose space)	18.4 ± 0.87^{c} (13)	16.0 ± 0.65^{e} (13)	$14.0 + 0.56$ (17)	15.4 ± 0.31^{e} (33)
Interstitial fluid (sucrose space minus plasma)	15.9	13.9	12.2	13.5
Total body water	73.2 ± 0.39^{f} (13)	70.3 ± 0.43^{d} (13)	71.4 ± 0.60 (17)	70.8 ± 0.41^{d} (40)
Intracellular fluid (total water minus sucrose space)	54.8	54.3	57.4	55.4

[a]From Thorson (1961). All volumes are expressed as percentage of body weight.
[b]Values are reported as means ± S.E., and numerals in parentheses indicate the number of individual determinations.
[c]$p < 0.001$.
[d]Not significant with respect to the corresponding value for the group of freshwater teleost species.
[e]$p < 0.05$.
[f]$p < 0.02$.

Table X Comparison of Inulin, Raffinose, and Sucrose Spaces of *Pseudoscarus guacamaia*[a]

Indicator	No.of specimens	Av body weight (g)	Volume (% body weight)	Range
Inulin	8	5451	11.4	9.2–14.5
Raffinose	4	4096	14.4	12.7–16.4
Sucrose	8	4696	16.6	14.3–18.9

[a]Data from Thorson (1961).

stages in the phylogeny of oxygen transport. In *Lampetra fuviatilis* the hemoglobin molecule contains a single peptide chain having an amino acid sequence more closely related to that of mammalian myoglobin than the α or β chains of mammalian hemoglobin (Braunitzer *et al.*, 1964). The molecular weight of this hemoglobin is 17,000 ; it is monomeric and it does not show any of the heme–heme interaction properties of the dimeric and tetrameric forms of hemoglobin. Furthermore, the molecule probably has an oxygen dissociation curve which is hyperbolic or sigmoidal with a flattened zone over the range of tissue and environmental oxygen partial pressures in the lamprey. Hemoglobin of this type would probably release relatively small amounts of oxygen over the range of oxygen partial pressures to which it was exposed. In contrast the heme–heme interaction properties of the tetrameric, and possibly the dimeric, forms of hemoglobin would result in sigmoidal oxygen dissociation curves with progressively smaller slopes over their intermediate ranges. These molecules would unload more oxygen per unit change in oxygen partial pressure. Therefore, compared to the number of monomeric hemoglobin molecules per unit volume of blood in the cyclostome, it would seem that fewer monomeric units of hemoglobin would be necessary to release the same volume of oxygen to the tissues if these units were arranged in dimers having heme-heme interaction properties. In the tetrameric form still fewer hemoglobin units would be required to release this volume of oxygen.

We feel therefore that a broader knowledge of the molecular forms of hemoglobin occurring in the fishes is necessary to fully explain the decreasing blood volumes in the more advanced forms. These studies should also include the characterization of the oxygen carrying and releasing properties of the hemoglobins at the tissue and environmental temperatures and at the oxygen tensions present in the tissues and environment of the organism.

3. Changes in the Extracellular Compartments of Euryhaline Species

Hatchery reared steelhead trout, *Salmo gairdneri*, maintained at seasonal temperatures and photoperiod, show significant declines in their plasma Ca^{2+},

Cl^-, and water concentrations and the concentrations of Ca^{2+} and Cl^- in muscle during the period of growth from 25 to 110 g body weight. This trend also occurred, but at a much slower rate, during the period of growth from 110 to 250 g body weight (Houston, 1959; Table XI). Estimation of the Cl^- space [Eq. (5)] in the muscle of these fish indicated that the extracellular volume per unit wet weight of muscle also declined quite markedly during the growth of the smaller trout and declined much less rapidly during the growth period of the larger individual fishes (Table XI). At the same time the relative intracellular volumes (total tissue water minus extracellular volume) of muscle samples from the smaller weight range of trout increased with increases in body weight, while in the larger fish the relative intracellular volume remained unchanged (Table XI).

The period of growth and development represented in the steelhead trout studied by Houston (1959) included the period of parr–smolt transformation in this species. This stage of development in all salmonid fishes is a period of profound physiological change and is associated with the preadaptation of the individual fishes to the marine environment (Huntsman and Hoar, 1939; Parry, 1958, 1961). Later studies by Houston (1960) and Houston and Threadgold (1963) were directed toward an elucidation of the changes occurring in the composition and distribution of body fluids during the parr–smolt transformation of the Atlantic salmon, *Salmo salar*. They found that the plasma Cl^- concentration of this species declined sharply with the onset of smoltification, but later recovered to a level somewhat higher than that observed in the nonsmolting parr. Some decrease was observed in the muscle Cl^- concentration at the onset of smoltification, but no secondary increase was observed as the meta-morphosis progressed. Changes in the Cl^- space of the Atlantic salmon suggest that the parr–smolt transformation process is characterized by a shift in the distribution of the body fluids in this species also. A sharp decline occurred in the extracellular volume (Cl^- space) of muscle during the silvery parr stage of development. This corresponded to the growth period from 20 to 35 g body weight. With the onset of the full smolting condition the extracellular volume of muscle stabilized at a value which was approximately 20% less than that of the presmolting fish. Concomitant with these changes in the muscle extracellular volume, increases in the intracellular volume (total muscle water minus Cl^- space) were observed.

Houston and Threadgold (1963) suggested that the changes in the compartmental distribution of electrolytes and water during the smoltification of the Atlantic salmon were consistent with possible changes in the pattern of renal excretion. We now know, at least in the trout, *Salmo gairdneri*, that this is indeed true. During the period of onset of smolting in this species the rates of urine flow and electrolyte excretion declined to approximately one-half of the presmolt values (Holmes and Stainer, 1966). This reduction was entirely attributable to a reduction in the glomerular filtration rate. If the trout were retained in freshwater until they eventually lost their overt characteristics of

Table XI Variation of Plasma and Tissue Chloride and Calcium, Plasma Water, Chloride Space, and Cellular Space (Tissue Water-Chloride Space) with Weight in Freshwater Steelhead Trout[a]

	June–July, 1957			February–March, 1958		
Parameter	*Sample size*	*Weight range (g)*	*Regression*[b] $(Y = a + bX)$	*Sample size*	*Weight range (g)*	*Regression*[b] $(Y = a + bX)$
Plasma chloride	39	25–70	$Y = 152.8 - 0.30\ X$	23	76–250	$Y = 141.0 - 0.04\ X$
Plasma water	39	25–70	$Y = 958.4 - 0.22.X$	19	76–250	$Y = 934.3 - 0.016\ X$
Plasma calcium	24	47–118	$Y = 2.71 - 0.004\ X$			
Tissue chloride	35	30–70	$Y = 31.9 - 0.18X$	23	80–250	$Y = 14.9 - 0.02\ X$
Tissue calcium	23	47–118	$Y = 3.24 - 0.015\ X$			
Chloride space	33	32–70	$Y = 217.1 - 1.28\ X$	19	80–250	$Y = 103.4 - 0.156\ X$
Cellular space	33	32–70	$Y = 628.3 + 0.43\ X$	19	80–250	$Y = 619.2 - 0.023\ X$

[a]Data from Houston (1959).
[b]$Y = a + bX$, where Y = plasma or tissue electrolyte and water concentrations (mEq/liter, mEq/kg wet weight, g H_2O/kg wet weight), a = the ordinate intercept, b = the slope, and X = body weight (g).

smolting, the glomerular filtration rate and the renal excretory pattern returned to those found in the presmolting fish (Holmes and Stainer, 1966). Therefore, since the smolting salmonid shows (1) a reduction in the extracellular volume and an increase in the intracellular volume of the muscle, (2) a reduction in the muscle water content, and (3) a reduction in the rate of water and electrolyte excretion via the kidneys, the new steady state must be presumed to be accompanied by a concomitant decrease in the extrarenal influx of water and electrolytes. Furthermore, it is possible that an increase in the extrarenal efflux of electrolytes also occurs at this time.

These data do not establish whether the changes in compartmental volumes occur as a direct result of the process of smoltification per se or whether they are merely manifestations of changes occurring due to the growth of the organism. Decreases in the extracellular compartment are characteristic of periods of rapid growth in several vertebrate species. Fellers *et al.* (1949) have demonstrated decreases of 79 and 57% in the thiocyanate and Na^+ spaces, respectively, in humans between infancy and maturity. Similar changes have also been reported for the rat and the chicken (Barlow and Manery, 1954; Medway and Kare, 1959).

Nevertheless, the physiological changes which occur in the smolting salmonid, whether they result from the smolting process per se or coincidental changes in growth rate, certainly predispose the individual fish to life in the marine environment (Gordon, 1959a,b; Houston, 1959, 1960, 1963; Parry, 1961). The reduced glomerular filtration and urine flow rates of the trout, *Salmo gairdneri* (Holmes and Stainer, 1966), may be interpreted as part of the preadaptation to a marine environment. Upon adaptation to seawater, this species shows even larger decreases in the rates of urine flow (R. M. Holmes, 1961) and glomerular filtration (Holmes and McBean, 1963). Furthermore, Houston (1960) was able to demonstrate that Atlantic salmon in the full smolting condition were able to adapt to seawater much more readily than fish in the early stages of the parr–smolt transformation. Nonsmolting parr, on the other hand, were invariably unable to withstand an abrupt transfer from freshwater to seawater.

Following the abrupt transfer of salmonid fishes from freshwater to seawater, a series of physiological changes take place. These responses may be divided into two phases. An acute adaptive phase occurs immediately following transfer of the fish to seawater, and this is followed by a chronic regulative phase which ultimately results in the establishment of a new steady state with respect to the tissue water and electrolyte composition of the fish. Several recent studies involving a variety of salmonid species have been devoted to an examination of the changes which occur in the inorganic ion and water composition of the fish during adaptation to seawater (e.g., Gordon, 1959a,b; Houston, 1959; Parry, 1961). Typical of these studies are the data derived from the trout, *Salmo gairdneri*, and included in Fig. 3 (W. N. Holmes, unpublished data). Of particular interest are the oscillatory changes which occur in some parameters between 24 and 72 hr and between

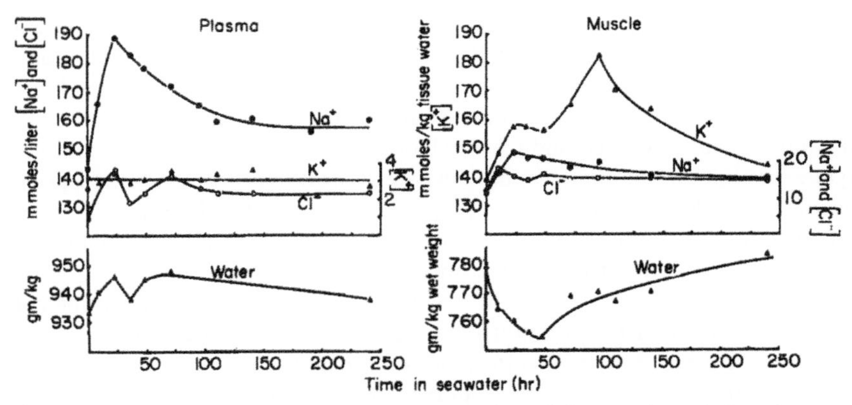

Fig. 3. Changes in the electrolyte and water concentrations of plasma and muscle from the trout, *Salmo gairdneri*, following abrupt transfer to 60% standard seawater (284 mM Na^+, 6 mM K^+) at approximately 5°C. The response may be divided into two phases: an acute adaptive phase immediately following transfer, followed by a chronic regulative phase which ultimately establishes a new steady state of electrolyte distribution (W. N. Holmes, unpublished data).

72 and 140 hr in the regulatory phase (Fig. 3). Similar oscillations were reported in this species by Houston (1959). Estimations of the Na^+ and Cl^- spaces [Eqs. (5) and (6), respectively] in the skeletal muscle of the freshwater trout before transfer to seawater indicated essentially similar values for the extracellular fluid volume (W. N. Holmes, unpublished data; Fig. 4) These values were 53.0 ± 1.8 and 50.8 ± 0.5 g/kg wet weight of muscle for the Na^+ and the Cl^- spaces, respectively. In the same species, Houston (1959) reported somewhat higher values for the muscle Cl^- space (63–72 g/kg wet weight) in fish of the same size range (200–250 g body weight). During the first 10 hr after transfer to seawater both the Cl^- and Na^+ spaces in muscle showed rapid increases to approximately 75 g/kg wet weight muscle (Fig. 3). After 10 hr, however, the Cl^- space commenced to decline but the Na^+ space continued to rise until at 24 hr after transfer it was almost twice the freshwater value (Fig. 2). Thereafter, the Na^+ space also declined and between 140 and 240 hr after transfer to seawater the Na^+ and Cl^- spaces were essentially constant at a level which was approximately 45% higher than the freshwater value. Between 72 and 140 hr after transfer, however, a secondary rise occurred in the Cl^- space of the muscle; a similar phenomenon was also reported by Houston (1963) for this species, but the reason for its occurrence remains obscure.

Clearly, the estimated volume of distribution in muscle of either Na^+ or Cl^- is not a reliable index of extracellular fluid volume during the early regulative phase of adaptation to seawater. Only the simultaneous estimations of inulin and/or sucrose space, together with measurements of the tissue water and electrolyte compositions, will elucidate this problem.

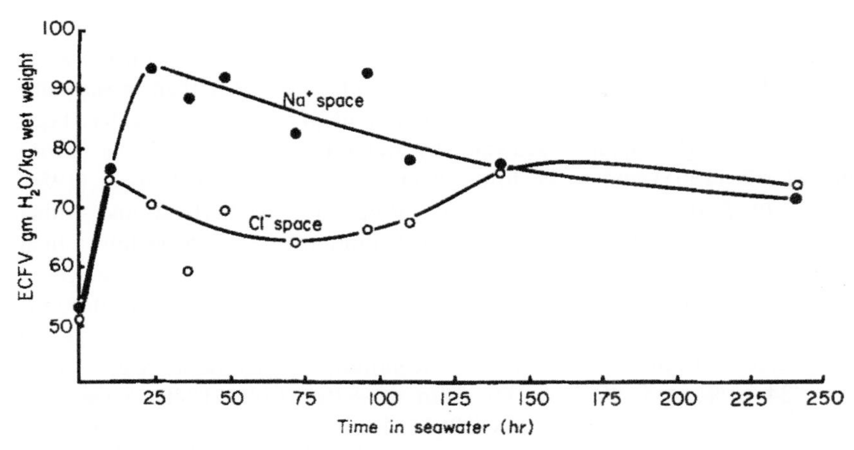

Fig. 4. Changes in the Na^+ and Cl^- spaces of skeletal muscle from trout (*Salmo gairdneri*) following abrupt transfer to 60% seawater (284 m*M* Na^+, 6 m*M* K^+) at approximately 5°C. (ECFV = extracellular fluid volume.) Cl^- space and Na^+ space as given in Eqs. (5) and (6). (From W. N. Holmes, unpublished data.)

V. ELECTROLYTE COMPOSITION

A. Class Agnatha

1. ORDER MYXINIFORMES

Early determinations of the blood freezing point of the myxinoids indicated that the blood was slightly hypertonic to seawater (Dekhuyzen, 1904; Greene, 1904). Borei (1935) found the plasma of *Myxine glutinosa* to be considerably hypotonic to seawater, but this finding has never been substantiated. In more recent studies the serum of *Myxine glutinosa* was found to be virtually isosmotic to seawater by Robertson (1954). The earlier findings of hypertonic blood were explained by McFarland and Munz (1958) when they discovered a relationship between serum osmotic pressure and the degree of handling which the fish had received. Slime production resulted in a 1–3% hypertonicity which lasted an hour or more. It is now generally agreed that the Myxiniformes are isosmotic with seawater (Morris, 1960; Chester-Jones *et al.*, 1962; Robertson, 1963; McFarland and Munz, 1965). Morris (1965), however, still maintains that the blood is slightly hypertonic with respect to the environmental seawater and that for this reason the animal produces a small amount of hypotonic urine.

Exposure of *Polistotrema stoutii* to 80% seawater resulted in rapid weight gain with a return to normal after 7 days; exposure to 122% seawater resulted in weight loss with no return to initial weight. These findings are explained by a high permeability to water and a lack of regulatory mechanisms for sodium chloride (McFarland and Munz, 1965).

Although the osmotic pressure of myxinoid plasma is very similar to that of the environment, its ionic composition is dissimilar. The Na^+ concentration in both *Myxine glutinosa* and *Polistotrema stoutii* is higher than in seawater (Table XII). The plasma Cl^- concentration in *Myxine glutinosa* has been found to be higher than in seawater by Bellamy and Chester-Jones (1961) and Morris (1965) and to be lower than in seawater by Robertson (1954, 1966). In *Polistotrema stoutii*, Urist (1963) and McFarland and Munz (1958, 1965) have found the plasma concentration of Cl^- to be lower than in seawater. The plasma concentration of K^+ is similar to that in seawater. The divalent ions Ca^{2+}, Mg^{2+}, and SO_4^{2-} are all found at lower concentrations in the plasma than in seawater. In *Myxine glutinosa* plasma Ca^{2+} ranges in concentration from 53–67% of its concentration in seawater, Mg^{2+} varies from 24–61%, and SO_4^{2-} varies from 12–87% (Cole, 1940; Robertson, 1954; Bellamy and Chester-Jones, 1961; Morris, 1965; Robertson, 1966). In contrast to a urea concentration of 60mmoles/ liter determined by Borei (1935) the serum urea content of *Myxine glutinosa* was found to be 4 mmoles/liter by Cole (1940) and 2–4 mmoles/kg of water by Robertson (1954, 1966); thus, urea plays no significant osmoregulatory role in the hagfish.

McFarland and Munz (1965) have shown that although there is probably no active transport of Na^+ across the gut, gills, or skin, the low Na^+ content of the slime secretion probably serves to maintain the high plasma Na^+ concentrations. In contrast to a low Na^+ content, the slime has a high content of Ca^{2+}, Mg^{2+}, and K^+. In addition, Mg^{2+}, K^+, SO_4^{2-}, and phosphate are secreted into the glomerular filtrate by the mesonephric duct cells and appear in the urine at higher concentrations than in the plasma (Munz and McFarland, 1964; McFarland and Munz, 1965). The lack of a renal mechanism for the reabsorption of sodium chloride or water supports the theory of the marine origin of the Myxiniformes.

2. *Order Petromyzontiformes*

Unlike the hagfish which have an exclusively marine habitat the lampreys invariably breed in freshwater. Lampreys spend the first part of the life cycle in freshwater as amoecoete larvae and after metamorphosis they either remain in freshwater (*Lampetra planerii*), migrate into estuarine or coastal waters (*Lampetra fluviatilis*), or migrate into the open sea (*Petromyzon marinus* and *Lampetra tridentata tridentata*). In the Great Lakes there is a potodramous population of *Petromyzon marinus* which gained access to their present habitat when the Welland Canal, which bypasses Niagara Falls, was opened in 1827 (Urist, 1963). This population spawns in the freshwater streams which drain into the Great Lakes.

Although the anadromous lampreys spend much of their life cycle in salt water the only analysis so far available from this habitat is that of Burian (1910) who showed the osmotic pressure of the blood of a single specimen of

Table XII Cyclostomata, Myxiniformes—Blood Chemistry

Fish or medium	Na	K	Mg	Ca	NH_4	Cl	HCO_3	PO_4	SO_4	Urea	Protein (g %)	Total ions	Units	Osmotic pressure (mOsm/liter)	Comments	Author	Date
Myxine glutinosa	—	—	—	—	—	—	—	—	—	—	—	—	—	962[a]	—	Dekhuysen	1904
Seawater	—	—	—	—	—	—	—	—	—	—	—	—	—	930[a]	—	—	—
Polistotrema stouti	—	—	—	—	—	—	—	—	—	—	—	—	—	1059[a]	—	Greene	1904
Seawater	—	—	—	—	—	—	—	—	—	—	—	—	—	1032[a]	—	—	—
Polistotrema stoutii	—	—	—	—	—	344	—	—	—	0	—	—	mmoles/liter	—	Blood chloride varies linearly with seawater chloride	Bond *et al.*	1932
Seawater diluted to 2.5% salinity	—	—	—	—	—	384	—	—	—	—	—	—	mmoles/liter	—	—	—	—
Polistotrema stoutii	—	—	—	—	—	414	—	—	—	0	—	—	mmoles/liter	—	—	—	—
Seawater diluted to 3.0% salinity	—	—	—	—	—	467	—	—	—	—	—	—	mmoles/liter	—	—	—	—
Polistotrema stoutii	—	—	—	—	—	471	—	—	—	0	—	—	mmoles/liter	—	—	—	—

Continued

Table XII Cyclostomata, Myxiniformes—Blood Chemistry—Cont'd

Fish or medium	Na	K	Mg	Ca	NH_4	Cl	HCO_3	PO_4	SO_4	Urea	Protein (g %)	Total ions	Units	Osmotic pressure (mOsm/liter)	Comments	Author	Date
Seawater	—	—	—	—	—	530	—	—	—	—	—	—	mmoles/liter	—	—	—	—
Polistotrema stoutii	—	—	—	—	—	570	—	—	—	0	—	—	mmoles/liter	—	—	—	—
Seawater concentrated to 4.0%	—	—	—	—	—	626	—	—	—		—	—	mmoles/liter	—	—	—	—
Myxine glutinosa	—	—	—	—	—	325	—	—	—	60	—	—	mmoles/liter	806[a]	Urea value has proved to be erroneous	Borei	1935
Seawater	—	—	—	—	—	520	—	—	—	—	—	—	—	1021[a]	—	—	—
Myxine glutinosa	402	9.1	22.5	5.3	—	448	3.7	—	6.0	4	—	901	mmoles/liter	—	—	Cole	1940
Seawater	416	9.1	50.2	9.4	—	483	2.2	—	30.4	—	—	1000	—	—	—	—	—
Myxine glutinosa	558	9.6	19.4	6.3	—	576	—	12.5[b]	6.7	3	—	1183	mmoles/kg	—	Serum is isotonic within 1%	Robertson	1954
Seawater	506	10.7	58	11.1	—	592	—	—	30.6	—	—	1209	—	—	—	—	—
Polistotrema stouti	370	—	—	—	—	408	—	—	—	—	—	—	mmoles/liter	859[a]	30 hr immersion	McFarland and Muns	1958
85% Seawater	338	—	—	—	—	444	—	—	—	—	—	—	—	852[a]	—	—	—

Polistotrema stouti	428	—	—	—	—	483	—	—	—	—	—	—	—	1022[a]	30 hr immersion	—	—
Seawater	450	—	—	—	—	498	—	—	—	—	—	—	—	1003[a]	—	—	—
Polistotrema stouti	505	—	—	—	—	522	—	—	—	—	—	—	—	1200[a]	30 hr immersion	—	—
116% Seawater	583	—	—	—	—	608	—	—	—	—	—	—	—	1165[a]	—	—	—
Myxine glutinosa	549	11.1	189	5.1	—	563	—	5.0	—	—	—	1152	mmoles/ liter	—	—	Bellamy and Chester-Jones	1961
Seawater	470	12.2	49.7	8.4	—	550	—	5.0	—	—	—	1067	mmoles/ liter	—	—	—	
Myxine glutinosa	535	9.1	—	—	—	—	—	—	—	—	—	—	mmoles/ liter	—	—	Chester-Jones *et al.*	1962
Seawater	489	10.2	—	—	—	—	—	—	—	—	—	—	—	—	—	—	—
Myxine gluiinosa	355	5.2	—	—	—	—	—	—	—	—	—	—	—	—	Serum sodium remains higher than seawater sodium	—	
60% Seawater	287	5.5	—	—	—	—	—	—	—	—	—	—	—	—	—		—
Myxine glutinosa	371	7.8	—	—	—	—	—	—	—	—	—	—	—	—	—	—	—
73% Seawater	344	7.7	—	—	—	—	—	—	—	—	—	—	—	—	—	—	—
Myxine glutinosa	1136	14.5	—	—	—	—	—	—	—	—	—	—	—	—	Water loss increases serum sodium	—	—

Continued

Table XII Cyclostomata, Myxiniformes—Blood Chemistry—Cont'd

Fish or medium	Na	K	Mg	Ca	NH_4	Cl	HCO_3	PO_4	SO_4	Urea	Protein (g %)	Total ions	Units	Osmotic pressure (mOsm/liter)	Comments	Author	Date
165% Seawater	776	14.8	—	—	—	—	—	—	—	—	—	—	—	—	—	—	—
Polistotrema stouti	544	7.7	10.4	5.4	—	446	5.2	1.0	4.4	—	4.2	1026.4	mmoles/liter	—	—	Urist	1963
Seawater	509	30.0	47.5	10.0	—	540	2.0	0.0	30.0	—	—	1168.5	—	—	—	—	—
Polistotrema stouti	570	7.0	12.0	4.5	—	547	—	3.7	0.9	—	—	—	mmoles/kg	1034	—	Munz and McFarland	1964
Seawater	496	10.3	51.6	10.9	—	543	—	0.0	25.7	—	—	—	—	1029	—		—
Polistotrema stouti	522	10.9	13.8	3.9	—	501	—	—	—	—	—	—	mmoles/kg	954	2 days' immersion	McFarland and Munz	1965
100% Seawater	456	11.3	50.8	8.9	—	528	—	—	—	—	—	—	—	953	—	—	—
Polistotrema stouti	405	10.3	12.8	4.4	—	371	—	—	—	—	—	—	—	743	2 days' immersion	—	—
75% Seawater	363	8.4	40.4	6.9	—	402	—	—	—	—	—	—	—	740	—	—	—
Polistotrema stouti	439	7.6	10.8	3.6	—	407	—	—	—	—	—	—	—	706	33 days' immersion	—	—
73% Seawater	336	7.9	39.6	7.9	—	371	—	—		—	—		—	700	—	—	—
Myxine glutinosa	529	10.4	25.6	6.4	—	534	—	—	18.3	—	—	1123.4	mmoles/kg	—	Running seawater	Morris	1965
Seawater	455	9.4	52.6	9.8	—	524	—	—	27.3	—	—	1078.1	—	—	—	—	—

Myxine glutinosa	471	12.1	50.7	4.9	—	500	—	—	19.7	—	—	1058.6	—	1038[a]	Still seawater high magnesium an effect of urethan		
Seawater	463	9.5	52.3	9.8	—	535	—	—	27.4	—	—	1096.7	—	1011[a]	—	—	—
Myxine glutinosa	486	8.2	11.9	5.1	—	508	7.2	2.1	3.0	2.8	—	1035	mmoles/kg	—	—	Robertson	1966
Seawater	439	9.3	50.0	9.6	—	513	2.2	—	26.4	—	—	1050	—	—	—	—	—

[a]Derived from freezing point depression.
[b]Expressed in mEq/kg water as and $H_2PO_4^{-}$– HPO_4^{2-} with a valency of 1.84.

Petromyzon marinus to be 317 mOsm/liter compared with 1236 mOsm/ liter for Mediterranean seawater. Thus, with respect to its osmotic concentration, the blood of the marine lamprey closely resembles that of the marine teleost and is approximately four times more dilute than that of the hagfish. Urist (1963) transferred metamorphosed poto-dramous lampreys, *Petromyzon marinus*, into artificial seawater and observed the changes in serum composition after 2 and 4 hr of immersion. In 4 hr the plasma Ca^{2+} concentration increased from 1.8 to 3.5 mmoles/ liter, inorganic phosphorous from 2.7 to 4.3 mmoles/liter, Cl^- from 105 to 173 mmoles/liter, and Na^+ from 124 to 173 mmoles/liter, but urea nitrogen remained constant at 0.17 to 0.15 mmoles/liter. Thus, there was an increase in inorganic ions during the 4-hr adaptation but no urea retention, suggesting that the mechanism of urea retention is of more recent origin than the cyclostomes. Other workers have captured anadomous lampreys during their spawning migration in freshwater and transferred them back into seawater in order to gain a knowledge of the ionic composition of their blood in saltwater (Fontaine, 1930; Galloway, 1933; Hardisty, 1956; Morris, 1956, 1958). It appears, however, that the osmoregulatory ability of the seawater lamprey is rapidly lost upon entry into freshwater, and therefore these experiments have been only partially successful.

Migrating *Petromyzon marinus* (Fontaine, 1930) and *Lampetra fluviatilis* (Galloway, 1933) (Table XIII) were found to show an increased blood osmotic pressure on transfer to dilute seawater. Immersion in full seawater was fatal in both cases. Morris (1958) caught maturing *Lampetra fluviatilis* at an early stage of their migration up the River Trent in England and transferred them to 50% seawater. Of the 18 animals tested only 4 were able to osmoregulate on the basis of maintaining a relatively constant body weight and a plasma osmotic pressure well below that of the environment. The mean osmotic pressure of these *Lampetra fuviatilis* was 306 mOsm/liter compared to 317 mOsm/liter for *Petromyzon marinus* captured in the Mediterranean, thus these two species may have similar blood concentrations in the marine habitat. Morris (1960) has proposed that there are three factors which contribute to the inability of maturing lampreys to osmoregulate in freshwater: First, an increase in the water permeability of the external surface which may lead them to migrate into water of lower salinity, second, a reduction in the swallowing rate related to a decrease in the diameter of the alimentary canal, and, third, a decrease in the abundance of Cl^- excretory cells in the gill epithelium.

A complete analysis of the plasma constituents of freshwater *Lampetra fuviatilis* by Robertson (1954) (Table XIII) showed it to be very similar in ionic composition to the plasma of the freshwater teleost, *Coregonus clupoides* (Table XVI). The only notable differences were that the Na^+, Cl^-, and HCO_3^- concentrations are lower in the lamprey. Urist (1963) and Urist and Van de Putte (1967) have analyzed the sera of the potodramous *Petromyzon marinus* in the Great Lakes and the anadromous *Petromyzon tridentata* tridentata of Oregon (Table XIII). The amoecoete larvae of *Petromyzon*

Table XIII Cyclostomata, Petromyzontiformes—Blood Chemistry

Fish or medium	Na	K	Mg	Ca	NH_4	Cl	HCO_3	PO_4	SO_4	Urea	Protein (g %)	Total ions	Units	Osmotic pressure (mOsm/ liter)	Comments	Author	Date
Lampetra fluviatilis	—	—	—	—	—	—	—	—	—	—	—	—	—	258[a]b	—	Dekhuyzen	1904
Petromyzon marinus	—	—	—	—	—	—	—	—	—	—	—	—	—	317[a]	In seawater	Burian	1910
Seawater	—	—	—	—	—	—	—	—	—	—	—	—	—	1236[a]	—	—	—
Petromyzon marinus	—	—	—	—	—	119	—	—	—	—	—	—	mmoles/ liter	290[a]	—	Fontaine	1930
Freshwater	—	—	—	—	—		—	—	—	—	—	—	—	11[a]	—	—	—
Petromyzon marinus	—	—	—	—	—	115	—	—	—	—	—	—	—	301[a]	—	—	—
Dilute seawater	—	—	—	—	—		—	—	—	—	—	—	—	263[a]	—	—	—
Petromyzon marinus	—	—	—	—	—	127	—	—	—	—	—	—	—	355[a]	—	—	—
Dilute seawater	—	—	—	—	—		—	—	—	—	—	—	—	435[a]	—	—	—
Petromyzon marinus	—	—	—	—	—	147	—	—	—	—	—	—	—	414[a]	—	—	—
Dilute seawater	—	—	—	—	—		—	—	—	—	—	—	—	537[a]	—	—	—
Petromyzon marinus	—	—	—	—	—	240	—	—	—	—	—	—	—	581[a]	—	—	—
Seawater	—	—	—	—	—		—	—	—	—	—	—	—	1054[a]	—	—	—
Lampetra fluviatilis	—	—	—	2.7	—	121	—	—	—	—	—	—	mmoles/ liter	247[a]	In tap water	Galloway	1933

Continued

Table XIII Cyclostomata, Petromyzontiformes—Blood Chemistry—Cont'd

Fish or medium	Na	K	Mg	Ca	NH_4	Cl	HCO_3	PO_4	SO_4	Urea	Protein (g %)	Total ions	Units	Osmotic pressure (mOsm/ liter)	Comments	Author	Date
Lampetra fluviatilis	—	—	—	—	—	—	—	—	—	—	—	—	—	280[a]	22 hr in seawater	—	—
One-third seawater	—	—	—	—	—	—	—	—	—	—	—	—	—	306[a]	—	—	—
Lampetra fluviatilis	—	—	—	—	—	—	—	—	—	—	—	—	—	301[a]	5.5 hr in seawater	—	—
One-half seawater	—	—	—	—	—	—	—	—	—	—	—	—	—	462[a]	—	—	—
Lampetra fluviatilis	—	—	—	—	—	—	—	—	—	—	—	—	—	403[a]	7 hr in seawater	—	—
Seawater	—	—	—	—	—	—	—	—	—	—	—	—	—	935[a]	—	—	—
Lampetra fluviatilis	—	—	—	—	—	—	—	—	—	—	—	—	—	253[a]	—	Wikgren	1953
Lampetra fluviatilis	119.6	3.2	2.1	2.0	0.4	95.9	6.4	12.8[b]	2.7	—	3.6	239	mmoles/ kg	—	—	Robertson	1954
Lampetra planeri	—	—	—	—	—	58	—	—	—	—	—	—	mmoles/ liter	220	Ammoecoete larva	Hardisty	1956
Lampetra planeri	—	—	—	—	—	101	—	—	—	—	—	—	—	226	Adult	—	—
Lampeira planeri	—	—	—	—	—	61.0	—	—	—	—	—	—	—	—	Ammeocoete control	—	—
Lampetra planeri	—	—	—	—	—	54.1	—	—	—	—	—	—	—	—	Ammoecoete 14 days' distilled water	—	—
Lampetra planeri	—	—	—	—	—	71.0	—	—	—	—	—	—	—	—	Ammoecoete 14 days' tap water	—	—

Lampetra fluviatilis	—	—	—	—	—	113	—	—	—	—	—	—	—	286	November adult	—	—
Lampeira fluviatilis	—	—	—	—	—	118	—	—	—	—	—	—	—	272	March adult	—	—
Lampetra fluviatilis	—	—	—	—	—	96.8	—	—	—	—	—	—	mmoles/ liter	—	November freshwater	Morris	1956
Lampetra fluviatilis	—	—	—	—	—	100.5	—	—	—	—	—	—	mmoles/ liter	—	January seawater	—	—
Lampetra fluviatilis	—	—	—	—	—	—	—	—	—	—	—	—	—	247[a]	Freshwater	Morris	1958
Lampetra fluviatilis	—	—	—	—	—	—	—	—	—	—	—	—	—	296[a]	In 33% seawater	—	
33% Seawater	—	—	—	—	—	—	—	—	—	—	—	—	—	355[a]	—	—	—
Lampetra fluviatilis	—	—	—	—	—	—	—	—	—	—	—	—	—	306[a]	In 50% seawater 4 out of 18 fresh run fish	—	—
50% seawater	—	—	—	—	—	—	—	—	—	—	—	—	—	522[a]	—	—	—
Lampetra fluviatilis	85	33	—	—	—	—	—	—	—	—	—	—	mmoles/ liter	—	Tap water	Bentley and Follett	1963
Lampetra fluviatilis	89	37	—	—	—	—	—	—	—	—	—	—	—	—	100 mmoles/liter NaCl	—	—
Petromyzon tridentata	87.0	6.1	1.4	2.8	—	80.1	3.1	4.7	0.5	—	3.6	186.1	mmoles/ liter	—	Migrating anadromous	Urist	1963
Petromyzon marinus	139.0	6.2	1.9	2.4	—	113.0	5.2	1.4	0.9	—	—	278.9	mmoles/ liter	—	Migrating potodramous	—	
Petromyzon marinus	136.0	5.1	1.8	2.7	—	112.0	5.2	1.3	0.7	—	3.9	273.9	mmoles/ liter	—	Spawning potodramous	—	—
Lake Huron water	0.02	0.05	0.25	0.9	—	0.05	1.75	0.003	2.3	—	—	5.32	mmoles/ liter	—	—	—	—

Continued

Table XIII Cyclostomata, Petromyzontiformes—Blood Chemistry—Cont'd

Fish or medium	Na	K	Mg	Ca	NH_4	Cl	HCO_3	PO_4	SO_4	Urea	Protein (g %)	Total ions	Units	Osmotic pressure (mOsm/ liter)	Comments	Author	Date
Lampetra planeri	99.4	6.4	—	—	—	81.2	—			—	—	—	mmoles/kg	227	November	Bull and Morris	1967
Ammoecoete larva	97.4	7.8	—	—	—	80.3	—	—	—	—	—	—	—	240	July	—	—
Petromyzon marinus	103.0	3.4	1.6	2.4	—	91.0	6.0	1.3	0.1	—	2.7	208.8	—	—	Ammoecoete larva	Urist and Van de Putte	1967
Petromyzon marinus	134.0	4.0	1.8	2.4	—	122.1	5.0	1.3	0.1	—	3.2	270.6	—		Metamorphosed downstream	—	—
Petromyzon marinus	137.0	3.3	2.0	2.2	—	122.1	5.0	1.4	0.1	—	3.5	272.1	—	—	Parasitic adult	—	—
Petromyzon marinus	124	—	—	1.8	—	105	—	2.7	—	0.17	—	—		—	Metamorphosed freshwater	—	—
Petromyzon marinus	184	—	—	3.1	—	166	—	3.3	—	0.16	—	—	—	—	Metamorphosed 2 hr in seawater	—	—
Petromyzon marinus	212	—	—	3.5	—	173	—	4.3	—	0.15		—	—	—	Metamorphosed 4 hr in seawater	—	—

[a]Converted from freezing point.
[b]Expressed in mEq/kg water present as H_2PO_4- and HPO_4^{2-} with a valence of 1.84.

marinus were found to have a low serum ionic concentration of only 209 mmoles/liter. After metamorphosis and downstream migration there was an increase in serum Na^+ and Cl^- concentrations to give a total ionic concentration of 271 mmoles/liter. During parasitic adult life, upstream migration, and spawning there was no significant change in the total ion concentration, but an increase in the K^+ and SO_4^{2-} concentrations occurred (Table XIII). Anadromous *P. tridentata tridentata* migrating up the Willamette River in Oregon had a much lower concentration of serum ions (186 mmoles/liter) than the potodramous *Petromyzon marinus* at the same stage in the life cycle. Urist (1963) and Urist and Van de Putte (1967) showed that this was the result of a loss of Na^+ and Cl^- in *P. tridentata tridentata* owing to a breakdown of osmoregulatory ability during its spawning migration.

Transfer of *Petromyzon marinus* from freshwater to Ca^{2+}-deficient freshwater for 2hr resulted in a small drop in serum Ca^{2+} concentration from 2.6 to 2.2 mmoles/liter and an increase in the serum inorganic phosphate concentration from 3.0 to 4.8 mmoles/ liter (Urist, 1963). The increase in inorganic phosphate is similar to the response in bony vertebrates with hypocalcemia. The finding of only a small decrease in Ca^{2+} concentration indicates that the lamprey is able to regulate Ca^{2+} by means of its gill membranes and mucosal skin despite the lack of a skeletal Ca^{2+} reservoir (Urist, 1963).

Hardisty (1956) recorded a seasonal variation in the total Cl^- content of the amoecoete larvae of *Lampetra planeri*. Bull and Morris (1967) have shown that this change was related to the nutritional status of the animal and that there is no difference in the ionic composition of the serum in animals sampled in July and November. The concentrations of Na^+ and Cl^- observed by Bull and Morris are lower than those found by Robertson (1954) in adult *Lampetra fluviatilis* but similar to those found by Urist and Van de Putte (1967) in the amoecoete larva of *Petromyzon marinus* indicating that low Na^+ and Cl^- concentrations may be a general characteristic of amoecoete larvae. The serum K^+ concentration obtained by Bull and Morris, on the other hand, was higher than those obtained by either Robertson (1954) or Urist and Van de Putte (1967) except in the case of migrating and spawning *Petromyzon marinus* (Table XIII).

B. Class Chondrichthyes

1. Subclass Elasmobranchit (Marine Species)

a. Plasma Composition. Although the marine Elasmobranchii resemble the Myxiniformes in having a plasma osmotic pressure similar to that of seawater they differ in that urea is responsible for a considerable portion of the plasma osmotic pressure (Table XIV). Urea was first identified in the blood of both the Rajiformes and Squaliformes by Staedeler and Frerichs (1858). Rodier (1899) found that the freezing point of elasmobranch blood was a little lower than that of seawater and that urea was responsible for approximately

Table XIV Chondrichthyes Blood Chemistry

Fish or medium	Na	K	Mg	Ca	Cl	HCO_3	CO_2	PO_4	SO_4	Urea	TMAO	Protein (g %)	Total ions	Units	Osmotic pressure (mOsm/ liter)	Comments	Author	Date
Elasmobranchii Marine																		
Mustelus canis	—	—	—	—	—	—	—	—	—	—	—	—	—	—	1011[a]	—	Garrey	1905
Seawater	—	—	—	—	—	—	—	—	—	—	—	—	—	—	978[a]	—	—	—
Raja stabuloforis	255	4.9	2.8	3.8	241	—	6.1	1.0	—	453	—	—	507	mmoles/ liter	—	—	Smith	1929a
Seawater	416	9.1	50.0	9.4	483	—	2.2	—	30.4	—	—	—	1090	—	—	—	—	—
Raja diaphenes	237	6.8	3.5	5.1	227	—	5.6	1.4	3.1	377	—	—	501	—	—	—	—	—
Seawater	206	4.6	22.6	4.5	225	—	2.0	—	13.2	—	—	—	504	—	—	—	—	—
Carcharias littoralis	267	5.5	2.3	5.5	235	—	10.3	3.0	0.5	381	—	—	544	—	—	—	—	—
Mustelus canis	270	5.0	3.0	5.5	234	—	11.6	3.5	2.2	381	—	—	541	—	—	—	—	—
Seawater	445	10.6	50.0	14.5	517	—	2.4	0.2	30.6	—	—	—	1162	—	—	—	—	—
Scyllium canicula	156	—	—	4.3	299	—	—	—	—	—	—	3.0	—	—	1107[a]	♂	Pora	1936b
	186	—	—	6.7	267	—	—	—	—	—	—	4.0	—	—	1123[a]	♂ Elevated oxygen 48 hr	—	—
	192	—	—	3.3	277	—	—	—	—	—	—	3.2	—	—	1098[a]	♀ —	—	—
	207	—	—	5.1	248	—	—	—	—	—	—	4.0	—	—	1124[a]	♀ Elevated oxygen 48 hr	—	—

Raja undulata	—	—	—	4.1	—	—	—	—	—	—	—	4.2	—	—	1097[a]	♂	Pora	1936c
	—	—	—	3.6	—	—	—	—	—	—	—	2.3	—	—	1125[a]	♀	—	—
Squatina angelus	—	—	—	4.9	255	—	—	—	—	—	—	5.3	—	—	1102[a]	♂ —	Pora	1936d
Torpedo marmorata	—	—	—	—	369	—	—	—	—	—	—	2.5	—	—	1098[a]	♂	—	—
Raja clavata	—	—	—	3.6	285	—	—	—	—	—	—	2.5	—	—	1095[a]	♀	—	—
Seawater	—	—	—	—	—	—	—	—	—	—	—	—	—	—	1074[a]	—	—	—
Raja erinacea	—	—	—	—	—	—	—	—	—	—	—	—	—	—	968[a]	Osmotic pressure is of whole blood	Chaisson and Friedman	1935
Seawater	—	—	—	—	—	—	—	—	—	—	—	—	—	—	925	—	—	—
Raja erinacea	254	8.0	2.5	6.0	255	—	—	—	—	320	—	—	—	mmoles/liter	—	—	Hartman *et al.*	1941
Rhinobatus percellens	143	12.8	1.0	3.7	144	—	—	—	—	349	—	—	—	mmoles/liter	—	—	Pereira and Sawaya	1957
Narcine brasiliensis	134	7.0	1.5	6.0	159	—	—	—	—	209	—	—	—	—	—	—	—	—
Seawater	354	9.8	37.6	8.0	—	—	—	—	—	—	—	—	—	—	—	—	—	—
Squalus acanthias	—	—	—	—	—	—	—	—	—	—	71	—	—	mmoles/liter	—	—	Cohen *et al.*	1958
Mustelus clavata	—	—	—	—	275	—	—	—	—	—	—	—	—	mmoles/kg	970	—	Davson and Grant	1960
Mustelus canis	288	8	3	5	270	6	—	2	3	342	97	2.85	1037	mmoles/kg	962	—	Doolittle *el al.*	1960
Aprionodon isodon	238	7.0	—	—	252	—	—	—	—	—	—	—	—	mmoles/liter	—	—	Sulya *et al.*	1960

Continued

Table XIV Chondrichthyes Blood Chemistry—Cont'd

Fish or medium	Na	K	Mg	Ca	Cl	HCO_3	CO_2	PO_4	SO_4	Urea	TMAO	Protein (g %)	Total ions	Units	Osmotic pressure (mOsm/ liter)	Comments	Author	Date
Carcharhinus limbatus	258	10.0	—	—	241	—	—	—	—	—	—	—	—	—	—	—	—	—
Sphyrna tiburo	289	12.5	—	—	254	—	—	—	—	—	—	—	—	—	—	—	—	—
Mustelus canis	—	—	—	—	—	—	—	—	—	—	—	—	—	—	981.3	—	Bloete *et al.*	1961
Raja clavata	289	4.0	—	—	311	—	—	—	—	444	—	—	—	mmoles/ kg	995[a]	—	Murray and Potts	1961
Seawater	—	—	—	—	—	—	—	—	—	—	—	—	—	—	973[a]	—	—	—
Platyrnoidoidis triseriata	234	11.4	4.6	5.3	208	3.2	—	2.1	—	—	—	—	—	mmoles/ liter	—	—	Urist	1961
Carcharhinus leucas leucus	223.4	9.0	2.9	4.5	236	5.1	—	2.0	0.6	333	—	2.4	484	mmoles/ liter	—	—	Urist	1962a
Squalus acanthias	255	6.6	—	—	239	—	—	—	—	—	—	—	—	mmoles/ liter	973/kg		Maren	1962a
Triakus semifasciatus	235	10.0	3.0	5.0	230	5.0	—	1.2	0.5	333	—	3.5	—	mmoles/ liter	—	—	Urist	1962b
Seawater	509	30.0	47.5	10.0	540	2.0	—	0.0	30.0	—	—	—	—	—	—	—	—	—
Raja stabuliforis	182	4.2	3–5	3–5	220	—	—	—	1	—	—	—	—	mmoles/ liter	958/kg	—	Maren *et al.*	1963
Raja ocellata	285	3.5	3–5	3–5	255	—	—	—	1	—	—	—	—	—	928	—	—	—
Raja erinacea	260	3.5	3–5	3–5	253	—	—	—	1	285	—	—	—	—	917	—	—	—
Raja clavata	285	4.0	—	—	240	—	—	—	—	—	—	—	—	—	—	—	Eager	1964
Squalus acanthias	240	3.6	—	—	259	—	—	—	—	—	—	—	—	mmoles/ liter	—	—	Robin *et al.*	1964

Squalus acanthias	—	—	—	—	234.6	—	—	—	—	—	—	—	—	mmoles/liter	997	—	Burger	1965
Squalus acanthias	—	—	—	—	242.6	—	—	—	—	—	—	—	—	mmoles/liter	1000	Rectal gland removed 21 days	—	—
Seawater	—	—	50	9	500	—	—	—	—	—	—	—	—	—	932	—	—	—
Dasyatis americana	251	18.8	1.9	11.6	256	—	—	—	—	351	—	3.55	—	mmoles/liter	864	—	Bernard *et al.*	1966
Dasyatis saj	256	20.6	1.6	9.8	262	—	—	—	—	382	—	2.60	—	mmoles/liter	840	—	—	—
Negaprion brevirostris	307	5.5	—	—	277	—	5.9	—	—	—	—	—	—	mmoles/liter	—	—	Oppelt *et al.*	1966
Ginglymostoma cirratum	291	4.2	—	—	287	—	7.8	—	—	—	—	—	—	—	—	—	—	—
Squalus acanthias	263	4.1	3.1	6.6	249	6	—	—	—	357	—	4.0	—	—	1007	—	Murdaugh and Robin	1967
Raja eglanteria	—	—	—	—	195	—	—	—	—	320	—	—	—	mmoles/liter	737[a]	24 hr immersion	Price and Creaser	1967
Dilute seawater	—	—	—	—	291	—	—	—	—	—	—	—	—	—	608[a]	—	—	—
Raja eglanteria	—	—	—	—	211	—	—	—	—	344	—	—	—	—	823[a]	24 hr immersion	—	—
Dilute seawater	—	—	—	—	378	—	—	—	—	—	—	—	—	—	703[a]	—	—	—
Raja eglanteria	—	—	—		222	—	—	—	—	368	—	—	—	—	844[a]	—	—	—
Seawater	—	—	—	—	421	—	—	—	—	—	—	—	—	—	817[a]	—	—	—
Raja eglanteria	243	11	—	—	249	—	—	—	—	366	—	—	—	mmoles/liter	—	—	Price	1967
Heterodontus triseriata	235	10.0	3.0	5.0	230	5.0	—	1.2	0.5	338	—	—	—	mmoles/liter	—	—	Urist and Van de Putte	1967

Continued

Table XIV Chondrichthyes Blood Chemistry—Cont'd

Fish or medium	Na	K	Mg	Ca	Cl	HCO_3	CO_2	PO_4	SO_4	Urea	TMAO	Protein (g %)	Total ions	Units	Osmotic pressure (mOsm/ liter)	Comments	Author	Date
Elasmobranchii Freshwater																		
Pristis microdon	—	—	—	—	170	—	—	3.1	—	130	—	—	—	mmoles/ liter	548[a]	—	Smith	1931
Dasyatis uarnak	—	—	—	—	212	—	—	—	—	104	—	—	—	—	548[a]	—	—	—
Carcharhinus melanop	—	—	—	—	158	—	—	—	—	103	—	—	—	—	484[a]	—	—	—
Hypolopus sephen	—	—	—	—	146	—	—	—	—	81	—	—	—	—	—	—	—	—
C. Leucas nicaraguensis	200.1	8.2	2.0	3.0	180.5	6.0	—	4.0	0.5	132	—	3.4	404.3	mmoles/ liter	—	—	Urist	1961
Freshwater San Juan River	0.7	0.1	0.2	0.8	0.8	1.7	—	0.001	0.7	—	—	—	5.1	—	—	—	—	—
Pristis perotteti	216.6	6.5	0.9	4.2	193.1	—	—	—	—	—	—	3.7	—	mmoles/ liter	—	—	T. B. Thorson	1967
Carcharhinus leucas	245.8	6.4	1.6	4.5	219.3	7.0	—	4.4	0.7	180	—	3.0	—	—	—	—	—	—
Holocephali																		
Hydrolagus colliei	—	—	—		—	—	—	—	—	—	—	—	—	—	801[a]	—	Nicol	1950
Chimaera monstrosa	363	10.2	—	—	380	—	—	—	—	266	—	—	—	mmoles/ liter	—	—	Fänge and Fugelli	1962
Hydrolagus colliei	268	6.9	1.5	4.8	272	1.7	—	—	0.6	303	—	—	557.7	mmoles/ liter	—	—	Urist	1966

[a]Value converted from freezing point.

one-third of its osmotic pressure. These findings were subsequently confirmed by Fredericq (1904), Garrey (1905), and Bottazzi (1906).

The first thorough investigation of the inorganic constituents of marine elasmobranch plasma was carried out by Smith (1929a) who examined the plasma of the sharks, *Carcharias littoralis* and *Mustelus canis*, and the skates, *Raja stabuliforis* and *Raja diaphenes* (Table XIV). No striking differences were apparent between the concentrations of ions and urea in the sharks and in the skates. The mean Na^+ concentration in plasma was 257 mmoles/liter, while K^+ was 5.5 mmoles/liter, Ca^{2+} 5.0 mmoles/liter, and Cl^- 234 mmoles/liter. The concentrations observed by Smith (1929a) were similar to those of McCallum (1926) who examined the sera of *Acanthias vulgaris* and *Carcharias littoralis* except that Smith observed a greater variability in the K^+ and Ca^{2+} levels. McCallum (1926) concluded that the high plasma urea concentrations which he observed in sharks were a result of an inability of the kidney to remove them; whereas Smith (1929a, 1936), noting that both urea and trimethylamine oxide (TMAO) are present in the urine at a lower concentration than in the plasma and that the uremia persists even when the fish is in a state of inanition, argued that urea and trimethylamine oxide are actively reabsorbed from the glomerular filtrate. Thus, urea and trimethylamine oxide form a significant component of the plasma osmotic solutes, the gills and integument being relatively impermeable. Hartman *et al.* (1941) extended the findings of Smith (1936) on the reciprocal relationship between plasma Na^+ and urea concentrations. During inanition there was a decrease in the plasma urea concentration and an increased plasma Na^+ concentration in Raja erinacea. This change was partially offset by feeding. Intramuscular injection of urea resulted in an increased plasma urea concentration and a decreased plasma Na^+ concentration. Neither interrenalectomy nor removal of the rectal gland was found to have a significant effect on the plasma constituents of the skate (Hartman *et al.*, 1944; Burger, 1965). Since the investigations of Smith (1929a) and Hartman *et al.* (1941), several workers have examined the serum composition of other elasmobranchs, and these data are listed in Table XIV. Urist and Van de Putte (1967) report ranges for marine elasmobranchs of 480-490 mmoles/liter for total serum ion concentration, 225-250 mmoles/liter for Na^+, 4.0–5.0 mmole/liter for Ca^{2+}, and 250–330 mmoles/liter for urea.

Pora (1936a,b) investigated the effect of hyperoxygenation on *Scyllium canicula* and found increases in the serum Na^+, Ca^{2+}, protein, and total osmotic concentrations and a decrease in the serum Cl^- concentration of both males and females (Table XIV). Pora (1936c) also observed a sex difference in the serum composition of *Raja undulata*, the osmotic pressure being higher in the female and serum protein and Ca^{2+} concentrations higher in the male (Table XIV). In *Scyllium canicula* the male was found to have the higher serum osmotic pressure. In contrast to the findings of a sex difference in serum Ca^{2+} concentration by Pora (1936c), Hess *et al.* (1928) reported that there was no sex difference for serum Ca^{2+} concentration in the dogfish; the concentration was high in both male and female. This was confirmed by Smith (1929a) and Hartman *et al.* (1941, 1944). Urist (1961) examined the serum Ca^{2+} concentration in several species of elasmobranch both male and female, mature and immature, gravid and nongravid, and found it to be high in all cases (5 mmoles/liter), a situation quite different from that in the teleost (Hess *et al.*, 1928).

The effects of immersion in dilute seawater on the plasma Cl^{-}, urea, and osmotic concentrations of the skate, *Raja eglanteria*, were investigated by Price and Creaser (1967) (Table XIV). Immersion in water of low salinity resulted in a loss of serum Cl^{-} and urea. Introduction of skates adapted to low salinity water (21‰) into high salinity water (up to 31‰) resulted in an increase in serum urea and Cl^{-}. The time taken to reach osmotic equilibrium in *Raja eglanteria* was 48 hr after a change of 2.4‰ and 70 hr after a change of 10.0‰ in the external salinity. Thus, skate in living in esturine conditions where the salinity changes with each tide probably never reach osmotic equilibrium. Although Price and Creaser (1967) found a lowering of serum Cl^{-} concentration in low salinity water in the laboratory, Price (1967) was unable to show a significant relationship between serum Cl^{-} concentration and external salinity in captured skate.

b. Composition of Other Fluids. i. Pericardial and perivisceral fluids. Smith (1929a) found the pH of the pericardial fluid to be lower than that of the plasma with a correspondingly lower concentration of HCO_3^{-}. The Ca^{2+} concentration was also found to be lower while the K^{+} concentration was higher (Table XV). Perivisceral fluid was also found to be more acidic than plasma and to have a lower Ca^{2+} concentration. Unlike the pericardial fluid the concentration of K^{+} in the perivisceral fluid was similar to that in the serum and the concentrations of SO_4^{2-} and Mg^{2+} were higher than those in the serum. Pericardial and perivisceral fluid urea concentrations were the same or lower than the plasma (Smith, 1929a) (Table XV).

In *Raja erinacea* the perivisceral fluid was found to have a higher concentration of Mg^{2+}, K^{+}, Cl^{-}, and urea and a lower concentration of Ca^{2+} than the plasma (Hartman *et al.*, 1941) (Table XV). Rodnan *et al.* (1962) and Murdaugh and Robin (1967) compared plasma and coelomic fluid concentrations in *Squalus acanthias* and found the Na^{+}, Mg^{2+}, H^{+}, and Cl^{-} concentrations to be higher and the Ca^{2+}, HCO_3^{-}, and protein concentrations to be significantly lower in the coelomic fluid (Table XV). Although NH_4^{+} was not measured, Rodnan *et al.* (1962) suggest that it may make up some of the cation deficit in the coelomic fluid.

The most recent measurements of batoid pericardial and perivisceral fluids have been carried out on the stingray, *Dasyatis americana*, by Bernard *et al.* (1966) (Table XV); Ca^{2+}, Mg^{2+}, and protein were less concentrated, and the osmotic pressure was lower in both pericardial and perivisceral fluid. In pericardial fluid K^{+} was less concentrated, and H^{+}, Cl^{-}, and NH_4^{+} were elevated in both pericardial and perivisceral fluids. These data were interpreted by

Table XV Chondrichthyes— Chemistry of Other Body Fluids

Fish and fluid	Na	K	Mg	Ca	Cl	HCO_3	CO_2	PO_4	SO_4	Urea	TMAO	Protein (g %)	Total ions	Units	Osmotic pressure (mOsm/ liter)	Comments	Author	Date
Elasmobranchii Marine																		
Raja stabuloforis	—	—	—	—	—	—	—	—	—	—	—	—	—	—	—	—	Smith	1929a
Cranial fluid	264	4.1	0.9	4.9	263	—	5.55	1.0	—	436	—	—	—	mmoles/ liter	—	—	—	—
Perivisceral fluid	283	5.8	17.9	2.4	309	—	0.35	0.7	7.2	450	—	—	—	mmoles/ liter	—	—	—	—
Raja stabuloforis	—	—	—	—	—	—	—	—	—	—	—	—	—	—	—	—	—	—
Pericardial fluid	335	20.2	2.5	0.6	369.5	—	0.4	0.7	—	364	—	—	—	mmoles/ liter	—	—	—	—
Raja diaphenes	—	—	—	—	—	—	—	—	—	—	—	—	—	—	—	—	—	—
Cranial fluid	—	—	—	—	—	—	4.6	0.9	0.4	198	—	—	—	mmoles/ liter	—	—	—	—
Perivisceral fluid	155	6.2	14.0	2.1	188	—	—	0.7	8.0	254	—	—	—	mmoles/ liter	—	—	—	—
Pericardial fluid	—	—	—	—	—	—	—	—	—	—	—	—	—	—	—	—	—	—
Carcharias littoralis	—	—	—	—	—	—	—	—	—	—	—	—	—	—	—	—	—	—
Cranial fluid	—	—	—	—	—	—	10.5	—	—	317	—	—	—	mmoles/ liter	—	—	—	—
Perivisceral fluid	276	8.0	21.4	5.1	306	—	—	0.6	1.27	355	—	—	—	mmoles/ liter	—	—	—	—
Pericardial fluid	290	9.3	2.7	2.8	—	—	—	—	—	—	—	—	—	—	—	—	—	—
Mustelus canis	—	—	—	—	—	—	—	—	—	—	—	—	—	—	—	—	—	—

Continued

Table XV Chondrichthyes— Chemistry of Other Body Fluids—Cont'd

Fish and fluid	Na	K	Mg	Ca	Cl	HCO_3	CO_2	PO_4	SO_4	Urea	TMAO	Protein (g %)	Total ions	Units	Osmotic pressure (mOsm/ liter)	Comments	Author	Date
Cranial fluid	270	6.0	1.0	4.0	260	—	8.8	1.7	1.0	300	—	—	—	mmoles/ liter	—	—	—	—
Perivisceral fluid	—	—	—	—	—	—	—	—	—	—	—	—	—	—	—	—	—	—
Pericardial fluid	—	—	—	—	—	—	—	—	—	—	—	—	—	—	—	—	—	—
Raja erinacea	—	—	—	—	—	—	—	—	—	—	—	—	—	—	—	—	Hartman *et al.*	1944
Perivisceral fluid	—	10	21	—	300	—	3	—	—	—	—	—	—	mmoles/ liter	—	—	—	—
Acanthias vulgaris	—	—	—	—	—	—	—	—	—	—	—	—	—	—		—	Jensen and Vilstrup	1954
Endolymph	280	60	—	—	—	—	—	—	—	—	—	—	—	—	—	—	—	—
Perilymph	247	37	—	—	—	—	—	—	—	—	—	—	—	—	—	Potassium value erroneous	—	—
Squalus acanthias	—	—	—	—	—	—	—	—	—	—	—	—	—	—	—	—	Maren and Frederick	1958
Plasma	—	—	—	—	—	—	7.4	—	—	—	—	—	—	mmoles/ liter	—	—	—	—
Aqueous humor	—	—	—	—	—	—	8.9	—	—	—	—	—	—	mmoles/ liter	—	Control	—	—
Aqueous humor	—	—	—	—	—	—	5.3	—	—	—	—	—	—	mmoles/ liter	—	Acetazolamide treated	—	—
Mustelus canis	—	—	—	—	—	—	—	—	—	—	—	—	—	—	—	—	Davson and Grant	1960
Aqueous humor	—	—	—	—	245	—	—	—	—	—	—	—	—	mmoles/ kg	947	Hypotonic to plasma	—	—

Cerebrospinal fluid	—	—	—	—	282	—	—	—	—	—	—	—	—	—	964	Isotonic to plasma.	—	—
Subdural fluid	—	—	—	—	273	—	—	—	—	—	—	—	—	—	944	Similar to aqueous humor	—	—
Mustelus canis	—	—	—	—	—	—	—	—	—	—	—	—	—	—	—	—	Doolittle *et al.*	1960
Aqueous humor	279	7	3	3	256	15	—	1	4	320	85	22	980	mmoles/kg	935	—	—	—
Mustelus canis	—	—	—	—	—	—	—	—	—	—	—	—	—	—	—	—	Bloete *et al.*	1961
Anterior aqueous humor	—	—	—	—	—	—	—	—	—	—	—	—	—	—	960.7/kg	—	—	—
Posterior aqueous humor	—	—	—	—	—	—	—	—	—	—	—	—	—	—	952.8	—	—	—
Vitreous humor	—	—	—	—	—	—	—	—	—	—	—	—	—	—	968	—	—	—
Raja clavata	—	—	—	—	—	—	—	—	—	—	—	—	—	—	—	—	Murray and Potts	1961
Cranial fluid	280	3.3	—	—	311	—	—	—	—	437	—	—	—	mmoles/kg	—	—	—	—
Perilymph	281	3.5	—	—	321	—	—	—	—	447	—	—	—	—		—	—	—
Endolymph jelly	295	63.4	—	—	391	—	—	—	—	381	—	—	—	—	—	—	—	—
Lorenzini jelly	443	12.5	—	—	581	—	—	—	—	75	—	—	—	—	—	—	—	—
Squalus acanthias	—	—	—	—	—	—	—	—	—	—	—	—	—	—	—	—	Maren	1962a
Cerebrospinal fluid	271	7.8	—	—	264	—	—	—	—	—	—	—	—	mmoles/liter	986/kg	—	—	—

Continued

Table XV Chondrichthyes— Chemistry of Other Body Fluids—Cont'd

Fish and fluid	Na	K	Mg	Ca	Cl	HCO_3	CO_2	PO_4	SO_4	Urea	TMAO	Protein (g %)	Total ions	Units	Osmotic pressure (mOsm/ liter)	Comments	Author	Date
Aqueous humor	276	7.5	—	—	251	—	—	—	—	—	—	—	—	—	979	—	—	—
Squalus acanthias	—	—	—	—	—	—	—	—	—	—	—	—	—	—	—	—	—	—
Coelomic fluid	296	4.4	3.7	2.2	328	0	—	—	—	415	—	0.02	—	mmoles/ liter	1005	—	Rodnan *et al.*	1962
Raja clavata	—	—	—	—	—	—	—	—	—	—	—	—	—	—	—	—	Enger	1964
Cranial fluid	286	4.6	—	—	255	—	—	—	—	—	—	—	—	mmoles/ liter	—	—	—	—
Endolymph	287	58.7	—	—	322	—	—	—	—	—	—	—	—	—	—	—	—	—
Cetorhinus maximus	—	—	—	—	—	—	—	—	—	—	—	—	—	—	—	—	—	—
Cranial fluid	248	3.9	—	—	238	—	—	—	—	—	—	—	—	—	—	—	—	—
Endolymph (saccular cavity)	276	56.0	—	—	319	—	—	—	—	—	—	—	—	—	—	—	—	—
Endolymph semicircular canal	286	86.9	—	—	378	—	—	—	—	—	—	—	—	—	—	—	—	—
Perilymph	259	3.1	—	—	237	—	—	—	—	—	—	—	—	—	—	—	—	—
Dasyatis americana	—	—	—	—	—	—	—	—	—	—	—	—	—	—	—	—	Bernard *et al.*	1966
Perivisceral fluid	255	20.4	0.3	2.6	310	—	—	—	—	365	—	0.111	—	mmoles/ liter	827	—	—	—

Pericardial fluid	262	11.9	0.7	0.9	308	—	—	—	—	340	—	0.031	—	mmoles/ liter	828	—	—	—
Dasyatis say	—	—	—	—	—	—	—	—	—	—	—	—	—	mmoles/ liter	—	—	—	—
Perivisceral fluid	236	20.5	1.5	3.9	289	—	—	—	—	356	—	0.153	—	mmoles/ liter	813	—	—	—
Pericardial fluid	258	11.7	0.8	1.9	295	—	—	—	—	354	—	0.040	—	mmoles/ liter	810	—	—	—
Seawater	346	21.4	30	20.1	399	—	—	—	—	—	—	—	—	mmoles/ liter	709	—	—	—
Nagaprion brevirostris	—	—	—	—	—	—	—	—	—	—	—	—	—	mmoles/ liter	—	—	Oppelt *et al.*	1966
Ventricular fluid	317	5.2	—	—	285	—	8.4	—	—	—	—	—	—	—	—	—	—	—
Extrabrain fluid	313	5.3	—	—	290	—	9.4	—	—	—	—	—	—	—	—	—	—	—
Ginglymostoma cirratum	—	—	—	—	—	—	—	—	—	—	—	—	—	—	—	—	—	—
Ventricular fluid	332	5.5	—	—	310	—	8.3	—	—	—	—	—	—	—	—	—	—	—
Squalis acanthias	—	—	—	—	—	—	—	—	—	—	—	—	—	—	—	—	Cserr and Rall	1967
Plasma	—	4.1	—	—	—	—	—	—	—	—	—	—	—	—	—	—	—	—
Plasma	—	6.0	—	—	—	—	—	—	—	—	—	—	—	—	—	Anoxic 15 min	—	—
Cerebrospinal fluid	—	3.5	—	—	—	—	—	—	—	—	—	—	—	—	—	—	—	—
Cerebrospinal fluid	—	5.8	—	—	—	—	—	—	—	—	—	—	—	—	—	Anoxic 15 min	—	—
Extradural fluid	—	3.4	—	—	—	—	—	—	—	—	—	—	—	—	—	—	—	—
Extradural fluid	—	3.2	—	—	—	—	—	—	—	—	—	—	—	—	—	Anoxic 15 min	—	—

Continued

Table XV Chondrichthyes— Chemistry of Other Body Fluids—Cont'd

Fish and fluid	Na	K	Mg	Ca	Cl	HCO_3	CO_2	PO_4	SO_4	Urea	TMAO	Protein (g %)	Total ions	Units	Osmotic pressure (mOsm/ liter)	Comments	Author	Date
Elasmobranchii Freshwater																		
Pristis microdon	—	—	—	—	—	—	—	—	—	—	—	—	—	mmoles/ liter	—	—	Smith	1931
Coelemic fluid	—	—	—	—	204	—	—	—	—	150	—	—	—	—	538	—	—	—
Pericardial fluid	—	—	—	—	201	—	—	—	—	98		—	—	—	—	—	—	—
Dasyatis uarnak	—	—	—	—	—	—	—	—	—	—	—	—	—	—	—	—	—	—
Coelemic fluid	—	—	—	—	278	—	—		—	—		—	—	—	—	—	—	—
Pericardial fluid	—	—	—	—	216	—	—	—	—	—	—	—	—	—	—	—	—	—
Carcharhinus leucas	—	—	—	—	—	—	—	—	—	—	—	—	—	—	—	—	Murdaugh and Robin	1967
Cranial fluid	247.4	4.7	1.8	3.0	219.3	6.6	—	0.18	0.5	—	—	1.3	—	mmoles/ liter	—	—	—	—
Perivisceral fluid	211.0	5.8	2.2	0.8	217.4	—	—	—	0.025	—	—	Trace	—	—	—	—	—	—
Pericardial fluid	249.6	10.9	1.8	1.4	254.7	—	—	—	Trace	—	—	Trace	—	—	—	—	—	—

Bernard *et al.* (1966) as indicating that communication between the perivisceral fluid and seawater via the abdominal pores and between the pericardial cavity and the abdominal cavity via the pericardioperitoneal canals must be nonexistent or limited.

ii. Cerebrospinal and cranial fluids. The composition of the cranial fluid of several elasmobranchs was found by Smith (1929a) to approximate that of a plasma dialysate (Table XV). Davson and Grant (1960) compared subdural fluid (cranial fluid) to true cerebrospinal fluid (CSF) in *Mustelus canis* and found CSF to be isosmolar and similar in Cl^- concentration to plasma while the subdural fluid had a lower osmotic pressure but a similar Cl^- concentration (Table XV). In *Raja clavata* cranial fluid had a significantly lower concentration of K^+ than the plasma; concentrations of sodium chloride and urea were not different (Murray and Potts, 1961) (Table XV). Maren (1926a) examined the true cerebrospinal fluid of 200 *Squalus acanthias* and found the Cl^- concentration to be 7% higher than in the plasma. In the CSF, Na^+ and K^+ were slightly or questionably higher (Table XV). The carbonic anhydrase inhibitor acetazolamide abolished the normal Cl^- excess in the CSF indicating that Cl^- is being actively secreted into the CSF (Maren, 1926b). Acetazolamide also changed the CSF-plasma ratio for carbon dioxide from 1.15 to 1.72 suggesting that carbonic anhydrase may be responsible for the removal of metabolic carbon dioxide from the central nervous system (Maren and Frederick, 1958). Cranial fluid and plasma were found to have similar concentrations of Cl^-, Na^+, and K^+ in *Raja clavata* and *Cetorhinus maximus* by Enger (1964) (Table XV). Oppelt *et al.* (1966) (Table XV) have compared the composition of CSF and extrabrain fluid (cranial fluid) in the lemon shark, *Nagaprion brevirostris*. Sodium chloride and total carbon dioxide concentrations were higher in CSF than in plasma. Extrabrain fluid was closer in composition to CSF than to plasma even though it has been regarded as being similar to plasma in inorganic composition. No connection has been found between the CSF and the extrabrain fluid, the former fluid having a relatively rapid turnover rate compared to the latter (Oppelt *et al.*, 1966).

Cserr and Rall (1967) have questioned the finding by Maren (1962a) and Oppelt *et al.* (1966) that the K^+ concentrations in CSF and plasma are similar. Rapidly sampled *Squalus acanthias* CSF had a K^+ concentration of 3.5 mmoles/kg (Table XV) which was relatively independent of the plasma K^+ concentration and similar to the concentration observed in mammalian CSF.

iii. Ear fluids and Lorenzini jelly. In the mammal the perilymph is regarded as being similar in composition to the cerebrospinal fluid, while the endolymph contains a high concentration of K^+ balanced by a low Na^+ concentration (Potts and Parry, 1964). The first analyses of these fluids in elasmobranchs were carried out by Kaieda (1930) (*Scolioidontus laticandus*) and Jensen and Vilstrup (1954) (*Acanthias vulgaris*), and in both cases the K^+ concentration in the endolymph was less than twice that in the perilymph. More recently, Murray and Potts (1961) (Table XV) have examined *Raja clavata* and found the perilymph to be similar in composition to the serum while the endolymph K^+ concentration was 19 times as concentrated as that in the perilymph. This compares to a 30-fold differential reported for the mammal. Also, Na^+ and Cl^- were higher in the endolymph while urea was present at a lower concentration. The jelly occupying the tubes and ampullae of Lorenzini was found to be 5% hypertonic to seawater and to contain more Na^+, K^+, and Cl^- and considerably less urea than plasma (Murray and Potts, 1961) (Table XV). Also, K^+ and Cl^- were found to be present at higher concentrations in endolymph in *Raja clavata* than in plasma by Enger (1964), but in this report no difference was shown in the Na^+ concentrations (Table XV).

iv. Eye fluids. A complete analysis of aqueous humor and plasma in the smooth dogfish, *Mustelus canis*, has been published by Doolittle *et al.* (1960) (Table XV). Urea, trimethylamine oxide, Na^+, Cl^-, protein, and osmotic concentrations were lower in the aqueous humor while HCO_3^- was present at a concentration of 15 mmoles/kg compared to 6 mmoles/kg in the plasma. Davson and Grant (1960) also found low osmotic and Cl^- concentrations in the aqueous humor of this fish. Maren and Frederick (1958) found the ratio between carbon dioxide concentration in aqueous humor and plasma to be 1.2. This was lowered to 0.7 after acetazolamide and carbonic anhydrase activity was located in the ciliary process, iris, and retina. Thus, as in the mammalia, carbonic anhydrase may be responsible for the transfer of carbon dioxide into the aqueous humor. Bloete *et al.* (1961) measured the osmotic concentrations of the eye fluids of *Mustelus canis* and found them to be in the order arterial plasma > anterior aqueous > posterior aqueous > vitreous. This sequence suggests that water probably does not pass through the cornea from the hypotonic seawater.

In contrast to the earlier work, Maren (1962a) (Table XV) found the carbon dioxide equilibrium and osmotic pressure to be the same in the plasma and aqueous humor of *Squalis acanthias*. The observation that Na^+ was definitely more concentrated and that K^+ was slightly more concentrated in the aqueous humor led Maren (1962a) to suggest that Na^+ secretion may be the major factor in aqueous humor formation. It is interesting to note, however, that Maren (1967) recently referred to his 1962 data (Table XV) as indicating a slightly greater amount of HCO_3^- in aqueous humor than in plasma in *Squalus acanthias*.

2. *Subclass Elasmobranchin (Freshwater Species)*

a. Plasma Composition. Four Malaysian elasmobranchs collected in freshwater were examined by Smith (1931). These were the shark, *Carcharhinus melanopterus*, the sawfish, *Pristis microdon*, and two rays, *Dasyatis uarnak* and *Hypolophus sephen*. Most of the experiments were carried out on *Pristis microdon* caught from the Perak River at least 25 miles upstream from the last traces of seawater. Plasma osmotic concentration was 548 mOsm/ liter in

Pristis microdon. This value is much lower than that found in marine elasmobranchs but higher than that found in freshwater teleosts. The blood urea concentration was only 30%, and the Cl^- concentration only 75% of the concentrations observed in marine elasmobranchs (Table XV).

Although elasmobranchs are widely distributed in freshwater (Herre, 1955), recent studies have been confined to those occurring in Lake Nicaragua and the Rio San Juan which connects it to the ocean (Urist, 1961; Thorson, 1967) (Table XIV). Urist (1961) compared the serum composition of *Carcharhinus leucas nicaraguensis* from Rio San Juan with those of the teleost, *Megalops atlanticus*, from the same location and *Carcharhinus leucas leucas* from a marine habitat. The total ion con centration of the serum in the freshwater *Carcharhinus* was 83%, the Ca^{2+} 66%, and the urea 30% of the values observed in the marine *Carcharhinus* (Table XIV). The serum of *Megalops atlanticus* had a quite different composition having a much lower Na^+ concentration and a minimal concentration of urea (Table XVI). The freshwater sharks and sawfish of the Rio San Juan and Lake Nicaragua are now believed to have migrated freely from the Atlantic Ocean and thus to be identical to those occurring in the ocean and capable of osmoregulating both in the seawater and in freshwater (Thorson, 1967). Serum concentrations in *Pristis perotteti* and *Carcharhinus leucas* from Lake Nicaragua were similar to those observed for *Carcharhinus leucas* nicaraguens*is* (Urist, 1961) except that the Na^+ and urea concentrations were somewhat higher in the *Carcharhinus leucas* from Lake Nicaragua (Thorson, 1967) (Table XIV).

b. Composition of Other Fluids. Early work by Smith (1931) showed that urea is present in both the perivisceral and pericardial fluids of the freshwater elasmobranchs, the concentration being lowest in the pericardial fluid as in the marine elasmobranchs (Table XV). The pericardial fluid of *Carcharhinus leucas* from Nicaragua has the expected high K^+ concentration; however, the perivisceral fluid of this fish does not have the elevated Cl^- concentration found in some marine elasmobranchs (Thorson, 1967) (Table XV). The cranial fluid of *Carcharhinus leucas* from Lake Nicaragua is similar in composition to the serum except that it has a lower phosphate and protein concentration (Thorson, 1967) (Table XV).

3. *Subclass Holocephali*

The first determination of the osmotic pressure of the blood of the ratfish, *Hydrolagus colliei*, indicated that it was slightly hypoosmotic to seawater (801 mOsm/liter) (Nicol, 1950). Fänge and Fugelli (1962) examined the blood *of Chimaera monstrosa* and concluded that it was probably isosmotic with seawater. It differed from the marine elasmobranchs in having higher Na^+ and Cl^- concentrations and a lower urea concentration (Table XIV). The observations of Fänge and Fugelli have since been confirmed by Urist (1966) who observed a similar situation with respect to Na^+, Cl^-, and urea in *Hydrolagus collei* (Table XIV).

Table XVI Osteichthyes—Blood Chemistry

Fish or medium	Na	K	Mg	Ca	NH_4	Cl	HCO_3	CO_2	PO_4	SO_4	Urea	Protein (g %)	Total ions	Units	Osmotic pressure (mOsm/liter)	Comments	Author	Date
Sarcopterygii																		
Coelacanthiformes																		
Latimeria chalumnae	181	51.3	14.4	3.5	—	199	4.7	—	—	—	355	5.1	—	mmoles/liter	1181	Fish frozen, blood hemolyzed	Pickford and Grant	1967
Seawater	—	—	—	—	—	—	—	—	—	—	—	—	—	—	1090	—	—	—
Dipteriformes																		
Proptopterus aethiopicus	99.0	8.2	Trace	2.1	—	44.1	—	35.0	1.0	Trace	0.6	—	—	mmoles/liter	238	Nonestivating male	Smith	1930
Actinopterygii																		
Acipenseriformes																		
Acipenser stellatus	—	—	—	—	—	96	—	—	—	—	3.5	—	—	mmoles/liter	—	Seawater	Korjuev	1938
Acipenser stellatus	—	—	—	—	—	92	—	—	—	—	3.4	—	—	—	—	Freshwater	—	—
Acipenser sturio	—	—	—	—		92	—	—	—	—	3.3	—	—	—	—	Seawater	—	—
Acipenser sturio	—	—	—	—	—	93	—	—	—	—	3.5	—	—	—	—	Freshwater	—	—
Acipenser oxyrhynchus	150.6	2.67	0.9	1.9	—	112.9	—	—	3.1	—	—	—	—	mmoles/liter	—	Young fish freshwater	Magnin	1962
Acipenser oxyrhynchus	164.9	2.84	1.3	1.5	—	132.9	—	—	2.5	—	—	—	—	—	—	Young fish seawater	—	—
Acipenser sturio	163.6	4.65	1.57	2.1	—	126.4	—	—	4.6	0.5	—	—	—	—	343[a]	Adult seawater	—	—
Acipenser sturio	155.8	4.3	1.47	2.3	—	119.7	—	—	5.0	0.7	—	—	—	—	318[a]	Adult freshwater	—	—

Acipenser fulvescens	143	4.2	1.25	1.28	—	107	—	—	2.6	—	—	—	—	—	—	Ottawa River	—	—
Acipenser fulvescens	143	3.56	0.85	1.25	—	104	—	—	2.4	—	—	—	—	—	—	St. Lawrence River	—	—
Acipenser fulvescens	148	4.46	1.6	1.4	—	113	—	—	2.7	—	—	—	—	—	—	Lake Nipissing (mature)	—	—
Acipenser transmontanus	130	2.5	2.1	1.7	—	115.1	5.2	—	2.9	0.4	1.0	2.6	259.9	mmoles/ liter	—	Seawater males	Urist and Van de Putte	1967
Acipenser transmontanus	129	2.7	2.0	1.8	—	111.0	6.0	—	3.3	0.5	0.9	2.5	256.3	—	—	Freshwater males and females	—	—
Acipenser transmontanus	123	2.0	1.1	4.6	—	116.0	5.0	—	4.1	0.2	1.1	4.0	256.0	—	—	Freshwater mature females	—	—
Amiiformes																		
Amiatus calva	132.5	2.0	0.4	5.3	—	119.5	—	4.3	4.1	2.2	—	—	—	mmoles/ liter	—	—	Smith	1929b
Lepisosteiformes																		
Lepisosteus osseus	140	2.7	0.3	6.1	—	118	—	9.1	3.9	2.4	—	—	—	mmoles/ liter	—	—	Smith	1929b
Lepisosteus productus	149	5.0	—	—	—	141	—	—	—	—	—	—	—	mmoles/ liter	—	Low salinity species	Sulya *et al.*	1960
Lepisosteus spatula	153	6.2	—	—	—	129	—	—	—	—	—	—	—	—	—	Euryhaline species	—	—
Lepisosteus osseus	159	4.2	—	—	—	133	—	—	—	—	—	—	—	—	—	Low salinity species	—	—
Teleostei—Marine																		
Syngnathus	212	16.5	9.7	3.4	—	192	—	—	4.7	—	20.5	—	—	mmoles/ liter	—	—	Edwards and Condorelli	1928
Hippocampus	162	—	—	—	—	146	—	—	—	—	44.1	—	—	mmoles/ liter	—	—	—	—

Continued

Table XVI Osteichthyes—Blood Chemistry—Cont'd

Fish or medium	Na	K	Mg	Ca	NH_4	Cl	HCO_3	CO_2	PO_4	SO_4	Urea	Protein (g %)	Total ions	Units	Osmotic pressure (mOsm/liter)	Comments	Author	Date
Lophius	229	6.6	3.7	2.3	—	162	—	—	8.0	—	9.4	—	—	mmoles/liter	—	—	—	—
Mureana	253	—	—	3.9	—	177	—	—	3.1	—	14.9	—	—	—	—	—	—	—
Lophius piscatorius	198	7.4	5.8	2.9	—	186	—	6.6	5.3	2.7	—	—	—	mmoles/liter	—	—	Smith	1929b
Gadus callarias	180	4.9	3.8	5.0	—	158	—	4.8	3.1	—	—	—	—	mmoles/liter	—	—	—	—
Spheroides maculatus	—	3.2	4.5	8.3	—	162	—	8.0	5.3	1.0	—	—	—	mmoles/liter	—	—	—	—
Conger vulgaris	—	—	—	—	—	186	—	—	2.4	—	—	4.5	—	mmoles/liter	430[a]	—	Boucher-Firley	1934
Mureana helena	—	—	—	—	—	187	—	—	2.7	—	—	5.6	—	mmoles/liter	441[a]	—	—	—
Conger vulgaris	—	—	—	—	—	161	—	—	—	—	—	5.6	—	—	425[a]	—	Pora	1936d
Lophius piscatorius	180	5.1	2.5	2.8	—	196	—	—	9.8	—	0.3	3.9	—	mmoles/liter	452[a]	—	Brull and Nizet	1953
Mureana helena	211.8	1.95	2.43	3.87	2.07	188.4	8.0	—	9.6[b]	5.7	—	8.0	—	mmoles/kg	—	—	Robertson	1954
Seawater	564	12.0	64.4	12.4	—	659	1.4	—	—	34.0	—	—	—		—	—	—	—
Lophius americanus	198	3.4	0.8	2.2	—	177	3.6	—	5.2	1.2	—	4.3	396	mmoles/liter	—	Average of fish 11 and 12	Forster and Berglund	1956

Scomberomorus maculatus	188	9.8	—		—	167	—	—	—	—	—	3.5	—	mmoles/ liter	386	—	Becker *et al.*	1958
Thunnus thynnus	190	26.8	—	—	—	181	—	—	—	—	—	6.5	—	—	437	—	—	—
Mycteroperca venenosa	190	6.4	—	—	—	181	—	—	—	—	—	2.3	—	—	467	—	—	—
Sphyraena barracuda	215	6.4	—	—	—	189	—	—	—	—	—	3.3	—	—	476	—	—	—
Mycteroperca bonasi	228	7.9	—	—	—	208	—	—	—	—	—	2.8	—	—	461	—	—	—
Promicrops itaiara	200	5.8	—	—	—	166	—	—	—	—	—	5.6	—	—	384	—	—	—
Seawater	445	9.8	—	—	—	537	—	—	—	—	—	—	—	—	1070	—	—	—
Brevoortia paironus	178	11.8	—	—	—	154	—	—	—	—	—	—	—	mmoles/ liter	—	Euryhaline species	Sulya *et al.*	1960
Dorosoma petenese	166	14.3	—	—	—	163	—	—	—	—	—	—	—	—	—	Low salinity species	—	—
Dorosoma cepedianum	163	2.8	—	—	—	—	—		—	—		—	—	—	—	Low salinity species	—	—
ElopS saurus	176	9.8	—	—		106	—	—	—	—	—	—	—	—	—	Euryhaline species	—	—
Galeichthys felis	187	10.0	—	—		178	—	—	—	—	—	—	—	—	—	Euryhaline species	—	—
Bagre marina	185	10.8	—	—	—	189	—	—	—	—	—	—	—	—	—	Euryhaline species	—	—
Mugil cephalus	177	6.3	—	—	—	153									—	Euryhaline species	—	—
Mugil curema	179	5.6	—	—	—	151	—	—	—	—	—	—	—	—	—	Euryhaline species	—	—

Continued

Table XVI Osteichthyes—Blood Chemistry—Cont'd

Fish or medium	Na	K	Mg	Ca	NH_4	Cl	HCO_3	CO_2	PO_4	SO_4	Urea	Protein (g %)	Total ions	Units	Osmotic pressure (mOsm/liter)	Comments	Author	Date
Oligophites saurus	182	7.0	—	—	—	167	—	—	—	—	—	—	—	—	—	High salinity species	—	—
Caranx hippos	201	3.0	—	—	—	170	—	—	—	—	—	—	—	—	—	Euryhaline species	—	—
Coryphaena hippurus	192	9.2	—	—	—	170	—	—	—	—	—	—	—	—	—	High salinity species	—	—
Cynoscion nebulosus	195	7.8	—	—	—	251	—	—	—	—	—	—	—	—	—	Euryhaline species	—	—
Cynoscion arenarius	219	5.8	—	—	—	—	—	—	—	—	—	—	—	—	—	Euryhaline species	—	—
Micropogon undulatus	179	7.6	—	—	—	—	—	—	—	—	—	—	—	—	—	Euryhaline species	—	—
Sciaenops ocellata	189	8.6	—	—	—	160	—	—	—	—	—	—	—	—	—	Euryhaline species	—	—
Leiostomus xanthurus	176	3.1	—	—	—	—	—	—	—	—	—	—	—	—	—	Euryhaline species	—	—
Pogonias cromis	156	6.8	—	—	—	—	—	—	—	—	—	—	—	—	—	Euryhaline species	—	—
Lagodon rhomboides	183	8.3	—	—	—	174	—	—	—	—	—	—	—	—	—	Euryhaline species	—	—
Archosargus probatocephalus	163	7.6	—	—	—	157	—	—	—	—	—	—	—	—	—	Euryhaline species	—	—

Scomberomorus maculatus	181	8.8	—	—	—	178	—	—	—	—	—	—	—	—	—	High salinity species	—	—
Paralabrax clathratus	180	5.0	1.5	3.0	—	147	10.0	—	1.3	0.5	0.1	3.5	—	mmoles/liter	—	—	Urist	1962b
Gadus morhua	185	3.7	—	—	—	157	—	—	—	—	—	—	—	mmoles/liter	—	—	Enger	1964
Cottus scorpius	184	3.8	—	—	—	173	—	—	—	—	—	—	—	—	—	—	—	—
Salmo irideus	170	1.4	—	—	—	135	—	—	—	—	—	—	—	—	—	—	—	—
Serranus scriba	189	—	—	—	—	159	—	—	—	—	—	—	—	mmoles/liter	—	Seawater	Motais *et al.*	1965
Serranus scriba	136	—	—	—	—	109	—	—	—	—	—	—	—	—	—	Freshwater 3 hr	—	—
Sphaeroides maculatus	98	17.1	—	3.4	—	185	—	—	—	—	—	—	—	mmoles/liter	—	Control	Eisler and Edmunds	1966
Sphaeroides maculatus	107	23.4	—	3.6	—	183	—	—	—	—	—	—	—	mmoles/liter	—	1 ppm Endrin 96 hr	—	—
Seawater	175	6.8	—	7.3	—	406	—	—	—	—	—	—	—	mmoles/liter	—	—	—	—
Teleostei—Freshwater																		
Tinca vulgaris	—	—	—	—	—	106	—	—	—	—	—	—	—	—	280[a]	—	Keys and Hill	1934
Esox lucius	—	—	—	—	—	—	—	—	—	—	—	—	—	—	274[a]	—	—	—
Coregonus clupoides	140.9	3.81	1.69	2.67	0.33	116.8	10.58	—	8.7[b]	2.29	0.7	4.2	—	mmoles/liter	—	—	Robertson	1954
Micropterus dolomieue	128	2.5	—	—	—	123	—	—	1.68	—	—	2.6	—	mmoles/liter	—	June 24	Shell	1959
	128	2.3	—	3.4	—	111	—	—	1.32	—	—	2.2	—	—	—	July 24	—	—
	140	3.9	—	2.5	—	128	—	—	1.28	—	—	2.3	—	—	—	August 27	—	—
	140	2.1	—	3.0	—	116	—	—	0.99	—	—	2.7	—	—	—	October 7	—	—

Continued

Table XVI Osteichthyes—Blood Chemistry—Cont'd

Fish or medium	Na	K	Mg	Ca	NH_4	Cl	HCO_3	CO_2	PO_4	SO_4	Urea	Protein (g %)	Total ions	Units	Osmotic pressure (mOsm/liter)	Comments	Author	Date
	140	2.3	—	4.9	—	110	—	—	0.97	—	—	4.5	—	—	—	November 18	—	—
Megalops atlanticus	101	6.2	1.4	2.5	—	140	10.1	—	5.0	0.40	6.0	6.0	266.6	mmoles/liter	—	Anadromous taken in freshwater	Urist	1961
Salmo gairdnerii irideus	139.2	4.7	—	2.2	—	128.1	—	—	1.0	—	—	3.7		mmoles/liter	—	1-4 days after capture	Huggel *et al.*	1963
	115.5	5.1	—	1.8	—	98.1	—	—	1.2	—	—	4.3	—	—	—	12 days after capture	—	—
Squalus cephalus	140.7	4.2	—	2.2	—	121.1	—	—	0.8	—	—	4.5	—	—	—	Caudal plasma	—	—
Carassius auratus	—	—	—	—	—	—	—	—	—	—	—	—	—	—	289[a]	—	Levine and Musallam	1964
Salvelinus namaycush	—	—	—	—	—	117.3	6.40	6.65	—	—	—	—	—	mmoles/kg	298	Control	Hoffert and Fromm	1966
	—	—	—	—	—	111.5	8.47	8.96	—	—	—	—	—	—	—	Diamox	—	—
Cyprinus carpio	130	2.93	1.23	2.12	—	125.2	—	—	0.39	—	—	2.2	—	mmoles/liter	274	17°C	Houston and Madden	1968

[a]Derived from freezing-point depression.
[b]Expressed in mEq/kg water present as $H_2PO_4^-$ and HPO_4^{2-} with a valency of 1.84.

C. Class Osteichthyes

1. Subclass Sarcopterygi

a. Order Coelacanthiformes. Morphological studies have shown that the coelacanths evolved from a group of fish lying close to the ancestral stock of gnathostomes, the Osteolepids (Rhipidistia), which gave rise to the lungfishes (Dipnoi), and also to the ancestral amphibia (Young, 1950). The Chondrichthyes, Sarcopterigii, and Amphibia differ biochemically from the Actinopterygii in that the three former groups are able to form urea by the ornithine urea cycle (Brown and Brown, 1967). Although the blood chemistry of only one specimen of the living coelacanth, *Latimeria chalumnae*, has been examined it is evident that these fish possess a blood urea concentration within the range found in the marine elasmobranchs and that the serum osmolarity is close to that of seawater (Pickford and Grant, 1967) (Table XVI). Blood from the renal and hepatic portal veins was hyperosmotic to seawater while blood from the heart was hyposmotic. If the value for heart blood turns out to be the correct one then *Latimeria* must drink seawater and excrete the excess ions ingested. But, if the blood is indeed hyperosmotic then *Latimeria* resembles an elasmobranch. The serum Na^+ concentration was lower than that of marine elasmobranchs and similar to that of marine teleosts. The measured values for blood K^+ and Mg^{2+} were both high owing to hemolysis. The Mg^{2+} concentration was also high in the aqueous humor (Table XVII), but in the elasmobranchs the Mg^{2+} concentration of this fluid is similar to that of plasma. Similar high Mg^{2+} concentrations had previously only been found in stressed *Lophius piscatorius* (Brull and Cuypers 1954). In contrast to the normal situation in teleosts, *Latimeria* had a blood Cl^-concentration higher than that for Na^+. This may, however, be explained by the complete hemolysis which occurred in the blood sample.

b. Order Dipteriformes. The lungfishes normally have a very low concentration of urea in their plasma (Table XVI). During the dry season, the shallow freshwaters which they inhabit dry up and the fish estivate in the underlying mud. During this period of water and nutritional deprivation the fish relies on tissue protein as an energy source and the amino acid nitrogen is metabolized to urea. To conserve the water, which would be necessary for urea excretion, the urea is allowed to accumulate in the blood and muscle until the end of the estivation period (Smith, 1930; Janssens, 1964). The inorganic composition of the blood of the normal lungfish was determined by Smith (1930). The plasma Cl^- concentration which he observed appears to be very low when compared to that in teleosts.

2. Subclass Actinopterygii

a. Order Acipenseriformes. Although the sturgeons have an almost wholly cartilaginous skeleton and bear other resemblances to the elasmo-branchs, it is now accepted that they are of actinopterygian descent (Young, 1950).

Table XVII Osteichthyes—Blood Chemistry

Fish or medium	Na	K	Mg	Ca	NH_4	Cl	HCO_3	CO_2	PO_4	SO_4	Urea	TMAO
						Teleostei—Euryhaline—General						
Alosa pseudoharengus	—	—	—	—	—	168	—	—	—	—	—	—
Alosa pseudoharengus	—	—	—	—	—	151	—	—	—	—	—	—
Fundulus kansae	160	3.9	—	2.5	—	—	—	—	—	—	—	—
Fundulus kansae	231	2	—	4.6	—	—	—	—	—	—	—	—
Fundulus kansae	157	1.2	—	3.0	—	—	—	—	—	—	—	—
Seawater	440	14	—	10.5	—	—	—	—	—	—	—	—
Pleuronectes flesus	193.7	5.4	—	—	—	166.1	—	—	—	—	—	0.7
Pleuronectes flesus	156.8	5.1	—	—	—	113.7	—	—	—	—	—	0.4
Platichthyes flesus	168	—	—	—	—	156	—	—	—	—	—	—
Platichthyes flesus	162	—	—	—	—	148	—	—	—	—	—	—
Platichthyes flesus	141.7	3.4	—	3.3	—	168.1	—	—	—	—	—	—
Platichthyes flesus	123.9	2.9	—	2.7	—	131.7	—	—	—	—	—	—

Fish	Na	K	Mg	Ca	NH_4	Cl	HCO_3	CO_2	PO_4	SO_4	Urea	Cholester (mg %)
						Teleostei—Euryhaline—Salmonoidei						
Oncorhynchus keta	—	—	—	1.55	—	89.5	—	—	—	—	5.33	—
Oncorhynchus keta	—	—	—	1.50	—	86.8	—	—	—	—	4.06	—
Oncorhynchus keta	—	—	—	1.17	—	80.8	—	—	—	—	20.9	—
Oncorhynchus keta	—	—	—	1.03	—	71.8	—	—	—	—	11.4	—
Oncorhynchus gorbushka	—	—	—	1.50	—	84.0	—	—	—	—	3.46	—
Oncorhynchus gorbushka	—	—	—	1.65	—	76.3	—	—	—	—	2.7	—
Oncorhynchus gorbushka	—	—	—	1.40	—	77.0	—	—	—	—	12.5	—
Oncorhynchus gorbushka	—	—	—	1.36	—	72.2	—	—	—	—	6.33	—
Oncorhynchus nerka	—	—	0.33	2.68	—	—	—	—	—	1.9	2.14	570

Protein (g %)	Total ions	Units	Osmotic pressure (mOsm/liter)	Comments	Author	Date
	Teleostei—Euryhaline—General					
5.9	—	mmoles/ liter	—	Prespawners seawater	Sindermann and Mairs	1961
5.3	—	—		Poatspawners freshwater	—	—
—	—	mmoles/ liter	181	Freshwater	Stanley and Fleming	1964
—	—	—	242	Seawater 3 days	—	—
—	—	—	190	Seawater 20 days	—	—
—	—	—	—	—	—	—
—	—	mmoles/ liter	364	Seawater 10 days	Lange and Fugelli	1965
—	—	—	304	Freshwater 10 days	—	—
—	—	mmoles/ liter	—	Seawater	Motais *et al.*	1965
—	—	—	—	Freshwater 3 hr	—	—
—	—		297	Seawater	Lahlou	1967
—	—	—	240	Freshwater	—	—

Protein (g %)	Total ions	Units	Osmotic pressure (mOsm/liter)	Comments	Author	Date
	Teleostei—Euryhaline—Salmonoidei					
—		mmoles/ liter		Male Estuary	Lysaya	1951
—	—	—	—	Female Estuary	—	—
—	—	—	—	Male Spawning grounds	—	—
—	—	—	—	Female Spawning grounds	—	—
—	—	—	—	Male Estuary	—	—
—	—	—	—	Female Estuary	—	—
—	—	—	—	Male Spawning grounds	—	—
—	—	—	—	Female Spawning grounds	—	—
—	—	mmoles/ liter	—	Male Lummi Island	Idler and Tsuyuki	1968

Continued

Table XVII Osteichthyes—Blood Chemistry—Cont'd

Fish	Na	K	Mg	Ca	NH_4	Cl	HCO_3	CO_2	PO_4	SO_4	Urea	Choleste (mg %)
						Teleostei—Euryhaline—Salmonoidei						
Oncorhynchus nerka	—	—	0.18	3.41	—	—	—	—	—	2.0	2.21	572
Oncorhynchus nerka	—	—	0.77	2.48	—	—	—	—	—	0.86	1.53	436
Oncorhynchus nerka	—	—	0.59	2.97	—	—	—.	—	—	0.61	1.32	499
Oncorhynchus nerka	—	—	0.42	1.56	—	—	—	—	—	1.5	1.74	394
Oncorhynchus nerka	—	—	0.45	1.94	—	—	—	—	—	1.53	1.5	202
Salmo trutta	155	5.1	0.45	1.55	—	121	—	—	1.5	0.04	—	—
Salmo trutta (brown trout)	150	2.7	—	—	—	133	—.	—			—	—
Salmo trutta (brown trout)	—	—	—	—	—	185	—	—	—	—	—	—
Salmo trutta (brown trout)	163	3.1	—	—	—	150	—	—			—	—
Salmo trutta (sea trout)	155	5.3	—	—	—	124	—	—	—	—	—	—
Salmo trutta (sea trout)	205	0.9		—	—	191	—	—	—	—	—	—
Salmo trutta (sea trout)	166	3.5	—	—	—	138		—	—	—	—	—
Salmo trutta (sea trout)	144	>10	—	—	—	122	—	—	—	—	—	—
Oncorhynchus masou	—	—	—	—	—	108	—	—	—	—	—	—
Oncorhynchus masou	—	—	—	—	—	109	—	—	—	—	—	—
Oncorhynchus masou	—	—	—	—	—	101		—	—	—	—	—
Oncorhynchus tschawytscha	187	1.8	—	—	—	—	—	—	—	—	2.7	501
Oncorhynchus tschawytscha	162	1.0	—	—	—	—	—	—	—	—	—	530
Oncorhynchus tschawytscha	160	0.8	—	—	—	—	—	—	—	—	3.1	754
Oncorhynchus tschawytscha	145	0.7	—	—	—	—	—	—	—	—	2.9	365
Oncorhynchus tschawytscha	165	2.0	—	—	—	—	—	—	—	—	3.1	126
Salmo salar—parrs	117	2.19	—	2.33	—	130	—	—	—	—	—	—
Salmo salar—smolts	131	3.03	—	2.00	—	183	7.3	—	—	—	—	—

Protein (g %)	Total ions	Units	Osmotic pressure (mOsm/liter)	Comments	Author	Date
	Teleostei—Euryhaline—Salmonoidei					
—	—	—	—	Female Lummi Island	—	—
—	—	—	—	Male Lillooet	—	—
—	—	—	—	Female Lillooet	—	—
—	—	—	—	Male Forfar Creek	—	—
—	—	—	—	Female Forfar Creek	—	—
—	—	mmoles/ liter	—	Freshwater	Phillips and Brockway	1958
—		mmoles/ liter	326	Freshwater	Gordon	1959a
—	—	—	335	½ seawater 240 hr, Beawater 48 hr	—	—
—	—	—	342	½ seawater 240 hr, seawater 64 days	—	—
—	—		351	Freshwater	—	—
—	—	—	430	Seawater 24 hr	—	—
—	—	—	356	Seawater 5 months	—	—
—	—		358	Brackish indefinite	—	—
—	—	mmoles/ liter	387	Inshore or estuarine waters	Kubo	1961
—	—	—	387	Lower reaches	—	—
—	—	—	349	Upper reaches	—	—
6.6	—	—		Sea	Robertson *et al.*	1961
4.6	—	—	—	Migrating fall run	—	—
5.8	—	—	—	Migrating spring run	—	—
4.1	—	—	—	Spawning fall run	—	—
1.7		—	—	Spawning spring run	—	—
—	—	mmoles/ kg	328	Freshwater	Parry	1961
—	—	—	312	Freshwater	—	—

Continued

Table XVII Osteichthyes—Blood Chemistry—Cont'd

Fish	Na	K	Mg	Ca	NH_4	Cl	HCO_3	CO_2	PO_4	SO_4	Urea	Choleste (mg %)
						Teleostei—Euryhaline—Salmonoidei						
Salmo salar—post-smolts	156	3.28	—	—	—	133	—	—	—	—	—	—
Salmo salar—smolts	159	3.62	—	—	—	166	—	—	—	—	—	—
Salmo salar—adults	212	3.15	—	3.43	—	157	11.0	—	—	—	—	—
Fish.	Na	K	Mg	Ca	NH_4	Cl	HCO_3	CO_2	PO_4	SO_4	Urea	Cholester (mg %)
Salmo salar—adults spawning	176	3.03	—	3.45	—	172	—	—	—	—	—	—
Salmo salar—adult kelts	294	3.44	—	3.98	—	—	—	—	—	—	—	—
Salmo gairdnerii	145	2.33	—	—	—	—	—	—	—	—	—	—
Salmo gairdnerii	160	2.42	—	—	—	—	—	—	—	—	—	—
Salmo gairdnerii	144	6.0	—	2.65	—	151	—	—	—	—	—	—
Salmo gairdnerii	162	3.81	—	2.02	—	124.6	—	—	1.27	—	—	—
Oncorhynchus kisutch	146	—	—	—	—	132	—	—	—	—	—	—
Oncorhynchus kisutch	147	—	—	—	—	131	—	—	—	—	—	—
Oncorhynchus kisutch	182	—	—	—	—	149	—	—	—	—	—	—
Oncorhynchus kisutch	205	—	—	—	—	182	—	—	—	—	—	—
Oncorhynchus tschawytscha	159	0.4	1.8	2.9	—	112	4.0	—	4.5	0.3	13.6	—
Oncorhynchus tschawytscha	161	0.3	1.2	2.7	—	114	4.1	—	4.0	0.5	10.3	—
Oncorhynchus tschawytscha	160	1.3	1.0	2.3	—	137	4.7	—	4.7	0.3	7.5	—
Oncorhynchus tschawytscha	179	1.0	0.9	1.0	—	139	5.0	—	4.1	0.3	7.1	—
Fish	Na	K	Mg	Ca	NH_4	Cl	HCO_3	CO_2	PO_4	SO_4	Urea	TMAC
						Teleostei—Euryhaline—Anguilloidei						
Anguilla anguilla	143.5	1.49	—	3.32	—	—	—	—	—	—	—	—
Anguilla anguilla	112	4.13	—	7.65	—	—	—	—	—	—	—	—
Anguilla anguilla	150	2.75	—	—	—	105	—	—	—	—	—	—

Protein (g %)	Total ions	Units	Osmotic pressure (mOsm/liter)	Comments	Author	Date
	Teleostei—Euryhaline—Salmonoidei					
—	—		349	Freshwater	—	—
—	—	—	—	Seawater 2 weeks	—	—
—	—	—	344	Seawater	—	—
Protein (g %)	Total ions	Units	Osmotic pressure (mOsm/liter)	Comments	Author	Date
—	—	—	328	Freshwater		
—	—	—	371	Freshwater + 2 days seawater	—	—
—	—	—	—	Freshwater	Holmes and McBean	1963
—	—	—	—	80% seawater 10 days	—	—
—	—	mmoles/liter	—	Freshwater	Fromm	1963
—	—	mmoles/liter	—	—	Hickman *et al.*	1964
—	—	mmoles/liter	295	Freshwater—control	Conte	1965
—	—	—	291	Freshwater—X ray	—	—
—	—	—	331	Seawater—control	—	—
—	—	—	372	Seawater—X ray	—	—
6.7	280.9	mmoles/liter	—	Spawning freshwater—male	Urist and Van de Putte	1967
5.1	287.8	—	—	Spawning freshwater—female	—	—
4.9	311.4	—	—	Marine habitat—male	—	—
4.7	331.9	—	—	Marine habitat—female	—	—
Protein (g %)	Total ions	Units	Osmotic pressure (mOsm/liter)	Comments	Author	Date
Teleostei—Euryhaline—Anguilloidei						
—	—	mmoles/liter	—	Control	Fontaine	1964
—	—	—	—	Stanniectomized	—	—
—	—	mmoles/liter	328	Silver eel—freshwater	Sharratt *et al.*	1964

Continued

Table XVII Osteichthyes—Blood Chemistry—Cont'd

Fish	Na	K	Mg	Ca	NH_4	Cl	HCO_3	CO_2	PO_4	SO_4	Urea	TMAC
					Teleostei—Euryhaline—Anguilloidei							
Anguilla anguilla	175	3.15	—	—	—	154.5	—	—	—	—	—	—
Anguilla anguilla	143	3.4	—	—	—	70	—	—	—	—	—	—
Anguilla anguilla	181	2.8	—	—	—	148	—	—	—	—	—	—
Anguilla anguilla	98.8	6.2	—	—	—	—	—	—	—	—	—	—
Anguilla anguilla	75.7	9.0	—	—	—	—	—	—	—	—	—	—
Anguilla anguilla	128.0	3.5	—	—	—	—	—	—	—	—	—	—
Anguilla anguilla	149.4	3.52	—	—	—	96.6	—	—	—	—	—	—
Anguilla anguilla	49.1	2.77	—	—	—	47.3	—	—	—	—	—	—
Anguilla anguilla	162.8	3.49	—	—	—	107.1	—	—	—	—	—	—
Anguilla anguilla	159.6	4.50	—	—	—	134.0	—	—	—	—	—	—
Anguilla anguilla	178.0	4.2	—	—	—	141.0	—	—	—	—	—	—
Anguilla anguilla	143.2	2.26	3.04	2.31	—	—	—	—	1.30	—	—	—
Anguilla anguilla	150.1	1.75	2.13	2.29	—	88.25	—	—	1.81	—	—	—
Anguilla anguilla	164,2	3.35	4.18	2.75	—	—	—	—	2.02	—	—	—
Anguilla anguilla	183.3	3.22	3.74	2.37	—	139.6	—	—	1.75	—	—	—
Anguilla anguilla	133.8	2.75	4.08	2.90	—	—	—	—	1.86	—	—	—
Anguilla anguilla	111.7	2.38	2.11	2.89	—	—	—	—		—	—	—
Anguilla anguilla	110.4	3.34	2.48	3.53	—	—	—	—	2.10	—	—	—
Anguilla anguilla	124.7	3.47	2.69	3.30	—	55.9	—	—	1.81	—	—	—
Anguilla anguilla	185.0	3.45	7.93	3.73	—	—	—	—	2.30	—	—	—
Anguilla anguilla	194.0	4.33	9.63	4.03	—	207.7	—	—	1.50	—	—	—

Protein (g %)	Total ions	Units	Osmotic pressure (mOsm/liter)	Comments	Author	Date
			Teleostei—Euryhaline—Anguilloidei			
—	—	—	377	Silver eel—seawater	—	—
—	—	—	—	Yellow eel—freshwater	—	—
—	—	—	—	Yellow eel—seawater	—	—
—	—	mmoles/liter	—	Distilled water—6 weeks	Chester-Jones *et al.*	1965
—	—	—	—	Distilled water—12 weeks	—	—
—	—	—	—	Aldactone—6 weeks	—	—
—	—	mmoles/liter	—	Freshwater controls (silver)	Butler	1966
—	—	—	—	Freshwater hypox—3 weeks (silver)	—	—
—	—	—	—	Freshwater controls (yellow)	—	—
—	—	—	—	Seawater controls (silver)	—	—
—	—	—	—	Seawater hypox—3 weeks (silver)	—	—
—		mmoles/liter	—	Control—yellow eel—freshwater	Chan *et al.*	1967
—	—	—	—	Control—silver eel—freshwater	—	—
—	—	—	—	Control—yellow eel—4-5 weeks in seawater	—	—
—	—	—	—	Control—silver eel—4-5 weeks in seawater	—	—
—	—	—	—	Control—yellow eel—4 weeks in distilled water	—	—
—	—	—	—	Control—silver eel—4 weeks in distilled water	—	—
—	—	—	—	Stanniectomized—yellow eel—3 weeks freshwater	—	—
—	—	—	—	Stanniectomized—silver eel—3 weeks freshwater	—	—
—	—	—	—	Stanniectomized—yellow eel—3 weeks seawater	—	—
—	—	—	—	Stanniectomized—silver eel—3 weeks seawater	—	—

Continued

Table XVII Osteichthyes—Blood Chemistry—Cont'd

Fish	Na	K	Mg	Ca	NH_4	Cl	HCO_3	CO_2	PO_4	SO_4	Urea	TMAO
Teleostei—Euryhaline—Anguilloidei												
Anguilla anguilla	66.6	3.64	2.74	1.64	—	—	—	—	—	—	—	—
Anguilla anguilla	119.7	2.35	1.69	2.33	—	—	—	—	—	—	—	—
Anguilla anguilla	122.6	1.82	1.88	1.79	—	—	—	—	1.94	—	—	—
Anguilla anguilla	195.4	3.15	10.31	3.05	—	—	—	—	2.23	—	—	—
Anguilla anguilla	222.4	3.24	10.27	4.23	—	214.4	—	—	1.93	—	—	—
Anguilla anguilla	111.5	3.10	1.78	3.47	—	—	—	—	—	—	—	—
Anguilla anguilla	84.9	4.05	—	2.57	—	53.0	—	—	—	—	—	—
Anguilla anguilla	205.1	4.08	10.75	4.65	—	181.4	—	—	—	—	—	—

Fish	Na	K	Mg	Ca	NH_4	Cl	HCO_3	CO_2	PO_4	SO_4	Urea	TMA
Anguilla anguilla	147.1	—	—	—	—	130.4	—	—	—	—	—	—
Anguilla anguilla	181.5	—	—	—	—	167.9	—	—	—	—	—	—
Anguilla anguilla	139.8	—	—	—	—	137.3	—	—	—	—	—	—
Anguilla anguilla	187.7	—	—	—	—	175.0	—	—	—	—	—	—
Anguilla anguilla	136.6	—	—	—	—	133.4	—	—	—	—	—	—
Anguilla anguilla	136.3	—	—	—	—	130	—	—	—	—	—	—
Anguilla anguilla	136.7	—	—	—	—	121	—	—	—	—	—	—
Anguilla rostrata	153.4	5.0	—	—	—	103.6	—	—	—	—	—	—
Anguilla rostrata	146	5.3	—	—	—	103.2	—	—	—	—	—	—

Protein (g %)	Total ions	Units	Osmotic pressure (mOsm/liter)	Comments	Author	Date
			Teleostei—Euryhaline—Anguilloidei			
—	—	—	—	Stanniectomized—yellow eel—3 weeks distilled water	—	—
—	—	—	—	Adrenal insufficient—yellow eel—3 weeks freshwater	—	—
—	—	—	—	Adrenal insufficient—silver eel—3 weeks freshwater	—	—
—	—	—	—	Adrenal insufficient—yellow eel—3 weeks seawater	—	—
—	—	—	—	Adrenal insufficient—silver eel—3 weeks seawater	—	—
—	—	—	—	Stanniectomized + adrenal insufficient—yellow eel—1 week freshwater	—	—
—	—	—	—	Stanniectomized + adrenal insufficient—silver eel—2 weeks freshwater	—	—
—	—	—	—	Stanniectomized + adrenal insufficient—silver eel—2 weeks seawater	—	—

Protein (g %)	Total ions	Units	Osmotic pressure (mOsm/liter)	Comments	Author	Date
—	—	mmoles/liter	—	Seawater—sham	Mayer *et al.*	1967
—	—	—	—	Seawater—adx.	—	—
—	—	—	—	2 hr freshwater—sham + cortisol	—	—
—	—	—	—	2 hr freshwater—adx.	—	—
—	—	—	—	2 hr freshwater—adx. + cortisol	—	—
—	—	—	—	12 hr seawater—sham	—	—
—	—	—	—	12 hr seawater—adx.	—	—
—	—	mmoles/liter	—	Controls	Butler and Langford	1967
—	—	—	—	Mock adx. 3 weeks steel sutures	—	—

Continued

Table XVII Osteichthyes—Blood Chemistry—Cont'd

Fish	Na	K	Mg	Ca	NH_4	Cl	HCO_3	CO_2	PO_4	SO_4	Urea	TMA
Anguilla rostrata	132.5	4.1	—	—	—	72.8	—	—	—	—	—	—
Anguilla rostrata	147.1	4.7	—	—	—	89.8	—	—	—	—	—	—
Anguilla rostrata	153.6	4.74	—	—	—	113.7	—	—	—	—	—	—
Anguilla rostrata	143.4	6.00	—	—	—	103.3	—	—	—	—	—	—
Anguilla rostrata	137.0	6.10	—	—	—	98.4	—	—	—	—	—	—
Anguilla rostrata	138.1	2.83	1.26	3.33	—	70.48	—	—	—	—	—	—
Anguilla rostrata	138.0	2.95	1.23	3.29	—	67.31	—	—	—	—	—	—
Anguilla rostrata	132.0	2.94	0.89	3.26	—	70.94	—	—	—	—	—	—
Anguilla rostrata	151.3	2.23	1.28	3.00	—	109.60	—	—	—	—	—	—
Anguilla rostrata	144.0	2.39	1.40	2.98	—	85.54	—	—	—	—	—	—
Anguilla rostrata	135.4	2.65	1.20	2.72	—	80.20	—	—	—	—	—	—
Anguilla rostrata	150.7	2.77	1.21	3.19	—	111.11	—	—	—	—	—	—
Anguilla rostrata	143.3	2.60	1.22	3.11	—	77.35	—	—	—	—	—	—
Anguilla rostrata	131.6	2.90	1.07	2.80	—	82.88	—	—	—	—	—	—

Protein (g %)	Total ions	Units	Osmotic pressure (mOsm/liter)	Comments	Author	Date
—	—	—	—	Adx. 3 weeks silk sutures	—	—
—	—	—	—	Adx. 3 weeks steel sutures	—	—
—	—	mmoles/ liter	—	Intact	Butler	1967
—	—	—	—	Hypox—3 weeks controls	—	—
—	—	—	—	Hypox—3 weeks prolactin	—	—
—	—	mmoles/ liter	271.6	Freshwater silver intact controls (March)	Butler *et al.*	1969b
—	—	—	266.7	Freshwater silver—sham adx. 3 weeks (March)	—	—
—	—	—	265.6	Freshwater silver—adx. 3 weeks (March)	—	—
—	—	—	307.9	Freshwater yellow intact controls (June)	—	—
—	—	—	287.4	Freshwater yellow—sham adx. 3 weeks (June)	—	—
—	—	—	271.6	Freshwater yellow—adx. 3 weeks (June)	—	—
—	—	—	297.4	Freshwater yellow intact controls (July)	—	—
—	—	—	275.8	Freshwater yellow—sham adx. 3 weeks (July)	—	—
—	—	—	251.8	Freshwater yellow—adx. 3 weeks (July)	—	—

Physiological uremia does not occur in *Acipenser* either in freshwater or in saltwater (Korjuev, 1938) (Table XVI). An investigation has been carried out into the changes in blood chemistry which occur during the migration of young sturgeon and spawned adults from freshwater into saltwater and of the migration of prespawning adults in the reverse direction by Magnin (1962). When immature *Acipenser oxyrhynchus* migrated from freshwater to saltwater there were increases in the plasma concentrations of Na^+, Mg^{2+}, and Cl^- and decreases in the concentrations of Ca^{2+} and phosphate (Table XVI). Migration of prespawning *Acipenser sturio* from the sea into freshwater resulted in a significantly lower plasma osmolar, Na^+ and Cl^- concentration. Plasma SO_4^{2-} was elevated in freshwater. Magnin (1962) noted a greater degree of variability in the ionic composition of the mature fish and concluded that gonadal muturation results in a perturbation of the metabolism of water and electrolytes. In the third species which he investigated, *Acipenser fulvescens*, Magnin compared specimens captured in the Ottawa and St. Lawrence rivers with those from Lake Nipissing. The mean plasma concentrations of Na^+, Cl^-, and Mg^{2+} were higher in the lake fish, a consequence of the majority of these fish being sexually mature (Table XVI). Magnin concluded that the homeostatic mechanisms of the sturgeon like those of the salmon did not completely regulate the osmotic pressure and ionic composition of the blood when the fish passed from one osmotic environment to another. Instead there were small but significant changes in the plasma concentrations of the ions.

In a recent investigation, the plasma of *Acipenser transmontanus* was found to have only a slightly lower total concentration of ions in freshwater than in seawater (Urist and Van de Putte, 1967) (Table XVI) indicating the presence of a very effective homeostatic mechanism in this species. The low concentration of Ca^{2+} in the plasma of this and other species of sturgeon (Table XVI) is unique and may be related to the progressive loss of bone during the evolution of the Acipenseriformes (Urist and Van de Putte, 1967). An exception to this generalized hypocalcemia are the mature female *Acipenser transmontanus* which show the hypercalcemia and hyperphosphotemia characteristic of mature female teleosts.

b. Orders Amiformes and Lepisosteiformes. Representatives of each of these two holostean orders are found in the Great Lakes. Of the two, the long-nosed gar, *Lepidosteus osseus*, is the more primitive, while *Amiatus calva* diverged more recently from the teleostean stock (Young, 1950). Blood analyses by Smith (1929b), however, revealed no major differences between the two species or between them and the freshwater teleosts except that the two holosteans had approximately double the Ca^{2+} and one quarter the Mg^{2+} found in *Coregonus clupoides* (Robertson, 1954) (Table XVI). Three species of gars including *Lepidosteus osseus* examined from the Gulf of Mexico (Sulya *et al.*, 1960) (Table XVI) had higher concentrations of Na^+, K^+, and Cl^- in the plasma than were reported by Smith, but this may be explained by the presumably higher environmental salinity.

3. Group "Teleosti"

a. Blood Chemistry. Although many analyses of the blood chemistry of marine teleosts were carried out in the first half of this century, Robertson (1954) appears to be the first worker to have carried out an analysis of the environmental seawater for comparison. In his studies, Na^+ comprised 92% and K^+ less than 1% of the cations in plasma while Cl^- accounted for 87% of the anions. Only HCO_3^- and phosphate ions were present at higher concentrations in the blood plasma than in seawater. The abundances of the other ions in plasma with respect to their concentrations in seawater were in the following order: $Na^+ > Ca^{2+} > Cl^- > SO_4^{2-} > K^+ > Mg^{2+}$, where plasma Na^+ was 38% of the Na^+concentration in seawater and plasma Mg^{2+} was 4% of the concentration of this ion in seawater. The total concentration of ions plus urea was 32% of the seawater ionic concentration (Robertson, 1954) (Table XVI). In an earlier analysis of the plasma of *Mureana* and three other teleosts, erroneously high urea concentrations were obtained (Edwards and Condorelli, 1928) (Table XVI).

Analysis of the plasma composition of the freshwater teleost, *Coregonus clupoides*, showed it to have a total ion concentration which was only 66% of that in *Mureana helena*, thus the freshwater teleost does not maintain ionic concentrations as high as those in the marine teleost. All ions except K^+ and HCO_3^- were lower in *Coregonus* with the major ions of sodium and chloride showing the greatest differences (Robertson, 1954) (Table XVI). Apart from the data of Robertson other complete analyses have been presented for the marine teleosts, *Lophius piscatorius* (Brull and Nizet, 1953), *Lophius americanus* (Forster and Berglund, 1956), and *Paralabrax clathratus* (Urist, 1962b) and for the teleost *Megalops atlanticus* (Urist, 1961) captured in freshwater (Table XVI). Other investigators have covered a greater number of different species but measured only Na^+, K^+, and Cl^- (Becker *et al.*, 1958; Sulya *et al.*, 1960) (Table XVI). One observation of note from the former paper is the high plasma K^+ concentration in the bluefin tuna, *Thunnus thynnus* (Table XVI). However since only one fish was examined it is not possible to determine whether this is a characteristic of the species as a whole. The only other plasma K^+ concentration which approached that in *Thunnus thynnus* was reported by Eisler and Edmunds (1966) for *Sphaeroides maculatus* treated with the insecticide Endrin (Table XVI).

An interesting study was carried out by Shell (1959) to determine whether seasonal changes occur in the electrolyte concentrations of the freshwater teleost, *Micropterus dolomieue*. Although the investigation was only carried out between June and November in one year, some seasonal differences are apparent (Table XVI). This observation may indicate the presence of some cyclic metabolic patterns. These cycles were related by Shell to a presumed change in the endocrine status of the bass during the period of the study. Experiments have shown that the stress of capture and transportation on *Salmo gairdnerii irideus* resulted in increased plasma Na^+ and Cl^-

concentrations and decreased plasma K^+ and protein concentrations. These concentrations returned to normal after 12 days (Huggel *et al.*, 1963) (Table XVI).

b. Euryhaline Species. Many teleosts are stenohaline and thus limited to habitats having only a narrow range of osmolarity such as freshwater rivers and lakes or the open sea. Other species are euryhaline and can accommodate themselves by means of homeostatic mechanisms to changes in external ionic concentrations. The euryhaline species may be divided into two groups, one of which inhabits the interface between the true freshwater and the true seawater environments, e.g., in an estuary. The second group of euryhaline fishes spend part of their life cycle in freshwater and part of it in seawater and may be divided into catadromous species which migrate downstream to spawn in the sea, e.g., *Anguilla rostrata*, and anadromous species which migrate upstream from the sea to spawn in freshwater, e.g., *Oncorhynchus nerka*. Much scientific effort has been expended in investigations of the changes and the reasons for the changes which occur in the plasma composition of these euryhaline species. We will first deal with some of the coastal euryhaline species and then, in more detail, some members of the orders *Anguilliformes* and *Salmoniformes*.

Intertidal blennies have been studied by Raffy (1949), House (1963), Gordon *et al.* (1965), and Evans (1967a). The plasma Na^+ and Cl^- concentrations in *Xiphister atropurpureus*, which has a salinity tolerance of from 10 to 100% seawater, were found to be stable over the range of 31 to 100% seawater but to fall by about 15% when transferred to salinities equivalent to between 31 and 10% seawater. Mature fish showed lower concentrations of Na^+ and Cl^- than the nonreproductive fish in 10% sea-water. Inability to lower its permeability to Cl^- may determine the lower end of the salinity range of this fish (Evans, 1967b). Transfer of the Hlounder, *Pleuronectes flesus*, from seawater to freshwater results in a decrease in plasma osmolarity and in the concentration of Na^+, K^+, and Cl^-, and, to a lesser extent, trimethylamine oxide (Lange and Fugelli, 1965) (Table XVI). This lowering of the plasma osmotic pressure would be expected to result in an increase in the volume of the cells from the osmotic influx of water. However, a loss of intracellular free ninhydrinpositive substances and trimethylamine oxide occurs, and consequently the size of the cells remain constant. When the euryhaline, *Platichthys flesus*, together with the stenohaline marine fish, *Serranus scriba*, were transferred from seawater to freshwater for 3 hr it was found that there was only a small drop in the plasma Na^+ and Cl^- concentrations in the former fish but a considerable drop, leading to death, occurred in the plasma concentrations of these ions in the stenohaline species (Table XVI). Transfer of the flounder to freshwater was found to result in an immediate 85% decrease in the rate of Na^+ and Cl^- efflux. This did not occur in the stenohaline fish (Motais *et al.*, 1965). Hickman (1959) transferred the starry flounder, *Platichthys stellatus*, from 25% seawater to 5.4% seawater and freshwater. The plasma osmolarity was reduced, reaching equilibrium in less than 24 hr in the former medium and between 1 and 3 days in the latter medium. Recently, the changes observed in the plasma composition of *Platichthys flesus* after transfer from seawater to freshwater have been reconfirmed by Lahlou (1967) (Table XVI).

Transfer of the euryhaline mud skipper, *Periophthalmus sobrinus*, from 1000 mOsm/ liter seawater to 200 mOsm/liter seawater resulted in no significant change in the plasma osmotic, Na^+ or Cl^- concentrations. Direct transfer to freshwater, however, resulted in a sharp drop in these three parameters and a high rate of mortality. Fish adapted to 200 mOsm/ liter seawater over a period of 6 days on the other hand were able to withstand transfer to freshwater (Gordon *et al.*, 1965).

An attempt to show an effect of the steroids hydrocortisone, 9α-fluoro hydrocortisone and *dl*-aldosterone on the changes in plasma electrolyte concentrations accompanying the transfer of the stenohaline wrasse, *Thalassoma dupperey*, the partially euryhaline mullet, *Mugil cephalus*, and the euryhaline tilapia, *Tilapia mossambica*, did not succeed (Edelman *et al.*, 1960).

On transfer from freshwater to seawater plasma Na^+, Ca^{2+}, and osmolar concentrations increased during the first 3 days and then returned to near normal levels after 20 days in *Fundulus kansae*. The plasma K^+ concentration on the other hand was lower after 3 and 20 days in seawater (Stanley and Fleming, 1964). A remarkable observation in this fish was the finding of hypertonic urine between the second and tenth days after transfer to seawater (Stanley and Fleming, 1964; Fleming and Stanley, 1965). In relation to these findings it has been shown recently that the increase in permeability to Na^+ to Cl^- after transfer of *Fundulus heteroclitus* from freshwater to saltwater takes many hours. In contrast, the reverse transfer results in a very rapid decrease in permeability (Potts and Evans, 1967). The plasma composition of an andromous species, the alewife, *Alosa pseudoharengus*, has been investigated in prespawning fish in saltwater and postspawning fish in freshwater. The latter group had lower serum protein and Cl^- concentrations than the former; however, it was not possible to distinguish between reproductive and migratory effects (Sindermann and Mairs, 1961).

c. Order Anguilliformes. The changes in the plasma electrolyte composition of eels in relation to their catadromous migration have been the subject of intensive study in the last 5 years. The major part of the life cycle of *Anguilla anguilla* is spent as a yellow eel in freshwater. When the eel becomes sexually differentiated prior to its catadromous migration, it stops feeding and the ventral surface turns a silvery white color. Transfer of silver eels from freshwater to seawater resulted in a 15% increase in the serum osmotic concentration. This change was mainly accounted for by increases in serum Na^+ and Cl^-. Freshwater yellow eels had a serum Cl^- concentration noticeably lower than the freshwater silver eels (Sharratt *et al.*, 1964, Table XVII). Removal of the corpuscles of Stannius from silver eels resulted in a significant decrease in the serum Na^+ concentration and significant increases in the serum Ca^{2+} and K^+ concentrations. Injection of a lyophilized preparation of the corpuscles into stanniectomized eels tended to lower serum Ca^{2+} and K^+ concentrations and raise serum Na^+ concentration, indicating a possible osmoregulatory role for the corpuscles of Stannius (Fontaine, 1964) (Table XVII). Recently, the effects of stanniectomy have been studied in the goldfish, *Carassius auratus* (Ogawa, 1968). Removal of the corpuscles in the goldfish maintained in fresh-water resulted in a statistically nonsignificant fall in plasma osmolarity and sodium concentration, which was reversed by injection of angiotensin II. Maintenance of freshwater silver eels in distilled water for 6 or 12 weeks resulted in a progressive decrease in serum Na^+ and a progressive increase in serum K^+ concentrations. Treatment of the distilled-water eels over the 6-week period, with the antialdosterone compound, Aldactone, resulted in a higher serum Na^+ and a lower serum K^+ concentration (Chester-Jones *et al.*, 1965) (Table XVII). Hypophysectomy of freshwater silver eels, *Anguilla anguilla*, resulted in significant decreases in the serum Na^+, K^+, and Cl^- concentrations. Intact yellow eels had higher serum Na^+ and Cl^- concentrations than intact silver eels in the same environment. Hypophysectomized silver eels maintained in seawater had higher serum concentrations of Na^+ and Cl^- than intact silver eels in seawater (Butler, 1966) (Table XVII). Thus, the pituitary gland plays a role in serum electrolyte homeostasis in *Anguilla anguilla*. Hypophysectomy of freshwater Anguilla rostrata also resulted in decreased serum Na^+ and Cl^- concentrations, but the changes were not reversed by injection of ovine prolactin (Butler, 1967) (Table XVII). This lack of response may have resulted from the amount and dosage regimen used, or may have indicated that other pituitary factors play a major osmoregulatory role in this fish. Contrary to the findings of Butler (1966), Chan *et al.* (1967) (Table XVII) noted lower serum Na^+ and higher serum K^+ concentrations in freshwater yellow eels than in freshwater silver eels. Adaptation of silver and yellow eels to seawater resulted in increases in serum Na^+, K^+, Cl^-, Mg^{2+}, and phosphate concentrations (Table XVII). Stanniectomy of freshwater eels resulted in a decrease in serum $Na^+$and Cl^- and an increase in serum $K^+$and Ca^{2+} concentrations, thus corroborating the findings of Fontaine (1964). Stanniectomized seawater eels, also analyzed 3 weeks post-operatively, had increased serum concentrations of Na^+, Ca^{2+}, and Mg^{2+}. In addition, silver eels showed an increase in serum K^+ concentration (Table XVII). Adrenalectomy was followed by a lowering of serum Na^+, Ca^{2+} and Mg^{2+} concentrations in freshwater eels. Adrenalectomy of seawater silver eels caused an increase in the serum concentrations of Na^+, Ca^+, and Mg^{2+}, but it must be emphasized that data for sham-adrenalectomized eels were not presented. After both stanniectomy and adrenalectomy, freshwater yellow eels had lower serum Na^+ and Mg^{2+} concentrations, while freshwater silver eels had lower serum Na^+ and Cl^- concentrations and increased serum K^+ concentrations (Table XVII). Injection of cortisol lowered the serum Na^+,

K^+, and phosphate concentrations in the freshwater yellow eels and lowered serum Na^+ concentration in freshwater silver eels. Aldosterone lowered the serum K^+ concentrations in both yellow and silver freshwater eels. Injection of the adrenal inhibitor metyrapone plus betamethasone caused a lowering of the serum Na^+ concentrations. Adrenocorticotropic hormone (mammalian) had no effect on serum electrolyte concentrations even when injected at the rate of 2IU/ day for 10 days (Chan *et al.*, 1967). These authors draw attention to the fact that while there are similarities between the effects of adrenalectomy and stanniectomy there are also differences (Table XVII). The changes following adrenalectomy were interpreted by Chan *et* al. as being a result of the removal of cortisol and possibly also an unidentified mineralocorticoid. The reasons for the changes following stanniectomy, however, remain obscure (Chan *et al.*, 1967). Additional work on *Anguilla anguilla* has further confirmed that cortisol is probably the Na^+ excreting factor in this fish (Mayer *et al.*, 1967) (Table XVII).

Recently, Butler and Langford (1967) (Table XVII) showed that while adrenalectomized freshwater yellow eels with silk-wound closure had significantly lower serum Na^+ and Cl^- concentrations than intact control eels, fish with stainless steel wire-wound closure did not show serum electrolyte changes. These findings were interpreted as indicating that the silk-sutured fish had lost ions by diffusion owing to poor wound healing and that partial adrenalectomy as carried out in the experiment did not affect serum electrolyte composition. Since adrenalectomy by the above technique did not measurably lower the plasma cortisol concentration (Butler and Donaldson, 1967) the operation has been improved in such a way as to remove a greater portion of the interrenal tissue, and this we have found to result in plasma cortisol concentrations similar to those observed in hypophysectomized eels (Butler *et al.*, 1969a,b). Adrenalectomy by the improved technique resulted in a lower plasma Mg^{2+} concentration in freshwater silver eels when compared to sham-operated fish. There was no significant difference between the serum composition of adrenalectomized and sham-adrenalectomized freshwater yellow eels in the June experiment. In the July experiment there were significant decreases in the serum Na^+, Ca^{2+}, and osmotic concentrations of adrenalectomized freshwater yellow eels relative to the shamadrenalectomized fish (Butler *et al.*, 1969b) (Table XVII). These changes observed in the July experiment are explained by the fact that additional interrenal tissue was removed from these fish. Although the removal of further interrenal tissue did not result in a detectable decrease in the plasma cortisol concentration it may have lowered the cortisol concentration below a critical threshold for electrolyte homeostasis. Experiments by Chan *et al.* (1968) point to a role for both cortisol and prolactin in the endocrine control of the electrolyte composition of the freshwater European eel, *Anguilla anguilla*.

d. Order Salmoniformes. Early work dealing with the changes in blood chemistry which occur in migrating salmon has been reviewed by Black

(1951). A 2-year study of blood composition in the pink salmon, *Oncorhynchus nerka*, and chum salmon, *Oncorhynchus gorbushka*, of the Amur Estuary and Amgun River (Lysaya, 1951) showed the expected decrease in plasma Cl^- concentrations when male and female fish migrated from the estuary to the spawning grounds. The blood Ca^{2+} concentrations observed in this study, however, did not show the usual sex difference observed in other maturing fish (refer to Section V, C, 3, f below). The increase in the blood urea concentration during the anadromous migration of these two species was considerable (Table XVII) and remained unconfirmed prior to the recent work of Urist and Van de Putte (1967) on *Oncorhynchus tschawytscha* (Table XVII); however, the change was not as great in this latter study. Lysaya considered the increasing urea concentration during the anadromous migration to be a homeostatic mechanism which would tend to maintain a constant blood osmotic pressure in the face of the observed electrolyte depletion and that this uremia eventually leads to the death of the salmon. This thesis has yet to be substantiated. Idler and Tsuyuki (1958) in the course of their investigation of the changes in the blood of sockeye salmon, *Oncorhynchus nerka*, during its anadromous migration in the Fraser River system also measured plasma urea, but rather than finding an increase in urea concentration during the migration they actually observed a slight decline. These authors also noted a decline in the cholesterol, Mg^{2+}, and Ca^{2+} concentrations of the plasma as the fish migrated up the river (Table XVII). At Lummi Island at the start of the upstream migration the female fish had a higher plasma Ca^{2+} concentration, but the sex difference was not apparent during the migration, indicating that either the major mobilization of the calcium phosphoprotein occurred prior to the fluvial migration or that the rate of utilization of this material by the ovary during the migration was sufficient to prevent an increase in the plasma concentration.

Transfer of *Oncorhynchus nerka* at the end of their spawning migration up the Dalnee River back into full seawater or even isotonic or somewhat hypotonic media resulted in an increase in plasma osmolarity, loss of coordination and death after 4-5 hr (Zaks and Sokolova, 1961). In contrast, *Oncorhynchus nerka* collected at the end of their anadromous migration to Great Central Lake while showing poor survival when transferred directly to full seawater showed very good survival if the salinity was increased gradually (Donaldson, 1968). Retention of downstream migrating *Oncorhynchus nerka* in freshwater resulted in a slight decrease in plasma osmolarity after 5 days, while the plasma albumin concentration fell from its initial level of 8.7% to 5.7% after 24 days. These changes together with changes in kidney function were interpreted as indicating a tendency toward "freshwater" characteristics (Zaks and Sokolova, 1965). Rapid transfer of adult *Oncorhynchus nerka* from saltwater to freshwater resulted in a decline in plasma Na^+ concentration of 22 mmoles/ liter and osmolarity of 63 mOsm/ liter. As in the earlier study by these workers, transfer of fish at the end of their upstream migration back into seawater with a freezing-point depression greater than −0.6° resulted in high mortality (Zaks and Sokolova, 1965).

An investigation of physiological changes in the blood of migrating adult *Oncorhynchus tschawytscha* (Robertson *et al.*, 1961) (Table XVII) showed that plasma Na^+ and especially K^+ concentrations fell during the upstream migration. The hypokalemia was especially noticeable in the fall run fish. The plasma cholesterol concentration rose during the migratory period and then fell in the spawning fish to levels which were lower than the concentration in the sea salmon. In this study, unlike those of Lysaya (1951) and Urist and Van de Putte (1967), there was no increase in plasma urea during the anadromous migration. Studies of the inshore migration of *Oncorhynchus masou* showed an increase in the blood osmotic and Cl^- concentration in the late spring at the beginning of the inshore migration followed by a decrease as the fish approached the estuary (Kubo, 1960). As the fish migrated up the river there was a decline in the blood osmolar and Cl^- concentration followed by a temporary rise during preparation for spawning and finally a marked drop just prior to spawning (Kubo, 1961) (Table XVII). Kubo has also investigated the changes occurring in the blood during the downstream migration. At first it seemed that the osmotic pressure of the blood became higher after the parr–smolt transformation when compared to parr or dark parr fish (Kubo, 1953). Later studies on the blood of *Oncorhynchus masou*, however, showed that when the smolts are about to descend into the sea the osmotic pressure of the blood became lower again (Kubo, 1955). Examination of *Oncorhynchus tschawytscha* in the marine habitat showed that they had a blood chemistry similar to that of marine teleosts. On the other hand, specimens of these fish sampled in freshwater showed a noticeable hypokalemia which was related to the prolonged starvation which the fish had undergone (Urist and Van de Putte, 1967) (Table XVII). Urist also pointed out that the plasma Ca^{2+} in the female salmon fell from a high of 6.0 mmoles/ liter down to 2.5 mmoles/ liter when the ova were fully developed. Ionizing radiation was found to cause an increase in the plasma Cl^-, Na^+, and osmotic concentrations of juvenile coho salmon, *Oncorhynchus kisutch*, kept in saltwater but not in coho kept in freshwater. The irradiation appeared to have interfered with the extrarenal osmoregulatory systems (Conte, 1965) (Table XVII). The plasma osmotic concentration of premigratory juvenile *Oncorhynchus kisutch* increased from 295 to 400 mOsm/liter immediately after transfer to seawater but then tended to decline with time. In no case did the osmotic concentration exceed 415 mOsm/ liter. Migratory and postmigratory fish were also able to osmoregulate satisfactorily (Conte *et al.*, 1966). These findings for the coho were in marked contrast to those for the steelhead trout, *Salmo gairdnerii*, which showed a regression of osmoregulatory activity after the migration period had passed (Conte and Wagner, 1965).

The first fairly complete analysis of the plasma of a fish from the genus *Salmo* was carried out on *Salmo trutta* (Phillips and Brockway, 1958)

(Table XVII). In general the concentrations were similar to those determined in the carp by Field *et al.* (1943). A comparison of the osmoregulatory abilities of the brown trout and sea trout forms of *Salmo trutta* in the Aberdeen area indicated that they were virtually identical in this respect (Gordon, 1959a) (Table XVII) and also similar to American hatchery raised *Salmo trutta* (Gordon, 1959b). Plasma ionic concentrations were either unchanged or increased by less than 10% after the fish had been transferred from freshwater to seawater and acclimated (Gordon, 1959a) (Table XVII). A complete study of the major plasma ions at various points in the life cycle of the Atlantic salmon, *Salmo salar*, has been published by Parry (1961). There was an increase in the osmotic pressure of the plasma during development from the fry to the smolt stage in the life cycle. This was followed by a decline during the downstream migration of the smolt. This decrease in *Salmo salar* was not significant owing to the individual variability of the fish. Kubo (1955), however, attached more significance to his observed decrease in plasma osmolarity at this stage in *Oncorhynchus masou*. *Salmo salar* smolts which had been in seawater for 2 weeks had an elevated plasma osmotic concentration. Postsmolts which had been confined to freshwater for a year after they would normally have migrated also showed the same response. Adult prespawning fish in seawater had lower plasma osmotic pressures than the smolts in seawater and the osmotic concentration decreased further after entry into freshwater on the anadromous migration. Postspawned salmon (kelts) which had been in seawater for 2 days had the highest plasma osmotic pressure (Parry, 1961) (Table XVII). The hypokalemia noted in spawning *Oncorhynchus tschawytscha* (Urist and Van de Putte, 1967) was not present in the spawning *Salmo salar* (Table XVII) indicating that it may be related to the inevitable postspawning death of the Pacific salmon. Hickman *et al.* (1964) determined the inorganic composition of plasma from *Salmo gairdnerii* acclimated to 16° (Table XVII) and of trout transferred gradually to water at 6°. The most notable changes observed were the immediate significant declines in the plasma Cl^- and phosphate concentrations. Fromm (1963) has also reported on the inorganic composition of *Salmo gairdneri* plasma (Table XVII).

An investigation of the role of adrenocorticosteroids in the osmoregulatory processes of *Salmo gairdnerii* has been carried out by Holmes and his co-workers. The injection of corticosteroids into saline-loaded freshwater *Salmo gairdnerii* resulted in increases in extrarenal Na^+ excretion of 61% (deoxycorticosterone) and 89% (cortisol) (Holmes, 1959). Injection of cortisol, corticosterone, or aldosterone into nonsmolting freshwater rainbow trout resulted in a decrease in the plasma Na^+ concentration, the response to aldosterone being the most dramatic. Plasma K^+ concentrations showed an initial rise after the first injection followed by a decline. These findings and those of Chester-Jones (1956) on *Salmo trutta* are in contrast to the normal pattern of response to corticosteroids in the mammal, i.e., an increase in the Na–K ratio (Holmes and Butler, 1963). As a part of a study devoted to determining the glomerular filtration rate of *Salmo gairdnerii*, Holmes and McBean (1963) measured plasma Na^+ and K^+ concentrations in freshwater-adapted fish and in fish after they had been transferred to seawater. A significant increase in the plasma Na^+ concentration ($p < 0.001$) occurred, but no change in the K^+ concentration was observed (Table XVII). Although differences have been observed in the urine electrolyte composition of *Salmo gairdnerii* smolts, presmolts and postsmolts maintained in freshwater, no significant differences have been observed in their plasma electrolyte composition (Holmes and Stainer, 1966). At the present time we have a fairly consistent picture of the changes in blood chemistry which occur in members of the Salmonoidei during their downstream and anadromous migrations, but our knowledge of the hormonal control of osmoregulation in these fish is incomplete. For example, there is some evidence that *Salmo gairdnerii* is able to survive in freshwater after hypophysectomy (Donaldson and McBride, 1967) longer than many other fish (Schreibman and Kallman, 1966) indicating that prolactin may not be obligatory for the maintenance of plasma osmolarity in these fish. On the other hand, we have evidence for the presence of prolactin in the pituitary of *Oncorhynchus tschawytscha* collected in freshwater (Donaldson *et al.*, 1968), and it may be that the 3-month freshwater survival of some hypophysectomized *Salmo gairdnerii* was a result of only a gradual loss of Na^+ similar to that which occurs in the goldfish, which also has a long, but not indefinite, survival period in freshwater following hypophysectomy (Yamazaki and Donaldson, 1968) and in the eel (Maetz *et al.*, 1967).

e. Effect of Low Temperature on Plasma Constituents. The first experiment to determine the effect of cold on teleost plasma was that of Platner (1950) who observed a 159% increase in serum Mg^{2+} in goldfish after 60 hr at 0–1°C. Anoxic goldfish at the same temperature showed an increase in plasma Mg^{2+} concentration of 627%. The source of the increased Mg^{2+} was probably the red blood cells. Arctic teleosts living in deep water have plasma freezing points in the range $-0.91°$ to $-1.01°$. The water in which they live has a year round temperature of $-1.73°$, and the plasma is thus in a supercooled state. Arctic teleosts living in surface waters on the other hand have plasma freezing points of $-0.80°$ to $-0.81°$ in the summer and $-1.47°$ to $-1.50°$ in the winter (Scholander *et al.*, 1957) (Table XVIII). Thus the plasma osmotic pressure of the surface species, which are subject to ice crystal seeding, is raised in the winter to prevent the plasma from freezing, while those species living in deeper water are not subject to the occurrence of ice seeding. Concentrations of Na^+, K^+, and Cl^- were found to be elevated in the plasma of the cod, *Gadus morhua*, caught in water below 2° in the winter, but this was not observed in cod caught below this temperature during the summer months. The data were interpreted to indicate an osmotic imbalance in those fish living at low temperatures during the winter months (Woodhead and Woodhead, 1958, 1959). Furthermore, these authors proposed that the increased plasma osmotic concentration of the arctic surface species examined

Table XVIII Teleostei—the Effect of Low Temperature on Blood Chemistry

												Freezing point depression					
Fish	*Na*	*K*	*Mg*	*Ca*	*Cl*	*PO_4*	*NPN*	*Urea*	*Amino N*	*Protein (g %)*	*Units*	*Plasma*	*Seawater*	*Water temp.*	*Comments*	*Author*	*Date*
Carassius auratus	—	—	0.8	—	—	—	—	—	—	—	mmoles/ liter	—	—	20–25	Control	Platner	1950
Carassius auratus	—	—	3.0	—	—	—	—	—	—	—	—	—	—	0–1	Nonaerated	—	—
Gadassius auratus	—	—	1.5	—	—	—	—	—	—	—	—	—	—	0–l	Aerated	—	—
Carus ogac	—	—	—	—	—	—	—	—	—	—	—	1.47	—	−1.73	Winter, depth 2–8 meters	Scholander *el al.*	1957
Gadus ogac	—	—	—	—	—	—	—	—	—	—	—	0.80	—	4–7	Summer, depth 2–6 meters	—	—
Myxocephalus scorpius	—	—	—	—	—	—	—	—	—	—	—	1.50	—	−1.73	Winter, depth 2–8 meters	—	—
Myxocephalus scorpius	—	—	—	—	—	—	—	—	—	—	—	0.80	—	4–7	Summer, depth 2–6 meters	—	—
Salvelinus alpinus	—	—	—	—	—	—	—	—	—	—	—	0.81	—	4–7	Summer, depth 2–6 meters	—	—
Boreogadus saida	—	—	—	—	—	—	—	—	—	—	—	1.02	—	−1.73	Temperature constant depth 100–200 meters	—	—
Lycodes turneri	—	—	—	—	—	—	—	—	—	—	—	0.97	—	−1.73	100–200 meters	—	—
Liparis koefoedi	—	—	—	—	—	—	—	—	—	—	—	0.91	—	−1.73	100–200 meters	—	—

Gymnacanthus tricuspis	—	—	—	—	—	—	—	—	—	—	—	0.93	—	−1.73	100–200 meters	—	—
Icelus spatula	—	—	—	—	—	—	—	—	—	—	—	0.96	—	−1.73	100–200 meters	—	—
Gadus callarias	—	—	—	—	188	—	—	—	—	—	mmoles/kg	0.797	—	−1.5	—	Eliassen *et al.*	1960
Gadus callarias	—	—	—	—	178	—	—	—	—	—	—	0.693	—	4.5	—	—	—
Gadius callarias	—	—	—	—	162	—	—	—	—	—	—	0.64	—	15.0	—	—	—
Cyclopterus lumpus	—	—	—	—	218	—	—	—	—	—	—	0.88	—	−1.5	—	—	—
Cyclopterus lumpus	—	—	—	—	165	—	—	—	—	—	—	0.57	—	10.0	—	—	—
Anarhichas minor	—	—	—	—	195	—	—	—	—	—	—	0.80	—	−1.5	—	—	—
Drepanopsetta platessoides	—	—	—	—	210	—	—	—	—	—	—	0.93	—	−1.5	—	—	—
Pseudopleuronectes americana	—	—	—	—	—	—	—	—	—	—	—	1.15	—	—	Winter, [NaCl] = 57% of osmotic pressure	Pearcy	1961
Pseudopleuronectes americana	—	—	—	—	—	—	—	—	—	—	—	0.63	—	—	Summer, [NaCl] = 83% of osmotic pressure	—	—
Myxocephalus scorpius	216	4.3			234	0.55	92.5	14.3	10.0	—	mEq/liter	1.25[a]	−1.75	−1.7	Labrador spring	Gordon *et al.*	1962
Myxocephalus scorpius	276	6.4	—	—	184	—	121.5	—	—	—	—	0.80[a]	—	—	New Brunswick spring	—	—
Gadus ogac	216	5.5	—	—	243	0.72	289	24.9	17.8	—	—	0.94[a]	—	—	Labrador spring	—	—
Micragadus tomcod	246	8.3	—	—	166	—	92.5	—	—	—	—	0.98[a]	—	—	New Brunswick spring	—	—

Continued

Table XVIII Teleostei—the Effect of Low Temperature on Blood Chemistry—Cont'd

												Freezing point depression					
Fish	*Na*	*K*	*Mg*	*Ca*	*Cl*	*PO_4*	*NPN*	*Urea*	*Amino N*	*Protein (g %)*	*Units*	*Plasma*	*Seawater*	*Water temp.*	*Comments*	*Author*	*Date*
Microgadus tomcod	231	5.1	—		142	—	71.5	—	—	—	—	0.82[a]	—	—	New Brunswick spring	—	—
Carassius auratus	—	—	—	—	123	—	—	—	—	—	mmoles/ liter	—	—	21	—	Houston	1962
Carassius auratus	—	—	—	—	90	—	—	—	—	—	—	—	—	3	800 min immersion	—	—
Cyprinus carpio	127.9	2.74	1.19	1.94	115.6	0.39	—	—	—	1.8	—	0.515[a]	—	4	3 weeks immersion	Houston and Madden	1968

[a]Derived from osmotic pressure.

by Scholander *et al.* (1957) may also have been in a state of osmotic imbalance. *Gadus callarias* caught at −1.5° in the Bering Sea had a plasma freezing point of 0.797°, a value which, although higher than the freezing point of plasma in these fish at 4.5°, still indicated a considerable degree of supercooling similar to that found by Scholander *et al.* (1957) (Eliassen *et al.*, 1960) (Table XVIII). The plasma freezing point of *Cottus scorpius* was reduced from −0.60° at a water temperature of 10° to −0.80° to −0.90° after being kept 6−8 weeks at −1.5° regardless of the season (Eliassen *et al.*, 1960). These authors question the theory that there was a definite temperature threshold below which the fish enter a state of osmotic imbalance and regarded the uniform response to cold which they observed in each group of fish as indicative of an adjustment to a new state of balance (Eliassen *et al.*, 1960).

The winter flounder, *Pseudopleuronectes americanus*, also showed a lowered serum freezing point in cold water and, like the cod, Cl^- accounted for a lower percentage of the total plasma osmolarity during the winter months than during the summer months (Pearcy, 1961) (Table XVIII). Thus, organic components in addition to inorganic components may be responsible for the lowering of plasma freezing point at low temperatures. With this in mind Gordon *et al.* (1962) returned to Hebron Fjord (Scholander *et al.*, 1957) to reinvestigate the problem. After examining many plasma components (Table XVII), these workers concluded that the antifreeze components of *Gadus ogac* may be part of the nonprotein nitrogen (NPN) fraction, but this was not the case for the sculpin, *Myxocephalus scorpius.*

f. Calcium in the Female Teleost. The first data showing a sex difference for plasma calcium concentration in the teleost were provided by Hess *et al.* (1928). These workers found a range of 9–12.5 mg% in the male cod and 12.7–29 mg% in the female cod. The females showing the highest concentrations were those with large mature gonads. A similar situation was found in the puffer fish. Pora (1935, 1936e) examined the serum of male and female *Cyprinus carpio* and *Labrus bergylta*, confirmed the findings of Hess *et al.*, and in addition noted a somewhat elevated protein concentration in the female. Estrogen treatment of male or female goldfish resulted in an increased serum concentration of nonultrafilterable calcium and phosphorus and in total protein. Some of the increased calcium was present as colloidal calcium phosphate, the remainder was bound to the serum phosphoprotein, vitellin (Bailey, 1957). Estrogenization of maturing sockeye salmon, *Oncorhynchus nerka*, resulted in increased serum levels of calcium, protein phosphorous, and total protein. Untreated maturing female salmon had higher serum concentrations of calcium and protein phosphorous than maturing male salmon (Ho and Vanstone, 1961). Female brook trout, *Salvelinus fontinalis*, examined between July and May showed a markedly increased total serum protein and calcium concentration during the spawning period of October to November. Male *Salvelinus* did not show similar changes (Booke, 1964). Oguri and Takada (1966) noted a fivefold increase in the serum calcium concentration of the snake headfish, *Channa argus*, after injection of estradiol. They also noted an increase in serum calcium in the goldfish after injection of estradiol and were able to relate the serum calcium concentration in normal female goldfish to the gonadosomatic index (Oguri and Takada, 1967). The target tissue in the maturing female fish which responds to estrogen is the liver which synthesizes a calcium phosphoprotein–phospholipid glycolipoprotein complex which is then transported via the bloodstream to the developing ova (Urist, 1964).

g. Chemistry of Other Body Fluids. i. Perivisceral fuid. The chemical composition of this fluid in the Osteichthyes appears only to have been examined by Smith (1929b, 1930). In *Lophius piscatorius* and *Gadus callarias* there were higher carbon dioxide concentrations in perivisceral fluid than in plasma indicating a greater alkalinity. In the elasmobranchs the reverse was true. Other ions were present in similar concentrations to those in the plasma, and only a trace of protein was measured. In the lungfish, *Protopterus aethiopicus*, Smith (1930) reported no differences between plasma and perivisceral fluid (Table XIX).

ii. Pericardial fluid. This fluid was also found to have a higher carbon dioxide content than plasma by Smith (1929b) (Table XIX). The ionic composition of the pericardial fluid of *Lophius americanus* showed no differences from that of plasma which cannot be explained in terms of Gibbs-Donnan effect and experimental error (Forster and Berglund, 1956) (Table XIX).

iii. Cranial fluid. This fluid which is collected from the brain case was found to be similar in ionic composition to the plasma in *Lophius piscatorius* by Smith (1929b) (Table XIX). Enger (1964) examined the cranial fluid of three species of teleost and compared it to plasma and endolymph. The cranial fluid was similar in composition to the plasma in the two marine teleosts examined, but in *Salmo irideus* there was a considerably higher K^+ concentration in the cranial fluid (Table XIX). The composition of the cerebrospinal fluid does not appear to have been investigated in the Osteichthyes.

iv. Endolymph. This fluid has been isolated from the ear and analyzed in three teleosts. In all cases the fluid showed the characteristically high K^+ concentration found throughout the vertebrates. The endolymph to cranial fluid ratio for Na^+ in these teleosts was 0.7 and for Cl^- was 1.0 , whereas in the elasmobranchs examined the ratios were 1.05 and 1.3, respectively (Enger, 1964) (Tables XV and XVIII).

v. Eye fluids. The vitreous body of *Lophius piscatorius* was found to have a lower concentration of Cl^- than that of plasma by Derrien (1937), while Davson and Grant (1960) reported that a combination of aqueous and vitreous humor is hypoosmotic with respect to plasma in both *Stenotomus versicolor* and *Tautoga onitis* (Table XIX). These findings have recently been substantiated by Hoffert and Fromm (1966) who found both the osmotic and Cl^- concentrations were lower in the aqueous humor. Treatment with the carbonic anhydrase inhibitor Diamox increased the Cl^- and lowered the total carbon dioxide and HCO_3^- concentrations in the aqueous humor, but the exact role

Table XIX Osteichthyes—Chemistry of Other Body Fluids

Fish and fluid	Na	K	Mg	Ca	Cl	HCO_3	CO_2	PO_4	SO_4	Urea	TMAO and TMA	Protein (g %)	Total ions	Units	Osmotic pressure (mOsm/ liter)	Comments	Author	Date
Lophius piscatorius	—	—	—	—	—	—	—	—	—	—	—	—	—	mmoles/ liter	—	—	Smith	1929b
Spinal fluid	194	6.4	6.0	2.7	184	—	7.2	4.7	3.1	—	—	—	—	mmoles/ liter	—	—	—	—
Pericardial fluid	197	5.9	8.2	2.6	197	—	9.0	—	3.2	—	—	—	—	mmoles/ liter	—	—	—	—
Perivisceral fluid	195	6.1	6.8	3.0	192	—	8.8	4.4	4.1	—	—	—	—	mmoles/ liter	—	—	—	—
Gadus callarias	—	—	—	—	—	—	—	—	—	—	—	—	—	mmoles/ liter	—	—	—	—
Perivisceral fluid	165	4.6	3.1	5.1	151	—	11.0	3.3	—	—	—	—	—	mmoles/ liter	—	—	—	—
Pericardial fluid	163	4.7	1.0	5.1	169	—	7.2	—	—	—	—	—	—	mmoles/ liter	—	—	—	—
Protopterus aethiopicus	—	—	—	—	—	—	—	—	—	—	—	—	—	mmoles/ liter	—	—	Smith	1930
Perivisceral fluid	91.0	4.2	Trace	2.1	46	—	36.0	1.6	—	—	—	—	—	mmoles/ liter	222	—	—	—
Pericardial fluid	97.0	3.2	Trace	1.9	34	—	—	—	—	—	—	—	—	—	208	—	—	—
Lophius piscatorius	—	—	—	—	—	—	—	—	—	—	—	—	—	—	—	—	Derrin	1937

Continued

Table XIX Osteichthyes—Chemistry of Other Body Fluids—Cont'd

Fish and fluid	Na	K	Mg	Ca	Cl	HCO_3	CO_2	PO_4	SO_4	Urea	TMAO and TMA	Protein (g %)	Total ions	Units	Osmotic pressure (mOsm/ liter)	Comments	Author	Date
Vitreous body	—	—	—	—	152	—	—	—	—	—	—	—	—	mmoles/ liter	—	Plasma Cl = 183	—	—
Lophius americanus	—	—	—	—	—	—	—	—	—	—	—	—	—	mmoles/ liter	—	—	Forster and Berglund	1956
Pericardial fluid	200	4.3	1.7	2.3	187	—	—	4.5	1.1	—	—	—	—	—	—	Mean of Nos. 1,9, and 15	—	—
Stenotomus versicolor	—	—	—	—	—	—	—	—	—	—	—	—	—	—	—	—	Davson and Grant	1960
Aqueous and vitreous humor	—	—	—	—	—	—	—	—	—	—	—	—	—	—	364	Plasma osmotic pressure – 397	—	—
Tautoga onitis	—	—	—	—	—	—	—	—	—	—	—	—	—	—	—	—	—	—
Aqueous and vitreous humor	—	—	—	—	—	—	—	—	—	—	—	—	—	—	321	Plasma osmotic pressure = 360	—	—
Gadus morhua	—	—	—	—	—	—	—	—	—	—	—	—	—	mmoles/ liter	—	—	Enger	1964
Cranial fluid	193	3.2	—	—	167	—	—	—	—	—	—	—	—	—	—	—	—	—
Endolymph	136	72.9	—	—	166	—	—	—	—	—	—	—	—	—	—	—	—	—

Cottus scorpius	—	—	—	—	—	—	—	—	—	—	—	—	—	—	—	—	—	—
Cranial fluid	191	3.6	—	—	176	—	—	—	—	—	—	—	—	—	—	—	—	—
Endolymph	152	50.8	—	—	177	—	—	—	—	—	—	—	—	—	—	—	—	—
Salmo irrideus	—	—	—	—	—	—	—	—	—	—	—	—	—	—	—	—	—	—
Cranial fluid	151	5.0	—	—	143	—	—	—	—	—	—	—	—	—	—	—	—	—
Endolymph	104	72.9	—	—	—	—	—	—	—	—	—	—	—	—	—	—	—	—
Salvelinus namaycush	—	—	—	—	—	—	—	—	—	—	—	—	—	—	—	—	Hoffert and Fromm	1966
Aqueous humor	—	—	—	—	93.5	6.19	6.4	—	—	—	—	—	—	mmoles/ liter	244	Control	—	—
Aqueous humor	—	—	—	—	127.4	4.79	4.96	—	—	—	—	—	—	—	—	Diamox	—	—
Latimeria chalumnae	—	—	—	—	—	—	—	—	—	—	—	—	—	—	—	—	Pickford and Grant	1967
Aqueous humor	—	—	3.7	—	—	—	—	—	—	303	—	—	—	mmoles/ liter	952	—	—	—

of carbonic anhydrase in aqueous humor formation in the teleost remains to be clarified (Hoffert and Fromm, 1966; Maren, 1967b). The aqueous humor of the coelacanth, *Latimeria* chalumnae, appears to have a lower urea concentration and a higher osmolarity than heart blood from the same animal (Pickford and Grant, 1967) (Table XIX).

REFERENCES

Armin, J., Grant, R. T., Pels, H., and Reeve, E. B. (1952). The plasma cell and blood volumes of albino rabbits as estimated by the dye T 1824 and P-32 marked cell methods. *J. Physiol. (London)* **116**, 59–73.

Bailey, R. E. (1957). The effect of estradiol on serum calcium, phosphorus, and protein of goldfish. *J. Exptl. Zool.* **136**, 455–469.

Barlow, J. S., and Manery, J. F. (1954). The changes in electrolytes, particularly chloride, which accompany growth in chick muscle. *J. Cellular Comp. Physiol.* **43**, 165.

Becker, E. L., Bird, R., Kelly, J. W., Schilling, J., Solomon, S., and Young, N., (1958). Physiology of marine teleosts. I. Ionic composition of tissue. *Physiol. Zool.* **31**, 224–227.

Bellamy, D., and Chester-Jones, I. (1961). Studies on *Myxine glutinosa*. I. The chemical composition of the tissues. *Comp. Biochem. Physiol.* **3**, 175–183.

Bentley, P. J., and Follet, B. K. (1963). Kidney function in a primitive vertebrate, the cyclostome *Lampetra fluviatilis*. *J. Physiol. (London)* **169**, 902–918.

Bernard, G. R., Wynn, R. A., and Wynn, G. G. (1966). Chemical anatomy of the pericardial and perivisceral fluids of the stingray *Dasyatis americana*. *Biol. Bull.* **130**,18–27.

Black, V. S. (1951). Some aspects of the physiology of fish. II. Osmotic regulation in teleost fishes. *Publ. Ontario Fisheries Res. Lab.* **71**, 53–89.

Bloete, M., Naumann, D. C., Frazier, H. S., Leaf, A., and Stone, W., Jr. (1961). Comparison of the osmotic activity of ocular fluids with that of arterial plasma in the *dogfish*. *Biol. Bull.* **121**, 383–384.

Bond, R. M., Cary, M. K., and Hutchinson, G. E. (1932). A note on the blood of *the hagfish Polistotrema stouti* (Lockington). *J. Exptl. Biol.* **9**, 12–14.

Booke, H. E. (1964). Blood serum protein and calcium levels in yearling brook trout. *Progressive Fish Culturist* **26**, 107–110.

Borei, H. (1935). Uber die Zusammensetzung der Korperflussigkeiten von *Myxine glutinosa* L. (Composition of blood of cyclostome, Myxine.), *Arkiv Zool.* **28B**, 1–5.

Bottazzi, F. (1906). Sulla regulazione della pressione osmotica negli organismi animali. *Arch. Fisiol.* **3**, 416–446.

Boucher-Firley, S. (1934). Sur quelques constituants chimiques du san de Congre et de Murene. *Bull. Inst. Oceanog.* **651**, 1–6.

Braunitzer, G., Hilse, K., Rudloff, V., and Hilschmann, N. (1964). *Advan. Protein Chem.* **19**, 1–65.

Brodie, B. B. (1951). *Methods Med. Res.* **4**, 31.

Brown, G. W., and Brown, S. G. (1967). Urea and its formation in Coelacanth liver. *Science* **155**, 570–573.

Brull, L., and Cuypers, Y. (1954). Quelques characteristiques biologiques de, *Lophius piscatorius L. Arch. Intern. Physiol.* **62**, 70–75.

Brull, L., and Nizet, E. (1953). Blood and urine constituents of *Lophius piscatorius*. *J. Marine Biol. Assoc. U. K.* **32**, 321–328.

Bull, J. M., and Morris, R. (1967). Studies on fresh water osmoregulation in the ammocoete larva of *Lampetra planeri* (Bloch). I. Ionic constituents, fluid compartments, ionic compartments and water balance. *J. Exptl. Biol.* **47**, 485–494.

Burger, J. W. (1965). Roles of the rectal gland and the kidneys in salt and water excretion in the spiny dogfish. *Physiol. Zool.* **38**, 191–196.

Burian, R. (1910). Funktion der Nierenglomeruli und Ultrafiltration. *Arch. Ges. Physiol.* **136**, 741–760.

Butler, D. G. (1966). Effect of hypophysectomy on osmoregulation in the European eel (*Anguilla anguilla* L.). *Comp. Biochem. Physiol.* **18**, 773–781.

Butler, D. G. (1967). Effect of ovine prolactin on tissue electrolyte composition of hypophysectomized fresh water eels *(Anguilla rostrata). J. Fisheries Res. Board Can.* **24**, 1823–1826.

Butler, D. G., and Donaldson, E. M. (1967). Unpublished data.

Butler, D. G., and Langford, R. W. (1967). Tissue electrolyte composition of the fresh water eel (*Anguilla rostrata*) following partial surgical adrenalectomy. *Comp. Biochem. Physiol.* **22**, 309–312.

Butler, D. G., Donaldson, E. M., and Clarke, W. C. (1969a). Physiological evidence for a pituitary adrenocortical feedback mechanism in the eel (*Anguilla rostrata*). *Gen. Comp. Endocrinol.* **12**, 173–176.

Butler, D. G., Clarke, W. C., Donaldson, E. M., and Langford, R. W. (1969b). Surgical adrenalectomy of a teleost fish (*Anguilla rostrata Lesueur*): Effect on plasma cortisol and tissue electrolyte and carbohydrate concentrations. *Gen. Comp. Endocrinol.* **12**, 502–514.

Chaisson, A. F., and Freidman, M. H. F. (1935). The effect of histamine, adrenaline and destruction of the spinal cord on the osmotic pressure of the blood in the skate. *Proc. Nova Scotian Inst. Sci.* **18**, 240–244.

Chan, D. K. O., Chester-Jones, I., Henderson, I. W., and Rankin, J. C. (1967). Studies on the experimental alteration of water and electrolyte composition of the eel (*Anguilla anguilla* L.). *J. Endocrinol.* **37**, 297–317.

Chan, D. K. O., Chester-Jones, I., and Mosley, W. (1968). Pituitary and adrenocortical factors in the control of the water and electrolyte composition of the freshwater European eel (*Anguilla anguilla* L.). *J. Endocrinol.* **42**, 91–98.

Chester-Jones, I. (1956). The role of the adrenal cortex in the control of water and salt electrolyte metabolism in vertebrates. *Mem. Soc. Endocrinol.* **5**, 102–119.

Chester-Jones, I., Phillips, J. G., and Bellamy, D. (1962). Studies on water and electrolytes in cyclostomes and teleosts with special reference to *Myxine glutinosa* L. (the Hagfish) and *Anguilla anguilla* L., (the Atlantic Eel). *Gen. Comp. Endocrinol.* Suppl. **1**, 36–47.

Chester-Jones, I., Henderson, I. W., and Butler, D. G. (1965). Water and electrolyte flux in the European eel (*Anguilla anguilla*). *Arch. Anat. Microscop. Morphol. Exptl.* **54**, 453–468.

Cohen, J. J., Krupp, M. A., and Chidsey, C. A., III (1958). Renal conservation of trimethylamine oxide by the spiny dogfish, *Squalus acanthias. Am. J. Physiol.* **194**, 229–235.

Cole, W. H. (1940). The composition of fluids and sera of some marine animals and of the sea water in which they live. *J. Gen. Physiol.* **23**, 575–584.

Conte, F. P. (1965). Effects of ionizing radiation on osmoregulation in fish *Oncorhynchus kisutch. Comp. Biochem. Physiol.* **15**, 292–302.

Conte, F. P., and Wagner, H. H. (1965). Development of osmotic and ionic regulation in juvenile steelhead trout *Salmo gairdneri. Comp. Biochem. Physiol.* **14**, 603–620.

Conte, F. P., Wagner, H. H., Fessler, J., and Gnose, C. (1966). Development of osmotic and ionic regulation in juvenile coho salmon *Oncorhynchus kisutch. Comp. Biochem. Physiol.* **18**, 1–15.

Cotlove, E., and Hogben, C. A. M. (1962). *Mineral Metab.* **2**, Part B, 109–173.

Cserr, H., and Rall, D. P. (1967). Regulation of cerebrospinal fluid [K^+] in the spiny dogfish *Squalus acanthias*. *Comp. Biochem. Physiol.* **21**, 431–434.

Davson, H., and Grant, C. T. (1960). Osmolarities of some body fluids in the elasmobranch and teleost. *Biol. Bull.* **119**, 293.

Dekhuyzen, M. C. (1904). Ergebnisse von osmotischen Studien, nametlich bei Knochenfischen, und der biologischen Station des Bergenser Museums. *Bergens Museums Arbok. Naturv. Rekke* **8**, 3–7.

Derrickson, M. B., and Amberson, W. R. (1934). Determination of blood volume in the lower vertebrates by direct method. *Biol. Bull.* **67**, 329 (abstr.).

Derrien, Y. (1937). Repartition du chlorure de sodium et du glucose entre le plasma sanguin et le corps vitre de *Lophius piscatorium* L. *Compt. Rend. Soc. Biol.* **126**, 943–945.

Donaldson, E. M. (1968). Unpublished observations.

Donaldson, E. M., and McBride, J. R. (1967). The effects of hypophysectomy in the rainbow trout *Salmo gairdnerii* (Rich.) with special reference to the Pituitary Interrenal axis. *Gen. Comp. Endocrinol.* **9**, 93–101.

Donaldson, E. M., Yamazaki, F, and Clarke, W. C. (1968). Effect of hypophysectomy on plasma osmolarity in Goldfish and its reversal by ovine prolactin and a preparation of salmon pituitary "Prolactin." *J. Fisheries Res. Board Can.* **25**, 1497–1500.

Doolittle, R. F., Thomas, C., and Stone, W., Jr. (1960). Osmotic pressure and aqueous humor formation in dogfish. *Science* **132**, 36–37.

Edelman, I. S., Young, H. L., and Harris, J. B. (1960). Effects of corticosteroids on electrolyte metabolism during osmoregulation in teleosts. *Am. J. Physiol.* **199**, 666–670.

Edwards, J. G., and Condorelli, L. (1928). Electrolytes in blood and urine of fish. *Am. J. Physiol.* **86**, 383–398.

Eisler, R., and Edmunds, P. H. (1966). Effects of endrin on blood and tissue chemistry of a marine fish. *Trans. Am. Fisheries Soc.* **95**, 153–159.

Eliassen, E., Leivestad, H., and Moller, D. (1960). Effect of low temperature on the freezing point of plasma and on the potassium: Sodium ratio in the muscle of some boreal and subartic fishes. *Arbok. Univ. Bergen, Mat.-Nat. Ser.* No. **14**, 1–24.

Elkinton, J. R., and Danowski, T. S. (1955). "The Body Fluids." Williams and Wilkins, Baltimore, Maryland.

Enger, P. S. (1964). Ionic composition of the cranial and labyrinthine fluids and saccular D.C. potentials in fish. *Comp. Biochem. Physiol.* **11**, 131–137.

Evans, D. H. (1967a). Sodium, chloride and water balance of the intertidal teleost, *Xiphister atropurpureus*. 1. Regulation of plasma concentration and body water content. *J. Exptl. Biol.* **47**, 513–518.

Evans, D. H. (1967b). Sodium, chloride and water balance of the intertidal teleost, *Xiphister atropurpureus*. 2. The role of the kidney and the gut. *J. Exptl. Biol.* **47**, 519–534.

Fänge, R., and Fugelli, K. (1962). Osmoregulation in chimaeroid fishes. *Nature* **196**, 689.

Fellers, F. X., Barnett, H. L., Hare, K., and McNamara, H. (1949). Changes in thiocyanate and $sodium^{24}$ spaces during growth. *Pediatrics* **3**, 622.

Field, J. B., Elveljem, C. A., and Juday, C. (1943). A study of blood constituents of carp and trout. *J. Biol. Chem.* **148**, 261–269.

Fleming, W. R., and Stanley, J. G. (1965). Effects of rapid changes in salinity on the renal function of a euryhaline teleost. *Am. J. Physiol.* **209**, 1025–1030.

Fontaine, M. (1930). Recherches sur le milieu intérieur de la lamproie marine (*Petromyzon marinus*). Ses variations en fonction de celles du milieu extérieur. *Compt. Rend.* **191**, 680–682.

Fontaine, M. (1964). Corpuscules de Stannius et regulation ionique (Ca,K,Na) du milieu intérieur de l'Anguille (*Anguilla anguilla* L.). *Compt. Rend.* **259**, 875–878.

Forster, R. P, and Berglund, F. (1956). Osmotic diuresis and its effect on total electrolyte distribution in plasma and urine of the aglomerular teleost, *Lophius americanus. J. Gen. Physiol.* **39**, 349–359.

Fredericq, L. (1904). Sur la concentration moléculaire du sang et des tissues chez les animaux aquatiques. *Arch. Biol. (Liege)* **20**, 709–730.

Fromm, P. O. (1963). Studies on renal and extra-renal excretion in a freshwater teleost, *Salmo gairdneri. Comp. Biochem. Physiol.* **10,** 121–128.

Galloway, T. M. (1933). Osmotic concentration of blood of Cyclostome *Petromyzon. J. Exptl. Biol.* **10**, 313–316.

Garrey, W. E. (1905). The osmotic pressure of sea water and of the blood of marine animals. *Biol. Bull.* **8**, 257–270.

Gordon, M. S. (1959a). Osmotic and ionic regulation in Scottish brown trout and sea trout (*Salmo trutta* L.). *J. Exptl. Biol.* **36**, 253–260.

Gordon, M. S. (1959b). Ionic regulation in the brown trout (*Salmo trutta* L.). *J. Exptl. Biol.* **36**, 227–252.

Gordon, M. S., Amdur, B. H., and Scholander, P. F. (1962). Freezing resistance in some northern fishes. Biol. Bull. **122**, 52–56.

Gordon, M. S., Boettius, J., Boettius, I., Evans, D. H., McCarthy, R., and Oglesby, L. C. (1965). Salinity adaptation in the mudskipper fish *Periophthalmus sobrinus. Hvalradets Skrifter Norske Videnskaps. Akad. Oslo* **48**, 85–93.

Gray, S. J., and Sterling, K. (1950). The tagging of red cells and plasma proteins with radioactive chromium. *J. Clin. Invest.* **29**, 1604–1613.

Greene, C. W. (1904). Physiological studies of the chinook salmon. *U.S. Bur. Fisheries, Bull.* **24**, 431–456.

Gregersen, M. I., and Rawson, R. A. (1959). Blood volume. *Physiol. Rev.* **39**, 307–342.

Hardisty, M. W. (1956). Some aspects of osmotic regulation in lampreys. *J. Exptl. Biol.* **33**, 431–447.

Hartman, F. A., Lewis, L. A., Brownell, K. A., Sheldon, F. F., and Walther, R. F. (1941). Some blood constituents of the normal skate. *Physiol. Zool.* **14**, 476–486.

Hartman, F. A., Lewis, L. A., Brownell, K. A., Angerer, K. A., and Sheldon, F. F. (1944). Effect of interrenalectomy on some blood constituents in the skate. *Physiol. Zool.* **17**, 228–238.

Herre, A. W. C. T. (1955). Sharks in fresh water. *Science* **122**, 417.

Hess, A. F., Bills, C. E., Weinstock, M., and Rivkin, H. (1928). Difference in calcium level of the blood between the male and female cod. *Proc. Soc. Exptl. Biol. Med.* **25,** 349–350.

Hickman, C. P., Jr. (1959). The osmoregulatory role of the thyroid gland in the starry flounder, *Platichthys stellatus. Can. J. Zool.* **37**, 997–1060.

Hickman, C. P., Jr., McNabb, R. A., Nelson, J. S., Van Breemen, E. O., and Comfort, D. (1964). Effect of cold acclimation on electrolyte distribution in rainbow trout (*Salmo gairdnerii*). *Can. J. Zool.* **42**, 577–597.

Ho, F. C. W., and Vanstone, W. E. (1961). Effect of estradiol monobenzoate on some serum constituents of maturing sockeye salmon (*Oncorhynchus nerka*). *J. Fisheries Res. Board Can.* **18**, 859–864.

Hoffert, J. R., and Fromm, P. O. (1966). Effect of carbonic anhydrase inhibition on aqueous humor and blood bicarbonate ion in the teleost *(Salvelinus namaycush*). *Comp. Biochem. Physiol.* **18**, 333–340.

Holmes, R. M. (1961). Kidney function in migrating salmonids. *Rept. Challenger Soc. (Cambridge)* **3**, No, **13**, 23.

Holmes, W. N. (1959). Studies on the hormonal control of sodium metabolism in the rainbow trout (*Salmo gairdnerii*). *Acta Endocrinol.* **31**, 587–602.

Holmes, W. N., and Butler, D. G. (1963). The effect of adrenocortical steroids on the tissue electrolyte composition of the fresh water rainbow trout (*Salmo gairdneri*). *J. Endocrinol.* **25**, 457–464.

Holmes, W. N., and McBean, R. L. (1963). Studies on the glomerular filtration rate of rainbow trout (*Salmo gairdneri*). *J. Exptl. Biol.* **40**, 335–341.

Holmes, W. N., and Stainer, I. M. (1966). Studies on the renal excretion of electrolytes by the trout (*Salmo gairdneri*). *J. Exptl. Biol.* **44**, 33–46.

Holmes, W.N. Unpublished data.

House, C. R. (1963). Osmotic regulation in the brackish water teleost, *Blennius pholis*. *J. Exptl. Biol.* **40**, 87–104.

Houston, A. H. (1959). Osmoregulatory adaptation of steelhead trout (*Salmo gairdneri* Richardson) to sea water. *Can. J. Zool.* **37**, 729–748.

Houston, A. H. (1960). Variations in the plasma level of chloride in hatchery reared yearling Atlantic salmon during parr–smolt transformation and following transfer into sea water. *Nature* **185**, 632–633.

Houston, A. H. (1962). Some observations on water balance in the goldfish *Carassius auratus* L. during cold death. *Can. J. Zool.* **40**, 1169–1174.

Houston, A. H., and Madden, J. A. (1968). Environmental temperature and plasma electrolyte regulation in the carp *Cyprinus carpio*. *Nature* **217**, 969–970.

Houston, A. H., and Threadgold, L. T. (1963). Body fluid regulation in smolting Atlantic salmon. *J. Fisheries Res. Board Can.* **20**, 1355–1367.

Huggel, H., Kleinhaus, A., and Hamzehpour, M. (1963). The blood composition of *Salmo gairdneri irideus* and *Squalius cephalus* (Teleostei, Pisces). *Rev. Suisse Zool.* **70**, 286–290.

Huntsman, A. G., and Hoar, W. S. (1939). Resistance of Atlantic salmon sea water. *J. Fisheries Res. Board Can.* **4**, 409.

Idler, D. R., and Tsuyuki, H. (1958). Biochemical studies on sockeye salmon during spawning migration. I. Physical measurements, plasma cholesterol, and electrolyte levels. *Can. J. Biochem. Physiol.* **36**, 783–791.

Janssens, P. A. (1964). The metabolism of the aestivating African lungfish. *Comp. Biochem. Physiol.* **11**, 105–117.

Jensen, C. E., and Vilstrup, T. (1954). Determination of some inorganic substances in the labyrinthine fluid. *Acta Chem. Scand.* **8**, 697–698.

Johansen, K., Fänge, R., and Johannessen, M. W. (1962). Relations between blood, sinus fluid and lymph in *Myxine glutinosa* L. *Comp. Biochem. Physiol.* **7**, 23–28.

Kaieda, J. (1930). Biochemische Untersuchungen des Labyrinthwassers und der cerebrospinal Flussigkeit der Haifische. *Z. Physiol. Chem.* **188**, 193–202.

Keys, A., and Hill, R. M. (1934). The osmotic pressure of the colloids in fish sera. *J. Exptl. Biol.* **11**, 28–33.

Korjuev, P. A. (1938). Urea and chlorides of the blood of sea ganoids. *Bull. Biol. Med. Exptl. URSS* 6, 158–159.

Kubo, T. (1953). On the blood of salmonid fishes of Japan during migration. I. Freezing point of blood. *Bull. Fac. Fisheries, Hokkaido Univ.* 4, 138–148.

Kubo, T. (1955). Changes of some characteristics of the blood of the smolts of Oncorhynchus masou during seaward migration. *Bull. Fac. Fisheries, Hokkaido Univ.* 6, 201–207.

Kubo, T. (1960). Notes on the blood of Masou salmon during inshore migration with special reference to the osmoconcentration. *Bull. Fac. Fisheries, Hokkaido Univ.* **11**,15–19.

Kubo, T. (1961). Notes on the blood of Masou salmon (*Oncorhynchus masou*) during upstream migration for spawning with special reference to the osmoconcentration. *Bull. Fac. Fisheries, Hokkaido Univ.* **12**, 189–195.

Lahlou, B. (1967). Excretion renale chez un poisson euryhaline, le flet (*Platichthys flexus* L.): Characteristiques de l'urine normale en eau douce et en eau de mer et effects des changements de milieu. *Comp. Biochem. Physiol*, **20**, 925–938.

Lange, R., and Fugelli, K. (1965). The osmotic adjustment in the euryhaline teleosts, the flounder, *Pleuronectes flesus* L. and the three-spined stickleback, Gasterosteus aculeatus L. *Comp. Biochem. Physiol.* **15**, 283–292.

Lehninger, A. L. (1965). "The Mitochondrion." Benjamin, New York.

Lehninger, A. L., Rossi, C. S., and Greenwalt, J. W. (1963). Respiration-dependent accumulation of inorganic phosphate and Ca^{++} by rat liver mitochondria. *Biochem. Biophys. Res. Commun.* **10**, 444.

Lennon, R. E. (1954). Feeding mechanism of the sea lamprey and its effect on host fishes. *U.S. Fish Wildlife Serv., Fishery Bull.* **56**, 247–293.

Levine, L., and Musallam, D. A. (1964). A one-drop cryoscope: The tonicity of frog and goldfish sera. *Experientia* **20**, 508.

Lysaya, N. M. (1951). Changes in the blood composition of salmon during the spawning migration. *Izv. Tikhookeansk. Nauchn.-Issled. Inst. Rybn. Khoz. i* Okeanogr. **35**, 47–60.

McCallum, A. B. (1926). Paleochemistry of body fluids and tissues. *Physiol. Rev.* **6**, 316–357.

McCarthy, J. E. (1967). Vascular and extravascular fluid volumes in the Pacific hagfish *(Eptatretus stoutii, Lockington)*. M.A. Thesis, Oregon State University.

McCarthy, J. E., and Conte, F. P. (1966). Determination of the volume of vascular and extravascular fluids in the Pacific Hagfish, *Eptatretus stoutii. Am. Zoologist* **6**, 605 (abstr.).

McFarland, W. N., and Munz, F. W. (1958). A re-examination of the osmotic properties of the Pacific hagfish (*Polistotrema stouti*). *Biol. Bull.* **114**, 348–356.

McFarland, W. N, and Munz, F. W. (1965). Regulation of body weight and serum composition by hagfish in various media. *Comp. Biochem. Physiol.* **14**, 383–398.

Maetz, J., Mayer, N., and Chartier-Baraduc, M. M. (1967). La balance minérale du sodium chez *Anguilla anguilla* en eau de mer, en eau douce et au cours de transfert d'un milieu a l'autre: Effets de l'hypophysectomie et de la prolactine. *Gen. Comp. Endocrinol.* **8**, 177–188.

Magnin, E. (1962). Recherches sur la systématique et la biologie des acepenserides *Acipenser sturio, Acipenser oxyrhynchus, et Acipenser fulvescens.* Ch. 4. Quelques aspects de l'équilibre hydrominéral. *Ann. Stat. Centr. Hydrob. Appl.* **9**, 170–242.

Manery, J. F. (1954). Water and electrolyte metabolism. *Physiol. Rev.* **34**, 334–417.

Maren, T. H. (1962a). Ionic composition of cerebrospinal fluid and aqueous humor of the dogfish (*Squalus acanthias*). I. Normal values, *Comp. Biochem. Physiol.* **5**,193–200.

Maren, T. H. (1962b). Ionic composition of cerebrospinal fluid and aqueous humor of the dogfish *Squalus acanthias.* II. Carbonic anhydrase activity and inhibition. *Comp. Biochem. Physiol.* **5**, 201–215.

Maren, T. H. (1967a). Special body fluids of the elasmobranch. *In* "Sharks, Skates and Rays" (P. W. Gilbert, R. F. Mathewson, and D. P. Rall, eds.), pp. 287–292. Johns Hopkins Press, Baltimore, Maryland.

Maren, T. H. (1967b). Carbonic anhydrase: Chemistry, physiology, and inhibition. *Physiol. Rev.* **47**, 595–781.

Maren, T. H., and Frederick, A. (1958). Carbonic anhydrase inhibition in the elasmobranch: Effect on aqueous humor and cerebrospinal fluid CO_2 *Federation Proc.* **17**, 391.

Maren, T. H., Rawls, J. A., Burger, J. W., and Myers, A. C. (1963). The alkaline (Marshall's) gland of the skate. *Comp. Biochem. Physiol.* **10**, 1–16.

Martin, A. W. (1950). Some remarks on the blood volume of fish. *In* "Studies Honoring Trevor Kincaid," pp. 125–140. Univ. of Washington Press, Seattle, Washington.

Mayer, N., Maetz, J., Chan, D. K. O., Forster, M., and Chester-Jones, I. (1967). Cortisol: A sodium excreting factor in the eel (*Anguilla anguilla* L.) adapted to sea water. *Nature* **214**, 1118–1120.

Medway, W., and Kare, M. R. (1959). Thiocyanate space in growing domestic fowl. *Am. J. Physiol.* 196, 783.

Morris, R. (1956). The osmoregulatory ability of the lamprey *Lampetra fluviatilis* in sea water during the course of its spawning migration. *J. Exptl. Biol.* **33**, 235–248.

Morris, R. (1958). The mechanism of marine osmoregulation in the Lampern (*Lampetra fluviatilis* L.) and the causes of its breakdown during the spawning migration. *J. Exptl. Biol.* **35**, 649–664.

Morris, R. (1960). General problems of osmoregulation with special reference to cyclostomes. *Symp. Zool. Soc. London* **1**, 1–16.

Morris, R. (1965). Studies on salt and water balance in *Myxine glutinosa* (L.). J. *Exptl. Biol.* **42**, 359–371.

Motais, R., Romeu, F. G, and Maetz, J. (1965). Mechanism of eurhalinity-comparative investigation of *Platichthys* and *Serranus* following their transfer to fresh water. *Compt. Rend.* **261**, 801–804.

Munz, F. W., and McFarland, W. N. (1964). Regulatory function of a primitive vertebrate kidney. *Comp. Biochem. Physiol.* **13**, 381–400.

Murdaugh, H. V., and Robin, E. D. (1967). Acid-base metabolism in the dogfish shark. *In* "Sharks, Skates and Rays" (P. W. Gilbert, R. F. Mathewson, and D. P. Rall, eds.), pp. 249–264. Johns Hopkins Press, Baltimore, Maryland.

Murray, R. W., and Potts, W. T. W. (1961). Composition of the endolymph, perilymph, and other body fluids of elasmobranchs. *Comp. Biochem. Physiol.* **2**, 65–76.

Nicol, J. A. C. (1960). The autonomic nervous system of the Chimaeroid fish *Hydrolagus colliei. Quart. J. Microscop. Sci.* **91**, 379–400.

Ogawa, M. (1968). Osmotic and ionic regulation in goldfish following removal of the corpuscles of Stannius or the pituitary gland. *Can. J. Zool,* **46**, 669–676.

Oguri, M., and Takada, N. (1966). Effects of some hormonic substances on the urinary and serum calcium levels of the snake headfish *Channa argus. Bull. Japan. Soc. Sci. Fisheries* **32**, 28–31.

Oguri, M., and Takada, N. (1967). Serum calcium and magnesium levels of goldfish, with special reference to the gonadal maturation. *Bull. Japan. Soc. Sct. Fisheries* **33**,161–166.

Oppelt, W. W., Adamson, R. H., Zubrod, C. G., and Rall, D. P. (1966). Further observations on the physiology and pharmacology of elasmobranch ventricular fluid. *Comp. Biochem. Physiol.* **17**, 857–866.

Parry, G. (1958). Size and osmoregulation in salmonid fishes. *Nature* **181**, 1218.

Parry, G. (1961). Osmotic and ionic changes in blood and muscle of migrating salmonids. *J. Exptl. Biol.* **38**, 411–427.

Pearcy, W. G. (1961). Seasonal changes in osmotic pressure of flounder sera. *Science* **134**, 193–194.

Pereira, R. S, and Sawaya, P. (1957). Contributions to the study of the chemical composition of the blood of certain selachians of Brazil. *Univ. Sao Paulo, Fac. Filsof., Cienc. Letras, Bol. Zool.* **21**, 85–92.

Phillips, A. M., Jr., and Brockway, D. R. (1958). The inorganic composition of brown trout blood. *Progressive Fish Culturist* **20**, 58–61.

Pickford, G. E., and Grant, F. B. (1967). Serum osmolarity in the coelacanth *Latimeria* chalumnae urea retention and ion regulation. Science **155**, 568–570.

Platner, W. S. (1950). Effects of low temperature on Mg content of blood, body fluid and tissue of goldfish and turtle. *Am. J. Physiol.* **161**, 399–405.

Pora, E. A. (1935). Differences minerales dans la composition du sang suivant le sexe, chez *Cyprinus carpio. Compt. Rend. Soc. Biol.* **119**, 373–375.

Pora, E. A. (1936a). De l'influence de l'oxgénation du milieu extérieur sur la composition du sang, chez (*Scyllium canicula*). *Compt. Rend. Soc. Biol.* **121**, 194–196.

Pora, E. A. (1936b). Influence des fortes oxygénations du milieu extérieur sur la composition du sang de *Scyllium canicula. Ann. Physiol. Physicochim. Biol.* **12**, 238.

Pora, E. A. (1936c). Sur les différences chimiques et physico-chimiques du sang des deux sexes des Sélaciens. *Compt. Rend. Soc. Biol.* **121**, 105–107.

Pora, E. A. (1936d). Quelques données analytiques sur la composition chimique et physico-chimique du sang de quelques invertebrés et vertebrés marins. *Compt. Rend.* **121**, 291–293.

Pora, E. A. (1936e). Sur les différences chimiques et physico-chimiques du sang, suivant les sexes chez *Labrus bergylta. Compt. Rend. Soc. Biol.* **121**, 102–105.

Potts, W. T. W., and Evans, D. H. (1967). Sodium and chloride balance in the killifish *Fundulus heteroclitus. Biol. Bull.* **133**, 411–425.

Potts, W. T. W., and Parry, G. (1964). "Osmotic and Ionic Regulation in Animals." Pergamon Press, Oxford.

Price, K. S. (1967). Fluctuations in two osmoregulatory components, urea and sodium chloride, of the clearnose skate *Raja eglanteria* Bosc 1802. II. Upon natural variation of the salinity of the external medium. *Comp. Biochem. Physiol.* **23**; 77–82.

Price, K. S., and Creaser, E. P. (1967). Fluctuations in two osmoregulatory components, urea and sodium chloride, of the clearnose skate *Raja eglanteria* Bosc 1802. I. Upon laboratory modification of external salinities. *Comp. Biochem. Physiol.* **23,** 65–76.

Prosser, C. L., and Weinstein, S. J. F. (1950). Comparison of blood volume in animals with open and closed circulatory systems. *Physiol. Zool.* **23**, 113–124.

Raffy, A. (1949). L'euryhalinité de *Blennius pholis* L. *Comp. Rend. Soc. Biol.* **143**, 1575–1576.

Robertson, J. D. (1954). The chemical composition of the blood of some aquatic chordates including members of the *Tunicata, Cyclostomata* and *Osteichthyes. J. Exptl. Biol.* **31**, 424–442.

Robertson, J. D. (1963). Osmoregulation and Ionic composition of cells and tissues. *In* "Biology of Myxine" (A. Brodal and R. Fänge, eds.), pp. 504–515. Oslo Univ. Press, Oslo.

Robertson, J. D. (1966). Osmotic constituents of the blood plasma and parietal muscle of *Myxine glutinosa* L. *In* "Some Contemporary Studies in Marine Science" (H. Barnes, ed.), pp. 631–644. Allen & Unwin, London.

Robertson, O. H., Krupp, M. A., Favour, C. B., Hane, S., and Thomas, S. F. (1961). Physiological changes occurring in the blood of the Pacific Salmon (*Oncorhynchus tshawytscha*) accompanying sexual maturation and spawning. *Endocrinology* **68**, 733–746.

Robin, E. D., Murdaugh, H. V., Jr., and Weiss, E. (1964). Acid-base, fluid and electrolyte metabolism in the elasmobranch. 1. Ionic composition of erythrocytes, muscle and brain. *J. Cellular Comp. Physiol.* **64**, 409–418.

Rodier, E. (1899). Observations et expériences comparatives sur l'eaux de mer, le sang et les liquides internes des animaux marins. *Trav. Lab. Soc. Sci. Stat. Zool. Arachon* pp. 103–123.

Rodnan, G. P., Robin, E. D., and Andrus, M. H. (1962). Dogfish coelomic fluid. 1. Chemical anatomy. *Bull. Mt. Desert Isl. Biol. Lab.* **4**, 69–70.

Schiffman, R. H., and Fromm, P. O. (1959). Measurement of some physiological parameters in rainbow trout (*Salmo gairdnerii*). *Can J. Zool.* **37**, 25–32.

Scholander, P. F., Van Dam, L., Kanwisher, J. W., Hammel, H. T., and Gordon, M. S. (1957). Supercooling and osmoregulation in arctic fish. *J. Cellular Comp. Physiol.* **49**, 5–24.

Schreibman, M. P., and Kallman, K. D. (1966). Endocrine control of fresh water tolerance in teleosts. *Gen. Comp. Endocrinol.* **6**, 144–155.

Sharratt, B. M., Chester Jones, I., and Bellamy, D. (1964). Water and electrolyte composition of the body and renal function of the eel (*Anguilla anguilla*). *Comp. Biochem. Physiol.* **11**, 9–18.

Shell, E. W. (1959). Chemical composition of the blood of small mouth bass. Ph.D. Thesis, Cornell University.

Sindermann, C. J., and Mairs, D. F. (1961). Blood properties of pre-spawning and post-spawning anadromous alewives (*Alosa pseudoharengus*). *U.S. Fish Wildlife Serv., Fishery Bull.* **183**, 145–151.

Smith, H. W. (1929a). The composition of the body fluids of elasmobranchs. *J. Biol. Chem.* **81**, 407–419.

Smith, H. W. (1929b). The composition of the body fluids of the goose fish (*Lophius piscatorius*). *J. Biol. Chem.* **82**, 71–75.

Smith, H. W. (1930). Metabolism of the lungfish *Protopterus aethiopicus*. *J. Biol. Chem.* **88**, 97–130.

Smith, H. W. (1931). The absorption and excretion of water and salts by the elasmobranch fishes. I. Fresh-water elasmobranchs. *Am. J. Physiol.* **98**, 279.

Smith, H. W. (1936). The retention and physiological role of urea in the *Elasmobranchii*. *Biol. Rev.* **11**, 49–92.

Soberman, R. J. (1950). Use of antipyrine in measurement of the total body water in animals. *Proc. Soc. Exptl. Biol. Med.*, **74**, 789–792.

Staedeler, G., and Frerichs, F. T. (1858). Uber das Vorkommen von Harnstoff' Taurin und Scyllit in den Organen der Plagiostomen. *J. Prakt. Chem.* **73**, 48–55.

Stanley, J. G., and Fleming, W. R. (1964). Excretion of hypertonic urine by a teleost. *Science* **144**, 63–64.

Sterling, K., and Gray, S. J. (1950). Determination of the circulating red cell volume in man by radioactive chromium. *J. Clin. Invest.* **29**, 1614–1619.

Sulya, L. L., Box, B. E., and Gunther, G. (1960). Distribution of some blood constituents in fish from the Gulf of Mexico. *Am. J. Physiol.* **199**, 1177–1180.

Thorson, T. (1958). Measurement of the fluid compartments of four species of marine Chondrichthyes. *Physiol. Zool.* **31**, 16–23.

Thorson, T. (1959). Partitioning of the body water in sea lamprey. *Science* **130**, 99–100.

Thorson, T. (1961). The partitioning of body water in Osteichthyes: Phylogenetic and ecological implications in aquatic vertebates. *Biol. Bull.* **120**, 238–254.

Thorson, T. (1962). Partitioning of body tluids in the Lake Nicaragua shark and three marine sharks. *Science* **138**, 688–690.

Thorson, T. B. (1967). Osmoregulation in fresh water elasmobranchs, *In* "Sharks, Skates and Rays" (P. W. Gilbert, R. F. Mathewson, and D. P. Rall, eds.), pp. 265–270. Johns Hopkins Press, Baltimore, Maryland.

Urist, M. R. (1961). Calcium and phosphorus in the blood and skeleton of the elasmobranchii. *Endocrinology* **69**, 778–801.

Urist, M. R. (1962a). Calcium and other ions in blood and skeleton of Nicaraguan freshwater Shark. *Science* **137**, 984–986.

Urist, M. R. (1962b). The bone-body tluid continuum: Calcium and phosphorous in the skeleton of extinct and living vertebrates. *Perspectives. Biol. Med.* **6**, 75–115.

Urist, M. R. (1963). The regulation of calcium and other ions in the serums of hagfish and lamprey. *Ann. N.Y. Acad. Sci.* **109**, 294–311.

Urist, M. R. (1964). Further observations bearing on the bone-body fluid continuum: Composition of the skeleton and serums of cyclostomes, elasmobranchs and bony vertebrates. *In* "Monographs on Bone Biodynamics" (H. M. Frost, ed.), pp. 151–179. Little, Brown, Boston, Massachusetts.

Urist, M. R. (1966). Calcium and electrolyte control mechanisms in lower vertebrates. *In* "Phylogenetic Approach to Immunity" (R. T. Smith, R. A. Good, and P. A. Miescher, eds.), pp. 18–28. Univ, of Florida Press, Gainesville, Florida.

Urist, M. R., and Van de Putte, K. A. (1967). Comparative biochemistry of the blood of fishes. *In* "Sharks, Skates and Rays" (P. W. Gilbert, R. F. Mathewson, and D. P. Rall, eds.), pp. 271–292. Johns Hopkins Press, Baltimore, Maryland.

Vasington, F. D., and Murphy, J. V. (1962). Ca^{++} uptake by rat kidney mitochondria and its dependence on respiration and phosphorylation. *J. Biol. Chem.* **237**, 2670.

Welcker, H. (1858). Bestimmungen der Menge des Korperblutes und der Blutfarbkraft, sowie Bestimmungen von Zahl, Mass, Oberfläche und Volumn des einzelnen Blutkörperchen bei Thieren und bei Menschen. Z. *Rationelle Med.* **4**, 145.

Wikgren, B. (1953). Osmotic regulation in some aquatic animals with special reference to the influence of temperature. *Acta Zool. Fennica* **71**, 1–102.

Winkler, A. W., Elkinton, J. R., and Eisman, A. J. (1943). Comparison of sulfocyanate with radioactive chloride and sodium in the measurement of extracellular fluid. *Am. J. Physiol.* **139**, 239.

Woodhead, P. M. J., and Woodhead, A. D. (1958). Effects of low temperature on the osmoregulatory ability of the cod (*Gadus callarias*) in arctic waters. *Proc. Linnean Soc. London* **169**, 63–64.

Woodhead, P. M. J., and Woodhead, A. D. (1959). Effects of low temperature on the physiology and distribution of the cod, *Gadus morhua* L. in the Barent Sea. *Proc. Zool. Soc. London* **133**, 181–199.

Yamazaki, F., and Donaldson, E. M. (1968). Unpublished observations.

Young, J. Z. (1950). "The Life of Vertebrates." Oxford Univ. Press, London and New York.

Zaks, M. G., and Sokolova, M. M. (1961). O mekhanismakh adaptatsii k izmeneniian solenosti vody u nerki *Oncorhynchus nerka* (Walb). *Vopr. Ikhtiol.* 1, 333–346 (*Fisheries Res. Board Can.* Transl. No. 372).

Zaks, M. G., and Sokolova, M. M. (1965). Ismenenie tipa osmoregulizatsii v raynye periody migratsionnogo tsikla u nerki *Oncorhynchus nerka* (Walb.). *Vopr. Ikhtiol.* **5**, 331–337 (*Fisheries Res. Board Can.* Transl. No. 773).

Chapter 9

Stimulation of a framework for future acid–base regulation studies in fish

David J. Randall* and Colin J. Brauner
University of British Columbia, Vancouver, BC, Canada
**Corresponding author: e-mail: randall@zoology.ubc.ca*

David J. Randall and Colin J. Brauner discuss the C. Albers chapter "Acid-Base Balance" in *Fish Physiology*, Volume 4, published in 1970. The next chapter in this volume is the re-published version of that chapter.

In 1970, relatively little was known about acid–base regulation in fishes. Claus Albers was a physiologist studying acid–base regulation in humans and had some experience working with fish. DJR asked C. Albers to write a review of acid–base regulation which was of course heavily influenced by mammalian models, with the goal of stimulating a frame-work for studies to investigate the mechanisms of acid–base regulation in fish. Since this chapter, our understanding of the mechanisms of fish acid–base regulation has greatly evolved, and this body of knowledge has contributed to our understanding of the mechanisms through which fish are impacted by environmental change, and in some cases, mitigations that can be employed to reduce this impact.

Fifty years ago, interest in acid–base regulation in fish was sporadic but growing. DJR invited Claus Albers to contribute a chapter to "Fish Physiology" because, although his interests were in mammalian (human) acid–base regulation, he also worked with fish. He was asked to provide a general review of acid–base regulation in vertebrates with comment on the studies in fish, which at the time was quite limited. At the time, the field of comparative physiology was evolving and instrumentation for measuring gases was becoming available, enhancing experimental capacity. The overall goal was to stimulate the development of a framework from which to advance our understanding of the mechanisms of acid–base regulation in fishes; a goal that was achieved by this chapter.

In his chapter, Albers concluded that there were two ways of regulating acid–base balance; the "metabolic" way and the "respiratory" way. The respiratory way concerned regulation of CO_2 through changes in ventilation, the metabolic way concerned buffering and cellular regulation of acid–base status. While the chapter was excellent in stimulating interest in subsequent study of mechanisms of acid–base regulation, it was limited in the identification of the important

Fish Physiology, Vol. 40A. https://doi.org/10.1016/bs.fp.2023.08.006

differences between terrestrial and aquatic systems and the wide variety of possibilities in fish that we understand today. To address the former, DJR was invited to spend summers at the Max Planck Institute for Experimental Medicine in Göttingen, Germany to assist in development of comparative studies in fish. DJR met and worked with Norbert Heisler both in Göttingen and at the Naples Marine Laboratory during the time that the Albers chapter was written. At that time, Norbert Heisler and DJR were working together on the effects of environmental CO_2 increases on dogfish. The Albers chapter was written to inform "fish" people about acid–base regulation in mammals. The chapter promoted comparison between the fish and mammalian systems. At a meeting in the late 60's, dominated by mammalian acid base researchers, DJR pointed out that there was no evidence of ventilation-controlled CO_2/pH regulation in aquatic animals. Many were surprised. In 1976, Norbert Heisler joined DJR in the Amazon, he was amazed at the ability of water breathing animals to regulate pH via the "metabolic" and not the "respiratory" way. Subsequent studies of acid–base regulation in fish showed that fish live in many complex environments with a range of ion, oxygen, and temperature levels and blood pH was managed without modulation of blood or body CO_2 levels.

Rather than fishes having a primitive version of terrestrial acid–base regulation, as implied in the chapter, we are seeing the aquatic origins of the extant terrestrial system, as well as recognizing the complexity of acid–base regulation in fish. For example, there is very limited ventilatory regulation of blood CO_2 in water-breathing fish (Randall and Tsui, 2006) and so acid–base regulation is dominated by the "metabolic" way (Evans et al., 2005; Heisler, 1984). Water contains much less dissolved oxygen than an equal volume of air. Ventilation/perfusion ratios are thus an order of magnitude greater in water-breathing fish than in air-breathing mammals. Moreover, the gills are a counter-current gas exchanger and very efficient at removing oxygen from the water. These requirements for oxygen uptake ensure ventilation is more than sufficient for CO_2 excretion, and blood CO_2 and bicarbonate levels are low in fish compared with that of mammals (Evans et al., 2005). Although exposure to high environmental CO_2 (i.e., $1\% \approx 7$ mmHg $\approx 10{,}000\,\mu$atm CO_2) increases ventilation in fish, there is no evidence that changes in ventilation alter arterial blood CO_2, and therefore, blood pH levels in water-breathing vertebrates at these high CO_2 levels (Randall and Tsui, 2006). Changes in environmental PCO_2 levels cause similar changes in fish arterial PCO_2 and any subsequent changes in pH are largely associated with changes in plasma [HCO_3^-] (in exchange for [Cl^-]), the "metabolic" way (Heisler, 1984).

CO_2 excretion in fish is limited by the rate of HCO_3^- entry into the red blood cell (Perry, 1986) and carbonic anhydrase rapidly catalyzes the conversion of $HCO_3^- + H^+$ to CO_2 within the RBC and is not available to plasma flowing through the gills of most teleost fish (Randall et al., 2014). Furthermore, in many fishes, CO_2 excretion is tightly coupled with O_2 uptake, where CO_2 excretion rates are enhanced by Hb-O_2 binding at the gills (Brauner and Randall, 1998). Ice fish lacking red blood cells are an exception, having carbonic anhydrase available to plasma in the gills (Harter et al., 2018). Carbonic anhydrase is also available to plasma flowing through the gills of elasmobranchs and both icefish and elasmobranchs can excrete bicarbonate directly from the plasma without passage through the red blood cells (Nikinmaa et al., 2019). What is clear is that there can be no "respiratory" regulation of acid–base status in teleost fish because blood CO_2 excretion is limited HCO_3^- entry into the red blood cell (Perry, 1986). Thus, acid–base status is regulated at the cellular level coupled to ion transfer via the gills and kidney. There are significant movements of water, protons and ions between the environment and the body of fish, mostly across the gills (Evans et al., 2005; Heisler, 1984).

During exposure to elevated environmental CO_2, fish rapidly compensate the ensuing reduction in blood pH (pHe) through a relatively rapid (24–96 h) elevation of plasma HCO_3^- in exchange for Cl^-, through processes at the gills. However, there is an upper CO_2 level beyond which this "metabolic" way cannot compensate pHe. In places like the Amazon, fish can experience very high CO_2 levels, beyond the capacity to compensate blood pHe, and some fishes have the ability to rapidly and tightly regulate intracellular pH (pHi; Heisler, 1982), despite large sustained, reductions in pHe (preferential pHi regulation). While Albers only briefly mentioned aspects related to intracellular pH values and regulation, it is now clear that preferential pHi regulation may be a relatively common characteristic among water and air-breathing fishes (Brauner and Baker, 2009; Shartau et al., 2020). From this work, it has been concluded that preferential pHi regulation may be an exaptation for air-breathing in vertebrates, and represent another important adaptation in the transition of vertebrates from water to land (Shartau et al., 2016).

Acid–base balance in fishes is influenced by the excretion of toxic ammonia, however only a few studies of interactions between ammonia and acid–base regulation in fish have been conducted. Most fish excrete ammonia across their gills as mentioned by Albers. The gills are much more permeable to ammonia than ammonium ions and ammonia excretion tends to raise the pH of blood. The ammonia efflux can be as high as one third of the CO_2 efflux. The aquatic environment is variable, adding to the complexity of acid–base regulation in fish. For example, under alkaline conditions fish ammonia excretion is limited. In terrestrial vertebrates, ammonia is converted to less toxic urea, via the ornithine urea cycle, and then excreted. A few teleost fish also have an active ornithine urea cycle excreting urea to reduce ammonia toxicity (Randall et al., 1989) but the majority use less expensive ammonia excretion across the gills. The evolution of the ornithine urea cycle in aquatic vertebrates paved the way for the movement of vertebrates onto land. None of this was discussed in the Albers chapter.

Another environmental variable that influences acid–base status is water temperature. The gills are an excellent heat exchanger so arterial blood leaving the gills is at the same temperature as the water leaving the gills; only local increases in tissue temperature are possible in some fishes. In fishes, blood pH increases with a decrease in temperature as reviewed by Albers. Blood pH remains midway between the ammonia/ammonium and CO_2/bicarbonate reactions (Randall and Wright, 1989), maintaining fluxes of both CO_2 and ammonia with changes in temperature. It was postulated (alpha-stat hypothesis) that the increase in pH offset the effects of temperature on protein function. Considerations of the effect of temperature on ammonia excretion, and thus implications for blood pH changes, were not considered by this mammalian centric view of acid base relations but remain an exciting area for further research.

The goal of Albers chapter was to stimulate a framework for further investigation of the mechanisms of acid–base regulation in fish. In the following 50 years, a great deal has been learned about the specific acid–base regulatory mechanisms involved at the gills, kidney and gastro-intestinal system in fish that is central to their success in a tremendous diversity of environments (Brauner et al., 2019; Claiborne et al., 2002). However, these discoveries have also allowed us to understand the mechanisms through which human induced changes of the environment impact fish. This ranges from addressing the impacts of acid rain in the 1960s–70s, and more recently environmental acidification due to elevated atmospheric CO_2 levels in seawater (Ocean acidification; OA) and freshwater systems, the latter of which has received only limited attention (Ou et al., 2015).

Understanding the mechanisms associated with acid-rain toxicity, for example, permitted the development of mitigative measures that greatly reduced the impact of acid-rain on aquatic species (Wood, 2022). In terms of OA, the early work on how fish respond to elevated environmental CO_2 has been central in informing OA impacts in fish. This early work, including that by Albers, involved exposing fish to high environmental CO_2 levels (1% CO_2) to induce an acid–base disturbance and monitor recovery. From this work, it is clear that most fish have a high capacity for correcting such acid–base disturbances in both freshwater and seawater (Toews et al., 1983; Tovey and Brauner, 2018), and initially it was assumed that fish may not be severely impacted by the much lower CO_2 levels associated with OA (i.e., 0.1% $\approx$ 0.7 mmHg $\approx$ 1000 μatm CO_2 by the end of the century). However, many studies now show that this is not the case, particularly in early life stages where behavioral responses are greatly affected (Munday et al., 2019). These behavioral responses are thought to be linked to altered $GABA_A$ neurotransmitter responses (Heuer et al., 2019), where the elevation in plasma HCO_3^- and reduction in plasma Cl^- associated with the "metabolic" way of acid–base regulation, renders some $GABA_A$ receptors as excitatory rather than inhibitory (Nilsson et al., 2012), resulting in a range of behavioral abnormalities (Munday et al., 2019).

While Claus Albers was a mammalian, mechanistic physiologist at heart, his chapter was important in stimulating tremendous advances in fish acid–base balance in the subsequent half a century. Findings that have spanned from elucidating basic mechanisms, to shedding insight into evolutionary adaptations, to understanding mechanisms (and in some cases mitigation) of human induced environmental impacts.

References

Brauner, C.J., Baker, D.W., 2009. Patterns of acid-base regulation during exposure to hypercarbia in fishes. In: Glass, M.G., Wood, S.C. (Eds.), Cardio-Respiratory Control in Vertebrates. Springer-Verlag, Berlin, pp. 43–63.

Brauner, C.J., Randall, D.J., 1998. The linkage between oxygen and carbon dioxide transport. In: Perry, S.F., Tufts, B. (Eds.), Fish Physiology; Fish Respiration. Academic Press, New York, pp. 283–319.

Brauner, C.J., Shartau, R.B., Damsgaard, C., Esbaugh, A.J., Wilson, R.W., Grosell, M., 2019. 3 –Acid-base physiology and CO_2 homeostasis: Regulation and compensation in response to elevated environmental CO_2. In: Grosell, M., Munday, P.L., Farrell, A.P., Brauner, C.J. (Eds.), Fish Physiology. Academic Press, pp. 69–132.

Claiborne, J., Edwards, S., Morrison-Shetlar, A., 2002. Acid-base regulation in fishes: cellular and molecular mechanisms. J. Exp. Zool. 293, 302–319.

Evans, D.H., Piermarini, P.M., Choe, K.P., 2005. The multifunctional fish gill: dominant site of gas exchange, osmoregulation, acid-base regulation, and excretion of nitrogenous waste. Physiol. Rev. 85, 97–177.

Harter, T.S., Sackville, M.A., Wilson, J.M., Metzger, D.C.H., Egginton, S., Esbaugh, A.J., Farrell, A.P., Brauner, C.J., 2018. A solution to Nature's haemoglobin knockout: a plasma-accessible carbonic anhydrase catalyses CO_2 excretion in Antarctic icefish gills. J. Exp. Biol. 221.

Heisler, N., 1982. Intracellular and extracellular acid-base regulation in the tropical fresh-water teleost fish Synbranchus marmoratus in response to the transition from water breathing to air breathing. J. Exp. Biol. 99, 9–28.

Heisler, N., 1984. Acid-base regulation in fishes. In: Hoar, W.S., Randall, D.J. (Eds.), Fish Physiology. Academic Press, San Diego, pp. 315–401.

Heuer, R.M., Hamilton, T.J., Nilsson, G.E., 2019. 5 – The physiology of behavioral impacts of high CO_2. In: Grosell, M., Munday, P.L., Farrell, A.P., Brauner, C.J. (Eds.), Fish Physiology. Academic Press, pp. 161–194.

Munday, P.L., Jarrold, M.D., Nagelkerken, I., 2019. 9 – Ecological effects of elevated CO_2 on marine and freshwater fishes: From individual to community effects. In: Grosell, M., Munday, P.L., Farrell, A.P., Brauner, C.J. (Eds.), Fish Physiology. Academic Press, pp. 323–368.

Nikinmaa, M., Berenbrink, M., Brauner, C.J., 2019. Regulation of erythrocyte function: multiple evolutionary solutions for respiratory gas transport and its regulation in fish. Acta Physiol. 227, e13299.

Nilsson, G.E., Dixson, D.L., Domenici, P., Mccormick, M.I., Sorensen, C., Watson, S.-A., Munday, P.L., 2012. Near-future carbon dioxide levels alter fish behaviour by interfering with neurotransmitter function. Nat. Clim. Change 2, 201–204.

Ou, M., Hamilton, T.J., Eom, J., Lyall, E.M., Gallup, J., Jiang, A., Lee, J., Close, D.A., Yun, S.-S., Brauner, C.J., 2015. Responses of pink salmon to CO_2-induced aquatic acidification. Nat. Clim. Change 5, 950–955.

Perry, S.F., 1986. Carbon dioxide excretion in fishes. Can. J. Zool. 64, 565–572.

Randall, D.J., Rummer, J.L., Wilson, J.M., Wang, S., Brauner, C.J., 2014. A unique mode of tissue oxygenation and the adaptive radiation of teleost fishes. J. Exp. Biol. 217, 1205–1214.

Randall, D.J., Tsui, T.K.N., 2006. Tribute to R. G. Boutilier: acid–base transfer across fish gills. J. Exp. Biol. 209, 1179–1184.

Randall, D.J., Wood, C.M., Perry, S.F., Bergman, H., Maloiy, G.M.O., Mommsen, T.P., Wright, P.A., 1989. Urea excretion as a strategy for survival in a fish living in a very alkaline environment. Nature (London) 337, 165–166.

Randall, D., Wright, P., 1989. The interaction between carbon dioxide and ammonia excretion and water pH in fish. Can. J. Zool. 67, 2936–2942.

Shartau, R.B., Baker, D.W., Crossley, D.A., Brauner, C.J., 2016. Preferential intracellular pH regulation: hypotheses and perspectives. J. Exp. Biol. 219, 2235–2244.

Shartau, R.B., Baker, D.W., Harter, T.S., Aboagye, D.L., Allen, P.J., Val, A.L., Crossley, D.A., Kohl, Z.F., Hedrick, M.S., Damsgaard, C., Brauner, C.J., 2020. Preferential intracellular pH regulation is a common trait amongst fishes exposed to high environmental CO_2. J. Exp. Biol. 223, jeb208868.

Toews, D.P., Holeton, G.F., Heisler, N., 1983. Regulation of the acid-base status during environmental hypercapnia in the marine teleost fish Conger conger. J. Exp. Biol. 107, 9–20.

Tovey, K.J., Brauner, C.J., 2018. Effects of water ionic composition on acid–base regulation in rainbow trout, during hypercarbia at rest and during sustained exercise. J. Comp. Physiol. B 188, 295–304.

Wood, C.M., 2022. Conservation aspects of osmotic, acid-base, and nitrogen homeostasis in fish. In: Cooke, S.J., Fangue, N.A., Farrell, A.P., Brauner, C.J., Eliason, E.J. (Eds.), Fish Physiology. Academic Press, pp. 321–388.

Chapter 10

ACID–BASE BALANCE☆

C. ALBERS

Chapter Outline

I. INTRODUCTION

This chapter deals with the chemical reactions and the physiological mechanisms affecting the concentration of hydrogen ions in the various fluid compartments of the body. Since acids are substances capable of delivering hydrogen ions, the concentration of hydrogen ions $[H^+]$depends primarily on the amount of the various acids present in the body fluids. The most important of these is carbonic acid which is derived from carbon dioxide (CO_2), one of the major end products of metabolism. Hence the equilibrium equations of carbonic acid and its reactions with the so–called buffer substances form the chemical basis of CO_2 transport and acid–base balance. The concentration of CO_2 and the chemical composition of the body fluids are therefore the principal factors governing acid–base balance. The former is controlled by ventilation, the latter by the action of various excretory mechanisms.

In the term "acid–base balance," acid stands for the sum of all anions except the hydroxyl ion and base stands for the sum of all cations except

☆This is a reproduction of a previously published chapter in the Fish Physiology series, "1970 (Vol. 4)/Acid-Base Balance: ISBN: 978-0-12-350404-3; ISSN: 1546-5098".

Fish Physiology, Vol. 40A. https://doi.org/10.1016/bs.fp.2023.08.002

the hydrogen ion. This "medical" definition of acids and bases does not seem to be in accord with the proper chemical definition. This disagreement, however, turns out to be formal rather than substantial, as pointed out by Siggaard–Andersen (1965). The medical definition was adopted several decades ago by van Slyke and his group who contributed most to our basic knowledge of the acid–base balance in mammals. Because much more is known about the acid–base balance in warm-blooded animals, some basic facts about the acid–base balance in mammals will be included in this chapter. The question of whether or not these facts are pertinent to the acid–base balance of fish has to be left open in many cases. It is the hope of the author that his description of the acid–base balance will prompt other investigators to fill in the gaps in this really fascinating field of comparative physiology and biochemistry.

II. BASIC CONCEPTS OF PHYSICAL CHEMISTRY

A. The Dissociation of Water and the Definition of pH

Water molecules are dissociated into hydrogen ions and hydroxyl ions to a very small extent. The ionic product of water

$$K_W = [H^+][OH^-] \tag{1}$$

is of the order of 10^{-14}, depending on the temperature and the ionic strength of the solution. The hydrogen ion is normally present in the hydrated form H_3O, but for the sake of simplicity we shall stick throughout this chapter to the more convenient form of denoting the hydrogen ion as H^+. Since in pure water hydrogen ions equal hydroxyl ions the concentration of H^+ is ½ K_w or about 10^{-7} mole/liter. If $[H^+]$ is changed by adding acids or bases, the concentration of hydroxyl ions $[OH^-]$ changes inversely according to Eq. (1). For instance, if we have a 0.1 *N* HCl solution, $[H^+]=10^{-1}$ and $[OH^-]=10^{-13}$ (for $K_w=10^{-14}$).

The classic laws of thermodynamics describing the reactions of ions are valid only for ideal, infinitely diluted solutions. Very dilute solutions may be regarded as nearly ideal solutions. In all practical cases, however, the thermodynamic laws are valid only if activities rather than concentrations are used. The activity is obtained by multiplying the concentration by the so-called activity coefficient *f*. For hydrogen ions we have $a_{H^+}=[H^+]\ f_{H^+}$. Although for practical purposes the distinction between the concentration and the activity of an ion is important, we shall neglect the difference between $[H^+]$ and a_{11^+} in the following derivations. This is also justified because all methods for the determination of $[H^+]$ actually give information about a_{H^+}.

Sometimes it is more convenient to use logarithmic units. A logarithmic scale is also justified on a physicochemical basis, since the chemical potential or the energy associated with the activity of an ion is related to the logarithm of the activity. This leads to the concept of pH, which was introduced first by Sorensen (1909) as the negative logarithm of the concentration of hydrogen ions. The definition of pH commonly accepted now is

$$pH = -\log a_{H^+} \tag{2}$$

From this definition it is easy to see that an increase in $[H^+]$ (or a_{H^+}) is denoted by a decrease in pH and vice versa. The pH scale covers a large concentration range, one unit being equivalent to a tenfold change in $[H^+]$. When the pH scale is used, the actual changes in $[H^+]$ are often underrated. Actually, a change of 0.3 pH units does not look too impressive, but it really means a doubling of $[H^+]$. One should not forget this fact when speaking about the relative constancy of arterial pH for instance.

B. Dissociation of Weak Acids

Let us consider an acid AH which dissociates into hydrogen ions and the anion A^-. According to the law of mass action we may write

$$\frac{[H^+][A^-]}{[AH]} = K \tag{3}$$

If the equilibrium constant K is very large, the concentration of [AH] is negligible and the dissociation may be regarded as virtually complete. This is the case with strong acids like HCl or HNO_3. If K is very small, only a fraction of the acid is dissociated and the acid is a weak acid. Most organic acids belong to this group. As a thermodynamic constant K depends on the temperature but not on the ionic strength. If we convert to activities in Eq. (3), assuming that $a_{AH} = [AH]$ and $a_{A^-} = [A^-]f_{A^-}$, then

$$\frac{a_{H^+}[A^-]}{[AH]} \cdot f_{A^-} = K = K' \cdot f_{A^-} \tag{4}$$

K' now depends on the temperature and on the ionic strength. Its negative logarithm is written pK'. If we take the logarithms on both sides of Eq. (4) and multiply by minus one, we obtain the important equation

$$pH = pK' + \log \frac{[A^-]}{[AH]} \tag{5}$$

which is known as the Henderson–Hasselbalch equation.

For a given concentration [AH] we obtain from Eq. (4) $[H^+]\,[A^-] = \text{const}$, which can be interpreted in a similar manner as Eq. (1). Therefore, if $[H^+]$ increases $[A^-]$ must decrease. If we add a strong acid to a weak acid, hydrogen ions recombine with the anion A^- to form the undissociated acid AH. Thus, the dissociation of AH is dependent on $[H^+]$. If h is the fractional extent to which a weak acid is dissociated, then

$$h = \frac{[A^-]}{[A^-] + [AH]} \tag{6}$$

Since we have from Eq. (3)

$$[AH] = \frac{[H^+] \cdot [A^-]}{K}$$

we obtain

$$h = \frac{K}{K + [H^+]} \tag{7a}$$

or in the logarithmic form

$$h = 1/\left(1 + 10^{pK-pH}\right) \tag{7b}$$

Therefore, if $[H^+]$ is very large with respect to K, h approaches zero. If $[H^+]$ is very small, h approaches unity. If $[H^+]=K$, h is 0.5, that means exactly one–half of the weak acid is dissociated when pH equals pK'. A plot of h vs. pH according to Eq. (7b) is shown in Fig. 1 for lactic acid (p$K'=3.9$ at 25°C) and for boric acid (p$K'=9.2$ at 25°C). Exactly the same curve is obtained if we titrate a weak acid with a strong base and plot the fractional extent of neutralization vs. pH. We may call this curve the standardized titration curve, whereas the ordinary titration curve is obtained by plotting the amount of base consumed (e.g., in milliliter of 0.1 N NaOH) versus pH. This results in a sigmoidshaped curve too. It is customary to determine the pK'

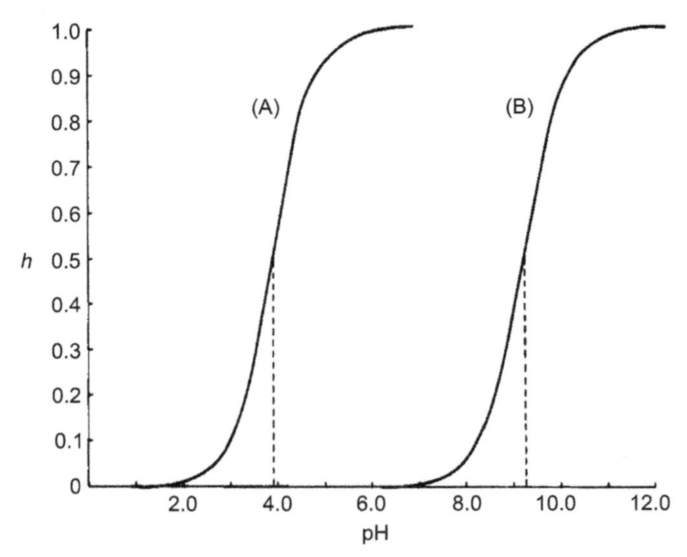

Fig. 1. Fractional dissociation h of lactic acid (A) and of boric acid (B) as a function of pH.

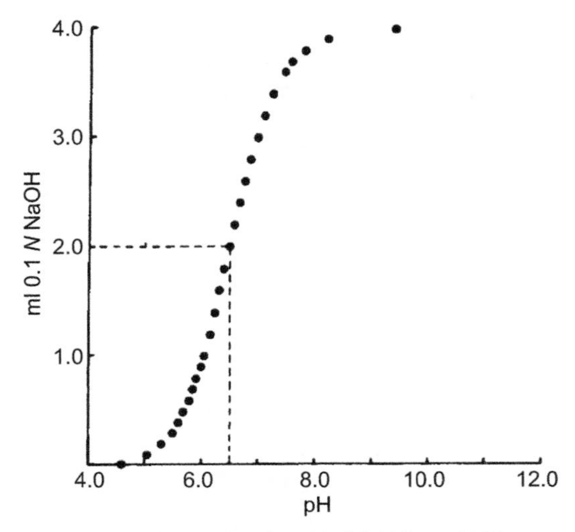

Fig. 2. Titration curve of 4ml of 0.1 *N* DMO at 19.6°C (Albers, 1966).

value of an acid by reading the pH of half neutralization from a titration curve. Figure 2 shows the titration of the weak acid DMO (dimethyloxazoledinedione), which is used for the indirect determination of the intracellular pH (see below). From Fig. 2 we read for half neutralization pH=6.48; pK' of DMO is therefore 6.48 at that temperature.

C. Carbonic Acid

The amount of gaseous CO_2 physically dissolved in water or electrolyte solutions is proportional to the partial pressure of CO_2 according to Henry's law

$$[CO_2] = Sp_{CO_2} \tag{8}$$

where S is the solubility coefficient in millimoles per liter per torr and p_{CO_2} the partial pressure of CO_2 in torr. Tables of S commonly refer to the sum of physically dissolved CO_2 and carbonic acid according to Eq. (9a). If a liquid has been equilibrated with a gas mixture containing CO_2 until the number of CO_2 molecules escaping from the liquid into the gas phase equals the number of CO_2 molecules entering the liquid from the gas phase, the net exchange between the two phases is zero. For this state of equilibrium it is said that the liquid phase has the same partial pressure of CO_2 as has the gas phase, whether or not the liquid is still in contact with the gas phase. Even if a liquid has never been in contact with a gas phase, gaseous components derived from chemical reactions exert a partial pressure which is linked to the amount of dissolved gas by the solubility coefficient S by Eq. (8). It is possible to calculate the partial pressure of CO. from analytical data as well as to measure it directly with membrane–covered glass electrodes.

CO_2 dissolved in water reacts to form carbonic acid:

$$CO_2 + H_2O \rightleftharpoons H_2CO_3 \tag{9a}$$

The carbonic acid as a dibasic acid dissociates into bicarbonate ions and carbonate ions

$$H_2CO_3 \rightleftharpoons H^+ + HCO_3^- \tag{9b}$$

$$HCO_3^- \rightleftharpoons H^+ + CO_3^{2-} \tag{9c}$$

From the law of mass action we have

$$[CO_2] = L[H_2CO_3] \tag{10a}$$

$$\frac{[H^+][HCO_3^-]}{[H_2CO_3]} = K_1 \tag{10b}$$

$$\frac{[H^+][CO_3^{2-}]}{[HCO_3^-]} = K_2 \tag{10c}$$

The equilibrium constant L is very large, less than 0.5% of the dissolved CO_2 being transformed into H_2CO_3. K_1 is the true dissociation constant of carbonic acid and about 20 times greater than that of acetic acid. But because of the large value of L in Eq. (10a), dissolved CO_2 acts as a much weaker acid. This may be seen if we combine Eqs. (10a) and (10b):

$$[H^+] = \frac{K_1}{L} \cdot \frac{[CO_2]}{[HCO_3^-]} \tag{11}$$

$K_1' = K_1/L$ is the so–called apparent first dissociation constant of carbonic acid and in the order of 4×10^{-7} at 20°C. Correspondingly, pK_1' of carbonic acid is about 6.4 at 20°C.

At the high body temperature of mammals the concentration of carbonate ions is practically zero. In fish, however, a small fraction of the chemically bound CO_2 is present as carbonate. Therefore, we also have to consider reactions (9c) and (10c). If we denote the total concentration of CO_2 as C_T, we have

$$C_T = Sp_{CO_2} + [HCO_3^-] + [CO_3^{2-}]$$

Solving Eqs. (10b) and (10c) for $[HCO_3^-]$ and $[CO_3^{2-}]$, we finally arrive at the rather complex equation

$$pH = pK_1' + \log \frac{C_T - Sp_{CO_2}}{Sp_{CO_2}(1 + K_2/[H^+])} \tag{12a}$$

which obviously cannot be solved for pH because $[H^+]$ appears in the denominator on the right side of the equation. After rearranging Eq. (12a) into the form

$$pH = pK_1' - \log\left(1 + \frac{K_2}{[H^+]}\right) + \log\left(\frac{C_T}{Sp_{CO_2}} - 1\right) \quad (12b)$$

and if

$$pK_1'' = pK_1' - \log\left(1 + \frac{K_2}{[H^+]}\right) \quad (13)$$

then we return to the familiar form of the Henderson–Hasselbalch equation

$$pH = pK_1'' + \log\left(\frac{C_T}{Sp_{CO_2}} - 1\right) \quad (12c)$$

which is equivalent to Eq. (5) and where pK_1'' now appears to depend not only on the temperature but also on the pH. The first reason for the interaction between pK_1'' and pH is the incorporation of the second dissociation constant K_2 of carbonic acid according to Eq. (13). A second reason is the formation of the ion $NaCO_3$ – from the reaction $Na^+ + CO_3^{2-} = HCO_3^-$ (Siggaard–Andersen, 1965).

It is possible to determine the chemically bound $CO_2[HCO_3^-] + [CO_3^{2-}]$ by titration. If this quantity is introduced in the derivations above, another pK'' is obtained. Likewise some authors prefer the use of the activity of carbonic acid

$$a_{H_2CO_3} = a_0 a_{H_2O} p_{CO_2}$$

where p_{CO} is the partial pressure of CO_2, a_0 the solubility of CO_2 in pure water, and a_{H_2O} the activity of water obtained from the freezing point depression. This leads to another definition of pK_1' and hence pK_1''. There are several ways to define pK_1', depending on the use of concentration or activity of one or more members participating in the equilibrium equation. Therefore, when using tabulated values of pK_1', the basic assumptions underlying such tables must not be neglected. With these reservations in mind we shall simply write pK_1' rather than pK_1'' in the following paragraphs.

For practical purposes, pK_1' is obtained by simultaneous determination of C_T and pH in samples of plasma equilibrated at a known p_{CO_2}. From the analytical data pK_1' can be calculated. Figure 3 shows a line chart for reading pK_1' as a function of pH and temperature. This line chart is valid for mammalian plasma and may also be used for most teleost fish, because ionic strength is similiar in both cases (see Table I). In Fig. 4 pK_1' values of dogfish plasma are shown which differ from the values of mammalian plasma owing to the higher ionic strength of elasmobranch blood. The reader should keep in mind that pK_1' defined here is not a constant nor does it have a thermodynamic meaning. It is nothing but an operational figure serving to fit three measurable quantities (C_T, pH, and p_{CO_2}) into an equation. This equation has proved to be most useful and its importance is beyond any doubt. These simplifications lead Homer W. Smith (1956) to call this equation "a most useful monument to human laziness."

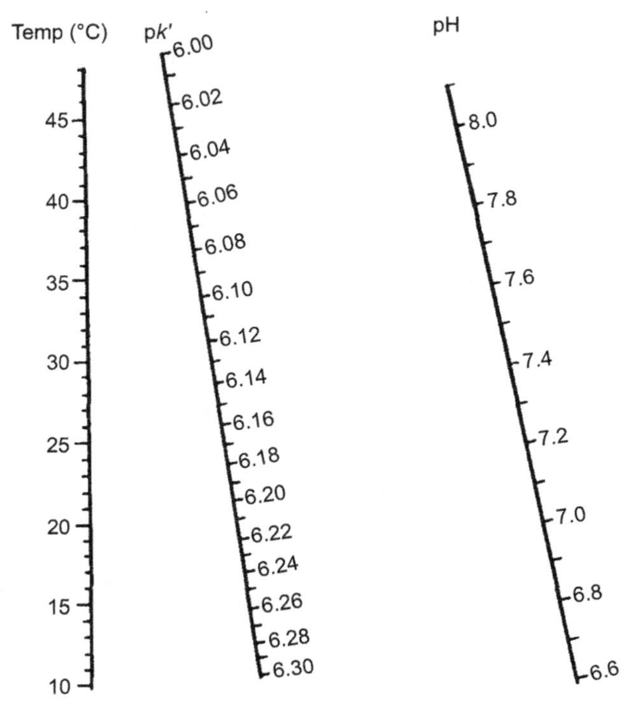

Fig. 3. Operational pK' of carbonic acid as a function of pH and temperature. From Severinghaus *et al.* (1956). Reprinted by permission of *J. Appl. Physiol.*

D. Buffer Action and Its Mathematical Description

Close inspection of the titration curves (Figs. 1 and 2) reveals an important relationship. Both the top and the bottom part of the curves indicate large changes in pH if small amounts of the base are added. On the other hand, in the neighborhood of the inflection point even large amounts of base result in only relatively small changes in pH. At the inflection point half of the acid is neutralized and present as salt. Thus, such a mixture of a weak acid and its salt with a strong base tends to maintain its pH when other acids or bases are added and is called, therefore, a buffer solution. As seen from Figs. 1 and 2 the buffer action is most effective if salt and acid are present in equal amounts. The buffer action can be described quantitatively by the slope of the curve, $d\mathrm{B}/d\mathrm{pH}$, where $d\mathrm{B}$ is an infinitesimal amount of base added to the solution. If we start with a pure acid AH at a concentration C_r and add B moles of the base per liter, it follows $[A^-]=B$ and $[AH]= C_P-B$. Substitution of these values into Eq. (5) and logarithmic differentiation finally gives

Table I Ionic Composition of Fish Plasma and Calculated Values of the Ionic Strength μ (all concentrations in mmole /kg H_2O^2)

Species	Na	K	Ca	Mg	Cl	SO_4	HCO_2[b]	HPO_4	μ	Reference
Myxine	558	9.6	12.5	38.8	576	13.3	2	?	0.70	Robertson (1954)
Lampetra fluviatilis	125	3.3	4.0	4.4	100	5.6	6.7	?	0.15	Robertson (1954)
Raia stabuluforis	277	6.5	5.1	3.0	252	1	8.1	2.0	0.29	Smith (1929)
Raia erinacea	270	8.6	12.9	4.1	275	(4)	29.6	?	0.33	Hartman *et al.* (1941)
Narcine	144	7.5	17.2	3.2	171	(4)	13.3	?	0.22	Pereira and Sawaya (1957)
Rhinobatus	154	13.7	7.8	2.1	155	(4)	24.5	?	0.20	Pereira and Sawaya (1957)
Cyprinus carpio	130	6.3	2.9	1.4	127	(4)	(10)	?	0.15	Field *et al.* (1943)
Corregonus	141	3.8	5.3	3.4	117	4.6	10.6	?	0.16	Robertson (1954)
Muraena	212	2.0	7.7	4.9	188	11.4	8.0	?	0.25	Robertson (1954)
Thunnus	190	26.8	(4)	(3)	181	(4)	41.8	?	0.24	Becker *et al.* (1958)

[a]Water content of plasma was assumed to be 930 g/ liter if not stated otherwise in the references. Assumed values in parentheses.
[b]Bicarbonate values if not given in the reference are calculated as the difference between the sum of anions and the sum of cations.

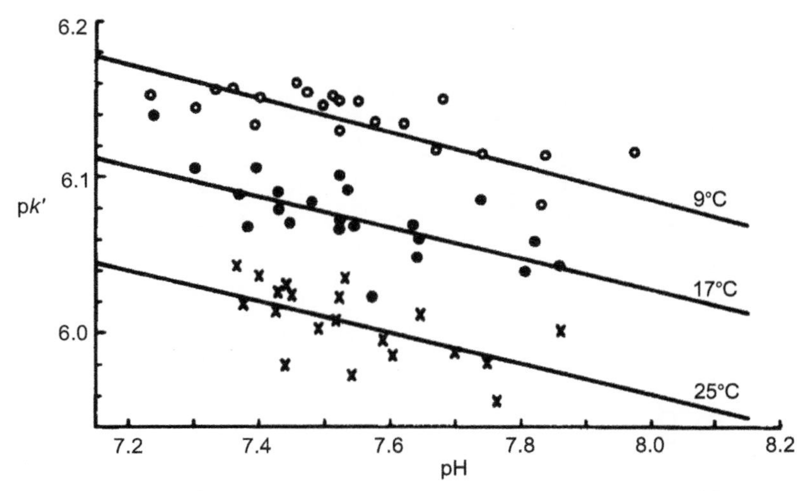

Fig. 4. Operational pK' of carbonic acid in dogfish plasma as a function of pH and temperature. S_{CO_2} = 73 % CO_2 solubility in pure water. From Albers and Pleschka (1967). Reprinted by permission of North Holland Publishing Co.

$$\frac{dB}{dpH} = 2.3C_P \frac{K[H^+]}{(K + [H^+])^2} \tag{14}$$

which apparently has its maximum, if $[H^+] = K$ and $dB/dpH = 0.575\ C_P$. This formula was derived by van Slyke, who called dB/dpH the buffer capacity of a solution. Obviously the buffer capacity depends on (1) the concentration C_P of the buffer substance, (2) the dissociation constant K of the buffer substance, and (3) the concentration of hydrogen ions. Buffer systems where $C_P = [A^-] + [AH]$ is constant are called "homogeneous systems." Biological examples are tissues and blood passing through tissues. The buffer substances in the blood are the plasma proteins and especially hemoglobin in the red cells. Because of their ampholytic dissociation, plasma proteins and hemoglobin can be considered to be much weaker acids than carbonic acid. They are able therefore to buffer the hydrogen ions arising from the carbonic acid. In the tissues, proteins are less important for buffering which is effected chiefly by anorganic and organic phosphate compounds. A homogeneous buffer system is completely described by Eq. (14). If a narrow range of $[H^+]$is considered, dB/dpH is nearly constant and the relationship between B and pH can be approximated by a straight line. This fact is used for the determination of the so–called buffer lines of true plasma (see below).

Quite another type of buffering is realized in systems where [AH] is kept constant rather than C_P. In the case of a bicarbonate buffer [AH] is kept constant if the partial pressure of CO_2, and hence the product Sp_{CO_2}, remains unchanged. Since addition of an acid to a bicarbonate buffer would primarily increase the concentration of carbonic acid [Eq. (9b)] and of CO_2 [Eq. (9a)], there must be another system linked to the buffer which takes up the excess of CO_2. An example of such a buffer is seawater which is in equilibrium with the p_{CO_2} of the atmosphere. If an acid is added to seawater, CO_2 escapes into the atmosphere; conversely, if a base is added, CO_2 is taken up from the atmosphere until p_{CO_2} is restored to its initial value. Since we have two systems linked to each other, such buffering is said to occur in a heterogeneous system. The quantitative behavior of such a heterogeneous system is obtained from Eq. (12c) where the denominator is kept constant while the numerator is changed. Two curves corresponding to $Sp_{CO_2}=0.01$ and 0.1mmole/ liter and $pK'=6.4$ are shown in Fig. 5. In contrast to the sigmoid–shaped curves of Figs. 1 and 2, we obtain curves with a slope and thus with a buffer capacity increasing continuously with the pH. For $Sp_{CO_2}=$ const the buffer capacity is found by differentiation of Eq. (12c) to be $dB/dpH=2.3[HCO_3^-]$. The absolute value of pH depends on p_{CO_2}, a tenfold change in p_{CO_2} results in a change of pH by one unit. The maintenance of a constant p_{CO_2} therefore is the major factor controlling the pH in such a heterogeneous system. The pH of the blood passing through the capillaries of the lungs is chiefly regulated by the proper adjustment of the ventilation to yield an almost constant arterial CO_2 tension.

We should keep in mind that the basic mechanisms of buffering in a homogeneous (or closed) system and in a heterogeneous (or open) system are fundamentally different. In a homogeneous system the buffering is a purely chemical process, depending on the nature and the concentration of the participating buffer substances. In a heterogeneous system the buffering depends on physical and physiological mechanisms involved in the maintenance of a constant CO_2 tension. As we have seen, acid–base regulation takes advantage of both types of buffering. Within the tissue acids entering the

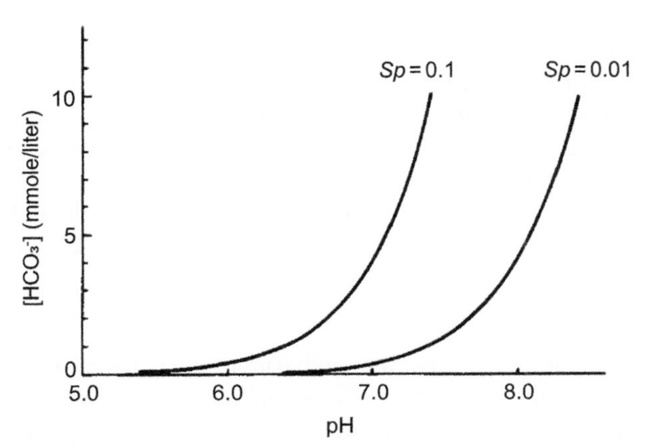

Fig. 5. Relationship between bicarbonate concentration $[HCO_3^-]$ and pH for a heterogeneous buffering system. Physically dissolved CO_2 0.1 and 0.01 mmole/liter, respectively.

blood are buffered by chemical processes according to a homogeneous system, whereas in the lungs or gills the blood, being in proximity to the environment, shows the typical properties of a heterogeneous buffer system.

E. Effects of Ionic Strength and Temperature

1. Effect of Ionic Strength

All biological fluids contain salts in an appreciable concentration, resulting in an ionic interaction which has to be taken into account. When investigating the acid–base balance, the ionic interaction is especially important for polyvalent ions. The total ionic interaction can be expressed as the ionic strength μ, which is half the sum of the products $c_i z_i^2$, where c_i is the concentration of the ion species i and z_i is the corresponding valency. The ionic strength of human plasma is 0.167 at 38°C. Table I shows the ionic composition of plasma of various freshwater and marine fish and the ionic strength calculated from these data. Some freshwater fish such as *Cyprinus carpio* and *Lampetra fluviatilis* have about the same ionic strength as mammals. In many species the ionic strength is higher, reaching about 0.3 for *Raia* and 0.7 for *Myxine.* In solutions having such an ionic strength the law of mass action can be applied only in a much more sophisticated form. Since the quantitative approach to the ionic interaction is beyond the scope of this chapter, the reader is referred to any good textbook on physical chemistry for details. Briefly summarized, an increase in ionic strength has two consequences: (1) it decreases the solubility of gases and (2) it increases the equilibrium constants for most reactions. The latter becomes evident if we recall from Eq. (4) that the operational constant K' is related to the thermodynamic constant K by the activity coefficient $K' = K/f$. Since f is lowered by an increase in ionic strength, K' must increase (or pK' decrease) together with μ. Examples are seen in Table II where K_1' of boric acid and of carbonic acid as well as K_2' of carbonic acid are listed as a function of temperature and chlorinity of seawater. Values for the solubility S of CO_2 are also given. Obviously at any given temperature S is highest for pure water (chlorinity $=0$) and decreases if the salt content and hence the ionic strength increase. The dissociation constants show the opposite effect.

In the same way the ionic strength increases the ionic product of water K_w. If we pass from concentrations to activities we obtain from Eq. (1)

$$K_W = \frac{a_{H^+} \cdot a_{OH^-}}{a_{H_2O}} \tag{15}$$

where $a_{H^+} \cdot a_{OH^-}$ is called the thermodynamic dissociation product. Since K_W and a_{H_2O} are affected by the ionic strength in an opposite direction, the effect of the ionic strength on the thermodynamic dissociation product is small. It seems reasonable, therefore, to derive the definition of neutrality from the thermodynamic dissociation product rather than from the ionic product K_w.

Table II Effect of Temperature and Chlorinity on the Dissociation Constants of Boric Acid and Carbonic Acid and the Solubility of CO_2[a]

t (°C)	Chlorinity (‰)	Boric acid $K' \times 10^9$	Carbonic acid		S_{CO_2} (mmole/liter atm)
			$K'_1 \times 10^6$	$K'_2 \times 10^9$	
0	0	0.40	0.26	0.023	77.0
	15	1.10	0.58	0.43	67.4
	20	1.29	0.63	0.57	64.0
10	0	0.50	0.34	0.032	53.6
	15	1.35	0.74	0.60	47.2
	20	1.58	0.80	0.80	45.2
20	0	0.60	0.40	0.042	39.4
	15	1.62	0.89	0.76	35.1
	20	1.95	0.97	1.02	33.7
30	0	0.72	0.45	0.051	29.9
	15	1.95	1.01	0.93	27.0
	20	2.29	1.10	1.26	26.0

[a]From Harvey (1966). Reprinted by permission of Cambridge University Press.

2. Effect of Temperature

There is no constant in the whole realm of physical chemistry which does not depend on temperature. As a general rule the dissociation of a molecule into ions is favored by an increase in temperature. As a consequence most values of K' increase together with the temperature. This can be seen from Table II which also shows that the effect of temperature on the second dissociation constant for carbonic acid is even greater than that on the first.

Of particular interest is the dissociation of water. Since with increasing temperature K_W increases, the pH indicating neutrality decreases. It is only at 25°C that pure water has a pH of 7.0, whereas at 5°C it has a pH of 7.36 and at 50°C it has a pH of 6.65. That is to say, the "meaning" of pH depends on the temperature. A pH of 7.0 "means" a neutral solution at 25°C, a slightly acid solution at 5°C, and a slightly alkaline solution at 50°C. To avoid the difficulty of interpreting a pH value, Winterstein (1954) suggested the use of the ratio $[OH^-]/[H^+]$ which of course is unity for a neutral solution, smaller than unity for an acid solution, and larger than unity for an alkaline solution. For similar reasons Rahn (1967) introduced the term "relative alkalinity" which is defined by $[H^+]_N/[H^+]$, where $[H^+]_N$ is the hydrogen ion concentration at neutrality and $[H^+]$ the actual hydrogen ion concentration. Since $[H^+]_N/[H^+] = \sqrt{K_W}/[H^+]$ and $[OH^-]/[H^+] = K_w/[H^+]^2$, the relative alkalinity of Rahn is the square root of Winterstein's term $[OH^-]/[H^+]$. A line chart for both terms as a function of temperature and pH is shown in Fig. 6.

In a similar way temperature affects the dissociation constants of the substances involved in the buffering processes. This has two consequences: (1) if a blood sample is cooled anaerobically, its pH will increase by roughly 0.015 per each degree centigrade; (2) if blood is equilibrated with CO_2, not only the physically dissolved CO_2 but also the chemically bound CO_2 increases if the temperature is lowered (see Fig. 12). This may be explained as follows: If the blood is cooled, the K' of the buffer substances decreases more than the K' of carbonic acid. Therefore, hydrogen ions recombine with the buffer to increase the undissociated moiety and more cations are available to form the dissociated salts of the carbonic acid: The chemically bound CO_2 increases. In addition, the physically dissolved CO_2 is increased, too, because the solubility of gases varies inversely with the temperature (Table II). Since the effect of temperature on the physically dissolved CO_2 is more pronounced than that on the chemically bound CO_2, the denominator in Eq. (12c) decreases more than does the numerator. Therefore, the pH must increase.

The effect of temperature on the dissociation constants of carbonic acid can be seen in Table II. From the data in Table II it is obvious that pK_2' is affected more than pK_1' especially for salt solutions. Although these effects are important for the physical chemistry of freshwater and seawater, they are of minor importance for the temperature effect on the acid–base balance of the blood when compared with the effects of temperature on the solubility of CO_2 and on the dissociation constants of the buffer substances.

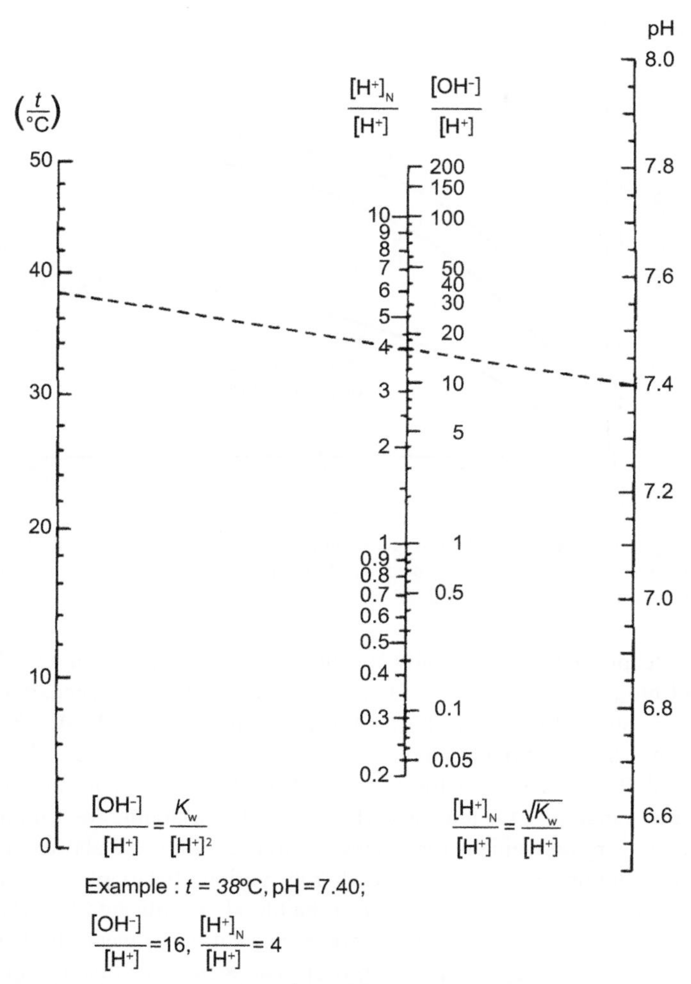

Fig. 6. Relative alkalinity and $[OH^-]/[H^+]$as a function of pH and temperature. Adopted from Albers (1962).

III. THE TRANSPORT OF CO_2 IN THE BLOOD

A. The CO_2 Combining Curve of the Blood

The curve relating the total CO_2 content of the blood to the CO_2 tension is commonly called the CO_2 dissociation curve of the blood. Since total CO_2 comprises physically dissolved CO_2 as well as chemically bound CO_2, the term "CO_2 combining curve" seems to be more appropriate. Figure 7 shows the CO_2 combining curves of oxygenated and deoxygenated blood of the salmon and the dogfish. In the region of low CO_2 tensions, the curves

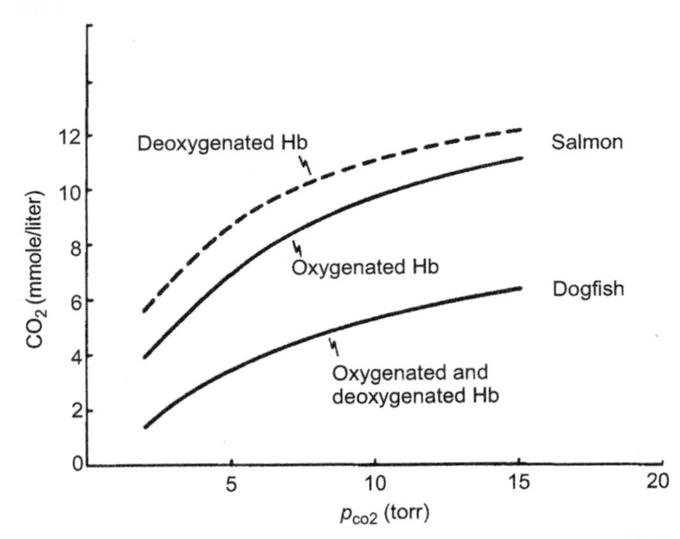

Fig. 7. CO_2 combining curves of the Altantic salmon, *Salmo salar*, and the landlocked salmon, *Salmo salar sebago*, at 5°C and of the dogfish, *Mustelus canis*, at 22°C. Data from Black *et al.* (1966b,c) and from Ferguson *et al.* (1938).

increase steeply but then gradually flatten out. In the trout the curves of deoxygenated blood show a higher CO_2 content than those of oxygenated blood, whereas no such effect could be demonstrated in the dogfish. As shown by Henderson (1932) for mammalian blood and by Ferguson *et al.* (1938) and Albers and Pleschka (1967) for fish blood, such curves yield straight lines in a limited range of CO_2 tensions when plotted in a double logarithmic system. This is very convenient for practical purposes. Figure 8 shows examples of CO_2 combining curves of the blood from various fishes plotted this way. Obviously, the CO_2 combining power of the blood, as indicated by these parameters, varies considerably. There is some correlation between the hemoglobin content and the CO_2 combining power. The reason for this as well as for the effects of oxygenation and deoxygenation will be discussed below.

It seems necessary to stress the narrow range of CO_2 tensions encountered under physiological conditions. Generally the arterial p_{CO_2} is about 1–2 torr only and the venous p_{CO_2} rarely exceeds 10 torr. *In vitro* investigations of the CO_2 transport should concentrate largely on this range. Equilibrium of fish blood with CO_2 tensions as high as 70 torr as done in the early papers in this field does not seem to provide useful information.

1. Physically Dissolved CO_2

If the partial pressure of CO_2 is known, the amount of physically dissolved CO_2 can be calculated with the aid of the solubility coefficient S [see Eq. (8)]. Whereas values of S are well established for mammalian blood, there

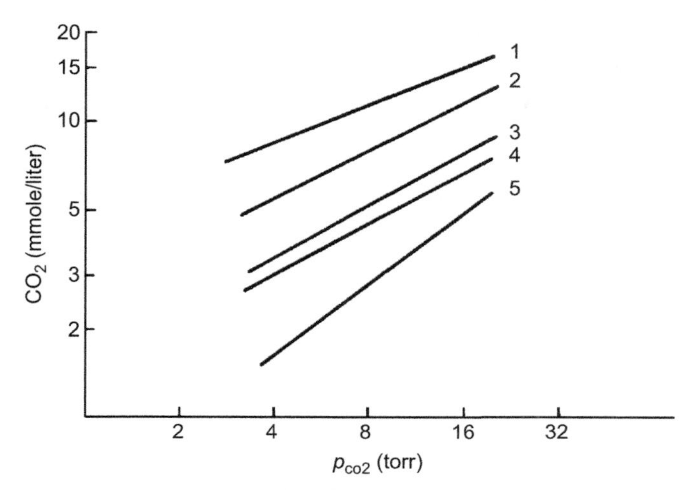

Fig. 8. CO_2 combining curves of various fishes plotted in a double log system. 1, *Cyprinus carpio* (15°C) (Ferguson and Black, 1941); 2, *Raia oscillata* (10°C) (Dill *et al.*, 1932); 3, *Opanus tau* (20°C) (Root, 1931); 4, *Mustelus canis* (22°C) (Ferguson *et al.*, 1938); 5, *Pterodorus* (28°C) (Willmer, 1934, quoted by Fry, 1957).

are almost no data reported on fish blood. In the case of marine fish some authors simply use the solubility of CO_2 in the surrounding seawater. Since the ionic composition and the ionic strength of fish plasma displays a fairly large variability, it is highly desirable to obtain reliable data for S especially if the Henderson–Hasselbalch equation is to be used. If approximations of *S* are applied, the results of such calculations are jeopardized by substantial errors.

Reaction (9a)

$$CO_2 + H_2O \rightleftharpoons H_2CO_3$$

as a molecular reaction proceeds very slowly. The time needed for full equilibrium is about 200 times longer than the time spent by the blood in the gills or in the lungs. The red cells of all vertebrates, however, contain the enzyme carbonic anhydrase which speeds up the formation and splitting of carbonic acid. The concentration of the enzyme in the red cells is correlated to the metabolic activity of each species. Although the concentration of the enzyme in fish blood is lower in most cases than that in mammals (Table III) there is evidence that the concentration of the enzyme is always in excess of the physiological demands (Maren, 1962; Larimer and Schmidt–Nielsen, 1960). If the enzyme is blocked by sulfanilamide a respiratory acidosis ensues with an elevation of the arterial CO_2 tension and a decrease in pH (Table IV). The increase of p_{CO_2} is not only seen in the blood but also in the swim bladder, cerebrospinal fluid, and aqueous humor.

Table III Concentration of Carbonic Anhydrase in Red Blood Cells of Various Species [a]

Species	Carbonic anhydrase (en/g)
Squalus acanthias	40
Raia ocellata	32
Ameiurus nebulosus	300
Lophius piscatorius	300
Perca fluviatilis	2400
Dog	1400
Rat	2400
Cat	2000

[a]From Maren (1967).

Table IV Effect of Blocking Carbonic Anhydrase with Diamox on p_{CO_2}, pH, and CO_2 Content of Blood, Cerebrospinal Fluid, and Aqueous Humor

Specimen		p_{CO_2} (torr)	pH	C_{CO_2} (mmoles/liter)
		Squalus acanthias[a]		
Venous blood	Control	9	7.56	7.96
	Diamox	16	7.47	12.1
Cerebrospinal fluid	Control	8	7.66	8.57
	Diamox	18	7.56	17.1
Specimen		p_{CO_2} (torr)	pH	C_{CO_2} (mmoles/kg H_2O)
		Salvelinus namaycush[b]		
Venous blood	Control	4.4	7.61	6.65
	Diamox	9.3	7.38	8.96
Aqueous humor	Control	4.2	7.65	6.40
	Diamox	8.4	7.22	4.96
Swim bladder	Control	4.6	–	–
	Diamox	10.2	–	–

[a]From Maren (1962).
[b]From Hoffert and Fromm (1966) and Hoffert (1966).

2. Chemically Bound CO_2

After carbonic acid has formed, the following reactions take place:

$$H_2CO_3 + BPr \rightleftharpoons BHCO_3 + HPr$$
$$H_2CO_3 + BO_2Hb \rightleftharpoons BHCO_3 + HO_2Hb$$
$$H_2CO_3 + BHb \rightleftharpoons BHCO_3 + HHb$$

where B denotes a cation bound either to plasma proteins (Pr) or to oxygenated or deoxygenated hemoglobin (O_2Hb and Hb, respectively). Since carbonic anhydrase is not present in the plasma, the first reaction is of minor importance. The other two reactions take place within the red cell. From the law of mass action it is apparent that the higher the concentration of hemoglobin, the higher the concentration of bicarbonate; hence, the CO_2 combining power of the blood. At the low body temperatures of most fish a small fraction of the bicarbonate further dissociates into carbonate ions and hydrogen ions. Because of the presence of ion pairs like $NaCO_3^-$ (Siggaard–Andersen, 1965) it is impossible at the present time to obtain quantitative information about the carbonate concentration from the few data available. The cations for the formation of bicarbonate are made available by the ampholytic dissociation of the protein component of the hemoglobin. The functional group involved is the imidazole, the dissociation of which is strongly affected by the oxygenation of the hemoglobin (see below).

Another chemical reaction between CO_2 and hemoglobin depends on some free α–amino groups, which at the alkaline side of the isoelectric point form so–called carbamino hemoglobin according to the equation

$$HbNH_2 + CO_2 \rightleftharpoons HbNCOO^- + H^+$$

Since an increase in the concentration of CO_2 on the left side of the equation is always associated with an increase in the concentration of H^+at the right side of the equation, the equilibrium of the reaction is hardly affected by the CO_2 tension. It is, however, strongly dependent on the oxygenation of the hemoglobin. Although in mammals the carbamino hemoglobin is only a small fraction of the absolute amount of the total CO_2, it plays an important role in the changes of the CO_2 concentration associated with the delivery of CO_2 from the tissues to the capillary blood. This may be seen from the following example: In human beings arterial blood contains an average 22 mmoles/ liter of total CO_2, which is made up of 1.25 mmoles/ liter of physically dissolved CO_2 (or 5.7% of the total CO_2), 19.65 mmoles/liter of bicarbonate (or 89.3% of the total), and 1.1 mmoles/liter of carbamino hemoglobin (or 5.0% of the total CO_2). When this blood passes through a capillary it takes up 2.2 mmoles/ liter of CO_2, the venous blood having 24.2 mmoles/ liter of total CO_2. The arteriovenous difference of 2.2 mmoles/liter is partitioned into 0.18 mmoles/ liter of physically dissolved CO_2 (or 8.2% of the arteriovenous difference), 1.39 mmoles/liter of bicarbonate (or 63% of the arteriovenous difference), and 0.63 mmoles/liter of carbamino hemoglobin (or 28.8% of the arteriovenous difference). The changes in carbamino hemoglobin therefore account for almost one–third of the total change in CO_2 when blood passes from an artery to a vein (Roughton, 1964).

Whether or not carbamino compounds are formed in fish blood is not known. Indirect evidence seems to rule out the presence of carbamino hemoglobin in the dogfish, *Mustelus* (Ferguson *et al.*, 1938), and in the carp (Ferguson and Black, 1941), whereas in the trout some caramino hemoglobin may be present (Ferguson and Black).

From the distribution of CO_2 between cells and plasma in the dogfish, Ferguson *et al.* (1938) conclude that substances other than hemoglobin might be involved in the buffering of CO_2 within the red cell. Apart from the so–called Y–bound CO_2 (an association of protein and bicarbonate), they point out that not yet identified substances in the nuclei may participate in the buffering process. This seems to be an interesting phenomenon which also could be of importance for the buffering in the tissues. During heavy exercise it is claimed that creatinine diffuses out of the muscles and increases the buffering capacity of the blood in the trout (Black *et al.*, 1959).

3. Interaction between Red Blood Cells and Plasma

Mammalian red blood cells are practically impermeable to cations but are freely permeable to anions. Because of the negative charges of the nondiffusing protein inside the red blood cell, a Donnan equilibrium exists across the cell membrane. The distribution of the ions is characterized by the ratio r according to

$$r = \frac{[\mathrm{Cl}^-]_i}{[\mathrm{Cl}^-]_e} = \frac{[\mathrm{HCO_3}^-]_i}{[\mathrm{HCO_3}^-]_e} = \frac{[\mathrm{A}^-]_i}{[\mathrm{A}^-]_e} = \frac{[\mathrm{H}^+]_e}{[\mathrm{H}^+]_i} \tag{15a}$$

where $[A^-]$ is the concentration of all monovalent anions except bicarbonate. The subscripts refer to the red cells (i) and plasma (e). The concentrations have to be expressed per kilogram of cell water and of plasma water, respectively, not per liter of plasma or red cells. As a corollary of the Donnan equilibrium, the increase of bicarbonate within the red cell leads to a redistribution of anions. Freshly formed bicarbonate diffuses out of the cells into the plasma in exchange for chloride, until Eq. (15a) is fulfilled.

Since the bicarbonate formed increases the number of osmotic particles in the cell water, there is also a redistribution of water, which enters the cells and causes a small but measurable increase in cell volume.

The two predominant consequences of these processes in mammalian blood are:

(1) From the total CO_2 in whole blood one–third only is found in the red cells and two–thirds in the plasma although the volume of red cells in man is about 45% of the blood. The CO_2 content of the plasma therefore exceeds that of whole blood (see Fig. 9).

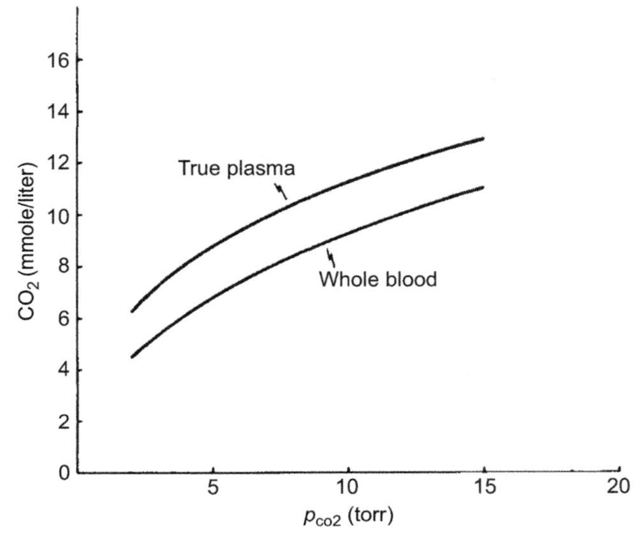

Fig. 9. CO_2 combining curve of true plasma and of whole blood in the rainbow trout, *Salmo gairdneri* Richardson. Data from Ferguson and Black (1941).

(2) When blood passing through a capillary takes up CO_2, almost all bicarbonate is formed within the red cells but only one–half stays in the cells and the other half is exchanged for chloride.

There are almost no data available on the Donnan equilibrium of fish erythrocytes. In the dogfish, Ferguson *et al.* (1938) determined the Donnan ratio for chloride to be 0.49–0.61, whereas for CO_2 the ratio was 1.03–1.97. These authors stress the experimental difficulty of such determinations, especially in fish blood. They state, ". . . it cannot be concluded from these data that there is no shift in water or chloride when CO_2 is added to dogfish blood." Albers *et al.* (1969), however, demonstrated in the dogfish, *Scyliorhinus canicula,* a decrease in the chloride ratio when the CO_2 tension was elevated. According to Ferguson and Black (1941), in the rainbow trout there is a higher Donnan ratio for CO_2 than for chloride, whereas in the carp both ratios are of the same order of magnitude. With increasing CO_2 tensions the red cells of the trout display an unusual increase in volume together with an appreciable decrease in plasma chloride. It is assumed that acids other than carbonic acid are buffered within the red cells and are also exchanged for chloride (Ferguson and Black, 1941).

4. *True Plasma versus Separated Plasma*

Because of the interaction between cells and plasma it is necessary to introduce the distinction between true plasma and separated plasma. If red cells are spun down, the remaining plasma may be equilibrated with various CO_2 tensions. The resulting CO_2 combining curve is, as a rule, flat and shows

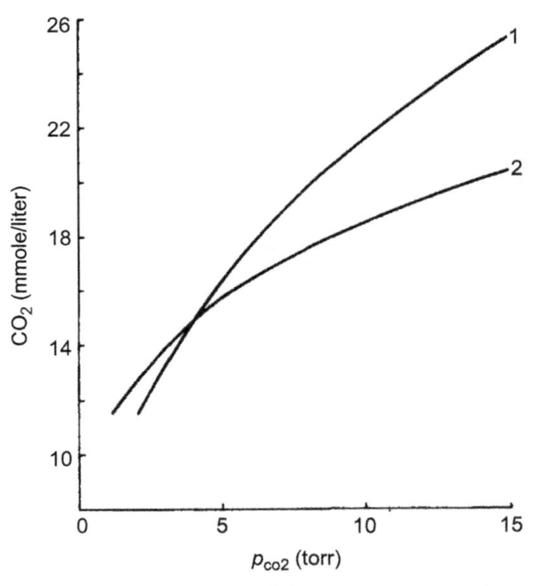

Fig. 10. CO_2 combining curves of true plasma (1) and of separated plasma (2) in the dogfish, *Scyliorhinus stellaris*. Separation at $p_{CO_2}=4$ torr (Albers and Pleschka, 1967).

a smaller CO_2 combining power than does whole blood. Since the only buffer substances present are the plasma proteins, the buffer capacity (see below) of such plasma is very low. Plasma equilibrated with CO_2 without red cells present is called "separated plasma." If we equilibrate whole blood with various CO_2 tensions and spin the red cells under paraffin oil, analyze the plasma for CO_2, and plot the CO_2 concentrations against the p_{CO_2}, we obtain a steep CO_2 combining curve with a CO_2 combining power greater than that of whole blood and with a much greater buffer capacity (Fig. 9).

Plasma which has been equilibrated in the presence of red cells is called "true plasma." It should be noted that the curves of true plasma and of separated plasma have one point in common: This is the point of the CO_2 tension where the separation of red cells and plasma has taken place. Examples of the different CO_2 combining curves of true and separated plasma are given in Fig. 10 for the dogfish, *Scyliorhinus stellaris*. Differences between true plasma and separated plasma have been reported for the carp and the trout (Black *et al.*, 1959). No marked differences in the buffer capacity of true and separated plasma are found in the spiny dogfish (Lenfant and Johansen, 1966).

B. The pH of the Blood as Related to CO_2

1. *THE pH — LOG p_{CO_2} DIAGRAM*

The foregoing sections dealt with the transport of CO_2 by the blood. In this section we shall discuss how CO_2 affects the blood pH. The concentrations of physically dissolved and chemically bound CO_2 are related to the pH by Eq. (12c)

$$pH = pK_1{}' + \log\left(\frac{C_T}{Sp_{CO_2}} - 1\right)$$

where pK' is the operational constant which depends on temperature as well as on pH and ionic strength. Total $CO_2(C_T)$ and p_{CO_2} are linked together by the CO_2 combining curve. It immediately becomes apparent then that for a given p_{CO_2} the pH will be different according to the position of the CO_2 combining curve, i.e., the value of the total CO_2. When calculating the pH from the CO_2 combining curve, the relationship between $[H^+]$ and p_{CO_2} turns out to be a linear one, at least for a limited range of p_{CO_2}. This is valid for mammalian blood (Henderson, 1932) and was demonstrated to be true for fish blood by Root (1931). To avoid the conversion from pH into $[H^+]$ it is common practice to plot pH vs. log p_{CO_2}, which, of course, also results in straight lines. Examples are given in Fig. 11. These lines can be obtained experimentally and are widely used for the indirect determination of p_{CO_2} in blood samples. Two facts are evident from Fig. 11:

(1) As mentioned above, for a given p_{CO_2} the pH may differ from species to species according to the position of the CO_2 combining curve.

(2) There are differences in the slope of the lines. For a given increase in p_{CO_2} some species show a larger decrease in pH indicating a poor buffer capacity, whereas other species show a smaller decrease in pH, indicating a higher buffer capacity.

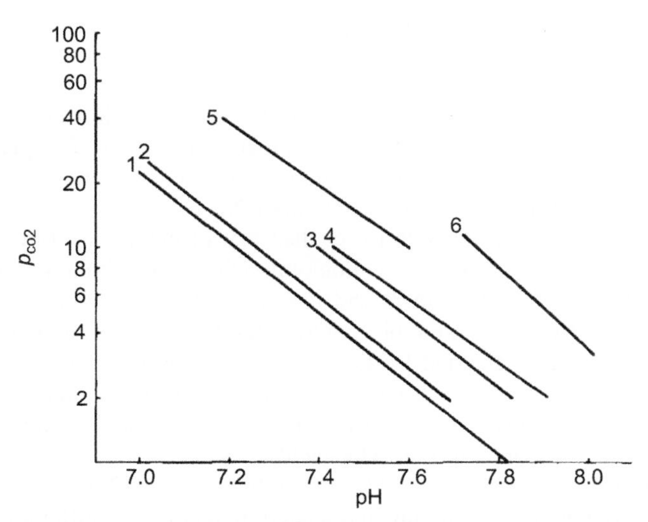

Fig. 11. pH–log p_{CO_2} lines in various fishes and in man. I, *Opsanus tau* (20°C); 2, *Prionotus carolinus* (20°C); 3, *Scyliorhinus stellaris* (17°C); 4, *Cyprinus carpio* (15°C); 5, *Scomber* (20° C); 6, man(15°C). 1, 2, and 5 from Root (1931); 3 and 6 from Albers and Pleschka (1967); 4 from Ferguson and Black (1941).

2. *The Buffer Capacity of Plasma and Blood*

The buffer capacity of a buffer solution is quantitatively described by Eq. (14). The chemical reactions described above can be considered as typical examples for buffer reactions if we keep in mind that carbonic acid is the "strong acid" which has to be buffered by the plasma proteins and by hemoglobin. At the physiological pH these proteins act as weak acids which are partly present as dissociated salts thus forming a typical buffer solution. If carbonic acid is added to this system, hydrogen ions recombine with the proteins to form the undissociated "weak acid," leaving the cation together with the bicarbonate ion as fully dissociated salt. Metaphorically speaking, carbonic acid and proteins compete with each other for the available cations to form salts. The resulting compromise depends on the dissociation constant of the buffer substance: the lower the dissociation constant (i.e., the weaker the acid), the more hydrogen ions are bound to the buffer and the more cations are left for the strong acid to form a salt. We shall refer to this relationship when we consider the effect of the oxygenation of hemoglobin on the buffer capacity (see below).

For a quantitative comparison of the buffer action the buffer capacity as described by Eq. (14) has to be modified: Since the concentration of the buffer substances can be different, the buffer capacity often is expressed per unit weight of the buffer substance and denoted:

$$\beta = \frac{1}{C_P} \frac{d\mathrm{B}}{d\mathrm{pH}} \tag{16a}$$

$$\beta = 2.3 \frac{K[\mathrm{H}^+]}{(K + [\mathrm{H}^+])^2} \tag{16b}$$

Values of β and of $d\mathrm{B}/d\mathrm{pH}$ are listed in Table V. Although β of plasma proteins in some elasmobranchs is higher than that of mammals, $d\mathrm{B}/d\mathrm{pH}$ of separated plasma in these species is about the same as in mammals owing to the lower concentration of plasma proteins (Dill *et al.*, 1932). In most fish, $d\mathrm{B}/d\mathrm{pH}$ of whole blood is much lower than that of mammals. The main reason is the lower concentration of hemoglobin, which seems to have the same β in the dogfish as in man (Albers and Pleschka, 1967). However, there may exist specific differences in the physiochemical properties of the participating buffer substances, as claimed by Lenfant and Johansen (1966). There is an open field for further research which seems to be of great interest since the buffer capacity is considered to be of vital importance and a limiting factor for exercise performance in fish (Hochachka, 1961).

The sum of the cations available for the buffering of CO_2 equals the sum of chemically bound CO_2 and the dissociated moiety of the buffer and is often

Table V Buffer Capacity of Whole Blood (d B/dpH) and Estimated Buffer Capacity per Gram Hemoglolin (β) for Various Fish

Species	t (°C)	Hb (g/liter)	dB/ dpH	β	References
Opsanus tau	20	46	6.7	0.147	Root (1931)
Scomber	20	118	14.8	0.127	Root (1931)
Prionotus	20	57	6.7	0.118	Root (1931)
Cyprinus	15	92	—	0.160	Ferguson and Black (1941)
Salmo gairdneri	15	74	—	0.230	Ferguson and Black (1941)
Mustelus canis	22	30	10–20	—	Ferguson *et al.* (1938)
Squalus acanthias	11	30	9	0.117	Lenfant and Johansen (1966)
Sculiorhinus	17	31	10	0.190	Albers and Pleschka (1967)

referred to in the literature as the concentration of "buffer base." The buffer base can be considered as constant as long as no strong acids other than carbonic acid are entering the blood. However, if for instance lactic acid enters the blood, an equivalent amount of cations is converted from the buffer base to the so–called "fixed base." The amount of cations available for the reversible buffer action is thus diminished.

As a consequence, for a given p_{CO_2} less CO_2 can be bound chemically and the CO_2 combining curve will be lowered by an equivalent amount. Likewise the position of the pH $-$ log p_{CO_2} line will be shifted to lower pH values for a given p_{CO_2}. In the medical literature a loss in buffer base is called a "metabolic acidosis" and an increase in buffer base is called a "metabolic alkalosis."

A typical example for a metabolic acidosis in fish is the accumulation of lactic acid in blood and tissues after severe exercise (Black, 1958; Black *et al.*, 1959, 1962). In the rainbow trout, for example, Black *et al.* (1966a) observed an increase in blood lactate from 5.1 to 69.7 mg% after 15 min of strenuous exercise which would correspond to a decrease in buffer base of 7 mEq/ liter.

3. *The Effects of Oxygenation of Hemoglobin*

As seen in Fig. 7 the CO_2 combining power of oxygenated blood is less than that of reduced blood. The following two factors contribute to this change in the CO_2 combining curve:

(1) Because of the difference in the dissociation constant between oxyhemoglobin and hemoglobin (see below) more bicarbonate is formed at a given p_{CO_2} in a solution containing reduced hemoglobin than in a solution containing oxygenated hemoglobin.

(2) The formation of carbamino hemoglobin is more pronounced with reduced hemoglobin than with oxygenated hemoglobin.

Both factors participate almost equally in the change of the CO_2 combining curve of mammals, although there is still some controversy about the exact quantitative relationship. The effect of the oxygenation on the CO_2 combining curve is one aspect of the so–called Haldane effect. The other and more important aspect is the effect on the buffer capacity and the pH – log p_{CO_2} line. From Eqs. (14) and (16b) it becomes immediately evident that a decrease in K must increase β or dB/dpH. The oxygenation of hemoglobin now increases its dissociation constant appreciably. In horse hemoglobin for instance, K is changed by a factor of about 30($pK_{O_2Hb} = 6.68, pK_{Hb} = 7.95$). Similar effects are obtained in the blood of other mammals and also in some fish. Reduced hemoglobin, therefore, having a lower K and a higher pK, respectively, must be a stronger buffer substance. Since at the tissues the transfer of CO_2 into the blood takes place when O_2 is released from the oxyhemoglobin, there is a simultaneous increase in the buffer capacity. The importance of this effect is obvious: If the exchange ratio R of CO_2 and O_2 is 0.7, it can be assumed with the figures of pK given above that the CO_2 entering the blood will not cause any change in pH because of the Haldane effect. If $R = 0.8$, about 85% of the hydrogen ions from the carbonic acid are mopped up by the concomitant change in the oxygenation and only 15% have to be buffered according to Eq. (14). As a result the arteriovenous pH difference is low, at least in mammals, where the pH of the mixed venous blood from the right heart has a pH which is about 0.03–0.05 lower than that of the arterial blood. In fish the difference may be higher because of the lower buffer capacity. For instance, in the dogfish the difference in **pH** between arterial and venous blood has been found to be 0.07 (Baumgarten–Schumann and Piiper, 1969).

4. *The Effect of Temperature*

As pointed out earlier, temperature affects the dissociation constants of the buffer substances as well as the solubility of CO_2. As a result, the CO_2 combining curves of blood are shifted upward if the temperature is lowered. This is observed in mammalian blood (Harms and Bartels, 1961), in the blood of the skate (Dill *et al.*, 1932), and in the dogfish (Fig. 12a). Since the observed shift is much greater than can be accounted for by the change in solubility, the chemically bound CO_2 must also have been altered by temperature (Fig. 12b).

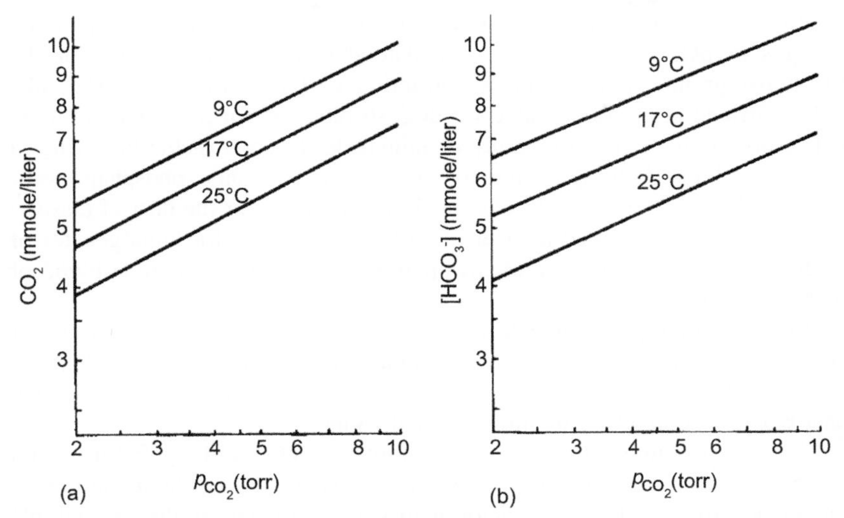

Fig. 12. Effect of temperature on the CO_2 combining curve in the dogfish, *Scyliorhinus stellaris*. (a) Total CO_2 content of whole blood plotted against p_{CO_2}. (b) Chemically bound CO_2 of true plasma (designated as HCO_3^-) plotted against p_{CO_2}. From Albers and Pleschka (1967). Reprinted by permission of North Holland Publishing Co.

Wherease a change in the CO_2 combining curve due to changes in the oxygenation of the hemoglobin inevitably affects the relationship between p_{CO_2} and pH, this does not necessarily happen if the CO_2 combining curve is altered by temperature. If in Eq. (12c)

$$\text{pH} = \text{p}K_1' + \log\left(\frac{C_T}{Sp_{CO_2}} - 1\right)$$

total CO_2, (C_T), and S are altered by temperature in the same proportion, no change of the pH – log p_{CO_2} line would occur. This is the case in human blood at physiological CO_2 tensions in the range between 37° and 26°C (Brewin *et al.*, 1955). However, as these authors point out, this constancy of the pH – log p_{CO_2} line is purely fortuitous: Temperature affects the pH – log p_{CO_2} line of both plasma and red cells but in an opposite direction. In man, at the prevailing quantitative relationship between red cell volume and plasma volume, both changes cancel out each other. Therefore, if the hematocrit is changed, the pH – log p_{CO_2} line does not remain unchanged. However, at very low CO_2 tensions and at temperatures between 25° and 9°C the temperature effect on the pH – log p_{CO_2} line of human blood is marked (Albers and Pleschka, 1967). In fish, data are available only for elasmobranchs. In dogfish blood an increase in temperature decreases the pH for a given p_{CO}, as shown by Albers and Pleschka (1967). These authors were unable to detect a significant change of dB/dpH with temperature in dogfish blood. This is in accord with the findings in mammalian blood.

From the changes in the CO_2 combining curve and the accompanying changes in pH it is possible to calculate the heat of dissociation of the functional groups involved in the buffering of CO_2. Albers and Pleschka (1967) arrived at the same value for dogfish blood as for mammalian blood, indicating the participation mainly of imidazole. In contrast the buffering in tissues is almost entirely owing to inorganic and organic phosphate compounds (Netter, 1959), resulting in much lower values for the heat of dissociation. This was confirmed by Mersch (1964) who found for homogenates of rat liver a value of 500 cal/mole as opposed to the value in human blood of 6300 cal/mole.

Because of the profound effects of temperature and the great variability of body temperatures in fish of various habitats, data on the acid–base balance and the CO_2 transport should always include the temperature. Otherwise such data would lose much of its informational value.

A most revealing effect of temperature was observed by Rahn (1967) when he acclimatized carps, turtles, and frogs to various temperatures in the range from 5° to 30°C. With increasing temperature the arterial pH fell in parallel to the pH of neutrality in all species. Thus the difference $pH - pN$ was maintained constant. Since $pH - pN$ measures the relative alkalinity (see Section II, E, 2), and moreover, since the same $pH - pN$ is found in many warm–blooded animals (Fig. 13), Rahn (1967) claims the relative alkalinity to be one of the most important quantities for the evolution of respiration.

5. *Summary*

CO_2 transport may now be summarized briefly as follows:

(1) CO_2 formed by metabolic processes in the tissues diffuses into the blood and reacts with water to form carbonic acid. The reaction is slow in the plasma. Within the red cells it is speeded up by the enzyme carbonic anhydrase.

(2) Carbonic acid dissociates into hydrogen ions, bicarbonate ions, and to a small extent into carbonate ions.

(3) The hydrogen ions recombine with plasma proteins and hemoglobin which represent the buffer substances. Proteins and hemoglobin can be considered as weak acids which are partly present as fully dissociated salts.

(4) Bicarbonate leaves the red cells in exchange for chloride according to Donnan's law.

(5) Carbonic acid also reacts with hemoglobin to form carbamino hemoglobin.

(6) Oxygenation of the hemoglobin decreases the buffer capacity as well as the ability to form carbamino hemoglobin (Haldane effect) in teleosts but not in elasmobranchs.

(7) In the lungs or gills the above–mentioned reactions occur in the opposite direction.

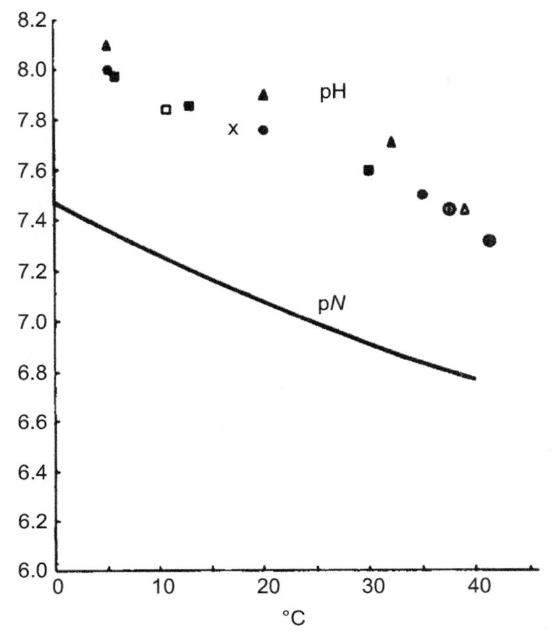

Fig. 13. Arterial pH of various vertebrate species as a function of temperature: (■) carp, (▲) frog, (●) turtle (all data from Rahn, 1967), (□) skate, (×) dogfish, (⊙)man, (Δ) dog, (⊕) duck (from various references). p*N* = pH of pure water.

IV. THE INTRACELLULAR pH

The rather complex processes of the CO_2 transport and the buffering of carbonic acid and other acids in the blood are fairly simple when compared with the analogous processes within the tissues. A comparative study of tissue CO_2 in several vertebrates has recently been published by Haning and Thompson (1965). In contrast to the blood, proteins are much less important as buffer substances. Some CO_2 within the tissues seems to be bound in a carbaminolike fashion and cannot be precipitated by barium hydroxide (Conway and Fearon, 1944). There is still some controversy about the chemical meaning of the so–called "barium soluble fraction" of CO_2. Data on this fraction are given for several vertebrate tissues by Haning and Thompson (1965). In the muscle of the channel catfish, *Ictalurus punctatus*, these authors found 70% of the tissue CO_2 to be barium soluble as compared with 77% in rat muscle and 72% in frog muscle. As recently shown by Butler *et al.* (1967) the "barium soluble CO_2 " does not result from a carbaminolike compound but from the inhibitory effect of proteins on the precipitation of barium carbonate. As mentioned earlier the buffer capacity of tissues is made up predominantly of inorganic and organic phosphate compounds. The total buffer capacity of tissues can well compete with that of the blood in some organs of mammals (e.g., liver tissue of rats) although in most organs the buffer capacity is lower than in blood. The buffer processes in the interior of a cell are closely linked to so many biochemical reactions that a theoretical approach seems to be impossible.

The intracellular pH has been determined using a variety of methods and has been found to be always lower than that of blood. Chambers *et al.* (quoted by Netter, 1959) reported for the nerve cells of fish a pH of 6.8–6.9, based on a colorimetric indicator method. Electrometric pH measurements have also been employed, yielding results comparable with those of indicator methods or with methods applying the Henderson–Hasselbalch equation for tissue CO_2.

An interesting approach to the intracellular pH was made by Waddell and Butler (1959); it is based on the weak acid DMO which can permeate the cell membrane only as an undissociated acid. Thus, extracellular and intracellular concentrations of the undissociated acid are equal. If the total concentration of DMO is determined in the extracellular fluid (plasma) as well as in the intracellular fluid, the undissociated moiety can be calculated by means of the pK' of DMO and the plasma pH. In the next step of the calculation the dissociated moiety is obtained by simple subtraction of the undissociated fraction from the total concentration. Finally, the intracellular pH is obtained by again applying the Henderson–Hasselbalch equation. Using the glass electrode as well as the DMO method, Robin (1961) determined the pH of the pericardial tluid of the dogfish (*Squalus acanthias*) and found good agreement between both methods. The average pH of the pericardial fluid was about 5.6. Robin *et al.* (1964) determined the intracellular pH with the DMO method in the same species and reported 7.41 as average for the brain and 6.95 for the muscle.

V. CONTROLLING MECHANISMS OF THE ACID–BASE BALANCE

It is not the aim of this section to describe in detail the operation of the controlling mechanisms of the acid–base balance. The reader is referred to the pertinent chapters of these volumes. Instead we shall briefly summarize the basic principles of the regulatory mechanisms involved as can be deduced from the previous sections. All quantities used in the many equations of this chapter can be divided into three groups:

(1) The first group consists of pure physicochemical constants, such as the ionic product of water K_w, the dissociation constants K of the participating acids and buffer substances, the solubility of CO_2, etc. These constants depend on temperature and ionic strength. The organism has to cope with these quantities which are completely independent of biological regulation.

(2) The second group is made up of the concentrations of the substances involved in the acid–base balance except the physically dissolved CO_2 (and the bicarbonate). Among these quantities are the various electrolytes, the plasma proteins, the hemoglobin, and other buffer substances. This group is subjected to regulatory mechanisms of several organs, especially to the function of excretory organs like the kidney. The "buffer base" can be changed by a redistribution of electrolytes between tissues, extracellular fluid, and plasma. In addition to the regulatory mechanisms in mammals, in fish there is also an exchange possible between the plasma and the surrounding seawater. Robin *et al.* (1966) and Dejours (1966) point to the possibility of an active excretion of bicarbonate by the gills. The gills also excrete ammonia (Dejours *et al.*, 1968), which in part is exchanged for sodium taken up by the gills (Garcia Romeu and Motais, 1966). Long lasting exposure to asphyxic conditions can produce an increase in hemoglobin, which is not only observed in mammals but also in fish. All changes of variables belonging to the second group are recognized by the evaluation of the CO_2 combining curve or by the $\mathrm{pH} - \log p_{CO_2}$ line. Generally such changes are long–term.

(3) There is only one quantity in this last group. This is the partial pressure of CO_2 which regulates the physically dissolved CO_2 and, hence, according to the CO_2 combining curve, the pH of the arterial blood. The p_{CO_2} is adjusted by the effective irrigation of the gills since for a given CO_2 production the p_{CO_2} varies inversely with it. Ventilatory changes of p_{CO_2} and pH are almost instantaneous and provide the basis for short–term adaptations required, e.g., during muscular exercise. Changes of the arterial p_{CO_2} can be measured directly.

To put it in another way, we may also say that there are two principally different ways of regulating the acid–base balance. There is the metabolic way, dealing chiefly with the buffer processes of the homogeneous type, and there is the respiratory way, dealing chiefly with the buffer processes of the heterogeneous type. As an open system the fish like all other animals is defended twofold against changes of the acidity of its *milieu intérieur*. This defense, however, appears less powerful in fish than in mammals. Whether this is considered as a lower stage of the phylogenetic development or as an adaptation to the ecological demands of the aquatic life is a question the reader may decide.

REFERENCES

Albers, C. (1962). Die ventilatorische Kontrolle des Säure–Basen–Gleichgewichts in Hypothermie. *Anaesthesist* 11, 43–51.

Albers, C. (1966). Unpublished data.

Albers, C., and Pleschka, K. (1967). Effect of temperature on CO_2 transport in elasmobranch blood. *Respiration Physiol.* **2**, 261–273.

Albers, C., and Pleschka, K. (1967). Unpublished data.

Albers, C., Pleschka, K., and Spaich, P. (1969). Chloride distribution between red blood cells and plasma in the dogfish (Scyliorhinus canicula). *Respiration Physiol.* **7**, 295–299.

Bartels, H., and Wrbitzky, R. (1960). Bestimmung des CO_2–Absorptionskoeffizienten zwischen 15 und 38°C in Wasser und Plasma. *Arch. Ges. Physiol.* **271**, 162–168.

Baumgarten–Schumann, D., and Piiper, J. (1969). Gas exchange in the gills of resting unanesthetized dogfish (Sciliorhinus stellaris). *Respiration Physiol.* **5**, 317–325.

Becker, E. L., Bird, R., Kelly, J. W., Schilling, J., Solomon, S., and Young, N. (1958). Physiology of marine teleosts. I. Ionic composition of tissue. *Physiol. Zool.* **31**, 224–227.

Black, E. C. (1958). Hyperactivity as a lethal factor in fish. *J. Fisheries Res. Board Can.* **15**, 573–586.

Black, E. C., Chiu, W., Forbes, F. D., and Hanslip, A. (1959). Changes in pH, carbonate and lactate of the blood of yearling Kamloops Trout (Salmo gairdnerii) during and following severe muscular activity. *J. Fisheries Res. Board Can.* **16**, 391–402.

Black, E. C., Robertson Connor, A., Lam, K., and Chiu, W. (1962). Changes in glycogen, pyruvate and lactate in rainbow trout (Salmo gairdnerii) during and following muscular activity. *J. Fisheries Res. Board Can.* **19**, 409–436.

Black, E. C., Manning, G. T., and Hayashi, K. (1966a). Changes in levels of hemoglobin, oxygen, carbon dioxide, pyruvate, and lactate in venous blood of rainbow trout (Salmo gairdnerii) during and following severe muscular activity. *J. Fisheries Res. Board Can.* **23**, 783–795.

Black, E. C., Kirkpatrick, D., and Tucker, H. H. (1966b). Oxygen dissociation curves of the blood of Landlocked Salmon (Salmo salar sebago) acclimated to summer and winter temperatures. *J. Fisheries Res. Board Can.* **23**, 1581–1586.

Black, E. C., Tucker, H. H., and Kirkpatrick, D. (1966c). Oxygen dissociation curves of the blood of Atlantic Salmon (Salmo salar) acclimated to summer and winter temperatures. *J. Fisheries Res. Board Can.* **23**, 1187–1195.

Brewin, E. G., Gould, R. P., Nashat, F. S., and Neil, E. (1955). An investigation of problems of acid–base equilibrium in hypothermia. *Guy's Hosp. Rept.* **104**, 177–214.

Butler, T. C., Poole, D. T., and Waddell, W. J. (1967). Acid–labile carbon dioxide in muscle: Its nature and relationship to intracellular pH. *Proc. Soc. Exper. Biol. Med.* **125**, 972–974.

Conway, E. J., and Fearon, P. J. (1944). The acid–labile CO_2 in mammalian muscle and the pH of the muscle fibre. *J. Physiol. (London)* **103**, 274–289.

Dejours, P. (1966). Respiratory gas exchange of aquatic animals during confinement. *J. Physiol. (London)* **186**, 126–127.

Dejours, P., Armand, J., and Verriest, G. (1968). Carbon dioxide dissociation curves of water and gas exchange of water–breathers. *Respiration Physiol.* **5**, 23–33.

Dill, D. B., Edwards, H. T., and Florkin, M. (1932). Properties of the blood of the skate (Raia oscillata). *Biol. Bull.* **62**, 23–36.

Ferguson, J. K. W., and Black, E. C. (1941). The transport of CO_2 in the blood of certain fresh water fishes. *Biol. Bull.* **80**, 139–152.

Ferguson, J. K. W., Horvath, S. M., and Pappenheimer, J. R. (1938). The transport of carbon dioxide by erythrocytes and plasma in dogfish blood. *Biol. Bull.* **75**, 381–388.

Field, J. B., Elvehjem, C. A., and Juday, C. (1943). A study of the blood constituents of carp and trout. *J. Biol. Chem.* **148**, 261–269.

Fry, F. E. J. (1957). The aҫuatic respiration of fish. *In* "The Physiology of Fishes" (M. E. Brown, ed.), Vol. 1, pp. 1–63. Academic Press, New York.

Garcia Romeu, F., and Motais, R. (1966). Mise en évidence d'échanges Na^+/NH_{4^+} chez l'anguille d'eau douce. *Comp. Biochem. Physiol.* **17**, 1201–1204.

Haning, Q. C., and Thompson A. M. (1965). A comparative study of tissue carbon dioxide in vertebrates. *Comp. Biochem. Physiol.* **15**, 17–26.

Harms, H., and Bartels, H. (1961). CO_2–Dissoziationskurven des menschlichen Blutes bei Temperaturen von 5 – 37°C und unterschiedlicher O_2–Sättigung. *Arch. Ges. Physiol.* **272**, 384–392.

Hartman, F. A., Lewis, L. A., Brownell, K. A., Shelden, F. F., and Walther, R. W. (1941). Some blood constituents of the normal skate. *Physiol. Zool.* **14**, 476–486.

Harvey, H. W. (1966). "The Chemistry and Fertility of Sea Waters." Cambridge Univ. Press, London and New York.

Henderson, L. J. (1932). "Blut" (German transl. by M. Tennenbaum). Steinkopff, Dresden.

Hochachka, P. W. (1961). The effect of physical training on oxygen debt and glycogen reserves in trout. *Can. J. Zool.* **39**, 767–776.

Hoffert, J. R. (1966). Observations on ocular fluid dynamics and carbonic anhydrase in tissues of lake trout (Salvelinus Namaycush). *Comp. Biochem. Physiol.* **17**, 107–114.

Hoffert, J. R., and Fromm, P. O. (1966). Effect of carbonic anhydrase inhibition on aqueous humor and blood bicarbonate ion in the teleost (Salvelinus Namaycush). *Comp. Biochem. Physiol.* **18**, 333–340.

Larimer, J. L., and Schmidt–Nielsen, K. (1960). A comparison of blood carbonic anhydrase of various mammals. *Comp. Biochem. Physiol.* **1**, 19–23.

Lenfant, C., and Johansen, K. (1966). Respiratory function in the elasmobranch Squalus suckleyi. *Respiration Physiol.* **1**, 13–29.

Maren, T. H. (1962). Ionic composition of cerebrospinal fluid and aqueous humor of the dogfish, Squalus acanthias. II. Carbonic anhydrase activity and inhibition. *Comp. Biochem. Physiol.* **5**, 201–215.

Maren, T. H. (1967). Carbonic anhydrase: Chemistry, physiology and inhibition. *Physiol. Rev.* **47**, 595–781.

Mersch, F. D. (1964). Der Temperatureinfluss auf die CO_2– Dissoziationskurven von Rattenleberhomogenaten. Inaugural Dissertation, Justus Liebig University, Giessen.

Netter, H. (1959). "Theoretische Biochemie." Springer, Berlin.

Pereira, R. S., and Sawaya, P. (1957). Contribution à l'étude de la composition chimique du sang de certains sélaciens du Brésil. *Univ. de Sao Paulo, Fac. Filosof., Cienc. Letras, Zool.* **21**, 85–92.

Prosser, C. L., and Brown, F. A. (1961). "Comparative Animal Physiology." Saunders, Philadelphia, Pennsylvania.

Rahn, H. (1967). Gas transport from the external environment to the cell. *Ciba Found. Symp. Develop. Lung.* pp. 3–29.

Robertson, J. D. (1954). The chemical composition of the blood of some aquatic chordates, including members of the Tunicata, Cyclostomata and Osteichthyes. *J. Exptl. Biol.* **31**, 424–442.

Robin, E. D. (1961). Of men and mitochondria–intracellular and subcellular acid–base relations. *New Engl. J. Med.* **265**, 780–785.

Robin, E. D., Murdaugh, H. V., and Weiss, E. (1964). Acid–base, fluid and electrolyte metabolism in the elasmobranch. I. Ionic composition of erythrocytes, muscle and brain. *J. Cellular Comp. Physiol.* **64**, 409–422.

Robin, E. D., Murdaugh, H. V., and Millen, J. E. (1966). Acid–base, fluid and electrolyte metabolism in the elasmobranch. III. Oxygen, CO_2, bicarbonate and lactate exchange across the gill. *J. Cellular Physiol.* **67**, 93–100.

Root, R. W. (1931). The respiratory function of the blood of marine fishes. *Biol. Bull.* **61**, 427–466.

Roughton, F. (1964). Transport of oxygen and carbon dioxide. In "Handbook of Physiology" (Am. Physiol. Soc., J. Field, ed.), Sect. 3, Vol. I, p. 767. Williams & Wilkins, Baltimore, Maryland.

Severinghaus, J. W., Stupfel, M., and Bradley, A. F. (1956). Variations of serum carbonic acid p*K'* with pH and temperature. *J. Appl. Physiol.* **9**, 197–200.

Siggaard–Andersen, O. (1965). "The Acid–Base Status of the Blood." Munksgaard, Copenhagen.

Smith, H. W. (1929). Body fluids of elasmobranchs. *J. Biol. Chem.* **81**, 407–419.

Smith, H. W. (1956). "Principles of Renal Physiology." Oxford Univ. Press, New York.

Waddell, W. J., and Butler, T. C. (1959). Calculation of intracellular pH from the distribution of 5,5–dimethyl –2,4–oxazolidinedione (DMO). Application to skeletal muscle of the dog. *J. Clin. Invest.* **38**, 720–729.

Winterstein, H. (1954). Der Einfluss der Körpertemperatur auf das Säure–Basen–Gleichgewicht im Blut. *Arch. Exptl. Pathol. Pharmokol.* **223**, 1–18.

Chapter 11

The lasting impact of Toki-o Yamamoto's pioneering chapter on fish sex determination and differentiation: A retrospective analysis of its contributions to reproductive biology and influences on aquaculture and fisheries sciences

J. Adam Luckenbach[a,b,*], Kiyoshi Kikuchi[c], Takashi Iwamatsu[d], Yoshitaka Nagahama[e,f], and Robert H. Devlin[g]

[a]*Environmental Physiology Program, Northwest Fisheries Science Center, National Marine Fisheries Service, National Oceanic and Atmospheric Administration, Seattle, WA, United States*
[b]*Center for Reproductive Biology, Washington State University, Pullman, WA, United States*
[c]*Fisheries Laboratory, University of Tokyo, Shizuoka, Japan*
[d]*Aichi University of Education, Kariya, Aichi, Japan*
[e]*National Institute for Basic Biology, Okazaki, Japan*
[f]*College of Science and Engineering, Kanazawa University, Kanazawa, Japan*
[g]*Fisheries and Oceans Canada, West Vancouver, BC, Canada*
**Corresponding author: e-mail: adam.luckenbach@noaa.gov*

J. Adam Luckenbach, Kiyoshi Kikuchi, Takashi Iwamatsu, Yoshitaka Nagahama and Robert H. Devlin discuss the T. Yamamoto chapter "Sex Differentiation" in *Fish Physiology*, Volume 3, published in 1969. The next chapter in this volume is the re-published version of that chapter.

In 1969, Professor Toki-o Yamamoto published a landmark review on sex differentiation in fishes, which provided the first comprehensive synthesis of the various influences on sexual phenotype, clarified the nomenclature, and sought to define the

Fish Physiology, Vol. 40A. https://doi.org/10.1016/bs.fp.2023.08.003

mechanisms that determine sex. Yamamoto's review highlighted the genetic basis of sex determination and the influence of sex steroids on gonadal sex differentiation. He argued that genetic sex determination was influenced by the net effect of sex chromosomes, when present, and contributions of female- and male-influencing alleles. Yamamoto also demonstrated that sex steroid treatments could override underlying genetic controls, resulting in sex reversal (e.g., XX-genotype males or XY-genotype females) in gonochoristic species. His comprehensive review has since stimulated a wide range of fields including aquaculture and fisheries sciences, genetics, and aquatic toxicology. This retrospective discusses Yamamoto's scientific legacy and remarkable life, which was devoted to studying the reproductive biology of fishes.

The process of sexual differentiation has fascinated scientists and scholars for centuries. From Aristotle to modern-day geneticists, physiologists, and environmental biologists, researchers have sought to understand the intricate factors that determine sex and drive the differentiation of testes or ovaries in different species. In 1969, Professor Toki-o Yamamoto published a landmark paper that provided the first comprehensive synthesis of the various influences involved in fish sex differentiation. Prior to Yamamoto's work, the field was complicated by conflicting hypotheses and a lack of consensus. Yamamoto drew on a wide range of research and presented a clear assessment of the validity of the existing theories. The paper established a firm foundation for subsequent researchers to build upon and remains a seminal work.

1 A brief overview of Yamamoto (1969)

Yamamoto (1969) defined the sex of an individual based on the obvious production of eggs or sperm, and further described the major sexual systems in fishes: gonochorism, which is reflected by two distinct sexes, female and male (the case with most fish species), or hermaphroditism, which has synchronous or consecutive forms (Fig. 1). Synchronous (also known as simultaneous) hermaphroditic species can produce eggs and sperm simultaneously, which, remarkably, allows for self-fertilization in some species (Tatarenkov et al., 2009). Consecutive (or sequential) hermaphroditic species can be further divided into protandrous hermaphrodites that mature first as males then change to females, or protogynous hermaphrodites that mature first as females then change to males (Fig. 1). Yamamoto noted that both protogynous and protandrous species start gonadal development containing testis and ovarian rudiments, suggesting that individuals do not switch sex, but rather emphasize the development and maturation of eggs or sperm at different phases of their reproductive life history. He prophetically hypothesized (see more below in present section) that the endogenous production of male vs. female sex hormones plays a role in controlling whether male or female gonadal development ensues.

During gonadal development in some gonochoristic fishes and other vertebrates, the gonads first develop as "undifferentiated" forms, which resemble ovaries, then secondarily the gonads in half of the individuals become testes.

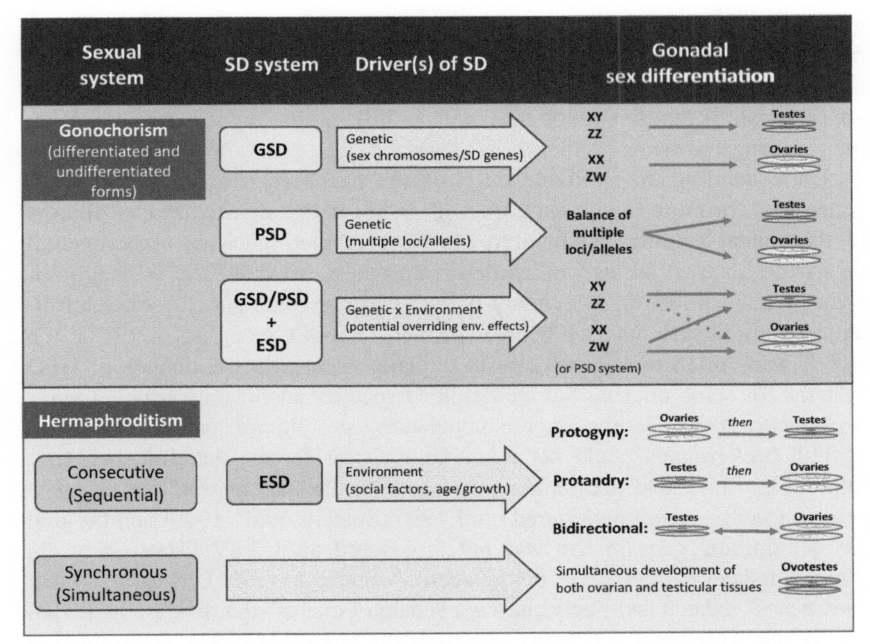

FIG. 1 Schematic diagram of the diverse sexual systems and mechanisms/drivers of sex determination (SD) and gonadal sex differentiation in fishes. Sex determination systems shown are genetic sex determination (GSD), polygenic sex determination (PSD), environmental sex determination (ESD), and combinations thereof. In gonochoristic species, naturally occurring instances of masculinization of genotypic females is much more common (arrow with solid line) than feminization of genotypic males (arrow with dashed line). *Drawing based on Yamamoto, Y., Luckenbach, J.A., accepted. Sex determination and gonadal sex differentiation. In: Alderman, S., Gillis, T. (Eds.), Encyclopedia of Fish Physiology, second ed. Elsevier.*

In differentiated forms, the gonads directly develop into either testes or ovaries. Yamamoto (1969) noted at that time that there were few known documented differentiated gonochorists (e.g., medaka (*Oryzias latipes*) and platyfish (*Xiphophorus maculatus*)), whereas today both undifferentiated and differentiated gonochorists are prevalent. Interestingly, undifferentiated forms are considered more labile and subject to hermaphroditism.

Yamamoto (1969) described clearly and at length the curious case of all-female (monosex) species, where mothers produce only daughters. Some females of the genus *Poecilia* are able to modify meiosis to generate ova (all of female genotype) and utilize sperm from other related species to activate their ova while, remarkably, preventing retention of the paternal genome. Similarly generated all-female progeny are also found in goldfish (genus *Carassius*). In another monosex female species group, *Poeciliopsis*, females can produce diploid ova (likely through a failure of polar body extrusion) which can be fertilized by a male from a related species. In this case, the male genome is retained in the resulting triploid progeny but excluded during meiosis to yield all-female ova in the next generation. Yamamoto's inclusion of this peculiar mode of reproduction provides the reader with an enhanced appreciation of the breadth of mechanisms influencing reproductive development among fishes.

Understanding the heritable basis of sex has been a major objective of geneticists for more than a century, with fishes being significant contributors to that understanding. Yamamoto (1969) presented previous experimental data from his own lab as well as that from others. At that time, sex chromosomes had been identified in many species and two main genetic sex determination (GSD) systems had been described, XX-XY and WZ-ZZ (Fig. 1), which Yamamoto took great care to explain, especially the nuance of GSD systems for some unusual but well-studied species, such as swordtails (genus *Xiphophorus*). Depending on the population, sex chromosomes in different genetic backgrounds could act as either male or female determinants. It is important to note that the human gene encoding the *sex determining region Y protein* (*SRY*) was not discovered until 1990 (Sinclair et al., 1990) and the first sex-determining gene in fish was not discovered until 2002 (Matsuda et al., 2002; Nanda et al., 2002; see below). Still, Yamamoto (1969) emphasized that "sex genes" existed and likely caused a "chain of events" that lead to the sexual phenotype. This quite accurately describes the processes of sex determination and differentiation in gonochoristic species as we know it today (Luckenbach and Yamamoto, 2018; Nagahama et al., 2021).

It is also noteworthy that Yamamoto (1969) included a section on polygenic sex determination (PSD), which is influenced by epistatic sex genes on sex chromosomes and/or autosomes. Yamamoto felt that sex is determined by the sum of genetic tendencies to induce each sex phenotype, with sex chromosomes (e.g., Y or W) possessing the strongest influences relative to those from autosomes or the second allosome. This polygenic view, also presented by Winge (1934), explained examples of apparent sex reversal simply as occurrences of genetic deviations from the normal balance of male or female genetic determinants. Yamamoto (1969) provided experimental data with medaka showing that sex-determining factors can be subject to selection such that males can develop in an otherwise XX genetic background, consistent with Aida's (1936) earlier findings. More recently, PSD has been found to be more widespread, including in zebrafish which possess family effects with multiple loci involved, and cichlids in which multiple sex chromosomes and sex determination systems co-exist (Bachtrog et al., 2014; Nagahama et al., 2021). Changes in genetic systems arising from captive breeding are also thought to play an important role in sex-determination systems in zebrafish (Wilson et al., 2014).

Yamamoto (1969) highlighted a number of examples of species that are "characterized by diverse sex ratios." Although environmental sex determination (ESD; Fig. 1) was not characterized in fishes until more than a decade later (Conover and Kynard, 1981), Yamamoto's appreciation of anomalous sex ratios provided the basis for subsequent research investigating the high degree of sexual plasticity observed in some fishes. We of course now know that ESD, and temperature-dependent sex determination (TSD) in particular, is widespread (50+ fish species; Luckenbach and Yamamoto, 2018), especially when reared in captivity under unnatural conditions which can trigger sex reversal and give rise to skewed sex ratios. The stress endocrine axis and epigenetic modifications have more recently been demonstrated to play important roles in mediating the ESD response (Yamamoto and Luckenbach, accepted).

It is clear when reading Yamamoto (1969) that sex control research (manipulation of sex with exogenous treatments) was one of his greatest scientific passions; note that the thought-provoking quote at the outset of the *Control of Sex Differentiation* section was one of his own: "Science unfolds and controls nature." Arguably, one of Professor Yamamoto's most impactful areas of research was on hormonal sex control using medaka as a model species (Fig. 2). He discovered that any steroid downstream of androstenedione can induce sex reversal. Through sex reversal experiments with medaka, he was able to generate XX-genotype male and YY-genotype female medaka and explore outcomes from a series of breeding crosses demonstrating that sex steroid treatment can override the underlying GSD mechanisms. His own research summarized in the chapter was noteworthy for a number of reasons, including demonstrating that (1) exogenous androgen treatment can drive testicular differentiation in genetically-female fish, (2) exogenous estrogen treatment can drive ovarian differentiation in genetically-male fish, (3) sex reversal can be induced with relatively low steroid dosages, and (4) monosex populations can be generated indirectly through sex reversal in the parental

FIG. 2 Medaka (*Oryzias latipes*) d-rR strain developed and used for research by Professor Toki-o Yamamoto. Top: orange-red strain (male shown; Bottom: white strain (female shown). Note the color difference along the dorsal surface between morphs. *Photo credit: Takashi Iwamatsu.*

line followed by targeted breeding crosses (see Section 3.2). Yamamoto thus posited that natural sex steroids play a critical role in sex differentiation. These were important conclusions that are supported by today's science and have been a topic in numerous subsequent reviews (e.g., Devlin and Nagahama, 2002; Donaldson, 1996; Luckenbach and Yamamoto, 2018; Nagahama et al., 2021; Pandian and Sheela, 1995; Piferrer, 2001a; Piferrer and Guiguen, 2008). He also proposed that the balance between androgens and estrogens is important in determining whether male or female differentiation occurs. As discussed further below, we now know that estrogens play a pivotal role in fish ovarian differentiation, development, and maintenance (Guiguen et al., 2010; Nagahama et al., 2021; Nakamoto et al., 2018; Zhou et al., 2021). However, the role of androgens in testicular differentiation is less clear, with some species potentially dependent on androgen production and signaling, and others not (Piferrer and Guiguen, 2008).

Yamamoto (1969) emphasized the criticality of the timing and dosage of steroid treatment application for successful induction of "complete" sex reversal. He noted that both sexes are responsive to steroid treatment and could result in at least partial transformation (e.g., induction of ovo-testis and testis-ova, referred to as "intersex") in fish treated at later stages of development. Treatment of undifferentiated gonochorists with androgens or estrogens during early development can induce sex reversal. However, differentiated species such as medaka may not be capable of complete sex reversal if treatments are applied after the fry stage and gonadal differentiation has begun. To overcome this difficulty, ovoviviparous adult females could be treated with sex steroids at a time when embryos they carry still possessed undifferentiated gonads, resulting in fully sex-reversed progeny. The key point was that steroid administration must be initiated "at the stage of the indifferent gonad and continue through the stage of gonadal sex differentiation" (Yamamoto, 1969).

The nature of "sex inducers" was discussed by Yamamoto (1969) where he presented how they may function in controlling sex differentiation. Prior work by other authors suggested that non-sex steroids were capable of modulating phenotypic sex, and that some androgens in high doses can cause feminization. These works were consistent with a non-sex steroidal mechanism of control. In contrast, working with medaka, Yamamoto noted that androgens and estrogens were highly specific in their effects on sex differentiation (e.g., no paradoxical feminization) and, as mentioned above, they required very low doses to induce sex reversal. Based on these observations, and that of Hishida (1962) who localized labeled sex steroids in juvenile gonads, Yamamoto (1969) concluded that a sex steroid-based control of sex differentiation was likely operating in fishes, but also noted that research on the specific steroidal pathways utilized was at a very early stage of what should be "promising areas of research." As you will see in Section 3.2 below, his prediction was correct.

Although the main focus of Yamamoto (1969) was determination and differentiation of the fish gonad ("primary sex differentiation"), his chapter closes with an overview of the differentiation of secondary sex characters with an emphasis on influences of gonadectomy or sex steroid treatment. Male-specific secondary characteristics (most common) can typically be induced by androgen treatment and suppressed by castration, and female-specific characteristics can be stimulated by estrogen treatment and suppressed by ovariectomy. However, in fewer cases, removal of the ovaries can stimulate the appearance of male-specific characteristics in females, consistent with an inhibitory control of the trait arising from the ovaries.

Perhaps one of the most important take-away messages from Yamamoto (1969) is his demonstration that sex determination and differentiation are highly complex and diverse processes among fishes. Indeed, this conclusion has been found to be true based on subsequent research (see recent reviews, Luckenbach and Yamamoto, 2018; Nagahama et al., 2021; Yamamoto and Luckenbach, accepted).

2 The life and career of Toki-o Yamamoto

Toki-o Yamamoto (Fig. 3) was born February 12, 1906 in a small village in the Akita prefecture, the northern part of Japan's main island of Honshu. He grew up surrounded by nature, which developed his keen interest in insects and fishes, and motivated him to begin studying zoology in 1926 at the University of Tokyo, Department of Zoology. He received his Ph.D. in 1936 and became an Assistant Professor immediately in the same department. During his term in Tokyo, he carried out numerous physiological studies on fertilization and activation of ova using several fishes, most notably medaka. Later, Yamamoto postulated that an invisible "fertilization wave" was propagated ahead of the wave of the breakdown of cortical alveoli within the ova (Yamamoto, 1944, 1961).

In 1942, the school of science was established in Nagoya Imperial University (later, Nagoya University) where Yamamoto was promoted to Associate Professor and subsequently to Professor in 1943. Yamamoto's laboratory was reduced to ashes during the Second World War and he lost almost all of his research data and materials. However, Yamamoto's strong passion for research and observational skills allowed him to rebuild and initiate new research projects, including his important work on fish sex differentiation (reviewed in Yamamoto, 1969). Using a new d-rR strain of medaka that he developed (Fig. 2), Yamamoto successfully induced artificial sex reversal for the first time ever in medaka (Yamamoto, 1953).

FIG. 3 Professor Toki-o Yamamoto in his aquarium facility at the Biological Institute, Nagoya University, Nagoya, Japan (1962). He stands among tanks used to hold medaka for his research. *Photo credit: Takashi Iwamatsu.*

During his 27-year term at Nagoya University, Yamamoto expertly instructed many students and researchers. While his love of nature was especially noteworthy, he was also passionate about drinking sake (rice wine), collecting stones and seashells, and listening to classical music. He frequented a classical music café near the university where he wrote and revised manuscripts with his students, including Takashi Iwamatsu, Yamamoto's last direct successor and co-author of the present article. When Yamamoto socialized at banquets, he often sang the self-composed tune, "The Song of Medaka."

After his retirement at the age of 63 in 1969, he continued to serve as a Professor at the Faculty of Agriculture, Meijo University, Nagoya. Despite being diagnosed with cancer in April of that year, he continued to conduct research for the next 8 years while enduring the illness. Yamamoto died at the age of 71 in Nagoya on April 5, 1977. During his illustrious career, he received several honors including, the 1950 Zoological Society of Japan

Award, 1957 Genetics Society of Japan Prize, 1976 Imperial Prize of the Japan Academy, 1976 the Order of the Rising Sun, Gold Rays with Neck Ribbon, and 1977 Junior Third Rank.

3 The legacy of Yamamoto (1969)

Yamamoto's (1969) chapter in the enduring book series *Fish Physiology* has been cited over 800 times and has influenced a variety of scientific fields including reproductive biology, aquaculture, fisheries science, genetics, aquatic toxicology, and behavioral biology (Fig. 4). Importantly, Yamamoto's chapter provided the first comprehensive overview of the research questions, hypotheses, and discoveries from the previous 100 years, many of which were complicated by conflicting hypotheses and a lack of consensus. It must be emphasized that this was nothing like writing a review article today—running literature searches and perusing downloaded PDF files on one's desktop computer. A significant amount of the prior literature was published in other languages, such as German and French, thus it was surely painstaking work to obtain the reference articles/books and translate the text. However, through his effort, Yamamoto provided a review that has stood the test of time and greatly benefitted the subsequent basic and applied fish sex determination

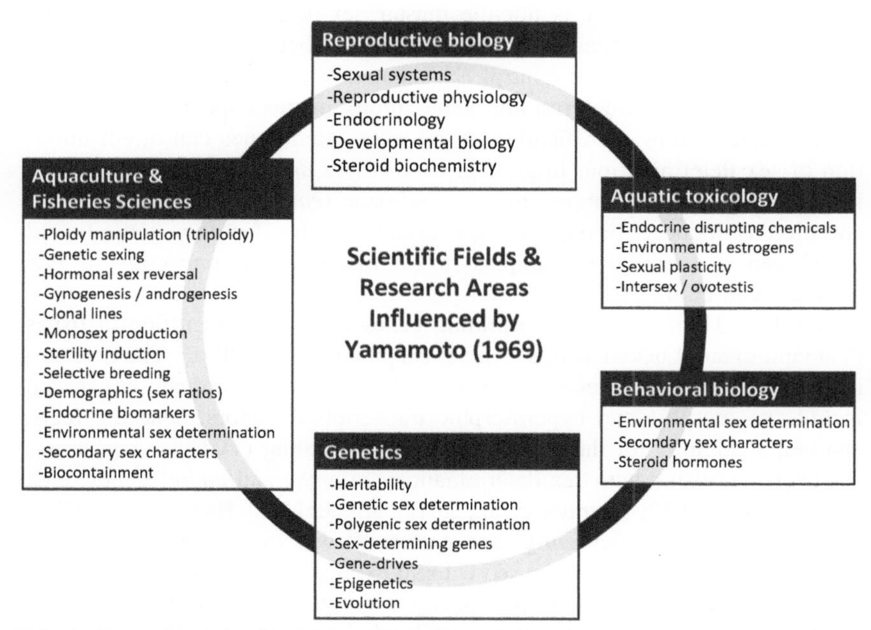

FIG. 4 Some of the scientific fields and areas of research influenced by Yamamoto (1969) over the past 50+ years.

and differentiation work described below. We were limited to mostly citing major review articles but encourage readers to further explore the primary literature on basic and applied aspects of fish sex determination and differentiation influenced by Yamamoto's chapter.

3.1 Basic discoveries stemming from Yamamoto (1969)

A recurrent theme in Yamamoto (1969) is that sex determination is largely genetically controlled in gonochoristic species. This conclusion was based on segregation patterns of sex-linked phenotypes, as well as sex ratios of progeny derived from hormonally-treated, sex-reversed parents, which confirmed that a chromosomal basis of sex determination is operating in many species. With the advent of molecular biology and genomics, Y-chromosome-linked sequences were identified in fishes, and subsequently specific genes were located in the sex-determining region of sex chromosomes. The first master sex-determining gene in a fish, identified in medaka (*DMY*, Matsuda et al., 2002; *dmrt1by* Nanda et al., 2002), was found to have high homology to *dmrt1*, a transcription factor involved in testicular differentiation in vertebrate and invertebrate species. While gain- and loss-of-function experiments clearly demonstrated the masculinizing abilities of *DMY/dmrt1by* in medaka, its specific role in inducing male development remains unknown. It is also interesting that *DMY/dmrt1by* is not the master sex-determining gene in some other members of the genus *Oryzias*, but rather, other sex-determining genes (i.e., *gsdf* and *sox3*) are in control (Nagahama et al., 2021). This lack of a conserved master sex-determining gene among medaka species is a revealing example of how modification of sex-associated genes can direct initiation of sex determination. In another case where *dmrt1* is essential to male sex determination, in tongue sole, *Cynoglossus semilaevis*, testicular differentiation is induced through dosage effects of the Z-chromosome-linked *dmrt1* gene present in two copies in ZZ males versus one copy in WZ females (Cui et al., 2017).

To date, 10 master sex-determining genes have been identified in fishes (Yamamoto and Luckenbach, accepted; Nagahama et al., 2021). The mechanisms by which these genes regulate sex differentiation are not conserved, and include neomorphic and hypermorphic transcription factors, dosage effects, and cell signaling. For the latter, altered cell signaling (TGF-β superfamily) has been associated with sex determination via *amhy* and *amhr2* genes (e.g., Hattori et al., 2012; Kamiya et al., 2012; Kikuchi and Hamaguchi, 2013). In contrast, a male determining variant of *gsdf*^{Y} was found to possess altered regulatory sequences which caused ectopic expression of *gsdf*^{Y} and induction of male differentiation in multiple species (e.g., *O. luzonensis* (Myosho et al., 2012); sablefish *Anoplopoma fimbria* (Herpin et al., 2021)). The gene *sox3* in *Oryzias dancena* has also been found to possess modified regulatory sequences that increase *sox3*^{Y} expression, which in turn elevates *gsdf* and drives testicular differentiation (Takehana et al., 2014). Uniquely, sex is determined in salmonids by *sdY*, a gene that originated not by modification of other sex-determining genes, but rather by modification of a completely novel gene (a truncated immune-related gene) not previously associated with sex determination/differentiation pathways (Bertho et al., 2018).

The above studies have also shown the clear interplay among known sex-determining genes as well as their roles in controlling sex differentiation pathways (e.g., *foxl2*, *cyp19a1a*, and *hsd17b1* in females, and *dmrt1*, *sox*, *gsdf*, and *amh* in males, among others; Nagahama et al., 2021). The genes *gsdf* and *DMY/dmrt1by* are co-expressed at the same developmental stage, consistent with *DMY/dmrt1by* being able to bind to the *gsdf* promoter to increase its expression. These analyses, as well as other studies, have clearly shown the direct control of the male path of sex differentiation by *DMY/dmrt1by* directed by *gsdf*. Similarly, functional studies with *amh* in other species have found this gene to be necessary and sufficient to stimulate testicular differentiation (Nagahama et al., 2021). In females, *cyp19a1/cyp19a1a* is essential for gonadal estradiol-17β synthesis and is one of the earliest markers of ovarian differentiation (Piferrer and Guiguen, 2008). Similarly, *foxl2* seems to play a supportive role in estrogen production. Germ cells are also critical for ovarian development in some species, as sterile fish, lacking germ cells, exhibit elevated androgens and decreased estrogen (Tanaka, 2019).

As mentioned earlier, Yamamoto (1969) hypothesized that sex steroids were endogenous sex inducers in fish embryos, which significantly catalyzed research on their involvement in the early stages of gonadal development. This has led to an accumulation of knowledge on the timing of gene expression and enzyme activity related to steroid production (e.g., *cyp19a1* and *cyp11c*) in fish gonads (Piferrer and Guiguen, 2008). Furthermore, evidence derived from functional analyses using steroid inhibitors and steroid-deficient fish largely corroborates the model that estrogens play indispensable roles in initiating and maintaining the process of ovarian differentiation in many fishes (Guiguen et al., 2010; Zhou et al., 2021). Interestingly, depletion of estrogens can also induce functional female-to-male sex reversal in sexually-mature female gonochoristic species, including medaka, demonstrating that gonochorists maintain their sexual plasticity until adulthood and estrogens play a critical role in maintaining the female phenotype (Nagahama et al., 2021).

In conjunction with identification of the diverse master sex-determining genes, it can reasonably be surmised that estrogen synthesis is situated downstream of these master genes (Nagahama et al., 2021). However, a direct link between the master genes and estrogens had not been elucidated until recently. The first link was reported in rainbow trout (*Oncorhynchus mykiss*); its sex-determining protein, SdY, directly suppresses *cyp19a1* expression in cooperation with cofactors (Foxl2 and Nr5a1 proteins), resulting in depletion of estrogens in the male genotype (XY) (Bertho et al., 2018). A more direct link was reported in amberjack fish including *Seriola dumerili* whose sex-determining locus encodes Hsd17b1 (Koyama et al., 2019). In these fish, a Z-linked variant of *hsd17b1* has been found to have a reduced ability to produce estradiol-17β and testosterone compared to the W-linked allele, which supports high levels of steroid production in the female genotype (ZW) (Koyama et al., 2019). These studies clearly support the steroidal basis of sex differentiation in fishes as advanced by Yamamoto (1969).

It is worth noting that the role of estrogens in the onset of ovarian differentiation in medaka remains controversial; the manifestation of apparent sex-reversal/intersex characteristics in the female genotype (XX) occurs only during the later stages of development subsequent to pharmacological and genetic suppression of estrogen synthesis (e.g., Nakamoto et al., 2018). Dr. Takashi Iwamatsu is continuing research on this issue.

In protandrous and protogynous fishes (Fig. 1), the shift in dominance of testicular vs. ovarian tissues is correlated with expression of masculinizing and feminizing genes (see above, and Avise and Mank, 2009; Nagahama et al., 2021). This is not simply a consequence of natural shifts in ovarian and testicular tissue, but rather is supported by functional studies diminishing masculinizing influences (e.g., via *dmrt1* knockout strains that lack male germ cells, resulting in promotion of ovarian differentiation; Wu et al., 2012). Masculinization of the gonad is possible since undifferentiated germ cells reside in the ovary that can be recruited to begin testis development during sex change in protogynous fishes. Indeed, even some gonochorists have been found to possess undifferentiated germ cells in ovaries. For simultaneous hermaphrodites, the role of the brain and pituitary has been found key, mediated via expression of gonadotropins (Lh and Fsh).

As noted earlier, ESD has now also been described in many fishes. Most often, elevated temperature induces testicular differentiation, but multiple additional factors can also affect sex differentiation, including growth rate, photoperiod, exposure to particular wavelengths of light, salinity, pH, and breeding behavior. ESD-based masculinization is associated with downregulation of female pathway genes such as *cyp19a1* and *foxl2*, and upregulation of male pathway genes like *dmrt1* and *amh* (Yamamoto and Luckenbach, accepted). Environmentally-induced masculinization has been associated with increased epigenetic modification (methylation) of the *cyp19a1* promoter in some species (Navarro-Martin et al., 2011; Piferrer, 2021b). These observations reveal how non-genetic influences are linked to, and can play important roles, in sex determination.

3.2 Practical applications stemming from Yamamoto (1969)

The case can certainly be made that the field of hormonal sex control in fishes was spawned from Yamamoto (1969). Although his primary goal was to examine effects of steroids on gonadal sex differentiation to unveil the underlying genetics and potential role of sex steroids as "natural inducers,"

Yamamoto's work, and that summarized in his chapter, was foundational to a number of applied sex control technologies that have greatly influenced aquaculture and environmental sciences (Fig. 4).

One of the most important practical sex control techniques used in aquaculture and fisheries science, foreshadowed by Yamamoto (1969), is that of genetic sex identification (Devlin et al., 1991; Luckenbach and Guzmán, 2022; Piferrer, 2001a). Being able to modify the phenotypic potential of an individual, independent of its genotypic sex, allows development of populations of fish that possess aquaculture production benefits relative to normal mixed-sex populations (Martínez et al., 2014). Yamamoto showed that medaka can be sex reversed (via appropriate steroid treatment timing and dosage) to generate XX males that can be mated with regular XX females to sire all-female populations (referred to nowadays as "indirect feminization"; Fig. 5). Thus, for some aquacultured species where performance differs between the sexes (e.g., sablefish, salmon, carp, tilapia), restricting production

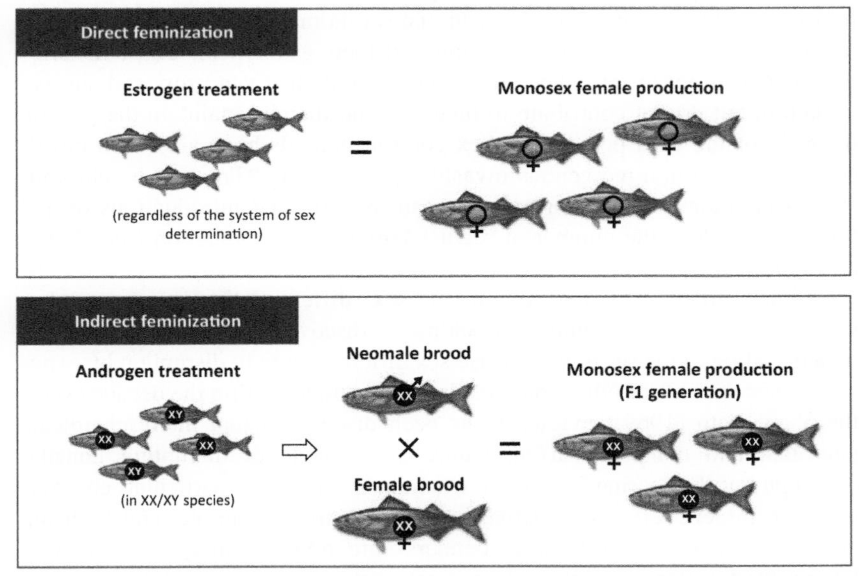

FIG. 5 Direct and indirect feminization approaches successfully used to produce monosex female populations of sablefish (*Anoplopoma fimbria*) for aquaculture. Direct feminization may be effective regardless of the sex determination system utilized by a species, whereas indirect feminization is streamlined in XX/XY species, such as sablefish, and can produce monosex female populations in the F1 generation. Genetic sexing methods can greatly facilitate sex control work by allowing researchers to determine the underlying genetic sex for individual fish with only DNA isolated from a fin clip. *Schematic diagram reproduced from Luckenbach, J.A., Fairgrieve, W.T., Hayman, E.S., 2017. Establishment of monosex female production of sablefish (*Anoplopoma fimbria*) through direct and indirect sex control. Aquaculture 479, 285–296. doi:10.1016/j.aquaculture.2017.05.037.*

to the better performing sex using "monosex technology" can be highly advantageous (Luckenbach et al., 2017; Martínez et al., 2014; Piferrer, 2001a). For this scheme, the genetic sex of XX and XY males can be distinguished by using test crossing (two generations) or molecular genetic assays (one generation). Depending on the degree to which sex is determined by sex chromosomes versus polygenic or environmental influences, monosex strains can be very stable.

Monosex approaches also have benefits in reducing impacts arising from introduced species by combining monosex and chromosome-set manipulation methods. For example, in rainbow trout, triploid females fail to produce functional ovaries, whereas triploid males develop testes and produce functional, albeit aneuploid sperm that is capable of fertilization but produces nonviable embryos (Lincoln and Scott, 1984). To enhance numbers of fish for commercial and sportfishing harvest, introduction of triploid fish has been used to reduce genetic introgression into wild populations by preventing interbreeding between introduced and wild fish (Piferrer et al., 2009). Releasing triploid males has the potential to modify the genetic architecture of the population, for example by competition with wild males (via mate selection) and females (affecting numbers of viable offspring and their genotypes). Thus, fisheries managers can opt to introduce all-female triploids that can grow and survive in nature, but do not contribute to breeding and thereby maintain the genetic integrity of the wild population. Sex control methods have also been developed to control non-indigenous invasive species using "Trojan" sex chromosome technology, which, for example can use repeated introductions of YY females to reduce the number of regular females in a population (see Cotton and Wedekind, 2007).

Yamamoto's lucid description of sex differentiation processes has provided background for numerous additional discoveries based on an elevated understanding of the sensitivity of fish sex differentiation to disruption by exposure to sex steroids or other compounds. For example, during the decades since the Yamamoto (1969) review, it has been discovered that many chemicals (steroidal and non-steroidal) in nature are capable of disrupting gonadal development (Endocrine Disrupting Chemicals; EDCs), in particular, environmental estrogens released to nature by human activity (Sumpter and Johnson, 2008). These EDCs or "gender benders" are now routinely monitored in human wastewater in an effort to mitigate potential impacts on aquatic ecosystems and human health. Furthermore, in fisheries science, skewed sex ratios in natural populations can be analyzed with sex-specific DNA markers to determine whether such deviations arose from sex reversal or by natural demographic fluctuations. Other examples of fish DNA sex technology now in use include determination of the sex of prey in scat samples (e.g., preferential selection of Chinook salmon by Orca whales), eDNA estimates of sex ratios, and determination of sex of salmon in Indigenous People's ancient middens (Ford and Ellis, 2006; Royle et al., 2018).

At the end of the 1960s, when Yamamoto published his review, little was known regarding the physiological and genetic pathways responsible for fish sex determination and differentiation. However, he was clearly aware of the importance of this area as he noted to the reader that "Research along these lines should be fruitful." Indeed, Yamamoto was correct in that prophecy, such that today, in the era of molecular biology and genomics, an abundance of genes involved in fish sex determination/differentiation are now known (Nagahama et al., 2021). In addition to providing a deep understanding of how these genes function to influence sex, they have also provided opportunities in applied science to control the reproductive or survival capabilities of fishes. Similar to the above objectives for sex control, transgenic and CRISPR-Cas9 based approaches have been developed to conditionally control gene constructs capable of supporting reproduction or inducing sterility (Xu et al., 2023). A variety of genetic or physiological pathways are being explored including the *dead end* gene (affecting migration and identity of PGCs), thiaminase (inducing vitamin dependency), Lh (control of gonadal maturation), and Zp3:caspase constructs affecting egg development. In addition, efforts are underway to generate CRISPR-Cas9 gene-drive systems (e.g., daughterless based) that could be used to control invasive/unwanted populations in nature (e.g., lamprey; Ferreira-Martins et al., 2021). None of these approaches are yet deployed in fishes in nature or aquaculture, and, while significant debate surrounds their use, they do hold significant promise to control unwanted reproduction.

4 In closing

In the closing section of his chapter, Yamamoto (1969) states that "sex is a phenotypic expression" that of course results from diverse and complex mechanisms. Although the genetic sex for most fishes is established at the time of fertilization, a high degree of plasticity remains whereby minor genetic, epigenetic, and/or environmental factors, as well as exogenous factors (e.g., EDCs, water temperature) may influence the ultimate sexual phenotype of the individual. Yamamoto again concludes that endogenous sex steroids likely play a key role in the natural process of gonadal sex differentiation, but also notes that "artificial control of sex differentiation may be one of the key projects in biology," thus showing his appreciation for how the ability to control sex might influence aquaculture, genetics, and fisheries science into the future. As discussed above, these were profound conclusions from his own life's work and effort towards developing Yamamoto (1969).

We can all learn from Professor Yamamoto's enduring nature and passion for research. After his aquatic facilities and laboratory were destroyed in World War II, despite the significant losses, he rebuilt and carried on with his work. The same occurred after he was first diagnosed with cancer; he was unstoppable until the end. In his quest to review the body of literature on fish sex differentiation, Yamamoto fearlessly challenged prior findings and ideas (e.g., see pg. 129 Oka) and clearly sought to identify common, unifying threads in the research up to 1969, identify spurious published results or conclusions, and establish consistent nomenclature for the field. He minced no words in his chapter and was willing to hold colleagues to task in order to find the truth regarding the mechanisms of sex differentiation. Overall, Yamamoto's contributions greatly advanced our understanding of sex differentiation and serve as a model for scientific rigor in research.

Acknowledgments

Each of the authors of this retrospective review benefited from Yamamoto-sensei's pioneering empirical and theoretical works on sex determination and differentiation, for which we are most grateful. The authors thank Su Kim (NWFSC), Visual Information Specialist, for her suggestions on graphic design of the figures. The diagram presented in Fig. 5 was modified from that previously published in Aquaculture, Volume 479, J.A. Luckenbach, W.T. Fairgrieve, and E.S. Hayman, Establishment of monosex female production of sablefish (*A. fimbria*) through direct and indirect sex control, pages 285–296, Copyright Elsevier (2017). The authors also sincerely appreciate comments by Dr. Yuzo Yanagitsuru (NWFSC) on an earlier draft of the manuscript.

References

Aida, T., 1936. Sex reversal in *Aplocheilus latipes* and a new explanation of sex differentiation. Genetics 21, 136–153. https://doi.org/10.1093/genetics/21.2.136.

Avise, J.C., Mank, J.E., 2009. Evolutionary perspectives on hermaphroditism in fishes. Sex. Dev. 3, 152–163. https://doi.org/10.1159/000223079.

Bachtrog, D., Mank, J.E., Peichel, C.L., Kirkpatrick, M., Otto, S.P., Ashman, T.L., Hahn, M.W., Kitano, J., Mayrose, I., Ming, R., Perrin, N., Ross, L., Valenzuela, N., Vamosi, J.C., 2014. Tree of Sex Consortium. Sex determination: why so many ways of doing it? PLoS Biol. 12, e1001899. https://doi.org/10.1371/journal.pbio.1001899.

Bertho, S., Herpin, A., Branthonne, A., Jouanno, E., Yano, A., Nicolb, B., Mullerc, T., Pannetierd, M., Pailhouxd, E., Miwae, M., Yoshizaki, G., Schartl, M., Guiguen, Y., 2018. The unusual rainbow trout sex determination gene hijacked the canonical vertebrate gonadal differentiation pathway. Proc. Natl. Acad. Sci. U. S. A. 115, 12781–12786. https://doi.org/10.1073/pnas.1803826115.

Conover, D.O., Kynard, B.E., 1981. Environmental sex determination–interaction of temperature and genotype in a fish. Science 213, 577–579. https://doi.org/10.1126/science.213.4507.57.

Cotton, S., Wedekind, C., 2007. Control of introduced species using Trojan sex chromosomes. Trends Ecol. Evol. 22, 441–443. https://doi.org/10.1016/j.tree.2007.06.0.

Cui, Z., Liu, Y., Wang, W., Wang, Q., Zhang, N., Lin, F., Wang, N., Shao, C., Dong, Z., Li, Y., Yang, Y., Hu, M., Li, H., Gao, F., Wei, Z., Meng, L., Liu, Y., Wei, M., Zhu, Y., Guo, H., Cheng, C.H., Schartl, M., Chen, S., 2017. Genome editing reveals *dmrt1* as an essential male sex-determining gene in Chinese tongue sole (*Cynoglossus semilaevis*). Sci. Rep. 7. https://doi.org/10.1038/srep42213.

Devlin, R.H., McNeil, B.K., Solar, I.I., Donalson, E.M., 1991. Isolation of a Y-chromosomal DNA probe capable of determining genetic sex in Chinook salmon (*Oncorhynchus tshawytscha*). Can. J. Fish. Aquat. Sci. 48, 1606–1612. https://doi.org/10.1139/f91-190.

Devlin, R.H., Nagahama, Y., 2002. Sex determination and sex differentiation in fish: an overview of genetic, physiological, and environmental influences. Aquaculture 208, 191–364. https://doi.org/10.1016/S0044-8486(02)00057-1.

Donaldson, E.M., 1996. Manipulation of reproduction in farmed fish. Anim. Reprod. Sci. 42, 381–392. https://doi.org/10.1016/0378-4320(96)01555-2.

Ferreira-Martins, D., Champer, J., McCauley, D.W., Zhang, Z., Docker, M.F., 2021. Genetic control of invasive sea lamprey in the Great Lakes. J. Great Lakes Res. Suppl. 1 (47), S764–S775. https://doi.org/10.1016/j.jglr.2021.10.018.

Ford, J.K.B., Ellis, G.M., 2006. Selective foraging by fish-eating killer whales *Orcinus orca* in British Columbia. Mar. Ecol. Prog. Ser. 316, 185–199. https://doi.org/10.3354/meps316185.

Guiguen, Y., Fostier, A., Piferrer, F., Chang, C.F., 2010. Ovarian aromatase and estrogens: a pivotal role for gonadal sex differentiation and sex change in fish. Gen. Comp. Endocrinol. 165, 352–366. https://doi.org/10.1016/j.ygcen.2009.03.002.

Hattori, R.S., Murai, Y., Oura, M., Masuda, S., Majhi, S.K., Sakamoto, T., Fernandino, J.I., Somoza, G.M., Yokota, M., Strüssmann, C.A., 2012. A Y-linked anti-Müllerian hormone duplication takes over a critical role in sex determination. Proc. Natl. Acad. Sci. U. S. A. 109, 2955–2959. https://doi.org/10.1073/pnas.1018392109.

Herpin, A., Schartl, M., Depincé, A., Guiguen, Y., Bobe, J., Hua-Van, A., Hayman, E.S., Octavera, A., Yoshizaki, G., Nichols, K.M., Goetz, G.W., Luckenbach, J.A., 2021. Allelic diversification after transposable element exaptation promoted Gsdf as the master sex determining gene of sablefish. Genome Res. 31, 1366–1380. www.genome.org/cgi/doi/10.1101/gr.274266.120.

Hishida, T., 1962. Accumulation of testosterone-4-C^{14} propionate in larval gonad of the medaka, *Oryzias latipes*. Embryol. Nagoya 7, 56–67. https://doi.org/10.1111/j.1440-169X.1962.tb00159.x.

Kamiya, T., Kai, W., Tasumi, S., Oka, A., Matsunaga, T., Mizuno, N., Fujita, M., Suetake, H., Suzuki, S., Hosoya, S., Tohari, S., Brenner, S., Miyadai, T., Venkatesh, B., Suzuki, Y., Kikuchi, K., 2012. A trans-species missense SNP in Amhr2 is associated with sex determination in the tiger pufferfish, *Takifugu rubripes* (fugu). PLoS Genet. 8 (7), e1002798. https://doi.org/10.1371/journal.pgen.1002798.

Kikuchi, K., Hamaguchi, S., 2013. Novel sex-determining genes in fish and sex chromosome evolution. Dev. Dyn. 242, 339–353. https://doi.org/10.1002/dvdy.23927.

Koyama, T., Nakamoto, M., Morishima, K., Yamashita, R., Yamashita, T., Sasaki, K., Kuruma, Y., Mizuno, N., Suzuki, M., Okada, Y., Ieda, R., Uchino, T., Tasumi, S., Hosoya, S., Uno, S., Koyama, J., Toyoda, A., Kikuchi, K., Sakamoto, T., 2019. A SNP in a steroidogenic enzyme is associated with phenotypic sex in *Seriola* fishes. Curr. Biol. 29, 1901–1909. https://doi.org/10.1016/j.cub.2019.04.069.

Lincoln, R.F., Scott, A.P., 1984. Sexual maturation in triploid rainbow trout *Salmo gairdneri* Richardson. J. Fish Biol. 25, 385–392. https://doi.org/10.1111/j.1095-8649.1984.tb04886.x.

Luckenbach, J.A., Guzmán, J.M., 2022. Reproduction. In: Midway, S.R., Hasler, C.T., Chakrabarty, P. (Eds.), Methods for Fish Biology, second ed. American Fisheries Society, Bethesda, Maryland, https://doi.org/10.47886/9781934874615. Chapter 12.

Luckenbach, J.A., Yamamoto, Y., 2018. Genetic & environmental sex determination in cold-blooded vertebrates: fishes, amphibians, and reptiles. In: Skinner, M.K. (Ed.), Encyclopedia of Reproduction, second ed. vol. 6. Elsevier Publishing, https://doi.org/10.1016/B978-0-12-809633-8.20553-0. Chapter 29.

Luckenbach, J.A., Fairgrieve, W.T., Hayman, E.S., 2017. Establishment of monosex female production of sablefish (*Anoplopoma fimbria*) through direct and indirect sex control. Aquaculture 479, 285–296. https://doi.org/10.1016/j.aquaculture.2017.05.037.

Martínez, P., Viñas, L., Sánchez, A.M., Díaz, L., Ribas, N., Piferrer, F., 2014. Genetic architecture of sex determination in fish: applications to sex ratio control in aquaculture. Front. Genet. 5, 340. https://doi.org/10.3389/fgene.2014.00340.

Matsuda, M., Nagahama, Y., Shinomiya, A., Sato, T., Matsuda, C., Kobayashi, T., Morrey, C.E., Shibata, N., Asakawa, S., Shimizu, N., Hori, H., Hamaguchi, S., Sakaizumi, M., 2002. *DMY* is a Y-specific DM-domain gene required for male development in the medaka fish. Nature 417, 559–563. https://doi.org/10.1038/nature751.

Myosho, T., Otake, H., Masuyama, H., Matsuda, M., Kuroki, Y., Fujiyama, A., Naruse, K., Hamaguchi, S., Sakaizumi, M., 2012. Tracing the emergence of a novel sex-determining gene in medaka, *Oryzias luzonensis*. Genetics 191, 163–170. https://doi.org/10.1534/genetics.111.137497.

Nagahama, Y., Chakraborty, T., Paul-Prasanth, B., Ohta, K., Nakamura, M., 2021. Sex determination, gonadal sex differentiation, and plasticity in vertebrate species. Physiol. Rev. 101 (3), 1237–1308. https://doi.org/10.1152/physrev.00044.2019.

Nakamoto, M., Shibata, Y., Ohno, K., Usami, T., Kamei, Y., Taniguchi, Y., Todo, T., Sakamoto, T., Young, G., Swanson, P., Naruse, K., Nagahama, Y., 2018. Ovarian aromatase loss-of-function mutant medaka undergo ovary degeneration and partial female-to-male sex reversal after puberty. Mol. Cell. Endocrinol. 460, 104–122. https://doi.org/10.1016/j.mce.2017.07.013.

Nanda, I., Kondo, M., Hornung, U., Asakawa, S., Winkler, C., Shimizu, A., Shan, Z., Haaf, T., Shimizu, N., Shima, A., Schmid, M., Schartl, M., 2002. A duplicated copy of DMRT1 in the sex-determining region of the Y chromosome of the medaka, *Oryzias latipes*. Proc. Natl. Acad. Sci. U. S. A. 99, 11778–11783. https://doi.org/10.1073/pnas.182314699.

Navarro-Martin, L., Vinas, J., Ribas, L., Diaz, N., Gutierrez, A., Di Croce, L., Piferrer, F., 2011. DNA methylation of the gonadal aromatase (*cyp19a*) promoter is involved in temperature-dependent sex ratio shifts in the European sea bass. PLoS Genet. 7 (12), e1002447. https://doi.org/10.1371/journal.pgen.1.

Pandian, T.J., Sheela, S.G., 1995. Hormonal induction of sex reversal in fish. Aquaculture 138, 1–22. https://doi.org/10.1016/0044-8486(95)01075-0.

Piferrer, F., 2001a. Endocrine sex control strategies for the feminization of teleost fish. Aquaculture 197, 229–281. https://doi.org/10.1016/S0044-8486(01)00589-0.

Piferrer, F., 2021b. Epigenetic mechanisms in sex determination and in the evolutionary transitions between sexual systems. Philos. Trans. R. Soc. Lond. Ser. B Biol. Sci. 376, 20200110. https://doi.org/10.1098/rstb.2020.0110.

Piferrer, F., Guiguen, Y., 2008. Fish gonadogenesis. Part II: molecular biology and genomics of sex differentiation. Rev. Fish. Sci. 16, 35–55. https://doi.org/10.1080/10641260802324644.

Piferrer, F., Beaumont, A., Falguiré, J.C., Flajšhans, M., Haffray, P., Colombo, L., 2009. Polyploid fish and shellfish: production, biology and applications to aquaculture for performance improvement and genetic containment. Aquaculture 293, 125–156. https://doi.org/10.1016/j.aquaculture.2009.04.036.

Royle, T.C.A., Sakhrani, D., Speller, C.F., Butler, V.L., Devlin, R.H., Cannon, A., Yang, D.Y., 2018. An efficient and reliable DNA-based sex identification method for archaeological Pacific salmonid (*Oncorhynchus* spp.) remains. PLoS One 13 (3), e0193212. https://doi.org/10.1371/journal.pone.0193212.

Sinclair, A.H., Berta, P., Palmer, M.S., Hawkins, J.R., Griffiths, B.L., Smith, M.J., Foster, J.W., Frischauf, A.M., Lovell-Badge, R., Goodfellow, P.N., 1990. A gene from the human sex-determining region encodes a protein with homology to a conserved DNA-binding motif. Nature 346, 240–244. https://doi.org/10.1038/346240a0.

Sumpter, J.P., Johnson, C., 2008. 10th Anniversary perspective: reflections on endocrine disruption in the aquatic environment: from known knowns to unknown unknowns (and many things in between). J. Environ. Monit. 10, 1476–1485.

Takehana, Y., Matsuda, M., Myosho, T., Suster, M.L., Kawakami, K., Shin-I, T., Kohara, Y., Kuroki, Y., Toyoda, A., Fujiyama, A., Hamaguchi, S., Sakaizumi, M., Naruse, K., 2014. Co-option of Sox3 as the male-determining factor on the Y chromosome in the fish *Oryzias dancena*. Nat. Commun. 5, 4157. https://doi.org/10.1038/ncomms5157.

Tanaka, M., 2019. Regulation of germ cell sex identity in medaka. Curr. Top. Dev. Biol. 134, 151–165. https://doi.org/10.1016/bs.ctdb.2019.01.010.

Tatarenkov, A., Lima, S.M., Taylor, D.S., Avise, J.C., 2009. Long-term retention of self-fertilization in a fish clade. Proc. Natl. Acad. Sci. U. S. A. 106, 14456–14459. https://doi.org/10.1073/pnas.0907852106.

Wilson, C.A., High, S.K., McCluskey, B.M., Amores, A., Yan, Y., Titus, T.A., Anderson, J.L., Batzel, P., Carvan, M., Schartl, M., Postlethwait, J.H., 2014. Wild sex in zebrafish: loss of the natural sex determinant in domesticated strains. Genetics 198, 1291–1308. https://doi.org/10.1534/genetics.114.16928.

Winge, O., 1934. The experimental alteration of sex chromosomes into autosomes and vice versa, as illustrated by *Lebistes*. Compt. Rend. Trav. Lab. Carlsberg. Ser. Physiol. 21, 1–49.

Wu, G.C., Chiu, P.C., Lin, C.J., Lyu, Y.S., Lan, D.S., Chang, C.F., 2012. Testicular *dmrt1* is involved in the sexual fate of the ovotestis in the protandrous black porgy. Biol. Reprod. 86, 1–11. https://doi.org/10.1095/biolreprod.111.095695.

Xu, L., Mingli, R., Hyung, J., Hayman, E.S., Fairgrieve, W.T., Zohar, Y., Luckenbach, J.A., Wong, T.T., 2023. Reproductive sterility in aquaculture: a review of induction methods and an emerging approach with application to Pacific Northwest finfish species. Rev. Aquac. 15, 220–241. https://doi.org/10.1111/raq.12712.

Yamamoto, T., 1944. Physiological studies on fertilization and activation of fish eggs. II. The conduction of the "fertilization wave" in the eggs of *Oryzias latipes*. Anno. Zool. Jp,. 22 (3), 126–136.

Yamamoto, T., 1953. Artificially induced sex-reversal in genotypic males of the medaka (*Oryzias latipes*). J. Exp. Zool. 123 (3), 571–594. https://doi.org/10.1002/jez.1401230309.

Yamamoto, T., 1961. Physiology of fertilization in fish eggs. Int. Rev. Cytol. 12, 361–405. https://doi.org/10.1016/s0074-7696(08)60545-8.

Yamamoto, T., 1969. Sex differentiation. In: Hoar, W.S., Randall, D.J. (Eds.), Fish Physiology. vol. III. Reproduction, Academic Press, New York, pp. 117–175. https://doi.org/10.1016/S1546-5098(08)60113-2.

Yamamoto, Y., Luckenbach, J.A., accepted. Sex determination and gonadal sex differentiation. In: Alderman, S., Gillis, T. (Eds.), Encyclopedia of Fish Physiology, second ed. Elsevier. https://doi.org/10.1016/B978-0-323-90801-6.00052-5.

Zhou, L., Li, M., Wang, D., 2021. Role of sex steroids in fish sex determination and differentiation as revealed by gene editing. Gen. Comp. Endocrinol. 313, 113893. https://doi.org/10.1016/j.ygcen.2021.113893.

Chapter 12

SEX DIFFERENTIATION☆

TOKI-O YAMAMOTO

Chapter Outline

It thus becomes of great interest to discover the mechanism by which sex is determined, and to find whether by any means we can bring it under our control.

Julian Huxley (1938)

☆This is a reproduction of a previously published chapter in the Fish Physiology series, "1969 (Vol. 3)/Sex Differentiation: ISBN: 978-0-12-350403-6; ISSN: 1546-5098".

Fish Physiology, Vol. 40A. https://doi.org/10.1016/bs.fp.2023.09.010

I. INTRODUCTION: SEXUALITY IN FISHES

Members belonging to the class Pisces exemplify an almost complete range of the various types of sexuality from synchronous hermaphroditism, protandrous and protogynous hermaphroditism, to gonochorism.

At the outset, it is necessary to clarify the concept of sex. The male and female can best be defined as sperm and egg producers, respectively. This definition, although self-evident, is needed to correct statements that appear frequently in literature, with respect to spontaneous sex reversal in gonochorists (Sections IV, B and C). The term "bisexuality" used to denote hermaphroditism is likely to lead to a misunderstanding. This term should be used to describe gonochorist.

The term "intersex" (Goldschmidt, 1915, 1927, 1931) is used in the present review to denote either sporadically appearing or experimentally produced hermaphroditic individuals of a species in which all or nearly all individuals are gonochoristic.

Whether hermaphroditism is the more primitive condition from which bisexuality or gonochorism may have arisen or a specialization derived from the more usual vertebrate gonochorism is a matter for debate. The solution of this intriguing problem will require a great deal more information in the future. Nevertheless, fishes provide excellent material to approach the problems of sex differentiation and of evolution of sex among animals.

In this study, stress is laid on sex differentiation as a process rather than examining sex phenotypes as they appear in adults. For the latter approach, Gordon's review (1957) may be consulted.

II. HERMAPHRODITISM

Unlike other vertebrates, a number of teleost fishes are hermaphrodites. Atz (1964), among others, defined the types of hermaphroditism. An individual is hermaphroditic if it bears recognizable male and female tissues. If all, or nearly all, individuals possess both ovarian and testicular tissues, that species is hermaphroditic. Synchronous (balanced) hermaphrodities are those in which the male and female sex cells ripen at the same time, regardless whether or not self-fertilization is possible. In consecutive (metagonous) hermaphrodites there are two types: protogynous hermaphrodites that function first as females and then transform into males and protandrous hermaphrodites that transform from males into females. Atz lists 13 families of teleosts, belonging to five orders, that include species of these types. The transformation may be accomplished in several ways, depending upon the arrangement of sexual tissues (Reinboth, 1962, 1967; Smith, 1965, cf. Fig. 3).

A. Synchronous Hermaphroditism

Dufossé (1854, 1856) found that *Serranus scriba* (Serranidae) is a synchronous, functional hermaphrodite. This was confirmed by van Oordt (1929).

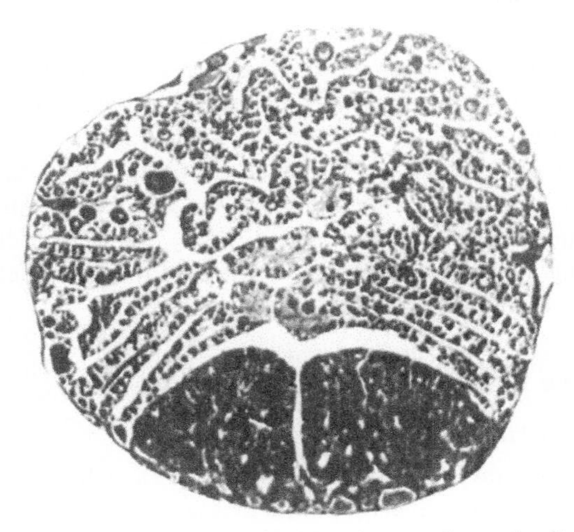

Fig. 1. Transverse section of the gonad of the synchronous hermaphrodite, *Serranus scriba* (Serranidae). Upper portion is the ovary and lower region the testis. After D'Ancona (1950).

D'Ancona's intensive studies (1949a,c,d, 1950) revealed that other Mediterranean serranids such as *S. cabrilla* and *Hepatus hepatus* belong to this category. The gonad of these fishes is separated into the ovarian and testicular areas (Fig. 1). D'Ancona (1949a,c,d, 1950) suggested the possibility of self-fertilization. Clark (1959, 1965) in *S. sub-ligerius*, a Florida serranid proved that self-fertilization and development are possible. In captivity, an isolated individual can emit sperm and fertilize its own eggs. In both nature and in the aquaria containing two or more fish, the fish may form spawning pairs. Immediately after one fish in the female phase spawns, the partner fertilizes the eggs. At this moment, the color pattern (vertical stripes) in the female phase changes to that of the male phase. Then, the first fish reverses its sexual role and acts as a male.

Of serranid fishes from Bermuda studied by Smith (1959), four species (one belonging to the genus *Hypoplectrus* and three of the genus *Prinodus*) are synchronous hermaphrodites while at least nine species are protogynous hermaphrodites (see Section II, B, 2).

Harrington (1961, 1965) demonstrated that most *Rivulus marmoratus*, an oviparous cyprinodont, are genuine hermaphrodites capable of internal self-fertilization. It is remarkable that over 10 uniparental generations have been propagated and each fish has been kept in lifelong isolation *ab ovo*. Tissue grafts between the parent and its offspring and among siblings were histocompatible, thus providing tight evidence of self-fertilization. According to Harrington (1967), low temperature (18°–20°C) tends to transform the hermaphrodites to males.

The order Perciformes includes four families (Serranidae, Sparidae, Centracanthidae, and Labridae) in which either synchronous or consecutive hermaphroditism occurs normally. In all these families, however, there are some gonochoristic species.

Mead (1960) and Gibbs (1960, cited in Atz, 1904) discovered that several fishes of four families belonging to the order Myctophiformes (also called Iniomi) are hermaphroditic (cf. Atz, 1964). Some of Mead's fish gave evidence of being synchronous hermaphrodites. However, at least five of the remaining dozen families in the order are gonochoristic.

B. Consecutive Hermaphroditism

1. Protandrous Hermaphrodites

Since the time of Syrski (1876) and Brock (1879), it has been known that the Mediterranean bream, *Sparus auratus*, is a protandrous hermaphrodite. Pasquali (1941) and D'Ancona (1941) described the precise process of gonadal differentiation from male to female phases. They demonstrated that its gonad consists of both testicular and ovarian areas from a very young stage.

In smaller fish the lateroventral testicular region predominates over the ovarian zone, and in larger fish the reverse is true (Fig. 2). Similar patterns have been reported by Syrski (1876), McLeod (1881), Hoeck (1891a,b), Williamson (1910), and van Oordt (1929). The latter author found in *Sargus (Diplodus) annularis* that there are some individuals in which the sexes are separated, Le Gall (1929) found that *Pagellus centrodontus* is a protandrous hermaphrodite.

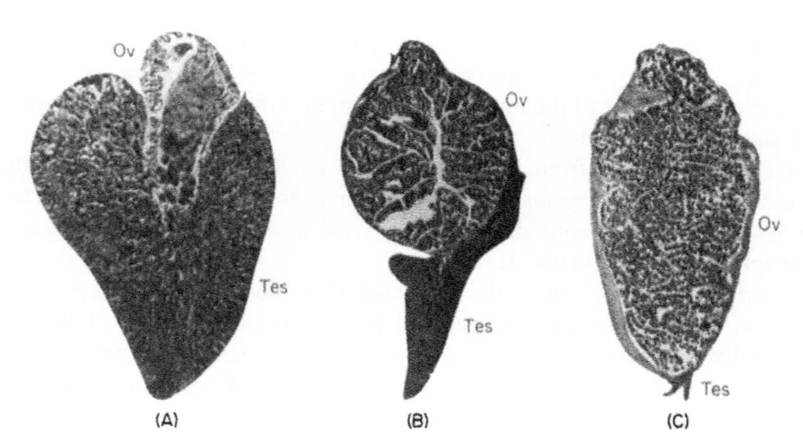

Fig. 2. Transverse sections of the gonad of the protandrous hermaphrodite, *Sparus auratus* (Sparidae). (A) Male phase, (B) transitory phase, and (C) female phase. Here Ov stands for ovary and Tes for testis. After D'Ancona (1950).

D'Ancona (1949a–d, 1950, 1956) showed that some sparids such as *Sparus auratus*, Sargus (*Diplodus*) *sargus*, and *Pagellus mormyrus* are protandrous hermaphrodites (see Fig. 2), while others such as *S. annularis S. vulgaris, Puntazzo* (*Charax*) *puntazzo, Boops boops, Oblada melanula*, and *Dentex dentex* are rudimentary hermaphrodites. *Pagellus acarne* is protandrous (Reinboth, 1962).

The classification of Japanese sparids is notoriously unsatisfactory. Here, the writer follows Dr. Abe's recent personal communication (1967). Kinoshita (1936) reported "sex reversal" from male to female in *Acanthopagrus schlegeli* (syn. *Sparus longispinis, Mylio macrocephalus*). He (1939) also reported sex reversal in *A.* (*Sparus, Mylio*) *latus* and *Sparus sarba* (*Sparus aries, Rabdosargus sarba*). He remarked that not all individuals transform from male to female, i.e., some males retain maleness even when they become large. Okada's histological study (1952b) indicates that an ambisexual organization is present in the early stages, so that the situation is similar to the reports of D'Ancona and others.

In the flat-head fish, *Inegocia* (*Cociella*) *crocodila* (Platycephalidae), Aoyama *et al.* (1963) examined a large number of individuals (>1000) caught by trawls in the East China Sea and the Yellow Sea. Small individuals had testes and medium-sized fish hermaphroditic gonads with functional testes, whereas large individuals had ovaries. Another flat-head fish, *Inegocia* (*Suggrundus*) *meerdervoort*, is also a protandrous hermaphrodite (Aoyama and Kitajima, 1966). Okada (1966) postulated that this form repeats the hermaphroditic state.

In *Gonostoma gracile* (Gonostomatidae), a deep-sea luminescent fish, Kawaguchi and Marumo (1967) found that individuals less than 7 cm are mostly males and those more than 9 cm are invariably females. Sex succession takes place in the medium-sized fish (7–9 cm), and hermaphrodites are frequently found in the specimens of 6–7 cm.

2. *Protogynous Hermaphrodites*

Certain fishes belonging to the Sparidae, i.e., *Pagellus erythrinus* (D'Ancona, 1950; Larrañeta, 1953, 1964) and *Spondyliosoma* (*Cantharus*) *cantharus* (D'Ancona, 1950, 1956; Reinboth, 1962) are protogynous hermaphrodites. Among Japanese sparids, *Dentex* (*Taius*) *tumifrons* seems to be protogynous. Inversion of sex is brought about by development of the testicular region and regression of the ovarian part.

In *Dentex* (*Taius*) *tumifrons* (now Sparidae), Aoyama (1955) found fishes with hermaphroditic gonads. He concluded that some of the females change into males. Of serranids from Bermuda four species were synchronous hermaphrodites, as already stated, while at least nine species belonging to the genera *Epinephelus, Mycteroperca, Alphes, Petrometopon*, and *Cephalopholis* were protogynous hermaphrodites (Smith, 1959). Smith pointed out that there

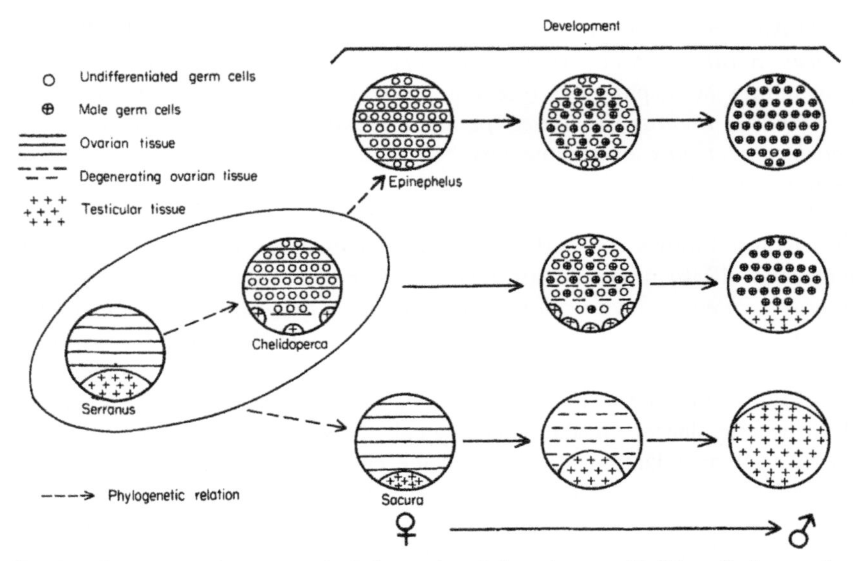

Fig. 3. Gonadal development and phylogenetic relations in serranid fishes. Redrawn after Reinboth (1967).

are three patterns of hermaphroditism in serranids: the *Serranus, Rypticus*, and *Epinephelus* type (Fig. 3). The patterns are essentially similar to those hermaphrodites reported by Reinboth (1967; cf. Fig. 3). *Epinephelus* and its allies have gonads in which the male tissue is present throughout the germinal epithelium lining the central lumen of the gonad. This male tissue becomes functional only after the female tissue has ceased to function. The genus *Rypticus* shows an intermediate type of gonad in which scanty male tissue is present in the lower part but is also found intermixed with the female tissue. The protogynous *Chelidoperca hirundinacea* is also of this type (Reinboth, 1967). The eastern Pacific *Paralabrax clathratus* seems to be secondarily gonochoristic. Here, we can make an inference on the process in evolution of gonochorism from hermaphroditism. Figure 3 illustrates the three types.

Kuroda (1931) postulated that larger and red *Sacura margaritacea* (Serranidae) and smaller and yellow *S. pulcher* are males and females, respectively, of the same species. He found some intermediate individuals. The protogynous hermaphroditism in this species was demonstrated by Reinboth (1963) and Okada (1965a,b). Protogynous sex reversal in this species seems to occur by degeneration of ovarian tissue after spawning. The Atlantic sea bass, *Centropristes striatus* (Serranidae), is also a protogynous hermaphrodite (Lavenda, 1949; Reinboth, 1965). In Centracanthidae (Maenidae), three species *Spicara smaris, S. chryselis*, and *S. maena*, of the genus *Spicara* (Maena) are known to occur in the Mediterranean. Zei (1949) has postulated that the two species *S. smaris* and *S. chryselis* are protogynous hermaphrodites. Lozano Cabo (1951, 1953) confirmed this in *S. smaris* and concluded that *S. smaris* and *S. alcedo* are two sexes of the same species. Lepori (1960) also concluded that *S. maena* and *S. chryselis* are protogynous species. Reinboth (1962) performed an elaborate study on *S. maena*.

For a long time the greenish and reddish Japanese wrasses (Labridae) have been regarded as different species; the former was called *Julis poecilopterus* and the latter *J. pyrrhogramma*. Jordan and Snyder (1902) suggested that the two forms might be sex variants of the same species, *Halichoeres poecilopterus*. Y. Kinoshita (1934, 1935, 1936) has verified this experimentally. He observed that the blue wrasse rapidly loses its secondary sex characters after castration, while the red wrasse is not affected by ovariotomy. Transplantation of a testis into the red wrasse transforms it into a blue wrasse. However, transplantation of an ovary into the blue wrasse causes no change in coloration of the recipients. Kinoshita found testis-ova in some individuals and postulated that this fish may be protogynous. Okada (1962) corroborated this. He (1964b) performed experiments on the effects of androgen and estrogen on sex reversal.

The Mediterranean labrid fishes *Coris giofredi* and *C. julis* have long been regarded as separate species. The two types differ in both size and color (Fig. 4). Smaller individuals (*C. giofredi*) are usually females and larger ones

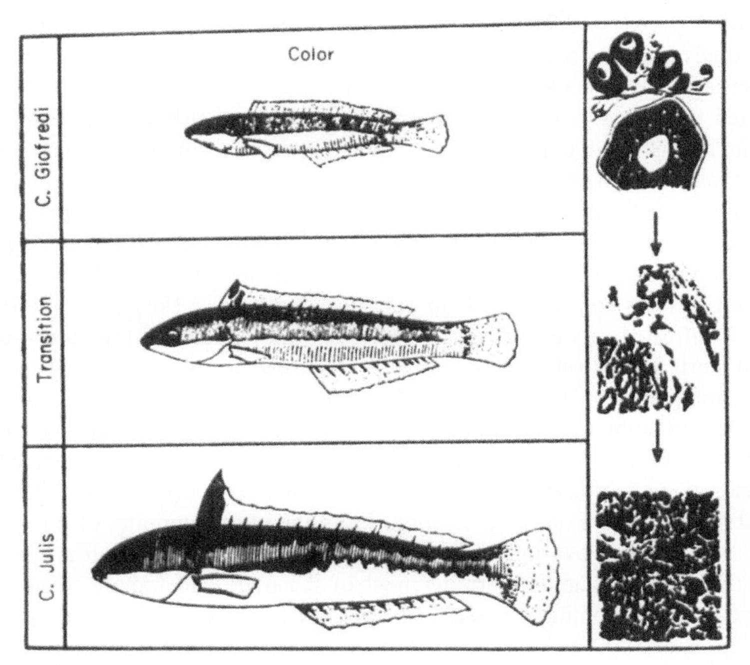

Fig. 4. Change of color and sex phase in the labrid, *Coris julis*. After Bacci, 1966: "Sex Determination," Fig. 7.1. Reprinted with permission from the author and Pergamon Press.

(*C. julis*) are males. Some medium-sized fish show intermediate color and have testes (Bacci and Razzauti, 1957, 1958). The two types are the same species, *C. julis*, a protogynous hermaphrodite. An interesting fact is that a few large fish with *C. giofredi* color are nevertheless males when their gonads are examined. Because each individual is at all times either a male or female, this labrid is termed a "false gonochorist."

The two species of labrids in the waters of the Mediterranean around Livorno, *Labrus turdus* and *L. merula*, are also protogynous species Sordi, 1962). In both species, all individuals less than 27 cm have ovaries. In the largest fishes of *L. turdus*, all individuals have testes. In *L. merula*, the largest individuals have a balanced sex ratio. This seems to indicate that while 50% of the fish change from female to male, 50% never change but remain females. In both *Coris julis* and a Caribbean labrid, *Thalassoma bifasciatum*, Roede (1965) observed 70% ♀♀+30% ♂♂ in the young and 100% ♂♂ in adults.

Reinboth (1962) pointed out the existence of two types of morphologically distinct males in labrids (*Coris julis* and *Haliochoeres poecilopterus*). The primary male looks like a female but remains a male throughout its life, and the secondary male that changes to a male from the female. However, according to Vandini (1965), there are no primary males that retain the *C. giofredi* color throughout their life cycle, viz., even the *C. giofredi* males eventually develop *C. julis* color. In the striped wrasse, *Labrus ossifagus*, there are some individuals with a red and others with a blue pattern. It was believed that the former is the female and the latter the male. Lönneberg and Gustafson (1937) reported that females greatly outnumber males in red fish and the reverse is true in blue-striped fish. They also found intersexual individuals changing from females to males. This seems to indicate that this form is protogynous and most of these fish, if not all, exhibit sex succession. Liu (1944) discovered that small individuals of the synbranchoid eel, *Monopterus albus* (*M. javanensis*), are females and the large ones males and offered a good case for protogynous hermaphroditism (cf. also Bullough, 1947). Liem (1963, 1985) confirmed this using nearly 1000 fish, and concluded that every individual starts its reproductive cycle as a functional female and then becomes a functional male. The sexuality of this fish has the following sequence: juvenile hermaphrodite → functional female → intersex → functional male.

In consecutive hermaphrodites such as Sparidae and Serranidae, either protandrous or protogynous, there is a basic feature in common. The juvenile gonad has an ambisexual organization. From the very beginning of differentiation of the gonad, ovarian and testicular rudiments are present in every fish, although the topographical arrangement of the male and female tissues and change of the dominant tissue in time differ from species to species (cf. Figs. 3 and 5).

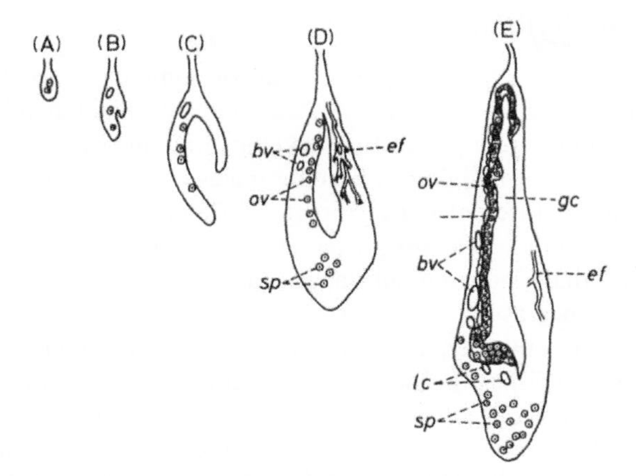

Fig. 5. Schematic transverse sections of early developmental stages (A → E) of the gonad in Sparidae, where ov stands for ovogonia; sp, spermatogonia; gc, gonocoelom; ef, efferent ducts; lc, lacunae; and bv, blood vessels. Redrawn from D'Ancona (1950).

A question arises as to whether all individuals in a consecutive hermaphroditic species inverse sex in sequence. In observations made on natural populations, occasional individuals are found smaller or larger than is usual for this sex. Smith (1959) noted unusually large females in some protogynous serranids, and Reinboth (1965) reported unexpectedly small males of *Centropristes striatus*, also a protogynous species. In *Dentex dentex* only some individuals undergo a transitional hermaphroditic stage (Lissia Frau, 1964). According to Larrañeta (1964) about 5% of the fish in a population of *Pagellus erythrinus* are males throughout their life, while 45% transform from female to male and 50% never transform but remain females. Rijavec and Zuvanovic (1965) obtained comparable results in the same species. Smith (1967) proposed a theory of hermaphroditism in which he postulated that sex reversal takes place in different individuals at different sizes and ages viz., sexual succession is a prolonged continuous process for the population. "During the first time interval following sexual maturity, only a part of the population changes sex, and during each time period a fraction of the remaining individuals transform." He interpreted Larrañeta's results as indicating that 5% of the individuals change before sexual maturity and that the life span is such that half the females do not live to change. Researches on consecutive hermaphrodites have been mostly based on observations on natural populations. There are only a few experimental or physiological strides and these are not satisfactorily documented. Hence, we know little about the cause of sex reversal. It seems that sex reversal may be caused by sex hormone imbalances. In the female phase the female hormone might dominate the male hormone and in the male phase the reverse might be true.

III. GONOCHORISM

As Witschi (1914a,b, 1930) has pointed out for amphibians so also in gonochoristic fishes there are "undifferentiated" and "differentiated" species. In the former, the indifferent gonad first develops into an ovary-like gonad and then about one-half of individuals become males and the other half females. In the latter, the indifferent gonad directly differentiates into either a testis or an ovary. In both types, sex differentiation seems to be brought about by male- and female-inducing substances (Section VI, A). It is natural that undifferentiated species are more unstable than differentiated ones in sexuality. Unfortunately, however, the two types have been studied embryologically in only a few fishes. In the absence of embryological evidence, it is as yet hazardous to correlate the two modes with the occurrence of sporadic intersexes which represent a remnant of the embryological condition manifested late in some adults. However, a few species in which undifferentiated and differentiated conditions are known provide evidence that spontaneous intersexes among gonochorists occur mostly in the "undifferentiated" species while in the "differentiated" ones the occurrence of intersex is rarely or never seen.

A. Undifferentiated Gonochorists

The lampreys and hagfishes (Agnatha) are undifferentiated species. Sexuality of the brook lamprey, *Lampetra lamottei* (*Entosphenus wilderi*), has been studied by Lubosch (1903) and more extensively by Okkelberg (1921). Ammocoete larvae up to 3.5 cm possess indifferent gonads. After this stage, some of the larval gonads contain both male and female germ cells. Of 15 adult males (sperm producers) undeveloped ova were found in the testis of seven.

In ammocoete larvae of *L. planeri* and *L. fluviatilis*, differentiation of the gonads is mostly completed at 5–6 cm in length. Hardisty (1960) recorded the sex ratio in animals longer than 6 cm. In *L. planeri* ammocoetes there is a small but significant excess of females. There are a number of transitional individuals with disintegrating oocytes and developing stroma, which may eventually become males. Among hagfish, Atlantic species, *Myxine glutinosa*, seems to be undifferentiated. Conel's (1917) and Schreiner's (1904, 1955) reports seem to indicate that this species is an intermediate type between hermaphroditism and gonochorism, in which juvenile hermaphrodites are common. The Japanese hagfish, *Eptatretus burgeri*, is also an undifferentiated gonochorist. Dean (cited by Conel, 1931) found only one male with an ovotestis out of 569 fish.

Turning to the class Pisces, there are many undifferentiated gonochorists which normally develop into either male or females. However, sporadic intersexes are known to occur in these fishes. It is beyond the scope of the present chapter to cite a vast number of instances; Freund (1923) and Atz (1964) have listed numerous cases.

Grassi (1919) in the eel, *Anguilla anguilla*, and Mršić (1923, 1930) in the rainbow trout, *Salmo gairdneri irideus*, have demonstrated that these species are undifferentiated. In the latter, accidental intersexes have been reported not infrequently. In Gadidae, sporadic intersexes have also been frequently reported. In the herring, *Clupea harengus*, there are several types of the intersexual gonad (Gäbler, 1930; Rudolf, 1931).

In the minnow, *Phoxinus laevis*, Bullough (1940) found 10 fish with intersexual gonads. In all cases the ovarian portion was suppressed while the testicular region was normal. He thought that these intersexes represented transitory stages changing from female to male. The effects of an androgen and an estrogen supported this postulation. However, a complete sex reversal was not achieved with these hormones.

These are three species of the paradise fishes: *Macropodus opercularis, M. chinensis*, and *M. concolor;* all three are undifferentiated species. Whereas the male index (♀♀/♀♀+♂♂) of *M. opercularis* and *M. chinensis* is 50%, that of *M. concolor* varies from 68% to 91%; this there is a preponderance of males (Schwier, 1939). The presence of multiple sex factors in *M. concolor* is apparent. While the offspring from *M. opercularis* × *M. chinensis* are fertile, males from *M. concolor* × *M. chinensis* are sterile.

The live-bearing cyprinodont *Poecilia* (formerly *Lebistes*) *reticulata* is an undifferentiated gonochorist (Goodrich *et al.*, 1934; Dildine, 1936). Blacher (1926a) and Winge (1927) found a few aged intersexes with an XX constitution. Eggert's report (1931, 1933) on intersexes of the mudskipper, *Periophthalmus vulgaris* (Gobiidae), seems to suggest that there are two geographical races in the same species. In the fish on the south coast of Java, the two sexes differentiate independently from the start, whereas on the north coast the testes of young have basically ovarian structure indicating that the male pass through the female phase. However, number and sizes of specimens examined by him were very limited.

B. Differentiated Gonochorists

Only a few fishes have been shown to be differentiated gonochorists. Sexuality of differentiated species is fairly stable among fishes. Wolf (1931) demonstrated that the "domesticated" platyfish, *Xiphophorus* (formerly *Platypoecilus*) *maculatus*, is a differentiated species. Bellamy and Queal (1951) stated that not a single sporadic intersex has ever been found among 50,000 specimens.

The situation is the same in the medaka, *Oryzias latipes* (Cyprinotontidae). No spontaneous, true intersex has ever been found among more than thirty thousand fish studied during 40 years. Yamamoto (1953) showed that this species is a differentiated gonochorist. In this connection, Oka's report (1931b) on the occurrence of oviform cells in three males of this species may be mentioned. His finding was based on fish which had been left to starve for 3 months. It is unfortunate and even strange that the title of this paper included the words "accidental hermaphroditism" since Oka carefully observed the cytological difference between these enlarged cells and genuine ovocytes, calling them "pseudo-ovocytes." It is probable that these cells are enlarged proto- or spermatogonia prior to degeneration, As to the so-called testis-ova in this species, induced by various agents, comments are given in Sections V, C and VI, A.

C. All-Female Species

The viviparous toothcarp, *Poecilia* (*Mollienisia*) *formosa*, which is thought to be a form of hybrid origin, inhabits northeastern Mexico and southwestern Texas. The "all-female" phenomenon in this form was first found by C. L. Hubbs and Hubbs (1932, 1946). In natural habitats, it propagates itself by mating with sympatric, related bisexual species, either *P. latipinna* or *P. sphenops*. However, paternal characters are not transmitted to offspring which are solely females. Subsequently, this fact has been confirmed by Meyer (1938), C. Hubbs *et al.* (1959), Haskins *et al.* (1960), C. Hubbs (1964), Kallman (1962, 1964a,b). C. L. Hubbs and Hubbs, and Kallman concluded that the monosexuality is the result of gynogenesis, the mate providing a stimulus to activate development of the ovum without syngamy. Nevertheless, there is strong evidence that the species is diploid (cf. C. Hubbs *et al.*, 1959; Kallman, 1962). In terms of DNA levels, *P. formosa* nucleus is the same as in the diploid spaces *P. sphenops* (Rasch *et al.*, 1965).

C. Hubbs *et al.* (1959) discovered a single wild male in the Brownville population, and Haskins *et al.* (1960) reported a single male in their laboratory stock. It proved fertile in mating to *P. formosa* siring all females. The appearance of extremely rare males is still a matter for debate. In this connection, Kallman's study (1964b) deserves attention. He found among thousands of offspring from gynogenetic *P. formosa* mated with males of *P. vitatta* or *P. sphenops* 14 exceptional fish of which 12 exhibited paternal patterns and possessed morphologically intermediate features while the remaining two had mosaic patterns with some areas patriclinous and others matriclinous. This shows that in rare cases syngamy occurs or a single chromosome from the male nucleus, governing pattern formation, may become accidentally incorporated into some or all cells of developing embryos giving rise to mosaic patterns.

Gynogenesis has been reported in certain natural populations of the crucian carp, called the "silver goldfish," *Carrassius auratus gibelio*, which produce all-female progeny (Lieder, 1955, among others). He regarded this phenomenon as a natural parthenogenesis. Artificial insemination revealed that a gynogenetic stimulus to the ova can be provided by the common goldfish, *C. auratus auratus*, or by the carp, *Cyprinus carpio*.

Spurway (1953) claimed the occurrence of "spontaneous parthenogenesis" in two anomalous guppy females. Later (Spurway, 1957) this postulation was withdrawn and she assumed that the phenomenon results from the self-fertility of "functional" intersexes. Subsequent to Spurway's study, Stolk (1958) reported pathological gynogenesis observed in the guppy and the swordtail. The ovary of each female was infested by a phycomycete fungus, *Ichthyophorus hoferi*, which apparently was responsible for the pathological activation of the ova. More than 64 broods were produced by these fishes; these without exception were daughters. Stolk criticized Spurway's finding by saying that her fish might also be infected by the parasite.

The mode of reproduction of the two unisexual "species" or strains of the genus *Poeciliopsis*, viviparous toothcarps of northwestern Mexico, is unique to vertebrates (Miller and Schultz, 1959; Schultz, 1961, 1966). At first, it was considered that each of the two undescribed species had both bisexual and unisexual strains. Bisexual strains were tentatively referred to as C and F and unisexual ones as Cx and Fx, respectively. Miller (1960) identified the C and Cx as *P. lucida*, but F and Fx are still undescribed. Later, Schultz (1966) showed that monosexual Cx is a form different from the bisexual *P. lucida* but closely allied to it. In nature, Cx propagates itself by mating with *P. lucida*. Hybridization experiments between Cx and males of two bisexual species *P. lucida* or *P. latidens* provided evidence that genetic factors of both parent combine to form the F_1 offspring. However, the entire male genome appears to be eliminated during ovogenesis, probably at meiosis. Hence, ova have no paternal genome in each generation.

Besides Cx, Schultz (1967) further reported two additional all-female stocks, Cy and Cz, which were previously thought to be Cx. *Poecilia lucida* provides sperm for these monosexual forms. While Cx and Cz are diploid expressing characteristics of both parents in the F_1, Cy is a triploid and in mating with males of various bisexual species, e.g., *P. latidens*, produces all-female, triploid offspring by gynogenesis devoid of paternal characters. Thus, the mode of reproduction of Cy is quite different from that of Cx and Cz in which paternal chromosomes of the F_1 generation are not transmitted through the ova; only those characteristics of the female germ line pass to the next generation.

In gynogenetic diploid forms, the diploid complement might be maintained by suppression of one of the meiotic divisions, reentry of the second polar body, or suppression of the first mitotic cleavage. The means by which the triploid Cy undergoes meiosis and produces fertile triploid eggs is obscure. At this point, it may be mentioned that Rasch *et al.* (1965) mated the gynogenetic, diploid *Poecilia formosa* to *P. vittata* and obtained offspring which were triploids as judged by their nuclear DNA. However, these triploids were sterile.

IV. GENETIC BASIS OF SEX DETERMINATION

> The genetic study of sex is important not merely because sex is instinctively our major interest, but it lies at the root of Mendelian heredity itself and is the major factor in evolution. It provides such admirable material for the study of gene interaction, of phenogenetics, that is, of developmental physiology.
>
> Abbreviated from H. J. Muller (1932)

The classic sex factor studies use notations such as FF = ♀, FM = ♂, where M>F in ♂ heterogametic species, and FM = ♀, MM = ♂, where F>M in ♀ heterogametic forms, which are oversimplified and are not generally found in nature, if the F and M symbolize single sex determiners (genes), Hartmann (1951, 1956) has proposed the formulas (A̸G) FF = ♀ and (AG̸) MF = ♂, where the M and F represent the sex determiners (realizers), the AG stands for ambisexual potencies and the symbols A̸ and G̸ represent the inhibiting effects exerted by the combinations of sex realizers. The supposed existence of the double system is also a formalization and has been a matter for debate since experiments on the localization of the AG complex have not been fruitful. Kosswig (1964) is also an opponent to these formulas.

A. XX-XY and WZ(Y)-ZZ(YY) Types

At this point, it may be appropriate to comment on the designation of sex chromosomes. The symbols in male heterogametic forms, that is, XX for female and XY (or XO) for male, are in universal agreement. Fishes which were found to have this type are listed in Table I. Among fishes, those which have this type are more numerous than those with female heterogamety.

On the other hand, in organisms with female heterogamety, some authors use the formula WZ(OZ) ♀–ZZ♂, while others, including outstanding geneticists, use the formula XY (or XO) ♀–XX♂. The latter denotation is not only perplexing but also is based on the misleading assumption that the Z=X and W=Y, which is in contradiction to the experimental facts revealed in fishes. This notation becomes seriously confusing when matings are made between male heterogametic and female heterogametic species or races. In Table II, symbols used for the female heterogametic platyfish by several authors are listed. As a result of the studies that follow, the symbol XY ♀–XX ♂ should be eliminated.

In connection with the Bellamy's experiment (1936) of the inter-specific matings between ♀ heterogametic *Xiphophorus maculatus* and ♂ heterogametic *X. variatus*, Castle (1936) suggested use of the terms XY for ♀ and YY for ♂ in the ♀ heterogametic species, where the Y is equivalent to the Z and the X to the W. In fact, Gordon (1946a,b, 1947a) proved experimentally in xiphophorin fishes that the Z = Y. However, since there was no evidence that W=X, he suggested that the formula WZ ♀-ZZ ♂ might have been better written as WY ♀-YY ♂. German authors continue to use the WZ-ZZ system, whereas American investigators, particularly Kallman

Table I Fishes with Male Heterogamety (XX ♀, XY♂) Based Mostly on Genetic Evidence

Species	Family	Author	Year
Oryzias latipes	Cyprinodontidae	Aida	1921
Poecilia[a] *reticulata*	Poeciliidae	Winge	1922
Poecilia[b] *nigrofasciata*	Poeciliidae	Breider	1935
Xiphophorus variatus	Poeciliidae	Kosswig	1935
X. xiphidium	Poeciliidae	Kosswig	1935
X. maculatus[c]	Poeciliidae	Gordon	1946
X. couchianus	Poeciliidae	Gordon and Smith	1938
X. milleri	Poeciliidae	Kallman	1965
X. montezumae cortezi	Poeciliidae	Kosswig	1959
Betta splendens	Anabantidae	Kaiser and Schmidt	1951
Mogrunda obscula[d]	Gobiidae	Nogusa	1955
Cottus pollux[d]	Cottidae	Nogusa	1957
Carassius auratus[e]	Cyprinidae	Yamamoto and Kajishima	1969

[a]Formerly *Lebistes*.
[b]Formerly *Limia*.
[c]Mexican populations.
[d]Cytologically established.
[e]Cultivated goldfish.

Table II Symbols of Sex Chromosomes in the ♀ Heterogametic Platyfish, *Xiphophorus maculatus*[a]

♀	♂	Author	Year	Stock
XY	XX	Bellamy	1922	Domesticated
		Bellamy and Queal	1951	Domesticated
WZ	ZZ	Gordon, Kosswig, Breider, Bellamy	1926	Domesticated
XY	YY	Castle	1936	Domesticated
WY	YY	Gordon	1947	Domesticated
WY	YY	Gordon	1950	Belize river
WY	YY	Kallman	1965	Rio Hondo

[a]From several authors, but chiefly Gordon (1952).

(1965b), stress that it is best to eliminate the symbol Z. The W of ♀ heterogametic forms is considered to have strong F gene(s), and the Y of ♂ heterogametic forms is considered to have strong M gene(s).

In this connection, it may be noted that in the interspecific mating between ♀ heterogametic *X. maculatus* (WZ-ZZ) and ♂ heterogametic *X. variatus* (XX-XY), Kosswig (1935a) stated that the Z of the former behaved like the Y of the latter, although at that time he considered that Z=X and W=Y.

Heterosomal sex-determining mechanisms have often been called "monogenic." This term is inadequate at least in fishes since sex is "determined" indirectly by the totality of sex genes in heterosomes and autosomes. Polygenic sex determination with or without heterosomes will be discussed later (Section IV, B).

Female heterogamety in the "domesticated" platyfish, *Xiphophorus* (*Platypoecilus*) *muculatus*, has been established by Bellamy (1922, 1928), Gordon (1927), Fraser and Gordon (1929), and Kosswig (1933). The dual mechanism of sex determination operating in the same species has been found by Gordon (1946a,b,c, 1947a, 1950, 1951a) who showed that, while the Mexican wild populations are male heterogametic, those originally from British Honduras are female heterogametic. The "domesticated" varieties are considered to have derived from British Honduras.

Kallman (1965a,b) extended Gordon's findings by studying the genetic behavior of populations from a vast area of Guatemala, intermediate between Mexico and British Honduras. The populations in this area contain both male and female heterogametic types even in a single pond. They interbreed with each other with the result that WX females were also collected. In view of evolution of the WY-YY and XX-XY types, this study is most interesting.

Tilapia mossambica (Cichlidae) is considered to be a superspecies or group of closely related forms. Crosses of Malayan females to African males result in progeny consisting only of males, or nearly so, whereas the reciprocal crosses produce offspring in a sex ratio of 1 ♀: 3♂ (Hickling, 1960). Whether the Malayan form has the XX-XY and the African form the WZ(Y)-ZZ(YY) system, analogous to the dual system in the platyfish, *X. maculatus*, has not yet been clarified.

B. Polygenic Sex Determination and So-called Genetic Sex Reversal

To grasp the real situation of sex determination, we must adopt the broader view originally advanced by Bridges (1922, 1925) for the fruitfly and elaborated by Winge (1934) for fish viz. that "a given property, the *sex included*, depends upon all the chromosomes, some of which pull in one direction and others in the other direction, some strongly and others faintly or not demonstrably at all" (Bridges, 1939).

In fishes in which the homogamety/heterogamety is established, exceptional XX ♂♂ or XY ♀♀ and WZ(Y) ♂♂ appear occasionally, although no exceptional ZZ(YY) ♀♀ are reported. These exceptions have been misleadingly called genetic "sex reversals" even by distinguished geneticists. This is based on the concept that only sex chromosomes are carriers of the sex determiners. Only a few geneticists—of whom Winge (1934) is the most celebrated—have grasped the intrinsic nature of sex differentiation. He has never referred to such exceptions as sex reversals because his theory is based on multiple sex factors with superior sex genes in the allosomes.

The genetic evidence indicates that the guppy, *Poecilia* (*Lebistes*) *reticulata*, has sex chromosomes, XX for ♀ and XY for ♂ (Winge, 1922b). However, minor or polygenic F and M genes are distributed throughout the autosomes. The Y has superior (epistatic) M gene(s) and the X is supposed to possess epistatic F gene(s) (Winge, 1934; Winge and Ditlevsen, 1947, 1948). In the majority of individuals, autosomal sex genes are more or less in balance. Consequently, sex in most individuals is determined by the heterosomal combination. However, by fortuitous combinations of autosomes or recombinations a few exceptions may appear in which $\Sigma M > \Sigma F$ in spite of an AAXX constitution. In *Mutatis mutandis*, exceptional XY females are considered to be individuals in which the totality of sex genes becomes $\Sigma F > \Sigma M$. These exceptions cannot be regarded as genetic "sex reversals" since they have a genetic basis to develop either into males or females.

Alda's breeding results (1930) in the medaka, *Oryzias* (*Aplocheilus*) *latipes*, can be interpreted by this postulation (Winge, 1930). Since polygenic sex genes are numerous, by selective breedings of XX-♂♂, the sex ratio of offspring can be varied. Aida (1936) established an XX-XX strain of the medaka and adopted a theory of polygenic sex differentiation which is somewhat different from that of Winge. He suggested that XX males may be the result of a lowering of the female-determining potency of the X chromosome. It is unfortunate that he referred to XX males as "sex reversals."

In our d-*rR* strain of the medaka, where normally X^rX^r are females and X^rY^R males about 0.5% of the progeny are exceptions (0.2% cross-overs and 0.3% of X^rX^r♂ plus X^rX^R♀, Yamamoto, 1959a, 1964a,b). These rare X^rX^r♂♂ and X^rX^R♀♀ are regarded as exceptions in which autosomal M genes and F genes, respectively, over-accumulate in such a way that the sum of autosomal sex genes outbalances the superior heterosomal sex factors. As pointed out earlier (Yamamoto, 1963), the number of possible autosomal combinations is 2^{46}, an astronomical number. Even if only some of autosomes are assumed to be sufficiently different in respect to M and F genes (or modifiers), the rest being more or less in even balance, the number of combinations would be enormous. Figure 6 is a graphical diagram illustrating the concept of the polyfactorial sex determination with epistatic sex genes in sex chromosomes, as illustrated in the guppy, the platyfish, and the medaka. For the sake of simplicity, it is taken for granted that the ratio of ΣM of AA: ΣF of AA of a

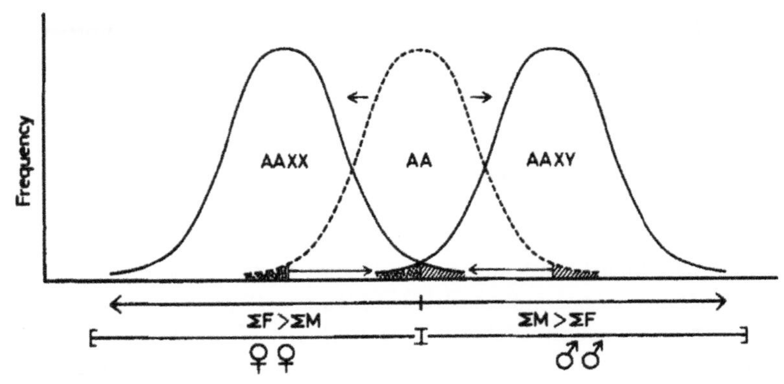

Fig. 6. A diagram of polyfactorial "sex determination" with epistatic sex genes in sex chromosomes.

population (where A stands for a set of autosomes) has a center (mean or mode) around which the individuals are fluctuating, thus forming a normal distribution curve (broken line).

In an AAXY constitution the AA curve is shifted in the male direction (right) because of epistatic M gene(s) in the Y, and in the AAXX constitution the curve is pulled to the female direction (left), perhaps because of the presence of epistatic F gene(s) in the X or absence of the Y. Exceptional XX males and XY females encountered in normal breeding may be regarded as individuals which fall in the two extreme regions of the AA curve (stippled and oblique hatched), i.e., the ΣM greatly overweighs the ΣF of AA or vice versa.

By inbreeding and systematic selections, the AA curve may be shifted in either direction. In fact, the concept of polygenic sex factors has been substantiated by selective breeding. Thus, Aida (1936) obtained a strain of the medaka, in which both female and male have the XX constitution. In Aida's stock, exceptional XX males were rare at first; by continued selection for XX males a breed was established in which XX males out-numbered XX females.

Formerly, Kosswig (1932, 1933, 1935a,b) and Breider (1935a) have asserted that sex of the platyfish is determined genotypically, while that of the swordtail is determined phenotypically. Kosswig (1939a,b, 1954) has maintained that the X in ♂ heterogametic species and the Z in ♀ heterogametic forms are devoid of sex factors and that ♀ determiners in the former and ♂ determiners in the latter are present in autosomes. Later, Kosswig and Öktay (1955) modified their earlier views on mechanisms of sex determinations in xiphophorin fishes (*X. maculatus*, *X. variatus*, *X. xiphidium*, and *X. helleri*). In this article, they maintained that in ♂ heterogametic (XX ♀-XY♂) fish the X contains the F gene which is weaker than the M gene of the Y, and that the W has a strong F gene which is stronger than the M gene of the Z in ♀ heterogametic fish. The x of *X. helleri* has a strong masculinizing potency,

Kosswig and his colleagues (Breider, 1942, 1949; Öktay, 1959, Anders and Anders, 1963; Peters, 1964; Dzwillo and Zander, 1967) eventually adopted the concept of polygenic sex determination (cf. Kosswig, 1964). This being so, exceptions are either prospective males or females from the very start of development in spite of their opposite sex chromosome formulas.

Öktay (1959) and Anders and Anders (1963) correctly attributed the occurrence of exceptional XX males and XY females, appearing in the Mexican platyfish, to the effects of autosomal sex genes. MacIntre (1961), working on the same animal, reported a not infrequent occurrence of phenotypic females with the XY constitution. When mated to normal XY males, "sex-reversed" females gave rise to XX daughters and XY and YY sons. It seems that autosomal F genes must have been accumulated in these strains by several generations of inbreeding.

In fishes with entirely polygenic sex determination, sex chromosomes are as yet not differentiated. They are characterized by diverse sex ratios. In crossing them with a species with a homogametic/heterogametic system, sex chromosomes of the latter usually become epistatic. The swordtail, *X. helleri*, is the most famous gonochorist with a polygenic system. Kosswig (1932, 1935b), Breider and Kosswig (1937), and Gordon (1948) symbolized a pair of chromosomes which are counterparts of sex chromosomes as xx or X′X′ in both sexes.

Sengün (1941) reported that the Wx class of F_1 (Mh) from the interspecific matings between *X. maculatus* (WZ) and *X. helleri* (xx) and of backcross (Mhh) to *X. helleri* are all ♀ ♀, while Xx class of F_1 (Mh) of the Mexican *X. muculatus* (XX) ♀ × *X. helleri* ♂ (xx) comprised both sexes. This means that the W is stronger than the X and Xx-♂♂ are the result of *X. helleri*-polygenic M genes. Öktay (1963) performed a hybridization between the swordtail, *X. helleri*, and ♂ heterogametic platyfish, *X. xiphidium*. The F_1 (Hx) from *X. helleri* ♀ (xx) × *X. xiphidium* ♂ (XY) were comprised of both ♀ ♀ (xX) and ♂♂(xY). In the backcross (Hxh) cross, the F_1 (Xh) from XX♀ × xx♂ consisted of xx ♀ ♀ and xX♂♂. This can be interpreted as indicating that the totality of sex genes in *X. helleri* ♀ is $\Sigma F > \Sigma M$, while that of *X. helleri* ♂ is $\Sigma M > \Sigma F$.

There are three species of the genus *Poecilia* (*Limia*) (Poeciliidae) in the Caribbean Islands, *P. nigrofasciata, P. caudofasciata*, and *P. vittata*. Of these, *P. nigrofasciata* possesses the XX-XY mechanism while the other two species show polygenic sex determination (Breider, 1935b, 1936b; Breider and Kosswig, 1937). The F_1 of XX-*P. nigrofasciata* ♀ × xx-*P. caudofasciata* ♂ are all females (Xx). In the reciprocal cross, xx-*P. caudofasciata* ♀ × XY-*P. nigrofasciata* ♂, the F_1 are in a ratio of 1 ♀ (xX):1♂(xY). The Y of *P. nigrofasciata* ♂ contains a strong M factor.

In Section III, A, Schwier's studies (1939) on the genus *Macropodus* (Anabantidae) are cited. Of three species, *M. opercularis, M. chinensis*, and

M. concolor, M. opercularis and *M. chinensis* seem to have a homogametic/heterogametic mechanism, because the sex ratio of the latter two species is 1♀ : 1♂. In the absence of any sex-linked character the heterogametic sex cannot be decided. In *M. concolor*, males outnumber females. Zander (1965) interpreted this fact as indicating that the X chromosome might be absent in this form, viz., sex formulas, being xx ♀-xY♂ type. Some males are xY and the others are xx males produced by autosomal, polygenic $\Sigma M > \Sigma F$ genes. Females are considered to be invariably xx constitution with autosomal $\Sigma F > \Sigma M$ genes.

In short, sex determination in gonochorists is polyfactorial with or without epistatic sex gene(s) in sex chromosomes, in both cases, a zygote in which $\Sigma M > \Sigma F$ differentiates into a male and that in which $\Sigma F > \Sigma M$ develops into a female.

In the experiments of Bellamy and Queal (1951) using the ♀ heterogametic platyfish, *X. maculatus*, a type of exceptional male appeared, which sired offspring in a ratio of 3♀ : 1♂ as if it had the genotype WZ(Y). About one-third of the females bred as though they were WW and produced only daughters. These exceptions were regarded as "sex reversals." If such were the case, we would expect to find some intersexes at the transitional stage. Confronted with the fact that not a single intersex has ever been found among 50,000 fishes handled in the past 28 years, the authors postulated that early sex inversion might have taken place. However, Wolf (1931) demonstrated that this species is a differentiated one. Hence, no juvenile hermaphroditism can occur.

In the platyfish, *X. maculatus*, with opposing sex-determining mechanisms, one XX-XY the other WY(Z)-YY(ZZ), exceptional XX or WY males and XY females appear occasionally. Besides Bellamy and Queal, Gordon (1946b,c, 1947a, 1951a), MacIntre (1961) and Kallman (1965b) regarded these as "sex reversals," notwithstanding the last-mentioned author's statement that sexuality of this fish is stable and that although more than 100,000 platyfish have been examined during the last 25 years, not a single female has ever changed into a male (fish with a testis). Consequently he defined a sex-reversed fish as one that is functionally one sex, but *genotypically* the other. However, if we take autosomal sex genes with various potencies into consideration, it is likely that it is genotypically also one sex and not the other.

If *we* accept these exceptions as real sex reversals, we are confronted with an array of paradoxical facts: (1) Both the medaka and the platyfish are sexually stable among fishes; (2) not a single sporadic intersex has ever been found, however, so-called sex reversals have been reported occasionally; and (3) no artificial induction of reversal in sex differentiation by any means, including sex hormones, has ever been accomplished in the platyfish, despite the occurrence of "spontaneous sex reversals." These puzzling facts can be clarified if we accept that they are genotypic males ($\Sigma M > \Sigma F$) and genotypic females ($\Sigma F > \Sigma M$), respectively. On the basis of the concept of multiple autosomal sex genes or modifiers these exceptions are not sex reversals at all. Breider (1942) described a WZ♂ of ♀ heterogametic *X. maculatus*. In mating this exception with a normal WZ♀, the offspring were in a ratio of 3♀ (1 WW ♀, 2 WZ ♀): 1 ♂ (ZZ). Breider did not detect the WW female. Bellamy and Queal (1951) detected the genotype by progeny tests.

To sum up, in strictly "differentiated" gonochorists such as the platyfish and the medaka the occurrence of spontaneous sex reversals is impossible.

C. "Spontaneous Sex Reversal" in the Swordtail

A male and female swordtail are illustrated in Fig. 7. Amateur breeders have repeatedly claimed instances of spontaneous sex reversal in some of their fishes, particularly in aged swordtail *X. helleri*. Biologists have also reported the same phenomenon (Harms, 1926a,b; Friess, 1933). Hild (1940) regarded sex reversal in this form as a regular phenomenon.

To accept a fish as a true functional sex reversal, it is necessary to ascertain that (1) the fish first functioned as a female and (2) later it turned out to be a functional male. Some reports on the matter have not ascertained these facts. Popoff (1929) and Sacks (1955) doubted that a true functional sex reversal occurred.

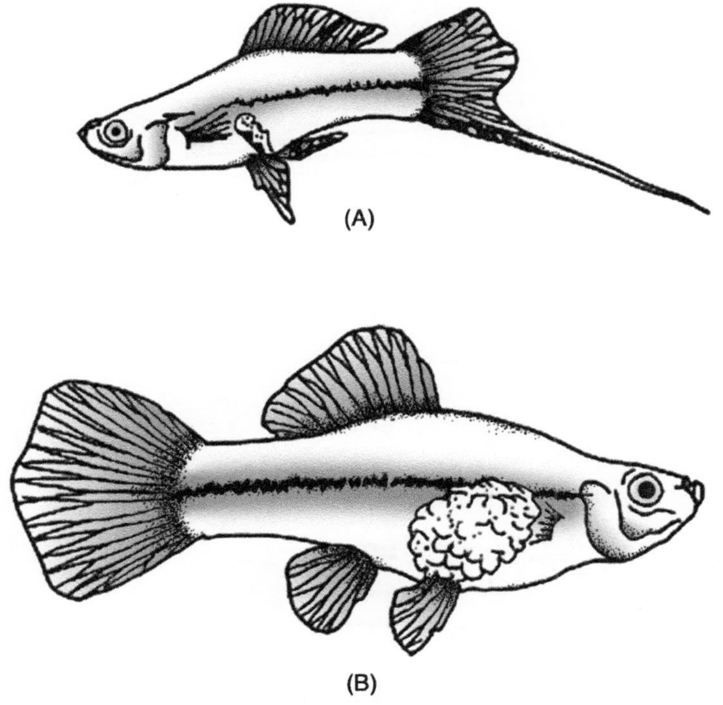

Fig. 7. (A) Male and (B) female of the swordtail, *Xiphorphorus helleri*. Redrawn by Dr. G. Eguchi after Gordon (1956).

Pathological masculinization in two guppy females was recorded by Wurmbach (1951). The fish developed a gonopodium. Histological study revealed that they were infected by a fungus, *Ichthyophorus hoferi*, which caused degeneration of peripheral ovocytes, followed by appearance of spermatogonia derived from ovarian-cavity epithelium. He pointed out that Fries' "rest bodies" in the masculinized gonad of the swordtail are not remnants of degenerated ovocytes but are cysts of the parasite.

Peters (1964) found that in *X. helleri* from Honduras there are two types of males: the normal or rapidly growing males (♂$_F$) and the slowly developing males (♂$_S$) which reach larger sizes and differentiate sexually more slowly. With this background, his criticism on reported sex-reversal merits special attention. Among some hundreds of females which he observed during several years, not a single sex reversal has ever been found, although he obtained some arrhenoid advanced-aged females (Fig. 8A). The outer appearance of ♂s looks like females in the immature stage, and male secondary sex

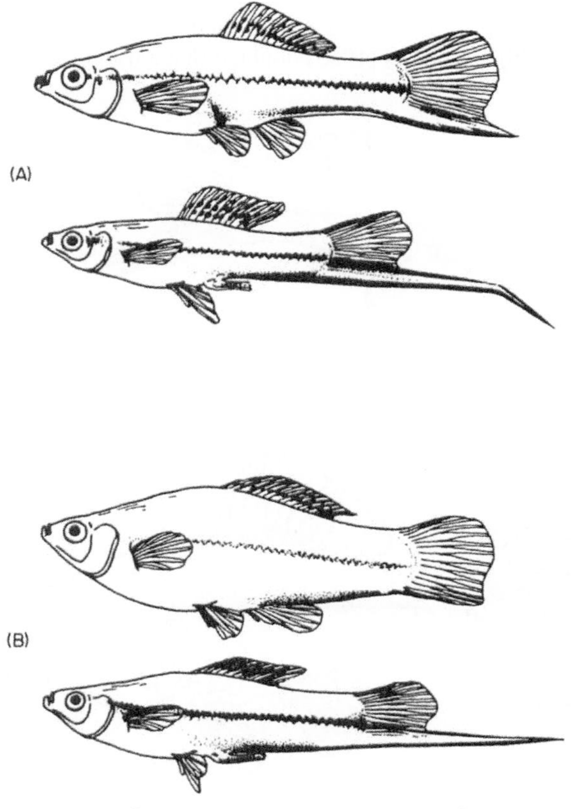

Fig. 8. (A) An arrhenoid ♀ (above) and a rapidly growing ♂$_F$. (B) A slowly growing ♂$_s$ (above) and a rapidly developing ♂$_F$ (below) of the swordtail, *Xiphophorus helleri*. Drawn by Dr. G. Eguchi from photos by Peters (1964).

characters manifest themselves only later (Fig. 8B). It is natural that slowly developing ♂s can fertilize ova. Peters considered that alleged sex reversals might not be males but arrhenoid females caused by hormonal upsets in old age, and unripe ♂s are likely to be taken as females. Philippi (1908) described arrhenoid females with male sex characters in *Glarichthyes* (Poeciliidae).

The most famous cases are in Essenberg's report (1926) on spontaneous, functional sex reversal in the swordtail. He claimed that two fish (B_{16} and C_3), after producing broods, turned out to be males and sired a few offspring in mating with "virgin females." This article has been cited in zoology textbooks without qualification. However, Gordon (1956, 1957), a celebrated poeciliid geneticist, expressed doubt as to the validity of Essenberg's conclusions. He pointed out that a functional swordtail male must have not only a complete testis but a functional gonopodium (modified anal fin) as well as its suspensorial skeleton. If the fish in question were first female and then turned out to be male, the female-type anal fin and suspensoria must have transformed into a perfect gonopodium. For him, it is inconceivable that such a drastic change should have occurred. Incomplete male secondary sexual characters engendered by hormonal upsets appear in aged swordtails. He questioned the virginity of the females used as mates of alleged sex reversals. It is well known that sperm from a single copulation can sire up to eight broods in the subsequent 8 months. Gordon (1956) states that "unconditional evidence is required before the reality of spontaneous, complete and functional sex reversal in a fish can be accepted."

V. CONTROL OF SEX DIFFERENTIATION

Science unfolds and controls nature.
Toki-o Yamamoto

A. Surgical Operation

Surgical castrations of female Siamese fighting fish, *Betta splendens*, have in rare instances produced fish in which the regenerated gonad was a functional testis. These experiments were first performed independently in the United States (Noble and Kumpf, 1937) and later in Germany (Kaiser and Schmidt, 1951). The American authors obtained only seven reversals out of 150 spayings, of which only three proved to be fertile, siring both sexes in mating with females. The German authors obtained three reversals, of which only one was fertile and fathered solely a total of 109 daughters. The latter experiment proved that the male is heterogametic.

B. Modification of Sex Differentiation by Sex Hormones

In the rainbow trout, *Salmo gairdneri irideus*, Padoa (1937, 1939a) reported that injection of a follicular hormone produced ova in the testes, while testosterone induced testicular tissue in the ovaries. Ashby (1952, 1956, 1959) reported the paradoxical similarity of action of estradiol and testosterone in the brown trout, *Salmo trutta*; both hormones retarded gonadal development and produced hypertrophy of the somatopleure. There was no evidence of sexual inversion. The result might be ascribed to the fact that he started hormone administration in alevins in which gonadal sex differentiation has already been established.

A number of attempts have been made in viviparous toothcarps to modify sex differentiation by administration of sex hormones, especially androgens, starting with newly born broods or adults. However, androgens mainly affect the secondary sexual characters and the action on the gonad has been either incomplete, pathological, or negative.

In the guppy, *Poecilia reticulata*, the effects of sex hormones have been studied by Witschi and Crown (1937), Berkowitz (1937, 1938, 1941), Régnier (1938), Eversole (1939, 1941), Gallien (1946, 1948), Hildeman (1954), Miyamori (1961), and Querner (1956). Berkowitz caused the production of ovo-testis by administering estrogens to young male guppies. He reported that the optimal dosage resulted in an almost complete reversal of testis to ovary. The same effect was observed by Querner. Miyamori produced ovo-testis by administration of androgen to young females. In the top minnow, *Gambusia holbrookii*, which shows "transitional intersexuality," Lepori (1942a,b, 1948) produced testis-ova in females by administration of an androgen and in males by an estrogen. Comparable results were obtained in *G. affinis* by Okada (1944). In the swordtail, *X. helleri*, Baldwin and Goldin (1940) reported that androgen administration to young females not only simulated the male secondary characters such as formation of a caudal sword and the gonopod but also induced masculinization of the gonad after degeneration of ovocytes. Querner (1956) showed that androgens produce ovo-testis in genetic females. Contrary to the cases in amphibians, corticoids have no effect on sex differentiation.

Vivian (1952a) reported that 11 incompletely hypophysectomized female swordtails developed an involuted ovary with so-called "rest bodies" and claimed that two or three of them showed partial masculinization. The fact that these rest bodies are not degenerated ovocytes but are cysts of a parasitic mycomycete has already been mentioned (Section III, C). This pathological change might be caused by infection of *Ichthyophorus hoferi*.

Failure to obtain complete reversal of sex differentiation in these studies may be because administration of hormones was started after the onset of gonadal sex differentiation. As stated before, although the guppy and swordtail are "undifferentiated" species, gonadal sex differentiation has already been established when broods are born (Essenberg, 1923, in the swordtail; Goodrich *et al.*, 1934; Dildine, 1936, in the guppy). Dzwillo's success (1962) in obtaining complete reversal in sex differentiation in the guppy will be discussed later.

In the oviparous toothcarp, *Oryzias latipes*, Okada (1943b, 1949, 1952a) claimed that the formation of testis-ova in adult males can be induced by administration of either estrogens or androgens. Treated males had gonads containing large oviform cells which look like ovocytes but which are not surrounded by follicle cells. The nature of these large-sized cells will be discussed later (Section VI, A). Okada (1964b) obtained a true testis-ova by estrogen administration starting in juveniles. In this case, ovocytes are genuine since they are accompanied by follicle cells.

C. Complete (Functional) Reversal of Sex Differentiation

In the swordtail (undifferentiated species), Dantschakoff (1941) claimed to have induced males by the administration of androgens to females. There have been some reservations regarding these observations (cf. Gordon, 1956, 1957, and Peters, 1964, in Section IV, D). If, however, this is true, it is possible that the induction was accomplished because this "undifferentiated" species is very labile in sexuality. As stated before (Section III, A) this species has no sex chromosomes. By treatment with an androgen, female germ cells including ovocytes might degenerate and protogonia might differentiate into male germ cells.

In the differentiated species such as the platyfish and the medaka, where neither spontaneous sex reversal nor even sporadic intersexes occurs, induction of complete reversal may be impossible if hormone treatment starts after the onset of gonadal sex differentiation. The rank of stability in sexuality seems to be platyfish $\overline{>}$ medaka $>$ guppy $>$ swordtail. No sex reversals have ever been successful by administration of sex hormones in the platies (Cohen, 1946; Tavolga, 1949; Laskowski, 1953) because heterologous sex hormones have inevitably been administered after birth when the gonadal sex differentiation has already been established.

The medaka, *Oryzias latipes* (Fig. 9), an oviparous toothcarp, is a strict gonochorist. Genetically, the sex-determining mechanism normal to this fish

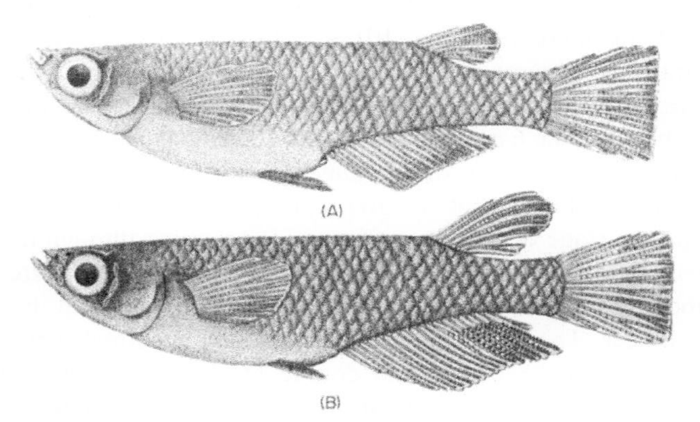

Fig. 9. (A) Female and (B) male of the medaka, *Oryzias latipes*.

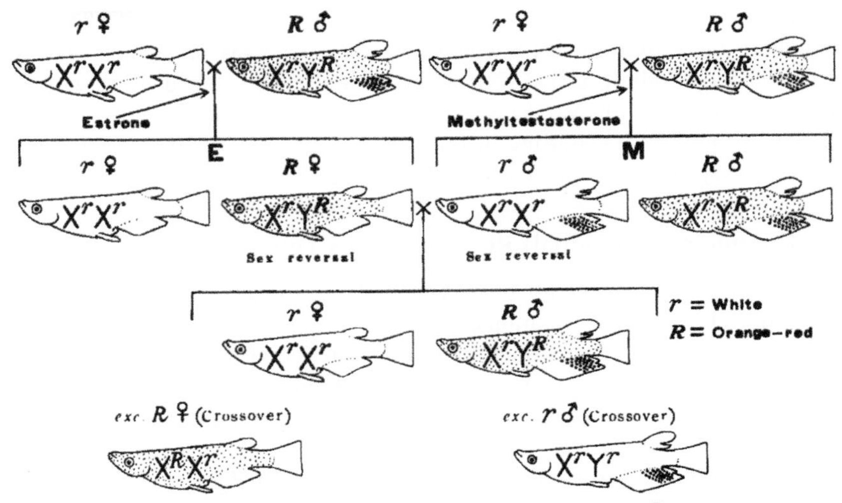

Fig. 10. A diagram illustrating the progeny of estrone-induced X^rY^R female mated with methyltestosterone-induced X^rX^r male (from Fig. 4, Yamamoto, 1961).

was established as XX for female and XY for male (Aida, 1921). Embryologically, it is a differentiated species (Yamamoto, 1953). Using our genetically analyzed strain (*d-rR*) of the medaka, in which white is female X^rX^r and orange-red is male X^rX^R, where *R* stands for xanthic pigmentation, functional reversal in sex differentiation has been accomplished in both directions by heterologous sex hormones (Yamamoto, 1953, 1958). Thus, it has become possible to control sex differentiation *ad libitum* either from genetic males (XY) to functional females or from genetic females (XX) to males. In other words, we are able to inverse the prospective fate of the fish, genetically destined to develop into one sex and direct it to differentiate into the other. Success in reversing sex differentiation in both directions in the medaka rendered it possible to mate estrone-induced XY females with methyltestosterone-induced XX males (Yamamoto, 1961, Fig. 10).

In contrast to viviparous poeciliids, the newly hatched fry (≅4.8 mm) has an indifferent gonad (gonad primordium) and gonadal sex differentiation takes place between the 6 and 11 mm stages. To accomplish complete reversal in sex differentiation in any differentiated gonochorists, certain conditions should be fulfilled. First, a heterologous *sex* hormone should be administered starting with the stage of indifferent gonads and passing through the stage of sex differentiation. In our experiments, heterologous hormones were administered by the oral route to newly hatched fry and continued until they reached 12 mm or longer. Thereafter, they were raised on a normal diet until they reached maturity. Second, adequate dosage levels of hormones should be used; these differ in potency with different sex hormones (Tables III and IV). Estrogen-induced X^rY^R Females (egg producers) were first obtained

Table III GD_{50} Dosage Levels of Estrogens Acting as Gynotermone (Gynogenin) to Induce 50% XY Females in the Medaka, *Oryzias latipes*

Estrogens	GD_{50} (μg/g diet)	Authors
Hexesterol	0.4	Yamamoto and Iwamatsu (unpublished data)
Euvestin[a]	0.8	Yamamoto, Hishida, and Uwa (unpublished data)
Ethynyl estradiol	1.7	Yamamoto, Noma, and Tsuzuki (unpublished data)
Estradiol-17β	5.8	Yamamoto and Matsuda (1963)
Stilbestrol	7.5	Yamamoto and Matsuda (1963)
Estrone	20	Yamamoto (1959b)
Estriol	130	Yamamoto (1965)

[a] *p,p*′-Dicarboethoxyoxy-*trans*-α,β-diethylstilbene.

Table IV AD_{50} Dosage Levels of Androgens Acting as Androtermone (Androgenin) to Induce 50% XX Males in the Medaka, *Oryzias latipes*

Androgens	AD_{50} (μg/g diet)	Authors
19-Nor-ethynyltestosterone	1.0	Yamamoto, Hishida, and Takeuchi (unpublished data)
Fluoxymesterone (Halotestin)	1.2	Yamamoto and Hishida (unpublished data)
17-Ethynyltestosterone (Pregneninolone, ethisterone)	3.4	Yamamoto and Uwa (unpublished data)
Methylandrostenediol	7.8	Yamamoto and Onitake (unpublished data)
Methyltestosterone	15	Yamamoto (1958)
Androstenedione	500	Yamamoto (1968)
Testosterone propionate	560	Yamamoto *et al.* (1968)
Androsterone	580	Yamamoto *et al.* (1968)
Dehydroepiandrosterone	>3200	Yamamoto and Oikawa (unpublished data)

Table V Production of Y^rY^r (White) Males and Estrone-induced Y^rY^r Females in the Medaka, *Oryzias latipes*[a,b]

P_1	$r(X^rX^r)$♀♀ (12) × $R(X^RY^r)$♂♂ (12)
	Normal diet |
I_c	$R(X^RX^r)$♀♀ (44) $r(X^rY^r)$♂♂ (40)
P_1	$r(X^rX^r)$♀♀ (12) × $R(X^RY^r)$♂♂ (12)
	Estrone diet |
I_e	$ER(X^RX^r)$♀♀ (38) $Er(X^rY^r)$♀♀ (24) $Er(X^rY^r)$♂♂ (2)[c]
P_2	$Er(X^rY^r)^d$♀♀ (24) × $R(X^RY^r)$♂♂ (12)
	Normal diet |
II_c Detected	$R(X^RX^r)$♀♀ (11) $R(X^RY^r)$♂♂ (6) $r\frac{(X^rY^r + Y^rY^r)}{5 + 4}$♂♂ (17)
P_2	$Er(X^rY^r)^d$♀♀ (24) × $R(X^RY^r)$♂♂ (12)
	Estrone diet |
II_e Detected	$ER\frac{(X^RX^r + X^RY^r)}{3 + 5}$♀♀ (28) $Er\frac{(X^rY^r + Y^rY^r)}{3 + 3}$♀♀ (13)

[a]From Yamamoto (1967).
[b]Subscripts c and e denote control and experimental, and I and II are the first and the second generations. Numbers enclosed in parentheses represent numbers of parents and offspring obtained. Animals designated by broken lines were submitted to testcrosses. Free figures are numbers of fish detected by testcrosses. Sex chromosome formulas following *r* or *R* are presumed sex genotypes. E denotes estronized fish; *r*, white, and *R*, orange-red.
[c]Nonreversals.
[d]In the original paper (1967) this formula was misprinted as X^rX^r.

by estrone and stilbestrol (Yamamoto, 1953, 1959a). The induced X^rY^R females in matings with normal X^rY^R males produced offspring in a ratio of 1♀:≅2♂ instead of 1♀ :3♂, indicating that viable Y^RY^R males are rare. However, two Y^RY^R males which sired all-male progeny were actually detected among 57 (X^rY^R+Y^RY^R) males singly tested by X^rX^r females (Yamamoto, 1955, 1959a). By using other mating systems (cf. Table V), it happened that Y^RY^r males produced by mating estrogen-induced X^rY^R females with X^rY^r males were all viable (Fig. 11). The Y^rY^r males yielded by mating induced X^RY^r or X^rY^r females with X^rY^r males also all survived. By administration of estrone in two consecutive generations, it was possible to invert sex differentiation in YY zygotes (Yamamoto, 1963, 1967). Although it does not relate directly to the main theme of this section, reference should be made to the intriguing problem of viability of YY zygotes. This problem has been fully discussed elsewhere (Yamamoto, 1964a,b). In short, the regular

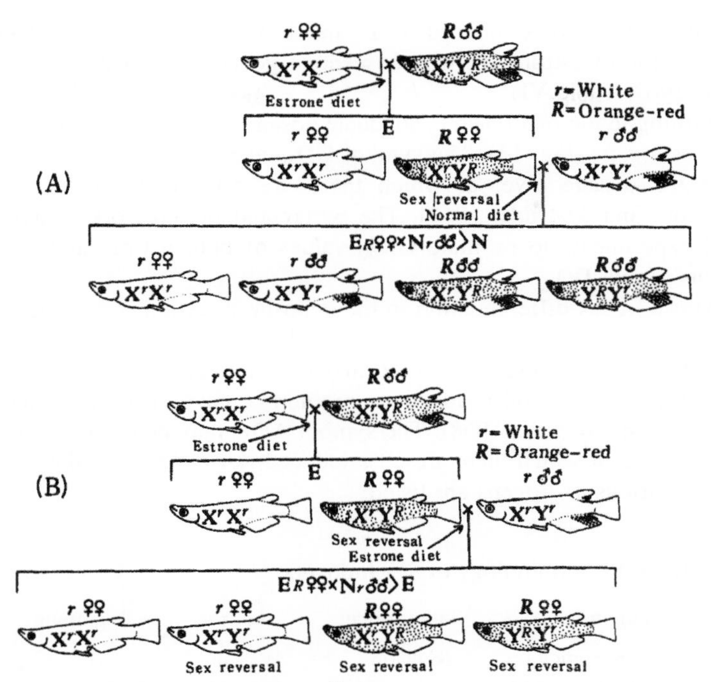

Fig. 11. (A) A diagram of production of Y^RY^r males (control of B) by normal feeding to offspring from estrone-induced X^rY^R females mated with normal X^rY^r males. (B) Induction of Y^RY^r females by estrone feeding to offspring from estrone-induced X^rY^R females mated to normal X^rY^r males (from Figs. 1 and 2, Yamamoto, 1963a).

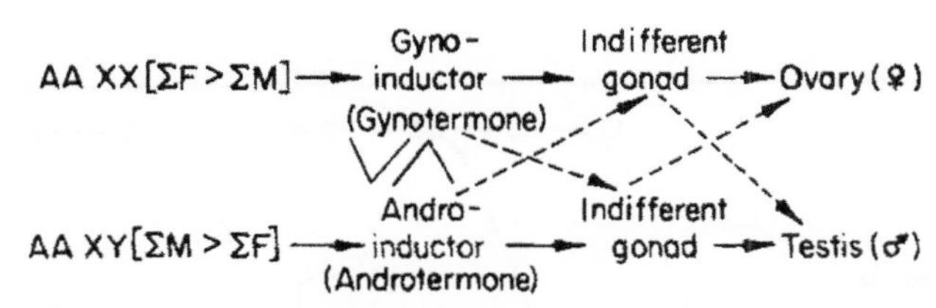

Fig. 12. A scheme of normal sex differentiation (solid lines) and reversal in sex differentiation by means of heterologous sex inductors (from Yamamoto, 1962).

Y^R chromosome contains an inert section (–) which in duplex condition ($Y^{R,-}Y^{R,-}$) renders the zygote nonviable, while both regular X^r and X^R possess the viability section (+). Rare survival X^RY^R zygotes are experimentally verified to have the $Y^{R,-}Y_c{}^{R,+}$ constitution the (+) of which is known to have been derived from $X^{r,+}$ by crossing over. This being so, it is quite rational that the Y^r is $Y^{r,+}$ in constitution and Y^RY^r is $Y^{R,-}Y^{r,+}$. It is hardly surprising that practically all Y^rY^r zygotes ($Y^r+Y^{r,+}$) are viable (Fig. 12). Using white females (X^rX^r) and orange-red (X^RY^r) males as parents in the first generation, estrone-induced Y^RY^r and Y^rY^r females were produced by administration of estrone in two consecutive generations (Yamamoto, 1963, Fig. 11; 1967, Table VI).

Returning to the main theme, adequate dosage levels are very different for various estrogens (Fig. 13). The symbol GD_{50} was suggested as the designation of the dosage level at which there are 50% induced XY females (Yamamoto and Matsuda, 1963). The writer and his co-workers have performed experiments to estimate GD_{50} values of both natural and synthetic estrogens (Table IV).

Reversal in sex differentiation in the opposite direction, viz., induction of X^rX^r male by methyltestosterone, has also been accomplished (Yamamoto, 1958); these males in mating with normal X^rX^r females sired all-female progenies. Androsterone and testosterone propionate have also a male-inducing action (Yamamoto *et al.*, 1968). The symbol AD_{50} may be used as a notation of dosage level at which there are 50% induced XX males. In Table X, AD_{50} values of various androgens are listed.

Table VI Terminology of Sex Inducers

Female inducer	Male inducer	Author	Beginning
Cortexin (corticin)	Medullarin	Witschi	1931
Gynogenin	Androgenin	D'Ancona	1949b
Gynotermone	Androtermone	Hartmann	1951

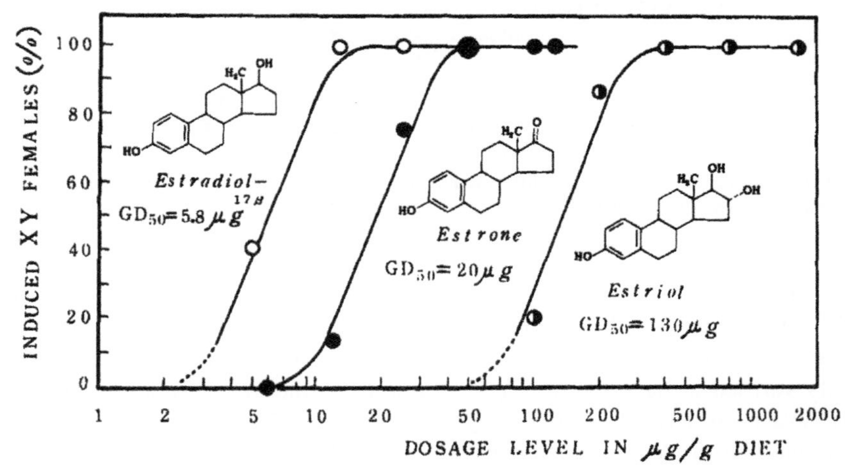

Fig. 13. Potencies of estradiol, estrone, and estriol in inducing XY females (from Fig. 2, Yamamoto, 1965b).

In the guppy, an undifferentiated species, all attempts to control sex by hormone administrations after birth when the gonadal sex differentiation is already established (cf. Goodrich *et al.*, 1934) failed until Dzwillo (1962) succeeded in getting functional XX males by androgen administration to gravid females containing embryos having indifferent gonads. Clemens (1965) was able to obtain androgen-induced XX males in *Tilapia mossambica* by administration starting with newly hatched fry having indifferent gonads. In the goldfish, *Carassius auratus*, Yamamoto and Kajishima (1969) obtained functional reversals in both directions by administering heterologous sex hormones for 2 months to newly hatched fry with indifferent gonads. By test-crosses of the hormone-treated fish, evidence for male heterogamety (XY) in the goldfish was first obtained.

In differentiated species, neither spontaneous intersex nor sporadic sex reversals can appear. Nevertheless, induction of complete reversal in sex differentiation can be achieved if the following necessary condition is fulfilled. Sexogens (estrogens and androgens) at suitable dosage levels should be administered consistently, beginning with the stage of the indifferent gonad and passing through the stage of sex differentiation.

In summary, by finding the correct dose of a suitable hormone for treating embryos or larvae and by starting medication at the indifferent gonadal stage, it is possible to convert an otherwise potential female into one that functions as a male or vice versa. This principle will hold true not only for "undifferentiated" species but also for "differentiated" ones.

VI. NATURE OF NATURAL SEX INDUCERS

A. Steroid versus Nonsteroid Theories

Witschi (1914, 1929) postulated the inducer theory of sex differentiation in amphibians. Although the sex genes are effective sex determiners, their action appears to be mediated by the sex-gene-controlled sex inducers as far as the vertebrate is concerned. Terminology of sex inducers differs among authors (Table VI).

The exact chain of events which leads from sex genes to sex differentiation has not yet been clarified but is outlined in Fig. 12 (Yamamoto, 1962). The important point is that sex genes are not direct sex determiners, but it is the sex-gene-controlled inducers that determine sex differentiation.

There are pros and cons for the steroid theory of sex inducers. In amphibians, Witschi (1942, 1950, 1955, 1957) and his followers are "cons" and postulated a protein nature of the sex inducers without evidence. Chieffi (1965) also stated that "the embryonic sex inducers ..., are in all probability different from sex hormones of the adults." The postulation is based *inter alia* on (1) the so-called paradoxical feminization of the gonad by high doses of androgens, (2) the nonspecificity of sex hormones as sex inducers, viz., non-sex-hormonic corticoids are believed to also have the potency of sex inducers, and (3) the ineffectiveness of androgens on some WZ amphibians. These statements are based on the sex ratio scored by a limited number of gonads of young, the sex genotype of which is unknown. Neither a single case of androgen-induced functional nor a corticoid-induced sex reversal has ever been produced. Furthermore, criteria for ovocytes are questionable. Experiments on the adult male medaka show that not only estrogens but androgens, corticoids, and any noxious treatments such as heat treatment and starvation result in the formation of large-sized cells in the testis (testis-ova), without characteristics of ovocytes such as the presence of follicle cells (cf. Egami, 1955a–e, 1956a,b). These oviform cells, which actually look like auxocytes, may be regarded as degenerating protogonia or spermatogonia enlarged prior to deterioration (Yamamoto, 1958). As to point (3) above we should not rely on negative results. Reported experiments have been made by immersion in hormone. It may well be that androgen immersion cannot simulate the natural condition surrounding the protogonia and induce them to differentiate into male germ cells. Inducers are thought to be densely produced only by cells surrounding the protogonia.

In the medaka the following occurs: (1) Estrogens act as female inducers and androgens function as male inducers; no paradoxical phenomenon are engendered. (2) Sex hormones act specifically as sex inducers; non-sex-hormonic corticoids have no effect. (3) Effective dosage levels of estrone and methyltestosterone are so low that it may be possible for juvenile fish to elaborate a small amount of sexogens (Yamamoto, 1959b, Hishida, 1965). (4) By administration of radioactive testosterone-16-^{14}C propionate and diethylstilbestrol-(monoethyl-1-^{14}C), Hishida (1962, 1965) showed that these are selectively incorporated into the juvenile gonad. Point (4) seems to indicate that the sex-differentiating gonad requires more sex hormones than other organs. On the basis of these results, we are inclined to the *sex* steroid theory of natural inducers.

The current view of pathways of biosynthesis of sexogens in adults is illustrated in Fig. 14 based on Meyers (1955a,b), Solomon *et al.* (1956), and others. Whether or not the pathways are valid for cells surrounding the indifferent gonad is not known. Of these naturally occurring steroids, pregnenolone, progesterone, and 17α-hydroxyprogesterone are found to be ineffective (Yamamoto and Matsuda, 1963; Yamamoto, 1968) while androstenedione and other sex steroids are effective (Tables III and IV).

B. Detection of Steroids and Relevant Enzymes in Fish Gonads

First of all, it is not known whether all the sex hormones of adult fishes are identical with those of mammals. The identity of some hormones has been established by Chester Jones and Phillips (1960), Chieffi and Lupo (1962, 1963), Lupo and Chieffi (1963), and Wotiz *et al.* (1960).

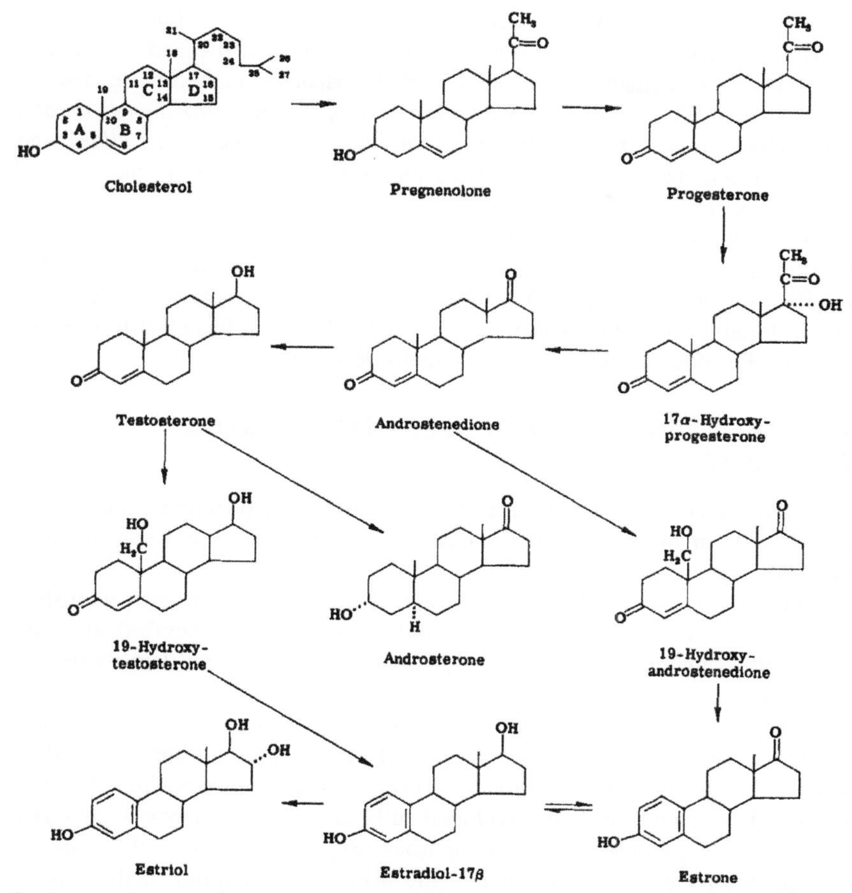

Fig. 14. A diagram of the current view of biosynthesis of key sex steroids (based on Meyers, 1955a; Solomon *et al*, 1956; and others).

Research on the detection of steroids in the larval gonad at the onset of sex differentiation by classic chemical methods is difficult because of the paucity of available material. Therefore, almost all approaches depend on histochemical and enzymic methods. In the elasmobranch *Scyliorhinus caniculas* Chieffi (1955) found a few sudanophilic granules in the gonadal medulla before sex differentiation. After the onset of sex differentiation, there appear many sudanophilic as well as Schultz-positive granules. The identification of enzymes involved in biosynthesis of sex steroids is another approach. Techniques have been developed for the histochemical demonstration of three relevant enzymes: Δ^5-3β-dehydroxysteroid dehydrogenase (Δ^5-3β-HSDH), Δ^5-3β-ketoisomerase (Δ^5-3β-KIM) and 17β-hydroxysteroid dehydrogenase (17β-HSDH) which is DPN and TPN dependent. Together with Δ^5-3β-KIM, Δ^5-3β-HSDH is involved in the pathway from pregnenolone to progesterone, one of the first steps of biosynthesis of sex steroids (cf. Fig. 14). Chieffi and co-workers (1961-1963) were unable to detect Δ^5-3β-HSDH in the embryonic gonads of the elasmobranchs, *Scyliorhinus caniculas*, *S. stellaris*, and *Torpedo marmorata*. The enzyme, however, was detected in the interrenal tissue of *S. stellaris* and *T. mumorada*. Bara (1966) showed the presence of 3β-HSDH in the testis of *Fundulus heteroclitus*.

Many questions remain unanswered. It is not known, for instance, where the supposed sex inducers are produced and in what way they influence the indifferent gonia (protogonia) to develop into either ovocytes or spermatocytes. Finally, the chemical nature of natural sex inducers still remains obscure. Research along these lines should be fruitful.

VII. DIFFERENTIATION OF SECONDARY SEXUAL CHARACTERS

Our primary concern is with gonadal or primary sex differentiation. References to the manifestation of secondary sexual characters are so numerous that only a brief account is presented here.

Within the last two decades some excellent reviews have been published (Hoar *et al.*, 1951; Hoar, 1955, 1957a,b; Dodd, 1955). Pickford and Atz (1957) presented a monograph on the piscine pituitary gland and its role in reproduction.

It is convenient to divide secondary sex characters into two categories: (1) temporary characteristics which normally appear only during the breeding season such as nuptial coloration, pearl organs, and the ovipositor of bitterlings; (2) permanent organs developed fully at the onset of sexual maturity such as the gonopodium of viviparous cyprinodonts and the papillar processes on the male anal fin and the urinogenital papilla in the female medaka, *Oryzias*.

It is also convenient to define male- and female-positive secondary sex characters (Yamamoto and Suzuki, 1955). The former is a character that is either specific to the male or is more developed in the male than in the female. The latter is the reverse. The mechanism by which these characters manifest themselves may only be elucidated by experiments.

Most of secondary sex characters of fishes are male-positive. Kopêĉ (1918, 1928) was the first to demonstrate in the minnow *Phoxinus laevis* that the nupital coloration is dependent on the testicular hormone. This has been confirmed in the sticklebacks, *Gasterosteus pungitius* and *G. aculeatus*, by van Oordt (1923, 1924), van Oordt and van der Mass (1927), Bock (1928), Craig-Bennett (1931), and Ikeda (1933); in the bitterlings by Tozawa (1929), Wünder (1931), and Glaser and Haempel (1932); and in the guppy by Blacher (1926b) and Samokhvalova (1933). Similar results were obtained in the medaka (Niwa, 1965a,b). Tozawa (1923) proved that the pearl organs of the goldfish, *Carassius auratus*, are controlled by testicular hormone. While castration results in the disappearance of these male-positive characters, ovariotomy causes no effect. This means that the absence of these in the female is not because of the inhibitory action of the ovary.

On the contrary, in the ganoid, *Amia calva*, Zahl and Davis (1932) showed that the gray-black pattern (male-positive) is absent in the female because of an inhibitory actiton of the ovary; ovariotomy produced this pattern. The male swordtail and the platyfish have complicated suspensorial skeletons including three hemal spines in the caudal vertebrae; these are necessary for proper functioning of the gonopodium. In the female, the three hemal spines are absent providing an undivided space required by the gravid female for the brood of embryos. Vivian (1952b) reported that spayed swordtails develop the three hemal spines as in the male, indicating that these are absent in the female by an inhibiting action of the ovary. Rubin and Gordon (1953) showed in the platyfish that α-estradiol benzoate induced the dissolution of one or two of these hemal spines in the male and methyltestosterone induced fusion of basal bony elements (mesonost and baseost) in the female, which are normally separated.

The secondary sex characters of the medaka, *Oryzias latipes*, have been described by Oka (1931a). Nagata (1934, 1936) demonstrated that castration results in a regression of both the shape and papillar processes of the male anal fin. Transplantation of a testis into the sprayed fish produces an almost perfect replica of the male characters. Administration of an androgen to females induces the male-type anal fin (Okada and Yamashita, 1944a). Other sexual differences such a*s* teeth, bones, and body shape were reported by Egami (1956b, 1959b) and Egami and Ishii (1956).

In the topminnow, *Gambusia holbrookii*, Dulzetto (1933) noted the correlation between testicular development and the gonopod formation. A host of experiments on gonopod formation has been performed by Turner (1942a,b, 1947, 1960) and Okada and Yamashita (1944b) in the topminnow *G. affinis*, by Grobstein (1940a,b, 1942a,b, 1948) in the platyfish *X. maculatus*, by Hopper (1949) and Hildemann (1954) in the guppy, and by Taylor (1948) in the guppy-swordtail hybrid. Some of the earlier studies of hormone-induced characters, aimed at inducing modifications of sex differentiation by administration of steroids in poeciliids, have already been discussed (Section V, B). Vivian (1950) reported that X-ray irradiation in the females induced formation of a malelike anal fin (incomplete gonopod).

At this point, it may be remarked that not all fins are sensitive to androgens. For example, androgens have a stimulating action on the anal and dorsal fins of the medaka but exert no effect on the caudal fin. We will refer to the pelvic fin later. In the gobiid fish, *Pterogobius zonoleucus*, Egami (1959c) stated that while the dorsal fin is elongated by administration of an androgen, other fins are not affected. Another male-positive character is the large tooth in the male medaka. Large teeth can be produced in the female by administration of methyltestosterone (Egami, 1957; Takeuchi, 1967).

Turner's extensive researches (1960) demonstrated that regional parts in the anal fin of *Gambusia* have a differential sensitivity to androgens. The 3-4-5 ray complex is the most sensitive. He among others suggested that the enzyme pattern may be different in each area of the fin.

Only a few female-positive secondary sex characters are present in teleosts. The urinogenital papilla (UGP) and the pelvic fin of the medaka and the ovipositor (elongated UGP) of the bitterlings are examples. Although the UGP is stimulated by both estrogen and androgen, it is far more sensitive to the former than the latter (Yamamoto and Suzuki, 1955). It appears that normally the UGP of the female is produced by an estrogen from the ovary.

The UGP of the Indian catfish, *Heteropneustes fossils*, is also female-positive. Kar and Ghosh (1950) reported that injections of either an estrogen or an androgen in females resulted in drastic hypoplasia of the ovaries, oviduct, and the UGP. They attributed this seemingly curious effect of the estrogen to the paucity of gonadotropin through suppression of its action by a large amount of exogenous hormone and concluded that the UGP is under the control of the ovarian hormone.

The pelvic fin of the female medaka is longer than that of the male (Oka, 1931a). Niwa (1965b) showed that this is because of an inhibitory action of the testicular hormone and cannot be attributed to a stimulating action of the ovarian hormone.

In the Japanese bitterling *Acheilognathus indermedia* Tozawa (1929), showed that ovariotomy inhibits ovipositor lengthening. Research on the ovipositor formation in the European species, *Rhodeus amarus*, has provided a very confusing and puzzling problem. Fleishmann and Kann (1932) and Ehrhardt and Kühn (1933) found that estrogen induces this reaction in the nonbreeding season. In 1935, American and German authors proclaimed that this response can be used for the diagnosis of pregnancy. It has been reported that urine from both men and pregnant women usually gives a positive reaction, while that from nonpregnant women may or may not cause this reaction. A male hormone in the urine was believed to be responsible. Banes *et al.* (1936) reported, however, that crude extracts of adrenal cortex gave a positive response while androsterone was negative. On the other hand, Kleiner *et al.* (1937) obtained a positive response by administration of androgens. De Wit (1939) confirmed this and further stated that progesterone and deoxy-corticosterone (DOCA) are far more active than an androgen. Bretschneider and de Wit (1940–1941, 1947) and de Wit (1940, 1941, 1955) developed a quantitative method and studied this phenomenon.

However, a survey of literature reveals that not only the above-mentioned substances but also adrenaline and anesthetics have been reported to be active in lengthening the oviposter. De Groot and de Wit (1949) reported that alcohol, heat shock, and strong light also causes ovipositor growth. This would seem to indicate that stresses in the sense of Selye also cause the reaction. Furthemore, the presence of mussels and/or males and an exhalent current from mussels have the influence on the cyclic ovipositor growth (de Wit, 1955). It appears that normally environmental factors exert their effects on the pituitary gland through intermediacy of the nervous activity evoked by sense organs. Further work must be done before these questions can be adequately answered.

VIII. SUMMARY

The need to clarify our thinking about sexuality in fishes is emphasized. First, it is unsound to consider that sexuality in all members of the class Pisces is labile, There are in fact many sex types: synchronous, protandrous, and protogynous hermaphrodites, as well as gonochorists. Among gonochorists, there are "undifferentiated" and "differentiated" species in the sense of Witschi. The latter are more stable than the former in sexuality. Evidently, there are graded levels in sexuality among fishes.

Second, sex is a phenotypic expression. In gonochorists, it is important to clarify the concept of sex to eliminate confusion. A male is a sperm producer and a female an egg producer. This definition, although selfevident, is particularly important in avoiding controversy in the problem of sex reversal. By definition, sex genotypes of zygotes before sex differentiation such as AAXX and AAXY are usually presumptive or prospective females and males, respectively. Although combinations of sex genes are determined at the time of fertilization, sex is actually determined by sex-gene-controlled sex inducers at a certain critical period of development, that is, sex genes act only indirectly upon sex differentiation. Their action is directly mediated by sex inducers.

Sex genes are present not only in the sex chromosomes but also distributed over a great number of autosomes. In view of multiple (polygenic) sex factors, exceptional AAXX males or AAXY females and WZ(Y) males are not "genetic sex reversals." These exceptions appear when the totality of sex genes become $\Sigma M > \Sigma F$ by fortuitous combinations of autosomes in spite of AAXX constitution and $\Sigma F > \Sigma M$ despite AAXY constitution. This concept may also be applied to exceptional WZ(Y) males in female heterogametic species. They have the genetic basis for either male or female. A diagram illustrates this concept (Fig. 6). In general, sex differentiation is induced by sex inducers controlled by sex genes with or without (in species without sex chromosomes) sex chromosomes. Sex chromosomes contain epistatic (superior) sex genes.

Reports on spontaneous sex reversals in gonochorists have been criticized. Such cases seldom, if ever occur in fishes. Most cases are masculinized females (with ovary) brought about by hormonal upsets that accompany advanced age.

Artificial control of sex differentiation may be one of the key projects in biology. One approach is to override the AAXX (in which $\Sigma F > \Sigma M$) or AAXY (in which $\Sigma M > \Sigma F$) constitution by artificial means. This has been successfully accomplished in both directions in the medaka and the goldfish by administration of heterologous sex hormones, i.e., estrogens to XY zygotes and androgens to XX zygotes. In the medaka, it is possible to invert sex differentiation of even YY zygotes into females by administration of estrone in two consecutive generations. Reversal in one direction, from the presumptive female to the functional male, was achieved in the guppy (Dzwillo, 1962) and *Tilapia* (Clemens, 1965).

It is of paramount importance that heterologous sex steroid administration must start at the stage of the indifferent gonad and continue through the stage of gonadal sex differentiation and that a suitable dosage should be used. GD_{50} of estrogens and AD_{50} of androgens which indicate relative potencies in inducing XY females and XX females in the medaka, respectively, are listed.

The chemical nature of the endogenous sex inducers are discussed. There are pros and cons of the steroid theory of the natural sex inducers. Because of (1) the specificity of sex steroids as exogenous sex inducers, (2) the very low effective dosage of natural sex steroids, and (3) the selective incorporation of sex steroids into the differentiating gonad, it is likely that endogenous sex inducers are allied to or identical with sex steroids.

A brief review on the manifestation of secondary sexual characters was given.

REFERENCES

Abe, T. (1967). Personal communication.

Aida, T. (1921). On the inheritance of color in a fresh-water fish *Aplocheilus latipes* Temminck and Schlegel, with special reference to the sex-linked inheritance. *Genetics* **6**, 554–573.

Aida, T. (1930). Further genetical studies of *Aplocheilus latipes*. *Genetics* **15**, 1–16.

Aida, T. (1936). Sex reversal in *Aplocheilus latipes* and a new explanation of sex differentiation. *Genetics* **21**, 136–153.

Anders, A., and Anders, F. (1963). Genetisch bedingte XX- und XY-♀ und XY- und YY-♂ beim wilden *Platypoecilus maculatus* aus Mexico. Z. *Vererbungslehre* 94, 1–18.

Aoyama, T. (1955). On the hermaphroditism in the yellow sea bream, *Taius tumi-frons*. *Japan. J. Ichthyol.* **4**, 119–129 (in Japanese with English résumé).

Aoyama, T., and Kitajima, T. (1966). Sex reversal in the flat-head fish, *Suggrundus meerdervoort* (Bleeker). Oral communication.

Aoyama, T., Kitajima, T., and Mizue, K. (1963). Study of the sex reversal of Inegochi, *Cociella crocodila* (Tisesius). *Bull. Seikai Regional Fisheries Res. Lab.* **29**, 11–33.

Ashby, K. R. (1952). Sviluppo del sistema riproduttivo di *Salmo trutta* L. in condizioni normali e sotto l'influenza di ormoni steroidi. *Riv. Biol.* (*Perugia*) **44**, 3–19.

Ashby, K. R. (1956). The effect of steroid hormones on the brown trout (*Salmo trutta* L.) during the period of gonadal differentiation. *J. Embryol. Erptl. Morphol.* **5**, 225–249.

Ashby, K. R. (1959). L'effetto di basse concentrazioni di estradiolo e di testosterone sul differenziamento gonadico di *Salmo trutta* L. (Nota preliminare). *Riv. Biol.* (*Perugia*) **51**, N. S. No. 11, 453–468.

Ashby, K. R. (1965). The effect of steroid hormones on the development of the reproductive system of *Salmo trutta* L. when administered at the commencement of spermatogenetic activity in the testis. *Riv. Biol.* (*Perugia*) **80**, N.S. No. 18, 139–169.

Atz, J. W. (1959). The relation of the pituitary to reproduction in fishes. *In* "The Physiology of the Pituitary Gland of Fishes" (G. E. Pickford and J. W. Atz, eds.), Part VII, pp. 178–270. N.Y. Zool. Soc., New York.

Atz, J. W. (1964). Intersexuality in fishes. *In* "Intersexuality in Vertebrates including Man" (C. N. Armstrong and A. J. Marshall, eds.), pp. 145–232. Academic Press, New York.

Bacci, G. (1966). "Sex Determination." Pergamon Press, Oxford.

Bacci, G., and Razzauti, A. (1957). Falso gonocorismo in *Coris julis* (L.). *Atti Accod. Nazl. Lincei, Rend., Classe Sci. Fis., Mat. Nat.* [8] **23**, 181–189.

Bacci, G., and Razzauti, A. (1958). Protogynous hermaphroditism in *Coris julis* L. *Nature* **181**, 432–433.

Baldwin, F. M., and Goldin, H. S. (1940). Effects of testosterone propionate of the female viviparous teleost, *Xiphophorus helleri*. *Proc. Soc. Exptl. Biol. Med.* **42**, 813–819.

Bara, G. (1966). 3β-Hydroxysteroid dehydrogenase activity in the testis of *Fundulus heteroclitus*. *Anat. Record* **152**, 449.

Barnes, B. O., Kanter, A. E., and Klaw, A. H. (1936). Bitterling ovipositor lengthening produced by adrenal extracts. *Science* **84**, 310.

Bellamy, A. W. (1922). Sex-linked inheritance in the teleost *Platypoecilus maculatus*. *Anat. Record* **24**, 419–420.

Bellamy, A. W. (1928). Bionomic studies on certain teleosts (Poeciliinae). II. Color pattern inheritance and sex in *Platypoecilus maculatus* Günth. *Genetics* 13, 226–232.

Bellamy, A. W. (1936). Interspecific hybrids in *Platypoecilus*: One species ZZ-WZ; the other XY-XX. *Proc. Natl. Acad. Sci. U.S.* **22**, 531–536.

Bellamy, A. W., and Queal, M. (1951). Heterosomal inheritance and sex determination in *Platypoecilus maculatus*. *Genetics* **36**, 93–107.

Berg, O., and Gordon, M. (1953). The relationship of atypical pigment cell growth to gonadal development in hybrid fishes. *In* "Pigment Cell Growth" (M. Gordon, ed.), pp. 43–72. Academic Press, New York.

Berkowitz, P. (1937). Effect of oestrogenic substances in *Lebistes reticulatus*. *Proc. Soc. Exptl. Biol. Med.* **36**, 416–418.

Berkowitz, P. (1938). The effect of oestrogenic substances in *Lebistes reticulatus*. (guppy). *Anat. Record* **71**, 161–175.

Berkowitz, P. (1941). The effect of oestrogenic substances in the fish (*Lebistesreticul atus*). *J. Exptl. Zool.* **87**, 233–243.

Blacher, L. J. (1926a). Fall von Hermaphroditism bei *Lebistes*. *Trans. Lab. Exptl. Biol. Zoo-Park Moscow* **1**, 90–95.

Blacher, L. J. (1926b). The dependence of secondary sex-characters upon testicular hormones in *Lebistes reticulatus*. *Biol. Bull.* **50**, 374–381.

Bock, F. (1928). Kastration und sekundäre Geschlechtscharactere bei Teleostiern. Z. *Wiss. Zool.* **130**, 453–468.

Breider, H. (1935a). Über Aussenfaktoren, die das Geschlechtsverhähtnis bei *Xiphophorus helleri* Heckel kontrollieren sollen. *Z. Wiss. Zool.* **A**, **146**, 383–416.

Breider, H. (1935b). Geschlechtsbestimmung und—differenzierung bei *Limia nigro-fascida, Caudofasciata, vittata* und deren Artbastarden. *Z. lnduktive Abstammungs-Vererbungslehre* **68**, 265–299.

Breider, H. (1936a). Eine Allelenserie von Genen verschiedener Arten. *Z. Induktive Abstammungs- Verebungslehre* **72**, 80–87.

Breider, H. (1936b). Weiteres über die Geschlechtsbestimmung und Geschlechts-differenzierung bei *Limia vittata* Poy. *Zool. Anz.* **114**, 113–119.

Breider, H. (1942). ZW-Männchen und WW-Weibchen bei *Platypoecilus maculatus*. *Biol. Zentr.* **62**, 187–195.

Breider, H. (1949). Der Geschlechtsbestimmungs-mechanismus in der Artdifferen-zierung. *Biol. Zentr.* **68,** 32–49.

Breider, H., and Kosswig, C. (1937). Neue Ergebnisse der Geschlechtsbestimmung-analyse bei Zahnkarpfen. *Verhandl. Deut. Zool. Ges.* 1937, pp. 275–284.

Bretschneider, L. H., and de Wit, J. J. D. (1940–1941). Histophysiologische Analyse der sexuell-endokrinen Organisation des Bitterlings-weibchens (*Rhodeus am-arus*). *Z. Zellforsch. Mikroskop. Anat.* **31**, 227–344.

Bretschneider, L. H., and de Wit, J. J. D. (1947). "Sexual Endocrinology of Non-mammalian Vertebrates." Elsevier, Amsterdam.

Bridges, C. B. (1922). The origin and variation in sexual and sex-linked characters. *Am. Naturalist* **56**, 51–63.

Bridges, C. B. (1925). Sex in relation to chromosomes and genes. *Am. Naturalist* **59**, 127–137.

Bridges, C. B. (1939). Cytological and genetic basis of sex. Sex and Internal Secretions (E. Allen, 2nd ed pp. 15–63. Williams & Wilkins, Baltimore, Maryland.

Brock, J. (1879). Beiträge zur Anatomie und Histologie der Geschlechtsorgane der Knochenfische. *Gegenbaurs Jahrb.* **4**, 505–572.

Bullough, W. S. (1940). A study of sex reversal in the minnow (*Phoxinus laevis* L.). *J. Exptl. Zool.* **85**, 475–494.

Bullough, W. S. (1947). Hermaphroditism in the lower vertebrates, *Nature* **160**, 9–11.

Castle, W. E. (1936). A simplified explanation of Bellamy's experiments concerning sex determination in tropical fishes. *Proc. Natl. Acad. Sci. U.S.* **22**, 679–682.

Chavin, W., and Gordon, M. (1951). Sex determination in *Platypoecilus maculatus*. I. *Zoologica* **36**, 135–145.

Chester Jones, I., and Phillips, J. G. (1960). "Adrenocorticosteroids in fish." *Symp. Zool. Soc. London* **1**, 17–32.

Chieffi, G. (1955). Nuove osservazioni sull'organsgenesi della mudulla gonade nei Vertebrati: Ricerche isto chemiche in *Rana esculenta, Bufo viridis* e *Scylliohinus canicula. Pubbl. staz. Zool. Napoli.* **27**, 62–72.

Chieffi, G. (1959). Sex differentiation and experimental sex reversal in elasmobranch fishes. *Arch. Anat. Microscop. Morphol. Exptl.* **48**, 21–36.

Chieffi, G. (1961). La luteogenesi nei Selaci ovo-vivipari. Richerche istologiche e istochimiche in *Torpedo marmorata* e *Torpedo ocellata. Pubbl. Staz. Zool. Napoli* **32**, 145–166.

Chieffi, G. (1965). Onset of steroidogenesis in the vertebrate embryonic gonads. *In* "Organogenesis" (R. L. De Haan and H. Urspring, eds.), pp. 653–471. Holt, New York.

Chieffi, G., and Botte, V. (1963). Histochemical reaction for steroid-3 βol-dehydrogenese in the interrenal and the corpuscles of Stannius of *Anguilla anguilla* and *Conger conger. Nature* **200**, 793–794.

Chieffi, G., and Lupo, C. (1961). Identification of estradiol-17 testosterone and its precursors from *Scylliorhinus stellaris* testes. *Nature* **190**, 169–170.

Chieffi, G., and Lupo, C. (1962). Analisi comparativa degli ormoni sessuali negli Ittiopsidi. *Excerpta Med. Intern. Congr. Ser.* **51**, 146–147.

Chieffi, G., and Lupo, C. (1963). Identification of sex hormones in the ovarian extracts of *Torpedo marmorata* and *Bufo vulgaris. Gen. Comp. Endocrinol.* **3**, 149–152.

Chieffi, G., Botte, V., and Visca, T. (1963). Attività steroide-3-deidrogenasica nell'interrende di alkuni Selacei. *Acta Med. Romana* **1**, 108–116.

Clark, E. (1959). Functional hermaphroditism in a serranid fish. *Science* **129**, 215–216.

Clark, E. (1965). Mating of groupers. New studies detected reversal of stripes in hermaphroditic fish. *Nat. Hist., N.Y.* **74**, 22–25.
Clemens, H. P. (1965). Sex direction in presumptive female, *Tilapia Mossambica.* "Intersexuality in Fishes." Abstr. Papers, Conf. Cape Haze Marine Lab., Sarasota, Florida.
Cohen, H. (1964). Effects of sex hormones on the development of the platyfish, *Platypoecilus maculatus. Zoologica* **31**, 121–127.
Conel, J. L. (1917). The urogenital system of Myxinoids. *J. Morphol.* **29**, 75–138.
Conel, J. L. (1931). The genital system of the Myxinoidea; a study based on notes and drawings of these organs in *Bdellostoma* made by Bashford Dean. *In* "The Bashford Dean Memorial Volume Archaic Fishes" (E. W. Gudger, ed.), pp. 63–110. Am. Museum Nat. Hist., New York.
Craig-Bennett, A. (1931). The reproductive cycle of the three spinned stickleback, *Gasterostetus aculeatus*, Linn. *Phil. Trans. Roy. Soc. London, Ser. B*, **219**, 197–278.
D'Ancona, U. (1941). Ulteriori osservazione e considerazioni sull'ermafroditismo dell'orata (*Sparus auratus* L.). *Pubbl. Staz. Zool. Napoli* **18**, 314–336.
D'Ancona, U. (1943). Nuove ricerche sulla determinazione sessuale dell'anguilla. Parte I. Lo sviluppo della gonade e i processi della gametogenesi. *Arch. Oceanog. Limnol.* (*Roma*) **3**, 159–269.
D'Ancona, U. (1945). L'ermafroditismo nel genere *Diplodus* (sin. Sargus; Teleostei). *Boll. Soc. Ital. Biol. Sper.* **20**, 153–155.
D'Ancona, U. (1948). Prime osservazioni sull'azioni deghi ormoni sessuali sulla gonade dell'-anguilla. *Atti. Accad. Nazl. Lincei, Rend., Classe Sci. Fis., Mat. Nat.* [8] **5**, 82–87.
D'Ancona, U. (1949a). Osservazioni sull'organizzazione nella gonade ermafrodita di alcuni Serranidi. *Nova Thalassia* **1**(5), 1–15.
D'Ancona, U. (1949b). Il differenziamento della gonade e l'ínversione sessuale degli Sparidi. *Arch. Oceanog. Limnol.* (*Roma*) **6**, 97–163.
D'Anmna, U. (1949c). Ermofroditismo ed intersexualità nei Teleostei. *Proc. 13th lntern. Congr. Zool., Park, 1948*, pp. 144–146.
D'Ancona, U. (1949d). Ermafroditismo ed intersessualità nei Teleostei. *Experientia* **5**, 381–389.
D'Ancona, U. (1950). Détermination e differenciation du sexe chez les poissons. *Arch. Anat. Microscop. Morphol. Exptl.* **39**, 274–294.
D'Ancona, U. (1956). Inversion spontanées et expérimentales dans les gonades des Teleostéens. *Annee Biol.* [3] **32**, 89–99.
Dantschakoff, V. (1941). "Der Aufbau des Geschlechts beim Köhren Wirbeltier." Fischer, Jena.
de Groot, B., and de Wit, J. J. D. (1949). On the artificial induction of ovipositor growth in the bitterling (*Rhodeus amarus* Bl.). II. Ovipositor growth caused by different chemical and physical agents. III. The relation between artificially induced ovipositor growth and the adaptation syndrome of Selye. *Acta Endocrinol.* **3**, 266–297.
de Wit, J. J. D. (1939). Ovipositor lengthening of the female bitterling produced by administration of progesterone. *Endocrinology* **24**, 580–582.
de Wit, J. J. D. (1940). A quantitative and qualitative test for steroid hormones based on the ovipositor reaction of the female bitterling (*Rhodeus amarus* Bloch). *J. Endocrinol*, **2**, 141–156.
de Wit, J. J. D. (1941). Eine neue Testmethode für endokrinologische Zweck (der Legeröhrentest am weiblichen Bitterling). *Arch. Ges. Physiol.* **245**, 282–299.
de Wit, J. J. D. (1955). Some results of investigations into European bitterling. *Rhodeus amarus Bloc. Japan. J. Ichthyol.* **9**, 94–104.
Dildine, G. C. (1936). Studies on teleostean reproduction. I. Embryonic hermaphroditism in *Lebistes reticulatus. J. Morphol.* **60**, 261–277.

Dodd, J. M. (1955). The hormones of sex and reproduction and their effects in fish and lower chordates. *Mem. Soc. Endocrinol.* **4**, 166–187.

Dufossé, M. (1854). Sur l'hermaphroditisme de certain Percoides. *L'Institut*, **22**, 394–395.

Dufossé, M. (1856). De l'hermaphroditisme chez le serran. *Ann. Sci. Nat.* [4] **5**, 295–332.

Dulzetto, F. (1933). La struttura del testicolo di *Gambusia holbrookii* e la sua evoluzioni in rapporto con lo sviluppo del gonopodio. *Arch. Zool. Ital.* **19**, 405–437.

Dzwillo, M. (1962). Über künstliche Erzeugung funktionellee Männchen weiblichen Genotypus bei *Lebistes reticulatus. Biol. Zentr.* **81**, 575–584.

Dzwillo, M., and Zander, C. D. (1967). Geschlechtsbestimmung und Geschlechtsumstimmung bei Zahnkarpfen (Pisces). *Mitt. Hamburgischen Zool. Museum Inst.* **64**, 147–162.

Egami, N. (1954a). Appearance of the male character in the regenerating and transplanted rays of the anal fin in females of the fish *Oryzias latipes*, following treatment with methyl dehydroxytestosterone. *J. Fac. Sci., Univ. Tokyo, Sect. IV* **7**, 271–280.

Egami, N. (1954b). Effects of hormonic steroids on the formation of male characteristics in female of the fish *Oryzias latipes*, kept in water containing testosterone propionate. *Annotationes Zool. Japan.* **27**, 122–127.

Egami, N. (1955a). Production of testis-ova in adult males of *Oryzias latipes.* I. Testis-ova in the fish receiving estrogens. *Japan. J. Zool.* **11**, 21–34.

Egami, N. (1955b). Production of testis-ova in adult males of *Oryzias latipes.* IV. Effect on testis-ova production of non-estrogenic steroids given singly or simultaneously with estradiol. *Japan. J. Zool.* **11**, 35–39.

Egami, N. (1955c). Production of testis-ova in adult males of *Oryzias latipes.* III. Testis-ova production in starved males. *J. Fac. Sci., Univ. Tokyo, Sect. IV* **7**, 421–428.

Egami, N. (1955d). Production of testis-ova in adult males of *Oryzias latipes.* IV. Effect of X-ray irradiation on testis-ovum production. *J. Fac. Sci., Univ. Tokyo, Sect. IV* **7**, 429–441.

Egami, N. (1955e). Production of testis-ova in adult males of *Oryzias latipes.* V. Note on testis-ovum production in transplanted testes. *Annotationes Zool. Japon.* **28**, 206–209.

Egami, N. (1956a). Production of testis-ova in adult males of *Oryzia latipes.* VI. Effect on testis-ova production of exposure to high temperature. *Annotationes Zool. Japon.* **29**, 11–18.

Egami, N. (1956b). Note on sexual difference in size of teeth of the fish, *Oryzias latipes. Japan. J. Zool.* **12**, 66–69.

Egami, N. (1957). Production of testis-ova in adult males of *Oryzias latipes.* VIII. Effect of administration of desiccated frog pituitary on testis-ovum production in males receiving estrone pellet. *J. Fac. Sci., Hokkaido Univ. Ser. VI* **13**, 369–372.

Egami, N. (1959a). Note on sexual difference in the shape of the body in the fish, *Oryzias latipes. Annotationes Zool. Japan.* **32**, 59–64.

Egami, N. (1959b). Note on sexual difference in the shape of the body in the fish, *Oryzias latipes. Annotationes Zool. Japan.* **32**, 59–64.

Egami, N. (1959c). Effect of testosterone on the sexual characteristics of the gobiid fish, *Pterogobius zonoleucus. Annotationes Zool. Japan.* **32**, 123–128.

Egami, N., and Ishii, S. (1956). Sexual differences in the shape of some bones in the fish, *Oryzias latipes. J. Fac.* Sci., *Univ. Tokyo, Sect. IV* **7**, 563–571.

Eggert, B. (1931). Die geschlechts Organe der Gobiiformes und Blenniformes. *Z. Wiss. Zool.* **189**, 249–558.

Eggert, B. (1933). Die Intersexualität bei Knochenfischen. *Z. Wiss. Zool.* **144**, 402–420.

Ehrhardt, K., and Kühn, K. (1938). Über einen neuen Test für weibliches Sexual hormon. *Monatsschr. Geburtshulfe Gynaekol.* **94**, 64.

Essenberg, J. M. (1923). Sex differentiation in the viviparous teleost *Xiphophorus helleri*. *Biol. Bull.* **45**, 46–97.

Essenberg, J. M. (1926). Complete sex-reversal in the viviparous teleost *Xiphophorus helleri*. *Biol. Bull.* **51**, 98–111.

Eversole, W. J. (1939). The effects of androgens upon the fish (*Lebistes reticulatus*). *Endocrinology* **25**, 328–330.

Eversole, W. J. (1941). Effect of pregneninolone and related steroids on sexual development in fish *Lebistes reticulatus*. *Endocrinology* **28**, 603–610.

Fleischmann, W., and Kann, S. (1932). Über eine Funktion des weiblichen Sexualhons bei Fischen (Wachstum der Legeröhre des Bitterlings). *Arch. Ges. Physiol.* **230**, 662–667.

Fraser, A. C., and Gordon, M. (1929). The genetics of *Platypoecilus*. II. The linkage of the two sex-linked characters. *Genetics* **14**, 160–179.

Freund, I. (1923). "Bibliographia Pathologiae Piscium." Prag.

Friedman, B., and Gordon, M. (1934). Chromosome numbers in xiphophorin fishes. *Am. Naturalist* **68**, 446–455.

Friess, E. (1934). Untersuchungen über die Geschlechtsumkehr bei *Xiphophorus helleri*. *Arch. Entwicklungsmech. Organ.* **129**, 255–355.

Gäbler, H. (1930). Zwei Fälle von Zwittergonaden bei *Clupea harangus* L. *Zool. Anz.* **91**, 72–75.

Gallien, L. (1946). Réactivité spécifique à la prégnéninolone chez *Lebistes reticulatus*, des gonades et des caractères sexuels somatique. *Compt. Rend.* **223**, 52–53.

Gallien, L. (1948). Sur la structure nageoire chez les femelles de *Lebistes reticulatus* Regan, masculinisées par pregneninolone. *Compt. Rend.* **226**, 1749–1751.

Glaser, E., and Haempel, O. (1932). Das experimentell hervorgerufene Hochzeitskleid des kastrierten Fisches als Stigme einer Test-und Standardisierungs method des männlichen Sexual hormones. *Arch. Ges. Physiol.* **229**, 1–14.

Goldschmidt, R. (1915). Vorläufige Mitteilung über weiteren Versuche zur Vererbung und Bestimmung des Geschlechts. *Biol. Zentr.* **35**, 565–570.

Goldschmidt, R. (1927). "Physiologische Theorie der Vererbung." Springer, Berlin.

Goldschmidt, H. (1931). "Die sexuelle Zwischenstufen." Springer, Berlin.

Goldschmidt, R. (1937) A critical review of some recent work in sex determination. I. Fishes. *Quart. Rev. Biol.* **12**, 426–439.

Goodrich, H. B., Dee, J. E., Flynn, B. M., and Mercer, R. N. (1934). Germ cells and sex-differentiation in *Lebistes reticulatus*. *Biol. Bull.* **67**, 83–96.

Gordon, M. (1926). Melanophores of *Platypoecilus*, the top minnow of geneticists. *Anat. Record* **34**, 138 (abstr.).

Gordon, M. (1927). The genetics of a viviparous top-minnow *Platypoecilus*; the inheritance of two kinds of melanophores. *Genetics* **22**, 376–392.

Gordon, M. (1946a). Genetics of *Xiphophorus maculatus*. Sex determining in two wild populations of the Mexican platyfish. *Genetics* **31**, 641 (abstr.).

Gordon, M. (1946b). Interchanging genetic mechanisms for sex-determination. J. *Heredity* **37**, 307–320.

Gordon, M. (1946c). Sexual transformation of a genetically constituted female fish (*Platypoecilus maculatus*) into a functional male. *Anat. Record* **96**, (abstr.).

Gordon, M. (1947a). Genetics of *Platypoecilus maculatus*. IV. The sex determining mechanism in two wild populations of Mexican platyfish. *Genetics* **32**, 8–17.

Gordon, M. (1947b). Speciation in fishes. Distribution in time and space of seven dominant multiple alleles in *Platypoecilus maculatus*. *Advan. Genet.* **1**, 95–132.

Gordon, M. (1948). Effects of five primary genes on the site of melanomas in fishes and the influence of two color genes on their pigmentation. *Spec. Publ. N.Y. Acad. Sci.* **4**, 216–268.

Gordon, M. (1950). Genetics and speciation in fishes. *Am. Phil. Soc., Year Book* pp. 158–159.

Gordon, M. (1951a). Genetics of *Platypoecilus maculatus.* V. Heterogametic sex-determining mechanism in females of a domesticated stock originally from British Honduras. *Zoologica* **36**, 127–134.

Gordon, M. (1951b). *Platypoecilus* now becomes *Xiphophorus Aquarium* **20**, 277–279.

Gordon, M. (1951c). Genetics and correlated studies of normal and atypical pigment cell growth. *Growth Symp.* **10**, 153–219.

Gordon, M. (1952). Sex-determination in *Xiphophorus* (*Platypoecilus*) *maculatus.* III. Differentiation of gonads in platyfish from broods having a sex ratio of the three females to one male. *Zoologica* **37**, 91–100.

Gordon, M. (1956). "Swordtail—The Care and Breeding of Swordtails." T. F. H., Publ. Inc., New Jersey.

Gordon, M. (1957). Physiological Genetics of fishes. *In* "The Physiology of Fishes" (M. E. Brown, ed.), Vol. 2, pp. 431–501. Academic Press, New York.

Gordon, M., and Aronowitz, O. (1951). Sex determination in *Platypoecilus maculatus.* II. History of a male platyfish that sired all-female broods. *Zoologica* **36**, 147–153.

Gordon, M., and Smith, G. M. (1938). The production of a melanotic neoplastic disease in fishes by selective matings. IV. Genetics of geographical species hybrids. *Am. J. Cancer* **34**, 543–565.

Gottfried, H., Hunt, S. V., and Wright, R. S. (1962). Sex hormones in fish. The estrogens of cod *Gadus callarias. J. Endocrinol.* **24**, 425–430.

Grassi, B. (1919). Nuove ricerche sulla storia naturale dell'*Anguilla. R. Comitato Talassografico Ital. M.* **67**, 1–141.

Grobstein, C. (1940a). Endocrine and developmental studies of gonopod differentiation in certain poeciliid fishes. I. The structure and development of the gonopod in *Platypoecilus maculatus. Univ. Calif. (Berkeley) Publ. Zool.* **47**, 1–22.

Grobstein, C. (1940b). Effect of testosterone propionate on regeneration of anal fin of adult *Platypoecilus maculatus* females. *Proc. Soc. Exptl. Biol. Med.* **45**, 484–486.

Grobstein, C. (1942a). Effect of various androgens on regenerating anal fin of adult *Platypoecilus maculatus* females. *Proc. Sac. Exptl. Biol. Med.* **49**, 477–478.

Grobstein, C. (1942b). Endocrine and developmental studies of gonopod differentiation in certain poeciliid fishes. II. Effect of testosterone propionate on the normal and regenerating anal fin of adult *Platypoecilus maculatus* females. *J. Exptl. Zool.* **89**, 305–328.

Grobstein, C. (1948). Optimum gonopodial morphogenesis in *Platypoecilus maculatus* coith constant dosage of methyl testosterone. *J. Exptl. Zool.* **109**, 215–236.

Hardisty, M. W. (1960). Sex ratio of ammocoetes. *Nature* **186**, 988–989.

Harms, J. W. (1926a). Beobachtungen über Geschlechtsumwandlungen reifer Tiere und deren F_1-Generation. *Zool. Anz.* **67**, 67–79.

Harms, W. (1926b). "Körper und Keimzellen." Springer, Berlin.

Harrington, R. W., Jr. (1961). Oviparous hermaphroditic fish with internd self-fertilization. *Science* **135**, 1749–1750.

Harrington, R. W., Jr. (1965). Intersexuality in *Rivulus marmoratus. In* "Intersexuality in Fishes," Abstr. Papers, Conf. Cape Haze Marine Lab., Sarasota, Florida.

Harrington, R. W., Jr., (1967). Environmentally controlled induction of primary male gonochorists from eggs of the self-fertilizing hermaphroditic fish, *Rivulus marmoratus* Poey. *Biol. Bull.* **132**, 174–199.

Hartmann, M. (1951). "Geschecht und Geschlechtsbestimmung in Tier- und Pflan-zenreich." de Gruyter, Berlin.

Hartmann, M. (1956). "Die Sexualität." Fischer, Stuttgart.

Haskins, C. P., Haskins, E. F., and Hewitt, R. E. (1960). Pseudogamy as an evolutionary factor in the poeciliid fish *Mollienisia formosa*. *Evolution* **14**, 473–483.

Hickling, C. F. (1960). The Malacca *Tilapia* hybrids. *J. Genet.* **57**, 1–10,

Hild, S. (1940). Versuche zur Geschlechtsdifferenzierung von *Xiphophorus helleri*. *Jena. Z. Naturw.* **73**, 135–143.

Hildemann, W. H. (1954). Effects of sex hormones on secondary sex characters of *Lebistes reticulatus*. *J. Exptl. Zool.* **126**, 1–15.

Hishida, T. (1962). Accumulation of testosterone-4-C^{14} propionate in larval gonad of the medaka, *Oryzias latipes*. *Embryologia Nagoya* **7**, 56–67.

Hishida, T. (1965). Accumulation of estrone-16-C^{14} and diethylstilbestrol (Monoethyl-l-C^{14}) in larval gonads of the medaka, *Oryzias latipes*, and determination of the minimum dosage of estrogen for sex reversal. *Gen. Comp. Endocrinol.* **5**, 137–144.

Hoar, W. S. (1955). Reproduction in teleost fish. *Mem. Soc. Endocrinol.* **4**, Part 1, 5–24.

Hoar, W. S. (1957a). Endocrine organs. *In* "The Physiology of Fishes" (M. E. Brown, ed.), Vol. 1, pp. 254–285. Academic Press, New York.

Hoar, W. S. (1957b). The gonads and reproduction. *In* "The Physiology of Fishes" (M. E. Brown, ed.), Vol. 1, pp. 287–321. Academic Press, New York.

Hoar, W. S. (1965). Comparative physiology: Hormones and reproduction in fishes. *Ann. Rev. Physiol.* **27**, 51–70.

Hoar, W. S., Black, V. S., and Black, E. C. (1951). Some aspects of the physiology of fish. I. Hormones in fish. *Publ. Ontario Fisheries Res. Lab.*, **61**, 1–52.

Hoek, P. P. C. (1891a). Hermaphroditisme bij visschen. Verhandl. *3rd Ned. Nat. Geneesk. Congr.* pp. 185–187.

Hoek, P. P. C. (1891b). Over het hermaphroditisme van de visschen uit de familien der Percidae en Sparidae. *Tijdschr. Ned. Dierk. Ver.* [2] **3**, 37–38.

Holly, M. (1930). Zur Geschlechtsumbildung bei Lebendgebärern. *Bl. Aquar.- Terrarienk.* **41**, 114–119.

Hopper, A. F. (1943). The early embryology of *Platypoecilus maculatus*. *Copeia* pp. 218–224.

Hopper, A. F. (1949). The effect of ethynyl testosterone on the intact and regenerating anal fins of normal and castrated females and normal males of *Lebistes reticulatus*. *J. Erptl. Zool.* **111**, 393–414.

Hubbs, C., Jr. (1964). Interaction between a bisexual fish species and its gynogenetic sexual parasite. *Bull. Trans. Mem. Museum* **8**, 1–72.

Hubbs, C., Drewry, G. E., and Warburton, B. (1959). Occurrence and morphology of a phenotypic male of a gynogenetic fish. *Science* **129**, 1227–1229.

Hubbs, C. L., and Hubbs, L. C. (1932). Apparent parthenogenesis in nature, in a form of fish of hybrid origin. *Science* **76**, 628–630.

Hubbs, C. L., and Hubbs, L. C. (1946). Breeding experiments with the invariable female, strictly matroclinous fish, *Mollienisia formosa*. *Rec. Genet. Soc. Am.* **14**, 48.

Huxley, J. (1938). *In* "Essays in Popular Science," p. 45. Penguin Books, Ltd., Harmondsworth, England.

Ikeda, K. (1933). Effect of castration on the sexual characters of anadromous three-spined stickle-head, *Gasterosteus aculeatus aculeatus* (L.). *Japan. J. Zool.* **5**, 135–157.

Jordan, D. S., and Snyder, J. O. (1902). A review of the labroid fishes and related forms found in the waters of Japan. *Proc. U.S. Natl. Museum* **24**, 595–662.

Kaiser, P., and Schmidt, E. (1951). Vollkommene Geschlechtsumwandlung beim weiblichen siamesischen Kampffish, *Betta splendens*. *Zool. Anz.* **146**, 66–73.

Kallman, K. D. (1962). Gynogenesis in the teleost, *Mollienesia formosa* (Girard), with a discussion of the detection of parthenogenesis in vertebrates by tissue transplantation. *J. Genet.* **58**, 7–24.

Kallman, K. D. (1964). Homozygosity in a gynogenetic fish. *Genetics* **50**, 260–261.

Kallman, K. D. (1965a). Sex determination in the teleost *Xiphophorus milleri*. *Am. Zoologist* **5**, 246–247.

Kallman, K. D. (1965b). Genetics and geography of sex determination in the poeciliid fish, *Xiphophorus maculatus*. *Zoologica* **50**, 151–190.

Kar, A. B., and Ghosh, A. (1950). Responses of the genital system and the urogenital papilla of the female catfish *Heteropneustes fossilis* (Block) to sex hormones. *Arch. Entwicklungsmech. Organ.* **144**, 257–264.

Kawaguchi, K., and Marumo, R. (1967). Biology of *Gonostoma gracile* (Gonosto-matidae). I. Morphology, life history and sex reversal. *Inform. Bull. Planktol. Japan., Commen. No. of Dr. Y. Matsui* pp. 53–67.

Kinoshita, J. (1936). On the conversion of sex in *Sparus longispinis* (Temminck et Schlegel) (Teleostei). *J. Sci. Hiroshima Univ.* **B4**, 69–89.

Kinoshita, J. (1939). Studies on the sexuality of genus *Sparus* (Teleostei). *J. Sci. Hiroshima Univ.* **B7**, 25–37.

Kinoshita, Y. (1934). On the differentiation of the male color pattern, and the sex ratio in *Halichoeres poecilopterus* (Temminck and Schlegel). *J. Sci. Hiroshima Univ.* **B3**, Div. 1, 65–76.

Kinoshita, Y. (1935). Effect of gonadectomies on the secondary sexual characters in *Halichoeres poecilopterus* (Temminck and Schlegel). *J. Sci. Hiroshima Univ.* **B4**, Div. 1, 1–74.

Kinoshita, Y. (1936). Testis-ova found in the Japanese wrasse, *Halichoeres poecilopterus*, and its sex reversal. *Botany & Zool. (Tokyo)* **4**, 35–38 (in Japanese).

Kleiner, I. S., Weisman, A., and Mishkind, D. I. (1937). The similarity of action of male hormones and adrenal extracts on the female bitterling. *Science* **85**, 75.

Kopéĉ, S. (1918). Contribution to the study of the development of the nuptial colors of fishes. *Sprawozdania Posiedzen Towarz. Nauk. Warszaw.* **3**, 11.

Kopéĉ, S. (1928). Experiment on depence of nuptial hue in the gonads in fish. *Biol. Generalis* **3**, 259.

Kosswig, C. (1932). Genotypische und phänotypische Geschlechtsbestimmung bei Zahnkarpfen und ihren Bastarden. I. Z. *lnduktive Abstammungs-Vererbungslehre* **62**, 23–34.

Kosswig, C. (1933). Genotypische und phänotypische Geschlechts bestimmung bei Zahnkarpfen und ihren Bastarden. III. *Arch. Entwicsklungsmech. Organ.* **128**, 393–446.

Kosswig, C. (1935a). Genotypische und Phänotypische Geschlechts bestimmung bei Zahnkarpfen. V. Ein X (Z)-Chromosom als Y-chromosom in fremdem Erbgut. *Arch. Entwicklungsmech. Organ.* **133**, 118–139.

Kosswig, C. (1935b). Genotypische und phänotypische Geschlechts bestimmung bei Zahnkarpfen. VI. *Arch. Entwicklungsmech. Organ.* **133**, 140–155.

Kosswig, C. (1939a). Geschlechtsbestimmungs analyse bei den Zahnkarpfen. *Istanbul Univ. Fen. Fak. Mecmuasi* **B4**, 239–270.

Kosswig, C., (1939b). Uber einen neuen Farbfaktor des *Platypoecilus maculatus*. *Istanbul Univ. Fen. Fak.. Mecmuasi* **B4**, 395–402.

Kosswig, C. (1954). Zur Geschlechtsbestimmungs analyse bei den Zahnkarpfen. *Istanbul Univ. Fen. Fak. Mecmuasi* **B19**, 187–190.

Kosswig, C. (1959). Beiträge zur genetischen Analyse xiphophoriner Zahnkarpfen. *Biol. Zentr.* **78**, 711–718.

Kosswig, C. (1964). Polygenic sex determination. *Experientia* **20**, 1–10,

Kosswig, C. (1965). Polygene Geschlechtsbestimmung. *Naturw. Rundschau* **18**, 392–401.

Kosswig, C., and Öktay, M. (1955). Die Geschlechtsbestimmung bei den Xiphophorini (Neue Tatsachen und neue Bedeutung). *Istanbul Univ. Fen. Fak. Hidrobiol.* **2**, 133–156.

Kuroda, N. (1931). Male and female of Sakura-dai and their scientific names. *Zool. Mag.* (*Tokyo*) **43**, 627–628 (in Japanese).

Larrañeta, M. G. (1953). Observaciones sobre la sexualidad de *Pagellus erythrinus* L. *Publ. Inst. Biol. Apl.* (*Barcelona*) **13**, 83–101.

Larrañeta, M. G. (1964). Sobre la biologia de *Pagellus erythrinus* (L.) especialmente del de costas de Castellon. *Invest. Pesquera* **27**, 121–146.

Laskowski, W. (1953). Reaktionen der primären und sekundären Geschlechtsmerkmale von *Platypoecilus variatus* (♂ Heterogamet) und *Platypoecilus maculatus* (♂ Homogamet) auf Sexual hormone. *Arch. Entwicklungsmech. Organ.* **146**, 137–182.

Lavenda, N. (1949). Sexual differences and normal protogynous hermaphroditism in the Atlantic sea bass, *Centropristes striatus*. *Copeia* **3**, 185–194.

Le Gall, J. (1929). Note sur la sexualitée de la dorade (*Pagellus* centrodontus) reported. *Rev. Trau. Office Peches Maritimes* **2**, 31–32.

Lepori, N. G. (1942a). Azione della follicolina sul sesso di *Gambusia holbrookii* (GRd.). *Atti Soc. Toscana Sci. Nat. Mem.* **51**, 1–19.

Lepori, N. G. (1942b). Inversione sperimentale delle gonadi femminili di *Gambusia holbrooki* Grd. trattala con testosterone. *Monit. Zool. Ital.* **53**, 117–131.

Lepori, N. G. (1945). Osservazioni sul determinismo dei carrateri sessuali nei pesci. *Rev. Biol.* (Perugia) **37**, 67–89.

Lepori, N. G. (1946). Effecti nulli degli ormone sessuali sulle creste genitali delle ceche. *Atti Soc. Toscana Sci. Nat. Mem.* **53**, 1–9.

Lepori, N. G. (1948). Osservazioni sulla "Intersessualita transitoria" nei pesci. *Boll. Pesca, Piscicolt. ldrobiol.* [N.S.] **11**, 3–19.

Lepori, N. G. (1960). Ermafroditismo Proteroginico in *Maena maena* (L.) ed in *Maena chryselis* (Cuv. e Val.) (Perciformes, Centracanthidae). *Boll. Pesca, Piscicolt. Idrobiol.* [N.S.] **14**, 155–165.

Lieder, U. (1955). Männchenmangel und natürlische Parthenogenese bei der Silber-karausche *Carassius auratus* gibelio (Vertebrate, Pisces). *Naturwissenschaften* **42**, 590.

Liem, K. F. (1963). Sex reversal as a natural process in the Synbranchiform fish Monopterus albus. *Copeia* No. 2, 303–312.

Liem, K. F. (1965). Intersexuality in symbranchoid eels. *In* "Intersexuality in Fishes," Abstr. Papers, Conf. Cape Haze Marine Lab., Sarasota, Florida.

Lissia Frau, A. M. (1964). Osservazioni sul differenziamento sessuale di *Denter dentex* (L.). (Teleostei, Sparidae). *Boll. Zool.* **31**, 783–747.

Liu, C. K. (1944). Rudimentary hermaphroditism in the symbranchoid eel, *Monopterus javanensis*. *Sinensia* **15**, 1–18.

Lönneberg, E., and Gustafson, G. (1937). Contributions to the life history of the striped wrasse, *Labrus ossifagus*. *Arkiv Zool.* **29A**, No. 7, 1–16.

Lozano Cabo, F. (1951). Nota preliminare sull' identificazione di *Spicaris smalis* L. e *Spicaris alcedo* C. V. come sessi differenti di una medisma con processo di inversione sessuale. *Atti. Accadi. Naz. Lincei, Rend., Classe Sci. Fis., Mat. Nat.* [8] **11**, 127–132.

Lozano Cabo, F. (1953). Monographia de los Centracandidos mediterraneos, con etude especial de la biometria, biologia y anatomia de *Spicaris smalis* L. *Mem. Acad. Cienc. Madrid* **17**, No. 2, 1–128.

Lubosch, W. (1903). Ueber die Geschlechtsdifferenzierung bei Ammocoetes. *Anat. Anz.* **23**, 66–74.

Lupo, C., and Chieffi, G. (1963). Oestrogens and progesterone in ovaries of the marine teleost *Conger conger. Nature* **197**, 596.

MacIntre, P. A. (1961). Spontaneous sex reversals of genotypic male in the platyfish *Xiphophorus maculatus. Genetics* **46**, 575–580.

McLeod, J. (1881). Recherches sur la structure et le développment de l'appareil reproducteur femelle des Téléostéens. *Arch. Biol.* (*Liege*) **2**, 497–530.

Mead, G. W. (1960). Hermaphroditism in archibenthic and pelagic fishes of the Order Iniomi. *Deep-Sea Res.* **6**, 234–235.

Meyer, H. (1938). Invertigation concerning the reproductive behavior of *Mollienisia fomosa. J. Genet.* **36**, 329–366.

Meyers, A. S. (1955a). 19-Hydroxylation of Δ^4-androstene-3, 17-dione and dehy-droepiandrosterone by bovine adrenals. *Experientia* **11**, 99–102.

Meyers, A. S. (1955b). Conversion of 19-Hydroxy-Δ^4-androstene-3, 17-dione to estrone by endocrine tissue, *Biochim. Biophys. Acta* **17**, 441–442.

Miller, R. R. (1960). Four new species of viviparous fishes, genus *Poeciliopsis*, from northwestem Mexico. *Occasional Papers Museum Zool., Univ. Mich.* **619**, 1–11.

Miller, R. R., and Schultz, R. J. (1959). All female strains of the teleost fishes the genus *Poesiliopsis. Science* **130**, 1655–1657.

Miyamori, H. (1961). Sex modification of *Lebistes reticulatus* induced by androgen administration. *Zool. Mag.* (*Tokyo*) **70**, 310–332 (in Japanese).

Mršić, W. (1923). Die Spätbefruchtung und deren Einfluss auf Entwicklung und Geschlechtsbildung, experimentell nachgefrüft an der Regenbogenforelle. *Arch. Mikroskop Anut. Entwicklungsmech.* **98**, 129–209.

Mršić, W. (1930). Uber das Auftreten intermediärer Stadien bei der Geschlechts-differenzierung der Forelle. *Arch. Entwiklungsmech. Organ.* **123**, 301–332.

Muller, H. J. (1932). Some genetic aspects of sex. *Am. Naturalist* **66**, 118–138.

Nagata, Y. (1934). Castration experiments on the medaka, *Oryzias latipes. Zool. Mag.* (*Tokyo*) **46**, 293–294 (in Japanese).

Nagata, Y. (1936). Transplantation of testis into ovariotomized *Oryzias latipes. Zool. Mag.* (*Tokyo*) **48**, 102–108 (in Japanese).

Niwa (Suzuki), H. (1965a). Effects of castration and administration of methy testosterone on the nupital coloration of the medaka (*Oryzias latipes*). *Embryologia* (*Nagoya*) **8**, 289–298.

Niwa (Suzuki), H. (1965b). Inhibition by estradiol of methyltestosterone-induced nupital coloration in the medaka (*Oryzia latipes*). *Embryologia* (*Nagoya*) **8**, 299–307.

Noble, G. K., and Kumpf, K. F. R. (1937). Sex reversal in the fighting fish, *Betta splendens. Anat. Record* **70**, 97.

Nogusa, S. (1955). Chromosome studies in Pisces. IV. The chromosome of *Mogrunda obscura* (Gobüdae), with evidence of male heteromamety. *Cytologia* (*Tokyo*) **20**, 11–18.

Nogusa, S. (1957). Chromosome studies in Pisces. VI. The X-Y chromosomes found in *Cottux pollux* Günth. (Cottidae). *J. Fac. Sci., Hokkaido Univ., Ser. VI* **13**, 289–292.

Oka, T. B. (1931a). On the processes on the fin rays of the male of *Oryzias latipes* and other sex characters of this fish, *J. Fac. Sci., Imp. Univ. Tokyo, Sect. IV* **2**, 209–218.

Oka, T. B. (1931b). On the accidental hermaphroditism in *Oryzias latipes. J. Fac. Sci., Imp. Univ. Tokyo, Sect. IV* **2**, 219–224.

Okada, Yô K. (1943a). A review of literature on secondary sexual characters in fish and their experimental consideration. *Ann. Rev. Exptl. Morphol.* **1**, 34–64 (in Japanese).

Okada, Yô K. (1943b). Production of testis-ova in *Oryzias latipes* by oestrogenic substances. *Proc. Imp. Acad. (Tokyo)* **19**, 501–504.

Okada, Yô K. (1943c). On the bipotentiality of the sex cells in poeciliid fish. *Proc. Imp. Acad. (Tokyo)* **20**, 244–250.

Okada, Yô K. (1944). On the bipotentiality of the sex cells in poeciliid fish. *Proc. Imp. Acad. (Tokyo)* **20**, 244–250.

Okada, Yô K. (1949). Experimental studies on the testis-ova in fish. *Zikkenkeitaigaku (Exptl. Morphol.)* **5**, 149–151 (in Japanese).

Okada, Yô K. (1952a). Experimental production of testis-ova in the fish *Oryzias latipes. Papers Coord. Comm. Res. Genet.* **3**, 139–142 (in Japanese).

Okada, Yô K. (1952b). Studies on the sex and sex separation in a fish, *Sparus longispinis. Papers Coord. Comm. Res. Genet.* **3**, 147–150 (in Japanese).

Okada, Yô K. (1962). Sex reversal in the Japanese wrasse, *Halichoeres poecilopterus. Proc. Japan Acad.* **38**, 508–513.

Okada, Yô K. (1964a). A further note on sex reversal in the wrasse, *Halichoeres poecilopterus. Proc. Japan Acad.* **40**, 533–535.

Okada, Yô K. (1964b). Effects of androgen and estrogen on sex reversal in the wrasse, *Halichoeres poecilopterus. Proc. Japan Acad.* **40**, 541–544.

Okada, Yô K. (1964c). A further note on testis-ova in the teleost, *Oryzias latipes. Proc. Japan Acad.* **40**, 753–756.

Okada, Yô K. (1965a). Sex reversal in the serranid fish, *Sacura masgaritacea*. I. Sex characters and changes in gonads during reversal. *Proc. Japan Acad.* **41**, 737–740.

Okada, Yô K. (1965b). Sex reversal in the serranid fish, *Sacura masgaritacea*. II. Seasonal variations in gonads in relation to sex reversal. *Proc. Japan Acad.* **41**, 741–745.

Okada, Yô K. (1966). Sex reversal in *Inegocia meerdervoort* with special reference to repitation of hermaphroditic state. *Psoc. Japan Acad.* **42**, 497–502.

Okada, Yô K., and Yamashita, H. (1944a). Experimental investigation of the manifestation of secondary sexual characters in fish, using the medaka, *Oryzias latipes* (Temminck and *Schlegel*) as material. *J. Fac. Sci., Imp. Univ. Tokyo, Sect. IV* **6**, 383–437.

Okada, Yô K., and Yamashita, H. (1944b). Experimental investigation of the sexual characters of poeciliid fish, *J. Fac. Sci., Imp. Univ. Tokyo, Sect. IV* **6**, 589–633.

Okkelberg, P. (1921). The early history of the germ cells in the brook lamprey, *Entosphenus wilderi*, up to and including the period of *sex* differentiation. *J. Morphol.* **35**, 1–152.

Öktay, M. (1959). Uber Ausnahmemaennchen bei *Platypoecilus maculatus* und eine neue Sippe mit XX-Maennchen und XX-Weibchen. *Istanbul Univ. Fen Fak. Mecmuasi* **B24**, 75–92.

Öktay, M. (1963). Die Rolle artfremder Gonosomen bei der Geschlechtsbestimmung von Bastarden mit *Platypoecilus xiphidium. Istanbul Univ. Fen Fak. Hidrobiol.* **B6**, 1–13.

Padoa, E. (1937). Differenziazone e inversione sessuale (feminizzazione) di avanotti di Trota (*Salmo irideus*) trattati con ormone follicolare. *Monit. Zool. Ital.* **48**, 195–203.

Padoa, E. (1939a). Observations ultérieures sur la differenciation du sexe, normale et modifiée par l'administration d'hormone folliculaire, chez la truite iridée (*Salmo irideus*). *Bio-Morphosis* **1**, 337–354.

Padoa, E. (1939b). Prime osservazioni sulle gonadi di un telesteo ermafrodita (*Hepatus hepatus* L.) trattati con ormone femmile. *Monit. Zoot. Ital.* **50**, 129–132.

Pasquali, A. (1941). Contributo allo studio dell'ermafroditismo e del differenziamento della gonade nell'orata (*Sparus auratus* L.). *Pubbl. Staz. Zool. Napoli* **18**, 282–312.

Peters, G. (1964). Vergleichende Untersuchungen an drei Subspecies von *Xiphorus helleri* Heckel (Pisces). Z. *Zool. Syst. Evolut.-Forsch.* **2**, 185–271.

Philippi, E. (1908). Fortpflanzungsgeschichte der viviparen Teleostier *Glaridichthyes januarius* und *Glaridichthyes decemmaculatus* in ihren Einfluss auf Lebensweise, makro- und mikroscopische Anatomie. *Zool. Jahrb., Abt. Anat. Ontog. Tiere* **27**, 1–94.

Pickford, G., and Atz, J. (1957). "Physiology of the Pituitary Gland in Fishes." N.Y. Zool. Soc., New York.

Popoff, W. (1929). Über Grösse des Schwertfishmännchens (*Xiphophorus helleri*), *Zool. Anz*, **86**, 159–160.

Querner, H. (1956). Der Einfluss von Steroidhormonen auf die Gonaden juveniler Poeciliiden. *Biol. Zentr.* **75**, 28–51.

Rasch, E. M., Darnell, R. M., Kallman, K. D., and Abramoff, P. (1965). Cytophotometric evidence for triploidy in hybrids of the gynogenetic fish, *Poecilia formosa. J. Exptl. Zool.* **160**, 155–170.

Régnier, M. T. (1938). Contribution à l'etude de la sexualité des cyprinodontes vivpares (*Xiphophorus helleri, Lebistes reticulatus*). *Bull. Biol. France Belg.* **72**, 385–493.

Reinboth, R. (1960). Natürliche Geschlechtsumwandlung bei adulten Teleosteern. *Verhandl. Deut. Zool. Ges. Bonn/Rhein* pp. 259–262.

Reinboth, R. (1962). Morphologische und funktionelle Zweiggeschlechtlichkeit bei marinen Teleostieren (Serranidae, Sparidae, Centracanthidae, Labridae). *Zool. Jahrb. Abt. Allgem.. Zool. Physiol. Tierre* **69**, 405–480.

Reinboth, R. (1963). Natürliche Geschlechtswechsel bei *Sacura margaritacea* (Hilgendorf) (Serranidae). *Annotationes Zool. Japan.* **36**, 173–178.

Reinboth, R. (1965). Sex reversal in the black sea bass *Centropristes striatus. Anat. Rec.* **151**, 403.

Reinboth, R. (1967). Protogynie bei *Chelidoperca hirundinacea* (Cuv. et Val.) (Serranidae). —Ein Diskussionsbeitrag zur Stammesgeschichte amphisexueller Fische. *Annotationes Zool. Japan.* **40**, 181–193.

Rijavec, L., and Zupanovic, S. (1965). A contribution to the knowledge of biology of *Pagellus erythrinus* L. in the middle Adriatic. *Rappt. Proces-Verbaux Reunions, Monaco* **18**, 195–200.

Roede, M. J. (1965). Intersexuality of Labrids. "Intersexuality in Fishes," Abstr. Papers, Conf. Cape Haze Marine Lab., Sarasota, Florida.

Rubin, A. A., and Gordon, M. (1953). Effects of alpha-estradiol benzoate and methyl testosterone upon the platyfish, *Xiphophorus maculatus* skeleton. *Proc. Soc. Exptl. Biol. Med.* **83**, 646–648.

Rudolf, H. (1931). Wieder ein Herringszwitter. *Zool. Anz.* **95**, 37–41.

Rust, W. (1941). Genetische Untersuchungen über die Geschlichtsbestimmungtypen bei Zahnkarpfen unter besonderer Berücksichtigung von Artkreuzungen mit *Platypoecilus variatus. Z. Induktive Abstammungs Vererbungslehre* **79**, 336–395.

Sacks, W. B. (1955). Über *Xiphophoms helleri* Heckel. *Deut. Aqua. Tern. Zeit.*, pp. 309–310.

Samokhvalova, G. V. (1933). Correlation in the development of the secondary sexual characters and the sex glands in *Lebistes reticulatus. Trans. Dyn. Develop.* **7**, 65–76.

Schreiner, K. E. (1904). Uber das Generations-organ von *Myxine glutinosa* (L.). *Zool. Anz.* **24**, 91–104.

Schreiner, K. E. (1955). Studies on the gonad of *Myxine glutinosa* L. *Univ. Bergen Arbok, Naturvitenskap. Rekke* **8**, 1–37.

Schultz, R. J. (1961). Reproductive mechanisms of unisexual and bisexual strains of the viviparous fish *Poeciliopsis. Evolution* **15**, 302–325.

Schultz, R. J. (1966). Hybridization experiments with an all-female fish of the genus *Poeciliopsis. Biol. Bull.* **130**, 415–429.

Schultz, R. J. (1967). Gynogenesis and triploidy in the viviparous fish *Poeciliopsis*. *Science* **157**, 1564–1567.

Schwier, H. (1939). Geschlechtsbestimmung und-differenzierung bei *Macropodus opercularis, concolor, Chinensis* und deren Artbastarden. Z. *Induktive Abstammungs Vererbungslehre* **77**, 291–435.

Sengün, A. (1941). Ein Beitrag zur Geschlechtsbestimmung bei *Platypoecilus maculatus* und *Xiphophorus helleri*. *Istanbul Univ. Fen. Fak. Mecmuasi* **B6**, 33–48.

Smith, C. L. (1959). Hermaphroditism in some serranid fishes from Bermuda. *Papers Mich. Acad. Sci.* **44**, 111–119.

Smith, C. L. (1965). The patterns of sexuality and the classification of serranid fishes. *Am. Museum Novitates* **2207**, 1–20.

Smith, C. L. (1967). Contribution to a theory of hermaphroditism. *J. Theoret. Biol.* **17**, 76–90.

Solomon, S., Wiese, R. V., and Lieberman, S. (1956). The *in vitro* synthesis of 17α- Hydroxyprogesterone and Δ^4-Androstene-3, 17-dione from progesterone by bovine ovarian tissue. *J. Am. Chem. Soc.* **78**, 5453–5454.

Sordi, M. (1962). Ermafroditismo proteroginico in *Labrus turdus* L. e in *L. merula*. *Monit. Zool. Ital.* 69–84.

Spurway, H. (1953). Spontaneous parthenogenesis in a fish. *Nature* **171**, 437.

Spurway, H. (1957). Hermaphroditism with self-fertilization, and the monthly extrusion of unfertilized eggs, in the viviparous fish *Lebistes reticulatus*. *Nature* **180**, 1248–1251.

Stolk, A. (1958). Pathological parthenogenesis in viviparous toothcarps. *Nature* **181**, 1660.

Syrski, S. (1876). Ergebnisse von Untersuchungen der Geschlechtsorgane von Knochenfischen (*De piscium osseorum organis genitalibus*). *Kosmos* (*Lwow*) **1** 418–455.

Takeuchi, K. (1967). Large tooth formation in female medaka, *Oryzias latipes*, given methyl testosterone. *J. Dental Res.* **46**, 750.

Tavolga, M. C. (1949). Differential effects of estradiol and pregneninolone on *Platypoecilu maculatus*. *Zoologica* **34**, 215–238.

Taylor, A. B. (1948). Experimental sex reversal in the red swordtail hybrid *Xiphophorus-Platypoecilus*. *Trans. Am. Microscop. Soc.* **67**, 155–164.

Tozawa, T. (1923). Studies on the pearl organ of the goldfish. *Annotationes Zool. Japan.* **10**, 253–263.

Tozawa, T. (1929). Experiments in the development of the nuptial coloration and pearl organs of the Japanese bitterling. *Folia Anat. Japan.* **7**, 407–417.

Turner, C. L. (1942). Morphogenesis of the gonopodial suspensorium in *Gambusia afinis* and the induction of male suspensorial characters in the female by androgenic hormone. *J. Exptl. Zool.* **91**, 167–193.

Tumer, C. L. (1947). The rate of morphogenesis and regeneration of the gonopodium in normal and castrated males of *Gambusia afinis*. *J. Exptl. Zool.* **106**, 125–143.

Tumer, C. L. (1967). The effects of steroid hormones on the development of some secondary sexual characters in cyprinodont fishes. *Trans. Am. Microscop. Soc.* **79**, 320–333.

Vandini, R. Z. (1965). Il problema dei maschi primari di *Coris Julis*. *Monit. Zool. Ital.* **83**, 102–111.

van Oordt, G. J. (1923). Secondary sex characters and testis of the ten-spined stickle-back (*Gasterosteus pungitius*). *Koninkl. Ned. Akad. Wetenschap., Proc. Sect. Sci.* **26**, 309–314.

van Oordt, G. J. (1924). Die Veränderung des Hodens während des Auftretens der sekundären Geschlechtsmerkmale bei Fischen, *Gasterosteus pungitius* L. *Arch. Mikroskop. Anat. Entwicklungsmech.* **102**, 379–405.

van Oordt, G. J. (1925). The relation between the development of the secondary sex characters and the structure of the testis in the teleost, *Xiphophorus helleri* Heckel. *Brit. J. Exptl. Biol.* **3**, 43–49.
van Oordt, G. J. (1928). The duration of life of the spermatozoa in the fertilized female of *Xiphophorus helleri*. *Tijdschr. Ned. Dierk. Ver.* **3**, 77–81.
van Oordt, G. J. (1929). Zur mikroskopischen Anatomie des Ovariotestes von *Serranus* und *Sargus* (Teleostei). *Z. Mikroskop.-Anat. Forsch.* **19**, 1–17.
van Oordt, G. J., and van der Mass, C. J. J. (1927). Castration and implantation of gonads in *Xiphophorus Helleri* Hechel. *Koninkl. Ned. Acad. Wetenschap., Proc.* **29**, 1172–1175.
Vivian, J. (1950). Masculinisation des femelles de Xiphophores par action des r ons X sur la gonade. *Compt. Rend. Acad. Sci., Paris* **231**, 1166–1168.
Vivian, J. (1952a). Rôle de l'hypophyse dans le déterminism de l'involution ovarienne et de l'inversion sexuelle chez le Xiphophores. *J. Physiol. (Paris)* **44**, 349–351.
Vivian, J. (1952b). Sur l'existence de charactères sexuels secondaires femelles et sur le de'terminisme de leur différenciation chez *Xiph. hell.* Heckel. *Compt. Rend. Acad. Sci., Paris* **235**, 1697–1699.
Williamson, H. C. (1910). Report on the reproductive organs of *Sparus centrodontus* Delaroche; *Sparus canthalus* L.; *Sebastes marinus* (L.) and *Sebastes dactyloptems* (Delaroche); and on the ripe eggs and larvae of *Sparus controdontus* and *Sebastes marinus*. *Rept. Fishery Board. Scotland Sci. Invest.* **1**, 1–35.
Winge, Ö. (1922a). A peculiar mode of inheritance and its cytological explanation. *J. Genet.* **12**, 187–144.
Winge, Ö. (1922b). One-sided masculine and sex-linked inheritance in *Lebistes reticulatus*. *J. Genet.* **12**, 145–162.
Winge, Ö. (1927). The location of eighteen genes in *Lebistes reticulatus*. *J. Genet.* **18**, 1–43.
Winge, Ö. (1930). On the occurrence of XX-males in *Lebistes* with some remarks on Aida's so-called "non-disjunctional" males in *Aplocheilus*. *J. Genet.* **23**, 69–76.
Winge, Ö. (1934). The experimental alteration of sex chromosomes into autosomes and vice versa, as illustrated by *Lebistes*. *Compt. Rend. Trav. Lab. Carlsberg, Ser. Physiol.* **21**, 1–49.
Winge, Ö., and Ditlevsen, E. (1947). Colour inheritance and sex determination in *Lebistes*. *Heredity* **1**, 65–83.
Winge, Ö., and Ditlevsen, E. (1948). Colour inheritance and sex determination in *Lebistes*. *Compt. Rend. Trav. Lab. Carlsberg, Ser. Physiol.* **24**, 227–248.
Witschi, E. (1914a). Experimentelle Untersuchungen über die Entwicklungsgeschichte der Keimdrüsen von *Rana temporaria*. *Arch. Mikroskop. Anat.* **85**, 9–113.
Witschi, E. (1914b). Studien über die Geschlechtsbestimmung bei Froschen. *Arch. Mikroskop. Anat.* **86**, 1–50.
Witschi, E. (1929). Bestimmung und Vererbung des Geschlechts bei Tieren. *Handbuch Vererbungswiss.* **2**, 1–115.
Witschi, E. (1930). Studies on sex differentiation and sex determination in Amphibians. IV. The geographical distribution of the sex races of the European grass frog (*Rana temporaria* L.). *J. Exptl. Biol.* Zool. **56**, 149–165.
Witschi, E. (1931). Studies on sex differentiation and sex determination in amphibians. V. Range of the cortex-medulla antagonism in parabiotic twins of Ranidae and Hyalidae. *J. Exptl. Zool.* **58**, 113–145.
Witschi, E. (1939). Modification of development of sex. Sex and Internal Secretions (E. Allen, 2nd ed pp. 145–226. Williams & Wilkins, Baltimore, Maryland.

Witschi, E. (1942). Hormonal regulation of development in lower vertebrates. *Cold Spring Harbor Symp. Quant. Biol.* **10**, 145–151.

Witschi, E. (1950). Génétique et physiologie de la différenciation du sexe. *Arch. Anat. Microscop. Morphol. Exptl.* **39**, 215–246.

Witschi, E. (1955). On morphogenic capacities of the estrogens. *J. Clin. Endocrinol. Metab.* **15**, 647–652.

Witschi, E. (1957). The inductor theory of sex differentiation. *J. Fac. Sci., Hokkaido Univ., Ser. VI* **13**, 428–439.

Witschi, E., and Crown, E. N. (1937). Hormones and sex determination in fishes and in frogs. *Anat. Record* **70**, 121–122.

Wolf, L. E. (1931). The history of the germ cells in the viviparous teleost *Platypoecilus maculatus*. *J. Morphol.* **52**, 115–154.

Wotiz, H. H., Botticelli, C. R., Hisaw, F. L., Jr., and Olsen, A. G. (1960). Estradiol-17β, estrone and progesterone in the ovaries of dogfish (*Squalus suckleyi*). *Proc. Natl. Acad. Sci. U.S.* **46**, 580–583.

Wünder, W. (1931). Experimentelle Erzeugung des Hochzeitskleides beim Bitterling (*Rhodeus amarus*) durch Einpritzung von Hormonen. *Z. Vergleich. Physiol.* **14**, 676–708.

Wurmbach, H. (1951). Geschlechtsumkehr bei Weibschen von *Lebistes reticulatus* bei Befall mit *Ichthyophonous hoferi* Plehn-Mulsow. *Arch. Entwicklungsmech. Organ.* **145**, 109–124.

Yamamoto, T. (1953). Artificially induced sex reversal in genotypic males of the medaka (*Oryzias latipes*). *J. Exptl. Zool.* **123**, 571–594.

Yamamoto, T. (1955). Progeny of artificially induced sex reversals of male genotype (XY) in the medaka (*Oryzias latipes*) with special reference to YY-male. *Genetics* **40**, 406–419.

Yamamoto, T. (1958). Artificial induction of functional sex-reversal in genotypic females of the medaka (*Oryzias latipes*). *J. Exptl. Zool.* **137**, 227–262.

Yamamoto, T. (1959a). A further study of induction of functional sex reversal in genotypic males of the medaka (*Oryzias latipes*) and progenies of sex reversals. *Genetics* **44**, 739–757.

Yamamoto, T. (1959b). The effects of estrone dosage level upon the percentage of sex-reversals in genetic male (XY) of the medaka (*Oryzias latipes*). *J. Exptl. Zool.* **141**, 133–154.

Yamamoto, T. (1961). Progenies of induced sex-reversal females mated with induced sex-reversal males in the medaka, *Oryzias latipes*. *J. Exptl. Zool.* **146**, 163–179.

Yamamoto, T. (1962). Hormonic factors affecting gonadal sex differentiation in fish. *Gen. Comp. Endocrinol.* Suppl. 1, 341–345.

Yamamoto, T. (1963). Induction of reversal in sex differentiation of YY zygotes in the medaka, *Oryzias latipes*. *Genetics* **48**, 293–306.

Yamamoto, T. (1964a). The problem of viability of YY zygotes in the medaka, *Oryzias latipes*. *Genetics* **50**, 45–58.

Yamamoto, T. (1964b). Linkage map of sex chromosomes in the medaka, *Oryzias latipes*. *Genetics* **50**, 59–64.

Yamamoto, T. (1965). Estriol-induced XY females of the medaka (*Oryzias latipes*) and their progenies. *Gen. Comp. Endocrinol.* **5**, 527–533.

Yamamoto, T. (1967). Estrone-induced white YY females and mass production of white YY males in the medaka, *Oryzias latipes*. *Genetics* **55**, 329–336.

Yamamoto, T. (1968). Effects of 17α-hydroxyprogesterone and androstenedione upon sex differentiation in the medaka, *Oryzias latipes*. *Gen. Comp. Endocrinol.* **10**, 8–13.

Yamamoto, T., and Kajishima, T. (1969). Sex-hormonic induction of reversal of sex differentiation in the goldfish and evidence for its male heterogamety. *J. Exptl. Zool.* **168**, 215–222.

Yamamoto, T., and Matsuda, N. (1963). Effects of estradiol, stilbestrol and some alkyl-carbonyl androstanes upon sex differentiation in the medaka, *Oryzias latipes. Gen. Comp. Endocrinol.* **3**, 101–110.

Yamamoto, T., and Suzuki, H. (1955). The manifestation of the urinogenital papillae of the medaka (*Oryzias laripes*) by sex hormones. *Embryologia* (*Nagoya*) **2**, 133–144.

Yamamoto, T., Takeuchi, K., and Takai, M. (1968). Male-inducing action of androsterone and testosterone propionate upon XX zygotes in the medaka, *Oryzias latipes. Embryologia* (*Nagoya*) **10**, 142–151.

Zahl, A. P., and Davis, D. D. (1932). Effects of gonadectomy on the secondary sexual characters in the ganoid fish, *Amia calva* Linnaeus. *J. Exptl. Zool.* **63**, 219–307.

Zander, C. D. (1965). Die Geschlechtsbestimmung bei *Xiphophorus montezumae cortezi* Rosen (Pisces). *Z. Vererbungslehre* **96**, 128–141.

Zei, M. (1949). Typical sex-reversal in teleosts. *Proc. Zool. Soc. London* **119**, 917– 921.

Chapter 13

Beginning with Blaxter—An early summary of embryonic and larval fish development

Casey A. Mueller*
Department of Biological Sciences, California State University San Marcos, San Marcos, CA, United States
Corresponding author: e-mail: cmueller@csusm.edu

Casey A. Mueller discusses the J.H.S. Blaxter chapter "Development: Eggs and Larvae" in *Fish Physiology*, Volume 3, published in 1969. The next chapter in this volume is the re-published version of that chapter.

The study of fish development is exciting and complicated and is central to understanding how phenotypes are formed. "Development: Eggs and Larvae" by Blaxter is a comprehensive review of dynamic developmental processes and serves as a useful gateway to the world of fish development. From the time of the chapter's publication to now, the focus on development has expanded from understanding fish recruitment, stock sizes, and taxonomy to more broadly examining the interplay between environment and physiology. Environment and development interactions are central to fish phenotypic development, and how young fish respond to the environment will often have long term consequences for the adult phenotype. Thus, studying development is vital for understanding the impacts of environmental change in the natural world and when rearing fish in aquaculture settings. Blaxter includes much information that is relevant and useful to larval fish research today, including developmental energetics, feeding, digestion and activity, physiological tolerance, and captive rearing. Blaxter's chapter has broad appeal and impact, as it highlights the importance of studying developmental stages across various fields of fish biology, from aquaculture to ecotoxicology to climate change research.

Fish development is phenotypically complex and fascinating, taking place in a wide range of timescales and thermal environments that include long, cold incubations (e.g., cold-water embryos that overwinter for 4–6 months in the North American Great Lakes) and very quick, warm incubations (e.g., a few days in many species, such as zebrafish (*Danio rerio*)). Fish egg size spans nearly 2 orders of magnitude from 0.28 mm diameter eggs of the viviparous shiner perch (*Cymatogaster aggregata*) to 2.5 cm diameter eggs of the mouth

Fish Physiology, Vol. 40A. https://doi.org/10.1016/bs.fp.2023.08.004

brooding gafftopsail catfish (*Bagre marinus*). Examples of amazing larval phenotypes include the delicate flat and transparent leptocephalus of the American eel (*Anguilla rostrata*), the spikey larval ocean sunfish (*Mola mola*), and the elaborate fin rays and external gut of the larval cusk eel (*Brotulotaenia nielseni*). Processes of embryonic and larval cellular differentiation and maturation interact with, and are influenced by, many aspects of the external environment. To study embryonic and larval processes is to focus on the formation of phenotypes, and it is exciting, complicated, and vital for our understanding of fish biology. Blaxter's chapter (Blaxter, 1969) is one of the original reviews that summarized these dynamic processes, and thus it serves as an excellent introduction to the world of development. This chapter, along with other reviews, chapters and books written and edited by Blaxter (e.g., Blaxter, 1988, 2012; Blaxter and Holliday, 1963) serve as the backbone for entry into the fish development literature.

I first read Blaxter's chapter as a PhD student. I was a "sometimes" fish physiologist in training, and my PhD research on embryonic energy use and gas exchange in Australian lungfish (*Neoceratodus forsteri*) (Mueller et al., 2011a, b) brought me to Blaxter. By the time I read the chapter, it had already been published for over 40 years, yet it was, and remains today, a seminal source of fish developmental physiology literature. It is reasonable to state that any physiologist working on developing fish is familiar with this chapter. A quick search of the literature citing this chapter indicates how useful this work was soon after it was published (A search on Google Scholar indicates 51 citations of the chapter from 1970 to 1975), and this activity of citation has continued to the present day (87 citations on Google Scholar from 2017 to 2022, over 900 total citations to date).

The original motivations for studying development have likely changed, with early studies on fish eggs and larvae driven by the desire to estimate the size of fish stocks, appraise stock-recruitment relationships, and understand fish taxonomy. However, the chapter's wide-ranging nature appeals to a broad readership and helped expand physiological and ecological study of developing fish. The broad nature of the chapter reflects Blaxter's own extensive interests in larval fish physiology, with published works on behavior (Blaxter, 1973, 1986), development and function of the swim bladder and sensory systems (Allen et al., 1976; Blaxter and Denton, 1976), and effects of temperature and salinity on physiological performance (Batty et al., 1993; Yin and Blaxter, 1987). The wide scope of this chapter is also evident in the type of literature that currently cite it, with recent literature on responses to temperature and salinity (Pacheco-Carlón et al., 2021; Skorupa et al., 2022), developmental plasticity (Dunn et al., 2020), larval dispersal (Swearer et al., 2019), ecotoxicology (Munnelly et al., 2021), and aquaculture protocols (Ribeiro et al., 2022). Such citations reflect the current motivations for larval fish research, with increasing emphasis on the interplay between environment and physiology, which is central to understanding fish development in the natural world and in an aquaculture setting.

The chapter begins with a description of the modes of incubation, fecundity, egg size, and the developmental processes from fertilization to metamorphosis. This is basic information on fish development that is vital for any researcher before embarking on physiological research. Details on metabolism and growth include a description of the all-important relationship between temperature and development rate, with data summarized from some of the earliest studies on Atlantic mackerel (*Scomber scombrus*), brook trout (*Salvelinus fontinalis*), brown trout (*Salmo trutta*), fourbeard rockling (*Enchelyopus cimbrius*), and lake whitefish (*Coregonus clupeaformis*). The concept of degree days, the product of incubation time and temperature, is discussed, including its limitations due to its linear nature attempting to describe a curvilinear relationship. Also included is the Q_{10} coefficient, describing the non-linear increase in development rate with increasing temperature. Blaxter cites two studies that calculated Q_{10}, with values between 6.5 and 1.5 for the temperature range of 5–23 °C in the fourbeard rockling (Battle, 1930) and 6.5–2.0 over the range of 3–18 °C in herring (*Clupea harengus*) (Blaxter, 1956). These two studies highlight how Q_{10} values are generally higher in developmental stages compared to adults, with the concept of Q_{10} coefficients also applied to other biological rate changes during development in response to temperature, such as oxygen consumption rate (Q_{10} values as high as 5) (Barrionuevo and Burggren, 1999; Mueller et al., 2011b).

The study of the energetics of fish embryos is particularly interesting as all sources of energy are contained within the egg. Yolk utilization efficiency (or yolk conversion efficiency), the relative mass of yolk that is converted to embryonic/larval tissue, the chemical composition of eggs, oxygen consumption rate, and growth are all measures that describe various aspects of endogenous energy use. Examples of these measures across development and at various temperatures are included for a number of species in this chapter, such as Atlantic salmon (*Salmo salar*), herring (*C. harengus*), killifish (*Fundulus heteroclitus*), Pacific sardine (*Sardinops caerulea*), and rainbow trout (*Oncorhynchus mykiss*). Blaxter discusses the challenges of studying energetics at different temperatures, as comparing a measure such as yolk utilization at common stages, for example, can be difficult because of the effect of temperature on development rate. In this regard, the chapter serves as a convenient summary of these measures and a how-to guide for investigating developmental energetics. There is still much to understand regarding energy substrate use during embryonic development, how energy is used and allocated under altered conditions, and whether changes in the environment may create energetic limitations during development. Thus, the field of developmental energetics remains a rewarding and essential area of study.

The important, but often perilous, transition from endogenously feeding embryo to exogenously feeding larvae, is marked by high mortality in some species. Blaxter and Hempel (1963) proposed the idea of the "point of no return" in herring (*C. harengus*), to describe when larvae do not exogenously feed (i.e., are starved) and ultimately become too weak to feed (i.e., show no feeding behavior). The "point of no return" is tied to food availability and the environmental conditions that govern it. While the idea of yolk absorption representing a "critical phase" for survival is discussed, so to is evidence that not all species experience big mortality events, with possible constant or steadily decreasing mortality rates. Whether a critical period of mortality is present prior to, during, or just after the shift from endogenous to exogenous feeding has been somewhat contentious (but see the review by Sifa and Mathias, 1987). The study of such responses is imperative for understanding how developmental processes are linked to the environment and ultimately population density and species success.

The concept of critical periods in developmental responses in Blaxter's overview of tolerance and optima is one I find to be most important, as it is of great interest to ecophysiologists such as myself. Blaxter highlights how developmental stages may show enhanced sensitivities to environmental variables such as temperature, salinity, oxygen level, pH, and radiation. Thus, an understanding of lethal limits of developmental stages is imperative if we wish to correctly understand the impacts of climate change. Of note is how tolerances may change across development, often with early stages showing greater sensitivity to the environment (e.g., thermal tolerance is generally higher in larval fish compared to embryonic fish; Dahlke et al., 2020). This is often attributed to the immature status of physiological mechanisms that regulate the internal environment. However, it is perhaps an oversimplification to state that sensitivity decreases, or tolerance increases, as development progresses (Mueller, 2018), and many species may have critical windows of sensitivity for various traits that we do not yet fully understand. For example, embryos of lake whitefish (*C. clupeaformis*) and rainbow trout (*O. mykiss*) demonstrate increased thermal sensitivity in traits such as survival, oxygen consumption, and heart rate during gastrulation and organogenesis (Eme et al., 2015; Melendez and Mueller, 2021; Mueller et al., 2015). Ecotoxicology studies also indicate "stage-specific responses" in which critical periods are linked to developmental events. For example, endocrine disruptors are particularly influential during periods of reproductive system plasticity in fathead minnow (*Pimephales promelas*), medaka (*Oryzias latipes*) and zebrafish (*D. rerio*) (Koger et al., 2000; Maack and Segner, 2004; van Aerle et al., 2002). Blaxter concludes this section by mentioning a trend of research examining responses to multiple environmental stimuli. Multi-stressor studies continue to be an important approach for understanding the numerous challenges imposed by climate change (Cominassi et al., 2019; Del Rio et al., 2019) and, when used in concert with approaches targeting critical windows, will provide the most comprehensive analysis of environmental responses during development. Such understanding of how the environment shapes development during critical periods influences the way we approach climate change and fisheries modeling, with studies on larval fish central to predicting abundance and distribution under changing conditions.

How young fish respond to and are influenced by the environment will often have long-term consequences for the adult phenotype. For example, higher incubation temperatures during organogenesis can program later life phenotypes, as demonstrated in the muscle of Atlantic salmon (*S. salar*) (Macqueen et al., 2008) and climbing perch (*Anabas testudineus*) (Ahammad et al., 2021). Likewise, hypoxia exposure during embryogenesis results in increased hypoxia tolerance (Robertson et al., 2014) and altered swimming performance and cardiac gene expression (Johnston et al., 2013) in larvae. Throughout the chapter, the importance of the environment is alluded to, and at times is the focus, of the discussion. This represents the groundwork for understanding the interactions between development and environment, which are at the forefront of physiological research today as we endeavor to understand how species will be impacted by global environmental change.

Other chapter sections of note include an overview of sense organs, activity, locomotion and buoyancy of larval stages. In discussing locomotion, Blaxter refers to the large residual yolk at hatching in many fish, and particularly salmonids, as a "hydrodynamic embarrassment to the larvae." As someone who has conducted research on hatchling rainbow trout (*O. mykiss*), I love this statement! The chapter concludes with an applied focus, discussing techniques for fish rearing and farming. This reflects the strong link between the field of basic fish physiology research and aquaculture, with research foci aimed at understanding the effects of rearing conditions on aquaculture produced larvae (Kuroki et al., 2019) and implications of using hatchery-reared fish for restocking wild populations (Araki and Schmid, 2010). Throughout, Blaxter guides the reader to other central literature that provide overviews of particular aspects of development (for example, reviews on the physiology of fertilization (Yamamoto, 1962), larval endocrine function (Pickford and Atz, 1957), and viviparity (Amoroso, 1960)), so that this work serves as both an overview and entryway to fish development literature.

The impact of Blaxter's chapter lies in its excellent overview of important findings and trends in larval physiology. Since its publication, the chapter has served as a foundation for studies that fall under the broad umbrella of "developmental fish physiology." Blaxter summarizes many important methodological approaches, such as calculating yolk conversion efficiencies and measuring oxygen consumption rate that are still relevant and important in today's research. However, layered on top of that foundation are current and future approaches that are expanding our understanding of embryonic and larval physiology. High throughput genetic and genomic analyses are allowing physiologists to explain whole organismal traits with underlying mechanisms. The emerging field of phenomics, high-dimensional organismal phenotyping (e.g., Fuentes et al., 2018; Li et al., 2021), is perhaps one of the most exciting areas of future research, as we strive to understand the holistic impacts of environmental change. Considerations such as critical windows of sensitivity, multi-stressor effects, and the long-term and multigenerational influence of the environment will continue to expand our understanding of how the environment and development interact. I urge anyone interested in development today to read this chapter. It will serve as a wonderful and informative place to begin when exploring the complexities of embryonic and larval fishes, as, to quote Blaxter "Fish eggs and larvae provide a relatively untapped source of biological material..." The study of fish eggs and larvae represents a fruitful endeavor that will ultimately expand our knowledge of fish physiology.

References

Ahammad, A.K.S., Asaduzzaman, M., Uddin Ahmed, M.B., Akter, S., Islam, M.S., Haque, M.M., Ceylan, H., Wong, L.L., 2021. Muscle cellularity, growth performance and growth-related gene expression of juvenile climbing perch *Anabas testudineus* in response to different eggs incubation temperature. J. Therm. Biol. 96, 102830.

Allen, J.M., Blaxter, J.H.S., Denton, E., 1976. The functional anatomy and development of the swimbladder-inner ear-lateral line system in herring and sprat. J. Mar. Biol. Assoc. UK 56, 471–486.

Amoroso, E., 1960. Viviparity in fishes. Symp. Zool. Soc. Lond. 1, 153–181.

Araki, H., Schmid, C., 2010. Is hatchery stocking a help or harm? Evidence, limitations and future directions in ecological and genetic surveys. Aquaculture 308, S2–S11.

Barrionuevo, W., Burggren, W., 1999. O_2 consumption and heart rate in developing zebrafish (*Danio rerio*): influence of temperature and ambient O_2. Am. J. Physiol. Reg. Int. Comp. Physiol. 276, R505–R513.

Battle, H.I., 1930. Effects of extreme temperatures and salinities on the development of *Enchelyopus cimbrius* (L.). Contrib. Can. Biol. Fish. 5, 107–192.

Batty, R.S., Blaxter, J.H.S., Fretwell, K., 1993. Effect of temperature on the escape responses of larval herring, *Clupea harengus*. Mar. Biol. 115, 523–528.

Blaxter, J.H.S., 1956. Herring rearing. II. The effect of temperature and other factors on development. Mark. Res. 5, 3–19.

Blaxter, J.H.S., 1969. Development: eggs and larvae. In: Hoar, W.S., Randall, D.J. (Eds.), Fish Physiology. vol. 3. Academic Press, San Diego, pp. 177–252.

Blaxter, J.H.S., 1973. Monitoring the vertical movements and light responses of herring and plaice larvae. J. Mar. Biol. Assoc. UK 53, 635–647.

Blaxter, J.H.S., 1986. Development of sense organs and behavior of teleost larvae with special reference to feeding and predator avoidance. Trans. Am. Fish. Soc. 115, 98–114.

Blaxter, J.H.S., 1988. Pattern and variety in development. In: Hoar, W.S., Randall, D.J. (Eds.), The Physiology of Developing Fish. Academic Press, San Diego, pp. 1–58.

Blaxter, J.H.S., 2012. The early life history of fish. In: The Proceedings of an International Symposium Held at the Dunstaffnage Marine Research Laboratory of the Scottish Marine Biological Association at Oban, Scotland, from May 17–23, 1973. Springer Science & Business Media.

Blaxter, J.H.S., Denton, E., 1976. Function of theswimbladder-inner ear-lateral line system of herring in the young stages. J. Mar. Biol. Assoc. UK 56, 487–502.

Blaxter, J.H.S., Hempel, G., 1963. The influence of egg size on herring larvae (*Clupea harengus* L.). ICES J. Mar. Sci. 28, 211–240.

Blaxter, J.H.S., Holliday, F., 1963. The behaviour and physiology of herring and other clupeids. Adv. Mar. Biol. 1, 261–394.

Cominassi, L., Moyano, M., Claireaux, G., Howald, S., Mark, F.C., Zambonino-Infante, J.-L., Le Bayon, N., Peck, M.A., 2019. Combined effects of ocean acidification and temperature on larval and juvenile growth, development and swimming performance of European sea bass (*Dicentrarchus labrax*). PLoS One 14, e0221283.

Dahlke, F.T., Wohlrab, S., Butzin, M., Pörtner, H.-O., 2020. Thermal bottlenecks in the life cycle define climate vulnerability of fish. Science 369, 65–70.

Del Rio, A.M., Davis, B.E., Fangue, N.A., Todgham, A.E., 2019. Combined effects of warming and hypoxia on early life stage *Chinook salmon* physiology and development. Conserv. Physiol. 7, coy078.

Dunn, N.R., O'Brien, L.K., Closs, G.P., 2020. Phenotypically induced intraspecific variation in the morphological development of wetland and stream *Galaxias gollumoides* McDowall and Chadderton. Diversity 12, 220.

Eme, J., Mueller, C.A., Manzon, R.G., Somers, C.M., Boreham, D.R., Wilson, J.Y., 2015. Critical windows in embryonic development: shifting incubation temperatures alter heart rate and oxygen consumption of Lake Whitefish (*Coregonus clupeaformis*) embryos and hatchlings. Comp. Biochem. Physiol. A 179, 71–80.

Fuentes, R., Letelier, J., Tajer, B., Valdivia, L.E., Mullins, M.C., 2018. Fishing forward and reverse: advances in zebrafish phenomics. Mech. Dev. 154, 296–308.

Johnston, E.F., Alderman, S.L., Gillis, T.E., 2013. Chronic hypoxia exposure of trout embryos alters swimming performance and cardiac gene expression in larvae. Physiol. Biochem. Zool. 86, 567–575.

Koger, C.S., Teh, S.J., Hinton, D.E., 2000. Determining the sensitive developmental stages of intersex induction in medaka (*Oryzias latipes*) exposed to 17β-estradiol or testosterone. Mar. Environ. Res. 50, 201–206.

Kuroki, M., Okamura, A., Yamada, Y., Hayasaka, S., Tsukamoto, K., 2019. Evaluation of optimum temperature for the early larval growth of Japanese eel in captivity. Fish. Sci. 85, 801–809.

Li, X., Dang, J., Li, Y., Wang, L., Li, N., Liu, K., Jin, M., 2021. Developmental neurotoxicity fingerprint of silica nanoparticles at environmentally relevant level on larval zebrafish using a neurobehavioral-phenomics-based biological warning method. Sci. Total Environ. 752, 141878.

Maack, G., Segner, H., 2004. Life-stage-dependent sensitivity of zebrafish (*Danio rerio*) to estrogen exposure. Comp. Biochem. Physiol. C 139, 47–55.

Macqueen, D.J., Robb, D.H., Olsen, T., Melstveit, L., Paxton, C.G., Johnston, I.A., 2008. Temperature until the 'eyed stage' of embryogenesis programmes the growth trajectory and muscle phenotype of adult Atlantic salmon. Biol. Lett. 4, 294–298.

Melendez, C.L., Mueller, C.A., 2021. Effect of increased embryonic temperature during developmental windows on survival, morphology and oxygen consumption of rainbow trout (*Oncorhynchus mykiss*). Comp. Biochem. Physiol. A 252, 110834.

Mueller, C.A., 2018. Critical windows in animal development: interactions between environment, phenotype and time. In: Burggren, W.W., Dubansky, B. (Eds.), Development, Physiology and Environment: A Synthesis. Springer, New York, pp. 60–91.

Mueller, C.A., Joss, J.M.P., Seymour, R.S., 2011a. Effects of environmental oxygen on development and respiration of Australian lungfish (*Neoceratodus forsteri*) embryos. J. Comp. Physiol. B. 181, 941–952.

Mueller, C.A., Joss, J.M.P., Seymour, R.S., 2011b. The energy cost of embryonic development in fishes and amphibians, with emphasis on new data from the Australian lungfish, *Neoceratodus forsteri*. J. Comp. Physiol. B. 181, 43–52.

Mueller, C.A., Eme, J., Manzon, R.G., Somers, C.M., Boreham, D.R., Wilson, J.Y., 2015. Embryonic critical windows: changes in incubation temperature alter survival, hatchling phenotype and cost of development in Lake whitefish (*Coregonus clupeaformis*). J. Comp. Physiol. B. 185, 315–331.

Munnelly, R.T., Windecker, C.C., Reeves, D.B., Rieucau, G., Portier, R.J., Chesney, E.J., 2021. Effects of short-duration oil exposure on bay anchovy (*Anchoa mitchilli*) embryos and larvae: mortality, malformation, and foraging. Aquat. Toxicol. 237, 105904.

Pacheco-Carlón, N., Guerrero-Tortolero, D.A., Cervantes-Montoya, L.B., Racotta, I.S., Campos-Ramos, R., 2021. The effects of constant and oscillating temperature on embryonic development and early larval morphology in longfin yellowtail (*Seriola rivoliana* Valenciennes). Aquac. Res. 52, 77–93.

Pickford, G.E., Atz, J.W., 1957. Physiology of the Pituitary Gland of Fishes. New York Zoological Society, New York.

Ribeiro, L., Hubert, F.N., Rodrigues, V., Rojas-Garcia, C., Dinis, M.T., 2022. Understanding fish larvae's feeding biology to improve aquaculture feeding protocols. In: Oceans. vol. 3. MDPI, pp. 94–113.

Robertson, C.E., Wright, P.A., Köblitz, L., Bernier, N.J., 2014. Hypoxia-inducible factor-1 mediates adaptive developmental plasticity of hypoxia tolerance in zebrafish, *Danio rerio*. Proc. R. Soc. Lond. B 281, 20140637.

Sifa, L., Mathias, J., 1987. The critical period of high mortality of larvae fish—a discussion based on current research. Chin. J. Oceanol. Limnol. 5, 80–96.

Skorupa, K., Mendonça, R.C., Araújo-Silva, S.L., Santana, D.d.S., Pinto, J.R.d.S., Tsuzuki, M.Y., 2022. The influence of salinity on egg incubation and early larval development of the flameback angelfish *Centropyge aurantonotus*. Aquac. Res. 53, 6616–6625.

Swearer, S.E., Treml, E.A., Shima, J.S., Hawkins, S.J., Allcock, A.L., Bates, A.E., Firth, L.B., Smith, I.P., Swearer, S.E., Todd, P.A., 2019. A review of biophysical models of marine larval dispersal. In: Oceanography and Marine Biology: An Annual Review. vol. 57. CRC Press, Boca Raton, pp. 325–356.

van Aerle, R., Pounds, N., Hutchinson, T.H., Maddix, S., Tyler, C.R., 2002. Window of sensitivity for the estrogenic effects of ethinylestradiol in early life-stages of fathead minnow, *Pimephales promelas*. Ecotoxicology 11, 423–434.

Yamamoto, T., 1962. Physiology of fertilization in fish eggs. Int. Rev. Cytol. 12, 361–405.

Yin, M., Blaxter, J.H.S., 1987. Temperature, salinity tolerance, and buoyancy during early development and starvation of Clyde and North Sea herring, cod, and flounder larvae. J. Exp. Mar. Biol. Ecol. 107, 279–290.

Chapter 14

DEVELOPMENT: EGGS AND LARVAE☆

J.H.S. BLAXTER

Chapter Outline

I. INTRODUCTION

Fish eggs and larvae provide a relatively untapped source of biological material, increased by the recent improvements in techniques for rearing marine species. Apart from their intrinsic interest, experimentally based information

☆This is a reproduction of a previously published chapter in the Fish Physiology series, "1969 (Vol. 3)/Development: Eggs and Larvae: ISBN: 978-0-12-350403-6; ISSN: 1546-5098".

Fish Physiology, Vol. 40A. https://doi.org/10.1016/bs.fp.2023.09.011

on these early stages is required for further progress in the advancing fields of fish culture and fisheries research. General textbooks on ichthyology such as those of Lagler *et al.* (1962), Nikolsky (1963), Norman (1963), and Marshall (1965) and on reproduction in fish by Breder and Rosen (1966) provide both general and some detailed information. Identification of eggs and larvae, apart from specialist papers, is possible from publications of Ehrenbaum (1909), D'Ancona *et al.* (1931–1933), through the current series of plankton sheets issued by the International Council for the Exploration of the Sea, and with the help of the extensive bibliographies by Dean (1916) and Mansueti (1954).

Most species of fish pass through a larval stage before assuming the adult form at metamorphosis. Sometimes the newly hatched fish is called a "prolarva" (or alevin in salmonids) until the yolk is resorbed, and then a "postlarva" (or fry), The term "larva" is used here for all stages to metamorphosis in marine fish, although alevin and fry may be used when referring to salmonids or other freshwater groups.

II. THE PARENTAL CONTRIBUTION

Apart from the more obvious genetical effects on differentiation, rate of development, body form, size, and behavior, the parents, and especially the female, have an important influence on the viability of the offspring both on a species and individual level, in terms of (a) the conditions for incubation, (b) fecundity, and (c) egg size.

A. Conditions for Incubation

Differences of spawning season and time and of spawning sites and substrate mean that incubation can take place in a great variety of conditions which influence the early development and physiology of the offspring.

1. *Eggs Single, with No Parental Care*

(a) Buoyant, planktonic—most marine fish, e.g., gadids, clupeids, flatfish, and deep-sea fish.

(b) Nonbuoyant, loose or attached to substrate—a common freshwater characteristic, e.g., cyprinids, pike *Esox*, or in littoral species, e.g., blenny *Blennius*, bullheads *Cottus*, sand eels *Ammodytes*; also found in some marine species, e.g., herring *Clupea harengus*, capelin *Mallotus villosus*, catfish *Anarhichas*, and American flounder *Pseudopleuronectes americanus*. Tendrils for attachment are found in many oviparous elasmobranchs, in the hagfish *Myxine*, smelt *Osmerus*, saury *Scomberesox*, and flying fish *Exocoetus*.

(c) Nonbuoyant, buried in sand or gravel—many salmonids, grunion *Leuresthes tenuis*, and lamprey *Petromyzon*; in peat or mud *Aphyosemion* and *Cynolebias* where the eggs undergo diapause during the dry season.

2. *Eggs Single, Special Environments*

The bitterling *Rhodeus amarus* lays eggs in the gills of the freshwater mussel and the lumpsucker *Careproctus* under the carapace of the Kamchatka crab.

3. *Eggs Single, with Parental Care*

(a) No nest, but eggs protected—found in many littoral forms, e.g., the bullheads Cottidae, blennies Blenniidae and gobies Gobiidae.
(b) Nests, often with parental protection and ventilation—also found in littoral species, e.g., blenny *Ictalurus*, sticklebacks *Gasterosteus*, and in other freshwater species such as sunfish Centrarcidae, bowfin *Amia*, lungfish *Protopterus* and *Lepidosiren*, and in the Cichlidae. Bubble nests giving good aeration are found in tropical or swamp species, e.g., Siamese fighting fish *Betta splendens*.
(c) Parents carrying eggs—sea horses *Hippocampus* and pipefish *Syngnathus* have brood pouches and the sheat fish *Platystacus* a specially modified area of "spongy" skin. Marine catfish Ariidae, cardinal fish Apogonidae and *Tilapia* are mouth brooders, and *Tachysurus* incubates the eggs intestinally.
(d) Ovovipiparity and viviparity (see Section IV, C)—elasmobranchs include picked dogfish *Acanthias*, smooth hound *Mustelus vulgaris*, electric ray *Torpedo*, stingray *Trygon*, and the nurse hound *Mustelus laevis*. Teleosts include redfish *Sebastes*, *Heterandria*, *Anableps*, poeciliids such as *Xiphophorus* and half beaks *Hemirhampus*.

4. *Eggs Massed*

Angler fish *Lophius* and yellow perch *Perca flavescens* have massed but unprotected eggs; in the lumpsucker *Cyclopterus* and butterfish *Blennius pholis* the eggs are protected by the male.

B. Fecundity and Egg Size

In higher latitudes the spawning season is often short and the eggs are liberated over a brief period of perhaps hours (clupeids) or over periods of some days, probably at certain times of the day or night (flatfish and gadids). Where the seasons are less marked spawning may occur over a much longer period or be intermittent throughout the year, especially where the time between generations is only a matter of weeks or months. Fecundity may be considered as the number of eggs produced in one year by a female although this may be very difficult to determine where spawning is protracted.

Some examples of fecundity, egg size, and length at hatching are given in Table I. In general, fecundity is high where the eggs are liberated into open marine waters; it is lower in freshwater species and where there is parental care. There is also a strong tendency for fecundity and egg size to be inversely related.

Table I Fecundity (Eggs/Female/Year), Egg Diameter, and Length at Hatching

Species	Common name	Fecundity	Diameter (mm)	Length at hatching (mm)
Molva molva	Ling	$20–30 \times 10^6$	1.0–1.1	3.0–3.5
Gadus morhua	Cod	$20–90 \times 10^5$	1.1–1.6	4.0
Melanogrammus aeglefinus	Haddock	$12 \times 10^3 - 30 \times 10^5$	1.2–1.7	4.0–5.0
Pleuronectes platessa	Plaice	$16 \times 10^3 - 35 \times 10^4$	1.7–2.2	6.0–7.0
Solea solea	Sole	15×10^4	1.0–1.5	3.2–3.7
Scomber scombrus	Mackerel	$35–45 \times 10^4$	1.0–1.4	3.0–4.0
Clupea harengus	Herring	$50 \times 10^2 - 20 \times 10^4$	0.9–1.7	5.0–8.0
Clupeonella delicatula	Kilka	$10–60 \times 10^3$	1.0	1.3–1.8
Salmo salar	Salmon	$10^3–10^4$	5.0–6.0	15.0–25.0
Osmerus eperlanus	Smelt	$50 \times 10^2 - 50 \times 10^3$	0.9	4.0–6.0
Acipenser sturio	Sturgeon	$80 \times 10^4 - 24 \times 10^5$	?	9.0
Cyprinus carpio	Carp	$18–53 \times 10^4$	0.9–1.6	4.8–6.2
Acanthurus triostegus	Convict surgeon fish	40×10^3[a]	0.7	1.7
Oryzias latipes	Medaka	20–40[a]	1.0–1.3	4.5–5.0
Scyliorhinus caniculus	Spotted dogfish	2–20	65.0 (length)	100.0
Lebistes reticulatus[b]	Guppy	10–50[a]	—	6.0–10.0
Zoarces viviparus[b]	Blenny	20–300	—	35.0–40.0
Sebastes viviparus[b]	Redfish	$12–30 \times 10^3$	—	5.0–8.0
Mustelus mustelus[b]	Smooth hound	10–30	—	250.0
Squalus acanthias[b]	Spur dogfish	2–7	24–32	240–310

[a]Number per spawning, which may be repeated often in one year.
[b]Viviparous or ovoviviparous.

Apart from enormous interspecific differences, there are also considerable variations of fecundity within a species. Many authors have found that fecundity increases with length, weight, or age (see Fig. 1) the relationship usually being of the form $F = aL^b$, where F is fecundity, L is length, and a and b are constants. Year-to-year differences resulting almost certainly from environmental effects are also well established. For instance, sea temperature was correlated by Rounsefell (1957) with the fecundity of pink salmon, *Oncorhynchus gorbuscha*, higher temperatures apparently resulting in lower fecundities. (Here the effect of temperature on growth is a complication.) Bagenal (1966) reported density-dependent factors operating in Scottish flatfish, high densities being correlated with low fecundity. Anokhina (1960) found that

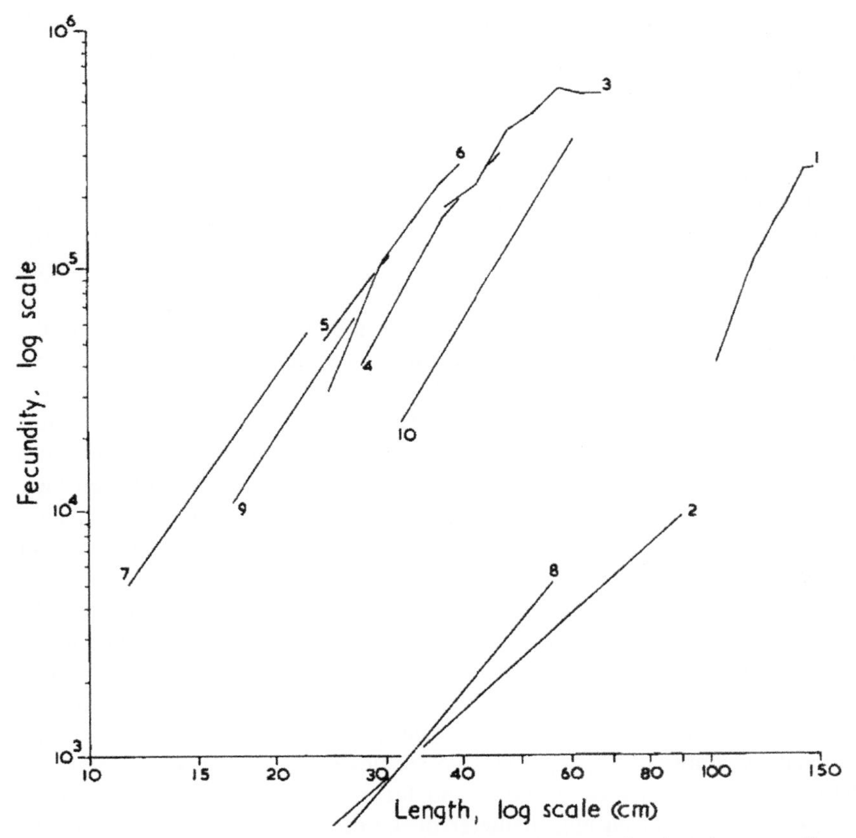

Fig. 1. The relationship between fecundity and length within a species. 1. *Acipenser stellatus* (Nikolsky, 1963); 2. *Salmo salar* (Pope *et al.*, 1961); 3. *Cyprinus carpio* (Nikolsky, 1963); 4. *Pleuronectes platessa* (Clyde) (Bagenal, 1966); 5. *Clupea harengus* (northern North Sea) (Baxter, 1959); 6. *Melanogrammus aeglefinus* (see Parrish, 1956); 7. *Osmerus eperlanus* (Lillelund, 1961); 8. *Salvelinus fontinalis* (Vladykov, 1956); 9. *Sardinops caerulea* (MacGregor, 1957); and 10. *Sebastes marinus* (Faroe Island) (Raitt and Hall, 1967).

fecundity in Baltic herring could be related to feeding conditions, high fat content of the female being related to high fecundity. Experimental studies by D. P. Scott (1961) indicated in rainbow trout, *Salmo gairdneri*, that an insufficient diet caused a reduction in egg number: in the guppy *Lebistes* fewer offspring were also produced when the females were kept on short rations (Hester, 1963). Extensions of this type of work are badly needed.

Differences within a species resulting from latitude, area, race, or season are no doubt interconnected. Considerable differences of this type were reported for plaice, *Pleuronectes platessa* (Bagenal, 1966), for herring (Baxter, 1959; Kändler and Dutt, 1958), for species of *Oncorhynchus* (Rounsefell, 1957), and for *Salmo salar* (Pope *et al.*, 1961). An interesting characteristic of certain species is a difference in fecundity of the left and right ovary. The left gonad contains more eggs in *Oncorhynchus* (Rounsefell, 1957), *Salmo* (Pope *et al.*, 1961), and smelt *Osmerus* (Lillelund, 1961). The significance of this is not clear.

Egg size varies at the interspecific level (see Table I), with larger eggs being especially associated with freshwater species like the salmonids or where fecundity is very low, as in many elasmobranchs. At the intraspecific level, differences of egg size as a result of area or river were noted by Rounsefell (1957) in *Oncorhynchus*, *Salmo*, *Cristivomer*, and *Salvelinus* species. In *Salmo salar* (Pope *et al.*, 1961) there are differences in egg diameter related to length, fecundity, and river. In *Tilapia* egg weight may increase 2–4 times or even more depending on the size of the female (Peters, 1963); in the flounder, *Platichthys flesus*, large females, or females from low salinities, have larger eggs (Solemdal, 1967). Larger eggs are also found in larger females of the spur dogfish *Acanthias* (Templeman, 1944). There are differences in average dry weight of the order of four times among the various races of herring (Hempel and Blaxter, 1967) and small differences between very young first spawners and repeat spawners. In two darter *Etheostoma* species with long spawning seasons, the egg diameter tends to be greater in the cooler winter months (Hubbs *et al.*, 1968).

The connection between egg size and fecundity in related species may be correlated with the conditions for incubation. For example, *Tilapia tholloni*, a substrate brooder, has 500–3000 eggs depending on length. On the other hand, *T. mossambica* and *T. macrocephala*, which are mouth brooders, have less than 500 eggs that are considerably greater in weight (Peters, 1963). Garnaud (quoted by Smith, 1957) reported two species of *Apogon*, one, *A. imberbis*, with an egg diameter of 0. 5 mm and fecundity of 22,000, and the other, *A. conspersus*, with an egg diameter of 4.5 mm and fecundity of 150. The relationship between egg size and egg weight *within* a species is shown for the different races of herring in Fig. 2. The winter–spring spawners have a low fecundity and large eggs, an adaptation to poor food for the young, but a low predator population. In summer–autumn conditions fecundity is high and egg size low, presumably an adaptation to good food supplies and many

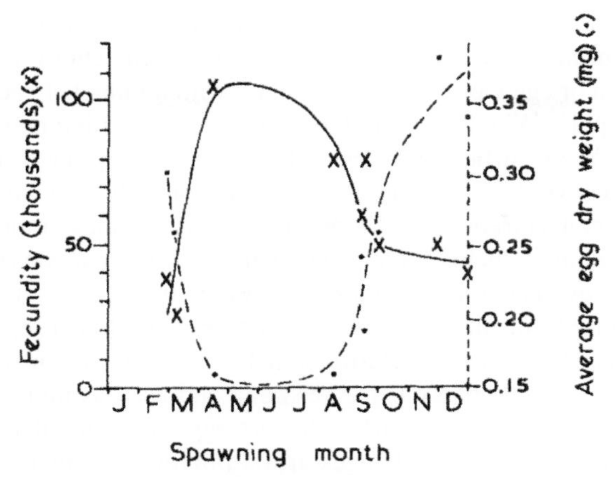

Fig. 2. Fecundity and average egg dry weight in different races of herring (data from Baxter, 1959; Hempel and Blaxter, 1967; Parrish and Saville, 1965; Kandler and Dutt, 1958).

predators. Intraspecific differences in fecundity and egg size deserve further study in terms of a link between ecological conditions and the physiology of maturation of the ovary.

A general pattern seems to emerge of marine fish with many, small buoyant eggs, a short incubation period, and vulnerable larvae. Freshwater fish have larger demersal eggs, a long incubation period and larger, less vulnerable larvae, while the littoral forms exhibit protective devices to prevent losses in this particularly difficult environment. The initial conditions of development determined by the genotype of the parent and reproductive behavior must have a considerable influence on the viability of the young and its physiology, in that environmental conditions such as temperature affect the speed of development, salinity presents problems of osmoregulation, and oxygen must be obtained for respiration. The egg is susceptible and yet cannot make defensive responses to mechanical shock, drift by current, toxins, light, and predators. The larva should live in conditions where food can be obtained and protective behavioral devices can be practiced.

III. EVENTS IN DEVELOPMENT

A. Fertilization

The physiology of fertilization in fish, with special reference to the extensive Japanese work, has been fully reviewed by Yamamoto (1961) from which much of the present account is taken.

There is some evidence for the action of gamones in *Lampetra* and in teleosts; these activate the sperm and serve as chemical attractants toward the egg, while other gamones are known to paralyze or agglutinate sperm. In the bitterling species *Acheilognathus* and *Rhodeus* sperm aggregation and activity have been noted in the micropyle region of the chorion.

The chorion or egg case is relatively tough with a funnel-shaped micropyle at the animal pole. Within the chorion a plasma or vitelline membrane [also called a pellicle or surface gel layer (Trinkaus, 1951)] surrounds the yolk and cytoplasm (ovoplasm) of the egg. Fertilization, which requires the presence of small concentrations of Ca or Mg ions, is normally monospermic in teleosts, the micropyle being too narrow to allow more than one sperm to pass at a time. The ovoplasm and chorion separate as the egg is activated by the sperm and a plug forms in the micropyle, further sperm being rejected. Where polyspermy occurs, as in some elasmobranchs, only one sperm fuses with the egg nucleus, the rest probably being resorbed and used as nutrient. Removal of the chorion permits polyspermy in teleost eggs; it seems that polyspermy is usually prevented by rapid changes at the micropyle, rather than over the egg cortex. In salmonids water activation (not to be confused with activation by sperm) takes place (e.g., see Prescott, 1955). When the egg is released into hypotonic solutions like river water the vitelline membrane becomes opaque and there are changes in its permeability. If sperm are not immediately available these changes may also affect fertilizability.

Following fertilization, the prominent alveoli in the egg cortex of salmonids, acipenserids, and lampreys disappear. In the medaka, *Oryzias latipes*, these alveoli break down progressively from the animal pole. The separation of the cortex from the chorion leads to the appearance of the perivitelline space. The chorion is permeable to water and small molecules, but larger molecules of a colloidal nature are retained in the perivitelline fluid. In *Oryzias* and *Lampetra* these colloids maintain an osmotically based tension within the chorion. It seems likely that the colloid is derived from polysaccharide material in the cortical alveoli so that the formation of the perivitelline space is partly owing to a decrease in volume of the ovoplasm, as the alveoli release colloid, and partly due to an osmotic distension of the chorion. According to Ginsburg (1961), polyspermy is blocked in sturgeon and trout eggs by the discharge of the cortical alveoli in the micropylar region. Chemicals, such as urethane, which cause polyspermy apparently retard the secretions of the alveoli, while removal of the perivitelline fluid in trout eggs allows the penetration of many sperm.

The chorion also hardens (see Section VIII) thus protecting the embryo in the early, more vulnerable stages. Probably the inner layer of glycoprotein is mainly responsible for this and it has been suggested that the alveolar colloid, Ca ions, phospholipids, or hardening enzymes also play a part. In salmonids, Zotin (1958) reported hardening of the chorion because of an enzyme in the perivitelline fluid; Ca ions affect the enzyme rather than the chorion itself. Ohtsuka (1960) considered that a phospholipid was liberated from the cortex (not from the alveoli) in *Oryzias* eggs. In *Fundulus* the chorion hardened with oxidizing agents but not with reducing agents. The soft chorion appeared to be impregnated with protein containing SH groups. Hardening resulted from oxidation of SH to SS groups by means of aldehydes produced from polysaccharides with α-glycol groups. Zotin distinguished between the initial enzyme action and subsequent hardening processes which last much longer and where the enzyme is no longer functioning. Thus the initial enzyme reaction is blocked when Ca ions are bound by citrate or oxalate or by the use of NaCl or other chlorides. Later hardening is not susceptible to many of these factors.

The eggs of *Oryzias*, *Gasterosteus*, and *Lampetra* can be activated by pricking. Other activating agents are surface-active chemicals and lipid solvents (perhaps emulsifying the cytoplasm at the animal pole) and thermal shock, electric fields, ultraviolet light, and high frequency vibrations. This type of artificial parthenogenesis usually leads to irregular cleavage, but stringent precautions are needed to prevent contamination with sperm in such experiments.

Of interest is the ability of eggs and sperm to retain their fertilizability after leaving the parent. According to Yamamoto (1961) fish eggs lose this capacity after a very short time, but this can be increased if they are retained in isotonic Ringer's solution. While this is true of some freshwater eggs, presumably as a result of water activation, in seawater the capacity for fertilization is retained for much longer—certainly for hours in the herring. Nikolsky (1963) states that sperm motility is short lived where spawning takes place in fast-flowing water, for example, 10–15 sec in *Oncorhynchus*. In slower flows, sturgeon sperm is motile for 230–290 sec, and in the sea herring sperm may be motile for hours or days. Observations of short-lived activity are difficult to make; thus, some of these figures must be considered approximate.

Storage of gametes is a useful technique in fish farming to allow controlled fertilizations in the laboratory and to obviate the need for transporting eggs in the susceptible pregastrulation stages. Salmonid gametes are best stored dry below 5°C (e.g., Barrett, 1951, Withler and Morley, 1968); those of the herring may be held in buffered egg-yolk diluents, but are also best kept dry at about 4°C (Blaxter, 1955). While this permits storage for, at most, a few days long-term techniques are also possible. Herring sperm, but not eggs, were kept for some months in a diluent consisting of 12.5% glycerol in 3% salt solution (diluted seawater) at −79°C and crosses made successfully between spring and autumn spawning races (Blaxter, 1953, 1955; Hempel and Blaxter, 1961). Sneed and Clemens (1956) also succeeded in holding out-of-season carp sperm immotile for 30 days at 3°–5°C in frog Ringer. Carp sperm could also be frozen and stored at −73°C in isotonic Ringer containing 6–12% glycerol with some survival when thawed after 60 hr of storage. More recently, Truscott *et al.* (1968) have shown that salmon sperm can be stored for 1–2 months at temperatures of −3° to −4.5°C using diluents such as 5% ethylene glycol or 5% dimethyl sulfoxide, retaining 70–80% fertility. Horton *et al.* (1967) obtained alevins from salmon eggs fertilized with sperm frozen in liquid nitrogen with dimethyl sulfoxide as a protecting agent, but the fertility rate was low. Hoyle and Idler (1968) have also obtained fertile salmon sperm after storage in liquid nitrogen using ethylene glycol with added lactose or serine. Slow freezing produced the best results. Mounib *et al.* (1968) succeeded in storing cod sperm for up to 60 days using 17–24% glycerol and temperatures of −79° and −196°C. Initial experiments suggest that faster cooling rates gave the better results with cod sperm.

B. Incubation (Fertilization to Hatching)

The progress of cleavage, formation of layers, and morphogenesis have been described in a number of standard textbooks such as Rudnick (1955), Waddington (1956) and Smith (1957), with Oppenheimer (1947) and Devillers (1961) stressing structural changes from the viewpoint of experimental embryology. More detailed information is limited mainly to freshwater species like the trout *Salmo trutta*, killifish *Fundulus*, medaka *Oryzias*, goldfish *Carassius*, and to the dogfish. New (1966) gives information on the problems of culturing *Fundulus*, *Oryzias*, and *Salmo* eggs for the purpose of experimental embryology.

Most fish eggs are round, although in the anchovy *Engraulis* and bitterling *Rhodeus* they are ovoid, and in certain gobies pear-shaped. Most species have telolecithal eggs with yolk more concentrated at the vegetative pole; some marine species have oil globules of varying size and number (see Simpson, 1956). Before fertilization the cytoplasm may be mixed with or separate from the yolk. The extent to which polarity exists at this stage has not been described in many species. After fertilization (as the cortical alveoli release colloid, the perivitelline space develops and the chorion hardens) cytoplasm migrates to the future blastodermal region, most of it arriving by the first cleavage. The remaining cytoplasm forms a "halo" or periblast under the blastoderm (see Fig. 3).

In lampreys cleavage is holoblastic but with the formation of micro- and macro-meres. In hagfish, elasmobranchs, and teleosts it is meroblastic. Other groups such as bowfin *Amia*, gar *Lepidosteus*, and sturgeon *Acipenser* have intermediate features. With some variation, the meroblastic group possess a blastodermal cap of cells at the animal pole after the initial stages of cleavage. In *Fundulus* the surface gel layer overlying the blastoderm, being sticky on its inner surface, serves to hold the outer blastomeres together (Trinkaus, 1951). Usually cleavage is not complete, and in the deeper layers the periblast becomes syncytial and is involved in mobilizing the yolk reserves. There are substantial cohesive forces between the developing blastomeres and the surrounding periblast which are important in the subsequent morphogenetic movements.

The blastoderm now commences to thin and overgrow the yolk (epiboly) and at the same time invaginate at its periphery (Fig. 3). The periblast seems closely connected with this spreading tendency of the blastoderm, which at the junction may be thickened to form a germ ring. The syncytial periblast

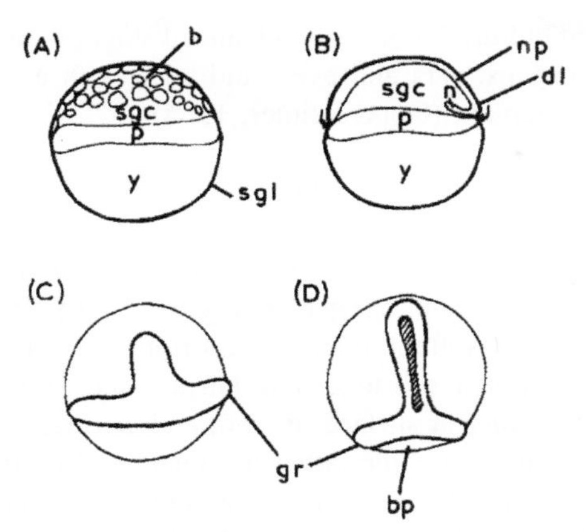

Fig. 3. (A) Transverse section of early blastula showing adhesion of blastomeres to the surface gel layer and attachment of blastoderm to the periblast at the periphery only (*Fundulus*, after Trinkaus, 1951). (B) Sagittal section of later blastula showing gastrulation; epiboly shown by arrows. (C) and (D) Surface view of eggs in later stages of gastrulation. Key: b, blastoderm; bp, blastopore; dl, dorsal lip of blastopore; gr, germ ring; n, notochord; np, neural plate; p, periblast; sgc, subgerminal cavity; sgl, surface gel layer; and y, yolk.

seems to have the property of autonomous spreading, and it is likely that cell proliferation in the blastoderm is relatively unimportant. Devillers (1961) suggests that the periblast acts as an intermediary between two "non-wettable" components—the blastoderm and yolk. As epiboly proceeds the blastopore contracts as does the surface gel layer over the yolk. In *Fundulus* this layer probably solates and passes inward (Trinkaus, 1951). The form of the developing germ then seems to be controlled by a balance of the forces of adhesion of the deeper blastomeres with each other and the syncytium, of tension in the surface (yolk) gel layer and of contractility of the periblast.

The embryonic axis is laid down by a process of convergence and concentration in relation to the dorsal lip of the blastopore, the quantity of yolk present having some influence on the time at which this event occurs. Presumptive areas have been mapped in the early gastrulae of some species and show considerable variation. In earlier stages the eggs seem to be of the regulatory type. In *Fundulus* the 2-cell and 4-cell stage can withstand a loss of half the number of blastomeres. The embryos of *Carassius* can be divided at the 8-cell stage, each part sometimes giving a normal embryo. Up to the 16-cell stage two embryos can be fused resulting either in twinning or an oversized single embryo. Removal of the yolk from the blastoderm before a critical stage is reached (8 cell stage in *Carassius*, 32 cell stage in *Fundulus*, blastula in *Salmo*) brings development to a halt. Incomplete removal of the yolk before the critical stage may, however, not prevent further development. It is likely that organizer substances rather than nutrient material diffusing from the yolk are more important. At later stages even the embryonic shield of *Fundulus* may be isolated and cultured to a fairly advanced stage, with the development of primitive axial organs, ears and eyes, and even with cardiac contractions and independent movement (Oppenheimer, 1964).

C. Hatching

The time to hatching is both a specifically and environmentally controlled character with temperature and oxygen supply exerting a considerable effect. Hatching results from a softening of the chorion (see Fig. 15) because of enzymic or other chemical substances which are secreted from ectodermal glands usually on the anterior surface or from endodermal glands in the pharynx. In the sturgeon *Acipenser* the latter are innervated by the palatine nerve (Ignat'eva, quoted by Deuchar, 1965). The activity of the larvae, which may be enhanced by increase of temperature or light intensity or by reduction of oxygen tension, assists in breaking through the chorion.

The biochemical aspects of hatching are dealt with by Hayes (1949), Smith (1957), and Deuchar (1965). The enzymes have apparently been identified in a number of species, but certainly in salmonids Hayes' work shows there is doubt about their mode of action. The chorion, which resists digestion by trypsin and pepsin, appears to be of "pseudo-keratin." Kaighn (1964) measured the amino acid and carbohydrate components in *Fundulus*. Cystine comprised only 1%, compared with 12% in keratin. It is therefore unlikely that disulfide links play a role in stabilizing chorionic protein as they probably do keratin. The hatching enzyme works best under alkaline conditions, pH 7.2–9.6 and temperatures of 14°–20°C having been reported as optima. Very little hydrolysis takes place and Hayes speculates that the enzyme may be a reducing agent which liquefies the chorion. In *Oryzias* the enzyme is probably a tryptase. In *Fundulus*, Kaighn (1964) obtained purified chorionase and concluded that digestion of the chorion was mainly a proteolytic process, although he could not duplicate its action with other proteases. Whatever the mode of action of the enzyme, it is likely that a considerable part of the nutrient material in the chorion can be utilized by the embryo via the perivitelline fluid and the losses at hatching may not be too serious (Smith, 1957).

D. The Larva

At hatching the larva is usually transparent with some pigment spots of unknown function. Notochord and myotomes are clear with usually little development of cartilage or ossification in the skeleton. A full complement of fins is rarely present, but a primordial fin fold is well developed in the sagittal plane. The mouth and jaws may not yet have appeared, and the gut is a straight tube. Although the heart functions for a considerable period before hatching, the blood is colorless in the majority of species and the circulation and respiratory systems poorly developed. The yolk sac is relatively enormous with, presumably, hydrodynamic disadvantages. Pigmentation of the eyes is very variable, but where the eye is not functioning at hatching it very soon develops. The kidney is usually pronephric with very few glomeruli. Very little is known about the endocrine glands, gonads, and other organs of the body cavity at such an early stage.

As the yolk is resorbed, the mouth begins to function, the gut and the eyes develop further, and the larva becomes fitted for transfer to sources of external food. One of the earlier systems to develop is that responsible for locomotion and support, the primordial fin being fairly soon replaced by median fins and the skeleton laid down. This is one of the better known aspects of later development because it is a system less easily damaged in such delicate organisms and because of the importance of meristic characters (see Section IX) in racial studies of fish. Branchial replaces cutaneous respiration as the gill arches and filaments appear. The swim bladder may or may not be present during the larval phase. It is possible that this and the eyes, which are potentially dangerous in making the transparent larva visible, are silvered in such a way as to render them inconspicuous.

E. Metamorphosis

A clear change or metamorphosis from the larval to adult form is to be found in many species. In others there may be a number of less marked metamorphoses, e.g., in salmonids and eels. The most obvious signs are the laying down of scales and other pigmentation and often the first appearance of hemoglobin in the circulation. The swim bladder and lateral line may also develop first at this stage. In flatfish there is rotation of the optic region of the skull and the change in the normal orientation of the body so that they eventually come to lie on one side. There are often concomitant changes in distribution and behavior such as schooling. Barrington (1961) gives detailed consideration to the physiological changes associated with metamorphosis in salmonids, eels, and the lamprey, especially from the aspect of thyroid activity and osmoregulatory functions.

The time to reach metamorphosis may be a matter of days in tropical species, a few weeks or months in the majority of fish from temperate latitudes, or periods of years in the sturgeon *Acipenser* and eel *Anguilla*. It is controlled not only genetically but also by temperature and food supply, which may affect the rate of growth, and possibly by social (hierarchical) factors as well.

F. Timing

To give some idea of the timing of the events just described, examples of the approximate duration of different stages under natural conditions are given in Table II. The modification of these times experimentally or by fluctuations in environmental conditions is discussed in the succeeding pages.

Table II Duration of Events in the Development of Some Species

Species	Weeks from fertilization to			Temp. range (°C)	Main author
	Hatch	*First feeding*	*Metamorphosis*		
Scomber scombrus (mackerel)	0.8–1.5	1.3–2.0	11–13	9–15	Sette (1943)
Roccus saxatilis (striped bass)	0.25	0.75	4–5	17	Mansueti (1958)
Osmerus eperlanus (smelt)	2.0–4.0	2.5–5.0	8–10	4–14	Lillelund (1961)
Acanthurus triostegus (convict surgeon fish)	0.15	0.7	?	26	Randall (1961)
Clupea harengus (Clyde herring)	2.5	3.5	16–18	7–10	Blaxter and Hempel (1963)
Pleuronectes platessa (plaice)	2.5	4.0	10–12	7–11	Ryland (1966)
Oryzias latipes (medaka)	1.5–2.0	2.0–2.4	Not clear cut	20–25	New (1966); T. Iwai, personal communication
Salmo salar (salmon)	20–22	26–28	Gradual	1–7	D. H. A. Marr, personal communication
Squalus acanthias (spur dogfish)	ca. 104	ca. 104	< 104	4–12	Templeman (1944)
Scyliorhinus caniculus (spotted dogfish)	24–32	28–36	28–36	4–12	Amoroso (1960)

IV. METABOLISM AND GROWTH

A. Rate of Development

Obvious specific differences in time to hatching may be masked by variations in ambient temperature, which is one of the most potent influences on rate of development. Detailed observations of temperature effects during development are scarce. Flüchter and Rosenthal (1965), however, showed between 3.5° and 9°C a doubling of heart rate in the embryos of the blue whiting, *Micromesistius poutassou*, and more rapid embryonic movements. The effect of temperature on time to hatching, a commonly used criterion, is shown in Fig. 4. Low temperatures retard hatching (see reviews by Battle, 1930; Hayes, 1949; Kinne and Kinne, 1962), and at a theoretical low temperature (the biological zero) the incubation period will be infinite. The product of incubation time (D) and temperature (T)—day-degrees—was originally thought to be constant, i.e.,

$$TD = k \tag{1}$$

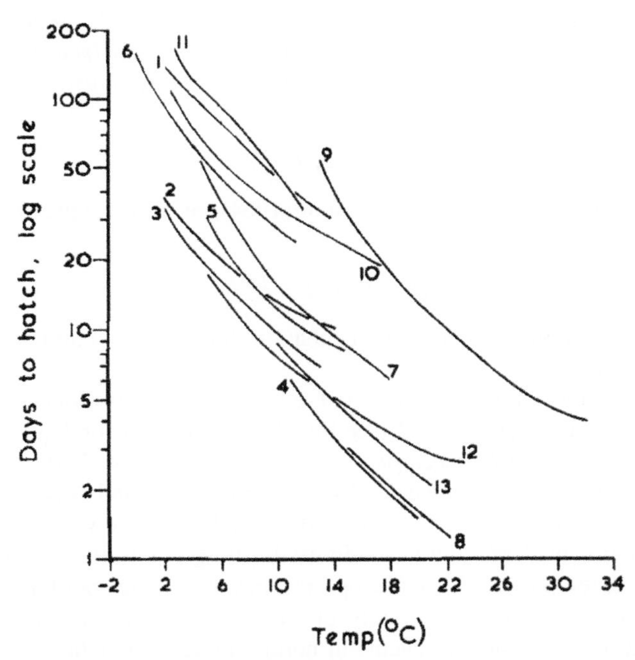

Fig. 4. Time from fertilization to hatching at different temperatures. 1. *Salvelinus fontinalis* (see Hayes, 1949); 2. *Pleuronectes platessa* (see Simpson, 1956); 3. *Gadus macrocephalus* (Forrester and Alderdice, 1966); 4. *Sardinops caerulea* (Lasker, 1964); 5. *Clupea harengus* (Blaxter and Hempel, 1963); 6. *Coregonus clupeaformis* (Price, 1940; Braum, 1964); 7. *Osmerus eperlanus* (Lillelund, 1961); 8. *Roccus saxatilis* (Mansueti, 1958); 9. *Cyprinodon macularius* (Kinne and Kinne, 1962); 10. *Salmo gairdneri* (*irideus*) (Garside, 1966); 11. *Salmo trutta* (*fario*) (Gray, 1928b); 12. *Enchelyopus cimbrius* (Battle, 1930); and 13. *Scomber scombrus* (Sette, 1943).

This was modified to use the temperature, not from zero, but from the biological zero (T_0), i.e.,

$$(T - T_0)D = k \tag{2}$$

There has been increasing criticism of the concept, for example, by Kinne and Kinne (1962) and Garside (1966), on various grounds. Plots of $1/D$ against T are *curvilinear* over wide ranges of temperature, simple linearity only applying over a narrow range. This invalidates the formulas given above. Furthermore, there may be inflections even of the curvilinear relationship at extreme temperatures. The biological zero, which is usually given between 0° and −2°C and most often around −1.5°C, may be below the freezing point of water or the body fluids themselves. In addition, abnormalities may occur at less extreme temperatures which are not necessarily lethal in the strict sense.

Improvements in describing the mathematical relation between D and T arise in later work. For example, Blaxter (1956) used the equation

$$(T - T_0)(D - D_0) = k \tag{3}$$

for development of *Clupea harengus*, where D_0 is the theoretical time to hatching at infinite temperature, not in itself a very satisfactory additional constant. Lasker (1964), working with the eggs of *Sardinops caerulea*, used the equation

$$D = aT^b \tag{4}$$

where a and b are constants, and Braum (1964), using the eggs of whitefish *Coregonus* and pike *Esox lucius*,

$$D = D_x + 1.26^{T_x - T} \tag{5}$$

where D_x is the minimum possible incubation time at the maximum permissible temperature T_x.

The van't Hoff values over different temperature ranges reflect the nonlinearity of plots of log D against T (see Fig. 4), the values being higher at lower temperatures. Thus in *Enchelyopus cimbrius* the Q_{10} varies between 6.5 and 1.5 over the temperature range 5°–23°C (Battle, 1930) and between 6.5 and 2.0 over the range 3°–18°C in herring (Blaxter, 1956). The value of this type of theoretical consideration may lie in establishing criteria for optimum conditions of development. Thus the optima may be where van't Hoff values lie between certain limits. Certainly the day-degree concept is useful as an approximation for predicting events in normal hatchery practice.

Other environmental factors influence the rate of development. Low salinities may accelerate or retard the time to hatching (see Kinne and Kinne, 1962; Forrester and Alderdice, 1966), while oxygen lack has a retarding effect on development, especially at higher temperatures (see Garside, 1966). Laale and McCallion (1968) found that the development of the zebra fish,

Brachydanio rerio, could be arrested before gastrulation by the use of the supernatant of homogenates produced from other zebra fish embryos. This arrest, which could be reversed, appeared to be an effect at the cellular level, the nuclei of the arrested embryos all being in interphase.

B. Yolk Utilization

The efficiency with which yolk is transformed to body tissue and the effect of the environment on utilization is important in that larger larvae may be expected to be stronger, better swimmers, less susceptible to damage, and less liable to predation. Efficiency at any time may be expressed as a percentage:

$$\frac{\text{dry weight increment of body}}{\text{dry weight decrement of yolk}} \times 100$$

More often efficiency is measured from fertilization to final yolk resorption (or to maximum weight attained on the yolk reserves). This is gross efficiency, i.e.,

$$\frac{\text{dry weight of final body}}{\text{dry weight of original yolk}} \times 100$$

or from fertilization to intermediate stages as

$$\frac{\text{dry weight of body}}{\text{dry weight of original yolk} - \text{dry weight of remaining yolk}} \times 100$$

or more precisely

$$\frac{\text{dry weight of body}}{\text{dry weight of body} + \text{dry weight of yolk used for maintenance}} \times 100$$

The difficulties of measuring efficiency by dry weight lie in the need for taking samples of an egg population at different stages with the accompanying problems of initial differences in egg weight. Utilization of material from the chorion or losses of excretory products are also difficult to allow for, as are the possibilities of uptake of organic matter from the environment (Pütter's theory, see Section V). Another serious problem when comparing, for example, environmental effects such as temperature on efficiency is the question of making dry weight measurements at "equivalent" stages (D. H. A. Marr, 1966). Both hatching and maximum weight (attained on the yolk) can be questioned for staging; hatching at different temperatures can result in larvae of quite different appearance, while full yolk utilization is often not complete when maximum weight is reached, some yolk remaining in the yolk sac (see Fig. 5). Furthermore, the disappearance of the yolk sac is no certain indication that all the yolk has been used as it may be present in storage spaces within the larval body, for example, in the subdermal spaces of cod and plaice larvae (Shelbourne, 1956). D. H. A. Marr (1966) adopted the ratio

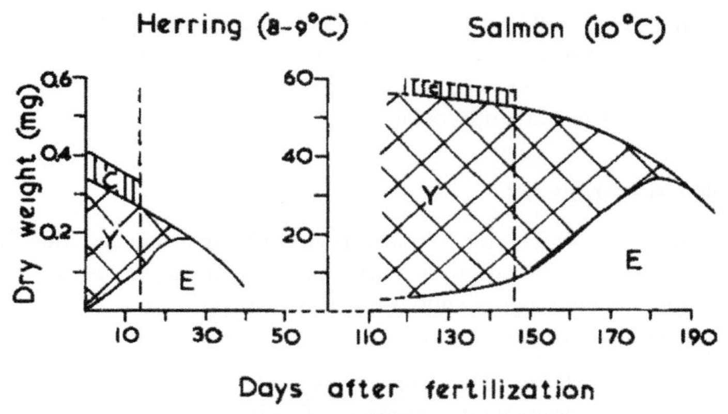

Fig. 5. The relative proportions of yolk (Y), embryo (E), and chorion (C) during the development of a small egg (herring: Blaxter and Hempel, 1963) and a large egg (salmon: D. H. A. Marr, 1966). The vertical dashed line represents hatching. Note the difference in the scale of the ordinates.

$$\frac{\text{dry weight of body}}{\text{dry weight of body} + \text{remaining yolk}} \times 100 \quad \text{(as percentage)}$$

as a criterion for *equivalent staging*, comparing in *S. salar* the efficiency of development at different temperatures between the 15% and 80% tages. Ryland and Nichols (1967) used the ratio

$$\frac{\text{rate of growth in length}}{\text{rate of yolk disappearance}} \times 100$$

for equivalent staging when comparing the efficiency of development at different temperatures during the yolk sac stage of *P. platessa.* The use of maximum weight as a stage for making comparisons still remains, however, a useful criterion and one with immediate meaning when deciding on the optimum conditions for hatchery practice.

Calculations of efficiency by various methods are given in Table III. Efficiency over the whole process of yolk utilization is mainly between 40% and 70% although clearly cumulative efficiency must decrease as growth proceeds and the maintenance requirements increase (Gray, 1928a). Experiments with temperature (see Fig. 6) indicate certain optima for maximum efficiency. Other influences on efficiency are the original egg weight at the intraspecific level in *C. harengus* (Blaxter and Hempel, 1966), and light conditions, contour of the substrate, and turnover of water in the photonegative alevin of *S. salar* living within the interstices of the spawning redd (D. H. A. Marr, 1965, 1967). Here highest efficiency is achieved under dark conditions, on a grooved substrate, with a rapid turnover of water.

A word of caution is required where larval feeding may occur well before final yolk resorption. High efficiency may result from low activity, a high proportion of yolk being used for growth; if this is reflected in low feeding activity it could be a very undesirable trait.

Table III Efficiency of Yolk Utilization

Species	Method	Stage	Temp. (°C)	Efficiency (%)	Author
Salmo trutta (trout)	Dry weights	Fertilization to 50–80 days	10	63	Gray (1926)
Salmo trutta (trout)	Dry weights	Fertilization to max. weight	15	56	Gray (1928a)
Salmo salar (salmon)	Dry weights	Hatching to 10 days after	0–16	42–59	Hayes and Pelluet (1945) (Fig. 6)
Salmo salar (salmon)	Calorific values	Fertilization to max. weight	10	41	Hayes (1949)
Salmo salar (salmon)	Special dry weight % index (see text)	15–80% (yolk sac stage)	7.6–14.3	64–70	D. H. A. Marr (1966) (Fig. 6)
Salmo gairdneri (rainbow trout)	Dry weight and metabolic criteria	Fertilization to max. weight	10	60	Smith (1957)
Fundulus heteroclitus (killifish)	Wet weights	Fertilization to max. weight	19.4–21.4	62	C. G. Scott and Kellicott (1916)
Silurus glanis (sheat fish)	(?) Wet weights	During yolk sac stage	?	66	Ivlev, quoted by Lasker (1962)
Sardinops caerulea (California sardine)	Calorific values and respiration	Fertilization to yolk resorption	14	79	Lasker (1962)
Clupea harengus (herring)	Dry weights	Fertilization to hatch	8–12	40–80	Blaxter and Hempel (1966) (Fig. 6)
		Fertilization to max. weight	8–12	50–65	
Pleuronectes platessa (plaice)	Rate of growth in length ÷ rate of yolk disappearance	During yolk sac stage	2.6–9.8	35–58	Ryland and Nichols (1967) (Fig. 6)

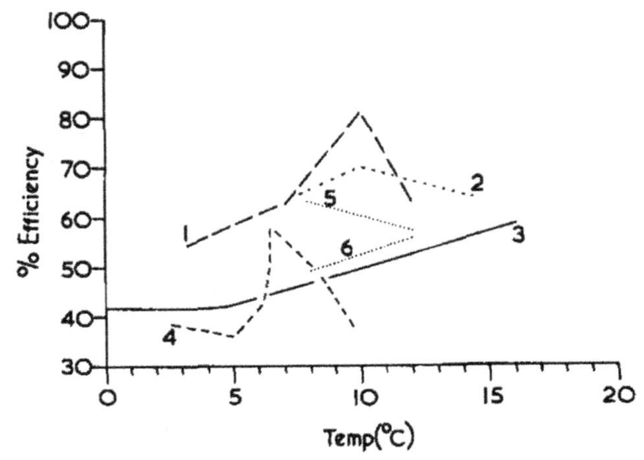

Fig. 6. Efficiency of development (see text and Table III), at different temperatures. 1. *Salmo trutta* (*furio*)—yolk sac period (see D. H. A. Marr, 1966); 2. *Salmo salar*—alevin stage (D. H. A. Marr, 1966); 3. *S. salar*—early alevin (Hayes and Pelluet, 1945); 4. *Pleuronectes platessa*—yolk sac larva (Ryland and Nichols, 1967); 5. *Clupea harengus*—small eggs, yolk sac period; 6. *Clupea harengus*—large eggs, yolk sac period (Blaxter and Hempel, 1966).

C. Viviparity

Amoroso (1960), who gives a comprehensive review of this subject, points out the rather indistinct barrier between ovoviviparity where the young develop within the female only on their yolk reserves, and viviparity, where the nutrient requirements are obtained from the mother. In the first place, there are a wide variety of methods of obtaining these nutrients: by absorption through simple external surfaces, by swallowing, or by "placental" connections. Perhaps all these may be considered as viviparous. In the second place, initial development may be ovoviviparous (on the yolk supply) with viviparity superimposed later. This is likely to be the case in the very early stages of *all* viviparous fish, but in the smooth dogfish, *Mustelus laevis*, and goodeid teleosts, for example, there is a change from one to the other rather later. Details of the functional morphology of viviparity are dealt with in the chapter by Hoar, this volume.

The changes of weight found during the development of various species (Amoroso, 1960) can be very striking. Some of the ovoviviparous ones, where maternal nutrients are scarce, show decreases of organic matter between the fertilized egg and final embryo, for example, of 23–34% in *Torpedo* spp. This gives an efficiency of 66–74% which puts this species very much in the same category as oviparous forms (Section IV, B) although it is rarely certain what nutrient is obtained from the mother. The long gestation period of 4–6 months with this order of efficiency suggests some nutrients are being absorbed from the oviduct. In other "ovoviviparous" species there may be gains in organic matter, e.g., of 356% in *Mustelus vulgaris* and 1628% in *Trygon*. In *Mustelus laevis*, sometimes more strictly called viviparous, the gain is 1064%, although Te Winkel (1963) reported a fall in weight of organic matter early in the development of a close relative, *M. canis*.

D. Biochemical Aspects

Much work has been done in the past on the larger salmonid eggs which can provide a greater bulk for analysis (see Hayes, 1949; Smith, 1957, 1958) or on whole ovaries (Lasker, 1962). Some more recent work which includes the use of isotopes, chromatography, and histochemistry is mentioned by Deuchar (1965) and Williams (1967).

1. *Water Relations*

The swelling of eggs, with the formation of the perivitelline space as a result of water uptake, is a general phenomenon signifying fertilization. The chorion at fertilization is permeable to water and also to urea, glucose, salts, and certain dyes. It seems likely that the colloidal material liberated from the cortical alveoli cannot escape through the chorion and creates an osmotic pressure which draws in water. This effect can be inhibited by high osmotic pressure in the outside medium. The chorion then hardens, a process taking a matter of a few hours (see Fig. 15) and the osmotic forces become matched by the resistance of the chorion. In hypertonic solutions the eggs of salmonids and of *Mullus barbatus* lose water only from the perivitelline space and not from the embryo (see Zotin, 1965). Some loss does, however, occur in acipenserids. The use of D_2O on water-activated and early fertilized eggs of *Oncorhynchus tshawytscha* seems to confirm the view that only the perivitelline space is penetrable by water (Prescott, 1955). Subsequent use of $^{3}H_2O$, $^{22}NaCl$, and $Na^{131}I$ on the water-activated eggs of *Salmo gairdneri* has shown a definite but limited permeability of the vitelline membrane to anions, cations, and water (Kalman, 1959). Unfortunately, this was not done on fertilized eggs. Recently, however, Potts and Rudy (1969) have confirmed with $^{3}H_2O$ that the vitelline membrane of fertilized eggs of *S. salar* has a high permeability before laying and during water hardening. Subsequently permeability is low until the eyed stage. The use of ^{24}Na (Rudy and Potts, 1969) showed that sodium exchange is confined initially to the perivitelline fluid but accumulation within the embryo occurs during the eyed stage. Terner (1968) reported that during the eyed stage of *Salmo gairdneri* external substrates such as ^{14}C-labeled pyruvate and acetate were apparently taken up and metabolized as judged by the presence of $^{14}CO_2$ in the respiratory CO_2. Wedemeyer (1968) also found that ^{65}Zn was taken up by developing eggs of coho salmon, *O. kisutch*. Almost all was bound to the chorion, but 26% was found in the perivitelline fluid, 2% in the yolk, and 1% in the embryo. Mounib and Eisan (1969) found that both ^{14}C-labeled pyruvate and glyoxylate could be utilized in the form of an exogenous substrate by salmon (*S. salar*) eggs. Lactate was produced and any carbon atom in these compounds could be incorporated by the eggs into organic acids, lipids, nucleic acids or proteins. The formation of ^{14}C-amino acids indicated the presence of an active transaminase system.

The use of freezing point measurements on the yolk of herring and plaice eggs (Holliday and Jones, 1965, 1967) gives an alternative picture. In the herring, which has a demersal egg, the yolk is not regulated osmotically until after gastrulation when it is covered by cells. However, in the pelagic floating egg of the plaice, regulation occurs from fertilization, indicating the ability of the vitelline membrane to osmoregulate. This seems necessary from the buoyancy aspect.

If Gray's (1926) measurements of wet and dry weights of the embryo and yolk of trout are generally true of salmonids, then the yolk as a high density nutrient (41% dry weight) requires considerable quantities of water for transformation to the relatively watery embryo (16% dry weight). Obtaining this water may be a problem if the permeability of the vitelline membrane remains limited throughout much of development. Smith (1957) suggests that growth may be retarded until hatching occurs and water becomes more readily available. It is likely that highly desiccated yolk is only feasible in the demersal egg; in *Sardinops caerulea* the water content of the larval yolk is about 91% (Lasker, 1962).

2. *Chemical Composition*

The chemical composition of the eggs of two species are given in Table IV. Further data are given by Phillips and Dumas (1959), who showed in particular that there was no difference in the constituents of different sized eggs of *Salmo trutta*.

There is considerable difficulty in deciding on the sequence of utilization of various materials for energy production. Analysis of the change in chemical components may be unreliable where substances like carbohydrates are being synthesized. Heat production and respiratory quotients are difficult to determine accurately in small eggs, and CO_2 liberation may be masked by a buffered external medium. Amberson and Armstrong (1933) found the RQ's of *Fundulus* eggs over the first 6 days of development were 0.90, 0.78, 0.77, 0.76, 0.72, and 0.72 suggesting very early carbohydrate metabolism. In *Oryzias*, Hishida and Nakano (1954) found 0.75 after fertilization, 0.92 at gastrulation, and 0.70 later. Another problem to overcome is the retention of nitrogenous excretory matter within the egg which gives values of protein metabolism which are too low. The original concept of Needham (1931) that materials were used up for energy in the sequence

carbohydrate — protein — fat

has to some extent been supported by Devillers (1965). However, Hayes (1949) was of the opinion that the sequence in salmon eggs was

hatch
↓
fat — protein — fat — protein

and Smith (1957, 1958) that it was

Table IV Analyses of Eggs[a]

Species	Dry weight (%)	Percent of wet weight					
		Protein	*Total fat*	*Oil*	*Phospholipid*	*Carbohydrate*	*Ash*
S. gairdneri (*irideus*) (rainbow trout)	33.8	20.2	—	3.6	3.8	0.2	1.3
Sardinops caerulea[b] (California sardine)	29.3	21.0	3.8	—	3.2	<0.3	2.1

[a]Data for trout from Smith (1957) and for sardine ovary from Lasker (1962).
[b]Calorific value of dry sardine yolk was 5.4 cal/mg.

hatch
↓
phospholipid — protein + carbohydrate — protein —
phospholipid — triglyceride fat

Further work is required in this field, especially of a comparative nature.

3. *Carbohydrate Metabolism*

Since carbohydrate is present in small quantities and synthesis is continuously occurring, its role is particularly difficult to ascertain by bulk analysis. Hayes (1949) reported a gradual increase in carbohydrate level during salmon development with a temporary fall in glucose level at hatching. Glycogen is probably synthesized near hatching and is present exclusively in the embryo, being stored in the liver near the end of the yolk sac stage. Glucose, in aqueous phase, is present in relatively greater quantities in the embryo than in the yolk due to its higher water content. Smith (1957) described falls in total carbohydrate after gastrulation (linked with the establishment of the circulatory system and therefore a higher potential for metabolism) and a further fall in glucose level at hatching, perhaps because of an interruption of glycogen synthesis. He did not report the general increase found by Hayes. Terner (1968) has more recently found that the ova of *Salmo gairdneri* have a free store of glucose which increases during development and only falls at hatching. Carbohydrate is certainly being synthesized in development for use in the final stages of energy production and for mechanical and osmoregulatory work.

4. *Protein*

Protein is the dominant raw material in the yolk and the main source for tissue formation. The use of labeled amino acids in *Oryzias* shows a slow passage from the yolk to a low molecular weight pool, a more rapid uptake occurring when the vitelline circulation is established (Monroy *et al.*, 1961). The proportion of yolk used for energy requirements is not easy to determine since the egg retains some nitrogenous metabolities (shown by an increase of nonprotein nitrogen with age) and only small amounts of ammonia are excreted. Of particular interest is the finding by Read (1968) of ornithine carbamoyltransferase and arginase in the embryos of *Squalus suckleyi* and *Raja binoculata*, suggesting an early functioning ornithine urea cycle with the retention of urea for osmoregulatory purposes. The use of protein for energy seems to decline after hatching, while the amount used over the whole of early development depends on whether calculations are made when some yolk remains or whether, at a later stage, the larva starts to consume its own tissues. While it has been suggested that 40% of the original protein may be metabolized, more recent detailed studies using isotopes (see Williams, 1967) on *Oryzias* indicate that all protein being resorbed from the yolk in the 9 days subsequent to gastrulation was being transformed into new tissue and none lost in combustion.

Regarding different types of protein it appears that in *Oryzias* 40% of phosphorus is incorporated as phospho-protein; it falls after gastrulation and the loss is very rapid at hatching, especially in *S. gairdneri* (*irideus*). Deuchar (1965), reviewing analyses of amino acids present in the same species, reports that aspartic and glutamic acid are always present, valine and leucine appearing after gastrulation.

5. Fat

There is some ambiguity about the role of fat in energy production during different phases of development. Fat is undoubtedly used as a fuel, possibly 70–80% eing consumed over the whole period of development. Hayes (1949) reported the main loss of fat to be in the fourth week after hatching in the salmon. It seems that the triglyceride fats are the last to be utilized before food is required from external sources. The energy requirements before hatching may also be met by fat, according to Hayes, or by fats in combination with protein (Smith, 1957).

According to Phillips and Dumas (1959) fat is synthesized from protein in the eyed stage of the eggs of brown trout *Salmo trutta*. Terner *et al.* (1968) found that ^{14}C-labeled acetate in the incubation medium of the eggs of *S. gairdneri*, *S. trutta*, and *Salvelinus fontinalis* was incorporated into free fatty acids and other lipids of the egg, which were probably the substrates for endogenous respiration. In addition, labeled acetate was found in the egg phospholipids, presumably as an intermediate stage in the synthesis of complex fats. These authors suggest that the lipids of the embryo are not transferred directly from the yolk but are resynthesized by the embryo after the breakdown of triglyceride fats in the lipid pool of the yolk.

E. Respiration

In the sense of obtaining oxygen, respiratory problems vary with the environment of the egg and larva, but they are probably rather rarely limiting with pelagic eggs. Eggs deposited on or in a substrate may act as an oxygen "sink" (Daykin, 1965) the oxygen level at the egg surface always being less than that of the surroundings even in high velocities of current. Hayes *et al.* (1951) calculated a flow of 1.2×10^{-2} $\mu l/cm^2/cm$ thickness/min through the chorion of salmon eggs. Sensitivity to low oxygen tensions or anoxia varies with both species and stage of development (see Devillers, 1965). In salmonids, cleavage may continue in anoxic conditions; perhaps this is possible because requirements are low and oxygen is stored within the egg, especially in the perivitelline fluid. Gastrulation is more easily blocked by lack of oxygen or by respiratory poisons such as cyanide or azide. Development can also be retarded by blocking oxidative phosphorylation with dinitrophenol. During oxygen lack there are often increases in lactic acid production but, in the shortterm, retardation of development to anoxia is reversible.

De Ciechomski (1965) succeeded in keeping the demersal eggs of *Austroatherina* for 17–18 days in "Vaseline oil." Development was at first normal, but no heart beat or movement was observed. Transfer to water resulted in the heart starting to beat and some movements occurred, but abnormal pigmentation developed, and no embryos hatched. Eggs of three pelagic species were killed by immersion in the oil.

Various adaptations are found which assist in obtaining oxygen. The spawning redd of the salmonid fish has a rapid current of water passing through it; parents guarding their eggs may ventilate them (e.g., the stickleback, *Gasterosteus aculeatus*) while the lungfish *Lepidosiren* has special pelvic "gills" for oxygenating the eggs in its nest; viviparous species obtain oxygen as well as nutrients from the female. After hatching functional gill filaments are often lacking; in herring they do not appear for some weeks at a length of about 20 mm (Harder, 1954). Oxygen is obtained over the body surface and almost certainly via internal body surfaces such as the pharynx and gut. The ultramicroscopic corrugations described by Jones *et al.* (1966) and Lasker and Threadgold (1968) on the epidermis might possibly be directing the flow of water over the body surface. In less favorable environments (see Nikolsky, 1963) external gills may be found, for example, in the bichir *Polypterus*, in the lungfish *Protopterus* and *Lepidosiren*, in the loach *Misgurnus fossilis*, and in *Gymnarchus niloticus*. In newly hatched trout, *Salmo trutta*, respiratory currents are produced by the pectoral fins which also keep the body clear of silt. After a short time mouth and gill movements replace those of the fins and then clogging of the gills by silt is prevented by a coughing reflex or aggregating the silt with mucus (Stuart, 1953).

Intake of oxygen has been measured in a number of species. In the egg stage this may be very low and variable before fertilization. In *Oryzias latipes* there is no sharp rise until a few hours after fertilization, although it seems likely that the respiratory rate becomes steadier at this stage (Nakano, 1953). Before any movements occur within the egg, relative values of oxygen uptake between different stages and species are quite valid, but subsequent to movement being possible the oxygen consumed can be dominated by bursts of activity. Oxygen uptake may be used to express metabolic functions, total or general metabolism consisting of two components, active metabolism (owing to activity) and basal metabolism (owing to maintenance functions). Standard metabolism applies to practically motionless fish under experimental conditions; in the fry of *Salmo salar*, for example, the basal metabolism was 64–69% the standard rate (Ivlev, 1960a). For active fish metabolism is proportional to KV^n, where V is velocity and K and n are constants. The change in oxygen uptake with varying velocity is shown in Fig. 7A. The function KV^n is called the "scope for activity" and has been estimated for a number of species (see Table V).

Some values of oxygen uptake expressed as Q_{O2} (μl/mg dry weight/hr) are given in Table VI, based where possible on subjective criteria of activity and

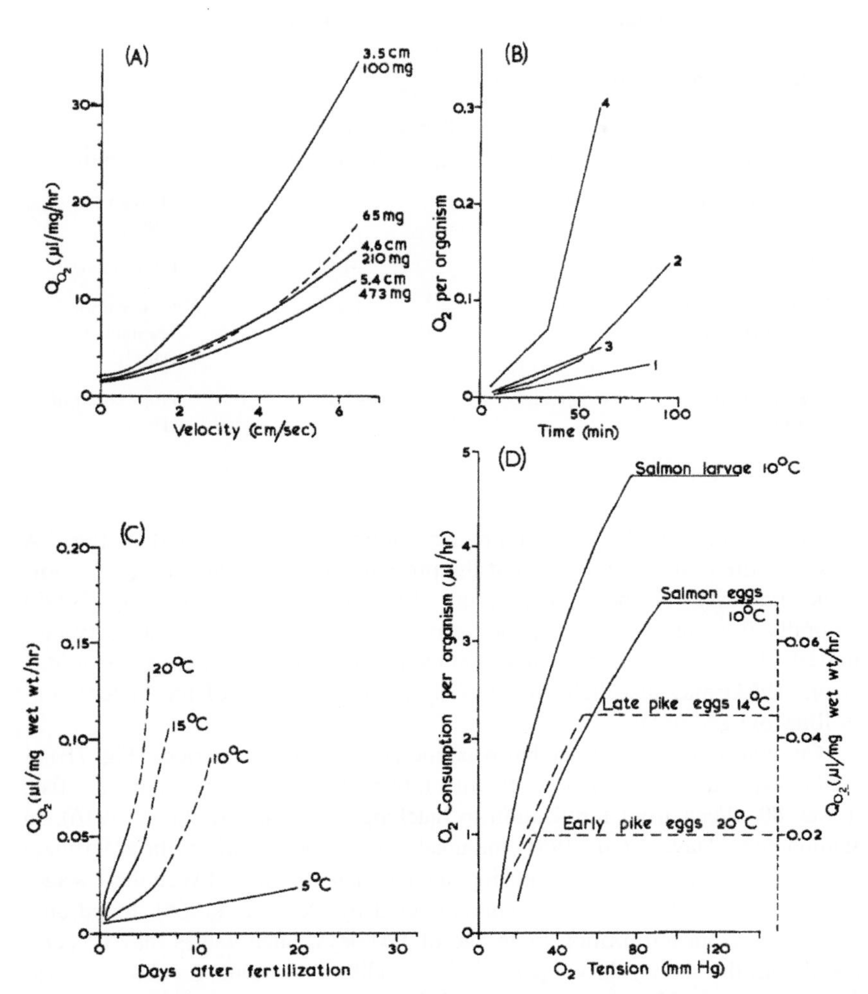

Fig. 7. (A) Oxygen uptake (Q_{o2}) for various sizes of young fish stemming different velocities of current at 20°C. (—) *S. salar* (Ivlev, 1960a); (—) *Alburnus alburnus* (Ivlev, 1960b). Wet weights of original data estimated as four times the dry weight. Estimated dry weight and the length given at the end of each curve. (B) Cumulative oxygen consumption in clupeid eggs and larvae. *Sardinops caerulea*: 1. 20 hr embryo at 14°C, 2. newly hatched active larvae at 14°C (Lasker and Theilacker, 1962); *Clupea harengus:* 3. newly hatched inactive larva at 8°C, 4. newly hatched active larva at 8°C (Holliday *et al*., 1964). (C) The effect of temperature on oxygen consumption (Q_{O2} based on *wet* weight) of different stages of the eggs of *Esox lucius;* hatching at about 11 days at 10°C (Lindroth, 1942). (D) The effect of oxygen tension on oxygen consumption. (—) and left-hand ordinate, eggs and larvae of salmon, *S. salar* (Hayes, 1949); (- - -) and right-hand ordinate, eggs of pike, *Esox lucius* (Lindroth, 1942).

Table V Scope for Activity

Species	Stage	Range of increase of Q_{02} owing to activity	Author
Salmo salar (salmon)	Eggs	3	Hayes *et al.* (1951)
Salmo salar (salmon)	Fry	7–14	Ivlev (1960a)
Sardinops caerulea (California sardine)	Larvae	3½	Lasker and Theilacker (1962)
Clupea harengus (herring)	Larvae	9–10	Holliday *et al.* (1964)

inactivity. Without the type of precise control as demonstrated in Fig. 7A these are the best available data at the present time. The standard Q_{02}'s show some similarity considering the variety of the material used. Active Q_{02}'s will naturally fluctuate widely, depending on the type and extent of activity permitted. The use of certain anesthetics to control activity (e.g., Holliday *et al.*, 1964) may also help in establishing at least some of the factors controlling oxygen uptake.

Cumulative oxygen uptake has been measured over short periods (Fig. 7B) as well as fairly long periods of development. In *Fundulus heteroclitus*, the killifish, it was 80 μl/egg from fertilization to hatching (Scott and Kellicott, 1916). In salmon eggs (Hayes *et al.*, 1951) the uptake was about 0.2 μl/egg/hr at fertilization rising to 3.4 μl/egg/hr at hatching, the cumulative total between these stages amounting to 1400 μl/egg. Privolniev, quoted by Devillers (1965), found only 850 μl/egg from fertilization to an age of 120 days. Such values may be compared with the short yolk sac period of the California sardine, *Sardinops caerulea*, where only 12 μl/larva were used from hatching to 180 hr subsequently (Lasker and Theilacker, 1962). Obviously cumulative uptake and uptake per unit time will vary with size, age, and temperature. Devillers (1965) considers that increases in oxygen uptake with age may be shown by the exponential equation

$$Q = ae^{kt} \tag{6}$$

where t is time and a and k are constants. Q_{02}'*s* will also vary with environmental conditions (see Fig. 7C). In herring eggs and larvae the Q_{10} for oxygen uptake (Q_{02}) was 2.0 between 5° and 14°C. Q_{02} also increased with a salinity shock but not with constant rearing salinities. There were only slight effects because of light intensity and starvation (Holliday *et al.*, 1964). In salmonids (see Section VIII) oxygen uptake falls at low ambient oxygen tensions (Fig. 7D), but the young stages of fish will resist oxygen lack quite well for a time.

Table VI Oxygen Uptake as Q_{O2} (μl/mg dry wt/hr)

Species	Stage	Temp. (°C)	Standard rate	Rate not defined	General (active) rate	Author
Salmo trutta (brown trout)	Egg, embryo only	10	0.55			Gray (1926)
S. salar (salmon)	Egg, embryo only[a]					
	19 days	10	0.9			Hayes *et al.* (1951)
	50 days	10		1.0		
S. salar (salmon)	Fry (no yolk)					
	3.5 mm	20	2.3		35.0	Ivlev (1960a)
	4.6 mm	20	2.0		15.3	
	5.4 mm	20	1.6		12.1	
Fundulus heteroclitus (killifish)	Egg, embryo only[b]	19–21	3.2			C. G. Scott and Kellicott (1916)
	6 day larvae (no yolk)[b]	19–21		3.9		
Esox lucius (pike)	Whole eggs[b] (range from fertilization to hatch)	5	0.02–?			Lindroth (1942)
		10	0.02–0.36			
		15	0.02–0.44			
		20	0.02–0.54			

Continued

Table VI Oxygen Uptake as Q_{02} (μl/mg dry wt/hr)—Cont'd

Species	Stage	Temp. (°C)	Standard rate	Rate not defined	General (active) rate	Author
Sardinops caerulea (California sardine)	Whole eggs	14	0.8		1.79	Lasker and Theilacker (1962)
	Whole larvae	14	1.33		2.68	
Clupea harengus (herring)	Whole eggs	8	1.5			Holliday *et al.* (1964)
		12	4.0			
	Larvae (less yolk)	5	2.0		2.5	
		14	3.5		5.0	

[a]Dry weight estimated at 16% of wet weight.
[b]Dry weight estimated at 25% of wet weight.

This raises the problem of oxygen debt in such experiments and indeed in any respiration measurements where activity is intense. The accumulation of such a debt can be tested by continuing respiration measurements after activity or anoxia. Hayes *et al.* (1951) found no oxygen debt after periods at reduced oxygen tensions in salmon eggs, suggesting there had been a reduction in metabolism under these conditions. Salmon fry, however (Ivlev, 1960a), showed an oxygen debt amounting to 48% of general metabolism after a period of activity. Fry of bleak, *Alburnus alburnus*, had a negligible debt when stemming a water velocity of 2 cm/sec but one of about 15% in 6.42 cm/sec (Ivlev, 1960b).

An overall review of respiration rates, including that of larval fish, allowed Winberg (1960) to deduce that the relationship

$$Q_s = 03\,W^{0.8} \tag{7}$$

represented the best fit for the data on standard metabolism in micro-liter per hour at 20°C for organisms of wet weight W mg. He provides conversion factors for other temperatures but stresses that such relationships are designed to evaluate experimental results rather than replace them. Blaxter (1966), recalculating the data of Holliday *et al.* (1964) for herring, concluded that in larvae of mean wet weight 1.53 mg respiration amounted to 17 μl O_2/day at 8°C. Using Winberg's relationship a value of 23 μl O_2/day is arrived at, fair agreement considering his relationship is operating at its limit.

Deductions can be made about the type of metabolism involved in respiration (Hayes, 1949; Ivlev, 1960b; Winberg, 1960). The consumption of 1000 μl O_2 corresponds to the utilization of different materials as follows:

$$1000\ \mu\text{l}\ O_2 \equiv \begin{array}{l} 1.05\ \text{mg protein} \\ 1.23\ \text{mg carbohydrate} \\ 0.50\ \text{mg fat} \end{array}$$

1 mg of protein yields 4.25 cal, 1 mg of carbohydrate yields 4.15 cal, and 1 mg of fat yields 9.45 cal;

therefore,

$$1000\,\mu\text{l}\ O_2 \equiv \begin{array}{rl} 4.5 & \text{cal from protein} \\ 5.1 & \text{cal from carbohydrate} \\ \underline{4.7} & \text{cal from fat} \\ 4.77 & \end{array}$$

The value 4.77 cal/1000 μl O_2 is sometimes called the oxycaloric coefficient. This makes it possible to calculate the requirements for growth and metabolism from measurements of growth rate and respiration. On the basis of 1000 μl O_2 being equivalent to 4.77 cal, and 1 mg wet weight of growth corresponding to 1 cal, Ivlev (1960b) concluded that young Baltic herring of

Table VII Calorific Intake in Some Species

Species	Wet weight (mg)	Daily intake (cal)	Temp. (°C)	Author
Engraulis japonicus (anchovy)	1500–6300	78–426	14–20	Takahashi and Hatanaka (1960)
Alburnus alburnus (bleak)	250	38.3	19–22	Ivlev (1960b)
Clupea harengus (herring)	66.5	16.2	(?) 15–17	Ivlev (1960b)
C. harengus (herring)	1.53	0.18	8	Blaxter (1966)

wet weight 66.5 mg required 16.2 cal/day for growth and maintenance. These and other data are given in Table VII.

Smith (1957, 1958) reviews the work on growth, heat production, and respiration. High respiration and high rates of growth appear to be correlated, at least at some phases of development. In particular, specific growth rate and heat produced show very similar trends at various stages of development of the rainbow trout, *Salmo gairdneri* (*irideus*). Up to hatching, plots of heat production on specific growth lie on a straight line with positive slope passing through the origin; after hatching the line passes through an intercept on the ordinate. There are two components in heat production—one resulting from growth and the other from maintenance. Temperature may affect the relative requirements of growth and maintenance thus giving differences in efficiency of development (see Section IV, B) with temperature optima.

F. Growth

Growth is influenced in the early stages by the ratio between embryo and yolk weight, especially by the problems of efficient yolk utilization (Section IV, B) and the need to mobilize the nutrient reserves. Higher maintenance requirements per unit weight in small organisms, surfacevolume ratios, change of diet with age, feeding and searching potential (Section VII, E) are also important. Growth has often been shown in terms of length but there is a need for uniformity, i.e., whether the caudal fin or fin fold is included, and length gives no information on the condition (fatness) of the organism or of the changes in body proportions with age; this is especially true in the yolk sac stage and at metamorphosis. Similarly, wet weights introduce the error of varying water content; thus, dry weight is the most satisfactory measure.

Examples of the relative quantities of yolk and embryo in a large and a small egg at different stages of development have already been given in Fig. 5. The increase of weight of the embryo and decrease of the yolk tend to be logarithmic in nature. It is clear (Section IV, B) that in the earlier stages relatively less yolk is being used for maintenance than for growth, giving a higher efficiency of utilization. Without external feeding the larval weight begins to fall before all the yolk is resorbed. Farris (1959) divided the growth in length of four species of pelagic larvae after hatching into three phases: an early rapid phase after hatching, a slow phase near the completion of resorption, and a subsequent negative phase if no food was available. This shrinkage in terms of length was also described by Lasker (1964) in the larvae of *Sardinops caerulea*.

Growth of salmonid and other freshwater fry in hatcheries has received considerable attention from an economic point of view, to provide maximum growth at minimum cost. Growth in marine conditions, where sampling of the same population is difficult owing to mortality, migration, or changes in net avoidance with age, is less adequately known, although examples are given in Fig. 8.

Some estimates of survival of fish larvae depend on a knowledge of growth rate and this is often inadequate or based on flimsy experimental data. Comparisons from tank experiments are difficult owing to mortalities, size-hierarchy effects, or other possible aquarium artifacts. Examples of size hierarchies in rearing experiments are given in Fig. 9, showing the enormous range of size which may be found from an initially fairly uniform stock. Thus Shelbourne (1964), after a series of rearing experiments on plaice, found the length ranged from 7.5 to 37.5 mm. In dense populations there were also

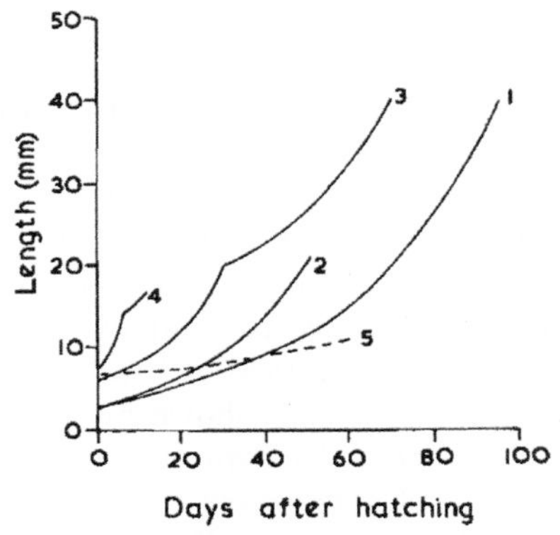

Fig. 8. Length of different species related to age under natural conditions. 1. *Melanogrammus aeglefinus*; 2. *Scomber scombrus*; 3. *Clupea harengus*; 4. *Esox lucius* (all from Sette, 1943); and 5. *Pleuronectes platessa* (Ryland, 1986).

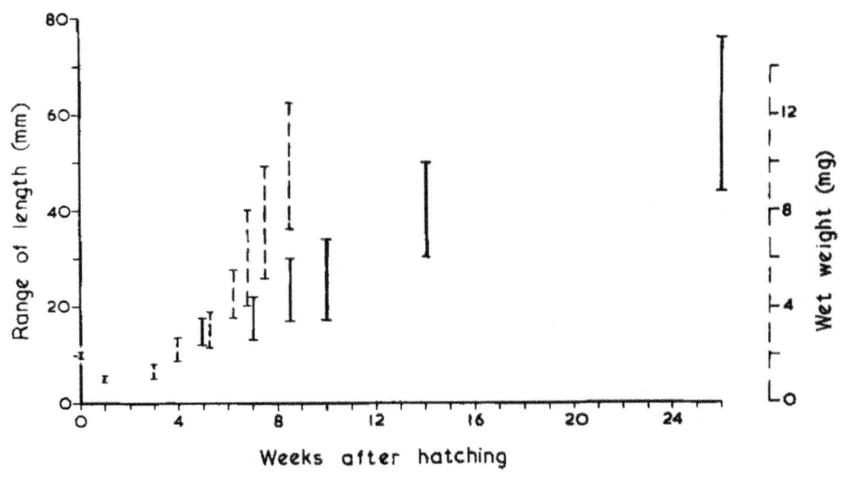

Fig. 9. Size hierarchies in rearing experiments: (—) *Clupea harengus* length (left-hand ordinate) (Blaxter, 1968a), and (- - -) *Pleuronectes platessa* wet weight (right-hand ordinate) (Ryland, 1966).

relatively larger numbers of small larvae. Magnuson (1962) tested the effect of food supply on *Oryzias* kept at high densities. As long as food was adequate, growth was not retarded, social hierarchies did not develop, and aggressive behavior was not manifested. Territorial behavior only occurred when food was limited spatially. Whether size hierarchies occur in natural conditions is not known. It seems they must stem from inherently different growth rates or from some type of dominance hierarchy—the first could occur in the wild, but the second is more likely to occur in crowded tank conditions.

Food supply and temperature are the most potent environmental factors in controlling growth in the later stages, and there are few data available on this except for salmonids (see chapter by Phillips, Volume I). In the egg stage low temperatures often produce longer larvae at hatching (see Table VIII), although Lasker (1964) found the maximum length of *Sardinops caerulea* larvae at intermediate temperatures. There are then probably temperature optima for yolk utilization which partially give rise to these effects, but often the range of temperature used may be insufficient to show them. On the assumption that high temperatures are nonoptimal other conditions producing smaller larvae, such as low oxygen tensions and current velocities for salmonids (Silver *et al.*, 1963; Shumway *et al.*, 1964), as well as high salinities in the desert minnow, *Cyprinodon macularius*, shown by Sweet and Kinne (1984) and in cod, *Gadus macrocephalus*, by Forrester and Alderdice (1966), may also be considered nonoptimal. Combinations of temperature and salinity suggest that the optimum conditions for growth in length *over the ranges used* were 26°C for 70‰ salinity, 28°C for 35‰ and 33°C for freshwater in *Cyprinodon*, and 6°C and 19‰ for *Gadus macrocephalus*. The mortality, length,

Table VIII Effect of Temperature on Size, Using Yolk Reserves Only

Species	Method	Temp. range (°C)	Corresponding size range (mm)	Author
Salmo trutta (brown trout)	Maximum wet weight	5–16	135–95 (mg)	Gray (1928b)
Coregonus clupeaformis (whitefish)	Hatching length	0.5–10	13–8.8	Price (1940)
Osmerus eperlanus (smelt)	Hatching length	12–18	5.2–4.6	Lillelund (1961)
Cyprinodon macularius (desert minnow)	Hatching length	28–35	5.3–3.7	Kinne and Kinne (1962)
			4.2–3.6	Sweet and Kinne (1964)
Coregonus wartmanni (whitefish)	Hatching length	1–7	11–9	Braum (1964)
Gadus macrocephalus (Pacific cod)	Hatching length	2–10	4.1–3.5	Forrester and Alderdice (1966)

and maximum weight of salmon alevins (on the yolk supply) was greatest when they were reared in the dark, on a grooved surface, where presumably activity was minimal (D. H. A. Marr, 1965).

G. Endocrines, Growth and Metamorphosis

Russian, Canadian, and other work on the role of the endocrine system in fish larvae is fully reviewed by Pickford and Atz (1957) and the role of the thyroid in metamorphosis by Barrington (1961). The main techniques used were histological, histochemical, or immersion in dilute solutions of thyroid hormone or antithyroid drugs. Evidence from a number of sources suggests that the pituitary is inactive until metamorphosis, for example, in herring *Clupea harengus*, eels *Anguilla anguilla*, bream *Abramis brama*, milkfish *Chanos chanos*, and sturgeon *Acipenser stellatus*.

The thyroid varies in its histological "activity" in the early larval stages but at metamorphosis in sturgeon, flatfish, eels, herring, pilchard *Sardina pilchardus*, and bone fish *Albula vulpes*, there are signs of secretion of colloid into the lumen of the thyroid follicles. Whether this is storage, or release into the circulation is taking place, can be satisfactorily tested only by measuring hormone levels in the blood. Earlier, thyroxine may retard growth and accelerate morphogenesis (e.g., in salmon, sturgeon, *Misgurnus fossilis* and *Lebistes*) at a concentration of 1 ppm or less. Rate of oxygen consumption may also be increased. In the developing eggs of *Scyliorhinus canicula* (see Dimond, 1963) the functioning of the thyroid seems highly dependent on temperature. At 8°C use of ^{131}I showed the thyroid concentrated and bound iodine, but there was no evidence of thyroid hormone formation. At room temperature thyroid hormone reached a measurable level. Thiourea at concentrations of about 100 ppm may delay hatching, retard yolk resorption, and inhibit growth in salmonids, inhibit morphogenesis in sturgeon larvae, and decrease oxygen consumption in salmon and sturgeon larvae by as much as 15%. Thiourea seems to improve the ability to utilize oxygen at low tensions, although it is not clear whether this is because of its antithyroid action. Thiourea might well improve survival in suboptimal conditions of oxygen concentration that might occur, for example, during transportation. The effect of thyroxine on promoting metamorphosis was tested on the eel (Vilter, 1946). There was an accelerating, but not a sudden, effect with an indication that an increasing threshold controls the sequence of events in metamorphosis. In other words, the later stages of metamorphosis require higher concentrations of thyroxine. Other experiments on lampreys showed no effect of thyroxine, thyroid extract, iodide, or iodine on metamorphosis (see Barrington, 1961).

It may be concluded that there is a suggestion of the thyroid playing a role in growth and differentiation in the larval stages but further work is required, especially at the important metamorphosis stage. If the pituitary is not functioning until metamorphosis the thyroid must be relatively independent of pituitary control in the early stages of life.

V. FEEDING, DIGESTION, AND STARVATION

Often the mouth is not completely formed at hatching, but rapid development in many marine fish larvae leads to the possibility of taking external food before the yolk is finally resorbed. Without success in feeding there is eventually self-metabolism and loss of weight. Yolk may be transported and stored within the body, especially in the base of the primordial fin and other subdermal spaces (Shelbourne, 1956), and fat may be stored in the mesenteries of the larval gut. Unlike demersal freshwater species there is no vitelline circulation in most pelagic larvae although *Scomberesox*, *Trachypterus*, and a few other species are known to be exceptions (Orton, 1957). Usually there is a yolk sac sinus in connection with the heart and with lateral branches to the subdermal spaces; in the species with a vascularized yolk sac these spaces are much less inflated.

After final resorption of the yolk the larvae retain their potential to feed for some days depending on species, egg size and temperature. The concept of a

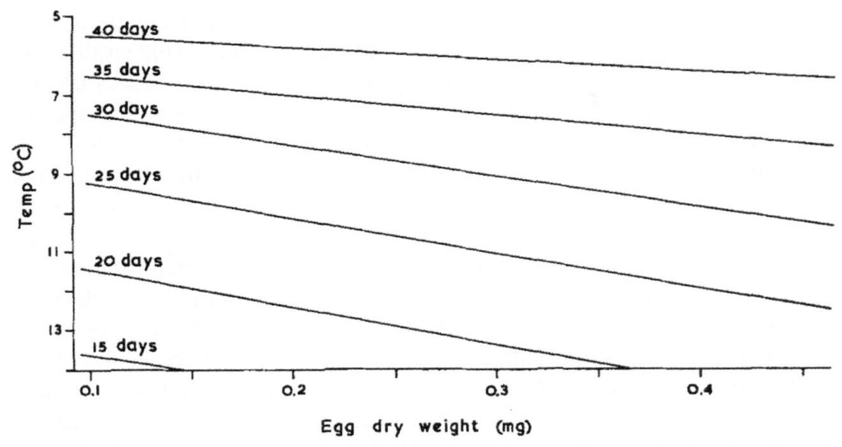

Fig. 10. The concept of "point of no return." The time from fertilization to the point where the larvae of *Clupea harengus* are too weak to feed is given in the form of a nomogram based on original egg dry weight of different races and the temperature (Blaxter and Hempel, 1963).

"point of no return" was introduced by Blaxter and Hempel (1963) for herring larvae (see Fig. 10). This point, after which the larvae are still living but too weak to feed, may vary from 15 days after fertilization for summer spawners in warmer water to 45 days for winter spawners. In the larval cisco, *Leuchichthys artedi*, it was 27–36 days after hatching at 3°–4°C (John and Hasler, 1956). The activity of larval *Solea solea* drops from about 70% of the time active at first feeding to 20% some 5–6 days later when, if feeding is unsuccessful, inanition occurs (Rosenthal, 1966). Information on the time to point of no return in relation to spawning and food supply may help to predict the probability of survival in different broods.

Most fish larvae are predatory with a large mouth and well-developed eyes. In herring the gape of the jaw increases by 50% during the yolk sac stage: the gape of 0.3–0.4 mm at first feeding depends on the length of the larvae and, therefore to some extent, on the original egg size (Blaxter, 1965). An elastic ligament at the articulation makes it possible for even larger organisms to be taken (Flüchter, 1962). Some species have very small gapes which is partly a function of their small size (e.g., lemon sole *Microstomus kitt*, sprat *Sprattus sprattus*, pilchard *Sardina pilchardus*, sardine *Sardinops caerulea*, and many gadoids). These must require very small food organisms for their early survival, a factor which is of considerably less importance for many freshwater fish.

The forward darting movements of fish larvae in order to engulf the prey are described in a number of species, for instance, in *Coregonus* and *Esox* by Braum (1964) and in sole, *Solea solea*, by Rosenthal (1966). Iwai (1964) found three types of feeding in the larvae of *Plecoglossus altivelis:* a predatory snapping, a respiratory current with a sieving action by the gill rakers, and a ciliary current associated with the olfactory pits.

The extent to which fish larvae are herbivorous and might make use of such currents is not clear. Green food in the gut may be incidental to the swallowing of water or due to fecal material from prey species, and its presence is no evidence of utilization. In the anchovy, *Engraulis anchoita*, however, phytoplankton in the guts of young stages can be correlated with the formation of gill rakers which presumably act as a sieve (De Ciechomski, 1967b). The utilization of dissolved organic matter (Putter's theory) has also never been adequately tested. Morris (1955), in a reappraisal of the theory, concluded from circumstantial evidence of a low incidence of feeding in larvae collected from nature, that it was likely they were utilizing at least fine particulate matter, which could be collected by mucous cells within the buccal cavity or during swallowing of seawater for osmoregulatory purposes. Morris estimated that the larvae of the night smelt, *Spirinchus starksi*, might pass about 1.15 ml of water per day across these mucous surfaces, possibly containing 0.011 mg of organic matter, a very small quantity even for a larva of 0.9 mg wet weight. The recent work of Terner (1968), mentioned in Section IV, D, showing that the ova of *Salmo gairdneri* can probably incorporate exogenous substrates into their endogenous reserves, may also be significant.

The guts of many pelagic marine larvae are often empty, a puzzling feature when one considers the fairly high food requirements in the early stages. Apart from a lack of suitable food this may result from the capture by net of the weaker feeders only, rapid digestion, defecation during capture, or diurnal feeding rhythms. Examples of such rhythms, which are given in Fig. 11, do indicate a decrease in feeding at night and an increase at dawn.

The role of vision in feeding has been tested experimentally and thresholds of light intensity measured (Table IX). The fall off in rate of feeding corresponds with the dusk and dawn periods. Some species can take food in the dark, e.g., cisco (John and Hasler, 1956), especially when it is present in high concentrations. Some, like sole, take food in the dark for most of the larval life and others, like plaice, only at later stages around metamorphosis (Blaxter, 1968b). Where vision is important the daily feeding period must vary considerably with season, latitude, clarity of water, and even average cloud cover. Ivlev (1960b) calculated, from the food requirements of young Baltic herring, their rate of feeding and the density of food, that they needed to feed 15 hr/day in August, in other words nearly all the hours of daylight. Blaxter (1966) estimated that much younger stages of herring had 10 hr/day to feed in the southern winter spawning groups and up to 24 hr/day in the more northern summer spawners. Variations in feeding time affect searching power, and therefore survival and growth, and these may be further influenced by the temperatures prevailing in different seasons.

Success in early feeding, feeding drives, and learning factors have been studied. At very early stages Braum (1964) found only 3–8% of feeding movements in *Coregonus* were successful, but 30% in *Esox*. In the yolk sac stage, herring larvae take food successfully in 3–10% of their feeding

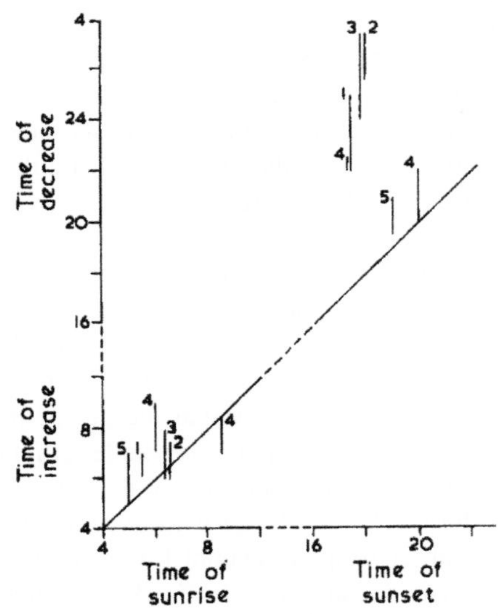

Fig. 11. The time period over which the gut contents of various species of larvae increase and decrease related to the time of sunrise and sunset. Note the lag at sunset caused, presumably, by the time taken to digest food taken earlier. 1. *Pleuronectes platessa* (Shelbourne, 1953); 2. *Pleuronectes platessa*; 3. *Ammodytes* (Ryland, 1964); 4. *Clupea harengus* (see Blaxter, 1965); and 5. *Salmo salar parr* (Hoar, 1942).

Table IX Main Range of Light Intensity over which Feeding Becomes Reduced

Species	Range of light intensity (mc)	Author
Oncorhynchus keta (chum salmon)	10^1–10^{-3}	Ali (1959)
O. gorbuscha (pink salmon)	10^1–10^{-3}	Ali (1959)
O. nerka (sockeye salmon)	10^1–10^{-3}	Ali (1959)
O. kisutch (coho salmon)	10^0–10^{-4}	Ali (1959), Brett and Groot (1963)
Coregonus wartmanni (whitefish)	?–10^0	Braum (1964)
Esox lucius (pike)	?–10^{-1}	Braum (1964)
Clupea harengus (herring)	10^2–10^{-1}	Blaxter (1966)
Pleuronectes platessa (plaice)	10^2–10^{-2}[a]	Blaxter (1968b)

[a]Older stages feed in the dark.

movements but later in 80–90% (Rosenthal and Hempel, 1968; Blaxter and Staines, 1969a). In plaice, feeding is 60–80% successful throughout larval development. In herring larvae from 20 to 40 mm in length the feeding drive depends on factors such as satiation and especially on the activity of the prey (Blaxter, 1965). Live organisms, although more difficult to catch, seem to increase the feeding drive and are probably very important in the young stages, where the main visual process may centre round movement perception (Section VI, A). Larval cisco, however, showed a much higher rate of feeding on dead than live *Cyclops* (John and Hasler, 1956).

The gut is usually a straight or simple tube at first feeding and the food is often digested near the anus. This is an interesting reverse of the normal situation in adult vertebrates. It seems likely that the gut is unspecialized at this stage and digestive enzymes are secreted along its length, although no critical work has been done on this. Harder (1960) gives a detailed account of the changes in the gut during the early growth of clupeids and engraulids; Ryland (1966) gives a rather simpler account for the plaice. Movement along the gut is mainly by peristalsis, although Iwai (1964) found cilia in the gut of ayu, *Plecoglossus altivelis*, especially in positions posterior to the liver. Backwardly directed ciliary currents and forward peristalsis seemed to cause a circulation of the gut contents, perhaps to aid digestion. In further reports (Iwai, 1967b,c) use of the electron microscope showed intermingling of ciliated cells and columnar cells with microvilli in both *Plecoglossus* and *Hypomesus olidus*. Reduction of cilia in later stages suggested a transition from cilia to microvilli. Whether cilia are to be found in many species is still open to investigation.

Rates of digestion have been measured by rate of disappearance of the gut contents (the gut wall and body being transparent) or by interposing differently colored food in the normal diet and observing the first signs of colored feces. Kurata (1959), using herring 12 days old, found that the gut clearance rate ranged from 12 to 19 hr or more at 9°C depending on the extent of food intake. The rate of digestion varies from 9 hr at 7°C to 4 hr at 15°C judged by transparency of the gut contents in this species (Blaxter, 1965).

Reports of larvae in poor condition or dead in net hauls may be due to selective capture or poor washing and are not certain evidence for mortality. Shelbourne (1957) estimated the condition of plaice larvae in the North Sea and found more signs of inanition early in the year. Measurements of condition factor [weight/length^3 x 1000) in herring larvae (Hempel and Blaxter, 1963; Blaxter, 1969a) compared with starving larvae in tanks suggested that larvae in the sea were often very near the point of starvation although, over a number of years, it was not possible to correlate high condition factors with a good food supply. Ivlev (1961) looked at various aspects of starvation in young catfish, *Silurus glanis*, and bream, *Abramis brama*. With complete starvation the survival time was 34–46 days corresponding to a weight loss of 33–49%. With rations below the maintenance requirement survival was prolonged, but even on a full maintenance diet, the weight remaining constant, the fry died after 126–151 days, indicating that a "need" for growth may be characteristic of these young stages and that merely to maintain uniform weight is physiologically or developmentally inadequate. The current velocity stemmed by young carp *Cyprinus carpio* and roach *Rutilus rutilus* for 5 min was reduced by 10 times, from about 100 to 10 cm/sec, over 52 days of starvation. Age had a considerable influence on survival time without food. In catfish, roach, and bleak *Alburnus alburnus*, it ranged from 3 to 6 days at yolk resorption and from 109 to more than 180 days when 100 days old. The loss in weight and percentage survival of 12 species of larvae 25–30 days old during starvation is shown in Fig. 12. Losses of weight of 30–45% were possible before death, with survival times of 30–80 days. Starving fish were much more susceptible to unfavorable conditions (see Table X) and also to infection and predation.

VI. SENSE ORGANS

A. Vision

The newly hatched larvae of many species have unpigmented, and presumably nonfunctioning, eyes (e.g., sole *Solea solea*, mackerel *Scomber scombrus*, whiting *Merlangius merlangus*, pilchard *Sardina pilchardus*, and sardine *Sardinops caerulea*); others have pigmented eyes (e.g., plaice *Pleuronectes platessa*, cod *Gadus morhua*, herring *Clupea harengus*, and salmonids). Schwassmann (1965) examined sections of the eye and brain of larval *Sardinops caerulea* after hatching. On the first day there was no pigment and little differentiation, but by the second day pigment with a stratified retina

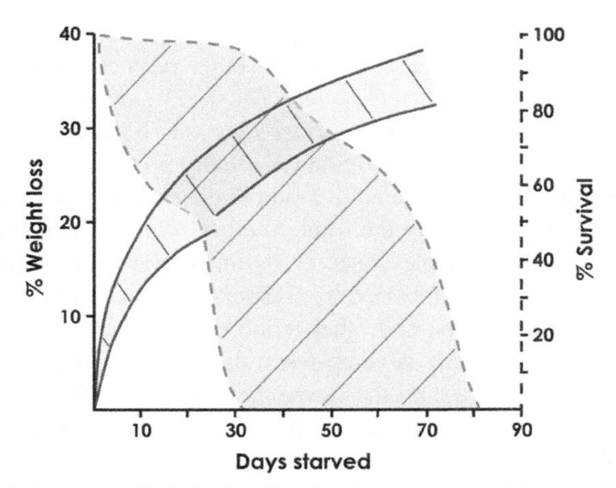

Fig. 12. Weight loss (— and left-hand ordinate) and percent survival (- - - and right-hand ordinate) as "envelopes" for 12 species of fish larvae during starvation (Ivlev, 1961).

Table X Lethal Limits of Certain Environmental Factors on Fry Before and After Starvation[a]

Species	Normal limit(s)	Starvation limit(s)	Days starved
	Salinity %		
Alburnus alburnus (bleak)	4.5	2.5	20
Caspialosa volgensis (Volga herring)	16.5	14.5	20
	pH		
Tinca tinca (tench)	4.0–10.8	5.7–8.3	50
Perca fluviatilis (perch)	5.3–8.2	6.3–7.8	30
	O_2 ppm		
Cyprinus carpio (carp)	0.7	2.1	50
Alburnus alburnus (bleak)	2.8	3.1	20

[a]From Ivlev (1961).

and visual cells had developed. At 5 days the optic nerve was myelinated and the optic chiasma showed some interdigitation. Considering there are only about 40 individual fibers in the optic nerve when feeding commences at 3 days, it seems probable that the eye at first feeding is only capable of coarse movement perception. The larval eye is also limited in other respects. In newly hatched *Lebistes* (Müller, 1952) and *Oncorhynchus* (Ali, 1959), and in *Clupea harengus* (Blaxter and Jones, 1967) and *Pleuronectes platessa* (Blaxter, 1968b) and in a number of other species (Blaxter and Staines, 1969b) up to metamorphosis, there is a pure-cone retina, although the adult has both cones and rods. Lyall (1957a,b) examined the developing eye of the trout, *Salmo trutta*, and concluded that the rods might develop from single cones or by an outward migration of bipolar cells, the latter view being supported by Blaxter and Jones (1967).

Retinal pigment migration and other retinomotor responses associated with dark and light adaptation only develop when the rods appear and are therefore absent in herring and plaice until the larvae metamorphose. This may be true of many other marine species. Thus the larvae are, visually, poorly equipped, with a single type of visual cell, no ability to dark or light adapt, and with a coarse retinal mosaic. *Oncorhynchus* species acquire rods and retinomotor responses very rapidly after hatching compared with the marine fish which have been examined. The range of light intensity over which light and dark adaptation takes place is then 10^1–10^{-1} meter-candles (mc), which corresponds fairly well with the thresholds for feeding (Table IX).

The acuity of the larvae is poor compared with the adult. Although the retinal cells are small and fairly closely packed, the eye is also small, as is the focal length of the lens. Baerends *et al.* (1960) found that young cichlids, *Aequidens portalegrensis*, could be trained to distinguish stripes 1.5 mm apart at 3 cm body length; this improved to 0.3 mm at 11 cm body length. Blaxter and Jones (1967) calculated the acuity of the eyes of larval herring (minimum separable angle) to be 200 minutes at 1 cm body length improving to 50 minutes at metamorphosis.

Thresholds and spectral sensitivity were measured by behavior techniques in herring, plaice, and sole larvae (Blaxter, 1968c, 1969b). Using a negative phototaxis as criterion the visual thresholds were about 10^{-5}–10^{-6} mc, the spectral sensitivity curves being plateaulike with peaks which might correspond to different cone populations.

Distances of perception and visual fields have been measured by a number of workers. These are given in Table XI.

B. Neuromast Organs

It has probably not been fully appreciated that many species of fish larvae have free neuromast organs. Iwai reviews earlier reports and his own work on *Tribolodon hakonensis*, *Tridentiger trigonocephalus*, and *Oryzias latipes* (1967a). These organs are usually situated in lateral rows along the body with the small hump of sensory cells being the main element visible especially in sea-caught or fixed specimens. Rearing, with the use of phase contrast microscopy, makes it possible to retain and see the cupulae which are relatively enormous—0.05 mm long in *Oryzias* but 0.1–0.4 mm long in some other species. In the adult it is likely that they are more often sunk into pits. The ability of larvae to avoid capture before the eyes fully develop is almost certainly a result of these neuromast organs.

VII. ACTIVITY AND DISTRIBUTION

A. Phototaxis and Activity

Much of the published work on phototaxis in fish larvae is anecdotal. Generalizations that marine fish larvae are photopositive and demersal freshwater larvae photonegative do not stand up to a detailed investigation using different temperatures, light intensities, species, or ages. Changes of the sign of the phototaxis may be related to temperature, especially where a sudden change is applied, e.g., in *Coregonus* (Braum, 1964). Herring larvae show a positive phototaxis at high light intensities which becomes negative below a certain threshold (Blaxter, 1968c).

Table XI Visual Fields and Perception Distances

Species	Size	Distance for locating food (mm)	Feeding distance (mm)	Visual field		Author
				Single eye	*Binocular*	
Coregonus (whitefish)	Early larvae	10	0.5–3	145°	45	Braum (1964)
Esox lucius (pike)	Early larvae	(?)10	3–4	150°	80°	Braum (1964)
Clupea harengus (herring)	12–14 mm	5	3	—	—	Blaxter (1966)
Clupea harengus (herring)	10–19 mm	10–26	—	—	—	Rosenthal and Hempel (1968)
Pleuronectes platessa (plaice)	6–8 mm	5	<5	—	—	Blaxter and Staines (1969a)

Activity levels are frequently associated with diurnal light changes and phototaxis. Hoar (1956, 1958), Ali (1959), and Heard (1964) review much work on activity, rheotaxis, schooling, and migration of young salmonids. Emergence of the alevins of sockeye salmon, *Oncorhynchus nerka*, seems to be controlled by incident light on the gravel and can be delayed by artificial light at night. The younger fry of pink salmon, *Oncorhynchus gorbuscha*, for example, are negatively phototactic and hide among the stones of their home stream by day. They become active at night and may rise to the surface and swim downstream or get displaced by the current. After schooling both pink and chum salmon *O. keta*, prefer lighted conditions and presumably substitute schooling for hiding as their protection against predators. They show then a positive phototaxis, stem the current by day displaying a cover reaction only with abrupt changes of light, and again become displaced downstream at night when the visually controlled rheotaxis phases out. Coho salmon fry, *O. kisutch*, are rather different in behavior; they are strongly territorial and generally more active, but their responses to light and their diurnal changes in activity are much less marked. The behavior of the fry of the different species of *Oncorhynchus* can, in fact, be related rather generally to their migratory habits. Pink and chum fry migrate to the sea quite soon after hatching, whereas sockeye fry tend to move into lakes, and coho remain in the home stream for a year or so. Loch trout, *Salmo trutta*, also show incipient territorial behavior at an early age (Stuart, 1953). Woodhead (1957), in a laboratory study of the larvae of *Salmo* species, related changes in photokinetic activity levels to ambient light intensity and age. Between 0.005 and 100 mc the photokinetic activity of *S. trutta* gradually increased above a basal dark level, and this species and *S. gairdneri* (*irideus*) and *S. salar* all showed increased activity with age. A positive photokinesis, together with a negative phototaxis, would work together to keep the larvae within the stones of their home stream. Woodhead and Woodhead (1955) found a similar increase in the activity of herring larvae with light intensity above 0.3 mc, and there was a basal level of activity in very low illumination as there was in the salmonids.

B. Vertical Distribution

Diurnal variations in distribution are linked to activity, phototaxis, and brightness discrimination. Some earlier work (see Simpson, 1956) on changes of vertical distribution did not take into account the need for opening and closing the plankton nets at defined depths and for preventing differential net avoidance by day and night. A general lack of larvae by day, especially larger specimens, partly results from this latter factor, as emphasized by Bridger (1956). With slow tow nets it can be difficult to estimate the relative effects of net avoidance and vertical distribution when comparing day and night catches. Strasburg (1960) used slow tow nets to catch tuna larvae. Some species such as skipjack, *Katsuwonus pelamis,* were rarely caught by day at all, while catches of all species were always much higher by night. He concluded that there was some net avoidance, but the main diurnal difference was owing to vertical migration. Stevenson (1962) used high speed nets for larvae of Pacific herring, *Clupea palasii.* The catches at night were far higher and the average size of the larvae was greater, from which he concluded that the larvae were concentrated at the surface by night and spread over a fairly wide stratum by day. There seemed to be net avoidance, even of high speed nets, by the larger larvae in the daytime. Ryland (1964) found a reverse distribution of larval plaice and sand eels *Ammodytes,* which were spread over a wide depth range from the surface to about 35 m by night and much more concentrated between 5 and 10 m during the day. The use of high speed nets in this study and in that of Colton (1965) on haddock larvae, *Melanogrammus aeglefinus,* prevented diurnal variation in size and presumably overcame the net avoidance factor.

C. Buoyancy and Pressure

There are a number of devices which provide buoyancy in pelagic fish eggs and larvae. In marine species, found in high salinities, maintenance of the body fluid concentration below that of the environment acts as a buoyancy mechanism. In the early stages lack of a skeleton also keeps the specific gravity fairly low. In freshwater the eggs are often demersal so that the much greater problem of maintaining buoyancy does not exist. In fact, in salmonid eggs the specific gravity may be very high in the early stages because of low water content of the yolk. It is sometimes said that a large perivitelline space is a mechanism for buoyancy. However, this normally contains water of the same concentration as the environment and is thus much more likely to act as a shock absorber, particularly in pelagic fish eggs subject to wave action. The embryo and yolk may regulate the body fluids from fertilization (e.g., in pelagic plaice eggs: Holliday and Jones, 1967) by means of the vitelline membrane. In the demersal herring egg there is no regulation until the yolk is covered by cellular tissue at gastrulation (Holliday and Jones, 1965). Many species such as pilchard *Sardina pilchardus,* mackerel *Scomber scombrus,* hake *Merluccius merluccius* and sole *Solea solea* have oil droplets, but these are not typical of pelagic eggs for they are absent in the sprat *Sprattus sprattus* and in *Gadus* and *Pleuronectes* species. Tendril-like outgrowths found, for example, in *Exocoetus* and *Scomberesox* eggs are as likely for attachment as for buoyancy.

There may be a mechanism to adjust specific gravity to the spawning medium. Solemdal (1967) found that variations in the osmotic pressure of the blood of female flounders, *Pleuronectes fleas,* at the spawning season depended on the salinity. This, in turn, caused changes in the specific gravity of the eggs when spawned. In water of low salinity the eggs were larger and their osmotic pressure lower. In the low saline water of the Baltic (6.5‰), fertilization and development were possible but the eggs rested on the bottom. In the sprat there is a correlation between low salinity and larger eggs (see Nikolsky, 1963) and also in herring (Holliday and Blaxter, 1960) and plaice (Holliday and Jones, 1967).

In the larval stages large subdermal spaces as in the plaice and cod (Shelbourne, 1956) will reduce specific gravity as long as the body fluids are maintained at an osmotic pressure below that of the environment. Other species have long processes to the dorsal or lateral fins, e.g., the angler fish *Lophius* and deal fish *Trachypterus*. It may be significant that Orton (1957) quotes the latter species as having only moderate fin folds and subdermal spaces, but whether the increased surface area assists buoyancy is open to question; these may be organs to reduce predation. In freshwater, the larvae of salmonids can afford to retain dense yolk reserves as they rest on the bottom. Others like carp and bream have cement organs for attaching themselves to weed.

The first filling of the swim bladder is well known in freshwater fish (see Tait, 1960). In the physostomatous salmonids like *S. trutta, S. gairdneri, Cristivomer namaycush,* and *Coregonus clupeaformis* the swim bladder is filled 1–3 months after hatching, depending on temperature or feeding. Air is taken in from the surface via the pneumatic duct and without access to the surface the swim bladder remains empty. The pneumatic duct remains open under such conditions and functions later. Lake trout, *Cristivomer namaycush,* were able to swim considerable vertical distances (280 m) with unfilled bladders suggesting that development in deep water presented no problems for filling the swim bladder. Physoclists like *Gasterosteus, Lebistes* (von Ledebur, 1928) and *Hippocampus* (Jacobs, 1938) also require access to air to fill the swim bladder but if deprived of this at the critical time the pneumatic duct closes and the swim bladder cannot be filled later.

Pressure effects on fish larvae were tested briefly by Bishai (1961). Newly hatched herring survived compression to and decompression from 4 atm and young plaice 37–50 mm long to and from 2 atm. Young salmonids could live at 5 atm to the end of the yolk sac stage, sometimes showing increased activity during compression or decompression. Older fry, about 20 days after yolk resorption, could withstand compression to 2 atm but had difficulty in dealing with air bubbles inside the body and in the swim bladder during decompression. Older fry still, about 60–140 days after yolk resorption, could not even withstand compression to 1.2 atm. At later stages, the ability to withstand compression increased but decompression was lethal. Qasim *et al.* (1983) observed changes in the vertical distribution of larval teleosts subjected to pressure changes. Young plaice larvae tended to swim upwards when the pressure was increased from 1 atm to 2 atm and sank when the pressure was returned to normal. Metamorphosing plaice did not show this response. Larvae of the blenny *Centronotus gunnellus* responded in a similar way to pressure changes equivalent to 25 cm of seawater and larvae of another blenny *Blennius pholis* to only 5 cm of seawater. Only the last of these species has a swim bladder. It is possible in such experiments that the larvae could have been responding to water currents produced by the pressure changes. If this is not so, one may ask how larvae without swim bladders can respond to pressure.

D. Locomotion and Schooling

Most species show signs of activity within the egg, and movement is often an important part of the hatching process. The relatively large yolk sac at hatching must be a hydrodynamic embarrassment to the larvae, but in salmonids, where it is especially large, there is little movement as the larvae grow in a rather passive way within the stream bed. The buoyant oil globule found in many marine pelagic larvae may also make equilibrium difficult. Two sorts of movement are found at yolk resorption-a serpentine eel-like mode of swimming in the long thin-bodied clupeoid type and a more maneuverable movement allowing "backing-up" with the use of the pectoral fin and marginal fin folds in the shorter flatfish type of larvae.

Swimming performance is shown in Fig. 13 in terms of burst speeds maintained for a few seconds, where the theoretical 10 body lengths/ sec (shown as dashed line) seems to fit quite well. Cruising speeds, of the order of 2–3 body lengths/sec (ideally the maximum sustainable speed without an oxygen debt accumulating), are shown in Table XII. Improvements in performance can be correlated with development and upturning of the caudal fin.

If schooling occurs it usually starts at metamorphosis. Presumably the larvae, being transparent, do not provide very good mutual stimuli for keeping schools together visually. Shaw (1960, 1961) analyzed the development of schooling in the silverside *Menidia* and found responses initiated between two fry at a body length of 8–9 mm with increasing participation as they grew. By 11–12 mm as many as 10 could be seen in parallel orientation. Fish reared

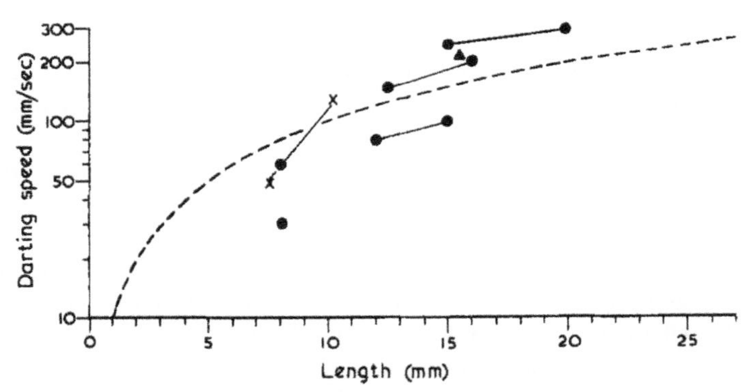

Fig. 13. Darting speeds of larvae of different length: (●—●) *Clupea harengus* 10°–15°C (Blaxter, 1962), (X—X) *Pleuronectes platessa* 6.5°–7.5° (Ryland, 1963), (▲) *Carassius auratus*, no temperature (Radakov, 1964), and (▼) *Perca fluviatilis*, no temperature (Radakov, 1964).

Table XII Cruising Speeds of Some Young Fish

Species	Length (mm)	Speed (mm/sec)	Temp. (°C)	Time maintained	Distance swum (m/hr)	Author
Pleuronectes platessa	7.5	12	6.5–7.5	A few sec		Ryland (1963)
(plaice)	9.6	27	6.5–7.5	A few sec		
Coregonus wartmanni	10–12	16	4	?	58	Braum (1964)
(whitefish)		29	16	?	103	
Rutilus rutilus (roach)	35–45	138	?	1 min		Aslanova quoted by Radakov (1964)
Abramis brama (bream)	45–55	126	?	1 min		
Cyprinus carpio (carp)	50–60	129	?	1 min		
Trachurus trachurus (horse mackerel)	30–40	136	?	1 min		
Mullus barbatus (mullet)	35–45	128	?	1 min		
Mugil sp. (mullet)	45–55	156	?	1 min		
Blicca bjorkna (white bream)	18–26	330	?	?		Radakov (1964)
Clupea hurengus (herring)	6.5–8	5.8	(?)14	45 min		Bishai (1960d)

Continued

Table XII Cruising Speeds of Some Young Fish—Cont'd

Species	Length (mm)	Speed (mm/sec)	Temp. (°C)	Time maintained	Distance swum (m/hr)	Author
Clupea harengus (herring)	12–14	10	9–10	60 min	36	Blaxter (1966)
Solea solea (sole)	4	6–9	15	Long periods intermittent swimming	17–19	Rosenthal (1966)
Alburnus alburnus (bleak)	0.26 g (wet wt)	29.8	20		107	Ivlev (1960b)
Micropterus dolomieui (smallmouth bass)	22	146–312	20–35			Larimore and Duever (1968)

in isolation in paraffin-coated bowls took a few hours to become integrated into existing schools. Generally speaking there was a considerable latency in schooling when "isolates" and "quasi-isolates" (isolated 5–7 days after hatching) were brought together at a size where schooling was normally well developed. Surprisingly, this initial latency was less among "isolates" than "quasi-isolates." In the newly emerged fry of *Oncorhynchus* species, schooling ceased in the dark and following subjection to light they needed 15 min or more to complete the school again, although in all species except coho this time became reduced with age (Ali, 1959). In herring (Rosenthal, 1968) schooling started about 2 months after hatching at a body length of 25–30 mm. Starving larvae tended to school more positively, but the school rapidly broke up when food was offered.

Rheotropic (optomotor) responses (orientation to currents or moving backgrounds) are often well developed in the larval stage as shown in some experiments on swimming performance and from the many observations on salmonid larvae by day and night (e.g., Hoar, 1956, 1958). This response seems to be maintained by either visual or tactile cues, and the sign may change under shock treatment. In young salmonids a temperature rise of 12°C changed the rheotaxis from positive to negative, presumably a defensive response (Keenleyside and Hoar, 1954).

E. Searching Ability

The ability to search for prey or, in the case f visual feeders. "optically filter" the environment, depends on cruising speed and the distance at which food is perceived. Perception distances (see Table XI) depend on the movement of the prey, its orientation in relation to the eye, contrast with the background, and illumination. Activity in terms of cruising speed may well be reduced after a period of feeding and is certainly influenced by feeding drives in older larvae. Starvation may also cause a reduction in activity. The volume searched per hour has been estimated in whitefish and herring larvae (see Table XIII), When multiplied up on a daily basis the volume searched becomes very dependent on day length, that is, on latitudinal and seasonal effects.

VIII. MORTALITY, TOLERANCE AND OPTIMA

From hatchery, rearing, and experimental studies it is well known that fish eggs and larvae go through periods of mortality. Susceptibility to mechanical shock or nonoptimal temperatures and salinities may vary with age (see Battle, 1944). Marine fish eggs seem especially delicate until the completion of gastrulation although this has not been systematically tested. Perhaps sensitive morphogenetic processes in the early stage or failure to osmoregulate are the cause. Similarly there are stages, such as the eyed stage in salmonids, where eggs may be particularly resistant to damage. After hatching, further

Table XIII Volume Searched during Feeding

Species	Size (mm)	Volume searched (liter/hr)	Author
Coregonus wartmanni (whitefish)	(?) 10	14.6	Braum (1964)
Clupea hrengus (herring)	8–16	0.3–2.0	Blaxter (1966) Blaxter and Staines (1969a)
Chupea harengus (herring)	10	1.5–2	Rosenthal and Hempel (1968)
	13–14	6–8	
Sardina pilchardus (pilchard)	5–7	0.1–0.2	Blaxter and Staines (1969a)
Pleuronectes platessa (plaice)	6–10	0.1–1.8	Blaxter and Staines (1969a)

stages of mortality may be experienced which can sometimes by connected with the further development of certain organs and with such physiological changes as the transition from cutaneous to gill respiration or the development of the swim bladder. Rearing studies on marine fish (Section X) show a high mortality at the end of the yolk sac stage in species such as cod *G. morhua,* haddock *Melunogrammus aeglfinus*, pilchard *Sardina pilchardus,* herring *Clupea harengus,* and lemon sole *Microstomus kitt.* These "critical phases" are probably in part the result of unsuitable food being supplied, especially from the aspect of size. The question of whether critical phases at yolk resorption occur under natural conditions is a much vexed question, examined especially by fishery biologists looking for the factors controlling brood survival. J. C. Marr (1956), in reviewing this hypothesis, concluded in two species (mackerel *Scomber scombrus,* and Pacific sardine *Sardinops caerulea*) that mortality was more likely to be constant or steadily decreasing without the catastrophic periods advocated by earlier workers. Undoubtedly some of the five species mentioned have small mouths and the finding of suitable food in the sea at the end of the yolk sac stage before the point of no return (Section V) must be a serious problem.

The lethal levels of various environmental factors such as salinity, temperature, oxygen tension, pH, radiation, and mechanical stimuli have been assessed for fish eggs and larvae. Salinity tolerance is surprisingly wide, ranging in marine species which have been examined from about 5%, up to 45–50‰ with even a wider range over short periods of exposure (see Chapter by Holliday, Volume I). Predictably, the lethal temperature varies

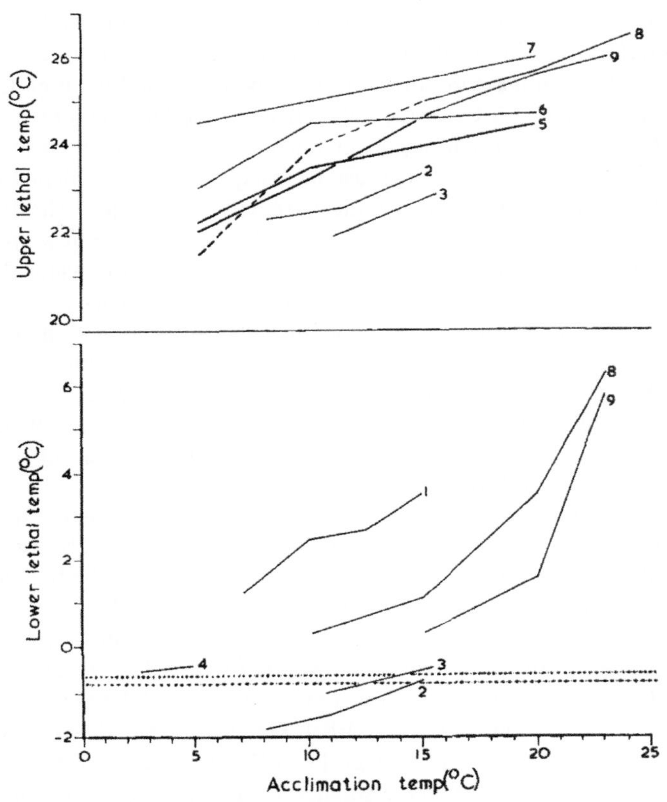

Fig. 14. Lower and upper lethal temperatures at different acclimation temperatures, criterion of lethal temperature being 50% survival after 24 hr; horizontal dashed lines give range of blood freezing point values. 1. *Brevoortia tyrannus*—in 24‰ salinity, acclimated for 12+ hr (Lewis, 1965). 2. *Clupea harengus* (spring spawned) and 3. *C. harengus* (autumn spawned)—in 34‰ salinity, acclimated for long period (Blaxter, 1960). 4. *Oncorhynchus* keta—in 28‰ salinity, acclimated for 21 days (Brett and Alderdice, 1958). 5. Salmo *solar,* 6. S. trutta, and 7. S. *trutta* (fario)-in freshwater, acclimated for 5 days (Bishai, 1960b). 8. Oncorhynchus *tshawytscha,* and 9. *O.nerka* (Brett, 1952).

with level of adaptation. A wide variety of criteria have, unfortunately, been used for defining lethal temperature, ranging from 50% mortality after one day (or longer) to the number of days to 50% mortality. Examples of high and low lethal temperatures where the criteria could be standardized are given in Fig. 14 related to temperature of acclimation. Temperature fluctuations are much damped down in the sea and may be negligible in deep water. In the tropics the seasonal fluctuations may also be slight, and this may lead to a narrow range of tolerance. In 10 species of fish larvae from the Indian Ocean, Kuthalingam (1959) found the range from lower to upper lethal temperature was only 3°–4°C (about 28°–32°C). Temperature tolerance may also vary with age. Combs (1965) measured long-term temperature thresholds over the period from fertilization to hatching in *Oncorhynchus* species. In *O. nerka,* for example, the low temperature threshold was about 5°C and the high temperature threshold about 13.5°C. At the 128-cell stage and later both *O. nerka* and Chinook *O. tshawytscha,* had a greater range of tolerances. In freshwater or high latitudes resistance to cold must be adequate or else adaptations are developed which prevent contact with extremely cold water. The flounder, *Pseudopleuronectes americanus* (Pearcy, 1962), is one of the few species of flatfish with a demersal egg and it spawns inshore in very cold water. Temperature optima may be judged by hatching success. For instance in *Coregonus clupeaformis* the highest percentage hatching was at 0.5°C with a very rapid fall off above 6°C (Price, 1940). In herring temperature optima vary with the race (Blaxter, 1956). In medaka *Oryzias* the best survival was obtained when the rearing temperature was alternated between 22° and 30°C every 12 hr (Lindsey and Ah, 1965). Generally, outside the optima not only does the percentage hatch fall but there are more abnormalities.

Oxygen is apparently only limiting at its low level. However, supersaturated solutions may be harmful if air bubbles are swallowed by the larvae. This "gas disease" can be fatal if the larvae cannot eliminate the bubbles from the gut (see Bishai, 1960a) as their buoyancy is disturbed. The eggs of the salmonids *S. salar,* rainbow trout *S. gairdneri,* Pacific salmon *O. tshawytscha* and *O. kisutch,* brook trout *Salvelinus fontinalis,* and lake trout *Salvelinus namaycush,* have been tested at low oxygen tensions (Hayes *et al.,* 1951; Silver *et al.,* 1963; Shumway *et al.,* 1964; Garside, 1959, 1966). Most were able to resist levels of 2.5 ppm or even less (25% saturation) although development was retarded and the larvae smaller at hatching. Reductions in current flow had a similar effect. Alderdice *et al.* (1958), using chum salmon eggs, measured the retardation in hatching resulting from low dissolved oxygen levels at different developmental stages. The intermediate stages, about one-third of the way from fertilization to hatching, were most susceptible. Hayes *et al.* defined the limiting tension not from its lethal aspect, but the tension at which normal oxygen consumption started to drop (see Fig. 7D). In S. *salar* eggs this was 3.0 ppm at 20 days, 7.5 ppm at 45 days, and 4.7 ppm at hatching. Other salmonid data on limiting tensions are summarized by Wicket (1954). In the egg of chum salmon, *O. keta,* the calculated limiting tension varied from 0.72 to 3.7 ppm increasing with age (temperatures from 0.1° to 8.2°C). In the eggs of pike, *Esox lucius* (Lindroth, 1942), it varied from 1.2 to 4.1 ppm depending on age and temperature. This type of finding is of value in assessing the survival chances of fish eggs buried in gravel. Wickett (1954)) for example, has shown that the flow rates and oxygen content of the water may be quite inadequate for chum salmon eggs, being even as low as 2 mm/hr and 0.2 ppm.

Marine and other fish larvae were studied by Bishai (1960a). Herring larvae at hatching were "restricted at 27–32% saturation and at 55–64%

saturation a few days later. Larvae of the lumpsucker, *Cyclopterus lumpus,* just tolerated 43% saturation 3 weeks after hatching and young plaice 4–5 cm in length just tolerated 14% saturation. He found that the larvae of the salmonids, *S. salar* and *S. trutta,* could survive very low levels of saturation, 3–10%, after hatching, but this ability decreased with age, the minimum level tolerated being 16–28%. A gain, criteria for defining lethal levels vary, and it is difficult to compare the results of different workers. To some extent acclimation to low oxygen increases resistance to hypoxia and there are also specific differences in tolerance.

Bishai (1960c) has also summarized work on the effects of pH on larval fish. Compared with seawater (pH about 8.0) which is fairly well buffered, freshwater has a pH which is more variable and more susceptible to change by effluents or run-off. Bishai found that herring larvae could survive values between 6.5 and 8.5. Lumpsucker larvae had a low tolerance level of about 6.9, plaice (4–5 cm) 6.2–6.5, while salmonid larvae had a range of tolerance from about 5.8 to 6.2 up to 9.0, or even 10.0 in sea trout alevins. To some extent the levels of tolerance depend on the substances used to change the pH. It is possible that CO_2 for reducing pH has an additional toxic or respiratory effect compared with the use of HCl. Certainly Alderdice and Wickett (1958) found high levels of CO_2 inhibited oxygen consumption in the eggs of *O. keta.*

Radiation effects on salmonid eggs and larvae are reviewed by Hamdorf (1960) and Eisler (1961). Strong visible light may cause early hatching, mortality, poor growth, and greater pigmentation. Exposure to light is less serious when the intensity is low or in the later stages, especially after the onset of pigmentation. It is, perhaps, not surprising that light is harmful to eggs and larvae which normally develop in complete or semidarkness among gravel or stones; increased activity probably contributes to its deleterious effect. Other species which develop in sand such as the grunion, *Leuresthes tenuis* (McHugh, 1954), or near the sea bed such as herring (Blaxter, 1956) may also show a poorer rate of hatching under lighted conditions. Careful controls are required in experiments since the light may cause other changes such as raising the temperature or increasing growth of phytoplankton in the water. The main effect on many other species is to accelerate hatching. Ultraviolet light has been tested on *Fundulus* and causes abnormalities of the skeleton, cylopia and twinning (see Eisler, 1961). In sockeye salmon, *O. nerka* (Bell and Hoar, 1950), it causes premature hatching, vertebral abnormalities, high mortality, and delays pigmentation. Marinaro and Bernard (1966) exposed the eggs of pelagic fish such as pilchard *Sardina pilchardus,* mullet *Mullus,* horse mackerel *Trachurus* and "sargue" *Diplodus annularis* to sunlight with and without ultraviolet filters, finding a lower rate of hatching with ultraviolet. The transparency of many fish eggs and larvae may be an adaptation to prevent absorption of light as well as acting as a means of camouflage. X rays have been tested, for example, by Solberg (1938) on *Fundulus* embryos. They were most sensitive after fertilization with a decrease of 10 times in sensitivity between early cleavage and 4–5 days later. Other work is reviewed by Eisler (1961) especially on salmonids. X rays have a number of effects such as higher mortality, decrease of growth, deformities, decrease in erythrocytes, and destruction of hematopoietic tissue and increased pigmentation. Generally, susceptibility decreases with age and in optimal conditions of temperature. Neyfakh (1959) used X rays for inactivating the nuclei in the developing eggs of the loach, *Misgurnus fossilis*. Irradiation during certain phases of development was followed by death at a very specific time, suggesting a periodical functioning of the nuclei.

Fish eggs in almost any environment are subject to mechanical stimuli which are potentially harmful-movement of gravel within a spawning redd, current borne objects on the sea bed, or wave action at the surface. The resistance of the egg may be measured by the maximum load it will take before the chorion bursts. Some results are shown in Fig. 15. They indicate an increase in the resistance of the chorion as it hardens subsequent to fertilization and a decrease in the strength of the chorion prior to hatching, presumably resulting from hatching enzymes. Changes in the chorion may in part account for periods of varying resistance to external conditions.

Other harmful influences may be hatching enzymes, especially for embryos which may be somewhat unhealthy in other respects, or where eggs are very crowded and the enzymes are concentrated in the water after hatching. Stuart (1953) observed sediments attached to the chorion of loch trout. These might well reduce oxygen intake. The alevins, however, seemed to have the ability to disperse such sediments by means of respiratory currents

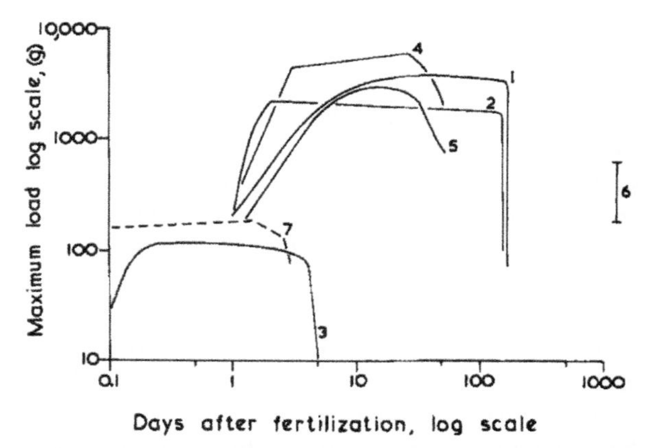

Fig. 15. The maximum load before bursting in eggs of different species during development. 1. *Salmo salar*, 2. *Coregonus lavaretus*, 3. *Acipenser* sp. (see Nikolsky, 1983); 4. *Salvelinus fontinalis* (?), 5. *Salmo salar* (see Hayes, 1949); 6. *Pleuronectes platessa* (range, varying age) (Shelbourne, 1963); and 7. *Engraulis anchoita* (De Ciechomski, 1967a).

passing over the body surface. With gill respiration sediments were aggregated by mucus and rendered less harmful.

A recent trend in research on tolerance is to use combinations of factors such as temperature, salinity, and oxygen concentration to see how they interact and to find out the optima. Kinne and Kinne (1962) used such combinations during the development of the desert minnow, *Cyprinodon mucularius* (i.e., temperature between 10° and 37°C, salinities from 0 to 85‰ and 70%, 100%, and 300% saturation of air in water). Mortality was least at near lethal temperatures when the salinity was held between one-half seawater and 35‰. The lethal temperature was lower in hypoxial conditions. It seemed that of the salinities used 35%, was the optimum for incubation. Lewis (1966)) using menhaden larvae, *Brevoortia tyrannus,* found that temperature tolerance was greatest at salinities of 10–15‰, suggesting that isotonic conditions were optimum for survival. Forrester and Alderdice (1966) calculated from experimental data that the optimum salinity and temperature for hatching eggs of cod, *Gadus macrocephalus,* was 19.4‰, and 53°C.

IX. MERISTIC CHARACTERS

Counts of vertebrae, myotomes, scales, gill rakers or fin rays have been used in racial studies. The variability in number and lability under different environmental conditions are striking examples of rather flexible raw material for evolutionary forces. Usually mean counts from samples of fish are required, the differences from race to race being inadequate for identification of individuals. While correlations between the counts in adults or young and the environmental conditions on the spawning ground first suggested that meristic characters were labile, confirmation has been obtained by experimental studies. They show that factors such as temperature, salinity, and oxygen content superimpose their effect on the range of meristic counts determined by the genotype. There is also an individual variation even in fish from the same parents reared under the same conditions. Earlier work, mainly on salmonid and inshore or estuarine species, has been reviewed by Tåning (1952) and Barlow (1961). Since it is necessary to keep the young until the meristic characters form there has been less work on species which require to be fed and reared for a considerable time after hatching.

The effects of some factors are given in Table XIV. Most work has been done on the effect of temperature on vertebral and fin ray counts, Generally a V-type relationship is found where the average vertebral count is minimum at an intermediate temperature. For example, in sea trout, *Salmo trutta,* Tåning found this was at about 6°C; in plaice it is at 8°C (Molander and Molander-Swedmark, 1957); at lower and higher temperatures the average counts were higher. With fin rays the V may tend to be asymmetrically inverted with counts highest at intermediate temperatures. In other instances the mean vertebral counts are inversely related to temperature, but the lack

Table XIV Environmental Effects on Meristic Characters by Experiment

Species	Character	Effects of increasing			Author
		Temperature	*Salinity*	*Oxygen*	
Lebistes reticulatus (guppy)	Fin rays	Increase			Schmidt (1917, 1919)[a]
Salmo trutta (sea trout)	Vertebrae	V relation			Schmidt (1921)[a]
Salmo trutta (sea trout)	Vertebrae	V relation		Decrease	Tåning (1952)
	Fin rays	A-relation (asymmetrical)			
Salmo gairdneri (rainbow trout)	Vertebrae	(?)Decrease		(?)Decrease	Mottley (1934, 1937)[a]
	Lateral scales	Decrease			
Oncorhynchus tshawytscha (Chinook salmon)	Vertebrae	V relation			Seymour (1959)
	Fin rays	A relation			
Fundulus heteroclitus (killifish)	Vertebrae	Decrease			Gabriel (1944)
Macropodus opercularis (paradise fish)	Vertebrae	V relation			Lindsey (1954)
	Fin rays	V relation			
Pleuronectes platessa (plaice)	Vertebrae	Decrease			Dannevig (1950)[a]
Pleuronectes platessa (plaice)	Vertebrae	V relation			Molander and Molander-Swedmark (1957)
	Fin rays	Increase			
Channa argus (snake-headed fish)	Vertebrae	V relation			Itazawa (1959)

Clupea harengus (herring)	Myotomes	Decrease	Increase		Hempel and Blaxter (1961)
	Vertebrae	Decrease			
Gasterosteus aculeatus (stickleback)	Vertebrae	V relation	See text		Lindsey (1962)
	Fin rays	Decrease			
Oryzias latipes (medaka)	Vertebrae	Decrease			Lindsey and Ali (1965)

[a]References in Tåning's (1952) review.

of an inflection may result from an insufficiently wide range of experimental temperatures being used.

Apart from the factors shown in Table XIV, both intensity and duration of light (McHugh, 1954; Lindsey, 1958) and CO_2, (Tåning, 1952) may have an effect. Higher intensities in the grunion, *Leuresthes tenuis,* reduce the number of vertebrae. A 16-hr day caused lower caudal vertebrae and possibly anal fin ray counts in *Oncorhynchus nerka,* while in sea trout, *Salm trutta,* increasing tension of CO, produced lower numbers of vertebrae.

The work of Heuts (1949) and Lindsey (1962) using different temperature-salinity combinations in *Gasterosteus aculeatus* were somewhat inconclusive resulting from the problem of getting sufficient numbers of survivors in an extensive series of experiments, some of which were in unfavorable conditions. Certainly it was shown that temperature effects vary with the salinity level. Using a freshwater and brackish water race, Heuts found the greatest effect of temperature on fin ray counts at the salinity of adaptation. Differential mortality and the unsolved problem of controlling, for example, oxygen and temperature independently have often led to difficulties in interpreting results.

It is the underlying reasons for meristic lability which is, perhaps, of most interest to the physiologist. Sensitive periods have been found during which lability of vertebral counts is at its greatest. In sea trout it is around gastrulation, with a further period when the last vertebrae are being preformed (Tåning, 1952). It is also before hatching in herring (Hempel and Blaxter, 1961) and killifish (Gabriel, 1944). However, in plaice (Molander and Molander-Swedmark, 1957) and paradise fish (Lindsey, 1954) meristic characters are still susceptible to modification after hatching. Generally speaking, fin ray counts are determined later than those of vertebrae. Later studies on *Oryzias* (Lindsey and Ali, 1965) showed that transfer to a higher temperature 2 days after fertilization produced fewer vertebrae, whereas transfer after 6 days resulted in a higher vertebral count. Regular diurnal alterations of temperature between 22° and 30°C gave vertebral counts intermediate between those resulting from sustained temperatures of 22° and 30°C.

Of the various hypotheses put forward to explain meristic lability those of Gabriel (1944) and Barlow (1961) fit the facts best. Often, but not always, environmental factors which delay hatching produce higher counts, Alternatively, higher counts may be considered to come from nonoptimum conditions. It is likely that the environmental factors change the relationship between growth and differentiation. Hayes *et al.* (1953), in their studies of morphogenesis in salmon at temperatures from 1° to 15°C, certainly found that the relative appearance of certain anatomical characters varied with temperature. If differentiation is late more tissue is available to be differentiated, leading to a higher count. Thus one may consider an interaction of the Q_{10} for growth and the Q_{10} for differentiation of the character; as the temperature is varied so the relationship between growth and differentiation varies. Where

V-shaped curves are obtained there are then points of inflection for the temperature coefficients. Garside and Fry (1959), using normal and reciprocal hybrid fry of speckled trout, *Sahelinus fontinalis,* and lake trout, *S. namaycush,* found that the mean myomere count was lower where the fish developed on the speckled trout yolks, which were smaller. There was also an inverse relationship between myomere count and the degree of twinning in Siamese twins of speckled trout, the counts being lower where there was less shared tissue.

X. REARING AND FARMING

A. Techniques

Three lines of approach are of interest at the present time:

(1) Tropical and temperate fish culture for food
(2) Rearing of salmonids and sturgeon to maintain population size after intensive fishing or hydroelectric schemes
(3) Temperate marine fish rearing and farming

A good general review on fish farming by Iversen (1968) is available and on tropical species by Hickling (1962). Some of the most popular species are the common carp *Cyprinus carpio,* mullets, *Tilapia,* milkfish *Chanos chanos,* the carp *Puntius javanicus,* and grass carp *Ctenopharyngodun idella.* Many of the species are reared in still water fertilized artificially or by sewage. The last four species mentioned are herbivorous which gives maximum efficiency of food utilization. Some of the species, e.g., *Tilapia* and *Chanos,* can be kept in salt or brackish water. In the case of *Tilapia* and the Israeli carp, breeding takes place in the holding ponds; in the European carp special breeding ponds are used, On the other hand, the fry of *Puntius, Chanos, Ctenopharyngodon,* and mullet must be caught in the wild and transferred to rearing ponds. Generally in tropical culture there are no serious problems of feeding the young which are of sufficient size to take the food available.

The extensive breeding of salmonids presents no problem of food supply for the young, which are relatively large when the plentiful yolk has been resorbed. Rearing is usually in running freshwater although acclimation of the young of *Salmo* species to marine conditions is now being developed, with improved growth and freedom from disease. The question of dietetics is well advanced in the group (see papers in *The Progressive Fish Culturist*).

Marine fish rearing (see reviews by Morris, 1956, and Atz, 1964) and fanning (review by Shelbourne, 1964), although attempted from time to time over the last hundred years, have only recently made significant advances. Starting with the liberation of millions of plaice and cod eggs or yolk sac larvae in coastal areas, only the later experiments by Dannevig (1963) seem convincingly exploitable. Very recently, *Mugil* and *Solea* have been established as self-perpetuating species in some of the Mediterranean coastal lagoons.

Following the earlier rearing in aquaria of some species to an advanced stage by highly empirical methods, a better experimental approach has now been adopted. Success in rearing *Pleuronectes platessa, Solea solea, Microstomus* kitt (Shelbourne, 1964; unpublished data) , *Clupea harengus* (see Blaxter, 1968a; Rosenthal, unpublished data), *Sardinops caerulea, Engraulis* and *Scomber* (Schumann, 1967; unpublished data), and *Sardinu pilchardus* (Blaxter, 1969c) has depended on some or all of the following factors:

(1) Food supply. The use of *Artemia* and *Balanus* nauplii, *Mytilus* trochophores, young *Tigriopus,* small nematodes, rotifers and oligochaetes and secondary foods like *Dunaliella* has improved survival rates, so also has the extensive collection and selection of natural plankton
(2) Antibiotics. Particularly with flatfish, dosages of mixed sodium penicillin (50 IU/ml) and streptomycin sulfate (0.05 mg/ml) have improved survival in the egg stage
(3) Black-walled tanks and uniform overhead illumination have helped feeding success by providing contrast of the food against the background and even distribution of food in the tank
(4) Tank hygiene

There are interesting physiological and behavioral aspects of rearing where the fish may be artificially crowded and conditions of stress set up (see Shelbourne, 1964; Riley and Thacker, 1963). Size hierarchy effects prevail (see Section IV, F), but growth seems to be density dependent only from the beginning of metamorphosis (Bowers, 1966a). Bitten fins are found more often in the smaller members of a tank community but are not always dependent on density. Perhaps more serious is the incidence of abnormal (albino or semi-albino) pigmentation in flatfish. It is generally worse in smaller fish and in very crowded tanks but may be reduced by using *spawning* stock acclimated to hatchery conditions (Shelbourne, 1964; unpublished data). Survival to metamorphosis is also better in larvae from acclimated spawning stock (Bowers, 1966b).

B. Sensory Deprivation

Hatchery-reared flatfish do not survive well when transferred to the sea, presumably because of high predation; hatchery-reared salmonids usually have a poorer swimming performance. In fact, the hatchery environment by preventing contact from predation, by often rather uniform physical and chemical conditions and lack of shelter may prevent the development of appropriate defensive responses and muscular systems. The need for high density may also encourage undesirable intraspecific relationships at the expense of the more desirable interspecific responses.

The more obvious shortcomings of the aquarium tank may be accompanied by sensory deprivation in a more general and less easily defined way. The uniformity of the aquarium environment may lead to insufficient sensory input. For instance, Qasim (1959), when rearing the young of *Blennius pholis,* obtained better survival with a day-night regime than with continuous light. Blaxter (1968a), rearing herring, found a hint of better survival when the light conditions over the tank were continually changed and when air jets playing on the surface kept up a continuous series of ripples. Hoar (1942) found that young salmon would feed on chopped earthworm in darkness during the day, but not at night, perhaps because of an activity (or retinomotor) rhythm. Such rhythms might eventually be suppressed by continuous light. Activity rhythms with periods of rest might play as important a part in the development of the sense organs and nervous system as sufficient sensory input during the active phases.

XI. CONCLUSIONS

The most substantial advances in recent years have been made in investigations of marine eggs and larvae. Further work is needed on the following:

(1) Factors determining spawning time and fecundity, especially of marine fish
(2) Permeability studies on the egg using isotope techniques, with associated studies on the fine structure of the membranes
(3) The order in which different substrates in the yolk are utilized and the interrelations between the organic materials in the yolk and embryo being used for growth and maintenance
(4) Differences in egg quality (of a subtle or less subtle nature) which may influence viability
(5) Feeding mechanisms and nutrition in larvae with special reference to the possible utilization of small particles or dissolved matter
(6) The respiratory and circulatory system in the larval stage where no functional gills or hemoglobin are present
(7) Differences in the sensory systems of larvae and adults
(8) The endocrine system in larvae, especially at metamorphosis
(9) Metamorphosis itself, particularly changes in the blood, laying down of scales and pigment, and behavior
(10) In rearing studies, so vital for investigations of larval stages and for techniques of fish culture, aspects of nutrition and stress need attention, especially the effect of tank conditions on growth, mortality, and the development of behavior

ACKNOWLEDGMENTS

The author is most grateful to Professor F. G. T. Holliday, University of Stirling, Dr. P. A. Orkin, University of Aberdeen, Dr. Sydney Smith, University of Cambridge, and to the editors, who read this chapter in draft.

REFERENCES

Alderdice, D. F., and Wickett, W. P. (1958). A note on the response of developing chum salmon eggs to free carbon dioxide in solution. *J. Fisheries Res. Board Can.* **15,** 797–799.

Alderdice, D. F., Wickett, W. P., and Brett, J. R. (1958). Some effects of temporary exposure to low dissolved oxygen levels on Pacific salmon eggs. *J. Fisheries Res. Board Can.* **15,** 229–249.

Ali, M. A. (1959). The ocular structure, retinomotor and photobehavioural responses of juvenile Pacific salmon. *Can. J. Zool.* **37,** 965–996.

Amberson, W. R., and Armstrong, P. B. (1933). The respiratory metabolism of *Fundulus heteroclitus* during embryonic development. *J. Cellular Comp. Physiol.* **2,** 381–397,

Amoroso, E. C. (1960). Viviparity in fishes, *Symp. Zool. Soc. London* **1,** 153–181.

Anokhina, L. E. (1960). Relationship between the fertility, fat content and the variations in the size of ova in the clupeid fishes. *Tr. Soveshch. Ikhtiolog. Komis. Akad. Nauk SSSR* **13,** 290–295 (in Russian).

Atz, J. W. (1964). A working bibliography on rearing larval marine fishes in the laboratory. *U.S. Fish Wildlife Serv. Bur. Sport Fishing Wildlife Res. Rept.* **63,** 95–102.

Baerends, G. P., Bennema, B. E., and Vogelzang, A. A. (1960). Über die Anderung der Sehschärfe mit dem Wachstum bei *Aeguidens portalegrensis* (Hensel) (Pisces, Cichlidae). *Zool. Jahrb.* **88,** 67–78.

Bagenal, T. B. (1966). The ecological and geographical aspects of the fecundity of plaice. *J. Marine Biol. Assoc. U.K.* **46,** 161–186.

Barlow, G. W. (1961). Causes and significance of morphological variation in fishes. *Systemutic Zool.* **10**, 105–117.

Barrett, I. (1951). Fertility of salmonid eggs and sperm after storage. *J. Fisheries Res. Board Can.* **8,** 125–133.

Barrington, E. J. W. (1961). Metamorphic processes in fishes and lampreys. *Am. Zoologist* **1,** 97–106.

Battle, H. I. (1930). Effects of extreme temperatures and salinities on the development of *Enchelyopus cimbrius* (L.). *Contrib. Can. Biol. Fisheries* [N.S.] **5,** 109–192.

Battle, H. I. (1944). The embryology of the Atlantic salmon (*Salmo salar* Linnaeus). *Can. J. Res.* **D22,** 105–125.

Baxter, I. G. (1959). Fecundity of winter-spring and summer-autumn herring spawners. *J. Conseil Conseil Perm. Intern. Exploration Mer* **25,** 73–80.

Bell, G. M., and Hoar, W. S. (1950). Some effects of ultra-violet radiation on sockeye salmon eggs and alevins. *Can. J. Res.* **D28,** 3543.

Bishai, H. M. (1960a). The effect of gas content of water on larval and young fish. z. *Wks. Zool.* **163,** 37–64.

Bishai, H. M. (1980b). Upper lethal temperatures for larval salmonids. *J. Conseil Conseil Perm. Intern. Exploration Mer* **25,** 129–133.

Bishai, H. M. (1960c). The effect of hydrogen ion concentration on the survival and distribution of larval and young fish. *Z. Wiss. Zool.* **164,** 107–118.

Bishai, H. M. (1960d). The effect of water currents on the survival and distribution of fish larvae. *J. Conseil Conseil Perm. Intern. Exploration Mer* **25,** 134–146.

Bishai, H. M. (1961). The effect of pressure on the survival and distribution of larval and young fish. *J. Conseil Conseil Perm. Intern. Exploration Mer* **26,** 292–31 1.

Blaxter, J. H. S. (1953). Sperm storage and cross fertilization of spring and autumn spawning herring. *Nature* **172,** 1189.

Blaxter, J. H. S. (1955). Herring rearing. I. The storage of herring gametes. *Marine Res. Scotland* No. **3,** 1–12.

Blaxter, J. H. S. (1956). Herring rearing. II. The effect of temperature and other factors on development. *Marine Res. Scotland* No. **5,** 1–19.

Blaxter, J. H. S. (1960). The effect of extremes of temperature on herring larvae. *J. Marine Biol. Assoc. U.K.* **39,** 605–608.

Blaxter, J. H. S. (1962). Herring rearing. IV. Rearing beyond the yolk-sac stage. *Marine Res. Scotland* No. **1,** 1–18.

Blaxter, J. H. S. (1965). The feeding of herring larvae and their ecology in relation to feeding, *Rept. Calif. Coop. Oceanog. Fisheries Invest.* **10,** 79–88.

Blaxter, J. H. S. (1966). The effect of light intensity on the feeding ecology of herring. *Symp. Brit.* Ecol. Soc. **6,** 393–409.

Blaxter, J. H. S. (1968a). Rearing herring larvae to metamorphosis and beyond. *J. Marine Biol. Assoc. U.K.* **48,** 17–28.

Blaxter, J. H. S. (1968b). Light intensity, vision and feeding in young plaice. *J. Exptl. Marine Biol. Ecol.* **2,** 293–307.

Blaxter, J. H. S. (1968c). Visual thresholds and spectral sensitivity of herring larvae. *J. Exptl. Biol.* **48,** 39–53.

Blaxter, J. H. S. (1969a). Feeding and condition of Clyde herring larvae. *Rapp Proces-Verbaux Reunions, Conseil Perm. Intern. Exphation Mer* (in press).

Blaxter, J. H. S. (1969b). Visual thresholds and spectral sensitivity of flatfish larvae. *J. Exptl. Biol.* **51,** 221–230.

Blaxter, J. H. S. (1969c). Experimental rearing of pilchard larvae, *Sardina pilchardus. J. Marine Biol. Assoc. U.K.* **49,** (in press).

Blaxter, J. H. S., and Hempel, G. (1963). The influence of egg size on herring larvae (*Clupea hurengus* L.). *J. Conseil Conseil Perm. Intern. Exploration Mer* **28,** 211–240.

Blaxter, J. H. S., and Hempel, G. (1966). Utilization of yolk by herring larvae. *J. Marine Biol. Assoc. U.K.* **461,** 219–234.

Blaxter, J. H. S., and Jones, M. P. (1967). The development of the retina and retinomotor responses in the herring. *J. Marine Biol. Assoc. U.K.* **47,** 677–697.

Blaxter, J. H. S., and Staines, M. (1969a). Food searching potential in marine fish larvae. *Proc. 4th European Marine Biol. Symp* (in press).

Blaxter, J. H. S., and Staines, M. (1969b). The pure-cone retina and retinomotor responses in teleost larvae. *J. Marine Biol. Assoc. U.K.* (submitted for publication).

Bowers, A. B. (1966a). Growth in hatchery-reared plaice. *Rept. Challenger Soc.* **3,** No. 18.

Bowers, A. B. (1966b). Marine fish culture in Britain. VI. The effect of the acclimatization of adult plaice to pond conditions on the viability of eggs and larvae. *J. Conseil Conseil Perm. Intern. Exploration Mer* **30,** 196–203.

Braum, E. (1964). Experimentelle Untersuchungen zur ersten Nahrungsaufnahme und Biologie an Jungfischen von Blaufelchen (*Coregonus wartmanni* Bloch), Weissfelchen (*Coregonus fera* Jurine) und Hechten (*Esox lucius* L.). *Arch. Hydrobiol.* **28,** 2/3, 183–244.

Breder, C. M., Jr., and Rosen, D. R. (1966). "Modes of Reproduction in Fishes." Natural History Press, Garden City, New York.

Brett, J. R. (1952). Temperature tolerance in young Pacific salmon Genus *Oncorhynchus. J. Fisheries Res. Board Can.* **9,** 265–323.

Brett, J. R., and Alderdice, D. F. (1958). The resistance of cultured young chum and sockeye salmon to temperatures below 0°C. *J. Fisheries Res. Board Can.* **15,** 805–813.

Brett, J. R., and Groot, C. (1963). Some aspects of olfactory and visual responses in Pacific salmon. *J. Fisheries Res. Board Can.* **20,** 287–303.

Bridger, J. P. (1956). On day and night variation in catches of fish larvae. *J. Conseil Conseil Perm. Intern. Exploration Mer* **22,** 42–57.

Colton, J. B., Jr. (1965). The distribution and behavior of pelagic and early demersal stages of haddock in ,relation to sampling techniques. *Spec. Publ. Intern. Comm. Northwest Atlantic Fisheries* **6,** 317–433.

Combs, B. D. (1965). Effect of temperature on development of salmon eggs. *Progressive* Fish *Culturist* **27,** 134–137.

D'Ancona, U., Sanzo, L., Sparta, A,, Bertolini, F., Ranzi, S., and Montalenti, G. (1931–1933). Uova, larve e stadi giovanili di Teleostei. "Fauna & Flora del Golfo di Napoli," Monograph 38, 4 parts. Publ. Naples Zool. Sta.

Dannevig, A. (1963). Artificial propagation of cod. *Rept. Norwegian Fishery Marine Invest.* **13,** 73–79.

Daykin, P. N. (1965). Application of mass transfer theory to the problem of respiration of fish eggs. *J. Fisheries Res. Board Can.* **22,** 159–171.

Dean, B. (1916). "A Bibliography of Fishes," 3 vols. New York Museum, Massachusetts Univ. Press, Cambridge, Massachusetts.

De Ciechomski, J. Dz. (1965). The development of fish embryos in a non-aqueous medium. *Acta Embryol Morphol. Exptl.* **8,** 183–188.

De Ciechomski, J. Dz. (1967a). Influence of some environmental factors upon embryonic development of the Argentine anchovy *Engraulis anchoita* (Hubbs, Marini). *Rept. Calif. Coop. Oceanog. Fisheries Invest.* **11,** 67–71.

De Ciechomski, J. Dz. (1967b). Investigations of food and feeding habits of larvae and juveniles of the Argentine anchovy *Engraulis anchoita. Rept. Calif. Coop. Oceanog. Fisheries Invest.* **11,** 72–81.

Deuchar, E. M. (1965). Biochemical patterns in early developmental stages of vertebrates. *In* "The Biochemistry of Animal Development" (R. Weber, ed.), Vol. 1, pp. 258–263. Academic Press, New York.

Devillers, C. (1961). Structural and dynamic aspects of the development of the teleostean egg. *Advan. Morphogenesis* **1,** 379–428.

Devillers, C. (1965). Respiration et morphogenèse dans I'oeuf des Téléostéens. *Ann. Biol.* **4,** 157–186.

Dimond, M. T. (1963). The relation of whole body I^{131} uptake to thyroid activity in the developing dogfish *Scyliorhinus canicula* (L.). *Biol. Bull.* **124,** 170–181.

Ehrenbaum, E. (1909). "Eier und Larven von Fischen des Nordischen Planktons." Lipsius & Tischer, Kiel and Leipzig.

Eisler, R. (1961). Effects of visible radiation on salmonid embryos and larvae. *Growth* **25,** 281–346.

Farris, D. A. (1959). A change in the early growth rates of four larval marine fishes. *Limnol. Oceanog.* **4,** 29–36.

Flüchter, J. (1962). Funktionsanatomische Untersuchungen am Kieferapparat der Heringslarven. *Kurze Mitt. Inst. Fischereibiol. Univ. Hamburg* **12,** 1–12.

Flüchter, J., and Rosenthal, H. (1965). Beobachtungen über das Vorkommen and Laichen des Blauen Wittlings (*Micromesistius poutassou* Risso) in der Deutschen Bucht. *Helgolaender Wiss. Meeresuntersuch.* **12,** 149–155.

Forrester, C. R., and Alderdice, D. F. (1966). Effects of salinity and temperature on embryonic development of the Pacific cod (*Gadus macrocephalus*), *J. Fisheries Res. Board Can.* **23,** 319–340.

Gabriel, M. L. (1944). Factors affecting the number and form of vertebrae in *Fundulus heteroclitus. J. Exptl. Zool.* **95,** 105–147.

Garside, E. T. (1959). Some effects of oxygen in relation to temperature on the development of lake trout embryos, *Can. J. Zool.* **D37,** 689–698.

Garside, E. T. (1966). Effects of oxygen in relation to temperature on the development of embryos of brook trout and rainbow trout. *J. Fisheries Res. Board Can.* **23,** 1121–1134.

Garside, E. T., and Fry, F. E. J. (1959). A possible relation between yolk size and differentiation in trout embryos. *Can. J. Zool.* **D37,** 383–386.

Ginsburg, A. A. (1961). The block to polyspermy in sturgeon and trout with special reference to the role of cortical granules (alveoli). *J. Embryol. Exptl. Morphol.* **9,** 173–190.

Gray, J. (1926). The growth of fish. I. The relationship between embryo and yolk in *Salmo fario. J. Exptl. Biol.* **4,** 215–225.

Gray, J. (1928a). The growth of fish. II. The growth-rate of the embryo of *Salmo fario. J. Exptl. Biot.* **6,** 110–124.

Gray, J. (1928b). The growth of fish. III. The effect of temperature on the development of the eggs of *Salmo fario. J. Exptl. Bid.* **6,** 125–130.

Hamdorf, K. (1960). Die Beeinflussung der Embryonal-und Larvalentwicklung der Regenbogen-forelle (*Salmo irideus* Gibb.) durch Strahlung im sichtbaren Bereich. Z. *Vergleich. Physiol.* **42,** 525–565.

Harder, W. (1954). Die Entwicklung der Respirationsorgane beim Hering (*Clupea harengus* L.), Z. *Anat. Entwicklungsgeschichte* **118,** 102–123.

Harder, W. (1960). Vergleichende Untersuchungen zur Morphologie des Darmes bei Clupeoidea. Z. *Wiss. Zool.* **163,** 65–167.

Hayes, F. R. (1949). The growth, general chemistry and temperature relations of salmonid eggs. *Quart. Rev. Biol.* **24,** 281–308.

Hayes, F. R., and Pelluet, D. (1945). The effect of temperature on the growth and efficiency of yolk conversion in the salmon embryo. *Can. J. Res.* **D23,** 7–15.

Hayes, F. R., Pelluet, D., and Gorham, E. (1953). Some effects of temperature on the embryonic development of the salmon (*Salmo salar*). *Can. J. Zool.* **D31,** 42–51.

Hayes, F. R., Wilmot, I. R., and Livingstone, D. A. (1951). The oxygen consumption of the salmon egg in relation to development and activity. *J. Exptl. Zool.* **116,** 377–495.

Heard, W. R. (1964). Phototactic behaviour of emerging sockeye salmon fry. *Animal Behaviour* **12,** 382–388.

Hempel, G., and Blaxter, J. H. *S.* (1961). The experimental modification of meristic characters in herring (*Clupea harengus* L.). *J. Conseil Conseil Perm. Intern. Exploration Mer* **26,** 336–346.

Hempel, G., and Blaxter, J. H. *S.* (1963). On the condition of herring larvae. *Rapp Proces-Verbaux Reunions, Conseil Perm. Intern. Exploration Mer* **154,** 35–40.

Hempel, G., and Blaxter, J. H. S. (1967). Egg weight in Atlantic herring (*Clupea harengus* L.). *J. Conseil Conseil Perm. Intern. Exploration Mer* **31,** 170–195.

Hester, F. J. (1963). Effects of food supply on fecundity in the female guppy *Lebistes reticulatus* (Peters). *J. Fisheries Res. Board Can.* **21,** 757–764.

Heuts, M. J. (1949). Racial divergence in fin ray variation patterns in *Gasterosteus aculeatus. J. Genet.* **49,** 183–191.

Hickling, C. F. (1962). "Fish Culture." Faber & Faber, London.

Hishida, T., and Nakano, E. (1954). Respiratory metabolism during fish development. *Embyologia (Nagoya)* **2,** 67–79.

Hoar, W. S. (1942). Diurnal variations in feeding activity of young salmon and trout. *J. Fisheries Res. Board Can.* **6,** 90–101.

Hoar, W. S. (1956). The behaviour of migrating pink and chum salmon fry. *J. Fisheries Res. Board Can.* **13,** 309–325.

Hoar, W. S. (1958). The evolution of migratory behaviour among juvenile salmon of the genus *Oncorhynchus. J. Fisheries Res. Board Can.* **15,** 391–428.

Holliday, F. G. T., and Blaxter, J. H. S. (1960). The effects of salinity on the developing eggs and larvae of the herring. *J. Marine Biol.* Assoc. *U.K.* **39,** 591–603.

Holliday, F. G. T., and Jones, M. P. (1965). Osmotic regulation in the embryo of the herring (*Clupea harengus*). *J. Marine Biol.* Assoc. *U.K.* **45,** 305–311.

Holliday, F. G. T., and Jones, M. P. (1967). Some effects of salinity on the developing eggs and larvae of the plaice (*Pleuronectes platessa*). *J. Marine Biol.* Assoc. *U.K.* **47**, 39–48.

Holliday, F. G. T., Blaxter, J. H. S., and Lasker, R. (1964). Oxygen uptake of developing eggs and larvae of the herring (*Clupeu harengus*). *J. Marine Bid.* Assoc. *U.K.* **44,** 711–723.

Horton, H. F., Graybill, J. R., and Wu, A. S. H. (1967). "Cryogenic Preservation of Viable Fish Sperm." Dept. Fisheries and Wildlife, Oregon State University, Corvallis, Oregon.

Hoyle, R. J., and Idler, D. R. (1968). Preliminary results in the fertilization of eggs with frozen sperm of Atlantic salmon (*Salmo salar*). *J. Fisheries Res. Board Can.* **25,** 1295–1297.

Hubbs, C., Stevenson, M. M., and Peden, A. E. (1968). Fecundity and egg size in two central Texas darter populations. *southwestern Naturalist* 13, 301–324.

Itazawa, Y. (1959). Influence of temperature on the number of vertebrae in fish. *Nature* **183,** 1408–1409.

Iversen, E. S. (1968). "Farming the Edge of the Sea." Fishing News (Books) Ltd., London,

Ivlev, V. S. (1960a). Active metabolic intensity in salmon fry (*Salmo salar* L.) at various rates of activity. Paper No. 213 (mimeo). Salmon & Trout Committee, Intern. Council for Exploration of the Sea, Copenhagen.

Ivlev, V. S. (1960b). On the utilization of food by planktonophage fishes. *Bull. Math. Biophys.* **22,** 371–389.

Ivlev, V. S. (1961). "Experimental Ecology of the Feeding of Fishes." Translation by Yale Univ. Press, New Haven, Connecticut.

Iwai, T. (1964). Feeding and ciliary conveying mechanisms in larvae of salmonid fish *Plecoglossus altivelis.* Temminck et Schlegel. *Physiol. Ecol.* (*Kyoto*) **12,** 38–44.

Iwai, T. (1967a). Structure and development of lateral line cupulae in teleost larvae. *In* "Lateral Line Detectors" (P. Cahn, ed.), pp. 27–44 . Indiana Univ. Press, Bloomington, Indiana.

Iwai, T. (1967b). The comparative study of the digestive tract of teleost larvae. I. Fine structure of the gut epithelium in larvae of Ayu. *Bull. Japan. Soc. Sci. Fisheries* **33,** 480–496.

Iwai, T. (1967c) The comparative study of the digestive tract of teleost larvae. II. Ciliated cells of the gut epithelium in pond smelt larvae. *Bull. Japan. Soc. Sci. Fisheries* **33,** 1116–1119.

Jacobs, W. (1938). Untersuchungen zur Physiologie der Schwimmblase der Fische. IV. Die erste Gasfüllung der Schwimmblase bei jungen Seepferdchen. Z. *Vergleich. Physiol.* **25,** 379–388.

John, K. R., and Hasler, A. D. (1956). Observations on some factors affecting the hatching of eggs and the survival of young shallow-water cisco *Leucichthys artedi* Le Sueur, in Lake Mendota, Wisconsin, *Limnol. Oceanog.* **1,** 176–194.

Jones, M. P., Holliday, F. G. T., and Dunn, A. E. G. (1966). The ultra-structure of the epidermis of larvae of the herring (*Clupea harengus*) in relation to the rearing salinity. *J. Marine Biol.* Assoc. *U.K.* 46, 235–239.

Kaighn, M. E. (1964). A biochemical study of the hatching process in *Fundulus heteroclitus. Develop. Biol.* **9,** 56–80.

Kalman, S. M. (1959). Sodium and water exchange in the trout egg. *J. Cellular Comp. Physiol.* **54,** 155–162.

Kändler, R., and Dutt, S. (1958). Fecundity of Baltic herring. *Rapp. Proces-Verbaux Reunions, Conseil Perm. Intern. Exploration Mer* **143,** 99–108.

Keenleyside, M. H. A,, and Hoar, W. S. (1954). Effects of temperature on the responses of young salmon to water currents. *Behaviour* **7,** 77–87.

Kinne, O., and Kinne, E. M. (1962). Rates of development in embryos of a cyprino-dont fish exposed to different temperature-salinity-oxygen combinations. *Can. J. Zool.* **D40,** 231–253.

Kurata, H. (1959). Preliminary report on the rearing of the herring larvae. *Bull. Hokkaido Reg. Fisheries Res. Lab.* **20,** 117–138.

Kuthalingam, M. D. K. (1959). Temperature tolerance of the larvae of ten species of marine fishes. *Current Sci. (India)* **28,** 75–76.

Laale, H. W., and McCallion, D. J. (1968). Reversible developmental arrest in the embryo of the Zebra fish *Bruchydunio rerio. J. Exptl. Zool.* **167,** 117–123.

Lagler, K. F., Bardach, J. E., and Miller, R. R. (1962). "Ichthyology." Wiley, New York.

Larimore, R. W., and Duever, M. J. (1968). Effects of temperature acclimation on the swimming ability of smallmouth bass fry. *Trans. Am. Fisheries Soc.* **97,** 175–184.

Lasker, R. (1962). Efficiency and rate of yolk utilization by developing embryos and larvae of the Pacific sardine *Sardinops caerulea* (Girard). *J. Fisheries Res. Board Can.* **19,** 867–875.

Lasker, R. (1964). *An* experimental study of the effect of temperature on the incubation time, development and growth of Pacific sardine embryos and larvae. *Copeia* No. **2,** 399–405.

Lasker, R., and Theilacker, G. H. (1962). Oxygen consumption and osmoregulation by single Pacific sardine eggs and larvae (*Sardinops caerulea* Girard). *J. Conseil Conseil Perm. Intern. Exploration Mer* **27,** 25–33.

Lasker, R., and Threadgold, L. T. (1968). "Chloride cells" in the skin of the larval sardine. *Exptl. Cell. Res.* **52,** 582–590.

Lewis, R. M. (1965). The effect of minimum temperature on the survival of larval Atlantic menhaden *Breuoortia tyrannus. Trans. Am. Fisheries Soc.* **94,** 409–412.

Lewis, R. M. (1966). Effects of salinity and temperature on survival and development of larval Atlantic menhaden *Breuoortia tyrannus. Trans. Am. Fisheries Soc.* **95,** 423–426.

Lillelund, K. (1961). Untersuchungen über die Biologie und Populations-dynamik des Stintes, *Osmerus eperlanus eperlanus* (Linnaeus 1758) der Elbe. *Arch. Fischereiwissenschaft* **12,** 1–128.

Lindroth, A. (1942). Sauerstoffverbrauch der Fische. II. Verschiedene Entwick-lungs-und Alter-stadien vom Lachs und Hecht. *Z. Vergleich. Physiol.* **29,** 583–594.

Lindsey, C. C. (1954) . Temperature-controlled meristic variation in the paradise fish. *Macropodus opercularis* (L.). *Can. J. Zool.* **D32,** 87–98.

Lindsey, C. C. (1958). Modification of meristic characters by light duration in kokanee *Oncorhynchus nerka. Copeia* No. **2,** 134–136.

Lindsey, C. C. (1962). Experimental study of meristic variation in a population of three spine sticklebacks *Gasterosteus aculeatus. Can. J. Zool.* **D40,** 271–312.

Lindsey, C. C., and Ali, M. Y. (1965). The effect of alternating temperature on vertebral count in the medaka (*Oryzias latipes*). *Can. J. Zool.* **D43,** 99–104.

Lyall, A. H. (1957a). The growth of the trout retina. *Quart. J. Microscop. Sci.* **98,** 101–110.

Lyall, A. H. (1957b). Cone arrangements in teleost retinae. *Quart. J. Microscop.* Sci. **98,** 189–201.

MacGregor, J. S. (1957). Fecundity of the Pacific sardine (*Sardinops caeruleu*). *US. Fish Wildlife Serv., Fishery Bull.* **57,** 427–449.

McHugh, J. L. (1954). The influence of light on the number of vertebrae in the grunion *Leuresthes tenuis. Copeia* No. **1,** 23–25.

Magnuson, J. J. (1962). An analysis of aggressive behaviour, growth and competition for food and space in medaka (*Oryzias latipes* (Pisces, Cyprinodontidae)). *Can. J.* **D40,** 313–363.

Mansueti, R. (1954). A partial bibliography of fish eggs, larvae and juveniles with particular reference to migratory and estuarine species of the Atlantic coast and supplemented by a check list and references to the early development of the fishes and fish-like chordates of

Maryland waters. Chesapeake Biol. Lab. (mimeo), Maryland Department of Research and Education.

Mansueti, R. (1958). Eggs, larvae and young of the striped bass *Roccus saxatilis*. *Contrib. Chesapeake Biol. Lab.* **112,** 1–35.

Marinaro, J. Y., and Bernard, M. (1966). Contribution à l'étude des oeufs et larves pélagiques de poissons méditerranéens. *Pelagos* **6,** 49–55.

Marr, D. H. A. (1965). Factors affecting the growth of salmon alevins and their survival and growth during the fry stage. *Yearbook Assoc. River Authorities* pp. 133–141.

Marr, D. H. A. (1966). Influence of temperature on the efficiency of growth of salmonid embryos. *Nature* **212,** 957–959.

Marr, D. H. A. (1967). Experiments on the artificial rearing of young Atlantic salmon (*Salmo salar* L.). Anadramous and Catadramous Fish Committee, Paper No. 25 (mimeo). Intern. Council for Exploration of the Sea, Copenhagen.

Marr, J. C. (1956). The "critical period" in the early life history of marine fishes, *J. Conseil Conseil Perm. Intern. Exploration Mer* **21,** 160–170.

Marshall, N. B. (1965). "The Life of Fishes." Weidenfeld & Nicolson, London.

Molander, A. R., and Molander-Swedmark, M. (1957). Experimental investigation on variation in plaice (*Pleuronectes platessa* Linné). *Inst. Marine Res., Lysekil, Ser. Biol. Rept.* **7,** 1–45.

Monroy, A., Ishida, M., and Nakano, E. (1961). The pattern of transfer of the yolk material to the embryo during the development of the teleostean fish *Oryzias latipes*. *Embryologia* (*Nagoya*) **6,** 151–158.

Morns, R. W. (1955). Some considerations regarding the nutrition of marine fish larvae. *J. Conseil Conseil Perm. Intern. Exploration Mer.* **20,** 255–265.

Morris, R. W. (1956). Some aspects of the problem of rearing marine fishes. *Bull. Inst. Oceanog.* **1082,** 1–61.

Mounib, M. S., and Eisan, J. S. (1969). Metabolism of pyruvate and glyoxylate by eggs of salmon (Salmo *salar*). *Comp. Biochem. Physiol.* **29,** 259–264.

Mounib, M. S., Hwang, P. C., and Idler, D. R. (1968). Cryogenic preservation of Atlantic cod (*Gadus morhua*) sperm. *J. Fisheries Res. Board Can.* **25,** 2623–2632.

Müller, H. (1952). Bau and Wachstum der Netzhaut des Guppy (*Lebistes reticulatus*). *Zool. Jahrb.* **63,** 275–324.

Nakano, E. (1953). Respiration during maturation and at fertilization of fish eggs. *Embryologia* (*Nagoya*) **9,** 21–31.

Needham, J. (1931), "Chemical Embryology," 3 vols. Cambridge Univ. Press, London and New York.

New, D. A. T. (1966). "The Culture of Vertebrate Embryos." Lagos Press, London.

Neyfakh, A. A. (1959). X-ray inactivation of nuclei as a method for studying their function in the early development of fishes. *J. Embryol Exptl. Morphol.* **7,** 173–192.

Nikolsky, G. V. (1963). "The Ecology of Fishes." Academic Press, New York.

Norman, J. R. (1963). "A History of Fishes" (P. H. Greenwood, ed.), 2nd ed. Ernest Benn, London.

Ohtsuka, E. (1960). On the hardening of the chorion of the fish egg after fertilization. III. The mechanism of chorion hardening in *Oryzias latipes*. *Biol. Bull.* **118,** 120–128.

Oppenheimer, J. M. (1947). Organization of the teleost blastoderm. *Quart. Rev. Biol.* **22,** 105–118.

Oppenheimer, J. M. (1964). The development of isolated *Fundulus* embryonic shields in salt solution. *Acta Embryol. Morphol. Exptl.* **7,** 143–154.

Orton, G. (1957). Embryology and evolution of the pelagic fish egg. *Copeia* No. **1,** 56–58.

Parrish, B. B. (1956). The cod, haddock and hake. *In* "Sea Fisheries" (M. Graham, ed.), pp. 251–331. Arnold, London.

Parrish, B. B., and Saville, A. (1965). The biology of the north-east Atlantic herring populations. *Oceanog. Marine Biol. Ann. Rev.* **3,** 323–373.

Pearcy, W. G. (1962). Distribution and origin of demersal eggs within the order Pleuronectiformes. *J. Conseil Conseil Perm. Intern. Exploration Mer* **27,** 232–235.

Peters, H. M. (1963). Eizahl, Eigewicht und Gelegeentwicklung in der Gattung *Tilapia* (Cichlidae, Teleostei). *Intern. Rev. Ces. Hydrobiol.* **48,** 547–576.

Phillips, A. M., Jr., and Dumas, R. F. (1959). The chemistry of developing brown trout eyed eggs and sac fry. *Progressive Fish Culturist* **91,** 161–164.

Pickford, G. E., and Atz, J. W. (1957). "The Physiology of the Pituitary Gland of Fishes." N.Y. Zool. Soc., New York.

Pope, J. A., Mills, D. H. and Shearer, W. M. (1961). The fecundity of Atlantic salmon (Salmo *solar* Linn). *Freshwater Salmon Fisheries Res., Scotland* **26,** 1–12.

Potts, W. T. W., and Rudy, P. P., Jr. (1969). Water balance in the eggs of the Atlantic salmon *Salmo salar. J. Exptl. Biol.* **50,** 223–237.

Prescott, D. M. (1955). Effect of activation on the water permeability of salmon eggs. *J. Cellular Comp. Physiol.* **45,** 1–12.

Price, J. W. (1940). Time-temperature relations in the incubation of the whitefish *Coregonus clupeaformis* (Mitchill). *J. Gen. Physiol.* **23,** 449–468.

Qasim, S. Z. (1959). Laboratory experiments on some factors affecting the survival of marine teleost larvae. *J. Marine Biol. Assoc. India.* **1,** 13–25.

Qasim, S. Z., Rice, A. L., and Knight-Jones, E. W. (1963). Sensitivity to pressure changes in teleosts lacking swim-bladders. *J. Marine Biol. Assoc. India* **5,** 289–293.

Radakov, D. V. (1964). "Velocities of Fish Swimming," pp. 4–28. Pamphlet from A. N. Severtsov Institute of Animal Morphology, Moscow, Nauka (in Russian).

Raitt, D. F. S., and Hall, W. B. (1967). On the fecundity of the redfish *Sebastes marinus* (L.). *J. Conseil Conseil Perm, Intern. Exploration Mer* **31,** 237–245.

Randall, J. E. (1961). A contribution to the biology of the convict surgeonfish of the Hawaiian Islands, *Acanthurus triostegus sandvicensis. Pacific Sci.* **15,** 215–272.

Read, L. J. (1968). Omithine-urea cycle enzymes in early embryos of the dogfish *Sqzbalus suckleyi* and the skate *Raja binoculata. Comp. Biochem. Physiol.* **24,** 669–674.

Riley, J. D., and Thacker, G. T. (1963). Marine fish culture in Britain. III. Plaice (*Pleuronectes platessa* (L.)) rearing in closed circulation at Lowestoft 1961. *J. Conseil Conseil Perm. Intern. Exploration Mer* **28,** 80–90.

Rosenthal, H. (1966). Beobachtungen uber das Verhalten der Seezungenbrut. *Helgobender Wiss. Meeresuntersuch.* **13,** 213–228.

Rosenthal, H. (1968). Beobachtungen uber die Entwicklung des Schwarmverhaltens bei den Larven des Herings *Clupea harengus. Marine Biol.* **2,** 73–76.

Rosenthal, H., and Hempel, G. (1968). Experimental Studies in Feeding and Food Requirements of Herring Larvae. Symposium on Food Chains. Univ. of Aarhus, Denmark.

Rounsefell, G. A. (1957). Fecundity of the North American Salmonidae. *U.S. Fish Wildlife Serv., Fishery Bull.* **57(122),** 451–468.

Rudnick, D. (1955). Teleosts and birds. *In* "Analysis of Development" (B. H. Willier, P. A. Weiss, and V. Hamburger, ed.), Saunders, Philadelphia, Pennsylvania.

Rudy, P. P., Jr., and Potts, W. T. W. (1969). Sodium balance in the eggs of the Atlantic salmon *Salmo salar. J. Exptl. Biol.* **50,** 239–246.

Ryland, J. S. (1963). The swimming speeds of plaice larvae. *J. Exptl. Biol.* **40,** 285–299.

Ryland, J. S. (1964). The feeding of plaice and sand-eel larvae in the southern North Sea. *J. Marine Biol. Assoc. U.K.* **44,** 343–364.

Ryland, J. S. (1966). Observations on the development of larvae of the plaice, *Pleuronectes platessa* L. in aquaria. I. *Conseil Conseil Perm. Intern. Exploration Mer* **30,** 177–195.

Ryland, J. S., and Nichols, J. H. (1967). Effect of temperature on the efficiency of growth of plaice proiarvae. *Nature* **214,** 529–530.

Schumann, G. O. (1967). La Jolla first to raise mackerel and sardines to advanced juvenile stage. *Corn. Fisheries Rev.* **29(2),** 24–25.

Schwassmann, H. O. (1965). Functional development of visual pathways in larval sardines and anchovies. *Rept. Calif. Coop. Oceanog. Fisheries Invest.* **10,** 64–70.

Scott, C. G., and Kellicott, W. E. (1916). The consumption of oxygen during the development of *Fundulus heteroclitus. Anat. Record* **11,** 531–533.

Scott, D. P. (1961). Effect of food quantity on fecundity of rainbow trout *Salmo gairdneri. J. Fisheries Res. Board Can.* **19,** 715–731.

Sette, O. (1943). Biology of the Atlantic mackerel (*Scomber scombrus*) of North America. Part 1. Early life history, including the growth, drift and mortality of the egg and larval populations. *US. Fish Wildlife Serv., Fishery Bull.* **50(38),** 149–237.

Seymour, A. (1959). Effects of temperature upon the formation of vertebrae and fin rays in young chinook salmon. *Trans. Am. Fisheries Soc.* **88,** 58–69.

Shaw, E. (1960). The development of schooling behaviour in fishes. *Physiol. Zool.* **33,** 79–86.

Shaw, E. (1961). The development of schooling in fishes, II, *Physiol. Zool.* **34,** 263–272.

Shelbourne, J. E. (1953). The feeding habits of plaice post-larvae in the Southern Bight, *J. Marine Biol. Assoc. U.K.* **32,** 149–160.

Shelbourne, J. E. (1956). The effect of water conservation on the structure of marine fish embryos and larvae. *J. Marine Biol. Assoc. U.K.* **35,** 275–286.

Shelbourne, J. E. (1957). The feeding and condition of plaice larvae in good and bad plankton patches. *J. Marine Biol. Assoc. U.K.* **36,** 539–552.

Shelbourne, J. E. (1963). Marine fish culture in Britain. II. A plaice rearing experiment at Port Erin, Isle of Man, during 1960 in open sea water circulation. *J. Conseil Conseil Perm. Intern. Exploration Mer* **28,** 70–79.

Shelbourne, J. E. (1964). The artificial propagation of marine fish. *Advan. Marine Biol.* **2,** 1–83.

Shumway, D. L., Warren, C. E., and Doudoroff, P. (1964). Influence of oxygen concentration and water movement on the growth of steelhead trout and coho salmon embryos. *Trans. Am. Fisheries* Soc. **93,** 342–356.

Silver, S. J., Warren, C. E. and Doudoroff, P. (1963). Dissolved oxygen requirements of developing steelhead trout and chinook salmon embryos at different water velocities. *Trans. Am. Fisheries Soc.* **92,** 327–343.

Simpson, A. C. (1956). The pelagic phase. *In* "Sea Fisheries" (M. Graham, ed.), pp. 207–250. Arnold, London.

Smith, S. (1957). Early development and hatching. *In* "The Physiology of Fishes" (M. E. Brown, ed.), Vol. Vol. 1, pp. 323–359. Academic Press, New York.

Smith, S. (1958). III. Yolk utilization in fishes. *In* "Embryonic Nutrition" (D. Rudnick, ed.), pp. 33–53. Univ. of Chicago Press, Chicago, Illinois.

Sneed, K. E., and Clemens, H. P. (1956). Survival of fish sperm after freezing and storage at low temperatures. *Progressive Fish Culturist* **18,** 99–103.

Solberg, A. N. (1938). The susceptibility of *Fundulus heteroclitus* embryos to X-radiation, *J. Exptl. Zool.* **78,** 441–469.

Solemdal, P. (1967). The effect of salinity on buoyancy, size and development of flounder eggs. *Sarsia* **29,** 431–442.

Stevenson, J. C. (1962). Distribution and survival of herring larvae (*Clupea pallasii* Valenciennes) in British Columbia waters. *J. Fisheries Res. Board Can.* **19,** 735–810.

Strasburg, D. W. (1960). Estimates of larval tuna abundance in the Central Pacific. *US. Fish Wildlife Serv. Fishery Bull.* **60,** 231–255.

Stuart, T. A. (1953). Spawning migration, reproduction and young stages of loch trout (*Salmo trutta* L.). *Freshwater Salmon Fisheries Res., Scotland* No. **5,** 1–39.

Sweet, J. G., and Kinne, O. (1964). The effects of various temperature-salinity combinations on the body form of newly hatched *Cyprinodon macularius* (Teleostei). *Helgolaender Wiss. Meeresuntersuch.* **11,** 49–69.

Tait, J. S. (1960). The first filling of the swim bladder in salmonids. *Can. J. Zool.* **D38,** 179–187.

Talbot, G. E., and Sykes, J. E. (1958). Atlantic coast migration of American shad. *US. Fish Wildlife Serv., Fishery Bull.* **58(142),** 473–490.

Takahashi, M., and Hatanaka, M. A. (1960). Experimental study on utilization of food by young anchovy, *Engraulis japonicus* Temminck et Schlegel. *Tohoku J. Agr. Res.* **11,** 161–170.

Tåning, A. V. (1952). Experimental study of meristic characters in fishes. *Biol. Rev. Cambridge Phil. Soc.* **27,** 169–193.

Templeman, W. (1944). The life history of the spur dogfish *Squalus acanthias* and the Vitamin A values of dogfish liver oil. *Res. Bull. Div. Fishery Resources, Newfoundland* No. **15,** 1–102.

Terner, C. (1968). Studies of metabolism in embryonic development. I. The oxidative metabolism of unfertilized and embryonated eggs of the rainbow trout. Comp. *Biochem. Physiol.* **24,** 933–940.

Terner, C., Kumar, L. A., and Choe, T. S. (1968). Studies of metabolism in embryonic development. II. Biosynthesis of lipids in embryonated trout ova. *Comp. Biochem. Physiol.* **24,** 941–950.

Te Winkel, L. E. (1963). Notes on the smooth dogfish *Mustelus canis* during the first three months of gestation. I and II. *J. Exptl.* Zool. **152,** 115–137.

Trinkaus, J. P. (1951). A study of the mechanism of epiboly in the egg of *Fundulus heteroclitus. J. Exptl. Zool.* **118,** 269–319.

Truscott, B., Idler, D. R., Hoyle, R. J., and Freeman, H. C. (1968). Sub-zero preservation of Atlantic salmon sperm. *J. Fisheries Res. Board Can.* **25,** 363–372.

Vilter, V. (1946). Action de la thyroxine sur la métamorphose larvaire de l'Anguille. *Compt. Rend. Soc. Biol.* **140,** 783–785.

Vladykov, V. D. (1956). Fecundity of wild speckled trout (*Salvelinus fontinalis*) in Quebec lakes. *J. Fisheries Res. Board Can.* **13,** 799–841.

von Ledebur, J. F. (1928). Beiträge zur Physiologie der Schwimmblase der Fische I. Z. *Vergleich. Physiol.* **8,** 44–86.

Waddington, C. H. (1956). "Principles of Development." Allen & Unwin, London.

Wedemeyer, G. (1968). Uptake and distribution of Zn^{65} in the coho salmon egg (*Oncorhynchus kisutch*). *Comp. Biochem. Physiol.* **25,** 271–279.

Wickett, W. P. (1954). The oxygen supply to salmon eggs in spawning beds. *J. Fisheries Res. Board Can.* **11,** 933–953.

Williams, J. (1967). Yolk Utilization. *In* "The Biochemistry of Animal Development" (R. Weber, ed.), Vol. 2, pp. 372–374. Academic Press, New York.

Winberg, G. *G.* (1960). Rate of metabolism and food requirements of fishes. *Fisheries Res. Board Can.* (*Trawl.*) 194.

Withler, F. C., and Morley, R. B. (1988). Effects of chilled storage on viability of stored ova and sperm of sockeye and pink salmon. *J. Fisheries Res. Board Can.* **25,** 2895–2699.

Woodhead, P. M. J. (1957). Reactions of salmonid larvae to light, *J. Exptl. Bid.* **34,** 402–416.

Woodhead, P. M. J., and Woodhead, A.D. (1955). Reactions of herring larvae to light: A mechanism of vertical migration. *Nature* **176,** 349–350.

Yamamoto, T. (1961). Physiology of fertilization in fish eggs. *Intern. Reo. Cytol.* **12,** 361–405.

Zotin, A. I. (1958). The mechanism of hardening of the salmonid egg membrane after fertilization or spontaneous activation. *J. Embryol. Exptl. Morphol.* **6,** 546–568.

Zotin, A. I. (1965). The uptake and movement of water in embryos. *Symp. Soc. Exptl. Biol.* **19,** 365–384.

Chapter 15

A foundational exploration of respiration in fish eggs and larvae

Daiani Kochhann[a,*] **and Lauren Chapman**[b,†]
[a]*Laboratory of Behavioural Ecophysiology, Center of Agrarian and Biological Sciences, Acaraú Valley State University, Sobral, CE, Brazil*
[b]*Department of Biology, McGill University, Montréal, QC, Canada*
[*]*Corresponding author: e-mail: daia.kochhann@gmail.com; lauren.chapman@mcgill.ca*

Daiani Kochhann and Lauren Chapman discuss the P.J. Rombough chapter "Respiratory Gas Exchange, Aerobic Metabolism, and Effects of Hypoxia during Early Life" in *Fish Physiology*, Volume 11A, published in 1988. The next chapter in this volume is the re-published version of that chapter.

Some contributions to science are so important to a field, it does not matter that time passes—they are classic and relevant. This is true for Peter Rombough's work entitled "Respiratory gas exchange, aerobic metabolism, and effects of hypoxia during early life" published in the Fish Physiology series in 1988. His herculean work of reviewing more than 500 studies in the field gave us the first conceptual framework on how respiratory gas exchange works in the early life stages of fish; how their metabolism is influenced by biotic and abiotic factors; and the impact of hypoxia exposure on these life stages. After 35 years since publication, we reviewed this foundational contribution, discussing its relevance to the field and some key advances in our knowledge of gas exchange and metabolism in the early life stages of fish.

Mechanisms of respiratory gas exchange are very diverse in fishes ranging from branchial and cutaneous respiration to facultative and obligatory air breathing. At the time when Peter Rombough wrote his seminal chapter in Fish Physiology, knowledge of gas exchange and aerobic metabolism of juvenile and adult fish was considered well advanced; but a conceptual framework for understanding oxygen uptake and supply in embryos and larval stages (ELS) of fishes was yet to emerge. This was despite the 500+ papers published on this subject in the 2 decades preceding the publication of

[†]Senior author.

Fish Physiology, Vol. 40A. https://doi.org/10.1016/bs.fp.2023.09.008

Rombough's chapter. Rombough's goal was to summarize and synthesize knowledge on respiratory gas exchange and energy use in early life stages, highlighting empirically driven insights and identifying key gaps to guide future research. After 35 years, the Rombough chapter has been cited almost 471 times, 100 times since 2015 (Google Scholar, June 11, 2023), clearly demonstrating the enduring influence of this contribution.

Although gas exchange is primarily cutaneous in embryos and larval fishes (branchial respiration becomes dominant toward the end of the larval period), Rombough treats them separately in his synthesis because of the major impact of the egg capsule during embryonic life. Like the layers of an onion, Rombough unfolds the egg capsule in his review to show the major resistances to respiratory gas exchange for fish embryos (boundary layer, egg capsule, perivitelline fluid). Although the egg capsule had been traditionally viewed as the major barrier to diffusive gas exchange during embryonic life, Rombough's integration of a growing body of studies demonstrated a strong impact of the perivitelline fluid on oxygen uptake, and the intriguing possibility that stirring of the perivitelline may be necessary to meet the metabolic demands on the embryo, with hatching regarded as a response to physiological hypoxia. Rombough proceeded to delve into mechanisms of cutaneous respiration noting that studies had focused on the morphology of assumed respiratory structures such as well-developed vascular networks forming under the skin during organogenesis. A summary of diffusion distances suggested that the cutaneous diffusion distances are only slightly greater than lamellar diffusion distances in juveniles and adult fishes. An important gap at the time of Rombough's review was the cardiovascular physiology of teleost embryos and larvae, an area that has sparked a suite of studies including a review of piscine cardiovascular development published by Rombough in 1998 (Rombough, 1998b), and a more recent review of cardiovascular development in embryonic and larval fishes by Burggren and colleagues published in 2017 (Burggren et al., 2017). Increasing interest in the effects of environmental stressors (e.g., hypoxia, contaminants) on fish ELS has led to studies of phenotypic plasticity in cardiovascular development. Advances in technology such as microtechniques that enable measurements of pressure and electrical signals of hearts in fish ELS, and functional assays that use image analyses of circulation have facilitated exciting advances in this area.

Rombough's review of respiratory pigments in the early life stages of fishes drew on several earlier studies that provided evidence for structurally and functionally distinct hemoglobins between early life stages in fishes and later stages (juveniles and adults). The suggestion of pigments other than hemoglobin such as myoglobin and carotenoids had been proposed by the time of Rombough's review, but evidence for an adaptive role was controversial. Although hemoglobins are critical to oxygen transport, Rombough's synthesis clearly pointed to the importance of water movement past the embryos post-hatch in enhancing embryonic gas exchange. This element of respiratory gas exchange in fish early life stages is particularly interesting because it can depend on the behavior of the parents and/or the larvae. Examples noted by Rombough included nest-fanning, mouth-brooding, and behavioral adaptations of the larvae. This is an area that continued to gain momentum in the decades following Rombough's synthesis with several species-specific studies that together demonstrate a diversity of strategies (reviewed in Chapman and McKenzie, 2009).

The larval stage in fishes is characterized by a shift in the site of respiratory gas exchange from the skin to the gills. At the time of Rombough's review, there was little known about the physiology of this important transition. However, it was clear that the shift to branchial respiration in larval fish is highly variable in terms of stage of timing of the transition. Rombough's review of branchial respiration highlighted not only this variation, but also the significance of the heart rate in regulating blood flow to the gills. Particularly interesting was the observation that ventilatory movements may be more coordinated and stronger under hypoxic conditions. Effects of hypoxia on respiratory gas exchange in fish embryos and larvae have gained momentum over the last few decades, driven in part by the global hypoxia crisis, and the potential interactions between climate warming and hypoxia. Rombough continued to publish articles on the skin-branchial transition including technological innovations (e.g., Rombough, 1998a).

There are clearly many factors affecting gas exchange and rate of oxygen uptake in ELS of fishes including body size. Understanding the effect of body mass on resting metabolic rates has been an active area of research for several decades. The allometric equation that relates body size to routine metabolism is $VO_2 = aM^b$, where b is the slope of the relationship between resting metabolic rate and body mass. Rombough compiled data for the metabolic mass exponents of fish larvae, noting that high mass exponents (approaching or exceeding 1) during early life of fishes appear to be widespread; but with exceptions (e.g., 0.82 and 0.65) for herring and plaice larvae (de Silva and Tytler, 1973). Rombough noted that mechanisms contributing to such variation in "b" remained unresolved. The mass/metabolism literature on larval fishes continued to grow after Rombough's review, synthesized for marine larvae in 2016 by Peck and Moyano. Although there was some evidence for an effect of thermal habitat (tropical, warm-temperate, cold-temperature, sub-polar), with highest "b" values in sub-polar species, Peck and Moyano pointed out that there are still very few data for some thermal groups. Peck and Moyano (2016) also noted the contributions of Rombough's compilation, with a 25-year gap between their review of respiration in marine larvae and Rombough's chapter during which time studies continued to explore important research gaps.

When Rombough wrote his review, the effect of temperature on metabolism of early life stages of fishes had been investigated in a few temperate species. At that time, there was evidence that water temperature had a stronger effect on embryos and larvae than in adults, with a Q10 for metabolism and development averaging about 3 in embryos (see Table IV in Rombough, 1988a) and 2 in adults (Fry, 1971). Rombough suggested that this difference probably reflected the fact that ELS tend to be more stenothermal than juveniles and adults. Although there were few data at that time, Rombough noted that at least some species of fish ELS were not capable of thermal acclimation or metabolic compensation. The subject of thermal acclimation has become increasingly significant in the face of climate change. Recent reviews have focused on upper thermal limits and the thermal acclimation capacity of fishes and other ectothermic vertebrates, with the recognition that the ability for fishes to adjust their thermal limits via plasticity may be an important mechanism for tolerating increases in mean temperature and temperature extremes (Campos et al., 2021; Chapman et al., 2022; Comte and Olden, 2017; Morley et al., 2019). There remains a bias toward studies of juvenile and adult fish. However, the few studies of ELS do show divergent patterns with the ELS of some species of fishes capable of thermal acclimation and metabolic compensation (Del Rio et al., 2019; McPhee et al., 2023; Pan and von Herbing, 2017) and others showing no evidence for thermal compensation or acclimation (Illing et al., 2020; Killen et al., 2008).

In exploring the relationship between metabolic rate and dissolved oxygen tension in early life stages of fishes, Rombough found that embryos and larvae generally acted as oxyregulators, meaning that their metabolic rate is independent of aquatic oxygen levels until reaching the critical oxygen tension (Pcrit) below which the metabolic rate becomes dependent on ambient oxygen concentration. This is a pattern seen in many juvenile and adult fishes as well. Rombough stressed the importance of both water temperature and the activity of ELS on Pcrit due to alteration of oxygen demands. From data on lethal dissolved oxygen levels, the consensus at that time was that ELS are very sensitive to low levels of oxygen and that sensitivity tends to increase as development proceeds, with the larval period being the most sensitive. Rombough linked larval sensitivity to hypoxia to changes in the site and the efficiency of respiratory gas exchange (de Silva and Tytler, 1973; Spoor, 1977) with a decline in sensitivity when larvae were able to use gills as a site of gas exchange. Rombough noted various challenges of estimating Pcrit in ELS, including the instability of polarographic sensors. Despite advances in respirometry since 1988, and many studies estimating Pcrit in fishes, the methods for estimating Pcrit and the relevance of the metric are still debated (Ultsch and Regan, 2019; Wood, 2018).

Aerobic scope, defined as the difference between maximum and standard metabolic rate in fishes, is a subject of great interest to physiologists because it represents the amount of energy that a fish has to cover the costs of various essential activities (growth, reproduction). Although empirical data for ELS was limited at the time, Rombough discussed the effect of temperature on aerobic scope. He noted that relationships appear to vary among species. However, for rainbow trout, larval fish show a pattern similar to juveniles and adults, whereby both active and routine metabolism increase with temperature, but active metabolism increases faster, driving an increase in aerobic scope. A decrease in aerobic scope is expected when approaching the thermal limit of species as oxygen availability limits active metabolism at high environmental temperature, a pattern discussed by Fry (1971) in his classic paper. At the time of Rombough's review, there was a clear deficiency of data on tropical species, which are the ones facing the highest temperatures and widespread natural and anthropogenically-induced hypoxia. Moreover, studies were unable to manipulate more than one factor at that time, making it harder to understand the relationship between temperature and hypoxia on metabolism and survival of ELS.

With the acceleration of studies exploring thermal tolerance of species, mechanisms underlying variation in thermal limits, and climate-driven range shifts in fishes, there are a number of recent meta-analyses and reviews that look for emergent and generalizable trends. The oxygen capacity limiting thermal tolerance (OCLTT) hypothesis has gained a great deal of attention from the scientific community as a possible mechanism explaining the upper thermal tolerance of ectotherms. The concept was proposed by Pörtner (2001) and relies on earlier studies of Fry (1947, 1971). It predicts that under warming temperatures, the maximum rate of oxygen consumption is predicted to exceed the capacity of the cardio-respiratory system to uptake and supply oxygen to tissues, manifested as a reduction in aerobic scope and a decline in critical thermal maximum when AS is reduced. Much work has been done to assess the value of OCLTT as a unifying mechanism of thermal tolerance in juvenile and adult fish, and more recently, some studies are investigating the potential of OCLTT in explaining thermal limitations in ELS of fishes. As mentioned by Rombough, the ELS of fishes are the most sensitive developmental stage when it comes to changes in temperature, and Pörtner et al. (2006) proposed that the narrow thermal tolerance in ELS was due to developmental constraints and insufficient capacity of central organs in larvae, mainly related to oxygen transport limitations. The limitation of oxygen is a reliable explanation for thermal tolerance in ELS of fishes, and the OCLTT concept is applied to explain the observed narrow thermal tolerance in ELS (Dahlke et al., 2020; Pörtner and Peck, 2010). Recent empirical work confirmed Rombough's idea that hypoxia and temperature have a synergistic effect on ELS of fishes (e.g., Del Rio et al., 2019, 2021; McPhee et al., 2023; Pan and von Herbing, 2017). Despite accelerating interest in aerobic scope in fishes, aerobic scope measurements are still challenging in ELS due to the difficulties of measuring maximum metabolic rate. In addition, the bias toward temperate species remains; this is, particularly important given the extraordinary diversity of fishes in many tropical regions and the relatively low thermal safety margin characteristic of tropical fishes (Campos et al., 2021).

The effects of pH and carbon dioxide concentration on ELS had received very little attention by the time of Rombough's data collation; however,

Rombough recognized that these are potentially important factors affecting ELS metabolic rate. There was only one study with pH (Rombough, 1988b) in steelhead embryos and one with carbon dioxide on chum salmon embryos (Alderdice and Wickett, 1958), and in both studies, metabolic rate was insensitive to the tested concentrations. With the advent of global climate changes, the question on how ELS of fishes will be affected by the increase in CO_2 concentrations and the consequent drop of pH has become an urgent topic of investigation, especially for marine species. Baumann (2019) reviewed the effects of elevated CO_2 on marine life and concluded that the sensitivity of fish ELS to CO_2 is highly variable and not necessarily negative. However, when associated with hypoxia or high temperature a few studies of fish ELS have observed a synergistic and deleterious effect (Di Santo, 2015; Flynn et al., 2015; Schwemmer et al., 2020). Clearly, there is still a need for additional work in this area.

Citing Blaxter (1969), Rombough highlighted activity as the most important single factor affecting metabolic rate of larvae, a pattern also noted for juveniles and adults (Brett, 1970). However, because of the particularities of larval development and metamorphosis, Rombough suspected that energetic relationships would probably be different from those of older fish. The idea was hard to test due to the challenges of inducing high levels of swimming activity in fish larvae and difficulties of measuring routine activity of larvae stage. Measuring the cost of swimming in fish larvae is still a "herculean challenge" as reviewed by Peck and Moyano (2016), and only a few studies were able to take up this challenge. Nilsson et al. (2007) and Killen et al. (2007) with different protocols and different fish larvae observed an increase in metabolic rate around twofold after exercising fish. Nilsson et al. (2007) came up with an extraordinary record of maximum oxygen uptake for a cold-blooded vertebrate when measuring the metabolic rate from larvae of two species of damselfish, *Chromis atripectoralis* and *Pomacentrus ambionensis*. The cost of routine swimming and daily rhythm has been rarely investigated in fish larvae. Blanco et al. (2020) observed no changes in routine metabolic rate under light and darkness in larval Atlantic bluefin tuna, which suggests no difference in larval activity; however, activity was not measured. In their study of *Gadus morhua* larvae Ruzicka and Gallager (2006), the cost of activity was estimated to be around 80% of the total metabolic cost for *G. morhua* larvae. Owing to advances in oxygen sensors and the increasing interest in the effects of environmental change on physiological traits of fishes, research on fish larvae is the subject of a growing number of studies. In their review of the challenges and advances of measuring respiration rates in larval marine fishes, Peck and Moyano (2016) reviewed 59 studies examining 53 fish species, with 35% of these studies carried out since 2000. They also noted the strong bias toward temperate fishes that still persists in this area of research.

Rombough's collation and synthesis of respiratory gas exchange, metabolism, and effects of hypoxia on fish embryos and larvae provided a rigorous exploration of a plethora of studies on the topic prior to 1988. Most importantly, Rombough was able to identify emergent trends, highlight controversial topics, and identify key research gaps, some of which have sparked new and exciting studies, often fueled by technological advances. His thorough exploration of the role of the gill in respiration and branchial development was an important precursor to later works examining the role of the larval gill in ion transport (e.g., Fu et al., 2010; Rombough, 1999). Rombough stressed the need for expansion of studies to better represent the diversity of fishes. Some fishes show far reaching adaptations to hypoxia (reviewed in Chapman, 2015), and comparative studies of physiological and anatomical traits that facilitate persistence in natural chronic or seasonal hypoxia may be very useful in understanding potential responses of fishes to anthropogenically-induced hypoxia. The global hypoxia crisis in coastal and estuarine waters (Diaz and Rosenberg, 2008) has fueled many studies of fish responses to hypoxia, and this will no doubt be an area that continues to flourish over the next few decades. Rombough also argued for more in-depth studies of the transition from cutaneous to branchial respiration as well as studies that explore the interactions between factors (biotic and abiotic) that affect gas exchange. The latter has led to a productive and expanding field of multi-stressor studies in fishes. The field of fish physiology has greatly benefitted from Rombough's timely synthesis of respiratory gas exchange and metabolism in the early life stages of fishes, particularly as scientists face an increasingly complex milieu of environmental stressors in aquatic systems, which may demand plastic and/or heritable change in embryonic and larval traits to facilitate persistence.

References

Alderdice, D.F., Wickett, W.P., 1958. A note on the response of developing chum salmon eggs to free carbon dioxide in solution. J. Fish. Res. B. Can. 15 (5), 797–799. https://doi.org/10.1139/f58-043.

Baumann, H., 2019. Experimental assessments of marine species sensitivities to ocean acidification and co-stressors: how far have we come? Can. J. Zool. 97, 399–408. https://doi.org/10.1139/cjz-2018-0198.

Blanco, E., Reglero, P., Ortega, A., Folkvord, A., de la Gandara, F., Hernandez de Rojas, A., Moyano, M., 2020. First estimates of metabolic rate in Atlantic bluefin tuna larvae. J. Fish Biol. https://doi.org/10.1111/jfb.14473.

Blaxter, J.H.S., 1969. Development: eggs and larvae. In: Hoar, W.S., Randall, D.J. (Eds.), Fish Physiology. 3. Academic Press, New York, pp. 178–241.

Brett, J.R., 1970. Fish-the energy cost of living. In: McNeil, W.J. (Ed.), Marine Aquaculture. Oregon State Univ Press, Corvallis, pp. 37–52.

Burggren, W.W., Dubansky, B., Bautista, N.M., 2017. Cardiovascular development in embryonic and larval fishes. In: Gamperl, A.K., Gillis, T.E., Farrell, A.P., Brauner, C.J. (Eds.), Fish Physiol. Vol 36 Part B. Academic Press, Cambridge, pp. 107–184.

Campos, D.R., Amanajás, R.D., Almeida-Val, V.M.F., Val, A.L., 2021. Climate vulnerability of South American freshwater fish: thermal tolerance and acclimation. J. Exp. Zool. A. 335, 723–734.

Chapman, L.J., 2015. Low-oxygen lifestyles in extremophilic fishes. In: Reisch, R., Plath, M., Tobler, M. (Eds.), Extremophile Fishes - Ecology and Evolution of Teleosts in Extreme Environments. Springer, Heidelberg, pp. 9–31.

Chapman, L.J., McKenzie, D., 2009. Behavioural responses and ecological consequences. In: Richards, J.G., Farrell, A.P., Brauner, C.J. (Eds.), Fish Physiology. vol. 27. Elsevier, San Diego, pp. 26–77.

Chapman, L.J., Nyboer, E.A., Fugère, V., 2022. Fish response to environmental stressors in the Lake Victoria Basin ecoregion. In: Fangue, N., Cooke, S. (Eds.), Fish Physiology. 39. Elsevier, pp. 273–324.

Comte, L., Olden, J., 2017. Climatic vulnerability of the world's freshwater and marine fishes. Nat. Clim. Change 7, 718–723.

Dahlke, F.T., Wohlrab, S., Butzin, M., Pörtner, H., 2020. Thermal bottlenecks in the life cycle define climate vulnerability of fish. Science 369, 65–70.

de Silva, C.D., Tytler, P., 1973. The influence of reduced environmental oxygen on the metabolism and survival of herring and plaice larvae. Neth. J. Sea Res. 7, 345–362. https://doi.org/10.1016/0077-7579(73)90057-4.

Del Rio, A.M., Davis, B.E., Fangue, N.A., Todgham, A.E., 2019. Combined effects of warming and hypoxia on early life stage Chinook salmon physiology and development. Cons. Physiol. 7 (1), coy078. https://doi.org/10.1093/conphys/coy078.

Del Rio, A.M., Mukai, G.N., Martin, B.T., Johnson, R.C., Fangue, N.A., Israel, J.A., Todgham, A.E., 2021. Differential sensitivity to warming and hypoxia during development and long-term effects of developmental exposure in early life stage Chinook salmon. Cons. Phys. 9 (1), 1–17. https://doi.org/10.1093/conphys/coab054.

Di Santo, V., 2015. Ocean acidification exacerbates the impacts of global warming on embryonic little skate, *Leucoraja erinacea* (Mitchill). J. Exp. Mar. Biol. Ecol. 473, 72–78. https://doi.org/10.1016/j.jembe.2014.11.006.

Diaz, J.R., Rosenberg, R., 2008. Spreading dead zones and consequences of marine ecosystems. Science 321, 926–929.

Flynn, E.E., Bjelde, B.E., Miller, N.A., Todgham, A.E., 2015. Ocean acidification exerts negative effects during warming conditions in a developing Antarctic fish. Cons. Phys. 3 (1), cov033. https://doi.org/10.1093/conphys/cov033.

Fry, F.E.J., 1947. Effects of the environment on animal activity. Ontario Fisheries Research Laboratory. 68, pp. 1–52.

Fry, F.E.J., 1971. The effect of environmental factors on the physiology of fish. In: Hoar, W.S., Randall, D.J. (Eds.), Fish Physiology. vol. 6. Academic Press, New York, pp. 1–98.

Fu, C., Wilson, J.M., Rombough, P.J., Brauner, C.J., 2010. Ions first: Na^+ uptake shifts from the skin to the gills before O_2 uptake in developing rainbow trout, *Oncorhynchus mykiss*. Proc. R. Soc. B 277, 1553–1560.

Illing, B., Downie, A.T., Beghin, M., Rummer, J.L., 2020. Critical thermal maxima of early life stages of three tropical fishes: effects of rearing temperature and experimental heating rate. J. Therm. Biol. 90, 102582. https://doi.org/10.1016/j.jtherbio.2020.102582.

Killen, S.S., Costa, I., Brown, J.A., Gamperl, A.K., 2007. Little left in the tank: metabolic scaling in marine teleosts and its implications for aerobic scope. Proc. R. Soc. B 274, 431–438. https://doi.org/10.1098/rspb.2006.3741.

Killen, S.S., Brown, J.A., Gamperl, A.K., 2008. Lack of metabolic thermal compensation during the early life stages of ocean pout *Zoarces americanus* (Bloch & Schneider): a benthic, cold-water marine species. J. Fish Biol. 72 (3), 763–772. https://doi.org/10.1111/j.1095-8649.2007.01735.x.

McPhee, D., Watson, J.R., Harding, D.J., Prior, A., Fawcett, J.H., Franklin, C.E., Cramp, R.L., 2023. Body size dictates physiological and behavioural responses to hypoxia and elevated water temperatures in Murray cod (*Maccullochella peelii*). Cons. Phys. 11 (1), coac087. https://doi.org/10.1093/conphys/coac087.

Morley, S.A., Peck, L.S., Sunday, J.M., Heiser, S., Bates, A.E., 2019. Physiological acclimation and persistence of ectothermic species under extreme heat events. Glob. Ecol. Biogeogr. 28, 1018–1037.

Nilsson, G.E., Östlund-Nilsson, S., Penfold, R., Grutter, A.S., 2007. From record performance to hypoxia tolerance: respiratory transition in damselfish larvae settling on a coral reef. Proc. R. Soc. B 274, 79–85. https://doi.org/10.1098/rspb.2006.3706.

Pan, T.C.F., von Herbing, I.H., 2017. Metabolic plasticity in development: synergistic responses to high temperature and hypoxia in zebrafish, *Danio rerio*. J. Exp. Zool. Part A - Ecol. Integr. Physiol. 327 (4), 189–199. https://doi.org/10.1002/jez.2092.

Peck, M.S., Moyano, M., 2016. Measuring respiration rates in marine fish larvae: challenges and advances. J. Fish Biol. 88, 173–205.

Pörtner, H.O., 2001. Climate change and temperature-dependent biogeography: oxygen limitation of thermal tolerance in animals. Naturwissenschaften 88, 137–146. https://doi.org/10.1007/s001140100216.

Pörtner, H.O., Peck, M.A., 2010. Climate change effects on fishes and fisheries: towards a cause-and-effect understanding. J. Fish Biol. 77, 1745–1779.

Pörtner, H.O., Bennett, A.F., Bozinovic, F., Clarke, A., Lardies, M.A., Lucassen, M., Pelster, B., Schiemer, F., Stillman, J.H., 2006. Trade-offs in thermal adaptation: the need for a molecular to ecological integration. Physiol. Biochem. Zool. 79, 295–313. https://doi.org/10.1086/499986.

Rombough, P.J., 1988a. Respiratory gas exchange, aerobic metabolism, and effects of hypoxia during early life. In: Fish Physiology. Vol. XIA. Academic Press, New York, pp. 59–161.

Rombough, P.J., 1988b. Growth, aerobic metabolism and dissolved oxygen requirements of embryos and alevins of the steelhead trout, *Salmo gairdneri*. Can. J. Zool. 66, 651–660.

Rombough, P.J., 1998a. Partitioning of oxygen uptake between the gills and skin in fish larvae: a novel method for estimating cutaneous oxygen uptake. J. Exp. Biol. 201, 1763–1769.

Rombough, P.J., 1998b. Piscine cardiovascular development. In: Burggren, W.W., Keller, B. (Eds.), Development of Cardiovascular Systems: Molecules to Organisms. Cambridge University Press, Cambridge, pp. 145–165.

Rombough, P.J., 1999. The gill of fish larvae. Is it primarily a respiratory or an iono-regulatory structure? J. Fish Biol. 55, 186–204. https://doi.org/10.1111/j.1095-8649.1999.tb01055.x.

Ruzicka, J.J., Gallager, S.M., 2006. The importance of the cost of swimming to the foraging behavior and ecology of larval cod (*Gadus morhua*) on Georges Bank. Deep-Sea Res. II Top. Stud. Oceanogr. 53, 2708–2734. https://doi.org/10.1016/j.dsr2.2006.08.014.

Schwemmer, T.G., Baumann, H., Murray, C.S., Molina, A.I., Nye, J.A., 2020. Acidification and hypoxia interactively affect metabolism in embryos, but not larvae, of the coastal forage fish *Menidia menidia*. J. Exp. Biol. 223 (22), jeb228015. https://doi.org/10.1242/jeb.228015. PMID: 33046569.

Spoor, W.A., 1977. Oxygen requirements of embryos and larvae of the largemouth bass *Micropterus salmoides*. J. Fish Biol. 11, 77–86.

Ultsch, G.R., Regan, M.D., 2019. The utility and determination of Pcrit in fishes. J. Exp. Biol. 222 (22), jeb203646.

Wood, C.M., 2018. The fallacy of the Pcrit—are there more useful alternatives? J. Exp. Biol. 221 (22), jeb163717.

Chapter 16

RESPIRATORY GAS EXCHANGE, AEROBIC METABOLISM, AND EFFECTS OF HYPOXIA DURING EARLY LIFE☆

PETER J. ROMBOUGH
Zoology Department, Brandon University, Brandon, Manitoba, Canada R7A 6A9

Chapter Outline

I. INTRODUCTION

The basic mechanisms involved in respiratory gas exchange in juvenile and adult fish are fairly well established (see reviews by Jones and Randall, 1978; Randall, 1982; Randall *et al.*, 1982; Randall and Daxboeck, 1984). The study of oxygen metabolism in older fish, similarly, is well advanced (reviewed by Fry, 1957, 1971; Beamish, 1978; Brett and Groves, 1979; Tytler and Calow, 1985). In contrast, relatively little is known of respiratory gas exchange and energy usage during early life. This arises not so much from

☆This is a reproduction of a previously published chapter in the Fish Physiology series, "1988 (Vol. 11A)/Respiratory Gas Exchange, Aerobic Metabolism, and Effects of Hypoxia During Early Life: ISBN: 978-0-12-350433-3; ISSN: 1546-5098".

Fish Physiology, Vol. 40A. https://doi.org/10.1016/bs.fp.2023.09.007

a lack of effort on the part of researchers—in excess of 500 papers dealing with various aspects of oxygen supply and demand during early life have been published in the last 20 years—but rather from the lack of a systematic approach to the problem. The aim of this review is to collate the large amounts of data that are currently available and to fit it into a conceptual framework that can be used as the basis for future investigations.

II. RESPIRATORY GAS EXCHANGE

Analytical models provide a useful framework for the study of respiratory gas exchange. The cascade model, in particular, has been used widely to describe various aspects of vertebrate respiratory function (Dejours, 1981; Piiper, 1982; Weibel, 1984; DiPrampero, 1985). In this model, respiratory gases are viewed as passing through a series of resistances, each of which is correlated with a specific process or structure. The overall resistance of the system is the sum of the individual resistances and, under steady-state conditions, overall flow through the system is equal to the flow through each of the elements. The model is especially useful in helping to define the nature and magnitude of the various resistances in the respiratory pathway and the partial pressure gradients necessary to overcome them. Such analysis is complicated for early life stages because of changes in respiratory rate and the nature and relative importance of resistances during development. These problems, while formidable, are not insurmountable, as evidenced by the successful application of the cascade model to the study of gas transport in mammalian (Dejours, 1981) and avian embryos (Dejours, 1981; Piiper and Scheid, 1984). Unfortunately, there is currently insufficient information to apply the model rigorously to the study of gas exchange during teleost ontogeny. However, enough is known of gas exchange in developing fish to use the cascade model in a more general way, that is, as a guide to help identify and describe the major resistances during development.

In many ways gas exchange is very similar in embryos and larvae, particularly once organogenesis is complete. In both stages gas exchange is primarily cutaneous. Branchial exchange typically becomes dominant only near the end of the larval period. However, in spite of this similarity, it is often convenient to treat embryos and larvae separately because of the major impact the egg capsule (zona radiata) has on gas exchange during embryonic life. The egg capsule, in addition to acting as a significant barrier in its own right, creates two other barriers, the external boundary layer and the perivitelline fluid, that together have an equal or greater impact (Fig. 1).

A. The Boundary Layer

The laminar boundary layer is a semistagnant region of water adjacent to the egg surface where oxygen is depleted and metabolic wastes accumulate. The

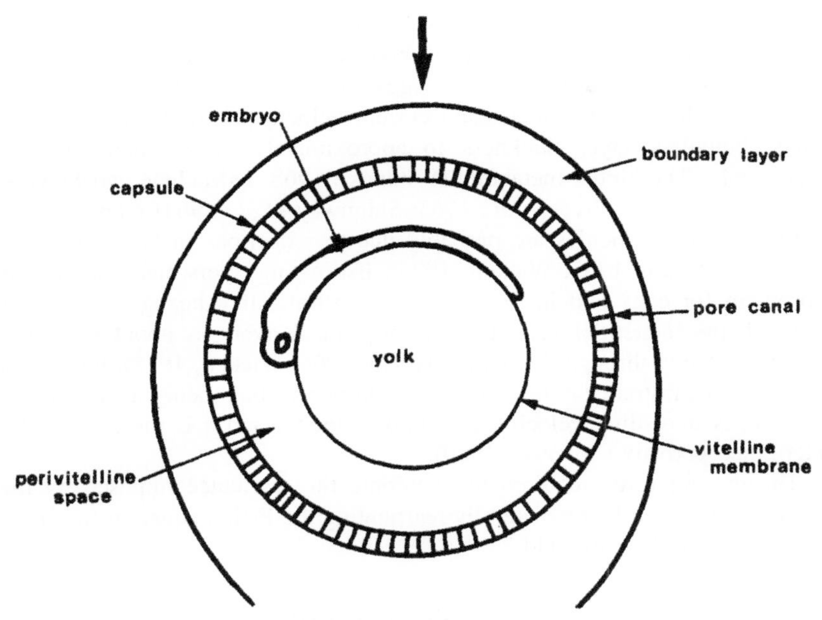

Fig. 1. Schematic diagram showing the major resistances to respiratory gas exchange for fish embryos (not in proportion). The heavy arrow indicates the direction of water flow. The actual shape of the trailing edge of the boundary layer will depend on water velocity and egg size.

boundary layer actually has no outer limit but for practical purposes is usually defined as the distance from the egg surface where local conditions are equivalent to 99% of free-stream conditions (Vogel, 1981). For the laminar flow regimes that eggs are typically exposed to (Reynolds numbers < 75; Johnson, 1980), the thickness of the boundary layer is proportional to egg size and inversely proportional to water velocity. Daykin (1965) indicated that the thickness of the solute boundary layer, d_s, is less than the thickness of the velocity boundary layer and can be estimated from the Sherwood number (Sh) and egg diameter (d) using the equation

$$d_s = d\left(\mathrm{Sh}^{-1}\right) \tag{1}$$

The Sherwood number is a dimensionless number dependent on two other dimensionless numbers, the Reynolds number (Re) and the Schmidt number (Sc). For spherical eggs, Sh can be estimated as

$$\mathrm{Sh} = 2.0 + 0.8(B \cdot \mathrm{Re})^{1/2}\ \mathrm{Sc}^{1/3} \tag{2}$$

where $\mathrm{Re} = \mu d v^{-1}$, $\mathrm{Sc} = vD^{-1}$, and B is the ratio of the interstitial and bulk velocities (Daykin, 1965; Johnson, 1980). Here μ is bulk water velocity, d is egg diameter, v is kinematic viscosity, and D is the diffusion coefficient.

Interstitial velocity is equal to bulk velocity for isolated eggs. According to Eq. (2), the thickness of the oxygen boundary layer would be about 0.02 and 0.05 cm, respectively, for single eggs with diameters of 0.1 and 0.5 cm in a 100 cm h^{-1} current. Increasing current velocity to 1000 cm h^{-1} would reduce boundary layer thickness to approximately 0.0008 and 0.02 cm, respectively. The higher metabolic (Winnicki, 1968; DiMichele and Powers, 1984a) and growth (Silver *et al.*, 1963; Shumway *et al.*, 1964) rates reported at higher water velocities are probably due to reductions in boundary-layer thickness (Daykin, 1965; Wickett, 1975). Estimating interstitial velocities is a problem for eggs laid in masses or in substrate. If a hexagonal array is assumed, the interstitial velocity in an egg mass averages about 9.1 times the bulk velocity through the mass (Daykin, 1965; Wickett, 1975). Interstitial velocities in substrate can be estimated from porous bed theory as $\mu_i = \mu\varepsilon^{-1}$, where μ_i is interstitial velocity, μ is bulk velocity, and ε is the empirically determined porosity (Johnson, 1980).

The driving force required to overcome the resistance imposed by the boundary layer can be predicted by rearranging the Fick's equation for diffusion through a plane to yield

$$C_1 - C_0 = \dot{V}O_2 d_s \left(4\pi r^2 D\right)^{-1} \tag{3}$$

where C_1 is the free-stream oxygen concentration, C_0 is the oxygen concentration at the egg surface, $\dot{V}O_2$ is the rate of oxygen consumption, and r is the egg radius. This equation is equivalent to the more widely used mass transport equation (Daykin, 1965; Wickett, 1975; Johnson, 1980) in which the reciprocal of the mass transport constant, k^{-1}, replaces $d_s D^{-1}$. The value of k is estimated as $k = (\mathrm{Sh})Dd^{-1}$. Equation (3) indicates that the driving force required to meet metabolic demands is directly proportional to the rate of oxygen consumption and thus increases more or less steadily throughout embryonic development. Daykin (1965) estimated that a partial pressure gradient across the boundary layer of about 52 mm Hg was required to meet the oxygen requirements of chum (*Oncorhynchus keta*) eggs ($r = 0.37$ cm) near hatch at 10°C and a flow rate of 85 cm h^{-1}. Smaller pressure differences are required for smaller eggs because of generally lower metabolic rates and thinner boundary layers. For example, Wickett (1975) estimated that cod (*Gadus macrocephalus*) eggs ($r = 0.05$ cm) incubated at 5°C in a current of 170 cm h^{-1} would require a pressure difference across the boundary layer of only about 17 mm Hg to fully satisfy their oxygen requirements.

Until recently it was assumed that forced convection (i.e., bulk water flow) was the major means of supplying oxygen to eggs. It now appears that under certain circumstances natural (free) convection may be important as well. Embryonic metabolism gives rise to solute concentration

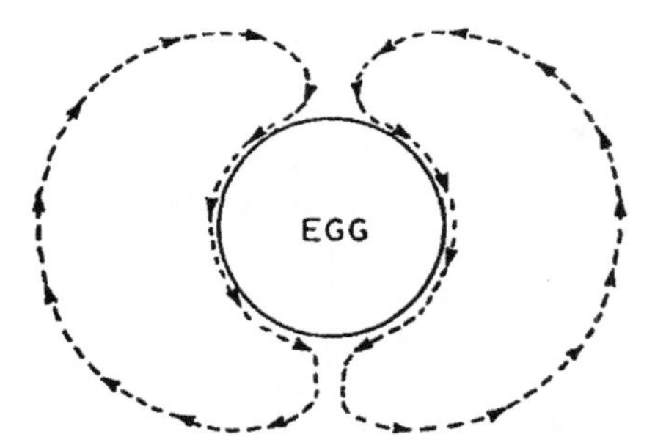

Fig. 2. Stylized depiction of the toroidal flow of water set up around a respiring egg as a result of natural convection in the absence of bulk water movements. [From O'Brien *et al.* (1978).]

gradients across the boundary layer. O'Brien *et al.* (1978) have shown that the oxygen-depleted, carbon dioxide-rich water immediately adjacent to the egg is denser than the well oxygenated, low CO_2 water in the free stream. In still water, this sets up a toroidal flow as the denser solution adjacent to the egg sinks (Fig. 2). Water velocities in the toroid can be relatively high. For example, O'Brien *et al.* (1978) observed an average velocity of 72 cm h^{-1} in the toroid set up by eyed eggs (400 degree-days, 10°C) of coho salmon (*O. kisutch*). At this velocity, natural convection would be about 150 times as effective as simple diffusion in supplying oxygen under "static" conditions. The effectiveness of natural convection can be expected to increase as metabolic rate increases, due to greater depletion of oxygen and accumulation of carbon dioxide in the boundary layer. Thus, natural convection may act as a homeostatic mechanism helping to balance oxygen supply and demand.

Whether natural convection plays a significant role in nature will depend on bulk water velocity and the orientation of the egg mass. Analysis of mixed regimes is complex but, in general, natural convection will play a greater role than forced convection at low bulk velocities [see Vogel (1981, pp. 178–195) on how the ratio of the Grashof number and the square of the Reynolds number can be used to estimate the relative importance of the two forces]. In nature, bulk velocities are often low. For example, the bulk flow in many spawning beds is less than the toroidal velocity set up by coho eggs in still water (O'Brien *et al.*, 1978). Similarly, interstitial velocities in egg masses of species such as lumpfish (*Cyclopterus lumpus*) and lingcod (*Ophiodon elongatus*) can be expected to be rather low. It has been suggested, but not demonstrated, that voids in such masses may provide avenues for natural convection to supply oxygenated water to the interior (O'Brien *et al.*, 1978; Giorgi and Congleton, 1984). Natural convection will be most effective if eggs are oriented so that the heavier oxygen-depleted water can sink and its movement is not opposed by forced convection (O'Brien *et al.*, 1978; Johnson, 1980).

B. The Egg Capsule

The egg capsule traditionally has been viewed as the major barrier to diffusive gas exchange during embryonic life. This may seem obvious, but the empirical evidence supporting this viewpoint is actually rather scanty. Strongest support comes from observations that incipient limiting oxygen tensions (P_c) drop significantly on removal of the capsule (Hayes *et al.*, 1951; Rombough, 1986, 1987). If metabolic rate and the drop in P_c are known, the diffusion coefficient of the capsule (D_c) can be calculated using Fick's equation for diffusion through a plane [Eq. (3)]. Several investigators have done this (Hayes *et al.*, 1951; Alderdice *et al.*, 1958; Daykin, 1965; Wickett, 1975), but the value most often cited, 0.18×10^{-5} cm^2 s^{-1}, is that worked out by Wickett (first presented in Daykin, 1965) using data provided by Hayes *et al.* (1951) for Atlantic salmon (*Salmo salar*). Wickett (1975) recognized that his estimate was based on rather sketchy data and indicated that standard diffusion tests should be conducted to check its validity. Unfortunately, this has not been done for teleosts, although Diez and Davenport (1987) recently used a Krogh-type diffusion chamber to estimate the oxygen diffusion coefficient of the egg case of the dogfish, *Scyliorhinus canicula*. Interestingly, the value they arrived at, 0.285×10^{-5} cm^2 s^{-1}, is similar to Wickett's value for Atlantic salmon. However, given the structural differences between the dogfish egg case and the salmon capsule, this cannot be taken as confirmation of Wickett's value.

Wickett's value for D_c is about one-tenth that of water. Wickett (1975) pointed out that the surface area of the radial pores that penetrate the capsule is similarly about one-tenth the total surface area of the capsule and speculated that this may indicate that diffusion takes place primarily through the pore canals rather than through the capsule matrix. If this is true, doubt is cast on the validity of the practice of some investigators (e.g., Daykin, 1965; Wickett, 1975; Kamler, 1976; DiMichele and Powers, 1984) applying diffusion coefficients, calculated for one species, to another unrelated species. There is considerable variation among teleosts in capsule structure (Lønning, 1972; Stehr and Hawkes, 1979; Groot and Alderdice, 1985). Even in closely related salmonids, pore area can vary anywhere between 7% and 30% of the total surface area (Groot and Alderdice, 1985).

Recent evidence suggests that the capsule may not be as great a barrier to diffusive gas exchange as was supposed previously. Berezovsky *et al.* (1979) used microelectrodes to measure the dissolved oxygen profile across the capsule, perivitelline fluid, and vitelline membrane of the recently fertilized loach eggs (*Misgurnis fossilis*). Surprisingly, they recorded very little drop in oxygen concentration across the capsule. In contrast, there was a gradual decline in oxygen tension across the perivitelline fluid and a sharp drop across the vitelline membrane (Fig. 3). Sushko (1982) reported a similar oxygen profile for loach eggs incubated in helium–oxygen and nitrogen–oxygen gas mixtures. Observations made by Alderdice *et al.* (1984) may provide an explanation for why the capsule

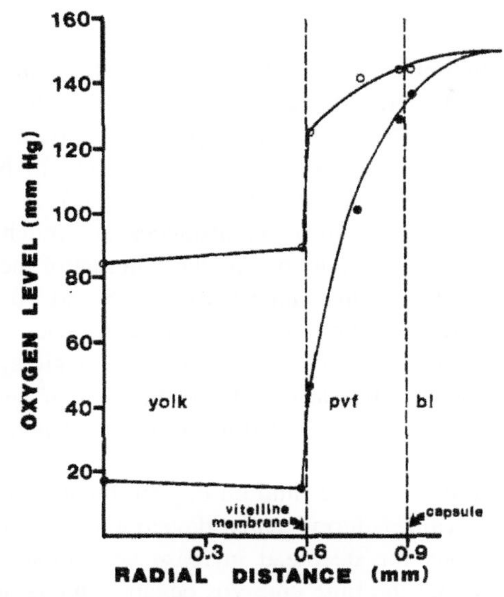

Fig. 3. Oxygen profiles across recently fertilized roach eggs measured using platinum microelectrodes. Open symbols, normal eggs; closed symbols, eggs in which rate of oxygen uptake was stimulated 3.3-fold using 10^{-4} *M* 2,4-dinitrophenol; pvf, perivitelline fluid; bl, boundary layer. [After Berezovsky *et al.* (1979).]

appears to offer relatively little resistance to gas exchange. Alderdice *et al.* (1984) observed that the hydrostatic pressure exerted on the capsule of steelhead (*S. gairdneri*) was considerably less than the osmotic pressure of the perivitelline fluid and calculated an effective filtration pressure of about −62 mm Hg driving water into the perivitelline space. They reasoned that according to Starling's hypothesis this should lead to the movement of water into the perivitelline space. Since the egg is in volume equilibrium, this must be balanced by an equal outflow by filtration. Exactly how this would be accomplished is not clear, since the capsule is not a linear structure like a capillary. Alderdice *et al.* (1984) suggested that the capsule may act like a balloon with micro-sieves in its wall. Increasing internal pressure would be accompanied by volume expansion. As the capsule expanded the pores would enlarge and more water would flow out. Volume and tension increases thus would be self-limiting. Alderdice *et al.* (1984) proposed that this process would tend to facilitate respiratory gas exchange and point to radiotracer studies (Potts and Rudy, 1969; Loeffler and Lovtrup, 1970; Loeffler, 1971) indicating an exchange of water across the capsule equivalent to the volume of the perivitelline fluid every 1–4 min. A connective flux of this magnitude would add to the diffusive flux of oxygen across the capsule but, perhaps more importantly, the currents generated would tend to prevent the establishment of a large concentration gradient across the capsule.

C. The Perivitelline Fluid

As noted earlier, microelectrode studies (Berezowsky *et al.*, 1979; Sushko, 1982) indicated that, at least for early embryos, much of the resistance to gas exchange previously attributed to the capsule actually resides in the perivitelline fluid. The perivitelline fluid can be expected to have an oxygen diffusion coefficient similar to that for water ($\simeq 2.5 \times 10^{-5}$ cm^2 s^{-1}) (Dejours, 1981), but because diffusion distances are much larger the net impact of the perivitelline fluid may be greater than that of the capsule. While capsule thicknesses in salmonids range between 15 and 70 μm (Groot and Alderdice, 1985), the distance across the perivitelline space can be in excess of 500 μm. Absolute distances tend to be smaller in pelagic eggs, but because of a thinner capsule and comparatively larger amount of perivitelline fluid, the relative impact of the perivitelline fluid on gas exchange would be even larger than in salmonids.

The impact of the perivitelline fluid on oxygen uptake can be seen in the rapid increase in P_c values during early development (Rombough, 1987). Critical oxygen tensions for steelhead increase very rapidly in relation to metabolic rate until about the time embryos began to move and stir the perivitelline fluid. Thereafter, P_c values increase more slowly and in direct proportion to metabolic rate (Fig. 4). Thus in a teleological sense, stirring of the perivitelline appears to be necessary if metabolic oxygen demands are to be met. Reznichenko *et al.* (1977) used a modified polarographic electrode to model oxygen exchange across the egg capsule and perivitelline fluid. He found that when the analog of the perivitelline fluid was stirred, the PO_2 at the body (electrode) surface increased while that under the capsule (membrane) decreased. This had the effect of increasing the steepness of the concentration gradient across the capsule and enhancing net oxygen transport. Peterson and Martin-Robichaud (1983) observed that Atlantic salmon embryos began to stir the perivitelline fluid fairly early in development. Trunk movements began abruptly at about 200 degree-days of development with an initial frequency of about 60–120 flexures h^{-1}. Dye studies indicated that trunk flexures resulted in rapid water movement along the trunk and from one side of the perivitelline fluid to the other. These movements were apparently of a respiratory nature, since an unexpected water failure leading to hypoxia resulted in an increase in frequency of trunk flexures. Trunk movements normally decline rather abruptly to a frequency of only 1 every 2–4 h by 350–400 degree-days in Atlantic salmon (Peterson and Martin-Robichaud, 1983). However, by this time the embryo had begun to move its pectoral fins rapidly at a rate of 40–150 min^{-1}. This rate was maintained until hatch. Dye studies indicated that these movements generated a rapid water flow ($\simeq$300 cm h^{-1}) in the immediate area of the pectoral fins but were not as effective as trunk flexures in completely mixing the perivitelline fluid. Complete mixing was accomplished by periodic trunk flexures.

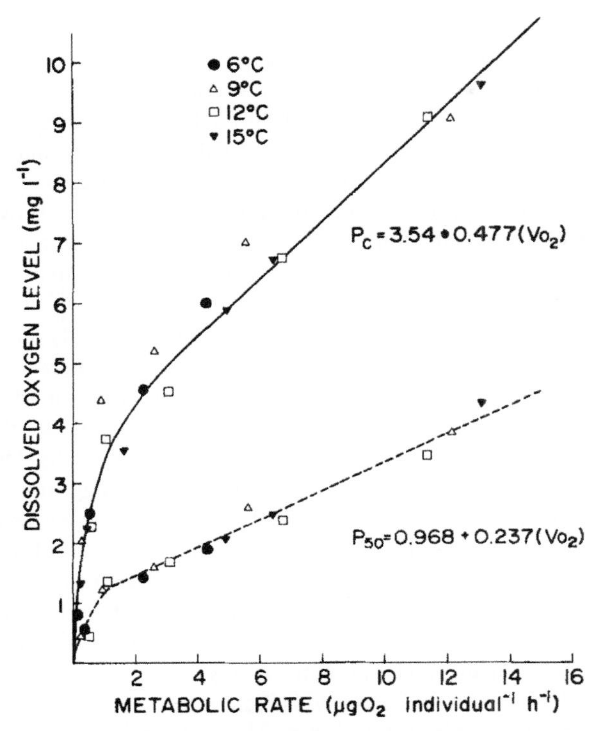

Fig. 4. Relationships between critical dissolved oxygen levels (P_c) and routine metabolic rate and P_{50} values (oxygen concentration at which $\dot{V}O_2 = 0.5\ r\dot{V}O_2$) and routine metabolic rate for steelhead embryos incubated at constant temperatures. Equations are for the linear portions of the curves. [From Rombough (1987).]

Recent studies of amphibians suggest that stirring of the perivitelline fluid may facilitate oxygen transport within an egg mass as well as within individual eggs. Burggren (1985) noted that oxygen partial pressures were higher and carbon dioxide partial pressures were lower in the interior of the egg mass of the frog *Rana palustris* than would be expected if simple diffusion was the only process involved. Frog embryos use cilia to stir the perivitelline fluid. Burggren (1985) suggested that the currents generated by movement of the cilia could lead to oxygen being transported to the interior of the egg mass by convection as well as diffusion. Oxygen would diffuse across the capsule at the surface of the egg closest to the outside of the egg mass. This mass of oxygen-rich water would then be moved by ciliary action to the opposite side of the egg where oxygen then would diffuse outward across the capsule toward the center of the egg mass. An oxygen molecule thus could be passed from egg to egg in a manner somewhat analogous to water being passed along a bucket brigade. Carbon dioxide would pass in the opposite direction.

This appears to be a plausible mechanism for supplying oxygen to the interior of teleost as well as amphibian egg masses. Many teleosts lay large, compact masses of eggs. For example, the egg mass of lingcod may be up to 5 liters in volume (Giorgi and Congleton, 1984). At present, there is not enough information on oxygen and current profiles within such egg masses to adequately test the "bucket-brigade" hypothesis. However, there is some circumstantial evidence that processes in addition to bulk water flows may be involved. Giorgi and Congleton (1984) noted that while oxygen concentrations in the center of a lingcod egg mass declined rather sharply following cessation of current flow, levels did not decline to zero as expected but stabilized at about 10% air saturation. Davenport (1983) similarly indicated that oxygen levels in the egg mass of lumpfish declined more slowly than expected when aeration ceased. These observations are intriguing but say little about the mechanisms involved. As discussed previously, these observations can be explained equally as well by natural convection as by the "bucket-brigade" hypothesis. They do, however, suggest that egg masses do not depend solely on forced convection to meet metabolic oxygen demands.

Braum (1973) noted that the deeper eggs in the egg mass of species such as herring (*Clupea harengus*) are threatened with asphyxia as a result of poor water circulation. He suggested that the perivitelline fluid could function as an oxygen reservoir to tide embryos over short periods of anoxia. This is unlikely to be of much significance. The amount of oxygen in the perivitelline fluid, assuming it is 100% saturated, is only sufficient to meet the oxygen requirements of advanced embryos for 1–2 min. This calculation assumes that oxygen would not diffuse back out of the capsule—which of course it would under hypoxic conditions—and that there is no convective exchange between the perivitelline fluid and the surrounding water—which is likely.

The perivitelline fluid provides the immediate environment for the developing embryo, and it is the gas concentration in the perivitelline fluid—not the surrounding water—that is of physiological significance. As predicted by the mass transport equation, Eq. (3), oxygen concentrations in the perivitelline fluid decline progressively as development proceeds. Assuming relatively constant capsule conductance and ambient PO_2, the only way the rising metabolic demands associated with tissue growth can be met is by an increase in the driving force across the capsule. This necessitates a reduction in the PO_2 of the perivitelline fluid. Berezovsky *et al.* (1979) demonstrated such a drop in perivitelline fluid PO_2 when the metabolic rate of loach embryos was stimulated by low concentrations of dinitrophenol (Fig. 3). A decline in perivitelline fluid PO_2 is also implied by the gradual increase in P_c that was seen during the course of steelhead development (Fig. 4; Rombough, 1987). Recently, Diez and Davenport (1987) showed that the PO_2 of the fluid in the egg case of the dogfish declined as development proceeded. Finally, similar declines in PO_2 have been well documented for reptilian and avian eggs (Dejours, 1981). Bird eggs in particular have been studied extensively, and since many of the structures in bird and fish eggs can be considered analogous, the type of relationships seen in bird eggs probably apply to fish eggs as well. For example in the hen egg, oxygen levels in the air space, which is analogous to the perivitelline fluid, decrease as metabolic rate increases (Wangensteen, 1972). This increases the diffusion gradient across the shell, which like the teleost capsule is pierced by tiny pores, and automatically ensures a greater rate of diffusive flux. It does so, however, at the expense of arterial PO_2 levels which gradually decline as development proceeds. Blood gas relationships have not been examined in fish embryos, but if the analogy with bird eggs holds, they probably follow a similar pattern.

A decrease in perivitelline fluid PO_2 late in embryonic development appears to be the trigger that initiates hatching in at least some teleosts. If advanced embryos are placed in hypoxic water, premature hatching occurs (Yamagami, 1981; DiMichele and Powers, 1984a; Ishida, 1985). Conversely, hatching can be delayed more or less indefinitely under hyperoxic conditions (Taylor *et al.*, 1977; DiMichele and Taylor, 1980; Ishida, 1985). Low oxygen levels do not appear to act directly on the hatching glands. Studies involving various anesthetics suggest the response is mediated by the central nervous system (Ishida, 1985). The location of the oxygen sensor is not known.

Hatching can be regarded as an adaptive response to physiological hypoxia. Escape from the confines of the egg capsule reduces the ambient oxygen level required to meet metabolic requirements by 30–50 mm Hg (Rombough, 1987). However, removal of the capsule does not alter the basic mechanisms involved in respiratory gas exchange.

D. Cutaneous Gas Exchange

Respiratory gas exchange in fish, and indeed in all vertebrates, is initially cutaneous. As development proceeds there is a gradual increase in the relative importance of gills, although in many species the skin remains the major site of gas exchange throughout the embryonic and larval periods. Recent evidence indicates that even in adults the skin may persist as an important site for respiratory gas exchange (Kirsch and Nonnotte, 1977; Lomholt and Johansen, 1979; Steffenson and Lomholt, 1985; Feder and Burggren, 1985).

Studies of gas exchange during the early life stages of teleosts have tended to be descriptive. As a result, most of what we know of respiratory mechanisms has been inferred from studies of the morphology of what are assumed to be respiratory structures. Morphological adaptations to facilitate gas exchange appear early in development. Boulekbache and Devillers (1977) suggested that the function of the microvilli present on the outer surfaces of blastomeres of rainbow trout (*S. gairdneri*) was to increase the surface area for respiratory gas exchange. In many species well-developed vascular networks form just under the skin during early organogenesis (Fig. 5). These capillary beds are often associated with specialized cutaneous structures, such as

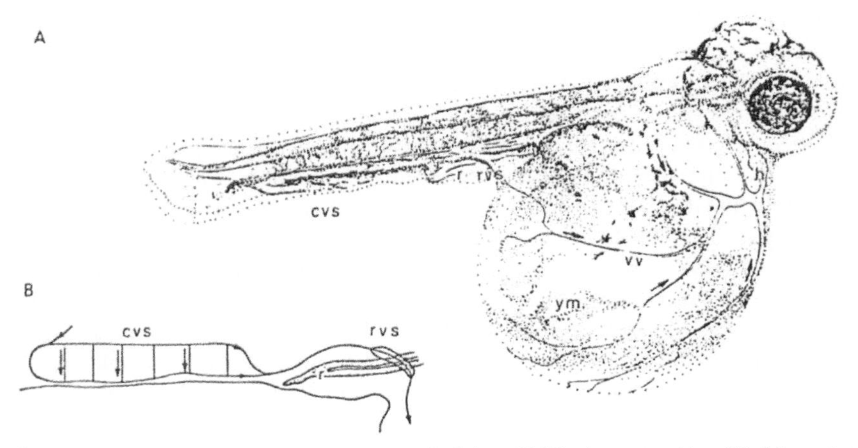

Fig. 5. (A) Cutaneous gas exchange structures in 5-day-old *Tilapia mossambica*. (B) Schematic diagram showing blood flow through caudal and rectal vascular systems: cvs, caudal vascular system; h, heart; r, rectum; rvs, rectal vascular system; vv, vitelline vein; ym, yolk mass. [From Lanzing (1976).]

an enlarged yolk sac, expansive medial finfolds, or enlarged pectoral fins, that greatly increase the surface area available for gas exchange. Detailed descriptions of such specialized structures are provided by Taylor (1913), Sawaya (1942), Wu and Liu (1942), Kryzanowsky (1934), Smirnov (1953), Soin (1966), Balon (1975), Lanzing (1976), McElman and Balon (1979), Liem (1981) and Hughes *et al.* (1986), among others.

The degree to which embryonic and larval respiratory structure are elaborated varies widely among species. Several authors have suggested that this reflects variations in oxygen levels in spawning habitat and can be used as the basis for a functional classification system (Kryzanowsky, 1934; Soin, 1966; Balon, 1975). It is beyond the scope of the current discussion to examine the merits of such systems, but attention will be drawn to one characteristic that these authors have considered particularly important, that is, whether the species lays pelagic or demersal eggs. In the marine and temperate freshwater environments, pelagic embryos tend to have relatively poorly developed capillary plexes near the body surface. Respiratory pigments (hemoglobin, myoglobin, perhaps carotenoids) usually do not appear until late in development, and gills may not become functional until near the end of the larval period (Balon, 1975). Pelagic eggs tend to be small and are normally found in well-oxygenated waters. As a result, it has been suggested that oxygen is not normally a limiting factor and thus specialized respiratory structures are not necessary (Hempel, 1979). In contrast, demersal eggs are usually larger and often are exposed to relatively low oxygen concentrations for extended periods. According to Balon (1975), this has resulted in selection for extensive vascularization of the body surface and the elaboration of specialized cutaneous gas exchange structures. Such structures tend to develop early and often persist throughout the larval period. Respiratory pigments appear early, and gills become functional soon after hatch. It should be recognized, though, that as with most generalizations in zoology, there are exceptions. For example, the Indian air-breather *Anabas testudineus* lays small pelagic eggs, but the body surface is well vascularized, and respiratory pigments appear early in development (Hughes *et al.*, 1986). These adaptations are not particularly surprising when one realizes that the eggs are laid in the very oxygen-poor waters of tropical swamps, and that it is only in the surface layer that oxygen levels are high enough to sustain embryonic development.

The effective surface area, the length of the diffusive pathway, the magnitude of the partial pressure gradient between the water and blood, the amount of blood perfusing the structure, and the convective movement of water past the structure are among the most important factors influencing the performance of respiratory gas exchangers. On the basis of these characteristics, cutaneous gas exchange in fish embryos and larvae would appear to be highly efficient. As mentioned previously, specialized exchange structures comprise a relatively large fraction of total body surface area in many species. For example, the well-vascularized medial and paired fins of larval herring (*C. harengus*) and plaice (*Pleuronectes platessa*) account for about 40% of total surface area at hatch (DeSilva and Tytler, 1973). In addition, the surface/volume ratio of most larvae is large because of their small absolute size. Total surface area for a 1.6 mg carp (*Cyprinus carpio*) larvae is in the order of 12,000 $mm^2\ g^{-1}$. In contrast, total surface area of a 350 mg juvenile is only about 1400 $mm^2\ g^{-1}$ (Oikawa and Itazawa, 1985).

Cutaneous diffusion distances have been estimated for only a few species, but the available evidence indicates that distances are only slightly greater than lamellar diffusion distances in juveniles and adults. The skin is only two cells thick over most of the body surface in young larvae (Lasker, 1962; Jones *et al.*, 1966; Roberts *et al.*, 1973). Lasker (1962) reported that the skin thickness of larval sardine (*Sardinops caerislea*) ranged from 1.7 μm on the finfold to 3.0 μm on the lateral portion of the trunk. Jones *et al.* (1966) estimated a minimum skin thickness in larval herring (*C. harengus*) of about 2.3 μm. The actual length of the diffusive pathway is somewhat greater. Many, particularly pelagic, species have a relatively thick fluid layer between the dermis and epidermis: 5.0 μm in the case of larval plaice (*P. platessa*; Roberts *et al.*, 1973). In addition, distances associated with diffusion across capillary walls and through the plasma should be taken into account. Even when this is done, distances remain relatively small. Webb and Brett (1972a) measured a mean distance of 4.7 μm from the surface of the skin to the center of blood capillaries in embryos of two species of viviparous seaperch (*Rhacohilus vacca* and *Embioteca lateralis*). Liem (1981) estimated 8–15 μm for the total thickness of the water–blood barrier in larval *Monopterus*. These distances are considerably less than the cutaneous diffusion distances of larval amphibians. Burggren and Mwalukoma (1983) estimated a blood–water barrier of 20–50 μm for larval bullfrog (*R. catesbeiana*). In this species up to 60% of total gas exchange takes place across the skin (Burggren and West, 1982). The skin of larval teleosts is probably at least as effective as an organ of gas exchange, given the shorter diffusion distances and, in many cases, more elaborate vascularization.

Cutaneous gas exchange in other vertebrates frequently suffers from a relatively small partial pressure gradient across the skin as a result of central mixing of oxygenated and deoxygenated blood prior to transit to the skin (Burggren, 1984). This problem appears to be minimized in many teleosts. Cutaneous gas exchange structures—for example, the caudal and rectal vascular systems of larval tilapia (Fig. 5) or the vitelline circulation of salmonids—typically receive blood that has already passed through at least a portion of the systemic circulation. Although no in situ measurements have been made, the blood entering the exchange structures should be comparatively poor in oxygen thus maximizing the partial pressure gradient between the blood and the water.

Extremely little is known of the cardiovascular physiology of teleost embryos. Cutaneous exchange obviously would be more effective if blood flow could be regulated. McElman and Balon (1979) noted that the amount of blood passing through cutaneous capillary beds varied during the development of walleye (*Stizostedion vitreum*) and suggested that this implied a mechanism for shunting blood so as to optimize gas exchange. Vascular recruitment as a result of higher systemic blood pressure caused by increased heart rate during periods of physiological hypoxia was proposed. However, there may be differences between ontogenetic changes in blood flow patterns and compensatory changes due to transient alterations in oxygen supply or demand. Reflex control of the amount of blood perfusing cutaneous gas exchange structures may well occur, capillary recruitment in response to hypoxia has been demonstrated in amphibian larvae (Burggren, 1984), but it has yet to be demonstrated in teleost embryos or larvae.

E. Respiratory Pigments

As mentioned previously, the stage at which hemoglobin first appears is highly variable. In many demersal species, such as salmonids, large numbers of pigmented erythrocytes are evident well before hatch. This is thought to be an adaptation to hypoxic conditions (Balon, 1975). In contrast, hemoglobin may not appear in the circulation of pelagic species, such as herring, until after metamorphosis (De Silva, 1974). Lack of hemoglobin has been proposed as a mechanism to limit predation by making pelagic larvae less conspicuous. However, it simply may be that hemoglobin is not required in well-oxygenated waters. Holeton (1971) reported that rainbow trout larvae showed little distress when their hemoglobin was poisoned by carbon monoxide. Similarly, Iuchi (1985) reported that rainbow trout larvae survived to the fry stage after having their erythrocytes destroyed by treatment with phenylhydrazine. Indeed, it may not even be necessary for small larvae to have a functioning circulatory system. Burggren (1984) points out that the so-called "cardiac lethal" larval mutant of the amphibian *Ambystoma,* in which the heart forms but fails to beat, is able to survive after hatching for many days in well-oxygenated water.

The evidence to date indicates that embryonic and larval hemoglobins are structurally and functionally distinct from those of juveniles and adults. Iuchi and Yamagami (1969) reported a gradual change in the electrophoretic banding pattern for the hemoglobins of rainbow trout during the period between hatch and gravel emergence. Similar shifts in electrophoretic banding patterns have been observed in Homasu salmon (*O. rhodurus*) and brook trout (*Salvelinus fontenalis*; Iuchi *et al.*, 1975) and in coho salmon (*O. kisutch*; Giles and Vanstone, 1976). Distinct embryonic and adult hemoglobins also have been reported for several viviparous species (Ingermann and Terwilliger, 1981a,b, 1982, 1984; Ingermann *et al.*, 1984; Weber and Hartvig, 1984; Hartvig and Weber, 1984).

Iuchi (1973b) compared the chemical and physiological properties of larval and adult hemoglobins of rainbow trout. Both were tetrameric, but larval hemoglobin displayed a higher oxygen affinity, less of a Bohr effect, virtually no Root effect, and greater cooperativity at physiological pH than adult hemoglobin. Larval and adult hemoglobins had P_{50} values of 31 mm Hg and 57.5 mm Hg, respectively, at pH 7.2 and 25°C. The Bohr effect (Δ log P_{50}/pH) was 0.023 for larval hemoglobin but 0.64 for adult hemoglobin. The oxygen capacity of larval hemoglobin was virtually unaffected by pH, while a drop to pH 6.5 reduced the oxygen capacity of adult hemoglobins to 50% of that at pH 8.0. Slopes of Hill plots were $n = 2.33$ and $n = 1.62$, respectively, at pH 7.2 for larval and adult hemoglobins. The high oxygen affinity and pH independence of larval hemoglobins are clearly advantageous to embryos and larvae exposed to the oxygen-poor, low-pH and high-CO_2. environments of the perivitelline fluid and interstices of the redd.

The shift in electrophoretic banding patterns suggests that the greater oxygen affinity of embryonic and larval blood is due primarily to intrinsic differences in the structure of the hemoglobins rather than to changes in concentrations of modulators of hemoglobin (Hb) affinity. This was shown to be the case for the viviparous eelpout *Zoarces viviparous* (Weber and Hartvig, 1984; Hartvig and Weber, 1984). Weber and Hartvig (1984) reported that fetal hemoglobin had a higher O_2 affinity (P_{50} values of 9 mm Hg and 23 mm Hg, respectively, at pH 7.5 and 10°C), reduced Bohr effect, and greater cooperativity than adult hemoglobin in nucleoside triphosphate-free preparations. Measurement of intraerythrocyte nucleoside triphosphate (NTP) concentrations revealed no significant difference in the NTP/Hb ratios of fetal and adult blood. Differences in modulator concentrations, though, do appear to be important in some species. Ingermann and Terwilliger (1981b, 1982, 1984) reported that part of the reason for the higher O_2 affinity of the hemoglobin of fetal seaperch *E. lateralis* was a lower NTP/Hb ratio. ATP was the most abundant modulator (82% total NTP), but there was also a significant amount of GTP (18% of total NTP) present within fetal erythrocytes. Indirect evidence suggests cofactors may modulate the O_2 affinity of larval hemoglobins in some oviparous species as well. DiMichele and Powers (1982) attributed differences in hatching times of different lactate dehydrogenase (LDH) genotypes of *Fundulus heteroclitus* to the ability to deliver oxygen to tissues. The LDH genotype (LDH B^aB^a) that hatched earliest was also the genotype that had the highest concentration of ATP in their erythrocytes as adults, and presumably as embryos. Increased ATP concentrations would lower hemoglobin O_2 affinity and thus reduce O_2 delivery to tissues near hatch when PO_2 levels in the perivitelline fluid are low. This would trigger release of the hatching enzyme and lead to early hatch.

Iuchi and Yamamoto (1983) demonstrated that the shift from embryonic–larval hemoglobins to juvenile hemoglobins in rainbow trout was the result of erythrocyte replacement. Erythrocytes containing embryonic hemoglobins were formed in the extraembryonic blood islands and intermediate cell mass beginning about one-third through embryonic development. These hemopoietic centers ceased production shortly after hatch, and centers in the kidney and spleen began to produce morphologically and antigenically distinct erythrocytes containing juvenile-type hemoglobins. Similar shifts in the site of hemopoiesis have been reported for Atlantic salmon (Vernidub, 1966) and anglefish *Pterophyllum scalare* (Al-Adhami and Kunz, 1976). Hemoglobin switching based on erythrocyte replacement may turn out to be widespread in lower vertebrates. Kobel and Wolff (1983) recently reported that the shift from embryonic to larval type hemoglobins in the amphibian *Xenopus borealis* also was associated with a shift in the site of erythropoiesis.

It has been suggested that pigments other than hemoglobin may be involved in respiratory gas exchange. De Silva (1974) noted that myoglobin was formed prior to hemoglobin in herring larvae and indicated that this may reflect a respiratory role during early development. This hypothesis has not been tested. Other authors have proposed a respiratory role for the carotenoid pigments (Smirnov, 1953; Volodin, 1956; Balon, 1975, 1984; Mikulin and Soin, 1975; Soin, 1977; Czeczuga, 1979). This hypothesis is based largely on circumstantial evidence, namely, that eggs of species containing high concentrations of carotenoids are often relatively large and, thus, are faced with relatively large diffusion distances. The problem of gas exchange is compounded for some of these species by the fact that they must develop in comparatively low oxygen environments. Unfortunately, there is little experimental evidence to indicate that carotenoids actually aid in oxygen transport under such conditions. In fact, there is now considerable evidence indicating that low carotenoid levels in eggs, normally rich in carotenoids, do not significantly reduce survival (Steven, 1949; Craik, 1985; Craik and Harvey, 1986; Tveranger, 1986). Craik (1985) speculates that carotenoids may play some minor, as yet undefined, role in oxygen transport but that present evidence does not warrant acceptance of carotenoid-based respiration as an established fact, as Balon (1975, 1984) would suggest.

We have seen how the movement of water past the egg and stirring of the perivitelline fluid enhance embryonic gas exchange. Adequate ventilation of body surfaces is similarly necessary for efficient gas exchange after hatch. Fish have adopted a number of tactics to ensure that this occurs. Some, such as nest-fanning, mouth-brooding, and wriggler-hanging, involve the parents. Others involve behavioral and physiological adaptations on the part of the larvae.

Hunter (1972) noted that newly hatched anchovy (*Engralis mordax*) exhibited regular bouts of swimming that did not appear to be associated with feeding or predator avoidance. He suggested that swimming might have a respiratory significance, presumably by enhancing ventilation of the body surface. This hypothesis was examined in some detail by Weihs (1980, 1981). Weihs (1980) pointed out that because of their small size, anchovy larvae existed in a viscous environment (i.e., low Reynolds number, Re). Consequently, both the larva and its immediate surroundings tend to be transported together by oceanic currents. A nonswimming larva thus would remain in the same mass of water and gradually deplete the available oxygen. Weihs (1980) developed a mathematical model for diffusive oxygen uptake by the larvae and estimated that oxygen would become limiting for a stationary day-old larva at concentrations below 63% air saturation (ASV). He then tested this prediction by observing larval swimming behavior at various oxygen concentrations. As predicted, both the frequency and duration of swimming significantly increased at oxygen levels below 60% ASV (Fig. 6).

Weihs' (1980) study indicates that the limiting step in cutaneous gas exchange, at least in anchovy larvae, is the convective flow of water past the exchange surface. It should be remembered that anchovy larvae live in a comparatively oxygen-rich environment. It would thus appear to be even more important for larvae inhabiting hypoxic environments to be able to generate large convective flows. This is accomplished in larvae of the Austrailian lungfish (*Neoceratodes forsteri*) by means of cilia that direct a water current posteriorly along the body surface (Whiting and Bone, 1980). Larvae of many warm-water teleosts use their pectoral fins to create a similar water flow. Liem (1981) reported that larvae of the air-breathing fish *Monopterus albus* use large, well-vascularized pectoral fins to direct a flow of relatively oxygen-rich surface water backward along their body surface. The yolk sac and caudal region of *Monopterus* larvae are also extensively vascularized. Microscopic observations indicated that surficial blood flow in these regions ran in the opposite direction to the flow of water generated by the pectoral fins. When larvae were placed in a tube with water flowing in the normally anterior to posterior direction the oxygen extraction efficiency was about

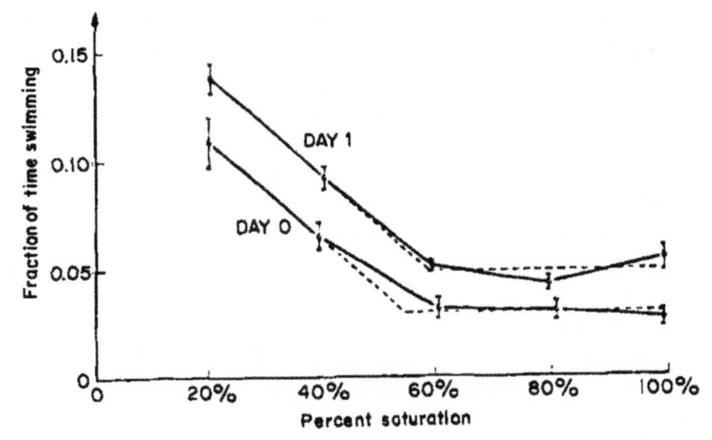

Fig. 6. Fraction of time spent actively swimming as a function of ambient oxygen concentration for newly hatched (day 0) and 24-h-old (day 1) northern anchovy larvae. [From Weihs (1980).]

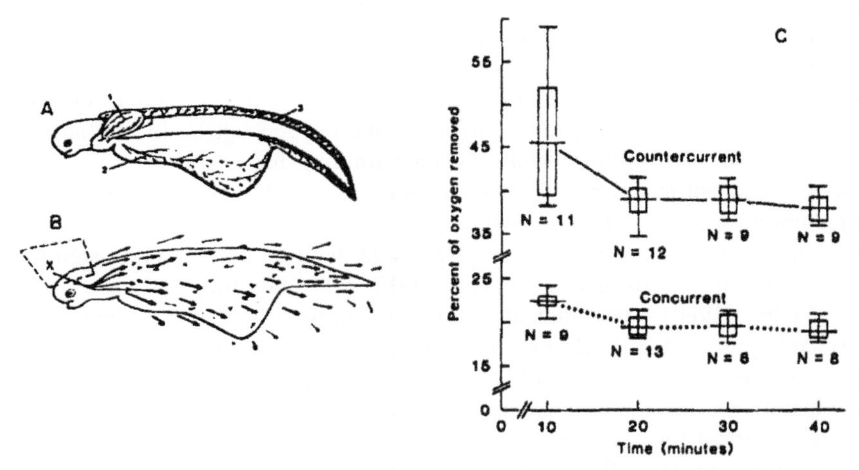

Fig. 7. (A) Larval *Monopterus albus* (4 days old) showing (1) large vascular pectoral fins, (2) well-developed capillary network in the yolk sac, and (3) well-vascularized unpaired medial fin. (B) Schematic representation of water currents generated by movement of pectoral fins. The stippled area shows the region from which water is drawn by the pectoral fins. (C) The effectiveness of oxygen extraction from water flowing in an anterior to posterior direction (countercurrent to blood flow) or in a posterior to anterior direction (concurrent to blood flow). The flow rate was 130 ml l^{-1}. (Reprinted with permission from Liem, 1981; copyright © AAAS.)

41%. When the direction of water flow was reversed, extraction efficiency dropped to about 20% (Fig. 7). Liem (1981) pointed out that on this basis the whole larva could be considered a functional analog of the gill lamellae of adult fish. He noted that similar elaborate vascular networks and mobile pectoral fins were common in larvae of other species inhabiting hypoxic waters and speculated that such countercurrent flow mechanisms might be widespread. It should be noted, though, that pectoral fin movements are not always associated with ventilating cutaneous gas exchange surfaces. Peterson (1975) found that, contrary to his expectations, the rapid pectoral fin movements of Atlantic salmon alevins did not direct water over the well-vascularized yolk sac. Instead, they appeared to be involved in drawing water over the gills.

F. Branchial Gas Exchange

The larval period is characterized by a shift in the site of respiratory gas exchange from the skin to the gills. Unfortunately, relatively little is known of the physiology of this transition. It is known that species vary in the stage at which gills first appear and the speed with which they are elaborated. For example, gill arches begin to form in rainbow trout shortly after gastrulation and by hatch are complete with functional filaments and secondary lamellae (Morgan, 1974a,b). Arctic char (*Salvelinus alpinus*) also possess filaments at hatch but secondary lamellae do not begin to form until about 8 days posthatch (at 6.5°C; McDonald and McMahon, 1977). In smallmouth bass (*Micopterus dolomieu*), filaments do not appear until about 7 days post-hatch (at 16°C; Coughlan and Gloss, 1984). Lamellae begin to form about 4 days later, about the time the larvae becomes free-swimming. Filaments and secondary lamellae first appear on gill arches of herring and plaice larvae about midway through the larval period, at body lengths of 10 and 8 mm, respectively (De Silva, 1974).

Such description provides relatively little information about the relative importance of the gills in larval gas exchange. This question can be answered best by directly measuring gas fluxes across the gills and skin. This has been done for amphibian larvae (Burggren and West, 1982; Burggren, 1984) but not for teleost larvae, even though the techniques developed for amphibians would be relatively easy to apply to some of the larger fish larvae. In the absence of such partitioning studies, the relative importance of gills and skin in fish larvae must be inferred from the rather limited amount of information available on the morphometrics of the two structures. As mentioned previously, the factors of particular importance in determining the efficiency of gas exchange structures are total surface area, diffusion distance, partial pressure gradients, and the convective supplies of blood and water.

De Silva (1974) reported that, at hatch, the gills of both herring and plaice accounted for an insignificant portion (<0.5%) of total body surface area. The surface area of the gills expanded rapidly during larval development; gill area weight exponents were 3.36 and 1.59, respectively. However, even at the end of the larval period, gill surface area was only a relatively small fraction (≃10% for herring and <5% for plaice) of the total surface area over which diffusive gas exchange could take place. Similarly, the surface area of the gills of newly hatched carp larvae accounted for < 0.2% of the total area available for gas exchange (gills, fins, body surface) (Oikawa and Itazawa,

1985). Gills grew rapidly, with allometric mass exponents of 7.066 between 1.6 and 3.0 mg and 1.222 between 3.0 and 200 mg. However, even for a 140 mg postlarva, the surface area of the gills remained less than one-third of the total potential exchange area. Thus, morphometric analysis would suggest that in these species the skin remains the most important organ for gas exchange throughout larval development. Gills appear to be relatively more important in salmonid larvae, but even here, cutaneous exchange remains significant until long after swim-up. McDonald and McMahon (1977) reported that early development of Arctic char gills was characterized by a rapid increase in filament number and size followed by an increase in the number and size of the secondary lamellae. At 15 days posthatch lamellae only accounted for about 10% of gill surface area, while by 47 days posthatch (at 6.5°C) they comprised 23% of total gill area. However, this was only about half the lamellar contribution (45% total gill area) in yearling char. McDonald and McMahon (1977) calculated that on the basis of body weight Arctic char larvae would require a lamellar surface area about four times that actually measured 47 days posthatch to meet metabolic oxygen demands if the gills were the only site of gas exchange.

Only a few measurements have been made of the thickness of the blood-water diffusion barrier in the gill of larval fish. Morgan (1974b) reported that the thickness of the blood–water pathway in rainbow trout decreased gradually from 11.1 μm at hatch to 6.8 μm 102 days posthatch. This decrease was due primarily to a shift from a multilayer respiratory epithelium at hatch to one that was only two cells thick. There was also a modest decrease in the thickness of the basement membrane. A reduction in the thickness of the diffusion barrier is probably typical of obligate water-breathing larvae. In contrast, lamellar diffusion distances increase as development proceeds in bimodal breathers. In *Anabas testudineus*, diffusion distances increased from 1–2 μm prior to air breathing to 4–8 μm after the start of air breathing (Mishra and Singh, 1979). In *Channa striatus*, diffusion distances increased similarly from 4 to 9 μm during the transition to air breathing (Prasad and Prasad, 1984). This increase in diffusion distances is probably an adaptation to prevent lamellar collapse while the gills are out of water.

Development of branchial blood vessels has been described for many species (Solewski, 1949; Markiewicz, 1960; Morgan, 1974a,b; McElman and Balon, 1979; Paine and Balon, 1984, among others). Much less is known about the functioning of the branchial circulation. Morgan (1974b) suggested that rainbow trout larvae might not be able to regulate blood flow through specific secondary lamellae. This was based on histological observations that pillar cells did not appear to be contractile. Overall flow to the gills presumably can be regulated by alterations in heart rate. Holeton (1971) and McDonald and McMahon (1977) noted that larvae responded to moderate hypoxia by increasing heart rate.

Holeton (1971) noted that ventilatory movements of newly hatched rainbow trout were weak and poorly coordinated and suggested that the respiratory pump was not fully functional at hatch. Peterson (1975), similarly, observed that opercular movements were infrequent and irregular in newly hatched Atlantic salmon. In contrast, Morgan (1974b) reported that the buccal and opercular pumps of rainbow trout were fully functional at hatch. The cause of these apparently contradictory results may lie in the oxygen concentrations at which the observations were made. McDonald and McMahon (1977) noted that the buccal and opercular movements of newly hatched Arctic char were infrequent and poorly coordinated in normoxic water, but if the larvae were placed in hypoxic water, ventilatory movements become stronger and fully coordinated. McDonald and McMahon (1977) proposed that this indicated that the larvae possessed the necessary neural and musculoskeletal mechanisms to irrigate their gills effectively but did not do so unless oxygen was limiting. In well-oxygenated water, oxygen requirements could be satisfied by cutaneous exchange at less cost. This raises the question of whether branchial gas exchange is cost-effective for very small larvae. Small larvae reside in a viscous environment (low Re). As a result, fluid accelerated by the buccal and opercular pumps would rapidly come to halt between cycles. Irrigating the gills using the buccal and opercular pumps thus can be expected to be relatively expensive and will be avoided by small larvae in favor of less expensive cutaneous exchange under normoxic conditions. A low Reynolds number environment may be the reason Atlantic salmon larvae use pectoral fin movements to assist in moving water over the gills. Pectoral fin movements are rapid ($\simeq$100 cycles min^{-1}; Peterson, 1975) and can be expected to produce a more or less constant flow, which at low Re may be more efficient than a pulsatile flow.

It has been suggested that larvae experience respiratory distress during the transition from cutaneous to branchial gas exchange. Iwai and Hughes (1977) reported that a preliminary study of the morphometrics of gill development in the black sea bream (*Acanthopagrus schlegi*) indicated a decline in the surface/volume ratio of the gills 5–8 days after hatch. They speculated that this would produce a situation unfavorable to gas exchange and might be the reason for high larval mortality during the so-called "critical period." McElman and Balon (1979) suggested that walleye larvae similarly may experience respiratory distress leading to mortality during the period when blood flow through the vitelline circulation is in the process of being reduced but before the gills are fully functional. Additional support for respiratory involvement in critical-period mortality in some species is provided by observations that critical oxygen levels rise (DeSilva and Tytler, 1973; Davenport, 1983) and resistance to hypoxia decreases (DeSilva and Tytler, 1973; Spoor, 1977, 1984) midway through the larval period. These observations will be discussed later in more detail.

III. AEROBIC METABOLISM

A. Measurement Techniques

Many factors influence the metabolic rate of embryos and larvae. It is virtually impossible to control all the factors when measuring metabolic rates and, as a result, the measurement technique itself often significantly affects the rate that is recorded. Three general techniques have been used to measure metabolic rates: indirect calorimetry, bioenergetic analysis, and direct calorimetry. Each of these techniques will be examined briefly with regard to those aspects that are most pertinent to their use with fish embryos and larvae.

1. *Indirect Calorimetry*

Indirect calorimetry is by far the most widely used technique. In virtually all cases the rate of energy production is estimated by measuring the rate of oxygen consumption. It is possible theoretically to estimate energy production by measuring the rate of carbon dioxide production, but because of practical difficulties this approach is rarely used. The few instances in which CO_2 production has been measured have been in conjunction with measurements of oxygen consumption. This allows calculation of respiratory quotients and provides some insight into energy sources during development (see Devillers, 1965; Kamler, 1976; Gnaiger, 1983a).

A variety of indirect methods have been used with fish embryos and larvae. The most widely used method seems to be the manometric technique using Warburg- or Gilson-type respirometers. The method is relatively easy to master and because of the standard design of the respirometers, results of different studies can be compared directly. However, the technique has a number of disadvantages, the most serious being the fact that respirometers are agitated. This disturbs the animals and may seriously affect the rate of oxygen consumption (Gnaiger, 1983b). It is usually assumed that shaking elevates metabolic rate (e.g., Eldridge *et al.*, 1977) but several investigators have speculated that in some species shaking may depress rather than enhance larval metabolism (Hunter and Kimbrell, 1980; Theilacker and Dorsey, 1980; Solberg and Tilseth, 1984). The impact of shaking can be reduced by only agitating flasks just prior to readings (e.g., DeSilva *et al.*, 1986), but if this is done, continuous readings of oxygen uptake cannot be made and much information is lost. Another problem with the method is the fairly long time required for equilibrium to occur, and as a result early responses may be unreliable. It is often suggested that because the system is closed, metabolic waste products accumulate and may influence oxygen uptake. Given the relative insensitivity of fish embryos and larvae to metabolic byproducts, this is probably not a major problem.

In a few instances, Cartesian divers have been used to measure oxygen consumption. The technique is very sensitive, and accuracies in the order of

$\pm 10^{-5}$ μl h^{-1} have been reported (Davenport, 1976; Gnaiger, 1983b). The problem is that divers are not readily available and require a good deal of skill to operate. As a result, their application has been limited. In addition, measurements involving active larvae are difficult because of the relatively small size of the divers.

The Winkler method has been used frequently in the past to measure oxygen utilization. It has the advantages of being inexpensive, simple, and reasonably accurate. It allows for a variety of respirometer designs and can be applied to either flow-through or closed systems. The major disadvantage of the technique is that it does not permit continuous monitoring of oxygen levels and in recent years has been largely superceeded by the use of polarographic oxygen sensors.

The general use of polarographic oxygen sensors to measure rates of oxygen consumption in the aquatic environment has been reviewed thoroughly (Hitchman, 1978; Forstner, 1983; Gnaiger, 1983b). Davenport (1976) reported that with care, accuracies in the order of ± 0.02 μl h^{-1} are possible. Polarographic sensors can be used in conjunction with virtually any type of respirometer. This flexability allows respirometers to be designed to suit the particular circumstances. For example, Dabrowski (1986) recently designed a circular respirometer to measure active metabolism in larvae. A banded paper drum moving around the respirometer induces activity, which is monitored visually. Polarographic electrodes allow oxygen uptake to be monitored continuously and correlated with activity. Both closed and flow-through systems have been used to measure oxygen uptake by fish embryos and larvae. As Forstner (1983) points out, each has advantages. Closed systems are easier to construct and accuracies are usually greater because only oxygen concentration has to be monitored (it is assumed that time is recorded in all cases). In flow-through tests both oxygen concentrations and flow must be monitored and accuracy may suffer. On the other hand, flow-through systems provide a more constant environment and eliminate the possibility of metabolic waste products influencing metabolic rate. In addition, long-term trends in oxygen uptake can be investigated. Gnaiger (1983b) points out that the choice of respirometer design requires a compromise between factors such as sensitivity, time resolution, a more or a less disturbed environment, and ease of construction. Forstner (1983) described an intermittent-flow system suitable for fish eggs or larvae that combines attributes of both flow-through and closed systems.

Plastics are frequently used in the construction of respirometers. Researchers should be aware that plastics vary considerably in their permeability to gases, and some, such as silastic [$P = 60 \times 10^{-9}$ cc cm (s cm^2 cm $Hg)^{-1}$] and acrylic [$P = 27 \times 10^{-9}$ cc cm (s cm^2 cm $Hg)^{-1}$], are relatively permeable to oxygen (see the Cole-Parmer 1987–1988 catologue for a listing of N_2, O_2, and CO_2 permeability coefficients for the more common plastics). As a result, significant amounts of oxygen may diffuse into or out of the system if the walls of the respirometer are thin. This is particularly a problem with plastic tubing because of the high surface-to-volume ratio. Plastics can be a problem even when chamber walls are relatively thick, since they have a tendency to act as "sponges" and thus dampen response times. Eriksen and Feldmeth (1967) indicated that the permeability of plastics could be reduced by coating them with water-soluble silicone (e.g., Siliclad, Clay-Adams Co.). The efficacy of this procedure has not been documented, but even if it does not significantly reduce gas permeability, it may still be worthwhile because of its effectiveness in inhibiting bubble formation. When all is said, probably the best procedure is to follow Forstner's (1983) recommendation and use glass or stainless steel instead of plastic wherever possible.

Control of flow is often overlooked when respirometers are designed. As we have seen, low flows reduce oxygen uptake in eggs by effectively increasing the thickness of the boundary layer (Winnicki, 1968; Wickett, 1975; DiMichele and Powers, 1984a,b). In addition, adequate flow is required to ensure rapid mixing between the organisms and the oxygen sensor to avoid the problem of excessive lag. Flow also influences oxygen uptake by larvae, but the situation is more complex than is the case for eggs. Low flows can reduce uptake by effectively increasing boundary layer thickness, while high flows may elevate rates of oxygen consumption by increasing activity. In flow-through respirometers, flows can be regulated by simply adjusting the head of the water entering the system. In closed systems, some type of pump is required. The choice of a suitable pump is not a trivial task. Centrifugal pumps often experience cavitation, which can lead to bubble formation. Reciprocating or peristaltic pumps cause pressure changes in the system that can influence the operation of the oxygen sensor. Low-speed gear pumps avoid these problems and would appear to be one of the most effective ways of regulating flows in closed-system respirometers (Rombough, 1987).

Dalla Via (1983) addresses the problem of bacterial contamination of respirometers as a source of error when measuring low rates of oxygen uptake such as those typical of fish eggs. He points out that the use of antibiotics is not particularly effective because of their potential toxicity to the test organism and the fact that many take an appreciable length of time to reduce bacterial populations. In some instances if bacteria are eliminated, fungal growth merely increases. Dalla Via (1983) suggests that perhaps the most effective way to deal with the problem is to run blank controls and subtract bacterial uptake from that of the test animal. This procedure compensates for microbes growing on the respirometer and in the water, but it does not take into account epibiotic contamination of the test organisms. Eggs in particular can harbor large numbers of bacteria (Yoshimizu *et al.*, 1980). Giorgi and Congleton (1984) tried to account for oxygen consumed by microbes growing on lingcod eggs by excising the capsule and measuring its rate of oxygen uptake separately. This technique indicated that between 10 and 24% of total oxygen consumption was due to epibiotic contamination.

2. *Bioenergetic Analysis*

Rates of oxygen consumption can be estimated from energetic analysis of embryonic and larval growth on the basis of the relationship

$$A = G + R + U \tag{4}$$

where A is the amount of energy assimilated, G is the energy equivalent of tissue elaborated, R is the energy equivalent of the amount of oxygen consumed, and U is the amount of energy lost in excretory products. The rate of oxygen consumption is estimated by rearranging the equation to yield $R = A - (G + U)$. Most investigators have used the dry-weight method of Toetz (1966) and Laurence (1969). Caloric equivalents are given by Elliot and Davison (1975), Brett and Groves (1979), and Gnaiger (1983a). Balanced energy budgets are rarely used as the sole means of estimating rates of oxygen consumption. Their main use has been to check the accuracy of metabolic rates measured by other methods (Gray, 1926; Smith, 1947; Laurence, 1969, 1973, 1978; Johns and Howell, 1980; Gruber and Wieser, 1983; Houde and Schekter, 1983; Rombough, 1987). Investigators cannot be sure that measured rates of oxygen consumption are representative of those in nature. Fry (1957) showed that simply placing fish in a respirometer significantly elevated their metabolic rate for a considerable length of time. However, if energy budgets can be shown to balance when measured rates of oxygen are substituted for R in Eq. (4), the investigator can be reasonably assured that measured rates are in fact representative of those in the natural environment (or at least in the laboratory rearing system).

Direct calorimetry has been used to estimate metabolic rates of fish embryos and larvae on only a few occasions (Smith, 1947; Gnaiger, 1979; Gnaiger *et al.*, 1981). The technique measures total heat production, but this can be converted to oxygen consumption using oxycaloric equivalents and assuming aerobic metabolism. The technique has the advantage of allowing the anaerobic contribution to total metabolism to be estimated if oxygen uptake is measured simultaneously (Gnaiger, 1983a). Descriptions of modern continuous-flow microcalorimeters are provided by Gnaiger (1983a), Knudsen *et al.* (1983), and Pamatmat (1983). According to Gnaiger (1983a), the sensitivity of such calorimeters is in the order of 2 μW, equivalent to about 0.35 μl O_2 h^{-1} assuming aerobic metabolism. This would appear to be adequate for measuring heat production of individual late embryos and larvae. Simultaneous direct and indirect calorimetry holds much promise for future investigations, since measurement of oxygen uptake alone fails to reflect total metabolic activity, particularly during periods of activity or hypoxia. Both these areas are of current interest.

3. *Oxygen Uptake during Early Development*

The early literature concerning the respiratory metabolism of fish embryos and larvae has been reviewed by Needham (1931, 1942), Smith (1957, 1958),

Devillers (1965), and Blaxter (1969). More recent but less comprehensive reviews are provided by Kamler (1976) and Hempel (1979). According to Needham (1931), the measurement of respiratory rates during early development began in 1896 with the publication of a monograph by Bataillon. Since then, hundreds of studies have been conducted, yet our understanding of many of the more basic aspects of aerobic metabolism in fish embryos and larvae is still rudimentary. For instance, controversy persists as to the manner in which metabolic rates vary during development and the relationship between metabolic rate and body size. Activity is known to have a profound effect on the rate of oxygen consumption, particularly in larvae, yet only recently have there been attempts to quantify the relationship. Fish vary considerably in their spawning habitats. Temperature and dissolved oxygen are probably the two most important environmental variables, yet their effects on metabolism during early life have been documented in only a few species. A host of other factors, such as carbon dioxide concentration, pH, salinity, light level, group size, and food intake, may also significantly affect metabolic rates, but our knowledge in these areas is extremely limited. The aim of this section of the review is to summarize what is known about the effects of the more important biotic and abiotic factors on the rate of aerobic metabolism of fish embryos and larvae.

Before beginning this examination, it would be useful to make a few general comments. Literature reports of oxygen uptake during early life are most often expressed on an individual basis. This is perfectly adequate for some purposes, such as illustrating general trends or calculating the total amount of oxygen consumed during specific phases of development. It is not particularly appropriate, though, if the aim is to compare species, different studies with the same species, or even different treatments within the same study, because of the overwhelming effect body mass has on oxygen consumption. The effect of body size can be largely eliminated by expressing oxygen uptake on a mass-specific basis ($\dot{V}O_2/M$; also termed metabolic intensity). Unfortunately, relatively few investigators have done so. Even in those cases in which metabolic rate is purported to be expressed on a mass-specific basis, care must be taken in interpreting results. Many investigators have used total mass (i.e., mass of the embryo plus yolk or in some cases mass of the embryo, yolk, egg capsule, and perivitelline fluid) instead of just the mass of formed tissue to calculate metabolic intensity. This results in considerable underestimation of metabolic intensities early in development when the ratio of nonrespiring yolk to metabolically active tissue is large. Care also should be taken to distinguish between values calculated on the basis of wet and dry masses. Another problem in interpreting the literature is uncertainty as to the level of metabolism measured. Rates of oxygen uptake can vary 3- to 10-fold during early life, depending on the level of activity. Activity levels are rarely given in the literature.

B. Biotic Factors

1. Stage of Development

a. Metabolic Rate. The most important factor affecting metabolic rate is the mass of actively metabolizing tissue, which under given conditions is a direct reflection of the stage of development. Rates of oxygen consumption ($\dot{V}O_2$) shortly after fertilization are low, in the range of 3.7 ng h^{-1} (*Gadus morhua*; Davenport *et al.*, 1979) to 70 ng h^{-1} (*S. gairdneri*; Czihak *et al.*, 1979), depending on species and water temperature. These values increase on average about 20-fold as a result of tissue growth during the period from fertilization to hatch, although the literature suggests a great deal of variation among species. Kaushik *et al.* (1982) indicated that there was only about a 3-fold increase in $\dot{V}O_2$ during embryonic development of carp (*Cyprinus carpio*), while Lukina (1973) reported a 50-fold increase for chum salmon (*O. keta*). Relative increases in $\dot{V}O_2$ tend to be greater for larger eggs and at higher temperatures. Dabrowski *et al.* (1984) indicated that oxygen consumption at hatch can be predicted if initial egg mass is known using the equation

$$\dot{V}O_2 = 0.1334M^{0.6634}T^{0.7143} \quad (5)$$

in which $\dot{V}O_2$ is μl O_2 h^{-1}, M is mg dry mass egg^{-1}, and T is °C. This equation was based on data for 11 species. It is open to question whether this relationship can be applied generally, but it is interesting that $\dot{V}O_2$ at hatch was not directly proportional to egg mass (i.e., metabolic mass exponent, $b < 1.0$). Dabrowski *et al.* (1984) attributed this to a higher proportion of metabolically inactive yolk in larger eggs.

The rate of oxygen consumption continues to increase after hatch—for fed larvae in a roughly exponential fashion (Laurence, 1973; Kamler, 1976). There are few reports of relative increases during the larval period, but they would appear to be of roughly the same order of magnitude as during the embryonic period. DeSilva and Tytler (1973) indicated that the rate of oxygen uptake of plaice and herring at metamorphosis was about 12.5 and 35 times, respectively, the rate of hatch. In many studies the larvae are not fed. If this is the case, larval $\dot{V}O_2$ typically increases about 3-fold, although again there is considerable species variation, before endogenous food supplies become limiting and metabolic rates decline. For unfed larvae, a parabolic model rather than an exponential model usually provides the best fit (up to maximum $\dot{V}O_2$) for data relating $\dot{V}O_2$ and developmental period (Hayes *et al.*, 1951; Rombough, 1987).

The general trend of increasing metabolic rates is well established, but some controversy remains as to whether there are significant deviations from this trend associated with specific developmental events. The events that have been considered most likely to sharply alter the rate at which oxygen is taken up are fertilization, gastrulation, the formation of the embryonic circulation, hatching, and the transition from cutaneous to branchial gas exchange. Early investigators suggested that there was a sharp increase in $\dot{V}O_2$ associated with fertilization (Boyd, 1928; Needham, 1931). More recently studies indicate that this is not the case and that $\dot{V}O_2$ actually increases rather gradually after fertilization (Nakano, 1953; Hishida and Nakano, 1954; Kamler, 1972, 1976; Czihak *et al.*, 1979). Trifonova (1937) reported that there were large fluctuations in $\dot{V}O_2$ during gastrulation and neurulation in the perch (*Perea fluviatus*). The reliability of these observations was later disputed (Hayes *et al.*, 1951; Winnicki, 1968). However, several more recent studies have also shown fluctuations in $\dot{V}O_2$ during this period, although not of the magnitude reported by Trifonova (Stelzer *et al.*, 1971; Kamler, 1972, 1976; Hamor and Garside, 1976). These studies have also tended to support early observations (Hyman, 1921; Amberson and Armstrong, 1933) that $\dot{V}O_2$ increases sharply as soon as the embryonic circulation is established. Many studies have indicated a leveling off or even a decrease in $\dot{V}O_2$ shortly before hatch (Amberson and Armstrong, 1933; Hayes *et al.*, 1951; Hamdorf, 1961; Stelzer *et al.*, 1971; Braum, 1973; DiMichele and Powers, 1984a,b). This is usually attributed to the capsule limiting the rate at which oxygen can be delivered to the embryo. Conversely, sharp increases in metabolic rate immediately after hatch (Eldridge *et al.*, 1977; Davenport and Lønning, 1980; Cetta and Capuzzo, 1982; Davenport, 1983; Solberg and Tilseth, 1984) have been attributed to the fact that the capsule no longer restricts oxygen supply or activity. The magnitude of the increases in $\dot{V}O_2$ associated with removal of the capsule is highly variable. Eldridge *et al.* (1977) reported a roughly 10-fold increase in Pacific herring (*Clupea harengus pallasi*). Davenport and Lønning (1980) observed only about a 40–60% increase in cod. They suggested that part of the reason for the relatively small increase was that the egg capsule of cod is comparatively thin ($\simeq 7\ \mu m$) and thus is less of a barrier to diffusion than the capsules of other species. The capsule normally does not appear to restrict oxygen consumption to all in some species. Robertson's (1974) data indicate little change in $\dot{V}O_2$ of southern pigfish (*Congiopodus leucopaecilus*) on hatch. Kamler (1976) similarly failed to observe an appreciable change in $\dot{V}O_2$ on hatch in carp (*Cyprinus carpio*). Kaushik *et al.* (1982) and Dabrowski *et al.* (1984) actually reported a small decline in $\dot{V}O_2$ immediately after hatch in carp and whitefish (*Coregonus lavaretus*), respectively. During larval development, the transition from cutaneous to branchial gas exchange has been linked to both an increase (Lukina, 1973) and a decrease (Kamler, 1972, 1976) in $\dot{V}O_2$.

b. Metabolic Intensity. Attempts to link changes in $\dot{V}O_2$ to particular developmental events have, on the whole, proved inconclusive because of

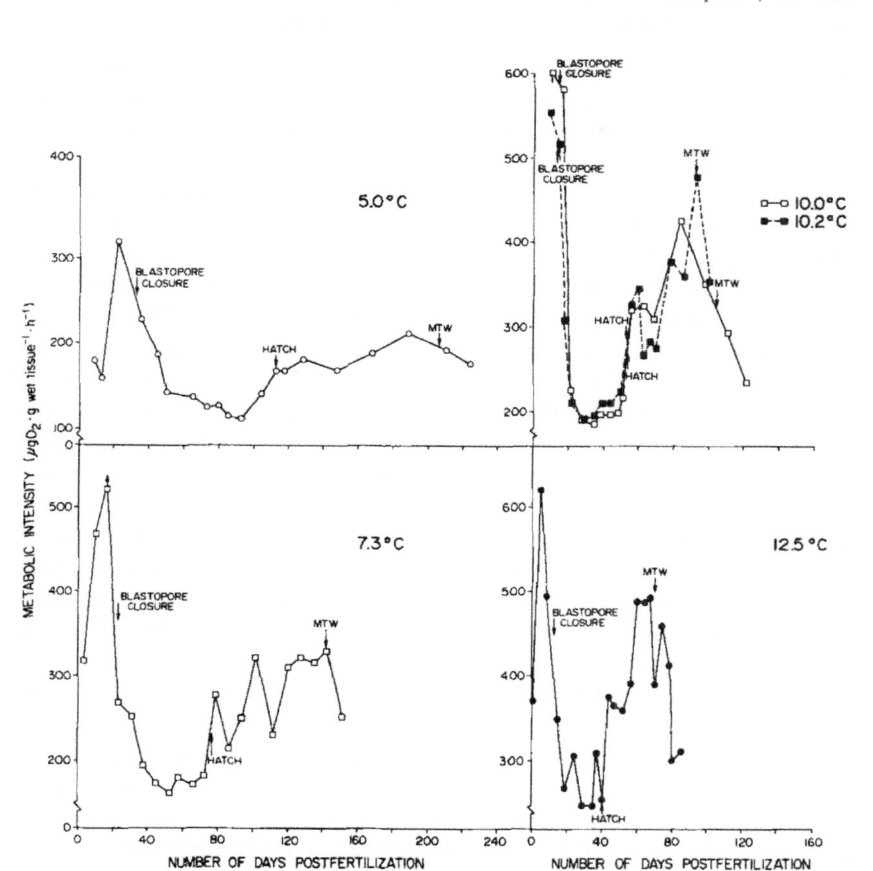

Fig. 8. Metabolic intensities of chinook (*Oncorhynchus tshawytscha*) embryos and larvae incubated at constant temperatures.

the overwhelming effect tissue mass has on oxygen consumption. As pointed out previously, changing tissue mass can be largely compensated for by expressing oxygen consumption on a mass-specific basis. The few studies in which this has been done indicate a more or less consistent pattern during embryonic development. It appears that values increase following fertilization to reach a maxima sometime prior to blastopore closure (Trifonova, 1937; Smith, 1957, 1958) (see Fig. 8). Trifonova (1937) indicated that there were two peaks during early development: one during cleavage and one during epiboly. Most other researchers have reported only a single peak during epiboly, but this may be simply a reflection of insufficient measurements during early development. Metabolic intensity declines following blastopore closure, to reach a minimum about midway through organogenesis (Hayes *et al.*, 1951; Smith, 1957, 1958; Alderdice *et al.*, 1958; Gruber and Wieser, 1983).

Thereafter there is a gradual increase to hatch. It is interesting that metabolic intensity continues to increase near hatch while $\dot{V}O_2$values are either stable or declining. This would suggest that oxygen supplies are maintained to existing tissues and that embryos compensate for the fact that the capsule limits oxygen uptake by reducing growth. Smith (1957, 1958) demonstrated that in both brown trout and Atlantic salmon metabolic intensity was directly correlated with specific growth rate.

There is typically a sharp increase in metabolic intensity on hatching. As was the case for $\dot{V}O_2$, this increase can be attributed to removal of the capsule and perivitelline fluid as barriers to diffusion and to enhanced activity (Holliday *et al.*, 1964; Davenport and Lønning, 1980; Gruber and Wieser, 1983; Davenport, 1983). Patterns after hatch are variable. Wieser and Forstner (1986) reported that $\dot{V}O_2/M$ (routine) began to decline exponentially almost immediately in three species of cyprinids (*Rutilis rutilis, Scardinius erythrophthalmus, Leuciscus cephalus*). DeSilva *et al.* (1986) noted a similar decline in the nile tilapia (*Oreochromis niloticus*). A more common pattern is for specific oxygen consumption to continue to increase until about midway through the period of endogenous feeding. This pattern has been reported for a number of salmonids (Smith, 1957, 1958; Gruber and Wieser, 1983; Wieser and Forstner, 1986; Rombough, 1987), sardine (Lasker and Theilacker, 1962), herring (Holliday *et al.*, 1964), cod (Davenport and Lønning, 1980), and lumpfish (Davenport, 1983). In most of these studies the larvae were not fed, so that $\dot{V}O_2/M$ declined rapidly once endogenous food supplies became limiting. For fed larvae, specific oxygen consumption may continue to rise until metamorphosis (Forstner *et al.*, 1983), remain relatively stable (Holeton, 1973), or decline gradually (DeSilva and Tytler, 1973), depending on the species.

2. *Mass Relationships*

Some investigators, rather than reporting mass-specific rates of oxygen consumption, have chosen to express the relationship between $\dot{V}O_2$ and body mass (M) in terms of the allometric power function

$$\dot{V}O_2 = aM^b \tag{6}$$

where a and b are constants. Such allometric relationships have long interested physiologists, although the underlying biological principles remain rather obscure. Expressing oxygen consumption in this fashion has the disadvantage that potentially important fluctuations in $\dot{V}O_2$ tend to be obscured by the overall trend. Nonetheless, the relationship is widely used, particularly for larvae.

The metabolic mass exponents [b, Eq. (6)] calculated for larvae are often compared to Winberg's (1956) generalized value of $b = 0.80$. However, Winberg's value was based primarily on studies involving juvenile and adult fish, and there is no reason to assume that the same exponent applies to earlier life stages. Indeed, ontogenic variations in the value of the metabolic mass exponent are widespread throughout the animal kingdom (Zeuthen, 1950, 1970; Adolph, 1983; Wieser, 1984). In virtually all cases, values decrease as development proceeds. This generalization appears to hold for many fish species. Kamler (1976) recognized four periods during the ontogeny of carp on the basis of different metabolic mass exponents: embryos, prefeeding larvae, feeding larvae, and postlarvae. Values were not calculated for embryos and prefeeding larvae, but graphic presentation of the data indicates that the mass exponents were significantly greater than 1.0. The mass exponent of feeding larvae was approximately unity ($b = 0.97$) while that of postlarvae was typical of juvenile and adult fish ($b = 0.80$).

High mass exponents during early life appear to be widespread, although not universal, among teleosts. Table I lists 16 species for which the larval mass exponent approached or exceeded unity. In addition, mass exponents close to unity can be inferred for larvae of brown trout (Gray, 1926; Wood, 1932), Pacific sardine (*Sardinops caerule*; Lasker and Theilacker, 1962), cod (Davenport and Lønning, 1980), Pacific mackerel (*Scomber japonicus*; Hunter and Kimbrell, 1980), and largemouth bass (*Micropterus salmoides*; Laurence, 1969) from the way in which metabolic intensity varies during development or from graphs showing the relationship between $\dot{V}O_2$ and M. Many investigators, however, have reported metabolic mass exponents for larvae that are not significantly different from those expected for juvenile and adult fish (Table I). Not all such values, though, can be taken as accurately representing the relationship between $\dot{V}O_2$ and M during much of early life. In most cases no attempt was made to differentiate between data collected for prefeeding and feeding larvae. Cetta and Capuzzo (1982) point out that this practice tends to mask any stage-specific differences in the value of the mass exponent. It will probably turn out that in many cases larval mass exponents are actually considerably higher prior to exogenous feeding than literature reports would suggest. For example, careful examination of data presented by Laurence (1978) indicates that $\dot{V}O_2$ tended to increase more rapidly with tissue mass for very young cod and haddock larvae than it did for older larvae. In the one case in which calculations were made using only data for small larvae (haddock, 4°C), it was found that the mass exponent was significantly greater than unity ($b = 1.267$). In contrast, when the data for all larval stages were combined, mass exponents were found to be typical of those seen in much older fish ($b = 0.55$–0.78). It should be recognized, however, that mass exponents are not always high, even for very young larvae. For example, a careful study by DeSilva and Tytler (1973) showed the metabolic mass exponents of both herring and plaice larvae to be considerably less than unity (0.82 and 0.65, respectively). The reason why some species have such relatively low mass exponents as larvae has not been established.

Table I Metabolic Rate–Mass Relationships of Fish Larvae[a]

Species	Temp. (°C)	Stage	Size range	Allometric relationship ($\dot{V}O_2 = aM^b$)				Comments	Reference[c]
				a	b	$\dot{V}O_2$ units	*M units*		
					$b \gtrsim 1.0$				
Cyprinus carpio	20	L	< 674 mg	0.64	0.97	$\mu l\ h^{-1}$	mg wet	Slope significantly different for small and large larvae	1
		F	1.2–4.6 g	1.27	0.80				1
Cyprinus carpio	20	L	< 1 g	0.60	0.95	$\mu l\ h^{-1}$	mg wet		2
Cyprinus carpio	20	L	< 1 g	0.60	0.98	$\mu l\ h^{-1}$	mg wet		3
			> 1 g	1.27	0.80				3
Cyprinus carpio[b]	23	L	2–70 mg	1.56	–0.054	$mg\ g^{-1}\ h^{-1}$	mg wet	$\dot{V}O_2/M$ recalculated	4
Abramis brama	20	L	1.6–40 mg	0.45	0.93	$ml\ h^{-1}$	g wet		5
Anabas testudineus	28	L	5–50 mg	1.00	0.95	$mg\ h^{-1}$	g wet	Recalculated	6
Channa punctatus	28	L	1–10 mg	1.00	0.91	$mg\ h^{-1}$	g wet	Recalculated	7
Coregonus sp.[b]	10	L	6–65 mg	14.8	0.12	$\mu mol\ h^{-1}\ g^{-1}$	mg wet	$\dot{V}O_2/M$, to metamorphosis	8

Acipenser baeri	20	L	11–28 mg	2.4	1.31	mg h^{-1}	g wet	Prolarvae	9
		F	11–390 mg	0.35	0.85			Feeding larvae	
Salvelinus alpinus	2	L	50–124 mg	2.45	1.09	mg h^{-1}	g wet	Yolk-sac larvae	10
Salmo gairdneri	4	L–F	80 mg–7 g	3.1	0.96	μmol h^{-1}	g wet	r$\dot{V}O_2$	11
			80 mg–7 g	11.0	1.11			a$\dot{V}O_2$	
Salmo gairdneri	12	L–F	80 mg–7 g	6.8	0.93	μmol h^{-1}	g wet	r$\dot{V}O_2$	11
				19.9	1.14			a$\dot{V}O_2$	
Clupea harengus	8	L			1.1	μl h^{-1}	mg dry	Hatch to MLDW	12
Clupea harengus	8	L	0.1–0.7 mg dry	1.19	0.74	μl h^{-1}	mg dry		13
	13	L	0.1–0.7 mg dry	3.63	1.33				
	18	L	0.1–0.7 mg dry	3.33	0.87				
Pleuronectes platessa	8	L	0.7–1.2 mg dry	1.21	0.78	μl h^{-1}	mg dry		13
	13	L	0.07–1.2 mg dry	2.20	0.96				
	18	L	0.07–1.2 mg dry	3.29	0.74				
Pseudopleuronectes americanus	7	L	7–10 μg protein (dry)	0.006	1.03	μl h^{-1}	μg protein (dry)		14
Melanogrammus aeglefinus	4	L	70–200 mg dry	0.0053	1.27	μl h^{-1}	μg dry	Small larvae only	15
	7	L	50–1000 mg dry	0.071	0.68				
	9	L	50–1000 mg dry	0.179	0.55				

Continued

Table I Metabolic Rate–Mass Relationships of Fish Larvae—Cont'd

Species	Temp. (°C)	Stage	Size range	Allometric relationship ($\dot{V}O_2=aM^b$)				Comments	Reference
				a	b	$\dot{V}O_2$ units	*M units*		
Anchoa mitchilli	26	L	8.9–424 μg dry	0.0077	0.98	μl h^{-1}	μg dry	Feeding larvae	16
Achirus lineatus	26	L	14.3–248 μg dry	0.014	0.94	μl h^{-1}	μg dry	Feeding larvae	16
Hypophthalmichthys molitrix	20	L	1–50 mg	0.33	1.05	μl h^{-1}	mg wet	Recalculated	17
Sparus aurata	19	L	28–1000 μg dry	7.09	1.00	μg h^{-1}	mg dry		18
	24	L	28–1000 μg dry	6.47	1.03				
$b \lesssim 0.8$									
Stizostedion lucioperca	20	L	0.8–79 mg	0.31	0.82	ml h^{-1}	g wet		5
Perca fluviatilis	20	L	1.5–32 mg	0.29	0.78	ml h^{-1}	g wet		5
Rutilis rutilis	20	L	1.6–62 mg	0.29	0.82	ml h^{-1}	g wet		5
Heteropneustes fossilis	28	L	2-20 mg	1.00	0.88	mg h^{-1}	g wet		19
Morone saxatalis	18	E–L	50–1500 μg	0.028	0.72	μl h^{-1}	μg dry		20
Lebeo calbasu	28	L	100–300 mg	0.50	0.84	mg h^{-1}	g wet	Recalculated	21
Oreochromis niloticus	30	F	1–10 mg?	9.68	0.42	μl h^{-1}	mg dry		22
Clupea harengus	10	L	0.1–10 mg	1.88	0.82	μl h^{-1}	mg dry	Recalculated	23

Pleuronectes platessa	10	L	0.5–10 mg	1.67	0.65	μl h^{-1}	mg dry		23
Pseudopleuronectes americanus	7	F	8–150 μg protein	0.016	0.78	μl h^{-1}	μg protein dry		14
Gadus morhua	4	L	50–1000 μg	0.018	0.71	μl h^{-1}	μg dry		15
	7	L	50–1000 μg	0.017	0.78				15
	10	L	50–1000 μg	0.054	0.69				15
Archosargus rhomboidalis	26	L	18–66 μg	0.018	0.84	μl h^{-1}	μg dry		16

[a]Abbreviations: L, larvae; F, fry; MLDW, maximum larval dry weight.
[b]Regression describes relationship between metabolic intensity (VO_2/M) and tissue mass.
[c]References: (1) Kamler, 1976; (2) Winberg and Khartova, 1953 (cited in Kamler, 1976) (3) Kamler, 1972; (4) Kaushik and Dabrowski, 1983; (5) Kudrinskaya, 1969; (6) Mishra and Singh, 1979; (7) Singh *et al.*, 1982; (8) Forstner *et al.*, 1983; (9) Khakimullin, 1985; (10) Holeton, 1973; (11) Wieser, 1985; (12) Blaxter and Hempel, 1966; (13) Almatar, 1984; (14) Cetta and Capuzzo, 1982; (15) Laurence, 1978; (16) Houde and Schekter, 1983; (17) Mukhamedova, 1977; (18) Quantz and Tandler, 1982; (19) Sheel and Singh, 1981; (20) Eldridge *et al.*, 1982; (21) Durve and Sharma, 1977; (22) DeSilva *et al.*, 1986; (23) De Silva and Tytler, 1973.

Several reasons have been advanced to explain high metabolic mass exponents during early development. Kamler (1976) suggested that the transition from the high larval value to the lower juvenile value in carp was the result of morphological changes affecting gas exchange. This idea was expanded upon by Pauly (1981). Pauly (1981) proposed that the underlying factor limiting metabolic rate and hence growth in fish was the availability of oxygen, rather than food or some intrinsic control mechanism. He reasoned that if oxygen was not limiting, metabolic rate should be directly proportional to body mass. To support his contention that oxygen is not limiting during early life, he noted that gill surface area of larval herring and plaice increased at a much faster rate than body mass (gill area mass exponents of 3.36 and 1.59, respectively; De Silva, 1974). This was rather an unfortunate choice of species, since in both herring and plaice the larval metabolic mass exponent is significantly less than unity (0.82 and 0.65, respectively; De Silva and Tytler, 1973). This does not necessarily disprove Pauly's basic hypothesis (i.e., that the ability to supply oxygen to tissues limits metabolic rate) but indicates that more sophisticated analysis of total exchange capacity, perhaps similar to that done by Ultsch (1973) for lungless salamanders, is required to adequately test the hypothesis. As discussed earlier, cutaneous gas exchange is extremely important in larvae and must be included in any analysis of diffusing capacity. In both herring and plaice cutaneous surface area expands at a slower rate than body mass (mass exponents of 0.58 and 0.50, respectively; De Silva, 1974). Thus, the total diffusing capacity of these larvae, in fact, may be expanding at about the same rate as metabolism. It must be pointed out that not all investigators agree with this hypothesis. In particular, Oikawa and Itazawa (1985) recently presented data for carp that they contend show no direct relationship between respiratory surface area and resting metabolism.

Forstner *et al.* (1983) attributed the high mass exponent of coregonid larvae to a preponderence of red muscle and concommitant high activity of oxidative enzymes. The decline in the value of the mass exponent following metamorphosis was attributed to a major reorganization of metabolism in which glycolytic activity becomes progressively more important (Forstner *et al.*, 1983; Hinterleitner *et al.*, 1987). There are, however, major differences among species in the developmental trajectories of oxidative and glycolytic enzymes (Hinterleitner *et al.*, 1987), and it is difficult to believe that high oxidative enzyme activity per se is sufficient to explain high mass exponents early in life. For example, salmonids generally display high mass exponents up to at least the end of the yolksac stage (Table I, Wieser and Forstner, 1986; Rombough, 1987), yet activity is borne almost entirely by white muscle (Forstner *et al.*, 1983), and levels of oxidative enzymes are very low (Hinterleitner *et al.*, 1987).

Quantz and Tandler (1982) attributed the high metabolic mass exponent of larval gilthead seabream (*Sparus aurata*) to their high feeding rate. They speculated that since the larvae were feeding more or less continuously, they should have a high specific dynamic action (heat increment). Noting that juvenile and adult fish had a mass exponent near unity for active metabolism (Brett and Groves, 1979), they suggested that it was not surprising that mass exponents for larvae were of a similar magnitude.

The allometric equations relating $\dot{V}O_2$ and M frequently have been used to compare metabolic levels of different species or ecological groupings. Konstantinov (1980) points out that such comparisons are only valid if weight exponents are equal. This severely restricts the use of such equations. Metabolic weight exponents are seldom the same for different species. Even in the same species, values for b vary during development and are influenced by factors such as temperature (Laurence, 1978; Konstantinov, 1980; Almatar, 1984) and activity (De Silva and Tytler, 1973; Wieser, 1985).

3. *Activity*

As for juveniles and adults (Brett, 1970), activity is the single most important factor influencing the metabolic intensity of larvae (Blaxter, 1969). The relationship between activity and oxygen consumption has been investigated in some detail for juvenile and adult fish (Beamish, 1964, 1978; Brett, 1964, 1970, 1972; Brett and Groves, 1979; Fry, 1971). Yet only recently has there been much interest in the earlier life stages, although there are compelling reasons to suspect that energetic relationships may be significantly different from those of older fish. The body musculature and supporting metabolic machinery develop gradually during the embryonic and larval periods and in many species do not assume typical juvenile patterns until after metamorphosis (Johnston, 1982; Forstner *et al.*, 1983; Batty, 1984; Wieser *et al.*, 1985). In addition, juveniles and larvae are exposed to different hydrodynamic regimes at routine swimming speeds. In larvae, routine activities occur in an intermediate hydrodynamic environment where both resistive and inertial forces are important (Webb and Weihs, 1986). As larvae grow or during burst swimming they move into the adult hydrodynamic regime, where inertial forces dominate. In many species this transition is accompanied by a change in swimming style from continuous movement of body and tail to a beat and glide pattern (Hunter, 1972, 1981; Weihs, 1980; Batty, 1984). To further confound matters, there are species-specific differences in development of muscle types and biochemical pathways (Wieser *et al.*, 1985; Hinterleitner *et al.*, 1987).

The study of the physiological energetics of the early life stages has been hampered by the lack of clear definitions for the various levels of metabolism associated with activity. In juvenile and adult fish, energy expenditure is normally described in terms of standard $(s\dot{V}O_2)$, routine $(r\dot{V}O_2)$ and active $(a\dot{V}O_2)$ metabolism (see Brett and Groves, 1979, for definitions). These terms have been applied to the early stages, but it should be noted that conditions are somewhat different during early life. In older fish standard metabolism refers to the postabsorptive state, a condition that is obviously not met during endogenous feeding. In addition, the standard metabolic rate of embryos and larvae includes a sizeable growth component so that, unlike in older fish, metabolic rates can be depressed considerably below the so-called standard level without affecting survival. What most investigators consider standard metabolism during early life is actually simply the metabolic rate under conditions of minimal neuromuscular activity. The vast majority of investigators have attempted to measure the embryonic and larval equivalent of routine metabolism. This is a rather nebulous term, since, as Wieser (1985) points out, it can vary at least twofold in response to a variety of intrinsic and extrinsic factors. For embryos and larvae, routine metabolism probably can be defined best as the average rate of aerobic metabolism under normal rearing conditions. Active metabolism in older fish usually refers to sustained activity. The young of many species, however, are not capable of sustained activity. As a result, most estimates of active metabolism during early life are based on burst activity. It is a moot point whether such results are comparable with values for juveniles and adults.

Absolute aerobic scope, defined as the difference between $a\dot{V}O_2$ and $s\dot{V}O_2$, is of particular interest to physiologists because it represents the amount of energy available to a fish to cover the cost of various biological activities. Unfortunately, it has proved difficult to estimate absolute scope during early life, although a new type of metabolic chamber recently described by Dabrowski (1986) may make such determinations easier in the future. It has proved somewhat easier to estimate selected portions of absolute scope, in particular the difference between $a\dot{V}O_2$ and $r\dot{V}O_2$ (termed the relative scope; Wieser, 1985) and the difference between $r\dot{V}O_2$ and $s\dot{V}O_2$ (termed the routine scope; Beamish, 1964). It thus is possible to obtain rough estimates of absolute scope indirectly by adding values reported for routine and relative scopes.

Some investigators have used the ratios of metabolic rates at the various activity levels, instead of their differences, as a way of expressing scope. The ratios $a\dot{V}O_2/s\dot{V}O_2$, $a\dot{V}O_2/r\dot{V}O_2$, and $r\dot{V}O_2/s\dot{V}O_2$ are referred to, respectively, as absolute factorial scope, relative factorial scope, and routine factorial scope. This is convenient for comparative purposes but does not tell the whole tale, since it gives no indication of absolute costs, which are of profound ecological importance. As will be discussed later, scopes and factorial scopes can change in response to various intrinsic and extrinsic factors. It should be recognized, however, that because of the manner in which they are calculated, they may not always change in the same direction. For instance, if $a\dot{V}O_2$ and $s\dot{V}O_2$ decline in parallel, absolute scope will remain constant but absolute factorial scope will increase.

Absolute aerobic scope $(a\dot{V}O_2 - s\dot{V}O_2)$ is routinely estimated for juvenile and adult fish by forcing them to swim against a current at progressively faster speeds. The oxygen uptake at maximum sustained swimming speed is taken to represent a$\dot{V}O_2$, while extrapolation of the power–performance curve back to zero swimming speed yields s$\dot{V}O_2$. Ivlev (1960a) appears to have been the only investigator to have applied this technique to very young fish. He reported that young Atlantic salmon (400 mg wet mass) were capable of sustaining a rate of oxygen uptake 22.5 times their standard rate. This value is sometimes cited as the degree of metabolic expansibility that can be expected for fish larvae (e.g., Blaxter, 1969). However, more recent evidence suggests that this is a gross overestimation and that absolute factorial scopes actually range from about 2.5 to 10, depending on species, fish size, and temperature (Table II).

Dabrowski (1986) recently used a modification of the standard power-performance procedure to estimate aerobic scope for salmon alevins and coregonid fry. The fish were induced to swim at progressively faster speeds by making use of their optomotor reaction to a moving background. Swimming speeds were monitored and correlated with rates of oxygen uptake. For young Atlantic salmon, the maximum $\dot{V}O_2$ recorded was about 3.3 times the value estimated for zero activity. The absolute factorial scope of *Coregonus schinizi* fry was somewhat lower, about 2.5.

A less common way of estimating aerobic scope in juvenile and adult fish is to take the difference between the maximum and minimum $\dot{V}O_2$ recorded while fish are held in a respirometer for an extended period. The value obtained is correctly termed the aerobic scope for spontaneous activity but, at least in some species, it appears to approximate absolute scope (Ultsch *et al.,* 1980). Several investigators have applied this technique to fish larvae. Holliday *et al.* (1964) indicated that there was approximately a 10-fold difference between the minimum and maximum $\dot{V}O_2$ of herring larvae. For sardine larvae the maximum difference in $\dot{V}O_2$ recorded was about 3.5-fold (Lasker and Theilacker, 1962), while in winter founder (*Pseudopleuronecter americanus*) the maximum difference was only about 3.0-fold (Cetta and Capuzzo, 1982). It is not clear why the apparent scope of herring should be so much greater than that of the other species, but it is interesting that larvae of Pacific herring (*Clupea pallasi*) also appear capable of increasing their metabolic rate about 10-fold (Eldridge *et al.,* 1977). In the case of the Pacific herring, it was stress caused by the shaking of the respirometer, rather than spontaneous activity, that led to elevated metabolic rates.

As mentioned previously, it has proved difficult to induce larvae to swim in a respirometer and, as a consequence, other techniques have had to be developed to estimate s$\dot{V}O_2$ and a$\dot{V}O_2$. The most common method for estimating s$\dot{V}O_2$ has been to anesthetize the larvae (Holliday *et al.,* 1964; DeSilva and Tytler, 1973; Davenport and Lønning, 1980; DeSilva *et al.,* 1986). This technique indicates routine factorial scopes $\left(r\dot{V}O_2/s\dot{V}O_2\right)$ in the range of 1.4–3.3, depending on species and stage of development (Table II). Another method has been to assume that the capsule severely restricts activity

Table II Scope for Activity

Species	Stage	Temp. (°C)	Scope	Technique	Reference
Routine factorial scope ($r\dot{V}O_2/s\dot{V}O_2$)					
Clupea harengus	E	8	2.1	$r\dot{V}O_2$/anesthetized $\dot{V}O_2$	Holliday *et al.* (1964)
	L	8	1.4	$r\dot{V}O_2$/anesthetized $\dot{V}O_2$	Holliday *et al.* (1964)
	L	8	2.5	Posthatch $\dot{V}O_2$/prehatch $\dot{V}O_2$	Holliday *et al.* (1964)
	L	10	1.9	$r\dot{V}O_2$/anesthetized $\dot{V}O_2$	De Silva and Tytler (1973)
Pleuronectes plattesa	L	10	1.6	$r\dot{V}O_2$/anesthetized $\dot{V}O_2$	De Silva and Tytler (1973)
Gadus morhua	L	5	2.2	$r\dot{V}O_2$/anesthetized $\dot{V}O_2$	Davenport and Lonning (1980)
	L	5	2.0	Posthatch $\dot{V}O_2$/prehatch $\dot{V}O_2$	Davenport and Lonning (1980)
Gadus morhua	L	5	1.6	$\dot{V}O_2$ in light/$\dot{V}O_2$ in dark	Solberg and Tilseth (1984)
Cyclopterus lumpus	L	5	2.0	Posthatch $\dot{V}O_2$/prehatch $\dot{V}O_2$	Davenport (1983)
Oreochromis niloticus	L	30	3.5	$r\dot{V}O_2$/anesthetized $\dot{V}O_2$	De Silva *et al.* (1986)
Salmo salar	L	20	1.5	Extrapolated from power–performance curve	Ivlev (1960a)

Salvelinus alpinus	L	4	2.6	Posthatch $\dot{V}O_2$/Prehatch $\dot{V}O_2$	Gruber and Wieser (1983)
Relative factorial scope ($a\dot{V}O_2/r\dot{V}O_2$)					
Alburnus alburnus	L	20	9.7[a]	Power–performance relationship	Ivlev (1960b)
Salmo gairdneri	L	4	2.7	Forced bursts	Wieser *et al.* (1985)
	L	12	1.9	Forced bursts	Wieser *et al.* (1985)
	L	20	~1.8	Forced bursts	Wieser *et al.* (1985)
Cyprinids (3 species)	L	12	2.4	Forced bursts	Wieser and Forstner (1986)
		20	2.1	Forced bursts	Wieser and Forstner (1986)
Spontaneous factorial scope					
Sardinops caerulea	E	14	2.8	Observed max. $\dot{V}O_2$/min. $\dot{V}O_2$	Lasker and Theilacker (1962)
	L	14	3.5	Observed max. $\dot{V}O_2$/min. $\dot{V}O_2$	Lasker and Theilacker (1962)
Pseudopleuronectes americanus	L	7	3.0	Observed max. $\dot{V}O_2$/min. $\dot{V}O_2$	Cetta and Capuzzo (1982)
Clupea harengus	L	8	10.4	Observed max. $\dot{V}O_2$/min. $\dot{V}O_2$	Holliday *et al.* (1964)
Clupea harengus pallasi	L	12.5	10.0	Posthatch max. $\dot{V}O_2$/prehatch $\dot{V}O_2$	Eldridge *et al.* (1977)

Continued

Table II Scope for Activity—Cont'd

Species	Stage	Temp. (°C)	Scope	Technique	Reference
Salmo salar	E	10.0	3.0	Observed max. $\dot{V}O_2$/min. $\dot{V}O_2$	Hayes *et al.* (1951)
Salvelinus alpinus	L	8	5.0	Posthatch max. $\dot{V}O_2$/prehatch $\dot{V}O_2$	Gruber and Wieser (1983)
Absolute factorial scope ($a\dot{V}O_2/s\dot{V}O_2$)					
Salmo salar	L	20	22.5[a]	Power–performance relationship	Ivlev (1960a).
Salmo salar	L	22	3.3	Power–performance relationship	Dabrowski (1986)
Coregonus sp.	F	14	2.5	Power–performance relationship	Dabrowski (1986)

[a]Unrealistically high, see text.

just before hatch (Davenport and Lønning, 1980; Davenport, 1983). Davenport and Lønning (1980) demonstrated that, at least in cod, the metabolic rate of embryos just before hatch was not significantly different from that of anaesthetized larvae shortly after hatch. Assuming that the metabolic rate of unanesthetized larvae shortly after hatch is representative of $r\dot{V}O_2$, this method gives routine factorial scopes for cod (Davenport and Lønning, 1980) and lumpfish (Davenport, 1983) of about 2.0.

The evidence is rather sketchy, but it appears that the difference between $r\dot{V}O_2$ and $s\dot{V}O_2$ (routine scope) tends to decrease as larvae mature (DeSilva and Tytler, 1973; Davenport and Lønning, 1980; De Silva *et al.*, 1986). For example, the routine scope of larval herring and plaice at metamorphosis, measured as the difference between unanaesthetized and anaesthetized $\dot{V}O_2$, was only about 25% of that at hatch (DeSilva and Tytler, 1973). The decrease in routine scope reflected a more rapid decline in $r\dot{V}O_2$ than in $s\dot{V}O_2$.

Wieser and his co-workers (Wieser, 1985; Wieser *et al.*, 1985; Wieser and Forstner, 1986) have attempted to obtain estimates of active metabolism by forcing larvae (using electrical or mechanical stimulation) to swim at burst speeds for short periods (30–60 s). The main driving force behind such activity was shown to be anaerobic (Wieser *et al.*, 1985), but it was felt that the maximum rate of oxygen uptake during activity or the first few minutes of recovery (the response time of the system was not fast enough to say precisely when $\dot{V}O_2$ was maximal) approached $a\dot{V}O_2$. Using the average rate of oxygen uptake prior to activity for $r\dot{V}O_2$, this technique gave estimates of 1.9–2.7 and 2.4–2.9 for the relative factorial scope $\left(a\dot{V}O_2/r\dot{V}O_2\right)$ of young rainbow trout (Wieser *et al.*, 1985) and larval cyprinids (Wieser and Forstner, 1986), respectively. Assuming that $a\dot{V}O_2$ is about twice $s\dot{V}O_2$, absolute factorial scopes would appear to range from about 4 to 6.

In juvenile and adult fish, aerobic scope tends to increase as the fish grows (Brett and Glass, 1973; Wieser, 1985). The pattern is not as obvious for younger fish. The metabolic expansibility of young rainbow trout increased with body mass, more or less as expected (Wieser, 1985; Wieser *et al.*, 1985). For example at 4°C, relative factorial scope increased from about 2.7 for yolk-sac larvae (80–120 mg) to about 5.2 for fry weighing between 3 and 10 g (Wieser *et al.*, 1985). On the other hand, in cyprinids, relative factorial scope was independent of body mass between 1 and 400 mg (Wieser and Forstner, 1986). Wieser and Forstner (1986) suggested that this may reflect the need of very small larvae to avoid the constraints small size normally places on aerobic scope if they are to escape predation.

The influence of temperature on aerobic scope appears to vary depending on the species. As was the case involving the effect of body mass, young rainbow trout follow a pattern similar to that seen in juveniles and adults (Wieser, 1985; Wieser *et al.*, 1985). Routine metabolic rate increases steadily with temperature between 4°C and 20°C (Fig. 9). Up to about 12°C, active

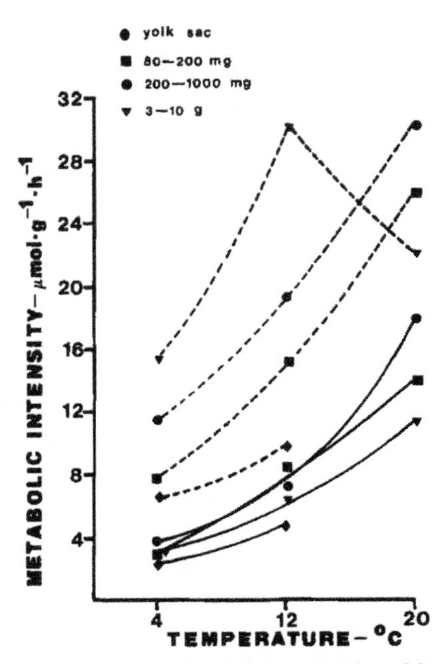

Fig. 9. Temperature dependence of rates of oxygen consumption of four size classes of rainbow trout *Salmo gairdneri*. Solid lines, routine rate; dashed lines, maximum rate during or immediately following forced burst activity. [After Wieser *et al.* (1985).]

metabolism also increases with temperature, but at a faster rate than routine metabolism, so that relative scope increases. At temperatures above 12°C, however, there is a decrease in the rate at which active metabolism expands so that relative scope $\left(a\dot{V}O_2/r\dot{V}O_2\right)$ remains constant or even declines. The reasons for this decline have not been demonstrated, but it may be that, as in adults (Fry, 1971), oxygen limits active metabolism at high ambient temperatures. Relative factorial scope in cyprinids, unlike in rainbow trout, appears to be relatively independent of temperature between 12°C and 24°C (Wieser and Forstner, 1986). This relationship, however, should probably be checked since Wieser and Forstner (1986) used data pooled from three species (*Rutilus rutilus, Leuciscus cephalis, Scardinius erythrophthalmus*), and the apparent temperature independence of aerobic scope in cyprinids may simply reflect compensating differences in temperature optima.

It is usually assumed that $a\dot{V}O_2$ is more or less fixed for a given temperature and size of fish. Thus factors that increase $r\dot{V}O_2$ would be expected to reduce relative scope. Wieser (1985) indicated that this may not be the case. He noted that, within a given size class of rainbow trout, those individuals with a high $r\dot{V}O_2$ also had a high $a\dot{V}O_2$, while individuals with a low $r\dot{V}O_2$ had a low $a\dot{V}O_2$. Wieser (1985) speculated that as yet undefined controlling factors (hormones?) may influence individual metabolism in the same way temperature does. Thus, an increase in r$\dot{V}O_2$ would result in an increase in a$\dot{V}O_2$ and a concomitant increase in relative scope. This hypothesis appears to be worth testing because if true, fish have a greater flexibility in their metabolic response to stress than previously assumed.

4. *Endogenous Rhythms*

It is recognized that diurnal variation in rates of oxygen consumption can be a major source of error in energetic analyses of early growth (Houde and Schekter, 1983). Empirical estimates of the magnitude of such variations during early life, though, are scanty. Ryzhkov (1965) reported oxygen uptake by eggs of the Sevan trout (*Salmo ischan*) peaked at about 0700 h and again at about 2100 h. Minimum and maximum values for $\dot{V}O_2$ differed 2- to 3-fold. Holliday *et al.* (1964) present data showing that the $\dot{V}O_2$ of both anesthetized and unanesthetized herring larvae was lower during the "night" than during the "day" even though all measurements were taken under identical indoor lighting conditions. This pattern was confirmed by De Silva and Tytler (1973). Oxygen uptake by both anesthetized and unanesthetized herring larvae declined at "dusk" and then gradually rose to typical daylight levels by "dawn." Again, all tests were conducted under indoor lighting. Minimun night-time values were about 50% of typical daytime values. De Silva *et al.* (1986) recently reported diurnal variation in $\dot{V}O_2$ from larvae and fry of the nile tilapia (*Oreochronis niloticus*). Unlike in herring, $\dot{V}O_2$ peaked twice in the course of a day. For newly hatched larvae, the peaks occurred just after sunset and just before dawn. The $\dot{V}O_2$ varied about 3-fold, with average nighttime rates being somewhat higher than average daytime rates. For older larvae (5 days) and fry (3 weeks), $\dot{V}O_2$ peaked at sunrise and then again at about noon. Rates varied 2- to 3-fold, with average daytime rates being somewhat higher than average nighttime rates. Geffen (1983) examined the effect of various light–dark regimes on oxygen uptake by Atlantic salmon embryos. During a 24-h period, $\dot{V}O_2$ rose to a single peak at about 1200 h when embryos were held in constant darkness. Under a 12:12 light-dark cycle, $\dot{V}O_2$ peaked toward the end of the dark period (1000 h) at about 0600 h. Under a 6:6 light–dark regime (two cycles per day), $\dot{V}O_2$ peaked in the middle of the first dark period, again at about 0600 h. There was no peak during the second dark period. Under all light regimes peaks were rather sharp, with $\dot{V}O_2$ remaining elevated for only 2–6 h. Peak $\dot{V}O_2$ values were from 2 to 4 times the average nonpeak $\dot{V}O_2$.

5. *Group Effect*

Metabolic rates of juvenile and adult fish can vary 20–50% depending on the number of individuals in the group (Brett, 1970; Itazawa *et al.*, 1978; Kanda and Itazawa, 1981). The possibility of a similar "group effect" during the early life stages has received considerable attention from Soviet researchers. Most of the earlier studies indicated that metabolic rates were significantly higher for isolated individuals than for groups (Ryzhkov, 1968; Grigor'yeva, 1967; Malyukina and Konchin, 1969; Kudrinskaya, 1969; Korwin Kossakowski *et al.,* 1981). This appeared to hold true for embryos as well as larvae and for schooling and nonschooling species. Reduced oxygen uptake was attributed to reduced motor activity, chemicals released into the water by other fish, or reduction in territoriality. However, Konchin (1971, 1981, 1982) has recently reexamined the "group effect" in several schooling and nonschooling species and concluded that earlier reports of reduced oxygen consumption by individuals in groups were simply the result of oxygen levels in respirometers dropping below the critical level and hence limiting gas exchange. He found that if fish to respirometer volume ratios were kept constant there was no significant difference in oxygen consumption of eggs and larvae of the summer bakhtah (*Salmo ischan*), roach (*Rutilus rutilus*), or pike (*Esox lucius*) as groups or as individuals. The rate of oxygen uptake of isolated individuals was not affected by visual contact with larvae of that particular species. Similarly, "conditioned" water (water in which larvae were previously held) did not affect oxygen uptake if it was aerated.

C. Abiotic Factors

1. Temperature

Next to stage of development and activity, temperature has the greatest influence on the metabolic rates of embryos and larvae. In spite of this, temperature relationships have been examined in detail in only a few species. In some ways thermal relationships during early life appear to be similar to those in later life. For example, when the logarithms of the routine metabolic rates of embryos and larvae of the various species listed in Table III are plotted against temperature, the resulting curve assumes the convex shape typical of that for juveniles and adults (Krogh's standard curve) (Fig. 10). In other ways, however, there are significant differences. In general, temperature changes have a more profound effect during early life. Literature values for the Q_{10} for metabolic rate in embryos and larvae range from about 1.5 (*Coregonus lavaretus*; Prokes, 1973) to 4.9 (*Salvelinus alpinus,* Gruber and Wieser, 1983), with an average value of about 3.0 (Table IV). In juvenile and adult fish, Q_{10} values average about 2.0 (Fry, 1971). The higher values during early life may reflect the fact that embryos and larvae tend to be more stenothermal than juveniles and adults. It is interesting that Q_{10} values for the rate of development are often similar to those for metabolism. Johns and Howell (1980) suggested that this similarity may explain why growth efficiency remains relatively constant over a relatively broad temperature range in many species.

The evidence is rather sketchy, but it appears that at least some species of fish may not be capable of thermal acclimation during early life. Clements

Table III Routine Metabolic Intensities during Early Life[a]

Species	Stage	Temp. (°C)	Metabolic intensity [μg(g wet wt.)$^{-1}$ h^{-1}]	Technique	Comments	Reference[b]
			Freshwater salmonid			
Salmo salar	L, newly hatched	4.0	164	W		1
	L, newly hatched	14	257	W		1
	E, 19–50 dpf	10.0	205	HG	Mean, $n = 11$, tissue wt. only	2
	L, 110 mg	9.0	354	W	At yolk absorption	3
Salmo gairdneri	E, latter half	10.0	442	DC	Average	4
	L, 120–200 mg	4.0	96	POS	Tissue wt. only, start free-swimming	5
	L, 120–200 mg	12.0	269	POS		5
	L, 120–200 mg	20.0	448	POS		5
	E, 16.4 mg	10.0	280	W		6
	E–L, 23.2 mg	6.0	200	POS	Mean fertilization to MLWW	7
	E–L, 27.8 mg	9.0	311	POS	Mean	7
	E–L, 24.9 mg	12.0	405	POS	Mean	7
	E–L, 21.6 mg	15.0	548	POS	Mean	7

Continued

Table III Routine Metabolic Intensities during Early Life—Cont'd

Species	Stage	Temp. (°C)	Metabolic intensity [μg(g wet wt.)$^{-1}$ h^{-1}]	Technique	Comments	Reference
Salmo trutta	L, 70 mg	10.0	571	W		8
	L, 70 mg	12.0	486	W		8
	L, 70 mg	14.0	514	W		8
	L, 70 mg	16.0	571	W		8
	L, 70 mg	18.0	786	W		8
	E–L	3.0	291	M	Mean, just before hatch to MLWW	9
	E–L	7.0	383	M		9
	E–L	12.0	836	M		9
	L, 45–70 dph	10.0	190	M		10
Salvelinus alpinus	L, newly hatched	8.0	131	POS	DW, 5.1 mg dry wt.	11
	E	4.0	64	POS	DW, tissue wt. only near hatch	12
	E	8.0	107	POS	DW, tissue wt. only, near hatch	12
Oncorhynchus gorbuscha	L, 245 mg	3.6	110	W	At emergence	13
	L, MLWW	8	929	POS		14

Oncorhynchus keta	E, 2.3–29 mg	10	231	W	Mean	15
Oncorhynchus tshawytscha	E–L, 1–500 mg wet	5	182	POS	Mean	16
	E–L, 1–500 mg	7.5	248	POS	Mean	16
	E–L, 1–500 mg	10.0	303	POS	Mean	16
	E–L, 1–500 mg	10.2	339	POS	Mean	16
	E–L, 1–500 mg	12.5	366	POS	Mean	16
Freshwater, nonsalmonid						
Cyprinus carpio	L, 100 mg	20	796	M	Estimated from allometric equation	17
	L, 100 mg	20	782	M	Estimated from allometric equation	18
	L, < 1000 mg	10	350	?	Estimated from allometric equation	19
	L, < 100 mg	20	829	?	Estimated from allometric equation	19
	L, 22–24 dph	20	639	M	Mean, stock differences	20
	L, 5.5–15.8 mg	22	727	M		21
	L, 2.1–11.1	28	1700	M		21
	L, 2.0	23	857	M		22
	L, 2–70 mg	23	1205		Mean	23

Continued

Table III Routine Metabolic Intensities during Early Life—Cont'd

Species	Stage	Temp. (°C)	Metabolic intensity [μg(g wet wt.)$^{-1}$ h^{-1}]	Technique	Comments	Reference
Cyprinids (three species combined)	L, 40–400 mg	12	208	POS	Mean	24
	L, 40–400 mg	16	384	POS	Mean	24
	L, 40–400 mg	20	509	POS	Mean	24
	L, 40–400 mg	24	733	POS	Mean	24
Rutilus rutilus	L	21	859	W	Swimbladder filled	25
	L, 2.6–3.6 mg	20	1134	?		26
Abramis bramis	L, 2.6–2.9	20	876	?		26
Stizostedion lucioperca	L, 2.3–4.6	20	1214	?		26
Perca fluviatis	L, 2.3–3.4	20	1486	?		26
	E, 5–11 dpf	22	1905	W		27
Esox lucius	L	21.5	892	W	Start exogenous feeding	25
Micropterus dolomieui	L, 1–12 dph	20	847	POS	Mean	28
Micropterus salmoides	L, 18–148 g	19	796	W	Mean	29
	L, 1.4–2.4 mg	20	1320	DOS	Mean 4–9 dpf	30

Morone saxatilis	L, 7 dpf	18	1333	M	DW, start feeding	31
Anguilla rostrata	L, 300 mg	23	737	POS	Elver	32
Hyophthalmichthys molitrix	L, 3 mg	20	706	W	Start exogenous feeding	33
Channa punctatus	L, 4.7 mg	28	1846	POS	Bimodal breather	34
Heteopneustes fossilis	L, 5.4 mg	28	1046	POS	Bimodal breather	35
Anabas testudineus	L, < 18.5 mg	28	1255	POS	Bimodal breather	36
Etroplus maculatus	L, ?	27	573	M	DW, prior to free-swimming	37
Lebeo calbasu	L, 100–260 mg	28	680	W		38
Coregonus sp.	L, 10–100 mg	10	640	POS		24
Coregonus lavaretus	L, 8 mg	5	588	W		39
	L, 8 mg	10	650	W		39
	L, 8 mg	15	875	W		39
Coregonus peled	L, 4.5 mg	5	667	W		39
	L, 4.5 mg	10	778	W		39
	L, 4.7	15	1894	W		39
Coregonus lavaretus	E, near hatch	12	331	POS	DW	40
	L, 15 dph	12.8	734	POS	DW	40
Coregonus schinzi	E, near hatch	12.5	429	POS	DW	40
Coregonus sp.	L, 15 mg	10	685	POS	At yolk absorption	41

Continued

Table III Routine Metabolic Intensities during Early Life—Cont'd

Species	Stage	Temp. (°C)	Metabolic intensity [μg(g wet wt.)$^{-1}$ h^{-1}]	Technique	Comments	Reference
Coregonus schinzi	L, ?	14.5	485	POS		42
Coregonus nasus	L, 6.8 mg	12	1057	W		43
Acipenser baeri	L, 20 mg	20	717	?		44
Oreochromis nilotis	L, feeding	30	1213	M	DW	45
			Marine			
Clupea harengus	L, yolk sac	10	499	M	DW	46
	L, first-feeding	8	282	M	DW, 2–10 days post yolk absorption	47
		13	571	M	DW	47
		18	724	M	DW	47
Clupea harengus	L, newly hatched	6	476	M	DW, tissue wt. only	48
	L, newly hatched	8	591	M	DW, tissue wt. only	48
	L, newly hatched	11	800	M	DW, tissue wt. only	48
	L, newly hatched	14	952	M	DW, tissue wt. only	48
	L, sac larvae	10	438	W	DW	49

Clupea harengus pallasi	L, yolk sac	12.5	6343	M	DW, near yolk absorption	50
Gadus morhua	L, yolk absorbed	5	286	POS	DW	51
	L, yolk absorbed	5	278	POS	DW	52
	L, 55 μg dry	5	362	M	DW, at MTW	53
Pleuronectes platessa	L, 1 dph	5	381	POS	DW	51
	L, 5 dph	10	785	M	DW	46
	L, first-feeding	8	265	M	DW	47
	L, first-feeding	13	436	M	DW	47
	L, first-feeding	18	724	M	DW	47
Sardinops caerulea	L, 70–180 hph	14	476	D	DW, estimated	54
	L	14	253	D	DW, inactive	55
Sardinops sagax	L, 1.8 dph	20	3293	POS	DW(?)	56
Cheilopogon unicolor	L, 4 dph	28	1215	POS	DW(?)	56
Sparus aurata	L, 50–150 μg dry	19	1499	POS	DW	57
	L, 50–150 μg dry	24	1670	POS	DW	57
Engrualis sp.	L	18	857	M(?)	DW	58

Continued

Table III Routine Metabolic Intensities during Early Life—Cont'd

Species	Stage	Temp. (°C)	Metabolic intensity [μg(g wet wt.)$^{-1}$ h^{-1}]	Technique	Comments	Reference
Scomber japonicus	L	18	1162	M(?)	DW	58
	L, 3–5 dph	18	1162	W	DW	59
	L, 3–5 dph	22	2171	W	DW, estimated from graph	59
Pseudopleuronectes americanus	L, 1000 μg	2	343	w	DW, estimated from graph	60
	L, 1000 μg	5	495	W	DW	60
	L, 1000 μg	8	743	W	DW	60
Parophrys retulus	E, near hatch	6	560	W	61	
Hirundichthys marginatus	L, 1.4 mg	22	1608	POS	DW(?), end endogenous feeding	56
Anchoa mitchilli	L, feeding	26	1352	POS	DW	62
Archosargus rhomboidalis	L, feeding	26	1886	POS	DW	62
Achiras lineatus	L, feeding	28	2248	POS	DW	62
Tautoga onitus	L, 10.9 mg dry	16	5371	M	DW, 5% yolk remaining	63
	L, 11 mg dry	19	4838	M	DW, 2.7% yolk remaining	63
	L, 11 mg dry	22	7810	M	DW, 10% yolk remaining	63

Lagodon rhomboides	L, 25 mg	15	500	M		64
Leistomus xanthurus	L, 42 mg	15	500	M		64
Congiopodus leucopaecilus	L, 18 dph	11.5	541	M		65
Brevoortia tyrannus	L, 47–55 mg	14	780	POS		66
	L, 47–55 mg	19	1000	POS		66
	L, 47–55 mg	24	1555	POS		66
Viviparous						
Zoarces viviparus	254 mg	5	92	POS	10–27 days prepartum	67
	180–380 mg	11	134	M		68
Clinus superciliosus	46.3 mg	16	676	M		69
Rhacochilus racca	389 mg	10	103	W		70
	3.8 g	18	222	W	At parturition	70

[a]Abbreviations: L, larva; E, embryo; dpf, days postfertilization; dph, days posthatch; hph, hours posthatch; W, Winkler; DC, direct calorimetry; POS, polarographic oxygen sensor; M, manometric; D, diver; Hg, dropping mecury electrode; DW, original data expressed on dry weight basis, converted to wet weight assuming dry weight = 13.3% wet weight; MLWW, maximum larval wet weight; MTW, maximum tissue weight.

[b]References: (1) Lindroth, 1942 (cited in Hayes *et al.*, 1951); (2) Hayes *et al.*, 1951; (3) Tamarin and Komarova, 1970 (4) Smith, 1958; (5) Wieser *et al.*, 1985; (6) Hamor, 1967 (cited in Hamor and Garside, 1975); (7) Rombough, 1987; (8) Penaz and Prokes, 1973; (9) Wood, 1932; (10) Gray, 1926; (11) Gnaiger, 1983b; (12) Gruber and Wieser, 1983; (13) Bailey *et al.*, 1980; (14) Storozhyk and Smirnov, 1982; (15) Alderdice *et al.*, 1958; (16) P. J. Rombough, unpublished data; (17) Kamler, 1976; (18) Winberg and Hartov, 1953 (cited in Kamler, 1972); (19) Kamler, 1972; (20) Jitariu *et al.*, 1971; (21) Korwin-Kossakowski *et al.*, 1981; (22) Kamler *et al.*, 1974; (23) Kaushik and Dabrowski, 1983; (24) Wieser and Forstner, 1986; (25) Konchin, 1981; (26) Kudrinskaya, 1969 (27) Trifonova 1937 (28) Spoor, 1984; (29) Laurence, 1969; (30) Spoor, 1977; (31) Eldridge *et al.*, 1982; (32) Gallagher *et al.*, 1984 (33) Mukhamedova, 1977; (34) Singh *et al.*, 1982; (35) Sheel and Singh, 1981; (36) Mishra and Singh, 1979; (37) Zoran and Ward, 1983; (38) Durve and Sharma, 1977; (39) Prokes, 1973; (40) Dabrowski *et al.*, 1984; (41) Forstner *et al.*, 1983; (42) Dabrowski and Kaushik, 1984; (43) Chernikova, 1964; (44) Khakimullin, 1985; (45) DeSilva *et al.*, 1986; (46) DeSilva and Tytler, 1973; (47) Almatar, 1984; (48) Holliday *et al.*, 1964; (49) Marshall *et al.*, 1937; (50) Eldridge *et al.*, 1977; (51) Davenport and Lonning, 1980; (52) Davenport *et al.*, 1979: (53) Solberg and Tilseth, 1984; (54) Lasker, 1962; (55) Lasker and Theilacker, 1962; (56) Klekowski *et al.*, 1980; (57) Quantz and Tandler, 1982; (58) Hunter, 1981; (59) Hunter and Kimbrell, 1980; (60) Laurence, 1975; (61) Alderdice and Forrester, 1971; (62) Houde and Schekter, 1983; (63) Laurence, 1973; (64) Kjelson and Johnson, 1976; (65) Robertson, 1974; (66) Hettler, 1976; (67) Broberg and Kristofferson, 1983; (68) Korsgaard and Andersen, 1985; (69) Veith, 1979; (70) Webb and Brett, 1972b.

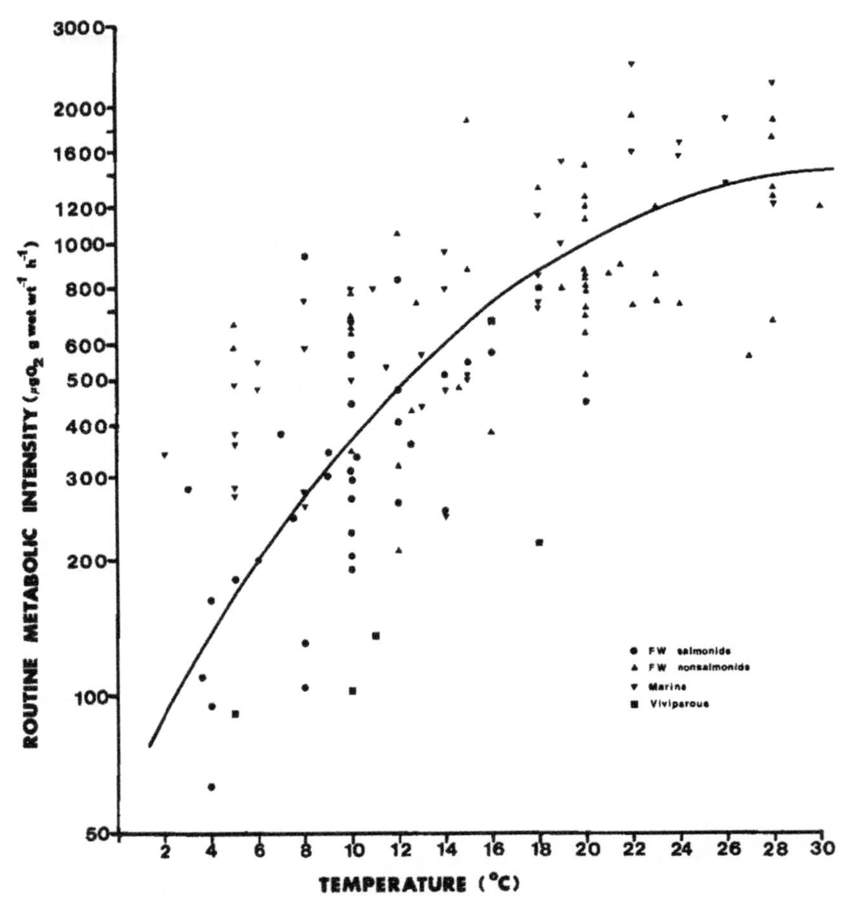

Fig. 10. Relationship between routine metabolic intensities (log scale) and temperature for fish embryos and larvae. Data were taken from Table III.

and Hoss (1977) monitored rates of oxygen consumption of larval flounder (*Paralichthys dentatus* and *P. lethostigwa*) transferred directly from environmental temperatures of 10–12°C to constant temperatures of 10 and 15°C. Rates of oxygen uptake, measured daily for 4 days, did not vary following transfer. This would suggest that thermal acclimation occurs either very rapidly (<1 day) or not at all. The latter is more likely. Hinterleitner *et al.* (1987) found no evidence of metabolic temperature compensation in larval roach or chub (*Levciscus cephalis*). Preliminary analysis of data indicates that steelhead embryos similarly do not show any signs of thermal acclimation upon reciprocal transfer between 5 and 10°C (P. J. Rombough, unpublished data).

Table IV Q_{10} Values for Embryos and Larvae

Species	Temp, range (°C)	Q_{10}	Reference
Salmo gairdneri	4–12	2.90	Wieser *et al.* (1985)
	12–20	1.80	Wieser *et al.* (1985)
Salmo gairdneri	6–9	4.37	Rombough (1987)
	9–12	2.41	Rombough (1987)
	12–15	2.74	Rombough (1987)
Salmo trutta	3–7	1.98	Wood (1932)
	7–12	4.76	Wood (1932)
Salmo trutta	10–20	1.87	Penaz and Prokes (1973)
Salmo salar	5–10	3.67	Hamor and Garside (1977)
Salvelinus alpinus	4–8	4.9	Gruber and Wieser (1983)
Onchorynchus tshaywtscha	5–7.5	3.45	P. J. Rombough (unpublished)
	7.5–10	2.23	P. J. Rombough (unpublished)
	10–12.5	2.12	P. J. Rombough (unpublished)
Cyprinus carpio	10–20	2.37	Kamler (1972)
Cyprinids	12–20	3.09	Wieser and Forstner (1986)
Coregonus lavaretus	5–15	1.49	Prokes (1973)
Coregonus peled	5–15	2.84	Prokes (1973)
"Chalcalburnus"	19–23	3.00	Karpenko and Proskurina (1970)
"Taran"	15.5–19	4.68	Karpenko and Proskurina (1970)
Oreochromis niloticus	25–30	2.42	DeSilva *et al.* (1986)
	30–35	2.09	
Clupea harengus	6–8	2.93	Holliday *et al.* (1964)
	8–11	2.75	Holliday *et al.* (1964)
	11–14	1.79	Holliday *et al.* (1964)

Continued

Table IV Q_{10} Values for Embryos and Larvae—Cont'd

Species	Temp, range (°C)	Q_{10}	Reference
Clupea harengus	8–13	1.9	Almatar (1984)
	13–18	3.0	Almatar (1984)
Pleuronectes platessa	8–13	1.8	Almatar (1984)
	13–18	6.4	Almatar (1984)
Pseudopleuronectes americanus	2–5	3.40	Laurence (1975)
	5–8	3.86	Laurence (1975)
Brevoortia tyrannus	14–19	1.64	Hettler (1976)
	19–24	2.40	Hettler (1976)
Tautoga onitus	16–22	1.87	Laurence (1975)
Scomber japonicus	18–22	4.77	Hunter and Kimbrell (1980)

The effects of temperature change on VO_2 are sometimes difficult to interpret because of changes in activity levels associated with the temperature change. For example, Gruber and Wieser (1983) attributed the high Q_{10} (4.9) calculated for Arctic char larvae held at 4 and 8°C to what they termed "warm stimulation" of activity at the higher temperature. Unfortunately, activity levels were not measured. Hettler (1976) did measure changes in activity levels of larval menhaden (*Brevoortia tyrannus*) transferred to different temperatures. He found that activity increased significantly as the temperature was raised but, interestingly, this was not reflected in a large increase in $\dot{V}O_2$ ($Q_{10} = 2.1$).

2. Dissolved Oxygen

Dissolved oxygen concentrations obviously greatly influence metabolic rate. The relationship between $\dot{V}O_2$ and oxygen concentration, however, is not simple. Most, if not all, fish can be classified as metabolic regulators on the basis of their standard $\dot{V}O_2$ as juveniles and adults (Beamish, 1964; Fry, 1971; Ultsch *et al.*, 1981). As mentioned previously, it is difficult to measure $s\dot{V}O_2$ during the early life stages, but if $r\dot{V}O_2$ is used as the basis for classification instead of $s\dot{V}O_2$, it appears that embryos and larvae, for the most part, also behave as metabolic regulators. This means that at high oxygen concentrations their metabolic rate is independent of the ambient oxygen concentration, but if oxygen levels are gradually reduced, a point is eventually reached below which metabolic rate becomes dependent on the ambient oxygen concentration. This point, termed the critical oxygen tension (P_c), defines the oxygen concentration required to maintain a particular level of metabolism. It is important to recognize that P_c is not fixed but varies in response to a variety of intrinsic and extrinsic factors. In juvenile and adult fish the two most important factors influencing P_c are activity and temperature (Beamish, 1964; Fry, 1971; Ott *et al.*, 1980). Combined high activity and temperature can result in P_c values near 100% air saturation (Brett, 1970). Activity (Broberg and Kristofferson, 1983) and temperature (Rombough 1986, 1987; Diez and Davenport, 1987) are also important factors influencing P_c during early life. In addition, the stage of development (Lindroth, 1942; Hayes *et al.*, 1951; Rombough, 1986, 1987) and the water flow (Fry, 1971) have profound effects. Temperature and activity influence P_c by altering oxygen demands. Stage of development affects both oxygen demand and supply, while water velocity primarily affects oxygen supply.

Routine P_c (the P_c associated with r$\dot{V}O_2$) is directly dependent on the stage of development and temperature. Values increase more or less steadily throughout embryonic development (Lindroth, 1942; Hayes *et al.*, 1951; Davenport, 1983; Rombough, 1987) and at any given stage of development are greater at higher temperatures (Rombough, 1987) (Fig. 11). The effect of temperature is an indirect one resulting from higher metabolic rates at higher temperatures, as indicated by the fact that when P_c is plotted against $\dot{V}O_2$ all points fall on the same line regardless of incubation temperature (Rombough, 1987) (Fig. 4). At high temperatures, routine P_c for large eggs, such as those of salmonids, may approach 100% air saturation near hatch. Such higher P_c values have led some investigators (e.g., Davenport, 1983; Gruber and Wieser, 1983) to classify teleost embryos as oxyconformers. However, this apparent conformity is simply a consequence of supply problems associated with the presence of the capsule and not intrinsic to the embryo itself. Hatching or artificial removal of the capsule results in an abrupt drop in P_c (Hayes *et al.*, 1951; Gnaiger, 1983b; Gruber and Wieser, 1983; Rombough, 1987).

It must be noted that not all studies have shown fish embryos to be metabolic regulators. Hamor and Garside (1979) noted that $\dot{V}O_2$ was lower at 30% and 50% air saturation than at 100% air saturation at *all* stages during the embryonic development of Atlantic salmon. This is difficult to explain since P_c values should have been well below 50% ASV during early development (Hayes *et al.*, 1951). It may be that metabolic response to chronic hypoxia is not the same as that to acute hypoxia. Other studies have indicated that embryos behave as neither true conformers nor true regulators (Davenport, 1983; Gruber and Wieser, 1983). Such inconclusive results may arise, at least in part, from practical difficulties associated with determining P_c (and $\dot{V}O_2$).

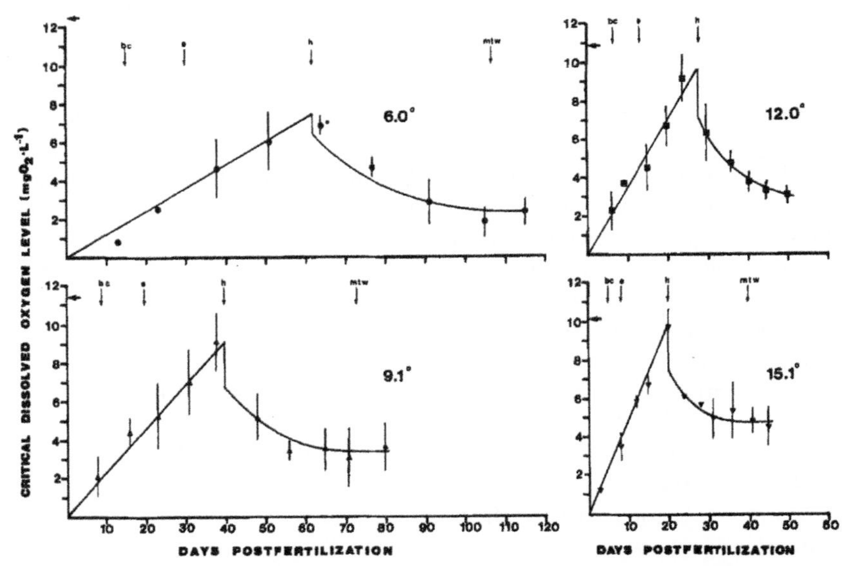

Fig. 11. Critical dissolved oxygen concentrations for steelhead embryos and larvae reared at constant temperatures of 6.0, 9.1, 12.0, and 15.1°C. The horizontal arrow indicates 100% air saturation. The point indicated by an asterisk represents the P_c for unhatched embryos at about 80% hatch; bc, blastopore closure; e, well eyed; h, hatch; mtw, maximum tissue weight. [From Rombough (1987).]

As we have seen, $\dot{V}O_2$ and hence P_c are dependent on many factors. Excitement and activity in particular are difficult to control. Simply placing fish in a respirometer can significantly elevate their metabolic rate (Fry, 1957). Typically, $\dot{V}O_2$ is high when the fish is first placed in the respirometer and declines as it adapts to the system. Associated with this gradual decline in $\dot{V}O_2$ is a gradual decline in P_c. Metabolic rates should eventually stabilize at r$\dot{V}O_2$, but if the respirometer is operated as a closed system, oxygen levels may drop below routine P_c before this can occur. If this occurs, critical levels will not be apparent and the animal may be classified as an oxyconformer. Even if metabolic rates stabilize before oxygen levels drop below routine P_c, the transition from oxygen independence to dependence may be obscured. Davenport (1983) presents data illustrating this point. Lumpfish eggs or larvae were placed in closed respirometers and $\dot{V}O_2$ was monitored until all oxygen was exhausted. Embryonic $\dot{V}O_2$ declined rapidly at first but, after a while, $\dot{V}O_2$ tended to stabilize even though ambient oxygen levels continued to decline. Oxygen uptake during this period probably approximated r$\dot{V}O_2$. Eventually oxygen levels in the respirometer dropped below routine P_c, and $\dot{V}O_2$ again declined. The problem is that the period of stable $\dot{V}O_2$ is not always easily recognized, particularly later in development when routine P_c is relatively high. Compensatory adjustments, such as changes in activity level or the efficiency of gas exchange, on the part of the embryo or larvae may also make it difficult to discern routine P_c.

Additional problems in estimating P_c using closed respirometers arise from response lags associated with such systems and instability of polarographic sensors caused by pressure changes when the system is closed. Both have the effect of appearing to increase initial $\dot{V}O_2$. Rombough (1987) was able to minimize these problems, and that of initial excitement on being placed in the respirometer, by allowing the fish sufficient time to adapt to the system before closing the respirometer and, in the case of advanced embryos, initiating tests at moderate levels of hyperoxia (110–160% air saturation). This ensured that apparent $\dot{V}O_2$ stablized well before oxygen levels in the respirometer reached routine P_c, allowing P_c to be easily estimated. The technique appears to have been effective since energy budgets calculated on the basis of $\dot{V}O_2$ at the estimated routine P_c balanced [[mean($G + R$)]A^{-1} = 102%]. Some of the problems associated with using "closed-system" respirometers to estimate P_c can be avoided by using flow-through systems, but this is very time-consuming, and in rapidly developing species routine P_c may have changed before the tests are completed.

The metabolic response of larvae to declining oxygen levels appears to be more variable than in embryos. As expected, activity is more of a problem, but even taking this into account there appear to be significant differences among species. All species can be expected to show an abrupt decrease in routine P_c at hatch. Rombough (1987) noted that the routine P_c for steelhead larvae continued to decline after hatch up to about midway through yolk absorption. The P_c then remained relatively stable until the end of yolk absorption, when experiments were terminated. Routine P_c was directly dependent on temperature. Levels during the latter half of the period of endogenous feeding ranged from 2.3 mg l^{-1} at 6°C to 4.8 mg l^{-1} at 15°C. Values were not determined for feeding larvae, but they probably continued to decline slowly since routine P_c values during endogenous feeding were somewhat higher than those reported for juvenile salmonids. De Silva and Tytler (1973) reported a different pattern for herring and plaice larvae. In both species P_c levels were lower for younger larvae than for older larvae. In fact, older larvae behaved as oxyconformers up to close to 100% air saturation. After metamorphosis, P_c levels dropped and both species again behaved as metabolic regulators. Davenport (1983) presented data indicating that P_c levels for lumpfish larvae similarly increased during the first 26 days of the larval period (at 5°C). The different patterns reported for steelhead and for herring, plaice, and lumpfish may reflect differences in the rate of transition from cutaneous to branchial gas exchange. The gills of salmonids are fairly well developed at hatch and probably assume an increasingly important role in respiratory gas exchange as development proceeds. In contrast, the gills of larval herring and plaice appear to be of little functional significance until close to metamorphosis. The skin remains the primary site of gas exchange, and because of an increasingly unfavorable surface/volume ratio the ability to extract sufficient oxygen from the environment may become limiting. The reduction in P_c after metamorphosis can be attributed to rapid elaboration of the gills and the appearance of hemoglobin in the blood.

3. Salinity

The influence of salinity on the rate of oxygen consumption of embryos and larvae has been investigated in only a few species, but results to date suggest that net ionoregulatory costs are negligible once acclimation has occurred. It is not clear whether this is because costs remain fairly constant over a broad range of salinities or because increases in costs are paralleled by proportional decreases in other metabolic processes. Evidence that net ionoregulatory costs are small comes from several sources. Lasker and Theilacker (1962) reported no significant difference in the $r\dot{V}O_2$ of sardine embryos held in half-, full-, and double-strength seawater. Holliday *et al.* (1964) similarly observed no significant difference in standard metabolic rates of anesthetized herring embryos and newly hatched larvae incubated at constant salinities between 5‰ and 50‰. DeSilva *et al.* (1986) recently reported no significant difference in either routine or standard (anesthetized) metabolic intensities of larvae of the eurhyaline species *Oreochromis niloticus* reared at salinities between 0 and 18‰.

Abrupt changes in salinity do affect rates of oxygen consumption. Holliday *et al.* (1964) reported up to a 10-fold increase in oxygen uptake by anesthetized herring embryos and larvae abruptly transferred from 35‰ to 5‰. Rates gradually declined to the pretransfer level over a period of 6–8 h. Holliday *et al.* (1964) indicated that the period of elevated metabolism coincided with the length of time required to restore osmotic imbalance brought about by the abrupt change in salinity. The pattern of oxygen uptake in unanaesthetized larvae was quite different. Transfer from 35‰ to both 15‰ and 5‰ resulted in a decrease in metabolic rate followed by a very gradual increase over the next 24 h to values typical of constant exposure to 35‰. Holliday *et al.* (1964) attributed the initial reduction in $\dot{V}O_2$ to reduced activity associated with buoyancy changes. Almatar (1984) similarly reported a reduction in oxygen uptake rates of herring and plaice yolk-sac larvae transferred from seawater (32‰) to low salinities (5 and 12.7‰). In feeding larvae, however, metabolic rates were elevated on transfer to 12.7‰ compared with control larvae (constant 32‰) and with larvae transferred to 5‰ and 40‰. Almatar (1984) attributed the apparent increase in $\dot{V}O_2$ to the fact that the larvae used in the 12.7‰ test were somewhat smaller than those used in the other tests, although why this should make a difference is not made clear.

4. Light

Within limits, light exerts a direct tonic effect on metabolic rate (MacCrimmon and Kwain, 1969; Hamor and Garside, 1975; Konchin, 1982a; Solberg and Tilseth, 1984). Konchin (1982a) indicated that the $\dot{V}O_2$ of roach (*Rutilus rutilus*) larvae was 18–25% higher in the "light" than in the dark. Solberg and Tilseth (1984) reported that the $\dot{V}O_2$ of larval cod was 25–70% higher in the "light" than in the dark, apparently because of greater activity. Light intensities were not specified in either study. Hamor and Garside (1975) found that $\dot{V}O_2$ of Atlantic salmon embryos only increased in response to increasing light intensities up to about 200 lux; at 250 lux, $\dot{V}O_2$ began to decline.

5. Other Factors

Very little is known of how potentially important factors such as pH, carbon dioxide concentration, and ration level influence aerobic metabolism during the early life stages. Oxygen uptake by steelhead embryos was not significantly affected by low pH until levels dropped below pH 4.0 (Rombough, 1987). Alderdice and Wickett (1958) reported that the metabolic rate of chum salmon embryos was independent of ambient CO_2 levels below 125 mg l^{-1} (PCO_2 not given). The relative insensitivity of embryos to pH and carbon dioxide is not surprising in light of the weak Bohr and Root effects seen in larval hemoglobins (Iuchi, 1973b). As mentioned previously, Quantz and Tandler (1982) suggested that a high specific dynamic action (SDA) may account for the high routine metabolic rate of larval gilthead seabream. Kaushik and Dabrowski (1983) attempted to measure the SDA of young carp but were not able to detect a significant postprandial increase in $\dot{V}O_2$ until the carp had attained a mass greater than 1 g. Dabrowski (1986) has since developed a more sensitive system for detecting changes in $\dot{V}O_2$ and has used it to estimate the specific dynamic action of Atlantic salmon alevins. In this species, an increase in metabolic rate equivalent to about 30% of standard metabolism and peaking 2.5–3 h postfeeding appears to be attributible to SDA. One of the reasons SDA has been difficult to detect is that it is partially compensated for by a decrease in active metabolism.

The effects of water-borne pollutants on the early stages of fish are dealt with elsewhere in this volume (von Westernhagen; Chapter 4), but it would be apropos to comment briefly on the effect of such substances on embryonic and larval $\dot{V}O_2$. A wide variety of substances, including pesticides (Kamler *et al.*, 1974; Klekowski *et al.*, 1977), metals (Storozhk and Smirnov, 1982; Akberali and Earnshaw, 1984), and hydrocarbons (Eldridge *et al.*, 1977; Davenport *et al.*, 1979; Hose and Puffer, 1984), have been tested for their effect on $\dot{V}O_2$. If a generalization is to be made, it is that results are highly variable and often difficult to interpret. A particular pollutant may have significant effects on one stage but not on another (Davenport *et al.*, 1979). A high concentration of the toxicant may have no effect, while a much lower concentration of the toxicant will produce a significant effect. For instance, oxygen uptake by embryonic grunion (*Leuresthes tenuis*) was not affected by a high body burden of benzo[a]pyrene but was significantly elevated in response to a low body burden (Hose and Puffer, 1984). Elevated $\dot{V}O_2$ at low exposure levels was attributed to hormesis (overcompensation to inhibitory challenge). Benzo[a]pyrene appears to be somewhat of an exception, and normally a relatively high concentration of a pollutant is required to produce a significant change in $\dot{V}O_2$. Davenport *et al.* (1979) estimated that because of normal variability in $\dot{V}O_2$ a change of about 40–50% r$\dot{V}O_2$ would be required to be statistically significant. For this reason, $\dot{V}O_2$ would not appear to be a particularly useful indicator of sublethal toxicity for most pollutants, especially since other more sensitive and easier to determine indicators are available (e.g., growth).

IV. EFFECT OF HYPOXIA

A. Environmental Hypoxia

Exposure to low levels of dissolved oxygen during early life can elicit a wide variety of responses, some compensatory, others clearly pathological. The particular response and its magnitude depends on species, stage of development, level of hypoxia, and duration of exposure. Other factors, such as temperature and water flow, also may be important. The most obvious place to begin an examination of the effects of environmental hypoxia is to describe the zone of tolerance—that is, to what levels oxygen can fall before mortality occurs.

1. *Lethal Levels*

Many studies have shown that fish are extremely sensitive to low levels of dissolved oxygen during early life (see reviews by Doudoroff and Shumway, 1970; European Inland Fisheries Advisory Commission, 1973; Davis, 1975; Alabaster and Lloyd, 1980; Chapman, 1986). Incipient lethal levels, however, remain poorly defined. There are two major reasons for this. The first arises from the failure of most investigators to follow standard bioassay procedures. Experimental conditions are often not adequately controlled (e.g., variable temperature regimes). Sufficient numbers of test levels to obtain anything more than the roughest estimate of lethal levels are rarely used. Control mortality is often not taken into account, or even reported. Levels of significance are seldom given, and comparisons are frequently made on the basis of variable exposure periods. If past performance is any indicator, future investigators would do well to review basic toxicological methods (e.g., Shepard, 1955; Sprague, 1973) before proceeding. The second major reason why incipient lethal levels remain poorly defined is inherent to the early stages themselves. Gottwald (1965) pointed out some of the problems: the slow response of embryos to hypoxia and the difficulty in assessing mortality during the early stages. The most significant difficulty, though, is that the organisms are undergoing profound developmental changes that alter their sensitivity to hypoxia even as it is being assessed. This makes it extremely difficult to determine precise response thresholds.

It is fairly well established that sensitivity to hypoxia tends to increase as development proceeds. Maximum sensitivity occurs during the larval period, with the precise stage depending on the particular species. Increasing sensitivity is implied by mortality patterns during chronic exposures to low levels of dissolved oxygen. For example, larval mortality is often recorded at oxygen levels that permitted survival to hatch (Brungs, 1971; Eddy, 1972; Siefert *et al.*, 1973, 1974; Dudley and Eipper, 1975). Mortality patterns are not always valid indicators of stage sensitivity. Rosenthal and Alderdice (1976) noted that with many toxicants, injury may be sustained at an early stage but not manifested until later in development. In the case of hypoxia, however, studies of acute toxicity also indicate increasing sensitivity as development proceeds. Alderdice *et al.* (1958) noted that the 7-day LC_{50} for chum embryos (*O. keta*) increased from 0.4 mg l^{-1} shortly after blastopore closure to 1.0–1.4 mg l^{-1} shortly before hatch (Table V). Likewise, Gottwald (1965) reported that 3-day LC_{50} values for rainbow trout increased from <0.9 mg l^{-1} at blastopore closure to 0.9–2.7 mg l^{-1} near hatch. Studies that have compared sensitivities shortly before and after hatch indicate that the newly hatched larvae are significantly more sensitive (Peterka and Kent, 1976; Spoor, 1977). This is somewhat surprising given the fact that the embryos are effectively exposed to a much lower ambient concentration because of the presence of the capsule. The particular stage at which larvae are most sensitive is highly variable. Bishai (1960) indicated that the sensitivity of young Atlantic salmon and brook trout continued to increase up to at least 80 and 127 days posthatch (at 5°C), respectively. Tamarin and Komarova (1972) reported that the "threshold level" (asphyxiation level in a closed container) of Atlantic salmon increased steadily to a maximum 42–60 days posthatch (at 8°C) and then declined to reach typical juvenile and adult levels 240 days posthatch. De Silva and Tytler (1973) reported that 12-h LC_{50} values for herring larvae increased from 2.8 mg l^{-1} shortly after hatch to reach a peak of 5.1 mg l^{-1} after 5–6 weeks of feeding before declining to 3.1 mg l^{-1} at metamorphosis. Smallmouth bass larvae (*Micropterus dolomieui*) were least resistant to severe hypoxia (1.0 mg l^{-1}) at about the start of exogenous feeding (9 days posthatch at 20°C; Spoor, 1984). In contrast, large-mouth bass larvae (*M. salmoides*) were least resistant shortly after hatch while still feeding endogenously (3 days posthatch at 20°C; Spoor, 1977). Larval plaice were similarly most sensitive shortly after hatch (De Silva and Tytler, 1973).

Changes in larval sensitivity have been linked to changes in the site and efficiency of respiratory gas exchange (DeSilva and Tytler, 1973; Spoor, 1977). The argument is basically as follows. The skin is the major site of gas exchange during much of early life. Young larvae with relatively low metabolic rates and large surface/volume ratios require a relatively small partial pressure gradient to meet oxygen demands. As the larvae grows its metabolic rate increases, but expansion of the area for cutaneous gas exchange fails to

Table V Lower Lethal Levels of Dissolved Oxygen for Teleost Embryos and Larvae[a]

Species	Temperature (°C)	Stage tested	Duration of test	LC_{50} (mg l^{-1})	No-effect level (mg l^{-1})	Comment	Reference
Chronic exposure, freshwater salmonids							
Salmo gairdneri	10.4	E	f–h (~30 days)	<2.8	3.0–4.5		Shumway *et al.* (1964)
Salmo gairdneri	10.0	E	f–h (~36 days)	1.6–2.5	1.6–2.5		Silver *et al.* (1963)
Oncorhynchus tshawytscha	10	E	f–h (~45 days)	1.6–2.5	1.6–2.5		
Oncorhynchus kisutch	9.5	E	f–h (~50 days)	<2.5	2.8–4.1		Shumway *et al.* (1964)
Salmo salar	5	E	f–h (77 days)	<3.7	6.8–11.9		Hamor and Garside (1976)
	10	E	f–h (43 days)	<3.9	5.8–11.6		
Oncorhynchus kisutch	8.5	E–L	119 days	1.4–2.9	1.4–2.9		Siefert and Spoor (1974)
Salvelinus fontinalis	8	E–L	133 days	2.3–2.9	2.3–2.9		
Salvelinus namaycush	7	E–L	131 days	2.4–4.3	4.3–6.0		Carlson and Siefert (1974)
	10	E–L	108 days	5.6–10.5	5.6–10.5	Near upper lethal temp.	

Salvelinus alpinus	4	E–L	93 days	<2.6	<2.6		Gruber and Wieser (1983)
	8	E–L	77 days	≃3.6	3.6–5.9	Near upper lethal temp.	
			Chronic exposure, freshwater nonsalmonids				
Stizostedion vitreum	12.3	E	f–h (~22 days)	<3.0	3.0–5.0	Walleye	Oseid and Smith (1971a)
Catostomus commersoni	12.3	E	f–h (~15 days)	<3.0	<3.0	White sucker	Oseid and Smith (1971b)
Morone chyrops	16	E	f–h (4 days)	<1.8	<1.8		Siefert *et al.* (1974)
Fundulus heteroclitus	20	E	f–h (20 days)	2.4–4.5	4.5–7.5		Voyer and Hennekey (1972)
Morone saxatilis	20	E	f–h (4 days)		>5.0		Turner and Farley (1971)
Coregonus artedii	2	E	f–h (166 days)	<1.0	1.0	Lake herring	Brooke and Colby (1980)
	4	E	f–h (122 days)	<1.0	1.0–2.0		
	6	E	f–h (84 days)	1.0–2.0	1.0–2.0		
	8	E	f–h (54 days)	2.0–4.0	2.0–4.0		
Micropterus salmoides	15	E	f–h		>2.4	Largemouth bass	Dudley and Eipper (1975)
	20	E	f–h		>2.4		
	25	E	f–h		>2.4		

Continued

Table V Lower Lethal Levels of Dissolved Oxygen for Teleost Embryos and Larvae—Cont'd

Species	Temperature (°C)	Stage tested	Duration of test	LC_{50} (mg l^{-1})	No-effect level (mg l^{-1})	Comment	Reference
Cyprinus carpio	25	E	f–h (70 h)	3.0–6.0	6.0–9.0	Carp	Kaur and Toor (1978)
Stizostedion vitreum	17	E–L	20 days	3.4–4.8	3.4–4.8	Walleye	Siefert and Spoor (1974)
Catostomus commersoni	18	E–L	22 days	1.2–2.5	1.2–2.5	White sucker	
Poxomis nigromaculatus	20	E–L	?	<2.7	<2.7	Black crappie	Siefert and Herman (1977)
Prosopium williamsoni	4	E–L	193 days	3.3–4.6	4.6–6.5	Mountain whitefish	Siefert *et al.* (1974)
	7	E–L	158 days	3.1–6.0	3.1–6.0		
Micropterus dolomieui	20	E–L	14 days	2.5–4.4	4.4–8.7	Smallmouth bass	
Morone chrysops	16	E–L	11 days	<1.8	1.8–3.4	White bass	
Micropterus salmoides	20	E–L	20 days	1.7–3.1	3.1–4.5	Largemouth bass	Carlson and Siefert (1974)
	23	E–L	20 days	1.7–3.0	1.7–4.2		

Ictalurus punctatus	25	E-L	19 days	2.4–4.2	4.2–5.0	Channel catfish	Carlson *et al.* (1974)
	29	E–L	19 days	2.3–3.8	3.8–4.6		
Esox lucius	15	E-L	20 days	2.6–4.9	2.6–4.9	Northern pike	
Pimephales promelas	24	L	30 days	4.02	4.02–5.01		Brungs (1971)
Acute exposure, freshwater species							
Oncorhynchus keta	10	E	7 days	0.4		12 dpf	Alderdice *et al.* (1958)
			7 days	0.6		22 dpf	
			7 days	0.6		32 dpf	
			7 days	1.0–1.4		48 dpf	
Salmo gairdneri	11.5	E	3 days	<0.9	0.9–1.7	Blastopore closure	Gottwald (1965)
			3 days	~0.9	1.7–2.7	Start circulation	
			3 days	0.9–2.7	2.7–4.3	Near hatch	
Salmo salar	5	L	120 h		0.4	Newly hatched	Bishai (1960)
			72 h		0.7	10 dph	
			48 h		0.7	40 dph	
			72 h		2.0	54 dph	
			72 h		3.1	80 dph	
			48 h		2.8	135 dph	

Continued

Table V Lower Lethal Levels of Dissolved Oxygen for Teleost Embryos and Larvae—Cont'd

Species	Temperature (°C)	Stage tested	Duration of test	LC_{50} (mg l^{-1})	No-effect level (mg l^{-1})	Comment	Reference
Salmo trutta	5	L	120 h		0.4	Newly hatched	
			72 h		0.7	10 dph	
			72 h		2.0	54 dph	
			72 h		2.4	127 dph	
			48 h		2.3	180 dph	
Coregonus sp.	6.5	L	?	3.3	6.0		Einsele (1965)
	11.5	L	?	3.7	6.9		
	19.0	L	?	5.5	7.5		
Esox lucius	10	E	8h		<0.6	3 days prehatch	Peterka and Kent (1976)
		L	8h	~0.3	0.3–1.8	1 dph (yolk-sac larvae)	
		L	8h	1.6–3.5	1.6–3.5	Feeding larvae	
Micropterus dolomieui	21.5	E	6 h	0.5	0.5–1.8	2 d prehatch	
		L	6 h		0.5–2.2	Yolk-sac larvae	
Lepomis macrochirus	26.5	E	4 h		<0.5		
		L	4 h	0.5–3.7	0.5–3.7	1–4 dph (yolk sac)	

Marine species							
Clupea harengus	14	L	24 h		2.2–2.6	Newly hatched	Bishai (1960)
			24 h		4.4	3 dph	
			24 h		5.1	4 dph	
Clupea harengus	8–13	L	12 h	2.8		Yolk-sac larvae	DeSilva and Tytler (1973)
			12 h	4.4		2–3 weeks feeding	
			12 h	5.1		5–6 weeks feeding	
			12 h	4.2		7–8 weeks feeding	
			12 h	3.1		Metamorphosis	
Pleuronectes platessa	8–13	L	12 h	3.9		Yolk-sac larvae	DeSilva and Tytler (1973)
			12 h	3.8		2–3 weeks feeding	
			12 h	3.6		6–7 weeks feeding	
			12 h	2.4		3–4 weeks metamorphosis	

Continued

Table V Lower Lethal Levels of Dissolved Oxygen for Teleost Embryos and Larvae—Cont'd

Species	Temperature (°C)	Stage tested	Duration of test	LC_{50} (mg l^{-1})	No-effect level (mg l^{-1})	Comment	Reference
Cyclopterus lumpus	12.7	L	24 h	3.6		24 dph	Bishai (1960)
Ophiodon elongatus	8.6	E	96 h	3.0	4.6–7.7	Lingcod	Giorgi and Congleton (1984)
Chasmodes bosquianus	20.8	L	24 h	2.5		Striped blenny, newly hatched	Saksena and Joseph (1972)
Gobiosoma bosci	20.8	L	24 h	1.3		Naked goby	
Gobiesox strumosus	20.5	L	24 h	0.7–1.2		Skillet fish	
Mugil cephalus	20	E	48 h	4.5–5.0	4.5–5.0	Near hatch, striped mullet	Sylvester *et al.* (1975)
		L	48 h	4.8–5.4	4.8–5.4	Newly hatched	
		L	96 h	6.4–7.9	6.4–7.9	Newly hatched	
Pterosmaris axillaris	13(?)	L	24 h	3.5	4.0 (LC_{10})	First-feeding static test	Brownell (1980)

Pachymetopon blochi		L	24 h	3.2	3.5 (LC_{10})	First-feeding static test	
Sparidae (5 species)		L	24 h	3.6	3.9 (LC_{10})	First-feeding static test	
Trulla capenis		L	24 h	<2.3	—	First-feeding static test	
Congiopodus spinifen		L	24 h	3.6	4.2 (LC_{10})	First-feeding static test	
Gaidropsarus capensis		L	24 h	2.5	2.9 (LC_{10})	First-feeding static test	
				2.1	2.7 (LC_{10})	First-feeding flow-through test	
Heteromycteris capensis		L	24 h	6.9	1.2 (LC_{10})	First-feeding flow-through test	

[a]Abbreviations: E, embryos; L, larvae; dph, days posthatch; f–h, fertilization to hatch.

keep pace. If metabolic demands are to be met, the partial pressure gradient across the skin must increase (recall $\dot{V}O_2 = GO_2PO_2$). The result is an increase in P_c (standard) and a reduction in the ability of the larva to tolerate hypoxia. As development proceeds, the gills become progressively more important as a site of respiratory gas exchange. Gills are assumed to be a more efficient organ of gas exchange and, as a result, P_c (standard) values gradually decline. This is reflected in a gradual decrease in sensitivity to hypoxia. Species vary in the timing of the transition from cutaneous to branchial gas exchange, and thus it is not surprising that they vary in the particular stage at which they are most sensitive to hypoxia. De Silva and Tytler (1973) indicated that the stage at which hemoglobin first appears also may be important. Spoor (1984) linked increased resistance of smallmouth bass larvae near the start of exogenous feeding to inflation of the swimbladder. It is unlikely that this represents a true increase in tolerance to hypoxia but more likely reflects the fact that the gas in the swimbladder can act as a temporary reservoir when oxygen is in short supply.

It is generally assumed that the ability of fish to resist hypoxia decreases with increasing temperature (European Inland Fisheries Advisory Commission, 1973; Alabaster and Lloyd, 1980; Chapman, 1986), although Chapman (1986) points out that the data supporting this contention is rather spotty. There is some evidence indicating that, within limits, hypoxic sensitivity may actually vary relatively little with temperature. Davis (1975) found that incipient sublethal response thresholds of adult fish were insensitive to temperature when expressed as a concentration (i.e., mg l^{-1}). This still represents an increase in the driving force required to meet oxygen demands, since, at a given concentration, partial pressures are greater at higher temperatures (because of decreased capacitance). Ultsch *et al.* (1978), on the other hand, reported that asphyxiation levels were actually lower at 20°C than at 10°C, even when expressed as partial pressures, in adults of six species of darters (*Etheostoma*).

Few studies have specifically examined the effect of temperature on the ability of embryos and larvae to tolerate a lack of oxygen. Temperature variations within a narrow range (3–5°C) appear to have little effect in many species (Siefert *et al.*, 1974; Carlson and Siefert, 1974; Carlson *et al.*, 1974; Hamor and Garside, 1976). There are some exceptions though. Increasing the temperature from 7 to 10°C resulted in a significant increase in the sensitivity of young lake trout (*Salvelinus namaycush*) (Carlson and Siefert, 1974). Similarly, Arctic char embryos and larvae were less tolerant of hypoxia at 8 than at 4°C (Gruber and Wieser, 1983). Both species are stenothermal, and in both studies the higher temperature was near the upper limit of the zone of tolerance. Lethal levels for embryonic (Brooke and Colby, 1980) and larval (Einsele, 1965) coregonids increased with temperature, but again increases were greatest at the higher temperatures. These results suggest that lethal oxygen *concentrations* may be relatively independent of temperature within the normal temperature range of a particular species but that at high temperatures, there is a strong likelihood of an additive or synergistic interaction.

High flow rates can be expected to increase the hypoxic tolerance of embryos by reducing the thickness of the boundary layer. The interaction between flow rate and dissolved oxygen level has been well documented using sublethal indicators such as growth (Silver *et al.*, 1963; Shumway *et al.*, 1964). There is less information on its effect on lethality. Shumway *et al.* (1964) reported that mortality among coho and steelhead embryos exposed to low oxygen levels tended to be greater at low flows. Miller (1972) reported that mortality among "unfanned" largemouth bass embryos was greater than among fanned embryos. High embryonic mortality in nature is often associated with low flows (Wickett, 1954; Coble, 1961; Phillips and Campbell, 1961; McNeil, 1966; Taylor, 1971; Hempel and Hempel, 1971; Giorgi, 1981). In most cases, however, flow rate and oxygen concentration varied simultaneously so it is difficult to separate effects due to low oxygen from those due to low flow.

Salinity, like temperature, appears to have little effect on hypoxic tolerance within normal limits. Alderdice and Forrester (1971) indicated that viable hatch of Pacific cod (*Gadus macrocephalus*) was largely independent of oxygen, provided levels were above 2–3 mg l^{-1} within the optimal range of temperatures (3–4.5°C) and salinities (17–23‰). Oxygen requirements tended to increase at higher salinities. Similarly, embryonic survival of pilchard (*Sardinops ocellata*) was largely independent of oxygen levels greater than 2.1 mg l^{-1} within the optimal range of temperature (16–21°C) and salinity (33–36‰) (King, 1977).

There are significant differences in the abilities of different species to tolerate hypoxia. There does not appear to be any phylogenetic pattern, but rather the sensitivity of a particular species seems to reflect oxygen levels in its normal habitat. For example, Spoor (1984) linked differences in habitat selection by larvae of smallmouth bass and largemouth bass to differences in their sensitivity to hypoxia. Chapman (1986) pointed out that while salmonids are more sensitive to hypoxia than most nonsalmonid freshwater fish as juveniles, they are less sensitive than many other groups as embryos and larvae. These observations lend some support to Balon's (1975) concept of reproductive guilds. Data in Table V indicate a mean embryonic–larval LC_{50} of 2.7 mg l^{-1} for salmonids at normal temperatures. The nonsalmonid freshwater species fall into two groups: white sucker, black crappie, white bass, and largemouth bass, with a mean embryonic–larval LC_{50} of 2.3 mg l^{-1}, and walleye, mountain whitefish, smallmouth bass, and pike with a mean embryonic-larval LC_{50} of 3.8 mg l^{-1}. Marine species are generally less tolerant than freshwater species. Among marine species, De Silva and Tytler (1973) attributed the greater resistance of newly hatched herring larvae compared with newly hatched plaice larvae to the fact that the former hatch from demersal eggs and are thus more likely to encounter low oxygen levels.

A glaring deficiency in the literature is the absence of any data on the hypoxia tolerance of tropical freshwater species, some of which spawn in virtually anoxic water.

Durborow and Avault (1985) reported significant differences among full-sib families of channel catfish (*Ictalurus punctatus*) in larval resistance to hypoxia. This raises the possibility of selecting strains that are resistant to hypoxia for use in aquaculture.

Shepard (1955) demonstrated that acclimation to low oxygen levels increased the ability of juvenile brook trout to tolerate hypoxia. There are limited data indicating that acclimation increases at least the resistance of larvae. McDonald and McMahon (1977) reported that chronic exposure of Arctic char larvae to low oxygen (2.5 mg l^{-1}) resulted in a 2.5-fold increase in median resistance time in 1.1 mg l^{-1} compared with normoxic-reared larvae.

Moderate levels of hyperoxia have been reported to enhance survival of embryos (Gulidov, 1969, 1974; Gulidov and Popova, 1978) and larvae (Sylvester *et al.*, 1975). At higher concentrations (>300% air saturation), though, oxygen becomes toxic. Species vary in their ability to tolerate hyperoxia. Gulidov (1969) reported significant mortality in pike (*Esox lucius*) eggs incubated in 36.4 mg l^{-1} (336% ASV). No pike embryos hatched in 45.3 mg l^{-1} (418% ASV), apparently because of suppressed neuromuscular activity. Embryos of verkhovka (*Leucospuis delineatus*) (Gulidov, 1974) and roach (*Rutilis rutilis*) (Gulidov and Popova, 1978), on the other hand, both hatched successfully in about 40 mg l^{-1}. In *L. delineatus,* hatching was delayed at high oxygen concentrations. Newly hatched larvae tended to have more body segments than normal, and no red blood cells were present. The initial phase of erythropoeisis was not inhibited, and red blood cells appeared in the circulation at about the normal time. The number of erythrocytes later declined, and red cells were absent at hatch. Absence of red blood cells did not adversely affect survival, which is not surprising given the high ambient oxygen concentration. Gulidov (1974) linked the high tolerance of *L. delineatus* and roach to hyperoxia to the fact that their eggs are frequently laid on vegetation and may thus be exposed to high oxygen concentrations in their natural habitat.

2. SUBLETHAL RESPONSES

Hypoxia elicits a broad spectrum of sublethal responses in embryos and larvae. These include reduced rates of growth and development, morphological changes, behavioral alterations, and a wide variety of metabolic and physiological adjustments. In general, sublethal response thresholds are even more poorly defined than lethal thresholds. The reasons for this are basically the same as those discussed previously: deficient experimental design and changes in the intrinsic sensitivity of the organism. The situation is actually somewhat worse than for studies of lethality. Sublethal response thresholds have tended to be higher than anticipated. Many investigators have chosen inappropriate experimental levels (i.e., too low) and, as a result, have been able to define response thresholds only very broadly (typically as lying somewhere between 30–50 and 100% air saturation), if at all. The following discussion, therefore, will be restricted for the most part to a qualitative description of the sublethal effects of hypoxia.

a. Development and Growth. Developmental velocity and early growth are highly sensitive to reductions in ambient oxygen levels. In many species incipient limiting levels appear to be close to, or even in excess of, 100% air saturation (Silver *et al.*, 1963; Shumway *et al.*, 1964; Eddy, 1972; Gulidov, 1974; Gulidov and Popova, 1978). Reduced growth and delays in development under moderate hypoxia should be regarded as compensatory responses whereby the animal adjusts metabolic demands to match available supply. Growth and development are normally closely linked, although there is some suggestion that developmental rates are less plastic than growth rates (Silver *et al.*, 1963).

Embryonic development is progressively retarded by continuous exposure to low levels of dissolved oxygen (Garside, 1959, 1966; Winnicki, 1968). Delays are typically insignificant during early development at moderate levels (> 20–30% ASV) of hypoxia. During later development, times to defined stages become progressively more delayed compared with normoxia. As expected, delays are more pronounced at lower oxygen levels.

Most investigators have not monitored developmental velocities closely. Instead they have looked at the effect of hypoxia on times to easily identifiable events such as hatch, emergence, or the onset of feeding. The problem with this approach is that times to some of these events are not indicative of the overall effect of hypoxia on developmental rate. For example, time to hatch is widely used as an indicator of sublethal hypoxic stress. Low oxygen levels elicit two responses that have opposing effects on time to hatch. Hypoxia reduces the overall rate of development and thus tends to delay hatching. However, once embryos have reached a certain stage, low oxygen levels initiate the release of hatching enzyme (DiMichele and Taylor, 1980; DiMichele and Powers, 1982; Yamagami *et al.*, 1984) and thus tend to reduce time to hatch. Which effect dominates depends on the particular species. Continuous exposure of coho, brook trout (Siefert and Spoor, 1974), walleye (Oseid and Smith, 1971a), mountain whitefish Siefert *et al.*, 1974), mummichog (*Fundulus heteroclitus*) (Voyer and Hennekey, 1972), lake trout (Garside, 1959; Carlson and Siefert, 1974), and lake herring (Brooke and Colby, 1980) to low oxygen levels resulted in significant delays in time to hatch. Continuous exposure of white sucker (Oseid and Smith, 1971b; Siefert and Spoor, 1974) and sockeye (Brannon, 1965) had no significant effect, while hypoxic incubation of smallmouth bass (Siefert *et al.*, 1974) and largemouth bass (Carlson and Siefert, 1974) caused premature hatch. These results suggest that if embryonic development is relatively rapid, as in smallmouth and largemouth bass, the premature release of hatching enzyme outweighs any general delay in development. The effect of low oxygen on time to emergence is similarly complicated by the fact that moderate levels of hypoxia can act as a directive factor and induce the larvae to leave the substrate prematurely. As a result, hypoxia has been reported variously to delay (Phillips *et al.*, 1966; Mason, 1969) or advance (Witzel and MacCrimmon, 1981; Bailey *et al.*, 1980; Bams, 1983) emergence.

Growth has been the most widely used indicator of hypoxic stress during early life (Silver *et al.*, 1963; Shumway *et al.*, 1964; Oseid and Smith, 1971a, b; Carlson *et al.*, 1974; Siefert *et al.*, 1973, 1974; Siefert and Spoor, 1974; Carlson and Siefert, 1974; Hamor and Garside, 1977; Gruber and Wieser, 1983; Florez, 1972). As mentioned earlier, response thresholds are high. Embryos appear to be more sensitive than larvae, although this has not been well documented. Embryonic growth, at least, is more severely inhibited by low oxygen at higher temperatures (Hamor and Garside, 1977; Eddy, 1972; Gruber and Wieser, 1983), as would be predicted from the fact that P_c values tend to increase with temperature. Also, as expected, low flow rates accentuate the limiting effect of low oxygen on embryonic growth (Silver *et al.*, 1963; Shumway *et al.*, 1964; Hamor and Garside, 1977) (Fig. 12). The effect of low flow appears to be more pronounced for larger eggs than for smaller eggs (Silver *et al.*, 1963; Shumway *et al.*, 1964; Brannon, 1965). Apparently the more favorable surface/volume ratios of small eggs make variations in the thickness of the boundary layer less important. High flow rates during the larval period can have the opposite effect and reduce growth efficiency, apparently as a result of enhanced activity levels (Brannon, 1965).

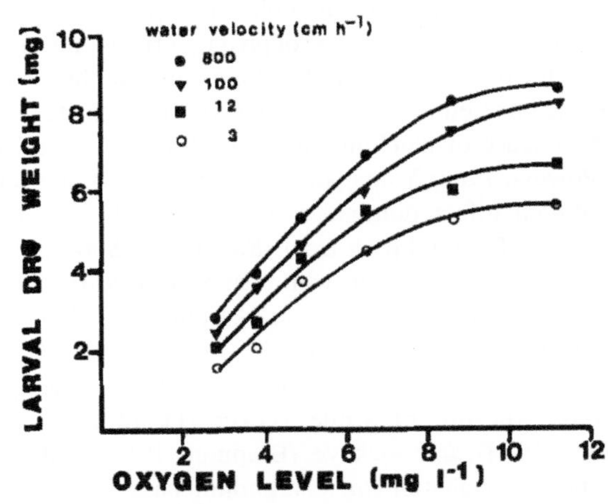

Fig. 12. Mean dry weights of newly hatched coho salmon (*Oncorhynchus kisutch*) larvae reared at various constant oxygen concentrations and bulk water velocities. [After Shumway *et al.* (1964).]

There is some question as to the significance of the reductions in growth noted at moderate levels of hypoxia. Eddy (1972) points out that although low oxygen levels result in smaller salmonid larvae at hatch and greatly extend the period of endogenous feeding, there is relatively little difference in the final size achieved. Growth efficiencies of Atlantic salmon (Hamor and Garside, 1977) and Arctic char (Gruber and Wieser, 1983) embryos were little affected by oxygen levels as low as 20–30% air saturation. Analysis of larval growth is more complicated. The growth efficiency of Atlantic salmon (Hamor and Garside, 1977) and Arctic char (Gruber and Wieser, 1983) larvae was significantly reduced at 20–30% ASV compared with at 100% ASV (Fig. 13). In both species, however, growth efficiency was significantly higher at 50% ASV than at normoxia. This was attributed to reduced locomotor activity under moderate hypoxia (Hamor and Garside, 1977; Gruber and Wieser, 1983). There is another possibility that serves to illustrate the difficulties in comparing early growth efficiencies. It arises from the fact that the larvae were not all at equivalent stages when the comparisons were made. For example, Hamor and Garside (1977) terminated their tests when between 80 and 111%(?) of the yolk present at hatch was consumed. Yolk conversion efficiencies decline sharply as yolk reserves near exhaustion. In steelhead, instantaneous yolk conversion efficiencies fall from 45% to 0% between 80 and 100% yolk utilization (Rombough, 1987). It would require only a relatively minor difference in the stages that are compared to produce an apparently significant difference in growth efficiency.

Even if growth efficiency is significantly reduced at low oxygen levels, the environmental significance is not always clear. It is generally assumed that smaller larvae are at a competitive disadvantage. Smaller larvae display slower absolute swimming speeds and hence can search a smaller volume for prey in a given time and may themselves be more susceptible to predation (Giorgi, 1981). Smaller larvae are also less effective at competing for territories (Mason, 1969). However, territories may not always be fully occupied or food always limiting. Mason (1969) observed that coho incubated under hypoxic conditions were smaller at emergence and could not compete successfully for territory with larger normoxic larvae. As a result, they were forced to migrate from the vicinity of the redd but subsequently did well providing the stream was not heavily populated. Bams (1983) reported that chum larvae from Japanese-style incubators were smaller and emerged earier than larvae incubated in upwelling gravel boxes, apparently because of hypoxic conditions in the Japanese-style incubators. These alevins, though smaller at emergence, subsequently grew faster than the alevins from upwelling boxes because of earlier feeding. The hatchery situation may be unusual because of the abundance of food, but, as Bams (1983) points out, it emphasizes that fitness components must be defined carefully and must be viewed in relation to the particular set of conditions.

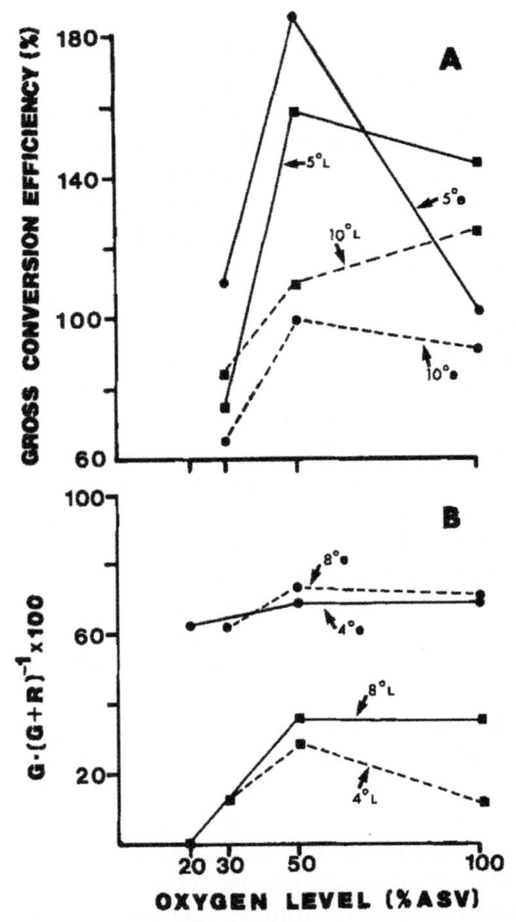

Fig. 13. Efficiency of early growth under conditions of chronic hypoxia. (ASV, air saturation.) (A) Gross conversion efficiencies (weight of tissue produced/weight of yolk consumed during a given period) of embryos (e) and larvae (L) of Atlantic salmon (*Salmo salar*) reared at 5 and 10°C. [After Hamor and Garside (1977).] (B) Growth efficiencies (energy content of tissue elaborated/energy expended on growth and metabolism during a given period) of embryos (e) and larvae (L) of Arctic char (*Salvelinus alpinus*) reared at 4 and 8°C. [after Gruber and Wieser (1983).]

b. Morphology. Teratogenic effects have been observed at oxygen concentrations close to the lethal level (Silver *et al.*, 1963; Shumway *et al.*, 1964; Garside, 1959; Alderdice *et al.*, 1958; Braum, 1973; Brooke and Colby, 1980). Deformities of the axial skeleton, jaws, and vitelline circulation appear most common. The changes in the vitelline circulation appear to be adaptive. Blood vessels become more finely divided and cover a greater proportion of the yolk. This can be expected to enhance respiratory gas exchange. Burggren and Mwalukoma (1983) noted that chronic exposure to low oxygen levels caused a similar increase in the density of capillaries supplying the skin of larval amphibians. This was associated with a reduction in the thickness of the blood–water diffusion barrier. Diffusion distances have not been reported for fish embryos or larvae chronically exposed to low oxygen, but observations that the yolk sac of hypoxic embryos ruptures readily (Garside, 1959; Brooke and Colby, 1980) suggest that this might be the case. Chronic hypoxia also brings about morphological adjustments in the gills that can be expected to enhance gas exchange. McDonald and McMahon (1977) observed lamellar hypertrophy in Arctic char larvae reared in 2.6 mg l^{-1} (at 6.5°C). Branchial development overall was inhibited as a consequence of a general inhibition of growth, so that by 47 days posthatch hypoxic larvae had 22% fewer filaments and 40% fewer lamellae than normoxic larvae. The surface area of individual lamellae, however, was significantly larger in the hypoxic larvae to the extent that there was no significant difference in total lamellar surface. The fact that the hypoxic larvae were more resistant to lethal oxygen concentrations led McDonald and McMahon (1977) to suggest that other adjustments, such as reduced blood–water distances, increases in the area of lamellar blood spaces, and increased lamellar perfusion, were also involved. Pinder and Burggren (1983) suggested that such morphological adjustments leading to increased conductance of gas-exchange organs was typical of the early stages of lower vertebrates. They contrasted this with the usual adult response to hypoxia of increased blood-carrying capacity and increased hemoglobin oxygen affinity.

c. Behavior and Physiology. Larvae respond to acute hypoxia with an increase in random movements (Spoor, 1977, 1984). The level at which they respond appears to be directly related to the lethal level, suggesting they are responding to hypoxemia rather than ambient oxygen levels. Spoor (1984) noted that the oxygen concentration at which smallmouth bass began to become agitated increased steadily during the first 8–10 days after hatch. During this period the length of time the larvae would resist acutely lethal oxygen concentrations also declined progressively. About 12 days posthatch the larvae became less responsive to low oxygen levels. This coincided with increased resistance.

The physiological responses of fish embryos and larvae to acute hypoxia are significantly different from those of juveniles and adults. Most notable is the absence of reflex bradycardia (Fisher, 1942; Holeton, 1971; McDonald and McMahon, 1977). In this regard they are similar to amphibian larvae (West and Burggren, 1982; Feder, 1983a,b; Quinn and Burggren, 1983).

Physiological and behavioral responses designed to facilitate embryonic and larval gas exchange were discussed in a previous section. Not unexpectedly, such responses tend to intensify during hypoxia. Peterson and Martin-Robichaud

(1983) noted that the frequency of trunk movements of Atlantic salmon embryos increased during hypoxia. Holeton (1971) noted that young rainbow trout larvae (1–8 days old) responded to acute exposure to moderate levels of hypoxia with increased pectoral fin movements, increases in rate and amplitude of breathing, and tachycardia. Only when oxygen levels dropped below about 40 mm Hg (at 10°C) did heart rate and breathing rate decline (Fig. 14). Heart rate and breathing rate recovered slowly in restoration of normoxia, unlike the quick recovery seen in adult fish. Peterson (1975) reported that the ventilatory rate also increased in Atlantic salmon larvae in response to acute hypoxia (30–80 mm Hg at 9°C). Increased opercular movements were accompanied by increases in the frequency of pectoral-fin movements. Movements of pectoral fins appear to help draw water across the gills, since ablation of the fins in normoxia results in a significant increase in opercular rate. McDonald and McMahon (1977) indicated that ventilatory and cardiovascular responses during chronic hypoxia were similar to those during acute hypoxia. Heart and circulatory rates of Arctic char larvae remained

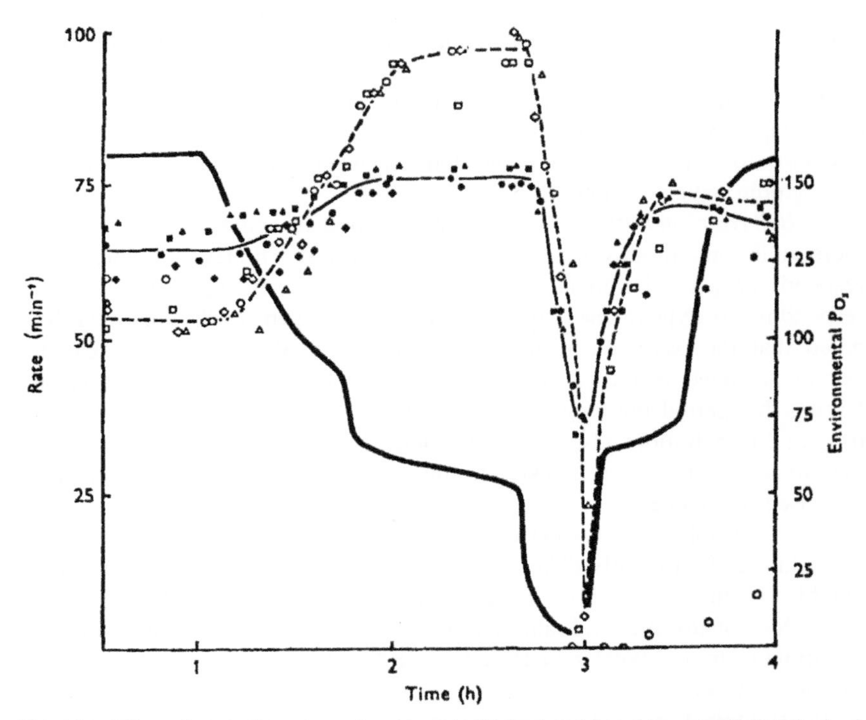

Fig. 14. Effect of acute hypoxia on heart rate (solid line) and breathing rate (dotted line) of rainbow trout larvae 8 days after hatch in 10°C water. Heavy line indicates PO_2 of the water. [From Holeton (1971).]

elevated in 33 mm Hg (at 6.5°C) for at least 47 days after hatch. As in Atlantic salmon (Peterson, 1975), ventilatory movements of newly hatched Arctic char were infrequent and uncoordinated. In contrast, rhythmic and rapid (60 min^{-1}) ventilatory movements were observed in hypoxic larvae within 1 h of hatch. McDonald and McMahon (1977) speculated that cutaneous respiration could provide sufficient oxygen to meet the respiratory demands of newly hatched larvae under normoxic conditions but not during hypoxia. The larvae compensate for a reduced driving force during hypoxia by branchial recruitment, effectively increasing the surface area for respiratory gas exchange (i.e., increasing conductance). Neither Holeton (1971) nor McDonald and McMahon (1977) observed the reflex bradycardia typical of juvenile and adult fish. Holeton (1971) speculated that the basic response to hypoxemia was increased breathing and heart rates and that the bradycardia observed in older fish was a superimposed mechanism designed to balance ventilation–perfusion rates for efficient delivery of oxygen to tissues. McDonald and McMahon (1977) argued that tachycardia was a more appropriate response during early life, since higher cardiac output would lead to opening of vascular channels within the gills and effectively increase the surface area for gas exchange.

d. Metabolism. During severe hypoxia, physiological adjustments may not be sufficient to allow the organism to meet its energy requirements aerobically. Partial compensation can be achieved by channeling energy through glycolytic pathways. Arctic char embryos are able to supply energy at 11–23% the aerobic rate $\left(r\dot{V}O_2\right)$ using anaerobic pathways (Gnaiger, 1979; Gnaiger *et al.*, 1981). The ability to produce a relatively high proportion of normal energy requirements anaerobically probably accounts for the high anoxic resistance of the embryonic stages, especially when it is remembered that a large proportion of normal metabolism goes toward growth rather than maintenance (Smith, 1957, 1958). Gnaiger *et al.* (1981) indicated that lactic acid was the predominant end product in Arctic char embryos, although small amounts of succinic acid were also produced. There was a transient increase in alanine concentration but no change in volitile fatty acid concentration. It thus appears that the high anoxic tolerance is due to high tolerance of lactic acid, rather than activation of more efficient anaerobic pathways. Recovery from anoxia is slow. Gnaiger (1979) observed that it took up to 48 h after restoration of normoxia for the metabolic rate of Arctic char embryos to return to typical aerobic levels.

The metabolic responses of larvae to acute hypoxia are complex and apparently involve both activation of anaerobic pathways and physiological or biochemical changes leading to more efficient utilization of oxygen. Gnaiger (1983a) monitored changes in total and aerobic energy production of Arctic char larvae following abrupt transfer from 100% to 48% air saturation using simultaneous direct and indirect calorimetry. Heat production dropped rapidly to about half the original rate but over the next 24 h gradually increased again to typical normoxic rates. During this period, total heat production measured directly was greater than that calculated on the basis of oxygen consumption, indicating a significant anaerobic contribution (≃20% midway through recovery). Gnaiger (1983a) termed this period the phase of anoxic compensation, although compensatory changes involving aerobic metabolism were also taking place as evidenced by the gradual increase in the rate of oxygen consumption. After about 24 h, the rate of oxygen consumption stabilized at approximately normoxic levels but total heat production began to decline so that *apparent* aerobic energy production was greater than total energy production. Gnaiger (1983a) termed this the phase of conservative compensation and suggested that it reflected coupled glyconeogenesis utilizing lactic acid accumulated during the period of anoxic compensation. Young salmonids apparently either oxidize or reconvert most of the lactate accumulated during periods of oxygen debt back to glycogen (Wieser *et al.*, 1985). This may not be true in all species. Broberg and Kristoffersson (1983) reported that young of the viviparous species *Zoarces viviparous* excrete a large proportion of accumulated lactate.

B. Physiological Hypoxia

Even under normoxic conditions, young fish may be subjected to hypoxemia as a consequence of developmental events or activity. Several investigators have attempted to link physiological hypoxia, as evidenced by increased lactic acid concentrations, with specific developmental events. It appears that in general there is a rise in lactic acid levels during gastrulation, a drop during early organogenesis, and a gradual rise toward hatch (Hishida and Nakano, 1954; Kamler, 1976; Boulekbache, 1981). Kamler (1976) reported that lactic acid levels continued to rise in carp larvae until near the end of endogenous feeding. The ontogeny of energy metabolism is beyond the scope of this review—recent reviews of this area are provided by Terner (1979) and Boulekbache (1981)—but it is important to recognize that many of the responses seen as a result of environmental hypoxia also may be elicited during normal development.

Wieser *et al.* (1985) recently examined the metabolic responses of young rainbow trout during physiological hypoxia induced by forced swimming. Anaerobic energy production (on a mass basis) was found to be independent of body mass and temperature. It was pointed out previously that maximum aerobic energy production of young rainbow trout increases with body size and temperature, up to a maximum at about 12°C (Wieser *et al.*, 1985; Wieser, 1985). The net result is that anaerobic sources are proportionally more important during larval activity than in older fish. In rainbow trout, anaerobic energy production during a 1-min burst of activity was 6–9 times the aerobic production in yolk sac larvae but only 2–6 times the aerobic production in free-swimming larvae and fry. Anaerobic energy production was derived entirely from depletion of phosphocreatine and ATP reserves during the first 30 s of activity. Glycolysis, as evidenced by increased lactic acid production, began to be important during the second 30 s of activity and during the period of recovery. AMP and ADP concentrations did not increase in response to ATP depletion, indicating the probable presence of a very active AMP deaminase. During recovery, phosphocreatine stores were the first to be replenished (67% control values within 5 min, 80% within 10 min). ATP levels remained low for the first 5 min of recovery but had regained 80% of the control value after 10 min. Wieser *et al.* (1985) compared the energy debts acquired during burst swimming (60 s) with energy liberated during recovery. For yolk-sac larvae the ATP required to replenish body energy stores balanced well with the amount of excess energy ($> r\dot{V}O_2$) supplied by anaerobic and aerobic processes during recovery. Larger fish, however, consumed progressively more energy during recovery than was required to repay the true oxygen debt. This was attributed to poststimulus excitement. Energy production during burst activity has not been reported for other species, but it would not be surprising if significant differences are found given the variations in patterns of muscle development and partitioning of energy between oxidative and glycolytic pathways that are known to occur (Forstner *et al.*, 1983; Wieser *et al.*, 1985; Hintleitner *et al.*, 1987).

V. CONCLUSIONS

In general, the ontogeny of respiration is not as well understood for teleosts as for higher vertebrate classes. However, there are signs that the situation is improving, thanks in large part to technological advances that have made it easier to work with such small and delicate organisms. Culture techniques have improved considerably during the past decade, allowing species that previously were difficult to obtain to be reared conveniently in the laboratory. Polarographic electrodes, including microelectrodes, have become readily available. Advances in electronics have permitted the construction of extremely sensitive microcalorimeters, while methods for assaying metabolite concentrations in small samples have improved greatly.

The last major review of this field was that of Blaxter (1969). Since then, a number of important advances have been made. It is now clearly estabished that the boundary layer represents a major barrier to diffusion (Daykin, 1965; Wickett, 1975; Vogel, 1981). A start has been made in recording PO_2 profiles within the respiring egg (Berezovsky *et al.*, 1979; Sushko, 1982; Diez and Davenport, 1987). Studies by Burgreen and co-workers (e.g., Burggren, 1985) with amphibians and Liem (1981) with fish larvae have brought about a greater appreciation of the sophistication of cutaneous gas-exchange structures. Researchers have begun to examine the transition from cutaneous to branchial gas exchange (e.g., De Silva, 1974; Morgan, 1974a,b;

McDonald and McMahon, 1977) and from water breathing to air breathing (Hughes *et al.*, 1986) and the possible ecological implications of these transactions (Iwai and Hughes, 1977). Iuchi (1985) and others have provided details concerning the ontogeny of respiratory pigments in fish. Information on energy partitioning during development is now available for a number of species (e.g., Smith, 1957; Laurence, 1969, 1975, 1977, 1978; Houde and Schekter, 1983; Gruber and Wieser, 1983; Rombough, 1987). Researchers have begun to examine the energetics of larval activity in detail, in terms both of functional morphology (Webb and Weihs, 1986) and of energy production (Wieser, 1985; Wieser *et al.*, 1985; Wieser and Forstner, 1986; Hinterleitner *et al.*, 1987). The physiological and morphological responses of the early stages to hypoxia have been shown to be quite different from those of adult fish (Holeton, 1973; McDonald and McMahon, 1977). The idea of using critical dissolved oxygen levels to predict oxygen requirements and sublethal response thresholds has been advanced (Rombough, 1986, 1987).

Much work, though, remains to be done. Most of what is known about respiration during early life is derived from studies of only a few species. Teleosts are a diverse group, and comparative studies of the physiological and anatomical adaptations that permit survival in hypoxic environments are a necessity. Other potentially productive areas of research include studies of how PO_2 profiles within the egg change during development, the problem of gas transfer in egg masses, and the functional aspects of the transition from cutaneous to branchial gas exchange. So far studies of this transition have been restricted to the structural aspects. The nature of the relationship between $\dot{V}O_2$ and tissue mass during early life remains controversial, and further investigations in this area would seem appropriate. Important aspects of aerobic metabolism in very young fish, such as power–performance relationships and the nature of endogenous rhythms, have received little attention. It would appear to be especially important to be able to relate metabolic rates measured in the laboratory with those that occur in the field. More research needs to be conducted on the physiological, morphological, and metabolic responses of embryos and larvae to chronic hypoxia and the ecological significance of such adjustments. Finally, there is the problem of factor interactions. The recent development of a general multivariate dose-response model (Schnute and Jensen, 1986; Jensen *et al.*, 1986) now would appear to make this problem more tractable.

REFERENCES

Adolph, E. F. (1983). Uptakes and uses of oxygen, from gametes to maturity: An overview. *Respir. Physiol.* **53**, 135–160.

Akberali, H. B., and Earnshaw, M. J. (1984). Copper-stimulated respiration in the unfertilized eggs of the Eurasian perch *Perca fluviatus*. *Comp. Biochem. Physiol. C* **78C**, 349–352.

Alabaster, J. S., and Lloyd, R. (1980). "Water Quality Criteria for Freshwater Fish. Dissolved Oxygen," pp. 127–142. Food Agric. Org., U. N./Butterworth, London.

Al-Adhami, M. A., and Kunz, Y. W. (1976). Haemopoietic centers in the developing angel fish, *Pterophyllum scalare* (Cuvier and Valenciennes). *Wilhelm Roux's Arch. Dev. Biol.* **179**, 393–401.

Alderdice, D. F., and Forrester, C. R. (1971). Effects of salinity, temperature and dissolved oxygen on early development of the Pacific cod *Gadus macrocephalus*. *J. Fish. Res. Board Can.* **28**, 883–902.

Alderdice, D. F., and Wickett, W. P. (1958). A note on the response of developing chum salmon eggs to free carbon dioxide in solution. *J. Fish. Res. Board Can.* **15**, 797–799.

Alderdice, D. F., Wickett, W. P., and Brett, J. R. (1958). Some effects of temporary exposure to low dissolved oxygen levels on Pacific salmon eggs. *J. Fish. Res. Board Can.* **15**, 229–249.

Alderdice, D. F., Jensen, J. O. T., and Velsen, F. P. J. (1984). Measurement of hydrostatic pressure in salmonid eggs. *Can. J. Zool.* **62**, 1977–1987.

Almatar, S. M. (1984). Effects of acute changes in temperature and salinity on the oxygen uptake of larvae of herring (*Clupea harengus*) and plaice (*Pleuronectes platessa*). *Mar. Biol. (Berlin)* **80**, 117–124.

Amberson, W. R., and Armstrong, P. B. (1933). The respiratory metabolism of *Fundulus heteroclitus* during embryonic development. *J. Cell. Comp. Physiol.* **2**, 381–397.

Bailey, J. E., Rice, S. D., Pella, J. J., and Taylor, S. G. (1980). Effects of seeding density of pink salmon *Oncorhynchus gorbuscha*, eggs on water chemistry, fry characteristics, and fry survival in gravel incubators. *Fish. Bull.* **78**, 649–658.

Balon, E. K. (1975). Reproductive guilds of fishes: A proposal and definition. *J. Fish. Res. Board Can.* **32**, 821–864.

Balon, E. K. (1984). Patterns in the evolution of reproductive styles in fishes. *In* "Fish Reproduction: Strategies and Tactics" (G. W. Potts and R. J. Wooton, eds.), pp. 35–53. Academic Press, London.

Bams, R. A. (1983). Early growth and quality of chum salmon (*Oncorhynchus keta*) produced in keeper channels and gravel incubators. *Can. J. Fish. Aquat. Sci.* **40**, 499–505.

Batty, R. S. (1984). Development of swimming movements and musculature of larval herring (*Clupea harengus*). *J. Exp. Biol.* **110**, 217–229.

Beamish, F. W. H. (1964). Respiration of fishes with special emphasis on standard oxygen consumption. III. Influence of oxygen. *Can. J. Zool.* **42**, 355–366.

Beamish, F. W. H. (1978). Swimming capacity. *In* "Fish Physiology" (W. S. Hoar and D. J. Randall, eds.), Vol. 7, pp. 101–187. Academic Press, New York.

Berezovsky, V. A., Goida, E. A., Mukalov, I. O., and Sushko, B. S. (1979). Experimental study of oxygen distribution in *Misgurnis fossilis* eggs. *Fiziol. Zh. (Kiev)* **25**(4), 379–389; *Can. Transl. Fish. Aquat. Sci.* No. 5209 (1986).

Bishai, H. M. (1960). The effect of gas content of water on larval and young fish. *Z. Wiss. Zool.* **163**, 37–64.

Blaxter, J. H. S. (1969). Development: Eggs and larvae. *In* "Fish Physiology" (W. S. Hoar and D. J. Randall, eds.), Vol. 3, pp. 178–241. Academic Press, New York.

Blaxter, J. H. S., and Hempel, G. (1966). Utilization of yolk by herring larvae. *J. Mar. Biol. Assoc. U.K.* **46**, 219–234.

Boulekbache, H. (1981). Energy metabolism in fish development. *Am. Zool.* **21**, 377–389.

Boulekbache, H., and Devillers, C. (1977). Etude par microscopie à balayage des modifications de la membrane des blastomères au cours des premiers stades du développement de l'oeuf de la truite (*Salmo irideus* Gibb). *C. R. Hebd. Seances Acad. Sci.* **285**, 917–920.

Boyd, M. (1928). A comparison of the oxygen consumption of unfertilized and fertilized eggs of *Fundulus heteroclitus*. *Biol. Bull.* (*Woods Hole, Mass.*) **55**, 92–100.

Brannon, E. L. (1965). The influence of physical factors on the development and weight of sockeye salmon embryos and alevins. *Int. Pac. Salmon Fish. Comm. Prog. Rep.* **12**, 1–26.

Braum, E. (1973). Einflüsse Chronischen Exogenen Saurstoffmangels auf die Embryogenese Des Herings (*Clupea harengus*). *Neth. J. Sea Res.* **7**, 363–375 (Engl. Abstr.).

Brett, J. R. (1964). The respiratory metabolism and swimming performance of young sockeye salmon. *J. Fish. Res. Board Can.* **21**, 1183–1226.

Brett, J. R. (1970). Fish–The energy cost of living. *In* "Marine Aquaculture" (W. J. McNeil, ed.), pp. 37–52. Oregon State Univ. Press, Corvallis.

Brett, J. R. (1972). The metabolic demand for oxygen in fish, particularly salmonids, and a comparison with other vertebrates. *Respir. Physiol.* **14**, 151–170.

Brett, J. R., and Glass, N. R. (1973). Metabolic rates and critical swimming speeds of sockeye salmon (*Oncorhynchus nerka*) in relation to size and temperature. *J. Fish. Res. Board Can.* **30**, 379–387.

Brett, J. R., and Groves, T. D. D. (1979). Physiological energetics. *In* "Fish Physiology" (W. S. Hoar, D. J. Randall, and J. R. Brett, eds.), Vol. 8, pp. 279–352. Academic Press, New York.

Broberg, S., and Kristoffersson, R. (1983). Oxygen consumption and lactate accumulation in the intraovarian embryos and young of the viviparous fish *Zoarces viviparus* in relation to decreasing water oxygen concentration. *Ann. Zool. Fenn.* **20**, 301–306.

Brooke, L. T., and Colby, P. J. (1980). Development and survival of embryos of lake herring *Coregonus artedii* at different constant oxygen concentrations and temperatures. *Prog. Fish-Cult.* **42**, 3–9.

Brownell, C. L. (1980). Water quality requirements for 1st feeding in marine fish larvae. 2. pH, oxygen and carbon dioxide. *J. Exp. Mar. Biol. Ecol.* **44**, 285–298.

Brummett, A. R., and Vernberg, W. B. (1972). Oxygen consumption in anterior vs. posterior embryonic shield of *Fundulus heteroclitus*. *Biol. Bull.* (*Woods Hole, Mass.*) **143**, 296–303.

Brungs, W. A. (1971). Chronic effects of low dissolved oxygen concentrations on the fathead minnow (*Pimephales promelas*). *J. Fish. Res. Board Can.* **28**, 1119–1123.

Burggren, W. (1984). Transition of respiratory processes during amphibian metamorphosis: From egg to adult. *In* "Respiration and Metabolism of Embryonic Vertebrates" (R. S. Seymour, ed.), pp. 31–53. Martinus Nijhoff/Dr. W. Junk Publishers, Dordrecht, The Netherlands.

Burggren, W. (1985). Gas exchange, metabolism, and "ventilation" in gelatinous frog egg masses. *Physiol. Zool.* **58**, 503–514.

Burggren, W., and Mwalukoma, A. (1983). Respiration during chronic hypoxia and hyperoxia in larvae and adult bullfrogs (*Rana catesbeiana*). I. Morphological responses of lungs, skin and gills. *J. Exp. Biol.* **105**, 191–203.

Burggren, W. W., and West, N. H. (1982). Changing respiratory importance of gills, lungs and skin during metamorphosis in the bullfrog *Rana catesbeiana*. *Respir. Physiol.* **47**, 151–164.

Carlson, A. R., and Siefert, R. E. (1974). Effects of reduced oxygen on the embryos and larvae of lake trout (*Salvelinus namaycush*) and largemouth bass (*Micropterus salmoides*). *J. Fish Res. Board Can.* **31**, 1393–1396.

Carlson, A. R., Siefert, R. E., and Herman, L. J. (1974). Effects of lowered dissolved oxygen concentration on channel catfish (*Ictalurus punctatus*) embryos and larvae. *Trans. Am. Fish. Soc.* **103**, 623–626.

Cetta, C. M., and Capuzzo, J. M. (1982). Physiological and biochemical aspects of embryonic and larval development of the winter flounder *Psuedopleuronectes americanus*. *Mar. Biol.* (*Berlin*) **71**, 327–337.

Chapman, G. (1986). "Ambient Water Quality Criteria for Dissolved Oxygen. Fresh-water Aquatic Life," Intern. Rep. U. S. Environmental Protection Agency, Washington, D.C.

Chernikova, V. V., (1964). Variations in the respiration rate of young *Coregonus nasus* (Pallos) with changes in the temperature and the oxygen and carbon dioxide concentration of the water. *In* "Fish Physiology in Acclimation and Breeding" Translated from Russian by the Israel Program for Scientific Translations, 1970 (T. I. Privol'nev, ed.), Keter Press, Jerusalem.

Clements, L. C., and Hoss, D. E. (1977). Effects of acclimation time on larval flounder *Paralichthys* sp. oxygen consumption. *Assoc. Southeast Biol., Bull.* **24**(2), 43.

Coble, D. W. (1961). Influence of water exchange and dissolved oxygen in redds on survival of steelhead trout embryos. *Trans. Am. Fish. Soc.* **90**, 469–474.

Coughlan, D. J., and Gloss, S. P. (1984). Early morphological development of gills in smallmouth bass (*Micropterus dolomieui*). *Can. J. Zool.* **62**, 951–958.

Craik, J. C. A. (1985). Egg quality and egg pigment content in salmonid fishes. *Aquaculture* **47**, 61–88.

Craik, J. C. A., and Harvey, S. M. (1986). The carotenoids of eggs of wild and farmed Atlantic salmon and their changes during development to the start of feeding. *J. Fish Biol.* **29**, 549–565.

Czeczuga, B. (1979). Carotenoids in fish. XIX. Carotenoids in the eggs of *Onchorhynchus keta* (Walbaum). *Hydrobiologia* **63**, 45–47.

Czihak, G., Peter, R., Puschendorf, B., and Grunicke, H. (1979). Some data on the basic metabolism of trout eggs. *J. Fish Biol.* **15**, 185–194.

Dabrowski, K. R. (1986). A new type of metabolism chamber for the determination of active and postprandial metabolism of fish, and consideration of results for coregonid and salmon juveniles. *J. Fish Biol.* **28**, 105–117.

Dabrowski, K., and Kaushik, S. J. (1984). Rearing of coregonid *Coregonus schinzi palea* (Cuv. et Val.) larvae using dry and live food. II. Oxygen consumption and nitrogen excretion. *Aquaculture* **41**, 333–344.

Dabrowski, K., Kaushik, S. J., and Luquet, P. (1984). Metabolic utilization of body stores during the early life of whitefish, *Coregonus lavaretus* L. *J. Fish Biol.* **24**, 721–729.

Dalla Via, G. J. (1983). Bacterial growth and antibiotics in animal respirometry. *In* "Polarographic Oxygen Sensors" (E. Gnaiger and H. Forstner, eds.), pp. 202–218. Springer-Verlag, Berlin and New York.

Davenport, J. (1976). A technique for the measurement of oxygen consumption in small aquatic organisms. *Lab. Pract.* **25**, 693–695.

Davenport, J. (1983). Oxygen and the developing eggs and larvae of the lumpfish, *Cyclopterus lumpus*. *J. Mar. Biol. Assoc. U.K.* **63**, 633–640.

Davenport, J., and Lønning, S. (1980). Oxygen uptake in developing eggs and larvae of the cod, *Gadus morhua*. *J. Fish Biol.* **16**, 249–256.

Davenport, J., Lønning, S., and Saethre, L. J. (1979). The effects of ekofisk oil extract upon oxygen uptake in eggs and larvae of the cod (*Gadus morhua*). *Astarte* **12**, 31–34.

Davis, J. C. (1975). Minimal dissolved oxygen requirements of aquatic life with emphasis on Canadian species. A review. *J. Fish. Res. Board Can.* **32**, 2295–2332.

Daykin, P. N. (1965). Application of mass transfer theory to the problem of respiration of fish eggs. *J. Fish. Res. Board Can.* **22**, 159–171.

Dejours, P. (1981). "Principles of Comparative Respiratory Physiology," 2nd ed. Elsevier/North-Holland Biomedical Press, Amsterdam.

De Silva, C. (1974). Development of the respiratory system in herring and plaice larvae. *In* "The Early Life History of Fish" (J. S. Blaxter, ed.), pp. 465–485. Springer-Verlag, Berlin and New York.

DeSilva, C. D., and Tytler, P. (1973). The influence of reduced environmental oxygen on the metabolism and survival of herring and plaice larvae. *Neth. J. Sea Res.* **7**, 345–362.

DeSilva, C. D., Premawansa, S., and Keemiyahetty, C. N. (1986). Oxygen consumption in *Oreochromis niloticus* (L.) in relation to development, salinity, temperature and time of day. *J. Fish Biol.* **29**, 267–277.

Devillers, C. (1965). Respiration and morphogenesis in the teleostean egg. *Fish Res. Board Can., Transl. Ser.* **3909**, 1–23.

Diez, J. M., and Davenport, J. (1987). Embryonic respiration in the dogfish (*Scyliorhinus canicula* L.) *J. Mar. Biol. Assoc. U.K.* **67**, 249–261.

DiMichele, L., and Powers, D. A. (1982). LDH-B genotype specific hatching times of *Fundulus heteroclitus* embryos. *Nature (London)* **296**, 563–564.

DiMichele, L., and Powers, D. A. (1984a). The relationship between oxygen consumption rate and hatching in *Fundulus heteroclitus. Physiol. Zool.* **57**, 46–51.

DiMichele, L., and Powers, D. A. (1984b). Developmental and oxygen consumption rate differences between lactate dehydrogenase-B genotypes of *Fundulus heteroclitus* and their effect on hatching time. *Physiol. Zool.* **57**, 52–56.

DiMichele, L., and Taylor, M. H. (1980). The mechanism of hatching in *Fundulus heteroclitus:* development and physiology. *J. Exp. Zool.* **217**, 73–79.

DiPrampero, P. E. (1985). Metabolic and circulatory limitations to VO_2 max at the whole animal level. *J. Exp. Biol.* **115**, 319–331.

Doudoroff, P., and Shumway, D. L. (1970). Dissolved oxygen requirements of freshwater fishes. *FAO Fish. Tech. Pap.* **86**, 1–291.

Dudley, R. G., and Eipper, A. W. (1975). Survival of largemouth bass embryos at low dissolved oxygen concentrations. *Trans. Am. Fish. Soc.* **104**, 122–128.

Durborow, R. M., and Avault, J. W., Jr. (1985). Differences in mortality among full-sib channel catfish families at low dissolved oxygen. *Prog. Fish-Cult.* **47**, 14–20.

Durve, V. S., and Sharma, M. S. (1977). Oxygen requirements of the early stages of the major carp *Lebeo calbasu. J. Anim. Morphol. Physiol.* **24**, 391–393.

Eddy, R. M. (1972). The influence of dissolved oxygen concentration and temperature on the survival and growth of chinook salmon embryos and fry. M.Sc. Thesis, Oregon State University, Corvallis.

Einsele, W. (1965). Problems of fish-larvae survival in nature and the rearing of economically important middle European freshwater fishes. *Calif. Coop. Oceanic. Fish. Invest. Rep.* **10**, 24–30.

Eldridge, M. B., Echeverria, T., and Whipple, J. A. (1977). Energetics of Pacific herring *Clupea harengus pallasi* embryos and larvae exposed to low concentrations of benzene, a mono aromatic component of crude oil. *Trans. Am. Fish. Soc.* **106**, 452–461.

Eldridge, M. B., Whipple, J. A., and Bowers, M. J. (1982). Bioenergetics and growth of striped bass, *Morone saxatilis,* embryos and larvae. *Fish. Bull.* **80**, 461–474.

Elliot, J. M., and Davison, W. (1975). Energy equivalents of oxygen consumption in animal energetics. *Oecologia* **19**, 195–201.

Eriksen, C. H., and Feldmeth, C. R. (1967). A water-current respirometer. *Hydrobiologia* **29**, 495–504.

European Inland Fisheries Advisory Commission (1973). Water quality criteria for European freshwater fish. Report on dissolved oxygen and inland fisheries. Prepared by European Inland Fisheries Advisory Commission Working Party on Water Quality Criteria for European Freshwater Fish. *EIFAC Tech. Pap.* **19**, 1–10.

Feder, M. E. (1983a). Response to acute aquatic hypoxia in larvae of the frog *Rana berlandieri. J. Exp. Biol.* **104**, 79–95.

Feder, M. E. (1983b). Effect of hypoxia and body size on the energy metabolism of lungless tadpoles, *Bufo woodhousei*, and air-breathing anuran larvae. *J. Exp. Zool.* **228**, 11–19.

Feder, M. E., and Burggren, W. W. (1985). Cutaneous gas exchange in vertebrates: Design, patterns, control and implications. *Biol. Rev. Cambridge Philos. Soc.* **60**, 1–45.

Fisher, K. C. (1942). The effect of temperature on the critical oxygen pressure for heart beat frequency in embryos of Atlantic salmon and speckled trout. *Can. J. Res., Sect. D* **20**, 1–12.

Florez, F. (1972). Influence of oxygen concentration on growth and survival of larvae and juveniles of the ide *Idus idus*. *Rep. Inst. Freshwater Res., Drottningholm* **52**, 65–73.

Forstner, H. (1983). An automated multiple-chamber intermittent-flow respirometer. *In* "Polarographic Oxygen Sensors" (E. Gnaiger and H. Forstner, eds.), pp. 111–126. Springer-Verlag, Berlin and New York.

Forstner, H., Hinterleitner, S., Mahr, K., and Wieser, W. (1983). Towards a better definition of "metamorphosis" in *Coregonus* sp: Biochemical, histological and physiological data. *Can. J. Fish. Aquat. Sci.* **40**, 1224–1232.

Fry, F. E. J. (1957). The aquatic respiration of fish. *In* "The Physiology of Fishes" (M. E. Brown, ed.), Vol. 1, pp. 1–64. Academic Press, New York.

Fry, F. E. J. (1971). The effect of environmental factors on the physiology of fish. *In* "Fish Physiology" (W. S. Hoar and D. J. Randall, eds.), Vol. 6, pp. 1–98. Academic Press, New York.

Gallagher, M. L., Kane, E., and Courtney, J. (1984). Differences in oxygen consumption and ammonia production among American elvers (*Anguilla rostrata*). *Aquaculture* **40**, 183–187.

Garside, E. T. (1959). Some effects of oxygen in relation to temperature on the development of lake trout embryos. *Can. J. Zool.* **37**, 689–698.

Garside, E. T. (1966). Effects of oxygen in relation to temperature on the development of embryos of brook trout and rainbow trout. *J. Fish. Res. Board Can.* **23**, 1121–1134.

Geffen, A. J. (1983). The deposition of otolith rings in Atlantic salmon, *Salmo salar* L., embryos. *J. Fish Biol.* **23**, 467–474.

Giles, M. A., and Vanstone, W. E. (1976). Ontogenetic variation in the multiple hemoglobins of coho salmon *Oncorhynchus kisutch* and effect of environmental factors on their expression. *J. Fish. Res. Board Can.* **33**, 1144–1149.

Giorgi, A. E. (1981). "The Environmental Biology of the Embryos, Egg Masses and Nesting Sites of the Lingcod, *Ophiodon elongatus*," NWAFC Processed Rep. 81–06. Northwest and Alaska Fisheries Center, National Marine Fisheries Service, U.S. Department of Commerce, Washington, D.C.

Giorgi, A. E., and Congleton, J. L. (1984). Effect of current velocity on development and survival of lingcod, *Ophiodon elongatus*, embryos. *Environ. Biol. Fishes* **10**, 15–28.

Gnaiger, E. (1979). Direct calorimetry in ecological energetics. Long term monitoring of aquatic animals. *Experientia, Suppl.* **37**, 155–165.

Gnaiger, E. (1983a). Calculation of energetic and biochemical equivalents of respiratory oxygen consumption. *In* "Polarographic Oxygen Sensors" (E. Gnaiger and R. Forstner, eds.), pp. 337–345. Springer-Verlag, Berlin and New York.

Gnaiger, E. (1983b). The twin-flow microrespirometer and simultaneous calorimetry. *In* "Polarographic Oxygen Sensors" (E. Gnaiger and R. Forstner, eds.), pp. 134–166. Springer-Verlag, Berlin and New York.

Gnaiger, E., Lackner, R., Ortner, M., Putzer, V., and Kaufmann, R. (1981). Physiological and biochemical parameters in anoxic and aerobic energy metabolism of embryonic salmonids, *Salvelinus alpinus*. *Eur. J. Physiol., Suppl.* **391**, R57 (abstr.).

Gottwald, St. (1965). The influence of temporary oxygen deprivation on embryonic development of rainbow trout (*Salmo gairdneri* Rich.). *Z. Fisch. Deren Hilfswiss.* **13**, 63–84.

Gray, J. (1926). The growth of fish. I. The relationship between embryo and yolk sac in *Salmo fario*. *J. Exp. Biol.* **4**, 215–225.

Grigor'yeva, M. B. (1967). The effect of schooling on the gas exchange of fishes. *In* "Fish Behavior and Reception," pp. 127–132. Nauka Press, Moscow.

Groot, E. P., and Alderdice, D. F. (1985). Fine structure of the external egg membrane of five species of Pacific salmon and steelhead trout. *Can. J. Tool.* **63**, 552–566.

Gruber, K., and Wieser, W. (1983). Energetics of development of the Alpine charr, *Salvelinus alpinus*, in relation to temperature and oxygen. *J. Comp. Physiol.* **149**, 485–493.

Gulidov, M. V. (1969). Embryonic development of the pike, *Esox lucius*, when incubated under different oxygen conditions. *Probl. Ichthyol.* **9**, 841–851.

Gulidov, M. V. (1974). The effect of different oxygen conditions during incubation on the survival and some of the developmental characteristics of the "Verkhova" (*Leucaspius delineatus*) in the embryonic period. *J. Ichthyol. (Engl. Transl.)* **14**(3), 393–397.

Gulidov, M. V., and Popova, K. S. (1978). The influence of increased O_2 concentrations on the survival and hatching of the embryos of the bream *Abramis brama*. *J. Ichthyol. (Engl. Transl.)* **17**, 174–177.

Hamdorf, K. (1961). The influence of the environmental factors (O_2 partial pressure and temperature) on the embryonic and larval development of the rainbow trout (*Salmo irideus* Gibb). *Z. Vergl. Physiol.* **44**, 523–549.

Hamor, T., and Garside, E. T. (1975). Regulation of oxygen consumption by incident illumination in embryonated ova of Atlantic salmon *Salmo salar*. *Comp. Biochem. Physiol. A* **52A**, 277–280.

Hamor, T., and Garside, E. T. (1976). Development rates of embryos of Atlantic salmon, *Salmo salar* L., in response to various levels of temperature, dissolved oxygen and water exchange. *Can. J. Zool.* **54**, 1912–1917.

Hamor, T., and Garside, E. T. (1977). Size relations and yolk utilization in embryonated ova and alevins of Atlantic salmon *Salmo salar* in various combinations of temperature and dissolved oxygen. *Can. J. Zool.* **55**, 1892–1898.

Hamor, T., and Garside, E. T. (1979). Hourly and total oxygen consumption by ova of Atlantic salmon, *Salmo salar* L., during embryogenesis, at two temperatures and three levels of dissolved oxygen. *Can. J. Zool.* **57**, 1196–1200.

Hartvig, M., and Weber, R. E. (1984). Blood adaptations for maternal–fetal oxygen transfer in the viviparous teleost, *Zoarces viviparus* L. *In* "Respiration and Metabolism of Embryonic Vertebrates" (R. S. Seymour, ed.), pp. 17–30. Martinus Nijhoff/Dr. W. Junk Publishers, Dordrect, The Netherlands.

Hayes, F. R., Wilmot, I. R., and Livingstone, D. A. (1951). The oxygen consumption of the salmon egg in relation to development and activity. *J. Exp. Zool.* **116**, 377–395.

Hempel, G. (1979). "Early Life History of Marine Fish. The Egg Stage," Washington Sea Grant Publ. Univ. of Washington Press, Seattle.

Hempel, I., and Hempel, G. (1971). An estimate of mortality in eggs of North Sea herring (*Clupea harengus* L.). *Rapp. P.-V. Reun., Cons. Int. Explor. Mer* **160**, 24–26.

Hettler, W. F. (1976). Influence of temperature and salinity on routine metabolic rate and growth of young Atlantic menhaden. *J. Fish Biol.* **8**, 55–65.

Hinterleitner, S., Platzer, U., and Wieser, W. (1987). Development of the activities of oxidative, glycolytic, and muscle enzymes during early larval life in three families of freshwater fish. *J. Fish Biol.* **30**, 315–316.

Hishida, T. O., and Nakano, E. (1954). Respiratory metabolism during fish development. *Embryologia* **2**, 67–79.

Hitchman, M. L. (1978). "Measurement of Dissolved Oxygen." Wiley, New York.

Holeton, G. F. (1971). Respiratory and circulatory responses of rainbow trout larvae to carbon monoxide and to hypoxia. *J. Exp. Biol.* **55**, 683–694.

Holeton, G. F. (1973). Respiration of Arctic char *Salvelinus alpinus* from a high Arctic lake. *J. Fish. Res. Board Can.* **30**, 717–723.

Holliday, F. G. T., Blaxter, J. H. S., and Lasker, R. (1964). Oxygen uptake of developing eggs and larvae of the herring (*Clupea harengus*). *J. Mar. Biol. Assoc. U.K.* **44**, 711–723.

Hose, J. E., and Puffer, H. W. (1984). Oxygen consumption of grunion *Leuresthes tenuis* embryos exposed to the petroleum hydrocarbon benzo-a-pyrene. *Environ. Res.* **35**, 413–420.

Houde, E. D., and Schekter, R. C. (1983). Oxygen uptake and comparative energetics among eggs and larvae of three subtropical marine fishes. *Mar. Biol. (Berlin)* **72**, 283–293.

Hughes, G. M., Datta Munshi, J. S., and Ojha, J. (1986). Post-embryonic development of water and air-breathing organs of *Anabas testudineus* (Bloch). *J. Fish. Biol.* **29**, 443–450.

Hunter, J. R. (1972). Swimming and feeding behavior of larval anchovy *Engraulis mordax*. *Fish. Bull.* **70**, 821–838.

Hunter, J. R. (1981). Feeding ecology and predation of marine fish larvae. *In* "Marine Fish Larvae: Morphology, Ecology and Relation to Fisheries" (R. Lasker, ed.), Washington Sea Grant Program, pp. 33–79. Univ. of Washington Press, Seattle.

Hunter, J. R., and Kimbrell, C. A. (1980). Early life history of Pacific mackerel, *Scomber japonicus*. *Fish. Bull.* **78**, 89–102.

Hyman, L. H. (1921). The metabolic gradients of vertebrate embryos. I. Teleost embryos. *Biol. Bull. (Woods Hole, Mass.)* **40**, 32–72.

Ingermann, R. L., and Terwilliger, R. C. (1981a). Oxygen affinities of maternal and fetal hemoglobins of the viviparous sea perch *Embiotoca lateralis*. *J. Comp Physiol. B* **142B**, 523–532.

Ingermann, R. L., and Terwilliger, R. C. (1981b). Intraerythrocytic organic phosphates of fetal and adult sea perch *Embiotoca lateralis*. Their role in maternal–fetal oxygen transport. *J. Comp. Physiol. B* **144B**, 253–260.

Ingermann, R. L., and Terwilliger, R. C. (1982). Blood parameters and facilitation of maternal-fetal oxygen transfer in a viviparous fish (*Embiotoca lateralis*). *Comp. Biochem. Physiol.* **73A**, 497–502.

Ingermann, R. L., and Terwilliger, R. C. (1984). Facilitation of maternal-fetal oxygen transfer in fishes: Anatomical and molecular specialization. *In* "Respiration and Metabolism of Embryonic Vertebrates" (R. S. Seymour, ed.), pp. 1–15. Martinus Nijhoff/Dr. W. Junk Publishers, Dordrecht, The Netherlands.

Ingermann, R. L., Terwilliger, R. C., and Roberts, M. S. (1984). Foetal and adult blood oxygen affinities of the viviparous Seaperch, *Embiotoca lateralis*. *J. Exp. Biol.* **108**, 453–457.

Ishida, J. (1985). Hatching enzyme: Past, present and future. *Zool. Sci.* **2**, 1–10.

Itazawa, Y., Matsumoto, T., and Kanda, T. (1978). Group effects on physiological and ecological phenomena in fish. Part 1. Group effect on the oxygen consumption of the rainbow trout and the medaka. *Bull. Jap. Soc. Sci. Fish.* **44**, 965–970.

Iuchi, I. (1973a). The post-hatching transition of erythrocytes from larval to adult type in the rainbow trout, *Salmo gairdneri irideus*. *J. Exp. Zool.* **184**, 383–396.

Iuchi, I. (1973b). Chemical and physiological properties of the larval and the adult hemoglobins in rainbow trout, *Salmo gairdneri irideus*. *Comp. Biochem. Physiol. B* **44B**, 1087–1101.

Iuchi, I. (1985). Cellular and molecular bases of the larval-adult shift of hemoglobins in fish. *Zool. Sci.* **2**, 11–23.

Iuchi, I., and Yamagami, K. (1969). Electrophoretic pattern of larval hemoglobins of the salmonid fish, *Salmo gairdneri irideus*. *Comp. Biochem. Physiol.* **28**, 977–979.

Iuchi, I., and Yamamoto, M. (1983). Erythropoiesis in the developing rainbow trout, *Salmo gardneri irideus:* Histological and immunochemical detection of erythropoietic organs. *J. Exp. Zool.* **226**, 409–417.

Iuchi, I., Suzuki, R., and Yamagami, K. (1975). Ontogenetic expression of larval and adult hemoglobin phenotypes in the intergeneric salmonid hybrids. *J. Exp. Zool.* **192**, 57–64.

Ivlev, V. S. (1960a). "Active Metabolic Intensity in Salmon Fry (*Salmo salar* L.) at Various Rates of Activity," Pap. No. 213 (mimeo). Salmon and Trout Comm., Int. Counc. Explor. Sea, Copenhagen.

Ivlev, V. S. (1960b). On the utilization of food by planktonophage fishes. *Bull. Math. Biophys.* **22**, 371–389.

Iwai, T., and Hughes, G. M. (1977). Preliminary morphometric study on gill development in black sea bream (*Acanthopagrus schlegeli*). *Bull. Jpn. Soc. Sci. Fish.* **43**, 929–934.

Jensen, J. O. T., Schnute, J., and Alderdice, D. F. (1986). Assessing juvenile salmonid response to gas supersaturation using a general multivariate dose-response model. *Can. J. Fish. Aquat. Sci.* **43**, 1694–1709.

Jitariu, P., Badilita, M., and Costea, E. (1971). Oxygen consumption during early stages in some selected breeds of carp in comparison with a local nonselected breed. *Stud. Cercet. Biol., Ser. Zool.* **23**, 213–217.

Johns, D. M., and Howell, W. H. (1980). Yolk utilization in summer flounder (*Paralichthys dentatus*) embryos and larvae reared at two temperatures. *Mar. Ecol.: Prog. Ser.* **2**, 1–8.

Johnson, R. A. (1980). Oxygen transport in salmon spawning gravel. *Can. J. Fish. Aquat. Sci.* **37**, 155–162.

Johnston, I. A. (1982). Physiology of muscle in hatchery raised fish. *Comp. Biochem. Physiol. B* **73B**, 105–124.

Jones, D. R., and Randall, D. J. (1978). The respiratory and circulatory systems during exercise. *In* "Fish Physiology" (W. S. Hoar and D. J. Randall, eds.), Vol. 7, pp. 425–501. Academic Press, New York.

Jones, M. P., Holliday, F. G. T., and Dunn, A. E. G. (1966). The ultrastructure of the epidermis of larvae of herring (*Clupea harengus*) in relation to rearing salinity. *J. Mar. Biol. Assoc. U.K.* **46**, 235–239.

Kamler, E. (1972). Respiration of carp in relation to body size and temperature. *Pol. Arch. Hydrobiol.* **19**, 325–331.

Kamler, E. (1976). Variability of respiration and body composition during early developmental stages of carp. *Pol. Arch. Hydrobiol.* **23**, 431–485.

Kamler, E., Matlak, O., and Srokosz, K. (1974). Further observations on the effect of sodium salt of 2,4-D on early developmental stages of carp. *Cyprinus carpio. Pol. Arch. Hydrobiol.* **21**, 481–502.

Kanda, T., and Itazawa, Y. (1981). Group effect on oxygen consumption and growth of catfish eel *Plotosus anguillaris. Bull. Jap Soc. Sci. Fish.* **47**, 341–346.

Karpenko, G. I., and Proskurina, E. S. (1970). Oxygen consumption rates of larvae and fry of *Cyprinidae* in relation to their ecology. *Proc. All-Union Res. Inst. Mar. Fish Oceanog.*, 1970. **4**, 7–13. Translated from Russian by *Fish Res. Board Can., Transl. Ser.* **2109**, 9.

Kaur, K., and Toor, H. S. (1978). Effect of dissolved oxygen on the survival and hatching of eggs of scale carp. *Prog. Fish Cult.* **40**, 35–37.

Kaushik, S. J., and Dabrowski, K. (1983). Postprandial metabolic changes in larval and juvenile carp (*Cyprinus carpio*). *Reprod. Nutr. Dev.* **23**, 223–234.

Kaushik, S. J., Dabrowski, K., and Luquet, P. (1982). Patterns of nitrogen excretion and oxygen consumption during ontogenesis of common carp (*Cyprinus carpio*). *Can. J. Fish. Aquat. Sci.* **39**, 1095–1105.

Khakimullin, A. A. (1985). Levels of standard and basal metabolism in young Siberian sturgeon, *Acipenser baeri* (*Acipenseridae*). *J. Ichthyol.* (*Engl. Transl.*) **24**(5), 29–33.

King, D. P. F. (1977). Influence of temperature, dissolved oxygen and salinity on incubation and early larval development of the south-west African pilchard (*Sardinops ocellata*). *S. Afr. Sea Fish Branch Invest. Rep.* **114**, 1–35.

Kirsch, R., and Nonnotte, G. (1977). Cutaneous respiration in three freshwater teleosts. *Respir. Physiol.* **29**, 339–354.

Kjelson, M. A., and Johnson, G. A. (1976). Further observations of the feeding ecology of post larval pinfish *Cagodon rhomboides* and spot *Leiostomus xanthurus*. *Fish. Bull.* **74**, 423–432.

Klekowski, R. Z., Korde, B., and Kaniuvska-Prus, M. (1977). The effect of sodium salt of 2,4-D on oxygen consumption of *Misgurnus fossilis* during early embryonal development. *Pol. Arch. Hydrobiol.* **24**, 413–422.

Klekowski, R. Z., Opalinski, K. W., and Gorbunova, N. N. (1980). Respiratory metabolism in early ontogeny of 2 Pacific flying fish species *Hirundichthys marginatus* and *Cheilopogon unicolor* and of Pacific sardine *Sardinaps sagax sagax*. *Pol. Arch. Hydrobiol.* **27**, 537–548.

Knudsen, J., Famme, P., and Hansen, E. S. (1983). A microcalorimeter system for continuous determination of the effect of oxygen on aerobic-anaerobic metabolism and metabolite exchange in small aquatic animals and cell preparations. *Comp. Biochem. Physiol. A* **74A**, 63–66.

Kobel, H. R., and Wolff, J. (1983). Two transitions of haemoglobin expression in *Xenopus:* From embryonic to larval and from larval to adult. *Differentiation* **24**, 24–26.

Konchin, V. V. (1971). The rate of oxygen consumption in the early ontogeny of the summer bakhtak [*Salmo ischchan* (Kessler)] when placed in respirometers in groups and singly. *J. Ichthyol.* (*Engl. Transl.*) **11**(6), 916–926.

Konchin, V. V. (1981). Analysis of the rate of gas exchange in young roach, *Rutilus rutilus*. *J. Ichthyol.* (*Engl. Transl.*) **21**(4), 122–132.

Konchin, V. V. (1982). Some characteristics of gaseous interchange in larvae of three species of freshwater fish with differing ecology under conditions of isolation and crowding. *Vestn. Mosk. Univ., Ser. 16: Biol.* **37**(2), 43–49.

Konstantinov, A. S. (1980). Comparative evaluation of the intensity of respiration in fishes. *J. Ichthyol.* (*Engl. Transl.*) **20**(1), 98–104.

Korsgaard, B., and Andersen, F. Q. (1985). Embryonic nutrition, growth and energetics in *Zoarces viviparus* L. as indication of a maternal–fetal trophic relationship. *J. Comp. Physiol. B* **155B**, 437–444.

Korwin-Kossakowski, M., Jowko, G., and Jerierska, B. (1981). The influence of group effect on oxygen consumption of carp (*Cyprinus carpio* L.) larvae. *Rocz. Nauk Roln., Ser. H* **99**, 49–62.

Kryzanowsky, S. G. (1934). Die Atmungsorgane der Fishlarven. *Zool. Jahrb., Zool. Jahrb., Abt. Anat. Ontog. Tiere* **58**, 21–61.

Kudrinskaya, O. I. (1969). Metabolic rate in the larvae of pike-perch, perch, carp-bream and roach. *Hydrobiol. J.* **5**(4), 68–72.

Lanzing, W. J. R. (1976). A temporary respiratory organ in the tail of *Tilapia mossambica* fry. *Copeia*, pp. 800–802.

Lasker, R. (1962). Efficiency and rate of yolk utilization by developing embryos and larvae of the Pacific sardine *Sardinops caerula* (Girard). *J. Fish. Res. Board Can.* **19**, 867–875.

Lasker, R., and Theilacker, G. H. (1962). Oxygen consumption and osmoregulation by single Pacific sardine eggs and larvae (*Sardinops caerulea,* Girard). *J. Cons. Int. Explor. Mer* **27**, 25–33.

Laurence, G. C. (1969). The energy expenditure of largemouth bass larvae, *Micropterus salmoides,* during yolk absorption. *Trans. Am. Fish. Soc.* **98**, 398–405.

Laurence, G. C. (1973). Influence of temperature on energy utilization of embryonic and prolarval tautog (*Tautoga onitis*) *J. Fish Res. Board Can.* **30**, 435–442.

Laurence, G. C. (1975). Laboratory growth and metabolism of the winter flounder *Pseudopleuronectes americanus* from hatching through metamorphosis at three temperatures. *Mar. Biol. (Berlin)* **32**, 223–229.

Laurence, G. C. (1977). A bioenergetic model for the analysis of feeding and survival potential of winter flounder, *Pseudopleuronectes americanus*, larvae during the period from hatching to metamorphosis. *Fish. Bull.* **75**, 529–546.

Laurence, G. C. (1978). Comparative growth, respiration and delayed feeding abilities of larval cod (*Gadus morhua*) and haddock (*Melanogrammus aeglefinus*) as influenced by temperature during laboratory studies. *Mar. Biol.* (*Berlin*) **50**, 1–7.

Liem, K. F. (1981). Larvae of air breathing fishes as countercurrent flow devices in hypoxic environments. *Science* **211**, 1177–1179.

Lindroth, A. (1942). Sauerstoffverbrauch der Fishe. II. Verschiedene Entwicklungs— und Alterstadien von Lach und Hecht. Z. *Vergl. Physiol.* **29**, 583–594.

Loeffler, C. A. (1971). Water exchange in the pike egg. *J. Exp. Biol.* **55**, 797–811.

Loeffler, C. A., and Lovtrup, S. (1970). Water balance in the salmon egg. *J. Exp. Biol.* **52**, 291–298.

Lomholt, J. P., and Johansen, K. (1979). Hypoxia acclimation in carp–how it affects O_2 uptake, ventilation and O_2 extraction from water. *Physiol. Zool.* **52**, 38–49.

Lønning, S. (1972). Comparative electron microscopic studies of teleostean eggs with special reference to the chorion. *Sarsia* **49**, 41–48.

Lukina, O. V. (1973). Respiratory rate of the North Okhotsk chum salmon (*Oncorhynchus keta* (Walb)). *J. Ichthyol.* (*Engl. Transl.*) **13**(3), 425–430.

MacCrimmon, H. R., and Kwain, W. H. (1969). Influence of light on early development and meristic characters in the rainbow trout, *Salmo gairdneri* Richardson. *Can. J. Zool.* **47**, 631–637.

McDonald, D. G., and McMahon, B. R. (1977). Respiratory development in Artic char *Salvelinus alpinus* under conditions of normoxia and chronic hypoxia. *Can. J. Zool.* **55**, 1461–1467.

McElman, J. F., and Balon, E. K. (1979). Early ontogeny of walleye (*Stizostedion vitreum*) with steps of saltatory development. *Environ. Biol. Fishes* **4**, 309–348.

McNeil, W. J. (1966). Effect of the spawning bed environment on reproduction of pink and chum salmon. *Fish. Bull.* **65**, 495–523.

Malyukina, G. A., and Konchin, V. V. (1969). Development of group effects during ontogenetic development of the Baltic salmon and the Sevan trout. *Probl. Ichthyol.* **9**, 292–297.

Markiewicz, F. (1960). Development of blood vessels in branchial arches of the pike (*Esox lucius* L.). *Acta Biol. Cracov. Zool. Ser.* **111**, 163–172.

Marshall, S. M., Nicholls, A. G., and Orr, A. P. (1937). On the growth and feeding of larval and post-larval stages of the Clyde herring. *J. Mar. Biol. Assoc. U. K.* **22**, 245–267.

Mason, J. C. (1969). Hypoxial stress prior to emergence and competition among coho salmon fry. *J. Fish. Res. Board Can.* **26**, 63–91.

Mikulin, A. Ye., and Soin, S. G. (1975). The functional significance of carotenoids in the embryonic development of teleosts. *J. Ichthyol.* (*Engl. Transl.*) **15**, 749–759.

Miller, R. W. (1972). Three methods for determining dissolved oxygen concentrations near fish embryos. *Prog. Fish-Cult.* **34**, 39–42.

Mishra, A. P., and Singh, B. R. (1979). Oxygen uptake through water during early life of *Anabas testudineus. Hydrobiologia* **66**, 129–134.

Morgan, M. (1974a). The development of gill arches and gill blood vessels of the rainbow trout, *Salmo gairdneri. J. Morphol.* **142**, 351–364.

Morgan, M. (1974b). Development of secondary lamellae of the gills of the trout *Salmo gairdneri* (Richardson). *Cell Tissue Res.* **151**, 509–523.

Mukhamedova, A. F. (1977). The level of standard metabolism of young silver carp, *Hypophthalmichthys molitrix*. *J. Ichthyol.* (*Engl. transl.*) **17**, 292–299.

Nakano, E. (1953). Respiration during maturation and at fertilization of fish eggs. *Embryologia* **2**, 21–31.

Needham, J. (1931). "Chemical Embryology." Cambridge Univ. Press, London and New York.

Needham, J. (1942). "Biochemistry and Morphogenesis." Cambridge Univ. Press, London and New York.

O'Brien, R. N., Visaisouk, S., Raine, R., and Alderdice, D. F. (1978). Natural convection: A mechanism for transporting oxygen to incubating salmon eggs. *J. Fish. Res. Board Can.* **35**, 1316–1321.

Oikawa, S., and Itazawa, Y. (1985). Gill and body surface areas of the carp in relation to body mass, with special reference to the metabolism–size relationship. *J. Exp. Biol.* **117**, 1–14.

Oseid, D. M., and L. L., Smith, Jr. (1971a). Survival and hatching of walleye eggs at various dissolved oxygen levels. *Prog. Fish-Cult.* **33**, 81–85.

Oseid, D. M., and L. L., Smith, Jr. (1971b). Survival and hatching of white sucker eggs at various dissolved oxygen levels. *Prog. Fish-Cult.* **33**, 158–159.

Ott, M. E., Heisler, N. H., and Ultsch, G. R. (1980). A re-evaluation of the relationship between temperature and the critical oxygen tension in freshwater fishes. *Comp. Biochem. Physiol. A* **67A**, 337–340.

Paine, M. D., and Balon, E. K. (1984). Early development of the northern logperch, *Percina caprodes semifasciata*, according to the theory of saltatory ontogeny. *Environ. Biol. Fishes* **11**, 173–190.

Pamatmat, M. M. (1983). Simultaneous direct and indirect calorimetry. *In* "Polarographic Oxygen Sensors" (E. Gnaiger and H. Forstner, eds.), pp. 167–175. Springer-Verlag, Berlin and New York.

Pauly, D. (1981). The relationship between gill surface area and growth performance in fish: A generalization of von Bertalanffy's theory of growth. *Meeresforschung* **28**, 251–282.

Penaz, M., and Prokes, M. (1973). Oxygen consumption in the brown trout *Salmo trutta*. *Zool. Listy* **22**, 181–188.

Peterka, J. J., and Kent, J. S. (1976). "Dissolved Oxygen, Temperature and Survival of Young at Fish Spawning Sites," EPA/600/3-76-113. Environ. Res. Lab., Duluth, Minnesota.

Peterson, R. H. (1975). Pectoral fin and opercular movements of Atlantic salmon (*Salmo salar*) alevins. *J. Fish. Res. Board Can.* **32**, 643–648.

Peterson, R. H., and Martin-Robichaud, D. J. (1983). Embryo movements of Atlantic salmon (*Salmo salar*) as influenced by pH, temperature and state of development. *Can. J. Fish. Aquat. Sci.* **40**, 777–782.

Phillips, R. W., and Campbell, H. J. (1961). The embryonic survival of coho salmon and steelhead trout as influenced by some environmental conditions in gravel beds. *Pac. Mar. Fish. Comm. Rep.* **14**, 60–73.

Phillips, R. W., Campbell, H. J., Hug, W. L., and Claire, E. W. (1966). "A Study of the Effects of Logging on Aquatic Resources, 1960–1966," Prog. Memo. Fish. No. 3. Res. Div., Oregon State Game Comm., Portland.

Piiper, J. (1982). Respiratory gas exchange at lungs, gills and tissues: Mechanisms and adjustments. *J. Exp. Biol.* **100**, 5–22.

Piiper, J., and Scheid, P. (1984). Respiratory gas transport systems: Similarities between avian embryos and lungless salamanders. *In* "Respiration and Metabolism of Embryonic

Vertebrates" (R. S. Seymour, ed.), pp. 181–191. Martinus Nijhoff/Dr. W. Junk Publishers, Dordrecht, The Netherlands.

Pinder, A., and Burggren, W. (1983). Respiration during chronic hypoxia and hyperoxia in larval and adult bullfrogs (*Rana catesbeiana*). II. Changes in respiratory properties of whole blood. *J. Exp. Biol.* **105**, 205–213.

Potts, W. T. W., and Rudy, P. P., Jr. (1969). Water balance in the eggs of the Atlantic salmon *Salmo salar*. *J. Exp. Biol.* **50**, 223–237.

Prasad, M. S., and Prasad, P. (1984). Changes in the water–blood diffusion barrier at the secondary gill lamellae during early life in *Channa striatus* synonym *Ophicephalus striatus*. *Folia Morphol.* **32**, 200–208.

Prokes, M. (1973). The oxygen consumption and respiratory surface area of the gills of *Coregonus lavaretus* and of *Coregonus peled*. *Zool. Listy* **22**, 375–384.

Quantz, G., and Tandler, A. (1982). "On the Oxygen Consumption of Hatchery-reared Larvae of the Gilthead Seabream (*Sparus auratus* L.)," Pap. F:7. Maricult. Comm., Int. Counc. Explor. Sea, Copenhagen.

Quinn, D., and Burggren, W. (1983). Lactate production, tissue distribution, and elimination following exhaustive exercise in larval and adult bullfrogs *Rana catesbeiana*. *Physiol. Zool.* **56**, 597–613.

Randall, D. J. (1982). The control of respiration and circulation in fish during exercise and hypoxia. *J. Exp. Biol.* **100**, 275–288.

Randall, D. J., and Daxboeck, C. (1984). Oxygen and carbon dioxide transfer across fish gills. *In* "Fish Physiology" (W. S. Hoar and D. J. Randall, eds.), Vol. 10, pp. 263–314. Academic Press, New York.

Randall, D. J., Perry, S. F., and Heming, T. A. (1982). Gas transfer and acid/base regulation in salmonids. *Comp. Biochem. Physiol. B* **73B**, 93–103.

Reznichenko, P. N., Solovev, L. G., and Gulidov, M. V. (1977). Simulation of the process of oxygen inflow to the respiratory surfaces of fish embryos by polarographic method. *In* "Metabolism and Biochemistry of Fishes" (G. S. Karzinkin, ed.), pp. 237–248. Indian Natl. Sci. Doc. Cent., New Delhi.

Roberts, R. J., Bell, M., and Young, H. (1973). Studies on the skin of plaice (*Pleuronectes platessa* L.). II. The development of larval plaice skin. *J. Fish Biol.* **5**, 103–108.

Robertson, D. A. (1974). Developmental energetics of the southern pigfish (Teleostei: Congiopodidae). *N. Z. J. Mar. Freshwater Res.* **8**, 611–620.

Rombough, P. J. (1985). Initial egg weight, time to maximum alevin wet weight and optimal ponding times for chinook salmon (*Oncorhynchus tshawytscha*). *Can. J. Fish. Aquat. Sci.* **42**, 287–291.

Rombough, P. J. (1986). Mathematical model predicting the dissolved oxygen requirements of steelhead (*Salmo gairdneri*) embryos and alevins in hatchery incubators. *Aquaculture* **59**, 119–137.

Rombough, P. J. (1987) Growth, aerobic metabolism and dissolved oxygen requirements of embryos and alevins of the steelhead trout, *Salmo gairdneri*. *Can. J. Zool.* (in press).

Rosenthal, H., and Alderdice, D. F. (1976). Sublethal effects of environmental stressors, natural and pollutional, on marine fish eggs and larvae. *J. Fish. Res. Board Can.* **33**, 2047–2065.

Ryzhkov, L. P. (1965). The circadian rhythm of oxygen consumption by roe of Sevan trout (*Salmo ishchan*). *Vopr. Ikhtiol.* **5**(2), 378–380.

Ryzhkov, L. P. (1968). The rate of gaseous exchange in the eggs, larvae and fry of the Sevan trout kept in groups or isolated. *Probl. Ichthyol.* **8**, 89–96.

Saksena, V. P., and Joseph, E. B. (1972). Dissolved oxygen requirements of newly hatched larvae of the striped blenny *Chasmodes bosquianus,* the naked goby *Gobiosama bosci* and the skilletfish *Gobiesox strumosus. Chesapeake Sci.* **13**, 23–28.

Sawaya, P. (1942). The tail of a fish larva as a respiratory organ. *Nature (London)* **149**, 169–170.

Schnute, J., and Jensen, J. O. T. (1986). A general multivariate dose-response model. *Can. J. Fish. Aquat. Sci.* **43**, 1684–1693.

Sheel, M., and Singh, B. R. (1981). Oxygen uptake through water during early life of *Heteropneustes fossilis. Hydrobiologia* **78**, 81–86.

Shepard, M. P. (1955). Resistance and tolerance of young speckled trout (*Salvelinus fontinalis*) to oxygen lack, with special reference to low oxygen acclimation. *J. Fish. Res. Board Can.* **12**, 387–433.

Shumway, D. L., Warren, C. E., and Doudoroff, P. (1964). Influece of oxygen concentration and water movement on the growth of steelhead trout and coho salmon embryos. *Trans. Am. Fish. Soc.* **93**, 342–356.

Siefert, R. E., and Herman, L. J. (1977). Spawning success of the black crappie (*Pomoxis nigromaculatus*) at reduced dissolved oxygen concentration. *Trans. Am. Fish. Soc.* **106**, 376–379.

Siefert, R. E., and Spoor, W. A. (1974). Effects of reduced oxygen on embryos and larvae of the white sucker, coho salmon, brook trout and walleye. *In* "The Early Life History of Fish" (J. H. S. Blaxter, ed.), pp. 487–495. Springer-Verlag, Berlin and New York.

Siefert, R. E., Spoor, W. A., and Syrett, R. L. (1973). Effects of reduced oxygen concentrations on northern pike (*Esox lucius*) embryos and larvae. *J. Fish. Res. Board Can.* **30**, 849–852.

Siefert, R. E., Carlson, A. R., and Herman, L. J. (1974). Effects of reduced oxygen concentrations on the early life stages of mountain whitefish, smallmouth bass, and white bass. *Prog. Fish-Cult.* **36**, 186–190.

Silver, S. J., Warren, C. E., and Doudoroff, P. (1963). Dissolved oxygen requirements of developing steelhead trout and chinook Salmon embryos at different water velocities. *Trans. Am. Fish. Soc.* **92**, 327–343.

Singh, R. P., Prasad, M. S., Mishra, A. P., and Singh, B. R. (1982). Oxygen uptake through water during early life in *Channa punctatus* Pisces *Ophicephaliformes. Hydrobiologia* **87**, 211–216.

Smirnov, A. I. (1953). Development of respiratory vessels in the skin of salmonid embryo. *Zool. Zh.* **32**, 787–790.

Smith, S. (1947). Studies in the development of the rainbow trout (*Salmo irideus*). I. The heat reproduction and nitrogenous excretion. *J. Exp. Biol.* **23**, 357–378.

Smith, S. (1957). Early development and hatching. *In* "The Physiology of Fishes" (M. E. Brown, ed.), Vol. 1, pp. 323–359. Academic Press, New York.

Smith, S. (1958). Yolk utilization in fishes. *In* "Embryonic Nutrition" (D. Rudnick, ed.), pp. 33–53. Univ. of Chicago Press, Chicago, Illinois.

Soin, S. G. (1966). Development, types of structure and phylogenesis of the vascular system of the vitelline sac in fish embryos, performing respiratory function. *Zool. Zh.* **45**(9), 1382–1397.

Soin, S. G. (1967). Ecological–morphological data on the relationship between carotenoids and the process of embryonal respiration in fish. *In* "Metabolism and Biochemistry of Fishes" (G. S. Karzinkin, ed.), pp. 536–552. Indian Natl. Sci. Docu. Cent., New Delhi, 1977.

Solberg, T., and Tilseth, S. (1984). Growth, energy consumption and prey density requirements in first feeding larvae of cod (*Gadus morhua* L.). *In* "The Propogation of Cod *Gadus morhua* L." (E. Dahl, D. P. Danielssen, E. Moksness, and P. Solemdal, eds.), Part 1, pp. 145–166. Inst. Mar. Res. Flødevigen Biol. Stn., Arendal, Norway.

Solewski, W. (1949). The development of the blood vessels of the gills of the sea trout, *Salmo trutta,* L.. *Bull. Acad. Sci. Crac.* **11**, 121–144.

Spoor, W. A. (1977). Oxygen requirements of embryos and larvae of the largemouth bass *Micropterus salmoides*. *J. Fish Biol.* **11**, 77–86.
Spoor, W. A. (1984). Oxygen requirements of larvae of the smallmouth bass *Micropterus dolomieui*. *J. Fish. Biol.* **25**, 587–592.
Sprague, J. B. (1973). The ABC's of pollutant bioassay using fish. *ASTM Spec. Tech. Publ.* **528**, 6–30.
Steffensen, J. F., and Lomholt, J. P. (1985). Cutaneous oxygen uptake and its relation to skin blood perfusion and ambient salinity in the plaice, *Pleuronectes platessa*. *Comp. Biochem. Physiol. A* **81A**, 373–375.
Stehr, C. M., and Hawkes, J. W. (1979). The comparative ultrastructure of the egg membrane and associated pore structures in the starry flounder, *Platichthys stellatus* (Pallas) and pink salmon, *Oncorhynchus gorbuscha* (Walbaum). *Cell Tissue Res.* **202**, 347–356.
Stelzer, R., Rosenthal, H., and Siebers, D. (1971). Influence of 2,4-dinitrophenol on respiration and concentration of some metabolites in embryos of the herring *Clupea harengus*. *Mar. Biol. (Berlin)* **11**, 369–378.
Steven, O. M. (1949). Studies on animal carotenoids. Carotenoids in the reproductive cycle of the brown trout. *J. Exp. Biol.* **26**, 295–302.
Storozhk, N. G., and Smirnov, B. P. (1982). Effect of mercury on the respiration rate of larval pink salmon, *Oncorhynchus gorbuscha*. *J. Ichthyol. (Engl. transl.)* **22**(5), 168–171.
Sushko, B. S. (1982). Microelectrode studies of oxygen tension and transport in loach *Misgurnus fossilis* eggs in helium–oxygen and nitrogen–oxygen media. *Fiziol. Zh. (Kiev)* **28**, 593–597, *Can. Transl. Fish. Aquat. Sci.* No. 5211 (1986).
Sylvester, J. R., Nash, C. E., and Emberson, C. R. (1975). Salinity and oxygen tolerances of eggs and larvae of Hawaiian striped mullet *Mugil cephalus*. *J. Fish Biol.* **7**, 621–630.
Tamarin, A. E., and Komarova, N. P. (1972). Some data on the respiration of Terek River salmon in relation to the bio-techniques of salmon culture. *Fish Res. Board Can., Transl. Ser.* **2055**, 1–11.
Taylor, F. H. C. (1971). Variation in hatching success in Pacific herring (*Clupea pallasi*) eggs with water depth, temperature, salinity, egg mass thickness. *Rapp. P.-V. Réun., Cons. Int. Explor. Mer* **160**, 34–41.
Taylor, M. (1913). The development of *Synbranchus marmoratus*. *Q.J. Microsc. Sci.* **59**, 1–51.
Taylor, M. H., DiMichele, L., and Leach, G. J. (1977). Egg stranding in the life cycle of the mummichog, *Fundulus heteroclitus*. *Copeia*, pp. 397–399.
Terner, C. (1979). Metabolism and energy conversion during early development. *In* "Fish Physiology" (W. S. Hoar, D. J. Randall, and J. R. Brett, eds.), Vol. 8, pp. 261–278. Academic Press, New York.
Theilacker, G. H., and Dorsey, K. (1980). Larval fish diversity, a summary of laboratory and field research. *In* "Workshop on the Effects of Environmental Variation on the Survival of Larval Pelagic Fishes" (G. D. Sharp, ed.), 10C Workshop Rep. 28, pp. 105–142. UNESCO, New York.
Toetz, D. W. (1966). The change from endogenous to exogenous sources of energy in bluegill sunfish larvae. *Invest. Indiana Lakes Streams* **8**(4), 115–146.
Trifonova, A. N. (1937). La physiologie de la différenciation et de la croissance. I. L'équilibre Pasteur-Meyerhof dans le développement des poissons. *Acta Zool. (Stockholm)* **18**, 375–445.
Turner, J. L., and Farley, T. C. (1971). Effects of temperature, salinity, and dissolved oxygen on the survival of striped bass eggs and larvae. *Calif. Fish Game* **57**, 268–273.
Tveranger, B. (1986). Effect of pigment content in broodstock diet on subsequent fertilization rate, survival and growth rate of rainbow trout (*Salmo gairdneri*) offspring. *Aquaculture* **53**, 85–93.

Tytler, P., and Calow, P. (1985). "Fish Energetics: New Perspectives." p. 349. Croom Helm, London.

Ultsch, G. R. (1973). A theoretical and experimental investigation of the relationships between metabolic rate, body size and oxygen exchange capacity. *Respir. Physiol.* **18**, 143–160.

Ultsch, G. R., Boschung, H., and Ross, M. J. (1978). Metabolism, critical oxygen tension and habitat selection in darters (*Etheostoma*). *Ecology* **59**, 99–107.

Ultsch, G. R., Ott, M. L., and Heisler, N. (1980). Standard metabolic rate, critical oxygen tension, and aerobic scope for spontaneous activity of trout (*Salmo gairdneri*) and carp (*Cyprinus carpio*) in acidified water. *Comp. Biochem. Physiol. A* **67A**, 329–335.

Ultsch, G. R., Jackson, D. C., and Moalli, R. (1981). Metabolic oxygen conformity among lower vertebrates: The toadfish revisited. *J. Comp. Physiol.* **142**, 439–443.

Veith, W. J. (1979). Reproduction in the live bearing teleost *Clinus superciliosus*. *S. Afr. J. Zool.* **14**, 208–211.

Vernidub, M. F. (1966). Composition of the red and white blood cells of embryos of Atlantic salmon and Baltic salmon *Salmo salar* L. and changes in this composition during growth of the organism. *Proc. Murmansk Inst. Mar. Biol.* **12–16**, 139–162, *Fish. Res. Board Can., Transl. Ser.* **1353**.

Vogel, S. (1981). "Life in Moving Fluids. The Physical Biology of Flow." Princeton Univ. Press, Princeton, New Jersey.

Volodin, V. M. (1956). Embryonic development of the autumn Baltic herring and their oxygen requirements during the course of development. *Voprosy Ikhtiol.* 7, 123–133. *Fish. Res. Board Can. Transl. Ser.* **252**.

Voyer, R. A., and Hennekey, R. J. (1972). Effects of dissolved oxygen on the two life stages of the mummichog. *Prog. Fish-Cult.* **34**, 222–225.

Wangensteen, O. D. (1972). Gas exchange by a bird's embryo. *Respir. Physiol.* **14**, 64–74.

Webb, P. W., and Brett, J. R. (1972a). Respiratory adaptations of prenatal young in the ovary of two species of viviparous seaperch, *Rhacohilus vacca* and *Embioteca lateralis*. *J. Fish. Res. Board Can.* **29**, 1525–1542.

Webb, P. W., and Brett, J. R. (1972b). Oxygen consumption of embryos and parents and oxygen transfer characteristics within the ovary of 2 species of viviparous seaperch *Rhacohilus vacca* and *Embioteca lateralis*. *J. Fish. Res. Board Can.* **29**, 1543–1553.

Webb, P. W., and Weihs, D. (1986). Locomotor functional morphology of early life history stages of fishes. *Trans. Am. Fish. Sci.* **115**, 115–127.

Weber, R. E., and Hartvig, M. (1984). Specific fetal hemoglobin underlies the fetal-maternal shift in blood oxygen affinity in a viviparous teleost *Zoarces viviparus*. *Mol. Physiol.* **6**, 27–32.

Weibel, E. R. (1984). "The Pathway for Oxygen." Harvard Univ. Press, Cambridge, Massachusetts.

Weihs, D. (1980). Respiration and depth control as possible reasons for swimming of northern anchovy *Engraulis mordax* yolk sac larvae. *Fish. Bull.* **78**, 109–117.

Weihs, D. (1981). Swimming of yolk sac larval anchovy *Engraulis mordax* as a respiratory mechanism. *Rapp. P.-V. Reun., Cons. Int. Explor. Mer* **178**, 327.

West, N. H., and Burggren, W. W. (1982). Gill and lung ventilatory responses to steady-state aquatic hypoxia and hyperoxia in the bullfrog tadpole. *Respir. Physiol.* **47**, 165–176.

Whiting, H. P., and Bone, Q. (1980). Ciliary cells in the epidermis of the larval Australian dipnoan, *Neoceratodus*. *J. Linn. Soc. London, Zool.* **68**, 125–137.

Wickett, W. P. (1954). The oxygen supply to salmon eggs in spawning beds. *J. Fish. Res. Board Can.* **11**, 933–953.

Wickett, W. P. (1975). Mass transfer theory and the culture of fish eggs. *In* "Chemistry and Physics of Aqueous Solutions" (W. A. Adams, ed.), pp. 419–434. Electrochem. Soc., Princeton, New Jersey.

Wieser, W. (1984). A distinction must be made between the ontogeny and the phylogeny of metabolism in order to understand the mass exponent of energy metabolism. *Respir. Physiol.* **55**, 1–9.

Wieser, W. (1985). Developmental and metabolic constraints of the scope for activity in young rainbow trout (*Salmo gairdneri*). *J. Exp. Biol.* **118**, 133–142.

Wieser, W., and Forstner, H. (1986). Effects of temperature and size on the routine rate of oxygen consumption and on the relative scope for activity in larval cyprinids. *J. Comp. Physiol.* **156**, 791–796.

Wieser, W., Platzer, U., and Hinterleitner, S. (1985). Anaerobic and aerobic energy production of young rainbow trout (*Salmo gairdneri*) during and after bursts of activity. *J. Comp. Physiol. B* **155B**, 483–492.

Winberg, G. G. (1956). "Rate of Metabolism and Food Requirements of Fishes" Nauch. Tr. Belorussk Gos Univ. Imeni V. I. Lenina, Minsk (Fish. Res. Board Can., Transl. Ser. No. 194, 1960).

Winberg, G. G., and Khartova, L. E. (1953). Rate of metabolism in carp fry. *Dokl. Akad. Nauk SSSR* **89**, 1119–1122.

Winnicki, A. (1968). Respiration of the embryos of *Salmo trutta* L. and *Salmo gairdneri* Rich, in media differing in gaseous diffusion rate. *Pol. Arch. Hydrobiol.* **15**, 23–28.

Witzel, L. D., and MacCrimmon, H. R. (1981). Role of gravel substrate on ova survival and alevin emergence of rainbow trout *Salmo gairdneri*. *Can. J. Zool.* **59**, 629–636.

Wood, A. H. (1932). The effect of temperature on the growth and respiration of fish embryos (*Salmo fario*). *J. Exp. Biol.* **9**, 271–276.

Wu, H. W., and Liu, C. K. (1942). On the breeding habits and the larval metamorphosis of *Monopterus javanensis*. *Sinensia* **13**, 1–13.

Yamagami, K. (1981). Mechanisms of hatching in fish: Secretion of hatching enzyme and enzymatic choriolysis. *Amer. Zool.* **21**, 459–471.

Yamagami, K., Yamamoto, M., Iuchi, I., and Taguchi, S. (1984). Retardation of maturation associated and secretion associated ultrastructural changes of hatching gland in the medaka embryos incubated in air. *Annot. Zool. Jap.* **56**, 266–274.

Yoshimizu, M., Kimura, T., and Sakai, M. (1980). Microflora of the embryo and fry of salmonids. *Bull. Jpn. Soc. Sci. Fish.* **46**, 967–975.

Zeuthen, E. (1950). Cartesian diver respirometer. *Biol. Bull.* (*Woods Hole, Mass.*) **98**, 303–318.

Zeuthen, E. (1970). Rate of living as related to body size in organisms. *Pol. Arch. Hydrobiol.* **17**, 21–30.

Zoran, M. J., and Ward, J. A. (1983). Parental egg care, behavior and farming activity for the orange chromide *Etroplus maculatus*. *Environ. Biol. Fishes* **8**, 301–310.

Chapter 17

Insights into the biophysical properties of electrogenesis and electroreception

Duncan Leitch*
Department of Integrative Biology and Physiology, University of California, Los Angeles, CA, United States
*Corresponding author: e-mail: dleitch@ucla.edu

Duncan Leitch discusses the M.V.L. Bennett chapter "Electric Organs" in *Fish Physiology*, Volume 5, published in 1971. The next chapter in this volume is the re-published version of that chapter.

Electrosensory systems, predominantly studied in aquatic species, enable the detection and generation of minuscule biogenic electric fields. Over 400 species, in South America and Africa, have independently evolved the ability emit specific electric signals to enhance communication, predation, and navigation. Despite research into electroreception commencing in the 1960s, a complete understanding of its mechanisms is yet to be fully deciphered. Through the use of African knifefish, the generation of electrical organ discharges, essential for "active electrolocation," was demonstrated. Detailed investigations by Michael Bennett and colleagues of the function of the electric organs and associated neural networks have illustrated their significance in sensory and evolutionary biology. In an intriguing display of convergent evolution, electric organs have independently emerged six times across vertebrate phylogeny, highlighting functional, genetic, and molecular similarities. Researching the neural control of the modulation of electric signals has revealed fundamental information on how the vertebrate nervous system detects salient external signals to produce diverse motor responses, with particular utility coming from study of the "jamming avoidance response" in weakly electric fish. Investigations into the molecular mechanisms initiating electrosensory transduction and the basis of diversity in electric organ signals are exciting testaments to the dynamic nature of scientific investigation in the field of electroreception.

Weak electric fields are ubiquitous throughout aquatic environments, generated both by inanimate geophysical sources or by animals (Keller, 2004). Fundamental physiological properties including respiration and cardiac function create fluctuating or uneven distribution of ions across permeable surfaces. Remarkably, a diverse range of vertebrates have neural adaptations to detect

Fish Physiology, Vol. 40A. https://doi.org/10.1016/bs.fp.2023.08.005

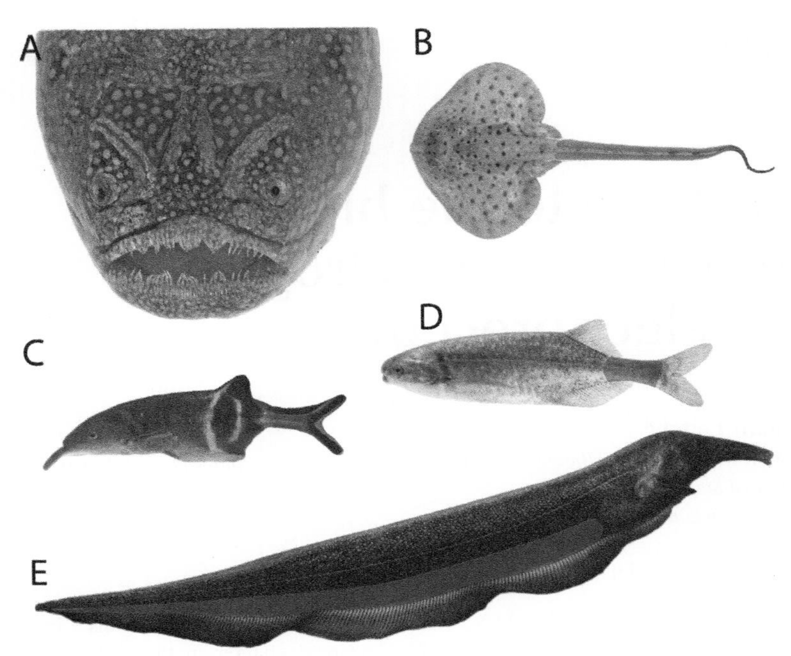

FIG. 1 Diversity of electric organs among electrogenic fishes. (A) The marine stargazer (*Astroscopus y-graecum*) has extraocular electrogenic organs (red shading); (B) The marine elasmobranch little skate (*Leucoraja erinacea*) has electric organs along the length of the tail; The African mormyrids elephant-nose fish (*Gnathonemus petersii*) (C) and *Brienomyrus brachyistius* (D), both pulse type emitting fish, have electric organs near the tail; (E) the South American electric knife fish (*Sternarchorhynchus retzeri*), a wave type emitting fish, has an electric organ on the ventrum along most of its length. *Photographs in (A and E) are used under a Creative Commons license (CC BY-ND 2.0). Photographs in (B–D) are by D.B. Leitch.*

these miniscule biogenic electric fields, often at the milli- to microvolt range, through their electrosensory system. Equally diverse is the range of fishes that have independently evolved the ability to produce specific electrical signals to facilitate communication, predation, and navigation (Fig. 1), including more than 200 South American species and 200 African species. This seemingly exotic sensory modality—electroreception—has captured the imagination of both scientists and the general public alike, yet fundamental understanding of the mechanisms by which the nervous systems produces and detects important electric signals has only been a topic of research since the early 1960s and 1970s (Phillips, 1988).

Despite historical identification of the "torporific" qualities from electric rays (*Torpedo*) (and their apparent utility in treating a wide range of human ailments) (Finger and Piccolino, 2011) to Alexander von Humboldt's striking descriptions of electric eels (*Electrophorus electricus*) defensively leaping from the water to shock horses (v. Humboldt, 1807), electroreception and the related study of the production of dedicated electrical signals (electrogenesis) have remained a unique scientific puzzle. In his *On the Origin of Species*, Charles Darwin reflected on the confusing, diminutive muscular anatomy of weakly electric fishes (African mormyrids and South American gymnotiforms, Fig. 1C–E) in comparison to the strongly electric eels that possessed large, elaborate electrogenic muscles, and considered these themes a challenge to his theory of natural selection (Darwin, 1859). Further progress in understanding the function and, later, neural mechanisms underlying electroreception remained elusive until almost a century later. British zoologist Hans Lissman demonstrated that African knifefish indeed produce a dedicated electrical signal (electric organ discharge; EOD) that could be readily amplified, visualized, and recorded (Lissmann, 1951). In a series of behavioral experiments, Lissman and colleagues recorded these EODs from a variety of mormyrid and gymnotiforms (Lissmann, 1958; Lissmann and Machin, 1958), noting their presence along with the movement of the fish as it navigated around objects of varying impedance, ultimately concluding that these weak electric signals mediate "active electrolocation." Presumably, the fish could electrically detect nearby objects based on their conductivity as they interfered with their self-generated EODs; consequently, these observations suggested that the fishes had specialized sensory systems adapted to the detection of these weak electric fields.

In this classic chapter (Bennett, 1971a), Michael Bennett surveyed the then current state of knowledge of the electrical signal producing organs and the neural networks related to their production. These discussions complement a subsequent chapter that focuses on the diversity of electroreceptors across fish taxa (Bennett, 1971b).

Although electric organs and electroreceptors have remained an important focus of study within the neurosciences, with notable textbooks covering their significance within sensory biology (including audition) (Bullock et al., 2005; Carlson et al., 2019) and evolutionary biology (Phillips, 1988), Bennett's thorough comparative discussion remains unique. More than a generalized "review," Bennett's wide-ranging discussion includes unpublished data related to his then recent experiences recording electric organ function during a series of unique research opportunities.

In particular, Bennett and colleagues (including the late Ted Bullock, widely considered one of the seminal figures in the history of neuroethology (Bullock, 1990; Zupanc and Zupanc, 2008)—the study of the neural basis of behavior) had the unique opportunity to conduct research on the Research Vessel (R.V.) Alpha Helix. Based at the Scripps Institution of Oceanography at the University of California in San Diego, with funding from the US National Science Foundation, the vessel was custom designed to host teams of experimental marine researchers with dedicate lab space (Day, 2009). Between 1966 and 1967, the R.V. Alpha Helix explored the Great Barrier Reef and Amazon River, with particular expeditions to the Rio Negro River system. In other realms of fish physiology, expeditions of the R.V. Alpha

Helix have become legendary, including pioneering studies related to the osmoregulatory and respiratory adaptations of diverse Amazonian fishes (Taylor and MacKinlay, 2002).

Bennett's experiences, particularly with the weakly electric gymnotiforms of the genus Sternarchus (the knifefishes) clearly influenced his discussion within the current chapter as well as subsequent further anatomical (Kristol et al., 1977; Waxman et al., 1972) and functional investigations of their electromotor systems. Some of the most notable aspects of the chapter include the detailed morphological and functional keys for distinguishing among the many gymnotid species and genera, with a similar key provided for the mormyrids (see Tables I and II in Bennett, 1971a). Indeed, the postscripts to Table II reveal personal research experience with the systematics of these fishes, describing a specific genus (*Sternarchorhynchus*) "considered to be monospecific, [although] two apparent species ... were found on the Rio Negro expedition of the Alpha Helix." These details bring a sense of discovery and adventure to the burgeoning field of understanding the natural history and physiology of electric fishes and underscore the dynamic nature of scientific investigation.

Underscoring the importance of these natural observations of biodiversity, subtle differences in EODs have remained an important tool in the systematics and identification of new species of electric fishes (Albert, 2001; Sullivan et al., 2000). Indeed, researchers have often remarked on the elegant simplicity of acoustically monitoring the amplified signals of these fishes' EODs, when transduced from an electric signal to an audible sound, from the hum of wave type gymnotids to the bursts of clicks from mormyrids, and the ability to "hear" a new species of weakly electric fish before ever seeing it.

Other notable passages from the chapter arise from Bennett's prescient predictions about the potential significance of the study of the electric organs. Beyond the domain of fish sciences, in the decades following the publishing of this chapter, Bennett established himself as a figure of stature within the neurosciences, as an expert in gap junctions—a particular form of an electrical synapse (Bennett, 1997, 2000). As specialized sites of neuronal interaction and communication, synapses have inspired heated historical debate surrounding their electrical or chemical basis. With his experience studying both electrical AND chemical nature of synapses using preparations derived from electric fishes, Bennett and his colleagues established many basic principles now widely appreciated within at the intersection of neurophysiology and biophysics including the chemical modulation of electrical synapses (Bennett et al., 1991, 2003).

The evolution of novelty is fundamental to understanding adaptation and the diversification of species, yet there are few examples of parallel innovation that are as clear as those posed by electric organs, which have independently emerged six times across vertebrate phylogeny. Although they are only distantly related, African mormyrids and South American gymnotiforms independently gained the ability to generate EODs via modified muscle systems, namely through the development of specialized "electrocyte" cells from myocytes (Crampton, 2019; Crampton et al., 2013). Specifically, mormyrids and gymnotiforms have co-opted homologs of a voltage-gated Na^+ channels, which typically facilitates transmembrane Na^+ currents in skeletal muscle, for functions in electrogenesis (Zakon et al., 2006). Following the whole genome duplication of fishes estimated to have occurred 226–316 million years ago, teleosts possessed two paralogs to the gene *Scn4a* (*Scn4aa* and *Scn4ab*), which encode alpha-subunits of the $Na_v1.4a$ and $Na_v1.4b$ channels. In electrocytes of electric organs of both mormyrids and gymnotiforms (despite 100 million years of independent trajectories of evolution), *Scn4aa* has shown particularly rapid sequence evolution, notably with substitutions of charged amino acids in key regions associated with voltage detection (Arnegard et al., 2010). More recent molecular genetic studies, which have examined both the genome of the electric eel as well as other weakly electric fishes have identified conserved patterns of differential gene expression in comparisons between myogenic skeletal muscle and electric organ electrocytes (Gallant et al., 2014). Namely, among electric fishes, transcription factors associated with skeletal "myogenic" profiles were downregulated whereas regulators of increased excitability, enhanced insulation, and signaling pathways associated with increased cell size were overexpressed in electrocytes. These remarkable findings underscore both the functional, genetic, and molecular similarities in the convergent evolution of electric organs.

Another fundamental theme that Bennett presciently highlighted to as an area for further research relates to neural control of the modulation of the EOD. He wrote, "the pathways of afferents to the command nuclei in variable frequency and intermittently discharging forms are by and large unknown ... the slight modifications of frequency in some high and constant frequency species also require investigation." Less than a decade prior, Watanabe and Takeda noted that providing weak electrical stimulation to a gymnotiform fish (*Eigenmannia*) caused a small shift in the fish's own EOD (Watanabe and Takeda, 1963), apparently representing an attempt to avoid "jamming" of the electrosensory signal. In other words, when two weakly electric fish producing EODs of similar frequency come within proximity of one another, the EOD of one fish shifts slightly higher than the original frequency while the EOD of the other shifts slightly lower. This phenomena was termed the "jamming avoidance response" (JAR) by Bullock (1969) and the basic hypothetical neural components of the circuit were described soon after by Bullock et al. (1972). In a landmark paper that further defined the deep convergent evolution of electrogenesis and electroreception, Walter Heiligenberg observed the JAR in the distantly-related African knifefish *Gynmnarchus niloticus*, showing that this behavior is not unique to only South American mormyrids but present among other taxa that independently evolved electrogenesis (Heiligenberg, 1975). It is difficult to overstate the significance of the study of the JAR of weakly electric fishes, with active research continuing to this day (Katz and Harris-Warrick, 1999; Metzner, 1999). With dozens of international labs studying the underlying circuitry producing these behavioral responses and their modulation, the JAR has become one of the first complex behaviours for which the entirety of its circuit has been delineated—from initial sensory transduction at specialized electroreceptors to multiple overlapping areas of sensory processing in the CNS to modulation of the motor output from the brainstem and cerebellum to electromotor neurons of the spinal cord. Bennett was certainly correct in predicting that more work on understanding the modulation of electrogenic signals would prove fruitful.

Apart from reading Bennett's work as an undergraduate student, my overlapping interests with the themes of this chapter arose through recent investigations of the molecular mechanisms initiating electrosensory transduction in elasmobranchs. This chapter along with the corresponding chapter on the electroreceptors provided an exceptional introduction to their respective fields, with discussions ranging from evolution to physiology to the current state of knowledge of the biophysical basis of these phenomena. Undoubtedly, Bennett's review of electric organs shaped our understanding of how to approach the study of the passive electroreceptors from the chain catshark (*Scyliorhinus rotifer*) in comparison to the tuning of receptors from the little skate (*Leucoraja erinacea*), which produces low frequency EODs (Bellono et al., 2017, 2018). Using a range of techniques from *de novo* transcriptomics, novel cellular physiology, and behavioral assays, we found that structural properties of discrete ion channels expressed in receptors (e.g., voltage gated sodium channels and potassium channels) appeared to selectively tune the sensitivity of sensory cells within electroreceptor organs to EOD-specific frequencies. The continued discussion that Bennett (1971a) provokes, cited more than 600 times at last count, even in the face of a sea change in methodology used in the study of electrogenesis, highlights its nature as a true classic in electroreception, and in the broader sensory neurobiological literature (Bennett, 1971b).

References

Albert, J.S., 2001. Species Diversity and Phylogenetic Systematics of American Knifefishes (Gymnotiformes, Teleostei). Museum of Zoology, University of Michigan.

Arnegard, M.E., Zwickl, D.J., Lu, Y., Zakon, H.H., 2010. Old gene duplication facilitates origin and diversification of an innovative communication system—twice. Proc. Natl. Acad. Sci. U. S. A. 107 (51), 22172–22177.

Bellono, N.W., Leitch, D.B., Julius, D., 2017. Molecular basis of ancestral vertebrate electroreception. Nature 543 (7645), 391–396.

Bellono, N.W., Leitch, D.B., Julius, D., 2018. Molecular tuning of electroreception in sharks and skates. Nature 558 (7708), 122–126.

Bennett, M., 1971a. Electric organs. Fish Physiol 5, 347–491. Elsevier.

Bennett, M., 1971b. Electroreception. Fish Physiol 5, 493–574. Elsevier.

Bennett, M.V.L., 1997. Gap junctions as electrical synapses. J. Neurocytol. 26 (6), 349–366.

Bennett, M.V., 2000. Electrical synapses, a personal perspective (or history). Brain Res. Rev. 32 (1), 16–28.

Bennett, M.V., Barrio, L., Bargiello, T., Spray, D., Hertzberg, E., Saez, J., 1991. Gap junctions: new tools, new answers, new questions. Neuron 6 (3), 305–320.

Bennett, M.V., Contreras, J.E., Bukauskas, F.F., Sáez, J.C., 2003. New roles for astrocytes: gap junction hemichannels have something to communicate. Trends Neurosci. 26 (11), 610–617.

Bullock, T.H., 1969. Species differences in effect of electroreceptor input on electric organ pacemakers and other aspects of behavior in electric fish; pp. 85–101. Brain Behav. Evol. 2 (2), 85–101.

Bullock, T.H., 1990. Goals of neuroethology. Bioscience 40 (4), 244–248.

Bullock, T.H., Hamstra, R.H., Scheich, H., 1972. The jamming avoidance response of high frequency electric fish. J. Comp. Physiol. 77 (1), 1–22.

Bullock, T.H., Fay, R.R., Hopkins, C.D., Popper, A.N., 2005. Electroreception. Springer Handbook of Auditory Research, Springer.

Carlson, B.A., Sisneros, J.A., Popper, A.N., Fay, R.R., 2019. Electroreception: Fundamental Insights from Comparative Approaches. Springer.

Crampton, W.G., 2019. Electroreception, electrogenesis and electric signal evolution. J. Fish Biol. 95 (1), 92–134.

Crampton, W.G.R., Rodriguez-Cattaneo, A., Lovejoy, N.R., Caputi, A.A., 2013. Proximate and ultimate causes of signal diversity in the electric fish Gymnotus. J. Exp. Biol. 216 (13), 2523–2541.

Darwin, C., 1859. On the Origin of Species. J. Murray, London.

Day, D., 2009. Alpha Helix Program Administrative History. Scripps Institution of Oceanography Archives.

Finger, S., Piccolino, M., 2011. The Shocking History of Electric Fishes: From Ancient Epochs to the Birth of Modern Neurophysiology. Oxford University Press.

Gallant, J.R., Traeger, L.L., Volkening, J.D., Moffett, H., Chen, P.-H., Novina, C.D., Phillips Jr., G.N., Anand, R., Wells, G.B., Pinch, M., 2014. Genomic basis for the convergent evolution of electric organs. Science 344 (6191), 1522–1525.

Heiligenberg, W., 1975. Electrolocation and jamming avoidance in the electric fish *Gymnarchus niloticus* (Gymnarchidae, Mormyriformes). J. Comp. Physiol. 103 (1), 55–67.

Katz, P.S., Harris-Warrick, R.M., 1999. The evolution of neuronal circuits underlying species-specific behavior. Curr. Opin. Neurobiol. 9 (5), 628–633.

Keller, C.H., 2004. Electroreception: strategies for separation of signals from noise. In: The Senses of Fish: Adaptations for the Reception of Natural Stimuli. Springer Netherlands, pp. 330–361.

Kristol, C., Akert, K., Sandri, C., Wyss, U.R., Bennett, M.V., Moor, H., 1977. The Ranvier nodes in the neurogenic electric organ of the knifefish Sternarchus: a freeze-etching study on the distribution of membrane-associated particles. Brain Res. 125 (2), 197–212.

Lissmann, H.W., 1951. Continuous electrical signals from the Tail of a Fish, *Gymnarchus niloticus* Cuv. Nature 167, 201–202.

Lissmann, H.W., 1958. On the function and evolution of electric organs in fish. J. Exp. Biol. 35 (1), 156–191.

Lissmann, H.W., Machin, K.E., 1958. The mechanism of object location in *Gymnarchus niloticus* and similar fish. J. Exp. Biol. 35 (2), 451–486.

Metzner, W., 1999. Neural circuitry for communication and jamming avoidance in gymnotiform electric fish. J. Exp. Biol. 202 (10), 1365–1375.

Phillips, R.E., 1988. Electroreception. In: Bullock, T.H., Heiligenberg, W. (Eds.), Wiley Series in Neurobiology. John Wiley & Sons, New York. 1986, Wiley Online Library.

Sullivan, J.P., Lavoue, S., Hopkins, C.D., 2000. Molecular systematics of the African electric fishes (Mormyroidea: Teleostei) and a model for the evolution of their electric organs. J. Exp. Biol. 203 (4), 665–683.

Taylor, T., MacKinlay, D., 2002. Physiological mechanisms: a tribute to Dave Randall. v. Humboldt, A. In: Proceedings of the International Congress on the Biology of Fish.

v. Humboldt, A., 1807. Jagd und kampf der electrischen aale mit pferden. Aus den reiseberichten des Hrn. Freiherrn Alexander v. Humboldt. Ann. Phys. 25 (1), 34–43.

Watanabe, A., Takeda, K., 1963. The change of discharge frequency by AC stimulus in a weak electric fish. J. Exp. Biol. 40 (1), 57–66.

Waxman, S., Pappas, G., Bennett, M., 1972. Morphological correlates of functional differentiation of nodes of Ranvier along single fibers in the neurogenic electric organ of the knife fish Sternarchus. J. Cell Biol. 53 (1), 210–224.

Zakon, H.H., Lu, Y., Zwickl, D.J., Hillis, D.M., 2006. Sodium channel genes and the evolution of diversity in communication signals of electric fishes: convergent molecular evolution. Proc. Natl. Acad. Sci. 103 (10), 3675–3680.

Zupanc, G., Zupanc, M., 2008. Theodore H. Bullock: pioneer of integrative and comparative neurobiology. J. Comp. Physiol. A. 194, 119–134.

Chapter 18

ELECTRIC ORGANS☆

M.V.L. BENNETT

Chapter Outline

I. INTRODUCTION

Electric organs are organs specialized for the production of an electric field outside the body. They are found only in fish but apparently have evolved independently in some six different groups (Fig. 1 and Table I). Electric fish are conveniently divided into two types, strongly and weakly electric. The discharge of a strongly electric fish is so large that the fish is painful to handle; the electric organ presumably or demonstrably functions as a weapon either defensively against predators or offensively in securing food. A weakly electric fish produces potentials that are too small to have value offensively or defensively; their organs function (at least in freshwater species) as part of an electrosensory system. The fish detects objects by means of the distortions they cause in the field set up by the electric organ. The sensing elements

☆This is a reproduction of a previously published chapter in the Fish Physiology series, "1971 (Vol. 5)/Electric Organs: ISBN: 978-0-12-350405-0; ISSN: 1546-5098".

Fish Physiology, Vol. 40A. https://doi.org/10.1016/bs.fp.2023.09.004

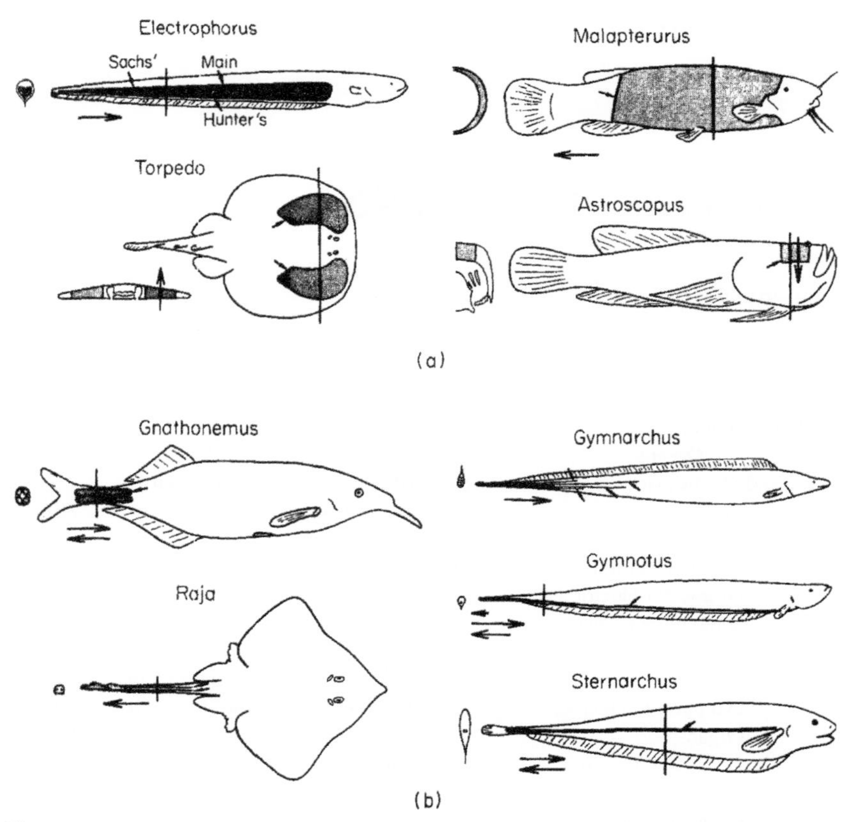

Fig. 1. Representative electric fish: (a) strongly electric and (b) weakly electric. All are shown from the side except *Torpedo* and *Raja*, which are shown from the top. Electric organs are stippled or solid and indicated by small arrows. Cross-sectional views through the organs are also shown at the levels indicated by the solid lines. The large arrows indicate the direction of active current flow through the organs; more than one arrow indicates successive phases of activity and the relative lengths are proportional to relative amplitudes of the phases. *Gnathonemus* is a mormyrid; other groups can be determined from family names in Table I. From Bennett (1968a).

are the electroreceptors which are described in the following chapter. Electrosensory systems can function in communication between fish. Some strongly electric fish also have weakly electric organs.

Strongly electric fish must have been known to primitive man, and *Torpedos* are said to have been used by the Romans in a primitive, and probably subconvulsant, form of shock therapy (Kellaway, 1946). Recognition that the discharge was electric came soon after the development of the Leyden jar allowed widespread study of electricity by scientists and experience of shocks. The *Torpedo* and electric eel were studied by many early physiologists, but the stargazer escaped the attention of the scientific community until just before Dahlgren and Silvester described its organ in 1906.

Table I Groups of Electric Fish[a]

Common name	Family	Genera and species	Strength of organ discharge	Distribution
Skates, ordinary rays	Rajidae	*Raja,* many species, a number of other genera not known to be electric	Weak	Marine, cosmopolitan
Electric rays, torpedos	Torpedinidae	A number of genera, many species	Strong, up to 60 V or 1 kW, some perhaps weak[b]	Marine, cosmopolitan
Mormyrids, elephant-nosed fish (many lack enlarged chin or snout)	Mormyridae	A number of genera, several with many species	Weak	Freshwater, Africa
Gymnarchus	Gymnarchidae	1 species, *G. niloticus*	Weak	Freshwater, Africa
Gymnotid eels; electric eel and knifefish	Electrophoridae	1 species, *Electrophorus electricus*	Strong, more than 500 V[c]	Freshwater, South America
	Gymnotidae	1 species, *Gymnotus carapo*	Weak	Freshwater, South America
	Sternopygidae	4 or 5 genera, a number of species	Weak	Freshwater, South America
	Rhamphichthyidae	2 monospecific genera	Weak	Freshwater, South America
	Sternarchidae	About 9 genera, a number of species	Weak	Freshwater, South America
Electric catfish	Malapteruridae	1 species, *Malapterurus electricus*	Strong, more than 300 V[d]	
Stargazers	Uranoscopidae	1 electric genus, *Astroscopus;* several species	Strong? about 5 V in air from small animals[e]	Marine, Western Atlantic

[a]Arranged phylogenetically; see Greenwood *et al.* (1966) for more detailed classifications. From Bennett (1970).
[b]From Bennett *et al.* (1961).
[c]From Albe-Fessard (1950a).
[d]From Remmler (1930).
[e]From Bennett and Grundfest (1961c).

Recognition of weakly electric fish was relatively delayed because of the imperceptibility of their discharges, and the electric activity of many was unknown until the 1950's (Lissmann, 1951, 1958; Coates *et al.*, 1954; Grundfest, 1957). However, all the groups had been recognized from morphological evidence by the early 1900's although their organs were sometimes classified as "pseudoelectric." Historical references are given in the earlier reviews by Grundfest (1957) and Keynes (1957), and the historical essays of Kellaway (1946) and Mauro (1969) are charming. Relatively brief reviews of recent work are given in Bennett (1970) and Grundfest (1967).

Electric fish have been studied in part because of their remarkable physiological abilities and the ease of detection of their discharges. Also, it has been hoped that they might reveal a great deal about normal function. The argument is that evolution may have exaggerated some aspect of organ or cell that makes a general phenomenon more understandable or easier to study. The giant axon of the squid is an outstanding example of a cell that has evolved in such a way—toward increased size—that makes feasible many kinds of experiments that are much more difficult in other tissues (Hodgkin, 1964).

The generating cells of electric organs are modified from muscle fibers except in the sternarchid family of the larger group, the gymnotids. In the sternarchids the myogenic part of the organ has been lost and the organ is modified from nerve fibers; that is, it is neurogenic. The evidence of origin will be discussed in Section II,G,l,f. In a number of species the generating cells are flattened and for this reason have been termed electroplaque(s), electroplate(s), or electroplax(es). However, in many of the more recently described organs, the generating cells have quite complex shapes. Although electroplaque still seems a natural term for flattened cells, the author has introduced the more general term electrocyte to refer to any generating cell whatever its shape (Bennett, 1970). Electric organs generally are rather gelatinous, and a large fraction of their volume is extracellular space. They contain a considerable amount of connective and other accessory tissues as well as blood vessels and motor nerves that control the discharge. As will be seen below the connective tissue can be important in channeling the flow of current.

Electrocytes work on the same general principles as ordinary nerve and muscle cells: potentials are generated across membranes. In all known cases the potentials result from selective permeability and passive movement of ions down their concentration gradients. But it is likely that some electrocytes will be found to have electrogenic pumping [movement of ions that involves net current flow and is linked to reactions such as adenosine triphosphate (ATP) hydrolysis (see Hodgkin, 1964; Albers, 1967)]. When the membranes on opposite faces of the generating cells are at the same potential that at rest is the resting potential, no current flows (Fig. 2A). When the membranes are at different potentials, current flows in a circuit that involves the two membranes, the cell cytoplasm, and the external medium (Fig. 2B). Figure 2C

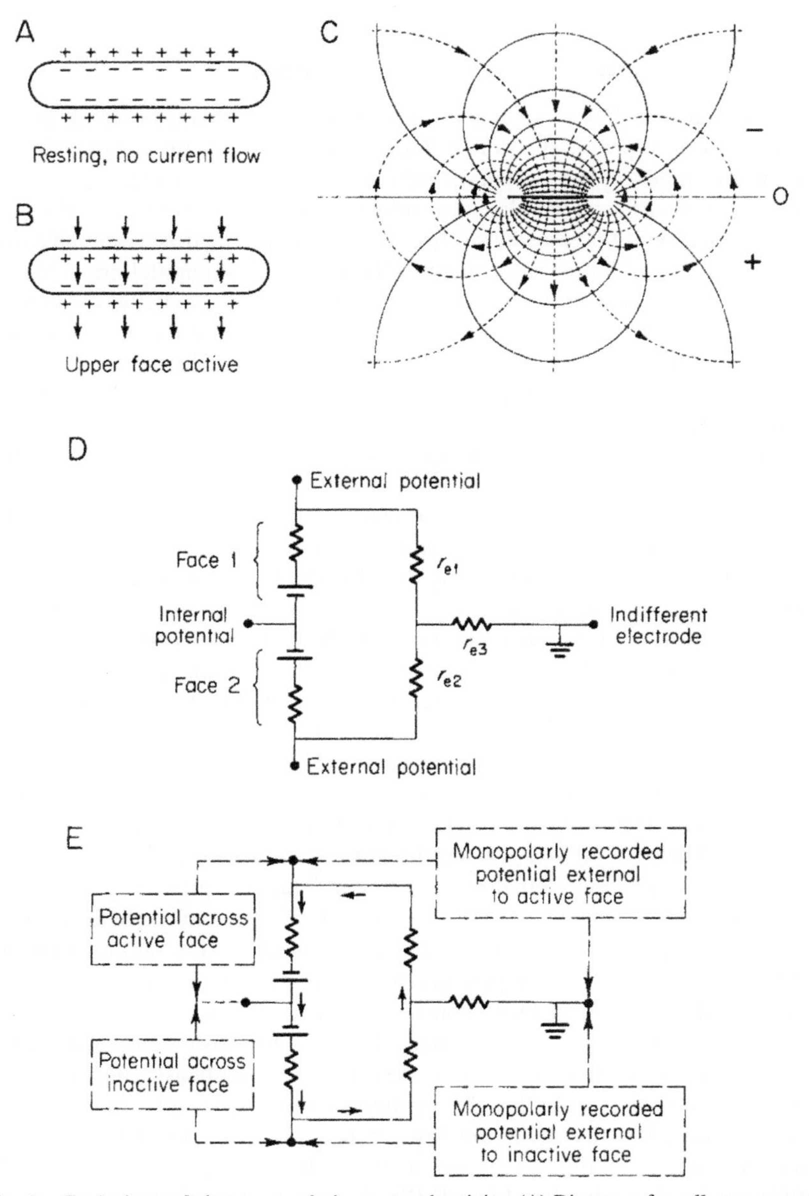

Fig. 2. Equivalents of electrocytes during rest and activity. (A) Diagram of a cell at rest; equal potentials are opposed. (B) When the upper face generates an overshooting action potential, two potentials act in the same direction and current flows as indicated by arrows. (C) Equipotentials (solid lines) and lines of current flow (dotted lines) around a thin electrocyte indicated by the heavy horizontal line. If the polarity of the cell corresponds to that in (B), the potential is negative above the thin horizontal line, which is the zero isopotential, and positive below it. For reasons of mathematical simplicity the diagram is for very long ribbon shaped cell of zero thickness. The membranes are assumed to act as current sources corresponding to a shell of magnetic dipoles. The isopotentials are separated by equal increments and meet at the two edges of the cell. (D) Electrical equivalent of a resting cell. The resistance of the external current path can be represented by r_{e1}, r_{e2} and r_{e3}. (E) Equivalent of an active cell. The small arrows indicate the direction of current flow.

represents current flow around an electrocyte when the two faces are at different, but uniform, potentials (dotted lines). These currents are associated with potential changes in the external medium, and surfaces at equal potentials (equipotential surfaces) are also diagrammed in Fig. 2C (solid lines). The current flow and field that an electric fish sets up around itself is essentially like that of Fig. 2C but larger since many cells in series and parallel are active at the same time. If an object such as a hand is present in the external medium, a potential difference is present across it and current flows through it. The current from a strongly electric fish is of sufficient magnitude to excite nerves, muscles, or receptors in the hand. An object distorts the electric field if it is of different conductivity than the medium, and the distortions, if large enough, can be detected by a fish's electrosensory system (see Chapter 11, this volume).

The relatively large size of external potentials generated by electric organs (Table I) as compared to other excitable tissues is not a result of larger membrane potentials although they may be slightly larger in a few instances. Rather the large outputs are a result of (a) arrangement of a cell's membranes in such a way as to maximize current outside the cell; (b) synchronous activity of many cells arranged in series and parallel; (c) to some degree, lower membrane resistances; and (d) accessory structures tending to channel current flow. The different kinds of adaptation will be discussed in connection with the various types of organ.

The types and patterns of electric organ discharge can be divided into several categories. The strongly electric organs all produce essentially monophasic pulses. They are all active intermittently, as indeed they must be, for the power outputs are so large that the fish can maintain them for only short periods. The organ is normally silent and is generally, if not exclusively, discharged in response to appropriate external stimuli. These stimuli may be tactile, chemical, electric, or perhaps visual. Responses are single pulses or trains of pulses usually of fairly constant size (Fig. 3A,A′).

Generally, weakly electric organs of freshwater fish continually emit pulses of rather constant size. The pulses may be monophasic, diphasic, triphasic, or even more complex. The patterns of emission fall into two categories. In one, the responses are brief pulses separated by long intervals. These species generally accelerate their discharges when presented with almost any kind of stimulus (Fig. 3B,B′). Acceleration results in an increased rate of testing the environment but may also represent a signal to another fish (Bullock, 1970; Black-Cleworth, 1970; Moller, 1970).

In the second category the duration of the pulses is as long or longer than the intervals between them (Fig. 3C,D). Generally pulses are emitted at a very constant frequency that can be very high. Recently, it has been found that in most species weak electric stimulation at a frequency close to that of the discharge causes small shifts in frequency (Watanabe and Takeda, 1963; Bullock, 1970). These changes apparently represent an attempt to avoid

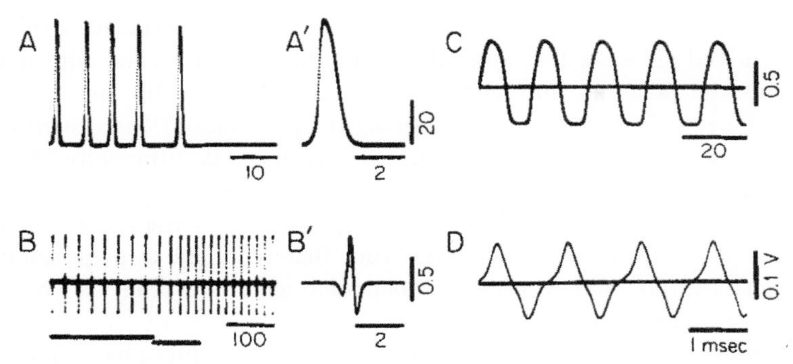

Fig. 3. Patterns of electric organ discharge. (A) An electric catfish 7 cm long. Potentials recorded between head and tail in a small volume of water with head negativity upward. Mechanical stimulation evokes a train of five pulses which attain a maximum frequency of 190/sec. (A′) Single pulses can also be evoked (faster sweep speed). (B–D) Weakly electric gymnotids immersed in water, discharges recorded between head and tail, head positivity upward. (A) A variable frequency gymnotid, *Gymnotus*; pulses are emitted at a basal frequency of approximately 35/sec. Tapping the side of the fish at the time indicated by the downward step in the lower trace causes an acceleration up to about 65/sec. The acceleration persists beyond the end of the sweep. The small changes in amplitude result from movement of the fish with respect to the recording electrodes. (B′) Faster sweep showing the pulse shape. (C) *Sternopygus*, a constant, low frequency gymnotid. The pulse frequency is about 55/sec. The horizontal line indicates the zero potential level. (D) *Sternarchus*, a constant, high frequency gymnotid. The frequency is about 800/sec. The horizontal line indicates the zero potential level. Calibrations in volts and milliseconds. From Bennett (1968a).

"jamming" of the electrosensory system by the applied signal. Other small changes of frequency may function in communication (Bullock, 1970; Black-Cleworth, 1970).

These two discharge patterns of weakly electric organs have been termed variable and constant frequency (although the terms now need a modifier such as relatively). They have also been called buzzers and hummers from the sound one perceives when the electric pulses are fed into a loudspeaker system. Variable frequency species include the mormyrids and many gymnotids. Constant frequency species include the remainder of the gymnotids and *Gymnarchus*. *Gymnarchus*, mormyrids, and several gymnotids are able to cease discharging for brief periods. This response appears to represent hiding by keeping quiet electrically and may also function in communication. Discharge patterns of weakly electric organs in marine fish are intermittent, but are poorly known under natural conditions.

II. ELECTRIC ORGANS AND ELECTROCYTES

A. Methods

An introduction to methods of establishing the properties of electrocytes may be helpful at this point. Microelectrodes are placed intracellularly and at various sites externally. Potentials at the tips can then be recorded by suitable amplifiers and associated equipment. A recording is termed monopolar if it represents the potential difference between the electrode tip and a distant grounded electrode in the vessel containing the preparation. The distant electrode is often called the indifferent electrode because it is sufficiently far from the active cell that moving it around does not detectably affect the recorded potential. One may use a differential amplifier to subtract the potential recorded through one microelectrode from that recorded through another; the result is differential recording. Monopolar recordings can also be subtracted after the experiment to obtain the same result.

In discussing properties of single electrocytes it is useful to have an equivalent circuit. As a first approximation one can consider an electrocyte to be a flattened cell with uniform potentials across each face that differ from each other during activity. Because of the uniformity of potential, each face can be given a simple equivalent circuit like those used for other membranes. (Of course a "naked" cell could not have a discontinuous change in potential at the edge of the cell for this would require an infinite current density. However, an electrocyte is often situated in an insulating connective tissue sheath that adheres closely to its edges, and the potentials over the faces are indeed quite uniform. In any case the nonuniformities are not important for most considerations of membrane properties.) Separate branches of the equivalent circuits for each membrane can be assigned to different ion species, each branch with a particular internal (equilibrium or Nernst) potential and a conductance that may or may not vary during activity. Alternatively, each membrane can be considered to consist of a single branch with a single internal potential and conductance, both of which can change. The membrane capacity, which may or may not be significant in electrocytes, is in parallel with the other branch or branches of the membrane equivalent. Generally, but not always, the resistance of the cytoplasm is negligible, and in the equivalent circuit of the entire cell the two membranes are directly connected together but oriented in the opposite direction. The circuit of the electrocyte can then be represented as a three terminal network, two terminals just external to each face and one inside the cell (Fig. 2D). The external terminals can be considered to be connected by a resistive path through the tissue and medium surrounding the cell. Another resistance leading to the indifferent electrode is required if monopolar recording is employed. This resistance is connected into the resistance of the external path thereby dividing it into two components. The placement of the connection depends on the symmetry with respect to the indifferent electrode of the material making up the external resistance. There are thus three resistances in the external path which are labeled r_{e1}, r_{e2}, and r_{e3} in Fig. 2D.

Consider what is recorded when one membrane of a cell generates a spike and the other membrane does not change its properties (Fig. 2E). Appropriate differential recordings show the potentials across each face and across the external medium. The recording across the active face gives the internal potential of the membrane less whatever internal voltage drop there is due to current flow through the opposed, inactive face and external path. Differential recording across the inactive face gives the passive voltage drop across this face that results from current flow. The potential across the external path is the voltage drop across the external resistances and also the difference between the potentials across the opposed membranes. Monopolar recording outside the two faces gives potentials of opposite sign; for a conventional depolarizing response the recording is negative going outside the active face and positive going outside the inactive face. The potentials are smaller than when differential recording is used because they represent voltage drops across part of the resistance around the external circuit (r_{e1} and r_{e2} in Fig. 2D). A monopolar recording by an intracellular electrode cannot distinguish between the two faces. The spike is smaller than that recorded across the active face by the voltage drop across r_{e1} in Fig. 2D. The spike is larger than the passive drop across the inactive face by the voltage drop across r_{e2}.

Obviously the same kind of inferences can be made from monopolar and differential recording since monopolar records can be subtracted to give the equivalent of differential recording. However, differential recordings often are more easily interpreted and electronic subtraction is much easier than the same process done graphically. An additional advantage is that interfering potentials resulting from activity of distant cells tend to be about the same size at different monopolar electrodes and are thus subtracted out by differential recording.

It is useful and often essential to pass stimulating currents through the faces. When current is passed from an intracellular electrode to an indifferent electrode, it flows outward through both faces in amounts depending on their resistances, capacities, and any responsiveness they exhibit as well as on the external resistances. Each face is depolarized and a response of one face may obscure a response of the other. For this reason it may be desirable to pass current between an external electrode outside one face and the indifferent electrode. (Use of two external electrodes, one outside each face, would be preferable in some ways but has not been carried out because of added technical complexity.) In this case part of the applied current flows inward through one face and outward through the other. Since this current polarizes the faces in opposite directions, it is generally simpler to evaluate the responsiveness of the two faces.

Electrocytes have been studied both *in situ* and in isolated tissue bathed in suitable physiological saline. For *in situ* work normal organ discharge can in many species be stopped by spinal section. Curare which blocks vertebrate neuromuscular transmission also blocks nerve-electrocyte transmission. It can be used to immobilize a fish as well as to block neurally controlled activity of myogenic electrocytes, but of course it prevents study of nerve-electrocyte transmission.

In a number of instances it is possible to apply currents to columns of cells or even to part of the intact animal. By means of a bridge circuit the potential resulting from voltage drops across fixed resistances in series with the active tissue can be subtracted from the records, and information similar to (and supplemental of) that obtained from single cells can be obtained (Albe-Fessard, 1950b; Bennett, 1961; Bennett *et al.*, 1961). The success of experiments of this kind is dependent on the series arrangement of the active membranes.

B. Membrane Properties

The surface membranes of electrocytes exhibit a number of different passive and active properties. In considering these properties, it is convenient to think of the cell surface as containing various kinds of sites where different and more less-specific ions can enter or leave the cell. These sites are intermixed in varying degrees and proportions to account for the different kinds of membrane activity. The membrane separating the sites has a bimolecular lipid core that presumably is of very high resistance and inactive in ion movement. Most of the membrane capacity can be assigned to this part of the surface. For most cells the specific capacitance is about 1 $\mu F/cm^2$ while membrane resistance—reflecting the number and nature of sites for ionic movement—can vary over a range of six or seven orders of magnitude. This concept of localized sites of ion movement has received strong support from recent experiments using artificial bimolecular lipid membranes. These membranes are similar to ordinary membranes in dimensions and capacitance but have an extraordinarily high resistivity. A number of compounds lower the resistance into the physiological range by providing sites for movement of ions or by acting as carrier molecules (e.g., Cass *et al.*, 1970).

In impulse responses of electrocytes there appear to be three types of sites that change their properties as a function of membrane potential. When the membrane is moderately depolarized, the conductance at sodium sites increases (sodium activation) allowing influx of sodium and further depolarization. This process is responsible for the active rising phase of the spike. If depolarization is maintained, the sodium conductance decreases again (sodium inactivation) and this change is a factor in return of the membrane potential to its resting level. At the resting potential the inactivated sodium sites gradually recover their ability to increase in conductance when the membrane is depolarized. (Any remaining activated sites rapidly return to their low, resting conductance at the resting potential and can be rapidly activated again without further delay.) One kind of potassium site in electrocytes increases in conductance when the membrane is depolarized (potassium activation or delayed rectification), but the change is delayed compared to sodium activation. The increased potassium conductance tends to restore the cell to the resting potential where the conductance gradually returns to normal. If the membrane is kept depolarized, the increase in potassium conductance can also reverse (potassium inactivation), a process which is generally much slower than either the reversal at the resting potential or sodium inactivation. So far these changes are like those of ordinary nerve (Hodgkin, 1964; Hille, 1970). Another kind of potassium site decreases in conductance when an outward current is passed through it. This change is very rapid in onset and reverses very rapidly when the outward current is reduced. It resembles and probably has the same mechanism as anomalous or outward rectification in muscle (Adrian *et al.*, 1970; cf. Bennett, 1970). It can also be considered a kind of inactivation (Bennett and Grundfest, 1966).

The resting membrane potential is apparently largely determined by potassium selective sites and the concentration gradient of potassium, although the cells may also be significantly permeable to chloride. In some electrocytes large hyperpolarizations cause the conductance of the resting membrane to decrease which can lead to what have been termed hyperpolarizing responses (Bennett and Grundfest, 1966). Since the conductance decreases by a large fraction of the resting value, these responses must involve a change in the potassium selective sites of the resting membrane. There is no known physiological significance of these responses.

In myogenic electrocytes another kind of a response is mediated by sites sensitive to the neurotransmitter acetylcholine released by the pre-synaptic nerve fibers. The transmitter increases the permeability at these sites, presumably to sodium and potassium ions, and the resulting current flow generates a postsynaptic potential (PSP). The change in conductance is virtually independent of the membrane potential, a property which has been termed electrical inexcitability (Grundfest, 1957; Bennett, 1964). An important consequence of there being an electrically inexcitable conductance increase is useful in establishing the chemical mediation of a postsynaptic response (Bennett, 1966). If the membrane is made sufficiently inside positive prior to the release of transmitter, the PSP current that would have depolarized the cell reverses and flows to make the inside of the cell less positive; the response is inverted. The potential at which the current reverses (loosely speaking the equilibrium potential by analogy with Nernst potentials) is generally slightly negative to the zero potential; this is the primary reason for believing that permeability is increased to both sodium and potassium (see N. Takeuchi, 1963).

Sites of the kinds described are combined in a number of ways in membranes comprising the faces of different electrocytes. These combinations are as follows:

(1) Spike generating membrane. This membrane contains sodium sites, and may or may not have potassium activation. Probably anomalous rectification is present in spike generating membranes of all myogenic organs. PSP generating sites may or may not be present.

(2) Membrane exhibiting delayed rectification without spike activity. This membrane appears to be spike generating membrane that has lost the sodium mechanism. PSP generating sites may or may not be present.

(3) Postsynaptic potential generating membrane. Postsynaptic potential generating sites may be the only responsive elements in a membrane as well as occurring intermixed in membrane of the preceding kinds.

(4) Low resistance inexcitable membrane. This membrane has a low resistance owing to a large potassium conductance, and it generates a resting potential; its properties do not change when the potential across it is changed. In some electrocytes there may be a significant Cl conductance.

(5) High capacity inexcitable membrane. This membrane has few sites for ionic movement, and its resistance is high. As a result, during activity virtually all of the current through it is capacitative. Its capacity per unit area of plasma membrane is probably similar to that of other membranes, that is, 1 $\mu F/cm^2$. Where membrane of this kind has been studied by electron microscopy, the surface is highly convoluted on an ultrastructural level, and therefore the capacity per unit area referred to macroscopic area is larger than the values for ordinary cells. [A similar situation obtains in muscle because of the transverse tubular system (cf. Gage and Eisenberg, 1969).] In electrocytes of the gymnotid *Sternopygus* one membrane has an effective capacity so large that it probably results from shift in ionic distributions on either side of the membrane (see Section II, C, 1, d).

In different electrocytes the two faces are made up of different combinations of the foregoing kinds of membrane. The equivalent circuits at rest and during activity are diagrammed in Fig. 4; they will be discussed in detail in sections concerning the different kinds of fish in which they occur.

There is good indirect evidence that sites mediating the different components of electrically excitable responses are independent of each other on a microscopic scale and that they are also separate from PSP generating sites (which often occur in the same general region of the cell). To be sure the spatial resolution of electrophysiological techniques may not allow measurement of the separation of different kinds of membrane responsiveness. Nonetheless, physicochemical considerations and separability of different kinds of electric activity under various electrical and pharmacological treatments strongly support the idea of separate channels mediating different response components (see, especially, Hille, 1970; but also Bennett, 1964; Grundfest, 1966). Another argument for separability of different components is the separate occurrence in different cells and in macroscopically different regions of a single cell. Electrocytes provide much of the comparative evidence of this kind.

A frequent characteristic of electrocytes is a kind of impedance matching between the faces (Bennett, 1961). The lower the membrane resistance the greater the electrical output of the cell, but also the greater the leakage at rest, and the greater the exchange between Na^+ and K^+ across spike generating membrane during activity. It would thus be inefficient to have one face of much lower resistance than the other (or of much lower resistance than the series resistances of the external medium and cytoplasm). One finds that a low resistance inexcitable face generally is of lower resistance than the

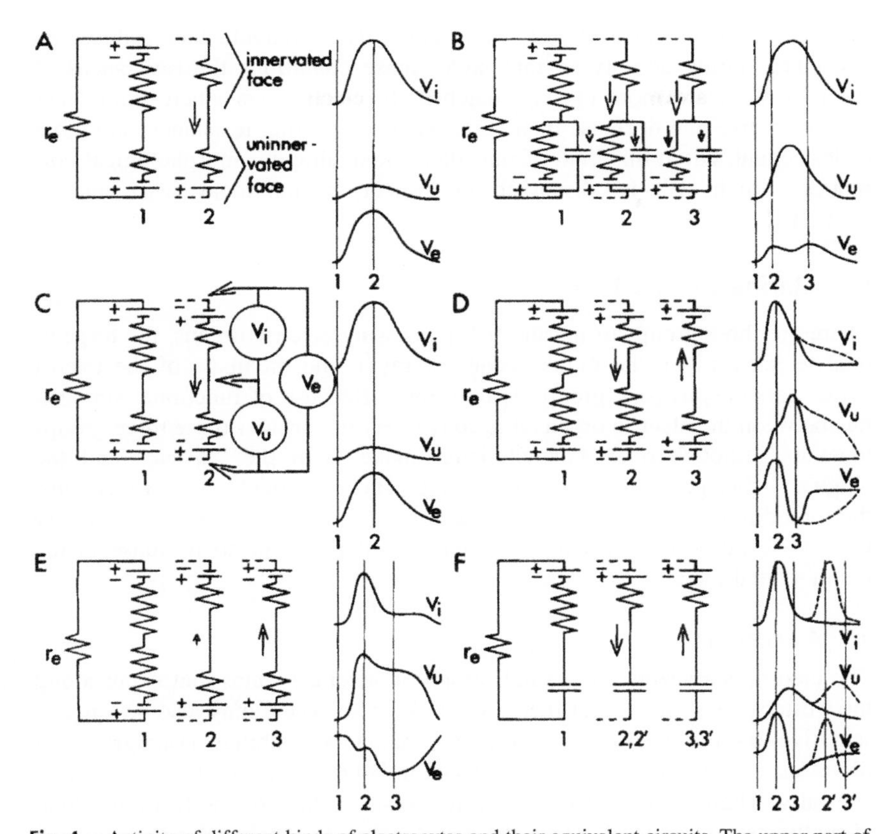

Fig. 4. Activity of different kinds of electrocytes and their equivalent circuits. The upper part of each circuit, labeled only in A, represents the innervated or stalk face; the lower part represents the uninnervated or nonstalk face. The resistance of the extracellular current path is represented by r_e. The potentials that are recorded differentially across the two faces (V_i and V_u) and across the entire cell (V_e) are drawn to the right of each circuit (intracellular positivity and positivity outside the uninnervated face shown upward; note that $V_i - V_u = V_e$). Placement of electrodes for recording these potentials is indicated in C. The successive changes in the membrane properties are shown by the numbered branches of the equivalent circuits, and their times of occurrence are indicated on the potentials. A lower membrane resistance is indicated by fewer zigzags in the symbol. Return to resting condition is omitted. (A) Electrocytes of strongly electric marine fish and disc-shaped electrocytes of rajids. The innervated face generates only a PSP. (B) Cup-shaped electrocytes of rajids. The innervated face generates a PSP and the uninnervated face exhibits delayed rectification. (C) Electrocytes of the electric eel and a few other gymnotids. The innervated face generates an overshooting spike. (D) Electrocytes of mormyrids and some gymnotids. Both faces generate a spike, and V_e is diphasic. In some the spike across the uninnervated face is longer lasting and the second phase of V_e predominates. These potentials are shown by dashes. (E) Electrocytes of the electric catfish. The stalk face is of higher threshold and generates a smaller spike (indicated by a smaller battery symbol) and the external potential is entirely negative on the nonstalk side distant from the stalk. (F) Electrocytes of *Gymnarchus*. The uninnervated face acts as a series capacity and the external response has no net current flow. The summation of a second response (2′,3′) on the first is shown by dashes. Electrocytes of sternarchids and *Eigenmannia* may operate similarly. From Bennett (1970).

opposed, excitable face at rest (Fig. 4A,C). The excitable face reduces its resistance during activity so that under these conditions the resistances of the two faces are more closely matched. In electrocytes where both faces become active and decrease their resistances, the resting resistances are more or less equal. Often this kind of impedance matching has morphological correlates, which will be discussed in respect to individual cases and in Section II, F.

C. Marine Electric Fish

There are three groups of marine fish possessing electric organs, the torpedinids (electric rays), the rajids (skates or rays), and members of the teleost genus *Astroscopus* (one group of stargazers). Because of functional similarities between the electric organs it is convenient to consider these three groups together, although other characteristics make them the "lowest" and the "highest" fish possessing electric organs. The electrocytes of *Astroscopus,* the torpedinids, and some of the rajids are the simplest known. There is only the one response component in the innervated face and no response in the uninnervated face.

1. *Astroscopus*

The electric stargazers are a small group of several species that occur along the western coast of the North and South Atlantic. There may also be a representative on the Pacific side of the Panamanian isthmus (Dahlgren and Silvester, 1906). These fishes are unusual in being somewhat flattened dorsoventrally. Their eyes are located on the dorsal surface of the head and look virtually straight upward, hence the name stargazer. Their habit is to burrow into sand leaving only their eyes protruding when they are practically invisible except for the two small black spots of their pupils. From this position they gulp down unwary minnows passing overhead. The electric organs lie just behind the eyes (Fig. 1), and they are in point of fact modified from extraocular muscles (Dahlgren, 1914, 1927). Other members of the same family, Urano-scopidae, are very similar in appearance but lack electric organs.

The electrocytes are large flattened cells that lie in the horizontal plane and are densely innervated on their dorsal surface by branches of the very large oculomotor nerves (Fig. 5). The ventral surface has many short processes or papilli that markedly increase its surface. This surface is further increased by many small invaginating tubules or canaliculi (Mathewson *et al.*, 1961; Wachtel, 1964). About 150–200 layers of cells are arranged in series, one above the other (Dahlgren and Silvester, 1906). Each layer of cells in parallel consists of about four large cells (approximately 5 mm in diameter in a 20-cm fish) surrounded by about 10 smaller ones. A single large cell can also spiral around to overlap itself to some extent.

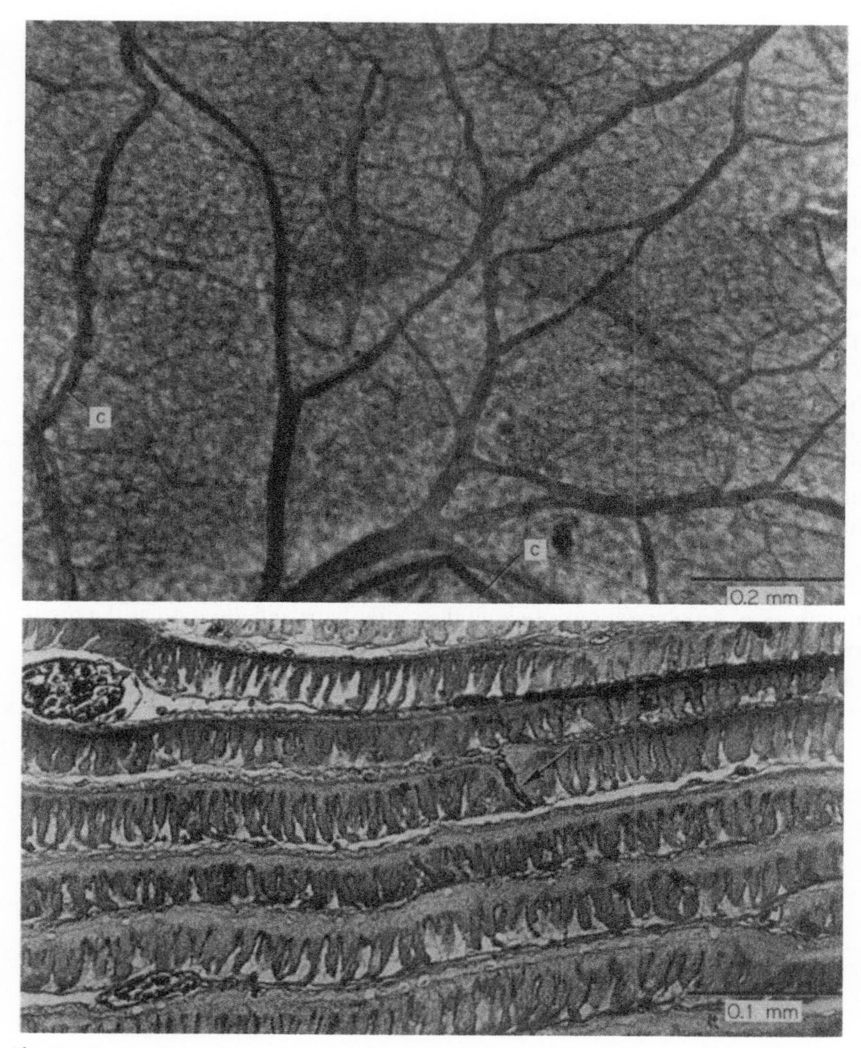

Fig. 5. Innervation and structure of electrocytes of the stargazer. The upper picture shows the dorsal surface of a single cell teased from formalin fixed material and stained with methylene blue. A nerve bundle enters from the lower left and forms profuse, sometimes anastomosing branches. Finer branches are not resolvable. Capillaries (c) are also seen. The lower micrograph is a vertical section through the organ, dorsal surface uppermost, following osmic acid fixation. The innervated, dorsal surfaces are smooth; the uninnervated, ventral surfaces have long papillae. Two nerve bundles are seen on the left. The edge of a cell between two others (lower right) illustrates the characteristic irregular layering of the cells. Just to the right of center is a vertical fissure (arrow) which may represent edge to edge apposition of two cells. From Bennett and Grundfest (1961b).

The organ discharges are pulses making the dorsal surface negative (and presumably also making the inside of the mouth positive, although this has not been directly verified). Responses to handling the fish range from single pulses up to trains of several tens of pulses at frequencies of about 50–100/sec. Trains of pulses are also emitted when the fish is capturing small minnows (Pickens and McFarland, 1964). The pulses are about 5 msec in duration. The amplitude is somewhat variable, particularly at the beginning of a train but is up to about 5 V recorded from a fish 20–30 inches in length with its dorsal surface in air (Fig. 6). The voltage across the organ is reduced somewhat if the discharge is evoked while the fish is immersed in seawater. The pulses are generated synchronously by the two organs.

The effectiveness of the discharge in aiding the capture of prey has not been demonstrated, but the unstimulated fish emits pulses rarely, if ever, at other times. Large fish can reach 40 cm in length, and the discharge is sufficiently strong that it is easily detected when the fish are handled (Dahlgren and Silvester, 1906). Even a small fish can cause mild discomfort if one's hands are quite wet and have a number of minor cuts (personal observation).

Only the innervated (dorsal) face of the electrocytes is active during organ discharges (Bennett and Grundfest, 1961b). The responses are PSPs and are monophasic depolarizations of about the same shape and duration as the organ pulses. The innervated membrane does not respond at all to depolarization but behaves linearly; it is electrically inexcitable. The response that is evoked by

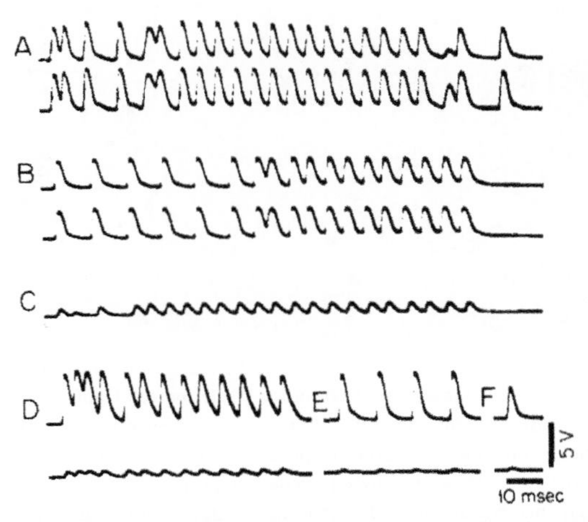

Fig. 6. Patterns of electric organ discharge of *Astroscopus*. (A,B) Simultaneous records from the two organs in each of two different fish. A probe electrode was placed on the skin over each organ with the dorsal surface of the fish in air. The reference electrode was on the ventral surface and grounded; dorsal negativity up. (C) Same fish as in B, but recording with the animal covered by seawater. (D–F) Simultaneous recordings as in A but from a fish which had had one of its electric organs denervated. The electrode on the denervated organ (lower traces) registered only a small pickup of the activity of the other organ.

stimuli applied directly to the organ is mediated by the presynaptic nerve fibers. The uninnervated face is of very low resistance and generates only a resting potential. (Presumably it is inexcitable, but this point has not been established critically because the membrane is of such low resistance that it may not have been depolarized sufficiently to excite it. However, one doubts that this membrane would have electrically excitable sites when they would be nonfunctional and when the innervated face lacks them.) The equivalent circuit during responses is shown in Fig. 4A. At rest the two faces generate equal potentials and no current flows. During activity the resting potential virtually disappears across the innervated face and the conductance of this face to (presumably) sodium and potassium greatly increases. Current flows inward across the innervated face and outward across the uninnervated face. [It is to be expected that the innervated face actually generates a potential of —10 to —20 mV (see N. Takeuchi, 1963). Note that the potential in the circuit is provided by what we have termed the inactive face.]

Monopolar recordings on which this description is based are shown in Fig. 7. (The straight horizontal line is a reference trace.) A brief stimulus is applied to the dorsal surface of the organ by means of a pair of small wire electrodes 1 msec after the start of the sweep. The recording trace goes off screen for about 0.2 msec during the stimulus and then returns more slowly toward its initial potential before the response begins with a latency of about 1 msec. As the electrode penetrates the cells a regular sequence of potential changes is observed that serves to identify electrode position as well as to characterize the responses. Immediately dorsal to the most superficial cell the response is a monophasic negativity of about 15 mV amplitude (A). The sign of the potential indicates that the underlying membrane is passing inward current. As the electrode is advanced the steady potential shifts about 90 mV negative which represents the resting potential across the innervated membrane (B). Simultaneously the response becomes a positive-going or depolarizing response of about 60 mV indicating that the electrode has crossed an

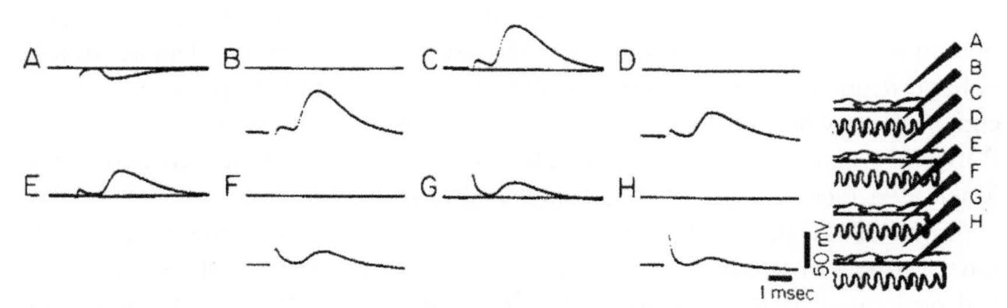

Fig. 7. Activity of a single electrocyte of *Astroscopus*. Monopolar recording *in vivo*. Stimuli are applied by a pair of fine wire electrodes close to the site of microelectrode penetration. The diagram on the right indicates positions of the recording electrode in successive records inferred from appearance and disappearance of resting potential and changes in response amplitude and sign. From Bennett and Grundfest (1961b).

active membrane. If it were differentially recorded across the innervated face (record B minus record A) the response would be about 75 mV in amplitude.

As the electrode is further advanced, the steady potential shifts back to its initial value signaling passage of the electrode through the cell into the underlying extracellular space (C). The resting potentials developed by the two faces are equal, and thus no current flows through the cell at rest. The response recorded outside the innervated face is virtually identical to that recorded in the cell. This indicates that the resistance of the uninnervated face is very low compared to the resistance in the external path. The response recorded differentially across the entire cell would of course be almost exactly like that recorded across the innervated face. When the electrode is further advanced, the steady potential again shifts negative indicating penetration of the second cell (D). Simultaneously the response amplitude decreases by about half indicating that the innervated membrane of this cell comprises a significant fraction of the resistance of the external path. Further advances show shifts in the steady potential as the electrode leaves and enters cells (E–H). The amplitude decreases as the electrode crosses innervated faces but not as it crosses uninnervated faces because of their relatively high and low resistances, respectively.

The neural mediation of these responses is indicated by the latency which remains about 1 msec even if very suprathreshold stimuli are used. If spike generating membrane were present, a much shorter latency could be obtained (an example is given in Fig. 18). Neural mediation is confirmed by pharmacological data and the effects of denervation (see below). Responses like those of Fig. 7 result from single nerve fibers for their amplitude varies in an all-or-none manner as stimulus strength is varied.

Other experiments involving passage of current during responses show that the resistance of the innervated faces is decreased during neurally mediated activity but is not affected by depolarization alone. Thus during activity the resistance of the innervated face moves toward that of the uninnervated face in the kind of impedance matching discussed in the preceding section. The resting resistances of the two faces correlate with the degree of surface elaboration seen in the fine structural studies; the lower resistance membrane is much more elaborated (Mathewson *et al.*, 1961). (The negligible voltage drop across the innervated faces need not obtain in responses of the entire organ when all the cells are active in series and much larger currents flow. In organ discharge there would also be a smaller potential across the innervated face because of its internal resistance.)

The PSPs of the electrocytes are cholinergic as they presumably are in the muscle fibers of origin. They are greatly reduced by the blocking agent curare and prolonged by the anticholinesterase eserine. The cells are depolarized by acetylcholine, presumably the actual transmitter, and the related compound carbamylcholine. Denervated cells retain their ability to respond to these drugs, but the response to electrical stimulation disappears confirming its neural mediation.

The microelectrode experiments demonstrate that unitary responses of the electrocytes are essentially of the same duration, shape, and sign as the organ discharges. If 150 cells in series give rise to a 5 V pulse about 30 mV per cell is required, a value exceeded somewhat in the microelectrode experiments. The discrepancy is explicable as a result of greater current flow during synchronous activity and failure of some of the cells to fire. Furthermore, cell counts, organ discharges, and PSP amplitudes may have varied in the different animals used for the measurements. Although most of the presynaptic fibers are active during a response (see Section III), some do fail to fire on occasion as indicated by the variable response amplitude, and a few inactive cells in series would markedly increase the internal resistance of the organ. In spite of the quantitative discrepancy the principles of series summation and synchronous activation are well illustrated by this electric organ.

2. *Torpedinids*

The electric rays are a cosmopolitan group of marine fishes. The electric organs of the genus *Torpedo* are diagrammed in Fig. 1. They are large flattened organs on each side of the head that extend completely through the "disc" from dorsal to ventral surfaces. The *Torpedos* are rather slow moving, and the use of the organ discharge in capture of much faster swimming prey has been studied in the European species, *Torpedo marmorata* (Belbenoit, 1970). When a fish comes near a *Torpedo* resting on the bottom, it swims forward and upward and then emits pulses. Small fishes can be stunned. The *Torpedo* drops back to the bottom over any immobilized prey. If it has been successful, it consumes the prey while continuing to emit pulses at a low frequency. The delay between movement and discharge indicates that the organ does not function in prey detection. The detection of prey appears to be mechanical or perhaps electrical. The effectiveness of the organs in predation is confirmed for the American species, *T. nobiliana,* by the presence of large and rapidly swimming fish in the stomach contents (Bennett *et al.*, 1961).

Each electric organ in *T. nobiliana* is made up of some 500–1000 closely packed and roughly circular columns of electrocytes that run from top to bottom of the organ (Fig. 8). The electrocytes are thin (10–30 μ) discs and have the same diameter as the columns; about 1000 are stacked one above another as in a role of coins. The columns can be quite large in diameter, 5–7 mm in a large *Torpedo* (1 meter across).

The cells are profusely innervated on their ventral surfaces. Nerves run in the interstices between columns and 5–7 nerve fibers enter the space between cells at roughly equal intervals around the periphery; each fiber innervates a sector of the cell (Fig. 9). The nerves arise from the electromotor lobe in the medulla (perhaps including parts of the seventh, ninth and tenth cranial nerves, Fig. 8). Electron microscopic examination reveals that the innervation is very dense (Fig. 10). The organ discharges of a large *Torpedo* are monophasic pulses, positive on the dorsal surface and about 5 msec in

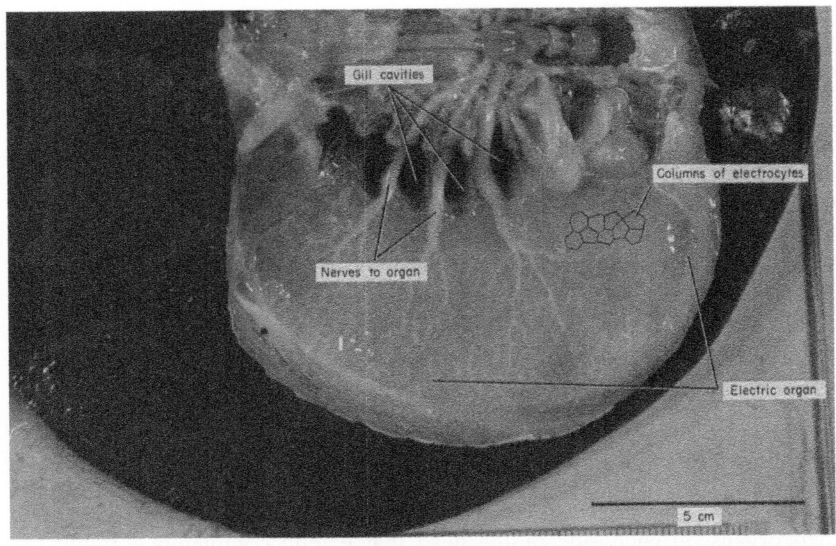

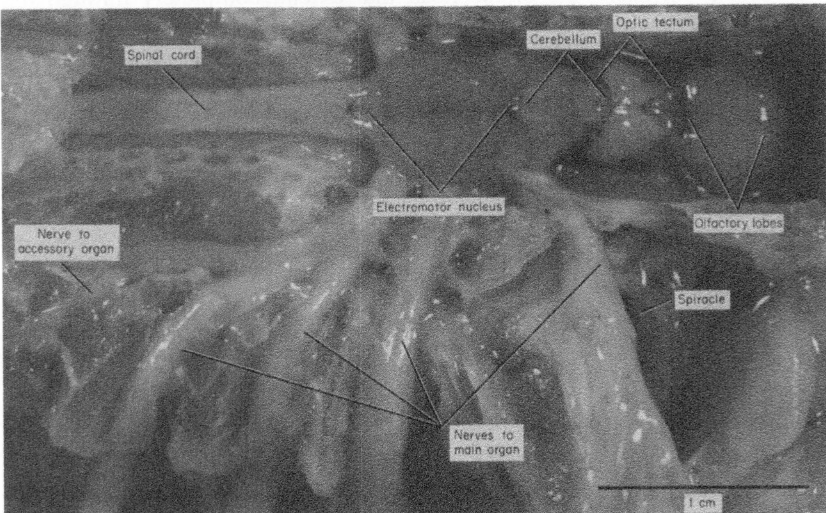

Fig. 8. Electric organ and brain of the torpedinid, *Narcine*. The nerves entering the organ were exposed by removal of the dorsal part of the organ. Separate columns of electrocytes are easily distinguished; a few are outlined. The brain and cranial nerves are shown at higher magnification in the lower picture.

duration and 50 V in amplitude recorded in air (Bennett *et al.*, 1961). The internal resistance of the organs is low, and the power output at the peak of the pulses can exceed 1 kW.

Because the electrocytes of *Torpedo* are very thin, it is difficult to use two or more electrodes for differential recording across the two faces and the

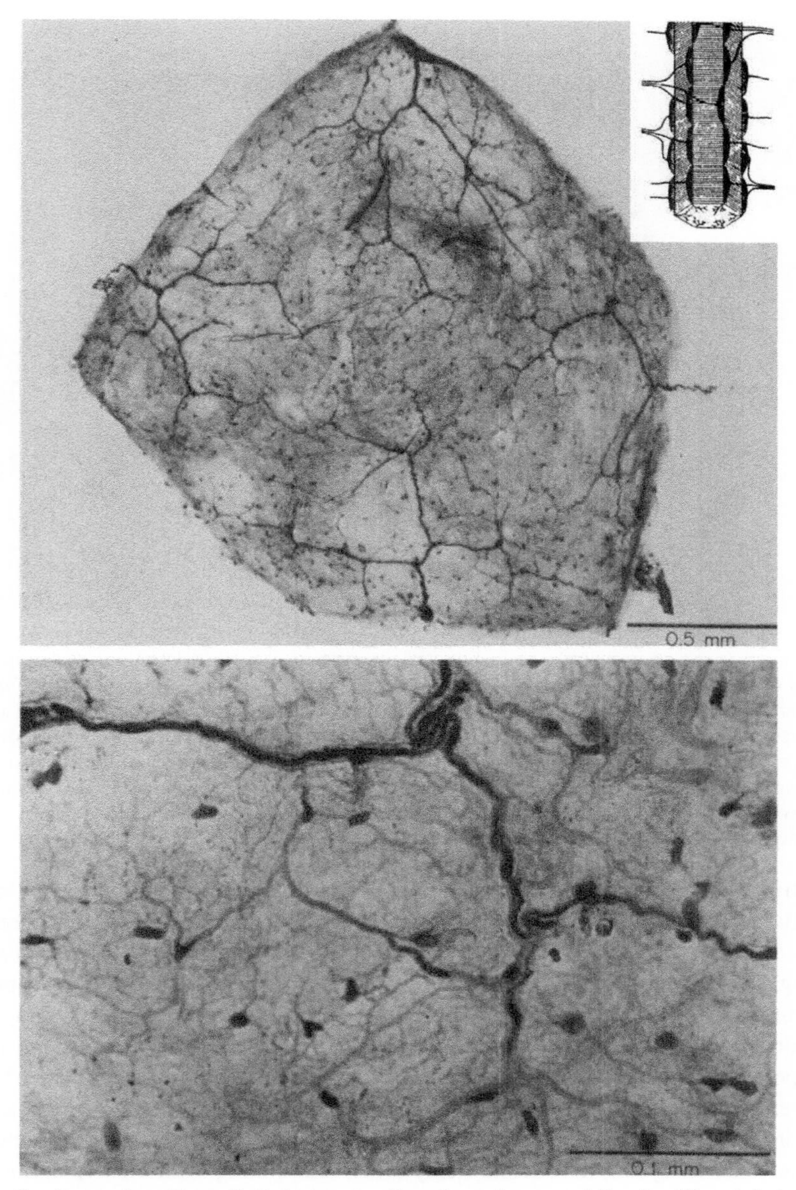

Fig. 9. Innervation of torpedinid electrocytes. From the main organ of *Narcine*. (Upper) A single cell teased out from formalin fixed material and stained with methylene blue. Four nerve fibers run across the surface from different points on the periphery. The fibers branch profusely, but each innervates a separate segment of the surface. The inset shows schematically the columnar arrangement of the electrocytes. (Lower) A region from the lower right of the isolated cell is shown at higher magnification. Nuclei of the electrocytes are stained as well as many fine nerve branches. Modified from Bennett and Grundfest (1961a) and Grundfest (1957).

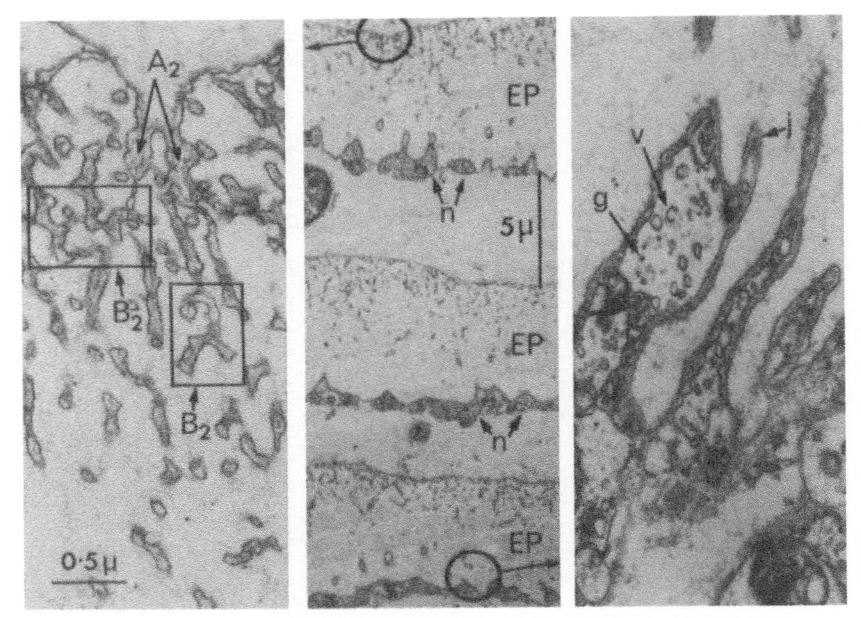

Fig. 10. Fine structure of torpedinid electrocytes (*Torpedo marmorata*). The central portion of the figure shows a perpendicular section through three electrocytes (EP), dorsal surface upwards. The ventral surface is densely covered with nerve endings (n). On the left is shown a region of the uninnervated surface at higher magnification. Several external openings of the many canaliculi are seen (A_2). The rectangles (B_2) indicate branch points of the canaliculi. Basement membrane material extends deeply into the canalicular network. On the right is shown a higher magnification of the innervated surface. Nerve profiles contain vesicles (v) as well as small granules (g), possibly glycogen, and lie embedded in the electrocyte. Occasion folds (j) extend quite deeply into the postsynaptic cell. From Sheridan *et al.* (1966).

entire cell. However, a single electrode can be advanced into a column of cells, and successive changes in resting potential and response can be used to determine electrode position as described in respect to *Astroscopus* (Fig. 7). The responses of the electrocytes of *Torpedo* are essentially the same as in *Astroscopus* (Bennett *et al.*, 1961). The active face is the innervated face, which however is on the ventral side. Again, this face does not respond to depolarization; it is electrically inexcitable. The response that is evoked by external stimulation is a monophasic depolarization, that is, a PSP mediated by activity of the presynaptic fibers. Its origin as a neurally mediated response is attested to by the irreducible delay when stimulating with closely applied electrodes and by the block of the response when curare is applied to the innervated face. In a column of tissue the response can be inverted by polarizing currents [the first tissue in which PSPs were shown to invert (Albe-Fessard, 1951; Bennett *et al.*, 1961)]. The uninnervated face is of very low resistance and generates only a resting potential. The area of this face is greatly increased by numerous invaginating canaliculi (Fig. 10).

The maximum response amplitude observed in a single cell is about 90 mV, which is somewhat larger than the recorded resting potentials. However, the cells are very thin and the canaliculi extend across at least half the cell thickness; a supposedly intracellular electrode might always be partly in the extracellular space outside the uninnervated face. Probably the full resting potentials are not recorded and like other PSPs those in *Torpedo* do not overshoot the resting potential. The duration of the PSPs is about 5 msec. Their latency with locally applied stimuli is 2–3 msec, an appreciable fraction of which may be conduction time in fine branches of the innervating fibers.

The PSPs in *Torpedo* electrocytes are cholinergic in that they are blocked by curare and dihydro-β-erythroidine and prolonged by the cholinesterase inhibitors eserine and physostigmine. Also, the cells are depolarized by actylcholine and carbamylcholine.

The synthesis of organ discharge from the responses of the single electrocytes simply requires synchronous activation as in *Astroscopus*. The duration of the PSPs is about the same as that of the organ discharge, and the amplitude of the PSPs is sufficient that the number of cells in series could produce the discharge amplitude.

Narcine is a smaller relative of *Torpedo*. It has a bilateral electric organ, the "main" organ, that resembles that of *Torpedo* both morphologically and physiologically (Bennett and Grundfest, 1961a). *Narcine* has an additional (bilateral) electric organ, the accessory organ, that lies at the posterior margin of the main organ (Mathewson *et al.*, 1958). This organ differs in one important respect from the main organ. The responses to low frequency stimuli are very small, but when the nerve is stimulated at up to 100/sec, the responses are considerably augmented and prolonged. However they are still somewhat smaller than in the main organ of *Torpedo*. Although the normal discharge of the accessory organ is unknown, these findings suggest that the discharge involves repetitive and fused responses of the single cells similar to what is observed in the rajids (see the next section). It seems likely that the accessory organ is used either in an active electrosensory system or in communication, although there is no direct evidence for this suggestion. The torpedinids like most other elasmobranchs have ampullae of Lorenzini, which are electroreceptors (see Chapter 11, this volume). A number of other small torpedinids are known, some of which live in the deep seas and are blind (Lissmann, 1958). Possibly their main electric organs are used in electrolocation. It is not known whether these species have accessory organs.

3. *Rajids*

The skates or ordinary rays are a large cosmopolitan group of marine fish comprising six or more genera and many species. They are weakly electric, but unlike freshwater species they emit pulses only infrequently. The discharges are sufficiently inconspicuous that a major taxonomic work on the group "Fishes of the Western North Atlantic" (Bigelow and Schroeder, 1953) makes no mention of the fact that these fish are electric. Other rays (suborder Myliobatoidea) apparently lack electric organs.

The electric organs of rajids are located in the tail in the center of the most lateral bundle of longitudinally running muscle fibers (Fig. 1). The organs are spindle-shaped and run most of the length of the tail. They are much greater in length than in diameter. The electrocytes are oriented anteroposteriorly and are innervated on their anterior faces. Each cell lies in a small connective tissue compartment. Two types of electrocytes have been described, the cup-shaped and the disc-shaped (Fig. 11). However, these terms are not particularly descriptive of the morphological differences. Cup-shaped cells lie at the anterior margin of their connective tissue chamber. Often they are convex posteriorly, which accounts for their name. Both faces are relatively smooth at the light microscopic level of resolution. Electron microscopy reveals a relatively small number of tubules invaginating into the innervated face and a somewhat greater number in the uninnervated face

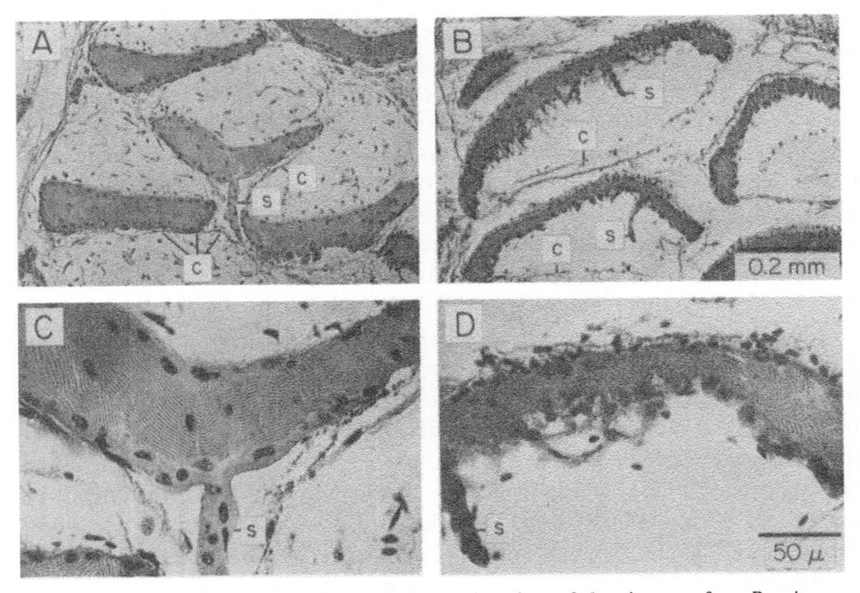

Fig. 11. Anatomy of rajid electrocytes. Longitudinal sections of electric organ from *R. erinacea*, which has cup-type cells (A,C) and from *R. eglantaria*, which has disc-type cells (B,D). Rostral surface up, hematoxylin and eosin stain, A,B, low power; C,D, high power. Cup-type cells are usually smooth on both surfaces (A) although there may be a few processes on the caudal face in some cells (A, lower right). The cells lie against the caudal walls of the connective tissue chambers (c) that contain them. The caudal face of disc-type cells usually has many processes although in some regions it may be relatively smooth (B, lower right). The cells lie anteriorly in their connective tissue chambers (c). Both types of cell contain striated material in a central area (C,D). There is often a short process or stalk (s) that is a remnant of the muscle fiber from which the cell develops (Ewart, 1892). From Bennett (1961).

(Mathewson *et al.*, 1961). Disc-shaped cells lie nearer the posterior of their chambers. The posterior, uninnervated faces have a large number of protuberances tens of microns in diameter and length (Fig. 11). Probably there are more invaginating tubules in these faces than in cupshaped cells. Both classes of cells contain striated filamentous material (Fig. 11, Wachtel, 1964) that reveals their myogenic origin.

The response properties of the two kinds of cell are somewhat different. The disc-shaped cells are physiologically similar to those of other marine electric fish, but the cup-shaped cells are more complex (Bennett, 1961). Some intergrading of the two extreme physiological types apparently does occur, but too few species have been studied to be sure of the extent of the correlation between form and physiological functioning. There is evidence that the differences between cup- and disc-shaped electrocytes are associated with other morphological characteristics and that the rajids can be divided into two groups (Ishiyama, 1958).

The organ discharges are monophasic and head negative. The fish can only sometimes be provoked into discharging by mechanical stimulation. Under these conditions the organ discharges are usually irregular and variable in size and duration (Fig. 12). The maximum pulse amplitude is of the order of a volt recorded in air, but tens of millivolts if the fish is in seawater (D). Sometimes more regular pulses are emitted (A), and preliminary results with long-term recording from animals in holding tanks suggest that "spontaneous" discharges or those evoked by light touch under these conditions are more constant

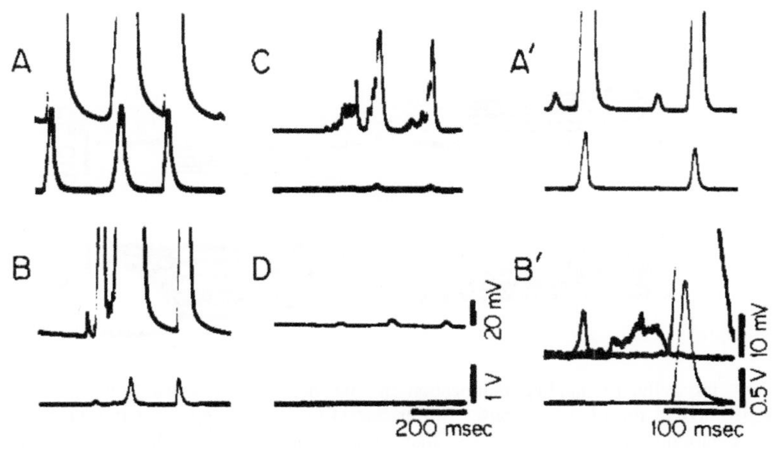

Fig. 12. Discharge of rajid electric organs. (A–D) *Raja erinacea*, which has cup-type electrocytes. (A′,B′) *Raja eglantaria*, which has disc-type electrocytes. Discharges are evoked by vigorous prodding and recorded differentially between tip and base of tail, caudal positivity up. Upper trace, higher gain. In both forms, discharge is asynchronous and variable in amplitude and duration. All records are in air except D, for which the animal is immersed in seawater while regularly responding as in B. The discharge is greatly reduced in amplitude. From Bennett (1961).

in amplitude (A. B. Steinbach, unpublished data). In any case the discharges that have been observed are long lasting compared to the responses of individual electrocytes, and the discharges most probably involve fused repetitive activity of many cells.

Unlike the electrocytes of strongly electric marine fish, cup-shaped electrocytes respond to depolarization. However, the response involves only delayed rectification. There is no sodium activation or other electrically excitable component leading to a regenerative response; the response to graded depolarizing pulses is graded. An example is shown in Fig. 13. For small depolarizing (outward) currents the cell behaves linearly with the same

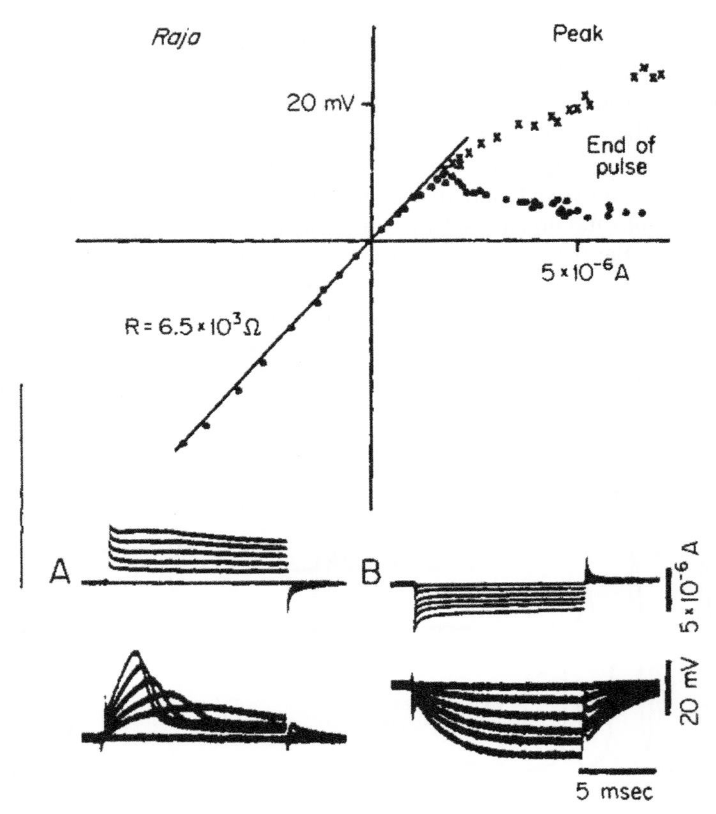

Fig. 13. Intracellularly applied polarization of cup-shaped electrocytes from *R. erinacea*. (A) Superimposed records of responses (lower traces) to depolarizing currents (upper traces). For larger currents, the voltage reaches an initial peak, then falls to a much lower steady value. (B) Records as in A in response to hyperpolarizing currents. The voltage change increases approximately exponentially toward a steady level. Graph: voltage–current relationship for the same cell as A and B, but using additional data. The relation is linear for hyperpolarizing current. For larger depolarizing currents, the initial peaks fall somewhat below the potentials they would have reached if the cell had the same resistance as for hyperpolarizing current. The potentials at the end of the pulses are much lower and continue to decrease as further current is applied. Modified from Bennett (1961).

resistance as for hyperpolarizing current. The resulting potentials rise slowly and more or less exponentially to a steady state value; the slowness of rise results from charging of the membrane capacity. For larger depolarizing currents there is an early peak of depolarization which then decreases as the rectification "turns on." The peak depolarization is always less than the steady state potential would be that corresponded to the resistance for small depolarizing and hyperpolarizing currents. If a current pulse is terminated on the rising phase of the initial depolarization, the potential immediately begins to return toward the base line; there is no tendency of the potential to continue in the depolarizing direction (Fig. 14). These findings lead to the conclusion that there is no regenerative component in the response. The increase in conductance is indicated not only by the reduction in potential during the current but also by the more rapid drop in potential following cessation of the current after the conductance increase has been produced (Fig. 14). The conductance increase caused by a brief stimulus lasts a few tenths of a second. It is associated with a small depolarization from the resting potential (Fig. 14D). The nature of the permeability change underlying the conductance increase is unclear. There is suggestive but incomplete evidence that it may be an increase in Cl permeability (Bennett, 1961; Grundfest, 1967).

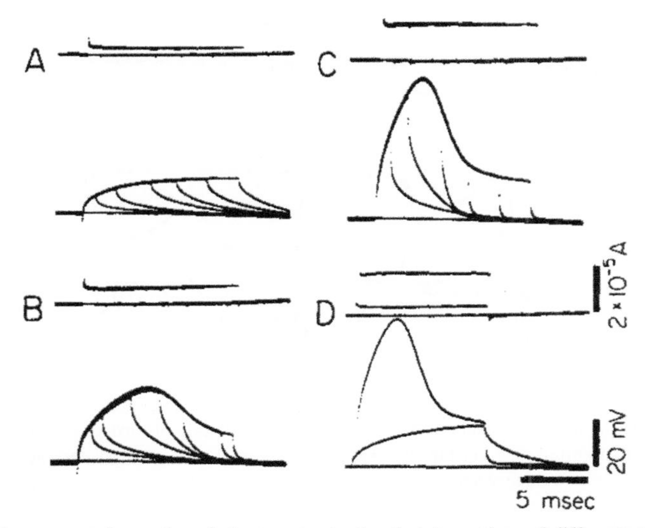

Fig. 14. Responses of cup-shaped electrocytes to depolarizing pulses of different strengths and durations (*R. erinacea*). Lower trace: intracellular voltage. Upper trace: current applied through second intracellular electrode. (A–C) Superimposed records of constant current pulses of different durations, magnitude increasing from A to C. (D) Superimposed records of a large and a small pulse of the same length; the traces cross following the end of the pulses. Whatever the current strength or duration, the voltage starts to decrease immediately on cessation of the current. Following the peaks, the voltages decrease more rapidly than if the pulses are stopped before them. The time course of the decrease deviates from exponential because the cells are not isopotential. In D a small depolarization is seen to be maintained following the larger stimulus. From Bennett (1961).

When the nerve supply to an electrocyte is stimulated, a PSP is produced that is graded in several all-or-none steps indicating that the cell is innervated by several fibers (Fig. 15C). The PSPs are cholinergic (Brock and Eccles, 1958). The larger PSPs activate the delayed rectification, and these PSPs decay more rapidly because of the shortened time constant of the cell.

In most cells the delayed rectification is found primarily in the uninnervated face; it thereby acts to increase external current flow in the medium around the electrocytes. At rest the resistance of innervated and uninnervated faces is relatively high. Depolarization by PSPs generated in the innervated face increases the conductance of the uninnervated face and thus allows more current to flow out through this face and around the cell. (Delayed rectification in the innervated face is maladaptive; it increases

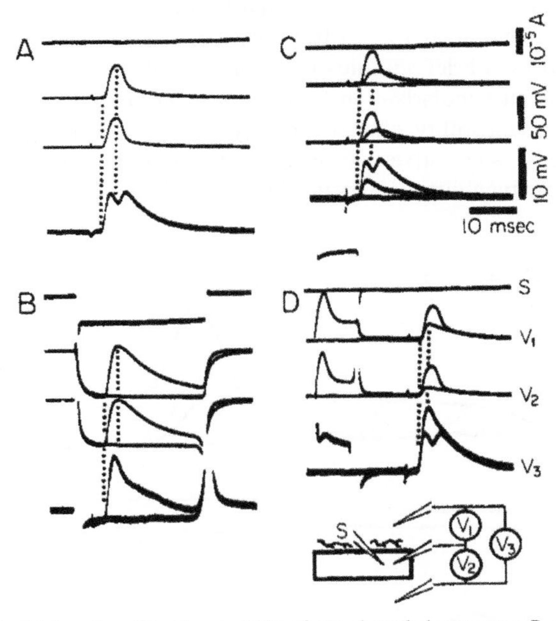

Fig. 15. Effect of delayed rectification on PSPs of cup-shaped electrocytes, *R. erinacea*. Recording and stimulating as in the diagram except that nerve stimulating electrodes are omitted. Upper trace: intracellular stimulating current. Second trace: potential across innervated face (V_1). Third trace: potential across uninnervated face (V_2). Fourth trace: potential across cell, posterior positivity up (V_3). A large PSP produces a posterior positive external potential that has two peaks (A). The second peak is blocked when the conductance increase is prevented by hyperpolarization (B) that also prolongs and increases the height of the response. Small PSPs that do not activate the delayed rectification also produce external potentials with a single peak (C, superimposed traces of large and small responses). When the delayed rectification is activated by a depolarizing pulse and a PSP is evoked, the external potential is increased and the potentials across the faces are reduced (D, superimposed responses with and without applied depolarizing pulse). The time constant of the uninnervated membrane is reduced so that all the potentials have the same time course and the peak of the external potential occurs later than the initial peak of responses when the delayed rectification has not been activated. From Bennett (1961).

microscopic current loops in this face and by loading the PSP generating membrane tends to reduce current flow around the cell.)

The effect of the delayed rectification on external potentials can be illustrated with reference to Figs. 4B and 15. The external response to a small PSP that does not turn on the delayed rectification is a single-peaked potential which is positive outside the uninnervated face (Fig. 15C). However, the peak of the external potential is earlier than that of the potential across the uninnervated face, a result that indicates that there is a capacitative component of the current. During the rising phase of the PSP the capacity of the uninnervated face is being charged and more capacitative current is flowing. At the peak of the potential no capacitative current is flowing and the current is entirely resistive. When a large PSP activates the delayed rectification, the initial part of the external record is of the same shape as that during small PSPs. However, after this period the delayed rectification turns on, outward current through the uninnervated face increases, and the external potential rises to a second peak (Fig. 15A,C). If the cell is hyperpolarized so that even a large PSP fails to activate the delayed rectification, the external response again has only a single peak that is earlier than the peaks of the transmembrane potentials (Fig. 15B). Note that in B the transmembrane potentials are slowly falling as they are for small PSPs. Moreover, hyperpolarization increases the amplitudes of the potentials because the PSP conductance change is unaffected while the driving force, the difference between the membrane potential and the PSP reversal potential, is augmented (see Bennett, 1961). If a large PSP is evoked after a depolarizing pulse that activates the delayed rectification, the external potential is enlarged, the voltage drop across the external response has only a single peak that occurs at nearly the same time as the peak of the transmembrane potentials (Fig. 15D). In addition, the external potential is enlarged, the voltage drop across the uninnervated face is reduced, and the PSP across the innervated face is also reduced because of the greater electrical load placed on it. A PSP that turns on the conductance increase causes the same changes in subsequent responses as does directly applied depolarization.

The utilization of the delayed rectification to reduce the resistance of the innervated face is an interesting adaptation. It allows the fish to maintain a high resting resistance but to achieve a large external current during activity. A possible disadvantage to this mechanism is that the increase in conductance is delayed, a property that would not appear to be very significant in an organ discharge that involves fused and presumably repetitive responses. The earliest PSPs would cause the conductance increase which would remain activated during the later PSPs. The double peak of the initial external responses of single cells is not seen in organ discharges although there may be an inflection on the rising phases (Fig. 12A). The difference is undoubtedly a question of synchronization because if electric stimuli are used to evoke synchronous PSPs in lengths of organ the external responses have the same shape as the responses of single cells.

D. Freshwater Electric Fish

1. GYMNOTIDS

The gymnotids are a diverse group of fish living in fresh waters of tropical South America. They are often divided into six families (Table I), but the relationship between the families is obvious from the similarities in body shape and in many other characteristics as well as from the possession of electric organs (Fig. 1). This group includes the electric eel, probably the best known electric fish, and a moderate number of other species that have only weakly electric organs. Because of their characteristic elongate shape, the weakly electric gymnotids are commonly called knife fish. The greater part of the length of the fish contains only muscle, spinal column, and electric organ. The viscera are located in the anterior end of the body, and the anus and genital papilla are just behind the chin. The anal fin begins slightly more posteriorly and runs to near the caudal end of the fish. The electric organ runs along most or all of the length of the body and is innervated by spinal nerves. A caudal fin is absent in all but the sternarchids, in which it is greatly reduced. There is little muscle caudal to the anal fin, and this region, sometimes called the *caudal* filament, contains mostly electric organ.

Gymnotids can regenerate new posterior regions if part is removed (Ellis, 1913). Often in fish taken from the wild the posterior is regenerated indicating that there were encounters with predators in which only the front end of the fish escaped. Ordinary swimming movements of the gymnotids are quite different from those of almost all other fish. The body is held straight, and waves along the anal fin provide the propulsive force. Most species appear to swim about equally well forward and backward and often investigate objects by swimming backward toward them (see *Gymnarchus,* Section II, C, 3). Most of the body musculature is not involved in movement of the anal fin and is probably used only in emergency movements when the body is rapidly flexed in the more usual fishlike manner.

The species of gymnotids are poorly described. Keys are available for a few areas of South America, but it is doubtful that these are complete. Tropical fish dealers, the usual source of supply of gymnotids, often bring in fish from other areas that are clearly new species. On the Rio Negro expedition of the R. V. Alpha Helix, perhaps one-third of the gymnotid species collected were undescribed taxomically. While there are difficulties in working on undescribed species, classification into genera is generally simple, and generic characteristics are apparently the most important ones from a physiological point of view. Nevertheless, it may be desirable to save specimens of a particular species studied for subsequent identification when the systematics of gymnotids is clarified.

Table II and Fig. 16 present a key to gymnotids that is modified from that of Ellis (1913). A second system has been added, one that is nondestructive and applicable to living specimens. Accurate classification sometimes

Table II Key to Families and Genera of Gymnotids

The gymnotids are recognizable from their elongate body and anal fin; the pectoral fins are small; the caudal is present only in the sternarchids in which it is very small; other fins are absent. There are two African species, *Notopterus ofer* and *Xenomystus nigri,* that are also commonly known as knife fish. They are similar in appearance to some weakly electric gymnotids, and confusion might arise if place of origin were unknown. African knife fish are easily distinguished by the short "tentacle" from the anterior naris and the anal fin that extends a short distance around the tip of the tail to form a false caudal fin (as is also true of the electric eel). *Notopterurus* has a small dorsal fin. In the following key the species name is given, if only a single species has been described.[a]

a. Head flattened dorsoventrally, lower jaw projects beyond upper so that the large mouth opens somewhat upwards

- b. Strongly electric; electric organ occupies much of the body caudal to the abdominal cavity; body not scaled; anal fin extends around the end of the tail to form a false caudal. Electrophoridae. *Electrophorus electricus*
- bb. Weakly electric; scaled; slender cylindrical tail extends beyond anal fin. Gymnotidae. *Gymnotus carapo*

aa. Head round in cross section or compressed laterally, lower jaw projecting little if any, mouth small or large

- c. Caudal fin absent; tail beyond anal fin slender and usually cylindrical; no dorsal filament
  - d. Snout short, Sternopygidae[b]
    - e. Teeth in both jaws, color uniform or with longitudinal stripes
      - f. Color uniformly dark except sometimes a lighter stripe along the posterior lateral line; body compressed laterally; orbital margin free, i.e., there is a distinct and deep cleft between the eye ball and adjacent skin; posterior air bladder long and conical; organ discharge more or less sinusoidal at a frequency of 50–150/sec. *Sternopygus*
      - ff. Color light; fairly transparent to pale yellow, often with several darker longitudinal stripes; body compressed laterally; eye covered by a thin membrane; posterior air bladder small, nearly spherical; organ discharge more or less sinusoidal at a frequency of 250–600/sec. *Eigenmannia*
      - fff. Very long caudal filament, dorsal profile of head concave. *Rabdolichops longicaudatus*[c]
    - ee. Teeth absent, color dark but patterned; brown to black mottled or with slightly diagonal banding; organ discharge consists of brief pulses separated by much longer intervals
      - g. Body compressed laterally, depth increases from posterior of head to shortly after beginning of anal fin, then decreases (Fig. 16); profile rounded; accessory organs in head region (Fig. 32); lies on side when resting on a flat surface. *Steatogenys elegans*[d]
      - gg. Body less compressed, depth near greatest at posterior of head (Fig. 16); profile shows a protruding mouth; no rostral accessory organs; rarely lies on side. *Hypopomus*
  - dd. Snout long and tubular. Rhamphichthyidae
      - g. Body entirely scaled. *Rhamphichthys rostratus*[c]
      - gg. Sides not scaled in anterior region. *Gymnorhamphichthys hypostomus*
- cc. Caudal fin present but quite small; dorsal filament (see Fig. 42) closely adherent to back but may be separated in fixed specimens; organ discharge frequency 700/sec or more. Sternarchidae[e]

Continued

Table II Key to Families and Genera of Gymnotids—Cont'd

h. Snout long, tubular, and down curving. *Sternarchorhynchus oxyrhynchus*[f]
hh. Snout long, tubular, horizontal and straight
  i. Mouth large, opens at least one-third the distance back to the level of the eye; upper profile of head markedly convex. *Sternarchorhamphus*
  ii. Mouth small, opens less than one-sixth the distance back to the level of the eye, upper profile quite straight. *Orthosternarchus tamandua*[c]
hhh. Snout not long unless mouth is very large
  j. Teeth present in both jaws; some located externally on what appear to be swollen lips. *Oedemognathus exodon*
  jj. Teeth present in both jaws but inside mouth
    k. Dorsal region of body virtually entirely scaled from head posteriorly to origin of dorsal filament
      l. Mouth large, its angle extending at least as far posteriorly as the anterior margin of the eye. *Sternarchus*
      ll. Mouth small, its angle extending no farther posteriorly than the posterior naris. *Sternarchella*
    kk. Scales absent from much of the body above the lateral line anterior to the dorsal filament (shaded in Fig. 16); scales near lateral line much larger than more dorsally. *Porotergus*
  jjj. Teeth absent from upper jaw
    m. Lower jaw toothless and with a distinct midline groove into which the beaklike upper jaw fits; middorsal region of body scaled anteriorly. *Adontosternarchus*
    mm. Lower jaw sometimes toothed and fitting into a midline groove in the upper jaw; middorsal region of body naked anteriorly. *Sternarchogiton*

[a]Modified from Ellis (1913), Eigenmann and Allen (1942), and Gery and Vu-Tân-Tuê (1964) who should be consulted for more extensive descriptions.
[b]*Sternopygus, Eigenmannia, Hypopomus,* and *Steatogenys* are normally grouped together as the family Sternopygidae. Properties of the electrocytes and their innervation indicates that *Hypopomus* and *Steatogenys* are more closely affiliated to the Rhamphich-thyidae. Greenwood *et al.* (1966) observe "Peculiar specializations in [the gymnotids] are many, and a complete study may alter the family arrangement accepted here."
[c]The author has not seen members of these genera nor are there any physiological data from them.
[d]Only one species is described in the taxonomic literature, but there is a second species described here that has somewhat different rostral electric organ.
[e]There is a proposed revision of Sternarchidae and *Sternarchus* to Apteronotidae and *Apteronotus* on grounds of priority (1800 vs. 1801). As *Sternarchus* has been used for a long time and provides the basis of many other generic names in the family, the author stands with Ellis (1913) who uses this name and derivatives.
[f]Although this genus has been considered to be monospecific, two apparent species of each were found on the Rio Negro expedition of the Alpha Helix.

requires dissection of the specimen, but a reasonable identification can usually be achieved without it, particularly if an oscilloscope is available to observe electric organ discharges.

a. The Electric Eel. The electric eel was the first electric fish for which the cellular mechanisms of the discharge were elucidated (Keynes and Martins-Ferreira, 1953; Altamirano *et al.*, 1953). The eel emits two classes of pulses, small pulses about 10 V in amplitude and large pulses some 500 V or more in amplitude. All the pulses are monophasic, head positive,

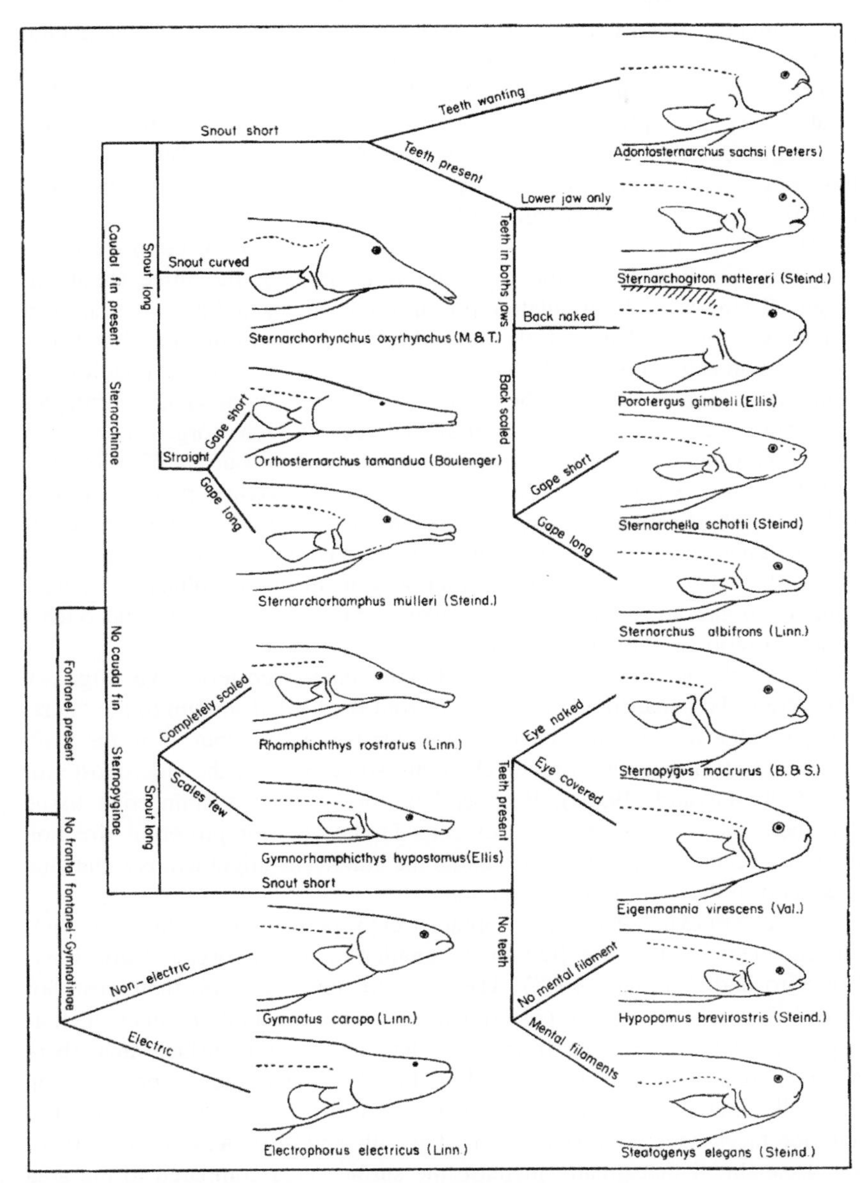

Fig. 16. Heads of representative gymnotids. Slightly modified from Ellis (1913).

and about 2 msec in duration. The voltage increases with the length of the fish, which can exceed 1.5 meters, but even a baby 7–10 cm long can emit close to 100 V. When the animal is resting, the small pulses are emitted at quite low frequencies down to a few per minute. When actively swimming, the pulse rate is usually increased to about 30 or more per second. Large pulses are emitted only in high frequency bursts at a frequency of several hundred per second. The first pulse in such a burst is small, but the amplitude increases to the maximum within 2 or 3 pulses (Albe-Fessard, 1950a). The weak pulses are implicated in the electrosensory system (see Chapter 11, this volume); the strong pulses have offensive or defensive value. The emission of moderate to high frequency pulses by one eel attracts other eels, and they also increase their rates of discharge (Bullock, 1970).

The electric tissue occupies much of the cross section of the body in the posterior three-fourths of the fish, and it is divided into three (bilateral) organs. There is axial musculature dorsal to the organs and a smaller amount of muscle ventrally that controls the anal fin (Fig. 17). In the dorsal and posterior region is the organ of Sachs, and in the ventral and anterior region is the main organ (Fig. 1). Beneath these organs is Hunter's organ. The electrocytes are more widely separated and somewhat larger in Sachs' organ than in the main organ, but otherwise the two organs are similar (Luft, 1957; Couceiro and Akerman, 1948). The main organ generates the greater part of the high voltage discharge used offensively or defensively. The organ of Sachs generates the greater part of the low voltage pulses involved in the electrosensory system but also contributes in a minor way to the high voltage discharge. Hunter's organ apparently functions like main organ anteriorly and Sachs' organ posteriorly (Albe-Fessard and Chagas, 1954).

The single electrocytes are flattened in the anterior posterior axis (Fig. 17). They are ribbon-shaped and tend to run from the medial septum to the lateral margin of the fish. In an adult there are dorsoventrally about 25 in the main organ and 10 in Hunter's organ. There are some 6000 in the anteroposterior axis (Albe-Fessard, 1950a). Each cell is contained in a connective tissue chamber, and successive layers of cells in the anterior posterior direction are fairly accurately aligned, one behind the other. The alignment is somewhat less good in the dorsoventral direction.

The electrocytes of the main organ occupy perhaps one-fifth to one-half the chamber volume; the fraction is smaller in Sachs' organ (Luft, 1957; Couceiro and Akerman, 1948). The cells are innervated on their posterior faces by spinal nerves. The posterior faces have a moderate number of short papilli or stalks protruding from them that increase the surface area; there are also some tubules in this face (Fig. 17). The innervation is primarily on the stalks and is much less dense than in electrocytes of torpedinids. The anterior faces have a large number of papilli and an extensive network of small canaliculi that greatly increase the surface area compared to the area that a simple planar membrane would have (Fig. 17, see also Mathewson *et al.*, 1961).

The properties of eel electrocytes are only slightly more complex than those of electrocytes of the stargazer and *Torpedo*. The innervated face responds to depolarization by generating a spike that overshoots the resting potential by about 50 mV. The uninnervated face is of very low resistance and does not become excited. The responses of a single cell are shown in

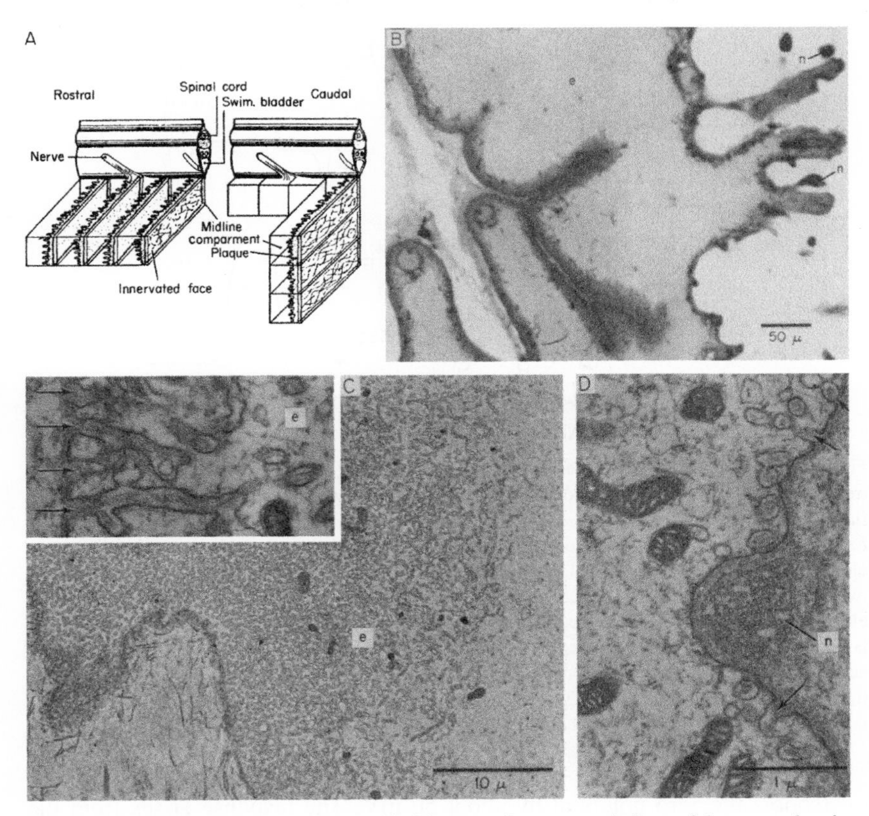

Fig. 17. Structure of eel electric organ. (A) Diagram of gross morphology of the organ showing series arrangement of electrocytes ("plaques") on the left and parallel arrangement on the right. From Altamirano *et al.* (1953). (B) Light micrograph of a section through a single electrocyte (e). The caudal, innervated surface (on the left) has some small processes from it. The innervation appears quite sparse and few nerve fibers (n) are visible. Relatively stout processes come off the anterior, uninnervated face and there is a pronounced layer of increased density associated with the membrane. (C) Low and high (inset) magnification electron micrographs of the uninnervated surface showing the extensive canalicular network responsible for the density associated with this face seen in B. The canaliculi open to the exterior (arrows) and branch profusely. Basement membrane material extends into the canaliculi, (D) High magnification electron micrograph of the innervated face showing a vesicle filled nerve terminal (n). The surface proliferation provided by canaliculi (openings indicated by arrows) is much smaller at this face. Magnification for inset in C same as that for D. Micrographs provided through the courtesy (and skill) or Drs. F. E. Bloom and R. Barnett.

Fig. 18. Current is applied by external electrodes in order to depolarize the innervated face, and differential recording between two microelectrodes is used. In Fig. 18A–C one electrode is advanced into the cell and then into the underlying extracellular space. When the "exploring" electrode is in the cell the internally negative resting potential of —90 mV is recorded across the innervated face (Fig. 18B). An adequate stimulus evokes a spike that

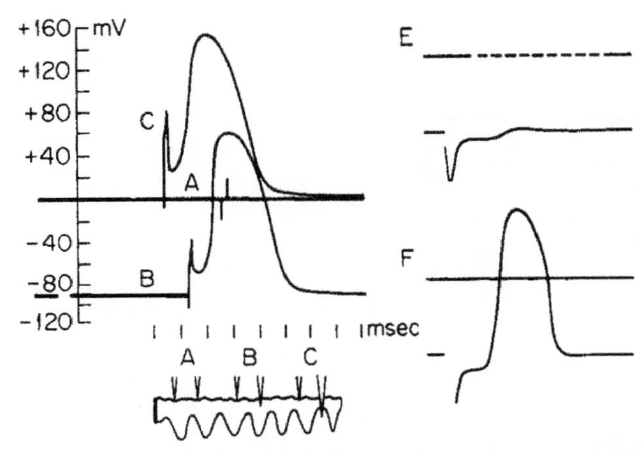

Fig. 18. Responses of electrocytes of the electric eel. Right: recording differentially as one electrode is advanced through a cell; positivity of this electrode shown upward. Current passed through the cell by external electrodes in order to depolarize the innervated face. (A) Both electrodes external to the innervated face; no response is seen (there is a brief diphasic stimulus artifact). (B) One electrode is advanced into the cell. The inside negative resting potential of about 90 mV and an overshooting action potential about 140 mV in amplitude are recorded. (C) When the exploring electrode is advanced to outside the uninnervated face, the resting potential disappears, but the spike is essentially unchanged. From Keynes and Martins-Ferreira (1953). (E–F) Differential recording across the innervated face, the upper trace shows the zero potential. Stimuli hyperpolarizing the innervated face can evoke PSPs (E) that arise after a latency of 1–2 msec and that if sufficiently large initiate a spike (F). From Altamirano *et al.* (1955). Calibrations are the same for A–F.

overshoots the zero potential by about 50 mV. The response arises from the depolarization produced by the stimulus; the delay can be made much shorter if a stronger stimulus is given. When the exploring electrode is advanced out through the uninnervated face of the cell, the resting potential disappears demonstrating that the resting potentials across the two faces are equal (Fig. 18C). The response amplitude however does not change. As discussed in respect to electrocytes of *Astroscopus* this finding indicates that the resistance of the uninnervated face is very low. Direct measurements using current pulses confirm its low resistance, which is much lower than that of the innervated face. The degree of surface elaboration seen cytologically correlates with the different resistances.

Stimulation of the nerve supply evokes PSPs that can depolarize the innervated membrane to the point where it generates a spike. Current applied as in Fig. 18A–C but in the opposite direction hyperpolarizes the innervated face and does not excite it. Such stimuli can excite nerve fibers in the tissue that then produce PSPs that arise after a delay of about 2 msec (Fig. 18D). If enough nerve fibers are stimulated the PSP initiates a spike (Fig. 18E).

The ionic basis of the action potential has been well studied in electrocytes of the eel. The cells are depolarized by high potassium solutions approximately as predicted by the Nernst relation, which indicates that at rest the potassium permeability of the cell predominates and the potential is largely determined by the intra- and extracellular concentrations of potassium (Higman *et al.*, 1964). The inward current responsible for the rising phase of the spike is sodium dependent (Keynes and Martins-Ferreira, 1953). Furthermore, it is eliminated by the pharmacological agent tetrodotoxin which is a specific blocking agent for the increase in sodium permeability produced by depolarization (Nakamura *et al.*, 1965).

Unlike the squid axon and many other tissues, the eel electrocytes lack K^+ activation or delayed rectification, but they do have anomalous rectification in the innervated face (Nakamura *et al.*, 1965). Actually, it makes sense for the cells to have anomalous rectification and not to have K^+ activation. For maximum effectiveness as an electric organ, the circuit for all the current carried inward by Na^+ should be completed by current in the external environment, not by local currents in the innervated face. The time constant of the cells is sufficiently short that the membrane rapidly returns to the resting potential without the restoring effect of delayed rectification (as is also very nearly true of myelinated nerve fibers, see Frankenhaeuser and Huxley, 1964). The anomalous rectification decreases the conductance enough to result in a severalfold decrease in eddy currents in the innervated face.

The excitability properties of the innervated face lead to an unusual sequence of conductance changes during the response (Morlock *et al.*, 1969). At the start of the spike, depolarization causes sodium activation and increased conductance and further depolarization; depolarization also decreases the conductance of anomalously rectifying channels, but the net result is increased conductance. Sodium inactivation ensues, and the total conductance decreases to below the resting value while the membrane potential falls from the spike peak; at a sufficiently low potential anomalous rectification reverses and conductance rises to the resting level. This diphasic sequence of conductance changes, increase followed by decrease, contrasts to the monophasic increase seen in nerve (Cole and Curtis, 1939; Tasaki and Freygang, 1955).

One might ask why the innervated membrane should not just have a high resting resistance instead of anomalous rectification. One possible reason is to provide a conductance in inactive cells for flow of current generated by active cells in series with them. As will be discussed further in Section III, A, the cells are not all active at the same time except in the largest discharges. The voltage of these responses is accounted for by synchronous activity of some 6000 cells in series each producing over 100 mV.

b. Gymnotus. This species is weakly electric; its organ discharge is a fraction of a volt recorded in water and not much more than a volt recorded in air. Its body shape is similar to that of the eel, but the electric organ is much smaller (Fig. 1). When undisturbed and resting it normally emits pulses at about 50/sec. The pulses are approximately triphasic, initially head negative, and about 1 msec in duration (Fig. 3B, B′). Mechanical stimuli, light, electric fields, and resistance changes can cause moderate transient accelerations of the discharge, and when the animal is feeding, the discharge frequency can briefly exceed 200/sec (Fig. 3, Lissmann, 1958; Bennett and Grundfest, 1959; Black-Cleworth, 1970). When swimming around its tank the fish maintains a fairly steady frequency somewhat above the resting level. *Gymnotus* is also capable of ceasing its discharge completely for brief periods, a response that may sometimes represent hiding or "listening." Both accelerations and cessations can be involved in communication between other members of the same species and a relatively potent stimulus for inducing cessation of firing is weak electric pulses at a frequency close to that of the organ discharge.

The electric organ runs longitudinally from just behind the chin to the tip of the caudal filament. The organ lies dorsal to the anal fin but extends beyond it rostrally to the cleithrum as well as caudally (Fig. 1). The organ on one side consists of about four longitudinal columns of drum-shaped cells, the flat faces of which are oriented anteroposteriorly (Fig. 19). The cells are about a millimeter in diameter and 300 μ thick in a fish 20 cm long. The number of columns is reduced anteriorly, and there may be several additional columns caudally. In each column there is one electrocyte per segment and there are about 90 segments in the animal. Each column of cells is enclosed in a connective tissue tube that is divided into chambers by loose septa between the cells. The cells of all but the most dorsal column are innervated by a number of fibers on their posterior faces. Aside from innervation the two faces are very similar. They are quite smooth with relatively few inpocketings and canaliculi (Schwartz *et al.*, 1971). The cells of the most dorsal column have their main innervation on the anterior face, but they have a few fibers ending on their posterior faces as well (Szabo, 1961d). As will be shown below, these cells behave physiologically like the more ventral cells but are oriented in the opposite direction. No obvious function of the posterior innervation was observed in the early physiological study of these cells (Bennett and Grundfest, 1959), but since the presence of the posterior innervation was not recognized at that time the question could well be reinvestigated.

The single electrocytes of *Gymnotus* generate external potentials that are diphasic (Fig. 20) in contrast to the monophasic discharges of the eel. This form of potential is produced because both faces generate spikes. The lower threshold face is the posterior, innervated face in all but the dorsal column of cells in which it is the anterior face. When the nerve supply is activated or when current is applied by an intracellular electrode, the lower threshold, innervated face fires first. Current flows inward through this face and outward through the opposite, uninnervated face which becomes excited, but with some delay with respect to firing of the innervated face. By this time the spike of the lower threshold (innervated) face is decreasing, and current flows in the reverse direction along the axis of the cell. This pattern of activity is indicated by the external recordings shown in Fig. 20. The monopolarly recorded

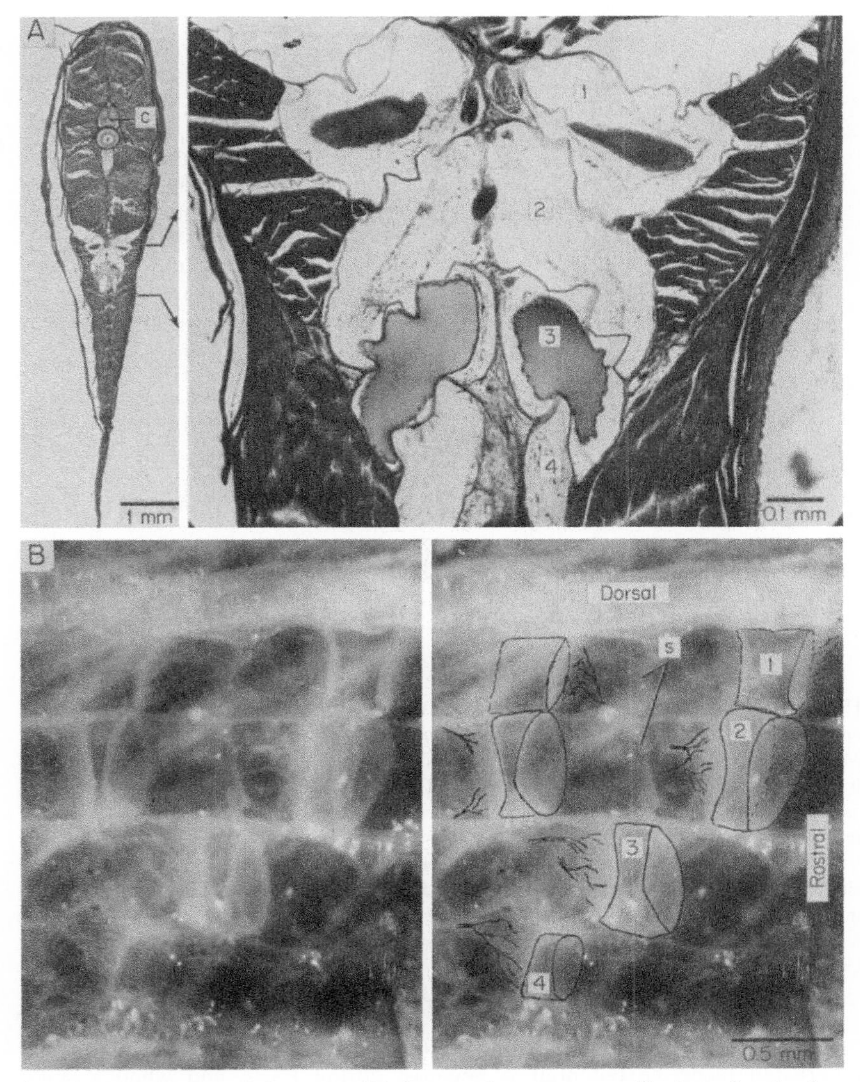

Fig. 19. Electric organ of *Gymnotus*. (A) Cross section at two magnifications of region 2 cm from the tip of the tail of a fish about 10 cm long; higher power view on right of the region indicated. The connective tissue tubes enclosing the four columns of electrocytes are numbered from dorsal to ventral; the section grazes electrocytes of column 1 and passes through nearly the maximum diameter of cells of column 3. The spaces dorsal to the tubes are fixation artifacts. At this level most of the body cross section is occupied by muscle. The vertebral column and spinal cord (c) are seen. (B) Photographs of a dissected unstained preparation. The cells and their main innervation are outlined on the right-hand copy. The caudal innervation of the most dorsal cells is also visible. Electrocytes of all four tubes are numbered. Modified from Bennett and Grundfest (1959).

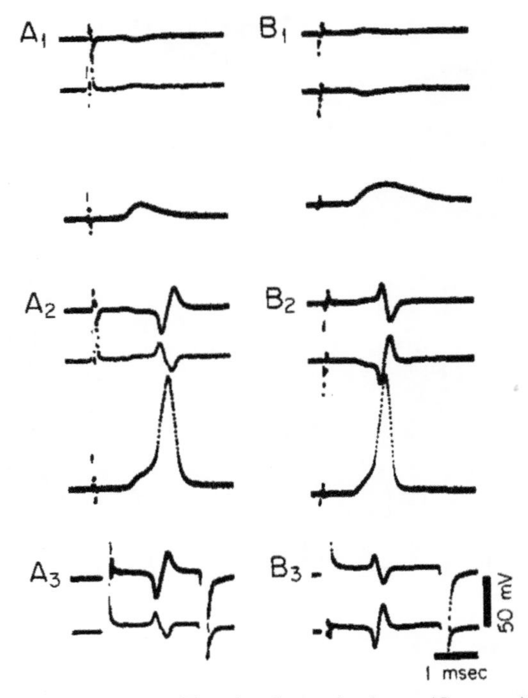

Fig. 20. Responses of electrocytes of dorsal and ventral columns (*Gymnotus*). (A) Records from an electrocyte of the most dorsal column. (B) Records from a cell of the third column ventrally. Monopolar recordings outside rostral face (upper traces), caudal face (middle traces), and intracellularly (lower traces). (A_1,B_1) Weak stimulation of the nerves evokes a depolarizing PSP which is associated with negativity external to the rostral face in A_1, and external to the caudal face in B_1, reflecting the different innervation of the two classes of cell. (A_2,B_2) Stronger stimuli evoke similar intracellular spikes in the two cells, but the diphasic external potential is initially positive outside the rostral face in A_2 and outside the caudal face in B_2. (A_3,B_3) Current passed through the intracellular electrode evokes an external response that is initially negative outside the innervated face in each case indicating that this is the lower threshold membrane. From Bennett and Grundfest (1959).

external potentials are of opposite sign outside the two faces. External to the innervated face, the potential is negative during PSPs, and when a spike arises, goes rapidly more negative. During the later part of the monopolarly recorded intracellular spike, the potential external to the innervated face reverses to go positive, indicating that the uninnervated face has a larger potential across it than the innervated face. The potential outside the uninnervated face has the same shape and about the same amplitude as that outside the innervated face but is opposite in sign. External to the edges of the cell the potentials are very small. This feature and the opposite polarity of potentials external to the two faces demonstrate the longitudinal orientation of the cell. (It should be noted that there will be some local circuit or eddy currents within each face that will not contribute significantly to the externally recorded potentials; for this reason the synapse in the innervated face could cause it to fire first even if it were not of lower threshold.) That the innervated face is indeed of lower threshold is indicated by the effect of intracellularly applied current, which depolarizes both faces equally if the external resistances are equal (see Fig. 2). This procedure always excites the innervated face first, even when the external resistance is somewhat greater on the innervated side.

As will be shown later with respect to *Gymnarchus* (Section II, C, 3), a diphasic external potential can arise if one face is inactive and behaves as a series capacitance. The excitability of both faces of *Gymnotus* electrocytes can be clearly demonstrated by stimulating with externally applied currents that run along the axis of the cell; these currents hyperpolarize one face and depolarize the other. An experiment of this kind is illustrated in Fig. 21. When a stimulating cathode is placed external to the uninnervated face, the applied current tends to depolarize this face and hyperpolarize the innervated face

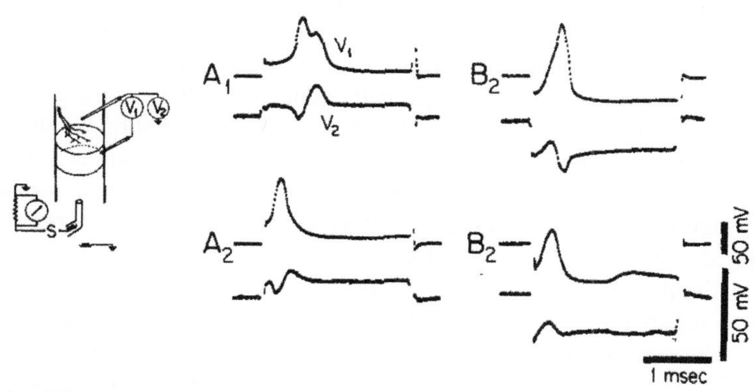

Fig. 21. Effects of axial stimulation on electrocytes of *Gymnotus*. Stimulation by rectangular current pulses and recording as indicated in inset except that the indifferent electrode is much farther away. Upper trace: differential recording across innervated face. Lower trace: monopolar recording external to innervated face. (A_1,A_2) Anodal stimulation external to the uninnervated face. (A_1) The stimulus is moderately above threshold and initiates a two component spike. The external record is initially negative indicating that the innervated face generates a spike first. (A_2) Stronger stimulation largely blocks the second spike component and reduces the positive phase in the external record; evidently the stimulus hyperpolarizes the uninnervated face sufficiently that it is only partially excited by the spike of the innervated face. (The residual positivity may represent local response in or capacitative currents through the uninnervated face, see Section II, D, 3.) (B_1,B_2) Cathodal stimulation external to the uninnervated face. (B_1) A two component spike is initiated but the external record is initially positive indicating that the uninnervated face fires first. (B_2) Stronger stimulation causes failure of the second spike component and a large reduction in the negative phase of the external record indicating failure of excitation of the innervated face. The small intracellular positivity and associated external negativity that develops after a latency of about 1 msec is a PSP resulting from stimulation of the nerve supply. Modified from Bennett and Grundfest (1959).

(Fig. $21B_1$). The uninnervated face fires first as indicated by initial positivity of the response recorded external to the innervated face. The spike recorded across the innervated face arises from a hyperpolarized level of membrane potential confirming that this face is not initiating the spike. The innervated face is excited by the activity of the uninnervated face, and negativity external to this face follows the positive phase. When the stimulating electrode external to the uninnervated face is an anode, the innervated face is depolarized and fires first; there is initial negativity outside this face (Fig. $21A_1$). Since the uninnervated face is hyperpolarized by the applied current, its firing is delayed compared to that evoked by neural or intracellular stimulation and the spike recorded across the innervated face has two quite distinct components. (The two spike components are only barely recognizable in the monopolar recordings of Fig. 20.)

Usually if stimulation only moderately above threshold is applied as in Fig. $21A_1$,B_1, the initially active face can excite the opposite face that is being hyperpolarized by the stimulus. However, strong stimuli can cause sufficient hyperpolarization of this face that the spike of the initially firing face excites it only partially if at all. Correspondingly, there is failure of the second spike component recorded intracellularly and disappearance or reduction of the second phase of the externally recorded responses (Fig. $21A_2$,B_2).

Although in *Gymnotus* the single cells generate diphasic external potentials, the overall organ discharge is triphasic. This configuration results from the opposite orientation and earlier firing of the most dorsal electrocytes (Fig. 22). These cells fire about ½ msec earlier than the more ventral cells, all of which fire synchronously. The activity of the anterior faces of the dorsal cells results in the initial head negativity. The activity of their uninnervated faces is simultaneous with the activity of the innervated faces of the ventral cells and summates with it to cause the second, head negative phase. Activity of the uninnervated faces of the ventral cells causes the final, head negative phase. Corresponding to the number of cells active, the initial phase is the smallest, the second phase is the largest, and the final phase is somewhat smaller than the second phase.

In *Gymnotus* the electric organ has at its rostral region a small number of modified cells that fire earlier than the main organ and apparently generate monophasic external potentials as do eel electrocytes. The detailed operation of this part of the organ has not been investigated. It resembles in its operation the rostral accessory organs of several other gymnotids (see below).

c. Hypopomus. At least three species of *Hypopomus* have been studied electrophysiologically, but the correspondence to the taxonomically named species is somewhat uncertain. Pulses are emitted at a basal frequency of 5–10/sec, and again there are large accelerations during swimming or when the fish is stimulated. One species can maintain its discharge rate quite constant at two or more levels, the higher ones generally associated with greater activity (Bullock, 1970). Cessation of discharge has also been observed (Bullock, 1970; Black-Cleworth, 1970). The pulses are about 2 msec in

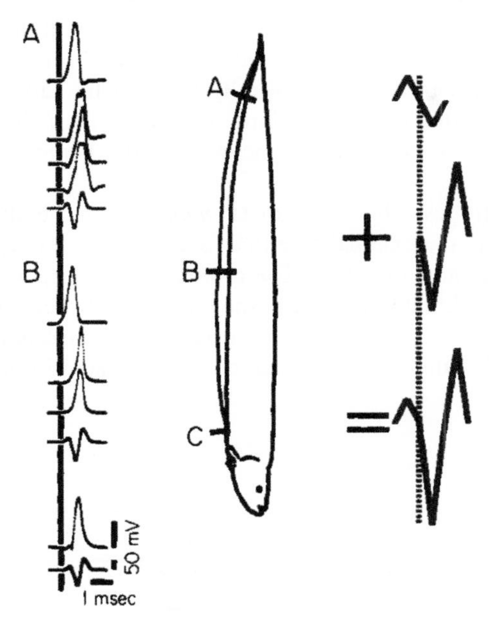

Fig. 22. Constitution of the organ discharge in terms of the responses of the different columns of electrocytes *(Gymnotus).* Left, intracellularly recorded responses of individual cells in different columns at the three levels of the fish indicated in the center diagram. The potentials are somewhat distorted by pickup from other active cells. In each case the lowest trace is the monopolarly recorded organ discharge at the tip of the tail which is used as a time reference; the vertical line indicates onset of organ discharge. (A) The electrocytes of the most dorsal column (upper traces) fire about 0.2 msec earlier than the cells of the three ventral columns (middle three traces). (B) At this level the electrocytes of the most dorsal column also fire earlier than those in the next two columns ventrally, but the discharge slightly precedes the firing of the dorsal column in A. (C) Only one column continues to this level. The electrocytes are caudally innervated and probably represent those in the third column. The response arises approximately at the same time as does activity in the caudally innervated electrocytes in A. Right, organ activity recorded at the tail as a summation of the potentials produced by the different columns. Broken vertical line indicates the start of discharge of the three columns of caudally innervated cells. The rostrally innervated cells contribute a smaller, earlier diphasic potential which is of opposite sign. From Bennett and Grundfest (1959).

duration and of the order of 1 V in amplitude; they are monophasic head positive in one of the species studied and diphasic initially head positive in another.

As in *Gymnotus* the electric organ runs from just behind the chin region to the tip of the caudal filament. The organ consists of 3–5 longitudinally running columns on each side in which there is one electrocyte per segment (Bennett, 1961). The cells are cylindrical, about 0.3 mm in diameter, and 0.2 mm thick in a 15-cm fish. All the electrocytes are innervated by a small number of nerve fibers on a short process or stalk from their posterior faces. In respect to innervation the electrocytes closely resemble those of the main organ of *Steatogenys* (Fig. 29).

In the species of *Hypopomus* (probably *H. artedi)* with the diphasic organ discharge, the potential is initially head positive when it is recorded at the head of the fish or some distance rostrally to it. At the tail and caudal to it, the potential is inverted (Fig. 23). Along the side of the fish the discharge is triphasic, probably because of asynchronous discharge of anterior and posterior cells.

The single cells resemble those of *Gymnotus* in that both faces generate spikes and the resulting external potentials are diphasic; however, all the cells are oriented in the same direction. The response of an electrocyte to intracellularly applied depolarizing current is shown in Fig. 24 recorded both monopolarly and differentially. The innervated (posterior) face is lower threshold because it fires first under these conditions as indicated by initial negativity

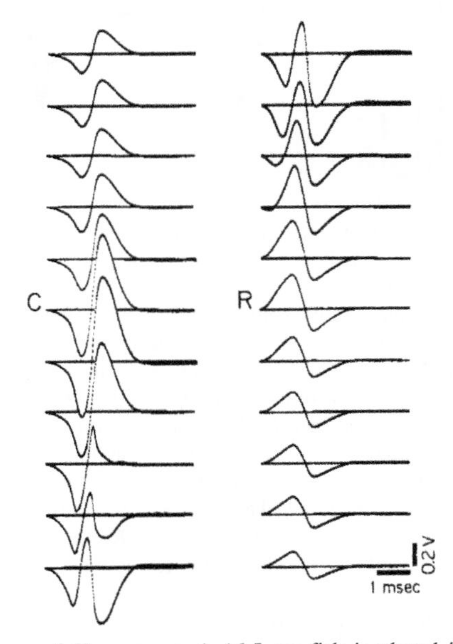

Fig. 23. Organ discharge of *Hypopomus.* A 16.5 cm fish is placed in a shallow plastic tray, 45 cm long by 24 cm wide by 5 cm deep, and held by a gauze tube lengthwise against the long side, midway between the surface and bottom. Three electrodes are used. One is fixed at the head, and one is moved at 1.5 cm intervals along the axis of the fish; these two electrodes record differentially against an electrode at the midpoint of the opposite side of the tray. The records are therefore somewhat larger but similar in form to what would be obtained by monopolar recording in a very large volume of water. The brief discharges occur at about 5/sec when the fish is unstimulated. The discharge at the rostral fixed electrode triggers the oscilloscope and is used to align the records in the figure obtained from the exploring electrode (positivity of this electrode up). At the tip of the tail (C) the potential is diphasic and initially tail negative. Moving the electrode farther caudally, it becomes smaller, but is similar in form (records above C). Moving rostrally along the body, the potential becomes triphasic and then diphasic, initially head positive, at the tip of the snout (R). The potential remains diphasic but decreases as the electrode is farther advanced (records below R). From Bennett (1961).

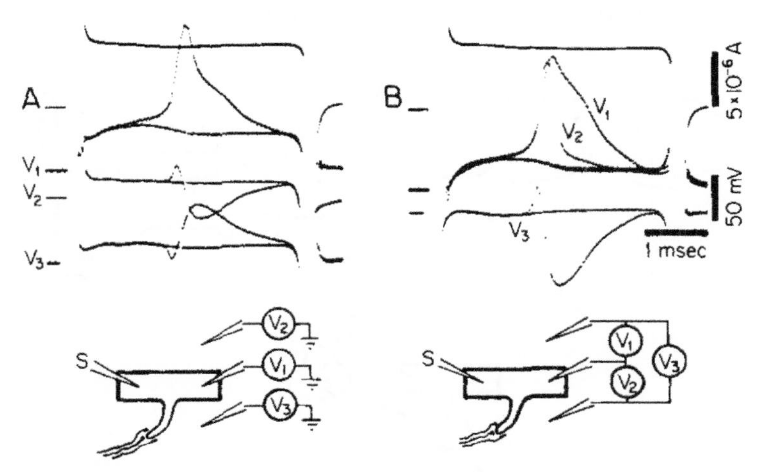

Fig. 24. Response of an electrocyte of *Hypopomus* to intracellularly applied depolarizing current. Superimposed threshold pulses that do and do not excite the cell. (A) Monopolar recording; (B) differential recording as in diagrams. Upper trace: stimulating current. The transmembrane potentials in B (V_1 and V_2) start from the same level. The external record in B (V_3) is shown with positivity external to the uninnervated face upward. Both faces are excited, the spike of the innervated face is lower threshold or faster rising and initially there is negativity external to this face. The spike of the caudal face is longer lasting and larger, and there is a longer lasting negativity external to this face. The characteristics of the spikes are more clearly seen from the differential recordings (B). The external records in A are mirror images, showing the longitudinal orientation of the cell. From Bennett (1961).

outside this face and initial positivity outside the uninnervated face. Differential recording across the two faces shows clearly that, although the spikes in each face are of about the same peak amplitude, the spike generated by the innervated face is considerably briefer than that generated by the uninnervated face. Thus, the head-negative phase of the external potential is larger than the head-positive phase.

External stimulation of the electrocytes also demonstrates that both faces generate spikes (Fig. 25). When current is passed in order to depolarize the uninnervated face and hyperpolarize the innervated face, the external record is initially head negative (Fig. 25A). The spike in the uninnervated face soon excites the briefer spike of the innervated face and the external record goes transiently more head positive. The longer lasting spike of the uninnervated face causes the external record to go head negative again. When current is passed that depolarizes the innervated face and hyperpolarizes the uninnervated face, the brief spike of the innervated face is evoked (Fig. 25B). However, this activity does not excite the long-lasting spike of the uninnervated face, which is apparently too hyperpolarized by the applied current. [A small head negativity occurs in the external record. This phase may represent a sub-threshold response of the uninnervated face or discharge of the capacity of this face (see *Gymnarchus,* Section II, D, 3).]

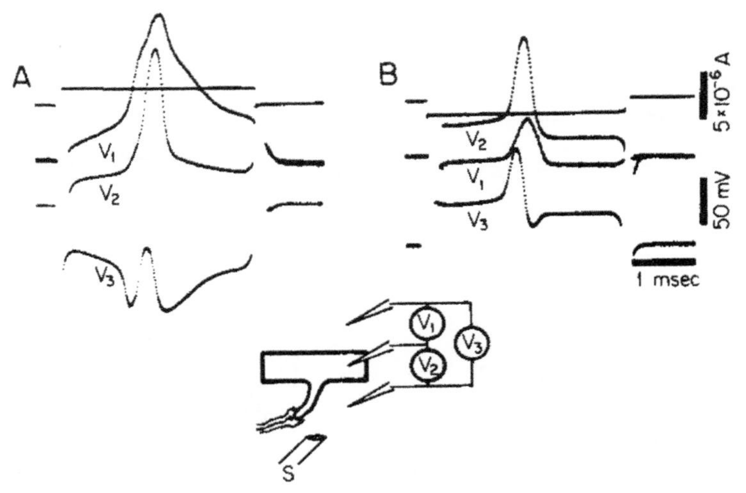

Fig. 25. Stimulation of an electrocyte by externally applied currents. *Hypopomus*, same cell as Fig. 24. Stimulation and recording as in diagram. Upper trace: stimulating current. Middle traces (V_1 and V_2): potentials across the two faces starting from the same level. Lower trace: potential between the external electrodes, rostral positivity up. When anodal current is passed (A), the caudal (stalk) face is hyperpolarized and the rostral (uninnervated) face is depolarized. The rostral face is excited first as indicated by initial rostral negativity externally. This activity excites the brief caudal spike which produces a rostral-positive phase in the external record which then becomes rostral-negative again. Oppositely directed currents (B) excite the caudal face first. The caudal spike fails to fire the rostral face, which is hyperpolarized by the stimulus. Only a small rostral negativity occurs in the external record. From Bennett (1961).

When the nerve supply is stimulated, excitation occurs first in the stalk and then propagates into the main part of the cell. The potentials in stalk and body can be quite different (Bennett, 1961) as is illustrated with respect to the species of *Hypopomus* with a monophasic discharge (Fig. 28).

The external potential generated by the electric organ of the monophasically discharging *Hypopomus* remains relatively constant in shape as an exploring electrode is moved along the fish and becomes very small just caudal to the center of the fish (Fig. 26). Evidently the fish behaves more like a dipole than the diphasically discharging *Hypopomus*, perhaps because of better synchronization between cells but also because the discharge of the individual cells is monophasic and the total output is therefore less sensitive to slight failures of synchronization.

The single cells have properties similar to those of the electric eel (Fig. 27). However, the uninnervated (anterior) faces have a resistance about half that of the innervated faces (Fig. 28D) and a spike in the innervated face causes an appreciable voltage drop across the uninnervated face. Still the external potentials differentially recorded across the cells are about 50 mV in amplitude. Corresponding to the different resistances the uninnervated face

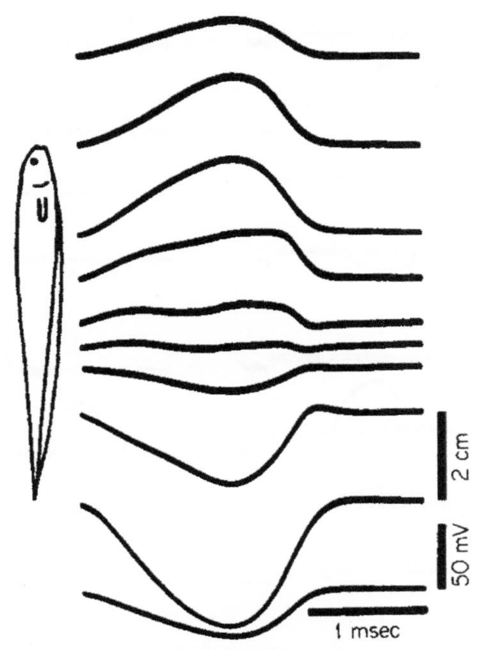

Fig. 26. Potentials generated by the monophasically discharging *Hypopomus*. Procedure as for Fig. 23, but recorded with the fish in the center of a somewhat smaller container. The distance calibration gives recording sites with respect to the fish referred to the level at the start of the records.

is more proliferated by tubules and canaliculi than is the innervated face (Schwartz *et al.*, 1971). The form of the recorded spike potentials depends on the form of the stimulus. A brief pulse near threshold produces a potential that rises more slowly than it falls, whereas the opposite is true of long-lasting stimuli. The response to brief pulses more closely resembles the externally recorded organ discharge and presumably corresponds more closely to activation by way of the synapses on the stalks.

During neural activation the impulse arises in the stalk and propagates into the cell body. An external stimulating electrode can be used both to evoke the neurally mediated response and to excite the cell directly. A response evoked by a brief anodal pulse applied by an electrode external to the uninnervated face is shown in Fig. 28A. Two small PSPs subthreshold for initiating a spike are seen in the stalk, but there is little spread to the main part of the cell. Stronger stimulation causes a larger PSP, and a spike is initiated in the stalk that is earlier, larger, and longer lasting than the spikes recorded across the faces of the cell. With a stronger and longer lasting stimulus the spike is initiated directly in the stalk (or posterior face) as indicated by the short latency of the response (Fig. 28C). At the time that the PSP occurs in

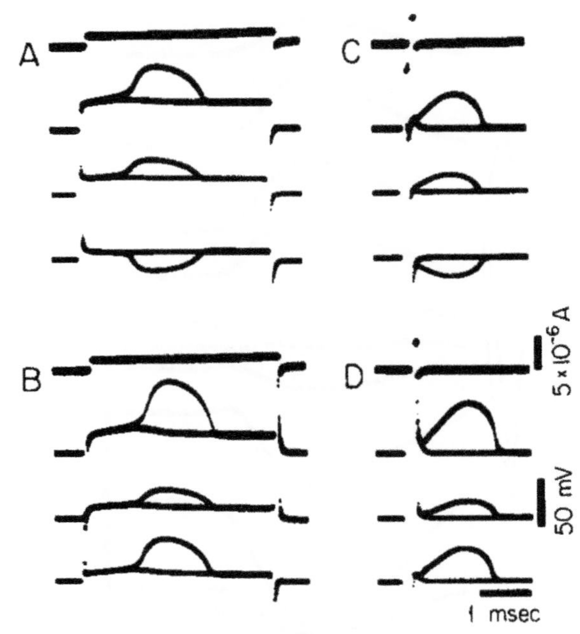

Fig. 27. Response of an electrocyte of monophasically discharging *Hypopomus* to intracellularly applied depolarization. Upper trace: stimulating current. (A,C) Monopolar recording. Second trace: intracellular; third trace: external to uninnervated face; fourth trace: external to innervated face. (B, D) Differential recording. Second trace: across innervated face; third trace: across uninnervated face; fourth trace: across entire cell. Superimposed traces of threshold stimulation with and without a response. The innervated face generates a spike, but there is only a small potential across the uninnervated face.

Fig. 28A there is a depression in the peak of the spike in the stalk. This response is the PSP that is inverted because the potential during the spike exceeds the reversal potential of the PSP (see del Castillo and Katz, 1956). If a cathodal stimulus is applied external to the uninnervated face a large depolarization of this face can be produced without exciting it, demonstrating its inexcitability (Fig. 28D). Stimulation of this polarity hyperpolarizes the innervated face, and if a prior brief stimulus initiates a neurally evoked response in the stalk (Fig. 28C), this activity can fail to invade the main part of the innervated face of the cell. In this event only a small potential is observed across the innervated face even though the spike recorded in the stalk remains large.

The role of stalks in electrocyte function remains obscure. It will be considered again in Section II, F.

One unidentified species of *Hypopomus* has been studied in which external recording suggests that electrocytes in about the anterior two-thirds of the body generate monophasic potentials and the posterior cells generate diphasic

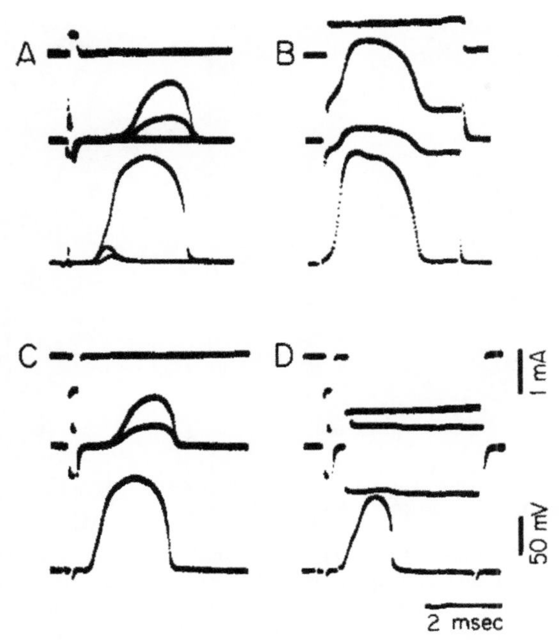

Fig. 28. Neural excitation and stimulation by axial currents of an electrocyte of monophasically discharging *Hypopomus*. First trace: stimulating current applied by an electrode external to the uninnervated face; next two traces starting from the same level: differential recording across the cell's faces; fourth trace: differential recording across the membrane of the stalk (one electrode external to the innervated face was used). (A) Three superimposed traces showing two subthreshold PSPs and one that initiated a spike. The spike in the stalk is larger than that across the main part of the innervated face of the cell. (B) A stronger and longer lasting stimulus excited the stalk and innervated face directly as indicated by the short latency. A PSP is still evoked and appears inverted on the peak of the spike in the stalk. (C) A brief cathodal stimulus causes a neurally mediated response to be initiated in the stalk (as indicated by latency and abrupt rise from the base line). (D) A strong cathodal stimulus prevents the stalk impulse from invading the body of the cell. A large amplitude but abbreviated spike remains in the stalk. The innervated face is depolarized by about 40 mV but fails to be excited.

potentials. This possibility requires exploration using microelectrode techniques. No species of *Hypopomus* has been observed to have a rostral accessory organ like those of *Gymnotus*, *Steatogenys*, and *Gymnorhamphicthys*.

d. Steatogenys. *Steatogenys* is quite similar in appearance to *Hypopomus* and is often confused with it by fish suppliers. The main organ of *Steatogenys* appears morphologically identical to that of the diphasically discharging *Hypopomus* (Fig. 29), and the form of discharge recorded near the tail or distant from the fish is similar (Fig. 30). The frequency of resting discharge is more like that of *Gymnotus,* and various stimuli cause moderate accelerations [although there is some dispute as to this point (see Bullock, 1970)].

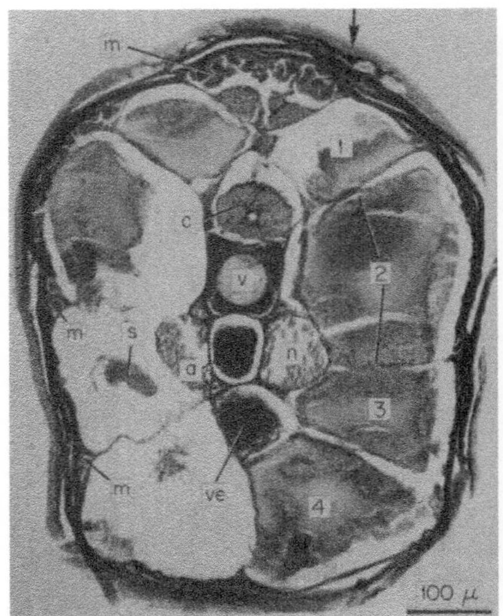

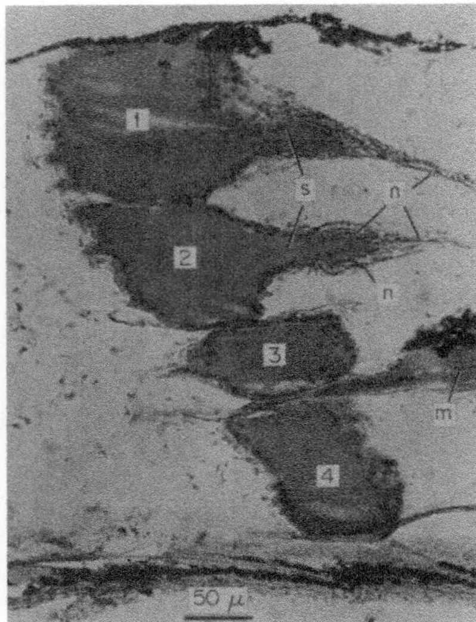

Fig. 29. Anatomy of the main electric organ of the small species of *Steatogenys*. Left: Cross section near the tip of the tail (dorsal up) stained with hematoxylin and eosin (5 cm fish). The four columns of electrocytes are numbered on the right. On the left the section passes through a stalk (s) of an electrocyte in column 3 and grazes the tip of a stalk in column 4. The spinal cord (c), vertebral column (v), dorsal aorta (a) and vein (ve), muscle (m) and electromotor nerves (n) are seen. Right: Paragittal section near the tip of the tail (dorsal up, rostral to the left) from a smaller fish at the level indicated by the arrow in the cross section, stained with silver, Romanes method. The stalks (s) of electrocytes 1 and 2 are seen with their innervating nerve fibers (n). Many nuclei lie just beneath the cell surface.

As in *Hypopomus* innervation occurs on a short stalk from the posterior face (Fig. 29). Both faces of the cells generate spikes and morphologically the two faces are similar although there is a somewhat greater proliferation of the uninnervated face (Schwartz *et al.*, 1971). Recordings across the faces and stalk of an electrocyte of the main organ are shown in Fig. 31. A brief stimulus is applied to the nerve supply and evokes a threshold synaptic potential. This PSP is much larger in the stalk than across either face of the cell. The impulse arises in the stalk and rapidly propagates into the main part of the cell. The external record is initially positive on the side of the uninnervated face, which indicates that the spike in the innervated (stalk) face rises more rapidly, although this is not clear in the figure. The impulse in the uninnervated face then becomes larger and a phase of head negativity appears in the external record. This latter phase is considerably larger than the initial, head-positive phase.

Steatogenys has an accessory electric organ in the chin region (Lowrey, 1913) which was called the *submental filament* by taxonomists (Fig. 32). In *Steatogenys elegans* (a relatively large species reaching 15–20 cm long) the submental organ consists of a single column of cells innervated on their

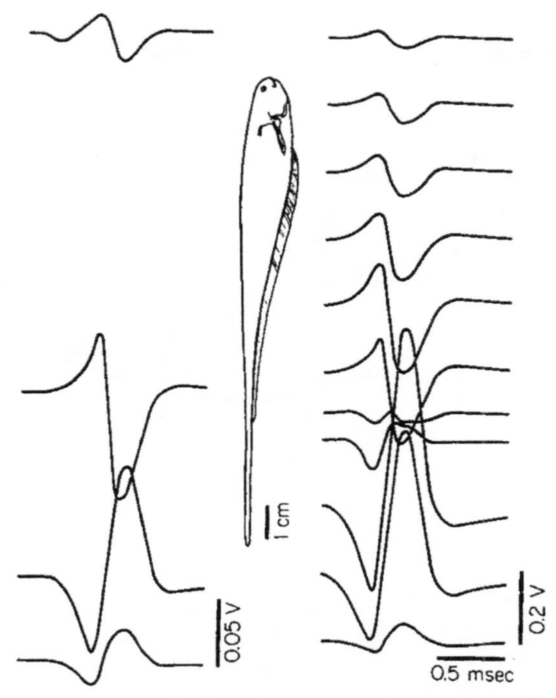

Fig. 30. Electric organ discharge of *Steatogenys elegans*. Recorded monopolarly along the side of the fish in a large dish of water. Time relations are taken from a simultaneous recording from a stationary electrode in the submental region. Records on the left are from the intact fish. The potentials are essentially diphasic in the tail region, but there is an early negative phase near the head (see Fig. 35). The early negative phase is lacking from the records on the right which were taken after removal of the rostral accessory organs. A minimum-sized potential is recorded 4 cm rostral to the tip of the tail and the diphasic potentials are of opposite sign on either side of this point.

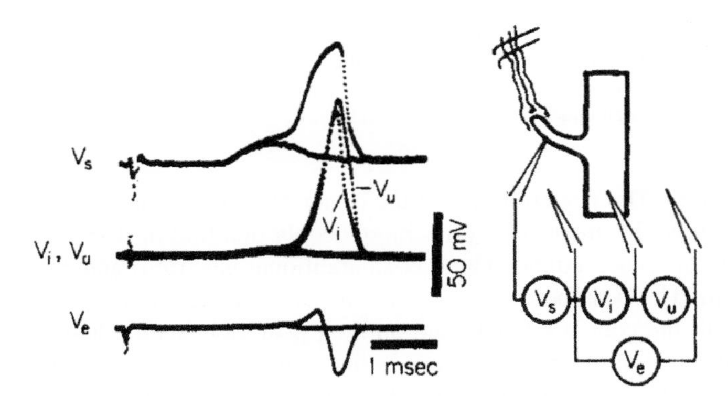

Fig. 31. Response of an electrocyte of the main organ of *Steatogenys elegans*. Recording and stimulating as indicated in the diagram. A brief stimulus is given to the nerve supply just after the beginning of the sweep and evokes a PSP just threshold for initiating a spike in the stalk (V_s, two superimposed sweeps). The spike of the innervated (stalk) face (V_1) rises more rapidly and is briefer than that of the uninnervated face (V_u) and the external potential (V_e) is diphasic, initially positive outside the uninnervated face. The PSP is considerably larger in the stalk than across either face of the cell and the spike in the stalk is longer lasting.

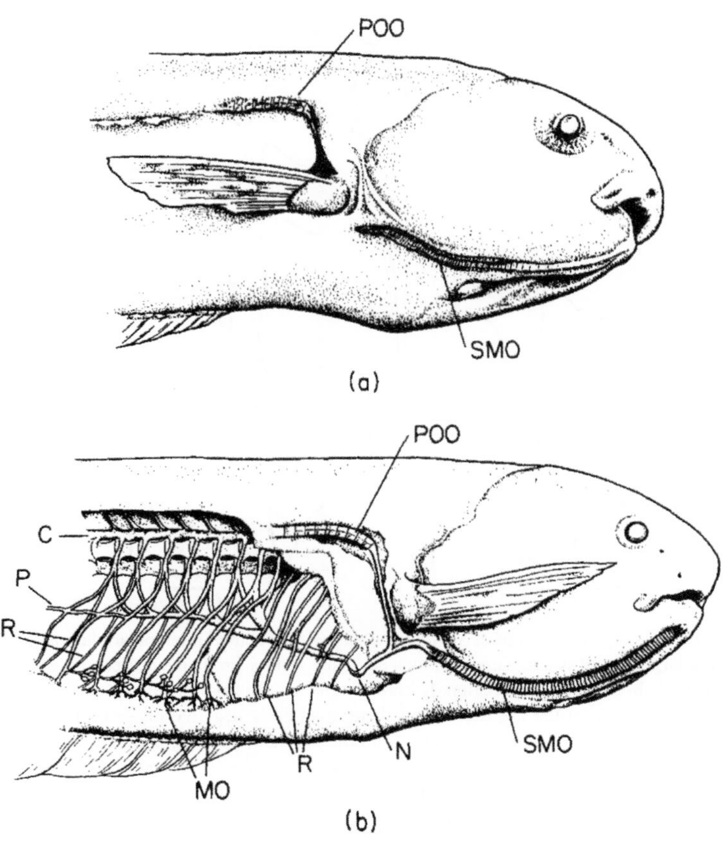

Fig. 32. Rostral accessory organs of *Steatogenys elegans*. (a) Appearance in intact fish; both postopercular organ (POO) and submental organ (SMO) are visible. (b) Overlying tissue is removed to show the full extent of the organs and the nerve (N) coming from spinal nerves to innervate them. The most rostral electrocytes of the main organ are also shown. Spinal cord, C; main organ, MO; longitudinal nerve plexus, P; and rib, R.

anterior faces. The posterior part of the organ can be seen through the overlying epidermis; the anterior part lies more deeply in a fold of dermis and is not visible in the intact animal. There is an additional electric organ behind each operculum. The postopercular organ is for part of its length a double column of cells; all the cells are innervated on their posterior faces. These accessory organs are innervated by spinal nerves that run rostrally from the most anterior level of the main organ (Fig. 32). Electrocytes of the accessory organs have several stalks instead of the single one found in cells of the main organ (Fig. 33). They are difficult to penetrate with microelectrodes because of the connective tissue surrounding the cells. However the isolated organs can be studied, and it can be concluded that they generate monophasic action

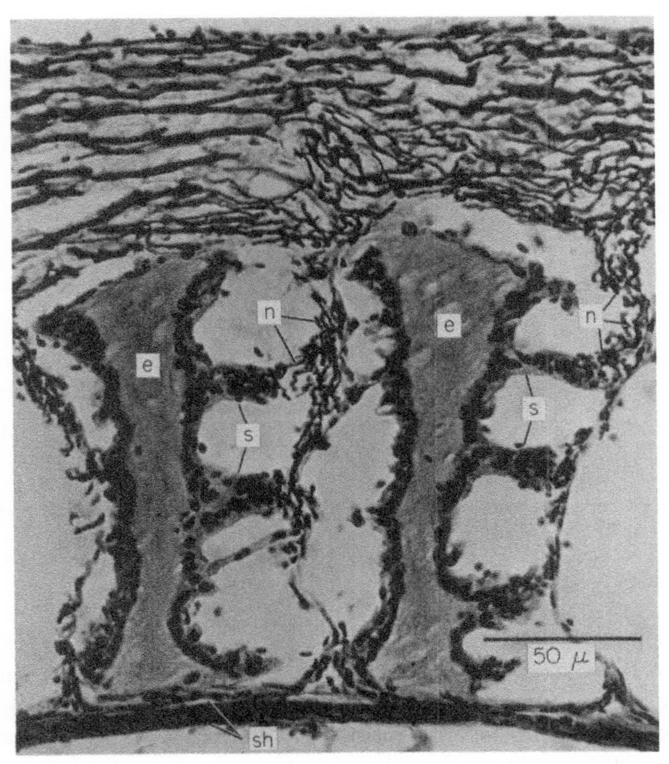

Fig. 33. Anatomy of submental electric organ of *Steatogenys elegans*. Romanes silver stain of paraffin embedded material. A longitudinal section, rostral to the right. Two electrocytes (e) are shown, each with several stalks (s) on the rostral face. Nerve fibers (n) come from the longitudinally running nerve trunk in the upper part of the figure to end on the stalks. The sheath (sh) around the organ is seen at the bottom.

potentials as do the cells of the eel and the monophasically discharging *Hypopomus*, and they have a similar marked proliferation of the uninnervated face (Schwartz *et al*., 1971).

When a column is stimulated by an axial current depolarizing the innervated faces, a monophasic spikelike potential is produced with a very short latency (about 0.1 msec in Fig. 34A). This response is positive with respect to the uninnervated face, and because of its short latency it can be identified as a directly excited spike. This potential is followed at a latency of about 1 msec by a smaller potential of similar sign that evidently represents PSPs and perhaps spike activity in the stalks. If a brief axial stimulus is given that hyperpolarizes the innervated face and depolarizes the uninnervated face, there is no short latency response indicating that the uninnervated face is inexcitable (Fig. 34C). This mode of stimulation does excite the nerve fibers; the delayed response is still present, but it becomes a full-sized spike since the

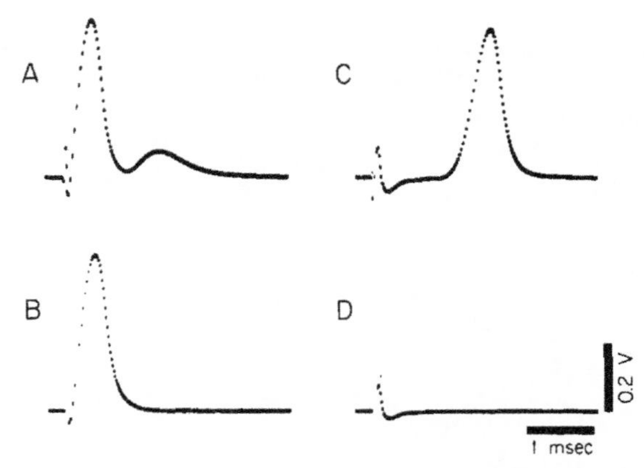

Fig. 34. Response of the submental organ of *Steatogenys elegans*. Recording in a bridge circuit (Bennett, 1961). Rostral positivity up. (A) A brief stimulus depolarizing the innervated (caudal) faces evokes a brief spikelike potential at short latency followed by a smaller potential. (B) Following curarization (about (0.1 m*M*) the brief spikelike potential is unaffected but the longer latency component is abolished. (C) Before curarization a brief stimulus hyperpolarizing the innervated faces evokes a response with a latency of about 1 msec. (D) Following curarization the same stimulus evokes no response.

innervated faces are not refractory as in Fig. 34A. The neural origin of the delayed response is confirmed by the effects of curare, a drug which blocks transmission at cholinergic synapses like those of nerve-electrocyte junctions. Following curare treatment the delayed response is absent for both polarities of stimulation, although the directly excited response persists unchanged (Fig. 34B,D).

The accessory organs fire about 1 msec before the main organ (Fig. 35). The submental electrocytes are innervated on their anterior faces, and the potential external to the anterior end of the organ is largely negative. The postopercular electrocytes are innervated on their posterior faces, and negativity is recorded external to the posterior end of this organ. At the same time the potential in the gill opening and in the mouth is positive. These electrocytes act to make the interior of the head positive. If the accessory organs are removed, the discharge in the head region becomes the simple diphasic potential of main organ (Fig. 30).

There is another much smaller species of *Steatogenys* (5–7 cm) which has submental electrocytes that are all innervated on a single stalk from their posterior faces. This species lacks a separate group of post-opercular electrocytes. These cells also have a monophasic discharge and function to make the interior of the head positive. The largest external potential is recorded at the posterior of the organ; there is very little potential at the anterior end of the organ under the chin. The direction of current flow with respect to the anterior posterior axis through this organ is opposite to that in the submental organs of

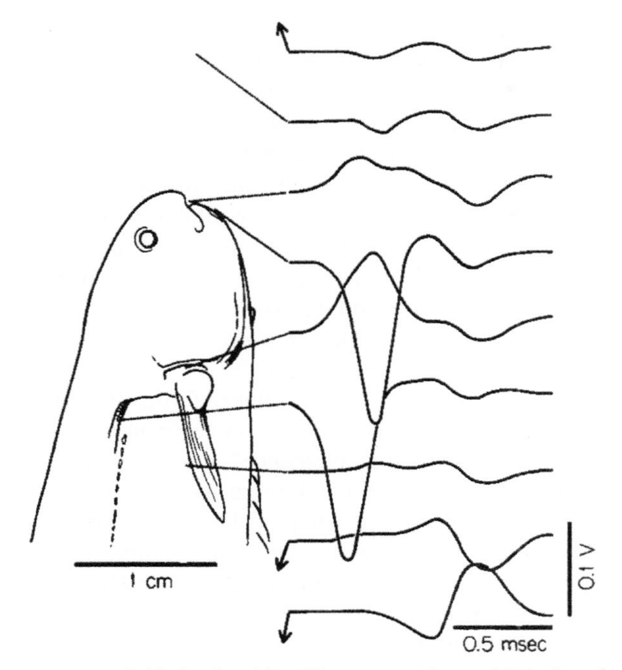

Fig. 35. Responses recorded in head region of *Steatogenys elegans*. Methods as for Fig. 30. The dotted lines indicate the sites of recording of the various traces. The three traces with arrows are recorded 2.5 cm rostral to the snout, 4.5 cm caudal to the snout and near the tip of the tail. There are large early negativities external to the rostral end of the submental organ and external to the caudal end of the postopercular organ. There are corresponding positivities just inside the mouth and gill opening.

Steatogenys elegans, but the transepidermal potentials are affected similarly because opposite ends of the organ are connected to the exterior and interior of the animal. The function of these accessory organs is presumably to increase sensitivity of the electrosensory system in the head region, but there are no experimental data.

e. Gymnorhamphichthys. Gymnorhamphichthys usually rests buried in sand during the day and emerges to swim around and feed at night. When in the sand it emits pulses at a quite constant rate of about 10–15/sec. When actively swimming it increases its discharge rate to a new and also quite constant frequency which it maintains for prolonged periods. Disturbances also cause increase in discharge rate. It has been shown to exhibit a circadian rhythm of motor activity and (perhaps secondarily) of discharge frequency under conditions of constant darkness (Lissmann and Schwassmann, 1965).

The main organ of *Gymnorhamphicthys* is cytologically and, as inferred by recording from the intact fish, electrophysiologically like that of *Steatogenys. Gymnorhamphichthys* also has submental electrocytes as does *Steatogenys.*

They are located beneath the dermis and thus are not obvious, which accounts for their not being seen previously. Each cell has multiple stalks from its posterior faces. External recording indicates that the submental cells fire before the main organ and generate monophasic action potentials. The related genus *Rhamphichthys* has not been studied with respect to its electric organs.

f. Sternopygus and Eigenmannia. Sternopygus and *Eigenmannia* are closely related and the organ discharges are so similar that they may be considered together. The discharges consist of head-positive pulses superimposed on a head-negative base line such that there is little dc component in the organ discharge; that is, averaged over one discharge cycle, there is little or no net current flow (Fig. 3C). The frequency in *Sternopygus* is about 50–100/sec, and in *Eigenmannia* it is about 250–600/sec.

The frequency is quite constant, and ordinary forms of stimulation do not affect it. Weak electric stimuli of nearly the same frequency as the animal's own discharge evoke the "jamming avoidance" response, that is, a small shift in frequency away from that of the interfering stimulus (Watanabe and Takeda, 1963; Bullock, 1970; see Chapter 11, this volume). *Sternopygus* can also stop discharging completely under these conditions.

The electric organ is located in the same position as in *Hypopomus* and there is no specialized region rostrally. The electrocytes are tubular to spindle-shaped with the long axis parallel to the anteroposterior axis of the fish. The cells are 1–2 mm long and 0.2 mm in diameter in a *Sternopygus* 15 cm long and somewhat larger in an *Eigenmannia* of similar size (Fig. 36). They are innervated on their posterior faces. There are about 6 columns on each side in *Eigenmannia* and about 10 in *Sternopygus*. The cells are accurately aligned in columns in *Eigenmannia*, but they are separated by only small spaces; the amount of extracellular space between cells is much smaller than the cells themselves. The alignment is less regular in *Sternopygus*, but the amount of extracellular space is similarly small.

It may not be obvious from records such as those of Fig. 3C that the discharge consists of head-positive pulses superimposed on a head-negative base line, and the distinction between pulse and interpulse interval may be even less obvious in the faster firing *Eigenmannia*. The distinction is readily established experimentally (Fig. 37). An appropriately timed stimulus to the spinal cord can cause the complete disappearance of a pulse (Fig. 37H) or cause a pulse to be generated earlier (Fig. 37C). The mechanism of block appears to be that the evoked volley in descending fibers finds the spinal neurons refractory from their immediately preceding response (initiated by the center in the medulla controlling the discharge, see Section III, A), and the evoked volley also collides with the next command volley from the medulla preventing it from reaching the spinal neurons. The stimulus however has no noticeable effect on the following pulse indicating that the organ is controlled rostral to the spinal cord and that the antidromic volley does not invade the controlling nucleus. [There is actually a very slight advance of subsequent pulses (see Bennett *et al.*, 1967c).]

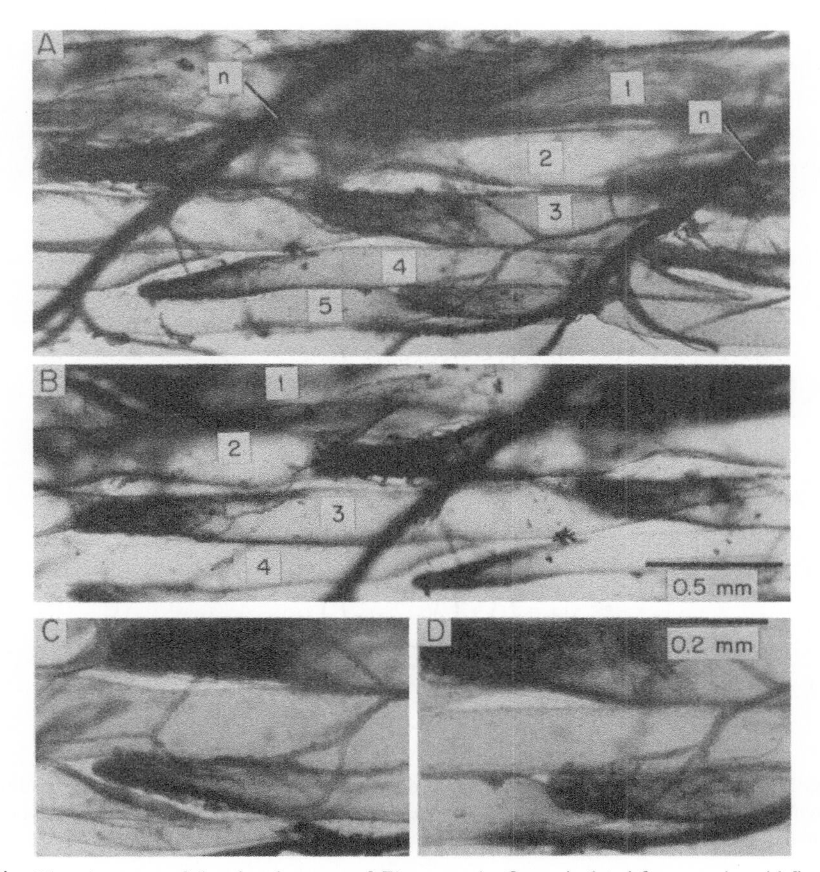

Fig. 36. Anatomy of the electric organ of *Eigenmannia*. Organ isolated from osmic acid fixed material. Dorsal up, rostral to the right viewed from the lateral surface. (A) Low power view. The five columns of electrocytes are numbered dorsoventrally. Two segmental nerves (n) run ventroposteriorly and give off branches to the electrocytes. A complete cell of column 4 is seen most clearly. The small spots on the cells are nuclei. The posterior, innervated ends of the cells are more darkly stained. There is little extracellular space between cells, and the cells usually overlap somewhat in the longitudinal axis. (B) The anterior end of a cell of column 3 lies lateral to the next cell anteriorly, and is similarly overlapped by the cell caudal to it. (C) Details of innervation of a cell from column 4. (D) Innervation of the anterior cell from column 5 in A. Magnifications the same in A and B and in C and D.

In *Sternopygus* the head-negative potential between responses can also be demonstrated by section of the spinal cord which blocks further pulse activity (Fig. 38). The head-negative potential remains immediately after spinal section but decays away over the next minute or so. Repetitive stimulation of the organ for some seconds causes at least a partial restoration of the head-negative potential which decays away again on cessation of stimulation [although somewhat more rapidly than following spinal section (Bennett, 1961)].

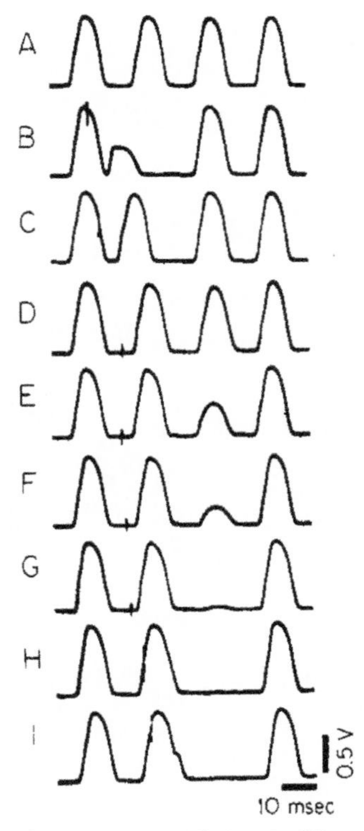

Fig. 37. Effects of spinal stimulation on organ discharge of *Sternopygus*. Potential recorded differentially along caudal portion of tail, rostral positivity up. (A) Normal discharge. (B–I) A brief stimulus to the spinal cord is given at successively later times in the discharge cycle.

The origin of these potentials has been elucidated by microelectrode studies of the electrocytes. The pulse component of the organ discharge is generated by the posterior end of the cell, and the head-negative component is generated by the anterior end. The cells are long compared to cells of weakly electric gymnotids discussed up to this point, and the potential can be quite different at the two ends of the cell. The spike and PSPs are larger at the posterior end of the cell (Fig. 39), and the threshold current is lower when applied through an electrode at this end of the cell. Longitudinal stimulation as in Figs. 21 and 25 has not been carried out. However, in the caudal filament there is little tissue other than the organ, and it can be stimulated by external electrodes. This procedure shows the posterior faces to be capable of generating a spike and the anterior faces to be inexcitable just as demonstrated for the submental organs of *Steatogenys* (Fig. 34).

In a cell that has not been stimulated for some time, the resting potential is the same at the two ends of the cell. Repetitive stimulation causes a

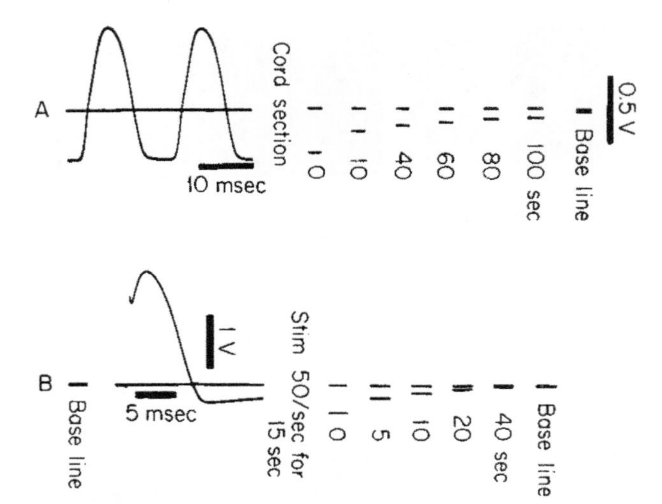

Fig. 38. Organ discharge of *Sternopygus*, effect of spinal section and repetitive stimulation. Potential recorded differentially along caudal portion of tail, rostral positivity up, zero level indicated by horizontal line and base line traces. Two preparations. (A) Two complete normal discharge cycles are shown on the left. The potential is head positive and head negative for approximately equal times. Spinal section abolishes the pulses, but a head-negative potential about equal to that during regular organ discharge remains (lower line at time 0). This potential decays over approximately 2 min (right). (B) Stimulation of organ inactivated by spinal section. Separate stimulating and recording electrodes, head positivity up. Brief pulses are passed through a condenser to give diphasic stimuli with no dc component. The pulses are oriented so that the initial brief phase excites the innervated faces. Their response appears superimposed on the slow opposite phase (left). Repetitive stimulation at 50/sec for 15 sec develops a head-negative potential that on cessation of stimulation decays over 15–20 sec (right). From Bennett (1961).

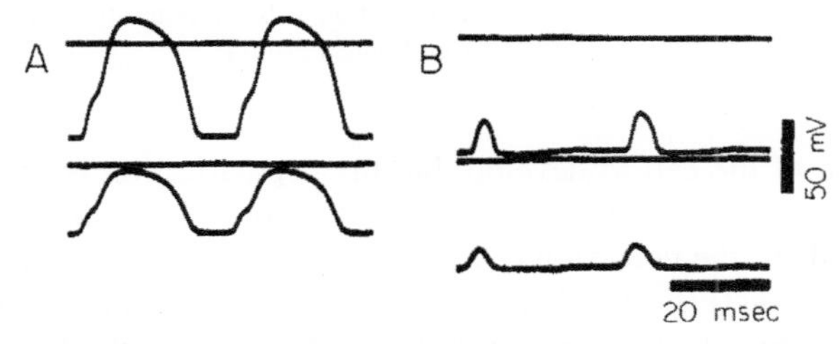

Fig. 39. Spikes and PSPs in an electrocyte of *Sternopygus* during organ discharge. Two intracellular recording electrodes are used, one 700 μ rostral to the other. Both record differentially against a closely applied external electrode to minimize pickup from other cells. The second trace gives the potential at the caudal electrode with the upper trace as its zero level. The fourth trace gives the potential at the rostral electrode with the third trace showing its zero level. When the cell is firing regularly, as in A, the spikes and the potential between spikes are greater at the caudal end. Repeated hyperpolarizing pulses are passed through a third intracellular electrode in order to block the spikes and, after some seconds, the records become as in B. The resting potential is equal at the two ends of the cell and only PSPs are produced, which are larger at the caudal end. From Bennett (1961).

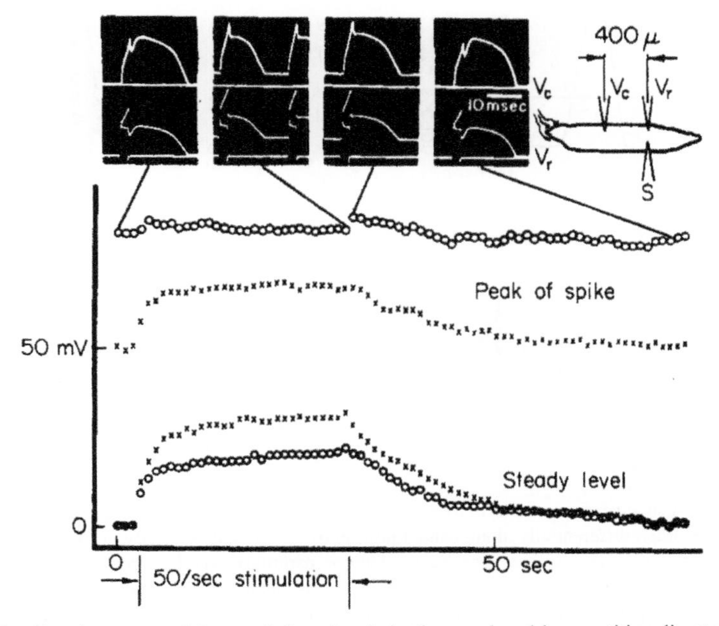

Fig. 40. Development and decay of slow depolarization produced by repetitive direct stimulation of a single cell of *Sternopygus*. Organ inactivated by spinal section. Three electrodes are used as in diagram. Brief pulses adequate to excite the cell are applied at 1/sec and 50/sec. Sample records are shown above graph with the connecting lines indicating when they are taken. Upper traces: potential at the caudal end of the cell and base line. Lower traces: potential at the rostral end of the cell and stimulating current serving also as a base line. Graph: data of complete experiment, resting potentials at the beginning taken as 0. Open circles: caudal. Crosses: rostral. Resting potentials determined at the end of the experiment are 64 mV at the rostral electrode and 67 mV at the caudal electrode. Repetitive stimulation at 50/sec causes a steady level of depolarization to develop, which is greater at the rostral end of the cell. On return to stimulation at 1/sec the depolarization subsides in about 20 sec. From Bennett (1961).

depolarization of the cell to develop, but the depolarization is greater at the anterior end of the cell (Fig. 40). On cessation of stimulation the potential of the two ends of the cells slowly equalizes again. Similar changes are seen in neural activation (Fig. 39). From these records it is clear that, during spikes, current flows from caudal to rostral in the interior of the cell, thus generating the head-positive phase of the organ discharge. When the cell has been active at the normal frequency for a sufficient period, current flows caudally in the cell in the interval between spikes thus generating the head-negative component of the overall discharge.

Insufficient experimental data are available to warrant a deailed discussion, but a possible explanation of the depolarization of the anterior face can be given. During the spike sodium current flows inward through the innervated face while outward current at the anterior end is carried by potassium ions. Potassium accumulates in the restricted extracellular space external to the anterior face and tends to depolarize it. This mechanism of ion accumulation provides an explanation of why there is so little extracellular space in this organ in contrast to those of other electric fish previously discussed. The question remains as to what ions carry the outward current through the innervated face during the head-negative phase of the organ discharge. It cannot be potassium if the concentration of this ion is the same outside the two faces as it would be except for cells at either end of the organ.

Spike generation in the electrocytes has an unusual feature: The resistance during the peak of the spike exceeds that at rest by a factor of 1.5 to 2. This property may be seen from the application of brief pulses between the spikes and at their peaks (Bennett, 1961) and is confirmed by ac impedance measurements from the cells of the caudal filament (Fig. 41). The sequence of conductance and permeability changes is probably as follows. During the rising phase of the spikes the conductance rises as a result of sodium activation. The conductance then decreases again at the spike peak because the conductance of anomalously rectifying membrane is a large fraction of the resting conductance and decreases more than enough to compensate for the increase in sodium conductance. Sodium inactivation ensues and the potential begins to fall. However, on the falling phase the conductance rises to above the maximum observed on the rising phase; this change is ascribable to potassium

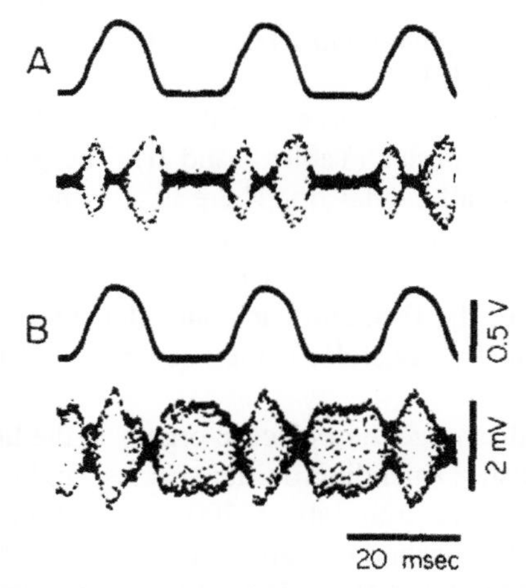

Fig. 41. Impedance changes during organ discharge of *Sternopygus*. The tip of the tail was placed in a bridge circuit. Upper trace: bridge signal after moderate amplification. Lower trace: bridge signal after filtering out low frequencies and much greater amplification. Amplitude and frequency of ac input 0.5 V and 5 kc. The bridge could be approximately balanced for both spike peaks and between spikes (A), but large imbalances occurred on the rising and falling phases. When the bridge was balanced for the falling phase (B), the imbalance at the spike peaks was somewhat greater than between spikes, and balance was approached during the rising phase. From Bennett (1961).

activation or delayed rectification. Except for the conductance increase on the falling phase this pattern of changes is like that in the electric eel for which the data quite adequately demonstrate the proposed mechanism (see Nakamura *et al.*, 1965; Morlock *et al.*, 1969). The similar conductances during spikes and at rest may be functional in maintaining an ac organ discharge. Exposure to media of different conductivities will load both phases of organ discharge similarly. Therefore less dc component will be associated with the new amplitude of organ discharge than there would be if the conductance during spikes were much lower than that between spikes.

Conductance changes associated with organ discharge are the same in *Eigenmannia* and in *Sternopygus*. Insufficient physiological data are available to distinguish whether the absence of a dc component in the discharge results from a mechanism like that in *Sternopygus* or whether the uninnervated face acts as a series capacity as in *Gymnarchus* and probably also the sternarchids (see Sections II, D, 3 and II, D, 1, g). The two different mechanisms might be characterized as a polarization capacity in contrast to a true membrane dielectric capacity. Ultra structural data suggest that *Eigenmannia* may employ a dielectric capacity. The uninnervated face of its electrocytes has many fine branching tubules or canaliculi that greatly increase the surface; the situation thus resembles that in *Gymnarchus* (Schwartz *et al.*, 1971). In *Sternopygus,* the innervated and uninnervated faces are similar and rather smooth, their areas being little increased compared to surface seen at the light microscope level of resolution. (A question here is why there is little extracellular space in the electric organ of *Eigenmannia.*) The significance of an ac discharge will be discussed in Section II, F.

g. Sternarchids. This is the largest of the gymnotid families in terms of numbers of genera and species (Tables I and II). The group is characterized by the presence of a small caudal fin. There is also the so-called dorsal filament that arises about the middle of the back and runs posteriorly for perhaps a third of the body length (Fig. 42). The dorsal filament is *in vivo* closely adherent to the back of the fish, and back and filament are contoured so that the filament is practically invisible; it may separate and become obvious in preserved specimens.

The sternarchids discharge their electric organs at the highest frequency of any known electric fish. Depending on the species, individual, and temperature the frequency ranges from about 700 to 1700 impulses per second (Steinbach, 1970). The discharge shape varies somewhat with the species, and will be treated separately below. The frequency is generally highly stable and is not affected by ordinary stimuli. Weak electric stimulation at closely neighboring frequencies does evoke small shifts in frequency, the "jamming avoidance" response (Larimer and MacDonald, 1968; Bullock, 1970; see Chapter 11, this volume). Several species have been observed to accelerate their discharge briefly, emitting a "chirp" (Bullock, 1970). These changes may function in intraspecific communication and in species recognition.

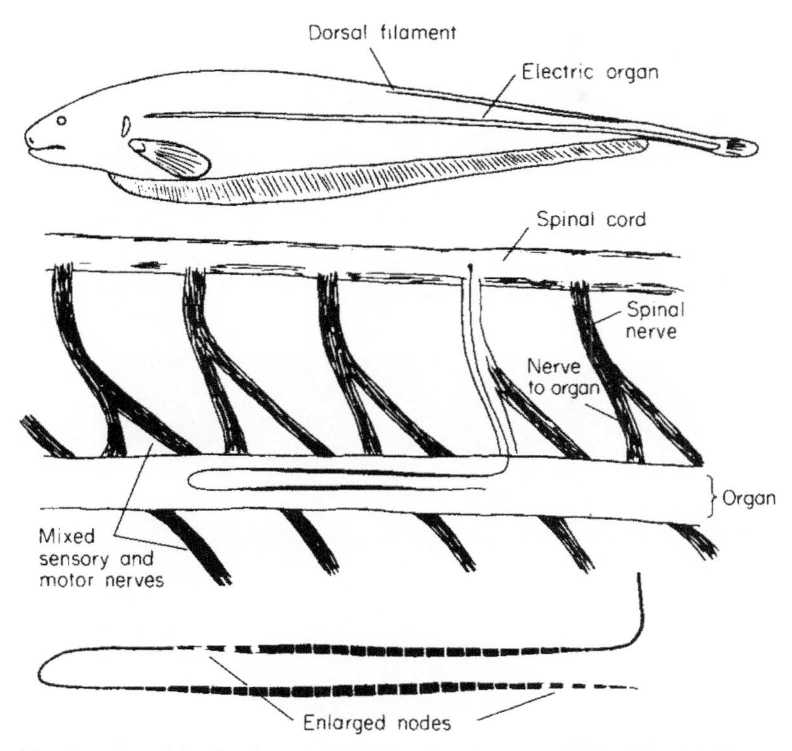

Fig. 42. Anatomy of the electric organ of *Sternarchus*. The upper diagram shows the location of the organ in the fish. The middle diagram shows the organ and nerves running to it and the course of a single fiber from its origin in the spinal cord. The lower diagram represents a single fiber and its nodal structure.

The electrocytes of the sternarchids turn out to be spinal neurons rather than cells of myogenic origin (de Oliveira Castro, 1955; Bennett, 1966, 1970). The best-studied species is *Sternarchus albifrons* because this is the one most commonly obtained by tropical fish dealers who call it the black ghost knife fish. The fish, electric organ, and electrocytes are shown diagrammatically in Fig. 42. The organ runs longitudinally just ventral to the spinal column over most of the length of the body. The axons of the spinal neurons descend from the cord and enter the electric organ (Fig. 43). They then run anteriorly for several segments, turn around and run posteriorly to end blindly at about the same level that they entered the organ. This course is established by dissection of single fibers from their point of entry into the organ until their termination. Evidently the organ has lost its myogenic electrocytes and enlarged the axons that formerly innervated them. This origin is confirmed by the existence of a nerve plexus in *Gymnotus, Hypopomus, Steatogenys,* and *Gymnorhamphichthys* in the same location as the sternarchid electric organ. In *Gymnotus* axons in this plexus have been traced to the (myogenic)

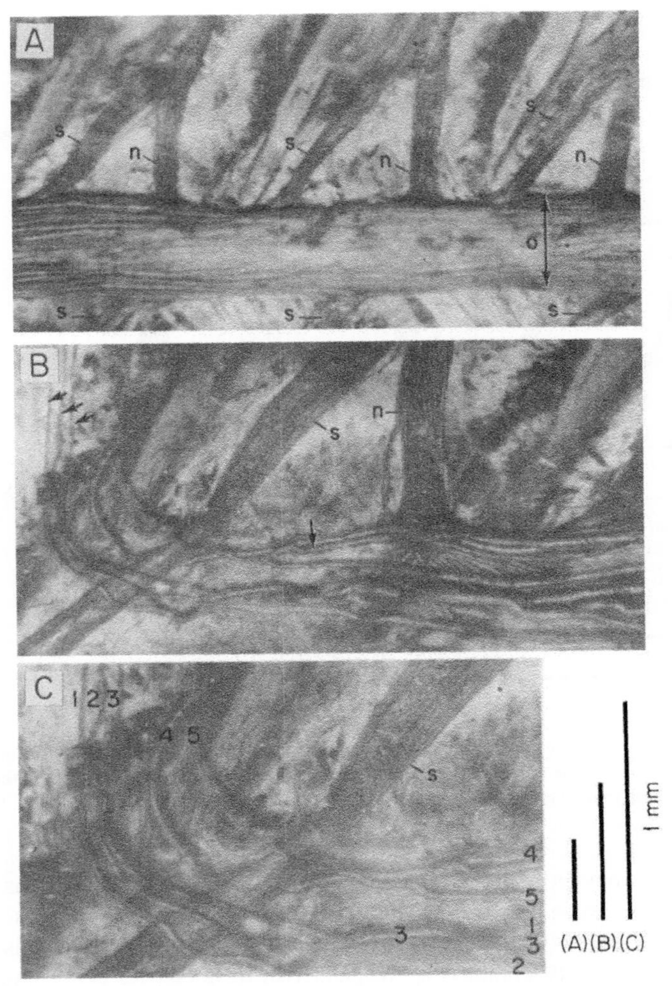

Fig. 43. Electric organ of *Sternarchus*. Formalin fixed material, dissected and stained with methylene blue. Dorsal to the top, anterior to the right. (A) The electric organ (o) following removal of its connective tissue sheath. Single longitudinally running electrocytes are visible. Nerves to the organ (n) and to more ventral tissues (s) are also seen. (B) Following dissection away of all electrocytes from nerves caudal to those shown and of most fibers entering the organ from the more caudal nerve in the figure. Three descending fibers of this nerve run approximately parallel in their origin position (arrows); two others are deflected anteriorly; the central part of a sixth is broken off shortly after entry into the organ (arrow). The more anterior nerve is undissected; its fibers diverge in the organ. Most of them lie medial to the fibers from the more caudal nerve. (C) Higher magnification of the left part of B. The five numbered fibers can be traced from the top left to the lower right. The smaller fibers of the ventral segmental nerve (s) are visible. Fiber 4 passes medial to this nerve. Calibrations on the lower right.

electrocytes. The plexus is absent in *Sternopygus, Eigenmannia,* and *Electrophorus,* a fact which has implications for the phylogenetic relations within the gymnotid group.

There are characteristic changes in the nerve fibers as they run along in the electric organ. In both its anteriorly and posteriorly running parts it becomes greatly dilated and can exceed 100 μ. in diameter. Before it enters the organ and where it turns around to run posteriorly, it is of the usual diameter for a large myelinated fiber, 10–20 μ. It tapers gradually before it terminates. The nodal structure also changes characteristically along the fiber (Fig. 44). The nodes become somewhat enlarged just after the fiber enters the organ. As the fiber dilates, the nodes become very narrow (although area is difficult to estimate because of the increased fiber diameter). As the fiber tapers near its most anterior part, there are several very long nodes. The nodes become of ordinary size again as the fiber turns around. This sequence repeats itself in the caudally running section of the fiber. There are first moderately enlarged nodes, then narrowed ones, then very enlarged ones.

The organ discharge of *Sternarchus* consists of diphasic pulses that are initially head positive (Figs. 3D and 45). As would be expected from the neural origin of the electric organ, the discharge is unaffected by a dose of curare that is sufficient to paralyze the animal completely.

The contribution of the single cells to the organ discharge is disclosed by microelectrode experiments. Just anterior to where the fiber enters the organ, a large impulse is recorded in the fiber during the head-positive phase of organ discharge (Fig. 46A). This impulse is smaller more anteriorly and the peak is delayed; thus, like the cells of *Sternopygus,* these cells are not isopotential. The response in this part of the fiber accounts for the head-positive phase of the discharge because current runs anteriorly along the interior of the cell at this time. In the posteriorly running section of the fiber the impulse is larger anteriorly (Fig. 46B). This activity occurs during the head-negative phase of the organ discharge and current flows in the cell in such a way as to give rise to the head-negative phase. The anatomy and recordings indicate that impulses coming from the spinal cord propagate along the electrocyte in the organ and that the activity of the anteriorly running part excites the posteriorly running part. Because the impulse is smaller at the distal end of each section of the fiber, it is likely that these regions are inexcitable like the uninnervated faces of eel electrocytes. The current paths and intracellular potentials during one discharge cycle are diagrammed in Fig. 47.

There is evidence that the inactive regions of the cells act as a series capacity rather than as a series resistance. The effects when one face of an electrocyte acts as a capacity are discussed in detail in respect to *Gymnarchus,* for which the experimental evidence is more complete (Section II, D, 3). The most important effect is that the organ discharge has no dc component At a constant frequency of activity current flows as follows. During a spike at one end of the cell the charge on the capacity of the other end is made more positive from

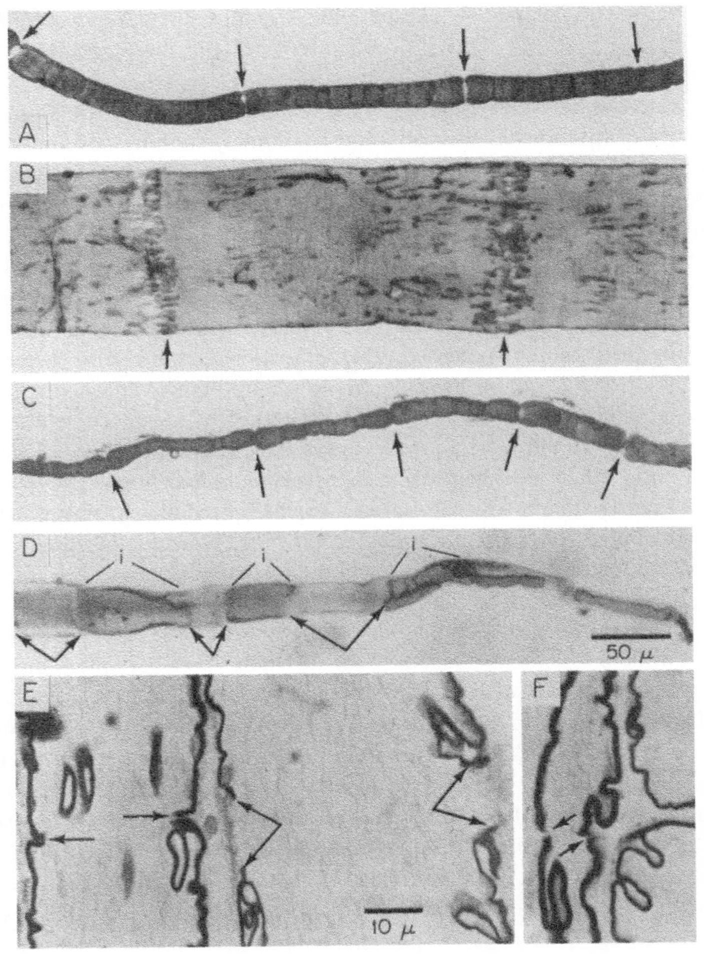

Fig. 44. Structure of *Sternarchus* electrocytes. (A–D) Photographs of a cell isolated by dissection following osmic acid fixation. (A) The fiber near its entry into the organ; 4 nodes (arrows) are visible; the 3 to the left appear somewhat enlarged. (B) A dilated part of the fiber in its longitudinally running course; 2 nodes of peculiar structure are seen (arrows). (C) The thin region of the fiber where it turns around to run posteriorly again; 5 nodes are seen; the 2 to the right appear somewhat enlarged. (D) The caudal termination of the fiber, at least three very large nodes (double arrows) and 3 small internodes (i) are seen. The nature of the adhering structures near the fiber termination is unclear. (E–F) Toluidine blue stained sections of osmic acid fixed Epon embedded material. (E) Arrows indicate a large node in the dilated fiber on the right and a small node in the dilated fiber on the left. (F) Arrows indicate a node in a narrow region of a fiber that is turning around at the anterior limit of its course. Note that the other fibers in this section are cut transversely. Magnifications the same in A–D and in E–F.

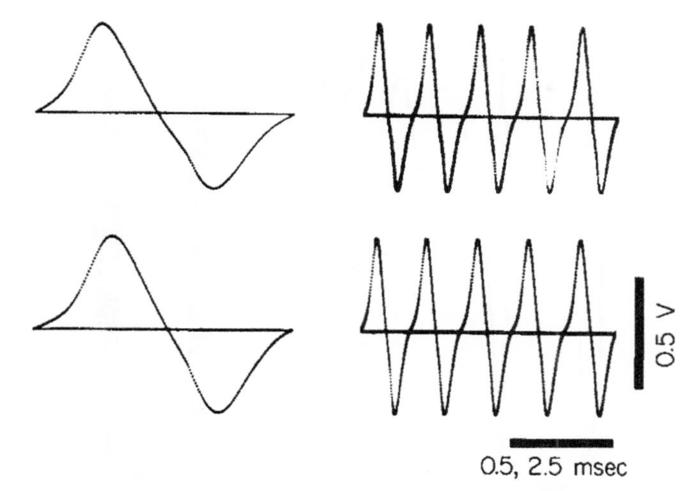

Fig. 45. Electric organ discharge of *Sternarchus*. Recorded differentially between head and tail at two sweep speeds. Upper records; prior to curarization. Lower records: after a dose of curare (10 mg/kg) that completely paralyzed the fish, organ discharge was unaffected.

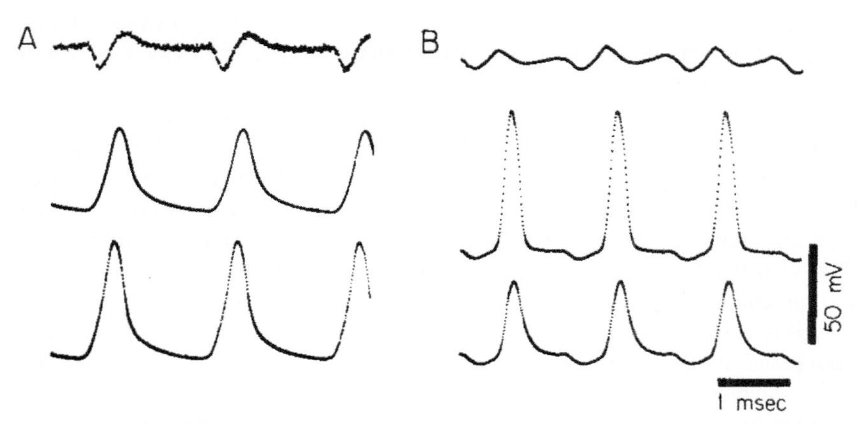

Fig. 46. Intracellular recordings from fibers in the electric organ of *Sternarchus*. Differential recording with respect to a closely applied external electrode to minimize contributions from other cells. Upper trace: monopolar recording of organ discharge. Middle trace: intracellular recording from anterior electrode. Lower trace: intracellular recording from an electrode in the same fiber several millimeters posteriorly. (A) Recording from an anteriorly running fiber segment as indicated by the larger size and earlier peak of the potential at the posterior electrode. The impulse occurs during the initial phase of organ discharge (negative going when monopolarly recorded at this level of the organ). (B) Recording from a caudally running fiber segment as indicated by the larger size and earlier peak of the potential at the anterior electrode. The impulse occurs during the second phase of organ discharge.

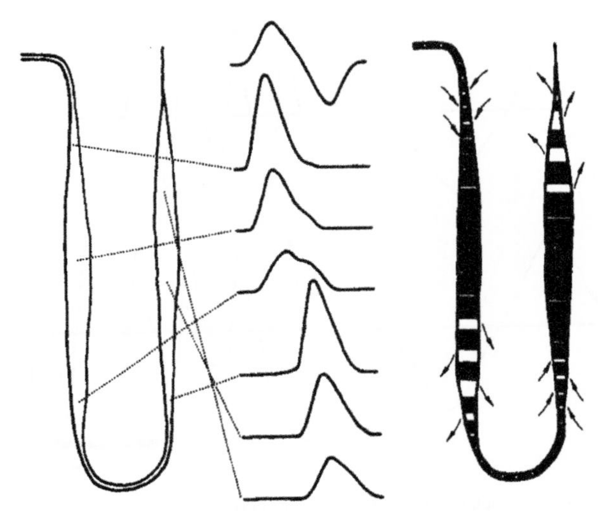

Fig. 47. Diagram of potentials and current flow at successive regions of the electrocytes during activity. The upper potential tracing represents an organ discharge. The directions of current flow during the spikes in the different regions are indicated by the arrows on the right.

the negative resting level; after the spike the capacity recharges to its original level, and no net current flows over the discharge cycle. The evidence for this property in sternarchids is largely comparative (M. V. L. Bennett and A. B. Steinbach, unpublished data). The organ discharge of all sternarchids has little or no dc component. In *Sternarchus* this could result, as in *Gymnotus,* from successive firing of opposite faces or ends of the cells. However, in *Sternarchorhamphus* the organ discharge consists of monophasic head-negative pulses superimposed on fairly level head-positive base line with no dc component (Fig. 48A,B). The electrocytes lack a rostrally running portion; the fibers turn and run caudally on entering the organ. The monophasic appearance of the discharge is consistent with activity of only the rostral regions of the fibers. The head positivity between impulses could be generated by a polarization capacity as in *Sternopygus*. It seems more likely that the mechanism is like that of *Gymnarchus,* and the head-negative phase results from charging of the dielectric capacity of the caudal portion of the fiber. That the potential between pulses is quite level only requires that the time constant of the system be long compared to the interval between pulses.

In some *Adontosternarchus* the main organ discharge has a small head-negative phase followed by a large head-positive phase separated by a more or less level base line. However, the base line is head positive so that again there is no net current flow during a discharge cycle (Fig. 48C,D). The electrocytes have both rostrally and caudally running portions, but the rostrally running section is shorter than the caudally running section. Thus, the electrocytes present an intermediate condition between those of *Sternarchus* and

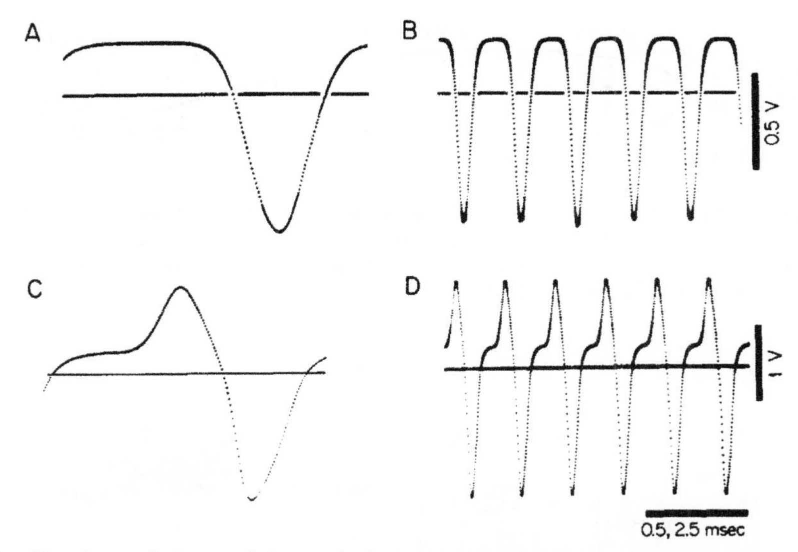

Fig. 48. Organ discharge of *Sternarchorhamphus* and *Adontosternarchus.* (A, B) Recorded monopolarly at the tail of *Sternarchorhamphus,* positivity down (equivalent to head positivity up). (C, D) Recorded differentially between head and tail of *Adontosternarchus,* head positivity up. Horizontal lines indicate zero potential levels. Faster sweep speed in A and C.

Sternarchorhamphus. The small head-positive phase of the pulse is ascribable to the small anteriorly running portion of the fiber; the large head-negative phase is ascribable to the large posteriorly running portion. The most likely way for the head-negative level between pulses to arise is for each portion of the fiber to have a series capacity.

These data do not require *Sternarchus* electrocytes to have a series capacity. However, if the fish is made anoxic the second phase of the organ discharge appears to fail. In spite of this change, the organ discharge still has no dc component. It seems probable then that both anteriorly and posteriorly running regions have a series capacity. In some individual *Sternarchus* the initial, head-positive phase is larger than the second, head-negative phase, although there is still no dc component. Correspondingly in some individuals, presumably the same ones, the rostrally running portion of the electrocytes is larger than the caudally running portion. Unfortunately, it is not possible to test these mechanisms by activating the organ over a wide range of frequencies. The cells continue to fire spontaneously after spinal section, and synchronous low frequency activity cannot be obtained.

One sternarchid, *Adontosternarchus,* has an accessory organ in the chin region (M. V. L. Bennett and A. B. Steinbach, unpublished data). This organ generates a potential of 20–25 mV amplitude outside the chin (Fig. 49). It is absent from the five or more other genera of sternarchids examined on the Rio Negro expedition of the R. V. Alpha Helix. It is made up of fibers that

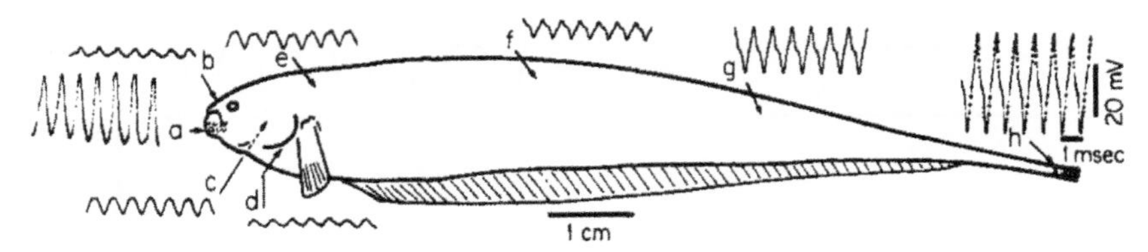

Fig. 49. Main and accessory organ discharges of *Adontosternarchus*. Recorded monopolarly with respect to a distant electrode in a large volume of water. There is a relatively large potential in the chin region (a) and at the tip of the tail (h). The potentials are smaller elsewhere over the head (b–e) and near the middle of the body (f, g).

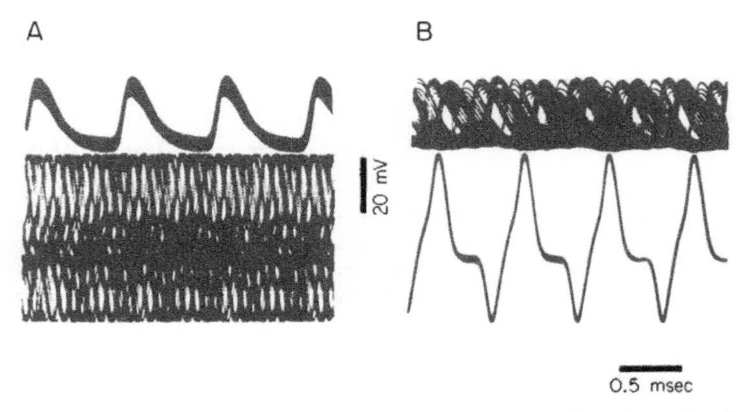

Fig. 50. Lack of synchronization of main and accessory organs of *Adontosternarchus*. Recording monopolarly from chin organ (upper trace) and tip of tail (lower trace). (A) When the sweep is triggered by the accessory organ discharge, repetitive superimposed sweeps show that the phase of main organ discharge changes. (B) When the sweep is triggered by the main organ discharge, the phase of the accessory organ discharge changes in different sweeps.

are probably modified from electrosensory nerve fibers for the fibers end in the skin in what appear to be modified electroreceptors. Consistent with its sensory origin the impulse frequency is set in the chin organ itself, and impulses proceed centripetally in the afferent fibers which can be sectioned without affecting organ discharge. Because of its peripheral origin the chin organ can fire at a somewhat different frequency from the main organ (Fig. 50).

Since electroreceptors themselves can, under certain conditions, generate maintained oscillations (see Chapter 11, this volume), it is necessary to distinguish the neurogenic potentials of the chin organ from potentials generated by electroreceptors that of course the fish also possesses. The most important difference is morphological. The fibers of the chin organs have very dilated myelin sheaths and peculiar nodal structures that closely resemble the characteristics of electrocytes in the main organ. Electroreceptor afferent fibers are ordinary myelinated fibers. Futhermore, if an electrode is advanced into the organ from the surface, the polarity of discharge does not invert until the electrode is deep into the tissue of the chin. In contrast, the oscillations at electroreceptors invert when the epidermis is crossed. Finally, the chin organ oscillations are little affected by external loading; the frequency is nearly constant whether the chin is in air or immersed in physiological saline. In contrast, electroreceptors do not oscillate when they are electrically loaded to only a small degree; generally, they will oscillate only when the skin is allowed to dry in air.

While measurements to date are inadequate, it is probable that discharge of the chin organ has little dc component. The function of the chin organ is obscure, but the potentials produced are large enough to activate receptors in the head region. Presumably the chin organ plays the same role as the rostral accessory organs of other gymnotids.

2. *The Electric Catfish*

The electric catfish, *Malapterurus electricus,* is the only silurid known to be electric. It is also distinguishable from other catfish by the absence of rays in its dorsal fin. The electric organ lies in the skin surrounding the body over most of the length of the fish (Fig. 1). It is innervated by two giant neurons, one on either side of the spinal cord in the first spinal segment (Fig. 72). Each neuron sends out a single axon that innervates the several million electrocytes on that side of the body. The electrocytes are shaped rather like lily pads. The main part of the cell is disc-shaped, about 1 mm in diameter and 20–40 μ. thick in a fish 15 cm long (Fig. 51). From a convoluted region in the center of the caudal face (called the rosette), a stalk protrudes that is about as long as the radius of the disc-shaped part. The cell is innervated on the tip of this stalk by a branch of the axon from the giant cell.

The electric organ of *Malapterurus* was long thought to be of glandular origin, in part because of its location and because the side opposite to the innervation of the electrocytes became negative. It thus violated the rule formulated by Pacini concerning innervation and polarity (who is better known for the corpuscles that bear his name) and generated potentials similar in sign to those known from some glandular tissue. It turns out that the discharge polarity is explicable in terms of ordinary mechanisms of excitability (Keynes *et al.*, 1961), and recent physiological and embryological data are consistent with a muscle origin for the organ (Johnels, 1956).

The organ discharges are primarily head-negative pulses 1–2 msec in duration (Fig. 3A,A′). The head-negative pulse is preceded by a very much smaller head-positive phase that was not observed in earlier studies (Fig. 52). This early phase results from activity of the stalks as will be discussed below. Thus, the organ really does not violate Pacini's rule. The pulses are emitted infrequently by an undisturbed fish. However, when the animal is feeding or mechanically stimulated a few pulses or long trains of pulses can be emitted.

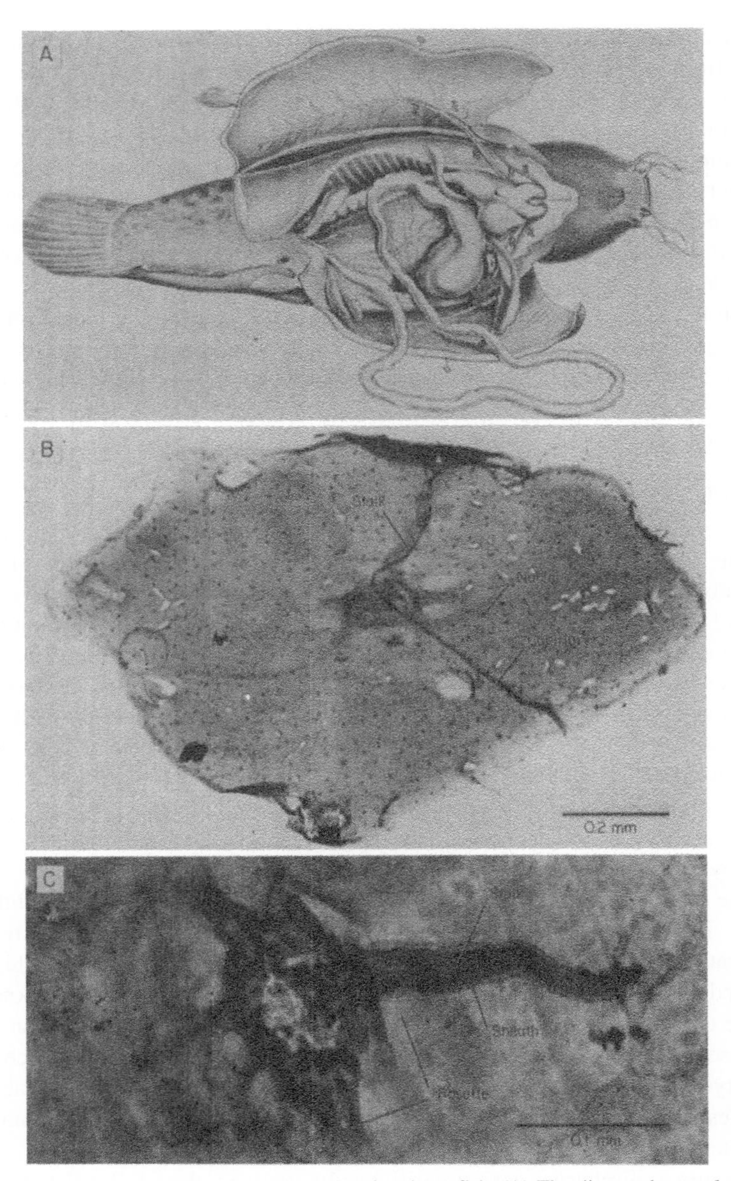

Fig. 51. Anatomy of the electric organ of the electric catfish. (A) The dissected ventral surface of the fish showing the organ (o) and the nerve (ne), artery (ae), and vein (ve) running to it (from Rosenberg, 1928). (B) The body of a single electrocyte dissected out following formalin fixation, and stained with methylene blue. (C) Higher magnification of the rosette region of another cell, Romanes' silver stain.

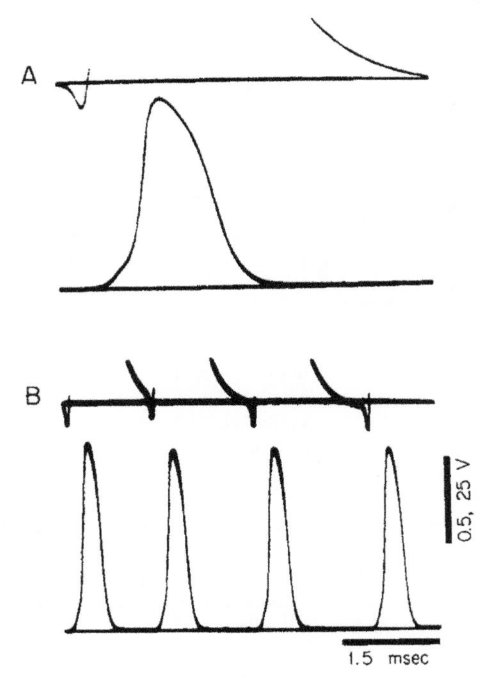

Fig. 52. Electric organ discharge of the electric catfish. Recorded between head and tail of a fish about 15 cm in length in a small container of aquarium water. High gain (upper traces) and low gain (lower traces), head negativity upwards, faster sweep in A. Two superimposed sweeps, one with and one without discharges to show base line. The primary head-negative discharge is preceded by a small head positivity. Single pulses (A) or trains (B) are emitted in response to touch.

Prey detection appears to precede pulse emission and the organ presumably functions in prey capture (Bauer, 1968). Inputs from taste receptors, which are found over much of the body surface, are powerful excitants. However, the effectiveness of the discharge in stunning small fish is minimal. The amplitude recorded in air is about 150 V from an animal 30 cm in length and even a small fish 5 cm long can generate 30 V (Keynes *et al.*, 1961). The amplitudes are considerably reduced in water but are still uncomfortably strong to someone trying to pick up the fish. The fish reaches a length of at least 50 cm, and 350 V discharges have been reported (Keynes, 1957).

It is interesting that the cells produce an almost entirely head-negative potential, although they are innervated on their posterior faces; in fact, the mechanism is not so different from that in the diphasicly discharging *Hypopomus* (Fig. 24). For the experiments of Figs. 53 and 54 a propagated impulse is set up by stimulation through a large monopolar electrode pressed against the main part of the cell distant from the recording site. A monopolar recording electrode detects a negative spike external to the nonstalk face (Fig. 53A) that is propagated in from the stimulation site. If this electrode is advanced into

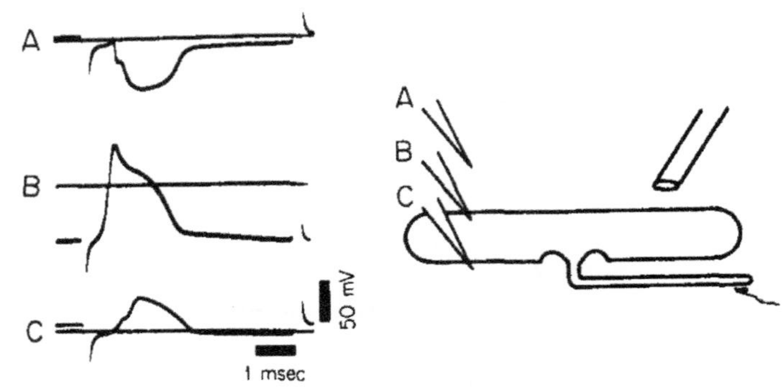

Fig. 53. Responses of a single electrocyte of the electric catfish. Monopolar recordings during advance of a microelectrode through the single cell. The horizontal trace shows the dc level. Stimulation by an external electrode distant from the recording site and recording as indicated in the inset diagram. A long-lasting pulse is used to minimize interference by stimulus artifacts. (A) Recorded outside the rostral face, the response is largely negative but is preceded by a small positive phase. (B) When the electrode is advanced into the cell, it records an internally negative resting potential and an overshooting spike with a shoulder on the falling phase. (C) When the electrode is further advanced to lie outside the rostral face, the resting potential disappears and a positive going response is recorded. Both external recordings have inflections on their rising phases at the time of the brief peak of the intracellular recording. From Keynes *et al.* (1961).

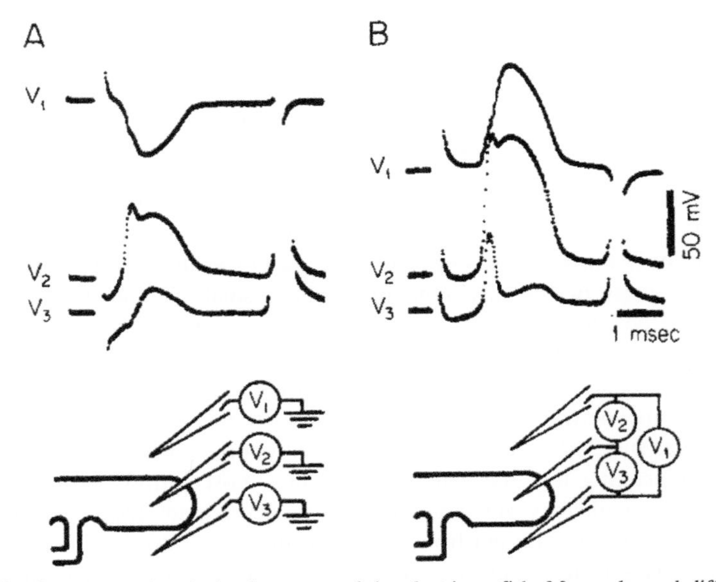

Fig. 54. Responses of a single electrocyte of the electric catfish. Monopolar and differential recordings from three separate electrodes as diagrammed, stimulation by a distant external electrode. (A) The monopolar recordings are similar to those in Fig. 53. (B) The differential recordings show the external response to be monophasic head negative with an inflection on the rising phase (V_1), the response of the rostral face to be a relatively long-lasting spike with two peaks (V_2), the response of the caudal face to be a brief spike with a small longer lasting and probably passive component (V_3). From Keynes *et al.* (1961).

the cell, it records an inside negative resting potential and an overshooting spike (Fig. 53B). Further advance of the electrode through the cell causes a loss of resting potential and recording of a positive external potential associated with the spike. Data of these kinds show the electrocyte activity to be of conventional polarity, but the shape of the external potentials is unlike that in other fish discussed up to this point.

The origin of the peculiar shape of the external potential becomes clearer in differential recordings (Fig. 54) from which it appears that the stalk face generates a small brief spike (followed by a much smaller hump) and the nonstalk face generates a larger and longer lasting spike. The sequence of potential changes during propagation of a spike thus is as follows (see Fig. 4E): outward current depolarizes each face, and the potential is initially positive outside each face as seen in the monopolar recordings of Fig. 53A,C. But negligible potential is produced across the cell because the positivities are about equal. The nonstalk face, being of lower threshold, begins to pass inward current that flows out through the stalk face depolarizing it further; the external potential becomes positive outside the stalk face and negative outside the nonstalk face. When the stalk face is excited, its activity opposes the spike of the nonstalk face, and, as a result, there is a reduction in the external potential or at least in its rate of rise. The spike in the stalk face then terminates while the spike in the nonstalk face continues. The external potential becomes much larger and then declines as the spike terminates. This sequence of firing also explains the shape of the spikes in the two faces. The spike in the nonstalk face often has two peaks; the larger potential generated initially is ascribable to reduced loading when the two faces fire together. Evidently this first maximum need not be accompanied by maximal sodium activation because the potential can rise to a second maximum after the brief spike of the stalk face. The residual potential across the stalk face following its brief spike is ascribable to voltage drop across this face produced by activity of the nonstalk face.

What is the significance of this sequence of activity? The resistance of the two faces at rest is about equal. The brief spike in the stalk face is apparently a rapid way to turn on delayed rectification in this membrane, and by increasing its conductance to increase external current flow. Electron microscopic observations show that both faces have a moderate number of branching tubules that increase their surface areas (Mathewson *et al.*, 1961). The proliferation is slightly greater in the nonstalk face. The similarity in areas is consistent with the suggestion that both surfaces increase their conductances during a response. Actually, the stalk face may not be capable of generating an all-or-none response. The response of this face to depolarization (applied by an external electrode) apparently is graded, and a propagated spike cannot be set up by external stimuli that depolarize this face and hyperpolarize the nonstalk face. It is even possible that this membrane only exhibits delayed rectification and totally lacks an inward current mechanism. The initial maximum seen in some external records would then be ascribable to capacitative current as in cup-type electrocytes of rajids (Fig. 15). The steep rise and fall of the potential across the nonstalk face makes this explanation unlikely, but it does not contradict available data.

From the location of the synapses and the observation that spikes can propagate along the body of the cell, it appears that a PSP sets up a spike in the stalk that propagates to invade the rest of the cell. As would be expected, neurally evoked responses in the body of the cells are similar in shape to directly evoked ones. Transmission at the synapses is cholinergic in that curare blocks neural excitation of the organ and a high concentration of cholinesterase is found histochemically at the synapses (Couteaux and Szabo, 1959).

Activity of the stalk can be recorded by electrodes placed (under visual control) close to the rosette, i.e., the site where the stalk joins the disc-shaped part of the cell. External stimulation in this region can excite the cell even if the polarity is such as to depolarize the stalk face (Fig. 55). External to the rosette

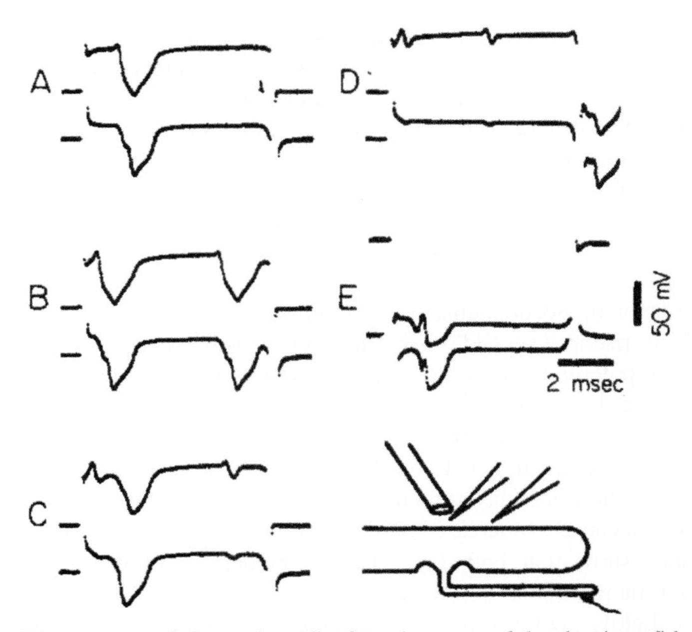

Fig. 55. Responses recorded near the stalk of an electrocyte of the electric catfish. External recording and stimulation as in the diagram, upper trace nearer the center of the rosette. Anodal stimuli outside the rostral face except in E. (A) A single largely negative spike is evoked that is preceded by a small but distinct positivity at the more central electrode. (B) A stronger stimulus evokes two responses. (C) A still stronger stimulus evokes two responses of the stalk but invasion of the body of the cell is delayed and blocked for the first and second responses, respectively. (D) A stronger stimulus evokes three responses of the stalk, but the first two fail to invade the body of the cell. The third occurs at the end of the stimulus and does invade the body of the cell. (E) Cathodal stimulation can excite the cell at this site, but the threshold is higher than for anodal stimulation.

opposite the stalk (upper trace) an initially positive potential is observed apparently resulting from activity of the stalk, because it is much smaller a short distance away (lower trace). If the stimulus strength is increased, repetitive firing is produced (Fig. 55B), which is never observed with stimulation of either polarity distant from the rosette. If the stimulus strength is increased further, the impulses fail to invade the body of the cell, but small biphasic external potentials remain that are initially positive opposite to the stalk (Fig. 55C,D). Evidently the hyperpolarization of the nonstalk face is so large that the stalk impulse cannot excite it. The predominance of the negative phases in the external records indicates that some excitation of membrane on the nonstalk side still occurs, although the activity does not propagate to the remainder of the cell. However, if the stalk impulse occurs at the end of the stimulus when hyperpolarization of the nonstalk face is terminated, the impulse in the stalk becomes able to invade the body of the cell as indicated by a large external response (Fig. 55D). A stimulus over the rosette that depolarizes the nonstalk face can also excite the cell, but the threshold is higher than for the opposite polarity of stimulation (Fig. 55E). When excited in this way the component of the external potential associated with firing of the stalk face is larger than when recorded distant from the rosette (Figs. 53 and 54).

One may question why the inflection that is seen on the rising phase of responses externally recorded close to single cells is generally not seen in the organ discharge. Since propagation time over the cell is an appreciable fraction of the duration of the spike of the stalk face, the activity of the stalk face is not synchronous in the different regions of the cell. It thus fails to produce a distinct component in the overall organ discharge which at any instant is an average of the contributions of all regions of all the cells.

3. *Gymnarchus*

The monospecific genus *Gymnarchus* is found in tropical Africa. It is closely related to the Mormyridae, but the mode of swimming is remarkably similar to that of gymnotids. The animal moves with a straight body by undulations of the dorsal fin. As far as locomotion is concerned, it is like an upside-down gymnotid (Fig. 1). Movement appears equally easy forward and backward, and often the animal seems to investigate strange objects with the tip of its tail. The electric organ pulses are emitted at a frequency of about 250/sec (Fig. 56). The frequency is not altered by ordinary kinds of stimuli, and "jamming avoidance" has not been observed (Bullock, 1970). Novel stimuli that perhaps startle the fish may evoke a sudden cessation and weak electric pulses may also be effective (Lissmann, 1958; Szabo and Suckling, 1964; Harder and Uhlemann, 1967).

The electric organ of *Gymnarchus* consists of four columns of electrocytes on each side of the body, one above the other. Each column runs to the tip of the caudal filament but their anterior extent varies (Fig. 1). The electrocytes are flattened cylinders innervated on their posterior faces by spinal nerves.

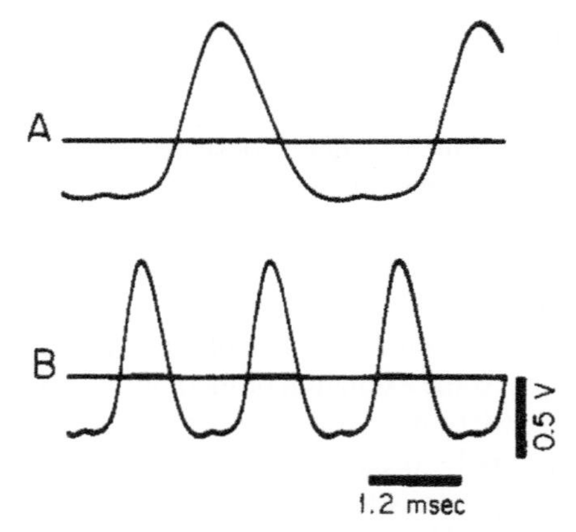

Fig. 56. Electric organ discharge of *Gymnarchus*. Recorded from an animal about 30 cm in length in a small container of aquarium water. Zero potential level indicated by the horizontal line. (A, B) Normal organ discharge at fast and slow sweep speeds.

Both faces are moderately convoluted. However, on a microscopic scale the uninnervated face has a large number of small canaliculi and processes that greatly increase its surface, while the innervated face is relatively smooth and has only a few canaliculi (Schwartz *et al.*, 1971).

The electric organ discharge consists of what appear to be head-positive pulses superimposed on an almost level base line (Fig. 56). However, the potential goes head-negative between pulses and as in *Sternopygus, Eigenmannia,* and sternarchids there is virtually no net current flow during one complete organ cycle. (There is a small head-positive bump between the large pulses; this component is apparently a result of nerve activity.) The absence of net current flow is a result of the properties of the uninnervated face of the electrocytes. The innervated face generates an ordinary spike when depolarized either synaptically or by applied currents. The uninnervated face has a large capacity and is of high resistance and inexcitable. Current flow through it is virtually entirely capacitative. During a spike in the innervated face the charge on the capacity of the uninnervated face is made more positive; between spikes the charge on the capacity tends to return to its initial value. The resulting external potential is diphasic and initially positive outside the uninnervated face.

The sequence of potential changes and the cell equivalent circuit are shown in Fig. 3E. The external medium behaves like an ohmic resistance (at the frequencies of the electric organ discharge). The external voltage is then proportional to the amplitude of transmembrane current, and the integral over time of the external voltage gives a measure of total current flow. Since the charge on the capacity of the uninnervated face before and after a spike becomes the same during a steady frequency of firing, no net current can flow through the capacity and the time integrals of the positive and negative

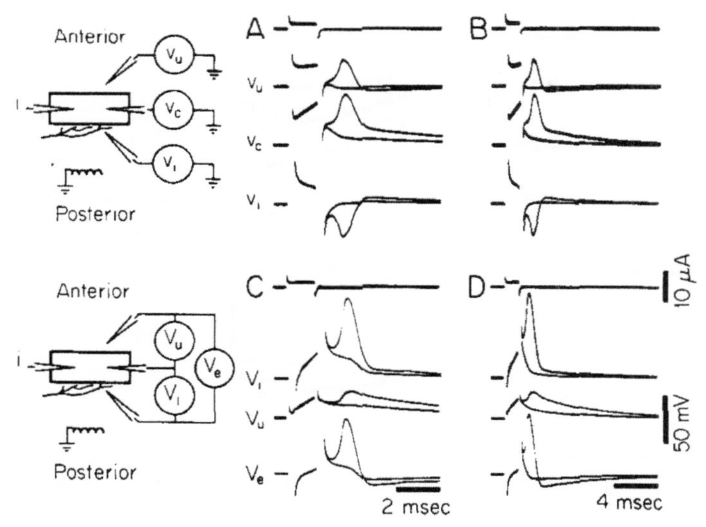

Fig. 57. Responses of a single electrolyte of *Gymnarchus*. Recording and stimulation as indicated in the diagrams. Faster sweep in A and C. (A, B) Outside the two faces monopolar recordings are diphasic and of opposite polarity. (C, D) Differentially recorded the response across the innervated face is a large monophasic spike. The potential across the uninnervated face V_u has a much smaller peak amplitude but decays away more slowly. The resulting external potential is diphasic initially head positive.

phases of the external potentials must be equal. This they are observed to be to within the 1 or 2% accuracy of the measurement.

The response of the single cells to depolarization applied by an intracellular electrode is shown in Fig. 57 recorded both monopolarly (Fig. 57A,B) and differentially (Fig. 57C,D). The monopolarly recorded potentials external to the two faces are diphasic and opposite in sign (Fig. 57A,B) establishing the longitudinal orientation of the cells. The potential outside the innervated face is initially negative indicating that this face becomes active. The differentially recorded potentials show a simple spike potential across the innervated face. The potential across the uninnervated face is also a monophasic depolarization, but its peak is much lower and somewhat later. During the falling phase, however, the potential across this face exceeds that across the innervated face, which accounts for the head-negative phase in the external records. It is not obvious from these data that the uninnervated face is purely passive and acts as a series capacity. These properties are established by passing longitudinal currents with an external stimulating electrode (Fig. 58). The results may be understood with respect to the equivalent circuit shown in Fig. 59.

If a long-lasting pulse of either sign is applied, there are transient potential changes across the innervated face at the onset and termination of the pulse, but the potential across this face is back very close to the resting value after 20 msec. After the capacity of the uninnervated face is charged to its new

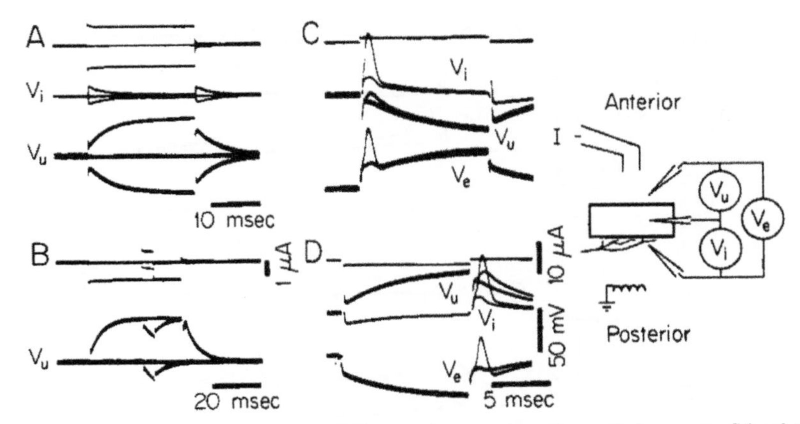

Fig. 58. Responses of an electrocyte of *Gymnarchus* to externally applied currents. Stimulation external to the uninnervated face and differential recording as indicated. (A) Oppositely directed current pulses of approximately equal amplitude produce approximately symmetrical potentials across the two faces. (B) When a brief pulse is superimposed on a long-lasting pulse that depolarizes the uninnervated face by about 50 mV, the brief pulse produces about the same change in potential as when it is given alone. Four superimposed sweeps with one, both, and no pulses. (C) The onset of an anodal stimulus depolarizes the innervated face; superimposed sweeps of threshold stimulation. (D) The termination of a cathodal stimulus depolarizes the innervated face; the threshold amplitude for stimulation by a long-lasting stimulus is the same as in C.

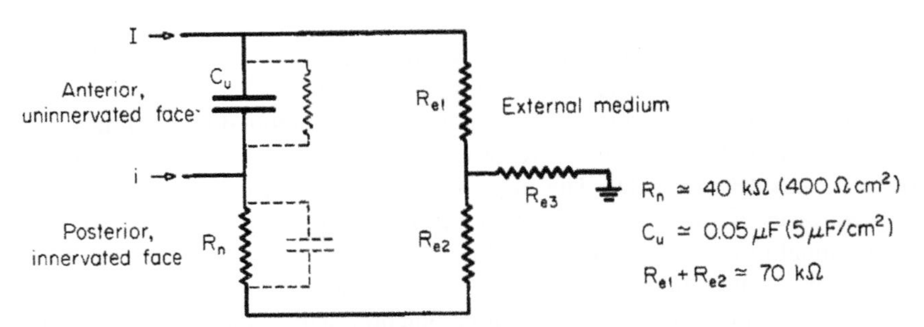

Fig. 59. Equivalent circuit of an electrocyte of *Gymnarchus*. Intracellularly applied current (i) and extracellularly applied current (I) are indicated. R_{e1}, R_{e2} and R_{e3} represent the resistance of the external medium. Negligible current flows through the resistance of the uninnervated face and the capacity of the innervated face and these elements are shown by dashed lines. The tentative values given for membrane parameters are referred to membrane area not corrected for surface convolutions.

value, no current flows through it and there is no voltage drop across the resistance of the innervated face (Fig. 58A). For pulses of this magnitude records for either direction of current are symmetrical. When depolarized by over 50 mV the uninnervated face is affected by a small pulse of current exactly as when it is at the resting potential (Fig. 58B). Both these results indicate the passivity of the uninnervated membrane.

Although no steady state voltage drop is produced across the innervated face by maintained current pulses applied externally, there are transients associated with charging and discharging the uninnervated face at onset and termination of the currents. If an anodal stimulus is applied outside the uninnervated face, the innervated face is depolarized by the initial surge and hyperpolarized following termination of the stimulus (Fig. 58C). If the stimulus is cathodal, the innervated face is hyperpolarized initially and depolarized after the termination of the stimulus (Fig. 58B). Provided the pulses are long lasting compared to the time constant of the system, the voltage across the innervated face must be identical at onset and termination of stimuli of equal amplitude but opposite polarity. This is experimentally true for *Gymnarchus* electrocytes; in Fig. 58 the threshold for firing at the onset of a cathodal stimulus (C) is the same as the threshold for firing at the termination of an anodal stimulus (D). Furthermore, if brief pulses are used to evoke responses of the innervated face during maintained current pulses, the amplitudes of the responses are the same as without maintained current, although they occur when the uninnervated membrane is either markedly depolarized or hyperpolarized.

One may ask if the membranes of the electrocytes have unique properties that give rise to the unusual function of the uninnervated face. From the equivalent circuit (Fig. 59) it is clear that intracellularly applied current and differential recording across the innervated face allow evaluation of the resistance of this face r_i. The resistances r_{e1} or r_{e2} can be evaluated by differentially recording across the cell while passing current through an electrode external to one or the other face. Where monopolar recording shows the external potentials outside the two faces to be equal, then $r_{e1} = r_{e2}$. The time constant of decay of potentials following either intra- or extracellular stimulation is given by $(r_{e1} + r_{e2} + r_i)$ C. Thus from the measured resistances and time constant one can obtain the capacity.

Although the data available are restricted, preliminary values are given in Fig. 59. These assume a uniform planar membrane on each surface. The resistance of the innervated face is quite reasonable. The capacity of the uninnervated face is if anything too small on the assumption that its capacity is the usual 1 $\mu F/cm^2$ and takes into account the great increase in membrane area resulting from the numerous tubules and processes of this face. The measurements indicate that the resistance of the uninnervated face is at least 50 times that of the innervated face. The great increase in area observed on a fine structural level means that the uninnervated membrane has a resistance of perhaps 250-500 times that of the innervated face. This is a high value, perhaps 200,000 Ω cm^2, but it is not unprecedentedly great (Bennett and Trinkaus, 1970). If the capacity of the innervated face were the usual 1 $\mu F/cm^2$ its time constant would be 0.4 msec. This value is much shorter than the time constant of the uninnervated face and is in reasonable agreement with the experimental observations. The significance of an organ discharge without net current flow is discussed in Section II, F.

4. Mormyrids

The mormyrids are found in freshwaters of tropical Africa. They are a large group of some 11 genera and large numbers of species. All that have been studied are weakly electric. A key to the genera is presented in Table III and Fig. 60. Identification as to species is generally unreliable with the possible exception of those areas for which keys are available (e.g., Greenwood, 1956; Pelligrin, 1923). Identification of species in our earlier work (Bennett and Grundfest, 1961c) was kindly provided by Dr. M. Poll of the Musée Royal du Congo Beige, Brussels, Belgium.

Table III Key to the Genera of Mormyrids

The mormyrids are identifiable as small scaled, ray finned fish without adipose fin (rayless fin caudal to the dorsal), without spines in front of the fins, and without barbels. The caudal fin is small and generally on a peduncle (Fig. 60). The mouth is small and there are teeth in both jaws. The eye is covered by a thin epithelium.[a]

- a. Anal very short compared to dorsal (ca. one-third the length, Fig. 60A). *Mormyrus* (many species)
- aa. Anal very long (5 times the length of dorsal, Fig. 60B). *Hyperopisus*
- aaa. Anal about same length as dorsal (no more than 30% difference)
  - b. Ventral closer to anal than pectoral; body elongate (Fig. 60C). *Isichthys* (one species, *I. henryi*)
  - bb. Ventral closer to pectoral than anal or equidistant between them
    - c. Teeth in several rows, villiform; terminal mouth with a chin appendage (Fig. 60D,E). *Genyomyrus* (one species, *G. donnyi*)
    - cc. Teeth in a single row in both jaws, 10–36 in each row; no chin appendage (Fig. 60G,H)
      - d. Nares farther from the eye than each other; mouth terminal (at the front of the head, Fig. 60F) or inferior. *Mormyrops* (many species)
      - dd. Nares close together and close to the eye; mouth always inferior (Fig. 601). *Petrocephalus* (many species)
    - ccc. Teeth in a single row in both jaws, 3–10 in each
      - e. One naris close to the corner of the mouth (Fig. 60J). *Stomatorhinus* (many species)
      - ee. Both nares distant from the corner of the mouth
        - f. Dentition of upper and lower jaw similar
          - g. Mouth inferior or subinferior without a chin appendage or elongated mouth (Fig. 60K). *Marcusenius* (many species)
          - gg. Mouth terminal (Fig. 60L) or on elongated jaws, sometimes greatly so (Fig. 60N); sometimes with short or long chin appendage (Fig. 60M,N). *Gnathonemus* (many species)
        - ff. A large pair of incisors in the lower jaw, without a chin appendage
          - h. Teeth of upper jaw fine conical, anal rather shorter than dorsal (Fig. 60P). *Myomyrus* (two species)
          - hh. Teeth of upper jaw obtuse or bicuspid, anal and dorsal of same length. *Paramyomyrus* (one species, *P. aequipinnis*)

[a]Modified from Poll (1957).

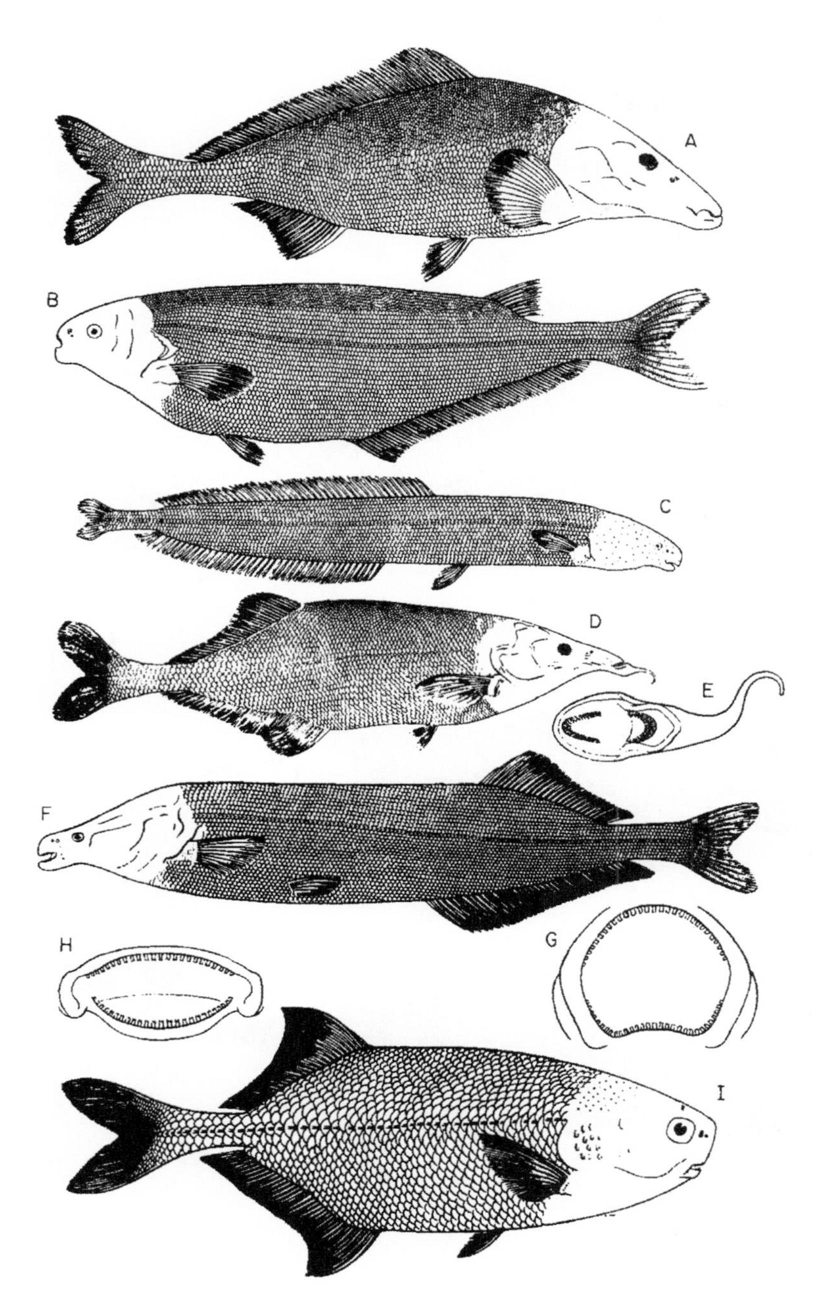

Fig. 60. Characteristics of representative mormyrid genera. (A) *Mormyrus caballus*. (B) *Hyperopisus bebe*. (C) *Isichthys henryi*. (D,E) *Genyomyrus donnyi*; side view and chin process. (F,G) *Mormyrops deliciosus*, side view and dentition. (H,I) *Petrocephalus sauvagei*, side view and dentition. (J) *Stomatorhinus corneti*. (K) *Marcusenius plagiostoma*. (L) *Gnathonemus leopoldianus*. (M) *Gnathonemus petersii*. (N) *Gnathonemus numenius*. (O,P) *Myomyrus macrodon*, side view and dentition. From Poll (1957).

(Continued)

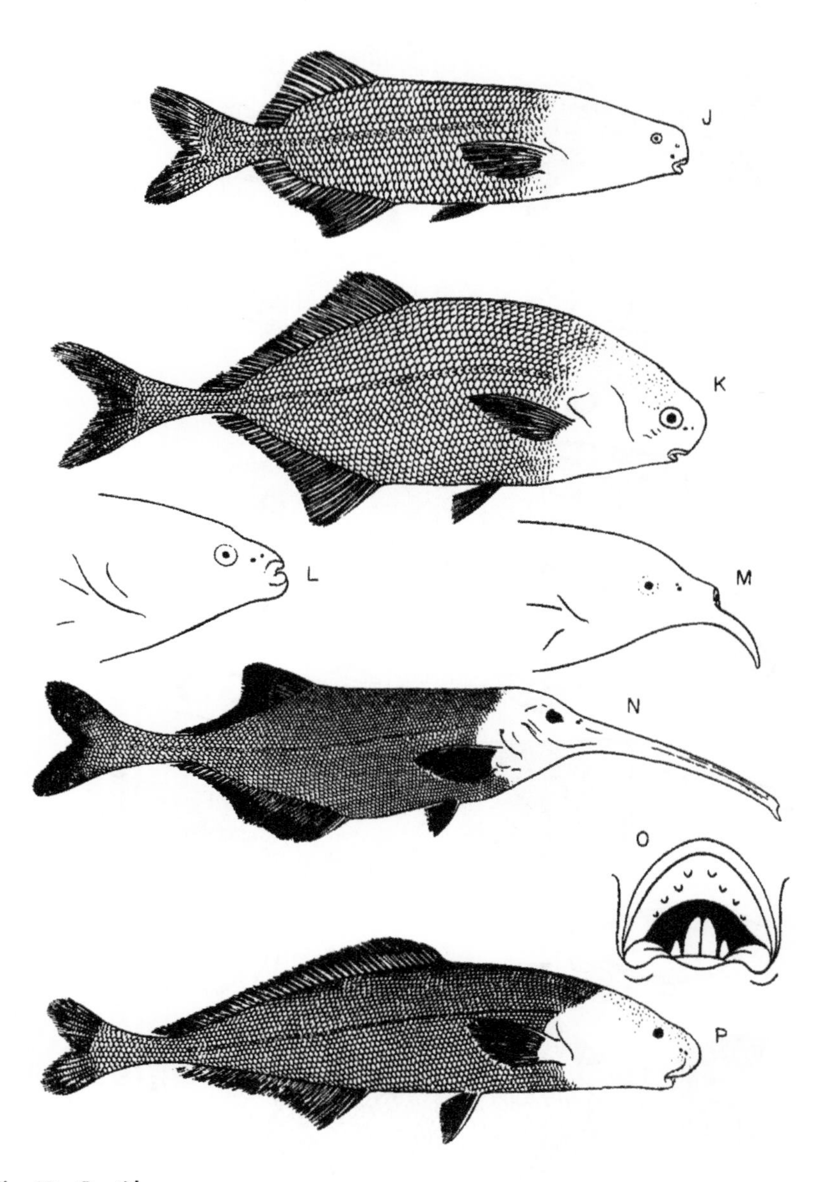

Fig. 60—Cont'd

The organ discharges are brief pulses that are emitted somewhat irregularly at a few per second when the animal is resting, but the discharge can accelerate to 40 or more per second during swimming (Lissmann, 1958). Accelerations are evoked by most modalities of stimulation including resistance changes and weak electric pulses (Szabo and Fessard, 1965; Moller, 1971). Accelerations can be classically conditioned as well as operantly conditioned in an avoidance paradigm (Mandriota *et al.*, 1965, 1968). Brief interruptions of discharge can also occur in response to novel stimuli and in interaction with other electric fish.

The electric organs lie just anterior to the caudal fin (Fig. 1). They are made up of four columns of cells and each column contains 100–200 cells in series. The cells are accurately aligned one behind the other. In many species the body narrows close to the tail to form a "caudal peduncle" and almost the entire cross section of the fish in this region is made up of electric organ. Each column of electrocytes occupies one quadrant and the only other structures are spinal column, skin, and tendons to the caudal fin.

The cells are innervated by spinal neurons that lie in three segments in the central region of the electric organ that extends over 8 or 10 segments. The nerve fibers end on stalks as in *Malapterurus* and some gymnotids, but the stalks are much more complex (Szabo, 1958, 1961a; Bennett and Grundfest, 1961c). A large number of stalks arise from the posterior faces (Fig. 61). These stalks fuse repeatedly (binarily) to form a greatly reduced number of stalks before they are innervated. In electrocytes of *Mormyrus* there may be more than 10 separate stalk systems, each with its own site of innervation. In *Mormyrops* there are one or two. In *Gnathonemus, Petrocephalus,* and *Marcusenius* there is only one.

There is a further complication in some species of *Gnathonemus*. After varying numbers of fusions the stalks turn anteriorly and pass through holes in the body of the cells (Fig. 61). They undergo their final stages of fusion on the anterior side of the cells. The degree of fusion before penetration depends on the species and correlates with the size of the initial head-negative phase of organ discharge as will be discussed below (Fig. 67). A similar penetrating stalk system occurs in *Mormyrops* (Grosse and Szabo, 1960) and in one of two specimens of a species of *Hyperopisus* that I have examined. As far as is known *Marcusenius* and *Petrocephalus* have only nonpenetrating stalk systems.

The form of organ discharge varies with the species. In *Gnathonemus, Marcusenius, Petrocephalus,* and *Hyperopisus* it is essentially diphasic, initially head-positive although there is a small initial head-negative phase if there is a penetrating stalk system (Figs. 62, 67). The discharges can be quite brief, 0.5 msec or less in duration. The discharge of *Mormyrops* is also essentially diphasic, but the initial large phase may be head-positive or head-negative. The head-negative discharges occur in specimens in which the innervation is on the posterior side, but the stalks are penetrating (Grosse and Szabo, 1960; Fig. 62B). In *Mormyrus* the discharge is also diphasic, but the second,

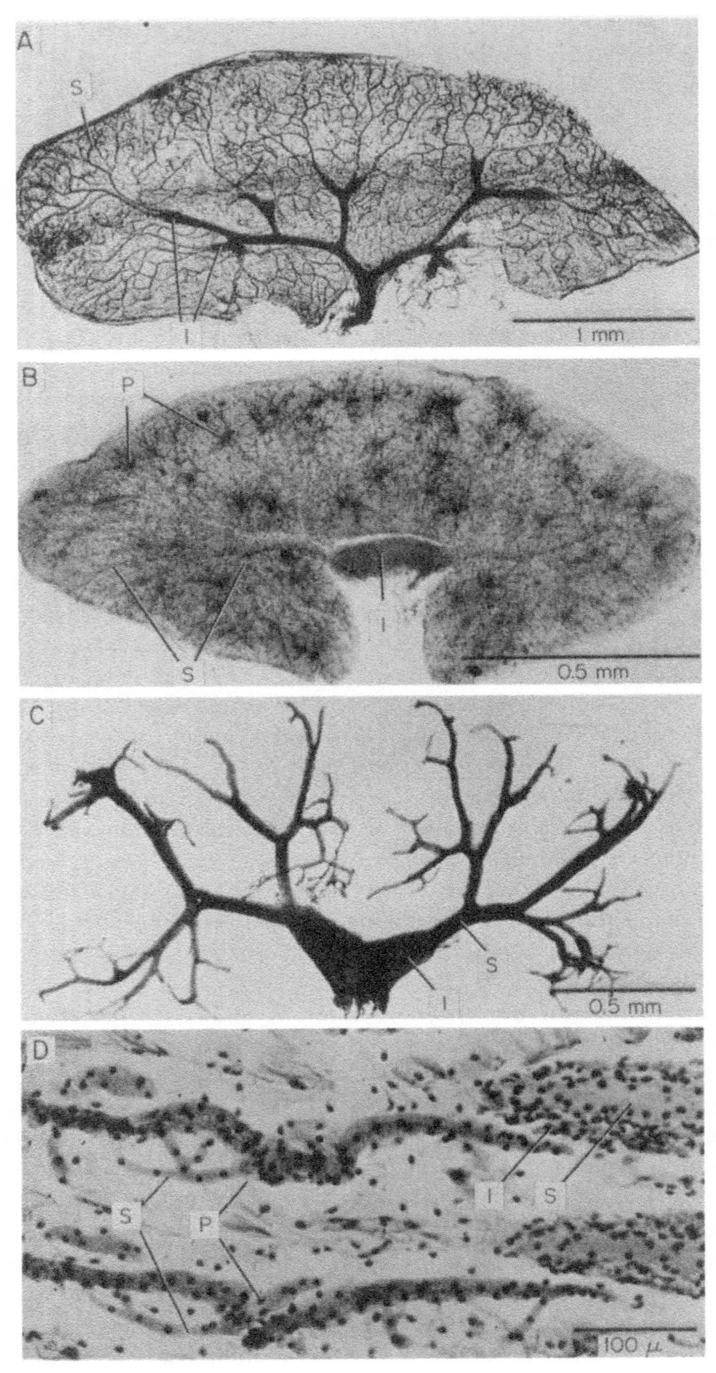

Fig. 61. See figure legend on opposite page.

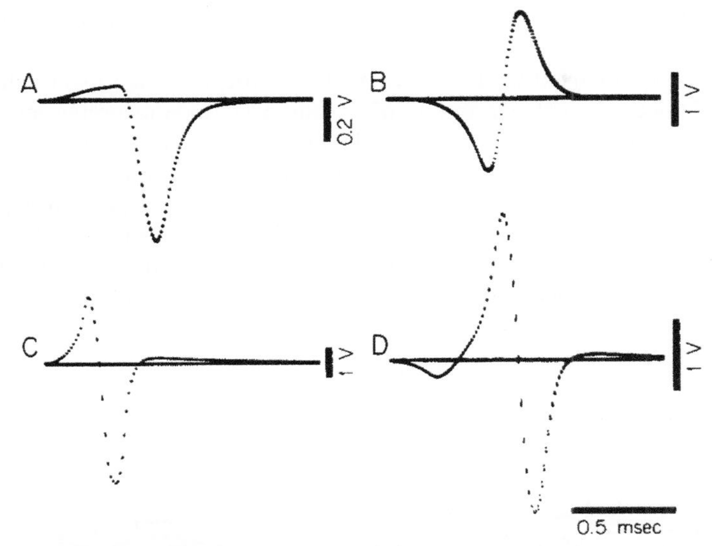

Fig. 62. Electric organ discharges of representative mormyrids. Recorded in a small volume of aquarium water between head and tail, head positivity up. (A) From *Mormyrus.* (B) From *Mormyrops.* The electrocytes of this species have penetrating stalks and are innervated on the caudal side. There are few penetrations and an initial head positive phase can be seen only with higher gain recording. (C) From *Petrocephalus.* (D) From *Hyperopisus.* The electrocytes from the specimen of this species with penetrating stalks were innervated anteriorly and had large numbers of penetrations; correspondingly the initial head-negative phase is quite large. Superimposed sweeps with (2 in D) and without organ discharges to show baseline.

head-negative phase is much larger than the initial phase (Fig. 24) and the entire response can be longer lasting (Bennett and Grundfest, 1961c; Szabo, 1961).

The single cells operate in a manner similar to those of electric fish already described (Bennett and Grundfest, 1961c). Both faces of the main part

---

Fig. 61. Anatomy of mormyrid electrocytes. (A–C) Single isolated cells stained with methylene blue. (A) Caudal surface of a cell from *Mormyrus rume.* The darkly stained, thick, branching nerve trunk innervates (i) at least eleven separate systems of stalks (s). (B) Rostral surface of an electrocyte of *Gnathonemus compressirostris* with penetrating stalks. The regions surrounding the penetrations (p) are darkly stained, elsewhere the stalks (s) and body of the cell are only lightly stained. A few stalks are seen running to the site where they are all fused and innervated (i). Under high magnification details of the stalk system could be seen, and the stalks on half the rostral surface of this cell are drawn in Fig. 67C. (C) Nonpenetrating stalk system torn off an electrocyte of *G. compressirostris* and heavily stained. The innervation (i) may be seen surrounding the stalk (s) at the central region. (D) Parasagittal section through the electric organ of *G. tamandua* stained with hematoxylin and eosin. Rostral surface up. Bodies of two electrocytes, about 30 μ thick and heavily stained, run horizontally across the figure. Fine stalks (s), about 10 μ in diameter, arise from the caudal surface, fuse, and pass through penetrations (p) in the bodies of the cells to the rostral surfaces where they leave the plane of section. The regions of innervation (i) were included in the section and lie rostral to the cell bodies. Darkly stained nuclei are seen in the stalk and in the bodies of the cells. From Bennett and Grundfest (1961c).

of the cells generate spikes. The external potentials are diphasic to nearly monophasic depending on the relations between the two faces.

The contributions of the two faces of an electrocyte from a fish producing a diphasic output are shown in Fig. 63. A microelectrode recording monopolarly is advanced through the cell from the rostral to caudal side (Fig. 63A–C). These records are then subtracted to give the equivalent of differential recording (Fig. 63B′,C′). Both rostral and caudal faces generate spikes. (In this particular experiment the separation between the two spikes is more marked than it often is, and external potentials that more closely resemble the organ discharges are usually obtained, Figs. 65 and 66.) Corresponding to the similar

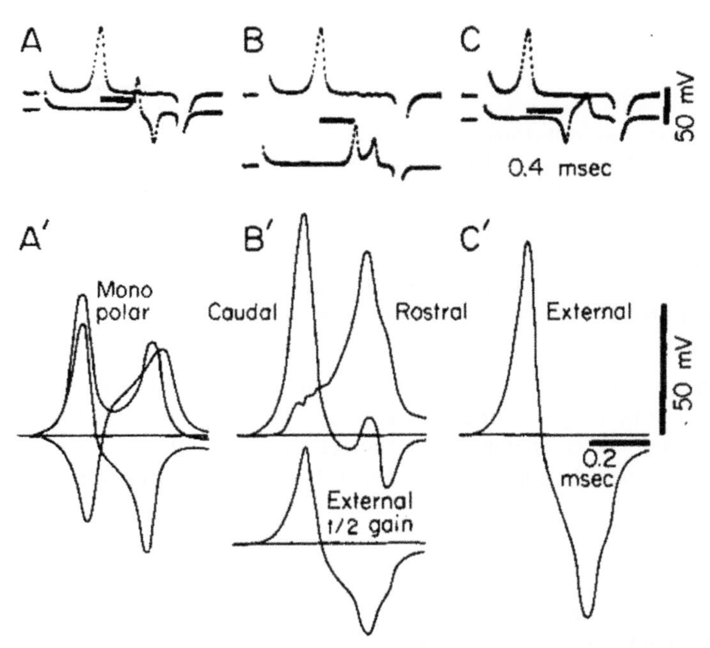

Fig. 63. Response of an electrocyte of *Gnathonemus*. Three electrodes are used, one for intracellular stimulation (1.6 msec pulse), another for intracellular recording near the stimulation site (A–C, upper trace), the third an exploring electrode about 2 mm distant (lower trace). (A) The exploring electrode is external to the rostral face and records a diphasic initially positive potential. (B) It is in the cell and shows the resting potential and a double-peaked spike. (C) It is advanced to lie outside the caudal face and records a diphasic response opposite to that in A. Time calibrations begin from the peak of the intracellular spike in the upper traces. (A′) Superimposed tracings of records made with the exploring electrode aligned with respect to the peak of the intracellularly recorded spike. (B′,C′) Potentials across the caudal and rostral faces and diphasic potential across the entire cell (rostral positivity up) obtained by graphical subtraction of the monopolar records. From Bennett and Grundfest (1961c).

responsiveness of the two faces, the surfaces appear similar electron microscopically; both surfaces are moderately increased by tubules, that of the anterior face slightly more so (Schwartz *et al.*, 1971).

In *Mormyrus rume* the organ discharge is predominantly head negative. Both faces of the cells appear to generate spikes, although it is difficult to be sure what fraction of responsiveness on the caudal side is to be assigned to the stalks (Fig. 64). The spike of the anterior face is longer lasting and the external potential is predominantly head negative. The initial slow head-positive phase of the organ discharge appears to be a result of activity in the stalks and can be observed with external stimulation of the single cells (Fig. 64A′,B′).

The location of the innervation on the stalks indicates that impulses arise in this region and propagate to involve the body of the cells. Action potentials can be recorded in the stalks, and the cell to which they go can be identified by intracellular stimulation in stalks or body of the cell. Impulses in stalks are

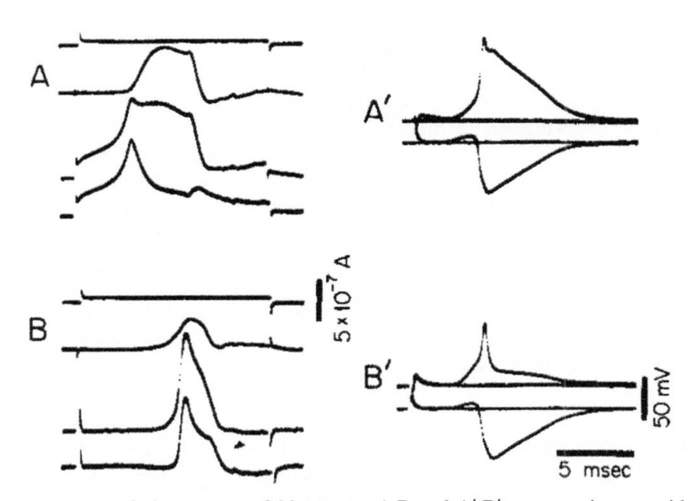

Fig. 64. Responses of electrocytes of *Mormyrus*. A,B and A′,B′ are experiments with different cells. (A) Three recording electrodes are close to the site of intracellular stimulation. Simultaneous traces show (from above down) stimulating current and differential recordings across the cell (rostral negativity up) and across the rostral and caudal membranes. The spike of the caudal face is briefer than that of the rostral face. The external response is largely rostral negative and lacks the initial rostral positivity of the organ discharge. (B) The same recording electrodes, but the response is evoked by intracellular stimulation about 1.5 mm from the recording site. The responses are somewhat shorter but are otherwise similar to those in A. The small potentials following the spikes result from excitation of neighboring electrocytes by activity of the penetrated cell. (A′,B′) Responses evoked by brief externally applied stimuli. Lower traces: recording across the cell, rostral positivity up. Upper traces: recording across the rostral face in A′ and the caudal face in B′. The transmembrane potentials are like those in A and B, but the external response much more closely resembles the organ discharge. Modified from Bennett and Grundfest (1961c).

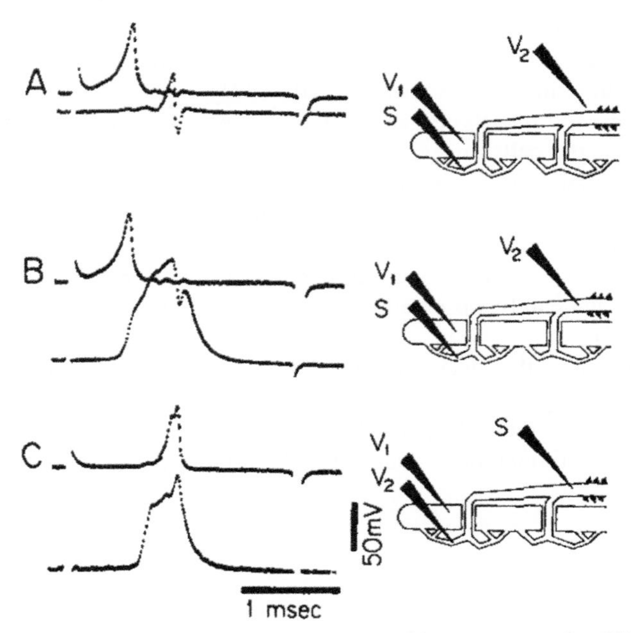

Fig. 65. Recording from an electrocyte stalk of *Gnathonemus tamandua*. The stalks are penetrating and the organ discharge is triphasic in this species. Three electrodes are used as in diagrams; the closely neighboring pair of electrodes are about 1 mm from the other one. V_1, upper trace; V_2, lower trace. (A,B) Before and after penetration of a stalk on the rostral side of the cell, stimulation in a stalk on the caudal side. Note the polarity of the triphasic external response in A. About the same potential appears to be superimposed on the broader intracellular spike in B. (C) Stimulation in the rostral stalk, recording in the body of the cell and in the stalk on the caudal side. The broad spike in the stalk has superimposed on it the response generated by the body of the cell. The early part of the brief response is obscure, but it clearly has a late positive phase. This polarity contrasts to the terminal negativity of the superimposed brief response on the opposite side of the cell in B. From Bennett and Grundfest (1961c).

longer lasting than those in the main part of the cells (Fig. 65). The PSPs that initiate the spike activity can be recorded with an appropriately placed electrode and are diagrammed in Fig. 73. The difference in spike shape, the recording of PSPs, and the fact that an electrode can enter and leave a stalk before penetrating the body of its cell allow the identification of the records as from stalks.

The discharges of *Gnathonemus* are very brief, and all regions of each face of the cells must fire quite synchronously. Direct measurement shows that a neurally evoked impulse reaches all regions of the posterior face of the cell highly synchronously. In contrast, stimulation in the body of the cell does not excite the cell synchronously. Because of the stalks an impulse initiated at one part of the body of the cell appears to be conducted along it with a very nonuniform velocity. Close to the site of stimulation (within 1 mm in the

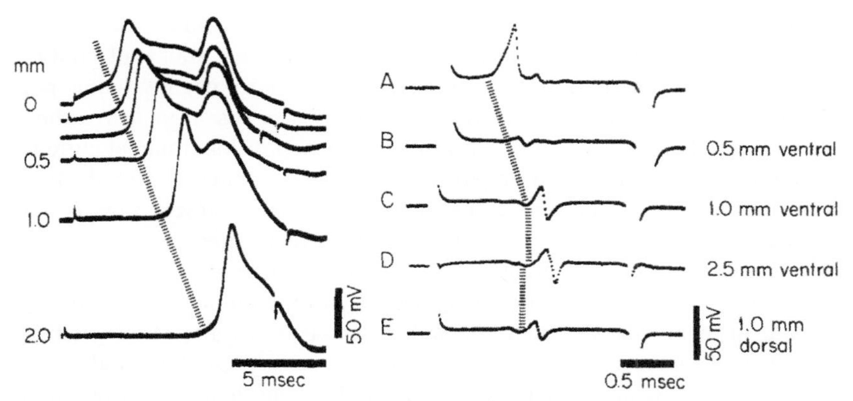

Fig. 66. Conduction along electrocytes of mormyrids. Three electrodes are used: one for intracellular stimulation, one for intracellular recording close to the stimulation site, and one exploring electrode that records at various distances along the edge of the cell. The records on the right are from *Mormyrus rume*. The response close to the site of stimulation is shown on the zero trace. Intracellular responses recorded by the other microelectrode as it is moved along the cell are below. The traces are displaced downward an amount proportional to the distances between the two recording electrodes. The spike recorded at the same time with the fixed electrode is used to position the traces in the horizontal (time) axis. Since the latency of the responses at the stimulation site varies, the stimulus artifacts do not align. The conduction velocity given by the slope of the broken line is 0.45 meter/sec. On the left are records from *Gnathonemus tamandua*. Procedure as for *Mormyrus* except that the exploring electrode records external to the rostral face. The external responses in this species are triphasic because of the large number of stalk penetrations (Fig. 67). The conduction velocity appears to be about 2 meters/sec up to a distance of 1 mm but is much faster between 1 mm and 2.5 mm. Modified from Bennett and Grundfest (1961c).

experiment of Fig. 66, right) the impulse propagates slowly, at a velocity of about 2 meters/sec. However, the impulse arrives at distant parts of the body of the cell nearly simultaneously. Evidently, the evoked impulse propagates antidromically (backward) up the stalks near the site of stimulation and then proceeds orthodromically to excite much of the cell in nearly the normal time relations. In cells of *Mormyrus* the conduction velocity along the cells is more or less uniform (Fig. 66, left). Not only is the stalk system divided into separate parts but also the stalks are much finer and the conduction velocity in them is likely to be lower.

There appears to be an evolutionary progression in going from the multiple innervation sites to the single site. *Mormyrus* is apparently the least complex stage after diffuse innervation (which is found in the related form *Gymnarchus*). In *Mormyrops* the number of sites is greatly reduced, and in *Gnathonemus* and most other genera it reaches its final limit of one. The organ discharge shows a parallel progression toward brevity, a fact that suggests more precise synchronization can be obtained when there are fewer sites of innervation. We will return to the question of synchronization in Section III, B.

The stalks play an additional roll in the form of the organ discharge. It is observed that fish possessing electrocytes with penetrating stalks all produce triphasic organ discharges; there is an initial phase of head negativity that precedes the predominantly diphasic response and that is absent in fish without penetrating stalks (Fig. 67). It is explicable as a result of longitudinal currents flowing along the organ due to impulses in the stalks passing through the bodies of the cells. The sequence of changes in direction of current flow is diagrammed in Fig. 68. In agreement with this explanation the size of the initial phase is greater the greater the number of penetrations (Fig. 67). No data are available on the functional significance of the initial head negativity. It would appear to be detectable in a fish where it is quite large (such as *Gnathonemus tamandua*), but in species where it is very small, it would seem to be of no significance at all. Behavioral studies may help to resolve these questions.

The embryological formation of the stalk system is an intriguing problem, particularly where the stalks penetrate the body of the cell. The cells are multinucleate and presumably arise by fusion of a number of cells (Szabo, 1961c). In any case the morphogenetic movements required and the control of cell fusion would appear to be highly involved. In spite of this apparent

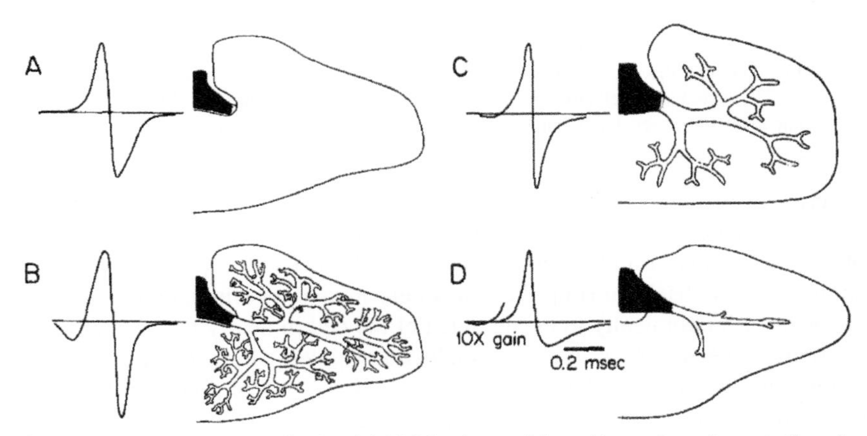

Fig. 67. Correlation of amplitude of initial head negativity with number of penetrations by stalks. Camera lucida drawings of one-half of a representative electrocyte from each of four species of *Gnathonemus* are shown with tracings of the discharge of their organs. Black area indicates zone of innervation. The potentials drawn at the same time scale but with amplitudes normalized to equal height of the head-positive phase. (A) *Gnathonemus compressirostris*, individuals which produce no initial head negativity do not have penetrating stalks. (B) The largest initial head-negative phase is in the discharge of *G. tamandua*, the electrocytes of which have many penetrations (175 in the drawing). (C) A specimen of *G. compressirostris* which has a small initial head-negative phase has electrocytes with a medium number of penetrations (25 in the drawing). (D) The organ discharge of a specimen of *G. moorii* shows initial head negativity only in high gain recordings. Its electrocytes have very few penetrations (5 in the drawing). From Bennett and Grundfest (1961c).

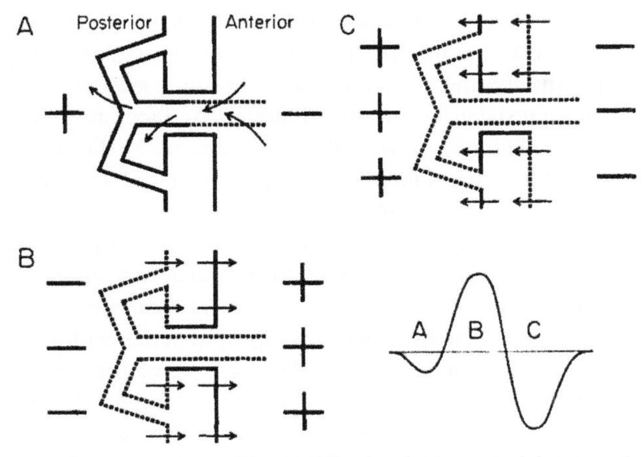

Fig. 68. Current flows generating triphasic pulses in electrocytes with penetrating stalks. Diagrams show a region near a single penetration during different stages of activity. Active membrane is indicated by dotted outlines. Arrows show direction of current flow. Resulting potential is shown in lower right. (A) Head negativity is produced when the stalk activity, initiated near the site of innervation, is passing through the penetration. (B) Head positivity results when the impulse in the stalk excites the caudal face. (C) Head negativity is again produced when the rostral face becomes active. From Bennett and Grundfest (1961c).

complexity, closely related species and perhaps even different individuals of the same species can have penetrating or nonpenetrating stalk systems (Grosse and Szabo, 1960; Bennett and Grundfest, 1961c).

E. Some Quantitative Considerations

The amplitudes of responses of individual electrocytes in several of the strongly electric fish tend to be somewhat larger than responses in other excitable cells. The PSPs in electrocytes of marine strongly electric fish are up to 90 mV in amplitude, and the peak is near zero resting potential in *Astroscopus*. This amplitude is larger than in any other known cell, but one reason is that most other cells are electrically excitable and their responses obscure the PSP. The PSP at the neuromuscular junction of frog twitch muscle fiber would also be very large; the reversal potential of the PSP is similar to that in the electric fish, and the conductance change is a substantial fraction of the resting conductance (Takeuchi and Takeuchi, 1959). In the electric eel the spike amplitude is about 150 mV, and the sodium equilibrium potential appears to be about +70 mV (Altamirano, 1955; Nakamura *et al.*, 1965). This compares with values of up to about 130 mV for other spike generating cells. Voltage amplitudes of membrane responses in electrocytes are otherwise unexceptional. Of course, the externally recorded responses of individual cells of even weakly electric organs are generally much larger than those of ordinary nerve and muscle cells.

The values of membrane resistance in electrocytes are far below the values for muscle [about 3000 Ω cm^2 for frog twitch muscle (Eisenberg and Gage, 1969)]. Correspondingly, the current outputs are much greater. In eel electrocytes the resting resistance of the innervated face is about 2–5 Ω cm^2 according to Nakamura *et al.* (1965); a value of 19 Ω cm^2 is reported by Cohen *et al.* (1969) using a different and perhaps more accurate technique. The resistance of the uninnervated face is about 0.2 Ω cm^2 (Keynes and Martins-Ferreira, 1953). During activity the resistance of the innervated face decreases to about 1 Ω cm^2 (calculated from Nakamura *et al.*, 1965), and the peak inward current is about 60 mA/cm^2. These values approach the resistance of the extracellular fluid and cytoplasm in series with the innervated face. Because of leakage and eddy currents, decrease of the membrane resistances to much below the series resistance would make the cells less efficient.

In *Torpedo* the resting resistance across an entire cell is 5-30 Ω cm^2 (Albe-Fessard, 1950b; Bennett *et al.*, 1961). The peak current during activity is about 75 mA/cm^2 and if the driving force for each cell is about 90 mV, the series resistance during activity is about 1.2 Ω cm^2 per cell.

Although the resting and active resistances of the electrocytes are far below those of muscle membranes, the values are comparable to those of the node of Ranvier [resting resistance, 30 Ω cm^2; active resistance, 2 Ω cm^2; peak inward current 10 mA/cm^2 (Dodge and Frankenhaeuser, 1959; Frankenhaeuser and Huxley, 1964)]. Furthermore, the figures for electrocytes are calculated from macroscopic areas disregarding the considerable surface convolutions, and a more realistic comparison would be referred to actual areas of plasma membrane. On the reasonable assumption that the plasma membranes of the electrocytes have a capacity of 1 μF/cm^2, one may estimate the true membrane area from the capacities measured per macroscopic area, which in the eel is about 15 μF/cm^2 (Cohen *et al.*, 1969) and in *Torpedo* is about 5 μF/cm^2 (Albe-Fessard, 1950b). Thus, the actual membrane parameters probably differ by factors of 5–15 from the reported ones (factors which appear consistent with the increase in surface seen morphologically, references below). Weakly electric organs tend to have some-what higher membrane resistances, several tens to several hundred Ω cm^2 referred to macroscopic surface. Insufficient data are available for an accurate comparison among weakly electric organs in terms of area of plasma membrane, but the membranes—with the exception of membranes acting as a series capacity—all have a shorter time constant than muscle and presumably are of lower resistivity.

The response characteristic in which electrocytes are most outstanding is frequency of firing. The electric organs of sternarchids discharge constantly at frequencies from 700 to 1700/sec, *Eigenmannia* operates at 250–600/sec, and *Gymnarchus* at about 250/sec. *Sternopygus* discharges at 50–100/sec, but it is like *Eigenmannia* on a 50% duty cycle in which the interval between pulses is about equal to the pulse duration. The higher values are unequaled for maintained frequency by muscle or other nerve. In fish other than the sternarchids the nerve-electrocyte synapse is chemically transmitting (and cholinergic) so that chemically mediated transmission can also operate at high frequencies. However, in gymnotids and *Gymnarchus* the synapses between the controlling neurons, which fire at the full organ frequency, may well all be electrically transmitting (see Section III).

The high frequency of maintained activity suggests that there could be a considerable power output per unit weight of organ. As yet no satisfactory data are available, but many of the cells contain large numbers of mitochondria suggesting a high metabolic rate (Schwartz *et al.*, 1971). The peak pulse power of the strongly electric organs is very large, but the pulses can be emitted at a high rate only for short times. The energy output presumably represents ions running down preexisting concentration gradients that are restored relatively slowly. Electrocytes of strongly electric organs do not have high densities of mitochondria (Mathewson *et al.*, 1961; Wachtel, 1964; Sheridan, 1965; Bloom and Barnett, 1966), and it would be predicted that the power output per gram of cell that could be maintained would be lower than for high frequency cells.

In marine electrocytes the site of production of electric energy is somewhat paradoxically the innervated face which is inexcitable in the strongly electric fish. It is across this face that the potential is found during organ discharge. In the eel the contribution of the two faces is about equal, as it is in many cells of which both faces generate spikes. For electrocytes in which one face acts as a capacitance the energy is produced entirely across the innervated face.

F. Adaptation and Convergent Evolution in Electric Organs

The evolutionary origin of electric organs will be discussed in the following chapter, because it appears intimately linked with the development of electrosensory systems. Ignoring the intermediate stages of evolution, one notes a number of convergences in electric organ systems that extend beyond simply the production of electricity (Table IV). Some of these convergences have obvious adaptive value. For example, it was recognized early that electric organs are relatively flattened in the two kinds of strongly electric marine fish and elongated in the two kinds of strongly electric freshwater fish. This difference is ascribable to the different resistances of the environment. The low resistance of seawater requires a high current, low voltage output; the high resistance of freshwater requires a low current, high voltage output.

Another striking difference between marine and freshwater electric fish is the absence of spike generating membrane in electrocytes of the three marine groups and its presence in those of the three freshwater groups. While it is natural to ascribe this difference to differences in loading by the environments, it is difficult to do so convincingly (Bennett, 1961, 1970) since each organ

Table IV Convergences in Evolution of Electric Organs[a]

	Elasmo-branchs		Teleosts									
			Related		*Gymnotids*							
	Rajids	*Torpedinids*	*Mormyrids*	Gymnarchus	*Electric eel*	Gymnotus	Hypopomus	Steatogenys	*Sternopygids*	*Sternarchids*	*Electric* catfish	*Stargazers*
No spikes, marine	X	X										X
Organ flattened		X										X
Spikes, fresh water			X	X	X	X	X	X	X	X	X	
Organ elongate	X		X	X	X	X	X	X	X	X	X	
Strongly electric organ		X			X						X	X
Weakly electric organ	X	X	X	X	X	X	X	X	X	X		
Accessory weak organ						X		X		X		
Intermittently active	X	X			X						X	X
Continuously active			X	X		X	X	X	X	X		
"Constant" frequency				X					X	X		
Variable frequency			X			X	X	X				
Can cease firing			X	X		X	X		X			
Monophasic, both faces respond	X										X	

Monophasic, one face series R	X	X			X	X[b]	X	X[b]				X
Diphasic, one face series C				X					X?	X		
Diphasic, both faces spike			X				X	X				
Triphasic			X			X						
Innervation on stalks			X		X?		X	X			X	

[a]Characteristics are shown along the left side; different groups are shown along the top. A particular group is indicated as having a particular characteristic if at least one member of the group has it; a question mark indicates that only suggestive evidence is available. Thicker vertical lines separate groups in which electric organs are virtually certain to be separately evolved. A convergence is indicated when a particular characteristic appears in more than one of the separately evolved groups. Other convergences, such as development of chin accessory organs, may occur within groups.
[b]In accessory organs.

consists of many elements in series-parallel array and the most energetically efficient kind of membrane would appear to be the best in both environments. Yet there is almost sure to be some advantage to PSP membrane for the marine environment because in each case the muscle giving rise to the electric organ does possess spike generating membrane (Bennett *et al.*, 1961; Grundfest and Bennett, 1961, and unpublished data).

An interesting convergence is in the development of stalks. These structures are most remarkable in mormyrids but are also prominent in catfish electrocytes. Stalks are present but seem of no great importance in gymnotids. Stalks do allow a smaller PSP current to excite the electrocytes since the input resistance of the stalks is higher, but the advantages of this are far from obvious if one considers marine electrocytes where the entire output consists of PSPs. The stalks act in synchronization of firing of mormyrid electrocytes, but the same output could be produced by diffusely innervated cells, at least in the species with longer lasting discharges and the more primitive stalk systems.

Another generalization concerns the presence of a dc component in the organ discharge. All strongly electric fish emit de pulses; perhaps this leads to more effective shocking of prey or predator. A few weakly electric fish that fire at low frequencies also emit pulses that are dc or have a large dc component, but in most weakly electric fish including all high frequency species, the organ discharge has little or no dc component. The absence of a dc component in the organ discharge allows the fish to have a dual electrosensory system in which one set of receptors is sensitive to low frequency signals arising in the environment and another set detects distortions in the high frequency field produced by the electric organ (see Chapter 11, this volume).

In *Gymnarchus,* the sternarchids, and perhaps *Eigenmannia* the absence of a dc component is achieved by modification of one electrocyte face to pass only capacitative current. In *Sternopygus* the uninnervated face apparently acts as a polarization capacity. In others such as *Gymnotus* and most mormyrids, the opposed faces of the electrocytes are active sequentially to achieve the same effect. Other convergences in terms of pulse shape and patterns of organ discharge can be found. Mormyrids are similar to the variable frequency gymnotids, and *Gymnarchus* is similar to the constant frequency gymnotids. The more striking differences are sure to be associated with differences in the operation of the sensory part of the system, and some of the subtler ones may be associated with species recognition.

The fine structure of electrocytes shows a few correlations with their functional properties. Where one face is low resistance and inexcitable and the other face generates PSPs or spikes, the surface of the low resistance face is greatly increased by projections or invaginations [torpedinids, electric eel, monophasically discharging *Hypopomus,* accessory organs of *Steatogenys* (Mathewson *et al.*, 1961; Sheridan, 1965; Bloom and Barnett, 1966; Schwartz *et al.*, 1971)]. The area of the active face may also be increased but to a smaller extent. The relative areas correlate with the changes in resistances during activity and are such that these resistances tend to become more nearly equal. The probable significance of this feature in terms of increased efficiency was noted in Section II, B. In *Gymnarchus* one face acts as a series capacity and this face is markedly increased in area; the morphological relations are similar in Eigenmannia (Schwartz *et al.*, 1971). In *Sternopygus* both uninnervated and innervated faces are quite smooth, which supports the hypothesis that the very large apparent capacity is a result of shifts of ionic concentrations rather than a dielectric capacity. Furthermore, the resistances during and between spikes are about equal.

A number of electrocytes in which both faces generate spikes have been studied at the fine structural level [*Gymnotus, Steatogenys* main organ, *Gnathonemus,* and *Malapterurus* (Mathewson *et al.*, 1961; Schwartz *et al.*, 1971)]. In each case there is some proliferation of both faces, but the degree is always less than in the inexcitable faces of monophasically responding cells. The proliferation in diphasically responding cells is somewhat greater in the uninnervated or nonstalk face; that in all but *Malapterurus* is the higher threshold, later firing face. The morphological difference may be an adaptation to the fact that the earlier firing face must operate through the series resistance of the later firing face at rest and that the latter should therefore have a low resting impedance. In *Gymnotus* at least the time constant of the later firing face is greater than that of the earlier firing face, and a significant fraction of the early outward current through the later firing face may be capacitative. The activity of the later firing face occurs during the phase of increased conductance owing to delayed rectification (potassium activation) in the earlier firing face, and the area of this face need not be so great. One would not be surprised to find that delayed rectification was more pronounced in the earlier firing face, but this possibility has not been investigated. Delayed rectification is more marked in *Gymnotus* cells than in those of *Steatogenys,* which may account for the smaller degree of proliferation of both surfaces in *Gymnotus* as compared to *Steatogenys.*

Although some nerve-electrocyte junctions are continuously active at very high rates, no morphologically distinctive features have been found (Schwartz *et al.*, 1971). All investigated do not appear significantly different from neuromuscular junctions.

It is a bit disappointing that current electron microscopic techniques have not revealed any further correlations with function. One cannot yet distinguish between a high resistance inexcitable membrane and a low resistance excitable membrane. This failure probably reflects the small fraction of membrane surface actually involved in ionic movements (cf. Hille, 1970). Perhaps freeze-cleaving techniques, which allow one to examine relatively large areas of membrane *en face,* will provide sufficient resolution. After all, it is to be expected that ionic channels, although themselves small, have associated with them protein molecules well within the range of resolution of the electron microscope.

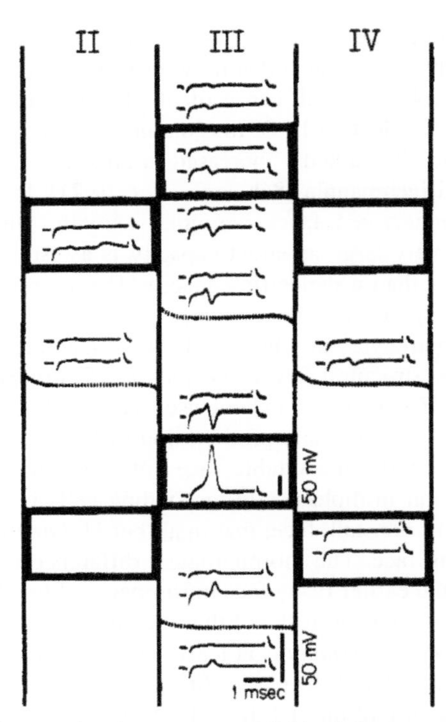

Fig. 69. Influence of connective tissue on external fields of electrocytes of *Gymnotus*. The diagram shows portions of the connective tissue tubes of the three ventral columns of electrocytes, two of which are indicated by heavy lines in each tube (see Fig. 19). Septa between chambers are shown as dotted lines. Rostral is toward the top. The lower cell in tube III was stimulated intracellularly and a third electrode explored the external field. Upper and lower records of each pair represent the response before and after penetrating the connective tissue tubes. The diphasic external response is initially positive on the rostral side. The external potentials are considerably larger inside tube III than elsewhere. The potentials decrement relatively little along the tube. A subthreshold PSP is evoked in the rostral cell of tube II. From Bennett and Grundfest (1959).

An important kind of adaptation in electric fish is the arrangement of connective tissue. In many electric organs, connective tissue appears to provide an insulating sheath that channels current flow along the axis of the organ. An example from *Gymnotus* is given in Fig. 69. In *Torpedo* the resistance of the skin on the dorsal and ventral surfaces of the organ is lower than that over the rest of the body (Bennett *et al.*, 1961). This difference tends to maximize current in the external medium and minimize current flowing through the animal's body.

An old but frequently recurring question is why strongly electric fish do not shock themselves. Their resistance to their own discharges as well as those of other fish is no doubt contributed to by the heavy fat and connective tissue layers that surround much of the nervous system. For example, the electromotor axons of *Malapterurus* are rather small, but they are surrounded by a sheath so that their apparent diameter is about 1 mm (in a fish about 15 cm long). This sheath becomes attenuated as the fiber branches, but it nevertheless extends beyond the synaptic terminals and part way down the electrocyte stalks (Mathewson *et al.*, 1961). One would not be surprised if the hearts of strongly electric fish were less easily driven into fibrillation by shock or if the electroconvulsive threshold were lower for their neural tissue, but no data have been obtained.

Another answer to the question of why electric fish do not shock themselves is that they do. Often an eel will twitch when it discharges its main organ. Self-stimulation becomes more prominent when the fish is out of water because the voltage developed across the animal's own tissue is increased. The electric catfish will normally not move when it discharges, but if a cut is made through the organ to the body wall each discharge will cause a twitch. Evidently opening of the inner lining of the organ allows more current to flow through the body. When mormyrids are placed in air, even they may show signs of selfstimulation. Organ discharges cause twitches of tail muscles and small after-discharges of the organ, probably because of electrical stimulation of the spinal cord (Bennett and Grundfest, 1961c).

G. Embryonic Origin and Development of Electric Organs

There is now little doubt that except in sternarchids all electric organs evolved as modifications of muscle. The muscle groups of origin can be located at any point along the body from eye muscles in *Astroscopus* to branchial muscles in torpedinids, to pectoral muscles in *Malapterurus,* to axial and tail muscles in the remainder. In electrocytes of *Astroscopus,* rajids, and mormyrids striations are easily seen under the light microscope. The orientation however is quite unrelated to the cell axis. Electron microscopy shows well-organized Z lines and thin filaments in both *Astroscopus* and rajids. *Gymnarchus* electrocytes have similar but less regular structures (Schwartz *et al.*, 1971). In the mormyrid *Gnathonemus* there are both thick and thin filaments and Z lines. The filaments are not very well aligned, however, and the cross bridges of normal muscle are certainly reduced in number if not absent altogether. Sarcoplasmic reticulum, associated in normal muscle with control of tension, is lacking in most regions or very sparse. It is not surprising then that the cells do not move, although it would be interesting to know the biochemical "deficiency," if any, in the contractile machinery.

Electrocytes of the other three evolutionary lines of electric fish, torpedinids, gymnotids, and *Malapterurus,* all have fine but disorganized filaments without Z lines (Wachtel, 1964; Schwartz *et al.*, 1971). The origin of these tissues from muscle is less obvious in the adult, although the development of the organ from myoblast-like tissue has been demonstrated histologically (Johnels, 1956; Keynes, 1961; see also Fritsch, 1890). The physiological and pharmacological properties strongly support derivation from muscle. Some histological aspects of electric organ development have also been described for *Gymnarchus, Astroscopus,* and *Mormyrus* (Dahlgren, 1914, 1927; Szabo, 1961c).

An interesting problem is the ontogeny of excitability in electrocytes of marine fish. Rajid electrocytes develop from rather normal appearing muscle fibers (Ewart, 1892). If they are normal, the spike generating membrane present in tail muscle of the adult (Grundfest and Bennett, 1961) would have to be lost. Embryological material is readily obtainable for this group unlike that for most other electric fish. Increase in the number of electrocytes during growth may also provide an opportunity for experimental analysis. Gymnotids regenerate their tails including electric organs (Ellis, 1913) and electric eels apparently add layers of electrocytes as they increase in length (Keynes, 1961).

The existence of a neurogenic accessory organ in the chin region of *Adontosternarchus* (Section II, D, 1, g) has developmental as well as evolutionary implications. It suggests a kind of preadaptation but one not involving a purposeful change toward a structure that has survival advantage only at some later time. A probable early stage in the chin organ development was maintained oscillatory activity of a group of electroreceptors (see Chapter 11, this volume) that served as a weakly electric organ for other receptors. The afferent fibers would then have become modified to resemble the fibers of the main electric organ. The independent evolution of rostral accessory organs in other gymnotids attests to the value of these organs. Is it to be supposed that the sensory fibers of the primitive accessory organ gradually enlarged over many generations to repeat the evolutionary sequence followed in the main organ? It seems much more likely that having evolved the developmental mechanism to make large generating nerve fibers in its main organ, the fish merely evolved the ability to apply the same mechanism to the chin fibers. In molecular terms, DNA coded information required to make the main organ fibers was also present in the DNA of the sensory neurons; the fish then acquired a way of expressing this information at the different site. In brief, the argument is that there is evolution of a mechanism of turning-on in a new part of the organism a previously evolved very complex developmental sequence. This process appears much more probable than evolution of a very similar complex sequence all over again.

Evolution of a structure in one part of the body and then the sudden appearance of a very similar structure in a quite different part of the body may not be uncommon in phylogenetic history. Another example from electric fish is Hunter's organ of the eel, which appears to be evolved from a different muscle group than the main and Sachs' organs, but which has very similar electrocytes (see Section II, D, 1, a). Other examples are discussed in Bennett (1970). In accessory electric organs of other gymnotids and *Narcine* the innervation suggests that the organs developed by migration of electrocytes from the main organ rather than by development from a different and local muscle group (see Sections II, C, 2 and II, D, 1, d).

This concept of preadaptation also raises the possibility that electric organs may have arisen in elasmobranchs only once. The argument for separate evolution is based on the muscles of origin being different in torpedinids and rajids. The organs could have originated in one muscle type in a common ancestor, then "jumped" to a second muscle type, and finally have been lost at the first site. The rajids seem the more generalized of the two, and a hypothetical electric common ancestor would probably have been rajid-like. The same argument appears inapplicable to separate evolution of electric organs in different teleost groups because of their much wider evolutionary separation.

H. Electrocytes as Experimental Material

Many electrocytes are very large cells that are of low resistance and easily studied by microelectrode techniques. Their greatest advantage over muscle is that they do not move when stimulated. However, the large size and low resistance is often a disadvantage because the membrane potential may not remain uniform over the cell when it is stimulated by a single intracellular microelectrode; that is, the cells are not "space-clamped" (cf. Bennett, 1970). Thus, qualitative results can be obtained using microelectrodes, but quantitative characterization of membranes in terms of Hodgkin–Huxley parameters as has been carried out for other tissues may not be possible (Hodgkin and Huxley, 1952; Frankenhaeuser and Huxley, 1964).

Because electrocytes have a uniform orientation, it is sometimes feasible to use external electrodes on columns of cells or even on the entire electric organ. This kind of preparation allows satisfactory impedance measurements which can be referred to the single cell (e.g., Fig. 41; see also Albe-Fessard, 1950b). Crude but useful current clamp measurements may also be possible where the properties of the individual cells are sufficiently alike that the current density through them is essentially uniform (e.g., Bennett *et al.*, 1961). The uniform orientation also has allowed measurements of thermal and optical changes associated with activity; most important, these changes can be studied under different degrees of electrical loading of the tissue (Cohen *et al.*, 1969; other references in Bennett, 1970).

Single eel electrocytes can be isolated. The single cell can then be placed between two baths and transcellular current restricted to a limited area by pressing the innervated face of the cell against a plastic sheet with a small hole through it. Current application by external electrodes is more or less uniform because the uninnervated face is inexcitable and of low resistance. This preparation has been used in voltage clamp and impedance measurements, although it is not clear that space clamping is possible even under these conditions (Nakamura *et al.*, 1965; Morlock *et al.*, 1969; cf. Bennett, 1970). As shown in respect to the squid axon, space clamping may fail if access resistance through the surrounding solution is too large compared to membrane resistance (K. S. Cole, 1968). It is difficult or impossible to clamp the squid axons showing the largest inward currents, and eel electrocytes pass considerably larger currents. Another problem is whether the small stalks on the innervated faces protrude far enough to be nonisopotential with neighboring membrane (as in several weakly electric gymnotids, Figs. 28 and 31). In the eel preparation the effects of membrane outside the edges of the window also must be evaluated. These factors require careful analysis as did voltage clamp of the squid giant axon, and it will be necessary to use exploring microelectrodes to verify space clamping. In spite of the possible or real shortcomings, gymnotid electrocytes allow voltage clamping superior in respect to temporal resolution to what has been possible with skeletal muscle fibers (Adrian *et al.*, 1970). One important measurement that has not been obtainable from muscle fibers is the high speed of onset and reversal of anomalous rectification (Bennett and Grundfest, 1966; Nakamura *et al.*, 1965; Morlock *et al.*, 1969). The primary factor is that the time constant of the excitable membrane is much lower in electrocytes and there is less interference from capacitative currents.

Isolated eel electrocytes can be of value in flux measurements (Higman *et al.*, 1964), although the spatial uniformity of concentration as well as potential becomes important and the conductances are so large that intracellular concentrations can change rapidly (Karlin, 1967). Still the large amounts of synaptic as well as spike generating membrane and the possibility of approximately controlling transmembrane potential during drug application and flux measurements are features not readily available in muscle.

Electrocytes do not appear as good as muscle for electrophysiological study of many aspects of synaptic transmission. The presynaptic fibers are no larger than those in muscle and usually do not end in a localized but accessible region. The low input resistance of the cells makes them less useful for study of miniature PSPs and actions of restricted synaptic areas. The low input resistance and wide distribution of synaptic membrane also impede studies using iontophoretic application of drugs, which remains the best method for study of kinetics of drug action. Nonetheless, through experimental simplicity, electrocytes have been and should continue to be useful in studies involving relatively gross application of drugs during current application and recording of PSPs (e.g., Karlin, 1969).

The kinds of problems where electrocytes are likely to be particularly useful are in the area of biochemistry. The organs are a rich source of acetylcholinesterase (Leuzinger and Baker, 1967), and much of the work characterizing the enzyme has been done on material of this origin. Moreover, the eel electric organ is probably the richest known source of the sodium–potassium transport ATPase and has been used in studying this enzyme (Albers, 1967; Post *et al.*, 1969). Eel electric organ does not have a particularly high concentration of mitochondria compared to a number of muscles, which suggests that its mean metabolic rate is not particularly high. Nevertheless, the only work it does is generation of electricity. Thus, it is reasonable that its sodium–potassium

ATPase levels should be very high. From the much greater concentration of mitochondria in electrocytes of repetitively active weakly electric fish (Schwartz *et al.*, 1971) one would expect there to be considerably higher levels of the transport ATPase. Of course, the weight of organ available is measured in grams rather than kilograms. When macromolecules of excitable membranes are to be isolated, electric organs are probably a good place to start. The ability to choose by choice of organ the type of excitability present in terms of PSP or spike generating membrane may be helpful in the future isolation procedures.

The electric organ of *Torpedo* is now being used for the isolation of vesicles containing acetylcholine (Israel *et al.*, 1968). Electric organs seem to be much better tissues than guinea pig brain in which to look for vesicles containing acetylcholine when only a very small fraction of interneuronal synapses can involve this transmitter. Another probably useful tissue for vesicle isolation is the electromotor lobe of the *Torpedo* brain. Available evidence indicates that synapses on the electromotor neurons are chemically transmitting, but the transmitter is unlikely to be acetylcholine (see Section III, B). The electromotor lobes provide a tissue sample of up to a gram of what appear to be neurons of a single type with a single class of synaptic ending on them. While the weight of tissue is small compared to the kilograms of electric organ, the size is very large compared to other neuronal groups of comparable homogeneity.

III. NEURAL CONTROL OF ELECTRIC ORGANS

For most electric fish it is of considerable importance that the firing of individual electrocytes be synchronous. Synchrony leads to an output that is larger in terms of both voltage and power because inactive cells act as a shunt or series resistance and thus reduce the amount of current that active cells produce outside the fish. When the discharge is diphasic or triphasic, synchronization is particularly important because slightly out of phase addition leads to cancellation. As has been seen in the section on electrocyte activity, organ discharge generally involves one highly synchronized response of each cell. Only weakly electric organs of marine fish appear to have discharges that involve fused and repetitive activity of the electrocytes.

There is another fundamental consideration in control of organ discharge; namely, that two spike generating membranes arranged in series tend to desynchronize each other's activity. Inward current generated by one excitable membrane tends to hyperpolarize the corresponding membrane of the next cell in series with it and thus prevent its firing (see Fig. 2). To achieve synchronous activity each cell in series must be separately innervated and its discharge controlled centrally. Of course, cell membranes in parallel may tend to excite one another as an action potential propagates along a single membrane. Still separate cells are always separately innervated. One exception to the rule of central control of organ discharge is known, the chin organ of *Adontosternarchus*. Since this organ consists of a single layer of parallel elements, the spontaneous activity of the cells tends to be synchronized and the discharge frequency is set in the organ itself without central control (see Section II, D, 1, g). As might be expected, this organ often operates at a somewhat different frequency from the main organ.

Control of electric organ discharge can be divided into two problems, how the fish "decides" to discharge its organ, and, having reached the decision, how it activates the different generating elements synchronously. The "decision" to discharge the organ is reached, depending on the kind of fish, by a small group of neurons in the higher spinal cord, medulla, or perhaps midbrain (Bennett *et al.*, 1967a,b,c; Bennett, 1968a). These cells make up the "command nucleus," and when they fire synchronously, the "command signal" to discharge the organ is initiated. This activity is then transmitted to the electrocytes either directly or through one or more neural relays.

Neurons of the command nucleus are probably spontaneously active in the continually discharging forms; that is, they are pacemaker neurons in a manner analogous to pacemaker cells in the heart. In species that discharge only intermittently, the command neurons receive excitatory inputs and perhaps inhibitory inputs as well. When these inputs reach threshold, the neurons "decide" that the organ will be discharged. A very significant feature of the pacemaker or command cells is that they are coupled to each other by means of "electrotonic synapses." The coupling is the basis of the highly synchronous firing that is observed in command neurons of the electric organ systems. The electrotonic synapses provide resistive pathways between cells that for this reason behave as if they are part of the same core conductor; potentials spread between the cells in the same way as they spread electrotonically along an axon. Two important properties of electrotonic synapses are relevant to their functioning in electric organ control. Current can flow in either direction across the synapses, and current begins to flow without delay when pre- and postsynaptic potentials differ. These properties differentiate electrotonic synapses from chemically transmitting synapses in which transmission is basically in one direction and postsynaptic current is delayed with respect to presynaptic impulses (by about ½ msec at room temperature). Electrotonically mediated PSPs are delayed with respect to presynaptic potentials because of the capacity of the postsynaptic cells, but generally the delay is very short compared to the delay at chemically transmitting synapses (Bennett, 1966).

The electrotonic synapses mediate rapid-acting positive feedback between cells; a relatively more depolarized cell tends to depolarize and excite its less depolarized neighbors and is itself made less depolarized and inhibited by them. Thus, the neurons tend to fire synchronously. The synapses are both excitatory and inhibitory, and it seems reasonable to call them synchronizing synapses (Bennett, 1968b). One may ask whether mutually excitatory, chemically transmitting synapses could also mediate synchronization. As will be seen below synchronization in most of the electric organ systems is so precise

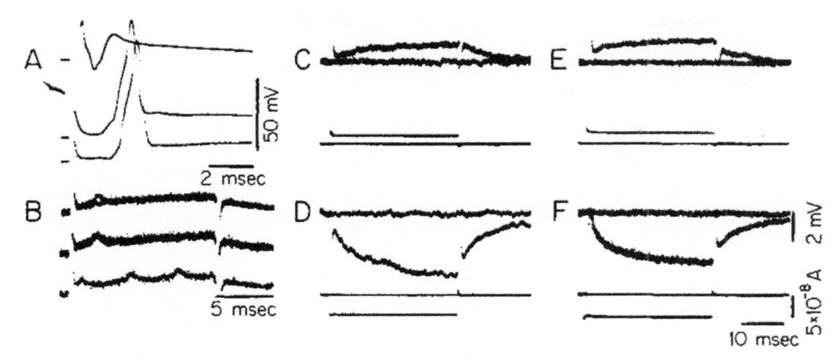

Fig. 70. Electrotonic coupling of oculomotor neurons innervating the electric organ of the stargazer *Astroscopus*. Two neighboring cells are penetrated by independently mounted electrodes. Their antidromic spikes are shown in A. The upper trace shows the antidromic volley recorded at the point of exit of the oculomotor nerve from the cranial cavity. When current is passed through the electrode in the cell of the upper trace both depolarization (C) and hyperpolarization (D) spread from cell to cell (superimposed sweeps with and without a pulse; current strength shown on the lower trace). When current is passed through the other electrode, depolarization (E) and hyperpolarization (F) also spread from cell to cell. When depolarization of the first cell is adequate to evoked spikes, there are corresponding small deflections in the second cell in addition to the maintained depolarization (B, increasing stimulus strength from top to bottom). The voltage gain is the same in B–F; the sweep speed is the same in C–F. From Bennett (1968b).

that the negligible delay of electrically mediated transmission is required; the delay associated with chemically mediated transmission would be too great.

The experimental demonstration and several properties of electrotonic transmission are illustrated by Fig. 70. In this experiment neighboring cells in the oculomotor nucleus of the stargazer are penetrated by microelectrodes; the antidromic spikes evoked by stimulation of the oculomotor nerve are shown in Fig. 70A. Depolarizing or hyperpolarizing current applied in either cell spreads to the other cell (Fig. 70C–F). If one or both electrodes are placed in a just extracellular position the recorded voltages are very greatly reduced. They do not disappear completely because the currents develop some voltage across the volume resistance of the neural tissue. It can be concluded that the voltages recorded intracellularly involve a special, junctional relation between cells and are not merely a result of proximity. Other less direct methods of demonstrating electrotonic transmission may be useful; for example, a PSP of very short latency can be assumed to be electrically mediated (cf. Bennett, 1966).

The morphological basis of electrotonic coupling has been investigated in a number of systems (Pappas and Bennett, 1966; Bennett *et al.*, 1967a,b,c; Kriebel *et al.*, 1969; Pappas *et al.*, 1971; these papers include other references). In every case close membrane appositions occur (Figs. 71 and 72) which are rare or absent in neighboring regions and which do not occur at synapses for which there is evidence that transmission is chemically mediated. These close appositions are believed to be the site of current

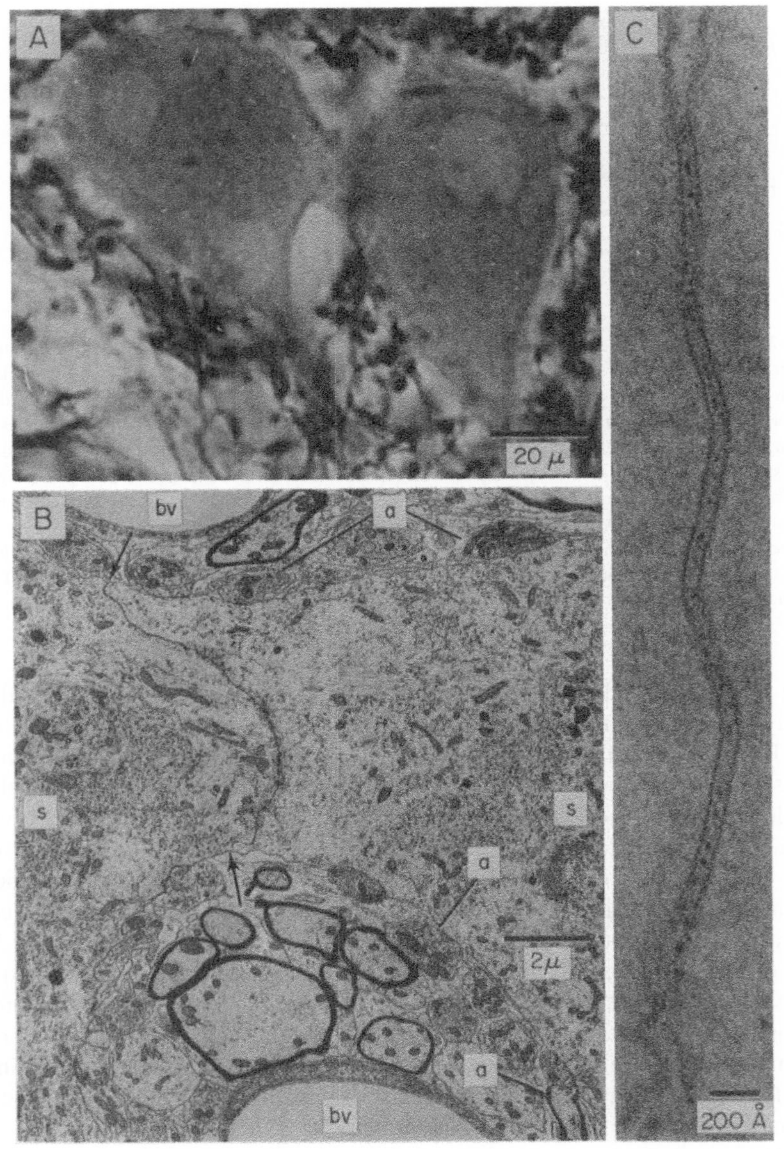

Fig. 71. Morphological basis of electrotonic coupling: dendrodendritic junctions between electromotor neurons of the mormyrid *Gnathonemus*. (A) In a silver stained preparation (Romanes' method) of the medullary relay nucleus, a thick bridge appears to connect the two cell bodies without there being any intervening membrane. (B) With the electron microscope such cells are seen not to have cytoplasmic continuity. The cell bodies (s) are separated by membranes the ends of which are indicated by the arrows. Blood vessels (bv) and myelinated nerve fibers are seen. Axon terminals (a) make contact with the cells; one on the lower right shows a portion of the myelin sheath. (C) At a similar region of apposition between spinal electromotor neurons, higher magnification shows large regions where the membranes appear fused. In this very thin section the central dark region appears as a series of dots. Osmic acid fixation. From Bennett *et al.* (1967a).

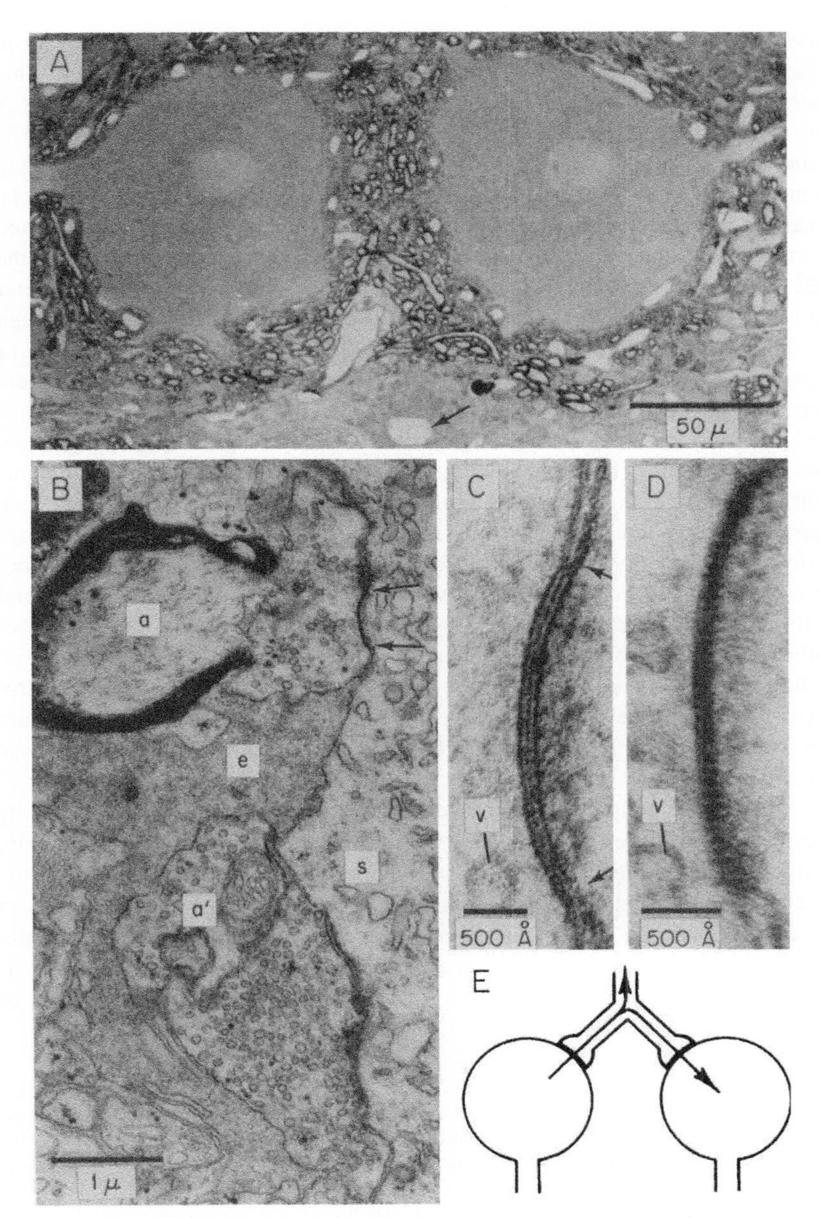

Fig. 72. Morphological basis of electrotonic coupling: axosomatic and axodendritic electrotonic synapses on giant electromotor neurons of the electric catfish. (A) Toluidine blue stained thick section that passes through the nuclei of the two giant cells. The central canal (arrow) is just ventral to the cells. A number of small dendrites come off the somata, but there is no apparent direct connection between the cells. (B) Electron micrograph of two axosomatic synapses (a,a′) on the cell. At the upper one (a) the myelin sheath is seen to terminate in the plane of section.

(Continued)

passage between cells. They are probably what have recently been termed *gap junctions* (Revel and Karnovsky, 1967; Brightman and Reese, 1969). This name arises from the appearance in perpendicular sections of a 20–30 Å gap between membranes following suitable fixation procedures. The gap is penetrable by marker substances applied in the extracellular space. The gap is not uniform but is made up of a more or less hexagonal lattice of channels. There is evidence that in the spaces outlined by the lattice, channels separated from extracellular space cross the junctional complex and interconnect the cell cytoplasms. The inter- cytoplasmic channels are perhaps 10 Å in diameter and provide sites for movement of ions and other small molecules between cytoplasms of the coupled cells (Payton *et al.*, 1969; Pappas *et al.*, 1971). This class of junctions is distinct from *tight junctions,* which are appositions where extracellular space appears completely occluded in perpendicular sections, and where there is no hexagonal structure in tangential sections (Brightman and Reese, 1969). (Both types were formerly called tight junctions. The current nomenclature in this area is confusing and will probably be revised when there is more general argreement as to morphological and physiological properties of the junctions and criteria for their identification.) These latter junctions generally occur in epithelia where they form complete rings around cells *(zonulae occludentes;* singular, *zonula occludens)* that prevent transepithelial leakage through intercellular clefts (Farquhar and Palade, 1963; Brightman and Reese, 1969). There is no evidence that *zonulae occludentes* form low resistance channels between cell cytoplasms because no cells are known to be both electronically coupled and joined exclusively by these junctions.

Electrotonic coupling between cells of the same kind can be mediated by electrotonic synapses between somata or dendrites (Fig. 71). Alternatively, cells can be coupled by way of presynaptic fibers that form electrotonic synapses on the cells; current then spreads from one post-synaptic cell to another through the presynaptic fibers (Fig. 72). (Since activity can normally be conducted in either direction across some of these synapses, pre- and post-synaptic denote the usual morphological relations rather than the direction of

Fig. 72—Cont'd There is a region of close apposition of axon and soma membranes at this synapse (between arrows), but probably not at the other one. Terminal a′ contains many vesicles, but there are relatively few in a. The former may be one of the chemically transmitting inhibitory synapses that occur on these cells. Near these endings there is a relatively large amount of extracellular space (e) filled with granular material. (C) Higher magnification of an axodendritic synapse. The axonal side is to the left and a presynaptic vesicle (v) is seen in it. A central dark line is formed at the close membrane apposition extending between the arrows. (D) An axodendritic synapse like that in C, but the close membrane apposition is cut more tangentially and striations appear with a periodicity of about 100Å. (E) Diagram of current flow where cells are coupled by way of presynaptic fibers. The cell on the left is more depolarized. All sections of osmic acid fixed material. Modified from Bennett *et al.* (1967b).

impulse propagation.) Coupling can also be mediated by both presynaptic fibers and dendrodendritic synapses in the same nucleus (Pappas and Bennett, 1966).

A. Pathways and Patterns of Neural Activity

The circuitry of electric organ command systems in teleosts that have been studied physiologically is diagrammed in Fig. 73. The physiological analysis has generally involved microelectrode recording from the entire animal paralyzed or anesthetized.

1. *The Electric Catfish*

The simplest control system is in the electric catfish (Bennett *et al.*, 1967b). Two neurons lie in the first spinal segment, one on either side (Fig. 72A). Each neuron innervates all the electrocytes on its side and each impulse fires the electrocytes once. The two neurons are closely coupled electrotonically, and hyperpolarization produced in one cell spreads to the other (Fig. 74C). The coupling is so close that an impulse initiated in one cell propagates into the other (Fig. 74B). The pathway of coupling is by way of presynaptic fibers (Fig. 72). Excitatory inputs gradedly depolarize the cells (Fig. 74A,D), and when one cell is excited the other must fire also. This explains why no stimulus can be found that excites one cell without exciting the other.

That the two cells comprise the command nucleus is indicated by the gradual rise in potential when organ discharge is evoked by cutaneous stimulation (Fig. 74D). When PSPs in one cell exceed threshold, both cells are depolarized since the presynaptic fibers end on both; once initiated the impulse rapidly propagates between the cells. The conduction requires less than 0.2 msec, which is more rapid than could be achieved by chemically mediated transmission (Fig. 74A). [The delay is greater when one cell is directly stimulated, for the unstimulated cell is much less depolarized when the impulse arises (Fig. 74B).]

2. *Weakly Electric Gymnotids*

In weakly electric gymnotids the command system has more than one neural level (Bennett *et al.*, 1967c). The organ frequency is set by what has been termed the pacemaker nucleus, which is in the medulla. (From a functional point of view it would be difficult for synchronous activity to arise within a group of neurons that are as widely separated as the spinal neurons innervating the electrocytes.) The pacemaker nucleus lies in the midline and contains some 30 to 200 cells depending on the species. It activates a relay nucleus also in the midline of the medulla, and this nucleus in turn activates the spinal neurons. There are 50 or so medullary relay neurons and some hundreds to thousands of spinal neurons. A single spike occurs at each level before each organ pulse (Fig. 75). In the pacemaker neurons there is a gradually rising

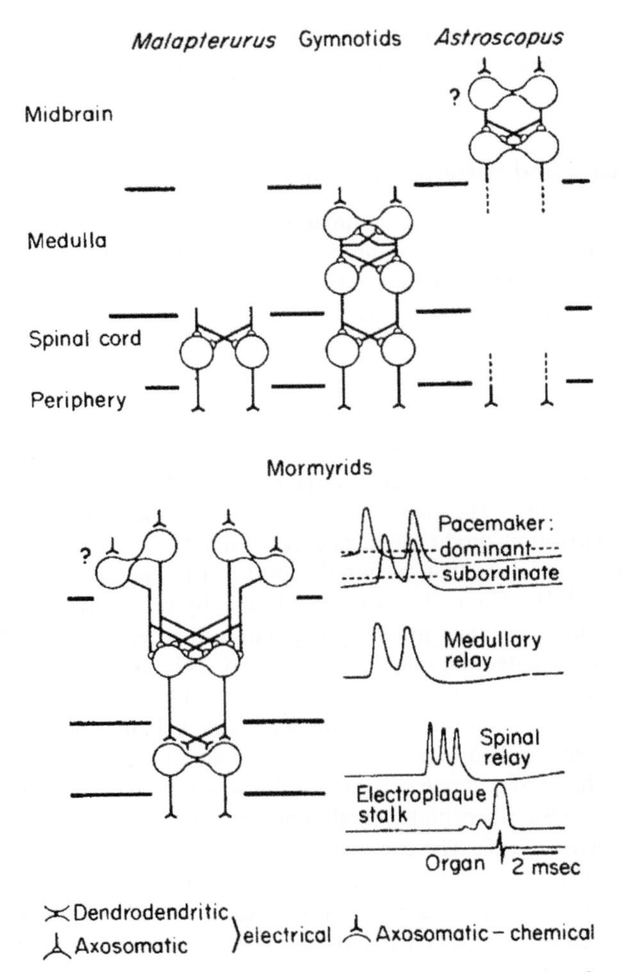

Fig. 73. Neural circuitry controlling electric organs of teleosts. The modes of transmission are diagrammed as shown in the key below. Where axosomatic synapses are indicated, axodendritic synapses are also found. The mode of transmission to the command nucleus is known only in the electric catfish, although it is indicated as chemically mediated in the others. Where there is a question mark, the cells have not been definitely localized, but several of their properties can be inferred. The command nucleus of the stargazer *Astroscopus* is now known to be in the medulla, not the midbrain. In *Malapterurus*, the gymnotids, and *Astroscopus*, a single command volley at each level precedes each organ discharge. In mormyrids, the activity is more complex and is diagrammed for each level. The dotted lines indicate the thresholds of the cells in the two pacemaker nuclei. The dominant pacemaker at any given time is the one receiving more excitatory inputs and firing before the other. From Bennett (1968a).

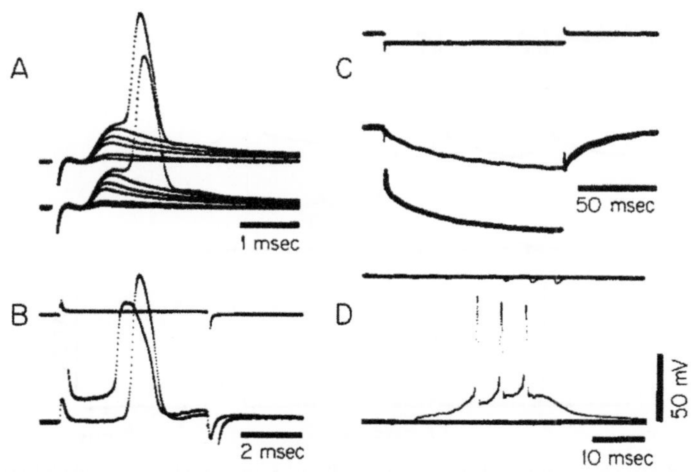

Fig. 74. Properties of the giant electromotor neurons of the electric catfish. (A) Upper and lower traces, recording from right and left cells, respectively. Brief stimuli of gradually increasing strength are applied to the nearby medulla (several superimposed sweeps, the stimulus artifact occurs near the beginning of the sweep). Depolarizations of successively increasing amplitude are evoked until in one sweep both cells generate spikes. (B) Two electrodes in the right cell, one for passing current (shown on the upper trace) and one for recording; one recording electrode in the left cell. The traces from the recording electrodes are the lower ones starting from the same base line. When an impulse is evoked in the right cell by a depolarizing pulse, the left cell also generates a spike after a short delay. (C) When a hyperpolarizing current is passed in the right cell, the left cell also becomes hyperpolarized, but more slowly and to a lesser degree (display as in **B**). **(D)** When organ discharge is evoked by irritating the skin, a depolarization gradually rises up to the threshold of the giant cell and initiates a burst of three spikes (lower traces, base line indicated by superimposed sweeps). Each spike produces a response in the organ (upper trace, recorded at high gain and greatly reduced in amplitude because curare is used to prevent movement). Modified from Bennett *et al.* (1967b).

depolarization between spikes (Fig. 75A,C). This depolarization is similar to pacemaker potentials in other tissues, and the cells appear to be spontaneously active. In the relay cells the potential between spikes is quite flat and the discharges arise abruptly from a level base line (Fig. 75B,D). These cells are clearly activated from a higher level which is in fact the pacemaker nucleus.

That the discharge frequency is set in the pacemaker nucleus is established by results like those in Fig. 76. If a hyperpolarizing pulse is applied in one cell, the next spike in that cell and each subsequent one is delayed (Fig. 76B,D). The descending volleys in the spinal cord are not changed in duration or desynchronized; they are reset in phase by the same amount as the spikes in the pacemaker neuron and follow the spikes at the normal interval. Thus, hyperpolarization in one cell affects the entire pacemaker nucleus and resets its phase of firing. Moreover, since hyperpolarization spreads between cells, the interaction between cells can be inferred to be electrotonic.

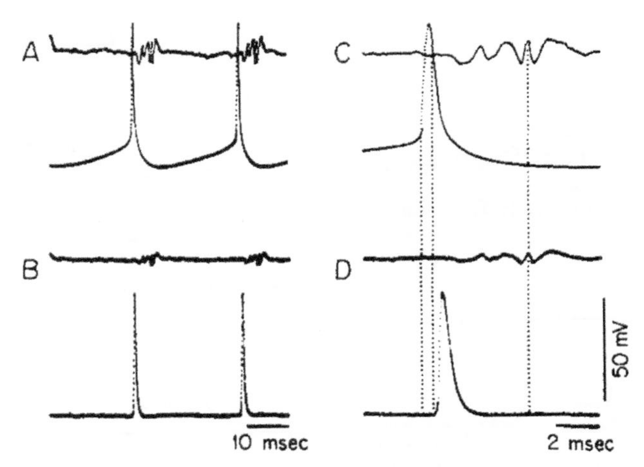

Fig. 75. Responses of pacemaker and relay cells in a weakly electric gymnotid *Gymnotus*. Upper traces: activity in the spinal cord and peripheral nerves leading to the electric organ (recorded by needle electrodes at high gain in a curarized animal). Lower traces: intracellular recordings in pacemaker (A, C) and relay (B, D) neurons. Faster sweep in C and D where the dotted lines indicate the times of firing of the cells in relation to the descending activity. From Bennett *et al.* (1967c).

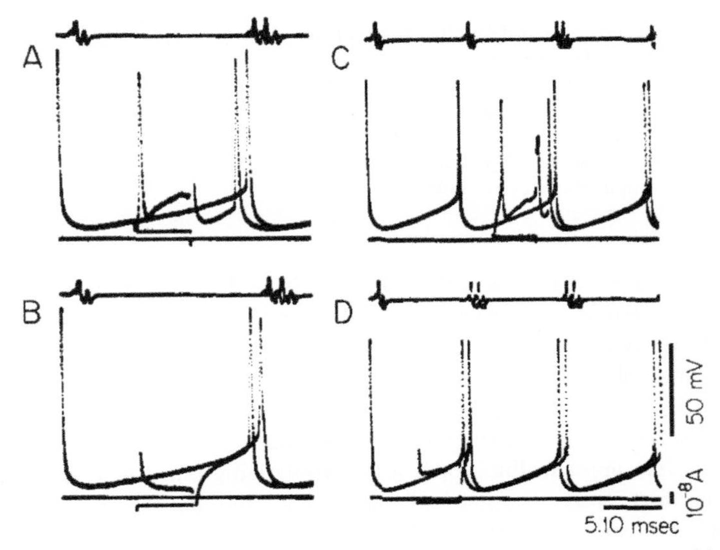

Fig. 76. Effect of polarization in a single pacemaker cell of a weakly electric gymnotid *Gymnotus*. Recording as in Fig. 75 except that current applied through the recording electrode is indicated on the lower trace. Two superimposed sweeps in each record, one with and one without applied current. The sweeps are triggered by the spike of the pacemaker cell. Faster sweep in A and B. (A,C) A depolarizing pulse that evokes a spike advances the next and subsequent spikes but does not desynchronize or itself cause any descending activity. (B,D) A hyperpolarizing pulse retards the next and subsequent spikes but does not desynchronize the descending activity. From Bennett *et al.* (1967c).

(This has been directly demonstrated also.) The coupling between pacemaker cells obviously leads to synchronization of their firing. Each cell is spontaneously active and would get out of step with the others if there were not coupling—or positive feedback—between them. The accuracy of synchronization is so great as to require the speed of electrotonic coupling; synchronization by mutually excitatory chemically transmitting synapses would involve too great a delay.

The coupling of gymnotid pacemaker neurons is not as close as in the catfish electromotor neurons and does not allow impulses to propagate from one cell to another. As shown in Fig. 76A,B an impulse in one pacemaker cell does not cause a descending volley in the spinal cord; it only advances the phase of firing. Thus, there is some spread of depolarization to the other cells but not a propagated impulse (unless the cells are very near to firing). The coupling is sufficiently close to synchronize the cells, and, when all the cells are firing in the vicinity of a cell penetrated by a microelectrode, it is extremely difficult to block that cell's activity by hyperpolarizing current. While a spike in any one cell affects its neighbors but little, many cells firing together can produce a large and very suprathreshold depolarization in an inactive cell.

As it happens, the medullary relay neurons are also electrotonically coupled in most gymnotids that have been studied. Coupling is by way of the pacemaker fibers afferent to them. At this level the coupling does not serve to keep the cells from firing out of phase, for each cell is always excited once per organ discharge by the large descending volley from the pacemaker nucleus. Presumably the coupling serves to synchronize relay cell firing, either because asynchrony was present in the initial pacemaker volley or because it has arisen in transmission from the pacemaker nucleus. The spinal relay neurons have been adequately studied only in the electric eel (Bennett *et al.*, 1964; Bennett, 1968a). These cells, too, are electrotonically coupled and by way of the descending fibers ending on them (Pappas and Bennett, 1966). Preliminary experiments indicate that the spinal relay neurons are similarly organized in *Gymnotus*. Again, coupling would tend to increase synchronization.

Sherrington (1906) called the motoneuron the final common path for muscle fiber activity. In the gymnotids the final common path for electric organ discharge extends three neurons back into the nervous system. Each level involves a number of neurons and a certain amount of signal shaping may go on in the relays, but these are only small extensions of the Sherrington concept.

3. *Astroscopus*

In the stargazer the command signal is initiated in what is probably a midline nucleus in the medulla and is relayed in the very large oculomotor nucleus (Bennett, 1968a, and unpublished data). Neurons are electrotonically coupled in both command and relay nuclei (Fig. 70). In the relay both dendrodendritic synapses and the presynaptic fibers mediate coupling. The organ discharges are somewhat variable in size and apparently do not always involve every oculomotor neuron (Fig. 6). The experimental evidence indicates that the command nucleus is not well synchronized. The oculomotor neurons do not simply relay the signal; they also "decide" whether or not a sufficiently large command volley is present to call for a discharge, and this decision is not always unanimous in spite of the coupling between cells.

Often firing of the command neurons is in pairs of spikes, and PSPs in the oculomotor neurons show two components. Apparently, a few oculomotor neurons fire twice in normal activity and give rise to the delayed component on the falling phase of some discharges (Fig. 6).

4. *Rajids*

In the skate the electric organ is probably controlled by a midline nucleus in the medulla since stimuli applied in this region evoke organ discharge (Szabo, 1955, 1961b). There is no evidence as to whether this nucleus is a relay or pacemaker, or whether the neurons are electrotonically coupled. This is a particularly interesting example because the discharge is repetitive and probably asynchronous. While it seems likely that positive feedback between neurons is involved in the decision to discharge the organ, the speed of electrical transmission is apparently not required. If feedback is still mediated electrically, it would suggest that electrical coupling may be involved in other slow systems as well.

5. *Mormyrids*

In the mormyrids control of organ discharge is more complex than in the fish discussed up to this point (Bennett *et al.*, 1967a; Bennett, 1968a). There is some evidence that the pacemaker nucleus is bilateral and that and organ discharge can be initiated by a command signal arising on either side (Fig. 73). This activity is relayed through a midline nucleus in the medulla to the spinal neurons and then to the electric organ. Although each pacemaker nucleus is able to initiate a command volley independently, either because it is spontaneously active or because it receives tonic excitatory inputs, pacemaker activity on the two sides is coordinated. An impulse arising in what may be considered the dominant nucleus at that time propagates from the medullary relay "antidromically" into the other pacemaker nucleus, the subordinate one. (Transmission at the pacemaker synapses is apparently electrotonic and allows "antidromic" propagation.) Thus both nuclei are excited, and the pace-maker potential of both nuclei is returned to the level of hyperpolarization that immediately follows spike activity; pacemaking in both nuclei is reset. The cells of the two nuclei then begin to depolarize again, but those in the dominant nucleus reach the firing level first, again resetting the activity of both nuclei.

In the mormyrids there is not simply a single impulse at each level of the electric organ control system (Fig. 73). PSPs from what is presumably a single spike in either pacemaker nucleus initiate a two-spike discharge in the medullary relay (the second spike propagates back to the dominant pacemaker and both spikes propagate to the subordinate pacemaker). Firing in "doublets" appears to be a property intrinsic to the cell membrane rather than a result of a long-lasting PSP because the cells give a two-spike discharge in response to a brief intracellularly applied current pulse. The two medullary relay volleys descend the spinal cord and cause two PSPs in spinal neurons that innervate the electrocytes. These PSPs are chemically mediated. The first PSP initiates a three-spike discharge. Firing in triplets appears intrinsic to the cell membrane of the spinal cells as does the doublet firing of the medullary cells. However, the third spike is somewhat labile and the second PSP from the medullary relay guarantees its occurrence. The three electromotor neuron spikes are propagated out to the synapses on the stalks of the electrocytes, but only a single postsynaptic spike is produced. The first PSP is very small, the second is greatly facilitated (increased in amplitude) but still subthreshold, and the third is facilitated sufficiently to reach threshold. The PSPs are not recorded external to the fish because the synapses are on stalks far from the body of the cells and there is little or no longitudinal current associated with them. (A similar firing pattern could be detected externally if the cells were diffusely innervated over one face.) The significance of this peculiar sequence of signal transformations is unknown. It also occurs in *Mormyrus* and *Mormyrops* in some species of which the discharge is longer lasting and the requirement for synchronization is not so great.

Both medullary and spinal neurons are closely coupled electrotonically, and an impulse in one cell propagates to all the other cells of that nucleus. (An antidromic impulse in an axon generally fails to invade the cell body, at least in part because of the low input resistance resulting from the close coupling of the cells.) The neurons are connected by thick dendrodendritic bridges in both nuclei (Fig. 71), and coupling by way of presynaptic fibers probably occurs in the medullary relay as well.

Interpretation of the peculiar multiple firing in the mormyrid control system is not likely to be aided by comparison to *Gymnarchus*. Light microscopy suggests that there are four interconnected nuclei in the medulla, but they have not been studied physiologically (Szabo, 1961b). All that is known is that a single volley descends the spinal cord, and a single PSP in the electrocytes initiates each organ discharge. An interesting aspect of the mormyrid control system is that the command system "informs" the sensory system that an organ discharge is coming (Bennett and Steinbach, 1969; see Chapter 11, this volume). The pathways mediating this activity and the effects on afferent volleys are just being explored.

6. *Torpedinids*

In *Torpedo* there is a command nucleus on each side that is located deep in the medulla (Szabo, 1954; Albe-Fessard and Buser, 1954). These nuclei activate the very large electromotor nuclei that lie on the dorsal surface of the medulla (Fig. 8). The electric organ discharges are synchronous on the two sides, and it is probable that either command nucleus can activate all the relay cells on both sides. No data are available about interaction between the two command nuclei, but some coordinating mechanism is likely to be found. Preliminary data indicate that the electromotor cells, which comprise a relay, are not coupled to each other and that transmission from the command nucleus is chemically mediated (cf. Saito, 1966). Single volleys from the command nucleus excite single volleys in the relay cells that produce the single PSPs in the electrocytes comprising the individual organ discharges.

7. *The Electric Eel*

The mechanism of control in the eel shows an interesting variation. All the electrocytes fire together during the large discharges, but only Sachs' organ and posterior part of Hunter's organ are active during the weak discharges. As in weakly electric gymnotids, the organs are controlled by a single midline relay nucleus in the medulla (Bennett *et al.*, 1964; Bennett, 1968a). The pacemaker nucleus has not yet been found in the eel. Each command volley from the medullary relay excites all the spinal neurons that innervate the electrocytes, and the command volley passes out ventral roots to anterior as well as posterior parts of the organ. At low frequencies PSPs in the main organ and the anterior part of Hunter's organ are small and do not excite the cells, but they are large enough to excite cells in Sachs' organ and in the posterior part of Hunter's organ (Albe-Fessard and Chagas, 1954). When the interval between command volleys becomes as small as several milli-seconds, the PSPs in the main organ and anterior part of Hunter's organ greatly increase in amplitude and become adequate to excite the cells of these organs. Thus the command to excite the main organ consists of high frequency activity in the same neurons that control Sachs' organ. Only a single bulbospinal relay system is required to control both organs. This increase in simplicity is obtained at the cost of several milliseconds' increase in the minimum latency at which the main organ can be activated. Probably the control system has evolved its dual functioning from an earlier stage in which there was only a single weakly electric organ. An intriguing problem is how pacemaker activity for the two organs is controlled.

B. Synchronization of Electrocyte Activity

Given that the command nucleus has reached a decision to fire the electric organ, that is, has itself fired, the individual electrocytes must be excited synchronously. This is a significant problem because the different parts of the electric organ may be quite far apart. Nerve tracts running directly to the different parts of the organ could not fire the cells synchronously unless they were much more rapidly conducting than any known nerves.

Two basic mechanisms are known to contribute to synchronization; both involve utilization of conduction time in nerve fibers or electrocyte stalks to equalize overall latency of the command signal in reaching the electrocytes. In one mechanism the nerve fibers run more or less directly to the more distant part of the organ, but take a more devious path to the nearer parts, thus tending to equalize path length and thereby conduction time (Fig. 77A). In the second mechanism shorter paths to nearer parts of the electric organ are of lower conduction velocity, again tending to equalize conduction time (Fig. 77B; fibers conducting more slowly are indicated as being of smaller diameter). Lower conduction velocity may also be found in only a portion of the path to the electrocytes; slowly conducting collaterals may branch off from the main rapidly conducting path to innervate the nearer parts of the organ (Fig. 77C).

Equalization of path length is an obvious feature of a number of electric organs where the different nerves enter the electric organ and run for some distance before giving off branches that return to end near the point of entry. Compensatory differences in conduction velocity apparently occur in the stalk system of mormyrid electrocytes where differences in conduction distance are

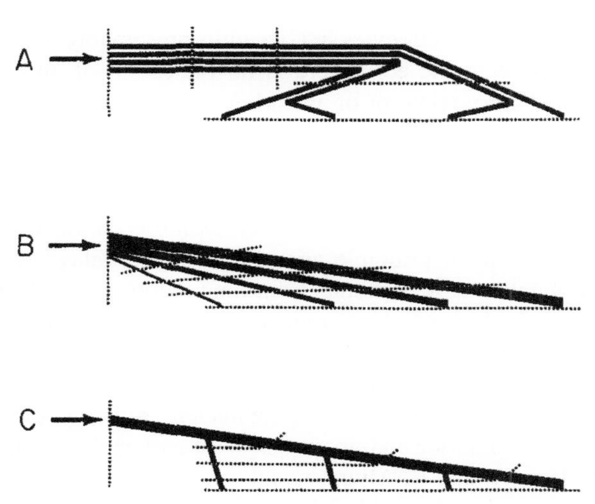

Fig. 77. Mechanisms of compensatory delay. Neural pathways are diagrammed leading to terminations in the periphery at different distances from a rostral command center. A command volley arises at the large arrow and the dotted lines represent times of arrival of impulses at equal time intervals afterward. (A) Equalization of path length. The paths to the nearer cells are made more devious so that all paths are of nearly equal length. (B) Compensatory differences in conduction velocity. The paths leading to the periphery are direct, but conduction is slower in the shorter, thinner paths. (C) Localized compensatory delays. Thin terminal branches in which conduction velocity is reduced are longer in the paths leading to the nearer parts of the periphery. From Bennett (1968a).

easily visualized and where synchronization between parts of the body of the cell is very precise (Fig. 61). As would be expected, the shorter paths involve stalks that are smaller in diameter. Probably most systems use a combination of the two mechanisms.

In the electric eel differences in conduction time down the spinal cord are compensated for by both increased delay at the spinal relays and increased delay from activity in ventral roots to spike initiation in the electrocytes (Albe-Fessard and Martins-Ferreira, 1953). Two lines of evidence indicate that the delay at the spinal relay arises in conduction time in collateral branches of descending axons. First, transmission from descending fibers to electromotor neurons is electrotonic and the PSPs are sufficiently rapidly rising that they can be delayed very little at the synapses themselves (Bennett, 1966). Second, action potentials at intermediate delays can be recorded in collaterals of the descending fibers. Preliminary morphological observations indicate that these collaterals are thinner at the anterior region of the spinal relay nucleus (R. M. Meszler and M. V. L. Bennett, unpublished data). The efferent axons entering the ventral roots are also thinner in the anterior regions of the spinal cord, and reduced conduction velocity probably contributes to the greater peripheral delay at the anterior of the organ.

The early firing of rostral accessory organs in a number of gymnotids and of the dorsal column of electrocytes in *Gymnotus* presumably involves similar mechanisms to those in the eel. There is no indication of an earlier firing pacemaker or medullary relay, or of an earlier firing group of spinal neurons. Apparently the relatively delayed firing of the majority of electrocytes is achieved peripherally. It may be that early firing of the accessory organs really represents delayed firing of the main organ.

At chemically transmitting synapses such as those on electrocytes, compensatory delays could in principle arise in "synaptic delay" [which at the neuromuscular junction at least results from time required for release of transmitter (Katz and Miledi, 1965a,b)]. Compensatory delays may occur at a number of chemically transmitting synapses in electromotor systems. One instance is in the spinal electromotor nucleus of the mormyrid *Gnathonemus* in which PSPs are synchronous at the two ends of the nucleus, although the descending volley arrives at the anterior end about 0.3 msec earlier than at the posterior end (M. V. L. Bennett and E. Aljure, unpublished data).

A further instance is provided by the electric organ of the electric catfish. A single large axon branches to innervate all the electrocytes on one side of the organ. There is some equalization of path length for the axon enters the organ somewhat posteriorly and runs both anteriorly and posteriorly from the site of entry. Stimulation of the axon at its point of entry leads to a discharge that is about the same duration as the responses of single cells (Fig. 78C). Stimulation of a small branch to a piece of organ from the head end of the fish produces a response that has a longer latency than stimulation of a similar preparation from the tail of the fish (Fig. 78B,B′). This difference

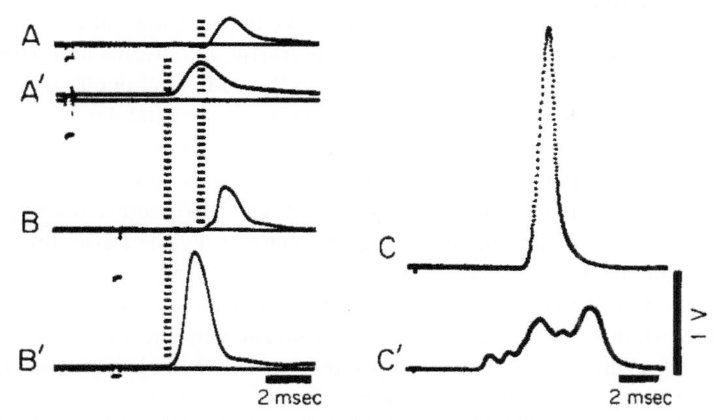

Fig. 78. Compensatory delays in tire catfish electric organ. (A,A′) Responses of a piece of electric organ dissected from near the tail of a fish about 20 cm long. The tissue has 10 cm of motor nerve attached to it and is stimulated at each end of this length of nerve. (B) Response of a piece of organ from same fish dissected near head, stimulated via the nerve close to point of entry into tissue. (B′) Response of another piece of organ, similarly stimulated, but taken from a position 10 cm nearer tail. (C) Response of the electric organ on one side from another large fish, stimulated orthodromically at the central end of the nerve. (C′) The response of the same organ, stimulated "antidromically" at the caudal end of the nerve. From Keynes *et al.* (1961).

in latency is just sufficient to compensate for the difference in conduction time down the main trunk of the axon (A,A′). The significance of the synaptic delay can be well illustrated by recording the output of one entire side of the organ when stimulating the nerve at different sites. If the nerve is stimulated at its caudal end rather than its site of entry into the organ, the differences in delay add to the differences in conduction time and a markedly asynchronous discharge results (Fig. 78C′).

The site of these compensatory delays is unknown; it could be in presynaptic nerve branches and terminals or at the synapses themselves. In the electric catfish it could also be in the stalks of electrocytes. However, no systematic differences in length of stalks have been observed in teased preparations, and the variability of stalk length is quite great in all parts of the organ. In both mormyrids and the electric catfish, morphological investigations may reveal that terminals are longer and thinner where delays are greater. In this case one would be less inclined to consider synaptic delay as a factor in the compensatory mechanism.

C. Organization of Electromotor Systems

As discussed here the control of electric organs is reasonably well described from one to several neural levels back into the nervous system. Provided one can accept the proposition that the pacemaker cells are autoactive, the analysis is reasonably complete for the very regular basal discharge frequency found in many species. Certainly one would like to know the ionic basis of the pacemaker activity, but in respect to impulse traffic, the origins and pathways are well defined. The pathways of afferents to the command nuclei in variable frequency and intermittently discharging forms are by and large unknown, although most if not all sensory modalities can be excitatory and responses can be of quite short latency (e.g., Fig. 3B). The slight modifications of frequency in some high and constant frequency species also require investigation. The relative fixity of jamming avoidance responses suggests fairly direct electrosensory inputs, while the brief modulations that perhaps have a signaling function probably involve much more complex pathways (Bullock, 1970). The involvement of higher centers is also indicated by the fact that changes in organ discharge rate can be operantly and respondently conditioned (Mandriota *et al.*, 1965, 1968) as well as serve for signaling (Black-Cleworth, 1970; Moller, 1971). An interesting problem is raised by the ability of *Gymnorhamphichthys* and some *Hypopomus* to maintain organ discharge frequency constant at low, high, and even intermediate levels. This ability suggests that there are neurons that maintain a tonic, asynchronous but quite constant excitatory input to the pacemaker nucleus. This input seems too constant to be due to casual stimulation of receptors resulting from movement, and an activating system that is itself quite stable would seem to be required to explain the data.

Pathways mediating the complete inhibition of discharge in *Gymnotus, Hypopomus,* and *Sternopygus* are also not worked out. The strong excitatory drive to the medullary relay in *Gymnotus* and probably also *Hypopomus* suggests that this nucleus is not involved. Probably the inhibition operates on the pacemaker nucleus. If so, it might be considered surprising that in *Gymnotus* the discharge always starts and stops abruptly with only a small degree of slowing below the basal rate before and after complete cessation. However, similar observations have been made during microelectrode recording from the pacemaker nucleus in which activity at this level did indeed cease when discharge ceased and restarted when renewed discharge was evoked by spinal stimulation. These sudden changes in frequency are consistent with calculations from the Hodgkin–Huxley equations, which predict that maintained firing of the squid axon in response to a steady current can occur only at rather high frequencies (Stein, 1967).

Unlike *Gymnotus, Sternopygus* does not start up again by emitting full-sized organ pulses; there is a gradual recovery of pulse amplitude and synchronization. While this observation might represent a difference in the pacemaker nucleus, it might also result from properties of lower level synapses. If the spinal cord is sectioned in *Gymnotus* a single brief stimulus to the cord will still cause a full-sized organ discharge. However, in *Sternopygus* moderately prolonged stimulation at about the normal organ frequency is required to restore the discharges to their full amplitude. Granted that transmission at the spinal relays is electrotonic, the most likely site of facilitation in *Sternopygus* is at the nerve-electrocyte synapses. The absence of cessation of discharge in the higher frequency species *Eigenmannia* and sternarchids is perhaps a result of the fact that discharge does not stop following spinal section but continues asynchronously.

The investigations to date have been largely restricted to elucidation of what may be considered the final common path for organ discharge. A number of modifications to the original concept of final common path appear necessary to describe control of electric organs. A group of neurons at a particular level can fire as a single neuron, and the common path may involve several levels of neurons. The highest level determines the frequency of discharge; the lower levels act only as relays and perhaps also do some signal shaping (or as in the mormyrids, transformation of impulse number). The command path may bifurcate (and then cease to be common). The command to a particular organ may be expressed in terms of frequency.

Several of these modifications are already required for ordinary motor systems. For example, some muscle fibers are innervated by more than one nerve fiber, which is analogous to paired command nuclei. If tension of a muscle fiber rather than any contraction at all is considered, the command may be coded in terms of frequency of firing. Furthermore, it is obvious that we do not understand all about the control of an electric organ or a muscle fiber when we know the final common path to it. We also need to know all the fibers afferent to neurons in that path. We have taken the decision to fire the organ as meaning the firing of the highest level nucleus of the final common path. The decision could also be defined as activity in any subset of fibers afferent to the highest nucleus that can cause the nucleus to fire. Similarly, the relevant output of a whole muscle, its tension, may be achieved (or coded for) by many different combinations of fibers active at different frequencies.

For freshwater weakly electric fish and for the electric catfish the neural circuitry guarantees that every electrocyte receives neural excitation in every discharge. The operation of the system is essentially all-or-none and an organ discharge is present or absent. It should be recognized that amplitude can vary minorly as a result of refractoriness (and perhaps facilitation), but these changes must be secondary to changes in frequency. The situation is similar in the eel except that the amplitude normally varies over a wide range. In all these fish and to a good approximation in *Torpedo* and the stargazers as well, the behavior of the animal with respect to its discharge can be characterized by a single series of time intervals, a simplicity which gives the system some attractiveness for quantitative study.

Two additional characteristics are frequently found in electric organ control systems. Where a group of neurons fire synchronously there is likely to be positive feedback between the cells. In command nuclei in which a highly synchronous volley arises, positive feedback must be present and must be electrically mediated because chemically mediated transmission is too slow.

In relay nuclei there is no absolute requirement for feedback, but cells are often coupled electrotonically, presumably to increase synchronization. Another characteristic often found in electric organ control systems is a tendency toward reduced numbers of cells at higher levels. Apparently the decision to discharge the electric organ is always reached in a group of cells that are few in number compared to the final generating cells. The most striking example is found in the electric catfish, in which two neurons control the several million electrocytes. Similar but less extreme pyramids of numbers are found in other electric fish. This aspect of neural organization may be denoted the committee principle. This term is chosen in analogy with decision making by committees which tends to be more rapid the smaller the committee.

One might question whether there could be a command "nucleus" containing a single cell, a committee of one. An example is provided by the Mauthner cells of lower vertebrates which can be considered single cell command systems for the axial musculature on either side of the body (Furukawa and Furshpan, 1963; Diamond and Yasargil, 1969).

One could propose a number of reasons for the existence of control systems containing more than one cell, such as protection against loss of neurons and production of enzymes for maintenance of synaptic transmission. At lower levels progressive increase in a number of neurons may be equivalent to progressive increase in size. The two giant neurons of the electric catfish are able to support a very small synaptic area on a large number of electrocytes, but it seems likely that a large number of neurons are required to provide the vast synaptic area in the electric organs of *Torpedo*. Phylogenetic or ontogenetic factors may also be responsible for the presence of the relays. Evolution may yet progress to single-celled command "nuclei" although in most instances the multicelled nuclei do about as well as required by the electric organ.

It is interesting to compare the function of relay nuclei to transmission of impulses along an axon. At a node of Ranvier an impulse that has been attenuated in electrotonic propagation from the preceding node triggers the generation of a new impulse that is of full amplitude. The node acts as a pulse restoring element, and decrementless conduction is thereby made possible. A relay nucleus potentially does more than this. The volley of impulses in the fibers efferent from the relay may be more synchronous than that in the afferent fibers, or all the efferents may become excited when only a fraction of the afferents are active (as appears to occur occasionally in *Astroscopus).* These functions do not *require* coupling of the cells; all that is necessary is that a number of afferent fibers converge on each relay cell.

Although midline nuclei have developed in a number of control systems, bilateral command nuclei may be fairly common because of the basically bilateral organization of the nervous system. If either of the two command nuclei can excite an entire effector system, synchronization between the two is not important. However, it would appear functional for each command to reset the phase of firing in both command nuclei. Probably this resetting occurs in the mormyrids by actual invasion of the impulses, but mutual inhibition could also be a mechanism in other systems. For example, crossed inhibition like that between Mauthner cells (Furukawa and Furshpan, 1963; A. A. Auerbach, unpublished data) would be equally effective in an electric organ system. In the hatchetfish each Mauthner fiber activates the muscles depressing both pectoral fins, and these cells thus constitute a bilateral command system for the depressor muscles (Auerbach and Bennett, 1969a,b). Contraction of these muscles causes the animals to dart upward in an escape reflex. Mutual excitation like that proposed for the mormyrid pacemaker nuclei would not work for the Mauthner cells, because each innervates the axial musculature on one side and near simultaneous excitation could lead to a counter-productive attempt to flip the tail to both sides simultaneously. Actually there is evidence that in the goldfish each Mauthner fiber excites motoneurons on one side and inhibits them on the other side but at a shorter latency (Diamond and Yasargil, 1969). Simultaneous activity of both Mauthner fibers excites no axial motoneurons. If one Mauthner fiber fires, it causes muscle contraction on one side and prevents the other Mauthner fiber from exciting the contralateral motoneurons for about 100 msec.

Recognition of the requirement for speed of transmission in synchronized systems has been useful in predicting sites where electrically mediated transmission has subsequently been found including neurons controlling sonic muscles (Bennett, 1966; Pappas and Bennett, 1966) and oculomotor neurons (Kriebel *et al.*, 1969). In mediating activity that is not very precisely synchronized, the speed of electrotonic synapses might not be required. Examples are the bursting behavior of respiratory neurons and various invertebrate cardiac ganglia. Electrotonic coupling does occur in a number of moderately synchronized systems in both vertebrates (cf. Bennett, 1968b) and invertebrates (Hagiwara *et al.*, 1959, Willows and Hoyle, 1969). The extent to which synchronization of slow systems is mediated by positive feedback and the degree to which the positive feedback is mediated electrically remain to be worked out. Electrotonical transmission appears to provide a simpler and perhaps more efficient means of positive feedback between cells (Bennett, 1968b).

One may question whether firing of an electric organ resembles other decision processes. The final common path concept of a decision requires extension even in some very simple systems. On the other hand, it is reasonable to think that the features observed in simple effector systems have some more general relevance. Negative feedback has been emphasized as a common property in neural control systems mediating homeostasis. It is also common in sensory and motor pathways where it may increase spatial resolution. Yet many phenomena involve recruitment of large numbers of cells in ways that appear to require mutual reinforcement of neuronal activities, that is, positive feedback.

IV. CONCLUSIONS AND PROSPECTS

The study of electric organs and electrocytes has illuminated many aspects of membrane physiology. Their primary usefulness is a result of evolution which has exaggerated different membrane functions in different electrocytes. For example, the concept of independent sites that pass different ions or groups of ions becomes more reasonable given that specific kinds of permeability can occur in isolation in different regions of a single cell or in different cells. Examples are the isolated occurrence of membrane generating postsynaptic potentials (in *Astroscopus* and torpedinids), of membrane exhibiting delayed rectification (in rajids), and of the electrically excitable sodium system without delayed rectification (in the electric eel). This macroscopic separation suggests microscopic separation, which in many single membranes can only be inferred from functional arguments or pharmacological data.

These specializations of different kinds of membrane cannot really be said to have been responsible for major advances in electrophysiological knowledge. Yet as new morphological, biochemical, and biophysical techniques become available, electrocytes may provide the best tissues for their evaluation.

Although one cannot determine electrophysiologically whether or not sodium and potassium channels of ordinary spike-generating membrane are separate, one can begin to think about isolating them biochemically. Electric organs would appear to provide a good starting material just as they are proving useful in the characterization of cholinesterase and sodium–potassium ATPase. By selection of cell type, one can select different kinds of membrane with relatively great degrees of purity compared to most other tissues.

The neural systems controlling electric organs have provided a large number of examples of electrically mediated transmission, which meets the functional requirement for rapid communication between cells. This mode of transmission also proves to be able to mediate many functions often considered as restricted to chemically mediated transmission. The correlation between morphologically close apposition and electrotonic coupling was considerably strengthened by the work on electromotor systems. This correlation helps to validate morphological identification of electrical transmission in other systems where electrophysiological analysis is not so simple.

It is not known whether there is any relevance to higher systems of the organizational principles deduced from electric organ systems. The next level of analysis of the electric organ systems may be no easier than study of less specialized systems that are of more general interest. Some knowledge is being obtained of afferent pathways from electroreceptors in weakly electric fish (see Chapter 11, this volume) which have important inputs to the electric organ control system. Both operant and respondent conditioning of the control system can be obtained and conditioned response latency can be very short. It is not unreasonable that the complete neural pathway of the conditioned response could be obtained in these cases. The central connections are minimally explored; one knows what goes in and one can go from the electric organ several synapses antidromically. The rewards for filling in the gap could be great, and prospects for at least some progress are bright.

ACKNOWLEDGMENTS

Some of the hitherto unpublished work was carried out in the laboratory of Neurophysiology, College of Physicians and Surgeons, Columbia University, with the support of Dr. H. Grundfest. Work on the Rio Negro expedition of the R. V. Alpha-Helix was in collaboration with A. B. Steinbach and allowed considerable advance in understanding of the sternarchids. The author is greatly indebted to Dr. T. H. Bullock for his assistance in that remote place. Supported in parts by grants from the National Institutes of Health (5P01 NB 07512 and HD-04248) and the National Science Foundation (GB-6880).

REFERENCES

Adrian, R. H., Chandler, W. K., and Hodgkin, A. L. (1970). Slow changes in potassium permeability in skeletal muscle. *J. Physiol. (London)* **208,** 645–668.

Albe-Fessard, D. (1950a). Les caractères de la décharge des poissons électriques. *Arch. Sci. Physiol.* **4,** 299–334.

Albe-Fessard, D. (1950b). Propriétés électriques passives du tissue électrogène des poissons électriques. *Arch. Sci. Physiol.* **4,** 413–434.

Albe-Fessard, D. (1951). Modification de l'activité des organes électriques par des courants de l'origine exterieure. *Arch. Sci. Physiol.* **5,** 43–73.

Albe-Fessard, D., and Buser, P. (1954). Analyse microphysiologique de la transmission réflexe au niveau du lobe électrique de la torpille *(Torpedo marmorata). J. Physiol. (Paris)* **46,** 923–946.

Albe-Fessard, D., and Chagas, C. (1954). Etude de la sommation a la jonction nerfélectroplaque chez le gymnote *(Electrophorus electricus). J. Physiol. (Paris)* **46,** 823–840.

Albe-Fessard, D., and Martins-Ferreira, H. (1953). Rôle de la commande nerveuse dans la synchronisation du functionnement des éléments de l'organe électrique du gymnote, *Electrophorus electricus* L. *J. Physiol. (Paris)* **45,** 533–546.

Albers, R. W. (1967). Biochemical aspects of active transport. *Ann. Rev. Biochem.* **36,** 727–756.

Altamirano, M. (1955). Electrical properties of the innervated membrane of the electroplax of electric eel. *J. Cell. Comp. Physiol.* **46,** 249–278.

Altamirano, M., Coates, C. W., Grundfest, H., and Nachmansohn, D. (1953). Mechanisms of bioelectric activity in electric tissue. I. The response to indirect and direct stimulation of electroplax of *Electrophorus electricus. J. Gen. Physiol.* **37,** 91–110.

Altamirano, M., Coates, C. W., and Grundfest, H. (1955). Mechanisms of direct and neural excitability in electroplaques of electric eel. *J. Gen. Physiol.* **38,** 319–360.

Auerbach, A. A., and Bennett, M. V. L. (1969a). Chemically mediated transmission at a giant fiber synapse in the central nervous system of a vertebrate. *J. Gen. Physiol.* **53,** 183–210.

Auerbach, A. A., and Bennett, M. V. L. (1969b). A rectifying synapse in the central nervous system of a vertebrate. *J. Gen. Physiol.* **53,** 211–237.

Bauer, R. (1968). Untersuchungen zur Entladungstätigkeit und zum Beutefangver-halten des Zitterwelses *Malapterurus electricus* Gmelin 1789 (Siluroidea, Malapteruridae, Lacep, 1803). Z. *Vergleich. Physiol.* **59,** 371–402.

Belbenoit, P. (1970). Comportement alimentaire et décharge associée chez *Torpedo marmorata* (Selacchii, Pisces). Z. *Vergleich. Physiol.* **67,** 205–216.

Bennett, M. V. L. (1961). Modes of operation of electric organs. *Ann. N. Y. Acad. Sci.* **94,** 458–509.

Bennett, M. V. L. (1964). Nervous function at the cellular level. *Ann. Rev. Physiol.* **26,** 289–340.

Bennett, M. V. L. (1966). Physiology of electrotonic junctions. *Ann. N. Y. Acad. Sci.* **137,** 509–539.

Bennett, M. V. L. (1968a). Neural control of electric organs. *In* "The Central Nervous System and Fish Behavior" (D. Ingle, ed.), pp. 147–169. Univ, of Chicago Press, Chicago, Illinois.

Bennett, M. V. L. (1968b). Similarities between chemically and electrically mediated transmission. *In* "Physiological and Biophysical Aspects of Nervous Integration" (F. D. Carlson, ed.), pp. 73–128. Prentice-Hall, Englewood Cliffs, New Jersey.

Bennett, M. V. L. (1970). Comparative physiology: Electric organs. *Ann. Rev. Physiol.* **32,** 471–528.

Bennett, M. V. L., and Grundfest, H. (1959). Electrophysiology of electric organ in *Gymnotus carapo. J. Gen. Physiol.* **42,** 1067–1104.

Bennett, M. V. L., and Grundfest, H. (1961a). The electrophysiology of electric organs of marine electric fishes. II. The electroplaques of main and accessory organs of *Narcine brasiliensis. J. Gen. Physiol.* **44,** 805–818.

Bennett, M. V. L., and Grundfest, H. (1961b). The electrophysiology of electric organs of marine electric fishes. III. The electroplaques of the stargazer, *Astro-scopus y-graecum. J. Gen. Physiol.* **44,** 819–843.

Bennett, M. V. L., and Grundfest, H. (1961c). Studies on morphology and electrophysiology of electric organs. III. Electrophysiology of electric organs in Mormyrids. *In* "Bioelectrogenesis" (C. Chagas and A. Paes de Carvalho, eds.), pp. 113–135. Elsevier, Amsterdam, 1961.

Bennett, M. V. L., and Grundfest, H. (1966). Analysis of depolarizing and hyperpolarizing inactivation responses in Gymnotid electroplaques. *J. Gen. Physiol.* **50,** 141–169.

Bennett, M. V. L., and Steinbach, A. B. (1969). Influence of electric organ control system on electrosensory afferent pathways in mormyrids. *In* "Neurobiology of Cerebellar Evolution and Development" (R. Llinás, ed.), pp. 207–214. Am. Med. Assoc., Chicago, Illinois.

Bennett, M. V. L., and Trinkaus, J. P. (1970). Electrical coupling between embryonic cells by way of extracellular space and specialized junctions. *J. Cell Biol.* **44,** 592–610.

Bennett, M. V. L., Würzel, M., and Grundfest, H. (1961). The electrophysiology of electric organs of marine electric fishes. I. Properties of electroplaques of *Torpedo nobiliana. J. Gen. Physiol.* **44,** 757–804.

Bennett, M. V. L., Giménez, M., Nakajima, Y., and Pappas, G. D. (1964). Spinal and medullary nuclei controlling electric organ in the eel, *Electrophorus. Biol. Bull.* **127,** 362.

Bennett, M. V. L., Pappas, G. D., Aljure, E., and Nakajima, Y. (1967a). Physiology and ultrastructure of electrotonic junctions. II. Spinal and medullary electromotor nuclei in Mormyrid fish. *J. Neurophysiol.* **30,** 180–208.

Bennett, M. V. L., Nakijima, Y., and Pappas, G. D. (1967b). Physiology and ultrastructure of electronic-junctions. III. Giant electromotor neurons of *Malapterurus electricus. J. Neurophysiol.* **30,** 209–235.

Bennett, M. V. L., Pappas, G. D., Giménez, M., and Nakajima, Y. (1967c). Physiology and ultrastructure of electrotonic junctions. IV. Medullary electromotor nuclei in gymnotid fish. *J. Neurophysiol.* **30,** 236–300.

Bigelow, H. B., and Schroeder, W. C. (1953). "Fishes of the Western North Atlantic, Part Two. Sawfishes, Guitarfishes, Skates and Rays." Sears Found. Marine Res., Yale University, New Haven, Connecticut.

Black-Cleworth, P. (1970). The role of electrical discharges in the non-reproductive social behavior of *Gymnotus carapo* (Gymnotidae, Pisces). *Animal Behavior Monographs* **3,** 3–77.

Bloom, F. E., and Barnett, R. (1966). Fine structural localization of acetylcholine-esterase in electroplaque of the electric eel. *J. Cell Biol.* **29,** 475–495.

Brightman, M. W., and Reese, T. S. (1969). Junctions between intimately apposed cell membranes in the vertebrate brain. *J. Cell Biol.* **40,** 648–677.

Brock, L. G., and Eccles, R. M. (1958). The membrane potentials during rest and activity of the ray electroplate. *J. Physiol. (London)* **142,** 251–274.

Bullock, T. H. (1970). Species differences in effect of electroreceptor input on electric organ pacemakers and other aspects of behavior in electric fish. *Brain, Behav. Evol.* **2,** 85–118.

Cass, A., Finkelstein, A., and Krespi, V. (1970). The ion permeability induced in thin lipid membranes by the polyene antibiotics nystatin and amphotericin B. *J. Gen. Physiol.* **56,** 100–124.

Coates, C. W., Altamirano, M., and Grundfest, H. (1954). Activity in electrogenic organs of knifefishes. *Science* **120,** 845–846.

Cohen, L. B., Hille, B., and Keynes, R. D. (1969). Light scattering and birefringence changes during activity in the electric organ of *Electrophorus electricus. J. Physiol. (London)* **203,** 489–509.

Cole, K. S. (1968). "Membranes, Ions and Impulses." Univ. of California Press, Berkeley, California.

Cole, K. S., and Curtis, H. J. (1939). Electric impedance of the squid giant axon during activity. *J. Gen. Physiol.* **22,** 649–670.

Couceiro, A., and Akerman, M. (1948). Sur quelque aspects du tissu électrique de l'*Electrophorus electricus* (Linnaeus). *Anais Acad. Brasil. Cien.* **20,** 383–395.

Couteaux, R., and Szabo, T. (1959). Siège de la jonction nerf-électroplaque dans les organes électriques à électroplaque pédiculées. *Compt. Rend.* **248,** 457–460.

Dahlgren, U. (1914). Origin of the electric tissues of *Gymnarchus niloticus. Carnegie Inst. Wash. Publ.* **183,** 161–194.

Dahlgren, U. (1927). The life history of the fish *Astroscopus* (the "Stargazer"). *Sci. Monthly* **24,** 348–365.

Dahlgren, U., and Silvester, C. F. (1906). The electric organ of the stargazer, *Astroscopus* (Brevoort). *Anat. Anz.* **15,** 387–403.

del Castillo, J., and Katz, B. (1956). Biophysical aspects of neuromuscular transmission. *Progr. Biophys. Biophys. Chem.* **6,** 121–170.

De Oliveira Castro, G. (1955). Differentiated nervous fibers that constitute the electric organ of *Sternachus albifrons* Linn. *Anais Acad. Brasil. Cien.* **27,** 557–560.

Diamond, J., and Yasargil, G. M. (1969). Synaptic function in the fish spinal cord: Dendritic integration. *Progr. Brain Res.* **31,** 201–209.

Dodge, F. A., and Frankenhaeuser, B. (1959). Sodium currents in the myelinated nerve fibre of *Xenopus laevis* investigated with the voltage clamp technique. *J. Physiol. (London)* **148,** 188–200.

Eigenmann, C. H., and Allen, W. R. (1942). "Fishes of Western South America." Univ. of Kentucky, Lexington, Kentucky.

Eisenberg, F. A., and Gage, P. W. (1969). Ionic conductances of the surface and tubular membranes of frog sartorius fibers. *J. Gen. Physiol.* **53,** 279–297.

Ellis, M. M. (1913). The gymnotid eels of tropical America. *Mem. Carnegie Museum* **6,** 109–204.

Ewart, J. C. (1892). The electric organ of the skate: Observations on the structure, relations, progressive development, and growth of the electric organ of the skate. *Phil. Trans. Roy. Soc. London* **B183,** 389–420.

Farquhar, M. G., and Palade, G. E. (1963). Junctional complexes in various epithelia. *J. Cell Biol.* **17,** 375–412.

Frankenhaeuser, B., and Huxley, A. F. (1964). The action potential in the myelinated nerve fibre of *Xenopus laevis* as computed on the basis of voltage clamp data. *J. Physiol.* (*London*) **171,** 302–315.

Fritsch, G. (1890). "Die elektrischen Fische. II. Die Torpedineen." Von Veit, Leipzig.

Furukawa, T., and Furshpan, E. J. (1963). Two inhibitory mechanisms in the Mauthner neurons of goldfish. *J. Neurophysiol.* **26,** 140–176.

Gage, P. W., and Eisenberg, R. S. Capacitance of the surface and transverse tubular membrane of frog sartorius muscle fibers. *J. Gen. Physiol.* **53,** 265–278.

Géry, J., and Vu-Tân-Tuê. (1964). *Gymnorhamphichthys hypostomus petiti,* ssp. nov. un curieux poisson gymnotoïde arénecole. *Vie Milieu,* Suppl. 17, 485–498.

Greenwood, P. H. (1956). The fishes of Uganda—II. *The Uganda Journal* **20,** No. 2, 129–143.

Greenwood, P. H., Rosen, D. E., Weitzman, S. H., and Myers, G. S. (1966). Phyletic studies of teleostean fishes, with a provisional classification of living forms. *Bull. Am. Museum Nat. Hist.* **131,** 341–455.

Grosse, J. P., and Szabo, T. (1960). Variation morphologique et fonctionelle de l'organe électrique dans une même espèce de Mormyrides (*Mormyrops deliciosus* Leach). *Compt. Rend.* **251,** 2791–2793.

Grundfest, H. (1957). The mechanisms of discharge of the electric organ in relation to general and comparative electrophysiology. *Progr. Biophys. Biophys. Chem.* **7,** 3–85.

Grundfest, H. (1966). Comparative electrobiology of excitable membranes. *Advan. Comp. Physiol. Biochem.* **2,** 1–116.

Grundfest, H. (1967). Comparative physiology of electric organs of elasmobranch fishes. *In* "Sharks, Skates and Rays" (P. W. Gilbert, R. F. Mathewson, and D. P. Rall, eds.), pp. 399–432. Johns Hopkins Press, Baltimore, Maryland.

Grundfest, H., and Bennett, M. V. L. (1961). Studies on the morphology and electrophysiology of electric organs. I. Electrophysiology of marine electric fishes. *In* "Bioelectrogenesis" (C. Chagas and A. Paes de Carvalho, eds.), pp. 57–95. Elsevier, Amsterdam.

Hagiwara, S., Watanabe, A., and Saito, N. (1959). Potential changes in synctial neurons of lobster cardiac ganglion. *J. Neurophysiol.* **22,** 554–572.

Harder, W., and Uhlemann, H. (1967). Zum Frequenzverhalten von *Gymnarchus niloticus* Cuv. (Mormyriformes, Teleostei). *Z. Vergleich. Physiol.* **54,** 85–88.

Higman, H. B., Podleski, T. R., and Bartels, E. (1964). Correlation of membrane potential and potassium flux in the electroplax of *Electrophorus. Biochim. Biophys. Acta* **79,** 138–150.

Hille, B. (1970). Ionic channels in nerve membranes. *Progr. Biophys. Mol. Biol.* **21,** 1–32.

Hodgkin, A. L. (1964). The Conduction of the Nervous Impulse." Liverpool Univ. Press, Liverpool.

Hodgkin, A. L., and Huxley, A. F. (1952). A quantitative description of membrane current and its application to conduction and excitation in nerve. *J. Physiol.* (*London*), **177,** 500–544.

Ishiyama, R. (1958). Studies on the rajid fishes (Rajidae) found in the waters around Japan. *J. Shimonoseki Coll. Fisheries* **7,** 189–394.

Israel, M., Gautron, J., and Lesbats, B. (1968). Isolement des vésicules synaptiques de l'organe électrique de la Torpille et localisation de l'acétylcholine à leur niveau. *Compt. Rend.* **D266,** 273–275.

Johnels, A. G. (1956). On the origin of the electric organ in *Malapterurus electricus. Quart. J. Microscop. Sci.* **97,** 455–464.

Karlin, A. (1967). Permeability and internal concentration of ions during depolarization of the electroplax. *Proc. Natl. Acad. Sci. U. S.* **58,** 1162–1167.

Karlin, A. (1969). Chemical modification of the active site of the acetylcholine receptor. *J. Gen. Physiol.* **54,** 245s–264s.

Katz, B., and Miledi, R. (1965a). The effect of temperature on the synaptic delay at the neuromuscular junction. *J. Physiol. (London)* **181,** 656–670.

Katz, B., and Miledi, R. (1965b). The measurement of synaptic delay, and the time course of acetylcholine release at the neuromuscular junction. *Proc. Roy. Soc.* **B161,** 483–495.

Kellaway, P. (1946). The part played by electric fish in the early history of bioelectricity and electrotherapy. *Bull. History Med.* **20,** 112–137.

Keynes, R. D. (1957). Electric organs. *In* "The Physiology of Fishes" (M. E. Brown, ed.), Vol. 2, pp. 323–343. Academic Press, New York.

Keynes, R. D. (1961). The development of the electric organ in *Electrophorus electricus* (L.). *In* "Bioelectrogenesis" (C. Chagas and A. Paes de Carvalho, eds.), pp. 14–18. Elsevier, Amsterdam.

Keynes, R. D., and Martins-Ferreira, H. (1953). Membrane potentials in the electroplates of the electric eel. *J. Physiol.* **119,** 315–351.

Keynes, R. D., Bennett, M. V. L., and Grundfest, H. (1961). Studies on morphology and electrophysiology of electric organs. II. Electrophysiology of electric organ of *Malapterurus electricus. In* "Bioelectrogenesis" (C. Chagas and A. Paes de Carvalho, eds.), pp. 102–112. Elsevier, Amsterdam.

Kriebel, M. E., Bennett, M. V. L., Waxman, S. G., and Pappas, G. D. (1969). Oculomotor neurons in fish: Electrotonic coupling and multiple sites of impulse initiation. *Science* **166,** 520–524.

Larimer, J. L., and MacDonald, J. A. (1968). Sensory feedback from electroreceptors to electromotor pacemaker centers in gymnotids. *Am. J. Physiol.* **214,** 1253–1261.

Leuzinger, W., and Baker, A. L. (1967). Acetylcholinesterase. I. Large-scale purification, homogeneity and amino acid analysis. *Proc. Natl. Acad. Sci. U.S.* **57,** 446–451.

Lissmann, H. W. (1951). Continuous electric signals from the tail of a fish, *Gymnarchus niloticus. Nature* **167,** 201.

Lissmann, H. W. (1958). On the function and evolution of electric organs in fish. *J. Exptl. Biol.* **35,** 156–191.

Lissmann, H. W., and Schwassman, H. O. (1965). Activity rhythm of an electric fish, *Gymnorhamphichthys hypostomus* Ellis. *Z. Vergleich. Physiol.* **51,** 153–171.

Lowrey, A. (1913). A study of the submental filaments considered as probable electric organs in the gymnotid eel, *Steatogenys elegans* (Steindachner). *J. Morphol.* **24,** 685–694.

Luft, J. H. (1957). The histology and cytology of the electric organ of the electric eel (*Electrophorus electricus* L.). *J. Morphol.* **100,** 113–140.

Mandriota, F. J., Thompson, R. L., and Bennett, M. V. L. (1965). Classical conditioning of electric organ discharge rate in mormyrids. *Science* **150,** 1740–1742.

Mandriota, F. J., Thompson, R. L., and Bennett, M. V. L. (1968). Avoidance conditioning of the rate of electric organ discharge in mormyrid fish. *Animal Behaviour* **16,** 448–455.

Mathewson, R., Mauro, A., Amatniek, E., and Grundfest, H. (1958). Morphology of main and accessory electric organs of *Narcine brasiliensis* (Olfers) and some correlations with their electrophysiological properties. *Biol. Bull.* **115,** 126–135.

Mathewson, R., Wachtel, A., and Grundfest, H. (1961). Fine structure of electroplaques *In* "Bioelectrogenesis" (C. Chagas and A. Paes de Carvalho, eds.), pp. 25–53. Elsevier, Amsterdam.

Mauro, A. (1969). The role of the Voltaic pile in the Galvani-Volta controversy concerning animal *vs*. metallic electricity. *J. History Med. Allied Sci.* **24,** 140–150.

Moller, P. (1970). "Communication" in weakly electric fish, *Gnathonemus niger* (Mormyridae). I. Variation of electric organ discharge (EOD) frequency elicited by controlled electric stimuli. *Animal Behaviour* **18,** 768–786.

Morlock, N. L., Benamy, D. A., and Grundfest, H. (1969). Analysis of spike electrogenesis of eel electroplaques with phase and impedance measurements. *J. Gen. Physiol.* **52,** 22–45.

Nakamura, Y., Nakajima, S., and Grundfest, H. (1965). Analysis of spike electrogenesis and depolarizing K inactivation in electroplaques of *Electrophorus electricus*, L. *J. Gen. Physiol.* **49,** 321–349.

Pappas, G. D., and Bennett, M. V. L. (1966). Specialized junctions involved in electrical transmission between neurons. *Ann. N. Y. Acad. Sci.* **137,** 495–508.

Pappas, G. D., Asada, Y., and Bennett, M. V. L. (1971). Morphological correlates of increased coupling resistance at an electrotonic synapse. *J. Cell Biol.* **49,** 173–188.

Payton, B. W., Bennett, M. V. L., and Pappas, G. D. (1969). Permeability and structure of junctional membranes at an electrotonic synapse. *Science* **166,** 1641–1643.

Pellegrin, J. (1923). "Les poissons des eaux douces de l'Afrique occidentale (du Sénégal au Niger)." Paris.

Pickens, P. E., and McFarland, W. N. (1964). Electric discharge and associated behaviour in the stargazer. *Animal Behaviour* **12,** 362–367.

Poll, M. (1957). "Les Genres des Poissons d'Eau Douce de l'Afrique." Publication de la Direction de l'Agriculture des Forêts et de l'Elévage, Brussels.

Post, R. L., Kume, S.. Tobin, T., Orcutt, B., and Sen, A. K. (1969). Flexibility of an active center in sodium-plus-potassium adenosine triphosphatase. *J. Gen. Physiol.* **54,** 306s–326s.

Remmler, W. (1930). Untersuchungen über die Abhängigkeit der nach aussen ableitbaren maximalen elektromotorischen Kraft des *Malopterurus elektricus* von der Temperatur mit Hilfe des Oszillographen. *Beitr. Physiol.* **4,** 151–178.

Revel, J. P., and Karnovsky, M. J. (1967). Hexagonal array of subunits in intercellular junctions of the mouse heart and liver. *J. Cell Biol.* **33,** C7–C12.

Rosenberg, H. (1928). Die elektrischen Organe. *In* "Handbuch normalischer pathologischen Physiologie," Vol. 8. pt. 2, pp. 876–925. Springer, Berlin.

Saito, N. (1966). Spike potentials of the electromotorneuron of the electric skate, *Narke japonica. Japan. J. Physiol.* **16,** 509–518.

Schwartz, I. R., Pappas, G. D., and Bennett, M. V. L. (1971). Fine structure of electrocytes in eight species of weakly electric teleosts (in preparation).

Sheridan, M. N. (1965). The fine structure of the electric organ of *Torpedo marmorata. J. Cell Biol.* **24,** 129–141.

Sheridan, M. N., Whittaker, V. P., and Israel, M. (1966). The subcellular fractionation of the electric organ of *Torpedo. Z. Zellforsch.* **74,** 291–307.

Sherrington, C. (1906). "The Integrative Action of the Nervous System." Cambridge Univ. Press, London and New York.

Stein, R. B. (1967). The frequency of nerve action potentials generated by applied currents. *Proc. Roy. Soc. London* **B167,** 64–86.

Steinbach, A. B. (1970). Diurnal movements and discharge characteristics of electric gymnotid fishes in the Rio Negro, Brazil. *Biol. Bull.* **138,** 200–210.

Szabo, T. (1954). Un relais dans le système des connexions du lobe électrique de la torpille. *Arch. Anat. Microscop. Morphol. Exptl.* **43,** 187–201.

Szabo, T. (1955). Quelques précisions sure le noyau de commande centrale de la décharge électrique chez la Raie (*Raja clavata*). *J. Physiol.* (*Paris*) **47,** 283–285.

Szabo, T. (1958). Structure intime de l'organ électrique de trois mormyrides. *Z. Zellforsch. Mikroskop. Anat.* **49,** 33–45.

Szabo, T. (1961a). Les organes électriques des Mormyrides. *In* "Bioelectrogenesis" (C. Chagas and A. Paes de Carvahlo, eds.), pp. 20–23. Elsevier, Amsterdam.

Szabo, T. (1961b). Anatomophysiologie de centres nerveux spécifiques de quelques organes électriques. *In* "Bioelectrogenesis" (C. Chagas and A. Paes del Carvahlo, eds.), pp. 185–201. Elsevier, Amsterdam.

Szabo, T. (1961c). Rapports ontogénetiques entre l'organ électrique, son innervation et sa commande enoéphalique. *Z. Zellforsch. Mikroskop. Anat.* **55,** 200–203.

Szabo, T. (1961d). Les organes électrique de *Gymnotus carapo*. *Koninkl. Ned. Akad. Wetenschap., Proc.* **C64,** 584–586.

Szabo, T., and Fessard, A. (1965). Le fonctionnement des électrorécepteurs ètudié chez les Mormyres. *J. Physiol.* (*Paris*) **57,** 343–360.

Szabo, T., and Suckling, E. E. (1964). L'arrêt occasionel de la décharge électrique continue du *Gymnarchus* est-il une réaction naturelle? *Naturwissenshaften* **51,** 92–94.

Takeuchi, A., and Takeuchi, N. (1959). Active phase of frog's end-plate potential. *J. Neurophysiol.* **22,** 395–411.

Takeuchi, N. (1963). Some properties of conductance changes at the end-plate membrane during the action of acetyl choline. *J. Physiol.* (*London*) 128–140.

Tasaki, I., and Freygang, W. H., Jr. (1955). The parallelism between the action potential, action current, and membrane resistance at a node of Ranvier. *J. Gen. Physiol.* **39,** 211–223.

Wachtel, A. W. (1964). The ultrastructural relationships of electric organs and muscle. I. Filamentous systems. *J. Morphol.* **114,** 325–360.

Watanabe, A., and Takeda, K. (1963). The change of discharge frequency by A. C. stimulus in a weak electric fish. *J. Exptl. Biol.* **40,** 57–66.

Willows, A. O. D., and Hoyle, G. (1969). Neuronal network triggering a fixed action pattern. *Science* **166,** 1549–1551.

Other volumes in the Fish Physiology series

Index

Note: Page numbers followed by "*f*" indicate figures and "*t*" indicate tables.

A

F

G

H

I

J

K

L

M

N

O

P

Q

R

S

Printed and bound by CPI Group (UK) Ltd, Croydon, CR0 4YY
08/06/2025
01896870-0012